The Elements

Element	Symbol	Atomic number	Atomic mass (amu)	Element	Symbol	Atomic number	Atomic mass (amu)
actinium	Ac	89	(227)	mendelevium	Md	101	(258)
aluminum	Al	13	26.98	mercury	Hg	80	200.6
americium	Am	95	(243)	molybdenum	Mo	42	95.94
antimony	Sb	51	121.8	neodymium	Nd	60	144.2
argon	Ar	18	39.95	neon	Ne	10	20.18
arsenic	As	33	74.92	neptunium	Np	93	237.0
astatine	At	85	(210)	nickel	Ni	28	58.69
barium	Ba	56	137.3	niobium	Nb	41	92.91
berkelium	Bk	97	(247)	nitrogen	N	7	14.01
beryllium	Be	4	9.012	nobelium	No	102	(259)
bismuth	Bi	83	209.0	osmium	Os	76	190.2
bohrium	Bh	107	(267)	oxygen	O	8	16.00
boron	B	5	10.81	palladium	Pd	46	106.4
bromine	Br	35	79.90	phosphorus	P	15	30.97
cadmium	Cd	48	112.4	platinum	Pt	78	195.1
calcium	Ca	20	40.08	plutonium	Pu	94	(244)
californium	Cf	98	(251)	polonium	Po	84	(209)
carbon	C	6	12.01	potassium	K	19	39.10
cerium	Ce	58	140.1	praseodymium	Pr	59	140.9
cesium	Cs	55	132.9	promethium	Pm	61	(145)
chlorine	Cl	17	35.45	protactinium	Pa	91	231.0
chromium	Cr	24	52.00	radium	Ra	88	(226)
cobalt	Co	27	58.93	radon	Rn	86	(222)
copper	Cu	29	63.55	rhenium	Re	75	186.2
curium	Cm	96	(247)	rhodium	Rh	45	102.9
darmstadtium	Ds	110	(269)	roentgenium	Rg	111	(272)
dubnium	Db	105	(262)	rubidium	Rb	37	85.47
dysprosium	Dy	66	162.5	ruthenium	Ru	44	101.1
einsteinium	Es	99	(252)	rutherfordium	Rf	104	(261)
erbium	Er	68	167.3	samarium	Sm	62	150.4
europium	Eu	63	152.0	scandium	Sc	21	44.96
fermium	Fm	100	(257)	seaborgium	Sg	106	(266)
fluorine	F	9	19.00	selenium	Se	34	78.96
francium	Fr	87	(223)	silicon	Si	14	28.09
gadolinium	Gd	64	157.3	silver	Ag	47	107.9
gallium	Ga	31	69.72	sodium	Na	11	22.99
germanium	Ge	32	72.40	strontium	Sr	38	87.62
gold	Au	79	197.0	sulfur	S	16	32.07
hafnium	Hf	72	178.5	tantalum	Ta	73	180.9
hassium	Hs	108	(277)	technetium	Tc	43	(98)
helium	He	2	4.003	tellurium	Te	52	127.6
holmium	Ho	67	164.9	terbium	Tb	65	158.9
hydrogen	H	1	1.008	thallium	Tl	81	204.4
indium	In	49	114.8	thorium	Th	90	232.0
iodine	I	53	126.9	thulium	Tm	69	168.9
iridium	Ir	77	192.2	tin	Sn	50	118.7
iron	Fe	26	55.85	titanium	Ti	22	47.87
krypton	Kr	36	83.80	tungsten	W	74	183.9
lanthanum	La	57	138.9	uranium	U	92	238.0
lawrencium	Lr	103	(260)	vanadium	V	23	50.94
lead	Pb	82	207.2	xenon	Xe	54	131.3
lithium	Li	3	6.941	ytterbium	Yb	70	173.0
lutetium	Lu	71	175.0	yttrium	Y	39	88.91
magnesium	Mg	12	24.31	zinc	Zn	30	65.41
manganese	Mn	25	54.94	zirconium	Zr	40	91.22
meitnerium	Mt	109	(268)				

Note: Parentheses ( ) denote the most stable isotope of a radioactive element.

P9-CFD-944

Essentials of General, Organic, and Biochemistry

Essentials of General, Organic, and Biochemistry

An Integrated Approach

Denise Guinn
The College of New Rochelle

Rebecca Brewer
Sandoz Inc.

W. H. FREEMAN • NEW YORK

Senior Development Editor: Susan Moran
Editorial Assistant: Brittany Murphy
Publisher: Clancy Marshall
Acquisitions Editor: Anthony Palmiotto
Market Development Manager: Kirsten Watrud
Marketing Director: John A. Britch
Executive Editor, Digital, Physical Sciences: Mark Santee
Media Editor: Dave Quinn
Supplements Editor: Kathryn Treadway
Project Editor: Jane O'Neill
Art Director: Diana Blume
Senior Illustration Coordinator: Bill Page
Illustrations: Network Graphics
Molecular Models: Denise Guinn and Gregory Williams
Photo Editor: Bianca Moscatelli
Photo Researcher: Donna Ranieri
Production Coordinator: Paul Rohloff
Composition and Layout: Black Dot
Manufacturing: RR Donnelley

The following figures first appeared in the text *General, Organic, and Biochemistry*, Second edition, by Ira Blei and George Odian (© 2009 W. H. Freeman and Company): Figures 2–7c, 4–5, 4–17, 4–18, 11–12, 12–19, 13–16, 13–19, 15–10, 15–21.

Library of Congress Control Number: 2009930489
ISBN-13: 978-0-7167-6121-1
ISBN-10: 0-7167-6121-1

ISBN-13: 978-1-4292-5103-7 (Preliminary Loose-leaf Edition)
ISBN-10: 1-4292-5103-4 (Preliminary Loose-leaf Edition)

© 2010 by W. H. Freeman and Company
All rights reserved
Printed in the United States of America
Third Printing

W. H. Freeman and Company
41 Madison Avenue
New York, NY 10010
Houndmills, Basingstoke RG21 6XS, England

www.whfreeman.com

Essentials of General, Organic, and Biochemistry

An Integrated Approach

Denise Guinn
The College of New Rochelle

Rebecca Brewer
Sandoz Inc.

W. H. FREEMAN • NEW YORK

Senior Development Editor: Susan Moran
Editorial Assistant: Brittany Murphy
Publisher: Clancy Marshall
Acquisitions Editor: Anthony Palmiotto
Market Development Manager: Kirsten Watrud
Marketing Director: John A. Britch
Executive Editor, Digital, Physical Sciences: Mark Santee
Media Editor: Dave Quinn
Supplements Editor: Kathryn Treadway
Project Editor: Jane O'Neill
Art Director: Diana Blume
Senior Illustration Coordinator: Bill Page
Illustrations: Network Graphics
Molecular Models: Denise Guinn and Gregory Williams
Photo Editor: Bianca Moscatelli
Photo Researcher: Donna Ranieri
Production Coordinator: Paul Rohloff
Composition and Layout: Black Dot
Manufacturing: RR Donnelley

The following figures first appeared in the text *General, Organic, and Biochemistry*, Second edition, by Ira Blei and George Odian (© 2009 W. H. Freeman and Company): Figures 2–7c, 4–5, 4–17, 4–18, 11–12, 12–19, 13–16, 13–19, 15–10, 15–21.

Library of Congress Control Number: 2009930489
ISBN-13: 978-0-7167-6121-1
ISBN-10: 0-7167-6121-1

ISBN-13: 978-1-4292-5103-7 (Preliminary Loose-leaf Edition)
ISBN-10: 1-4292-5103-4 (Preliminary Loose-leaf Edition)

© 2010 by W. H. Freeman and Company
All rights reserved
Printed in the United States of America
Third Printing

W. H. Freeman and Company
41 Madison Avenue
New York, NY 10010
Houndmills, Basingstoke RG21 6XS, England

www.whfreeman.com

To my sons, Charles and Scott,
my mother, Inge,
and brothers, Peter and Herb,
for their unfaltering support and inspiration

–Denise

To my husband, Mark,
my parents, Marion and Charles,
and my sisters, Debbe, Sandie, and Cindy
for their continued support and encouragement

–Becky

About the Authors

Jorge Madrigal, Nyack, NY

Denise Guinn received her B.A. in chemistry from the University of California at San Diego and her Ph.D. in organic chemistry from the University of Texas at Austin. She was a National Institutes of Health postdoctoral fellow at Harvard University before joining Abbott Laboratories as a Research Scientist in the Pharmaceutical Products Discovery Research Group. In 1992, Dr. Guinn joined the faculty at Regis University, in Denver, Colorado, as Clare Boothe Luce Professor, where she taught courses in general chemistry, organic chemistry, and the general, organic, and biochemistry course for nursing and allied health majors. Last year she moved to New York to join the chemistry department at The College of New Rochelle, where she teaches organic chemistry, biochemistry, and the general, organic, and biochemistry course for nursing students. Her area of research is synthetic organic chemistry; she has published in the Journal of Organic Chemistry, the Journal of the American Chemical Society, and the Journal of Medicinal Chemistry. She currently resides in Nyack, New York, with her sons, Charles and Scott.

Rebecca Brewer received her Ph.D. in chemistry from the University of Connecticut, Storrs, Connecticut, where her area of research was developing rapid analytical methods for field-testing contaminated groundwater and soil. She taught graduate-level courses in analytical chemistry at Sacred Heart University in Fairfield, Connecticut, part-time while working full-time at Cytec Industries, Stamford, Connecticut. More recently, she was an assistant professor of chemistry at Regis University in Denver, Colorado, where she taught the general, organic, and biochemistry course as well as analytical chemistry and general chemistry. Becky is currently doing pharmaceutical research and development for Sandoz in Broomfield, Colorado.

Brief Contents

Contents

Health Care Applications

Preface

As long-time instructors contemplating ways to improve our general, organic, and biochemistry course, we asked ourselves two questions: "Why do allied-health majors need to take chemistry?" and "How can chemistry help nursing and allied-health students understand human physiology and medicine so that they are better prepared for their careers?" We noticed that students seemed to have the most difficulty in the course when they couldn't relate the chemistry to their career interests and to their lives. We set out to write a text that made chemistry as relevant as possible *throughout* the course. We decided that the most effective way to prepare students for a health care career is to weave together general, organic, and biochemistry, while at the same time narrowing the focus to include the topics most motivating for students. The fundamentals of chemistry are therefore introduced in a context with which students can identify.

In developing *Essentials of General, Organic, and Biochemistry,* we listened to hundreds of instructors and students talk about their experiences with the general, organic, and biochemistry course. We asked instructors what worked and didn't work for them, and what they liked and didn't like about other books. We asked students how they learned difficult topics, and what they wanted from their textbook and their course. We also surveyed nursing faculty to see what outcomes they wanted from their students.

Their answers were remarkably similar, and provided us with three key goals. First, students need to see how chemistry relates to their everyday lives and to their careers, and why it is important. Second, students need a great deal of support in a variety of forms; they need the opportunity to review, practice, and *experience* the concepts in order to understand them and to be successful. And finally, we must set limits on the breadth of content we expect to cover in the course.

After six years of planning and consultation, we've addressed these needs in several ways, outlined below.

Integration of Organic and Biochemistry in Every Chapter

The allied-health chemistry course has traditionally been arranged linearly in three distinct areas—general chemistry, organic chemistry, and biochemistry. Often, allied-health students have trouble connecting general chemistry to their intended careers. The resulting early drop in student motivation is difficult to overcome, and many students perform poorly even after they reach the organic and biochemistry topics, which they perceive as more relevant.

We approach the course differently. In this textbook, organic and biochemistry concepts are included in every chapter, so that during every week of the course, students are engaging topics directly related to their careers. To make this integration possible, we have rearranged some topics; most notably, chemical reactions follow organic chemistry so that the chapter can include the chemical reactions of organic compounds, which are of particular interest to health care majors.

In addition, we often tailor the content to prepare students for their careers in health care. Two representative examples appear in Chapters 5 and 10:

- In examples of units of concentration in Chapter 5, we have chosen those that are most commonly used in medicine. As a consequence, we are able to write exercises that ask students to calculate concentrations and dosages, using common medications as the examples.

> Health care emphasized in Chapter 5, Solutions, Colloids, and Membranes

[Jose Luis Pelaez, Inc./Corbis]

Blood Test

John Doe

Blood Component	Value	Normal Range
Iron	125 µg/dL	40–150
Glucose	78 mg/dL	70–110
Chloride	104 mmol/L	98–109
CO_2	25 mmol/L	21–31

Integration of biochemistry in Chapter 8, Chemical Reaction Basics

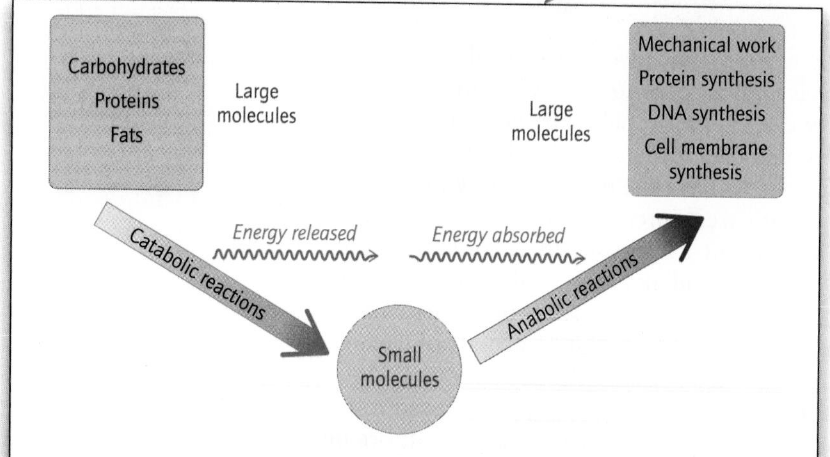

Figure 8-5 Overview of catabolic and anabolic reactions, which together constitute metabolism. The breakdown of large molecules in catabolism releases energy used to build large molecules in anabolism.

- Rather than introducing the traditional reactions such as displacement reactions between salts and redox reactions of metals, in Chapter 10 we use the five organic reactions performed by enzymes in the cell as examples. Thus, most of the redox reactions we selected are ones that occur in the cells of the body.

We have tried to make connections between chemical concepts and health care in the main text, rather than shunting this material aside in boxes. The student learns to see chemistry as integral to how the body functions and how medical treatments work.

Motivation increases significantly when students see the "why" of their efforts, and consistently motivated students are much more likely to succeed than those searching for relevance early in the course.

Meaningful, Supportive Learning Experiences

Some students can grasp a concept the first time they see it, but most require multiple interactions with the material in order to reach a useful understanding. In this textbook, we offer a progression of explanations and activities aimed at developing a meaningful learning experience, one that students value and can apply later in the course and in their careers. Our goal is to meet the students where they are, and take them where they need to go.

We believe that context and connection are crucial, so the first interactions with new concepts arrive in the form of vignettes that high-

Focus on biochemical reactions in Chapter 10, Reactions of Organic Functional Groups in Biochemistry

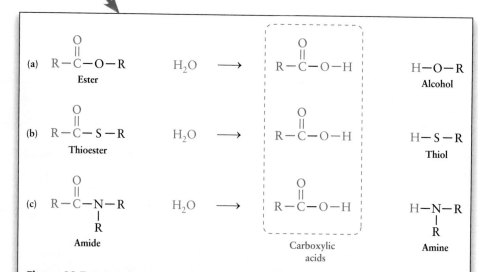

Figure 10-7 Hydrolysis reactions transfer an acyl group from a carboxylic acid derivative to a water molecule. The acyl group is shown in red.

light interesting, health-related topics connected to the concepts in the chapter. Explanations are reinforced by worked examples and by practice problems following each worked example. The unique Model Tool Kit exercises walk students through the process of building the molecules about which they are learning; we've found that this visual and hands-on interaction is effective, and that students begin to use the model kit on their own when solving problems. A small model kit, specifically designed for the exercises, is available with the book. Finally, these textbook elements are supported by online resources such as interactive tutorials, video walkthroughs of instructors solving problems, and an interactive version of the model kit.

Careful Selection of Concepts, Flexibility in Coverage

General, organic, and biochemistry are three important but potentially overwhelming subject areas to address in just one or two semesters. Instructors tell us that they often skip or skim over large sections of their textbook in order to meet time constraints while still delivering an effective overview of the chemistry students will need. We've written this textbook to focus the students on the most important and practical concepts for their careers. To ensure that we remained on track, we constantly asked ourselves and more than a hundred reviewers whether we were providing the proper depth and breadth.

Our reviews indicate that many instructors of the one-semester course do not cover some mathematically or otherwise challenging concepts, yet these same concepts are crucial to other instructors. We chose to cover such concepts within a clearly labeled **Extension Box.** When assigning chapters, instructors can easily indicate whether or not students will be responsible for material within the Extension Boxes, making it easier for instructors to tailor the course to their preferences. Examples of topics covered in these boxes include converting between different percent concentrations and molarity, reaction stoichiometry, and naming esters and amines.

Overall, everyone involved in the development of this textbook has focused on the best ways to foster a practical, memorable understanding of chemistry, so that students can apply their knowledge successfully to their chosen field.

Chapter Walkthrough

An Opening Vignette introduces the subject of each chapter and immediately immerses the student in a high-interest topic related to medicine or the human body.

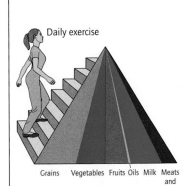

Grains   Vegetables Fruits Oils Milk Meats
and
beans

Figure 14-1 Current United States Government food pyramid.

The Role of Physical Activity in Weight Loss

In Chapter 8 you learned about the problems of starvation in the world. Now consider the millions of people who wrestle with the opposite problem—excess weight. Why does it seem so much harder to lose weight than it is to gain weight? Isn't it simply a matter of balancing the number of calories consumed (energy in) with the number of calories expended (energy out)? The biochemistry of metabolism—the subject of this chapter—can help explain some of the complexities involved in weight loss.

Why do diets that call for a severe reduction in calories usually not work? The problem with a drastic reduction in calories is that the brain needs a constant supply of glucose, in the amount of 70–110 mg per dL of blood. However, there are no metabolic pathways that produce glucose from stored fat. Once the body's limited stores of glycogen have been converted to glucose in response to a severe reduction in calories, the body begins to degrade proteins in muscle tissue in order to supply glucose to the brain. Furthermore, plummeting glucose levels make it difficult to stick to a weight loss plan that involves a severe reduction in calories. Low blood glucose levels signal the brain that it needs sugar, which initiates the desire to consume food.

Detailed Worked Exercises, paired with Practice Exercises throughout each chapter, give students a helpful roadmap for solving problems and utilizing the concepts they're learning in the text. Practice Exercises follow each worked exercise, giving students the immediate opportunity to check their understanding. Answers to the Practice Exercises appear at the end of each chapter.

WORKED EXERCISE 9-5   Acid–Base Neutralization Reactions

A base can be used to neutralize the effects of excess stomach acid. For example, magnesium hydroxide is a base commonly found in over-the-counter antacid medications, such as Phillips' Milk of Magnesia, used to treat heartburn caused by excess stomach acid. Write and balance the neutralization reaction of magnesium hydroxide, $Mg(OH)_2$, with stomach acid, HCl.

SOLUTION

$$\underset{\text{Base}}{Mg(OH)_2(s)} + \underset{\text{Acid}}{2\,HCl} \longrightarrow \underset{\text{Water}}{2\,H_2O} + \underset{\text{Salt}}{MgCl_2(aq)}$$

Since magnesium hydroxide yields two OH^- per formula unit, two HCl molecules are required to neutralize one formula unit of $Mg(OH)_2$. Thus, the coefficient 2 must be placed before HCl (yielding $2H^+$). Two water molecules are produced as well as the salt $MgCl_2$, which is electrically neutral.

Commercial antacids neutralize stomach acid. [© 1995 Michael Dalton, Fundamental Photographs]

PRACTICE EXERCISES

9.14 Write the balanced equation for the neutralization of hydrochloric acid with calcium hydroxide, $Ca(OH)_2$.

9.15 Write the balanced equation for the reaction of HNO_3 with the following:
   a. KOH   **b.** $Mg(OH)_2$ **c.** $Al(OH)_3$

9.16 Zoloft, primarily used to treat depression in adults, is sold as the hydrochloride salt of the amine. The structure of the neutral form of Zoloft is given below. Write the reaction of this molecule with HCl, showing the formation of the hydrochloride salt. Why is water not formed in this neutralization reaction?

9.17 **Critical Thinking Question:** Use Le Châtelier's principle to explain how certain bases having limited solubility in water, such as aluminum hydroxide $(Al(OH)_3)$ and magnesium hydroxide $(Mg(OH)_2)$, can still neutralize stomach acid.

Problem-Solving Tutorials, indicated by icons in the margin, reinforce many Worked Exercises. Students can reach the tutorials by going to www.whfreeman.com/gob. The tutorials provide additional explanation, a step-by-step walkthrough of a problem, and several interactive practice exercises. The tutorials are described in more detail below under "Premium Multimedia Resources."

The Model Tool 8-1 Balancing a Chemical Equation

I. Construction Exercise Part I

1. Obtain one black carbon atom, four red oxygen atoms, and four light-blue hydrogen atoms. Obtain four bent double bonds and four single bonds.

2. Construct a model of methane, CH_4, and two models of oxygen, O_2. Remember, an oxygen molecule contains an oxygen–oxygen double bond requiring two bent connectors. Imagine that these are the reactants for a chemical reaction.

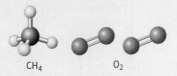

CH_4 O_2

3. How many C, O, and H atoms are present in the reactants?

II. Construction Exercise Part II

4. Simulate a chemical reaction by breaking *all* the bonds in your methane and oxygen molecules.

5. Using *only* these atoms and bonds, construct as many carbon dioxide, CO_2, and water, H_2O, molecules as you can from the atoms you have available from the reactants. Use the

Model Tool Exercises, using a model kit, provide a superb way for students to experience and readily understand the three-dimensional aspects of chemistry. Easy-to-follow steps are associated with practical questions, encouraging students to think at every step and build a deeper understanding of molecular structure. A model kit available with the text has been specifically created for the activities in this text.

Guidelines for step-by-step processes, such as naming organic compounds, are written to be used as both initial explanations and quick references when doing homework or other exercises. They are clearly set off in the text to emphasize their importance and to help students find them easily.

Guidelines for Predicting the Product of a Hydrolysis Reaction

Step 1: *Identify* **the type of carboxylic acid derivative in the reactant—ester, thioester, or amide.** From this information you can predict what functional groups will be produced in the product structures (right side of Figure 10-7). Remember, one of the products will always be a carboxylic acid, RCOOH (or a carboxylate ion, $RCOO^-$).

Step 2: *Break* **the carbonyl carbon–heteroatom bond.** Break the single bond between the carbonyl carbon and the O, S, or N atom, to give two partial structures:

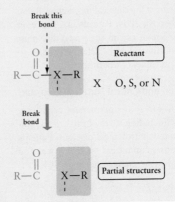

Extension Topic boxes provide flexibility for the instructor. While the mathematics integrated into the text focuses on performing basic calculations and unit conversions, more mathematically challenging treatments of stoichiometry and other topics are set off in these clearly labeled sections, allowing instructors to cover them according to their preferences.

Extension Topic 5-1 Converting Between Mass and Mole-Based Concentration Units

In this chapter, you have seen that some of the most common units of concentration in medicine are expressed as either mole-based or mass-based concentrations. Often it is more convenient to know the *number* of solute molecules or ions in a solution, expressed in moles, rather than the *mass* of the solute. In such cases, if you are given a concentration that is based on mass, you must perform a unit conversion to obtain the number of solute molecules or ions. The flow chart below illustrates the steps to take in converting a mass-based concentration to a mole-based concentration, and vice-versa.

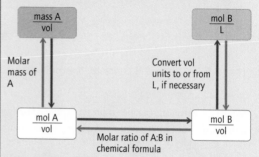

Suppose you need to know how many mol/L of Na^+ (B) there are in a physiological saline solution, which is 0.90% (m/v) NaCl (A).

Chemistry in Medicine boxes at the end of each chapter offer an in-depth look at how the chemical principles described in the chapter can be directly applied to a problem in health care.

Chemistry in Medicine Critical Needs for Human Calorimetry in Medicine

On May 27, 1995, Christopher Reeve, well known for his role as Superman in the 1978 blockbuster movie, was paralyzed from the neck down after being thrown from his horse and landing on his head. For the remainder of his life, Reeve was confined to a wheelchair and unable to breathe without the assistance of a mechanical respirator. Reeve passed away on October 10, 2004. Patients like Reeve on mechanical respirators cannot eat on their own and must have their caloric and nutritional needs determined through methods based on the principles of calorimetry.

Mechanical Respiration and Critical Nutrition Decisions

Several conditions can lead to the permanent need for a mechanical respirator, including severe brain injury, spinal-cord injury, and some neurological diseases. Patients on mechanical respirators cannot eat on their own, and a medical professional must manage feedings. It is crucial not to underfeed or overfeed these patients. Malnourishment caused by underfeeding is common in patients who require permanent mechanical respirators and, in severe cases, can lead to coma and death. Then again, overfeeding increases oxygen consumption and metabolic rate. The ventilator and lungs must work harder, possibly causing respiratory muscle fatigue or even respiratory failure.

How can a health care professional manage not to overfeed or underfeed such patients? It can be difficult to estimate the number of Calories that a patient on a mechanical respirator requires, especially without knowing how much energy that patient expends. Calorimetry provides a convenient method for determining a patient's energy expenditure. That information can then be used to calculate the proper number of Calories that the patient should consume per day.

Exceptionally plentiful Additional Exercises at the end of each chapter reinforce the concepts and skills presented in the text. Clearly labeled by section to help both students and instructors, these problems are also available in the textbook's accompanying online homework system. Answers to the odd-numbered exercises are available at the back of the book, and detailed solutions for all the exercises are available in the student solutions manual.

Chapter Descriptions

CHAPTER 1 Measurement, Atoms, and Elements

This chapter combines a primer on measurement with an introduction to atoms and the periodic table so that students can arrive quickly at more engaging material. Section 1.1 introduces the metric system, significant figures, unit conversions, and density measurements, using medically relevant examples such as ultrasounds, vitamin labels, and dosage calculations. The section introduces dimensional analysis as a way of solving problems involving calculations, a technique that is systematically used throughout the text. Section 1.2 presents the basic structure and properties of the atom, leading to a discussion of elements and isotopes. Section 1.3 teaches students how to navigate the periodic table. The final section describes the electron structure of atoms.

Integration of organic and biochemistry: A section on "Important Elements in Biochemistry and Medicine" classifies elements as building block elements, macro- and micronutrients, and radioisotopes used in medicine.

In the spirit of integration, certain topics often covered in a first chapter are moved to later chapters in order to (1) not overwhelm students with many introductory topics at once and (2) introduce those topics in a context in which they can be most effectively explained. These topics include (1) energy, heat, and temperature, which are introduced in Chapter 4 on states of matter and further described in Chapter 8 in discussing energy and chemical reactions; (2) states of matter, covered in Chapter 4; and (3) mixtures, covered in Chapter 5, in the section on solutions.

CHAPTER 2 Compounds

Chapter 2 shows how different atoms come together to make ionic and covalent compounds. Central to the chapter are discussions of how the electron structure of an atom determines the type of bond it forms. Students learn how to determine the formula unit for an ionic compound and how to interpret molecular formulas. They also learn how to write compounds as Lewis dot structures. In the final section, students learn how to calculate formula mass, molecular mass, and molar mass, and how to convert between grams and moles, using dimensional analysis.

Integration of organic and biochemistry: Organic compounds are used as examples to illustrate characteristics of compounds. Students learn to draw Lewis structures using organic compounds. A discussion of the effects of caffeine introduces students to receptors, and the *Chemistry in Medicine* box on blood chemistry introduces some biochemically important compounds such as urea and glucose.

The topics related to health care appear in sections on electrolytes, polyatomic atoms in health care (fluoride and tooth enamel, bleach), and interpreting molar quantities of important compounds in blood tests.

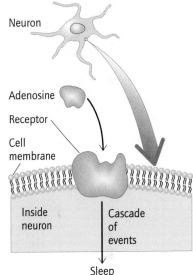

Figure 2-1 When adenosine binds to the adenosine receptor, it sets off a cascade of biochemical events in brain cells, leading to sleep.

CHAPTER 3 Shapes of Molecules and Their Interactions

This chapter connects molecular shape and intermolecular forces of attraction: molecular shape helps determine polarity, which in turn determines the types of intermolecular interactions a compound can participate in. Section 3.1 explains, using VSEPR, how electron geometry determines molecular shape, giving tetrahedral, trigonal planar, and linear electron geometries. Section 3.2 shows how molecular polarity is determined by the types of covalent bond present and the molecular shape. Section 3.3 introduces the basic types of intermolecular forces of attraction and shows how they determine certain physical properties of a molecule. Students learn how to interpret three-dimensional representations of molecules, including ball-and-stick models, space-filling models, and structural formulas.

Integration of organic and biochemistry: Examples are often simple organic molecules such as methane or simple nonorganic molecules important in biochemistry, such as H_2O, CO_2, and NH_3. A brief section introduces the role of hydrogen bonding in the structure of DNA.

CHAPTER 4 Solids, Liquids, and Gases

This chapter examines the factors that determine the physical state of a substance, and in particular introduces the central role of energy in determining that physical state. The concepts of energy, heat, and temperature are introduced in that context. Changes of state and the properties of each phase are explained in terms of the kinetic molecular theory. The second half of the chapter focuses on gases, including understanding pressure and vapor pressure and the gas laws.

Students learn why a steam burn causes more damage to skin than boiling water at the same temperature, how to interpret blood pressure readings, how gases in an anesthetic affect patient recovery time, and why hyperbaric oxygen is an effective treatment for the bends.

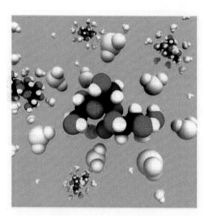

Figure 5-4 Sucrose, a molecule, dissolves in water because many water molecules surround each sucrose molecule and hydrogen bond with it.

CHAPTER 5 Solutions, Colloids, and Membranes

The first half of the chapter introduces the student to aqueous solutions, emphasizing their role in biology. The section on solution concentrations focuses on the units of concentration most often seen in a medical context. The student learns how to interpret the concentrations given in a blood test, and how to calculate concentrations and dosages of medicines. After an introduction to colloids and suspensions, the final section shows how osmosis, dialysis, and diffusion of gases are employed to control the movement of substances across a membrane.

Integration of organic and biochemistry: The chapter introduces cell membranes as semipermeable membranes and illustrates osmosis and dialysis using important molecules in biochemistry such as glucose and urea.

Topics of interest to health care majors include blood as a solution, colloid, and suspension; electrolytes; and drugs in solution or suspension.

CHAPTER 6 Hydrocarbons and Structure

This introduction to alkanes, alkenes, alkynes, and aromatic hydrocarbons explores the structure and physical properties of these compounds. Students learn how to recognize conformations, structural isomers, and geometric isomers. They also learn how to write compounds as condensed structural formulas and skeletal line structures. One section is devoted to the rules for naming hydrocarbons.

The simple examples that predominate in the chapter are supplemented with examples of compounds chosen from biology such as β-carotene, a neurotoxin, vanillin, and OTC analgesics with aromatic rings.

Integration of biochemistry: The section "cis–trans isomers in biochemistry" looks at an isomerization reaction in the chemistry of vision.

CHAPTER 7 Organic Functional Groups

This chapter introduces all the common organic functional groups, describing their structure, naming conventions, solubility properties, and ionic forms where appropriate. A section is devoted to derivatives of phosphoric acid because of their important role in biochemistry.

Integration of biochemistry: The chapter introduces several biochemically important structures, such as glucose, cholesterol, citric acid, triglycerides, and ATP. Also included are numerous structures of well-known medications such as aspirin, codeine, morphine, Prozac, and penicillin.

CHAPTER 8 Chemical Reaction Basics

Because organic functional groups have been introduced earlier, this chapter is able to teach chemical reactions using organic compounds of relevance to health care majors.

The chapter focuses on simple combustion reactions of organic compounds. The middle section of the chapter covers the basic principles of energy in chemical reactions, including exothermic and endothermic reactions, and calorimetry. The final section on reaction rates describes activation energy and factors that influence reaction rates.

Reaction stoichiometry is treated in an extension box to allow the flexibility of covering this material at the instructor's discretion.

Integration of organic and biochemistry: A brief, general overview of metabolism and energy demonstrates the relevance of chemical reactions to human health. A section on calorimetry shows students how to calculate the number of calories in the food they eat. Two paragraphs introduce enzymes as the catalysts of biochemical reactions, preparing students for Chapter 10.

Acetic acid

CHAPTER 9 Acids, Bases, pH, and Buffers

Acid–base reactions are the first type of reaction explored at length in the text. The first section presents the definitions and properties of acids and bases. Here the text introduces the concept of equilibrium and Le Châtelier's principle. Section 9.2 introduces the pH scale. Section 9.3 presents buffers. It includes a section on "buffers in the body" that looks at blood buffering and the consequences of acidosis and alkalosis.

Integration of organic and biochemistry: Because this chapter follows the chapter on organic functional groups, it is able to use organic molecules as examples: acetic acid, citric acid, and lactic acid as examples of weak acids; the ionization of pyruvate as an example of equilibrium and Le Châtelier's principle. A section on pH and the cell looks at the ubiquity of ionized forms in the body.

CHAPTER 10 Reactions of Organic Functional Groups in Biochemistry

This chapter introduces the student to several types of chemical reactions, focusing on types of reactions that are significant in biochemistry: oxidation–reduction reactions, hydration–dehydration, acyl group transfer reactions including hydrolysis, and phosphoryl group transfer reactions. Thus, by the time students see their first complete catabolic pathway in Chapter 12, they will be familiar with the kinds of reactions they see in the pathway. Simple examples are explored first; then more applied examples.

Integration of biochemistry: A number of the examples in the chapter are of key reactions in biochemistry, including the transfer of acetyl groups by acetyl CoA and the transfer of phosphoryl groups between ADP and ATP. The chapter introduces coenzymes $NAD^+/NADH$ and $FAD/FADH_2$ as key players in oxidation–reduction reactions in biochemical pathways. A brief section explains the role of ATP as the cell's "energy currency."

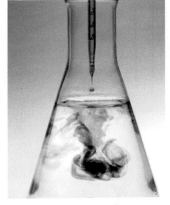

A chemical reaction taking place in an Erlenmeyer flask. Functional groups react in the same characteristic way whether the reaction is performed in a flask in the laboratory or in the aqueous medium of the cell. [© 1994 Paul Silverman, Fundamental Photographs, NYC]

CHAPTER 11 Proteins: Structure and Function

Chapter 11 introduces the structure of amino acids, how they join to make peptides by peptide bonds, and the four levels of protein structure. The concept of chirality and the role of enantiomers are introduced as applied to amino acids. A section introducing enzymes also describes competitive and noncompetitive inhibitors. The chapter emphasizes the relationship between protein structure and function and describes the importance of hydrophobic and hydrophilic interactions.

Integration of biochemistry: The opening vignette examines sickle-cell anemia, a disease caused by one wrong amino acid in the important protein hemoglobin.

CHAPTER 12 Carbohydrates: Structure and Function

The first half of the chapter covers the structures of the monosaccharides, dissaccharides, and complex carbohydrates, including chirality and diastereomers as applied to monosaccharides. A final section called "Oligosaccharides as cell markers" explains blood typing.

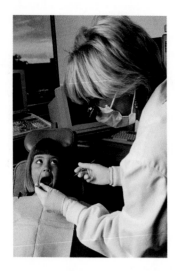

Cleanings by a dental hygienist help prevent tooth decay. [Scott T. Baxter/ Photodisc/Getty Images]

Integration of biochemistry: The chapter begins with a brief overview of the role of carbohydates in catabolism, then returns to biochemistry in the second half of the chapter with a section on the metabolism of carbohydrates. The section is in two parts: (1) the hydrolysis of carbohydrates as part of digestion and (2) glycolysis and fermentation. A brief section introduces the functions of four other glucose-processing pathways, without giving any individual steps: gluconeogenesis, the pentose-phosphate pathway, formation (glycogenesis) and degradation (glycogenolysis) of starch.

The text groups metabolic pathways with structure, in an attempt to avoid overwhelming students by presenting all of the biochemical pathways in one chapter.

CHAPTER 13 Lipids: Structure and Function

The first section sets out the structures of fatty acids and triglycerides, focusing on how double bonds determine the melting point of a triglyceride and thus its physical properties at room temperature. A second section presents the structures of the phospholipids and glycolipids, then goes on to explain how structure leads to the formation of cell membranes. The final section covers cholesterol and the steroid hormones, including bile acids and Vitamin D.

Integration of biochemistry: The third section focuses on the catabolism of lipids, their transport from the small intestine to fat and muscle cells, and the β-oxidation pathway in muscle cells. A section on lipoproteins introduces these particles and their roles.

CHAPTER 14 Metabolism and Bioenergetics

This chapter focuses on the third stage of catabolism, connecting it to the first two stages introduced for each of the biomolecules in the previous two chapters. The first half of the chapter describes the role of the citric acid cycle, the electron-transport chain, and oxidative phosphorylation to complete the story of catabolism begun in earlier chapters. This section ends with a "Summary of metabolism" that draws together all the biochemical pathways seen in this and the preceding chapters.

A final section on entropy and free energy, including coupled reactions, is for instructors who wish to include more advanced coverage of bioenergetics.

CHAPTER 15 Nucleic Acids: DNA and RNA

This succinct introduction to nucleic acids covers the structure of nucleotides and DNA and RNA, and the processes of DNA replication, transcription, and translation. Mutations are linked to their role in inherited diseases. The chapter ends with the role of nucleotides in the life cycle of the HIV virus.

CHAPTER 16 Nuclear Chemistry and Medicine

This chapter has been written as a stand-alone chapter—one that can be covered anytime in the course after Chapter 1. The first half of the chapter is a basic introduction to radioisotopes, balancing nuclear equations, and the types of radiation. The second half looks at the biological effects of nuclear radiation. The effects of ionizing radiation are analyzed in terms of its energy and penetrating power. The student also learns several units for measuring radiation, and sees how the amount of radiation can be used to predict the effects of radiation sickness. The chapter also covers medical uses of radiation in both treatment and diagnostic medicine, with particular attention to medical imaging technology.

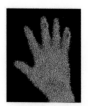
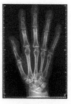

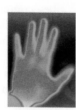

| Gamma ray image | X-ray image | Visible light image | Infrared image |

High energy ⟵——————⟶ Low energy

Student and Instructor Support

We believe a student needs to interact with a concept several times in a variety of scenarios in order to obtain a practical understanding. With that in mind, W. H. Freeman has developed the most comprehensive student learning package available.

Supplemental Materials
Instructor Ancillary Support

Test Bank by Rachel Jameton, Lewis-Clark State College
Printed ISBN: 1-4292-3467-9
CD-ROM ISBN: 1-4292-3133-5

The test bank offers over 1600 multiple-choice questions, and is designed to assess student knowledge at all levels of learning, from basic definitions to application and synthesis. Utilizing diagrams, figures, and structures, the test bank emphasizes visual understanding of an array of concepts, including applications to the health sciences. While the text bank is also available as a printed manual, the easy-to-use CD-ROM includes Windows and Macintosh versions of the widely used test generation software, which allows instructors to add, edit, and resequence questions to suit their testing needs.

Electronic Instructor Resources

Instructors can access valuable teaching tools through www.whfreeman.com/gob. These password-protected resources are designed to enhance lecture presentations, and include Textbook Images (available in .jpeg and PowerPoint format), the Online Quiz Gradebook, Clicker Questions, Lab Information, and more. There's even a set of Enhanced Lecture PowerPoints, which can contain complete class lecture content. Highlighting key chapter ideas and including worked examples, pre-class questions, and textbook figures and diagrams, these PowerPoints contain around 60 slides per chapter.

WebAssign Premium

For instructors interested in online homework management, W. H. Freeman and WebAssign have partnered to deliver WebAssign Premium—a comprehensive and flexible suite of resources for your course. Combining the most widely used online homework platform with a wealth of visualization and tutorial resources, WebAssign Premium extends and enhances the classroom experience for instructors and students by combining algorithmically generated versions of selected end-of-chapter questions with a fully interactive eBook at an affordable price. See below for more details, or visit www.whfreeman.com/gob to sign up for a faculty demo account.

For the Laboratory

Lab Manual by Sara Selfe, Edmonds Community College
ISBN: 1-4292-2433-9

The lab manual provides a wide variety of classic and innovative experiments covering the basic topics of general, organic, and biochemistry. These experiments emphasize biological implications of chemical concepts, often within the context of the health science field. Each experiment can easily be completed within a three-hour time frame, and is accompanied by data sheets and questions that guide students through the analysis of their data.

LabPartner Chemistry

W. H. Freeman's latest offering in custom lab manuals provides instructors with a diverse and extensive database of experiments published by W. H. Freeman and Hayden-McNeil Publishing—all in an easy-to-use, searchable online system. With the click of a button, instructors can choose from a variety of traditional and inquiry-based labs. LabPartner Chemistry sorts labs in a number of ways, from topic, title, and author, to page count, estimated completion time, and prerequisite knowledge level. Add content on lab techniques and safety, reorder the labs to fit your syllabus, and include your original experiments with ease. Wrap it all up in an array of bindings, formats, and designs. It's the next step in custom lab publishing—the perfect partner for your course.

Student Resources
Student Study Guide and Solutions Manual

by Rachel C. Lum, University of Colorado
ISBN: 1-4292-2432-0

The Student Study Guide and Solutions Manual provides students with a combined manual designed to help them avoid common mistakes and understand key concepts. After a brief review of each section's critical ideas, students are taken through stepped-out worked examples, try-it-yourself examples, and chapter quizzes, all structured to reinforce chapter objectives and build problem-solving techniques. The solutions manual includes detailed solutions to all odd-numbered exercises in the text.

Molecular Model Kit

ISBN: 1-4292-2687-0

Molecular modeling helps students understand the physical and chemical properties of molecules by providing a way to visualize the three-dimensional arrangement of atoms. This model set, created specifically for *Essentials of General, Organic, and Biochemistry*, uses polyhedra to represent atoms and plastic connectors to represent bonds (scaled to correct bond length).

Premium Multimedia Resources

Problem-Solving Tutorials reinforce the concepts presented in the classroom, providing students with self-paced explanation and practice. Each tutorial relates to a worked example from the text, and consists of the following four components:

- Conceptual Explanation: An easy-to-follow, thorough explanation of the topic.
- Worked Example: A step-by-step walkthrough of the problem-solving technique.
- Try It Yourself: An interactive version of the Worked Example that prompts students to complete the problem and supplies answer-specific feedback.
- Practice Problems: A series of problems designed to check understanding.

A key icon next to a Worked Exercise in the text indicates that students can find a Problem-Solving Tutorial related to the exercise at www.whfreeman.com/gob.

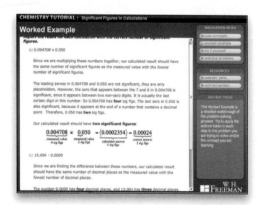

ChemCast Videos replicate the face-to-face experience of watching an instructor work a problem. Using a virtual whiteboard, the ChemCast tutors show students the steps involved in solving key worked examples, while explaining the concepts along the way. The worked examples were chosen with the input of general, organic, and biochemistry chemistry students. ChemCasts can be viewed online or downloaded to a portable media device, such as an iPod.

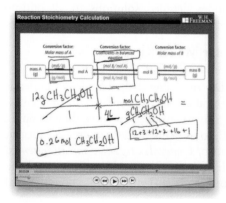

The Virtual Model Kit allows students to build molecular structures on the computer screen. Students select elements and bonds according to instructions in the text's Model Tool activities, as well as from additional exercises available within the associated Web site and homework system. After constructing the models, students are presented with automatically graded questions to probe their experience.

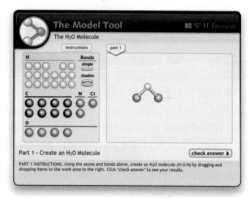

The multimedia-enhanced eBook contains the complete text with a wealth of helpful functions. All student multimedia, including the Problem-Solving Tutorials, ChemCasts, and Virtual Model Kit activities, are linked directly from the eBook pages. Students are thus able to access supporting resources when they need them—taking advantage of the "teachable moment" as students read. Customization functions include instructor and student notes, document linking, and editing capabilities.

Online Learning Environments

The above resources are available in two platforms. WebAssign Premium offers the most effective and widely used online homework system in the sciences, and is designed specifically for those instructors seeking graded homework management. The Student Companion Web site provides the student-oriented support materials independent of any homework system.

WebAssign Premium

For instructors interested in online homework management, WebAssign Premium features a time-tested, secure online environment already used by millions of students worldwide. Featuring algorithmic problem generation and supported by a wealth of chemistry-specific learning tools, WebAssign Premium for *Essentials of General, Organic, and Biochemistry* presents instructors with a powerful assignment manager and study environment. WebAssign Premium provides the following resources:

- Algorithmically generated problems: Students receive homework problems containing unique values for computation, encouraging them to work out the problems on their own.

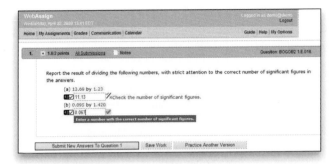

- Complete access to the interactive eBook is available from a live table of contents, as well as from relevant problem statements.
- Links to Problem-Solving Tutorials, ChemCasts, and other interactive tools are provided as hints and feedback to ensure a clearer understanding of the problems and the concepts they reinforce. (Tutorials and ChemCasts are described above.)
- Graded molecular drawing problems allow instructors to evaluate student understanding of molecular structure. The system evaluates virtually "drawn" molecular structures, returning a grade as well as helpful feedback for common errors.

- Labeling problems, reflecting a type of exercise found in the textbook, ask students to correctly identify functional groups and other aspects of molecular structures to ensure an understanding of molecular structure and characteristics.

Student Companion Web Site

The *Essentials of General, Organic, and Biochemistry* Book Companion Web site, accessed through www.whfreeman.com/gob, provides a range of tools for problem solving and chemical explorations. They include:

- Student self-quizzes
- Animations that allow students to visualize chemical events on a molecular level
- An interactive periodic table of the elements
- An English/Spanish Glossary
- Pre-drawn molecules for use in homework problems
- To access additional support, including the Problem-Solving Tutorials, ChemCasts, and Virtual Model Kit, students can upgrade their access through a direct subscription to the Premium component of the Web site.

Acknowledgments

We are grateful to the more than 100 instructors who throughout the development of this text have contributed their expertise and experience in providing thoughtful reviews of individual chapters. We are especially grateful to Joe Burnell of the University of Indianapolis and Valerie Keller of the University of Chicago, who reviewed the entire final manuscript for accuracy, and James Pazun of Pfeiffer University, who checked the first five chapters. Brian Shuch, M.D., of the University of California at Los Angeles School of Medicine, verified the accuracy of the medical information in the vignettes that open each chapter and the Chemistry in Medicine boxes, and Richard Burns, M.D, performed the same task for Chapter 16 on nuclear medicine. Valerie Keller, Colleen Craig of Seattle Central Community College, and Jason Ward of Norwell High School checked the solutions to all the exercises. Peter Krieger of Palm Beach Community College offered valuable suggestions for improving the figures. A special debt of gratitude goes to the instructors listed below who shared our vision for the book and agreed to class test the manuscript prior to publication. Together with the comments and support offered by instructors at focus groups, this feedback has been instrumental in creating the text you see here before you.

Class Testers
Mamta Agarwal, Chaffey College
George Bandik, University of Pittsburgh
Lois Bartsch, Graceland University
Martin Brock, Eastern Kentucky University
Stephen Dunham, Moravian College
Michelle Hatley, Sandhills Community College
Andrea Martin, Widener University
Phil McBride, Eastern Arizona University
Edmond O'Connell, Fairfield University
Anuhadha Pattanayak, Skyline College
Matthew Saderholm, Berea College

Jeffrey Sigman, Saint Mary's College
Tara Sirvent, Vanguard University
Carnetta Skipworth, Bowling Green Community College of Western Kentucky University
Lorraine Stetzel, Lehigh Carbon Community College
Christy Wheeler, College of Saint Catherine

Focus Group Participants
Loyd Bastin, Widener University
Scott Carr, Anderson University
Rosemarie Chinni, Alvernia College
Ana Ciereszko, Miami-Dade College

Mian Jiang, University of Downtown Houston
Booker Juma, Fayetteville State University
Annie Lee, Rockhurst University
Carol Libby, Moravian College
Samar Makhlouf, Lewis University
Andrea Martin, Widener University
Bryan May, Central Carolina Technical College
Elizabeth Pulliam, Tallahassee Community College
Rita Rhodes, University of Tulsa
Trineshia Sellars, Palm Beach Community College

Carnetta Skipworth, Bowling Green Community College of Western Kentucky University

Lee Ann Smith, Western Kentucky University

Reviewers

Ronald T. Amel, Viterbo University

Laura Anna, Millersville

Tasneem Ashraf, Cochise College

Theodore C. Baldwin, Olympic College

George Bandik, University of Pittsburgh

Thomas Barnard, Cardinal Stritch University

Bal Barot, Lake Michigan College

Lois M. Bartsch, Graceland University

Nick Benfaremo, St. Joseph's College of Maine

Jerry Bergman, Northwest State Community College

Chirag Bhagat, William Rainey College

John Blaha, Columbus State Community College

Carol E. Bonham, Pratt Community College

Martin Brock, Eastern Kentucky University

Diane M. Bunce, The Catholic University of America

Joe C. Burnell, University of Indianapolis

N. J. Calvanico, Lehigh Carbon Community College

David Canoy, Chemeketa Community College

Scott R. Carr, Anderson University

Stephen Cartier, Warren Wilson College

Lynne C. Cary, Bethel College

Amber Flynn Charlebois, Fairleigh Dickinson University

Ana A. Ciereszko, Miami-Dade College

Joana Ciurash, College of the Desert

Stuart C. Cohen, Horry-Georgetown Technical College

Jeannie T. B. Collins, University of Southern Indiana

Felicia Corsaro-Barbieri, Gwynedd-Mercy College

Brian Cox, Cochise College-Sierra Vista Campus

Milagros Delgado, Florida International University

Anthony B. Dribben, Tallahassee Community College

Stephen U. Dunham, Moravian College

Robert G. Dyer, Arkansas State University, Mountain Home

Eric Elisabeth, Johnson County Community College

Anne Felder, Centenary College

Francisco Fernandez, City University of New York, Hostos Community College

K. Thomas Finley, The College at Brockport

Hao Fong, South Dakota School of Mines and Technology

Karen Frindell, Santa Rosa Junior College

Priscilla J. Gannicott, Lynchburg College

Andreas Gebauer, California State University, Bakersfield

Louis A. Giacinti, Milwaukee Area Technical College

Eric Goll, Brookdale Community College

Nalin Goonesekere, University of Northern Iowa

Maralea Gourley, Henderson State University

Ernest Grisdale, Lord Fairfax Community College

Mehdi H. Hajiyani, University of the District of Columbia

Melanie Harvey, Johnson County Community College

Michelle L. Hatley, Sandhills Community College

Michael A. Hauser, St. Louis Community College–Meramec

John W. Havrilla, University of Pittsburgh–Johnstown

Jonathan Heath, Horry-Georgetown Technical College

Sherry Heidary, Union County College

Sara Hein, Winona State University

Steven R. Higgins, Wright State University

Jason A. Holland, University of Central Missouri

Byron Howell, Tyler Junior College

Michael O. Hurst, Georgia Southern University

T. G. Jackson, University of South Alabama

Rachel A. Jameton, Lewis-Clark State College

Mike Jezercak, University of Central Oklahoma

Matthew Johnston, Lewis-Clark State College

Booker Juma, Fayetteville State University

Cathie Keenan, Chaffey College

Mushtaq Khan, Union County College

Edward A. Kremer, Kansas City Kansas Community College

Bette A. Kreuz, The University of Michigan-Dearborn

Peter J. Krieger, Palm Beach Community College

Lida Latifzadeh Masoudipour, El Camino College

Annie Lee, Rockhurst University

Scott Luaders, Quincy University

Julie Lukesh, University of Wisconsin, Green Bay

Riham Mahfouz, Thomas Nelson Community College

Samar Makhlouf, Lewis University

Karen Marshall, Bridgewater College

Andrea Martin, Widener University

Christopher Massone, Molloy College

Bryan May, Central Carolina Technical College

Phil McBride, Eastern Arizona College

Ann H. McDonald, Concordia University Wisconsin

Patrick McKay, San Mateo County Community College

S. Ann Melber, Molloy College

Stephen Milczanowski, Florida Community College at Jacksonville

Michael N. Mimnaugh, Chicago State University

Luis D. Montes, University of Central Oklahoma

Peter P. Mullen, Florida Community College

Michael P. Myers, California State University, Long Beach

Grace M. Ndip, Shaw University

Nelson Nunez-Rodriguez, City University of New York, Hostos Community College

E. J. O'Connell, Fairfield University

Janice J. O'Donnell, Henderson State University

Elijah Okegbile, Pima Community College

C. Edward Osborne, Northeast State Community College

John D. Patton, Southwest Baptist University

Lynda R Peebles, Texas Woman's University

Paul Popieniek, Sullivan County Community College of SUNY

Jerry Poteat, Georgia Perimeter College

Ramin Radfar, Wofford College

Christina Ragain, University of Texas, Tyler

S. Ramaswamy, University of Iowa

Rita T. Rhodes, The University of Tulsa

Rosalie Richards, Georgia College and State University

Shashi Rishi, Greenville Technical College

Ghassan Saed, Oakland University

Steve P. Samuel, SUNY College at Old Westbury

Karen Sanchez, Florida Community College at Jacksonville
Shaun E. Schmidt, Washburn University
William Seagroves, New Hampshire Technical Institute
Sara Selfe, Edmonds Community College
Trineshia N. Sellars, Palm Beach Community College
Paul Seybold, Wright State University
Sonja Siewert, West Shore Community College
Jeffrey A. Sigman, Saint Mary's College of California
Nancy C. Simet, University of Northern Iowa

John W. Singer, Jackson Community College
Joseph F. Sinski, Bellarmine University
Carnetta Skipworth, Western Kentucky University
Robert Smith, Metropolitan Community College
Koni Stone, California State University, Stanislaus
Mary W. Stroud, Xavier University
K. Summerlin, Troy University, Montgomery Campus
Erach R. Talaty, Wichita State University
Eric R. Taylor, University of Louisiana–Lafayette

Jason R. Taylor, Roberts Wesleyan College
Rod Tracey, College of the Desert
Robert C. Vallari, St. Anselm University
Sarah Villa, University of California, Los Angeles
Maria Vogt, Bloomfield College
Linda Waldman, Cerritos College
Karen T. Welch, Georgia Southern University
Christy Wheeler West, College of St. Catherine
Ryan S. Winburn, Minot State University
Corbin Zea, Grand View College

This textbook benefits from the extensive effort of our supplements authors, to whom we extend our most heartfelt gratitude. Among them is an experienced laboratory text author, Sara Selfe of Edmonds Community College, who has created an impressive array of laboratory experiments uniquely designed to accompany this textbook. Rachel Lum of the University of Colorado prepared the solutions guide that accompanies the more than 1500 questions that appear at the end of each chapter. Martin Brock of Eastern Kentucky University, who shared our vision for the textbook early on, has created the comprehensive PowerPoint slides for instructors.

An exceptionally rich set of online resources accompanies the text. We wish to thank the following authors for their work creating these resources. Jonathan Bergmann and Aaron Sams, Woodland Park High School, Colorado, created the ChemCast problem-solving videos. We also thank the students from Widener University and Sandhills Community College for suggesting the topics to cover in video form. James Pazun, Pfeiffer University, created the online quizzes. Simon Bott, University of Houston, and Colleen Craig, Seattle Central Community College, wrote the Problem-Solving Tutorials, and Luis Montes, University of Central Oklahoma, and Jason Ward, Norwell High School, reviewed them.

This textbook would not have been possible were it not for the exceptional dedication and pool of talent provided by the editorial team at W. H. Freeman. Our deepest gratitude goes to Susan Moran, senior developmental editor at W. H. Freeman, whose attention to detail and considerable editorial talents were instrumental in achieving the clarity and readability of this text. Her enthusiasm and encouragement served as a beacon of light for the authors throughout the process. We are appreciative of the considerable skills brought to the project by publisher Clancy Marshall and acquisitions editor Anthony Palmiotto, who deftly managed the many facets of a modern textbook: ancillaries, media package, and all the associated activities that must be in place for a book to come to fruition. We thank the editorial staff at W. H. Freeman for sharing our vision for this text and Clancy Marshall for believing in our idea when first approached. We extend our appreciation to the many essential production staff members including project editor Jane O'Neill, who carefully shepherded the book through the proof stages; Margaret Comaskey, the copyeditor; Diana Blume, who created the elegant design; Bill Page, who guided the production of an entirely new illustration program; Paul Rohloff, who arranged the typesetting and printing; Donna Ranieri, who found all the photographs, and Bianca Moscatelli, photo editor.

Kirsten Watrud, market development manager, and her assistant Kerri Knipper made sure that we obtained valuable feedback from instructors from the very beginning of the project through surveys, class tests, and focus groups. John Britch, marketing director, has guided an impressive effort to bring this text to the attention of instructors all over the country.

We owe special thanks to Kathryn Treadway for ably guiding the development of the print supplements and to David Quinn for his excellent work assembling the impressive set of resources available on our book's Web site. In addition, Mark Santee managed the WebAssign integration.

The pedagogy of the book is greatly enhanced by the artwork produced by Network Graphics. This text is further enhanced and unique in the number and quality of authentic protein structures, which have all been rendered by Gregory Williams, who also produced many of the other molecular models, including the electron density models.

Finally, we would like to express our appreciation to our colleagues at various academic institutions who supported our efforts throughout this project with their encouragement and moral support. They are Steve Martin (The University of Texas at Austin), Pat Ladewig (Regis University), Paul Ewald (Regis University), Steve Doty (Regis University), Stephen Cartier (Warren Wilson College), Madeline Mignone (Dominican College), as well as our colleagues at The College of New Rochelle: Elvira Longordo, Tom Venanzi, Melanie Harasym, Lynn Petrullo, Faith Kostel-Hughes, and Dick Thompson.

Measurement, Atoms, and Elements

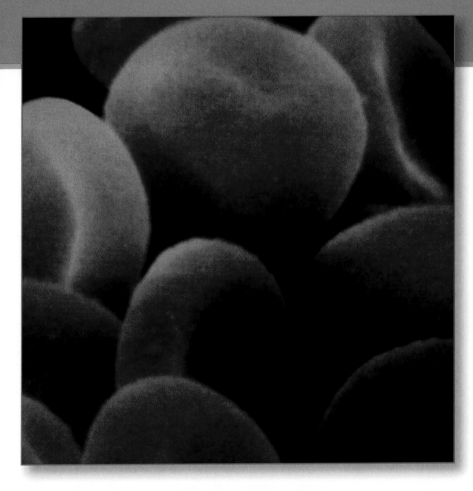

Scanning electron micrograph (SEM) of human red blood cells. Red blood cells carry oxygen to tissues throughout the body. The red pigment is due to the iron atoms in hemoglobin, which binds to oxygen. [Kenneth Eward/Photo Researchers, Inc.]

OUTLINE

This icon indicates that a **Problem-Solving Tutorial** is available at www.whfreeman.com/gob

What Happens When You Don't Get Enough Iron?

An iron deficiency is the most common nutritional deficiency in the world. When a person's iron reserves become depleted, a condition known as iron deficiency anemia (IDA) develops. **Anemia** is a general term for any condition in which there is a deficiency of red blood cells. People with IDA experience fatigue as well as some of the other symptoms listed in Table 1-1. IDA can also slow motor and mental development in children.

Table 1-1 Symptoms of Iron Deficiency Anemia
Inflammation or soreness of the tongue (glossitis)
Brittle nails
Frequent headaches
Fatigue and weakness
Lethargy
Shortness of breath
Pale skin
Taste disturbances
Dizziness

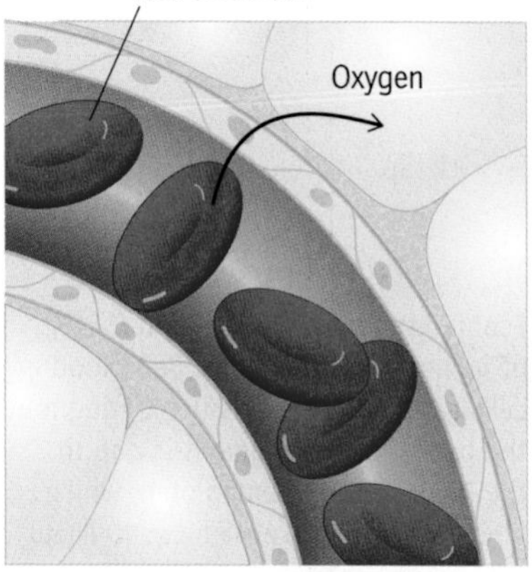

Every cell in the body requires iron, especially red blood cells. Red blood cells serve the important function of transporting oxygen to all parts of the body. The characteristic biconcave shape of a red blood cell exists to increase its surface area, thereby maximizing oxygen absorption.

Red blood cells constitute 40% of the volume of the blood. Each red blood cell contains about 250 million hemoglobin molecules, the oxygen-carrying substance of these cells. A molecule is composed of two or more atoms—the simplest components of all matter. In this chapter you will learn about atoms. Then, in Chapter 2 you will learn how atoms come together to form molecules. Hemoglobin is a large, complex molecule composed of 10,000 atoms. Four of these are iron atoms. In fact, iron is what makes your blood red in color.

Oxygen is continuously being breathed into the lungs from the air. Oxygen is a molecule, too, although a much smaller one than hemoglobin, containing only two atoms. Each of the four iron atoms in a hemoglobin molecule will bind to one oxygen molecule and carry it through the bloodstream to tissues that require oxygen, as illustrated in Figure 1-1. Every cell needs oxygen in order to produce energy.

When the body's supply of iron is depleted, as in IDA, hemoglobin can no longer be produced by the body. Consequently, fewer red blood cells are made, and tissues start to become starved of oxygen. Not surprisingly, fatigue is one of the symptoms of anemia. IDA is caused by poor absorption of iron, lack of iron in the diet, chronic blood loss, or pregnancy. Some medications can also reduce the body's ability to absorb iron.

The binding of oxygen to iron in your red blood cells is an example of how atoms and molecules interact—chemistry. The atomic world of chemistry is central to our understanding of health and disease. Chemistry also provides tools for treating disease. You will repeatedly see the central role chemistry plays in our understanding of health and disease in the coming chapters, as you build your knowledge of chemistry.

Figure 1-1 Red blood cells travel through the circulatory system delivering oxygen to cells.

Matter is anything that has mass and occupies volume. What is not matter?
• Energy
• Light
• An idea

Chemistry is the study of matter and changes in matter. ***Matter** is defined as anything that has mass and occupies volume.* Therefore, matter is all the "stuff" around you and in you. It includes matter that you can see as well as matter that is too small for you to see. In this chapter you will learn about the fundamental component of all matter, the atom. Then, in Chapter 2 you will see how different atoms come together to make compounds, substances composed of more than one type of atom. To understand matter, however, you must first be able to measure it. Hence, the chapter begins with measurement, an essential tool in all disciplines of science.

1.1 | Matter and Measurement

You can see a drop of blood on the head of a pin with the naked eye (Figure 1-2). If you look at this droplet of blood under a microscope, you will see that it is actually composed of millions of red blood cells, each with a diameter 1,000 times smaller than the width of the droplet of blood. Now imagine—because you can't see it with a microscope—peering inside one of these red blood cells. Among other things, you will "see" millions of hemoglobin molecules, a type of compound composed of many different atoms. Hemoglobin is the substance that carries oxygen from the lungs to the tissues throughout the body. Hemoglobin has a diameter 1,000 times smaller than a red blood cell and a million times smaller than the droplet of blood on the head of a pin. Nonetheless, a single hemoglobin molecule is composed of about 10,000 atoms. Zoom in another 10-fold and you will "see" one of four iron atoms in a hemoglobin molecule.

As you can see from Figure 1-2, matter can be described on different scales. The **macroscopic scale** includes all the material you can see, like the drop of blood. The **microscopic scale** includes matter such as red blood cells that cannot be seen by the naked eye, but can be seen with magnification through a microscope. The **nanoscale** and atomic scale describes matter such as a hemoglobin molecule or an atom that is too small to be seen even with a microscope. Modern imaging devices such as the scanning tunneling microscope (STM) have allowed scientists to create an image of the atom, such as the STM image of individual nickel atoms shown in the margin.

Chemistry helps us to understand the behavior of matter on the nanoscale, so that we can better explain the properties of matter that we observe on the macroscopic scale. Throughout this text you will see that an understanding of the *nanoscale* brings with it a better understanding of the *macroscale*—the things you can see—especially as it applies to health and disease. You will learn that many disease processes are caused by some malfunction at the atomic level.

Chemistry is an experimental science and as such requires that we be able to measure matter and changes in matter. Thus, we begin with measurement, one of the most important and fundamental tools of chemistry.

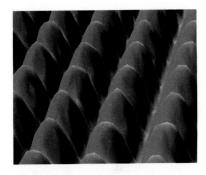

STM image of nickel atoms. A scanning tunneling microscope (STM) is a specialized instrument, very different from an optical microscope, that produces an image of an atom. [Don Eigler]

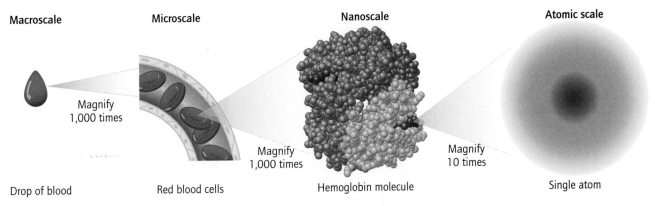

Macroscale **Microscale** **Nanoscale** **Atomic scale**

Magnify 1,000 times Magnify 1,000 times Magnify 10 times

Drop of blood Red blood cells Hemoglobin molecule Single atom

Figure 1-2 Matter at different scales ranging from the macroscale (a droplet of blood) to the microscale (red blood cells) to the nanoscale (a hemoglobin molecule) to the atomic scale (an iron atom).

Units and the Metric System

Medical professionals make measurements every day, whether obtaining a patient's weight, taking a patient's temperature or blood pressure, or administering a particular dose of medication. *Every measurement consists of both a number and a unit.* For example, a baby might weigh 10 lb, not just 10. Every measurement also has a margin of error associated with it, which can be conveyed in the number of digits reported in the number. Thus, the baby may weigh 10 lb, 10.0 lb, or 10.00 lb, depending on the balance used to weigh the baby.

Two systems of units that you will encounter in medicine are the English system and the metric or SI system. The metric system is the most widely used system of measurement in the world and it is the system of choice in the sciences. The metric system is based on a set of base units that measure a physical quantity, such as

- the meter (m) for length (about an arm's length),
- the gram (g) for mass (about the mass of a stick of gum),
- the second (s) for time (1/60 of a minute)

Other metric units are derived from the base units, such as

- the liter (L) for volume (about the volume of a bottled soda)
- density (*d*), the mass per volume of a substance

Prefixes in the Metric System A meter is an excellent unit of length for measuring a room but too short to measure a long distance conveniently and too long to measure a very short length. For measurements on different scales, the metric system creates larger or smaller units by adding a prefix before the base unit. Metric prefixes represent multiples of 10 of the base units. For example, the prefix "kilo" represents 1,000, so a **kilometer** is the equivalent of 1,000 meters and a **kilogram** is the equivalent of 1,000 grams. Metric prefixes are also created by dividing the base unit by a multiple of ten. For example, a **millimeter** is one-thousandth (1/1,000) of a meter. A micrometer is 1/1,000,000 of a meter. A given prefix can be applied to *any* base metric unit, and it will always refer to the same multiple. Thus, milli always represents 1/1,000 of the base unit, regardless of the base unit, as shown in the list below and in Table 1-2.

millimeter	mm	1/1,000 meter	a metric length
milligram	mg	1/1,000 gram	a metric mass
milliliter	mL	1/1,000 liter	a metric volume
millisecond	ms	1/1,000 second	a metric time interval

The international system of units **(SI)** was established by an international group of scientists for the purpose of setting a uniform set of units in the sciences. It is the predominant unit of science and commerce.

Table 1-2 Common Metric Prefixes

Scale	Prefix	Symbol	Factor	Factor in Scientific Notation
Macroscale	giga	G	1 000 000 000	10^9
	mega	M	1 000 000	10^6
	kilo	k	1 000	10^3
	Base unit		1	1
	deci	d	0.1	10^{-1}
	centi	c	0.01	10^{-2}
	milli	m	0.001	10^{-3}
Microscale	micro	μ	0.000 001	10^{-6}
Nanoscale	nano	n	0.000 000 001	10^{-9}
Atomic	pico	p	0.000 000 000 001	10^{-12}
Subatomic	femto	f	0.000 000 000 000 001	10^{-15}

The metric system is easy to use because you need only memorize the prefix symbol (column 3) and the associated multiple of ten (columns 4 and 5) that each prefix represents, as shown in Table 1-2.

The abbreviation for 1×10^x is simply 10^x. For more detailed guidelines on how to use **scientific notation,** consult the Math Appendix.

Length The meter is the metric base unit of length. Figure 1-3 shows that a decimeter, dm, is one-tenth (0.1) of a meter; a centimeter, cm, is one-hundredth (0.01) of a meter; and a millimeter, mm, is one-thousandth (0.001) of a meter. From Table 1-2 you can see the prefixes then start to go down in multiples of 1,000. A micrometer, μm, is one-millionth (0.000 001) of a meter; a nanometer, nm, is one-billionth (0.000 000 001) of meter; a picometer, pm, is one-trillionth of a meter, and so forth. Notice these prefixes all represent measurements smaller than the base unit. Table 1-2 also shows larger multiples of the base unit, represented by the prefixes kilo, mega, and giga; however, the kilometer is the only commonly used multiple of the meter for distance measurements.

Table 1-2 also shows that if you consider 1 mm the smallest length you can reasonably see with the naked eye, then every 1,000-fold decrease takes you into the range of another scale: macroscale (>mm) → microscale (μm) → nanoscale (nm) → atomic scale (pm). Figure 1-4 shows the range of metric lengths ranging from a woman measuring 1.65 m tall (the macroscale) to an iron atom with a diameter of 126 pm (the atomic scale).

Consider a technician performing an ultrasound scan of a developing fetus. He or she routinely measures the size of various parts of the fetus, including the crown–rump length, the biparietal diameter (distance between sides of the head), the femur length (thigh bone), and the abdominal circumference. These measurements of length provide assessments of gestational age, size, and growth of the fetus. For example, the biparietal diameter in a healthy fetus increases from about 2.4 cm at 13 weeks to about 9.5 cm at term. Structural abnormalities of the fetus such as spina bifida and cleft palate are reliably diagnosed using ultrasound measurements taken before 20 weeks. At some point in your career you will surely need to take measurements using the metric system.

Mass Mass is a measure of the amount of matter. For all practical purposes mass and weight are the same, although by definition the weight of an object depends on gravity. For example, the weight of an object will be different on the earth and on the moon, although its mass is the same. Mass is measured with a balance. A bathroom scale, for example, is a simple type of balance.

The base metric unit of mass is the gram (g). If you look at the label for any multivitamin, such as the one illustrated in the margin, you will see that the mass of each vitamin and mineral in the tablet is printed on the label. You may recognize several metric prefixes on the label, because the amount of

Scientific notation was developed to express extremely large or extremely small numbers in a convenient way that avoids having to write many zeros. For example, the number 3,000,000 can be expressed in scientific notation as 3×10^6 (3 million) The superscript, 6, following the "10" indicates that the decimal point is to be moved 6 places to the right, adding six zeros after the 3. A negative superscript indicates that the decimal point is to be moved to the left: $3 \times 10^{-6} = 0.000003$.

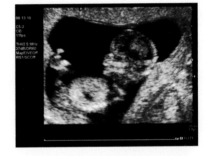

An ultrasound scan of a human fetus at 12 weeks shows the head (center right), abdomen (lower left), and legs (center left). At this stage in its development the fetus is about 8 centimeters long and weighs about 18 grams. Ultrasound scans are a useful way to monitor the health and growth of the fetus. [Gusto/ Photo Researchers, Inc.]

```
                                                  → 1 decimeter
|←————————————————————————————————————————→|    (1/10 meter)
|ꞁꞁꞁꞁꞁꞁꞁꞁꞁꞁꞁꞁꞁꞁꞁꞁꞁꞁꞁꞁꞁꞁꞁꞁꞁꞁꞁꞁꞁꞁꞁꞁꞁꞁꞁꞁꞁꞁꞁꞁꞁꞁꞁ|
   1     2     3     4     5     6     7     8     9    10

|←——→| 1 centimeter
|ꞁꞁꞁꞁꞁꞁ|   (1/100 meter)

→||← 1 millimeter
|ꞁ|    (1/1000 meter)
```

Figure 1-3 Some common measurements of length smaller than the meter: the decimeter (dm), centimeter (cm), and millimeter (mm), all shown at actual size.

Supplement Facts		
Serving Size: 2 Tablets		
Servings Per Container: 120		

(nutritional label table — Amount Per Serving / %DV)

A nutritional label for a multivitamin shows the mass of each vitamin present in 2 tablets. [Bianca Moscatelli]

each vitamin is significantly less than the base unit, the gram. For example, the prefix "m" in mg represents 1/1,000 (10^{-3}) of a gram, so 120 mg of vitamin C corresponds to 0.12 g of vitamin C. The microgram is represented by the Greek letter mu, "μ," placed before the base unit. Thus, 1 μg stands for 1/1,000,000 (10^{-6}) of a gram.

The microgram is also abbreviated mcg to avoid some of the confusion caused by the Greek letter μ. Thus, 80 mcg of vitamin K, shown on the label, corresponds to 0.000 080 g of vitamin K.

$$1\ \mu g = 1\ mcg$$

Volume Volume is a measure of how much three-dimensional space a substance, especially a gas or liquid, occupies. For example, 1 L of gasoline and 10 mL of cough syrup represent volume measurements. Volume units are derived from units of length. For example, the volume of a box can be obtained by multiplying the width, length, and depth of the box. Thus, a cube measuring 1 cm along each side has a volume of 1 cm × 1 cm × 1 cm = 1 cm^3, as shown in Figure 1-5. The volume of the box is written as 1 cm^3 or 1 cc. Notice in this calculation that the units are centimeters cubed (cm^3), which arises from the multiplication of the cm units. *Units are treated just like numbers for arithmetic operations like multiplication, division, addition, and subtraction.*

The base unit of volume in the metric system is the liter (L). One liter is the volume occupied by a cube measuring 10 cm on each side (10 cm × 10 cm × 10 cm = 1,000 cm^3), as shown in Figure 1-5. A milliliter is 1/1,000 of a liter, as expected from the prefix "milli." *The cubic centimeter is equivalent to one milliliter:*

$$1\ cm^3 = 1\ mL$$

Special labware is used to measure the volume of a liquid, such as the pipet, the graduated cylinder, and the syringe. Beakers also contain volume markings, but they are generally not used for precise volume measurements.

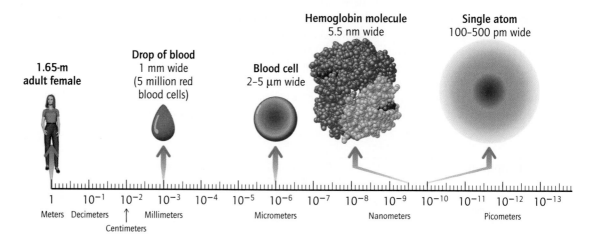

Figure 1-4 Metric measurements ranging from the macroscale to the atomic scale: woman → red blood cell in woman → hemoglobin in red blood cell → iron atom in hemoglobin.

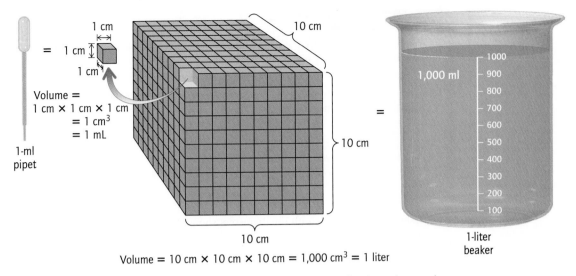

Figure 1-5 Volume is derived from length measurements. A cube measuring 1 cm along each side has a volume of 1 cm³ or 1 mL. A cube measuring 10 cm along each side has a volume of 1,000 mL or 1 L.

The volume of a solid object cannot be measured in the same way as that of a liquid. If it has a regular shape, its length or diameter can be measured and the volume calculated, as we saw for the volume of a cube calculated in Figure 1-5. Alternatively, and often the simplest way to measure the volume of a solid, is to determine how much water it displaces. For example, if a block of lead were fully submerged in a graduated cylinder filled with an initial volume of water (V_{initial}), the level of the water would rise by an amount corresponding to the volume of the block of lead:

$$V_{\text{object}} = V_{\text{final}} - V_{\text{initial}}$$

For example, you could determine your body's volume by measuring the volume of water you displace when submerged in the bathtub.

WORKED EXERCISES 1-1 Using Metric Units

1 Using Table 1-2, determine how many grams there are in one microgram. What fraction of one gram does the prefix μ in μg represent?
2 What is the volume in cubic centimeters of a cube measuring 5.00 cm a side? What does this volume correspond to in units of milliliters?
3 When a gold sphere is placed in a graduated cylinder containing 100.0 mL of water, the water level rises to 120.0 mL. What is the volume of the gold sphere in cm³?

SOLUTIONS

1 From Table 1-2 you see there is 1×10^{-6} g in 1 μg. The prefix μ in μg represents one millionth, or 1/1,000,000 of a gram.
2 The volume of the cube is 5.00 cm × 5.00 cm × 5.00 cm = 125 cm³. Remember to multiply the units as well as the numbers. Since 1 mL = 1 cm³, 125 cm³ corresponds to 125 mL.
3 The volume of the gold sphere is equal to the difference in the volume before and after the gold is submerged: V_{gold} = 120.0 mL − 100.0 mL = 20.0 mL, which corresponds to 20.0 cm³.

PRACTICE EXERCISES*

1.1 What fraction of a meter is 1 picometer? What fraction of a gram is 1 picogram? What does the prefix pico stand for?

1.2 What is the length, in centimeters, of the side of a cube having a volume of 27 cm³?

1.3 For each pair of measurements below, indicate which is the greater length. If the lengths are the same, state so.

    a. 1 μm or 1 mm   **b.** 1 km or $1 \times 3 \ 10^3$ m

1.4 What do the following prefixes stand for in terms of fractions or multiples of the base unit they precede?

    a. milli **b.** deci **c.** kilo

1.5 Which of the objects with the diameters shown below can be seen with the naked eye?

    a. 1 km **b.** 1 m **c.** 1 nm

1.6 For each pair below, indicate which measurement represents the greater mass. If the measurements in a pair are the same, state so.

    a. 1 mg or 1 μg   **b.** 1 ng or 1 μg   **c.** 10 mcg or 10 μg

1.7 What is the volume of an object in cubic centimeters if it is placed in 10 mL of water and causes the level of water to rise to 11 mL.

    *You can find the answers to the Practice Exercises at the end of each chapter.

Significant Figures

The numerical value obtained from any measurement always contains some degree of uncertainty. The uncertainty in a measurement depends on the accuracy and precision of the measuring device as well as the human error associated with reading any device—balance, thermometer, ruler, etc. *The number of digits reported conveys information about the error in a measurement.*

Precision and Accuracy Precision is a measure of how close together repeated measurements are to each other, while accuracy is a measure of how close the measurement is to the "true" value. Ideally, a measurement is both accurate and precise. However, measurements can be inaccurate and/or imprecise. Consider, for example, the weight of an infant measured three times on four different balances (Table 1-3). Assume the "true" weight of the infant is 7.5 lb, and that reasonable precision for the balance is ±0.1 lb.

Table 1-3 Three Readings of a 7.5-lb Infant Taken on Different Balances, A–D

Balance	Balance A	Balance B	Balance C	Balance D
Weight (lb)	7.5	8.1	8.0	9.0
	7.6	8.0	7.5	8.4
	7.4	8.2	7.0	7.9
Average weight	7.5	8.1	7.5	8.4
Type of error	Precise and accurate	Precise, but not accurate	Accurate, but not precise	Neither precise nor accurate

The average weight reported on balance A is 7.5 lb, the same as the "true" value, and all measurements are within 0.1 lb of the average; therefore, the measurements obtained on balance A are both precise and accurate. The values obtained using balance B are also within 0.1 lb of the average, but the average—8.1 lb—is not close to the "true" value; therefore, balance B is precise but not accurate. Balance C shows an average close to the "true" value, but two of the individual measurements differ from the average by 0.5 lb; therefore, this balance is accurate but not precise. The measurements taken on balance D are neither precise nor accurate, because the average is not near the "true" value and each measurement differs significantly from the average.

Even using balance A, you cannot know the infant's weight exactly, because of environmental factors at the time the measurement was taken, imperfections in the balance itself, when the balance was last calibrated, and inherent human error in taking any measurement. So, if you report a measurement of 7.5 lb, you may really mean the infant's weight is somewhere between 7.4 and 7.6 lb. When reporting measured values, the convention is to record all of the certain digits plus one uncertain digit to a best approximation. All the certain digits as well as the one uncertain digit are known as **significant figures.** In the previous example, "7" is the certain digit, and 4, 5, or 6 is the uncertain digit, so the correct way to report the number would be to write two significant figures: 7.5 lb. A scientist interpreting this measurement understands it to mean that the infant's weight is 7.5 ±0.1 lb, or somewhere between 7.4 and 7.6 lb.

The greater the precision of the measuring device, the more significant figures the measured reading may have. For example, a bathroom scale can be read to the tenths place (one place to the right of the decimal place), as in 5.0 lb. A top-loading balance, such as you might find in the chemical laboratory, can be read to the hundredths place, as in 1.66 g, and an analytical balance can be read to the ten thousandths place, as in 1.5598 g. Hence, more significant digits should be reported for a measurement made on an analytical balance (5 significant figures) than on a top-loading balance (3 significant figures) or a bathroom scale (2 significant figures).

By definition, the last digit (shown in red below) is understood to be the uncertain digit in the measurements listed.

5.0

5.00

5.000

5.0000

A top-loading balance (top) and an analytical balance (bottom) show different levels of precision: the digital display on the top-loading balance shows only three significant figures, while the display on the analytical balance shows six for the same sample. [Richard Megna/Fundamental Photographs]

These numbers have 2, 3, 4, and 5 significant figures respectively, and therefore, indicate measurements made on increasingly more precise instruments. Note that these measurements would normally also be followed by a unit.

In the laboratory and the clinic you will be taking measurements and recording them. You should always record the number of significant figures appropriate for the measuring device you are using. When interpreting measured numbers, consider all nonzero digits to be significant, and use the following guidelines to help you interpret whether zeros are significant or not.

Guidelines for Determining the Number of Significant Figures in a Measured Number

- All nonzero digits are significant.

    3.45   3 significant figures

- Zeros between nonzero digits are always significant regardless of whether or not there is a decimal point.

    3.05   3 significant figures

- Zeros following a nonzero digit in a number containing a decimal point are significant.

    0.400   3 significant figures
    4.0   2 significant figures

- If there is no decimal point, zeros following a nonzero digit are not significant; they are placeholders.

    6,000   1 significant figure

Note: A decimal point may be placed after the zeros to specify that the zeros are significant:

    6,000.   4 significant figures

- Zeros that appear *before* nonzero digits, whether or not there is a decimal point, are not significant. The zeros serve as placeholders.

    0.00040   2 significant figures

- All digits in a number expressed in scientific notation are significant.

    2.30×10^4 3 significant figures

Exact Numbers If you count 20 students in a class carefully, there is no uncertainty as to whether there are 19, 20, or 21 students. Exact numbers include numbers obtained by counting and numerical values in some definitions. *Exact numbers contain an infinite number of significant figures because they carry no uncertainty*. Some definitions are exact numbers, because they are not measured values. All the metric conversions in Table 1-2 are exact numbers. For example, in the definition: "60 minutes = 1 hour," both "60" and "1" are exact numbers, and therefore have an infinite number of significant figures—no uncertainty.

WORKED EXERCISE 1-2   Determining the Number of Significant Figures

Indicate the number of significant figures in the following measured values:

a. 4.507 cm
b. 0.00550 g
c. 2.0×10^5 m
d. 53,000. seconds
e. 189 students

SOLUTION

a. There are four significant figures in 4.507. Zeros between nonzero digits are significant, and all nonzero digits are significant.

b. There are three significant figures in 0.00550. The zeros before the nonzero digits are not significant; they are place holders. The zero after the five is significant.

c. There are two significant figures in 2.0 × 10⁵. Scientific notation is the only way to convey 2 significant figures in this measurement, because 20,000 has only one significant figure, while writing 20,000. specifies five significant figures, because of the decimal point.

d. There are five significant figures in 53,000. The decimal point after the zeros indicates that the three zeros are significant.

e. There are an infinite number of significant figures in this number, obtained by counting. There is no uncertainty in this number.

PRACTICE EXERCISES

1.8 How many significant figures are there in the measured values shown below? Explain your reasoning.

a. 0.007 m
b. 50 people
c. 23,000. seconds
d. 0.004050 mg

1.9 The measurements below were made on three different balances:

i. 5.5 g   ii. 5.51 g   iii. 5.5093 g

a. Which measurement was made on the most precise balance?
b. Which is the uncertain digit in each measurement?
c. A student takes a measurement on a balance where the display fluctuates between 5.50 g and 5.53 g. Which of the three measurements above would be the most correct number to report?

Significant Figures in Calculations

When calculations are performed on measured values, it is important to report the final answer with the correct number of significant digits. It is incorrect to simply report all the digits displayed on the calculator! The correct number of significant figures to report depends on whether the calculation involves multiplication and division or addition and subtraction.

Multiplying and Dividing Measured Values When multiplying or dividing measured values, the final calculated answer cannot have more significant figures than the measurement with the *fewest* number of significant figures. For example, if you multiplied the two measured lengths 3.55 m and 11.65 m to determine the area of a room, the answer displayed on the calculator would read many more digits than either of the measured values used in the calculation. If you reported all of these digits it would suggest the area of the room was more precise than either of the measurements that went into calculating the area, which is impossible. The answer must be rounded to three significant figures, because the measurement with the *fewest* significant figures, 3.55, has three significant figures:

3.55 m	×	11.65 m =	<41.3575 m²> =	41.4 m²
Measured value		measured value	calculator answer	correct answer
3 significant figures		4 significant figures	6 significant figures	3 significant figures

The correct answer to report for this calculation would be 41.4 m².

Review of Rounding Rules

After determining how many significant digits to include, look at the next digit to the right.

- Round up if this digit is greater than or equal to 5, and drop all digits that come after.

  (0. 846 862 becomes 0.847).

- Leave as is if this digit is less than 5, and drop all digits that come after.

  (0.846 324 becomes 0.846).

Adding and Subtracting Measured Values When adding or subtracting measured numbers, the final answer can have no more decimal places than the measured value with the fewest decimal places.

The sum of the numbers shown below should be reported as 19.3 cm, not 19.294 cm as displayed on the calculator, because the measurement with the fewest decimal places, 5.4, has its last digit in the tenths place.

$$
\begin{array}{r}
5.4 \ \ \text{cm}\\
6.55 \ \ \text{cm}\\
+\ \ 7.344 \ \text{cm}\\
\hline
<19.294 \ \text{cm}>
\end{array}
\qquad \text{round to} \qquad \textbf{19.3 cm}
$$

calculator answer   **correct answer**
thousandths place   tenths place

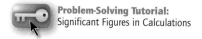

Problem-Solving Tutorial:
Significant Figures in Calculations

WORKED EXERCISE 1-3   Significant Figures in Calculations

Perform the following calculations and round the final answer to the correct number of significant figures:

a. $0.0022 \times 58.88 =$
b. $7.0 + 8.55 + 233 =$

SOLUTION

a. The calculation involves multiplication, so the final answer should have no more significant figures than the measured value with the fewest significant figures, which is 0.0022 with two significant figures. The calculator gives the following result: $0.0022 \times 58.88 = 0.129536$. Round the number that appears on the calculator display to two significant figures: **0.13.** The second digit is rounded up to 3 because the next digit to the right of 2 is 9—a number greater than 5.

b. The calculation involves addition, so the final answer should have no more decimal places than the measured value with the fewest decimal places. The calculator gives the sum 248.55. The number 233 has the fewest decimal places because its last decimal place is the ones place. Therefore, the final answer cannot have any numbers past the ones place. Thus, 248.55 should be rounded to 249. The 8 is rounded up, because the next digit to the right is 5—a number greater than or equal to 5.

$$
\begin{array}{r}
7.0\\
8.55\\
+\ 233.\\
\hline
<248.55>\\
\textbf{249}
\end{array}
$$

<248.55>   calculator answer
249   **final answer**

PRACTICE EXERCISES

1.10 Perform the following calculations. Report the answer to the correct number of significant figures.

  a. $56.50 \times 37.99 =$
  b. $5.999 + 6.001 + 3.2222 =$
  c. $57.200 \times \dfrac{67.55}{1.220} =$

1.11 **Critical Thinking Question:** $800 - 1723 =$

  Hint: Your final answer must show zeros (placeholders) after the hundreds place, because the tens and units place is uncertain in the number 800, and the arithmetic operation involves addition or subtraction.

Converting Between Metric Units

Medical professionals are routinely called upon to perform metric conversions as part of their daily routine. A metric **conversion** is the mathematical transformation of a value in one metric unit (the supplied unit) into another metric unit (the requested unit). To perform a conversion, a process known as dimensional analysis is used universally throughout the sciences. Dimensional analysis considers the units (the dimensions) to set-up the calculation. Dimensional analysis is a fail-safe method that insures numbers are not inadvertently multiplied when they should be divided and vice versa. As you can imagine, calculating the incorrect amount of medicine a patient receives can be catastrophic.

Guidelines for Performing Metric Conversions Using Dimensional Analysis.

1. **Step 1: Identify the Conversion(s).** Write the mathematical expression(s) that equate(s) the supplied unit and the requested unit. When there is no direct conversion, more than one conversion may be required. *For example, to convert 0.500 grams into milligrams, you need the conversion that equates grams, the supplied unit, and milligrams, the requested unit. From Table 1-2:*

$$1 \text{ mg} = 10^{-3} \text{ g}$$

2. **Step 2: Express the Conversion as Two Possible Conversion Factors.** A conversion can be turned into a **conversion factor** by expressing the equality in the form of a fraction. There are always two ways to write a conversion factor, which are obtained by inverting the numerator and the denominator. Be sure always to keep the number and its associated unit together. *For example, the conversion factors for the conversion in Step 1 are*

$$\frac{1 \text{ mg}}{10^{-3} \text{ g}} \quad \text{and} \quad \frac{10^{-3} \text{ g}}{1 \text{ mg}}$$

> Caution: When you memorize metric conversions, don't inadvertently mix up the numerical value and its associated unit. In other words
>
> $1 \text{ mg} = 10^{-3} \text{ g}$ is CORRECT, but $10^{-3} \text{ mg} = 1\text{g}$ is INCORRECT.
>
> Whenever you invert the numerator and denominator, be sure to keep the number together with its associated unit.

3. **Step 3: Set Up the Calculation by Multiplying the Supplied Unit by One or More Conversion Factors so that the Supplied Unit Cancels.** Multiply the supplied unit by the form of the conversion factor that causes the supplied units to cancel. As with identical *numbers*, identical *units* cancel algebraically, when they appear in both the numerator and the denominator of a fraction. The result is an answer in the requested unit.

$$\cancel{\text{Supplied unit}} \times \frac{\text{requested unit}}{\cancel{\text{supplied unit}}} = \text{requested unit}$$

For example, to convert 0.500 g into mg:

$$0.500 \cancel{\text{ g}} \times \frac{1 \text{ mg}}{10^{-3} \cancel{\text{ g}}} = 500. \text{ mg}$$

As a final step in any calculation, you should double check that you have reported the correct number of significant figures in the final numerical answer. Since multiplication and division are involved in metric conversions, the final answer should have no more significant figures than the measured value with the smallest number of significant figures. Note that metric conversions are exact numbers and therefore would not reduce the number of significant figures reported.

Problem-Solving Tutorial:
Using Unit Systems

WORKED EXERCISE 1-4 Metric Conversions

Convert 15 millimeters into units of centimeters using dimensional analysis and using the conversion factors shown in Table 1-2.

SOLUTION

The supplied unit is millimeters, mm, and the requested unit is centimeters, cm. There is no direct conversion between mm and cm, because neither is the base unit of length, the meter (m). Therefore, two conversion factors must be used: mm → m and m → cm.

Step 1: Identify the conversions needed. 1 mm = 10^{-3} m and 1 cm = 10^{-2} m.

Step 2: Express each conversion as two possible conversion factors.

$$\frac{1 \text{ mm}}{10^{-3} \text{ m}} \quad \text{or} \quad \frac{10^{-3} \text{ m}}{1 \text{ mm}} \quad \text{and} \quad \frac{1 \text{ cm}}{10^{-2} \text{ m}} \quad \text{or} \quad \frac{10^{-2} \text{ m}}{1 \text{ cm}}$$

Step 3: Set up the calculation by multiplying the supplied unit by the correct form of one or more conversion factors so that the supplied units cancel.

$$15 \text{ mm} \times \frac{10^{-3} \text{ m}}{1 \text{ mm}} \times \frac{1 \text{ cm}}{10^{-2} \text{ m}} = 1.5 \text{ cm}$$

Finally, check that the answer, 1.5 cm, has the correct number of significant figures. Here, the measured value is 15 mm, which has two significant figures; thus, the final answer should have only two significant figures.

PRACTICE EXERCISES

1.12 How many centimeters are there in 0.65 m?

1.13 How many grams are there in 7.7 kg?

1.14 How many liters are there in 561 mL of water?

1.15 How many picometers are there in 56 nm?

1.16 Hemoglobin has a spherical shape with a diameter of 5.5 nm. What is the diameter of a hemoglobin molecule in meters? What is it in millimeters?

1.17 One patient has a tumor 150. μm in diameter while another has a tumor 2 mm in diameter. Which patient has the larger tumor?

Converting Between Metric and English Units

In the United States and some other parts of the world, English units are still in common use, so you will find it necessary at times to convert from an English unit to a metric unit and vice versa. This type of calculation requires dimensional analysis using a metric-English conversion. Table 1-4 lists some of the common conversions between metric and English units for length, mass, and volume.

Table 1-4 Some Common English–Metric Conversions

Type of Measurement	Conversions
Length	1 inch = 2.54 cm (exact)
	1 m = 39.37 inches
Mass	1 lb = 453.5 g
	1 kg = 2.205 lb
Volume	1 L = 1.057 quarts

WORKED EXERCISES 1-5 Performing English-Metric Conversions

1 A patient weighs 125 lb. Calculate her weight in kilograms.
2 The Washington monument is 155 ft high. What is the height of the Washington monument in meters?

SOLUTIONS

1 The supplied unit is pounds, lb, and the requested unit is kilograms, kg.
Step 1: Identify the conversion needed. Table 1-4 shows that the conversion between pounds and kilograms is 1 kg = 2.205 lb.
Step 2: Express the conversion as two possible conversion factors:

$$\frac{1 \text{ kg}}{2.205 \text{ lb}} \quad \text{or} \quad \frac{2.205 \text{ lb}}{1 \text{ kg}}$$

Step 3: Set up the calculation by multiplying the supplied unit by one or more conversion factors so that the supplied unit cancels:

$$125 \cancel{\text{ lb}} \times \frac{1 \text{ kg}}{2.205 \cancel{\text{ lb}}} = 56.7 \text{ kg}$$

The measurement, 125, has three significant figures, so the answer must have three significant figures: 56.7.

2 In this exercise you are asked to convert from feet to meters.
Step 1: Identify the conversions needed. Table 1-4 does not show a direct conversion between feet and meters; however, it lists a conversion between inches and meters, and you know the conversion between feet and inches, so you will need to use two conversion factors: feet to inches and inches to meters.
Step 2: Express each conversion as two possible conversion factors:

$$\frac{1 \text{ m}}{39.37 \text{ in.}} \quad \text{or} \quad \frac{39.37 \text{ in.}}{1 \text{ m}} \quad \text{and} \quad \frac{12 \text{ in.}}{1 \text{ ft}} \quad \text{or} \quad \frac{1 \text{ ft}}{12 \text{ in}}$$

Step 3: Set up the calculation by multiplying the supplied unit by one or more conversion factors so that the supplied unit cancels.

$$155 \cancel{\text{ ft}} \times \frac{12 \cancel{\text{ in.}}}{1 \cancel{\text{ ft}}} \times \frac{1 \text{ m}}{39.37 \cancel{\text{ in.}}} = 47.2 \text{ m (3 significant figures)}$$

The measurement, 155 ft, has three significant figures. The feet-to-inch conversion factor involves exact numbers, but the inch-to-meter conversion factor involves a measured value (39.37) with 4 significant figures. Since the calculation requires multiplication and division, the final answer cannot have more than three significant figures, the number of significant figures in the measurement with the fewest number of significant figures.

PRACTICE EXERCISES

1.18 If you were to fill your 15-gal gas tank in Europe, how many liters of petrol (gasoline) would you need to purchase?
1.19 The Empire State Building in New York is 1,453 ft tall. How tall is the Empire State Building in meters?
1.20 A premature baby weighs 906 g. What is the weight of the baby in pounds?

Dosages Based on Patient's Weight For many medications, dosages must be adjusted for a patient's weight to ensure safety. This is especially true when giving medicine to children. Patients are often weighed in pounds, yet many drug handbooks recommend dosage per kilogram. From Table 1-4, you see there are 2.2 pounds in one kilogram. Converting pounds to kilograms is often the first step in calculating dose by weight. The dosage itself is given as a conversion factor between the mass of medicine and the weight of the patient; for example, 8 mg per kg body weight. The term "per" always indicates a ratio and can be expressed as a fraction in two forms using dimensional analysis:

$$\frac{8 \text{ mg}}{1 \text{ kg}} \quad \text{or} \quad \frac{1 \text{ kg}}{8 \text{ mg}}$$

Some common abbreviations used for ordering medications include "q.d" and "b.i.d," which are derived from the Latin phrases for administered "daily" and "twice daily," respectively. The number of times per day can also be expressed as a conversion factor:

$$\frac{1 \text{ time}}{\text{day}}, \quad \frac{2 \text{ times}}{\text{day}}$$

WORKED EXERCISE 1-6   Dosage Calculations

Tetracycline Elixir, an antibiotic, is ordered for a child weighing 52 lb at a dosage of 8.0 mg per kilogram of body weight q.d. How many milligrams should one dose for this child contain?

SOLUTION

Step 1: Identify the conversions. The conversion for the patient's weight from pounds to metric units is 1 kg = 2.205 lb.

Step 2: Express each conversion as two possible conversion factors. The English to metric conversion factors for the patient's weight are:

$$\frac{1 \text{ kg}}{2.205 \text{ lb}} \quad \text{or} \quad \frac{2.205 \text{ lb}}{1 \text{ kg}}$$

The dosage *is* a conversion factor between the mass of medicine in milligrams and the weight of the patient in kilograms per dose:

$$\frac{8.0 \text{ mg}}{1 \text{ kg}} \quad \text{or} \quad \frac{1 \text{ kg}}{8.0 \text{ mg}}$$

Since the dosage is marked q.d., the dose should be given once a day.

Step 3: Set up the calculation so that supplied units cancel. Only requested units remain.

$$52 \text{ lb} \times \frac{1 \text{ kg}}{2.205 \text{ lb}} \times \frac{8.0 \text{ mg}}{\text{kg}} = 190 \text{ mg given once a day (2 significant figures)}$$

This type of calculation may seem confusing, because the units of mass (kg and mg) appear in both the numerator and denominator of two of the conversion factors. The key to solving this type of problem is realizing the mass of the patient is different from the mass of the medication when setting up the conversion factors. Finally, check that the answer has no more significant figures than the supplied values: 52 lb and 8.0 mg/kg, in this example, have two significant figures.

PRACTICE EXERCISES

1.21 Quinidine is an antiarrhythmic agent. It is prescribed for an adult patient weighing 110 lb at a dosage of 25 mg per kilogram of body weight per day. How many milligrams should one dose contain?

1.22 Ampicillin, an antibiotic, is prescribed for a child weighing 63.0 lb at a dosage of 20.0 mg per kilogram of body weight per day, in four equally divided doses. How many milligrams should be given per dose?

Density

Osteoporosis is a disease characterized by low bone density. Density is a physical property of matter often used to characterize a substance or material, as seen in this example of bone. The **density** (d) of a material is defined as its mass (m) divided by its volume (V). Density is derived from measured values of mass and volume. Density typically has units of g/mL or g/cm^3.

$$d = \frac{m}{V}$$

The density of a substance is independent of the amount of substance. For example, the density of one drop of water is the same as that of a bathtub full of water. The density of some common substances is shown in Table 1-5.

Substances that are less dense than water, and which do not dissolve in water, will float on water, such as ice or oil.

While density represents a physical property, it can also be used as a conversion factor to calculate the mass of a substance whose volume and density are known, or the volume of a substance whose mass and density are known.

Table 1-5 Density of Some Elements and Compounds at 25°C

Element/ Compound	Density (g/mL or g/cm^3)
Water	1.00
Iron	7.9
Gold	19.32
Alcohol	0.79

WORKED EXERCISES 1-7 Calculations Involving Density

1 What is the density of a substance with a mass of 0.90 g and a volume of 1.2 mL?
2 Using Table 1-5, calculate the mass of a 5.0-cm^3 block of gold?

SOLUTIONS

1 Density is defined as $d = m/V$. To calculate density, substitute the numerical values supplied for mass and volume into this equation:

$$d = \frac{0.90 \text{ g}}{1.2 \text{ mL}} = 0.75 \text{ g/mL (2 significant figures)}$$

Since no units cancel, the final answer has units of g/mL, which are the typical units of density.

2 **Step 1: Identify the conversions.** The conversion factor is given in the physical definition of density, so skip to Step 2.

Step 2: Express each conversion as two possible conversion factors. Since volume is supplied in this problem, the density of gold (19.32 g/cm^3) becomes the conversion factor:

$$\frac{19.32 \text{ g}}{cm^3} \quad \text{or} \quad \frac{cm^3}{19.32 \text{ g}}$$

(density)

Step 3: Set up the calculation so that the supplied units cancel. Volume is the supplied unit, so use the form of the conversion factor that will cause volume to cancel, leaving units of mass:

$$5.0 \text{ } \cancel{cm^3} \times \frac{19.32 \text{ g}}{\cancel{cm^3}} = 97 \text{ g of gold}$$

conversion factor
(density)

Round the final answer to two significant figures, because the supplied value, 5.0 cm^3, has two significant figures.

PRACTICE EXERCISES

1.23 What is the density of a liquid that has a mass of 5.5 g and a volume of 5.0 mL?

1.24 Would a gold sphere with a diameter of 2 mm have a greater or a smaller mass than an iron sphere with the same diameter? *Hint:* No calculation required.

1.25 Would a gold sphere with a mass of 5 g have a greater diameter or a smaller diameter than a gold sphere with a mass of 10. g?

1.26 An object has a mass of 10.5 g. When it is submerged in a graduated cylinder initially containing 82.5 mL of water, the water level rises to 95.0 mL. What is the density of the object?

1.27 Using Table 1-5, what is the mass of 25 mL of water?

1.28 Using Table 1-5, what volume of water in mL would a 65-g piece of iron displace?

1.29 A piece of unknown metal was found to have a mass of 71.1 g and a volume of 9.0 cm³. Is this unknown metal gold or iron? Explain.

1.30 Would you expect a woman with osteoporosis to have a bone density greater than or less than normal? Explain.

1.31 Ice floats on liquid water. Does this mean ice is more or less dense than liquid water?

1.2 | Elements and the Structure of the Atom

The Greek philosophers were the first to propose that matter could not be cut into smaller and smaller pieces indefinitely; that there is a point after which you cannot cut any further. We now know this point to be the **atom,** the smallest intact component of all matter. Thus, if you were to take a piece of nickel and cut it into smaller and smaller pieces, the smallest indivisible piece of nickel would be one nickel atom, with the incredibly small diameter of 256 pm (256×10^{-12} m), shown in the STM image on page 3. In other words, the substance we identify as nickel is made up of an enormous number of nickel atoms.

There are 111 different types of atoms, differing in composition as explained below. Atoms are the fundamental component of the two types of matter: elements and compounds. An **element** is a substance composed of only one type of atom, which cannot be broken down into any simpler form of matter. Thus, nickel is an element composed of only nickel atoms, and gold is an element composed of only gold atoms. A **compound** is a substance composed of two or more different types of atoms held together in a unique proportion, which can be broken down into its elements through a chemical reaction. For example, water is a compound and not an element, because it is composed of both hydrogen and oxygen atoms in a 2:1 ratio and *can* be broken down into something simpler: the elements oxygen and hydrogen. In Chapter 2 you will learn how different atoms can come together to form compounds. Thus, the fundamental building blocks for all matter are the atoms that constitute the various elements.

The Parts of an Atom

An atom is a spherically shaped particle containing the following **subatomic particles:**

- **protons,**
- **neutrons,** and
- **electrons.**

The word **atom** comes from the Greek word for uncuttable.

Humankind has used its knowledge of the atom in many ways, from building the atom bomb to generating nuclear power. Nuclear medicine is another field that has emerged from our understanding of the atom, providing advanced treatments and diagnostic tools in medicine. To learn more about the role of the atom in nuclear medicine, see Chapter 16.

Some common elements are, reading clockwise, bromine (in the bottle), zinc, copper, calcium, sulfur, and iodine (center). [Richard Megna/Fundamental Photographs]

The subatomic particles differ in their charge, mass, and location within the atom, as summarized in Table 1-6.

Charge of the Subatomic Particles Two of the three subatomic particles have an electrical charge. A proton has a positive charge, $+1$; an electron has a negative charge, -1; and a neutron has no charge, 0. The net charge on an atom is zero, because atoms contain an equal number of electrons (negative charges) and protons (positive charges).

Location of the Subatomic Particles Within the Atom Both protons and neutrons are concentrated in an incredibly small volume at the center of the atom known as the **nucleus.** The nucleus is only about 10 fm in diameter. The electrons, by comparison, occupy a much larger volume of space and so define the size of an atom, ranging from 74 to 300 pm. To illustrate the size of the nucleus relative to the entire atom, imagine scaling up the nucleus of the atom to the size of a football. The size of the atom would be comparable to the size of the entire football stadium—bleachers and all.

Understanding the nature of electrons was one of the greatest scientific advances in physics in the early twentieth century. It was found that electrons do NOT orbit the nucleus, as scientists had first believed. Although the path travelled by the fast moving electron cannot be determined, the region of space occupied by an electron, known as an **electron orbital,** can be described. An orbital is a region of space in which the electron is *most likely* to be found. Figure 1-6 shows two representative orbitals, the $1s$ and the $2p_x$ orbital. The spherical $1s$ orbital describes the volume occupied by the two electrons closest to the nucleus of any atom (except hydrogen, which has only one electron). As you can see from the figure, orbitals are not always spherical in shape, as illustrated by the $2p_x$ orbital, which has the two-lobed shape shown. In atoms containing more electrons, the electrons occupy larger orbitals and therefore

In terms of size, a football is to a football stadium as the nucleus of an atom is to the atom. [Rick Doyle/Corbis]

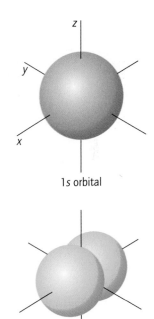

1s orbital

$2p_x$ orbital

Figure 1-6 The $1s$ and $2p_x$ electron orbitals. An s orbital is spherical in shape, but a p orbital has a two-lobed shape similar to a dumbbell.

Table 1-6 Distinguishing Characteristics of the Subatomic Particles

Subatomic Particle	Charge	Mass	Location Within the Atom
Proton	$+1$	1 amu	Nucleus
Neutron	0	1 amu	Nucleus
Electron	-1	.00055 amu	Orbitals

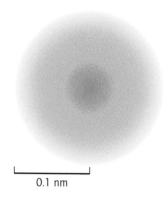

0.1 nm

Figure 1-7 Electron cloud for a hydrogen atom. The darker regions show where the electron is more likely to be found. The radius of a hydrogen atom is 0.1 nm.

are more likely to be found farther from the nucleus, making these atoms larger in size.

An orbital can also be represented as an **electron cloud,** analogous to a time-lapse photograph of an electron's positions over time, as shown for a hydrogen atom in Figure 1-7. The darker the area, the greater the probability of finding the electron in that region.

Mass of Subatomic Particles Protons and neutrons have an approximately equal mass, whereas electrons have practically no mass. Thus, the mass of an atom comes from its protons and neutrons.

The mass of a subatomic particle is measured in units called atomic mass units, abbreviated amu. The mass of a proton and a neutron is approximately 1 amu. The conversion between the amu and the metric unit for mass, the gram, is

$$1 \text{ amu} = 1.66 \times 10^{-24} \text{ g}$$

Atoms with more protons and neutrons have a greater mass (they are heavier) than atoms with fewer neutrons and protons. Moreover, since all of an atom's mass is concentrated in the tiny volume of the nucleus, the nucleus of the atom is extraordinarily dense. Continuing with the football stadium analogy, it would be as though the football was so massive, the entire football team could not lift it!

PRACTICE EXERCISES

1.32 Fill in the blanks with the name(s) of the correct subatomic particle(s) (*protons, neutrons, electrons*) that fit each of the following descriptions:

    a. _____ are located in the nucleus of the atom.
    b. _____ are the lightest subatomic particle.
    c. _____ have a positive charge.
    d. _____ have a mass approximately equal to 1 amu.

1.33 Which is more dense, the atom or the nucleus of the atom? Explain.

1.34 Where in the atom are the electrons found?

1.35 Describe the shape of a $2p_x$ orbital.

Atomic Number *The number of protons an atom contains is known as its* **atomic number.** For example, a helium atom contains two protons, and therefore has atomic number 2. *The identity of an element is determined by the number of protons in its atoms—its atomic number.* For example, the element carbon has atomic number 6; it is composed of atoms all containing six protons and six electrons. You may be familiar with some elements while others are probably new to you. Calcium, atomic number 20, for example, is a major constituent of bone and teeth; its atoms contain 20 protons. Iron atoms, the oxygen-binding atoms of hemoglobin in red blood cells, have 26 protons; the atomic number for iron is thus 26.

Every element has been assigned a one or two letter **element symbol.** The first letter is always a capital letter; for example, the element symbol for helium is He. Element symbols are one- or two-letter abbreviations derived from the English or Latin word for the element. The use of Latin accounts for unexpected element symbols like mercury, Hg, and iron, Fe.

Physical and Chemical Properties of the Elements

Every element has properties that we can observe on the macroscopic scale. These properties are divided into two categories:

- physical properties, and
- chemical properties.

Physical Properties The **physical properties** of an element are those properties we can observe or measure without considering how the element interacts with other substances. Examples of physical properties include color, hardness, density, and boiling point (the temperature at which the substance changes from a liquid to a gas). For example, some of the physical properties of gold (Au) are listed in Table 1-7.

Chemical Properties The **chemical properties** of an element refer to its behavior in the presence of *other* chemical substances. An element may undergo a chemical change with this substance, known as a **chemical reaction,** when it combines with another element or compound to form a new compound. For example, iron undergoes a chemical change in the presence of oxygen, transforming the iron and oxygen into the compound iron oxide (rust)—a substance different from both iron and oxygen. Iron oxide has its own unique physical and chemical properties, which are different from those of either iron or oxygen. Not all elements react with oxygen, and an element's tendency to react with oxygen is a characteristic *chemical property* of the element. For example, white phosphorus bursts into flames when exposed to the oxygen in air, while helium is completely unreactive in air—it is inert.

Isotopes of an Element

Although the number of protons and electrons is the same for all atoms of a given element, the number of neutrons varies. ***Isotopes*** *are atoms with the same number of protons, but a different number of neutrons.* For example, carbon is found naturally on earth as three different isotopes. All the isotopes of carbon contain 6 protons, by definition (atomic number 6). However, they differ in the number of neutrons they each contain: 6, 7, or 8 neutrons, as shown in Table 1-8.

Table 1-7	Physical Properties of Gold
Color	Shiny gold
Hardness	Most malleable and ductile metal
Density	19.32 g/mL
Boiling point	2856°C
Melting point	1064°C

White phosphorus bursts into flames upon contact with the oxygen in air.
[Richard Megna/Fundamental Photographs]

Table 1-8 Naturally Occurring Isotopes of Carbon, C

Atomic Number	Number of Protons	Number of Neutrons	Mass Number
6	6	6	12
6	6	7	13
6	6	8	14

Most elements have more than one naturally occurring isotope, and additional isotopes can sometimes be prepared artificially. For example, silicon, Si, has three naturally occurring isotopes, and seventeen man-made isotopes. All man-made isotopes are radioactive, a topic which will be explored in Chapter 16.

Since neutrons are neutral particles, their number does not influence the overall charge of an isotope. However, since the mass of a neutron is comparable to that of a proton, isotopes with more neutrons will have a greater mass. Isotopes are characterized by their **mass number,** the sum of the protons and neutrons of the isotope.

> Mass number = number of protons + number of neutrons

As you can see from the last column in Table 1-8, the three naturally occurring isotopes of carbon have the mass numbers 12, 13, and 14. The mass number for an isotope is often written as a **super**script to the left of the

Table 1-9 Some Common Representations of the Natural Isotopes of Carbon

Isotope Symbol	Alternative Representation
$^{12}_{6}C$	Carbon-12
$^{13}_{6}C$	Carbon-13
$^{14}_{6}C$	Carbon-14

element symbol, while the atomic number appears as a **sub**script to the left of the element symbol, as shown below for carbon-13.

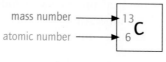

mass number = number of protons + number of neutrons
atomic number = number of protons

The second column in Table 1-9 shows another way to represent isotopes. The mass number can be written following a hyphen after the full name of the element. The atomic number is absent in this representation, but it can always be determined from the element symbol.

WORKED EXERCISES 1-8   Isotopes and Mass Number

1 How many protons and neutrons are there in radioactive strontium-90, which is the element with atomic number 38?
2 Write the symbol for the isotope of chlorine, Cl, with 17 protons and 20 neutrons.

SOLUTIONS

1 There are 38 protons in a strontium-90 atom. You can calculate the number of neutrons in an isotope by subtracting the number of protons (the atomic number) from the isotope's mass number (90):

For Sr: number of neutrons = mass number − atomic number
number of neutrons = 90 − 38 = 52 neutrons

2 The isotope is chlorine-37 or $^{37}_{17}Cl$. The sum of the protons and neutrons gives the mass number: 17 + 20 = 37.

PRACTICE EXERCISES

1.36 Fill in the blanks in the table below:

Isotope	Mass Number	Atomic Number	Number of Protons	Number of Neutrons
Chlorine-35		17		
Mercury-197		80		
Iodine-125			53	

1.37 Silicon has three naturally occurring isotopes: silicon-28, silicon-29, and silicon-30. Complete the table below:

Isotope	Mass Number	Atomic Number	Number of Protons	Number of Neutrons
Silicon-28		14		
Silicon-29			14	
Silicon-30				

a. What do all silicon isotopes have in common?
b. How are the isotopes of silicon different?
c. Which isotope of silicon has the greatest mass? Explain.

1.38 Only one stable isotope of gold exists and it contains 116 neutrons and 79 protons. Write this isotope using the element symbol with the atomic number and mass number in the conventional format.

Average Atomic Mass

What is the mass of a single atom? The mass of an atom is determined by the number of protons and neutrons that it contains. For example, a chlorine-35 atom has an approximate mass of 35 amu, because it contains 17 protons and 18 neutrons, each with a mass of 1 amu. Chlorine-37 has two more neutrons than chlorine-35, so its mass is approximately 37 amu. Since the natural abundance of chlorine on earth is 75% chlorine-35 and 25% chlorine-37, the *average* atomic mass of chlorine is 35.45 amu—a value that is in between the masses of the two isotopes but is closer to the mass of the most abundant of the two. An average that takes into account the greater contribution of some of its contributors is known as a *weighted* average. In this example, chlorine-35 makes a greater contribution to the average mass than chlorine-37, because there are three times as many chlorine-35 atoms as there are chlorine-37 atoms. The **average atomic mass** of an element is a weighted average of the mass of its isotopes based on their natural abundance. The term "average atomic mass" is often simplified to "atomic mass."

The average atomic mass of carbon is 12.007, although no single carbon atom has this mass; just as the average number of children per couple in the United States is 2.5, even though there is no such thing as a couple with half a child! It is just the nature of averages.

PRACTICE EXERCISE

1.39 Calcium has four naturally occurring isotopes: calcium-40, calcium-42, calcium-43, and calcium-44. The table below shows the natural abundance of each of these isotopes as a percentage of the total.

Calcium Isotope	Natural Abundance
Calcium-40	96.94%
Calcium-42	0.6479%
Calcium-43	0.1356%
Calcium-44	2.086%

a. Which of these isotopes has the greatest mass?
b. Which of these isotopes is the most abundant isotope of calcium on earth?
c. Which isotope has the greatest number of neutrons?
d. The average atomic mass of calcium is 40.08. Why is the average atomic mass most similar to the mass number for isotope calcium-40, yet not equal to 40?
e. How many more neutrons does calcium-44 have than calcium-40?
f. What are the units for average atomic mass?

1.3 The Periodic Table of Elements

To date, 111 different elements have been discovered. They are listed in the famous **Periodic Table of Elements,** shown in Figure 1-8 and on the inside front cover of this book. Each box in the periodic table provides information about a single element. In this section you will learn the significance of the layout of the periodic table of elements: the *rows* and *columns* on the table. Once you learn how to navigate the periodic table, it will become a useful tool when you build compounds from these elements in the next chapter. In Chapter 2 you will see that elements serve as the building blocks for all compounds. In other words, *all* matter is composed of atoms, including the life-saving drug given to a patient, the food you eat, and the hemoglobin in your blood cells.

How to Read the Periodic Table

Each box in the periodic table of elements displays the element symbol, atomic number, and average atomic mass of an element (Figure 1-8). Information

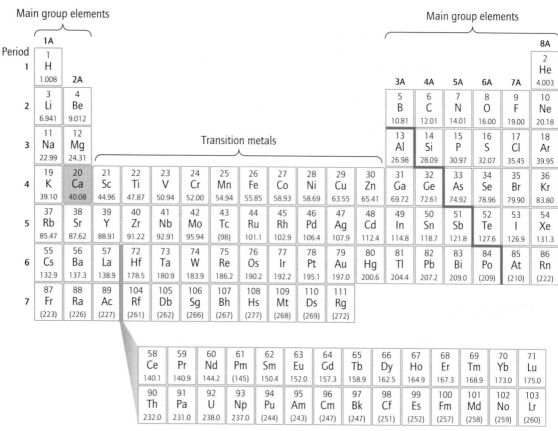

Figure 1-8 The periodic table of elements has 111 elements arranged in columns (called groups) and rows (called periods). The names of the elements are included on the inside front cover of the textbook.

about isotopes and their mass number does not appear in the periodic table, but must be obtained from other reference sources. Do not confuse the average atomic mass with the atomic number. They are readily distinguished, because the atomic number is always a whole number—the number of protons—while the atomic mass is a weighted average of mass that rarely is a whole number (see example below, which is also highlighted in the periodic table above).

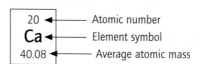

The elements are listed in the periodic table in order of their atomic number: from lowest atomic number to highest atomic number from left to right and from top to bottom across the table.

Problem-Solving Tutorial:
Reading the Periodic Table

WORKED EXERCISES 1-9 Using the Periodic Table to Obtain Information About an Element

1 Using the periodic table shown in Figure 1-8, determine the atomic number and element symbol for each of the following elements:
   a. carbon   **b.** gold

2 What are the element symbols and the full names for the elements with the following atomic numbers?
   a. atomic number = 7   **b.** atomic number = 80

3 How many protons and electrons do the following elements contain?

	number of protons	number of electrons
a. fluorine	_____	_____
b. sodium	_____	_____

SOLUTIONS

1 Locate the element symbol for the element in question on the periodic table. The atomic number will be listed in the box as well. It will always be a whole number:

    a. Carbon, C, atomic number = 6   b. Gold, Au, atomic number = 79

2 Locate the box with the indicated atomic number on the periodic table. The element symbol will be listed in the box as well. Atomic number 7 corresponds to nitrogen, N; atomic number 80 corresponds to mercury, Hg.

3 The number of protons for the atoms of an element is equal to the element's atomic number. An element is neutral so it will have an equal number of protons and electrons. Thus fluorine, F, has 9 protons and 9 electrons. Sodium, Na, has 11 protons and 11 electrons.

PRACTICE EXERCISES

1.40 What is the atomic number for each of the following elements?
    a. krypton   b. silver

1.41 What are the element symbols and the full names for the elements with the following atomic numbers?
    a. 10   b. 8

1.42 Which element has the fewest number of protons? What is its atomic number and element symbol?

1.43 Vanadium is found in some mineral supplements. How many protons and electrons does vanadium contain?

1.44 Which of the following are *not* elements? How can you tell?
    a. copper   b. bronze   c. gold   d. silver
    e. steel   f. uranium   g. tin   h. tungsten

Groups and Periods

The periodic table's unique array of 7 rows and 18 columns has special significance. Elements in the same column have similar physical and chemical properties—a property of the elements known as **periodicity.** A column of elements is known as a **group** or **family** of elements. Rows are known as **periods.**

Groups The 18 groups in the periodic table are numbered at the top of each column. The first two columns are numbered 1A and 2A, and the last six columns resume numbering from 3A to 8A, as shown in Figure 1-9. Elements in the first two and last six columns are known as **main group elements.** The middle section is numbered 1B to 8B, and contains the **transition metal elements.**

Although all groups can be identified by their group number, the four groups listed below also have common names:

Group 1A	**alkali metals**
Group 2A	**alkaline earth metals**
Group 7A	**halogens**
Group 8A	**inert gases** or **noble gases**

Elements within the same group exhibit similar chemical and physical properties. For example, two physical properties that the alkaline earth

Figure 1-9 Groups are represented in 18 columns on the periodic table. The groups representing the main group elements are shown in color.

Main group			Transition metals											Main group					

(Periodic table figure — groups labeled 1A–8A, 3B–2B, with alkali metals, alkaline earth metals, transition metals, halogens, noble gases indicated. Lanthanides and actinides shown separately below.)

Lanthanides

58 Ce	59 Pr	60 Nd	61 Pm	62 Sm	63 Eu	64 Gd	65 Tb	66 Dy	67 Ho	68 Er	69 Tm	70 Yb	71 Lu
90 Th	91 Pa	92 U	93 Np	94 Pu	95 Am	96 Cm	97 Bk	98 Cf	99 Es	100 Fm	101 Md	102 No	103 Lr

Actinides

The element magnesium is a malleable, shiny, silvery-white alkaline earth metal.
[Paul Silverman/Fundamental Photographs]

Magnesium reacts with oxygen when heated to produce magnesium oxide, released in the form of a white smoke.
[Richard Megna/Fundamental Photographs]

metals share are a silvery-white color and a soft consistency. Two readily observable chemical properties that the alkaline earth elements share are reactivity in air (except Be) and reactivity with water (except Be). If you were to examine another group of elements, you would observe a different set of physical and chemical characteristics, but similarities would be shared within the group.

Periods Each row in the periodic table is known as a *period*. There are seven periods in the table, as shown in Figure 1-10, which features the elements in period 4—the fourth row. Sections of periods 6 and 7 are always separated from the rest of the periodic table, and appear below the main table in two rows, each containing 14 elements. These two rows actually belong in the gap after the elements La and Ac, as indicated in the periodic table, and they are known as the **lanthanides** and **actinides,** respectively.

Periods

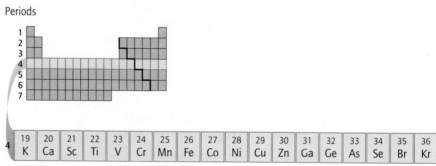

19 K	20 Ca	21 Sc	22 Ti	23 V	24 Cr	25 Mn	26 Fe	27 Co	28 Ni	29 Cu	30 Zn	31 Ga	32 Ge	33 As	34 Se	35 Br	36 Kr

Figure 1-10 Periods are represented in 7 rows on the periodic table. The elements in period 4 are featured here.

1 H																	2 He
3 Li	4 Be											5 B	6 C	7 N	8 O	9 F	10 Ne
11 Na	12 Mg											13 Al	14 Si	15 P	16 S	17 Cl	18 Ar
19 K	20 Ca	21 Sc	22 Ti	23 V	24 Cr	25 Mn	26 Fe	27 Co	28 Ni	29 Cu	30 Zn	31 Ga	32 Ge	33 As	34 Se	35 Br	36 Kr
37 Rb	38 Sr	39 Y	40 Zr	41 Nb	42 Mo	43 Tc	44 Ru	45 Rh	46 Pd	47 Ag	48 Cd	49 In	50 Sn	51 Sb	52 Te	53 I	54 Xe
55 Cs	56 Ba	57 La	72 Hf	73 Ta	74 W	75 Re	76 Os	77 Ir	78 Pt	79 Au	80 Hg	81 Tl	82 Pb	83 Bi	84 Po	85 At	86 Rn
87 Fr	88 Ra	89 Ac	104 Rf	105 Db	106 Sg	107 Bh	108 Hs	109 Mt	110 Ds	111 Rg							

Metals Metalloids Nonmetals

58 Ce	59 Pr	60 Nd	61 Pm	62 Sm	63 Eu	64 Gd	65 Tb	66 Dy	67 Ho	68 Er	69 Tm	70 Yb	71 Lu
90 Th	91 Pa	92 U	93 Np	94 Pu	95 Am	96 Cm	97 Bk	98 Cf	99 Es	100 Fm	101 Md	102 No	103 Lr

Figure 1-11 Metals (blue) and nonmetals (orange) are separated by the bold zigzag line running diagonally from boron to polonium. Metalloids (pink) appear along the bold zig-zag line.

Three representative halogens from group 7A. From left to right: chlorine, a pale yellow gas; bromine, a red brown liquid with a red vapor; and iodine, a purple solid with a purple vapor. A chemical property of the halogens is that their atoms are bonded together in pairs. [Chip Clark]

Metals, Nonmetals, and Metalloids

Another important distinction between elements in the periodic table is their classification as metals or nonmetals. **Metals** appear (in blue) to the left of the bold zigzag line that runs diagonally from boron to polonium, as shown in Figure 1-11. **Nonmetals** appear to the right of the bold line (in orange). Those elements located along the zigzag line (in pink in Figure 1-11) are called metalloids: they are boron, silicon, germanium, arsenic, antimony, tellurium, polonium, and astatine. **Metalloids** display characteristics of both metals and nonmetals.

Generally, metals and nonmetals have very different physical and chemical properties. Table 1-10 compares some of the physical properties of metals and nonmetals. In Chapter 2 you will learn that metals tend to lose electrons and nonmetals tend to gain electrons, which is an important chemical property that distinguishes metals from nonmetals.

Although hydrogen is placed at the top of the alkali metals group in the periodic table, because it has similar electron characteristics to the alkali metals, it is actually classified as a nonmetal.

All the elements that make up the important molecular compounds in your body are nonmetals. Many metals serve as important macronutrients and micronutrients in the body, as described in the next section.

Copper is a transition metal element with atomic number 29. It displays the shininess and malleability expected of a metal, and is also a good conductor of electricity. [Erich Schrempp/Fundamental Photographs]

Table 1-10 Physical Properties of Metals and Nonmetals

Metals	Nonmetals
Shiny	Dull
Exist as solids at room temperature (except mercury)	Exist as solids, liquids, and gases at room temperature
Good conductors of electricity	Poor conductors of electricity
Malleable (can be shaped)	Brittle, hard, or soft

WORKED EXERCISES 1-10 Groups and Periods

1 For the element potassium, indicate
   a. the family it belongs to,
   b. its group number,
   c. the period it belongs to, and
   d. whether it is a metal, nonmetal, or metalloid.

2 Indicate whether each of the following sentences describes a chemical property or a physical property:
   a. Cesium is a soft, shiny, gold-colored metal.
   b. Magnesium (Mg) reacts with hydrochloric acid to form magnesium chloride and hydrogen gas. Magnesium chloride is a compound different from elemental magnesium.

SOLUTIONS

1 a. alkali metal   b. group 1A   c. period 4   d. metal

2 a. These are physical properties of cesium, because they describe characteristics of the element that do not involve its interaction with another substance.
   b. This is a chemical property of magnesium, because it describes how magnesium behaves in the presence of hydrochloric acid, another substance, to produce yet other substances, magnesium chloride and hydrogen.

PRACTICE EXERCISES

1.45 State the group number and family name, if one exists, for the following elements:
      a. beryllium   b. iodine   c. argon   d. nitrogen

1.46 Which of the following elements are metals?
      a. sulfur   b. mercury   c. neon
      d. sodium   e. uranium   f. molybdenum

1.47 Provide the name of the element that fits the description listed in each case:
      a. A group 2A metal in period 5
      b. A halide in period 3
      c. A transition metal in period 4 with two more protons than calcium.

1.48 Indicate whether the following observations demonstrate chemical properties or physical properties of the element.
      a. When Cs is dissolved in water, it violently produces hydrogen gas, which ignites upon formation.
      b. Mercury is a dangerous metal, because it readily goes from the liquid state to the gas state, causing it to be readily inhaled. The maximum level of mercury in air allowed by the EPA (Environmental Protection Agency) is 0.1 mg/m^3.

Important Elements in Biochemistry and Medicine

Several elements are highlighted in the periodic table with a medical focus shown in Figure 1–12. These elements have the following important roles in biochemistry and medicine:

- Building block elements (C, H, N, O, P, and S), color-coded blue.
- Macronutrients (Na, K, Mg, Ca, P, S, Cl), color-coded orange.
- Micronutrients (V, Cr, Mn, Fe, Co, Cu, Zn, Mo, Si, Se, F, I), color-coded yellow.
- Radioisotopes in nuclear medicine (Ba, C, Ce, Cr, F, Ga, Au, I, Fe, Kr, Hg, P, Se, Sr, Te, Tl, Tc, Co), color-coded green.

The building block elements include carbon, hydrogen, nitrogen, oxygen, phosphorous, and sulfur. They are called building block elements because they are the elements that make up the chemical structure of the majority of molecular compounds found in living organisms.

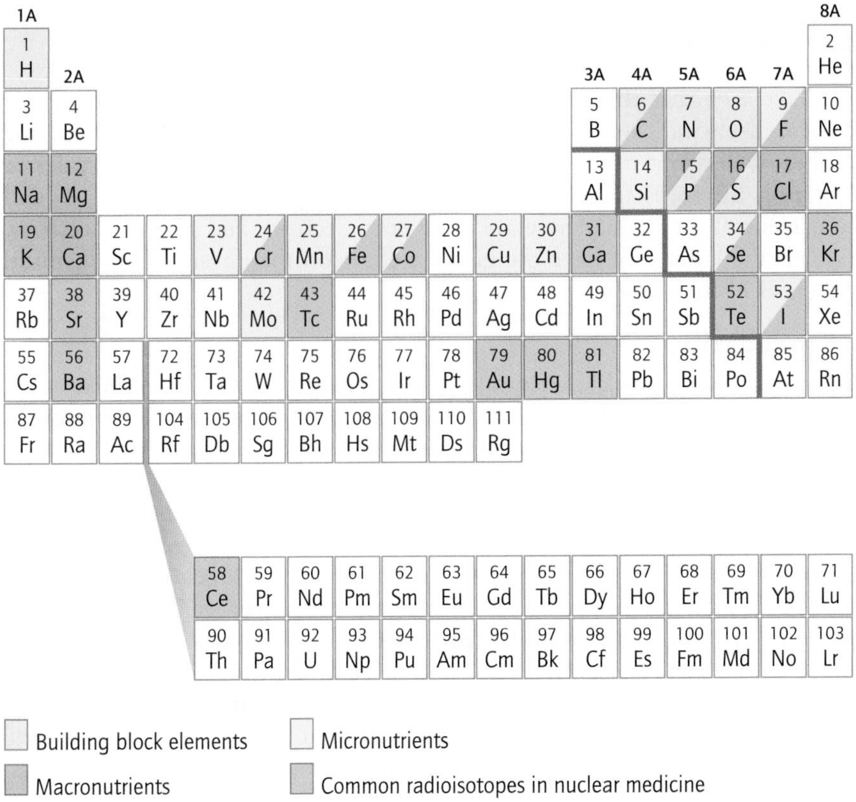

Figure 1-12 This periodic table of elements highlights the elements with particular medical interest.

☐ Building block elements ☐ Micronutrients

☐ Macronutrients ☐ Common radioisotopes in nuclear medicine

Essential nutrients supplied through the diet are divided into two categories: macronutrients and micronutrients. **Macronutrients** are elements required in large quantities—defined as more than 100 mg a day. They include sodium, potassium, magnesium, calcium, phosphorous, sulfur, and chlorine. **Micronutrients** are the 12 elements required in quantities of less than 100 mg a day, but nevertheless play a critical role in the functioning of your body. To learn more about the micronutrients, read *Chemistry in Medicine: The Micronutrients and Nutrition* at the end of the chapter.

Elements with important isotopes used in nuclear medicine are highlighted in green. These isotopes are used in the treatment or diagnosis of disease and are the subject of Chapter 16.

Micronutrients are also called **trace minerals**.

PRACTICE EXERCISES

1.49 Complete the table below for the six building block elements.

Element	Metal or Nonmetal	Group Number	Period Number	Atomic Number/ Number of Protons
Carbon				
Hydrogen				
Nitrogen				
Oxygen				
Phosphorus				
Sulfur				

1.50 What distinguishes a micronutrient from a macronutrient? How does your body obtain these nutrients?

1.51 How many of the seven macronutrients are main group elements?

1.52 How many of the building block elements are nonmetals?

1.4 Electrons

You have seen that the elements display a periodicity in their chemical and physical behavior. The main factor in determining this periodicity is the number of electrons in the outermost shell of an atom, known as the **valence electrons.** Electrons exist in shells that occur in layers, like an onion, around the nucleus of the atom. There are inner core electrons and valence electrons, and it is the latter that have the greatest effect on the physical and chemical properties of the element.

Electron Shells

Like all moving particles, electrons possess energy. Therefore, electrons can be characterized not only by the *shape* of their orbitals but by their *energy*. In general, electrons in shells closer to the nucleus have a lower energy than the valence electrons.

A peculiar property of electron energies is that only fixed energy levels are permitted. Imagine a car that could be driven only at fixed speeds of 25, 35, 55, and 85 miles per hour. The fixed electron **energy levels** are designated $n = 1, 2, 3$, etc., indicating increasingly higher electron energies. These energy levels are often referred to as **electron shells.** Electrons within an atom that belong to the same energy level are said to be in the same shell. Figure 1-13 shows how electron energy levels increase in increments (quanta): stepwise rather than continuously. From the discovery that electron energy levels are fixed (quantized) emerged the field of physics known as quantum mechanics, which revolutionized our understanding of the atom.

A given electron shell, n, can accommodate a maximum of $2n^2$ electrons, where $n = 1, 2, 3$, etc. Thus, the $n = 1$ shell can accommodate up to 2 electrons, the $n = 2$ shell, up to 8 electrons, and so forth, as shown in Table 1-11.

The lower the value of n, the closer the electrons are to the nucleus. The valence electrons are the electrons in the shell with the highest value of n. For example, a carbon atom contains 6 electrons. Two electrons are in the $n = 1$ shell, and 4 electrons are in the $n = 2$ shell. The valence electrons are therefore the 4 electrons in the $n = 2$ shell, while the other 2 electrons are inner core electrons, as shown in Figure 1-13. The inner core electrons are closest to the nucleus. The distance of shells from the nucleus increases with higher energy. Electrons in the outermost shell, with the highest n value, are the important valence electrons, which have the highest energy and the largest orbitals, and thus spend more time further from the nucleus.

Valence Electrons and Electron Shells

In the most stable form of an atom, its electrons will occupy the lowest energy levels, until the maximum number of electrons allowed in a shell has been reached.

Table 1-11 Maximum Number of Electrons Allowed in Each Electron Shell

Electron Shell	Maximum Number of Electrons, $2n^2$
1	2
2	8
3	18
4	32

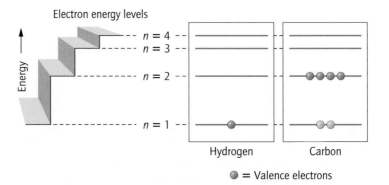

Figure 1-13 Electron orbital energy levels increase in a stepwise rather than continuous fashion. The single electron of hydrogen is in the lowest level, $n = 1$, whereas the six electrons of carbon are divided between two shells, with four valence electrons in the outermost $n = 2$ level.

Electron energy levels

Energy

$n = 4$
$n = 3$
$n = 2$
$n = 1$

Hydrogen   Carbon

● = Valence electrons

Table 1-12 Electron Arrangements for Elements Hydrogen to Argon
(Yellow cells denote valence electrons.)

Period	Element	Group Number	$n = 1$	$n = 2$	$n = 3$	$n = 4$
1	H	1	1			
	He	2	2			
2	Li	1	2	1		
	Be	2	2	2		
	B	3	2	3		
	C	4	2	4		
	N	5	2	5		
	O	6	2	6		
	F	7	2	7		
	Ne	8	2	8		
3	Na	1	2	8	1	
	Mg	2	2	8	2	
	Al	3	2	8	3	
	Si	4	2	8	4	
	P	5	2	8	5	
	S	6	2	8	6	
	Cl	7	2	8	7	
	Ar	8	2	8	8	

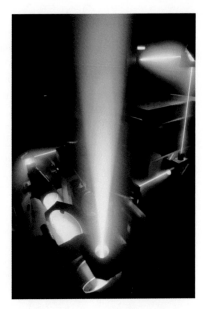

Green laser light emitted by the noble gas, argon. [Roger Ressmeyer/Corbis]

Consider the arrangement of electrons for elements with atomic numbers 1 through 18, shown in Table 1-12. Valence electrons are shown in yellow cells. The two elements in period 1 have electrons in only one shell, and therefore their only electrons are valence electrons. Hydrogen has 1 electron and helium has 2 electrons—enough to fill the $n = 1$ shell. Period 2 has eight elements, lithium through neon. These elements all have 2 electrons in the first shell, like the noble gas helium in period 1. The remaining electrons occupy the outermost $n = 2$ shell, as valence electrons. The $n = 2$ energy level can accommodate a maximum of 8 electrons. Lithium has 1 electron in the $n = 2$ shell, and therefore, 1 valence electron. Beryllium has 2 electrons in the $n = 2$ shell, and therefore, 2 valence electrons. Boron has 3 valence electrons, carbon has 4, nitrogen 5, oxygen 6, fluorine 7, and neon 8. Neon is the element in period 2 with a full $n = 2$ shell, because the $n = 2$ shell can accommodate a maximum of 8 electrons.

Every element within a group has the same number of valence electrons, which corresponds to the group number. For example, hydrogen, lithium, sodium, potassium, rubidium, cesium, and francium all have 1 valence electron, as illustrated for H, Li, and Na in Figure 1-14 in green, and for this reason they are grouped together in the periodic table as group 1A elements. The difference among the elements in group 1A is

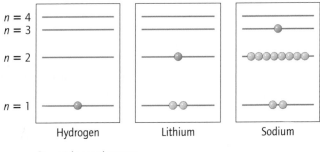

= Valence electrons

Figure 1-14 Electron energy levels for three elements from group 1A: hydrogen (H), lithium (Li), and sodium (Na). Each has 1 valence electron, found in increasingly higher energy levels: $n = 1, 2, 3$.

The Model Tool 1-1 The Building Block Elements

If you received a model tool kit with your textbook, your instructor may ask you to perform the following exercise.

1. Your model kit contains color coded atoms for five of the six building block elements. You will begin to use this kit regularly starting in Chapter 2. Here you are introduced to the atoms.

2. Fill in the table below for each building block element in your kit.

Element	C	N	O	Cl	H
(a) Color of the atom in your kit					
(b) Number of valence electrons					

which shell contains the valence electron: the energy level corresponding to the valence shell increases from hydrogen (period 1) to francium (period 7). Thus, hydrogen has its one electron in the $n = 1$ shell, while sodium has its one valence electron in an $n = 3$ shell, and francium has its one valence electron in an $n = 7$ shell.

The inert gases all contain a full shell of 8 valence electrons, except for helium which contains a full shell with 2 electrons. A full shell is a particularly stable electron arrangement, and accounts for the lack of chemical reactivity displayed by these elements—they are inert.

Elements in period 3 all have a full $n = 1$ and $n = 2$ shell. The elements sodium through argon have 1 through 8 valence electrons in the outermost $n = 3$ shell. Elements with atomic number greater than 20 have more complex electron arrangements; nevertheless, the number of valence electrons always corresponds to the group number for any main group element.

In general, atoms with valence electrons in higher energy levels will have a larger diameter than atoms with valence electrons in lower energy levels, because their valence electrons are in larger orbitals, which occupy a volume further from the nucleus. For example, a lithium atom is smaller than a sodium atom.

WORKED EXERCISES 1-11   Valence Electrons and Energy Levels

1 How many valence electrons do the following elements contain?
  a. bromine   **b.** thallium
2 Complete the table below.

Element	Symbol	Atomic Number	Group Number	Number of $n = 1$ Electrons	Number of $n = 2$ Electrons	Number of $n = 3$ Electrons
Carbon	C					
				2	8	2

SOLUTIONS

1 **a.** Bromine is a group 7A element so it has 7 valence electrons.
  b. Thallium is a group 3A element so it has 3 valence electrons.
2 Complete the table below.

Element	Symbol	Atomic Number	Group Number	Number of $n = 1$ Electrons	Number of $n = 2$ Electrons	Number of $n = 3$ Electrons
Carbon	C	6	4A	2	4	0
Magnesium	Mg	12	2A	2	8	2

PRACTICE EXERCISES

1.53 How many valence electrons do the following elements contain?

    a. barium **b.** oxygen

1.54 Complete the table below.

Element	Symbol	Atomic Number	Group Number	Number of $n = 1$ Electrons	Number of $n = 2$ Electrons	Number of $n = 3$ Electrons
Sulfur	S					
				2	8	8

1.55 Radioactive strontium is dangerous because, being similar to calcium, it can accumulate in bone. In terms of electrons, how is strontium similar to calcium? How is it different?

In this chapter you learned how scientists measure matter, entering the nanoscale world of the atom. You learned that elements are composed of one type of atom and that their physical and chemical properties are determined by their valence electrons. In Chapter 2 you will learn how atoms come together to form compounds, a tendency that is also determined by an atom's valence electrons.

Chemistry in Medicine The Micronutrients and Nutrition

100 mg is 0.1 g or 1/10 of a gram. This is about the mass of a few grains of dry rice.

There are 12 elements in the periodic table of elements that are classified as micronutrients, as shown in Figure 1-15. Micronutrients are elements that your body needs in only trace quantities, defined as less than 100 mg/day. Even though micronutrients are required in small quantities, they are essential for health, and must be supplied regularly through the diet.

Some micronutrients are found distributed throughout the body, while others are localized primarily in certain organs. For example, iodine is localized in the thyroid gland, as shown in Figure 1-16. The thyroid gland uses iodine to produce two important iodine-containing hormones that help regulate metabolism. They help control weight, heart rate, blood cholesterol, muscle strength, and skin condition.

Most people ingest about 100–200 µg/day of iodine, primarily from iodized salt. An iodine deficiency (<20 µg/day) can lead to an enlarged thyroid—called a goiter. Severe iodine deficiencies are associated with high infant mortality and mental retardation. It is estimated that 29% of the world's population lives in areas where the soils are depleted in iodine, causing locally grown foods to be deficient in this essential micronutrient. Iodine supplements and the fortification of food with iodine have been used to fight iodine deficiency disorders worldwide. A total of one teaspoon of iodine is all a person requires over a lifetime!

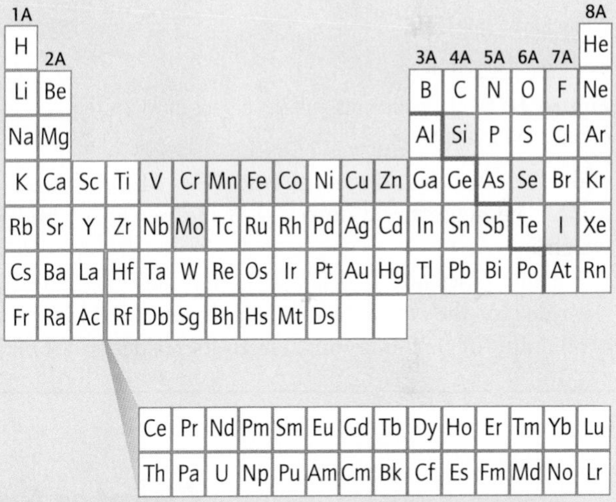

Figure 1-15 The micronutrients are highlighted in yellow in this periodic table. These are elements obtained from the diet and required in amounts less than 100 mg/day.

Some micronutrients, such as fluorine, serve a structural role in the body. Fluoride—a form of fluorine described in the next chapter—provides part of the structure of bone and teeth. Optimal fluoride intake has been shown to reduce dental cavities, because fluoride helps produce stronger tooth enamel. Fluoride is found in drinking water and tea. However, excessive intake of fluoride can cause yellowing of teeth, an enlarged thyroid gland,

continued

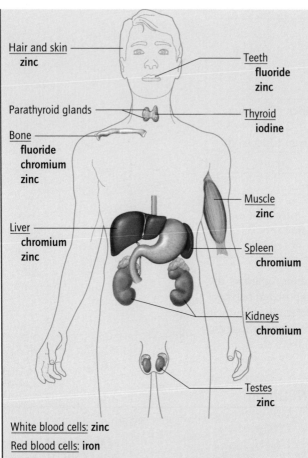

Hair and skin
zinc

Parathyroid glands

Bone
fluoride
chromium
zinc

Liver
chromium
zinc

Teeth
fluoride
zinc

Thyroid
iodine

Muscle
zinc

Spleen
chromium

Kidneys
chromium

Testes
zinc

White blood cells: **zinc**

Red blood cells: **iron**

Figure 1-16 Micronutrients and the organs in which they concentrate in the body.

and brittle bones and teeth. Large doses of fluoride are toxic. The acute toxic dose is 2–8 mg per kg of body weight.

Many cities add 0.7–1.2 parts per million (ppm) of fluoride to the local water supply to increase the availability of this micronutrient to its residents. 1 ppm fluoride means there is 1 mg of fluoride in every liter of water. Municipal water fluoridation has met with some controversy. Although fluoride is a naturally occurring mineral found in lakes, rivers, and all other water sources, some groups fear possible harmful health effects of artificial adjustment of the fluoride concentration in municipal water. Advocates of fluoridation argue that it is analogous to fortifying salt with iodine or adding vitamin D to milk.

Many of the micronutrients found among the transition metals in the periodic table are important for the proper functioning of enzymes, extremely important chemical substances that enhance the rate of chemical reactions in the body. For example, zinc is an element required by more than 100 different enzymes. Zinc is important in the immune system, and is often found in over-the-counter throat lozenges. The concentration of zinc is highest in bone, teeth, hair, skin, liver, muscle, white blood cells, and the testes, as shown in Figure 1-16.

30
Zn
65.38
Zinc

Adults usually require 8–12 mg/day of zinc. Good sources of zinc include oysters, breakfast cereal, beef, pork, chicken, yogurt, baked beans, and nuts. Zinc deficiencies can result in growth retardation, hair loss, diarrhea, skin lesions, and slow healing of wounds.

In the opening vignette you learned that iron is a part of hemoglobin, the molecule that transports oxygen in red blood cells. Iron is also found in many other oxygen transport molecules and enzymes that are involved in extracting energy from the foods you eat. Iron-rich foods include clams, pork, beef liver, iron-fortified cereal, and pumpkin seeds. Iron is believed to be absorbed more efficiently from the diet when meat, fish, and vitamin C are part of the diet; while coffee and tea hinder the absorption of iron.

26
Fe
55.85
Iron

Chapter Summary

Measurement and the Metric System

- Every measurement has a numerical value and a unit.
- The metric system consists of several base units and a number of prefixes that represent a multiple of ten of the base unit: giga, kilo, deci, centi, milli, micro, nano, pico.
- The meter is the base unit of length in the metric system and the gram is the base unit of mass in the metric system.
- The metric unit of volume is the liter. There is 1 milliliter in 1 cubic centimeter.
- The degree of uncertainly in the numerical value obtained from a measurement is indicated by the number of significant figures.
- Significant figures are all the nonzero digits and the zeros in a number that do not serve as place holders.

- Solutions to calculations that require multiplying and dividing measured numbers must be rounded so that there are no more significant figures than in the measured value with the least number of significant figures. Solutions to addition and subtraction calculations cannot have more decimal places than the measured value with the fewest decimal places.
- Metric conversions are performed using dimensional analysis, which involves multiplying a supplied unit by a conversion factor that equates the supplied and requested units and allows the supplied units to cancel algebraically.
- Density is a physical property of a substance defined as its mass per volume: $d = m/V$. Density can be used as a conversion factor to calculate mass and volume.

Elements and the Structure of the Atom

- The atom is the smallest stable component of matter. An atom is composed of protons, neutrons, and electrons.
- Protons have a positive charge, electrons a negative charge, and neutrons are neutral.
- The mass of the atom is the sum of the protons and neutrons, located in the nucleus.
- Electrons are located in orbitals and have a negligible mass. The size of the atom is determined by its electrons.
- An element is composed of one type of atom, and cannot be broken down into a simpler form of matter.
- The atomic number for an element corresponds to the number of protons its atoms contain.
- Isotopes are atoms with the same number of protons but a different number of neutrons.
- The mass number for an isotope is the sum of the protons and neutrons.
- The average atomic mass of an element is the weighted average of its isotopes, which depends on both the natural abundance of each isotope and the mass of each isotope.

The Periodic Table of Elements

- Elements in the same column in the periodic table belong to the same group or family. Elements in the same row in the periodic table belong to the same period.
- The main group elements are in groups 1A through 8A.
- Elements in the same group contain the same number of valence electrons, accounting for their similar chemical and physical properties.
- Metals and nonmetals are separated by the bold diagonal zigzag line on the right side of the periodic table. Metalloids are the elements along this line.
- Many of the elements have important biochemical and medical value as building block elements, micronutrients, macronutrients, and elements used in nuclear medicine.

Electrons

- Electrons occupy different energy levels, or shells, n, where $n = 1, 2, 3 \ldots$. The maximum number of electrons in a given shell is $2n^2$.
- Valence electrons are the outermost electrons of an atom, found in the shell with the highest n value. The other electrons are inner core electrons.
- Elements in the same group have the same number of valence electrons, which corresponds to the group number.

Key Words

Alkali metals The elements in group 1A.

Alkaline earth metals The elements in group 2A.

Atom The smallest component of an element that still displays the characteristics of the element.

Atomic number The number of protons in an atom; it defines the element.

Average atomic mass A weighted average of the mass of the isotopes of an element based on the natural abundance of each isotope and its mass.

Chemical properties Behavior of a substance in the presence of other substances. A chemical reaction transforms substances into new substances.

Conversion The mathematical expression equating two different units.

Conversion factor A relationship between units expressed as a fraction.

Density A physical property of a substance defined as its mass per unit volume: $d = m/V$.

Electron A subatomic particle of the atom with a negative charge and negligible mass, located in an orbital.

Electron orbital A region of space describing where the electron is most likely to be found.

Electron shell A fixed energy level of the electron, designated $n = 1, 2, 3$, etc. Higher numbered shells are higher in energy and larger in size, and they accommodate more electrons.

Element A substance composed of only one type of atom, which cannot be broken down into a simpler form of matter. There are 111 known elements.

Element symbol The one- or two-letter symbol used to identify an element.

Group A column in the periodic table.

Halogens The elements in group 7A.

Isotopes Atoms having the same number of protons but a different number of neutrons. They have a different mass.

Macronutrients Elements which must be supplied through the diet in a quantity greater than 100 mg/day.

Main group elements The elements in groups 1A through 8A.

Matter Anything that has mass and occupies volume.

Mass number The sum of the number of protons and neutrons in an isotope.

Metals The elements on the left side of the periodic table (left of the dark zigzag line).

Micronutrients Elements that must be obtained through the diet in quantities of less than 100 mg/day.

Neutron Subatomic particle with no charge and a mass of approximately 1 amu, located in the nucleus of the atom.

Noble gases The elements in Group 8A.

Nonmetals Elements on the right side of the periodic table (right of the dark zigzag line).

Nucleus The small dense center of the atom that contains the protons and neutrons.

Period A row in the periodic table.

Periodic table of elements A table of the 111 elements showing their symbol, atomic number and atomic mass, displayed in characteristic rows and columns, which reveals the periodicity of the elements.

Periodicity The repeating chemical and physical properties of elements, reflected in the layout of the periodic table.

Physical properties The properties of a substance that we can observe or measure that do not involve another substance.

Proton Subatomic particle with a positive charge, and a mass of 1 amu, located in the nucleus of the atom.

Scientific notation A number such as 3,000,000 with many zeros, expressed as 3×10^6, where the first number is a number between 1 and 10, and the superscript is a whole number, indicating how many places the decimal should be moved to the left or the right.

Significant figures All the certain digits and one uncertain digit in a measurement.

Subatomic particles The parts of an atom: protons, neutrons, and electrons.

Transition metal elements The metals in groups 1B–8B, positioned between groups 2A and 3A on the periodic table.

Valence electrons The outermost electrons of an atom, with the highest n value; equal to the group number.

Additional Exercises

Matter and Measurement

1.56 The fundamental component of all matter is the ____.

1.57 Classify the size of the following items as macroscopic, microscopic, or on the nanoscale:
- **a.** A skyscraper
- **b.** A skin cell
- **c.** DNA
- **d.** A red blood cell

1.58 Classify the size of the following items as being on macroscopic, microscopic, or on the nanoscale:
- **a.** A chromium atom
- **b.** The human body
- **c.** A grain of sand

1.59 Arrange the following metric prefixes in order of increasing size:
- **a.** nano
- **b.** kilo
- **c.** pico
- **d.** micro

1.60 Arrange the following metric prefixes in order of increasing size:
- **a.** femto
- **b.** centi
- **c.** milli
- **d.** deci

1.61 For each of the following pairs, which length is shorter? If the lengths are the same, state so.
- **a.** 1 m or 10 mm
- **b.** 1 cm or 10 mm
- **c.** 1 cm or 1 dm
- **d.** 15 cm or 1 dm

1.62 For each of the following pairs, which length is shorter? If the lengths are the same, state so.
- **a.** 10 m or 1 km
- **b.** 1 nm or 1022 m'
- **c.** 10^{-3} m or 1 mm
- **d.** 1 nm or 1 μm

1.63 A lead ball is dropped into a graduated cylinder containing 15.0 mL of water, causing the level of the water to rise to 16.5 mL. What is the volume of the lead ball?

1.64 An irregular shaped metal sample is placed in a graduated cylinder containing 200. mL of water. The water level increases to 203.5 mL. What is the volume of the metal sample?

1.65 For each of the following pairs, state which is the smaller number. If the numbers are the same, state so.
  a. 10^4 or 10^8
  b. 10^{-3} or 10^{-6}
  c. 3.7×10^4 or 3.7×10^{-4}

1.66 For each of the following pairs, state which is the smaller number. If the numbers are the same, state so.
  a. 1×10^3 or 1.5×10^3
  b. 1×10^2 or 0.1×10^3
  c. 2.5×10^{-3} or 25×10^{-2}

1.67 For each of the following pairs, state which measurement represents the smaller mass. If they have the same mass, state so.
  a. 1 ng or 1 mg
  b. 100 mg or 1 g
  c. 1000 mg or 1 g

1.68 For each of the following pairs, state which measurement represents the smaller mass. If they have the same mass, state so.
  a. 1 g or 1 kg
  b. 1 μg or 1 kg
  c. 1 g or 1×10^{-3} kg

1.69 Express the following numbers in scientific notation:
  a. 1,000,000,000,000,000,000
  b. 2,305,000,000
  c. 0.0000000000015
  d. 0.0208

1.70 Express the following numbers in scientific notation:
  a. 0.0000076   **c.** 10,000
  b. 0.001   **d.** 1400

1.71 Express the following numbers written in scientific notation in conventional form:
  a. 1×10^5 **c.** 1.65×10^2
  b. 2.4×10^{-3}

1.72 Which is the larger number: 4.5×10^{-2} or 4.5×10^2?

1.73 Answer TRUE or FALSE:
  a. _____ Only sloppy work leads to uncertainty in a measurement.
  b. _____ A balance must be calibrated in order to give accurate readings.
  c. _____ In experimental work, if a series of repeated measurements gives values very close to one another, but not close to the "true value" they are considered precise.
  d. _____ In experimental work, if a series of repeated measurements gives values very close to one another, but not close to the "true value" they are considered accurate.

1.74 Indicate the number of significant figures in each of the following measurements: (a) 57,000; (b) 4.60; (c) 0.00011; (d) 23,304.60.

1.75 Indicate the number of significant figures in each of the following measurements: (a) 304; (b) 5,000; (c) 5,110; (d) 0.000330.

1.76 Perform the following calculations. Show the correct number of significant figures in your answer, assuming each number is a measured value. Include units.
  a. 56.33 cm $\times$ 2.5 cm =
  b. 3.4 cm + 2.2 cm + 5.11 cm + 8.777 cm =
  c. 34.22 g/39.0 mL =

1.77 Perform the following calculations. Show the correct number of significant figures in your answer, assuming each number is a measured value.
  a. 33,000 + 910 =
  b. 0.333 g $\times$ 0.22 =
  c. (37.55 mL + 22.2 mL) $\times$ 56.66 =

1.78 Which of the following represent exact numbers, and therefore contain an infinite number of significant figures?
  a. 2.54 cm = 1 in
  b. a room 24 cm $\times$ 33 cm
  c. 55 students in a classroom

1.79 Which of the following represent exact numbers, and therefore contain an infinite number of significant figures?
  a. 1 mm = 10^{-3} m
  b. a mass of 56.7 kg
  c. a density of 1.2 g/mL

1.80 Carry out the following metric conversions. Remember to report your final answer to the correct number of significant figures.
  a. Convert 50,000. m into kilometers.
  b. How many micrograms are there in 0.66 g?

1.81 Carry out the following metric conversions. Remember to report your final answer to the correct number of significant figures.
  a. How many milliliters are there in 6.000 L?
  b. A sample of compact bone has a mass of 3.8 g and a volume of 2.0 cm³. What is the density of the sample?

1.82 What conversion factors would you use to convert 61,000. mm into picometers?

1.83 What conversion factor would you use to convert 4,000 m into kilometers?

1.84 How many milliliters are there in 75.6 μL?

1.85 Ibuprofen can be found in 200.-mg doses in over-the-counter analgesics like Advil and Motrin. How many grams of ibuprofen does an Advil tablet contain?

1.86 What is the volume of a cube measuring 24 cm a side?

1.87 How many seconds are there in 2.000 minutes?

1.88 Using dimensional analysis, calculate how many seconds there are in 5.2 years. Express the final answer in scientific notation.

1.89 What is the weight in pounds of an animal weighing 150. kg?

1.90 There are 5280 ft in a mile. How many kilometers are there in 68.2 miles?

1.91 There are 4 quarts in a gallon. How many liters are there in 86 gal?

1.92 There are 8 oz in a cup and 4 cups in a quart. How many milliliters are there in 2 oz?

1.93 The recommended dose of Ceclor, an antibiotic, is 20 mg/kg of body weight a day, in three equally divided doses, one given every 8 hours. How many milligrams per dose should a baby weighing 12 lb receive?

1.94 Tylenol is ordered for a child with a fever at a dose of 25 mg/kg of body weight a day. If the child weighs 34 lb, how much Tylenol should be given to the child per day?

1.95 Which object in the following pairs has the greater density? Explain why.
a. A loaf of bread or a brick
b. A bowling ball or a soccer ball
c. A bucket full of water or a bucket full of concrete

1.96 What is the density of an object with a volume of 2.2 mL and a mass of 3.0 g?

1.97 What is the mass of a gold sphere that displaces 2.3 mL of water? (Use Table 1-5.)

1.98 What substance has a density of 1.0 g/mL?

1.99 Using Table 1-5, calculate the mass of a gold cube with sides 2.20 cm in length.

1.100 Convert 500. mg of vitamin C into grams and micrograms (μg).

1.101 Rank from largest to smallest volume: 50.00 mL; 5,000 μL; 0.5000 L; 8.000 cm³.

1.102 A doctor must make an incision 2.5 cm long. How long will the incision be in meters?

1.103 Would a gold sphere with a mass of 15 g have a greater diameter or a smaller diameter than a gold sphere with a mass of 6 g? (*Hint:* No calculation required.)

Elements and the Structure of the Atom

1.104 What are the three subatomic particles in an atom? What are the charges on those subatomic particles?

1.105 Which is the lightest of the following subatomic particles: the proton, the electron, or the neutron?

1.106 Where are the protons, neutrons, and electrons located within an atom?

1.107 Gold has 79 electrons, while helium has 2. Which atom would you expect to have the smaller diameter? Explain.

1.108 What is the full name and atomic number of the following elements? (a) O; (b) Na; (c) Cu; (d) Sn; (e) Ru; (f) W; (g) Eu.

1.109 What is the full name and element symbol for the elements with the following atomic numbers? (a) 6; (b) 13; (c) 95; (d) 78; (e) 27.

1.110 How many protons and electrons are there in the following elements?
a. oxygen
b. chromium
c. phosphorus

1.111 How many protons and electrons are there in the following elements?
a. cesium
b. rhenium
c. manganese

1.112 Which element has 51 protons?

1.113 Which element has 33 protons?

1.114 Which element has 56 electrons?

1.115 Which element has 88 electrons?

1.116 Between which two elements in the periodic table is lead located?

1.117 Which atom do you expect to have a larger diameter, oxygen or selenium? Explain.

1.118 How is the atomic number of an element determined?

1.119 What is the mass number of an element?

1.120 What are isotopes of an element? How do isotopes of an element differ from each other? How are they the same?

1.121 Sulfur has four naturally occurring isotopes: sulfur-32, sulfur-33, sulfur-34, and sulfur-36. Complete the table below, which gives information about each of these isotopes.

Isotope	Mass Number	Atomic Number	Number of Protons	Number of Neutrons
Sulfur-32				
Sulfur-33				
Sulfur-34				
Sulfur-36				

1.122 What do all sulfur isotopes have in common?

1.123 Which sulfur isotope is the lightest? Explain.

1.124 In nature, 51% of bromine atoms are bromine-79 and the other 49% are bromine-81.
a. What is the atomic number for each of these isotopes?
b. What is the mass number for each of these isotopes?
c. What is the difference between these two isotopes?
d. Using the periodic table, look up the average atomic mass of bromine.
e. Why is the average atomic mass of bromine approximately 80 and not 79 or 81? Does this value for bromine's average atomic mass make sense given the relative abundance of the bromine isotopes?

1.125 Oxygen has three stable isotopes: oxygen-16, oxygen-17, and oxygen-18. Write each of these three isotopes using the symbol for the element along with the appropriate subscript and superscript.

1.126 Iron has four natural isotopes: iron-54, iron-56, iron-57, iron-58. The natural abundances of the isotopes are as follows:

Iron Isotope	Natural Abundance
Iron-54	5.80%
Iron-56	91.72%
Iron-57	2.20%
Iron-58	0.28%

a. Which isotope is the lightest?
b. Which isotope is the least abundant?
c. Which isotope is the most abundant?
d. Which isotope has the smallest number of neutrons?
e. Why is the average atomic mass of iron, 55.845, closest to the mass number of iron-56 but not exactly 56.000?

1.127 On earth, 75% of chlorine atoms are chlorine-35 and the other 25% are chlorine-37.
a. What is the mass number for each of these isotopes?
b. What is the atomic number for each of these isotopes?
c. How do these two isotopes differ from one another?

1.128 Technetium-99 is a radioactive isotope frequently used in medicine. How many protons, electrons, and neutrons does Tc-99 have? Write this isotope using the element symbol with the atomic number and mass number in the conventional format.

The Periodic Table of Elements

1.129 What is the difference between physical and chemical properties of an element?

1.130 What is a family or group of elements?

1.131 What groups constitute the main group elements? Where on the periodic table are the transition metal elements?

1.132 What is a row of elements in the periodic table called?

1.133 What is the difference between the alkali metals and the alkaline earth metals?

1.134 State the group number and the name of the family, if one exists, for the following elements: (a) calcium; (b) krypton; (c) bromine; (d) rubidium; (e) chlorine; (f) sodium.

1.135 Indicate whether the following observations demonstrate chemical properties or physical properties of the element.
 a. Nickel is a malleable, ductile, lustrous metal.
 b. Gallium melts and shrinks in volume when held in the hand.

1.136 What are the differences between nonmetals and metals?

1.137 Classify the following as *nonmetal, metal,* or *metalloid*: (a) oxygen; (b) germanium; (c) carbon.

1.138 Classify the following as *nonmetal, metal,* or *metalloid*: (a) tin; (b) beryllium; (c) helium; (d) silicon; (e) chromium; (f) phosphorus.

1.139 Provide the name of the element that fits the following critieria:
 a. A group 1A metal in period 4
 b. A noble gas in period 6

1.140 Provide the name of the element that fits the following critieria:
 a. A transition metal in period 5 with three more protons than technetium
 b. An actinide with two fewer protons than plutonium

1.141 What subatomic particle is most influential in determining the physical and chemical properties of an element?

1.142 How does the group number relate to the number of valence electrons in elements in that group?

1.143 Indicate the number of electrons in each energy level for the following elements and indicate the number of valence electrons: (a) boron; (b) phosphorus.

1.144 Indicate the number of electrons in each energy level for the following elements, and indicate the number of valence electrons: (a) oxygen; (b) beryllium.

1.145 What is the difference between an *s* and a *p* orbital?

1.146 Which family of elements has 7 valence electrons?

1.147 Which family of elements has 4 valence electrons?

1.148 Name three building block elements. Write the number of valence electrons for each of those elements. Why are they called building block elements?

1.149 Name two macronutrients and two micronutrients. What distinguishes micronutrients from macronutrients?

1.150 Are the building block elements metals or nonmetals?

Chemistry in Medicine

1.151 What makes an element a micronutrient?

1.152 What organ in the body requires the most iodine? What is the source of iodine for most people?

1.153 Why is iron an essential micronutrient?

1.154 What role does fluorine play in the body?

Answers to Practice Exercises

1.1 There is 1×10^{-12} m in 1 pm. There is 1×10^{-12} g in 1 pg. The prefix pico stands for 1×10^{-12} or 1/1,000,000,000,000 of the base unit.

1.2 3.0 cm

1.3 (a) 1 mm is longer than 1 μm; (b) 1 km and 10^3 m are the same length.

1.4 (a) milli = 1/1,000 or 10^{-3}; (b) deci = 1/10 or 10^{-1}; (c) kilo = 1,000 or 10^3.

1.5 1 km and 1m can be seen by the naked eye.

1.6 (a) 1 mg; (b) 1 μg; (c) 10 mcg is the same as 10 μg.

1.7 $V = 1 \text{ cm}^3$

1.8 (a) 1; the zeros are placeholders; (b) infinite (exact number) due to counting; (c) 5; zeros are significant because there is a decimal; (d) 4.

1.9 (a) iii; (b) The last digit in each number is uncertain; (c) ii.

1.10 (a) 2,146; (b) 15.222; (c) 3,167.

1.11 −900

1.12 $0.65 \text{ m} \times \dfrac{1 \text{ cm}}{10^{-2} \text{ m}} = 65 \text{ cm}$ (2 significant figures)

1.13 $7.7 \text{ kg} \times \dfrac{10^3 \text{ g}}{1 \text{ kg}} = 7.7 \times 10^3 \text{ g or } 7,700 \text{ g}$
(2 significant figures)

1.14 $561 \text{ mL} \times \dfrac{10^{-3} \text{ L}}{1 \text{ mL}} = 0.561 \text{ L}$ (3 significant figures)

1.15 $56 \text{ nm} \times \dfrac{10^{-9} \text{ m}}{1 \text{ nm}} \times \dfrac{1 \text{ pm}}{10^{-12} \text{ m}} = 5.6 \times 10^4 \text{ pm or}$
$56,000 \text{ pm}$

1.16 $5.5 \text{ nm} \times \dfrac{10^{-9} \text{ m}}{1 \text{ nm}} = 5.5 \times 10^{-9} \text{ m}$
$5.5 \times 10^{-9} \text{ m} \times \dfrac{1 \text{ mm}}{10^{-3} \text{ m}} = 5.5 \times 10^{-6} \text{ mm}$

1.17 $150. \text{ μm} \times \dfrac{1 \times 10^{-6} \text{ m}}{1 \text{ μm}} = .000150 \text{ m}$
(3 significant figures)
$2. \text{ mm} \times \dfrac{1 \times 10^{-3} \text{ m}}{1 \text{ mm}} = .002 \text{ m}$ (1 significant figures)
Since .002 m > 0.000150 m, the patient with the 2.-mm tumor has the larger tumor.

1.18 $15 \text{ gal} \times \dfrac{4 \text{ qt}}{1 \text{ gal}} \times \dfrac{1 \text{ L}}{1.057 \text{ qt}} = 57 \text{ L}$

1.19 $1,453 \text{ ft} \times \dfrac{12 \text{ in}}{1 \text{ ft}} \times \dfrac{1 \text{ m}}{39.37 \text{ in}} = 442.9 \text{ m}$

1.20 $906 \; \cancel{g} \times \dfrac{1 \; \cancel{kg}}{10^3 \; \cancel{g}} \times \dfrac{2.205 \; lb}{1 \; \cancel{kg}} = 2.00 \; lb$

1.21 $110 \; \cancel{lb} \times \dfrac{1 \; \cancel{kg}}{2.205 \; \cancel{lb}} \times \dfrac{25 \; mg}{\cancel{kg}} = 1.2 \times 10^3 \; mg$ or 1,200 mg given once a day

1.22 $63.0 \; \cancel{lb} \times \dfrac{1 \; \cancel{kg}}{2.205 \; \cancel{lb}} \times \dfrac{20.0 \; mg}{\cancel{kg} \cdot \cancel{day}} \times \dfrac{1 \; \cancel{day}}{4 \; doses} = 143 \; mg/dose$

1.23 $d = \dfrac{5.5 \; g}{5.0 \; mL} = 1.1 \; g/mL$

1.24 The gold sphere would have a greater mass because Table 1.5 indicates that gold is more dense than iron.

1.25 The 10-g sphere will have the larger diameter.

1.26 $V = 95.0 - 82.5 = 12.5 \; mL$; $d = m/V = 10.5 \; g/12.5 \; mL = 0.840 \; g/mL$

1.27 25 g (the density of water is 1.000 g/mL)

1.28 $65 \; \cancel{g} \times \dfrac{1 \; mL}{7.9 \; \cancel{g}} = 8.2 \; mL$ of water displaced

1.29 $\dfrac{71.1 \; g}{9.0 \; cm^3} = 7.9 \; g/cm^3$, which corresponds to the density of iron.

1.30 Someone with osteoporosis is losing bone *mass*, therefore their bone density (m/V) would be less than normal.

1.31 For ice to float on water, it must be less dense than liquid water.

1.32 (a) protons and neutrons; (b) electrons; (c) protons; (d) protons and neutrons.

1.33 The nucleus is more dense than the atom because they have the same mass but different volumes: the nucleus is a much smaller volume within the larger volume of the atom.

1.34 Electrons are located in orbitals.

1.35 A $2p_x$ orbital has a two-lobed dumbbell shape.

1.36

Isotope Symbol	Mass Number	Atomic Number	Number of Protons	Number of Neutrons
Chlorine-35	35	17	17	18
Mercury-197	197	80	80	117
Iodine-125	125	53	53	72

1.37

Isotope	Mass Number	Atomic Number	Number of Protons	Number of Neutrons
Silicon-28	28	14	14	14
Silicon-29	29	14	14	15
Silicon-30	30	14	14	16

(a) All silicon isotopes have 14 protons.

(b) The number of neutrons is different: 16, 15, and 14.

(c) Silicon-30 because it has more neutrons and neutrons have mass.

1.38 $^{195}_{79}\text{Au}$

1.39

Calcium Isotope	Natural Abundance
Calcium-40	96.94% (b) Most abundant on earth
Calcium-42	0.6479%
Calcium-43	0.1356%
Calcium-44	2.086% (a) Greatest mass
	(c) Greatest number of neutrons

(d) Average atomic mass of calcium, 40.08, is most similar to the mass number of calcium-40, because calcium-40 is the most abundant isotope, and atomic mass is a weighted average. It is not equal to 40 due to the contribution of the other isotopes to the average mass.

(e) Calcium-44 has four more neutrons than calcium-40.

(f) The units of atomic mass are amu—atomic mass units.

1.40 (a) 36; (b) 47.

1.41 (a) Ne, neon; (b) O, oxygen.

1.42 Hydrogen, H. It has the atomic number 1.

1.43 A vanadium atom has 23 protons and 23 electrons.

1.44 (b) bronze; (e) steel. You can tell they are not elements because they are not on the periodic table of elements.

1.45 (a) beryllium: Group 2A, alkaline earth; (b) iodine: Group 7A, halogens; (c) argon: Group 8A, noble gases or inert gases; (d) nitrogen: Group 5A.

1.46 (b) mercury; (d) sodium; (e) uranium; (f) molybdenum.

1.47 (a) strontium, Sr; (b) chlorine, Cl; (c) titanium, Ti.

1.48 (a) chemical property; (b) going from liquid to gas is a physical property, the danger it poses is a chemical interaction with the body—a chemical property.

1.49

Element	Metal or Nonmetal	Group Number	Period Number	Atomic Number/ Number of Protons
Carbon	Nonmetal	4A	2	6
Hydrogen	Nonmetal	1A	1	1
Nitrogen	Nonmetal	5A	2	7
Oxygen	Nonmetal	6A	2	8
Phosphorus	Nonmetal	5A	3	15
Sulfur	Nonmetal	6A	3	16

1.50 A micronutrient is required in amounts of less than 100 mg/day, while a macronutrient is required in amounts of greater than 100 mg/day. We obtain micronutrients and macronutrients from our diet.

1.51 All the macronutrients are main group elements.

1.52 All of the six building block elements are nonmetals.

1.53 (a) 2; (b) 6.

1.54

Element	Symbol	Atomic Number	Group Number	$n = 1$	$n = 2$	$n = 3$
Sulfur	S	16	6A	2	8	6
Argon	Ar	18	8A	2	8	8

1.55 Radioactive strontium is dangerous because it can accumulate in bone. It can replace calcium, because they are in the same family of elements, the alkaline earth metals, and thus have the same number of valence electrons, and therefore similar physical and chemical properties. They differ in which period they are in, period 4 for calcium and period 5 for strontium, which means they have a different number of inner core electrons.

CHAPTER 2

Compounds

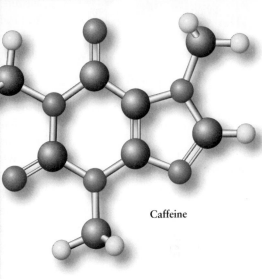

Caffeine

Coffee is a mixture that contains many chemical compounds, including caffeine and water. [Johann Helgason/Alamy]

OUTLINE

This icon indicates that a **Problem-Solving Tutorial** is available at www.whfreeman.com/gob

(a)

Neuron

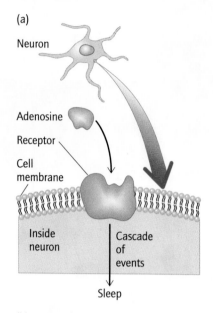

Adenosine

Receptor

Cell
membrane

Inside
neuron

Cascade
of
events

Sleep

(b)

Adenosine
cannot bind

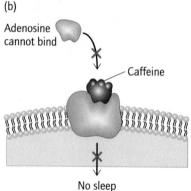

Caffeine

No sleep

Figure 2-1 When adenosine binds
to the adenosine receptor, it sets off a
cascade of biochemical events in brain
cells, leading to sleep.

What Is in Your Morning Coffee?

Do you start your day with a cup of tea or coffee? Do you drink soft drinks throughout
the day to give yourself a lift? If so, you may be among the 90% of Americans who
consume more than 300 mg of caffeine a day. What is caffeine?

Caffeine, the central nervous system stimulant in these beverages, is a *compound*.
In Chapter 1, you learned about elements, and in this chapter you will learn about
the most abundant type of matter: compounds. Compounds are substances composed
of more than one type of atom and held together by chemical bonds. Most beverages are actually composed of many different compounds and elements, known as a
mixture. The main substance in these beverages is water, a compound composed of
two hydrogen atoms and one oxygen atom held together by chemical bonds. Caffeine
is a compound more complex than water, composed of the following building block
elements: 8 carbon atoms, 10 hydrogen atoms, 4 nitrogen atoms, and 2 oxygen atoms.
Chemical bonds hold this assemblage of atoms together, creating what is known as a
molecule. A ball-and-stick model of a molecule of caffeine is illustrated on the opening
page of this chapter. In a ball-and-stick model each different colored sphere represents
a different type of atom: carbon is black, oxygen is red, nitrogen is blue, and hydrogen
is white. The "sticks" joining the atoms represent bonds.

What happens when caffeine enters your body and why does it keep you awake?
Within 15 minutes of ingestion, caffeine makes its way to the brain where it interacts
with brain cells. Caffeine works by blocking an important signal between brain cells
that prepares the brain for sleep. That signal is initiated when another compound
named adenosine is released. Adenosine signals brain cells to slow their activity and
prepare for sleep. It does this by binding with a huge molecule on the surface of brain
cells known as the adenosine receptor, as shown in Figure 2-1.

Cells have a variety of different receptors, often situated on the surface of the cell
membrane, which contain a site for a specific messenger molecule to bind—such as a
hormone, or adenosine in the case of the adenosine receptor. When a messenger molecule
binds to its receptor, a cascade of biological events is initiated inside the cell. In the presence of caffeine, the biological events needed for sleep are blocked because caffeine binds
to the adenosine receptor instead of adenosine, but it cannot initiate the subsequent cascade of events as adenosine does. So long as caffeine occupies the receptor site, instead of
feeling sleepy, the body feels alert—and that is perhaps why you are drinking that cup of
coffee in the first place!

Eventually, all this extra brain activity stimulates your pituitary and adrenal
glands to respond as they do in a crisis, releasing hormones that signal the fight-
or-flight response—the production of adrenaline so that you can be more alert during what the brain incorrectly perceives as a crisis. Adrenaline in turn stimulates the
release of dopamine, the well known "feel good" substance that activates the pleasure
centers in the brain. As you can see, a cascade of events occurs when caffeine blocks
the adenosine receptor.

In summary, caffeine makes you feel more awake by blocking the adenosine receptor,
which in turn makes you feel even more alert as a result of the release of adrenaline, which
in turn makes you feel good as a result of the release of dopamine. Eventually, the extra
adrenaline and dopamine wear off, and you begin to feel tired and possibly depressed. At
that point, you are probably heading for your next cup of coffee!

In order to understand the myriad of biochemical events in a healthy individual
as well as those involved in disease processes, you will need to understand what a
compound is at the atomic level. In this chapter you will learn about compounds, and
then in Chapter 5 you will learn how compounds and elements can be combined to
form mixtures, such as your morning cup of coffee.

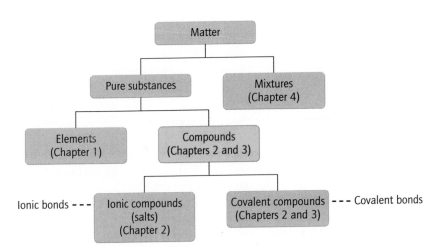

Figure 2-2 Classification of matter.

In Chapter 1 you learned that the smallest intact component of all matter is the atom. Each element is composed of one type of atom—defined by its atomic number. You saw that an element's valence electrons determine the location of the element on the periodic table as well as its physical and chemical properties. Valence electrons are at center stage again in this chapter when you will learn about one of the most important characteristics of atoms: their ability to combine with other atoms to form compounds. *A **compound** is a substance composed of two or more different atoms in a defined whole-number proportion.*

A substance composed of only one compound or element is classified as a **pure substance.** In other words, there are only two types of pure substances: elements and compounds, as shown in the flow chart in Figure 2-2. Water is an example of a **pure compound.** It is composed of hydrogen and oxygen atoms in a defined whole-number proportion: there are 2 hydrogen atoms for every 1 oxygen atom.

A few elements exist in nature in their elemental form as single atoms: these are the noble gases and some metals. Most atoms bond to other atoms to form compounds. Thus, elements are the building blocks for all compounds. There are two basic types of compounds:

- ionic compounds
- covalent compounds

The fundamental difference between an ionic compound and a covalent compound is the nature of the force of attraction that holds the atoms in the compound together—the chemical bonds. We first consider ionic compounds.

2.1 Ionic Compounds

A central theme in all science is that like charges repel and opposite charges attract. An **ionic compound** is formed when a metal atom transfers some or all of its valence electrons to a nonmetal atom, creating charged species known as *ions*. A positive and a negatively charged ion exhibit a strong force of attraction between them known as an **electrostatic attraction.** The electrostatic attraction between oppositely charged ions is the "glue" that holds ions together in an ionic compound. Ionic compounds are also called **salts.**

Ions

An **ion** is an atom that has lost or gained one or more valence electrons. It therefore has an unequal number of protons and electrons, which gives it either a positive (+) or a negative (−) charge. *The magnitude of the charge*

Sodium metal reacts violently with water (top), whereas the sodium ion in sodium chloride, table salt, is unreactive (bottom). [(Top) Andrew Lambert Photography/Photo Researchers, Inc.; (bottom) Timothy Lozinski/Alamy]

Recall that metals are the elements on the left side of the bold diagonal line on the periodic table and nonmetals are the elements to the right of this line.

Table 2-1 Subatomic Particles in Na and Na$^+$

Element/Ion	Number of Protons	Number of Electrons	Charge
Na (sodium)	11	11	0
Na$^+$ (sodium cation)	11	10	+1

on an ion is equal to the difference between the number of protons and electrons in the ion. For example, sodium (Na), a Group 1A metal, tends to lose its one valence electron to become a sodium ion (Na$^+$). An elemental sodium atom has 11 protons and 11 electrons. After losing one electron, the ion has 11 protons and 10 electrons, one less electron than the element, as shown in Table 2-1. When 11 protons, each positively charged, and 10 electrons, each negatively charged, exist together, a net +1 charge on the ion results.

When the number of electrons is greater than the number of protons, the ion is negatively charged, and is known as an **anion.** When the number of electrons is less than the number of protons, the ion is positively charged and is known as a **cation.** Keep in mind that ions are formed by the gain or loss of *electrons*; the number of *protons* never changes. Only in nuclear reactions will the number of protons or neutrons in an atom change (see Chapter 16). *As a general rule, metals tend to lose electrons to become cations, while nonmetals tend to gain electrons and become anions.*

Cation: the number of electrons is less than the number of protons

Anion: the number of electrons is greater than the number of protons

The symbol for an ion is the same as that for the element from which it was derived, because they have the same number of protons. A superscript is added to the right of the element symbol showing the magnitude and sign of the charge on the ion. For example, the symbol for a sodium ion is Na$^+$. An ion that has lost 2 electrons has a +2 charge, so the symbol for the magnesium ion, for example, is Mg^{2+}. The convention when writing the symbol of an ion is to place the sign of the charge (+ or −) *after* the number. If the charge is 1+ or 1−, the number "1" is implied and a + or − alone is used. The name of a cation is the same as the element followed by the word "ion."

Even though they have a similar symbol, an ion has significantly different physical and chemical properties from the element from which it was derived. For example, elemental sodium, Na, is a shiny metal, which is extremely flammable, exploding into a fireball upon contact with water. In contrast, the sodium cation, Na$^+$, is part of many ionic compounds, such as sodium chloride (NaCl), the familiar unreactive white crystalline substance used as table salt.

Main Group Metal Ions Why do metals tend to lose electrons—a process known as **ionization?** By losing all of its valence electrons, an ion derived from a main group element is left with a full outermost shell—an arrangement of electrons similar to the noble gas in the period above it. If you compare Figures 2-3b and 2-3c, you will see that Na$^+$ in period 3 has the same electron arrangement as the noble gas neon (Ne) in period 2. Recall from Chapter 1 that a full outermost shell is an especially stable electron arrangement.

For the same reason that group 1A elements lose 1 electron to form an ion, group 2A elements lose 2 electrons, and group 3A elements lose 3 electrons. For example, the group 2A element magnesium loses 2 electrons to form the magnesium cation, Mg^{2+}, giving it 12 protons and only 10 electrons, as shown in Figure 2-4b. Thus, a magnesium cation also has a full outermost shell, like neon. The tendency of the group 1A elements to lose 1 electron, and

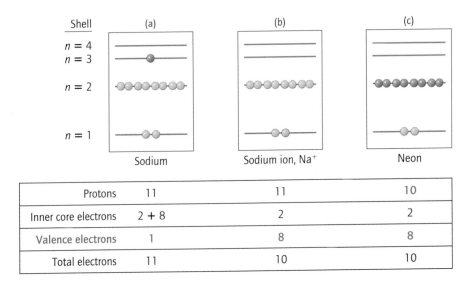

Shell	(a) Sodium	(b) Sodium ion, Na$^+$	(c) Neon
Protons	11	11	10
Inner core electrons	2 + 8	2	2
Valence electrons	1	8	8
Total electrons	11	10	10

Figure 2-3 Electron shell arrangements for (a) sodium, Na, (b) sodium ion, Na$^+$, and (c) neon, Ne. A sodium atom (a) loses the single electron in its outermost shell to become a sodium cation, Na$^+$ (b), which has the same electron arrangement as Ne (c).

group 2A elements to lose 2 electrons, and group 3A metals to lose 3 electrons is another example of the periodicity of the elements.

Transition Metal Ions Some metals, in particular many transition metal elements, can lose a variable number of electrons. For example, iron (Fe) can lose either 2 or 3 electrons to become Fe^{2+} or Fe^{3+}, respectively. All iron ions contain 26 protons, but Fe^{2+} has 24 electrons and Fe^{3+} has 23 electrons. Therefore, you cannot predict the number of electrons a transition metal (or a group 4A metal) loses simply from its group number. The common ionic forms of the transition metals are shown in Figure 2-5.

Lead (Pb) and mercury (Hg), along with other "heavy metals," form ions such as Hg^{2+} and Pb^{2+} that are poisonous. Lead ions are especially harmful to the developing brain of a child. Lead poisoning in children occurs when they are exposed to the dust from old lead-based paint, which was used in homes prior to 1978. Lead and mercury ions bind to the sulfur atoms in important proteins in the body, which cause them to stop functioning properly.

Nonmetal Ions Turn your attention now to the right side of the periodic table, the nonmetals. Nonmetal elements in groups 5A through 7A have a tendency to gain electrons and form anions. These elements gain the number of electrons

Heavy metal ions are poisonous, such as the lead ion (Pb^{2+}) used in some paints prior to 1978. [Jeff Albertson/Corbis]

Shell	(a) Magnesium	(b) Magnesium ion	(c) Neon
Protons	12	12	10
Inner core electrons	2 + 8	2	2
Valence electrons	2	8	8
Total electrons	12	10	10

Figure 2-4 Electron shell arrangements for (a) magnesium, Mg, (b) magnesium ion, Mg^{2+}, and (c) neon, Ne. A magnesium atom (a) loses the 2 electrons in its outer shell to become a magnesium cation, Mg^{2+} (b), which has the same electron arrangement as Ne (c).

Figure 2-5 Common ionic forms of some important ions.

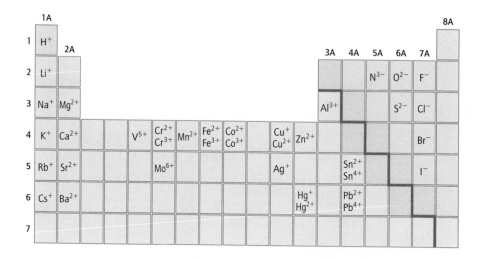

needed to achieve a full valence shell—typically 8 electrons (although hydrogen achieves a full valence shell with 2 electrons). For example, flourine has 7 valence electrons, just 1 electron short of the 8 electrons it needs to have a full valence shell, as shown in Figure 2-6a. The addition of 1 electron fills the shell and produces the fluoride ion, F^-, an anion with a -1 charge (Figure 2-6b). Indeed, all the group 7A elements form ions with a -1 charge: F^-, Cl^-, Br^-, and I^-. The name of an anion formed from an element is similar to the name of the element except that the ending is changed to -*ide*, as in fluor**ide.**

Similarly, the nonmetals in Group 6A need 2 electrons to achieve a full outermost shell, creating an anion with a -2 charge. For example, the sulfide ion, S^{2-}, is formed when sulfur, S, gains 2 electrons. Similarly, the group 5A nonmetals form -3 ions, such as N^{3-}.

The group 4A nonmetal elements have a half-filled valence shell and as such are unique in that they do not gain or lose electrons. Carbon, the most important group 4A element, seldom forms ions; instead it achieves stability by sharing electrons. Shared electrons are the basis for covalent bonds, and the subject of Section 2.2.

Most of the important nonmetal macronutrients and micronutrients are found in their anionic form in the body. See the *Chemistry in Medicine* box at the end of this chapter to learn more about the role of these ions in medicine and nutrition.

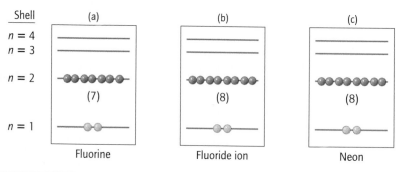

Figure 2-6 Electron shell arrangements for (a) fluorine, F, (b) fluoride ion, F^-, and (c) neon, Ne. A fluorine atom (a) gains 1 electron in its outermost shell to become a fluoride ion, F^- (b), which has the same electron arrangement as Ne (c).

	(a)	(b)	(c)
Protons	9	9	10
Inner core electrons	2	2	2
Valence electrons	7	8	8
Total electrons	9	10	10

WORKED EXERCISES 2-1   Predicting the Ionic Form of an Element and Naming Ions

1 Write the ions that can be formed from the following elements. Indicate whether the elements are metals or nonmetals:

   a. oxygen
   b. barium
   c. copper

2 Consider the ions Br^-, Sr^{2+}, and Co^{2+}.

   a. Indicate the number of protons and electrons in each.
   b. State whether the ion was formed from a metal or a nonmetal.
   c. Explain why the charge on the ion is as indicated.

SOLUTIONS

1 **a.** O^{2-}, oxide is derived from a nonmetal.
   b. Ba^{2+}, barium ion is derived from a metal.
   c. Cu^+ and Cu^{2+}, The two copper ions are derived from the transition metal copper, Cu. It is a transition metal with two forms.

2 **a.** Br^- has 35 protons and 36 electrons with a net charge of -1. Bromine is a group 7A nonmetal, with 7 valence electrons. Gaining 1 electron gives it a full outermost shell similar to the noble gas, krypton, Kr.
   b. Sr^{2+} has 38 protons and 36 electrons with a net charge of $+2$. Stronwtium is a group 2A metal, with 2 valence electrons. Losing its 2 valence electrons gives it a full valence shell similar to krypton, Kr.
   c. Co^{2+} has 27 protons and 25 electrons. Cobalt is a transition metal, and loses 1 or 2 electrons, as shown in Figure 2-5. The $+2$ charge indicates that it has lost 2 electrons.

PRACTICE EXERCISES

2.1 What is the difference between an element and a compound?

2.2 Write the symbol for the ions formed from the following elements, based on their position on the periodic table, if possible:

   a. selenium
   b. potassium
   c. arsenic
   d. zinc

2.3 Consider the ions I^-, N^{3-}, and Cr^{3+}

   a. Indicate the number of protons and electrons in each.
   b. State whether the ion was formed from a metal or nonmetal.
   c. Explain why the charge on the ion is as indicated.

2.4 Why do Group 6A elements tend to gain two electrons? What is the charge on these ions? Provide an example of a group 6A anion.

2.5 In the movie *Erin Brokovitch*, which is based on a true story, Brokovitch learns that the high incidence of cancer in a small town is the result of ground water contamination from chromium(VI) (Cr^{6+}). However, another form of the chromium ion, Cr^{3+}, is an important micronutrient. What is the difference between these two forms of chromium? How is it plausible that they would have such different effects in the body?

The Ionic Lattice

An ionic compound is formed as the result of the mutual attraction between cations and anions. Sodium chloride, commonly known as table salt, is perhaps the most familiar ionic compound. Table salt is composed of an equal number of sodium (Na^+) and chloride (Cl^-) ions held together by the electrostatic attraction between the positive charge on sodium cations and

(a)

(b)

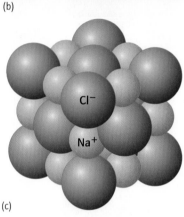

(c)

Figure 2-7 The ionic compound sodium chloride, NaCl, viewed on different scales. (a) On the macroscopic scale, NaCl is found in the familiar grains of salt. (b) On the microscopic scale, a crystal of NaCl. (c) On the atomic scale, NaCl is a lattice of ions. Na^+ is shown in yellow, Cl^- is shown in green. [(a) Charles D. Winters/Photo Researchers, Inc.; (b) Phototake Inc./Alamy]

the negative charge on chloride anions. The electrostatic attraction that holds ions together is also called an **ionic bond.**

Most ionic compounds are brittle solids at room temperature, like sodium chloride, shown on different scales in Figure 2-7a. When viewed under the scanning electron microscope, sodium chloride has the crystalline appearance shown in Figure 2-7b. On the atomic scale, sodium chloride exists in a crystalline lattice structure made up of an equal number of Na^+ and Cl^- ions, as shown in Figure 2-7c. In a **lattice structure,** every cation is surrounded by anions, and every anion is surrounded by cations. An extended lattice of ions goes on until the edge of the crystal. In fact, the only way to disrupt the lattice is to dissolve the compound in water or to heat the salt until it melts, and the latter requires very high temperatures.

Electrolytes When most ionic compounds are placed in water, the lattice structure falls apart and each cation and anion becomes surrounded by water molecules. The ions become separated from one another and are instead surrounded by water molecules; the salt is said to be *dissolved* in water. An ionic compound dissolved in water is considered a mixture. You will learn more about ionic compounds dissolved in water when you study mixtures in Chapter 5.

Ions dissolved in water are often referred to as **electrolytes.** The cell, composed mainly of water, contains many electrolytes that are critical to its function. The most important of these electrolytes are Na^+, K^+, Ca^{2+}, Mg^{2+}, and Cl^-. For example, sodium ion, Na^+, and potassium ion, K^+, are involved in regulating blood pressure and are important in the transmission of electrical signals in nerve cells. The most important electrolytes in the human body are the macronutrients and micronutrients shown in Figure 1-14. Indeed, many of the macronutrients and micronutrients exist in their ionic forms. For example, the iron in your blood is not elemental iron, Fe, but iron(II) ion Fe^{2+}.

Electrolytes are present in many popular sports drinks like Gatorade, Propel, and Vitaminwater. Electrolytes are also found in infant supplements such as Pedialyte. These drinks help to replenish essential electrolytes lost while perspiring—while engaging in sports—or from diarrhea.

Proper electrolyte balance is regulated by hormones in the body, with the kidneys playing an important role in eliminating excess electrolytes. A blood test measures the concentration of each of these electrolytes. When they are outside the normal range, it is often an indicator of organ malfunction. See the *Chemistry in Medicine* box at the end of this chapter to learn more about electrolytes and their significance.

The Formula Unit A pure ionic compound has a definite and unique composition, which is defined by its formula unit. The **formula unit** of an ionic compound represents the ratio of cation to anion in the lattice. For example, the formula unit for sodium chloride is NaCl, which indicates there are an equal number of Na^+ and Cl^- ions in the lattice. If you were to examine the lattice structure for Li_2O you would see that there are twice as many lithium ions as oxide ions, but every lithium ion is surrounded by oxide ions and every oxide ion by lithium ions; they are not individual Li_2O units. *Thus, an ionic compound is defined by the lowest whole number ratio of its component ions.*

The formula unit of an ionic compound is written according to the following set of rules:

- The chemical symbol of each ion in the compound is written, with the cation listed first, followed by the anion. The charges on the ions are not included in the formula unit.

- A numerical subscript following each chemical symbol is used to indicate the ratio of cation to anion in the compound. The subscript is understood to be "1" when none is shown. The lowest whole number ratio is used for the subscripts.
- *The sum of all the positive and negative charges of the individual ions of an ionic compound must always add up to zero, so that the ionic compound is neutral overall.* In other words, the positive and negative charges must cancel one another. Thus, for ionic compounds in which the cation and anion have the same magnitude of charge, the ratio of cation to anion is 1:1. For example:

Sports drinks replenish important electrolytes, needed by cells, that are lost during exercise. [Tommy Hindley/ Professional Sport/Topham/The Image Works]

$$\text{NaCl:} \quad \begin{array}{ccc} Na^+ & & Cl^- \\ (+1) & + & (-1) & = 0 \end{array}$$

$$\text{MgO:} \quad \begin{array}{ccc} Mg^{2+} & & O^{2-} \\ (+2) & + & (-2) & = 0 \end{array}$$

- When the magnitudes of the charges on the cation and the anion are different, subscripts must be added to indicate the ratio that results in an electrically neutral compound. For example, the ionic compound formed from Mg^{2+} and Cl^- has the formula unit $MgCl_2$, because **two** Cl^- ions are required to cancel the charge on **one** Mg^{2+} ion:

$$\text{MgCl}_2\text{:} \quad \begin{array}{ccc} Mg^{2+} & & 2 \times Cl^- \\ +2 & + & 2(-1) & = 0 \end{array}$$

General guidelines for determining the formula unit for an ionic compound are listed in the *Guidelines* box below.

Guidelines for Determining the Formula Unit for an Ionic Compound

Step 1: Determine the charge on the cation. For main group elements, the charge on the cation corresponds to its group number, obtained from the periodic table.

$$\begin{array}{ll} \text{Group 1A} & +1 \\ \text{Group 2A} & +2 \\ \text{Group 3A} & +3 \end{array}$$

If the cation is derived from a transition metal, there is usually more than one possible charge for the cation. You will need to determine which form it is in from other information provided.

For example, if the cation is iron, its charge is either +2 or +3.

Step 2: Determine the charge on the anion. The magnitude of the charge on the anion corresponds to the number of electrons needed to fill the outermost shell: group number − 8 = charge.

$$\begin{array}{ll} \text{Group 5A} & 5 - 8 = -3 \\ \text{Group 6A} & 6 - 8 = -2 \\ \text{Group 7A} & 7 - 8 = -1 \end{array}$$

For example, the fluoride ion has a charge of −1 because it is an anion derived from a group 7A nonmetal: 7 − 8 = −1

$$F \rightarrow F^-$$

Step 3: Insert subscripts. Write the cation followed by the anion without showing their charges. Insert subscripts so that the sum of the charges yields a net zero charge. If the subscript is one, it is implied and should not be written-in. One aid for determining subscripts is demonstrated below: The subscript on each ion is equal to the numerical value of the charge on the *other* ion.

For example, for calcium and fluoride, the formula of the salt can be obtained by performing the exercise shown:

$$Ca^{2+} \times Cl^- \longrightarrow CaCl_2$$

Step 4: If the subscripts can be divided by a common divisor, do so. The subscripts should be the lowest whole number representation of the ratio of cations to anions.

For example, in Mg_2O_2 both subscripts can be divided by 2, so the final formula is MgO:

$$Mg_2O_2 \xrightarrow[\text{by 2}]{\text{Divide subscripts}} MgO$$

Step 5: Double check that the formula unit represents a neutral compound. Do this by multiplying the charge on each ion by its subscript and adding your results together. The sum must equal zero to be a neutral compound:

(cation charge $\times$ cation subscript) + (anion charge $\times$ anion subscript) = **0**

In the case of CaF_2, there is one +2 cation and two −1 anions, giving a net charge of zero:

$$(+2) + 2(-1) =$$
$$(+2) + (-2) = \mathbf{0}$$

Naming Ionic Compounds

Note that the numerical subscripts of the formula unit do not appear anywhere in the name of an ionic compound.

To name an ionic compound, write the name of the cation first, followed by the name of the anion. For example, the compound with the formula unit $CaBr_2$ has the name calcium bromide.

A cation always has the same name as the element from which it is derived. For transition metals that have variable charged forms, the charge on the cation is indicated by placing a Roman numeral, corresponding to the magnitude of the charge, within parentheses immediately following the name of the cation.

You obtain the name of an anion by changing the ending on the element name to *-ide*. For example, $FeCl_2$ is named iron(II) chloride, where the (II) indicates that iron ion is in the +2 state and not in its alternative +3 state. Roman numerals should not be used for transition metals that exist in only one form, such as silver (Ag^+), or for any of the main group elements.

An older naming system still in use for ions with variable charge is shown below. The ending *-ous* is used for the ion with the lower charge and the ending *-ic* for the ion with the higher charge. For example:

Fe^{2+}	ferrous ion	Cu^+	cuprous ion	Pb^{2+}	plumbous
Fe^{3+}	ferric ion	Cu^{2+}	cupric ion	Pb^{4+}	plumbic

Using the older naming system, $FeCl_3$ would be called ferric chloride rather than iron(III) chloride.

Problem-Solving Tutorial:
IUPAC Naming of Ionic Compounds

WORKED EXERCISES 2-2   Practice with Formula Units

1 What is the formula unit for aluminum sulfide? Show your work.
2 What is the name of the compound with the formula unit Fe_2O_3?

SOLUTIONS

1 **Step 1: Determine the charge on the cation.** The charge on aluminum is $+3$, because it is a cation derived from a group 3A metal:

$$Al \longrightarrow Al^{3+}$$

Step 2: Determine the charge on the anion. The charge on sulfur is -2, because it is an anion derived from a group 6A nonmetal: $6 - 8 = -2$

$$S \longrightarrow S^{2-}$$

Step 3: Insert subscripts. Write the cation followed by the anion and insert subscripts such that the total positive charge cancels the total negative charge. Alternatively, use the algorithm shown below:

$$Al^{3+} \diagdown\!\!\!\!\diagup S^{2-} \longrightarrow Al_2S_3$$

Step 4: If the subscripts can be divided by a common divisor, do so. No common divisor exists for the subscripts 2 and 3.

Step 5: Double check that the formula unit represents a neutral compound.

(cation charge $\times$ cation subscript) + (anion charge $\times$ anion subscript) = 0

$$\begin{array}{ccccc} (+3 \times 2) & + & (-2 \times 3) & = & \\ (+6) & + & (-6) & = & 0 \end{array}$$

2 Determine the charge of the cation from the formula unit, Fe_2O_3, given that the overall formula must be neutral:

$$(x \times 2) + (-2 \times 3) = 0$$
$$(x \times 2) + (-6) = 0$$
$$x \times 2 = 6$$
$$x = +3$$

Construct the name by listing the cation first, followed by the anion: iron (III) oxide or ferric oxide. In the newer naming system, the Roman numeral three must be in parentheses immediately following the cation, indicating that the form of the iron cation is Fe^{3+} not Fe^{2+}.

PRACTICE EXERCISES

2.6 Answer the following questions for the lattice structures of NaCl and CsCl:

   a. How does the lattice structure confirm the formula unit given?
   b. Why is the ratio of cation to anion one to one?
   c. Why does the lattice show more ions than the formula unit does?
   d. What is holding the ions together in these lattice structures?
   e. What color are the spheres representing the chloride ion in each lattice?
   f. Which ion, Na^+ or Cl^-, has the smaller diameter?

2.7 Write the formula unit for the following ionic compounds:

   a. sodium sulfide   b. magnesium nitride
   c. copper (I) chloride   d. copper (II) chloride

NaCl

CsCl

2.8 Name the following ionic compounds:

    a. Li_2O **b.** MgI_2 **c.** KCl

2.9 From the formula unit Fe_2O_3, deduce whether the iron ion is in the +2 or +3 form.

2.2 | Covalent Compounds

The greatest variety of compounds are found among covalent compounds. Indeed, some of the most important substances in the cell are covalent compounds, including proteins, carbohydrates, lipids, DNA, hormones, and many others. While most elements are made up of single atoms, covalent compounds consist of *molecules*. A **molecule** *is a discrete entity—not a lattice—composed of two or more different nonmetal atoms held together by covalent bonds.* Although there are some elements that exist as molecules, most molecules are covalent compounds. Some molecules are quite large—for example, proteins can have over 10,000 atoms in their structure. Moreover, the variety of compounds that can be constructed from the building block elements is limitless. And scientists create new covalent compounds in the laboratory every day.

Covalent compounds can be distinguished from ionic compounds by the type of chemical bonds they contain. You have seen that ionic bonds arise from the attraction between oppositely charged ions: a metal cation and a nonmetal anion. In contrast, a **covalent bond** is formed when two *nonmetal* atoms come together and share some or all of their valence electrons. By sharing electrons, nonmetal atoms in a molecule achieve a full outermost shell of 8 electrons. One exception is hydrogen, which achieves a full outermost shell when it shares 2 electrons.

As a result of their different chemical bonds, ionic and covalent compounds have very different physical and chemical properties. For example, covalent molecules have lower melting and boiling points than ionic compounds. While most ionic compounds are solids at room temperature, covalent compounds can be found in all three states of matter at room temperature: solid, liquid, and gas, depending on the compound.

The Molecular Formula

A covalent compound has a definite and unique composition described by its **molecular formula.** In a molecular formula each atom type is listed, usually in alphabetical order, followed by a subscript indicating how many atoms of that type are in the molecule. For example, a water molecule contains two hydrogen atoms and one oxygen atom; hence its molecular formula is H_2O.

A molecular formula cannot be altered without also changing the identity of the compound. For example, hydrogen peroxide has the molecular formula H_2O_2, so it is composed of the same type of atoms as water, but in a different ratio, and therefore, it is an entirely different substance. A molecular formula gives limited information, however, because it does not necessarily indicate the arrangement of atoms and the type of covalent bonds present.

The Covalent Bond

Single Bonds The simplest example of a covalent bond is seen in the element hydrogen. Hydrogen is a molecule containing two hydrogen atoms, H_2. Molecules composed of two atoms are known as **diatomic molecules.** The diatomic elements include hydrogen, H_2, oxygen, O_2, nitrogen, N_2, fluorine, F_2, chlorine, Cl_2, bromine, Br_2, and iodine, I_2.

Hydrogen gas, H_2, is extremely flammable. The *Hindenburg* was an aircraft filled with hydrogen gas that exploded into a fireball in 1937, killing 35 of the 97 passengers and crew and one member of the ground crew. [Bettmann/Corbis]

Each of the hydrogen atoms in a hydrogen molecule shares its one valence electron with the other hydrogen atom, creating a covalent bond. The structure of a molecule can be depicted by showing the valence electrons as dots placed next to the element symbol. To indicate that two electrons are shared by both atoms, the electron dots are placed between the two atoms. To simplify the drawing process, a shared pair of electrons is usually represented as a line. In the case of hydrogen, the structure would be written as H—H.

A shared pair of electrons is known as a **covalent** bond. When only one pair of electrons is shared between two atoms, the bond is known as a **single bond.** The distance between two nuclei joined by a bond is known as the **bond length.** The bond length varies somewhat with different types of covalent bonds.

The sharing of electrons in a hydrogen molecule effectively gives each hydrogen atom two valence electrons, and so each atom achieves a full outermost shell of electrons like the noble gas helium (He) in period 1. The two shared electrons in H_2 spend their time around both nuclei. In other words, the two electrons reside in a molecular orbital that encompasses both hydrogen nuclei, as shown in Figure 2-8, and both electrons are attracted to the positive charge on each nucleus. Although the two electrons repel each other and the two nuclei likewise repel, these repulsions are more than offset by the strong

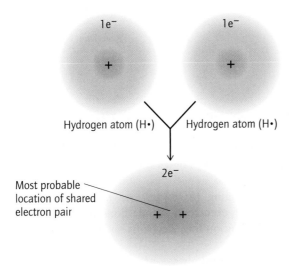

Figure 2-8 The orbitals of two isolated hydrogen atoms, each containing a single electron, come together to form a molecular orbital containing both electrons that encompass both hydrogen nuclei.

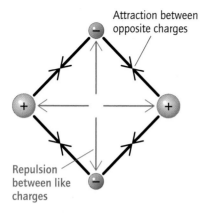

Figure 2-9 A covalent bond forms between atoms because the forces of attraction between opposite charges dominate the repulsive forces between like charges at the optimal distance between nuclei (the bond length).

attraction between the negatively charged electrons and the positively charged nuclei when the distance between the two nuclei is optimal (Figure 2-9).

Elements in period 2 share 1 to 4 valence electrons in order to achieve a full outermost shell of 8 electrons, similar to the arrangement of electrons in neon, the noble gas in period 2. Since the most common building block elements are in period 2, 8 electrons is the most common number of electrons in the outermost shell surrounding an atom in a molecule. *You can predict the structure of a molecule by arranging electrons so that each atom has a full outermost shell, known as the **octet rule**.* For example, a molecule of methane, CH_4 has the structure shown in the margin, where the carbon atom shares each of its 4 valence electrons with a single hydrogen atom, and each hydrogen atom shares its one valence electron with the carbon atom. Thus, in a molecule of methane, carbon has an octet of 8 electrons in its outermost shell, and each hydrogen atom has 2 electrons. Each pair of electrons can be represented with a single line. We say that there are four C—H bonds in a molecule of methane.

The structures we have been using to represent a pair of bonding electrons as a line or a pair of dots are called **Lewis dot structures.** Lewis dot structures serve as an electron bookkeeping tool that provides a two-dimensional representation of a molecule with valence electrons represented as dots. When an atom shares two electrons, the dots are placed between the two atoms to indicate that both atoms share the electrons, or the shared pair of electrons is represented as a line.

Methane, CH_4

Double and Triple Bonds Atoms within molecules are not limited to sharing only 2 electrons. In order for some atoms to achieve an octet, they must form multiple bonds: Two atoms may share 4 electrons to form a **double bond** or 6 electrons to form a **triple bond.** For example, ethylene has the chemical formula C_2H_4, and the only way for both carbon atoms to achieve an octet is if the two carbon atoms share four electrons as shown below. Similarly, the two carbon atoms in acetylene share six electrons.

Four shared electrons

Six shared electrons

Ethylene, C_2H_4

Acetylene, C_2H_2

When atoms share two or three pairs of electrons, two and three lines are drawn to depict a double and a triple bond, respectively.

Nonbonding Electrons In forming molecules, certain atoms do not share all of their valence electrons. For example, in a molecule of water, oxygen has 6 valence electrons, so it needs to share only 2 electrons. The other 4 valence electrons belonging to oxygen are not shared; these electrons are known as **nonbonding electrons.** While bonding electrons spend their time around both nuclei, nonbonding electrons spend their time around only one nucleus.

Nonbonding electrons are always represented in pairs, also known as **lone pairs,** written beside the atom they belong to, but with no second atom attached, as shown in red for the Lewis dot structure of a water molecule. The Lewis dot structure for water shows that oxygen is indeed surrounded by an octet of 8 electrons: 2 nonbonding pairs (2 × 2 = 4 electrons) and 2 bonding pairs (2 × 2 = 4).

Octet

Red:
nonbonding   H—Ö—H   Blue:
electrons   bonding
                           electrons

Water, H$_2$O

Both hydrogen atoms are surrounded by 2 electrons each, a duet. The total number of valence electrons in the molecule must always equal the sum of the valence electrons from each atom.

The periodicity of the elements is seen again in the number of bonds formed by elements within the same group. Halogens form 1 covalent bond, because they have 7 valence electrons—one short of an octet. The group 6A elements form 2 bonds, because they have 6 valence electrons—two short of an octet. Remember, two bonds can appear as two single bonds or one double bond. Similarly, group 5A elements form 3 bonds and group 4A elements form 4 bonds. The Lewis dot representations for the building block elements are shown in Figure 2-10. Detailed guidelines for writing Lewis dot structures are described in the next section, following the modeling exercise.

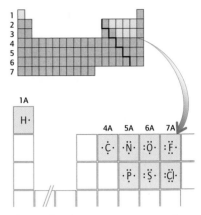

Figure 2-10 Lewis dot representations of the building block elements.

Problem-Solving Tutorial:
Ionic and Covalent Compounds

The Model Tool 2-1 Covalent Bonds in Molecules

You may have received a model tool kit with your textbook if your instructor requested it. Use this kit to perform the following exercise.

I. Construction of H$_2$, O$_2$, HCN, and CH$_4$

1. Obtain 1 black carbon atom, 7 light blue hydrogen atoms, 2 red oxygen atoms, and 1 blue nitrogen atom.

2. Your model kit has two types of bonds: straight and bent. Use one straight bond for any single bond between atoms. Use two bent bonds whenever you need to construct a double bond and three bent bonds whenever you need to construct a triple bond.

3. Make a model of a hydrogen molecule, H$_2$. How many electrons are shared? Why are electrons shared?

4. Make a model of O$_2$ similar to the ball-and-stick model shown at right. How many valence electrons does each oxygen atom have? Does each oxygen atom have an octet? Why does this molecule form a double bond? How many electrons are shared in a double bond? Write a Lewis dot structure for O$_2$.

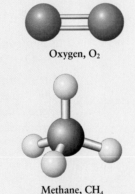

Oxygen, O$_2$

5. Make a model of CH$_4$ as shown at right. How many valence electrons does each atom have? Does the carbon atom have an octet? Does the hydrogen atom have a duet? Write a Lewis dot structure for CH$_4$.

Methane, CH$_4$

6. Make a model of HCN as shown at right. How many valence electrons does each atom have? Why is there one C—H bond and one carbon–nitrogen triple bond? Does each atom have an octet? Does the hydrogen atom have a duet?

Hydrogen cyanide, HCN

II. Inquiry Questions

7. Which of the three models represent diatomic molecules?

8. Which of the models has a double bond? How many shared electrons does a double bond represent?

9. Which of the models has a triple bond? How many shared electrons does a triple bond represent?

10. Which one of the models contains only single bonds? How many shared electrons does a single bond represent?

Table 2-2 Typical Number of Bonding and Nonbonding Pairs of Electrons in the Building Block Elements and the Halogens

Electrons	Carbon, C	Nitrogen, N; Phosphorus, P	Oxygen, O; Sulfur, S	Halogens: F, Cl, Br, I	Hydrogen, H
Bonding pairs (covalent bonds)	4	3	2	1	1
Nonbonding pairs	0	1	2	3	0

Writing Lewis Dot Structures

Knowing how to write and interpret Lewis dot structures is the foundation for learning organic chemistry and biochemistry. While determining the Lewis dot structure from the molecular formula alone is not always possible, especially for complex structures, you can determine the Lewis dot structure from a simple molecular formula relatively easily. If you become familiar with the number of bonds typically formed by the building block elements it is easier to predict the Lewis dot structure from a compound's molecular formula. Table 2-2 shows the number of bonds and nonbonding pairs of electrons typically found on the building block elements when they are part of a molecule. For example, Table 2-2 indicates oxygen usually forms two bonds, which means that oxygen can form either one double bond, $=O$, or two single bonds, $—O—$, when part of a molecule. Oxygen also typically has two nonbonding pairs of electrons. Using Table 2-2 together with the guidelines given below, you should be able to construct a Lewis dot structure from the molecular formula of most simple compounds.

Guidelines for Writing Lewis Dot Structures

1. **Determine which atom in the molecule is the central atom (or which two if there are two central atoms), and which are the surrounding atoms. Initially, place a single bond between each pair of atoms.** The central atom is usually the element positioned closest to the center of the periodic table, *because those elements form the most bonds. Use Table 2-2 to guide your decision. Except in the case of hydrogen peroxide, H—O—O—H, avoid making oxygen–oxygen bonds.*

   For example, in CCl_4, carbon is closer to the center of the periodic table, so it is the central atom; from Table 2-2, carbon forms 4 bonds while chlorine forms only 1 bond. Place a single bond between the carbon atom and each chlorine atom.

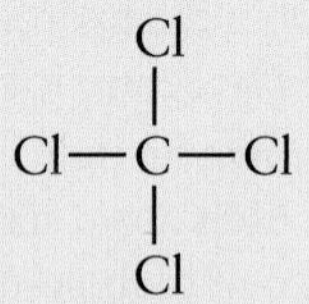

$$\begin{array}{c} Cl \\ | \\ Cl—C—Cl \\ | \\ Cl \end{array}$$

2. **Add up the total number of valence electrons and determine the number of valence electrons remaining to be distributed as nonbonding electrons.** Add up all the valence electrons of all atoms in the molecule and then subtract the bonding electrons already distributed in Step 1.

   In CCl_4 the total number of valence electrons is $4(7) + 4 = 32$ electrons: 7 for each chlorine atom and 4 from the carbon atom. Subtract each of the bonding pairs of electrons already distributed in the C—Cl single bonds: $32 − 4(2) = 24$. Thus, 24 valence electrons remain to be distributed.

3. **Perform a preliminary distribution of the remaining valence electrons as nonbonding pairs. Do this by placing pairs of dots around each atom so that each atom achieves an octet (8), except hydrogen, which should have a duet (2). Avoid placing nonbonding electrons on hydrogen or carbon atoms.**

$$
\begin{array}{c}
:\!\ddot{C}l\!: \\
| \\
:\!\ddot{C}l\!-\!C\!-\!\ddot{C}l\!: \\
| \\
:\!\ddot{C}l\!:
\end{array}
$$

In CCl_4, place 3 pairs of nonbonding electrons around each Cl atom to give each Cl atom an octet. This uses all remaining 24 valence electons. The carbon atom already has an octet from Step 1.

4. **Turn nonbonding electrons into multiple bonds if any atoms are short of an octet. If,** after distributing all the valence electrons, there are still some atoms short of an octet, turn a nonbonding pair of electrons on an adjacent atom into a multiple bond to the atom that is short, creating a double bond. If the atom is still short of an octet, a triple bond may be required. It may be formed by turning another nonbonding pair of electrons into a bonding pair.

No multiple bonds are required in this example, since every atom achieves an octet with the available valence electrons. See the worked example that follows for an example that requires this step to be applied.

5. **Double check each atom in the molecule against Table 2-2, to check that each atom is surrounded by the expected number of bonding and nonbonding electrons.**

In CCl_4, according to Table 2-2, the carbon atom should have 4 bonds and no lone pairs, and each chlorine atom should have 1 bond and 3 lone pairs. The Lewis dot structure drawn meets these criteria.

WORKED EXERCISE 2-3   Writing Lewis Dot Structures

Using the five steps in the guidelines provided above, write the Lewis dot structure for hydrogen cyanide, HCN.

Problem-Solving Tutorial:
Drawing Lewis Dot Structures

SOLUTION
Step 1: Determine which atom in the molecule is the central atom, and which atoms are the surrounding atoms. Place a single bond between each pair of atoms.

$$H-C-N$$

Carbon is the central atom because it is the atom closest to the center of the periodic table. H is on the end because it forms only one bond. Attach H and N to carbon since that is how it appears in the formula.

Step 2: Add up the total number of valence electrons and determine the number of electrons remaining to be distributed as nonbonding electrons. The total number of valence electrons is the sum of hydrogen, 1; carbon, 4; and nitrogen, 5: $1 + 4 + 5 = 10$ total valence electrons. Subtract the 4 electrons used in the H—C and C—N single bonds: $10 - 4 = 6$ electrons remain to be distributed. Remember that each line represents two shared electrons.

Step 3: Perform a preliminary distribution of the remaining electrons as nonbonding pairs.

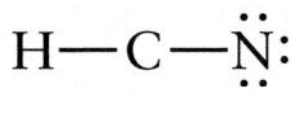

$$H-C-\ddot{\underset{..}{N}}:$$

Distribute the 6 remaining electrons around nitrogen in pairs, since Table 2-2 indicates that carbon doesn't have nonbonding electrons. Notice that the carbon atom does not have an octet; thus, a multiple bond is indicated.

Step 4: Turn nonbonding electrons into multiple bonds if any atoms are short of an octet.

$$H-C\equiv N:$$

Convert nonbonding pairs of electrons on nitrogen to C—N multiple bonds, which allows both carbon and nitrogen to be surrounded by more electrons. In order for both C and N to have an octet, a triple bond is required—6 electrons shared between carbon and nitrogen.

Step 5: Double check each atom in the molecule against Table 2-2. The carbon atom has four bonds (one single and one triple bond) and no lone pairs—the usual arrangement for carbon. The nitrogen atom has three bonds (one triple bond) and one nonbonding pair—the usual arrangement for nitrogen. The hydrogen atom has one single bond and no lone pairs, also as expected.

PRACTICE EXERCISES

2.10 Write the Lewis dot structures for the following compounds using the guidelines.
   a. ammonia, NH_3
   b. oxygen, O_2
   c. hydrogen sulfide, H_2S
   d. chlorine, Cl_2

2.11 Answer the following questions for the Lewis dot structure of carbon dioxide:

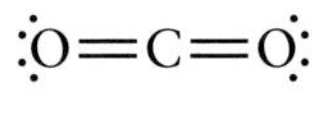

$$:\ddot{O}=C=\ddot{O}:$$

   a. How many nonbonding pairs of electrons and how many bonding pairs of electrons does each oxygen atom have?
   b. How many nonbonding electrons and how many bonding electrons does the carbon atom have?
   c. Does each atom have an octet of electrons?
   d. What types of covalent bonds are present in this molecule, single, double, or triple? How many electrons are shared between each carbon and oxygen atom in this molecule?

Expanded Octets

The key building block elements in period 2—carbon, nitrogen, and oxygen—will always have an octet in a molecule. Atoms in period 3 are sometimes found with an *expanded octet: surrounded by more than 8 electrons in the outermost shell.* Several important biomolecules contain a phosphorus atom in an expanded octet, including ATP, the energy currency of the cell, and DNA, the molecule containing the cell's genetic information.

Most of the biomolecules containing phosphorus are derivatives of phosphoric acid, H_3PO_4, whose Lewis dot structure is shown in Figure 2-11. Notice that the central phosphorus atom is surrounded by 10 electrons, rather than 8. Each oxygen atom in phosphoric acid follows the octet rule, as expected for a period 2 element.

In following the guidelines provided for drawing Lewis dot structures, you may find at the end that you have a surplus of electrons. In that case, a central atom with an expanded octet is probably present. Double check that the atom is in period 3 or higher, before placing the additional electrons around that atom.

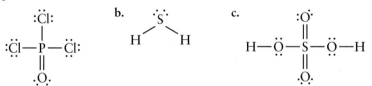

Figure 2-11 The Lewis dot structure of phosphoric acid, H_3PO_4, shows an expanded octet on P.

WORKED EXERCISE 2-4 Identifying Expanded Octets in Molecules

For each of the two phosphorus-containing molecules shown below, indicate whether it has an expanded octet, and how many electrons surround the central P atom. Is it the normal arrangement of electrons according to Table 2-2?

a.

PCl₃

b.

PCl₅

SOLUTION

a. PCl_3 contains a P atom with an octet (1 nonbonding pair and 3 bonding pairs). This is the normal arrangement of electrons for a group 5 element.
b. PCl_5 has an expanded octet with 10 electrons around P (5 bonding pairs and 0 nonbonding pairs). This is not the normal arrangement of electrons for a group 5 element, but because it is in period 3, it can have an expanded octet. Each Cl atom in both compounds has an octet (3 nonbonding pairs and 1 bonding pair).

PRACTICE EXERCISES

2.12 Evaluate each of the following molecules to determine if it contains an atom with an expanded octet. Indicate the number of electrons surrounding each expanded octet.

a.

b.

c.

2.13 Taurine, found in some high-energy drinks and cat food, has the structure shown below.

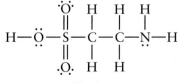

Taurine

a. Which element in taurine has an expanded octet? Is the element with an expanded octet in period 3? How many electrons surround this atom?
b. How many bonding electrons does each of the carbon atoms have?
c. How many bonding and nonbonding electrons does the nitrogen atom have?

Table 2-3 Prefixes Used to Indicate Multiple Atoms in a Binary Covalent Compound

Number of Atoms	Prefix
1	mono
2	di
3	tri
4	tetra
5	penta
6	hexa

Note: The rules for naming binary covalent compounds are different from the rules for naming ionic compounds, which do *not* include prefixes when a subscript is present.

Problem-Solving Tutorial:
Naming Covalent Compounds

Naming Simple Binary Covalent Compounds

To name covalent compounds containing several carbon atoms, a set of rules described in Chapters 6 and 7 is followed. However, for compounds containing only two different nonmetal elements—binary compounds—a simpler set of rules is used. Begin by naming the first element in the formula according to its element name, followed by the second element in the formula, but change the ending of the second element name to -*ide*. For example, NO is named nitrogen oxide.

If more than one atom of a nonmetal element is present, as indicated by subscripts, a prefix is inserted before the element name to indicate the number of atoms of that type present in the molecule. For example, N_2O has the name **di**nitrogen oxide, where the prefix *di-* indicates that there are two nitrogen atoms in the binary covalent compound. Dinitrogen oxide, more familiar as "laughing gas," is widely used in dentistry and other areas of medicine. Table 2-3 lists the common prefixes used to name binary compounds.

Sometimes, the prefix *mono-* is used ("mono" means "one"), but normally it is assumed that if an atom has no prefix, only one such atom is present. For example, CO is named carbon monoxide; but NO, with a similar formula, is named nitrogen oxide.

As a convention, when the two vowels **ao** or **oo** appear together, the first vowel is dropped. The binary compound N_2O_5 is named dinitrogen pent**ox**ide, not dinitrogen pent**ao**xide.

WORKED EXERCISES 2-5   Naming Binary Compounds

1 Name the compound PF_5.
2 What is the molecular formula for sulfur hexafluoride?
3 What is the difference between sulfur dioxide and sodium oxide?

SOLUTIONS

1 Phosphorus pentafluoride. There is 1 phosphorus atom, named after the element: phosphorus. There are 5 fluorine atoms, so the prefix *penta-* is inserted before the element name and the ending is changed to -*ide*.

2 SF_6. There is no prefix before sulfur, so there is only one sulfur atom. The prefix *hexa-* placed before "fluoride" indicates that there are 6 fluorine atoms.

3 Sulfur dioxide is a covalent compound with the formula SO_2. Since it is a covalent compound, the number of oxygen atoms in the formula appears in the name. The presence of the metal sodium is a clue that sodium oxide, Na_2O, is an ionic compound, so no prefixes are used in its name. The ending in both compounds is changed to -*ide*, per the convention.

PRACTICE EXERCISES

2.14 Name the following compounds:

  a. CO_2 b. N_2O_3

2.15 What are the molecular formulas for the following compounds?

  a. nitrogen trifluoride   b. carbon tetrachloride

2.16 What are the names of the following compounds? Indicate whether the compound is ionic or covalent.

  a. $AlCl_3$ b. CS_2 c. ZnO   d. N_2O_4

2.3 | Compounds Containing Polyatomic Ions

You have seen that ionic compounds are composed of metal cations and nonmetal anions joined by an electrostatic attraction and that covalent compounds are composed of nonmetal atoms sharing electrons. A third category of compounds exists that has characteristics of both ionic and covalent compounds. These are compounds containing one or more polyatomic ions.

Polyatomic Ions

An ion formed from a single atom, like those we have seen so far, is also called a **monatomic ion.** A **polyatomic ion** ("poly" means many) is an ion formed when a *molecule*, rather than a single *atom*, gains or loses 1, 2, or 3 electrons. Covalent bonds join the atoms, but an imbalance in the number of protons and electrons overall creates a net charge on the molecule, and hence, it is referred to as an ion. The charge can be *localized* on one atom or it can be *spread* over several atoms in the polyatomic ion. The charge is indicated as a superscript outside a pair of brackets that enclose the structure, or as a superscript following the formula unit of the polyatomic ion. For example, the hydroxide ion, HO^-, has the Lewis structure

$$\left[:\ddot{O}-H \right]^-$$

In this example, the oxygen and hydrogen atoms share a pair of electrons in a covalent single bond, yet there is a net -1 charge overall, because there is one more electron than protons in the two atoms combined. Note that polyatomic ions typically have a Lewis structure where one or more atoms do not correspond to the usual nonbonding and bonding arrangements described in Table 2-2.

Some common polyatomic ions are listed in Table 2-4. Each polyatomic ion has a unique name, which you should learn to associate with its formula unit and charge. For example, the carbonate ion has the formula unit CO_3^{2-}, a polyatomic ion with a -2 charge.

The Formula Unit and Naming

Like monatomic ions, polyatomic ions are part of a lattice structure and serve as either the cation or the anion component or both. As with monatomic ions, the ratio of cation to anion is given by the formula unit. The lowest whole number ratio of ions that gives an electrically neutral compound determines the subscripts.

Writing the formula unit of a compound containing a polyatomic ion is similar to the process described for monatomic ions, except that if there is a subscript following the polyatomic ion, the entire polyatomic ion must be enclosed in parentheses. For example, magnesium hydroxide has the formula unit $Mg(OH)_2$. Parentheses around the OH followed by the "2" indicate that there are 2 hydroxide (OH^-) ions for every 1 magnesium ion (Mg^{2+}).

When naming an ionic compound containing a polyatomic ion, name the cation followed by the anion. If either the cation or the anion is a polyatomic ion, use the name for the polyatomic ion given in Table 2-4. For example, Na_2CO_3 has the name sodium carbonate, because the name of the monatomic cation is sodium (Na^+) and the name of the polyatomic anion is carbonate, CO_3^{2-}. Note that the number of cations or anions in the formula unit does not appear as part of the name of the compound.

Table 2-4 Common Polyatomic Ions

Name	Formula
Anions	
Acetate	$CH_3CO_2^-$
Hydrogen carbonate (bicarbonate)	HCO_3^-
Carbonate	CO_3^{2-}
Cyanide	CN^-
Dihydrogen phosphate	$H_2PO_4^-$
Hydroxide	OH^-
Hyphochlorite	OCl^-
Nitrate	NO_3^-
Nitrite	NO_2^-
Phosphate	PO_4^{3-}
Hydrogen phosphate	HPO_4^{2-}
Sulfate	SO_4^{2-}
Sulfite	SO_3^{2-}
Cations	
Hydronium	H_3O^+
Ammonium	NH_4^+

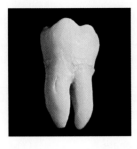

Tooth enamel is the top outer part of the molar. [Anna Blume/Alamy]

Problem-Solving Tutorial:
Ionic versus Covalent compounds

The active substance in bleach is sodium hypochlorite, NaOCl, a compound containing the polyatomic ion OCl^-.
[David Young-Wolff/Photo Edit]

Polyatomic Ions in Health Care

Tooth enamel is composed of the ionic compound calcium hydroxyapatite, which has the formula unit $Ca_5(PO_4)_3OH$. As you can tell from the formula unit, this ionic compound is made up of two different polyatomic anions: 3 phosphate ions (PO_4^{3-}) and 1 hydroxide ion (OH^-). One of the reasons fluoride ion, F^-, is added to municipal water systems, toothpastes, and mouth washes is that it replaces the hydroxide ion in calcium hydroxyapatite to form $Ca_5(PO_4)_3F$. This is a much stronger substance than hydroxyapatite and is therefore less likely to decay.

The active ingredient in bleach is the hypochlorite ion (OCl^-). This ion is a well known bacteriocide—a substance that kills bacteria. The preservatives sodium nitrate, $NaNO_3$, and sodium nitrite, $NaNO_2$, are compounds containing the polyatomic ions NO_3^- and NO_2^-, respectively, which are used to inhibit the growth of microorganisms in food.

WORKED EXERCISE 2-6   Polyatomic Ions in Ionic Compounds

Answer the questions below about the acetate ion.

$$\left[\begin{array}{c} \overset{\displaystyle H}{\underset{\displaystyle H}{\overset{|}{\underset{|}{H-C}}}} \overset{\displaystyle \cdot\cdot\overset{\displaystyle O}{}}{\overset{\|}{-C}} - \ddot{O}: \end{array} \right]^-$$

Acetate ion

a. The charge on the acetate ion is spread over both oxygen atoms. What is the total charge on the acetate ion?
b. How many covalent bonds does this polyatomic ion contain?
c. Why is acetate ion classified as a polyatomic ion and not a monatomic ion?
d. How is acetate ion different from a molecule?
e. What is the formula unit for sodium acetate?
f. What is the formula unit for calcium acetate?

SOLUTION

a. The charge on the acetate ion is -1, indicated outside the brackets.
b. There are seven covalent bonds, five single bonds and one double bond.
c. Acetate is a polyatomic ion because there are several atoms joined by covalent bonds and collectively they carry a net -1 charge.
d. Acetate is an ion not a molecule because there is a net charge.
e. $NaCH_3CO_2$ or $C_2H_3O_2^-$. Remember to write the cation first, followed by the anion.
f. $Ca(CH_3CO_2)_2$ or $Ca(C_2H_3O_2)_2$. Acetate must be enclosed in parentheses to indicate that there are two acetate ions for every calcium ion. The negative charge on two acetate ions $(2 \times -1) = -2$ is required to cancel the total positive charge on one calcium ion ($+2$).

PRACTICE EXERCISES

2.17 Answer the following questions about the ammonium ion.

$$\left[\begin{array}{c} H \\ | \\ H-N-H \\ | \\ H \end{array} \right]^+$$

a. The charge on the ammonium ion is localized on the nitrogen atom. What is the charge on the ammonium ion?

    b. How many covalent bonds does the ammonium ion contain?

    c. Why is ammonium considered a polyatomic ion and not a monatomic ion?

    d. Why is ammonium classified as an ion and not a molecule?

    e. What is the formula unit for ammonium chloride?

    f. What is the formula unit for ammonium sulfide?

2.18 Using Table 2-4, name the following ionic compounds:

    a. KNO_3 **b.** Na_2HPO_4 **c.** NaOH **d.** $AgNO_3$ **e.** $(NH_4)_2CO_3$

2.19 Using Table 2-4, write the formula unit for the following ionic compounds:

    a. lithium hydroxide **b.** strontium carbonate **c.** magnesium phosphate

2.20 Potassium cyanide is a poison. Write the formula unit for potassium cyanide.

2.21 Lithium carbonate is an antipsychotic drug used to treat bipolar disorder. What is the formula unit for lithium carbonate?

2.22 **Critical thinking question:** Most gems are finely polished minerals. A mineral is an ionic compound. Turquoise, used in jewelry, has the formula unit $CuAl_6(PO_4)_4(OH)_8$. Notice that there is more than one cation and anion; nevertheless, the ionic compound must be electrically neutral. Based on the charges on the polyatomic anions and the monatomic cations, what form of copper is present in turquoise, Cu^{2+} or Cu^+?

Turquoise. [TH Foto-Werbung/Photo Researchers, Inc.]

2.4 Molecular Mass, Formula Mass, and Molar Mass

You have considered the atomic scale of matter and know that an atom, an ion, and a molecule are incredibly small in size, so how can we ever know how many of these particles there are in a sample of matter? The answer is that if you know the mass of a single atom, you can determine the number of atoms present from the mass of a macroscopic sample of the pure element. Similarly, if you know the mass of a molecule or formula unit, you can determine the number of molecules or ions present. Your body mass, for example, is a result of the sum total mass of all the atoms, molecules, and ions that make up your body. In chemistry it is important to be able to keep track of the number of particles present in a given mass of a pure compound or element.

Formula Mass and Molecular Mass

In Chapter 1 you learned that the average atomic mass of an element can be found on the periodic table of elements and is given in atomic mass units (amu). What is the mass of one *molecule* constructed of atoms whose average atomic masses are known? *Quite simply, the mass of one molecule is the sum of the individual average atomic masses of its component atoms, known as its* **molecular mass,** *also given in amu.*

> *For covalent compounds*
> Molecular mass = sum of atomic masses of atoms in molecular formula (amu)

For example, a molecule of water, which has the chemical formula H_2O, has a molecular mass of 18.01 amu:

Atom type	no. of atoms	average atomic mass	total
H	2	$\times$ 1.008 amu/atom	= 2.016 amu/atom
O	1	$\times$ 16.00 amu/atom	= +16.00 amu/atom
			18.02 amu/molecule (molecular mass)

For an ionic compound, the process is identical, except that you use the mass of the ions in the formula unit to calculate the **formula mass** of the compound. Note that the mass of an ion is the same as the mass of the element from which it was derived. Since an ionic compound is made up of a lattice rather than individual molecules, *formula* mass instead of *molecular* mass is the term used to refer to the mass of the formula unit. Formula mass is also reported in atomic mass units (amu).

For ionic compounds
Formula mass = sum of atomic masses of ions in formula unit (amu)

WORKED EXERCISES 2-7 Calculating Molecular and Formula Mass

1 Calculate the molecular mass of caffeine, described in the opening vignette. The molecular formula for caffeine is $C_8H_{10}N_4O_2$. Is this a molecular compound or an ionic compound, and how can you tell?
2 Calculate the formula mass of calcium hydrogen carbonate, $Ca(HCO_3)_2$. Is this a molecular compound or an ionic compound, and how can you tell?

SOLUTIONS

1 $C_8H_{10}N_4O_2$

Atom type	no. of atoms		atomic mass (from periodic table)		total
C	8	×	12.01	=	96.08 amu
H	10	×	1.008	=	10.08 amu
N	4	×	14.01	=	56.04 amu
O	2	×	16.00	=	+ 32.00 amu
					194.20 amu/molecule

The molecular mass of caffeine is 194.20 amu.

This substance is a covalent compound, because it is composed of all nonmetals, so the terms molecular mass and molecular formula are used.

2 $Ca(HCO_3)_2$

Atom type	no. of atoms		atomic mass		total
C	2	×	12.01	=	24.02 amu
Ca	1	×	40.08	=	40.08 amu
H	2	×	1.008	=	2.016 amu
O	6	×	16.00	=	+ 96.00 amu
					162.12 amu/formula unit

The formula mass of calcium hydrogen carbonate is 162.12 amu. To avoid rounding errors, do not round the atomic mass values obtained from the periodic table until the end of the calculation.

This substance is an ionic compound, because it is composed of a metal and polyatomic ions, so the terms formula mass and formula unit are used. Remember, the atomic mass of atoms within parentheses must be multiplied by the subscript outside of the parentheses, as well as the subscript inside the parentheses. In this example, you see there are 6 oxygen atoms, as indicated by the subscript "3" following the symbol for oxygen, and the subscript "2" after the parentheses. The subscript "2" is also applied to the carbon and hydrogen atoms within the parentheses.

PRACTICE EXERCISES

2.23 Calculate the molecular mass of methane, CH_4. Is this a molecular compound or an ionic compound, and how can you tell?

2.24 Calculate the formula mass of aluminum hydroxide, $Al(OH)_3$. Is this a molecular compound or an ionic compound, and how can you tell?

2.25 Glucose—blood sugar—has the molecular formula $C_6H_{12}O_6$. Calculate the molecular mass of glucose. Is this a molecular compound or an ionic compound, and how can you tell?

The Mole

The mass of one atom, ion, or molecule is so small that it requires an enormous number of particles—on the order of 10^{23}—to arrive at a mass on the macroscale that can be conveniently seen and weighed. To go from the atomic scale to the macroscale, scientists scale up from one atom, ion, or molecule to **6.02 × 10²³** atoms, ions, or molecules, a value known as **Avogadro's number.** Avogadro's number represents a quantity known as a **mole** in the same way that the number 12 represents a quantity known as a dozen. In this way we can simply say we have 1 mole of a substance rather than trillions and trillions of atoms. Since *1 mole represents 6.02×10^{23} particles, it can be expressed as a conversion between number of particles and a mole:*

A mole of sodium chloride, water, and copper (reading clockwise from the top). [Richard Megna/Fundamental Photographs]

$$1 \text{ mol of atoms} = 6.02 \times 10^{23} \text{ atoms}$$

$$1 \text{ mol of molecules} = 6.02 \times 10^{23} \text{ molecules}$$

$$1 \text{ mol of formula units} = 6.02 \times 10^{23} \text{ formula units}$$

Although it is not much of a simplification, the abbreviation for the *mole* is *mol*. The photo in the margin shows 1 mole of the element copper, the ionic compound sodium chloride (NaCl), and the covalent compound water (H_2O). Since you can see a mole of matter, the mole represents a macroscopic amount of material.

We know why Avogadro's number needs to be such a large number, but why 6.02×10^{23}, and not just any large number? *By using Avogadro's number specifically, the numerical value for the mass of 1 mole of any element conveniently becomes the same numerical value as the average atomic mass of an atom of that element.* For example, one carbon atom has an atomic mass of 12.01 amu per atom, while one mole of carbon atoms has a mass of 12.01 grams per mole:

1 single carbon atom:	**12.01** amu/atom
1 mole of carbon atoms:	**12.01** g/mol

Likewise, the numerical value for the mass of 1 mole of any compound is the same numerical value as the molecular mass or formula mass of that compound. The difference lies only in the units: The mass of one mole of atoms, ions, or molecules is given in units of grams per mole, g/mol, while the mass of one atom, ion, or molecule is given in units of amu.

The mass of one mole of an element or compound is known as its **molar mass.** Since one mole of any element or compound always contains Avogadro's number of atoms, molecules, or ions, whenever you know the number of moles of a sample, you are keeping track of the number of atoms, molecules, or ions in that sample. *This is particularly important in chemistry because we rely on the mass of a substance to give us information about the number of particles of that substance.* In the next section you will learn how to perform calculations that convert between the units of mass and units of moles of an element or compound.

From your periodic table, you can see that one mole of sulfur, for example, has a mass of 32.07 g, while one mole of carbon has a mass of 12.01 g.

Three pineapples have a greater mass than three cherry tomatoes, because a single pineapple has a greater mass than a single cherry tomato. [Richard Megna/ Fundamental Photographs]

Lab Result Report

Jane Doe

Blood Component	Value	Normal Range
Chloride	104 mmol/L	98–109 mmol/L
CO_2	25 mmol/L	21–31 mmol/L

If you are comparing an equal number of atoms of both elements, why is one mole of sulfur atoms heavier than one mole of carbon atoms? Every sulfur atom has 16 protons in its nucleus plus some number of neutrons, while every carbon atom has only 6 protons in its nucleus. Since there are 6.02×10^{23} atoms in both cases, one mole of the heavier atoms will obviously weigh more—just as you would expect a dozen pineapples to weigh more than a dozen cherry tomatoes, because a single pineapple weighs more than a single cherry tomato. One important distinction between atoms and fruits, however, is that you could actually weigh an individual pineapple or cherry tomato if you had to. This is not possible with an individual atom, and therefore, the relationship between mass and number of particles is critical for doing scientific work.

In medicine you will frequently encounter units involving the mole. In a blood test, for example, the number of moles or millimoles in every liter of blood is typically reported for substances in the blood, used for diagnostic purposes. Jane Doe's blood test, shown in the margin, indicates that she has 104 millimoles of chloride ion (Cl^-) in every liter of her blood. Note the use of the prefix, milli, which you recall means 1/1000, so 104 mmoles is 0.104 mol. Thus, Jane Doe has 0.104 mol of chloride ion in every liter of her blood. In the column on the far right, you can compare this value to the normal range for chloride ion, which is 98 mmol to 109 mmol in every liter of blood. The test results show that Jane Doe's chloride levels are within the normal range.

To assess patient health, doctors and nurses essentially report the number of chloride ions in a sample of blood and compare it to the normal range of values. To determine the number of chloride ions that correspond to 0.104 mol requires a calculation using Avogadro's number, described in *Extension Topic 2-1: Converting Between Moles and Atoms, Ions, or Molecules*. You will learn more about units of millimoles per liter and other concentration units when you study solutions in Chapter 5. In general, the mole is used whenever the *number* of molecules, ions, or atoms is important.

WORKED EXERCISES 2-8   The Mole and Avogadro's Number

1 Which has the larger mass, 1 mol of gold or 1 mol of silver? Explain why.

2 What is the mass of 1 mol of silver? of 1 mol of gold? Be sure to include units in your answser.

3 How many silver atoms are there in 1 mol of silver? How many water molecules are there in 1 mol of water? How many Na^+ ions and Cl^- ions are there in 1 mol of NaCl?

4 What is the mass of half a mole of silver? How many silver atoms are there in half a mole of silver?

SOLUTIONS

1 A mole of gold has a greater mass than a mole of silver because the former has the greater average atomic mass.

2 Using the periodic table, a mole of gold has a mass of 197.0 g and a mole of silver has a mass of 107.9 g.

3 There are 6.02×10^{23} silver atoms in 1 mol of silver. There are 6.02×10^{23} water molecules in 1 mol of water. There are 6.02×10^{23} chloride ions and 6.02×10^{23} sodium ions in 1 mol of NaCl.

4 Half a mole of silver has a mass of 107.9 g/2 = 54.0 g. There are $6.02 \times 10^{23}/2 = 3.01 \times 10^{23}$ silver atoms in half a mole of silver.

PRACTICE EXERCISES

2.26 How many marbles are there in 1 mol of marbles?

2.27 Which has a greater mass, a mole of ping-pong balls or a mole of basketballs? Explain why.

2.28 What is the average atomic mass of one mercury atom? What is the mass of 1 mol of mercury atoms? How are these two numbers similar and different?

2.29 How many molecules are there in half a mole of ammonia (NH_3)?

2.30 What is Avogadro's number? Why is Avogadro's number so large?

2.31 Using the blood test results for Jane Doe, how many moles of carbon dioxide, CO_2, are there in every liter of Jane Doe's blood? Are Jane's blood CO_2 levels normal?

Converting Between Units of Mass and Moles of Any Substance: Grams → Moles or Moles → Grams

Converting between mass and moles and vice versa is a routine calculation in the chemical laboratory. Suppose you were asked to calculate how many grams there are in a particular number of moles of an element or compound. The key to solving this type of problem is recognizing that the molar mass of an element or compound can be used as a conversion factor between the mass and moles of that substance.

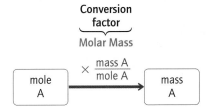

Alternatively, you might be asked to determine the number of moles of an element or compound in a given mass of that element or compound—the reverse process. The molar mass of the element also serves as the conversion factor in this calculation, but in its inverted form: mol/g.

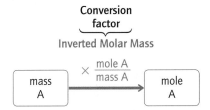

As with any calculation involving conversions, using conversion factors—described in Chapter 1—provides a reliable and straightforward method for arriving at the solution. Follow the guidelines below to convert between grams and moles or moles and grams of a substance.

Guidelines for Converting Between Units of Mass and Moles of an Element or Compound

Step 1: If the substance is a monatomic element, look up the molar mass of the element on the periodic table. If the substance is a compound or a diatomic element, calculate the molar mass from the molar mass of its constituent elements.

The molar mass of a monatomic element can be found on the periodic table—it is the same numerical value as the atomic mass, but the units are grams per mole. For a compound or a diatomic element, the molar mass is calculated from the sum of its constituent elements as described in the previous section.

For example, the molar mass of copper is 63.55 g/mol. The molar mass of water is 18.02 g/mol.

Step 2: Express the molar mass as two possible conversion factors.

Molar mass *is* a conversion factor, and as such can be used as is or in its inverted form: g/mol or mol/g.

For example, the molar mass of Cu is a conversion factor, and so is its inverted form:

$$\frac{63.55 \text{ g Cu}}{1 \text{ mol Cu}} \quad \text{or} \quad \frac{1 \text{ mol Cu}}{63.55 \text{ g Cu}}$$

(molar mass)

Step 3: Set up the calculation so that the supplied units cancel.

Next, multiply the supplied unit (grams or moles) by the form of the conversion factor that causes the supplied units to cancel, so that only the requested unit remains.

For example, to determine the number of moles in 50.0 g of copper, the second conversion factor above (inverted molar mass) would be used, because grams are in the denominator of this conversion factor.

$$50.0 \text{ g Cu} \times \frac{1 \text{ mol Cu}}{63.55 \text{ g Cu}} = 0.787 \text{ mol Cu (3 significant figures)}$$

Therefore, 50.0 g of Cu is equivalent to 0.787 mol of Cu atoms.

As with all calculations, double check that the final answer has been rounded to the correct number of significant figures.

WORKED EXERCISE 2-9   Converting From Mass to Moles

How many grams of sulfur are there in 5.600 mol of sulfur?

SOLUTION

Step 1: Look up the molar mass of the element on the periodic table. The substance is sulfur, an element, so the molar mass can be obtained from the periodic table: 32.07 g/mol.

Step 2: Express the molar mass as two possible conversion factors. The molar mass and its inverted form are the two possible conversion factors in any gram to mol or mol to gram calculation.

$$\frac{32.07 \text{ g}}{1 \text{ mol S}} \quad \text{or} \quad \frac{1 \text{ mol S}}{32.07 \text{ g}}$$

(molar mass)

Step 3: Set up the calculation so that the supplied units cancel. The calculation is similar to the one described in the guidelines, but the choice of conversion factor is different, because now the supplied unit is in moles and the requested unit is in grams. As always, multiply the supplied unit by the form of the conversion factor that allows the supplied unit (mole in this case) to cancel:

$$5.600 \text{ mol S} \times \frac{32.07 \text{ g S}}{1 \text{ mol S}} = 179.6 \text{ g of sulfur (4 significant figures)}$$

(conversion factor)

Therefore, 5.600 moles of sulfur have a mass of 179.6 g.

2.32 How many moles of mercury are there in 159 g of mercury?

2.33 How many moles of water are there in 45 g of water?

2.34 Calculate the number of moles of ibuprofen in a 200.-mg dose of Advil. The molecular formula for ibuprofen is $C_{13}H_{18}O_2$.

2.35 About 18 million people in the United States are diabetic. A fasting glucose ($C_6H_{12}O_6$) level of greater than 126 mg/dL of glucose is usually indicative of diabetes. How many moles of glucose in every deciliter (dL) of blood does this value correspond to?

Extension Topic 2-1 Converting Between Mass and Number of Particles

You have learned how to convert between grams and moles and moles and grams of a pure element or compound. Sometimes you may want to consider how many atoms, molecules, or ions are actually present in a given sample. *This type of calculation requires Avogadro's number, which always serves as a conversion factor between moles and the number of any particle.*

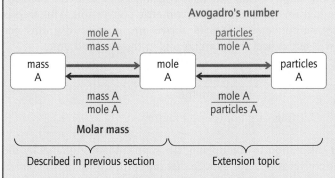

Whereas molar mass is the conversion factor between moles and mass, Avogadro's number is the conversion factor between *moles* and *number of particles*, whether they are atoms, ions, or molecules. Notice that the mole is at the center of both types of calculations. Thus, if you want to determine the number of atoms in a 5-g sample of copper, you must first convert the mass of copper into moles of copper, and then convert moles of copper to atoms of copper. In the first conversion, the inverted form of molar mass is used as the conversion factor, and in the second step Avogadro's number is used to create the conversion factor—as shown in the diagram.

WORKED EXERCISE E2-1 Converting from Number of Particles to Mass

What is the mass of 5.4×10^{20} water molecules?

SOLUTION

Steps 1 and 2: Determine the conversion factors needed. Avogadro's number will be used as a conversion factor:

$$\frac{6.02 \times 10^{23} \text{ molecules}}{1 \text{ mol}} \quad \text{or} \quad \frac{1 \text{ mol}}{6.02 \times 10^{23}}$$
(Avogadro's number)

The molar mass of water will be needed, which was calculated previously to be 18.01 g/mole. The molar mass, a conversion factor, can be expressed in two ways:

$$\frac{18.02 \text{ g}}{1 \text{ mol}} \quad \text{or} \quad \frac{1 \text{ mol}}{18.01 \text{ g}}$$
(molar mass)

Step 3: Set up the calculation so that the supplied units cancel. The calculation can be done in two sequential steps or in one step as shown:

$$5.4 \times 10^{20} \text{ molecules} \times \frac{1 \text{ mol}}{6.02 \times 10^{23} \text{ molecules}} \times$$
(inverted form of Avogadro's number)

$$\frac{18.01 \text{ g}}{1 \text{ mol}} = 0.016 \text{ g}$$
(molar mass   (2 significant
of water)   figures)

Thus, 5.4×10^{20} molecules of water are equivalent to a mass of 0.016 g of water.

PRACTICE EXERCISES

E2.1 How many copper atoms are there in 0.30 mol of copper?

E2.2 How many copper atoms are there in 5.0 g of copper?

E2.3 How many chloride ions are there in every liter of Jane Doe's blood?

E2.4 How many carbon dioxide molecules are there in 1 L of Jane Doe's blood?

E2.5 A normal dose of ibuprofen is 200. mg. How many ibuprofen molecules are there in a dose of ibuprofen? The molecular formula of ibuprofen is $C_{13}H_{18}O_2$.

E2.6 How many formula units of table salt are there in 1 g of sodium chloride? How many sodium ions is this? How many chloride ions is this?

WORKED EXERCISE 2-10   Converting From Mass to Moles

How many moles of caffeine, $C_8H_{10}N_4O_2$, are in 100. mg of caffeine?

SOLUTION

Step 1: For a compound, calculate the molar mass from its constituent atoms, using the periodic table.

Calculate the molar mass of caffeine, $C_8H_{10}N_4O_2$ from the atomic masses of its constituent atoms:

$$8\,C + 10\,H + 4\,N + 2\,O = (8 \times 12.01) + (10 \times 1.008) + (4 \times 14.01) + (2 \times 16.00) = 194.20 \text{ g/mol}$$

Step 2: Express the molar mass as two possible conversion factors.

The molar mass of caffeine and its inverted form are the two possible conversion factors:

$$\frac{194.19 \text{ g } C_8H_{10}N_4O_2}{1 \text{ mol } C_8H_{10}N_4O_2} \quad \text{or} \quad \frac{1 \text{ mol } C_8H_{10}N_4O_2}{194.19 \text{ g } C_8H_{10}N_4O_2}$$

(molar mass)

Step 3: Set up the calculation so that the supplied units cancel.

Extra Conversion Step: In this exercise, you must first convert the supplied unit from milligrams to grams, using a metric conversion. This step is necessary because the conversion factors above are in units of grams per mole and grams will not cancel with milligrams in the supplied unit.

$$100.\ \cancel{\text{mg caffeine}} \times \frac{10^{-3} \text{ g caffeine}}{1 \ \cancel{\text{mg caffeine}}} = 0.100 \text{ g caffeine}$$

(mg → g conversion factor)

Set up the next calculation by multiplying the supplied unit in grams by the correct form of the conversion factor. Notice that the correct form of the conversion factor has grams in the denominator so that it will cancel with grams from the supplied unit. The result will be an answer in moles, the requested unit.

$$0.100 \text{ g } \cancel{C_8H_{10}N_4O_2} \times \frac{1 \text{ mol } C_8H_{10}N_4O_2}{194.20 \text{ g } \cancel{C_8H_{10}N_4O_2}} = 5.15 \times 10^{-4} \text{ mol}$$

(conversion factor)   (3 significant figures)

PRACTICE EXERCISES

2.36 How many grams of mercury are there in 0.666 mol of mercury?

2.37 What is the mass of 1.5 mol of water?

2.38 How many grams of chloride ions, Cl^-, are there in 1 liter of Jane Doe's blood?

In this chapter you considered interactions of atoms *within* compounds: the electrostatic attractions between cations and anions that hold together ionic compounds and the covalent bonds that hold together covalent compounds. In learning about covalent compounds, you were introduced to the important skill of writing Lewis dot structures. Lewis dot structures are a form of electron bookkeeping that shows how atoms are connected to one another in a molecule. As such, Lewis dot structures are 2-dimensional representations of molecules. In the next chapter you will build upon your knowledge of Lewis dot structures to construct three-dimensional representations of molecules and consider how molecules interact with one another through *intermolecular forces of attraction.*

Chemistry in Medicine Blood Chemistry and the Diagnosis of Disease

Almost everyone has had a blood test taken at some point in his or her life. It is one of the most routine medical tests, is readily performed, and provides a wealth of information. A blood test is often called a **blood chemistry panel,** because it measures the concentration of some important atoms, ions, and molecules in the blood. Blood is a complex **mixture** of many substances.

Blood transports nutrients, gases, hormones, and other molecules to the tissues throughout the body, and at the same time it transports waste products to the kidneys, where they are filtered and eliminated from the body. Abnormal levels of certain ions and compounds in the blood are often the first sign of organ malfunction. Problems with the kidneys, liver, adrenal glands, heart, and other organs can be readily diagnosed with a simple blood test.

In a typical blood test, the concentrations of these substances are reported alongside a column showing the normal concentration range for a healthy individual. To understand a blood chemistry panel, first consider the composition of blood.

The Composition of Blood

When a sample of blood in a test tube is placed in a centrifuge and spun at high speed (5000 rpm), it separates into two layers as shown in Figure 2-12. The denser layer at the bottom of the test tube contains red blood cells, the cells that transport oxygen from the lungs to the tissues. The less dense layer at the top of the test tube is known as **blood plasma.** Although the number (hemocrit) and shape of the blood cells in the bottom layer provides important information, the plasma layer contains many of the dissolved ions, gases, and molecules of interest.

Blood plasma can be further separated into fibrinogen and serum. **Fibrinogen** is the substance in blood that causes it to clot. To prevent a blood sample from clotting, fibrinogen is often removed from the blood plasma, leaving what is known as blood **serum.** The focus here will be on blood serum chemistry.

$$Blood = plasma + blood\ cells$$
$$Plasma = serum + fibrinogen$$

Blood serum is a light yellow liquid that is composed of 90% water. Serum is not red, because it does not contain red blood cells. Hundreds of chemical substances are dissolved in the serum, and some of these are the substances of interest in a blood chemistry panel. These substances, grouped according to their chemistry, include:

- small molecules: glucose, amino acids, creatinine, urea, etc. These small molecules are all composed of building block elements.
- biomolecules (large biological molecules): proteins such as enzymes and albumin; lipids such as fats and cholesterol. These are much larger molecules, also composed of building block elements.
- electrolytes (Na^+, K^+, Ca^{2+}, Cl^-, HCO_3^-, etc.). These include most of the macronutrients as well as some of the important polyatomic ions.
- dissolved gases (CO_2, N_2, O_2). Although most of the oxygen in the blood is bound to hemoglobin, some is dissolved in the plasma. You will learn more about gases dissolved in liquids in Chapter 5.

Table 2-5 shows the normal concentration range for some of these substances, which are usually reported in a blood chemistry panel. Each concentration is expressed as a ratio of two units. The numerator gives the mass or

Table 2-5 Normal Range for Various Ions and Molecules in a Blood Chemistry Panel

Ion or Molecule	Normal Range
Creatinine	0.5–1.4 mg/dL
Urea	7–20 mg/dL
Albumin	3.9–5.0 g/dL
Total Protein	6.3–7.9 g/dL
Glucose	64–128 mg/dL
Calcium	8.5–10.9 mg/dL
Potassium	3.7–5.2 mEq/L
Chloride	101–111 mmol/L
Carbon dioxide	20–29 mmol/L
Cholesterol	100–240 mg/dL
Alkaline phosphatase enzyme	44–147 IU/L

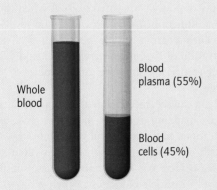

Whole blood

Blood plasma (55%)

Blood cells (45%)

Figure 2-12 Blood can be separated into two layers: the top layer is known as blood plasma, and the bottom layer contains blood cells.

continued

moles of the substance of interest, and the denominator represents the volume of blood containing the substance. Note, "equivalents" refers to the number of moles of charge, a term that applies specifically to electrolytes and will be described further in Chapter 5.

You will find that remembering your metric prefixes will be useful in interpreting Table 2-5. For example, the most common unit of volume used in a blood test is the deciliter, dL. Recall from Table 1-2 that the prefix *deci*- means $\frac{1}{10}$ or 0.1 of the base unit, which in this case is the liter, L. By comparison, the *milli*liter is $\frac{1}{1000}$ of a liter; therefore, 1 dL = 100 mL. You will learn more about the different units of concentration and their significance in Chapter 5.

Substances in a Blood Panel

Consider some of the important substances measured in a blood panel. *Urea* is a molecule with the formula CH_4N_2O produced from the breakdown of amino acids, and eliminated in the urine. Urea is often referred to as BUN, which stands for Blood Urea Nitrogen, because of the nitrogen in its structure. High serum BUN levels are often an indication of kidney problems.

Normal concentration: 7–20 mg/dL

$$H-N-C-N-H$$

Urea, CH_4N_2O

Glucose, commonly referred to as blood sugar, is measured to determine if a patient has hypoglycemia (low blood sugar), or hyperglycemia (high blood sugar). Since an individual's glucose concentration fluctuates throughout the day, depending on when and what was eaten last, this test usually must be taken after the person has been fasting for 12 hours. Fasting glucose levels above 126 mg/dL are usually an indicator of diabetes.

A glucometer is used to measure a diabetic's glucose levels throughout the day.

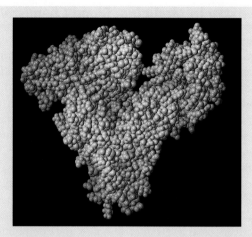

Figure 2-13 A single molecule of the blood protein albumin. Each sphere represents an atom. Proteins are large to extremely large biological molecules. [Courtesy of Pacific Northwest National Laboratory]

Diabetics must monitor their glucose levels throughout the day, typically with a drop of their blood, using a glucometer.

Because many proteins are produced in the liver, assessing the concentration of several key liver proteins in the blood is a common method for evaluating liver function. *Albumin*, for example, is the most abundant protein in the blood. Albumin regulates the amount of water in tissues and blood. Figure 2-13 shows a model of a molecule of albumin, which contains many building block atoms, most of which are carbon and hydrogen atoms. Each sphere in the figure represents an atom. The green spheres represent other molecules bound to albumin. Higher than normal concentrations of albumin are seen in some types of liver disease, shock, and dehydration; while low levels may suggest hemorrhage (bleeding), diarrhea, other liver diseases, or infection.

Potassium ion, K^+, is arguably the most important cation in the body, required by all cells. It is important for proper signaling in neurons and muscle cells. Abnormal potassium levels are often present when the adrenal glands malfunction. *Increased* potassium levels are also present during cardiac arrest (heart attack). The normal concentration of K^+ ranges between 3.7 and 5.2 mEq/L. An equivalent is a measure of the number of moles of charge.

As you can see, medicine is all about chemistry, as our bodies are made up of atoms, molecules, and ions whose concentrations must be carefully regulated by the body. Too much or too little of a substance is often associated with disease. Here you have seen that these substances are measured by a simple blood test, which provides a convenient window into the health of a patient.

Chapter Summary

Ionic Compounds

- Matter is composed of pure substances and mixtures. Pure substances include elements and compounds. There are two types of compounds: ionic and covalent.

- An ionic compound is formed as a result of the electrostatic attraction between a cation and an anion.

- A cation is formed when a metal atom loses electrons and an anion is formed when a nonmetal atom gains electrons.

- Main group ions typically have a complete outermost shell of electrons similar to the nearest noble gas on the periodic table.

- Transition metals lose varying numbers of electrons, and therefore, often exist in more than one ionic form.

- An ionic compound exists as a lattice structure wherein each ion is surrounded by ions with the opposite charge.

- Ions dissolved in water are known as electrolytes.

- The formula unit of an ionic compound indicates the lowest whole-number ratio of cation to anion that gives an electrically neutral compound.

- An ionic compound is named by writing the name of the cation first, followed by the name of the anion. If the cation has multiple forms, a Roman numeral is placed in parentheses immediately following the cation name.

Covalent Compounds

- Discrete entities known as molecules are formed by the sharing of electrons between nonmetal atoms, known as covalent bonds.

- Two shared electrons are known as a single bond and are written as a line between the two atoms. Multiple bonds occur when 4 or 6 electrons are shared between atoms.

- Molecules form in such a way that each atom achieves a stable octet— 8 valence electrons. However, hydrogen shares 2 electrons.

Writing Lewis Dot Structures

- Lewis dot structures are used to show how valence electrons are distributed among the atoms in a molecule.

- Bonding electrons are shown as a line, and nonbonding electrons are shown as a pair of dots.

- Hydrogen always forms one bond by sharing its one electron with an electron from another atom, to give a stable pair of bonding electrons.

- Elements in period 3 or higher are capable of forming expanded octets, a situation where there are more than 8 valence electrons on an atom.

Polyatomic Ions

- Polyatomic ions contain covalent bonds, but also have a net charge, so they are classified as ions.

- Polyatomic ions are characterized by their name and charge.

- Polyatomic ions form ionic compounds similarly to monatomic ions.

Molecular Mass, Formula Mass, and Molar Mass

- The molecular mass is the mass of an individual molecule, given in units of amu. It is calculated from the sum of the individual atomic masses of the atoms that make up the molecule.

- The formula mass of an ionic compound, given in units of amu, is calculated from the mass of the individual ions that make up the formula unit of the compound. This is the lowest whole-number ratio of ions in the lattice.
- The molar mass of a covalent compound is numerically equivalent to the molecular mass, and the molar mass of an ionic compound to the formula mass. However, the molar mass is reported in units of grams per mole.

Key Words

Anion A negatively charged ion resulting from a nonmetal atom gaining electrons, so that the number of electrons is greater than the number of protons.

Cation A positively charged ion resulting from the loss of electrons from a metal atom, so that the number of electrons is less than the number of protons.

Covalent bond The sharing of electrons between two nonmetal atoms in a molecule.

Covalent compound A molecule composed of more than one type of nonmetal atom joined by covalent bonds: H_2O, NH_3, etc.

Electrolyte An ion dissolved in water.

Electrostatic attraction The force of attraction between oppositely charged entities such as a proton and an electron or a cation and an anion.

Formula mass The mass of one formula unit of an ionic compound in units of amu.

Formula unit The symbols of the elements in an ionic compound. Subscripts indicate the lowest whole-number ratio of cations and anions.

Ion A positive or negatively charged atom. The charge results from an unequal number of protons and electrons, due to the gain or loss of electrons.

Ionic bond The electrostatic attraction that holds anions and cations together, such as the lattice structure of an ionic compound.

Ionic compound A compound composed of ions held together by electrostatic attractions.

Lattice A three-dimensional array of cations and anions forming an ionic compound. Cations are surrounded by anions and anions by cations.

Lewis dot structure The structure of a molecule showing how valence electrons are distributed around atoms and shared between atoms. Covalent bonds are drawn as lines and nonbonding electrons are drawn as dots.

Mixture A physical combination of two or more elements and/or compounds.

Molar mass The mass of one mole of an element or compound, given in units of g/mol.

Molecule A compound composed of more than one type of nonmetal atom, joined by covalent bonds.

Nonbonding electrons Valence electrons on an atom in a molecule that belong solely to that atom and are not shared. Represented as pairs of dots in a Lewis dot structure.

Octet rule The tendency for most atoms in a molecule to surround themselves with 8 valence electrons by bonding with other atoms.

Polyatomic ion Two or more atoms joined by one or more covalent bonds AND containing a charge on one or more atoms.

Pure compound A substance containing only one type of compound, ionic or covalent. Pure water contains only water molecules.

Additional Exercises

2.39 What is the difference between an element and a compound?

2.40 Ionic compounds are formed between _____ cations and _____ anions. [*metal* or *nonmetal*]

2.41 Covalent compounds are formed between _____ atoms. [*metal* or *nonmetal*]

Ionic Compounds

2.42 Is an ionic bond a strong or weak force of attraction?

2.43 Which of the following are repulsive interactions?
a. + and +   **b.** + and −   **c.** − and −

2.44 Write the common ions formed from the following elements:
a. lithium   **b.** phosphorus
c. iodine   **d.** vanadium

2.45 Write the common ions formed from the following elements:
a. calcium   **b.** chromium
c. nitrogen   **c.** silver

2.46 Define the following terms:
a. electrostatic attraction
b. ionization
c. an ion

2.47 What is the difference between a cation and an anion? What type of element forms cations?

2.48 What are the two types of ions? What type of element forms anions?

2.49 Why are ions not formed from group 8A elements?

2.50 Why are ions not formed from carbon and silicon?

2.51 Hydrogen can gain or lose one electron to form an ion. Write the symbol for these two possible ions derived from hydrogen. Offer a plausible explanation for why hydrogen can form both an anion and a cation.

2.52 Indicate the number of protons and electrons present in the following ions. Explain why the charge is as indicated. Name the ion.
a. Cs^+ **b.** Ag^+ **c.** Cr^{3+} **d.** Se^{2-} **e.** Br^-

2.53 Indicate the number of protons and electrons present in the following ions. Explain why the charge is as indicated. Name the ion.
a. Mg^{2+} **b.** Hg^{2+} **c.** Cl^- **d.** F^- **e.** O^{2-}

2.54 What is a salt?

2.55 What is an ionic compound?

2.56 What is the term for an ion dissolved in water?

2.57 What happens to the ionic lattice of NaCl when placed in water?

2.58 What is an electrolyte?

Writing Formula Units and Naming Ionic Compounds

2.59 What ions are present in the following formula units? Provide the name and the symbol of the ions:
a. KBr **b.** $MgCl_2$ **c.** KI **d.** $BaCl_2$ **e.** NaF

2.60 What ions are present in the following formula units? Provide the name and the symbol of the ions:
a. Na_2O **b.** CuO **c.** AgCl **d.** ZnS **e.** GaAs

2.61 Write the formula unit for the ionic compound formed from each of the following combinations. Indicate which ion is the cation and which ion is the anion.
a. lithium and iodine
b. rubidium and fluorine
c. calcium and bromine
d. barium and iodine
e. iron(II) and sulfur
f. aluminum and oxygen

2.62 Write the formula unit for the ionic compound formed from each of the following combinations. Indicate which ion is the cation and which ion is the anion.
a. strontium and chlorine
b. chromium(III) and oxygen
c. iron(II) and oxygen
d. cobalt(II) and chlorine
e. iron(II) and selenium
f. platinum(IV) and fluorine

2.63 Name the following compounds:
a. SrO **b.** KI **c.** Ga_2O_3 **d.** LiF
e. NaI **f.** Fe_2O_3

2.64 Name the following compounds:
a. $BaCl_2$ **b.** PtI_2 **c.** $HgCl_2$ **d.** $CoCl_2$ **e.** K_2S

2.65 Sodium ions are macronutrients that play an important role during nerve impulses. Write the symbol for the sodium ion.

2.66 Identify each of the compounds as either an ionic or a covalent compound:
a. $CH_3CH_2NH_2$ **b.** NH_4OH
c. $CaCO_3$

2.67 What is the charge on each of the ions in the following ionic compounds?
a. $Zn(OH)_2$ **b.** $CuCH_3CO_2$
c. $SnCl_4$ **d.** V_2O_5 **e.** CrO_3

Covalent Compounds

2.68 What is a molecule?

2.69 What is the difference between an ionic compound and a covalent compound?

2.70 What is the octet rule?

2.71 Which elements are diatomic?

2.72 How many electrons are shared in the following covalent bonds:
a. single bond **b.** double bond **c.** triple bond

2.73 Write a Lewis dot representation for the following atoms:
a. carbon **b.** hydrogen
c. oxygen **d.** phosphorus

2.74 Write a Lewis dot representation for the following atoms:
a. nitrogen **b.** chlorine
c. sulfur **d.** silicon

2.75 Answer the following questions for the Lewis dot structure of hydrogen cyanides shown below:

$$H-C\equiv N:$$

a. How many bonding electrons and how many nonbonding electrons does the carbon atom contain?
b. How many bonding electrons and how many nonbonding electrons does the nitrogen atom contain?
c. Does each atom have an octet of electrons?
d. What types of covalent bonds are present in this molecule: single, double, or triple bonds?
e. How many electrons are shared between the carbon atom and the nitrogen atom?

2.76 Answer the following questions for the Lewis dot structure of hydrogen fluoride, shown below:

$$H-\ddot{\underset{\cdot\cdot}{F}}:$$

a. How many bonding electrons and nonbonding electrons does the fluorine atom contain?
b. Does the fluorine atom have an octet?
c. How many electrons are shared between the fluorine atom and the hydrogen atom?
d. Is hydrogen fluoride a covalent or ionic compound?

2.77 Write a Lewis dot structure for Br_2. Is this molecule an element or a compound?

2.78 Write the typical number of bonds and nonbonding pairs of electrons found for the following elements:
a. carbon **b.** oxygen and sulfur
c. halogens (F, Cl, Br, I)
d. nitrogen and phosphorus
e. hydrogen

2.79 Write the Lewis dot structure for the following compounds:
a. ethane, C_2H_6 **b.** PH_3 **c.** CCl_4 **d.** CO_2
e. Critical Thinking Question: Methanol, CH_4O

2.80 Write the Lewis dot structure for the following compounds:
a. PF_3 **b.** SH_2
c. formaldehyde, CH_2O
d. propane, C_3H_8 **e.** HCl
f. Critical Thinking Question: Carbonic acid, H_2CO_3. (*Hint:* Hydrogen is bonded to the oxygen atoms.)

2.81 Part of the structure of ATP is shown below. Indicate which atoms have an expanded octet.

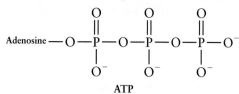

ATP

2.82 Draw the Lewis dot structure for acetonitrile, CH_3CN. (*Hint:* The atoms are connected as shown in the molecular formula.)

2.83 Name the compound SO_3.

2.84 Name the compound BCl_3.

2.85 What is the difference in the name for PCl_3 and the name for PCl_5?

2.86 Provide the names for the following binary compounds that contain nitrogen and oxygen:
a. NO   **b.** NO_2 **c.** N_2O **d.** N_2O_3
e. N_2O_4 **f.** N_2O_5

Compounds Containing Polyatomic Ions

2.87 Write the name of each of the following polyatomic ions:
a. HCO_3^- **b.** $CH_3CO_2^-$ **c.** HO^-

2.88 Write the name of each of the following polyatomic ions:
a. HPO_4^{2-} **b.** H_3O^+ **c.** NH_4^+

2.89 Write the formula and charge for the following polyatomic ions:
a. ammonium **b.** carbonate **c.** hydrogen phosphate

2.90 Write the formula and charge for the following polyatomic ions:
a. sulfate   **b.** hydronium   **c.** hydroxide

2.91 Answer the following questions for hydrogen carbonate, also known as bicarbonate ion.
a. What is the overall charge on this ion?
b. Why is this ion considered a polyatomic ion and not a monatomic ion?
c. What is the formula unit for sodium bicarbonate (baking soda)?
d. What is the formula unit for calcium hydrogen carbonate?

2.92 Answer the following questions for the carbonate ion.
a. What is the overall charge for this ion?
b. Write the formula unit for sodium carbonate.
c. How does the formula unit for sodium carbonate differ from that for sodium hydrogen carbonate? (See previous question.)
d. Write the formula unit for calcium carbonate (limestone).

2.93 Answer the following questions for the hydrogen phosphate ion.
a. What is the overall charge on this ion?
b. Why is this ion considered a polyatomic ion and not a monatomic ion?
c. What is the formula unit for potassium hydrogen phosphate?

d. What is the formula unit for magnesium hydrogen phosphate?

2.94 Identify the cations and anions present in the following compounds. What is the ratio of cations to anions for each compound and why?
a. NaOH   **b.** KCH_3CO_2 **c.** Li_2CO_3
d. NH_4OH **e.** $(NH_4)_2HPO_4$

2.95 Write the formula unit for the following compounds
a. sodium phosphate
b. ammonium chloride
c. magnesium hydroxide (milk of magnesia)

2.96 Lead accumulates in the body by forming lead phosphate, which accumulates in the bones. Write the formula unit for lead phosphate.

2.97 Calcium phosphate is one of the compounds present in kidney stones. What is the formula unit for calcium phosphate?

Molecular Mass, Formula Mass, and Molar Mass

2.98 How many M&Ms would there be in one mole of M&Ms?

2.99 What is Avogadro's number?

2.100 What is the mass of one mole of oxygen atoms?

2.101 What is the mass of one mole of nickel atoms?

2.102 Which has a smaller mass: a mole of feathers or a mole of chemistry books? Which is more: a mole of feathers or a mole of chemistry books?

2.103 Which is lighter: a mole of platinum or a mole of zirconium?

2.104 Which has more mass: a mole of hydrogen or a mole of zinc?

2.105 What is the molecular mass of butane, C_4H_{10}?

2.106 What is the formula mass of sodium hydroxide?

2.107 How many moles are there in 2.00 g of calcium chloride?

2.108 How many grams of hydrogen peroxide, H_2O_2, are there in:
a. 1 mole   **b.** 0.5 mole
c. 10 moles   **d.** 0.678 mole

2.109 How many moles of copper are there in 0.75 g of copper?

2.110 How many moles of molybdenum are there in 1.235 g of molybdenum?

2.111 How many grams are there in 3.5 mol of gallium?

2.112 How many atoms are there in a fourth of a mole of any type of atom?

2.113 Calculate the molar mass of the following elements:
a. aluminum, Al   **b.** hydrogen, H_2
c. calcium, Ca   **d.** nitrogen, N_2

2.114 Calculate the molar mass of the following compounds:
a. ethanol, C_2H_6O **b.** ammonia, NH_3

2.115 Calculate the molar mass of the following compounds
a. Li_2CO_3, a drug used to treat bipolar disorder
b. K_3PO_4

Chemistry in Medicine

2.116 When a blood sample is separated by a centrifuge, it separates into two layers. What are the two layers called?

2.117 Which of the following would you expect to be present in the larger amount in a blood sample: glucose or albumin?

2.118 What is BUN and what is it used to diagnose?

2.119 What is the molecular formula of urea? What is the molar mass of urea? Is urea a covalent compound or an ionic compound? How can you tell? How many moles of urea are present in 1 dL of serum?

Answers to Practice Exercises

2.1 An element is composed of only one type of atom. A compound is composed of two or more different types of atoms.

2.2 **a.** Se^{2-} **b.** K^+ **c.** As^{3-} **d.** Zn^{2+}

2.3 I^-
  a. 53 protons, 54 electrons
  b. a nonmetal
  c. gains 1 electron because it is a group 7A element that achieves an electron arrangement like the noble gas Xe when it has 8 valence electrons.
  N^{3-}
  a. 7 protons, 10 electrons
  b. a nonmetal
  c. gains 3 electrons because it is a group 5A element that achieves an electron arrangement like the noble gas Ne when it has 8 electrons.
  Cr^{3+}
  a. 24 protons, 21 electrons
  b. a metal
  c. loses three electrons, because this is one of the variable forms of this transition metal ion.

2.4 Group 6A elements gain 2 electrons because this gives them a full outermost shell. The charge on these ions is −2. Examples, include O^{2-}, S^{2-}, Se^{2-}.

2.5 Both ions have 24 protons. Cr^{3+} contains 21 electrons and Cr^{6+} contains 18 electrons. They are different ions, so it wouldn't be surprising to find that they have different physical properties and thus possibly different effects in the body.

2.6 **a.** The lattice structure shows a one to one ratio of cation to anion for both NaCl and CsCl.
  b. The ratio is one to one, because the charge on the cation is the same as on the anion.
  c. The formula unit is only a ratio, whereas the lattice structure is a continuous extended array of ions.
  d. Electrostatic forces of attraction hold the ions together.
  e. Chloride ions are shown in green.
  f. The sodium ion has a smaller diameter than a chloride ion.

2.7 **a.** Na_2S **b.** Mg_3N_2 **c.** $CuCl$ **d.** $CuCl_2$

2.8 **a.** lithium oxide
  b. magnesium iodide
  c. potassium chloride

2.9 Fe^{3+}

2.10 a. H—N̈—H   **b.** :Ö=Ö:
               |
               H

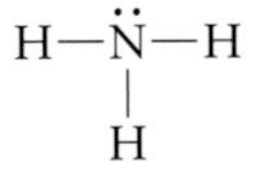

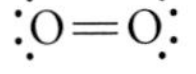

  c. H—S̈—H   **d.** :C̈l—C̈l:

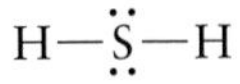

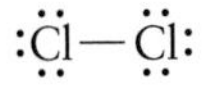

2.11 a. two bonding pairs of electrons and two nonbonding pairs of electrons.
  b. four bonding pairs of electrons and no nonbonding pairs of electrons
  c. Each atom has an octet—the sum of bonding and nonbonding pairs.
  d. Double bonds; Four electrons (2 pairs) are shared in a C=O bond.

2.12 a. The phosphorus atom contains an expanded octet: 10 electrons, 5 pairs.
  b. No expanded octets.
  c. Sulfur has an expanded octet: 12 electrons, 6 pairs.

2.13 a. Sulfur has an expanded octet. Sulfur is in period 3. There are 12 electrons, 6 pairs, surrounding the sulfur atom.
  b. Each carbon atom has pairs of bonding electrons.
  c. Nitrogen has 3 bonding and 1 nonbonding pair of electrons. This is the usual arrangement of electrons for N and C.

2.14 a. carbon dioxide **b.** dinitrogen trioxide

2.15 a. NF_3 **b.** CCl_4

2.16 a. aluminum chloride: ionic
  b. carbon disulfide: covalent
  c. zinc oxide: ionic
  d. dinitrogen tetroxide: covalent (if vowels *ao* appear together drop the first vowel)

2.17 a. +1 **b.** 4 bonds
  c. It contains more than one atom and there are one or more covalent bonds in the ion.
  d. Because it contains a charge, ammonium is classified as an ion.
  e. NH_4Cl **d.** $(NH_4)_2S$

2.18 a. potassium nitrate
  b. sodium hydrogen phosphate
  c. sodium hydroxide
  d. silver nitrate
  e. ammonium carbonate

2.19 a. $LiOH$ **b.** $SrCO_3$ **c.** $Mg_3(PO_4)_2$

2.20 KCN

2.21 Li_2CO_3

2.22 Cu^{2+} $x + 6(+3) + 4(-3) + 8(-1) = 0$. Therefore $x = +2$.

2.23 $12.01 + 4(1.008) = 16.04$ amu. It is a molecular compound because it is composed of all nonmetal atoms.

2.24 $3(16.00) + 3(1.008) + 26.98 = 78.00$ amu. It is an ionic compound because it contains the metal ion Al^{3+}.

2.25 $6(12.01) + 12(1.008) + 6(16.00) = 180.16$ amu. It is a molecular compound because it is composed of all nonmetal atoms.

2.26 6.02×10^{23} marbles

2.27 A mole of basketballs has a larger mass than a mole of ping-pong balls, because a single basketball has a greater mass than a single ping-pong ball.

2.28 One mercury atom: 200.6 amu. One mole of mercury atoms: 200.6 g/mol. They have the same numerical value, but different units.

2.29 3.01×10^{23} ammonia molecules

2.30 6.02×10^{23}. Avogadro's number is so large because atoms are small, so it takes many of them to obtain a macroscopic sample.

2.31 Jane Doe's level of CO_2 is 0.025 mol/L, which is in the normal range of 0.021 to 0.031 mol/L.

2.32 0.793 mol

2.33 2.5 mol

2.34 9.70×10^{-4} mol

2.35 6.99×10^{-4} mol

2.36 134 g

2.37 27 g

2.38 3.69 g

E2.1 1.8×10^{23} copper atoms

E2.2 4.7×10^{22} copper atoms

E2.3 6.26×10^{22} chloride ions

E2.4 1.5×10^{22} CO_2 molecules

E2.5 5.84×10^{20} ibuprofen molecules

E2.6 1×10^{22} formula units of NaCl, which is 1×10^{22} Na^+ ions and 1×10^{22} Cl^- ions

Shapes of Molecules and Their Interactions

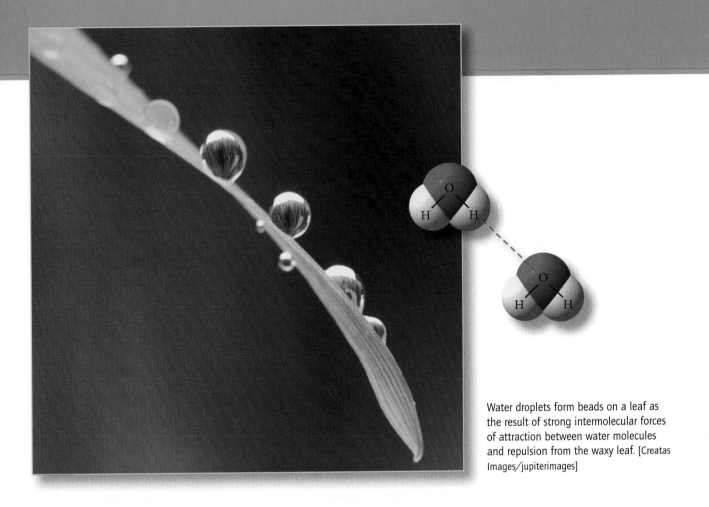

Water droplets form beads on a leaf as the result of strong intermolecular forces of attraction between water molecules and repulsion from the waxy leaf. [Creatas Images/jupiterimages]

OUTLINE

 This icon indicates that a **Problem-Solving Tutorial** is available at www.whfreeman.com/gob

Breast Cancer and the Shapes of Molecules

A woman born today has a one in eight chance of developing breast cancer in her lifetime. Breast cancer is a disease characterized by the uncontrolled growth of abnormal cells in breast tissue. Treatment may involve surgery and other therapies, including chemotherapy, radiation therapy (see Chapter 16), and hormone therapy. Here we will describe hormone therapy to demonstrate how the shapes of molecules play a pivotal role in biochemistry, and how with this knowledge, scientists have been able to develop drugs that prevent the progression of breast cancer. In living organisms molecules recognize each other in large part by shape and electrostatic interactions.

Hormone therapy for breast cancer is the long-term administration of a drug that blocks the effects of estrogen—such a drug is called an *antiestrogen*. Estrogens are a group of molecules that act as female hormones. The most important estrogen is *estradiol*, whose ball-and-stick model is shown in the margin. Estradiol is the main estrogen produced in women of child-bearing age, and it is also the main estrogen involved in breast cancer. Estradiol promotes the growth of cells found in bone, liver, uterine, brain, heart, and breast tissue. All these cells contain *estrogen receptors*, which are extremely large protein molecules. When estradiol binds to an estrogen receptor, it signals the cell to grow and reproduce. Since 75% of breast cancer cells also contain estrogen receptors, estradiol stimulates their growth as well. Breast cancer cells thrive on estradiol. The goal of hormone therapy is to prevent estradiol from binding to the estrogen receptor of cancer cells.

How does estradiol interact with the estrogen receptor, to promote the growth of breast cancer cells? The body contains hundreds of different types of receptors; many are located on the surface of the cell, while others are located inside the cell. Some, including the estrogen receptor, are located in the nucleus of the cell, where the genetic information for the cell is found. Each type of receptor has a characteristic shape and function. A receptor is activated when the right molecule (called the ligand) binds to the receptor at its binding site—a pocket or groove within the large receptor molecule. The "right" molecule is one that has a shape complementary to the binding site. When the estrogen receptor is activated, a cascade of biochemical events is set off, leading to cell growth and division.

In the case of the estrogen receptor, the molecule with the right shape is estradiol; however, similarly shaped man-made molecules can bind as well. Some of these man-made drugs are as good as or better than estradiol at binding to the receptor, but because they are somewhat different in shape, they can't activate the receptor—a characteristic of an antiestrogen drug. Figure 3-1 illustrates a drug or hormone molecule having a complementary shape to its receptor. Alternatively, some molecules and drugs that have a complementary shape to the binding site work by activating the receptor in the same way as the natural ligand.

In addition to having a complementary shape, a ligand is bound to its receptor binding site by electrostatic forces of attraction. These are not covalent bonding forces, but rather *intermolecular forces of attraction*—electrostatic forces that act *between* rather than *within* molecules. It is these forces of attraction that allow a ligand or drug molecule to temporarily "stick," or *bind*, to the receptor.

When a drug is developed to treat a particular disease, it is designed to have a shape similar to the natural ligand. It is designed either to have the same effects as the natural ligand or to block the effects of the ligand. In the case of antiestrogens, they are designed to bind to the estrogen receptor, but not to activate it. By occupying the receptor binding site, antiestrogens block estradiol from binding to the receptor, thus slowing the growth of breast cancer cells. This chapter's *Chemistry in Medicine* feature explores further the role of antiestrogens in breast cancer treatment.

In this chapter you will consider the details of molecular shape and intermolecular forces of attraction. Since cells recognize molecules by shape and electrostatic attractions, these chemical concepts are important for an understanding of health and disease.

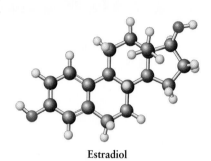

Estradiol

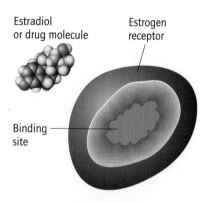

Estradiol or drug molecule

Estrogen receptor

Binding site

Figure 3-1 The interaction between a receptor and a hormone or drug molecule. Certain molecules are able to bind to the receptor binding site, in part because their shape is complementary to the shape of the site.

Cells rely on the shapes of molecules to communicate with other cells; hence, many molecules act as important messengers between cells in the body. Whether it is a hormone interacting with a receptor on the surface of a cell or a drug molecule interacting with other molecules in the cell, many biological molecules recognize one another largely by shape and also by forces of attraction that act between molecules.

In Chapter 2 you learned how to write Lewis dot structures, which show how atoms are joined in a molecule by using dots and/or lines to represent the valence electrons of the atoms. While Lewis dot structures do not show the actual three-dimensional shape of a molecule, they do provide the basis for predicting it. In this chapter you will learn how to use a Lewis dot structure to determine the three-dimensional shapes of simple molecules.

In Chapter 2 you also learned about the bonding forces of attraction between atoms *within* a molecule that form the covalent bonds. In this chapter, you will turn your attention to forces of attraction that exist *between* molecules—***inter***molecular forces of attraction. Then, in the next chapter you will learn how intermolecular forces of attraction influence the physical properties of a compound, such as whether it is a solid, liquid, or gas at a given temperature.

3.1 Three-Dimensional Shapes of Molecules

Most molecules have a three-dimensional shape; in other words, they are not two-dimensional (flat), as their Lewis dot structures might lead you to believe. Large molecules, in particular, can have quite complex and elaborate shapes. For example, the estrogen receptor described in the opening vignette has a complex shape, which is necessary for its specific function: binding estrogen molecules. However, small molecules have shapes that can be described in plain geometric terms.

The shape of a simple molecule containing one or two central atoms surrounded by two or more atoms is determined from the spatial arrangement of the *atoms* surrounding the central atom, which in turn is determined from the number and type (bonding or nonbonding pairs) of electrons on the central atom. The arrangement of *electrons* around the central atom is referred to as the molecule's **electron geometry,** while the arrangement of *atoms* around the central atom defines the **molecular shape** of the molecule.

Molecular Models

Molecular models are a valuable tool used to help visualize the three-dimensional shapes of molecules. Two types of molecular models are routinely used and will be seen throughout this text: the ball-and-stick model and the space-filling model.

Ball-and-stick models depict molecules by representing atoms as *balls* and bonds as *sticks*. Furthermore, the balls are color-coded to represent the various building block elements:

Carbon	black	Oxygen	red
Hydrogen	white	Nitrogen	blue

Ball-and-stick models are used primarily to show the bond angles in molecules, although they ignore the amount of space (volume) occupied by the atoms in a molecule.

A **space-filling model** is used show the overall amount of space occupied by the atoms in a molecule. However, this type of model obscures the bond

Ethanol, C_2H_5OH
(drinking alcohol)

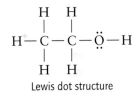

Lewis dot structure

Ball-and-stick model

Space-filling model

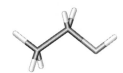

Tube model

angles in a molecule. Thus, different models are used depending on what structural aspect of the molecule needs to be visualized.

Occasionally, a tube model will be used in place of a ball-and-stick model, particularly when the molecule is large. In a **tube model** both bonds and atoms appear as part of a tube, where the end points represent the atoms, color-coded the same as ball-and-stick models. Each half of the bond bears the color of the atom it is attached to.

Using Lewis Dot Structures to Predict Electron Geometry

The process for determining the molecular shape of a molecule can be described in three steps.

Step 1: Write the Lewis dot structure for the molecule.
Step 2: Determine the electron geometry from the Lewis dot structure.
Step 3: Determine the molecular shape from the electron geometry.

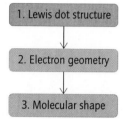

You learned how to perform the first step in Chapter 2. Next, we describe the second step, how to determine the electron geometry from the Lewis dot structure.

The Lewis dot structure of a molecule provides all the information needed to determine its electron geometry. The simplest theory for predicting electron geometry, known as **v**alence **s**hell **e**lectron **p**air **r**epulsion theory (**VSEPR**), is based on the premise that electrons around a central atom adopt a geometry that places them as far apart from one another as possible, while still remaining attached to the central atom. This theory makes scientific sense: since electrons have like charges, they will experience repulsion and therefore separate as far apart from one another as geometry will allow. In other words, groups of electrons will avoid other groups of electrons unless they are part of the same double or triple bond.

To determine the electron geometry around a central atom, count the groups of electrons surrounding the central atom in the Lewis structure. A *group* of electrons refers to one triple bond, one double bond, one single bond, or one nonbonding pair of electrons. The majority of molecules you will encounter in living systems have two-, three-, or four-electron groups surrounding the central atom. Exceptions are metals, which sometimes have five- or six-electron groups around the metal atom center.

From VSEPR theory we get the following three common *electron geometries*, which are derived from the number of electron groups around the central atom:

- 4 groups of electrons: **tetrahedral**
- 3 groups of electrons: **trigonal planar**
- 2 groups of electrons: **linear**

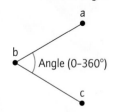

Recall from geometry that an angle is defined by three points. Point b is the central atom in a bond angle.

Each of the basic electron geometries is illustrated in Figure 3-2. The **linear geometry** is one-dimensional and has the shape of a straight line, with both groups of electrons directed 180° from each other. The trigonal planar geometry is two-dimensional: each of the three groups of electrons points to one of the three corners of an equilateral triangle, separated by 120°. The tetrahedral geometry is three-dimensional: each of the four groups of electrons points to one of the four corners of a tetrahedron, creating angles of 109.5°. As you can see, the fewer the number of electron groups, the farther apart the electrons can be placed in three-dimensional space, and the greater the **bond angle** between atoms in the molecule.

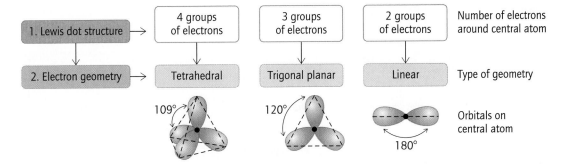

| 1. Lewis dot structure | → | 4 groups of electrons | 3 groups of electrons | 2 groups of electrons | Number of electrons around central atom |

Figure 3-2 Steps for determining the electron geometry of a molecule

---

WORKED EXERCISE 3-1 Predicting Electron Geometry from Lewis Dot Structures

Answer the following questions for carbon dioxide, CO_2.

  a. Write the Lewis dot structure.
  b. How many electron groups surround the central carbon atom? What are the groups: nonbonding electrons, single bonds, double bonds, or triple bonds?
  c. What is the electron geometry: linear, trigonal planar, or tetrahedral?
  d. What is the O—C—O bond angle?

SOLUTION

  a. :O=C=O:
  b. The central carbon atom is surrounded by *two* groups of electrons. Both groups are double bonds (C=O double bonds).
  c. The electron geometry is linear, because a linear geometry maximizes the distance between two groups of electrons surrounding the central carbon atom.
  d. A linear geometry has a bond angle of 180°.

PRACTICE EXERCISE

3.1 Predict the electron geometry and bond angles for the following compounds. Begin by writing the Lewis dot structure.
  a. CCl_4 **b.** HCN
  c. BF_3 (only 6 electrons surround the central boron atom—an exception to the octet rule).

Using Electron Geometries to Determine Molecular Shape

In the three-step process for determining the shape of a molecule, the last step uses the electron geometry as the basis for determining the shape of the overall molecule, as illustrated in Figure 3-3.

When the central atom in a molecule has only bonding electrons and no nonbonding electrons, its molecular shape is one and the same as its electron geometry. Thus, four bonding groups around a central atom produce a molecule with a tetrahedral shape, three bonding groups around a central atom produce a molecule with a trigonal planar shape, and two bonding groups of electrons around a central atom produce a molecule with a linear shape.

A molecule with a central atom containing one or more nonbonding pair of electrons will have a *molecular shape* that is different from its *electron geometry*, since shape is defined by the relative position of the *atoms*, and nonbonding electrons are not "seen." For example, you would not describe a water molecule as having a tetrahedral shape, even though it has a tetrahedral

Figure 3-3 Steps for determining the shape of a molecule from its electron geometry.

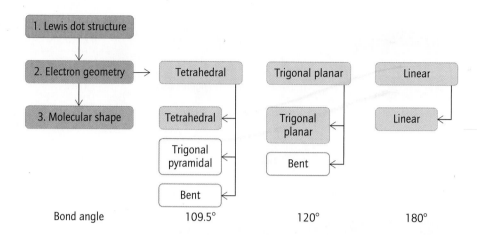

arrangement of the four groups of electrons around the central oxygen atom. What you *see* in a water molecule is simply three atoms with a bent (or angular) shape. In other words, the nonbonding electrons have a behind-the-scenes role in determining the shape of a molecule.

$$H - \overset{\cdot\cdot}{\underset{\cdot\cdot}{O}} - H$$

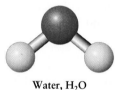

Water, H₂O

Shapes of Molecules with a Tetrahedral Electron Geometry
There are three different *molecular* shapes derived from a tetrahedral *electron* geometry, depending upon the number of nonbonding electrons on the central atom:

Tetrahedral Electron Geometry (4 groups of electrons)

Number of Bonding Groups	Number of Nonbonding Groups	Molecular Shape
4 bonding groups	0 nonbonding groups	tetrahedral
3 bonding groups	1 nonbonding group	trigonal pyramidal
2 bonding groups	2 nonbonding groups	bent

Since the tetrahedral, trigonal pyramidal, and bent shapes are all derived from a tetrahedral electron geometry, they all contain approximately the same bond angles: 109.5°.

Consider methane (CH₄), which has the Lewis dot structure shown in Figure 3-4a. Since the central carbon atom is surrounded by four groups of electrons, the electron geometry is tetrahedral (Figure 3-4b), the maximum separation for four groups of electrons. Each group of electrons around carbon is a single C—H bond. Since each electron group is a bonding group, the molecular shape is tetrahedral, like the electron geometry. (Figure 3-4c).

The tetrahedral molecular shape is the most common geometry seen in organic chemistry and biochemistry, and you will see this shape frequently throughout this text. Therefore, it is useful to know how to draw an accurate three-dimensional representation of a tetrahedral molecule. Chemists use a technique routinely used by artists: An atom projecting away from the viewer is written as a dashed line (simulates the appearance of getting smaller), and

Methane, CH₄

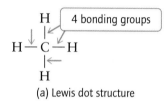

(a) Lewis dot structure

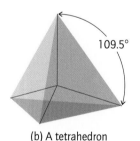

(b) A tetrahedron

(c) Tetrahedral electron geometry and molecular shape

Figure 3-4 (a) The Lewis dot structure for methane, CH₄. (b) The four bonding groups of electrons point to the four corners of a tetrahedron, creating 109.5° bond angles. (c) Methane has a tetrahedral electron geometry and molecular shape.

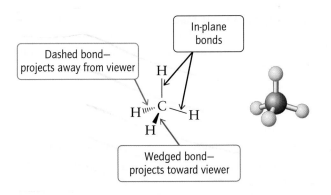

Figure 3-5 Using dashed lines and solid wedges to show the tetrahedral shape of methane in three dimensions.

an atom projecting toward the viewer is written as a wedged line (simulates the appearance of getting larger), as illustrated in Figure 3-5 for a molecule of methane. *The Model Tool 3-1: The Tetrahedral Electron Geometry* will give you hands-on familiarity with this important geometry.

The **trigonal pyramidal** shape arises when there are three bonding groups and one nonbonding pair of electrons around a central atom. Ammonia is an example of a molecule with a trigonal pyramidal shape. The Lewis dot structure for ammonia, shown in Figure 3-6a, has three N—H single bonds and one nonbonding pair of electrons on the nitrogen atom. Since there are four groups of electrons around the central nitrogen atom, ammonia has a tetrahedral electron geometry (Figure 3-6b). The molecular shape described by the atoms in the molecule, however, is trigonal pyramidal (Figure 3-6c).

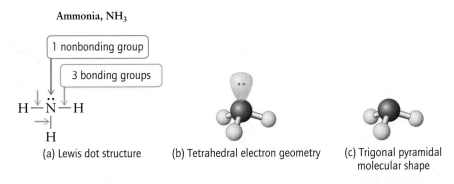

(a) Lewis dot structure
(b) Tetrahedral electron geometry
(c) Trigonal pyramidal molecular shape

Figure 3-6 (a) The Lewis dot structure for ammonia, NH_3. (b) The four electron groups of ammonia form a tetrahedral electron geometry. (c) Because one of the electron groups is nonbonding, the molecular shape of NH_3 is trigonal pyramidal. Bond angles are approximately 109.5°.

A **bent** (or angular) molecular geometry arises when there are two bonding and two nonbonding groups of electrons around a central atom. The Lewis dot structure for water has two O—H bonds and two nonbonding pairs of electrons on oxygen, shown in Figure 3-7a. The four groups of electrons on oxygen give it tetrahedral electron geometry (Figure 3-7b). The molecular shape, however, is defined by the two hydrogen atoms and the oxygen atom—a shape described as "bent" (Figure 3-7c).

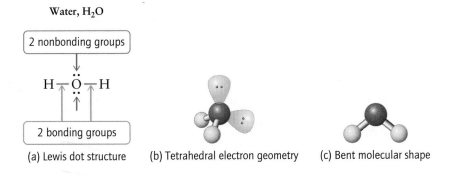

(a) Lewis dot structure
(b) Tetrahedral electron geometry
(c) Bent molecular shape

Figure 3-7 (a) The Lewis dot structure for water, H_2O. (b) The four electron groups on oxygen yield a tetrahedral electron geometry (b), but because two of the electron groups are nonbonding, the molecule has a bent shape (c). The three atoms form an approximately 109.5° bond angle.

Shapes of Molecules with a Trigonal Planar Electron Geometry In Figure 3-3 you learned that three electron groups around a central atom produces a trigonal planar electron geometry. There are two molecular shapes derived from a trigonal planar electron geometry.

Trigonal Planar Electron Geometry (3 groups of electrons)

Number of Bonding Groups	Number of Nonbonding Groups	Molecular Shape
3 bonding groups	0 nonbonding groups	trigonal planar
2 bonding groups	1 nonbonding group	bent

Remember to count the double bond in the Lewis dot structure as one group of electrons.

When all three groups of electrons around the central atom are bonding, then the molecular shape is trigonal planar. For example, formaldehyde has a trigonal planar electron geometry and molecular shape, shown in Figure 3-8. The trigonal planar electron geometry gives a molecule approximately 120° bond angles.

Figure 3-8 (a) The Lewis dot structure for formaldehyde, CH_2O. (b) The three electron groups around the carbon in formaldehyde point to the three corners of an equilateral triangle, and thus formaldehyde has a trigonal planar electron and molecular geometry (c). Bond angles are approximately 120°.

Formaldehyde, CH_2O

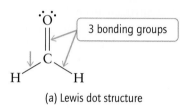

(a) Lewis dot structure

(b) Bond angles

(c) Trigonal planar molecular shape

If one of the electron groups in a molecule with a trigonal planar electron geometry is nonbonding, the molecular shape of the molecule is bent. An example of such a molecule is sulfur dioxide (SO_2), shown in Figure 3-9. The bent shape is similar in appearance to the bent shape described for water, except that the bond angle is approximately 120°. As you can see bond angles are determined from electron geometry.

The Linear Geometry When a molecule has a central atom surrounded by two groups of electrons, they will always be bonding groups; therefore, the molecule has both a linear electron geometry and a linear molecular shape. The bond angle will be 180°. For example, hydrogen cyanide, HCN, has two groups of electrons around the central atom, one triple bond and one single bond, so its shape is linear.

Hydrogen cyanide, HCN

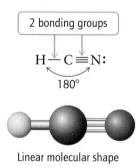

Linear molecular shape

Sulfur dioxide, SO_2

(a) Lewis dot structure

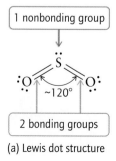

(b) Trigonal planar electron geometry

(c) Bent molecular shape

Figure 3-9 (a) The Lewis dot structure of sulfur dioxide, SO_2. Notice that the sulfur atom has an expanded octet, with 10 electrons surrounding it rather than 8. (b) The three electron groups of this molecule form a trigonal planar electron geometry, with approximately 120° bond angles. (c) Because one of the electron groups is nonbonding, the molecule has a bent shape.

Table 3-1 Common Molecular Shapes and Bond Angles Based on Number of Electron Groups Around a Central Atom

Total number of electron groups	2	3		4		
Number of bonding groups	2	3	2	4	3	2
Number of nonbonding groups	0	0	1	0	1	2
Bond angle	180°	120°		109.5°		
Electron geometry	Linear	Trigonal planar		Tetrahedral		
Molecular shape	Linear	Trigonal planar	Bent	Tetrahedral	Trigonal pyramidal	Bent

Table 3-1 summarizes the six common molecular shapes derived from the three common electron geometries.

WORKED EXERCISES 3-2   Predicting Electron Geometry and Molecular Shape from Lewis Dot Structures

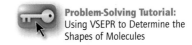
Problem-Solving Tutorial:
Using VSEPR to Determine the Shapes of Molecules

1 Ozone, O_3, is a form of elemental oxygen, which in the stratosphere protects us from harmful UV radiation. The Lewis dot structure for ozone is shown below (ignore the fact that two of the oxygen atoms do not have the usual two bonding and two nonbonding electron pairs). What is the electron geometry of this molecule? What is the molecular shape of the molecule? What is the O—O—O bond angle?

$$:\ddot{O}-\ddot{O}=\ddot{O}:$$

2 Answer the following questions for a molecule of PBr_3:
   a. Write the Lewis dot structure.
   b. How many electron groups surround the central phosphorus atom?
   c. What is the electron geometry? Explain.
   d. What is the molecular shape? Explain.
   e. What is the approximate Br—P—Br bond angle?

SOLUTIONS

1 The electron geometry of ozone is trigonal planar, because three groups of electrons surround the central oxygen atom: one double bond, one single bond, and one nonbonding pair. The molecular shape is bent with 120° bond angles, because one group of electrons is nonbonding, indicated by the pair of dots on the central oxygen.

2 **a.** 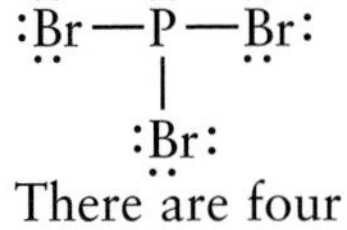 $:\ddot{B}r-\overset{\displaystyle |}{P}-\ddot{B}r:$
   $\qquad\quad :\ddot{B}r:$

   b. There are four electron groups around the central P atom: three single bonds and a nonbonding pair of electrons.
   c. The electron geometry is tetrahedral, because there are four groups of electrons around the central atom.
   d. The molecular shape is trigonal pyramidal, because there are three bonding groups and one nonbonding group.
   e. The Br—P—Br bond angle is close to 109.5°, because the electron geometry is tetrahedral.

PRACTICE EXERCISES

3.2 Write the Lewis dot structure for hydrogen sulfide, H_2S, the substance that gives rotten eggs their bad smell. What is the electron geometry of this molecule? What is the molecular shape? What is the H—S—H bond angle? Why is this molecule not linear?

3.3 Write the Lewis dot structure and then indicate the electron geometry, molecular shape, and bond angles for each of the following compounds:

   a. CH_2Cl_2
   b. OF_2
   c. $SiCl_4$
   d. PF_3

3.4 What do all the compounds in Exercise 3.3 have in common?

3.5 A ball-and-stick model of the hydronium ion is shown in the margin. What is the Lewis dot structure and molecular formula for this ion? What is the electron geometry of this molecule? What are the H—O—H bond angles? What is the molecular shape? Is this the shape shown in the ball-and-stick model?

Hydronium ion, H_3O^+

 ### The Model Tool 3-1   The Tetrahedral Electron Geometry

If a molecular modeling kit was provided with your textbook, you can use it to perform the following exercise.

I. Construction of Chloromethane, CH_3Cl

1. Obtain one black carbon atom, three light-blue hydrogen atoms, one green chlorine atom, and four straight bonds.
2. Using the Lewis dot structure as a guide, make a model of chloromethane.

$$
\begin{array}{c}
\text{H} \\
| \\
\text{H}-\text{C}-\ddot{\text{C}}\text{l}\!: \\
| \\
\text{H}
\end{array}
$$

3. Answer the questions under Part IV, Inquiry Questions.

II. Construction of Ammonia, NH_3

1. Obtain one blue nitrogen atom, three light-blue hydrogen atoms, and three straight bonds.
2. Using the Lewis dot structure as a guide, make a model of ammonia.

$$
\begin{array}{c}
\text{\textbf{..}} \\
\text{H}-\ddot{\text{N}}-\text{H} \\
| \\
\text{H}
\end{array}
$$

3. Answer the questions under Part IV, Inquiry Questions.

III. Construction of Water, H_2O

1. Obtain one oxygen atom, two light-blue hydrogen atoms, and two straight bonds.
2. Using the Lewis dot structure as a guide, make a model of water.
Answer the questions under Part IV, Inquiry Questions.

$$
\text{H}-\ddot{\text{O}}-\text{H}
$$

IV. Inquiry Questions

1. Examine your models. Why are the bond angles approximately the same in all three models?

2. What is the approximate bond angle in all these models?

3. Chloromethane: Does it matter where you place the Cl atom in the model? In other words, do you create different models depending on where you place the Cl atom?

4. Chloromethane: Is there any way to arrange the atoms around the central carbon atom that would place the atoms farther apart from each other? What is the advantage of adopting a tetrahedral geometry with 109.5° bond angles rather than a flat planar geometry with 90° bond angles?

5. Chloromethane: Provide a three-dimensional drawing of the molecule, using dashed and wedged bonds.

6. Ammonia: Compare your ammonia model to the model of methane. How is the electron geometry of ammonia similar to methane? How is the molecular shape of ammonia different from methane?

7. Ammonia: What prevents the bond angles from being 120° in this model?

8. Water: Compare your model of water to your models of methane and ammonia. How is the electron geometry of your water model similar to methane and ammonia? How is the molecular shape different?

9. How many different groups of electrons surround the central atom in each of your models: two, three, or four? Locate the appropriate cell in Table 3-1 into which each of your models falls.

10. How many groups of electrons are bonding electrons in each of your models: two, three, or four?

11. How many groups of electrons are nonbonding electrons in each of your models: zero, one, or two?

Shapes of Larger Molecules

The six different molecular shapes described here can also be used to evaluate the shapes of larger molecules having more than one central atom. Larger molecules can be viewed as a combination of the VSEPR molecular shapes of individual atom centers. For example, a molecule of methanol, CH_3OH, has two central atoms, one carbon atom and one oxygen atom. The arrangement of atoms around the carbon atom is tetrahedral and around the oxygen atom is bent, forming the molecular shape shown in the margin.

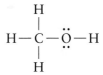

Methanol, CH_3OH

WORKED EXERCISES 3-3 Shapes of Larger Molecules

1 What is the molecular shape around each carbon atom in propane, C_3H_8? Describe the overall shape of propane.

2 Predict the shape of ethylene, C_2H_4.

SOLUTIONS

1 Propane is composed of three tetrahedral carbon atoms in a row, which gives the molecule an overall zigzag appearance.

2 To determine the shape of ethylene, begin by writing its Lewis dot structure:

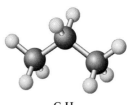

C_3H_8

Next, determine the shape based on the number of electrons around each carbon atom in the molecule. Both carbon atoms are surrounded by three groups of electrons and no nonbonding pairs, so they both have a flat, trigonal planar geometry, as shown in the ball-and-stick models below:

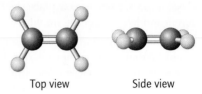

Top view   Side view

PRACTICE EXERCISE

3.6 Describe the overall three-dimensional shape of acetylene (C_2H_2) by writing the Lewis dot structure and analyzing the electron geometry and molecular shape around each carbon atom.

3.2 | Molecular Polarity

One reason why molecular shape is important is that it is one of the factors that determine whether a molecule is polar or nonpolar. A **polar molecule** has a separation of charge: it has one positive pole and one negative pole. The two oppositely charged poles create what is called a **dipole.** In a **nonpolar molecule** there is no such separation of charge. Polarity, a property that influences how molecules interact with one another, is determined by two factors:

- the types of covalent bonds in the molecule, whether polar or nonpolar, and
- the molecular shape of the molecule

The first factor, bond polarity, depends on the *electronegativities* of the atoms.

Electronegativity

A covalent bond is polar if the two atoms sharing the bonding electrons have a significantly different electronegativity. **Electronegativity** *is a measure of an atom's ability to draw electrons toward itself in a covalent bond.* The more electronegative an atom is, the more strongly it attracts electrons. In general, an atom's electronegativity is determined from its valence electron arrangement and the number of protons in its nucleus, which can be determined from its position on the periodic table.

The relative electronegativities for some representative elements is shown in the chart in Figure 3-10. Elements toward the bottom left portion of the periodic table, shown in blue, are the least electronegative; and elements toward the top right portion of the periodic table, shown in red, are the most electronegative. A value of 4.0 is found for the most electronegative element, fluorine, and values less than 1.5 are found among the least electronegative elements.

Several factors determine the electronegativity value for an element. One is the proximity of the valence electrons to the nucleus. The closer the valence electrons are to the positively charged nucleus, the more attracted to the nucleus they are and the more electronegative the atom is. Thus, within a group of elements, electronegativity increases as you go from the bottom to the top of a group. For example, fluorine is more electronegative than chlorine, which is more electronegative than bromine, and so forth. Within a period, electronegativity increases from left to right. For example, fluorine

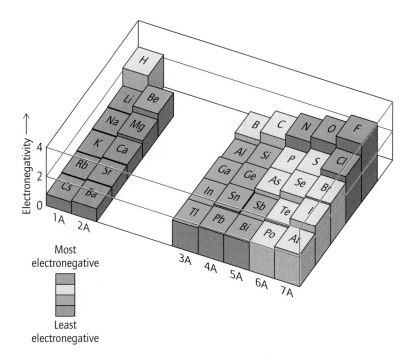

Figure 3-10 Chart of relative electronegativities for some main group elements.

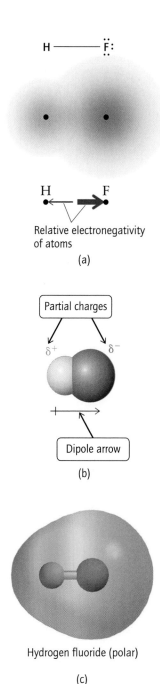

Figure 3-11 Polar covalent bond in hydrogen fluoride, HF. (a) The electron cloud probability diagram shows that electrons are more likely to be found around the fluorine atom, where the purple shading is more intense. (b) A dipole arrow points toward the more electronegative atom. Partial charges, δ^+ and δ^-, indicate separation of charge. (c) An electron density diagram is a space-filling model color coded to show partial positive charges as blue and partial negative charges as red. It is superimposed on a ball-and-stick model.

is more electronegative than oxygen, which is more electronegative than carbon, and so forth. Since noble gases generally do not form covalent bonds, they are usually excluded from this type of electronegativity table.

Bond Dipoles

In Chapter 2 you learned that a covalent bond is formed when two atoms share electrons. The electrons in a covalent bond were described as spending their time equally around both nuclei. Indeed, when two atoms are identical, the electron density is evenly distributed around both nuclei. This type of covalent bond is known as a **nonpolar covalent bond.** The diatomic elements, for example, all contain nonpolar covalent bonds and therefore, they are nonpolar molecules.

Covalent bonds formed between different atoms with comparable electronegativity values are also nonpolar. For example, a carbon–hydrogen bond is a relatively nonpolar bond, because the electronegativity of carbon is similar to that of hydrogen. The difference in their electronegativities is less than 0.5 on a scale of 0 to 4. Thus, all molecules containing only carbon and hydrogen atoms—hydrocarbons—are classified as nonpolar molecules.

When a covalent bond is formed between two atoms with significantly different electronegativities, the bonding electrons spend a greater amount of time around the more electronegative atom, as indicated in Figure 3-11a. Since electrons carry a negative charge, there is a partial negative charge (δ^-) on the more electronegative atom and a partial positive charge (δ^+) on the less electronegative atom. As a result, the bond will have a dipole, or separation of charge. This type of covalent bond is known as a **polar covalent bond.** For example, the single bond of hydrogen fluoride, H—F, is a polar covalent bond, because fluorine is much more electronegative than hydrogen.

The Greek symbols δ^+ and δ^- are used to indicate a partial positive and a partial negative charge on an atom, as illustrated in Figure 3-11b, as opposed to the full +1 or −1 carried by an ion. Alternatively, a **dipole arrow** can be placed alongside the bond with the tail of the arrow shown next to the less electronegative atom and the head of the arrow pointing toward the more electronegative atom (Figure 3-11b).

Figure 3-11c shows an electron density diagram of the HF molecule. An **electron density diagram** is a space-filling model that indicates regions of higher and lower concentrations of electrons in the molecule by color. Partial negative charges are indicated in red, partial positive charges in blue, and neutral regions in yellow or green.

You can think of ionic and covalent bonds as being part of a continuum that spans the range from ionic bonds at one extreme and nonpolar covalent bonds on the other extreme. Polar covalent bonds lie somewhere in the middle of the continuum:

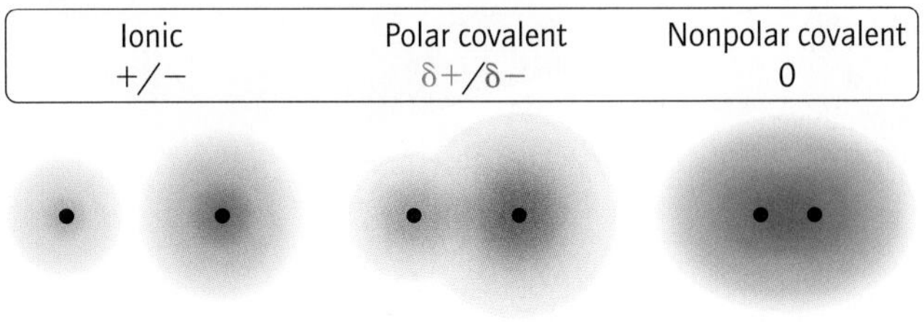

Ionic $+/-$	Polar covalent $\delta+/\delta-$	Nonpolar covalent 0

WORKED EXERCISES 3-4   Distinguishing Polar and Nonpolar Covalent Bonds

1 Using Figure 3-10, identify the more electronegative atom in each pair:

   a. carbon or oxygen
   b. sulfur or oxygen
   c. calcium or iodine

2 As a group, which are more electronegative: metals or nonmetals? Explain.

3 Indicate which of the following bonds are polar covalent bonds, by showing a bond dipole arrow alongside the bond.

   a. C—Cl **b.** N—H **c.** C—H

SOLUTIONS

1 **a.** Oxygen, because it is to the right of carbon in the same period.
   b. Oxygen, because it is above sulfur in the same group.
   c. Iodine, because it is much further to the right of calcium, even though iodine is in a lower period than Ca.

2 Nonmetals as a group are more electronegative than metals. Nonmetals are to the right of metals on the periodic table.

3 **a.** C—Cl
   b. N—H
   c. The bond is nonpolar because C and H have a similar electronegativity value.

PRACTICE EXERCISES

3.7 Using Figure 3-10, identify the more electronegative atom in each pair and explain your choice:

   a. carbon or nitrogen
   b. phosphorus or nitrogen
   c. lithium or chlorine
   d. silicon or nitrogen

3.8 As a group, which are more electronegative: metals or metalloids?

3.9 Indicate which of the following represent polar covalent bonds, by showing a bond dipole arrow.
  a. O—H
  b. C—O
  c. F—F

Polar Molecules

A polar molecule is a molecule with a charge separation; there is a positive pole and a negative pole within the molecule. When a molecule contains only one polar covalent bond, it is by definition a polar molecule. Thus, HF is a polar molecule, while nitrogen, N_2, is nonpolar.

If a molecule does not contain any polar bonds, it is nonpolar; the electrons are evenly distributed throughout the molecule. For example, hydrocarbons are nonpolar.

If a molecule contains more than one covalent bond, its molecular shape as well as the individual bond dipoles must be considered to determine if the overall molecule is polar. In other words, if bond dipoles counterbalance each other, the molecule will be nonpolar despite the presence of bond dipoles. For example, CO_2 is a nonpolar molecule even though it has two polar C=O bonds. This is because the molecule has a linear shape, and therefore the two bond dipoles are equal and in opposite directions, which causes them to cancel. Another way to understand why the molecule is nonpolar is to observe that the center of positive charge and the center of negative charge coincide so that there is no separation of charge in the molecule. In other words, if you add the two dipole arrows, they cancel.

$$\overset{\longleftarrow\;\;+}{\underset{\delta^-}{O}}=\overset{+}{\underset{\delta^+}{C}}=\overset{+\;\;\longrightarrow}{\underset{\delta^-}{O}}$$

Carbon dioxide
Nonpolar

The two O—H bonds of a water molecule, H_2O, are also polar covalent bonds. However, since water has a bent molecular shape, the two bond dipoles do not cancel. The center of negative charge is on the oxygen atom and the center of positive charge is located between the two hydrogen atoms, generating a net dipole, indicated by the dark black dipole arrow in Figure 3-12a. In other words, if you add the two blue bond dipole arrows, the result is the molecular dipole shown by the black dipole arrow. The electron density diagram in Figure 3-12b shows that oxygen has a partial negative charge and each hydrogen atom has a partial positive charge. Although nonbonding electrons play a role in polarity, you can ignore them when determining if the molecule is polar.

As a general rule, when the covalent bonds around a central atom are identical *and* the electron geometry is the same as the molecular shape, the individual bond dipoles will cancel and the molecule is nonpolar—as in the case of CO_2. Thus, tetrahedral, trigonal planar, and linear molecular shapes will be nonpolar if all the covalent bonds in the molecule are identical. In contrast, molecules with a trigonal pyramidal or bent shape will be polar if there are any polar covalent bonds at all present in the molecule—as in the case of water.

Now that you have learned that molecules can be defined by both their shape and their polarity, you are ready to consider how these characteristics influence how molecules interact with one another.

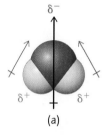

(a)

(b) Electron density diagram

Figure 3-12 Water is a polar molecule. (a) The sum of the individual bond dipoles, shown in blue, equals the net molecular dipole, shown in black. (b) The electron-density diagram of water shows that the oxygen atom has a partial negative charge, indicated in red, and the hydrogen atoms have a partial positive charge, indicated in blue.

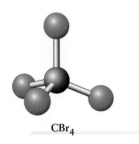

CBr₄

Problem-Solving Tutorial:
Molecular Polarity

WORKED EXERCISES 3-5   Identifying Polar and Nonpolar Molecules

1 Carbon tetrabromide, CBr_4, has four polar C—Br bonds, yet the molecule is nonpolar. Explain.

2 Explain why BF_3 is nonpolar, but NF_3 is polar. Note that B in BF_3 does not have an octet.

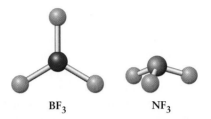

BF_3 NF_3

SOLUTIONS

1 C—Br bonds are polar covalent bonds because bromine is more electronegative than carbon. But, because each C—Br dipole is equivalent and they are directed to the four opposite corners of a tetrahedron, the dipoles cancel each other, causing the molecule to be nonpolar. In other words, the center of negative charge and the center of positive charge are both on carbon, so is no separation of charge.

2 Both BF_3 and NF_3 have three polar covalent bonds (B—F or N—F) because fluorine is more electronegative than boron and nitrogen. Since BF_3 has a trigonal planar electron geometry *and* a trigonal planar molecular shape, the three B—F bond dipoles cancel and the molecule is nonpolar. The center of both the positive charge and the negative charge in BF_3 is on the boron atom.

  NF_3 has a tetrahedral electron geometry and a trigonal pyramidal molecular shape, so the three N—F bond dipoles do not cancel and the molecule is polar. In other words, the center of positive charge is on the nitrogen atom, and the center of negative charge is in the center of the three fluorine atoms, so there is a separation of charge and the molecule is polar.

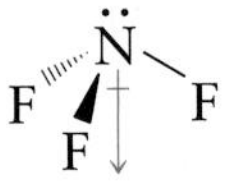

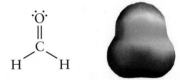

PRACTICE EXERCISES

3.10 Which of the following molecules are nonpolar? Explain why.
  a. Cl_2
  b. C_3H_8
  c. SO_2 (trigonal planar electron geometry; bent molecular shape)
  d. HCl
  e. CCl_4

3.11 A Lewis dot structure and an electron density diagram for formaldehyde, CH_2O, are shown below.

$$\ddot{O}$$
$$\|$$
$$C$$
$$H \quad H$$

  a. What is the electron geometry? What is the molecular shape? What are the bond angles?
  b. Is this molecule flat, i.e., two-dimensional?
  c. Is this molecule polar or nonpolar? If it is polar, label the positive and negative ends of the molecule with a net dipole arrow.
  d. What does the blue region in the electron density diagram signify? What does the red region signify?

3.12 An electron density diagram for estradiol, $C_{18}H_{24}O_2$, is shown below.

   a. What information is conveyed by the high percentage of green color in this electron density diagram?

   b. There are two OH groups in the molecule. Where are they and how did you identify them? Comment on their color.

3.3 Intermolecular Forces of Attraction

At room temperature (25°C), why is water (H_2O) a liquid and methane (CH_4) a gas? This question asks about a macroscopic property—a property that you can observe with the naked eye. Yet the answer depends in large part on how the molecules that make up these compounds interact with one another at the atomic level. You have already seen that atoms interact to form covalent bonds in molecules. In this section, you will see that molecules interact with one another through **inter**molecular forces of attraction—those that exist *between* molecules, rather than *within* molecules. These forces are not as strong as covalent bonds, but they play an important role in chemistry. **Intermolecular forces of attraction** determine the physical properties of a substance, such as boiling point, melting point, and solubility, which will be described in the next two chapters. Intermolecular forces of attraction also influence the three-dimensional shapes of large biomolecules such as proteins and DNA.

Methane, CH_4, is the familiar substance also known as natural gas. [Tetra images/Punchstock]

Similar to bonding forces, intermolecular forces arise from electrostatic interactions: the attraction between opposite charges and the repulsion between like charges. Partial charges resulting from dipoles in one molecule are attracted to the opposite partial charges in other molecules with dipoles. A dipole can be either a *permanent dipole*, such as those found in the polar molecules you learned about in Section 3.2, or a *temporary dipole*, those induced in a nonpolar molecule.

Intermolecular forces of attraction are much weaker than covalent bonding forces. Typically the strongest intermolecular forces of attraction are 5–10% of the strength of the average covalent bond, because the charges are much smaller (partial charges rather than full charges) and the distances between charges are greater. Although intermolecular forces of attraction are weaker than covalent bonds, they are extremely important in nature. Covalent bonds are found where relatively permanent connections between atoms are required, and intermolecular forces of attraction are found where *temporary* connections between molecules are required. For example, the two strands of a DNA molecule are held together by intermolecular forces of attraction, so that they are readily disrupted and the strands separated when the information on a strand must be read.

There are three basic types of intermolecular forces of attraction between molecules. In order of increasing strength, they are

- dispersion forces,
- dipole–dipole forces, and
- hydrogen bonding forces.

Figure 3-13 Guide to determining the type of intermolecular forces present in a covalent compound.

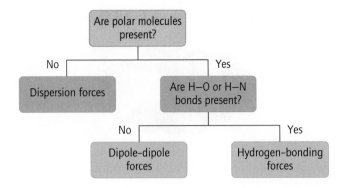

You can determine the strongest type of intermolecular force of attraction exerted by a molecule by evaluating the types of *individual* bond dipoles present in the molecule. Figure 3-13 provides a flow chart describing the steps in the evaluation process.

Dispersion Forces

All molecules and elements exhibit dispersion forces, but dispersion forces are the sole intermolecular force of attraction available to nonpolar molecules. Recall that nonpolar molecules do not have a dipole, because they have a uniform distribution of electrons. A uniform distribution of electrons refers to the *average* distribution of electrons over time (Figure 3-14a). Since electrons are in constant random motion, at any given *instant* in time electrons can temporarily shift toward one end of the molecule, creating a **temporary dipole** (Figure 3-14b). A temporary dipole *induces* a corresponding temporary dipole in adjacent molecules, causing them to shift their electrons in a way that brings opposite partial charges together (Figure 3-14c) and draws the molecules together. Induced temporary dipoles are transmitted throughout the sample. This type of electrostatic attraction is known as a **dispersion force,** because the intermolecular forces of attraction are "dispersed" throughout the entire sample. Since *induced* dipoles are short-lived, dispersion forces are the weakest of the three types of intermolecular forces of attraction.

Since many electrons are distributed throughout a molecule, there are many opportunities for temporary dipoles to form. It follows, therefore, that dispersion forces increase as

- the number of electrons in a molecule increases and
- the surface area of a molecule increases

Dispersion forces are also known as London forces.

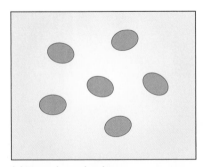

(a) Nonpolar molecules

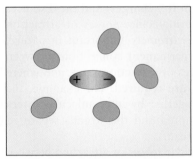

(b) Temporary dipole in one molecule

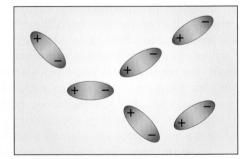

(c) Induced dipoles in neighboring molecules

Figure 3-14 Temporary dipoles change with each observation. (a) The distribution of electrons is uniform in nonpolar molecules. (b) One molecule develops a temporary dipole due to random electron movements. (c) A temporary dipole in one molecule induces temporary dipoles in neighboring molecules throughout the sample.

For example, hexane (C_6H_{14}) has more electrons and more surface area than ethane (C_2H_6). Thus, hexane exerts stronger dispersion forces than ethane.

The strength of a compound's dispersion forces determines many of its physical properties. For example, hexane molecules are held together by stronger dispersion forces than ethane; therefore, hexane is a liquid at room temperature, whereas ethane is a gas.

Dipole–Dipole Forces

In addition to dispersion forces, polar molecules are capable of interacting through **dipole–dipole forces,** which are electrostatic attractions between the permanent dipoles of polar molecules. These molecules arrange themselves in a way that brings together opposite charges, similar to the behavior of nonpolar molecules under the influence of dispersion forces. However, because the dipoles are permanent, dipole–dipole intermolecular forces of attraction are much stronger.

For example, formaldehyde has a permanent dipole as a result of its C=O bond (Figure 3-15a). Therefore, the partial positively charged end of one formaldehyde molecule will orient itself toward the partial negatively charged end of another formaldehyde molecule, and so forth throughout the sample, as shown in Figure 3-15b.

As a result of these stronger intermolecular forces of attraction, formaldehyde is a liquid rather than a gas at room temperature.

(a) Permanent dipole in a single
    formaldehyde molecule, three views

(b) Dipole–dipole interactions of many
    formaldehyde molecules, two views

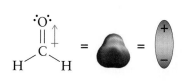

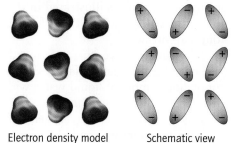

Electron density model   Schematic view

Figure 3-15 Dipole–dipole intermolecular forces of attraction in formaldehyde. Molecules arrange themselves so that the positively charged end of one molecule interacts with the negatively charged end of another, creating a dipole-dipole interaction.

Hydrogen Bonding Forces

If you have ever placed a water bottle in the freezer, you have experienced the effects of hydrogen bonding. As the water freezes it expands, possibly causing a full water bottle to crack. **Hydrogen bonding** is the strongest intermolecular force of attraction between molecules. Hydrogen bonding is *not* a covalent bond as the term "bonding" in the name might suggest. Rather, hydrogen bonding is a type of dipole–dipole interaction that exists when a molecule contains one of the three strongest polar covalent bonds:

- H—F
- H—O
- H—N

H—F is not found in the body, but O—H and N—H bonds are found in many large biomolecules, such as proteins, carbohydrates, and DNA. They are the two most polar covalent bonds in biological molecules, because they involve nonmetal atoms at opposite ends of the electronegativity scale: hydrogen and oxygen, and hydrogen and nitrogen. The hydrogen atom in these bonds has the greatest partial positive charge of any atom in a covalent bond, and nitrogen and oxygen have the greatest partial negative charge of any atom in a covalent bond found in biological molecules. It follows, therefore, that a molecule containing an O—H or N—H bond

Figure 3-16 Hydrogen bonding in ice. Each water molecule has hydrogen bonds to four other water molecules.

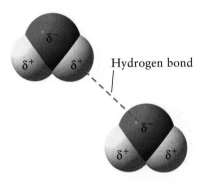

A dashed line is used to indicate hydrogen bonding, as for example, between a pair of water molecules.

Hydrogen bond

Ice floats on water because hydrogen bonding causes the molecules to occupy a greater volume in the solid state than in the liquid state. Therefore, the mass per volume (density) is less for the solid state than the liquid state. [Charles D. Winters/Photo Researchers Inc.]

should exhibit the strongest force of attraction between the positive pole—hydrogen—of one molecule and the negative pole—nitrogen or oxygen atom—in another molecule. A dashed line is used to designate a hydrogen bond between the hydrogen atom (δ^+) of one molecule and the nitrogen or oxygen atom (δ^-) of another molecule, as shown in the margin for a pair of water molecules.

As you can see in the five water molecules shown in Figure 3-16, there is more than one hydrogen bond made to any one water molecule. The network of hydrogen bonds extends throughout the liquid or solid sample of water, acting like a glue that holds the water molecules together. Thus, water molecules are held together strongly while nonpolar compounds exhibit a weaker interaction. Hydrogen bonding is responsible for water beading on a leaf, as illustrated in the photo on page 79. In contrast, non-polar molecules, such as those found in gasoline, do not form beads on a nonpolar surface.

In a sample of ice, there are four hydrogen bonds for every water molecule, creating a crystalline lattice that occupies a greater volume than the corresponding sample of liquid water with fewer hydrogen bonds per molecule. Hence, ice is less dense than water, causing it to float on water.

It is the existence of hydrogen bonding that causes water to be a liquid at room temperature, and the absence of hydrogen bonding that causes methane to be a gas at the same temperature. In order for liquid water to become a gas (steam), the hydrogen bonds must be broken. Because breaking hydrogen bonds requires energy, sufficient energy must be added to the water sample (by heat or microwaves) before the molecules can separate and enter the gas phase. In contrast, methane has no hydrogen bonds, so it is a gas at well below room temperature.

WORKED EXERCISES 3-6   Intermolecular Forces of Attraction

1 Which of the following molecules contain a permanent dipole?
    a. C_2H_6 **b.** H_2O

2 Which molecule in each pair exhibits stronger intermolecular forces of attraction with like molecules? Explain your choice. State the type of intermolecular force present.
    a. C_3H_8 or C_6H_{14} **b.** H_2O or H_2S

3 Which type of intermolecular force of attraction exists in all molecules? What types of molecules interact only through this intermolecular force of attraction?

SOLUTIONS

1 **a.** Water, H_2O, is a polar molecule and therefore it has a permanent dipole.

Problem-Solving Tutorial:
Intermolecular Forces

2 a. C_6H_{14} exerts stronger dispersion forces than C_3H_8, because there are more atoms in the molecule and therefore more electrons. Dispersion forces increase with the number of electrons in a molecule.

  b. H_2O exerts the stronger intermolecular force of attraction because it interacts with other water molecules through hydrogen bonding, the strongest intermolecular force of attraction. Water forms hydrogen bonds because it has two O—H bonds, one of the most polar of all covalent bonds. Sulfur is less electronegative than oxygen because it lies below oxygen on the periodic table, so its bond dipole to hydrogen is much weaker than the O—H bond dipole.

3 Dispersion forces exist among all molecules and elements. Dispersion forces are the only intermolecular force of attraction between nonpolar molecules and all nonmetal elements.

PRACTICE EXERCISES

3.13 Which type of intermolecular force of attraction is the weakest?
  a. hydrogen bonding
  b. dispersion forces
  c. dipole–dipole forces

3.14 What type of molecules exhibit only dispersion forces and why?

3.15 What is the difference between a temporary dipole and a permanent dipole?

3.16 Which of the following molecules contains a permanent dipole? Explain why.
  a. acetone (finger nail polish remover)

$$\begin{array}{ccccccc} & H & & O & & H & \\ & | & & \| & & | & \\ H- & C & - & C & - & C & -H \\ & | & & & & | & \\ & H & & & & H & \end{array}$$

  b. C_3H_8

3.17 Which molecule in each pair exhibits stronger intermolecular forces of attraction? Explain your choice. State the type of intermolecular force of attraction present.
  a. water, H_2O, or ammonia, NH_3

$$\begin{array}{ccccc} & H & & H & \\ & | & & | & \\ \textbf{b.}\ H- & C & - & C & -H \\ & | & & | & \\ & H & & H & \end{array} \quad \text{or} \quad \begin{array}{c} O \\ \| \\ C \\ H \qquad H \end{array}$$

  c. HCl or HF
  d. F_2 or HF

Hydrogen Bonding in DNA

Deoxyribonucleic acid (DNA) is an important biomolecule because it stores your genetic information (see Chapter 15). Basically, DNA is composed of two extremely large molecules—strands—wound together in the shape of a double helix. Hydrogen bonds hold the two strands of DNA together throughout the entire length of the helix.

Figure 3-17 shows a portion of the DNA double helix. Projecting perpendicular to each of the two strands shown in blue are the atoms that form hydrogen bonds. The hydrogen bonds are formed between the

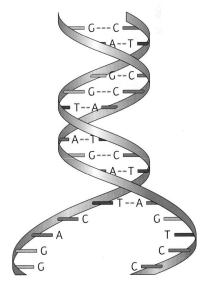

Figure 3-17 DNA has two strands bound together in the overall shape of a double helix. The two strands are held together by hydrogen bonds formed between atoms on adjacent strands, as shown by the dashed red lines.

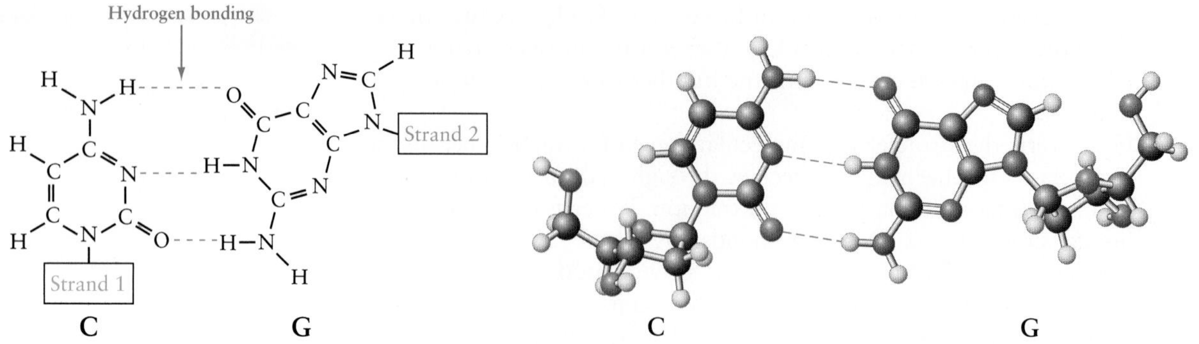

Figure 3-18 Molecular view of a section of two strands of DNA showing hydrogen bonding between a C and a G on opposite strands.

hydrogen of an N—H or O—H bond on one strand and either a nitrogen or an oxygen atom on the adjacent strand. Hydrogen bonding connects the ladder-like rungs in the structure of DNA shown in Figure 3-17. These bonds occur between four different nitrogen-containing groups seen throughout the DNA molecule and identified as C, A, T, and G. The dashed red lines represent the hydrogen bonds. If you were to zoom in on one rung of the ladder, for example, you would see two or three hydrogen bonds, as shown in Figure 3-18 for the interaction between a C and a G on adjacent strands of DNA.

When the time comes for DNA to be replicated or transcribed to make a protein, the hydrogen bonds are easily broken so that the two strands can come apart—like a zipper—making it possible for the information on one strand to be read. You will learn in Chapter 15 that reading the information from a portion of a strand of DNA known as a gene supplies the instructions for making a protein. The atoms that make up each DNA strand are held together by much stronger covalent bonds, which are necessary for maintaining the integrity of the information stored within the atoms of DNA's structure.

WORKED EXERCISE 3-7   Intermolecular Forces of Attraction

Hydrogen bonding between T and A of DNA is shown below.

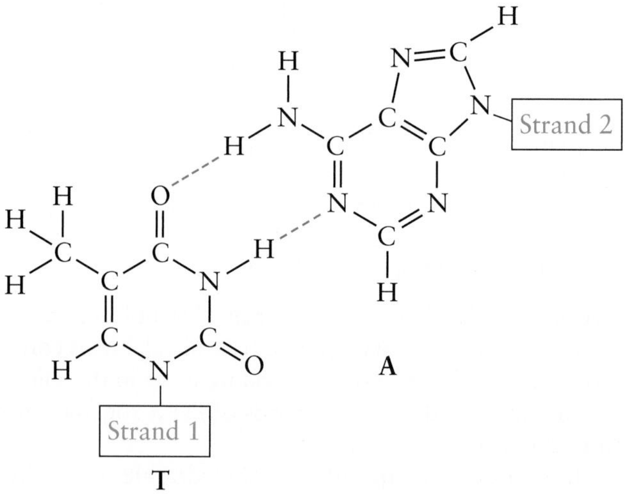

a. How many hydrogen bonds are shown and between what atoms?

b. Label the atoms as δ^+ or δ^-, appropriately.

c. What is the role of hydrogen bonding in the portion of DNA shown.

SOLUTION

a. Two hydrogen bonds are shown, one between C=O and H—N and one between N—H and N=C.

b.

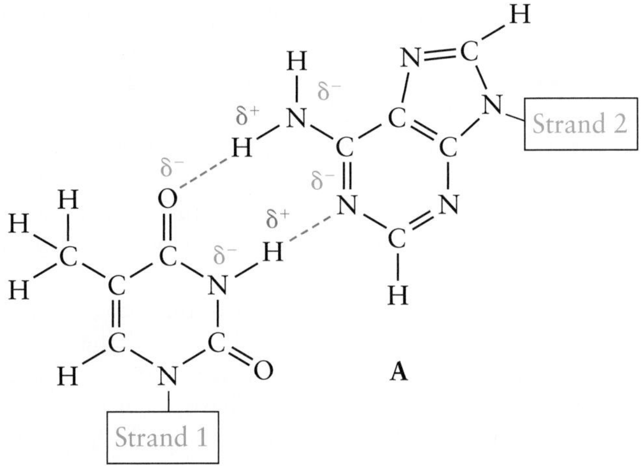

c. Hydrogen bonding holds the two strands of DNA together at this rung on the ladder until they need to be separated.

PRACTICE EXERCISES

3.18 What is DNA?

3.19 Parts of two strands of DNA are shown below, identified as C and G.

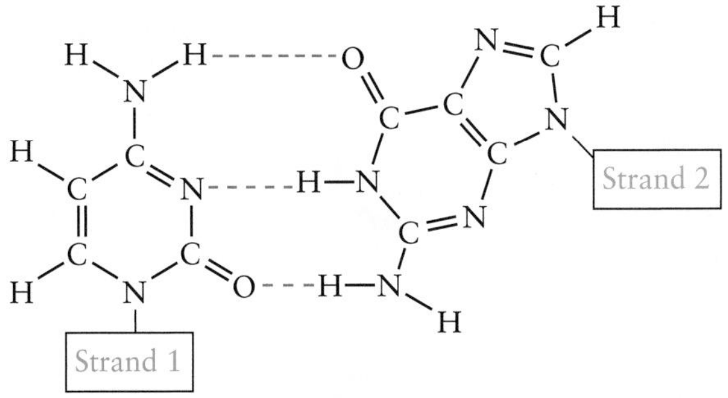

   a. How many hydrogen bonds are formed between C and G?

   b. Show the partial charges on all the atoms involved in hydrogen bonding.

   In this chapter you have seen the six basic molecular shapes of simple molecules. The shape of a molecule together with the individual bond dipoles determines the polarity of the overall molecule, which in turn determines how the molecule interacts with other molecules. Intermolecular forces impact the physical properties of a compound and play an important role in the structure of biological molecules such as DNA. In the next chapter you will apply your knowledge of intermolecular forces of attraction to see how they influence other physical properties of a compound.

Throughout a woman's life, estrogens cause normal breast and uterine cells to grow and divide, a process called *cell proliferation*. In the liver, estrogens promote the production of good cholesterol, and in bone, estrogens promote the growth of strong bones. These are normal and beneficial effects of estrogen. Scientists now have an understanding of how estrogens work on a molecular level: Estradiol binds to the estrogen receptor, a nuclear receptor that activates particular genes in these cells. In breast cells, gene activation produces the molecules required for cell proliferation.

Unfortunately, estrogen also stimulates the proliferation of breast cancer cells in breast tissue, and uterine cancer cells in the tissue lining the uterus, because most breast cancer cells also contain estrogen receptors. The challenge in designing a drug for the treatment of breast cancer is to find one that selectively blocks the estrogen receptor in breast and uterine cells, but not in liver and bone cells. In other words, an effective treatment for breast cancer is one that doesn't at the same time cause osteoporosis (loss of bone density) or cardiovascular disease to develop. Drugs that selectively block the estrogen receptor in one cell type

are known as **s**elective **e**strogen **r**eceptor **m**odulators (SERMs). SERMs have been used in the treatment of breast cancer as well as osteoporosis and other diseases affected by estrogen.

Tamoxifen, introduced in 1977, is the oldest SERM still in use today. Tamoxifen is often prescribed as a hormone therapy for up to 5 years following the surgical removal of breast cancer. Taking Tamoxifen has been found to reduce the risk of breast cancer by 49%. Unfortunately, one of the drug's side effects is that it stimulates the proliferation of uterine cells, increasing the likelihood of developing uterine cancer. Instead of blocking the effects of estrogen, as it does in breast cells, Tamoxifen acts like estrogen in uterine cells. Tamoxifen also promotes the growth of bone tissue, acting like estrogen in bone tissue.

How does Tamoxifen stop the progression of breast cancer? When estradiol binds to the estrogen receptor, the shape of the estrogen receptor changes. This change in shape allows other *specialized proteins* necessary for DNA activation to bind to the estrogen receptor. This cascade of events is summarized in Figure 3-19a.

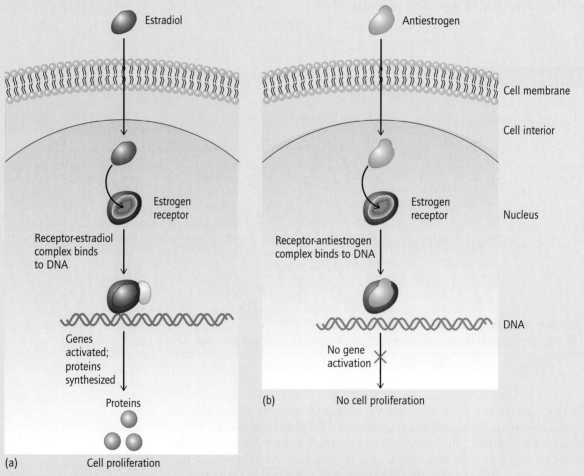

Figure 3-19 The actions of estrogens and antiestrogens. (a) Estradiol binds to the estrogen receptor, causing gene activation. (b) Antiestrogen binds to the estrogen receptor, preventing gene activation. As a result, proteins required for cell growth are not produced.

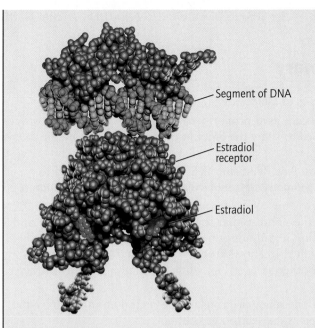

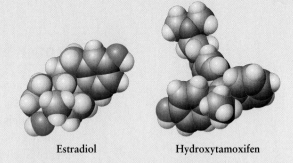

Estradiol   **Hydroxytamoxifen**

Figure 3-21 Space-filling models of estradiol and hydroxytamoxifen

Figure 3-20 The estrogen receptor showing a segment of DNA and bound estrogen.

When Tamoxifen enters the cell, the estrogen receptor is unable to distinguish it from the natural ligand, estradiol, so it binds to the receptor as estradiol does. However, the shape of Tamoxifen causes it to bind to the receptor in a slightly different way, causing the receptor to change its shape in a way that prevents the necessary specialized proteins from binding. Furthermore, Tamoxifen blocks estradiol from binding to the receptor and activating DNA. Consequently, cancer cells cease to grow and divide.

Intermolecular Forces of Attraction Between Estrogen and the Estrogen Receptor

The estrogen receptor is an extremely large protein molecule. You can see its three-dimensional shape in Figure 3-20. The receptor appears symmetrical because it is actually a pair of identical molecules, like a molecular twin, known as a *dimer*.

Estradiol binds to a cavity within the receptor, known as the estrogen binding site. Estradiol fits the cavity perfectly because it has a shape complementary to the binding site. In addition, intermolecular forces of attraction act between estradiol and atoms within the receptor binding

site. The two O—H groups in estradiol form hydrogen bonds to a nitrogen atom and an oxygen atom located on opposite ends of the estrogen receptor binding site. Nonpolar regions of the estrogen binding site are attracted to the largely nonpolar estradiol molecule through dispersion forces of attraction.

The Antiestrogen Tamoxifen

The drug Tamoxifen is transformed in the body into *hydroxytamoxifen,* which is the active drug molecule that binds to the estrogen receptor. Estradiol and hydroxytamoxifen have noticeably different overall shapes (Figure 3-21), yet they both fit the estrogen receptor binding site. However, they bind at different parts of the binding site, causing the receptor to change its shape in different ways, as shown in Figure 3-22. Consequently, the binding of hydroxytamoxifen prevents the activation of DNA in breast tissue. Paradoxically, Tamoxifen binding activates DNA in bone and uterine cells. For reasons not understood, it blocks the estrogen receptor only in breast tissue.

Clearly, shape and electrostatic interactions are central to molecular interactions. Tamoxifen binding to the estrogen receptor is just one example of one drug interacting with one nuclear receptor. It is estimated that there are over 150 different nuclear receptors which bind substances ranging from the thyroid hormones to vitamin D. Indeed, nuclear receptors are a major target for drug therapy. Efforts to design even better drugs than Tamoxifen continue. The goal for any drug therapy is to selectively prevent the progression of a disease, without introducing negative side effects.

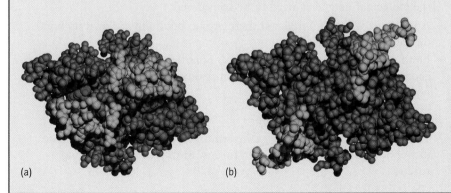

(a)   (b)

Figure 3-22 Estrogen receptor bound to (a) estradiol and (b) hydroxytamoxifen. Notice that the shape of the receptor changes depending on the ligand bound to the receptor.

Chapter Summary

Three-dimensional Shapes of Molecules

- Ball-and-stick models are used primarily to show bond angles, while space-filling models are used to show the amount of space occupied by the atoms in a molecule.
- Valence Shell Electron Pair Repulsion (VSEPR) theory can be used to predict the electron geometry of a molecule from the Lewis dot structure of a molecule.
- VSEPR is based on the premise that the central atom in a molecule and the atoms surrounding it will adopt a molecular shape that places all groups of electrons on the central atom, bonding and nonbonding, as far apart from each other as possible, while maintaining bonds to the central atom.
- A molecule with only bonding electrons on the central atom will have the same molecular shape as its electron geometry.
- Molecules with a tetrahedral electron geometry and one nonbonding pair of electrons will have a trigonal pyramidal molecular shape with approximately 109.5° bond angles. NH_3 is an example.
- Molecules with a tetrahedral electron geometry and two nonbonding pairs of electrons will have a bent molecular shape with approximately 109.5° bond angles. H_2O is an example.
- Molecules with a trigonal planar electron geometry and one nonbonding pair of electrons on the central atom will have a bent shape with approximately 120° bond angles.
- Molecules with a linear electron geometry will have a linear molecular shape and 180° bond angles. CO_2 is an example.
- The shapes of large molecules are a combination of the shapes of the individual atom centers in the larger molecule.

Molecular Polarity

- Electronegativity is a measure of an atom's ability to attract electrons toward itself in a covalent bond.
- Electronegativity increases as you go from the bottom to the top of a group in the periodic table. It also increases from left to right within a period. The most electronegative element is fluorine.
- A nonpolar covalent bond is formed between identical atoms and atoms with similar electronegativities.
- A polar covalent bond is formed between atoms with different electronegativities.
- In order to determine if a molecule is polar or nonpolar, bond dipoles and the molecular geometry need to be considered.
- A molecule will be nonpolar if there are no bond dipoles or if its bond dipoles cancel.
- Bond dipoles cancel in molecules with polar covalent bonds, if the electron geometry is the same as the molecular geometry AND all the covalent bonds are identical.

Intermolecular Forces of Attraction

- There are three basic types of intermolecular forces of attraction: dispersion forces, dipole–dipole forces, and hydrogen bonding forces.

- Dispersion forces are exerted by all molecules and nonmetal elements and are the only forces of attraction exhibited by nonpolar molecules.
- Dispersion forces are attractions between induced temporary dipoles.
- Dipole–dipole forces are attractions between the permanent dipoles of polar molecules. Polar molecules will arrange themselves so that like charges are kept apart and opposite charges are brought together.
- Hydrogen bonding is the strongest intermolecular force of attraction and occurs between the hydrogen atom in any of the bonds H—F, H—O, or H—N and an electronegative atom like N and O.
- All intermolecular forces of attraction are weaker than covalent bonds.
- Hydrogen bonding makes ice less dense than liquid water.
- Hydrogen bonding holds the two strands of DNA together.

Key Words

Ball-and-stick model A model of a molecule that represents atoms as colored spheres and covalent bonds as sticks. It is a tool for visualizing bond angles in a molecule.

Bent A molecular shape that arises from a tetrahedral electron geometry when there are two nonbonding pairs of electrons; or from a trigonal planar electron geometry, when there is one nonbonding pair of electrons.

Bond angle The angle created by a central atom and any two atoms bonded to the central atom.

Bond dipole A polar covalent bond with a positive end and a negative end—charge separation—is created when the two bonded atoms have differing electronegativities.

Dipole A charge separation in a bond or a molecule.

Dipole–dipole force Intermolecular force of attraction between the permanent dipoles of two polar molecules.

Dispersion force The only force of attraction between nonpolar molecules, created by temporary attractions between induced dipoles.

Electron density diagram A space-filling model color-coded to show the relative charge accumulation on the different atoms in the molecule. Red is used for δ^- and blue for δ^+.

Electron geometry The geometry created by the electrons around a central atom, including bonding and nonbonding groups of electrons.

Electronegativity A measure of an atom's ability to attract electrons toward itself in a molecule.

Hydrogen bonding forces The strongest intermolecular force of attraction that can exist between molecules. It typically occurs between the hydrogen atom in an N—H or O—H bond in one molecule and either an oxygen or nitrogen atom in another molecule. HF can also hydrogen bond, but it is not found in the body.

Intermolecular forces of attraction Forces of attraction that occur between molecules as a result of electrostatic attractions.

Linear geometry An electron geometry or molecular shape resembling a straight line, formed by two groups of electrons surrounding a central atom with a 180° bond angle.

Molecular shape The geometry of a molecule based on the relative positions of the atoms in the molecule.

Nonpolar covalent bond A bond formed between two atoms with the same or comparable electronegativities.

Nonpolar molecule A molecule with an even distribution of electrons; no separation of charge.

Polar covalent bond A covalent bond that contains a bond dipole. A polar covalent bond is formed when two atoms have significantly different electronegativities.

Polar molecule A molecule that has a separation of charge. A polar molecule is formed when the centers of positive and negative charge in a molecule do not coincide. The bond dipoles in the molecule do not cancel.

Space-filling model A visual representation of a molecule that shows the relative amount of space occupied by the atoms in the molecule.

Tetrahedral geometry An electron geometry or molecular shape resembling a tetrahedron. The electron groups around the carbon atom point to the four corners of a tetrahedron, and bond angles are 109.5°.

Trigonal planar geometry An electron geometry or molecular shape resembling an equilateral triangle. The electron groups around the central atom point to the three corners of the triangle, and bond angles are 120°.

Trigonal pyramidal geometry The geometry of a central atom with a tetrahedral electron geometry and one nonbonding pair of electrons. Bond angles are approximately 109.5°.

Valence Shell Electron Pair Repulsion Theory (VSEPR) A theory used to predict the shapes of simple molecules. VSEPR predicts the electron geometry from the number of electron groups (usually 2 to 4) around the central atom. Groups of electrons are positioned to achieve the maximum distance between them while maintaining a bond to the central atom.

Additional Exercises

Three-Dimensional Shapes of Molecules

3.20 What principle is VSEPR theory based upon?

3.21 Write the Lewis dot structure for the following molecules and then determine the electron geometry and the molecular shape. Indicate the bond angles in each one.
   a. CH_2Cl_2
   b. NI_3
   c. SeO_2

3.22 Write the Lewis dot structure for the following molecules and then determine the electron geometry and the molecular shape. Indicate the bond angles.
   a. PF_3
   b. SCl_2
   c. $SiCl_4$

3.23 If a molecule has only two groups of electrons surrounding the central atom, what is the shape of the molecule?

3.24 What is the electron geometry for a molecule with three groups of electrons around the central atom? What are the possible molecular shapes of the molecule? What is/are the bond angle(s)?

3.25 What is the electron geometry for a molecule with four groups of electrons around the central atom? What are the possible molecular shapes of the molecule? What is/are the bond angle(s)?

3.26 Which of the molecular shapes listed below are three-dimensional?
   a. linear
   b. trigonal planar
   c. tetrahedral

3.27 Which of the molecular shapes listed below are three-dimensional?
   a. trigonal pyramidal
   b. bent
   c. trigonal planar

3.28 What are the bond angles in a molecule that has a trigonal planar electron geometry?

3.29 What are the bond angles in a molecule that has a linear electron geometry?

3.30 What are the bond angles in a molecule that has a trigonal pyramidal molecular shape?

3.31 What are the bond angles in a molecule that has a tetrahedral molecular shape?

3.32 What are the two possible bond angles for a molecule with a bent shape? Why are they not the same.

3.33 What are the bond angles and what is the molecular shape of the following molecules?
   a. 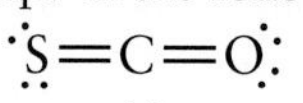 $\ddot{S}{=}C{=}\ddot{O}\!:$
   b. 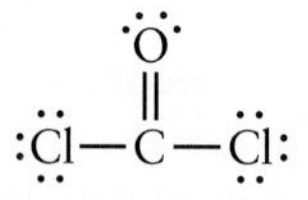
$$:\!\ddot{C}l\!-\!\underset{|}{\overset{:\ddot{O}:}{\overset{\|}{C}}}\!-\!\ddot{C}l\!:$$

3.34 What are the bond angle and the molecular shape of the following molecules?
   a. HCN
   b. 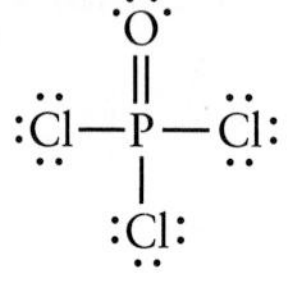
$$:\!\ddot{C}l\!-\!\underset{\underset{:\ddot{C}l:}{|}}{\overset{:\ddot{O}:}{\overset{\|}{P}}}\!-\!\ddot{C}l\!:$$

3.35 The Lewis dot structure for sulfate ion, $SO_4{}^{2-}$, is shown below. What is the geometry of this polyatomic ion? What is the O—S—O bond angle?
$$\left[\;:\!\ddot{O}\!-\!\underset{\underset{:\ddot{O}:}{|}}{\overset{:O:}{\overset{\|}{S}}}{=}\ddot{O}\;\right]^{2-}$$

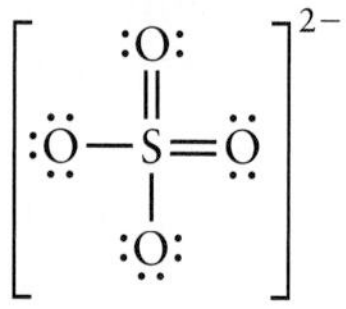

3.36 The Lewis dot structure for the nitrate ion, $NO_3{}^-$, is shown below. What is the geometry of this polyatomic ion? What is the O—N—O bond angle?
$$\left[\;:\!\ddot{O}\!-\!\overset{:O:}{\overset{\|}{N}}\!-\!\ddot{O}\!:\;\right]^-$$

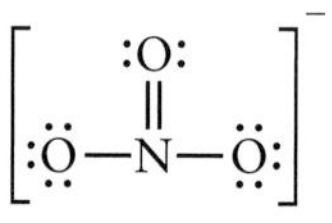

3.37 Critical Thinking Question: Why does the bond angle around the central atom decrease as the number of groups of electrons surrounding the central atom increases? What is the greatest bond angle? What is the smallest bond angle?

3.38 Write the Lewis dot structures of the following molecules and then indicate the molecular shape and the bond angles for each one.
   a. PH_3
   b. CCl_4
   c. BF_3 (boron is an exception to the octet rule and has six electrons around it)
   d. SF_2
   e. OCS

3.39 Write the Lewis dot structures for the following molecules and then indicate the molecular shape and the bond angles for each one.
   a. C_2H_4
   b. SiF_4
   c. C_2H_6
   d. C_2H_2

3.40 Fill in the blanks in the table below for the six common molecular shapes:

1. Total number of electron groups		3				
2. Number of bonding groups	2	3			3	
3. Number of nonbonding groups	0		1	0		2
4. Electron geometry		Trigonal planar				
5. Molecular shape					Trigonal pyramidal	
6. Bond angle					109.5°	

3.41 Ethanol has the following Lewis dot structure:

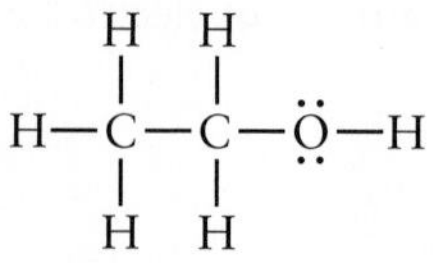

a. Provide a three-dimensional representation of ethanol using dashed and wedged bonds.
b. What is the bond angle around each carbon atom?
c. What is the bond angle around the oxygen atom?
d. What is the molecular shape around each carbon atom?
e. What is the molecular shape around the oxygen atom?

3.42 The Lewis dot structure for acetonitrile is shown below.

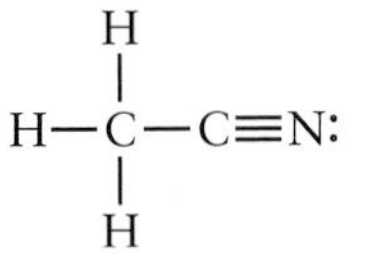

$$H—C—C≡N:$$

(with H above and H below the first carbon)

a. What is the molecular geometry around each carbon atom? Do both carbon atoms have the same molecular shape? Explain.
b. Is any part of this molecule flat (two-dimensional)?
c. What is the H—C—H bond angle?
d. What is the H—C—C bond angle?
e. What is the C—C—N bond angle?

3.43 What is the difference between a trigonal planar geometry and a trigonal pyramidal geometry? How are the bond angles different?

3.44 Why do two molecules with a bent geometry not necessarily have the same bond angle?

3.45 Diatomic molecules, molecules containing only two atoms, will always be linear. Explain why.

3.46 **Critical Thinking Question:** The Lewis dot structure of an epoxide is shown below. Based on VSEPR, what should the bond angle be for each carbon atom? Since this molecule has the shape of an equilateral triangle, based on geometry, what must these bond angles be? Why do you think this molecule is not commonly found in nature?

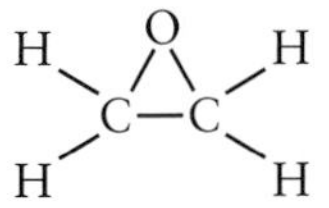

(epoxide structure with O at top, two C atoms, and H atoms)

Modeling Exercise

Exercises 3.47 and 3.48 require the use of a molecular modeling kit.

3.47 I Construction of Formaldehyde, H₂CO

1 Obtain one black carbon atom, two light-blue hydrogen atoms, one oxygen atom, two bent bonds, and two straight bonds.
2 Write the Lewis dot structure.
3 Make a model of H_2CO. *Hint:* Both hydrogen atoms are attached to the carbon atom.

II Observations of Formaldehyde Model

4 How many different groups of electrons surround the central carbon atom in your model: two, three, or four?
5 How many groups of electrons are bonding electrons: two, three, or four?
6 Are there any nonbonding electrons on the central atom?
7 Determine the electron geometry around the central carbon atom in formaldehyde.
8 What is the molecular shape of formaldehyde?
9 Does this molecule have a two-dimensional, flat shape or a three-dimensional shape?

10 What is the H—C—O bond angle? Are all the bond angles in the model approximately the same?
11 Why is the geometry of formaldehyde not a T-shape? In other words, what advantage does a trigonal planar geometry offer over a T-shape?

Modeling Exercise

3.48 I Construction of Carbon Dioxide, CO₂

1 Obtain one black carbon atom, two red oxygen atoms, and four bent bonds.
2 Using the Lewis dot structure as a guide, make a model of CO_2.

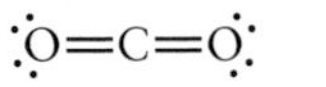

$$:O=C=O:$$

II Observations of Carbon Dioxide Model

3 How many different groups of electrons surround the central carbon atom in your model: two, three, or four?
4 Are these groups of electrons bonding or nonbonding?
5 Are there any nonbonding electrons around the central atom?
6 Determine the geometry around the central carbon atom in carbon dioxide.
7 Does carbon dioxide have a flat two-dimensional shape or a three-dimensional shape?
8 What is the O—C—O bond angle?

3.49 What structural information do ball-and-stick models convey? When the size of atoms is important, what type of model is often more suitable than a ball-and-stick model?

3.50 What is the Lewis dot structure represented by each model below? Indicate whether each model is a ball-and-stick or a space-filling model. For the ball-and-stick model indicate the bond angle(s).

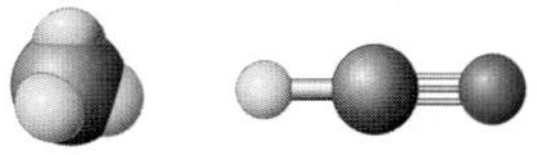

Molecular Polarity

3.51 Define the term *electronegativity*.

3.52 What is the most electronegative element found in biological molecules?

3.53 Indicate which element in each pair is more electronegative.
a. carbon or oxygen
b. sulfur or oxygen
c. lithium or fluorine

3.54 Indicate which element in each pair is more electronegative.
a. phosphorus or oxygen
b. carbon or chlorine (use your electronegativity table for this one)
c. sodium or chlorine

3.55 Does electronegativity increase or decrease as you move up in a group of elements?

3.56 Does electronegativity increase or decrease as you move from left to right within a period of elements?

3.57 Excepting the noble gas elements, which family of elements is the most electronegative?

3.58 Why are some covalent bonds polar?

3.59 How does the electronegativity of the atoms in a covalent bond determine whether it is a polar covalent bond or a nonpolar covalent bond?

3.60 Answer the following questions for a molecule of chloroform, $CHCl_3$.
   a. Write the Lewis dot structure for the molecule.
   b. Provide a three-dimensional representation of the molecule using dashed and wedged bonds.
   c. Are there any bond dipoles in the molecule? If so, which atom in each bond is more electronegative? Show all bond dipoles with a properly drawn dipole arrow.
   d. Is chloroform a polar molecule? Explain.

3.61 Which of the following molecules are polar? For polar molecules, show dipole arrows.
   a. H_2O

   b. ethanol,

   H—C—C—O̤—H with H atoms

   c. C_2H_4.

3.62 Which of the following molecules are polar? For polar molecules write a three-dimensional representation of the molecule along with dipole arrows.
   a. HF   **b.** Br_2 **c.** CO_2

3.63 Which of the following molecules are nonpolar? Explain.
   a. I_2 **b.** CH_4 **c.** HBr

3.64 Which of the following molecules are nonpolar? Explain.
   a. HCN   **b.** CS_2 **c.** CH_2F_2

Intermolecular Forces of Attraction

3.65 What is the difference between a covalent bond and an intermolecular force of attraction?

3.66 Which is stronger: a covalent bond or an intermolecular force of attraction?

3.67 What are the three types of intermolecular forces? Which one is the strongest? Which one is the weakest?

3.68 Nonpolar molecules are attracted to each other by what kind of intermolecular force of attraction?

3.69 Explain why dispersion forces are the weakest of the intermolecular forces of attraction.

3.70 Why are dipole–dipole interactions stronger than dispersion forces?

3.71 Draw a representation of how you might expect HBr molecules to arrange themselves. What type of intermolecular force exists between HBr molecules?

3.72 What three types of covalent bonds can be involved in hydrogen bonding?

3.73 What type of intermolecular forces would you expect each of the following molecules to exhibit?
   a. C_5H_{12}

   b. H—C—C—C—H, (with H, O, H above and H, H below)

   acetone (found in fingernail polish remover)
   c. water

3.74 What is the strongest intermolecular force you would expect each of the following molecules to exhibit?
   a.   H—C—O̤—H (with H atoms above and below C)

   b. HF
   c. C_3H_8

3.75 Why does ice float on water?

3.76 Why does water form beads on a leaf?

3.77 **Critical Thinking Question:** Draw the hydrogen bonding interaction that might occur between a water molecule, H_2O, and a methanol molecule,

   H—C—O̤—H (with H atoms above and below C)

3.78 **Critical Thinking Question:** Vinyl tablecloths are made of hydrocarbons. Explain why water spilled on a vinyl tablecloth will form beads instead of being absorbed.

3.79 What intermolecular force gives DNA a double helical shape?

3.80 **Critical Thinking Question:** DNA sequencing is used in crime scene investigations and also used to diagnose genetic disorders. The first step in sequencing DNA is to heat the DNA strand so that it separates into two strands. Explain on a molecular level why heat is needed to break apart the two strands of DNA.

Chemistry in Medicine

3.81 How does estrogen affect the growth of breast cancer cells?

3.82 How do antiestrogens prevent the growth of breast cancer cells?

3.83 What type of molecule is the estrogen receptor? Why does estradiol fit in the estrogen binding site?

3.84 Besides hydrogen bonding, what other intermolecular forces allow estradiol to bind to the estrogen receptor?

3.85 When Tamoxifen binds to the estrogen receptor in breast cancer cells, does gene activation occur? Why or why not?

3.86 **Critical Thinking Question:** Why does a decrease in circulating estrogen contribute to osteoporosis?

a. Provide a three-dimensional representation of ethanol using dashed and wedged bonds.
b. What is the bond angle around each carbon atom?
c. What is the bond angle around the oxygen atom?
d. What is the molecular shape around each carbon atom?
e. What is the molecular shape around the oxygen atom?

3.42 The Lewis dot structure for acetonitrile is shown below.

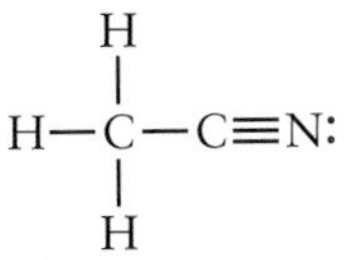

a. What is the molecular geometry around each carbon atom? Do both carbon atoms have the same molecular shape? Explain.
b. Is any part of this molecule flat (two-dimensional)?
c. What is the H—C—H bond angle?
d. What is the H—C—C bond angle?
e. What is the C—C—N bond angle?

3.43 What is the difference between a trigonal planar geometry and a trigonal pyramidal geometry? How are the bond angles different?

3.44 Why do two molecules with a bent geometry not necessarily have the same bond angle?

3.45 Diatomic molecules, molecules containing only two atoms, will always be linear. Explain why.

3.46 **Critical Thinking Question:** The Lewis dot structure of an epoxide is shown below. Based on VSEPR, what should the bond angle be for each carbon atom? Since this molecule has the shape of an equilateral triangle, based on geometry, what must these bond angles be? Why do you think this molecule is not commonly found in nature?

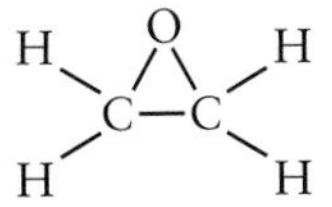

$$
\begin{array}{ccc}
H & O & H \\
 \diagdown & \diagup\diagdown & \diagup \\
 & C{-}C & \\
\diagup & & \diagdown \\
H & & H
\end{array}
$$

Modeling Exercise

Exercises 3.47 and 3.48 require the use of a molecular modeling kit.

3.47 I Construction of Formaldehyde, H_2CO

1 Obtain one black carbon atom, two light-blue hydrogen atoms, one oxygen atom, two bent bonds, and two straight bonds.
2 Write the Lewis dot structure.
3 Make a model of H_2CO. *Hint:* Both hydrogen atoms are attached to the carbon atom.

II Observations of Formaldehyde Model

4 How many different groups of electrons surround the central carbon atom in your model: two, three, or four?
5 How many groups of electrons are bonding electrons: two, three, or four?
6 Are there any nonbonding electrons on the central atom?
7 Determine the electron geometry around the central carbon atom in formaldehyde.
8 What is the molecular shape of formaldehyde?
9 Does this molecule have a two-dimensional, flat shape or a three-dimensional shape?

10 What is the H—C—O bond angle? Are all the bond angles in the model approximately the same?
11 Why is the geometry of formaldehyde not a T-shape? In other words, what advantage does a trigonal planar geometry offer over a T-shape?

Modeling Exercise

3.48 I Construction of Carbon Dioxide, CO_2

1 Obtain one black carbon atom, two red oxygen atoms, and four bent bonds.
2 Using the Lewis dot structure as a guide, make a model of CO_2.

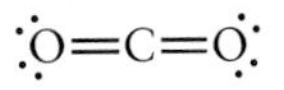

$$\ddot{\text{O}}{=}\text{C}{=}\ddot{\text{O}}$$

II Observations of Carbon Dioxide Model

3 How many different groups of electrons surround the central carbon atom in your model: two, three, or four?
4 Are these groups of electrons bonding or nonbonding?
5 Are there any nonbonding electrons around the central atom?
6 Determine the geometry around the central carbon atom in carbon dioxide.
7 Does carbon dioxide have a flat two-dimensional shape or a three-dimensional shape?
8 What is the O—C—O bond angle?

3.49 What structural information do ball-and-stick models convey? When the size of atoms is important, what type of model is often more suitable than a ball-and-stick model?

3.50 What is the Lewis dot structure represented by each model below? Indicate whether each model is a ball-and-stick or a space-filling model. For the ball-and-stick model indicate the bond angle(s).

Molecular Polarity

3.51 Define the term *electronegativity*.

3.52 What is the most electronegative element found in biological molecules?

3.53 Indicate which element in each pair is more electronegative.
a. carbon or oxygen
b. sulfur or oxygen
c. lithium or fluorine

3.54 Indicate which element in each pair is more electronegative.
a. phosphorus or oxygen
b. carbon or chlorine (use your electronegativity table for this one)
c. sodium or chlorine

3.55 Does electronegativity increase or decrease as you move up in a group of elements?

3.56 Does electronegativity increase or decrease as you move from left to right within a period of elements?

3.57 Excepting the noble gas elements, which family of elements is the most electronegative?

3.58 Why are some covalent bonds polar?

3.59 How does the electronegativity of the atoms in a covalent bond determine whether it is a polar covalent bond or a nonpolar covalent bond?

3.60 Answer the following questions for a molecule of chloroform, $CHCl_3$.
a. Write the Lewis dot structure for the molecule.
b. Provide a three-dimensional representation of the molecule using dashed and wedged bonds.
c. Are there any bond dipoles in the molecule? If so, which atom in each bond is more electronegative? Show all bond dipoles with a properly drawn dipole arrow.
d. Is chloroform a polar molecule? Explain.

3.61 Which of the following molecules are polar? For polar molecules, show dipole arrows.
a. H_2O

b. ethanol,
$$H-\overset{\overset{\displaystyle H}{|}}{\underset{\underset{\displaystyle H}{|}}{C}}-\overset{\overset{\displaystyle H}{|}}{\underset{\underset{\displaystyle H}{|}}{C}}-\ddot{\underset{..}{O}}-H$$

c. C_2H_4.

3.62 Which of the following molecules are polar? For polar molecules write a three-dimensional representation of the molecule along with dipole arrows.
a. HF **b.** Br_2 **c.** CO_2

3.63 Which of the following molecules are nonpolar? Explain.
a. I_2 **b.** CH_4 **c.** HBr

3.64 Which of the following molecules are nonpolar? Explain.
a. HCN **b.** CS_2 **c.** CH_2F_2

Intermolecular Forces of Attraction

3.65 What is the difference between a covalent bond and an intermolecular force of attraction?

3.66 Which is stronger: a covalent bond or an intermolecular force of attraction?

3.67 What are the three types of intermolecular forces? Which one is the strongest? Which one is the weakest?

3.68 Nonpolar molecules are attracted to each other by what kind of intermolecular force of attraction?

3.69 Explain why dispersion forces are the weakest of the intermolecular forces of attraction.

3.70 Why are dipole–dipole interactions stronger than dispersion forces?

3.71 Draw a representation of how you might expect HBr molecules to arrange themselves. What type of intermolecular force exists between HBr molecules?

3.72 What three types of covalent bonds can be involved in hydrogen bonding?

3.73 What type of intermolecular forces would you expect each of the following molecules to exhibit?
a. C_5H_{12}

b.
$$H-\overset{\overset{\displaystyle H}{|}}{\underset{\underset{\displaystyle H}{|}}{C}}-\overset{\overset{\displaystyle O}{\|}}{C}-\overset{\overset{\displaystyle H}{|}}{\underset{\underset{\displaystyle H}{|}}{C}}-H,$$
acetone (found in fingernail polish remover)
c. water

3.74 What is the strongest intermolecular force you would expect each of the following molecules to exhibit?
a.
$$H-\overset{\overset{\displaystyle H}{|}}{\underset{\underset{\displaystyle H}{|}}{C}}-\ddot{\underset{..}{O}}-H$$

b. HF
c. C_3H_8

3.75 Why does ice float on water?

3.76 Why does water form beads on a leaf?

3.77 **Critical Thinking Question:** Draw the hydrogen bonding interaction that might occur between a water molecule, H_2O, and a methanol molecule,
$$H-\overset{\overset{\displaystyle H}{|}}{\underset{\underset{\displaystyle H}{|}}{C}}-\ddot{\underset{..}{O}}-H \quad .$$

3.78 **Critical Thinking Question:** Vinyl tablecloths are made of hydrocarbons. Explain why water spilled on a vinyl tablecloth will form beads instead of being absorbed.

3.79 What intermolecular force gives DNA a double helical shape?

3.80 **Critical Thinking Question:** DNA sequencing is used in crime scene investigations and also used to diagnose genetic disorders. The first step in sequencing DNA is to heat the DNA strand so that it separates into two strands. Explain on a molecular level why heat is needed to break apart the two strands of DNA.

Chemistry in Medicine

3.81 How does estrogen affect the growth of breast cancer cells?

3.82 How do antiestrogens prevent the growth of breast cancer cells?

3.83 What type of molecule is the estrogen receptor? Why does estradiol fit in the estrogen binding site?

3.84 Besides hydrogen bonding, what other intermolecular forces allow estradiol to bind to the estrogen receptor?

3.85 When Tamoxifen binds to the estrogen receptor in breast cancer cells, does gene activation occur? Why or why not?

3.86 **Critical Thinking Question:** Why does a decrease in circulating estrogen contribute to osteoporosis?

Answers to Practice Exercises

3.1

a.

$$\begin{array}{c} :\ddot{Cl}: \\ | \\ :\ddot{Cl}-C-\ddot{Cl}: \\ | \\ :\ddot{Cl}: \end{array}$$

; tetrahedral electron geometry; 109.5° bond angles.

b. H—C≡N:; linear electron geometry; 180° bond angle.

c.

$$\begin{array}{c} :\ddot{F}: \\ | \\ B \\ :\ddot{F} \qquad \ddot{F}: \end{array}$$

, trigonal planar electron geometry, 120° bond angles.

3.2

$$\begin{array}{c} :\ddot{S}: \\ / \quad \backslash \\ H \qquad H \end{array}$$

; tetrahedral electron geometry; bent molecular shape; ~109.5° bond angle; the molecule is not linear because the nonbonding pairs of electrons demand as much space as bonding electrons (actually, demand more space), so all four groups of electrons around sulfur need to be as far apart as possible, which VSEPR predicts will be a tetrahedral arrangement of the electrons around sulfur.

3.3

a.

$$\begin{array}{c} H \\ | \\ :\ddot{Cl}-C-H \\ | \\ :\ddot{Cl}: \end{array}$$

; tetrahedral electron geometry; tetrahedral molecular shape; ~109.5° bond angles.

b. :F̈—Ö—F̈:; tetrahedral electron geometry; bent molecular shape; ~109.5° bond angles.

c.

$$\begin{array}{c} :\ddot{Cl}: \\ | \\ :\ddot{Cl}-Si-\ddot{Cl}: \\ | \\ :\ddot{Cl}: \end{array}$$

; tetrahedral electron geometry; tetrahedral molecular shape; 109.5° bond angles.

d.

$$\begin{array}{c} :\ddot{F}-\ddot{P}-\ddot{F}: \\ | \\ :\ddot{F}: \end{array}$$

; tetrahedral electron geometry; trigonal pyramidal molecular shape; ~109.5° bond angles.

3.4 The compounds in Exercise 3.3 all have four groups of electrons around the central atom: a tetrahedral electron geometry.

3.5

$$\left[\begin{array}{c} H-\ddot{O}-H \\ | \\ H \end{array} \right]^{+}$$

; H_3O^+; tetrahedral electron geometry; ~109.5° bond angles; trigonal pyramidal shape; this is indeed the shape shown in ball and stick model.

3.6 H—C≡C—H; the molecular shape around each carbon atom is linear, because each carbon atom is surrounded by two groups of electrons. All bond angles (H—C—C and C—C—H) are 180°, so the overall shape of the molecule is linear.

3.7

a. Nitrogen, because it is to the right of carbon in the same period on the periodic table.

b. Nitrogen, because it is above phosphorus and in the same group on the periodic table, which places the valence electrons closer to the nucleus.

c. Chlorine, because it is a nonmetal and much farther to the right in the periodic table.

d. Nitrogen, because it is to the right and above silicon, making it more electronegative on both counts (group and period).

3.8 Metalloids are more electronegative than metals, because they are farther to the right on the periodic table than metals.

3.9

a. $\overleftarrow{\text{O—H}}$

b. $\overrightarrow{\text{C—O}}$

c. nonpolar

3.10

a. Nonpolar, because both atoms (Cl) are identical.

b. Nonpolar, because it is a hydrocarbon; carbon and hydrogen have very similar electronegativities.

c. Polar

d. Polar

e. Nonpolar. All the C—Cl bond dipoles cancel, since the molecule is tetrahedral with four identical C—Cl bonds. The C—Cl bond dipoles are directed to the four corners of a tetrahedron, making the center of negative charge coincide with the center of positive charge (located on the carbon atom).

3.11

a. Trigonal planar electron geometry; trigonal planar molecular shape; ~120° bond angles

b. The molecule is flat—2 dimensional.

c. Polar.

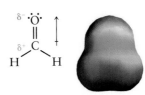

d. Red represents partial negative charge; blue represents partial positive charge

3.12 a. The green color represents nonpolar regions of the molecule. Since there is a lot of green color, it indicates estradiol is a relatively nonpolar molecule.

b. As positioned in the drawing, the OH groups are on the far left and far right side of the molecule, where the portion of red and blue appear after the large region of green. The oxygen atom contains a partial negative charge, indicated in red in an electron density diagram, and the hydrogen atom, a partial positive charge, indicated in blue in an electron density diagram.

3.13 b. Dispersion

3.14 Dispersion forces are the only forces of attraction in nonpolar molecules. Hydrocarbons, because they lack bond dipoles; elements that exist as molecules, because the atoms are identical; and molecules whose shapes cause their bond dipoles to cancel.

3.15 A permanent dipole exists in molecules at all times due to a permanent separation of charge, whereas a temporary dipole is induced and changes often.

3.16 Acetone due to the C=O polar covalent bond.

3.17 a. Hydrogen bonding in water, because the O—H dipole is stronger than the N—H dipole, because oxygen is more electronegative than nitrogen, and hydrogen is the same in both.

b. Dipole–dipole forces in second structure, because it contains the C=O bond, which is polar, because oxygen is more electronegative than carbon. The other structure only has dispersion forces, which are weaker.

c. Hydrogen bonding in H—F because fluorine is more electronegative than Cl and H is the same in both molecules. Dipole-dipole forces exist in HCl.

d. HF has hydrogen bonding, whereas F_2 only has dispersion forces.

3.18 DNA, deoxyribonucleic acid, is a biomolecule found in the nucleus of most cells, and it contains all of an organism's genetic information.

3.19 a. Three hydrogen bonds

b.

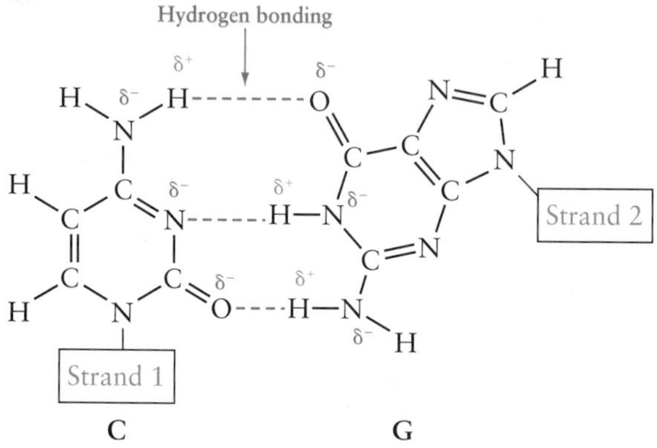

Hydrogen bonding

Strand 2

Strand 1

C   G

Solids, Liquids, and Gases

Matter exists in three different states—solid, liquid, and gas. Water exists in all three forms on the surface of the earth, as ice, liquid water in the sea, and water vapor in the air. [Michele Falzone/Getty Images]

OUTLINE

This icon indicates that a **Problem-Solving Tutorial** is available at www.whfreeman.com/gob

Scuba diving is a fun and safe activity when standard safety measures are followed. [Courtesy of Passions of Paradise, Australia]

Diver Rescue: "The Bends"

Every year thousands of people dive 60 feet or more beneath the surface of the sea, enticed by coral reefs, exotic fish, and famous shipwrecks. Although the majority of scuba divers are well-trained and avoid safety blunders, about 5 in every 100,000 divers experiences an accident. Consider the following scenario. On a diving trip, your companion surfaces looking confused and weak, complaining of a headache, itching, and joint pains. The diving instructor immediately radios for help, and a helicopter arrives, transporting your companion to the nearest medical facility for treatment. What happened? The short answer to this question is that your companion probably spent too much time at a certain depth, and ascended too quickly, causing a condition known as decompression sickness ("the bends").

The bends occurs when there is a sudden and significant change in external pressure on an individual. On the surface of the earth, the weight of the air above you creates a pressure on your body. At sea level, the measured value of that pressure is one atmosphere (atm). For every 30 feet you descend beneath the surface of the ocean, you experience an additional one atmosphere of pressure, as a result of the weight of the air and seawater above you. Air is a gas composed of 21% oxygen, 79% nitrogen, and small amounts of other gases. At increased pressures, a diver breathes in a greater number of nitrogen and oxygen molecules than at sea level.

What happens to the extra nitrogen and oxygen molecules that enter the diver's body? These extra molecules dissolve in the diver's blood. As a diver ascends to the surface, the pressure on the diver decreases, and the oxygen and nitrogen begin to diffuse out of the blood. If a diver ascends too quickly, the nitrogen in the blood will emerge too quickly, forming bubbles in the bloodstream. Think of how bubbles form in a bottle of a carbonated beverage when you twist open the cap. The unopened bottle is under pressure; when you twist open the cap, the pressure on the beverage suddenly decreases and gas molecules diffuse out and form bubbles. The bubbles formed in the diver's bloodstream can clog small blood vessels, disrupt circulation, and cause severe pain when they expand within the closed spaces of joints—all symptoms of the bends.

An important part of scuba training is learning how long you can stay at a particular depth without having to worry about ascending too rapidly. In the event that a case of the bends does occur, there is a treatment. See *Chemistry in Medicine: Hyperbaric Oxygen Therapy (HBOT)* at the end of this chapter to learn how the bends is treated. In this chapter, you will learn about solids, liquids, and gases; how energy and matter are closely linked; and the unique characteristics of gases—thus gaining a more in-depth understanding of why scuba diving can affect the body the way it does.

In the previous chapter you learned that intermolecular forces of attraction between molecules play a key role in determining their physical properties, such as whether they exist as a gas or a liquid at room temperature. In this chapter you will learn more about the different states of matter. All matter exists in one of three **physical states** or **phases**: solid, liquid, or gas. Examples of all three states are found in the body. The oxygen you breathe is in the gas state, making it possible to quickly fill the lungs with each breath you take. Blood is in the liquid state, so it can be pumped throughout the circulatory system, transporting important nutrients to cells. Skin and bone are in the solid state, providing structural integrity to the body.

In this chapter, you will examine the factors that determine the physical state of a substance, and you will be introduced to the central role energy plays in determining the physical state of a molecule. Since gases have unique but important characteristics, they are treated in detail in the last section of the chapter. By the end of this chapter, you will understand applications of this topic, such as why a steam burn causes more damage to skin than boiling water, how gases in an anesthetic affect patient recovery time, and why hyperbaric oxygen is an effective treatment for the bends.

The gasoline in this container exhibits potential energy by virtue of its composition. [Ron Chapple/Corbis]

4.1 | States of Matter

Consider first the macroscopic differences between solids, liquids, and gases. The three states of matter are usually distinguished in terms of shape and volume as illustrated in Figure 4-1:

- A **solid** has a definite *shape* and *volume*, which is independent of the shape of its container.
- A **liquid** occupies a definite *volume*, but does not have a definite *shape*; it conforms to the shape of its container.
- A **gas** has neither a definite *shape* nor a definite *volume*. A gas expands to fill its entire container.

What makes a gas take on the shape of its container while a solid has a fixed volume and shape? To answer these questions, it helps to imagine what is happening to the individual molecules at the atomic level. To understand the behavior of molecules in the solid, liquid, and gas state, we must first examine how energy and matter are intimately.

The balanced rock exhibits potential energy by virtue of its position. [Justin McCarthy]

Kinetic and Potential Energy

Central to the behavior of all molecules is energy, one of the most important concepts in science. **Energy,** *broadly defined, is the capacity to do work, where* **work** *is the act of moving an object.* Energy affects the physical states of matter, as well as physical and chemical *changes* of matter.

There are two fundamental forms of energy: kinetic energy and potential energy. **Kinetic energy** is the energy of *motion*, the energy a substance possesses as a result of the motion of its molecules or atoms. A moving car or a falling rock, for example, has kinetic energy. The kinetic energy of an object depends on both its mass, *m*, and its velocity, *v* (speed). Thus, faster moving molecules have greater kinetic energy than slower moving molecules, and heavier molecules have greater kinetic energy than lighter molecules.

Kinetic energy (KE)
Energy of motion

$$KE = \frac{1}{2}mv^2$$

A solid has a definite shape and volume.

A liquid occupies a definite volume but conforms to the shape of its container.

A gas conforms to the shape and volume of its container.

Figure 4-1
Macroscopic differences between a solid, a liquid, and a gas.

Potential energy
Stored energy of position, composition, or condition.

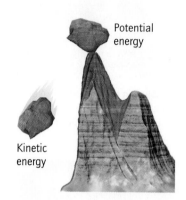

Potential energy

Kinetic energy

Figure 4-2 A rock on a precipice has potential energy, which is converted to kinetic energy when it falls.

Potential energy is stored energy, the energy a substance possesses as a result of the position, composition, and condition of its atoms. A rock poised at the top of a precipice, for example, possesses potential energy as a result of its *position*. When the rock falls, its potential energy is converted into kinetic energy (Figure 4-2). A gallon of gasoline has potential energy as a result of its composition—the chemical bonds in the molecules in gasoline. When gasoline is burned in a car engine, the potential energy in its chemical bonds is released as kinetic energy—used to perform the work of moving your car.

Heat is kinetic energy that is transferred from one object to another due to a difference in temperature. *Heat energy always flows from a hot object to a cold object.* For example, the reason you perceive ice to be cold is because heat from your body is transferred to the ice when you touch the molecules of ice, which are at a lower temperature than your body. Conversely, steam feels hot, because heat energy is transferred from steam molecules to your body.

It is important to note that temperature is *not* heat. While heat is a form of energy, **temperature** is a *measure* of the average kinetic energy of the molecules, ions, or atoms that make up a substance. Objects whose molecules have a higher average kinetic energy have a higher temperature than objects whose molecules have a lower average kinetic energy. As molecules move faster their kinetic energy increases, and the temperature of the substance increases. Rub your hands together quickly and vigorously. Do you feel them getting warmer? Your hands become warmer because the molecules on the top layer of skin are moving faster, causing them to have greater kinetic energy and a higher temperature. You detect the higher temperature as warmth because heat is being transferred to your hands.

Temperature Scales

Temperature can be measured using a thermometer and can be reported in one of three temperature scales: Celsius, Fahrenheit, and Kelvin. The Fahrenheit scale is used primarily in the United States. In science and medicine, the Celsius and Kelvin scales are the preferred temperature scales. The Kelvin scale is important in many applications in science because it represents an **absolute** scale, one that assigns 0 K to the temperature when all molecular motion has stopped. Although absolute zero has not been achieved, temperatures as low as 0.1 K have been reached. In contrast, the Celsius scale is a *relative* scale based on the freezing and boiling points of water, which are assigned the values 0 °C and 100 °C, respectively. The Celsius scale is divided into 100-degree increments between 0 °C and 100 °C. The Fahrenheit scale is also relative, and the corresponding freezing and boiling points of water are 32 °F and 212 °F. Figure 4-3 illustrates the differences between the three temperature scales.

The three units of temperature are listed in Table 4-1 along with the equations used to convert between temperature scales. Equations, not simply conversion factors, are required because the size of a degree Celsius is larger than the size of a degree Fahrenheit, and the two scales are offset from one another by 32 °F. On the other hand, a degree Celsius is the same size as a kelvin, but the scales are offset by 273 degrees. To convert from °F to K, it is easiest if you first convert from °F to °C and then from °C to K—a process requiring two steps.

A Kinetic Molecular View of the States of Matter

In Chapter 3 you learned that intermolecular forces of attraction draw molecules together. Intermolecular forces are a form of potential energy. Kinetic

Table 4-1 Temperature Unit Conversions

Temperature Units

Units (degrees)	Abbreviation
Celsius	°C
Fahrenheit	°F
kelvin*	K

Conversions

From °C to °F	$°F = \left(\dfrac{9}{5} \times °C\right) + 32$
From °F to °C	$°C = \dfrac{5}{9}(°F - 32)$
From °C to K	$K = °C + 273$
From K to °C	$°C = K - 273$

* Note that a degree sign (°) is not used in the Kelvin scale because it is an absolute scale.

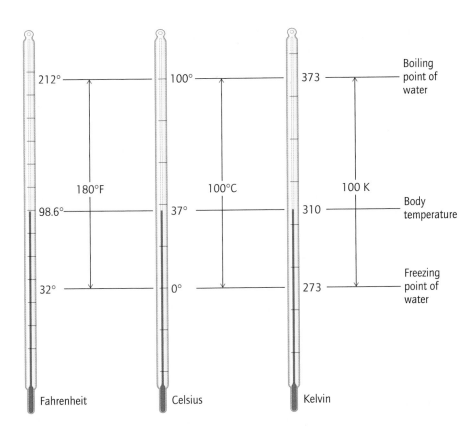

Figure 4-3 Comparison of Fahrenheit, Celsius, and Kelvin temperature scales.

energy, on the other hand, tends to pull molecules apart. These two opposing forces together determine the physical state of a substance: whether it is solid, liquid, or gas at a given temperature. The physical differences between the three states of matter can, therefore, be explained by what is known as the **kinetic molecular view** (Figure 4-4).

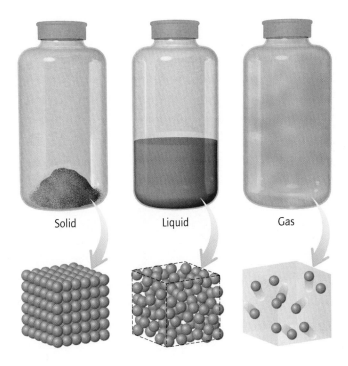

Figure 4-4 The macroscopic properties of solids, liquids, and gases correspond to differences at the atomic level.

In the *gas* phase, atoms or molecules have the highest kinetic energy. Atoms and molecules in the gas phase are moving faster than when they are in the liquid or solid phase and they are much farther apart. Since atoms or molecules are so widely spaced in the gas phase, intermolecular forces of attraction are practically nonexistent. In the gas state, kinetic forces dominate intermolecular forces of attraction.

In the *liquid* phase, these same atoms or molecules are much closer together, moving randomly and tumbling over one another. This is why liquids flow when poured. Intermolecular forces draw the particles together, and these forces are particularly strong when hydrogen-bonding forces of attraction are present. For example, water is a liquid at room temperature, because intermolecular forces of attraction dominate kinetic forces. The composition of a substance determines its polarity and influences what state it is in at a given temperature.

In the *solid* phase, atoms or molecules exist in a regular ordered pattern, much like the lattice structures you saw in Chapter 2, with intermolecular forces of attraction between atoms or molecules. Each H_2O molecule in ice (solid water), for example, is hydrogen bonded to four other water molecules in a rigid lattice structure, as shown in Figure 4-5. Molecules in the solid phase have less kinetic energy than in the liquid phase, so they remain in a fixed position with mainly vibrational motions. Intermolecular forces dominate kinetic forces in the solid phase.

Density of Ice Usually molecules or atoms in the solid phase are more closely packed than in the liquid phase; however, water is an exception. Hydrogen bonding between water molecules in the solid phase creates a rigid lattice structure that positions the molecules farther apart than molecules in the liquid phase. For this reason, ice is less dense than liquid water, causing it to float. All substances have the lowest density when they are in the gas phase, because they occupy a much greater volume than when in either the liquid or solid phase.

A water molecule

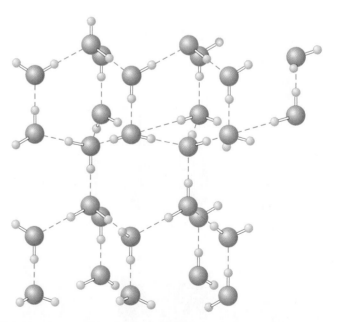

Figure 4-5 Lattice structure of ice showing hydrogen bonding as dashed lines.

Problem-Solving Tutorial:
Temperature Conversions

WORKED EXERCISES 4-1 **Temperature, Energy, and the States of Matter**

1 In which of the following states is the speed of molecules the slowest?
a. solid   **b.** liquid   **c.** gas

2 In which of the following states are there essentially no intermolecular forces of attraction between molecules? Explain why not.
a. solid   **b.** liquid   **c.** gas

3 While traveling in Europe, you feel ill and visit a health clinic. A nurse tells you that your temperature is 38.3 °C. Do you have a fever?

4 It is a comfortable summer day on the beach in southern California; the air temperature is 72 °F. Your friend who is visiting from Europe would like to know what 72 °F corresponds to in °C. What do you tell her? What is the Kelvin temperature on this summer day?

SOLUTIONS

1 (a) Molecules in the solid phase have the least kinetic energy, and therefore have the slowest speed.

2 (c) Few if any intermolecular forces of attraction exist in the gas phase, because molecules are too far apart to interact.

3 Yes, you have a fever of 101 °F. Locate the equation from Table 4-1 that shows °F isolated on one side of the equals sign:

$$°F = \left(\frac{9}{5} \times °C\right) + 32$$

Substitute 38.3 for the value of °C in the equation:

$$°F = \left(\frac{9}{5} \times 38.3\ °C\right) + 32 = 101\ °F$$

Always perform any mathematical operation enclosed in parenthesis before the other operations.

4 Convert °F to °C by using the equation from Table 4-1 that has °C isolated on one side of the equals sign:

$$°C = \frac{5}{9}\left(°F - 32\right)$$

Substitute 72 for °F in the equation and solve for °C:

$$°C = \frac{5}{9}\left(72 - 32\right) = 22\ °C$$

A temperature of 72 °F is equal to a temperature of 22 °C.
To find this temperature in kelvin, locate the equation from Table 4-1 used to convert °C to K:

$$K = °C + 273$$

Substitute 22 for °C and solve for K:

$$K = 22 + 273 = 295\ K$$

Thus, 72 °F = 22 °C = 295 K. All these values represent the same temperature; only the units are different.

PRACTICE EXERCISES

4.1 Describe the macroscopic differences between the solid, liquid, and gas states by comparing their shape and volume to the container they occupy.

4.2 Normal body temperature is 37.0 °C. Convert 37.0 °C into °F and kelvin. Show your work.

4.3 Many birds have a normal body temperature of 106 °F. Convert this temperature into kelvin.

4.4 The boiling point of helium is 4 K. What is this temperature in °C and °F?

4.5 Match the following descriptions to the state that describes it: *solid, liquid,* or *gas.*

    a. Molecules in this phase have the greatest kinetic energy.
    b. Molecules in this phase fill the entire volume of their container.
    c. Molecules in this phase are the most ordered.
    d. Hydrogen bonding between water molecules is present in these two states.

4.6 Indicate whether each of the following examples is a demonstration of potential energy or kinetic energy:

    a. a compressed spring
    b. a windmill turning
    c. hydrogen bonding
    d. molecules colliding with the walls of their container

4.2 | Changes of State

We have all seen water condense on an ice-cold drink on a warm, humid day, fog emerge from a block of dry ice, and an ice cube melt. These are all macroscopic examples of **changes of state**—the process of going from one physical state to another. Changes of state are classified as *physical changes,* not *chemical changes,* because the chemical bonds in the molecules do not change. That is, covalent bonds are not formed or broken. Thus, water is still composed of H_2O molecules in all three states: ice, liquid, and steam.

Several terms are used to describe the various changes of state, summarized in Figure 4-6. Many of these terms are probably already familiar to you. A change of state from the solid to the liquid phase is known as **melting;** the reverse change from liquid to solid is known as **freezing.** A change of state from the liquid to the gas phase is known as **vaporization,** while the reverse change from gas to liquid is known as **condensation.** A solid can also undergo a change of state directly to the gas phase without first entering the liquid phase, a process known as **sublimation.** For example, dry ice (CO_2) changes directly from the solid phase to the gas phase, at room temperature. The reverse process, changing from the gas phase to the solid phase, is known as **deposition.** An example of deposition is the formation of snow: water changing from a gas to a solid in the clouds.

Energy and Changes of State

We have seen that covalent bonds remain intact during a change of state. However, **inter**molecular forces of attraction, which are much weaker than covalent forces (see Chapter 3), *are* broken or formed during a change of state, particularly when the substance is changing to or from the gas phase.

To effect a change of state, energy must be added to or removed from a substance. Energy must be *added* to achieve *melting, vaporization,* and *sublimation,* and energy must be *removed* to achieve *freezing, condensation,* and *deposition.* For example, to convert a pot of liquid water into steam you must *add* energy to the water, perhaps using your stove. Heat energy from the stove is transferred to the molecules in the pot, increasing their kinetic energy. More energetic, faster-moving molecules are then able to break away from the intermolecular forces of attraction holding them together in the liquid phase, so that they can exist apart in the gas phase, as illustrated in Figure 4-7.

In a similar manner, when you step out of the shower, heat is transferred from your body to the water molecules on your skin. You know this because you feel cold. Heat from your body is being used to *evaporate* the water on your skin. Like the vaporization of water, evaporation of water occurs when liquid water changes to the gas state. Although evaporation may occur at any

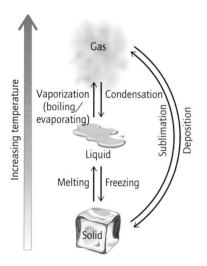

Figure 4-6 Changes of state.

Sweating during vigorous exercise helps to cool the body. [Jason Horowitz/zefa/ Corbis]

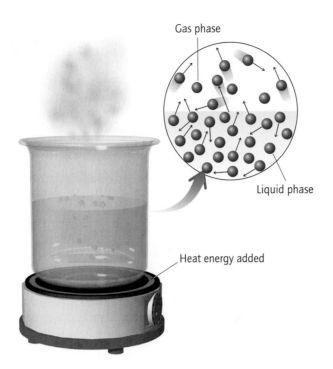

Gas phase

Liquid phase

Heat energy added

Figure 4-7 The process of vaporization in boiling water. A hot plate provides heat energy, which causes some water molecules to acquire enough kinetic energy to break away from the intermolecular forces of attraction that hold them in the liquid phase.

temperature, vaporization is defined as the change of state that occurs at the boiling point of the liquid. The cooling effect of evaporation is the reason the body produces sweat when it overheats. Thus, heat is removed from the surroundings during vaporization and evaporation.

The reverse process, condensation, requires that the *equivalent* amount of heat energy be *removed* from the gas phase. Removing energy reduces the kinetic energy of molecules, slowing their motion and allowing intermolecular forces of attraction to form as molecules come closer together to form the liquid phase. For example, if you place your hand over a pot of boiling water, the steam condenses on your much cooler skin. You can actually feel the heat transferred to your hand, as water molecules turn to liquid on your hand. Heat is transferred to the surroundings during condensation.

The **heat of vaporization** is the amount of energy that must be added to a liquid to transform it into a gas, or the amount of heat that must be removed to effect the reverse process. The *calorie* is a common unit of heat energy, defined as the amount of energy required to raise 1 g of liquid water 1 °C. For water, the heat of vaporization is 540 calories per gram (cal/g) of water. The **heat of fusion** is the amount of energy that must be added to turn a solid into a liquid. For water, the heat of fusion is 80 cal/g of water. The reverse process, freezing, requires exactly the same amount of heat to be *removed* per gram of water.

The heat of vaporization for water is a much higher value than the heat of fusion for water because the hydrogen bonds between water molecules must be broken when going from the liquid phase to the gas phase. In addition, a more significant separation of molecules is required between the liquid and gas phases than between the solid and liquid phases.

Melting and Boiling Points

A heating curve is a graph that shows how the temperature of a substance increases as heat energy is added to it. A cooling curve is a similar graph that shows the reverse process.

Consider the heating curve for water shown in Figure 4-8. As solid water (ice) is heated, its temperature increases as the kinetic energy of the water molecules increases. At a specific temperature unique for every substance, known as the **melting point,** the substance changes from the solid state to the liquid state. For

Figure 4-8 Heating curve for water. As heat is added, the temperature increases, except during a phase change, when the heat is used to effect a change of state. [Photos, left to right: David Arky/Corbis; SGM/Stock Connection; Martyn F. Chillmaid/ Photo Researchers, Inc.]

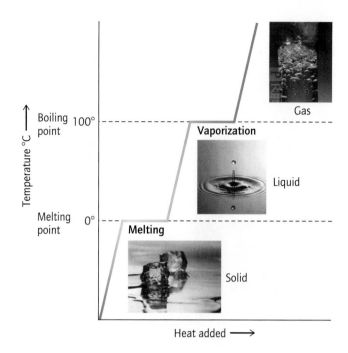

water, the melting point is 0 °C. At this temperature, heat will continue to be absorbed by the substance, but the temperature will not rise, as indicated by the lower horizontal line on the heating curve. Rather than increase the kinetic energy of the molecules, the added heat energy is used to break intermolecular forces of attraction—hydrogen bonding in the case of water; this is the heat of fusion.

Once the entire sample is in the liquid phase, additional heat once again causes the kinetic energy of the molecules to increase, and the temperature of the sample rises, as indicated by the second diagonal line on the heating curve. Another horizontal line in the curve occurs at the **boiling point** of the substance, when the substance changes from the liquid to the gas state. For water, the boiling point is 100 °C. As with the melting point, heat will continue to be absorbed, but the temperature will not rise until all the molecules are in the gas phase. At the temperature of the boiling point, heat energy is being used to break the remaining intermolecular forces of attraction; this is the heat of vaporization. After that, additional heat goes toward raising the temperature of the gas, as represented by the last diagonal line on the heating curve.

Each substance has a unique melting point and a unique boiling point, which depends on the strength and number of the intermolecular forces of attraction between the molecules of that substance. Substances that form hydrogen bonds require a greater amount of energy to change from the liquid to the gas phase. For example, liquid water must be heated to 100 °C to become a gas (steam). In contrast, propane (C_3H_8), commonly used as a fuel for barbecue grills, becomes a gas at a much lower temperature, −42 °C, because propane is a hydrocarbon and only dispersion forces hold the molecule together.

To understand that different substances have different capacities to absorb heat, perform the calculations in *Extension Topic 4-1: Specific Heat.*

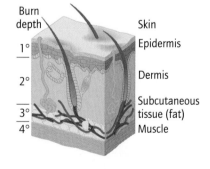

Skin burns. The severity of a burn depends on the depth of the skin tissue damage, classified as first degree through fourth degree. A first-degree burn affects the top layer of skin (epidermis) only. A second-degree burn affects the epidermis and some or all of the dermis. A third-degree burn extends into fat tissue, and a fourth-degree burn extends into the muscle. Steam burns cause more damage than burns from boiling water.

Steam Burns If boiling water and steam are at a temperature of 100 °C, why is a steam burn so much more severe than a burn from boiling water? It isn't that steam is at a higher temperature but rather that the amount of heat released as a result of the phase change upon contact with the cooler skin. Steam condenses when it comes into contact with the skin, a phase change that requires that heat be removed from the steam by an amount equivalent to the heat of vaporization of water. For 10 g of water, the phase change alone transfers 5,400 cal to the

Extension Topic 4-1 Specific Heat

In this chapter, you have learned that when heat is added to a substance, the temperature of that substance increases. **Specific heat** is the amount of heat required to raise the temperature of 1 g of a substance by 1 °C. The specific heat of a substance is a physical property and varies from one compound or element to another. It also depends on the state of matter. The specific heat of a substance can be calculated using the following equation:

$$\text{Specific heat} = \frac{\text{heat}}{\text{mass} \times \text{change in temperature } (\Delta T)}$$

Heat refers to the amount of heat energy transferred to the substance, and the change in temperature is the difference in the temperature before and after heat has been transferred. Heat is usually measured in calories and mass is measured in grams. Specific heat is commonly reported in units of cal/(g · °C). This equation can be used to calculate any one of the four variables, if the other three are known.

A substance with a large specific heat requires a greater input of heat energy for any increase in temperature. For example, consider walking on the beach. On a hot day, the sand under your feet feels very warm, so you step into the water, which is cool. At the same surrounding temperature, the sand feels much warmer than the water. Why is this so? The same amount of heat causes a much smaller temperature increase in the water than in the sand, as reflected in their different specific heats: 1.00 cal/g °C for water and 0.16 cal/g °C for sand. In fact, liquid water has one of the highest specific

heats of any substance as a result of hydrogen bonding. The large specific heat of water is important biologically because it allows the human body to maintain a steady temperature rather than varying with the temperature of the ambient air.

Table 4-2 lists the specific heat of some common substances.

WORKED EXERCISE E4.1   Calculations with Specific Heat

Using Table 4-2, calculate the number of calories that must be added to warm 22.1 g of each of the substances below from 25 °C to 42 °C. Which substance requires the most heat to raise its temperature?

a. water   **b.** paraffin wax   **c.** copper

SOLUTION

Rearrange the equation algebraically to solve for heat. You should obtain the following equation:

$$\text{Heat} = \text{specific heat} \left(\frac{\text{cal}}{\text{g °C}} \right) \times \text{mass (g)} \times \Delta T(\text{°C})$$

First, calculate the temperature difference, ΔT:

$$\Delta T = 42° - 25° = \textbf{17 °C}$$

a. For water: Amount of heat =

$$1.00 \, \frac{\text{cal}}{\text{g°C}} \times 22.1 \, \text{g} \times 17 \, \text{°C} = \textbf{380 cal}$$

b. For paraffin wax: Amount of heat =

$$0.60 \, \frac{\text{cal}}{\text{g°C}} \times 22.1 \, \text{g} \times 17 \, \text{°C} = \textbf{230 cal}$$

c. For copper: Amount of heat =

$$0.093 \, \frac{\text{cal}}{\text{g°C}} \times 22.1 \, \text{g} \times 17 \, \text{°C} = \textbf{35 cal}$$

Water requires the greatest amount of heat to raise the temperature by 17°.

PRACTICE EXERCISES

E4.1   Use Table 4-2 to calculate the number of calories required to raise 45 g of each of the following substances from 23.0 °C to 32.0 °C.

     a. sand   **b.** water   **c.** ethanol

E4.2   An unknown piece of metal weighing 5.3 g could be iron, aluminum, or lead. In going from a temperature of 26 °C to 38 °C, the metal absorbed 14 cal of heat. Use Table 4-2 to determine the identity of the metal.

E4.3   How is the specific heat of water important for the human body?

E4.4   Two houses are identical in all aspects except that one is made of brick and the other is made of wood. Which house will be cooler on the same hot summer day? Explain.

Table 4-2 Specific Heat of Some Common Substances

Substance	Specific Heat (cal/g · °C)
Water (liquid)	1.00
Water (gas)	0.497
Water (solid)	0.490
Paraffin wax	0.60
Ambient air	0.24
Ethanol	0.58
Brick	0.20
Sand	0.16
Wood	0.10
Copper	0.093
Iron	0.11
Aluminum	0.22
Lead	0.031

skin. This energy must be removed in order to form hydrogen bonds between liquid water molecules. Additional heat (630 cal) is transferred to the skin as the liquid cools from 100 °C to 37 °C (body temperature). In contrast, for 10 g of boiling liquid water, the heat removed corresponds only to 630 cal.

The transfer of heat from a colder to a hotter object is the principle behind the use of hot packs and cold packs. Cold packs are applied to injuries where heat must be removed from the tissue. Conversely, hot packs are applied where heat must be added to tissue.

PRACTICE EXERCISES

4.7 Identify the term that describes the changes of state listed below:
  a. gas → solid
  b. solid → liquid
  c. liquid → gas

4.8 Which of the phase changes in the question above involve(s) the transfer of heat energy to the surroundings?

4.9 What change of state occurs during the process of sublimation?

4.10 What change of state is the reverse of vaporization?

4.11 Explain why a steam burn at 100 °C causes greater damage to the skin than boiling water at 100 °C.

4.12 **Critical Thinking Question:** A microwave oven transfers energy to water molecules through a form of energy known as electromagnetic radiation. Microwaves are a type of electromagnetic radiation, similar to visible and ultraviolet light. Explain what must be happening to the water molecules to cause the temperature of the water, and therefore the temperature of the food, to increase.

4.13 **Critical Thinking Question:** When an infant has an extremely high fever, it is possible to reduce the fever by placing the infant in a bathtub of tepid water. The water temperature of the bathtub isn't really that important. Explain how evaporation of water from the infant's skin is involved in reducing the infant's body temperature, not the temperature of the bath.

4.3 Pressure

Have you ever had the misfortune of breaking a bone? If so, you know first hand the pain that excessive *pressure* can cause. When a bone breaks, a force applied to a given area of the bone exceeds the strength of the bone. **Pressure,** *P*, is a measure of the amount of force applied over a given area:

$$\text{Pressure } (P) = \frac{\text{force}}{\text{area}}$$

Pressure is an important factor influencing the states of matter, in particular the gas state. In this section you will learn about the units of pressure, which are used to report blood pressure and vapor pressure. Once you learn how pressure is measured and what information it provides, you will be able to see how it helps explain the unique properties of gases.

Pressure Units

Like other physical properties, pressure can be measured and may be reported in a variety of units. Some of these units are given in Table 4-3, along with their conversions to atmospheric pressure. The most common unit of pressure in medicine is millimeters of mercury, mmHg. This is the unit used for reporting blood pressure.

Molecules of air in the atmosphere press down on us as a result of gravity. This type of pressure is known as **atmospheric pressure** and represents the force exerted by the weight of air at any given place (area) on the earth (Figure 4-9).

Table 4-3 Some Common Units of Pressure and Their Conversions

Pressure Units	
Unit	Abbreviation
Atmospheres	atm
Torr	torr
Millimeters of mercury	mmHg
Pounds per square inch	psi or lb/in.^{2}
Pascal	Pa

Conversions
1 atm = 760 mmHg (exact)
1 atm = 760 torr (exact)
1 atm = 14.70 psi
1 atm = 1.013×10^5 Pa

Atmospheric pressure is commonly measured in units of *atmospheres* using a device called a *barometer*. At sea level, the atmospheric pressure due to the column of air above you is one atmosphere (1 atm). As you climb to higher altitudes, the atmospheric pressure is lower because there is less air above you. For example, at 14,000 ft, the atmospheric pressure is only 0.59 atm (Figure 4-10). Hikers often experience difficulty breathing at these altitudes, because the lower air pressure makes it harder for oxygen to reach the lungs and the brain. Sudden external changes in pressure on the human body can lead to a dangerous medical condition called decompression sickness. Some symptoms of decompression sickness include localized deep pain, dizziness, nausea, and shortness of breath. Deep-sea diving causes divers to experience significant *increases* in pressure, which can under certain circumstances lead to decompression sickness, the "bends." See this chapter's opening vignette for more information about the bends. Other situations that may lead to decompression sickness include:

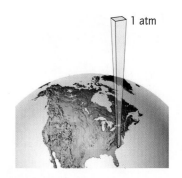

Figure 4-9 At sea level, a column of air above you applies a force per unit area that is equal to one atmosphere (1 atm).

- A worker coming out of a region that has been pressurized (a caisson or mine) to keep water out (caisson disease).
- An unpressurized aircraft increasing in altitude.
- The failure of the cabin pressurization system of an aircraft.
- A deep-sea diver flying in an aircraft shortly after diving.
- An astronaut exiting a spacecraft when his or her suit is not pressurized properly.

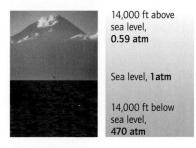

Figure 4-10 Pressure changes with distance above and below sea level.
[Colin Watmough/aliki image library/Alamy]

WORKED EXERCISE 4-2 Converting Between Units of Pressure

At an altitude of 14,000 ft, the atmospheric pressure is 0.59 atm. What is this pressure equivalent to in psi (lb/in.^{2})?

SOLUTION

Step 1: Identify the conversion. Use Table 4-2 to find the conversion between atm and psi: 1 atm = 14.70 psi.

Step 2: Express the conversion as a conversion factor.

$$\frac{14.70 \text{ psi}}{1 \text{ atm}} \quad \text{or} \quad \frac{1 \text{ atm}}{14.70 \text{ psi}}$$

Step 3: Set up the calculation so that the supplied units cancel. Use the conversion factor that allows atm units to cancel in order to obtain an answer in the requested unit, psi:

$$0.59 \text{ atm} \times \frac{14.70 \text{ psi}}{1 \text{ atm}} = \textbf{8.7 psi}$$

PRACTICE EXERCISES

4.14 Intraocular pressure is the pressure inside the eye. A typical intraocular pressure is 15 mmHg. What is the pressure in units of atmospheres?

4.15 What is the name of the device used to measure atmospheric pressure?

4.16 Does atmospheric pressure increase or decrease at higher altitudes? Explain.

4.17 Critical Thinking Question: When a golfer hits a shot off the tee in Denver it travels farther than it would in Florida. Explain this observation.

4.18 Critical Thinking Question: Tennis balls manufactured for use in communities at high altitude, such as Denver, are designed with less bounce. Why are different tennis balls made for communities at high altitude?

Problem-Solving Tutorial:
Pressure Conversions

Blood Pressure

Blood pressure is the pressure exerted by blood on the walls of blood vessels. Blood pressure is an important indication of an individual's overall health. As you know, the heart acts like a pump, contracting and relaxing, causing blood to

A patient getting her blood pressure measured with a sphygmomanometer.

flow throughout the circulatory system. When the heart muscle contracts, blood pressure against the arteries reaches a maximum as blood is forced through the arteries. This blood pressure maximum is known as the **systolic pressure.** When the heart muscle relaxes, blood pressure drops and the minimum blood pressure, known as the **diastolic pressure,** is reached. The normal range for systolic pressure is 90–130 mmHg, and the normal range for diastolic pressure is 60–80 mmHg. A typical blood pressure reading might be 111/86, read as "one-eleven over eighty-six."

A device called a *sphygmomanometer* is used to measure blood pressure. You undoubtedly are familiar with the cuff of the sphygmomanometer that is used with a stethoscope. In addition, the sphygmomanometer also contains a tube of mercury, the "manometer" of the sphygmo*manometer*, that measures pressure. The cuff is wrapped around the upper arm, and air inflates it until it cuts off the flow of blood through the brachial artery. The pressure of the cuff is then gradually reduced, while the stethoscope is used to listen for the sound of blood as it first begins to flow again. When the sound first appears, the pressure reading equals the systolic pressure. As the cuff deflates further, blood begins to flow freely through the artery, and eventually no sound can be heard. When the sound vanishes, the pressure reading equals the diastolic pressure.

Elevated blood pressure is a condition known as **hypertension.** One in three adults has hypertension. Untreated hypertension causes strain on the heart and may lead to a heart attack. Table 4-4 lists the range of blood pressures for different conditions ranging from hypotension (low blood pressure) to hypertension.

Table 4-4 Ranges of Blood Pressure

Condition	Systolic (mmHg)	Diastolic (mmHg)
Hypotension (low blood pressure)	<90	<50
Normal	90–130	50–90
Prehypertension	130–140	90–100
Hypertension (high blood pressure)	>140	>100

WORKED EXERCISE 4-3 Blood Pressure

Chronic hypertension (high blood pressure) is a serious disease that can damage blood vessels and cause hardening of the arteries. Suppose a patient has her blood pressure taken and the reading is 155/92.

a. Does the patient have a normal blood pressure? If not, what is her condition?
b. What are the units in this blood pressure measurement?
c. Which number is the systolic pressure? Which number is the diastolic pressure?
d. Covert the systolic pressure to units of atmospheres (atm).
e. Convert the diastolic pressure to units of torr.

SOLUTION

a. 155/92 is not considered a normal blood pressure. This patient suffers from hypertension.
b. 155 mmHg/92 mmHg
c. 155 is the systolic pressure, 92 is the diastolic pressure.

d. $155 \; \cancel{\text{mmHg}} \times \dfrac{1 \; \text{atm}}{760 \; \cancel{\text{mmHg}}} = 0.204 \; \text{atm}$

e. $92 \; \cancel{\text{mmHg}} \times \dfrac{760 \; \text{torr}}{760 \; \cancel{\text{mmHg}}} = 92 \; \text{torr}$

PRACTICE EXERCISES

4.19 A typical blood pressure reading for a healthy adult is 120/80 mmHg. Convert these pressures to units of pascals (Pa).

4.20 What is a sphygmomanometer?

4.21 Define systolic and diastolic pressure.

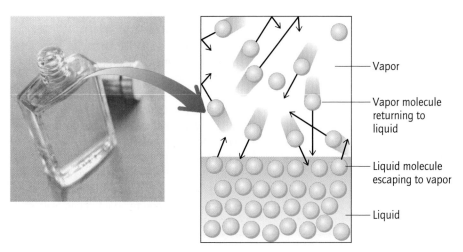

Figure 4-11 The movement of molecules between the liquid and gas phases is a result of a substance's vapor pressure. [Photo: BananaStock/Punchstock]

Vapor Pressure

While shopping, a friend passes you a bottle of cologne to sniff. Just by placing your nose above the open bottle, you are able to smell the fragrance. How is this possible, given that the cologne is a liquid and you did not actually touch any liquid to your nose? As you know, the atoms or molecules in a liquid are constantly colliding. At the surface of the liquid, these collisions cause some of the particles to gain sufficient kinetic energy to enter the gas phase, a process known as **evaporation.** Simultaneously, some of the particles in the gas phase directly above the liquid lose kinetic energy and return to the liquid phase (Figure 4-11). This movement of particles between the liquid and gas phases is a result of the vapor pressure of the liquid substance.

***Vapor pressure** is the pressure exerted by molecules in the gas phase in contact with molecules in the liquid phase.* Solids also have a vapor pressure, but it is usually much lower than the vapor pressure of liquids. Liquids with a high vapor pressure enter the gas phase more readily and are considered **volatile** liquids. The cologne you sniffed is volatile, so the liquid cologne molecules enter the gas phase readily. Since a gas expands to fill its container, the gas molecules reach your nose, where you can smell them.

The vapor pressure of a liquid increases with temperature. Figure 4-12 shows the vapor pressure of water increasing as the temperature increases. At the temperature when the vapor pressure of a substance equals the atmospheric pressure, bubbles appear in the liquid, indicating boiling. *The **boiling point** of a liquid is the temperature at which the vapor pressure of the liquid equals the atmospheric pressure.* The boiling point of a liquid at 1 atm is known as the **normal boiling point.** A substance with a *higher* vapor pressure will have a *lower* boiling point.

At sea level, atmospheric pressure is 760 mmHg and water boils at 100 °C. At higher altitudes, the atmospheric pressure is lower; hence, the boiling point of water is lower. For example, in Denver the atmospheric pressure is 560 mmHg; therefore, in Denver water boils at 92 °C rather than 100 °C.

Every substance has its own unique vapor pressure; it is one of the physical properties of a substance. It is especially important to know the vapor

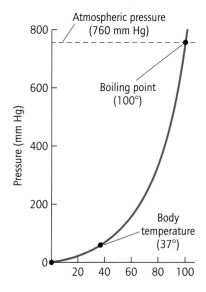

Figure 4-12 The vapor pressure of a substance increases as temperature increases.

Table 4-5 Vapor Pressures of Some Substances at 20 °C

Substance		Vapor Pressure (mmHg)	Common Uses/Notes
Acetone (C_3H_6O)	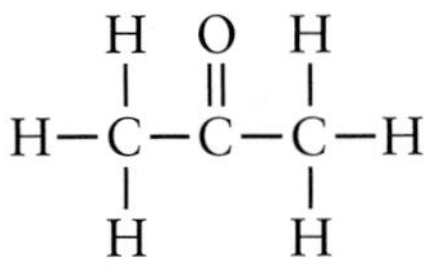 H O H | || | H—C—C—C—H | | H H	184	A common solvent used in the chemical and pharmaceutical industry; an ingredient in nail polish remover.
Mercury (Hg)		0.002	An element commonly used in dentistry, and also in thermometers and sphygmomanometers. Its vapor is toxic.
Methylene chloride (dichloromethane, CH_2Cl_2)	H | H—C—Cl | Cl	350	A common solvent used in the chemical industry, and widely used as a paint stripper and degreaser. Care must be taken when working with methylene chloride, as it is an acute inhalation hazard.
Phenol (C_6H_6O)	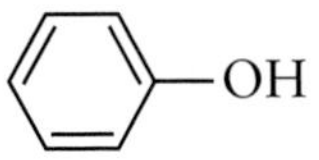 ⬡—OH	0.36	A compound commonly used in the purification of DNA. It's the active ingredient in anesthetic sprays such as Chloraseptic. Exposure of the skin to liquid phenol will cause severe burns.

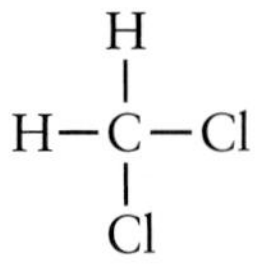

pressure of a toxic liquid, since the vapor from such a liquid can be hazardous if its concentration reaches a dangerously high level *and* people are exposed to it for a long period of time. The vapor pressures at 20 °C for some familiar commercial substances are listed in Table 4-5.

WORKED EXERCISE 4-4 Vapor Pressure

Consult Table 4-5 to answer the following questions:

a. Which has the higher vapor pressure, mercury or methylene chloride? Which is the more volatile substance?

b. Which would you expect to have a higher boiling point, phenol or acetone? Explain.

c. Convert the vapor pressure of acetone into atmospheres.

SOLUTION

a. Methylene chloride has a higher vapor pressure at 20 °C than mercury and is therefore more volatile.

b. Phenol has a higher boiling point than acetone because it has a lower vapor pressure at 20 °C.

c. $184 \text{ mmHg} \times \dfrac{1 \text{ atm}}{760 \text{ mmHg}} = 0.242 \text{ atm}$

PRACTICE EXERCISES

4.22 In what units of measurement is vapor pressure usually reported?

4.23 Why might it be important to know the vapor pressure of methylene chloride?

4.24 Define normal boiling point. Will the boiling point of a liquid be higher or lower at high altitude?

4.25 **Critical Thinking Question:** In an autoclave, water vapor is used to sterilize medical and laboratory equipment. Sterilization is a process that destroys bacteria, viruses, and other transmissable agents. One way of sterilizing something is by applying heat. The pressure in an autoclave is much greater than atmospheric pressure. Would water boil at a temperature greater than or less than 100 °C in an autoclave? Speculate on why items are sterilized so effectively in an autoclave.

4.26 How is the vapor pressure of a substance related to its boiling point?

4.27 Define vapor pressure.

4.28 The vapor pressure of water at normal body temperature (37 °C) is 47.07 torr. Convert this value to mmHg.

4.29 Skunks produce a compound with a very noxious odor; would you expect this compound to be volatile or not? To have a high or a low boiling point?

4.4 | Gases

Consider a cake baking in the oven. After a while, you notice an appetizing smell in the air. This wonderful smell is a result of odorous gas molecules from the cake diffusing throughout the air, filling the volume of your kitchen and reaching the olfactory receptors in your nose. From Section 4.1, you learned that gases are unique because they occupy the entire volume of the container they are in. Gases are also unique because they can be compressed into a much smaller volume. Because of this property, gases are more strongly influenced by temperature and pressure than are liquids or solids.

After centuries of careful observation and experimentation, a set of **gas laws** has been identified to describe the macroscopic behavior of gases in terms of four variables: pressure (P), volume (V), number of moles (n), and temperature (T). In this section, you will be introduced to several of the gas laws, but first let's consider the behavior of gas molecules.

Kinetic Molecular Theory

In Section 4.1 you learned that atoms or molecules in the gas phase have a greater kinetic energy than molecules in the liquid or solid phase. Since molecules and atoms are in constant motion, molecules and atoms in the gas phase are continuously colliding against the walls of their container, creating pressure. The pressure (P) of a gas is the force per area exerted by the *gas* particles colliding against those walls, illustrated in Figure 4-13. The more collisions with the container walls, the greater the pressure.

The characteristics of a gas can be described by the kinetic molecular view of gases as follows:

- **The particles of a gas are in constant, random motion.** Gas molecules move at high speeds in straight lines and in random directions, filling the entire volume of the container they occupy.

- **The total volume of all the gas particles in a container is negligible compared to the volume of the container.** The volume of the container consists mostly of empty space. This is why a gas is easily compressed.

- **The attractive forces among the particles of a gas are negligible.** Gas particles are far enough apart from one another that intermolecular forces of attraction between molecules are practically nonexistent.

- **The temperature of a gas depends on the average kinetic energy of the gas particles.** The faster gas particles move, the greater their kinetic energy and the higher the temperature of the gas.

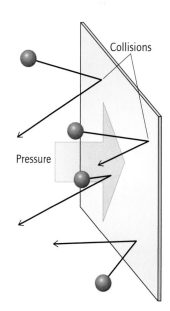

Figure 4-13 The pressure of a gas is the force per unit area exerted by the gas particles colliding with the walls of their container.

⚕ Why understanding gases is important in medicine

- Measurements of dissolved gases in blood, such as oxygen and carbon dioxide, are used to evaluate a patient's health.

- Understanding the use of oxygen tanks and respiratory therapies requires a knowledge of the special behavior of gases.

- Anesthesiology requires knowledge of the special behavior of gases.

- Understanding the threats posed by pressure changes at high altitudes (in the mountains or airplanes) or deep sea depths (scuba diving) requires a basic knowledge of gases.

WORKED EXERCISE 4-5 Kinetic Molecular Theory

Use kinetic molecular theory to explain the following:

a. The same number of oxygen molecules that would normally occupy a large volume of space in air can be compressed and contained in liquid form in an oxygen tank.
b. On a hot day, the pressure in the tires of your car increases.
c. Gasoline spilled in one part of a room is quickly detected at the other end of the room.

SOLUTION

a. Since the total volume of individual oxygen molecules is extremely small compared to the total volume of space they occupy, there is plenty of room in an oxygen tank to squeeze together the same number of oxygen molecules that normally occupy a large volume of space, turning it into a liquid.
b. On a hot day, the surrounding air has more kinetic energy, and some of that energy is transferred to the air in the tires, causing the oxygen and nitrogen molecules in the tires to move faster. The fast-moving gas particles collide more frequently against the walls of the container, resulting in a higher tire pressure.
c. Gas particles move at high speeds in all directions, filling the volume of the container they are in and causing them to eventually reach your nose.

PRACTICE EXERCISES

4.30 Identify the variable P, V, n, or T that is described by the following:
   a. Increases as the collisions of gas particles with the walls of the container increase.
   b. Increases as the kinetic energy of the gas particles increases.
   c. The space occupied by a gas.

4.31 **Critical Thinking Question:** Why might a balloon burst on a hot day?

STP and the Molar Volume of a Gas

To help make comparisons between different gases, the properties of a gas are described under a standard set of reference conditions. In chemistry, these conditions are known as **STP: S**tandard **T**emperature and **P**ressure.

   *Under the conditions of STP, one mole of any gas occupies a volume of 22.4 L, known as the **molar volume** of a gas* (Figure 4-14). The identity or mass of the gas does not matter; for example, one mole of helium gas and one mole of oxygen gas both have a volume of 22.4 L at STP. At STP, molar volume can be used as a conversion between the number of moles of a gas (n) and its volume (V):

$$22.4 \text{ L} = 1 \text{ mol at STP}$$

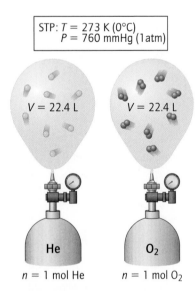

STP: T = 273 K (0°C)
      P = 760 mmHg (1 atm)

V = 22.4 L V = 22.4 L

He O_2

n = 1 mol He n = 1 mol O_2

Figure 4-14 At STP, one mole of any gas occupies a volume of 22.4 L. Thus, one mole of helium occupies the same volume as one mole of oxygen, 22.4 L.

WORKED EXERCISE 4-6 Performing Volume–Mole Calculations Using STP

What volume will be occupied by 3.2 mol of nitrogen at STP? Would the volume be different if the gas were argon?

SOLUTION

This exercise can be solved using dimensional analysis.

Step 1: Express the conversion as conversion factors. Since 22.4 L = 1 mol at STP,

$$\frac{22.4 \text{ L}}{1 \text{ mol}} \qquad \text{or} \qquad \frac{1 \text{ mol}}{22.4 \text{ L}}$$

Step 2: Set up the calculation so that the supplied units cancel.

$$3.2 \ \cancel{mol} \times \frac{22.4 \ L}{1 \ \cancel{mol} \ (STP)} = 72 \ L$$

Therefore, 3.2 mol of nitrogen at STP would occupy a volume of 72 L. The same number of moles of argon would also occupy 72 L, because the identity of the gas is not important.

PRACTICE EXERCISE

4.32 For each of the following, given the volume (V) or number of moles (n) of a gas at STP, calculate the missing variable, number of moles (n) or volume (V).

   a. Volume of xenon at STP = 34.9 L. n = ?
   b. Number of moles of chlorine at STP = 4.5 mol. V = ?
   c. Volume of a mixture of oxygen and nitrogen = 151 L. n = ?

Pressure–Volume Relationship of Gases

Suppose you took a syringe without the needle attached, depressed the plunger halfway, and covered the tip with your finger. The number of moles of gas, n, within the syringe is constant, and so is the temperature, T. You then depress the plunger slowly and steadily. You are compressing the gas within the syringe, causing the same number of moles of gas to occupy a smaller volume. What does this feel like on the finger depressing the plunger? The farther you depress the plunger, the harder it becomes to depress, because the pressure of the gas has increased.

Why has the pressure increased as the volume was decreased? Based on kinetic molecular theory, as the volume of the syringe decreases, there is less wall space for the gas molecules to collide with. Therefore, collisions are more frequent because the same number of gas molecules are colliding with less wall space, creating increased pressure. This demonstration shows that as the volume of a gas decreases, the pressure of the gas increases, and vice-versa, given constant T and n (Figure 4-15). In other words, pressure and volume are *inversely* related to one another—as one goes up, the other goes down. This relationship is known as **Boyle's law** and can be written as

$$P_1 V_1 = P_2 V_2 \ (n \ \text{and} \ T \ \text{are constant})$$

where the subscripts 1 and 2 refer to the initial and final values for P and V.

WORKED EXERCISES 4-7 The Pressure–Volume Relationship of Gases

1 An air bubble forms at the bottom of a lake, where the total pressure is 2.52 atm. At this pressure, the bubble has a volume of 3.31 mL. When the bubble rises to the surface, where the pressure is 1.00 atm, will the bubble have a larger volume or a smaller volume? Explain. Assume that the temperature and the amount of gas (n) within the bubble remain constant.

2 A nitrogen bubble forms in the left knee joint of a deep-sea diver as she ascends. At a depth of 52 ft the pressure is 3.15 atm and the bubble has a volume of 0.015 mL. Assuming constant temperature and constant number of moles of nitrogen in the bubble, what will be the volume of this bubble at the surface of the sea, where the pressure is 1.00 atm?

SOLUTIONS

1 The volume of the bubble will be larger after it rises to the surface because the pressure decreases from 2.52 atm at the bottom of the lake to 1.00 atm at the surface of the lake. Because volume and pressure are inversely related, the volume of the bubble will increase as the pressure decreases.

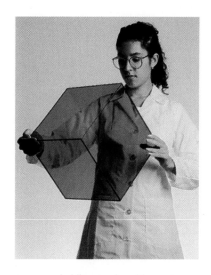

A woman holding a cube with a volume of 22.4 L—the molar volume occupied by a gas at STP. [W.H. Freeman. Photo by Ken Karp]

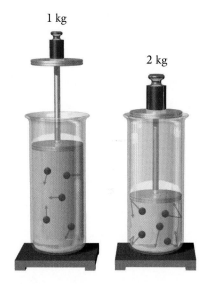

Figure 4-15 Boyle's law: As volume decreases, pressure increases.

Problem-Solving Tutorial:
Boyle's Law

2 Since both P and V are changing, while n and T are constant, we can use the following equation:

$$P_1 V_1 = P_2 V_2$$

Use the following steps to solve this problem:

Step 1: Define the variables and select the variable to solve for. The problem indicates that $P_1 = 3.15$ atm, $V_1 = 0.015$ mL, and $P_2 = 1.00$ atm; therefore you need to solve for V_2, the final volume.

Step 2: Algebraically isolate the unknown variable on one side of the equation. Use algebra to manipulate the equation above, so that V_2 is isolated:

$$V_2 = \frac{P_1 V_1}{P_2}$$

Step 3: Substitute the values from Step 1 into the equation in Step 2 and solve for the unknown variable. Substitute the values for P_1, V_1, and P_2 into the equation above and solve for V_2:

$$V_2 = \frac{3.15 \text{ atm} \times .015 \text{ mL}}{1.00 \text{ atm}} = 0.047 \text{ mL}$$

Consult Appendix A for a review of algebra.

PRACTICE EXERCISES

4.33 The pressure gauge on a patient's full 10.7-L oxygen tank reads 8.5 atm. At constant temperature, how many liters of oxygen can the patient's tank deliver at a pressure of 0.92 atm?

4.34 The average pair of human lungs can expand to hold about 6 L of air, although only a small amount of this capacity is used during normal breathing. Suppose at a pressure of 1 atm, air occupies 0.55 L in a pilot's lungs. If a pilot flying an airplane ascends rapidly to a pressure of 0.45 atm while holding his breath, could his lungs rupture?

4.35 Suppose you were 60 ft beneath the surface of the ocean scuba diving, and took a deep breath of air from your tank and held your breath. Why would it not be a good idea to start ascending without exhaling?

Breathing and the Pressure–Volume Relationship The pressure–volume relationship of gases is demonstrated every time you inhale and exhale (Figure 4-16). The lungs are an elastic structure contained in an airtight

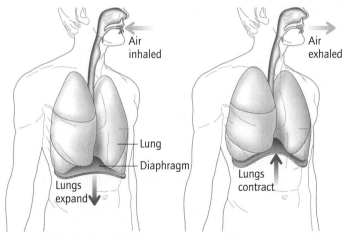

Figure 4-16 When you inhale, the lungs expand, lowering the pressure inside the lungs so more air can enter. When you exhale, the lungs contract, raising the pressure inside so that air is forced out.

As air pressure in lungs decreases, air flows in

As air pressure in lungs increases, air flows out

chamber. The diaphragm, also an elastic structure, is located at the base of this airtight chamber. When you inhale, your diaphragm moves downward and your lungs expand (V increases). As the volume of the lungs increases, the pressure within the lungs decreases, causing air to enter the lungs until the pressure within the lungs equals atmospheric pressure. Then exhalation begins. When you exhale, the diaphragm moves upward while the lungs contract in volume. As the volume within the lungs decreases, the pressure within the lungs increases and becomes greater than that of the atmosphere, so air flows out of the lungs.

In summary, upon inhalation, the pressure of the lungs *decreases* as volume *increases*, and upon exhalation, the pressure of the lungs *increases* as volume of the lungs *decreases*, while the atmospheric pressure is constant. These pressure changes determine whether air flows into or out of the lungs.

Pressure–Temperature Relationship of Gases

Consider the following scenario, which demonstrates the pressure–temperature relationship of gases. Suppose you took a syringe, depressed the plunger about halfway, locked it to a constant volume, and then blocked the tip with a small rubber stopper. If you placed this syringe in an oven and raised the temperature of the oven, what would happen to the rubber stopper at the tip of the syringe? As the temperature, T, rises, the pressure inside the fixed volume of the syringe rises. Eventually the pressure would be high enough to blow off the rubber stopper at the end of the syringe. This demonstration shows the relationship between pressure and temperature, when moles of gas and volume are constant. Under those conditions, as the temperature of a gas increases, the pressure of the gas increases.

This example illustrates that pressure and temperature are *directly* proportional to one another—as one goes up, so does the other (Figure 4-17). This relationship is known as **Gay-Lussac's law** and can be written as follows, where 1 and 2 refer to the initial and final values for P and T:

$$\frac{P_1}{T_1} = \frac{P_2}{T_2} \quad (n \text{ and } V \text{ are constant, } T \text{ must be in kelvin})$$

When performing calculations using any gas law equation, the temperature must be expressed in kelvin because the Kelvin scale is the only absolute scale.

2 kg   4 kg

$T = 200$ K $T = 400$ K
Greater kinetic energy

Figure 4-17 Gay-Lussac's law: An increase in temperature results in an increase in pressure, when V and n are constant.

WORKED EXERCISES 4-8 The Pressure–Temperature Relationship of Gases

1 Why is it a bad idea to throw an aerosol can into a fire?
2 An oxygen tank in a laboratory has a pressure of 12 atm at 25 °C. If the pressure inside the gas tank exceeds 25 atm, the tank will explode. If a fire occurs in the laboratory, raising the temperature of the gas inside the cylinder to 398 °C, will the tank explode?

SOLUTIONS

1 The contents of an aerosol can are under pressure. If the can is thrown into a fire, the temperature of the contents will increase and therefore so will the pressure. If the pressure of the contents exceeds the strength of the can, the can will explode.
2 Since both P and T are changing, and the volume and the number of moles of oxygen in the tank are constant, we can use Gay-Lusssac's law:

$$\frac{P_1}{T_1} = \frac{P_2}{T_2}$$

Use the following steps to solve this problem:

Step 1: Define the variables and select the variable to solve for. The problem indicates that $P_1 = 12$ atm, $T_1 = 25$ °C, and $T_2 = 398$ °C; therefore, you solve for P_2.

Step 2: Algebraically isolate the unknown variable on one side of the equation. Use algebra to manipulate the equation above, so that P_2 is isolated:

$$P_2 = T_2 \times \frac{P_1}{T_1}$$

Step 3: Substitute the values from Step 1 and solve for the unknown variable. Substitute the values for P_1, T_1, and T_2 into the equation above and solve for P_2. *Remember, you must first convert degrees celsius to kelvin!*

$$T_1 = 25 \text{ °C} + 273 = 298 \text{ K}$$
$$T_2 = 398 \text{ °C} + 273 = 671 \text{ K}$$
$$P_2 = 671 \cancel{K} \times \frac{12 \text{ atm}}{298 \cancel{K}} = 27 \text{ atm}$$

This value exceeds the 25-atm limit, so, yes, the tank will explode.

PRACTICE EXERCISES

4.36 Most bacteria will be killed after 15 min in an autoclave at a pressure of 103 kPa and a steam temperature of 121 °C. Prions are infectious agents composed of proteins, which must be exposed to steam at a temperature of 134 °C for 18 min in order to be destroyed. What will the pressure of the autoclave be at this temperature?

4.37 Your tire pressure at 25 °C reads 35.1 psi. After driving around for one hour on a summer day, your tire pressure reads 36.5 psi. What is the temperature in °C of the air in your tires?

4.38 **Critical Thinking Question:** A pressure cooker is a sealed vessel that does not permit air or liquids to escape or enter below a preset pressure. Mountain climbers often use pressure cookers to compensate for low atmospheric pressures at high altitudes. At high altitudes does water boil above or below 100 °C? What would be the effect on cooking at high altitudes? Explain why a pressure cooker is useful when cooking at high altitudes.

4.39 Use kinetic molecular theory to explain why, when V and n are constant, pressure will increase when temperature increases.

Autoclaves work on the principle of the pressure–temperature relationship of gases, where water in the form of steam is the gas. [Photodisc/Punchstock]

Volume–Temperature Relationship of Gases

Consider the following scenario, which demonstrates the volume–temperature relationship of gases. Suppose you took a syringe, partly filled it with air at room temperature and placed it in an oven at a higher temperature. The gas in the syringe will expand as it warms. Since you did not lock the syringe to a specific volume, the plunger will move up with the force exerted by the expanding gas, in order to maintain the pressure equal to atmospheric pressure. Therefore, as the temperature of a gas increases, the volume of the gas increases, if the number of moles of gas, n, and pressure, P, remain constant.

This example shows that volume and temperature are *directly* proportional to one another—as one goes up, so does the other (Figure 4-18). This relationship is known as **Charles' law** and can be written as follows, where 1 and 2 refer to the initial and final values for V and T:

$$\frac{V_1}{T_1} = \frac{V_2}{T_2} \quad (n \text{ and } P \text{ are constant, } T \text{ must be in kelvin})$$

2 kg

2 kg

$T = 200$ K $T = 400$ K

Figure 4-18 Charles' law: An increase in temperature results in an increase in volume, when P and n are constant.

WORKED EXERCISES 4-9 The Volume–Temperature Relationship of Gases

1 A balloon at room temperature is placed in a freezer. What happens to the volume of the balloon? Assume no gas escapes or enters the balloon. Explain.

2 A gas is warmed until its volume is 14.2 L. Originally the gas occupied a volume of 9.60 L at 68 °C. What is its final temperature, in °C, assuming P and n have remained constant?

SOLUTIONS

1 The balloon will shrink, because the pressure of the gas inside the balloon will decrease in the lower temperature of the freezer and there is a direct relationship between volume and temperature.

2 Assuming that P and n are constant, and V and T are changing, we can use Charles' law to predict the outcome:

$$\frac{V_1}{T_1} = \frac{V_2}{T_2}$$

Use the following steps to solve this problem:

Step 1: Define the variables and select the variable to solve for. The problem indicates that $V_1 = 9.60$ L, $T_1 = 68$ °C, and $V_2 = 14.2$ L; therefore you need to solve for T_2.

Step 2: Algebraically isolate the unknown variable on one side of the equation. Use algebra to manipulate the equation above, so that T_2 is isolated:

$$T_2 = T_1 \times \frac{V_2}{V_1}$$

Step 3: Substitute the values given in Step 1 and solve for the unknown variable. Substitute the values for P_1, T_1, and T_2 and solve for P_2. *Remember, you must first convert degrees celsius to kelvin,* even though you want the final answer in °C!

$$T_1 = 68 + 273 = 341 \text{ K}$$

$$T_2 = 341 \text{ K} \times \frac{14.2 \text{ \L}}{9.60 \text{ \L}} = 504 \text{ K}$$

Since the problem requests the temperature in °C, you must convert the temperature in kelvin back to °C:

$$°C = 504 - 273 = \textbf{231 °C}$$

PRACTICE EXERCISES

4.40 Use kinetic molecular theory to explain why, when P and n are constant, volume will increase when temperature increases.

4.41 Neon, a noble gas, is cooled from 76 °C to 38 °C. The original volume of neon gas was 19.5 L. Find the new volume in liters, assuming P and n have remained constant.

Combined Gas Law

A single mathematical relationship can be formed by combining the three previous relationships: pressure–volume, pressure–temperature, and volume–temperature. Using this single relationship, you can predict how changing two of the three variables (P, V, and T) affects the unknown variable, as long as the number of moles of gas molecules, n, is held constant. This combined relationship is known as the **combined gas law,** and it can be written as

$$\frac{P_1 V_1}{T_1} = \frac{P_2 V_2}{T_2} \quad (n \text{ is constant, } T \text{ must be in kelvin})$$

WORKED EXERCISE 4-10 The Combined Gas Law

A caisson is a pressurized, water-tight structure used in the construction of dams and bridges. A worker in a caisson inhales 125 mL of pressurized air at 1.8 atm and 27 °C. What is the pressure of the air in the worker's lungs if the gas expands to 212 mL at a body temperature of 37 °C?

SOLUTION

Since P, T, and V are changing and n is constant, we can use the following equation:

$$\frac{P_1 V_1}{T_1} = \frac{P_2 V_2}{T_2}$$

Use the following steps to solve this problem:

Step 1: Define the variables and select the variable to solve for. The problem indicates that $P_1 = 1.8$ atm, $V_1 = 125$ mL, $T_1 = 27$ °C, $V_2 = 212$ mL, and $T_2 = 37$ °C; therefore you need to solve for P_2.

Step 2: Isolate the unknown variable on one side of the equation. The exercise requests that the final pressure be determined given the initial values for P, V, and T and the final values for V and T. Manipulate the equation, so that P_2 is isolated:

$$P_2 = \frac{P_1 V_1 T_2}{T_1 V_2}$$

Step 3: Substitute the values given in Step 1 and solve for the unknown variable. Substitute the values for P_1, V_1, T_1, V_2, and T_2 and solve for P_2.

Remember you must first convert temperatures to kelvin!

$$T_1 = 27 + 273 = 300.\ \text{K}$$
$$T_2 = 37 + 273 = 310.\ \text{K}$$
$$P_2 = \frac{1.8\ \text{atm} \times 125\ \cancel{\text{mL}} \times 310.\ \cancel{\text{K}}}{300.\ \cancel{\text{K}} \times 212\ \cancel{\text{mL}}} = 1.1\ \text{atm}$$

PRACTICE EXERCISES

4.42 During ascent from a depth of 105 ft, a nitrogen bubble, having a volume of 0.010 mL, forms in the skin tissue of a scuba diver. The pressure at this depth is 4.2 atm and the temperature is 11 °C. What volume will this bubble have when the diver surfaces, and the pressure is 1.00 atm and the temperature is 29 °C?

4.43 A sample of oxygen gas has a volume of 9.75 L at a pressure of 1.33 atm and a temperature of 20.0 °C. Calculate the temperature of the gas in °C when the pressure and volume of the gas have changed to the following:
  a. 2.02 atm and 2.25 L
  b. 4.56 atm and 25.6 L
  c. 881 mmHg and 1561 mL

4.44 Critical Thinking Question: Avogadro's law is yet another gas law. It states that the volume and number of moles of a gas are directly related, as long as P and T are constant.
  a. Explain Avogadro's law based on kinetic molecular theory.
  b. Write an equation that expresses this relationship for initial and final values of V and n.

Partial Pressure and Gas Mixtures

You now have an understanding of how pressure, volume, and temperature affect gases. Next we explore what happens when we mix gases together. Consider the scenario depicted in Figure 4-19. Oxygen gas is confined to a cylinder at a pressure of 1.0 atm. A second tank of equal volume contains

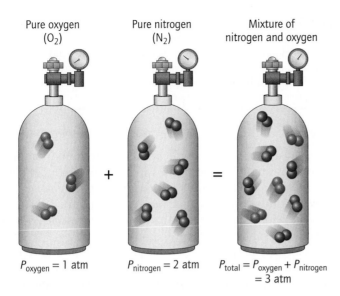

Pure oxygen
(O_2)

Pure nitrogen
(N_2)

Mixture of
nitrogen and oxygen

+

=

$P_{oxygen} = 1$ atm

$P_{nitrogen} = 2$ atm

$P_{total} = P_{oxygen} + P_{nitrogen}$
$= 3$ atm

Figure 4-19 When two gases are combined, the total pressure will equal the sum of the partial pressures exerted by each gas.

nitrogen gas at a pressure of 2.0 atm. Suppose that the contents of the oxygen tank are forced into the nitrogen tank. What would be the total pressure exerted by this new mixture of oxygen *and* nitrogen gases? The answer is that the pressure of the mixture is equal to the sum of the two individual gas pressures: 1.0 atm + 2.0 atm = 3.0 atm. *Dalton's law states that if you have a mixture of gases, each gas in the mixture will exert a pressure independent of the other gases present, and each gas will behave as if it alone occupied the total volume.*

The pressure exerted by one gas in a gas mixture is known as its **partial pressure,** P_N. **Dalton's law** states that the sum of the partial pressures of each gas present in the mixture equals the total pressure (P_{tot}):

$$P_{tot} = P_1 + P_2 + P_3 + \cdots + P_N$$

where P_1, P_2, P_3, $\cdots$, P_N represent the partial pressures of each gas in the mixture.

WORKED EXERCISE 4-11   Partial Pressures of Gases

An air sample in an average adult's lungs contains oxygen at a partial pressure of 0.18 atm, nitrogen at a partial pressure of 0.77 atm, carbon dioxide at a partial pressure of 0.06 atm, and water vapor at a partial pressure of 0.08 atm. What is the total pressure in this person's lungs?

SOLUTION
Since partial pressures are provided, use the following equation:

$$P_{tot} = P_1 + P_2 + P_3 + \cdots + P_N$$

Substitute the partial pressures for each gas into the equation and add the pressures:

$$P_{tot} = 0.18 \text{ atm} + 0.77 \text{ atm} + 0.06 \text{ atm} + 0.08 \text{ atm} = 1.09 \text{ atm}$$

PRACTICE EXERCISES

4.45 An air tank contains a mixture of nitrogen and oxygen at a total pressure of 3.00 atm. If the partial pressure of oxygen is 0.63 atm, what is the partial pressure of nitrogen inside the tank?

4.46 Often helium replaces nitrogen in scuba tanks prepared for dives greater than 150 ft below the surface of the ocean. A tank is prepared containing only oxygen and helium for a scuba diver who is going to descend 230 ft below the ocean surface. At that depth, the diver breathes a gas mixture that has a total pressure of 8.0 atm. If the partial pressure of oxygen in the tank at that depth is 1.6 atm, what is the partial pressure of helium?

Henry's Law

The gases oxygen and carbon dioxide are naturally found dissolved in the blood. The concentrations of these gases in the blood are commonly measured in medical tests. *Concentration* is a term used to report how much of a given substance (in this case, a gas) is mixed with another substance (in this case, water) *to form a solution* (blood). A solution is a uniform mixture of substances and is the subject of the next chapter. *The amount of gas dissolved in a solution is directly proportional to the partial pressure of that gas above the solution.* Thus, as the pressure of the gas above the solution increases, the concentration of the gas in the solution increases. This relationship is known as **Henry's law.** At constant temperature, Henry's law can be described mathematically by the equation

$$\text{Pressure} = k \times \text{concentration}$$
$$P = k\,C$$

where P = the partial pressure of the gas above the solution,
     k = a constant, which depends on the identity of the gas, and
     C = the concentration of the gas in solution.

Henry's law is especially useful in the field of anesthesiology when gaseous anesthetics are inhaled. Anesthetics have different Henry's constants, k, and therefore at a given pressure they have different concentrations. *The smaller the Henry's constant, the higher the concentration of anesthetic dissolved in blood.*

The anesthetic diethyl ether, for example, has a relatively small Henry's constant, k, allowing for relatively high concentrations of the anesthetic to be dissolved in blood. Because most of the diethyl ether is dissolved in the blood, it does not reach the brain as quickly as an anesthetic with a higher Henry's constant. For anesthetics with higher Henry's constants, the gas diffuses through tissues directly, allowing it to reach the brain more quickly. Therefore, a patient anesthetized with diethyl ether takes a long time to become anesthetized and a long time to regain consciousness. Desflurane, another anesthetic, has a larger Henry's constant. A patient anesthetized with desflurane becomes anesthetized quickly and regains consciousness quickly, once the source of desflurane has been removed.

> **WORKED EXERCISE 4-12**   Henry's Law
>
> At higher altitudes atmospheric pressure is lower, because there is less air. Considering Henry's law, would you expect the concentration of oxygen in your blood to be higher or lower compared to sea level if you were at an altitude of 14,000 ft above sea level?
>
> **SOLUTION**
>
> Henry's law shows that there is a direct relationship between the pressure and the concentration of a gas in solution; therefore, at lower atmospheric pressures you would have a corresponding lower concentration of oxygen in your blood. Hence, you would have a lower concentration of oxygen in your blood at 14,000 ft than at sea level. This is the reason that it is hard to breathe at high altitudes and that oxygen masks must be used at very high altitudes.

Scuba divers must be concerned with the consequences of Henry's law, also known as **Martini's law**. If too much nitrogen is dissolved in the diver's blood at deep depths, then for unknown reasons the diver may begin to feel and act drunk, and his or her judgment is clouded. This is definitely not a good situation when you're diving at great depths beneath the surface of the ocean! To help alleviate this problem, sometimes argon or helium is substituted for nitrogen in divers' tanks.

PRACTICE EXERCISES

4.47 At 30 ft beneath the surface of the ocean, would you expect the concentration of nitrogen in a person's blood to be higher or lower than the nitrogen concentration at the surface of the ocean? Explain using Henry's law.

4.48 Isoflurane, commonly used for veterinary anesthesia, has a larger Henry's constant (k) than diethyl ether. Which anesthetic would allow an anesthetized animal to regain consciousness more quickly, isoflurane or diethyl ether? Explain.

4.49 Two patients similar in build and weight were both anesthetized with the same amount of desflurane, but one patient was in Boston and one was in Denver. Considering Henry's law, which patient would you expect to become anesthetized in a shorter period of time? Explain.

4.50 Where would you expect a glass of champagne to have more bubbles, in London or at the top of the Swiss Alps?

Partial Pressures and Dissolved Gases in the Human Body As you know, the air we breathe is a mixture of different gases. Air is mostly nitrogen and oxygen, but it also includes small amounts of water vapor, carbon dioxide, and other gases. Therefore, the gases that we breathe have different partial pressures. Table 4-6 lists the partial pressures of the different gases in inhaled and exhaled air.

Table 4-6 Partial Pressures of Gases Involved in Breathing

	Nitrogen (mmHg)	Oxygen (mmHg)	Carbon Dioxide (mmHg)	Water Vapor (mmHg)	Total (mmHg)
Inhaled air	594.0	160.0	0.3	5.7	760.0
Exhaled air	569	116	28	47	760

Notice that the total pressure of the individual gases adds up to 760 mmHg, 1 atm, in accordance with Dalton's law. There is a higher partial pressure of oxygen in inhaled air because we inhale more oxygen than we exhale. Figure 4.20 shows the exchange of gases in the lungs. Oxygen from the lungs dissolves in the blood (Henry's law), diffusing into the cells, where some is used in biochemical reactions to generate energy in a process called cellular respiration. Carbon dioxide gas is produced as a by-product of cellular respiration in the cells and is carried by the blood to the lungs to be exhaled. Consequently, there is a higher partial pressure of carbon dioxide in exhaled air.

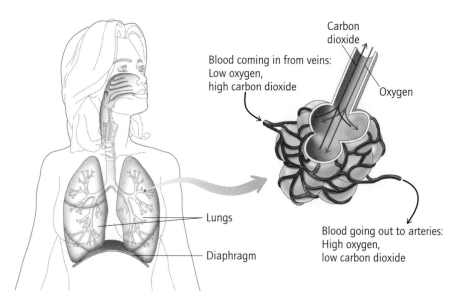

Figure 4-20 Gas exchange in the lungs. Inhaled air contains higher levels of oxygen than exhaled air. Exhaled air contains higher levels of carbon dioxide than inhaled air.

WORKED EXERCISE 4-13 Applying Gas Laws to the Human Body

When an asthmatic has an asthma attack, a lower volume of air enters the lungs, and therefore a lower volume of oxygen enters the bloodstream. How is the partial pressure of oxygen in the bloodstream affected during an attack? Explain.

SOLUTION

When an asthmatic has an attack, the partial pressure of oxygen in the bloodstream will decrease because the amount of oxygen decreases—a lower volume of air, and thus fewer oxygen molecules, enter the lungs.

PRACTICE EXERCISES

4.51 In the lungs, oxygen partial pressures should lie between 0.1 atm and 1.4 atm in order to avoid significant health problems.

   a. Which would you expect a person climbing Mount Everest to be concerned with: the partial pressure of oxygen in her lungs getting above 1.4 atm or below 0.1 atm? Explain.

   b. Consider a scuba diver at 150 ft below the surface of the sea breathing compressed air. Which would you expect the diver to be concerned with: the partial pressure of oxygen in his lungs getting above 1.4 atm or below 0.1 atm? Explain.

4-52 **Critical Thinking Question:** Why is the partial pressure of nitrogen not altered much in the process of breathing? That is, why is the partial pressure of nitrogen in inhaled air similar to its partial pressure in exhaled air?

Chemistry in Medicine Hyperbaric Oxygen Therapy (HBOT)

HyperBaric Oxygen Therapy (HBOT) is the use of high-pressure oxygen to treat medical conditions, including the "bends," which you learned about in the opening vignette. Some other conditions that warrant HBOT include carbon monoxide poisoning, diabetic wounds, and infections of necrotizing fasciitis (flesh-eating bacteria). Here we'll discuss the use of HBOT in the treatment of the bends, as well as carbon monoxide poisoning and diabetic wounds.

Treating The Bends

HBOT has been used to treat the bends in divers since the 1940s. The hyperbaric chamber is a clear plastic tube about 7 ft long, like the one shown in the photo. A diver experiencing the bends would be placed on a padded table that slides into the tube. The chamber would then gradually be filled with pure oxygen to a pressure 1.5–3 times that of normal atmospheric pressure (1 atm). This increase in pressure causes the nitrogen bubbles resulting from the bends to re-dissolve in the diver's blood, as expected from Henry's law. The chamber is then depressurized very slowly to allow nitrogen to emerge in bubbles small enough to circulate easily to the lungs and be safely exhaled.

You may be wondering why oxygen gas does not present the same problem. Nitrogen is an inert gas; it does not interact chemically with the body, while oxygen does. Although some of the inhaled oxygen simply dissolves in blood, most of the oxygen molecules bind to the iron-containing protein in blood, hemoglobin (Hb). One hemoglobin protein can bind up to four oxygen molecules. Because an oxygen molecule must be released from hemoglobin before it can diffuse out of blood, oxygen is released from the body more gradually than nitrogen. As a diver ascends to the surface, or as a hyperbaric chamber returns to atmospheric pressure, the excess oxygen that has dissolved in the blood comes out of solution more slowly than the nitrogen. This enables the diver to safely exhale the excess oxygen.

Treating Carbon Monoxide Poisoning Using HBOT

It has been estimated that more than 40,000 people each year in the United States seek medical attention for poisoning by carbon monoxide gas. Common sources of carbon monoxide poisoning include wood-burning stoves, motor vehicle exhaust, home furnaces that are not functioning properly, and fireplaces.

Carbon monoxide (CO) also binds to hemoglobin, producing carboxyhemoglobin (COHb). The attraction between hemoglobin and carbon monoxide is 240 times stronger than the attraction between hemoglobin and oxygen. When CO binds to hemoglobin, it replaces the oxygen in hemoglobin, and the level of oxygen available to tissues drops to dangerous levels. Such oxygen deprivation is called *hypoxia*. Because oxygen uptake by the brain is reduced, the victim may become unconscious, sustain brain damage, or even die.

A hyperbaric oxygen therapy chamber. [Jason Cohn/Reuters/Corbis]

Treating Diabetic Wounds Using HBOT

Of the 11 million Americans with diabetes, an estimated 25% will develop diabetic wounds on their feet. Diabetes is a disease that results in abnormally high blood sugar levels. It causes a number of complications, including neuropathy, a form of nerve damage that causes lack of sensation in the limbs. A diabetic patient with neuropathy is less likely to feel a wound occurring in the extremities, particularly the feet, and so is less likely to seek treatment. Moreover, a diabetic wound is more likely to become infected because bacteria will feed on the higher glucose levels found in a diabetic patient's blood. Infections restrict blood supply to the wound and surrounding tissue, and hence, oxygen levels are below normal in these tissues. Due to lack of oxygen and lack of treatment, the wound heals slowly.

Oxygen is essential to many cell functions, including tissue production. By oxygenating tissues near the wound, HBOT stimulates the formation of capillary blood vessels, enabling blood flow to hypoxic tissues as well as aiding tissue repair.

Hyperbaric oxygen therapy demonstrates a medical procedure in which gas pressure is used to alleviate acute problems, such as the bends and carbon monoxide poisoning, as well as chronic problems, such as diabetic wounds. Understanding how gases behave within the human body was critical in developing this distinctive therapy.

The human body naturally produces some carbon monoxide, but normal COHb levels in the body are less than 5% of total hemoglobin. Carbon monoxide is toxic at levels of COHb above 25%, and levels above 70% are often fatal. Carbon monoxide poisoning is treated either by administering pure oxygen at atmospheric pressure or by applying HBOT. With a COHb level of 10%, a hyperbaric treatment will reduce the time needed to drive out half the CO present in the blood from 5 hr to 30 min. Once COHb is eliminated, oxygen once again becomes available to tissues, including the brain.

In this chapter you learned about the different states of matter, including their physical properties and relationships that are useful in the field of medicine. You studied gases in more depth and discovered that having a basic knowledge of the special behavior of gases allows us to understand the relationship between breathing and important medical issues such as decompression sickness and anesthesia. In the next chapter—Solutions, Colloids, and Membranes—you will consider mixtures and their role in health and medicine.

Chapter Summary

States of Matter

- Matter exists in three physical states: solid, liquid, and gas.
- A solid has a definite shape and volume. A liquid has a definite volume but not a definite shape. A gas has neither a definite shape nor a definite volume.
- Energy is defined as the capacity to do work.
- There are two forms of energy: kinetic energy and potential energy. Kinetic energy is the energy of motion. Potential energy is stored energy.
- Heat is kinetic energy that is transferred from one object to another due to a difference in temperature.
- The temperature of a substance reflects the average kinetic energy of the molecules, ions, or atoms that make up that substance.

- Temperature can be measured in three temperature scales: Celsius, Fahrenheit, and Kelvin.
- Two competing forces determine the physical state of a substance: intermolecular forces of attraction (potential energy), which draw molecules together, and the motion of molecules (kinetic energy), which tends to pull them apart.
- The physical differences between the states of matter can be explained by a molecular view of the three states of matter. In the gas phase, molecules are far apart and moving rapidly and randomly. In the liquid phase, the atoms and molecules are close together and move randomly. In the solid phase, molecules or atoms are close together and are arranged in a regular ordered pattern with only vibrational motion.
- Most solids are more dense than liquids, except water, which occupies a greater volume in the solid phase due to hydrogen bonding. The density of a substance is lowest in the gas phase.

Changes of State

- The process of going from one state to another is called a change of state and involves a transfer of energy.
- Changes of state include: solid to liquid (melting); solid to gas (sublimation); liquid to gas (vaporization); liquid to solid (freezing or fusion); gas to liquid (condensation); and gas to solid (deposition).
- The heat of fusion is the amount of energy that must be added to turn a solid into a liquid; the same amount of energy must be removed to turn a liquid into a solid.
- The heat of vaporization is the amount of energy that must be added to change a liquid into a gas; the same amount of energy must be removed to change a gas into a liquid.
- A heating curve is a graph that shows how the temperature of a substance increases with added heat energy. A cooling curve is a graph that shows how the temperature of a substance decreases with removal of heat energy.

Pressure

- Pressure is the amount of force applied over a given area. Pressure has the greatest effect on the gas phase.
- Pressure can be measured in a variety of units: atmospheres, torr, millimeters of mercury (mmHg), pounds per square inch, and pascals.
- Molecules of air exert a pressure known as the atmospheric pressure, which can be measured by a barometer.
- Blood pressure is the pressure exerted by blood on the walls of blood vessels. A sphygmomanometer measures the systolic and diastolic blood pressure.
- Vapor pressure is the pressure exerted at a given temperature by molecules in the gas phase that are in contact with molecules in the liquid phase. The vapor pressure of a liquid increases with temperature. When the vapor pressure of a liquid equals the atmospheric pressure, the liquid will boil.

Gases

- Four variables describe the physical properties of a gas: pressure (P), volume (V), the amount of gas (n), and temperature (T).
- The kinetic molecular view of gases states that the particles of a gas are in constant, random motion; the attractive forces among the particles of a gas are negligible; the volume of a gas is mostly empty space; and the temperature of a gas depends on the kinetic energy of the gas particles.

- A standard set of reference conditions are used to make comparisons between different gases. Under standard temperature and pressure (STP), one mole of any gas occupies a volume of 22.4 L. STP is 1 atm and 0 °C.
- The pressure and the volume of a gas are inversely related to one another. (Boyle's law) $P_1 V_1 = P_2 V_2$
- The pressure and the temperature of a gas are directly proportional to one another. (Gay-Lussac's law) $\dfrac{P_1}{T_1} = \dfrac{P_2}{T_2}$
- The volume and the temperature of a gas are directly proportional to one another. (Charles' law) $\dfrac{V_1}{T_1} = \dfrac{V_2}{T_2}$
- The combined gas law combines the pressure–volume, pressure–temperature, and volume–temperature relationships into a single relationship.
  $$\frac{P_1 V_1}{T_1} = \frac{P_2 V_2}{T_2}$$
- The sum of the partial pressures of each gas present in a mixture equals the total pressure exerted by the mixture of gases. (Dalton's law)
  $P_{\text{Tot}} = P_1 + P_2 + P_3 + \dots P_n$
- The amount of gas dissolved in a solution is directly proportional to the partial pressure of that gas above that solution (Henry's law):
  $C = kP$.

Key Words

Absolute temperature scale The Kelvin scale. This temperature scale assigns a temperature of zero to the theoretical condition in which all molecular motion has stopped.

Atmospheric pressure The pressure exerted by the weight of air at any given place on the earth.

Blood pressure The pressure exerted by blood on the walls of the blood vessels.

Boiling point The temperature at which a pure liquid changes to the gas phase at atmospheric pressure.

Boyle's law The pressure and volume of a gas are inversely proportional to each other. $P_1/V_1 = P_2/V_2$

Change of state The process of going from one state of matter (s, l, g) to another.

Charles' law The volume and temperature of a gas are directly proportional to each other. $V_1/T_1 = V_2/T_2$

Combined gas law A gas law that combines Boyle's law, Gay-Lussac's law, and Charles' law into one equation. $P_1 V_1/T_1 = P_2 V_2/T_2$

Condensation The process of changing from the gas state to the liquid state.

Dalton's law The sum of the partial pressures of each gas present in a mixture equals the total pressure of the gas mixture. $P_T = P_1 + P_2 + \dots P_N$

Deposition The process of changing from the gas state to the solid state.

Diastolic pressure The temporary decrease in blood pressure that occurs when the heart muscle relaxes and there is less pressure against the arteries.

Energy The capacity to do work, where work is the act of moving an object.

Evaporation The process of changing from the liquid to the gas phase at any temperature. It occurs when atoms and/or molecules have enough kinetic energy to leave the surface of a liquid and enter the gas phase.

Freezing The process of changing from a liquid state to a solid state.

Fusion Freezing; the process of changing from a liquid state to a solid state.

Gas A state of matter that has neither a definite shape nor a definite volume. Atoms and molecules are moving rapidly and in random motion. There are no intermolecular interactions and there is mainly empty space between molecules. Kinetic energy of particles is higher than in other states.

Gay-Lussac's law The pressure and temperature of a gas are directly proportional to each other. $P_1 T_1 = P_2 V_2$

Heat Kinetic energy (molecular motion) that is transferred from one object to another due to a difference in temperature.

Heat of fusion The amount of energy that must be added to change a solid to a liquid. It is the same amount of energy that must be removed from a liquid during freezing.

Heat of vaporization The amount of energy that must be added to change a liquid to a gas. It is the same amount of energy that must be removed from a gas during condensation.

Henry's law The amount of gas dissolved in a solution is directly proportional to the partial pressure of that gas above the solution. $C = kP$

Hypertension A condition in which a person has elevated blood pressure.

Kinetic energy The energy of motion; the energy a substance possesses as a result of the motion of its atoms or molecules.

Kinetic molecular view An examination of the arrangement and motions of atoms and molecules at the subatomic level.

Liquid A state of matter that occupies a definite volume, but does not have a definite shape. Molecules have more kinetic energy in the liquid phase than in the solid phase, but less than in the gas phase. Intermolecular forces of attraction exist between molecules or atoms in the liquid phase.

Melting The process of changing from the solid state to the liquid state.

Melting point The temperature at which a substance changes from solid to liquid. At the melting point, a substance will absorb heat without a temperature change, because the energy is used to change state.

Molar volume of a gas The volume occupied by one mole of any gas at STP; that volume is 22.4 L for any gas.

Partial pressure The pressure exerted by one gas in a mixture of gasses.

Physical state or phase The existence of matter as a solid, liquid, or gas.

Potential energy Stored energy; the energy a substance possesses as a result of the position, composition, and condition of its atoms.

Pressure A measure of the amount of force applied over a given area.

Solid A state of matter that has a definite shape and volume, independent of its container. Molecules in the solid phase have the least amount of kinetic energy. Intermolecular forces of attraction hold the molecules or atoms in a lattice.

STP (standard temperature and pressure) A standard set of reference conditions for a gas: 273 K and 760 mmHg.

Sublimation The process of changing directly from the solid state to the gas state.

Systolic pressure The temporary increase in blood pressure that occurs when the heart contracts and forces blood through the arteries.

Temperature A measure of the average kinetic energy of the molecules, ions, or atoms that make up a substance. The three temperature scales are Celsius, Fahrenheit, and Kelvin.

Vapor pressure The pressure exerted by the gas phase molecules in contact with a liquid, at a given temperature.

Vaporization The process of changing from a liquid state to a gas state.

Volatile A substance with a high vapor pressure, which therefore readily enters the gas phase.

Additional Exercises

States of Matter

4.53 What are the three physical states of matter?

4.54 What two states of matter are influenced by the shape of the container that holds the matter?

4.55 What two states of matter have fixed volumes compared to the container?

4.56 Does a rock sitting at the top of a hill have potential energy or kinetic energy?

4.57 Does a rock rolling down a hill have potential energy or kinetic energy?

4.58 Indicate whether each of the following examples is a demonstration of potential energy or kinetic energy:
   a. Water flowing over a dam
   b. A skier at the top of a hill
   c. Water in a reservoir behind a dam
   d. Atoms bonded together within a crystal lattice

4.59 Indicate whether each of the following examples is a demonstration of potential energy or kinetic energy:
   a. A biker pedaling up hill
   b. A hiker standing at the top of a mountain
   c. Helium atoms in a balloon
   d. The wax in a candle

4.60 Which molecules with similar mass have more kinetic energy, faster moving ones or slower moving ones?

4.61 Which molecules with similar velocity have more kinetic energy, heavier ones or lighter ones?

4.62 Do molecules have more kinetic energy at higher temperatures or lower temperatures?

4.63 In which physical state do molecules have the least amount of kinetic energy?

4.64 In which physical state do molecules have the greatest amount of kinetic energy?

4.65 In which state do water molecules have the most kinetic energy: liquid water, steam, or ice?

4.66 While traveling in Europe, you notice that a thermometer reads 20. °C. Are you wearing winter clothes (coat, hat, scarf, etc.) or summer clothes? What is this temperature in kelvin and degrees Fahrenheit?

4.67 The thermometer in Christchurch, New Zealand, reads 31 °C. Are you wearing winter clothes or summer clothes? What is this temperature in degrees Fahrenheit and kelvin?

4.68 The thermometer in Tokyo, Japan, reads 11 °C. Are you wearing summer clothes or winter clothes? What is this temperature in degrees Fahrenheit and kelvin?

4.69 A normal temperature reading for the human body is 98.6 °F. What is this temperature reading in degrees Celsius?

4.70 While on vacation in Europe, you took your temperature and it was 38 °C. Do you have a fever?

4.71 The temperature in outer space is 2.7 K. What is this temperature in degrees Celsius and degrees Fahrenheit?

4.72 Liquid nitrogen is often used as a treatment to remove warts by freezing them off. Liquid nitrogen has a temperature of 77 K. What is this temperature in degrees Celsius and degrees Fahrenheit?

4.73 In what physical state are intermolecular forces the greatest?

4.74 Explain why, in terms of intermolecular forces and kinetic energy, you can pour water into a glass and it will take on the shape of the glass, but if you drop ice cubes into an empty glass, they retain their shape.

4.75 Would you expect a block of aluminum to float or sink in a container of molten aluminum?

4.76 Explain why ice floats in a glass of water.

Changes of State

4.77 Are bonding or intermolecular forces affected when a change of state occurs?

4.78 For which changes of state does energy need to be added to a substance?

4.79 For which changes of state does energy need to be removed from a substance?

4.80 Identify the changes of state in the following processes:
a. Melting snow
b. Dry ice (solid CO_2) evaporating
c. Formation of icicles
d. Formation of steam in your shower
e. Formation of water droplets on the outside of a glass of ice cold drink on a hot summer day
f. Fumes coming off the gas tank as you fill your car with gasoline
g. Formation of frost on a cold windshield overnight

4.81 Identify the changes of state in the following processes:
a. Your friend's liquid perfume produces an aroma that fills the room.
b. Formation of ice cream from milk in an ice cream maker
c. Wax dripping from a candle
d. Naphthalene (s) produces an odor coming from moth balls in the closet.
e. Your breath steams up your car windshield on a cold winter day.

4.82 Mercury is the only metal that is a liquid at room temperature. It is a toxic metal that is readily absorbed by the lungs. What phase change must be occurring to make this metal particularly dangerous?

4.83 Why is steam able to cause a skin burn?

4.84 Humans sweat as a way to regulate their body temperature. Why does the body cool down when sweat evaporates from the surface of the skin?

4.85 **Critical Thought Question:** Why is ethanol, CH_3CH_2OH, a liquid at room temperature, while carbon dioxide, CO_2, is a gas at room temperature?

Extension Topic 4-1: Specific Heat

4.86 Using Table 4-2, calculate in calories the heat that must be added to warm 13.4 g of each of the substances below from 29 °C to 51 °C. Which substance requires the greatest input of heat and why?
a. water
b. copper
c. sand

4.87 Using Table 4-2, calculate in calories the heat that must be added to warm 10.9 g of each of the substances below from 22 °C to 49 °C. Which substance requires the greatest input of heat and why?
a. brick
b. ethanol
c. wood

4.88 **Critical Thinking Question:** A metal object with a mass of 23 g is heated to 97 °C, then transferred to an insulated container containing 81 g of water at 19 °C. The water and the metal object reach a final temperature of 24 °C. What is the specific heat of this metal object? *Hint:* Assume that the heat lost by the metal object is equal to the heat gained by the water.

4.89 **Critical Thinking Question:** You can actually calculate the difference in the amount of heat absorbed by the skin as the result of a burn that occurred from exposure to boiling water as compared to a burn that occurred from exposure to steam. Perform the calculations for (a) and (b) below, and compare your answers for each. Why are steam burns so severe?
a. A burn caused by boiling water: Use the specific heat of water to calculate the amount of heat released when 25.0 g of liquid water at 100 °C hits the skin and cools to body temperature (37 °C).
b. A burn caused by steam: Assume that 25.0 g of steam hits the skin, condenses, and then cools to body temperature (37 °C). First use the heat of vaporization of water to calculate the energy released to condense gaseous water (steam) to liquid water on the skin. Add this value to the value obtained in (a)—the amount of heat released by the water (absorbed by the skin) in going from 100 °C to 37 °C.

Pressure

4.90 Define the term *pressure*.

4.91 What is the difference between the unit *atmosphere* and *atmospheric pressure*?

4.92 **Critical Thinking Question:** Why is the pressure at 14,000 ft above sea level 0.69 atm, while the pressure at 14,000 ft below sea level is 470 atm?

4.93 A patient had a high blood pressure reading of 160/110 mmHg.
a. Convert the systolic pressure into psi.
b. Convert the diastolic pressure into torr.

4.94 A patient with an intraocular pressure (the pressure inside the eye) greater than 21.5 mmHg has a greater risk of developing glaucoma. Convert 21.5 mmHg into atmospheres.

4.95 A patient comes into the ER with severe head trauma. His intracranial pressure (pressure from the brain on the intracranial space) is 30 mmHg. This pressure is high enough for the head trauma to be fatal. Convert this pressure into psi.

4.96 During the first stage of labor, a contraction is measured at 40.4 mmHg. Convert this pressure into atm and Pa.

4.97 A patient had a systolic pressure of 0.23 atm and a diastolic pressure of 2.17 psi. Does this patient have hypertension? What is the patient's blood pressure in mmHg?

4.98 Gasoline is flammable. The vapor pressure of gasoline is 6.9 psi. What is this pressure in mmHg? Why is it necessary to avoid electrical sparks when filling your car with gasoline, even if the liquid gasoline goes directly from the pump to your gasoline tank?

4.99 Using Table 4-5, predict which substance would have a higher boiling point, acetone or methylene chloride.

4.100 Using Table 4.5, predict which substance would have a lower boiling point, mercury or phenol.

4.101 Do you expect the boiling point of water at a base camp on Mt. Everest to be higher or lower than the boiling point of water on the beach in Honolulu? Explain using atmospheric pressure.

Gases

4.102 Use the kinetic molecular theory to explain the following:
a. Molecules in the gas phase cannot hydrogen bond.
b. On a cold day, the tire pressure in your car decreases.
c. The volume of a container filled with a gas contains mostly empty space.

4.103 What is the volume of 1 mol of carbon dioxide at STP?

4.104 How many moles of gas are present in 11.2 L at STP?

4.105 What volume will be occupied by 4.1 mol of helium at STP?

4.106 If argon occupies 15.3 L at STP, how many moles of argon are present?

4.107 How many moles of hydrogen occupy 220 L at STP?

4.108 What volume is occupied by 0.2 mol of neon at STP?

4.109 Imagine you are hiking in the Rocky Mountains near the Continental divide at 12,000 ft elevation. You drink all the water in your 1-L plastic water bottle, put the cap back on the bottle, and carry the bottle back to Denver (5280 ft in elevation). When you return to Denver the water bottle has collapsed in at the sides. Explain why the empty water bottle has collapsed.

4.110 When you buy bags of potato chips in Denver (5280 ft elevation), sometimes the bags look like filled balloons: the package has expanded as if it were ready to burst. These bags are packed at sea level. Explain why the bag expands in Denver, but not in Los Angeles. (Assume that the temperature is the same in the two cities.)

4.111 The pressure gauge on a patient's 2.24-L oxygen tank reads 142.8 atm. At a constant temperature, how many liters of oxygen can the patient's tank hold at a pressure of 0.84 atm?

4.112 A child holds a helium balloon with a volume of 1.1 L at a pressure of 0.91 atm. A little while later the volume of the balloon is 3.1 L. What is the pressure on the balloon? Did the child go up or down in altitude with the balloon?

4.113 Mark Harris holds the world record in free diving (diving underwater without the use of supplemental oxygen). He can free dive to a depth of 60 m. Assume that he takes a breath that fills his lungs to 3.6 L at the surface of the water (1 atm). To what volume does that correspond when he reaches the depth of 60 m under water (5.9 atm)?

4.114 Upon inhalation, the pressure of the lungs _____ as the volume _____. [increases or decreases]

4.115 Upon exhalation, the pressure of the lungs _____ as the volume _____. [increases or decreases]

4.116 Summarize how Boyle's law (*PV* relationship) describes breathing.

4.117 Why do the labels on cans of baked beans tell you not to heat the unopened can directly on the stove?

4.118 Why is cooking food with a pressure cooker faster than cooking food in an open pot on the stove?

4.119 If the tires on your car are filled to the same volume, will the tire pressure be higher in summer or winter?

4.120 In an autoclave, the pressure is increased to 103 kPa. What is the temperature of the steam inside the autoclave?

4.121 Your tire pressure reads 30.1 psi at 20.0 °C. What is the tire pressure after you drive for several hours and the temperature of the tires increases to 35.5 °C?

4.122 In a pressure cooker, the temperature of boiling water is 125 °C. At sea level, what is the pressure inside the pressure cooker compared to an open pot? Assume that the open pot and the pressure cooker have the same volume.

4.123 When you bake a cake, you use leavening agents that create pockets of carbon dioxide. Why does a cake rise when you put it in the oven to bake? (Assume that the external pressure has remained the same).

4.124 If a balloon filled with air is put into a container of liquid nitrogen (−195.8 °C), the size of the balloon will shrink. Explain why.

4.125 Nitrogen was heated from 12 °C to 34 °C. The original volume of nitrogen was 8.7 L. Find the new volume in liters assuming *P* and *n* remain the same.

4.126 Argon is cooled from 120. °C to 52 °C. The original volume of argon was 122 L. Find the new volume in liters, assuming *P* and *n* remain the same.

4.127 **Critical Thinking Question:** Most hot air balloons are flown in the early morning or evening when outside air temperatures are cooler. Explain why the hot air in a balloon makes the balloon float.

4.128 A gas occupies a volume of 18 L at a pressure of 4.3 atm and a temperature of 501 K. What will the volume expand or contract to if the pressure is raised to 10. atm and the temperature is changed to 250. K?

4.129 A sample of nitrogen is collected at 2.7 atm and 12.0 °C. It has a volume of 2.25 L. What would the volume of this gas be at 0.00 °C and 100. kPa?

4.130 A sample of oxygen has a pressure of 712 mmHg when its volume is 205 mL and its temperature is −44.0 °C. What would be the pressure of the oxygen if the volume is changed to 370. mL at 30. °C?

4.131 State Dalton's law.

4.132 An air tank contains a mixture of nitrogen and oxygen at a total pressure of 4.30 atm. If the partial pressure of nitrogen is 0.92 atm, what is the partial pressure of oxygen in the tank?

4.133 A mixture of neon and argon gases has a total pressure of 2.42 atm. The partial pressure of the neon alone is 1.81 atm. What is the partial pressure of the argon?

4.134 When a patient hyperventilates, she is breathing faster than normal and taking in too much oxygen and not eliminating enough carbon dioxide. A treatment for hyperventilation is to have the patient cover her nose and mouth with a small paper bag and breathe in and out slowly. This technique increases the amount of carbon dioxide in the bloodstream. Explain what happens to the partial pressure of carbon dioxide inside the bag as the patient breathes out into the bag.

4.135 Consider two people who have the same build and weight. One of them goes scuba diving and the other climbs Mt. Hood in Oregon. Considering Henry's law, which one of them has more oxygen in their blood?

4.136 Would you expect a glass of soda to have more bubbles on the beach or in the mountains?

4.137 Why would a patient anesthetized with desflurane regain consciousness more quickly than one anesthetized with diethyl ether?

Chemistry in Medicine

4.138 When would a patient need to use Hyperbaric Oxygen Therapy?

4.139 When a patient is treated for the bends with HBOT, the chamber is pressurized with pure oxygen. What does this increase in pressure do to the nitrogen bubbles in the blood? Using Henry's law, explain why the increase in pressure has this effect.

4.140 How does hemoglobin in the blood prevent oxygen gas from causing "the bends"?

4.141 Why is carbon monoxide poisonous? What does it do when it enters the bloodstream?

4.142 What is the advantage of using HBOT over administering oxygen at atmospheric pressure to treat carbon monoxide poisoning?

4.143 **Critical Thinking Question:** A scuba diver ascends too quickly and develops the bends.
a. A nitrogen bubble forms in the elbow. At a depth of 60 ft, where the pressure is 2.81 atm, the bubble has a volume of 0.021 mL. Assuming a constant temperature and a constant number of moles of nitrogen in the bubble, what volume will the bubble have at the surface of the sea, where the pressure is 1.00 atm?
b. The scuba diver is put into a hyperbaric oxygen chamber where the pressure is 2.25 atm. What is the volume of the nitrogen bubble once the patient is in the hyperbaric chamber?

4.144 **Critical Thinking Question:** In normal air, the partial pressure of oxygen is 0.18 atm, the partial pressure of nitrogen is 0.77 atm, and the partial pressure of carbon dioxide is 0.05 atm. Scuba divers often use tanks with an enhanced mixture of only oxygen and nitrogen. If the total pressure in the scuba tank is 3.83 atm and the partial pressure of oxygen is 1.38 atm, what is the partial pressure of nitrogen? Is the percentage of nitrogen in the tank more or less than what is found in normal air?

4.145 **Critical Thinking Question:** A patient undergoes laparoscopic surgery. During the procedure, her abdomen is filled with 5.58 L of carbon dioxide at 25 mmHg and 21 °C. How many moles of carbon dioxide have been added to her abdomen?

4.146 **Critical Thinking Question:** A 50:50 nitrous oxide/oxygen mixture is often used as a pain reliever and a sedative. If the partial pressure of oxygen is 0.50 atm, what is the partial pressure of nitrous oxide? What is the total pressure?

Answers to Practice Exercises

4.1 A solid has a definite shape and volume, independent of its container. A liquid occupies a definite volume, but does not have a definite shape—it conforms to the shape of the container. A gas has neither a definite volume nor a definite shape. A gas expands to fill its entire container.

4.2 99 °F and 310 K.

4.3 314 K

4.4 −269 °C and −452 °F

4.5 a. gas b. gas c. solid d. solid and liquid

4.6 a. potential b. kinetic c. potential d. kinetic

E4.1 a. 65 cal b. 4.1×10^2 cal c. 2.3×10^2 cal

E4.2 aluminum

E4.3 The large specific heat of water allows the human body to better regulate its temperature. If the specific heat were smaller, the temperature of the body would fluctuate with the temperature of the ambient air.

E4.4 The house made of brick will be cooler, since brick has a higher specific heat than wood.

4.7 a. deposition b. melting c. vaporization

4.8 deposition

4.9 solid → gas

4.10 gas → liquid (condensation)

4.11 When steam comes in contact with the skin, the amount of heat transferred to the skin includes both the heat of vaporization (heat removed for the steam condensing to water) and the heat removed to cool the water from 100 °C to 37 °C. When boiling water comes in contact with the skin, the heat transferred to your skin is only the amount of heat removed to cool the water from 100 °C to 37 °C.

4.12 When food is heated in the microwave, the energy of the individual water molecules increases and causes them move faster. This increase in kinetic energy of water molecules increases the temperature of the food.

4.13 Heat from the infant's body is used to evaporate water from his skin, causing his body temperature to decrease. This is a result of the relatively high heat of vaporization of water.

4.14 2.0×10^{-2} atm

4.15 a barometer

4.16 Atmospheric pressure decreases at higher altitudes. There is less air above you as you go higher in altitude.

4.17 Denver is one mile above sea level, while Florida is at sea level. The atmospheric pressure in Denver is therefore lower than it is in Florida. The lower atmospheric pressure allows the golf ball to have more "lift" and carry farther, since there is not as much pressure on the ball in Denver as there is in Florida.

4.18 Different tennis balls are made for communities at higher altitude because the balls will have more "lift" at the lower atmospheric pressure of higher altitudes.

4.19 120 mmHg = 1.6×10^4 Pa, 80 mmHg = 1.1×10^4 Pa.

4.20 A sphygmomanometer is a device used to measure blood pressure.

4.21 Systolic pressure is the temporary increase in blood pressure caused by the heart contracting and forcing blood through the arteries. Diastolic pressure is the temporary decrease in blood pressure as the heart muscle relaxes.

4.22 Vapor pressure is usually reported in units of mmHg.

4.23 Methylene chloride is an inhalation hazard. The vapor pressure indicates that methylene chloride is volatile and thus likely to be inhaled.

4.24 The normal boiling point of a liquid is the boiling point at 1 atm. Boiling points are lower at higher altitude because the atmospheric pressure is lower.

4.25 In an autoclave, the pressure is much greater than atmospheric pressure, so water boils at a temperature higher than 100 °C. The extra-hot water inactivates transmissible agents like bacteria, viruses, and spores.

4.26 A substance that has a low boiling point has a high vapor pressure.

4.27 Vapor pressure is the pressure exerted by the gas phase molecules in contact with the liquid, at a given temperature.

4.28 47.07 mmHg

4.29 This compound would be expected to be volatile, with a low boiling point.

4.30 **a.** The pressure of the gas **b.** the temperature of the gas **c.** the volume of the gas.

4.31 The balloon bursts due to the increased pressure, caused by the high temperature.

4.32 **a.** 1.56 mol **b.** 1×10^2 L **c.** 6.74 mol.

4.33 99 L

4.34 No, his lungs will not rupture. The air in the pilot's lungs will occupy a volume of 1.2 L at 0.45 atm.

4.35 As you ascend, the pressure decreases and the volume of air in your lungs would increase, potentially rupturing them.

4.36 106 kPa

4.37 37 °C

4.38 At high altitudes water boils below 100 °C, so food takes longer to cook. Using the pressure cooker, the boiling temperature of water can be increased to cook the food more quickly.

4.39 As the temperature increases, the kinetic energy of the molecules increases. The gas particles move faster and they collide more frequently with the walls of the container, causing pressure to increase.

4.40 As the temperature increases the gas particles will move faster. In order to keep the number of collisions with the walls the same (constant pressure), the gas expands and the volume increases.

4.41 17.4 L

4.42 0.045 mL

4.43 **a.** −170. °C **b.** 2370 °C **c.** −230. °C

4.44 **a.** At a constant temperature and pressure, the number of collisions with the walls of the container is constant. If the number of gas molecules increases, the space occupied by the gas molecules needs to increase to keep the number of collisions with the walls constant.

b. $\dfrac{V_1}{n_1} = \dfrac{V_2}{n_2}$

4.45 2.37 atm

4.46 6.4 atm

4.47 The concentration of nitrogen should be greater 30 ft below the surface than at the surface. The pressure is higher 30 ft below, therefore the concentration is higher.

4.48 Isoflurane. Since the Henry's constant is larger for isoflurane, the concentration of isoflurane in the bloodstream would be smaller.

4.49 The patient in Boston. The atmospheric pressure in Boston is greater than in Denver. The higher the pressure, the higher the concentration of anesthesia in the bloodstream.

4.50 London. The concentration of carbon dioxide (the bubbles in champagne) will be higher where the atmospheric pressure is higher. In the Swiss Alps, the CO_2 would escape more quickly once the bottle is opened and depressurized, causing the glass of champagne to go "flat" sooner.

4.51 **a.** Below 0.1 atm, as this partial pressure would indicate that the amount of oxygen in her lungs was decreasing. **b.** Above 1.4 atm, as this pressure would indicate that he was breathing in too much oxygen.

4.52 The nitrogen partial pressure does not change much between inhaled and exhaled air because nitrogen is not used in the biochemical reactions in the body.

Solutions, Colloids, and Membranes

A health care professional adjusts the drip rate of an intravenous (IV). An IV delivers a mixture of substances directly into the bloodstream. [Tetra Images/Punchstock]

OUTLINE

 This icon indicates that a **Problem-Solving Tutorial** is available at www.whfreeman.com/gob

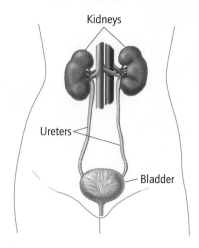

Figure 5-1 The kidneys are located in the middle of the back. Waste and excess water are filtered through the kidneys and flow into the bladder through the ureters.

Kidney Disease

In the United States alone, kidney disease affects more than 7.5 million people. There are approximately 61,000 people waiting for a donor kidney, according to the United Network of Organ Sharing (UNOS). People who have less than 15% kidney function must undergo *dialysis* until a kidney is found. Dialysis is a life support treatment that performs the functions that the damaged kidneys are no longer able to perform.

The kidneys are organs that filter about 200 L of blood daily to remove waste products (certain ions and small molecules) and about 2 L of water. The waste and surplus water become urine, which flows to the bladder through the ureters (Figure 5-1). Networks of arterial capillaries are present in the kidney. The walls of these capillaries are membranes that selectively allow certain compounds in the blood, such as urea, ions, glucose, and amino acids, to pass through the capillary walls directly into the kidney.

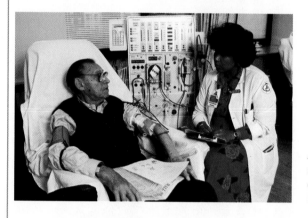

A nurse consults with a patient undergoing dialysis. [Hank Morgan/Photo Researchers, Inc.]

However, larger particles such as proteins do not pass through the capillary membranes, and therefore remain in the bloodstream. Some of the ions, small molecules, and water will be reabsorbed back into the bloodstream, while others such as urea, creatinine, and surplus water, will not. These are waste products that are excreted as urine.

Kidney disease can lead to complete kidney failure, called *renal failure*. Kidney failure may occur suddenly ("acute renal failure") from severe shock, dehydration, heart attack, or severe kidney infection. Or the onset of kidney disease may be gradual ("chronic renal failure"), as the result of diabetes, high blood pressure, or certain hereditary factors. When the kidneys stop working completely, the body retains water and waste products, causing swelling, which is especially noticeable in the hands and feet. Dialysis is the only way to remove the excess water and waste products from the body, and must be done frequently, on average three times a week or more. Although dialysis allows many people with kidney failure to function for long periods of time, it is uncomfortable and inconvenient—and commonly must be performed in a hospital. Kidney disease often cannot be cured, and kidney transplants become the only hope for a person's long-term survival.

In this chapter you will learn about the chemical properties of solutions and membranes, and how the body functions to regulate the concentrations of nutrients. By the end of the chapter you will be able to more fully understand the process of kidney dialysis, which is described in further detail in *Chemistry in Medicine: Kidney Dialysis*, at the end of this chapter.

In the previous chapters you studied pure substances in various states of matter. In this chapter you will study mixtures. *A **mixture** contains two or more elements or compounds in any proportion.* The body is composed largely of aqueous mixtures: mixtures in which water is the main component.

For example, blood is an aqueous mixture, and so are most of the other fluids in your body.

In this chapter, you will learn how scientists and medical professionals quantify the compounds and elements that make up a mixture. For example, a blood test is a report indicating the concentrations of oxygen, glucose, cholesterol, sodium, and other substances in your blood. You will learn how to interpret the concentrations given in a blood test, as well as how to calculate concentrations and dosages of medications. You will learn about colloids and suspensions and how to distinguish them from typical solutions. This is something your kidneys do naturally in filtering wastes and retaining necessary nutrients.

5.1 | Mixtures and Solutions

Matter is composed of both pure substances and mixtures. Recall from Chapter 2 that pure substances can be either elements or compounds. Whereas the atoms in a compound must be present in a definite proportion, the components of a mixture can be present in any proportion, and it still retains its identity. For example, premium and regular gasoline contain different proportions of the same compounds, but both are considered the mixture gasoline.

Mixtures are further distinguished from pure substances because a mixture *can* be separated into its pure components through physical separation techniques. Physical separations take advantage of the differences in physical properties between the various components of the mixture. For example, if you had a mixture of sugar and sand, you could separate the sugar from the sand by first adding water to the mixture. Sugar dissolves in water, while sand does not—a difference in physical properties. The mixture could then be filtered, separating the pure sand from the remaining solution. The water could be evaporated from the solution to isolate pure sugar. The various components of your blood are separated in a blood test using sophisticated instrumentation. Likewise, instruments are available to separate and identify illegal substances such as steroids, cocaine, or LSD that may be present in blood or urine. This type of analysis is routinely performed on athletes at most major athletic competitions such as the Olympics.

There are two general classes of mixtures: heterogeneous and homogenous. In a **heterogeneous mixture** the components are unevenly distributed throughout the mixture. Granite rock is an example of a heterogeneous mixture (Figure 5.2a). If you examine a sample of granite, you will see several different components, distinguished by color and texture, distributed unevenly throughout the rock. Conversely, the components of a **homogeneous mixture** are evenly distributed throughout the mixture. A homogeneous mixture is also known as a **solution.** A cup of coffee containing water, the coffee flavors, caffeine, and sugar would be considered a homogeneous mixture, or a solution (Figure 5.2b). The different classifications of matter are summarized in Figure 5-3.

(a) Heterogeneous mixture

(b) Homogeneous mixture

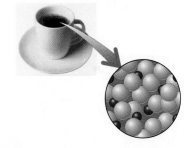

Figure 5-2 (a) Granite is a heterogeneous mixture. The components of heterogeneous mixtures are not evenly distributed. (b) Coffee is a homogeneous mixture. The components of homogenous mixtures are evenly distributed.
[(a) sciencephotos/Alamy; (b) D. Hurst/Alamy]

solute + solvent = solution

Solutions

A solution consists of a *solvent* and one or more *solutes*. A **solute** is a substance in the solution that is present in the lesser amount and the **solvent** is the substance present in the greatest amount. A solute is said to be *dissolved in* the solvent. The term "dissolved" means that the atoms or molecules of the solute are evenly distributed within the solvent. The solute molecules are no longer interacting with other solute molecules, but with the solvent molecules.

A solution may contain more than one solute. For example, if you dissolved a teaspoon of sugar and a teaspoon of salt in a full glass of water,

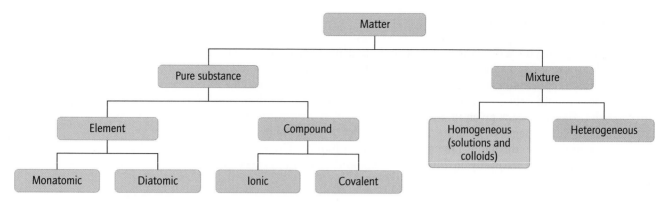

Figure 5-3 Classifications of matter.

(a)

(b)

(a) Sugar dissolves in water: sugar is the solute, water is the solvent. (b) In taffy, water dissolves in sugar: water is the solute, sugar is the solvent.

[(a) © 2006 Richard Megna/Fundamental Photographs; (b) © 2005 Mitch Wojnarowicz/ Amsterdam Recorder/The Image Works]

sugar and salt would be the solutes, and water would be the solvent, because less sugar and salt are present than water. If the amount of sugar exceeded the amount of water, then sugar would be the solvent and water the solute. This is the case for taffy, an extremely sweet candy.

Polar solvents will dissolve polar solutes, and nonpolar solvents will dissolve nonpolar solutes, as characterized in the well-known phrase *"like dissolves like."* In general, nonpolar molecules and polar molecules will not mix; they interact through different forces of attraction. Nonpolar molecules interact only through dispersion forces, whereas polar molecules are capable of dipole–dipole interactions and hydrogen bonding. Consider oil and vinegar as an example. Vinegar is a solution containing water as the solvent. Oil is insoluble in water because it is composed primarily of nonpolar molecules, which cannot hydrogen bond with water molecules. Therefore, when oil encounters water, the hydrogen bonding forces between water molecules exclude the oil molecules, "pushing" the less dense oil to the surface. When two liquids do not mix, they are said to be insoluble in one another, or **immiscible.** Most biological solutions consist of molecules or ions dissolved in water, known as **aqueous solutions,** which will be the focus of this chapter.

Table 5-1 | Solute/Solvent Combinations for Some Common Solutions

Solvent = Solid

Solvent	Solute	Common Solution
Cu(s)	Solid: Sn(s) and sometimes P(s)	Bronze, a metal alloy
Ag(s)	Liquid: Hg(l)	Amalgam used for dentistry

Solvent = Liquid

Solvent	Solute	Common Solution
Ethanol(l)	Solid: I_2(s)	Tincture of iodine
H_2O(l)	Liquid: Ethanol(l)	Alcoholic beverages
H_2O(l)	Gas: CO_2(g)	Carbonated beverages

Solvent = Gas

Solvent	Solute	Common Solution
Air(g)	Solid: Naphthalene(s)	The scent of mothballs
Air(g)	Liquid: Water droplets(l)	Humid air
N_2(g)	Gas: O_2(g)	Air

There is a limit to how much solute a solvent can dissolve. When a solution cannot dissolve any more solute it is called a **saturated solution** and the solute will physically separate from the solvent. You may see a separate phase, or even a precipitate, such as the undissolved sugar at the bottom of a glass of iced tea. Indeed, kidney stones and gout are the result of saturated solutions of calcium salts such as calcium phosphate ($Ca_3(PO_4)_2$), or calcium oxalate (CaC_2O_4) in the kidneys and joints, respectively. The calcium salts physically separate from solution.

Table 5-1 lists several familiar solutions. As you can see from the table, the solvent may be in any of the three possible states of matter—solid (s), liquid (l), or gas (g).

WORKED EXERCISES 5-1   Understanding Mixtures and Solutions

1 Classify each mixture below as a *homogeneous* or *heterogeneous* mixture.
   a. a chocolate chip cookie
   b. a cup of tea (tea bag removed)
   c. a bucket full of rocks and sand
2 If necessary, use Table 5-1 to identify the solute(s) and solvent in each solution.
   a. dental amalgam   b. a cola drink
   c. a saline solution for your contact lenses made from sodium chloride and water
   d. 1 tsp of sugar in an 8-oz glass of water

SOLUTIONS
1 a. Heterogeneous; the chocolate chips are not evenly distributed throughout the cookie.
   b. Homogeneous; the compounds that make up tea are uniformly distributed throughout the solvent, water.
   c. Heterogeneous; the rocks are not evenly distributed throughout the sand.
2 a. Solute: Hg, solvent: Ag.
   b. Solute: CO_2, solvent: water.
   c. Solute: sodium chloride, solvent: water.
   d. Solute: sugar, solvent: water.

PRACTICE EXERCISES

5.1 Explain the difference between a homogeneous and a heterogeneous mixture.
5.2 Identify the solute and solvent in each solution:
   a. 5 mL of ethanol and 25 mL of water
   b. 200 g of water containing 6 g of NaCl
   c. 0.005 L of CO_2 and 2 L of O_2
5.3 What does the phrase "like dissolves like" mean?
5.4 Why is vinegar immiscible in oil?
5.5 What is a saturated solution? How can you tell when a solution is saturated?
5.6 Based on the solubility characteristics of solutions, why do kidney stones form?

Three important types of solutes are present in biological solutions:
- molecules,
- ions, and
- gases.

(a)

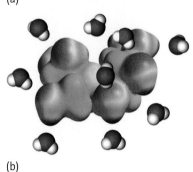

(b)

Figure 5-4 (a) Sucrose, a molecule, dissolves in water because many water molecules surround each sucrose molecule and hydrogen bond with it. (b) Water molecules surrounding a sucrose molecule, shown as an electron density diagram. The hydrogen atoms on the water molecules are attracted to the more "electron-rich" portions of the sucrose molecule (red areas). The oxygen atoms on the water molecules are attracted to the more "electron-poor" areas of the sucrose molecule (blue areas).

Molecules as Solutes

Recall that a molecule is a neutral compound or element containing atoms that are joined by covalent bonds. *When a molecule dissolves in a solvent, the covalent bonds remain intact.* When dissolved, the chemical structure of a molecule exists exactly as it did in its pure form, except that it is surrounded by solvent molecules rather than other solute molecules. For example, when sucrose ($C_{12}H_{22}O_{11}$) is dissolved in water, the sucrose molecules disperse *uniformly* throughout the aqueous solution, each molecule surrounded by many water molecules. During dissolution—the process of dissolving—several water molecules surround each sucrose molecule, as shown in Figure 5-4. Hydrogen bonding between sucrose and water aid the dissolution process. Because each solute molecule retains its structural integrity, dissolution is classified as a physical change rather than a chemical change.

Many solute molecules serve as nutrients. For example, the sugar glucose dissolved in the bloodstream provides a source of energy for cells in the body. Other solute molecules in blood include amino acids and the water-soluble vitamins.

Ions as Solutes

The crystal lattice of most ionic compounds—described in Chapter 2—dissolves readily in water to form individual ions surrounded by many water molecules. For example, sodium chloride, NaCl(s), familiar as table salt, dissolves in water to produce *separated* Na^+ ions and Cl^- ions, which are each surrounded by water molecules. The Na^+ and Cl^- ions separate from the solid NaCl lattice, and disperse uniformly throughout the aqueous solution. The partially negative oxygen atoms of water molecules attract the positively charged sodium cations, Na^+, drawing the cations into solution. Likewise, the partially positive hydrogen atoms of water molecules attract the negatively charged chloride anions, Cl^-, drawing the anions into solution. Figure 5-5 shows the dissolution process for sodium chloride. The solution will become saturated if too much NaCl is added, whereupon the lattice begins to re-form in a process called precipitation. Indeed, that is the case in kidney stones.

A substance that produces ions when dissolved in water is known as an **electrolyte.** Solutions containing electrolytes conduct an electric current, and this property is the origin of the term *electro*lyte. Some physiologically important electrolytes are listed in Table 5-2. The charge on an ion is a key part of what makes communication between nerve cells and muscle cells possible. Large electrolyte imblances in the body can cause dehydration and overhydration, which may lead to cardiac and neurological complications.

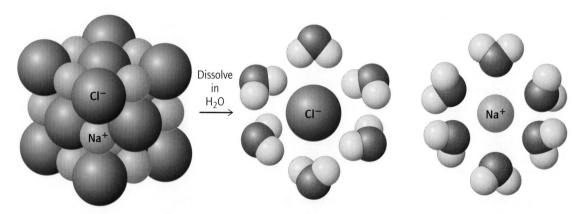

Dissolve in H_2O

Figure 5-5 Sodium chloride, NaCl, dissolves in water to produce Na^+ and Cl^- ions, both of which are surrounded by many water molecules.

The formula unit of an ionic substance allows you to determine which ions are produced and in what proportions when the ionic compound dissolves in water and forms an aqueous solution (aq). For example, the formula unit NaCl indicates that a 1:1 ratio of Na^+ and Cl^- ions is released into solution:

$$NaCl(s) \longrightarrow Na^+(aq) + Cl^-(aq)$$
$$\text{(1 sodium ion)} \qquad \text{(1 chloride ion)}$$

The formula unit for sodium phosphate (Na_3PO_4) indicates there are three sodium ions for every one polyatomic phosphate anion (PO_4^{3-}) released:

$$Na_3PO_4(s) \longrightarrow 3Na^+(aq) + PO_4^{3-}(aq)$$
$$\textbf{(3 sodium ions)} \qquad \textbf{(1 phosphate ion)}$$

Table 5-2 Physiologically Important Electrolytes
Sodium (Na^+)
Potassium (K^+)
Calcium (Ca^{2+})
Magnesium (Mg^{2+})
Chloride (Cl^-)
Hydrogen phosphate (HPO_4^{2-})
Hydrogen carbonate (HCO_3^-)

Gases as Solutes

Molecules in the gas phase can also dissolve in water. A familiar example of a gas solute dissolved in an aqueous solution is a carbonated beverage. Carbon dioxide is the solute, dissolved under pressure in water to produce soft drinks. When the bottle or can of soft drink is opened, the carbon dioxide gas comes out of solution—these are the bubbles you see as a result of the rapid depressurization. Blood contains the important dissolved gases oxygen, O_2, and carbon dioxide, CO_2.

WORKED EXERCISE 5-2 | Ions as Solutes

How many ions of each type will be produced when Na_3PO_4 is dissolved in aqueous solution? Would you describe the solutes as molecules, electrolytes, or gases? Explain.

SOLUTION

The formula unit for sodium phosphate (Na_3PO_4) indicates there are 3 sodium ions for every one polyatomic phosphate anion (PO_4^{3-}) released:

$$Na_3PO_4(s) \longrightarrow 3Na^+(aq) + PO_4^{3-}(aq)$$
$$\textbf{(3 sodium ions)} \qquad \textbf{(1 phosphate ion)}$$

Since Na_3PO_4 produces ions when dissolved in water, it is an electrolyte.

Recall from Chapter 2 that the sum of the charges is always zero.

PRACTICE EXERCISES

5.7 For the following aqueous solutions, indicate whether the solute(s) is(are) a molecule or electrolyte(s). If the solute is a molecule, indicate if it is a gas.

   a. glucose(aq), $C_6H_{12}O_6$
   b. KCl(aq)
   c. CO_2(aq)

5.8 For the following compounds, how many ions of each type will be produced per formula unit in aqueous solution?

   a. $MgCl_2$
   b. K_3PO_4
   c. $NaHCO_3$

5.9 Describe the formation of an aqueous solution of glucose, $C_6H_{12}O_6$, from pure, solid glucose.

5.10 Describe the formation of an aqueous solution of KI from pure, solid KI.

Figure 5-6 Solutions ranging from dilute to concentrated contain increasing amounts of the solute relative to the solvent. Here, the solute is a blue dye. [©2008 Richard Megna, Fundamental Photographs]

5.2 | Solution Concentrations

When a solution contains a large amount of solute, it is described as **concentrated.** When a solution contains a small amount of solute, it is described as **dilute.** Interpreting solution concentrations is an important part of a healthcare professional's responsibilities. For example, a solution with a concentration of 5 mg/mL is 1000 times more concentrated than a solution with a concentration of 5 μg/mL. As you can imagine, administering a solution to a patient that is 1000 times more concentrated *or* dilute than recommended can have fatal consequences.

Solutions can have a range of concentrations. You can *see* the differences in concentration of the solutions in Figure 5-6, which contain various amounts of a blue-colored solute. In the photo, lighter colored solutions contain less solute; they are more dilute. Darker colored solutions contain more solute; they are more concentrated. Some of the simplest ways to measure concentration are colorimetric measurements, which relate the intensity of the color of the solution to the concentration of solute. For example, the Bradford protein assay measures the amount of protein in a solution by adding a special colored dye that binds to the protein.

Measures of Solution Concentration

The **concentration** of a solution is a quantitative measure of how much solute is dissolved in a given quantity of solution. Concentration is always expressed as a ratio, where the amount of the solute is given in the numerator and the total amount of solution is given in the denominator. Note that the total amount of solution is defined as the amount of solute *plus* the amount of solvent.

$$\text{Concentration} = \frac{\text{amount of solute}}{\text{amount of solution}}$$

$$\text{Amount of solution} = \text{amount of solute} + \text{amount of solvent}$$

There are many different ways to express concentration, but most fall into one of two categories:

- concentrations based on mass of solute and
- concentrations based on moles of solute

These concentration units differ only in the way the solute is reported: in grams or in moles. The amount of solution—the denominator—is often reported as a volume unit, typically using the metric unit of the liter (L). The units of concentration most commonly encountered in the medical field are:

- mass/volume (m/v)   } units based on mass
- % mass/volume (%m/v)
- moles/volume   } units based on moles
- equivalents/volume

mass/volume When looking over the results of a blood test, you will notice that several important solutes are reported in concentrations of *mass* of solute per a given *volume* of blood. Medication dosages are also often expressed as a mass/volume concentration. The general form of the equation for determining a mass/volume concentration is

$$\frac{\text{mass}}{\text{volume}} = \frac{\text{mass of solute}}{\text{volume of solution}}$$

For example, consider a patient whose iron level is 125 μg/dL: 125 μg of iron (the solute) are dissolved in every deciliter of blood (the solution). Blood tests normally report glucose, cholesterol, and creatinine in units of mg/dL, and protein in units of g/dL. These are all units of mass/volume; they differ only in the amount of solute and therefore the metric prefix used.

Note that in many solution concentration problems, you will come across the word "per." In calculations, "per" can be translated into a conversion factor. Any unit *per* another unit can be expressed as a fraction where "per" indicates division of one unit by the other. For example, 60 miles per hour can be expressed as $\dfrac{60\ \text{mi}}{1\ \text{hr}}$ or $\dfrac{1\ \text{hr}}{60\ \text{mi}}$. Therefore, a blood iron concentration of 125 μg per deciliter can be expressed as $\dfrac{125\ \mu\text{g}}{1\ \text{dL}}$ or $\dfrac{1\ \text{dL}}{125\ \mu\text{g}}$.

[Jose Luis Pelaez, Inc./Corbis]

Blood Test

John Doe

Blood Component	Value	Normal Range
Iron	125 μg/dL	40–150
Glucose	78 mg/dL	70–110
Chloride	104 mmol/L	98–109
CO_2	25 mmol/L	21–31

WORKED EXERCISES 5-3 Calculating Mass/Volume Concentrations

1 The concentration of an intravenous (IV) levodopa solution, used to treat Parkinson's disease, is 3.0×10^2 mg levodopa in 2.5×10^2 mL of total solution. Express this concentration in milligrams per milliliter.

2 For a solution containing 137 μg/dL of Fe, how many grams of Fe are dissolved per milliliter of solution? *Note:* There are 100 milliliters in a deciliter.

SOLUTIONS

1 The supplied units are mg and mL, and the requested unit is mg/mL. In order to obtain the concentration in mg/mL, write the mass of solute in the numerator and the volume of the solution in the denominator. Then simply divide:

$$\frac{3.0 \times 10^2\ \text{mg}}{2.5 \times 10^2\ \text{mL}} = 1.2\ \text{mg/mL}$$

2 Recall from Chapter 1 how to use dimensional analysis to perform conversions. The same procedure works to solve solution concentration problems.

Step 1: Express conversions as conversion factors. In this problem, there are two conversions needed. The supplied unit is μg/dL and the requested unit is g/mL. Therefore, we need to convert μg to g and dL to mL in order to express this concentration in g/mL. Recall that there are always two ways to write a conversion factor, obtained by inverting the numerator and denominator.

$$\frac{1\ \mu\text{g}}{10^{-6}\text{g}} \quad \text{or} \quad \frac{10^{-6}\text{g}}{1\ \mu\text{g}}$$

and

$$\frac{1\ \text{dL}}{100\ \text{mL}} \quad \text{or} \quad \frac{100\ \text{mL}}{1\ \text{dL}}$$

Step 2: Set up the calculation so that the supplied units cancel. Begin with the supplied concentration. Choose the appropriate conversion factors for the numerator and the denominator so that the supplied units (μg/dL) cancel and you are left with the answer in the requested units (g/mL). This calculation can be performed in one step as shown:

$$\frac{137\ \cancel{\mu\text{g}}}{\cancel{\text{dL}}} \times \frac{10^{-6}\ \text{g}}{1\ \cancel{\mu\text{g}}} \times \frac{1\ \cancel{\text{dL}}}{100\ \text{mL}} = 1.37 \times 10^{-6}\ \text{g/mL}$$

PRACTICE EXERCISES

5.11 Lidocaine, a common local anesthetic and antiarrhythmic drug, comes in IV form in a concentration of 4 g/L. What is the concentration of this solution in milligrams per deciliter?

5.12 Isuprel, used to treat asthma, comes in IV form at a concentration of 4 mg in 500 mL. What is the concentration of this solution in grams per liter?

5.13 For a solution containing 99 mg/dL of glucose, how many micrograms of glucose are dissolved per liter of solution?

5.14 For a solution containing 122 mg/dL of Fe, how many grams are dissolved per liter of solution?

% mass/volume Many solutions used in **intravenous (IV)** therapy have their concentrations reported in units of percent (%) mass/volume. The % mass/volume is the ratio of mass of solute, in grams, to the volume of solution, in milliliters, reported as a percentage by multiplying by 100. Remember, % means per 100.

$$\%\frac{\text{mass}}{\text{volume}} = \frac{\text{g of solute}}{\text{mL of solution}} \times 100$$

The terms **ppm** (parts per million) and **ppb** (parts per billion) are analogous to % mass/volume, and are used when a very small amount of solute is present.

$$\%(\text{m/v}) = \frac{\text{g of solute}}{\text{mL of solution}} \times 100$$

$$\text{ppm(m/v)} = \frac{\text{g of solute}}{\text{mL of solution}} \times 1{,}000{,}000$$

$$\text{ppb(m/v)} = \frac{\text{g of solute}}{\text{mL of solution}} \times 1{,}000{,}000{,}000$$

For % m/v concentrations, the units must be in g/mL. All solutions described here will be aqueous solutions. Recall that the density of water is 1 g/mL; therefore 1 g of water = 1 mL of water. The use of g/mL gives us a unitless ratio to be multiplied by one hundred, for a true percentage. This is why the units g/mL are omitted from the concentration term and only the % is shown.

For example, the most commonly administered IV solution, physiological saline, has a concentration of 0.9% (m/v) of NaCl. That is, 0.9 g of the solute, NaCl, is dissolved in every 100 mL of total solution. Written as a fraction, the concentration is 0.9 g/100 mL.

Table 5-3 lists some common IV solutions reported in their standard form: % mass/volume. All are aqueous solutions used to treat dehydration and electrolyte imbalances. For example, the solute in a 5% dextrose solution is the sugar D-glucose. This solution is often given to patients at risk of having low blood sugar or an elevated sodium, Na^+, concentration.

D-glucose is also commonly known as dextrose.

Table 5-3 Some Common Solutions Used in IV Therapy

Solution Concentration	Common Name
0.9% NaCl	Normal saline
0.45% NaCl	Half-normal saline
5% dextrose	D5W
3.3% dextrose, 0.3% normal saline	2/3 & 1/3
0.6% NaCl, 0.31% sodium lactate, 0.03% KCl, 0.02% $CaCl_2$	Ringer's Lactate

WORKED EXERCISE 5-4 Calculating % m/v

A solution is prepared by dissolving 0.15 g of NaCl in enough water to give a final solution volume of 275 mL. What is the % m/v of the NaCl solution?

SOLUTION
The supplied units are g and mL. Recall,

$$\%\frac{\text{mass}}{\text{volume}} = \frac{\text{g of solute}}{\text{mL of solution}} \times 100$$

$$= \frac{0.15 \text{ g NaCl}}{275 \text{ mL solution}} \times 100 = \mathbf{0.055\% \ (m/v) \ NaCl}$$

PRACTICE EXERCISES

5.15 What is the % m/v of an I_2 solution prepared by dissolving 8.5 g of I_2 in enough ethanol to make 225 mL of solution?

5.16 Calculate the % m/v of sucrose in a carbonated beverage that contains 28 g of sucrose in 315 mL of beverage.

WORKED EXERCISE 5-5 Preparing IV Solutions

You are asked to prepare 2 L of a 5% dextrose solution, a common IV solution. Recall that a 5% solution contains 5 g of solute in every 100 mL of solution. How much of the solute do you need to weigh out (in grams)?

SOLUTION

Step 1: Express conversions as conversion factors. In this problem, the supplied information is 2L and 5% solution. The requested unit is grams of solute. You will need to use the concentration as a conversion factor between volume of solution (supplied unit) and the mass of solute (requested unit): 5% = 5 g/100 mL. You will also need the metric conversion between L and mL:

$$5\% = \frac{5 \text{ g}}{100 \text{ mL}} \text{ or } \frac{100 \text{ mL}}{5 \text{ g}}$$

$$\frac{1 \text{ mL}}{10^{-3} \text{ L}} \text{ or } \frac{10^{-3} \text{ L}}{1 \text{ mL}}$$

Step 2: Set up the calculation so that supplied units cancel.

$$2 \text{ L} \times \frac{1 \text{ mL}}{10^{-3} \text{ L}} \times \frac{5 \text{ g}}{100 \text{ mL}} = 100 \text{ g}$$

PRACTICE EXERCISES

5.17 You have been asked to prepare 1 L of a 0.45 % NaCl (% m/v) solution for IV therapy. How many grams of NaCl(s) should you weigh out?

5.18 You have been asked to prepare 5 L of a half-normal saline solution for IV therapy. Consulting Table 5-3, how many g of NaCl(s) should you weigh out?

You now have been introduced to the most common units of concentration encountered in medicine based on the *mass of solute:* m/v and % m/v. In the next section, you will be introduced to the most common units of concentration encountered in medicine based on *moles of solute:* moles/liter and equivalents/liter.

Molarity Abbreviated M, **molarity** is defined as the number of moles of solute in one liter of total solution: moles per liter.

$$\text{Molarity (M)} = \frac{\text{moles of solute}}{\text{L of solution}}$$

While the concentration units in the previous section describe the *mass* of the solute per unit volume, molar concentration is an indication of the absolute *number* of solute particles per unit volume, regardless of their mass. For

[Jose Luis Pelaez, Inc./Corbis]

Blood Test

John Doe

Blood Component	Value	Normal Range
Iron	125 µg/dL	40–150
Glucose	78 mg/dL	70–110
Chloride	104 mmol/L	98–109
CO_2	25 mmol/L	21–31

Recall that 1 mol = 6.02×10^{23} atoms, molecules, or ions.

example, a 1 molar solution will have 6.02×10^{23} solute particles per liter of solution regardless of the solute's identity or mass. Refer to *Extension Topic 5-1* to learn how to convert between % m/v and molarity, M.

Blood test results typically report calcium, sodium, potassium, chloride, and carbon dioxide concentrations in units of millimoles per liter, mmol/L. Remember, a mmol is 1/1000 of a mole. You may even encounter concentrations reported in μmol/L. The metric prefixes m and μ are often used when the concentration of solute is very small, in order to avoid many zeros in the numerical value. When performing calculations involving molarity, turn the capital M into units of mol/L.

WORKED EXERCISE 5-6 Calculating Molarity

A solution having a volume of 1.5 L contains 0.018 mol of CO_2 (carbon dioxide). What is the concentration of carbon dioxide in this solution in moles per liter, M?

SOLUTION

The supplied units are moles and liters, and the requested unit is mol/L. In order to obtain the concentration in mol/L, write the moles of solute in the numerator and the volume of the solution in the denominator. Remember, volume must be in units of liters. Then simply divide:

$$\frac{0.018 \text{ mol}}{1.5 \text{ L}} = 0.012 \text{ mol/L} = 0.012 \text{ M}$$

PRACTICE EXERCISES

5.19 A solution having a volume of 3.0 L contains 23 mmol of O_2 (oxygen). What is the molarity of this solution?

5.20 How many moles of potassium ions (K^+) are there in 5.0 L of 2.5 mM K_3PO_4?

5.21 How many millimoles of lithium ions (Li^+) are there in 2.5 L of 3.0 mM LiCl?

equivalents/liter Ion concentrations are often expressed in equivalents per liter, because this unit of concentration provides information about the number of charges per liter of solution. An **equivalent,** abbreviated **eq,** is the number of moles of charge. It is calculated by multiplying the number of moles of an ion by the charge on the ion.

Equivalents (eq) = moles of ion × charge on ion

For example, 1 mol of calcium ions (Ca^{2+}) represents 2 equivalents:

1 mol Ca^{2+} × 2 (magnitude of charge) = 2 eq Ca^{2+}

Table 5-4 lists some common ions, their charges, and the number of equivalents represented by 1 mol of ions.

Table 5-4 Common Ions and Equivalents

Charge	Common Ions	Equivalents in 1 mole
+1	Na^+, K^+, H^+, Li^+, NH_4^+	1
+2	Ca^{2+}, Mg^{2+}	2
+3	Fe^{3+}, Cr^{3+}, Al^{3+}	3
−1	Cl^-, F^-, $C_2H_3O_2^-$ (acetate)	1
−2	O^{2-}, S^{2-}, CO_3^{2-} (carbonate)	2
−3	PO_4^{3-} (phosphate), $C_6H_5O_7^{3-}$ (citrate)	3

Extension Topic 5-1 Converting Between Mass and Mole-Based Concentration Units

In this chapter, you have seen that some of the most common units of concentration in medicine are expressed as either mole-based or mass-based concentrations. Often it is more convenient to know the *number* of solute molecules or ions in a solution, expressed in moles, rather than the *mass* of the solute. In such cases, if you are given a concentration that is based on mass, you must perform a unit conversion to obtain the number of solute molecules or ions. The flow chart below illustrates the steps to take in converting a mass-based concentration to a mole-based concentration, and vice-versa.

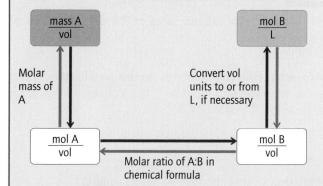

Suppose you need to know how many mol/L of Na^+ (B) there are in a physiological saline solution, which is 0.90% (m/v) NaCl (A).

Two conversions must be performed to solve this problem: Convert the numerator (0.90 g) from mass to moles, and the denominator (100 mL) from 100 mL to L. You can start with either conversion. The following begins with the conversion in the numerator.

Step 1. *Rewrite the % m/v concentration as a fraction, such that you have a specific mass divided by a specific volume.* For a 0.90% m/v solution of NaCl, you know that for every 100 mL of this solution, 0.90 g of NaCl are dissolved.

$$0.90\% \text{ (m/v) NaCl} = \frac{0.90 \text{ g NaCl}}{100 \text{ mL solution}} \times 100$$

Step 2. *Convert moles into mass.* Convert the numerator, 0.90 g of NaCl, into moles of NaCl. Use the molar mass of NaCl, which is 58.44 g/mol, as the conversion factor. Note that the 100 mL isn't being changed yet.

$$0.90 \text{ g NaCl} \times \frac{1 \text{ mol NaCl}}{58.44 \text{ g NaCl}} =$$

$$1.5 \times 10^{-2} \text{ mol NaCl}$$

Step 3. *Convert mass of the compound to moles of the requested ion.* Still focusing on the numerator, convert 1.5×10^{-2} mol of NaCl to mol of Na^+. Look at the subscripts in the chemical formula to determine the ratio of moles of ion to moles of compound. You know

from the chemical formula of NaCl there is 1 mol of Na^+ for every 1 mol of NaCl,

$$\frac{1 \text{ mol Na}^+}{1 \text{ mol NaCl}} \quad \frac{(B)}{(A)}$$

$$1.5 \times 10^{-2} \text{ mol NaCl} \times \frac{1 \text{ mol Na}^+}{1 \text{ mol NaCl}} =$$

$$1.5 \times 10^{-2} \text{ mol Na}^+$$

Note: Carry at least one more significant figure than is necessary until your final calculation; then round to the appropriate number of significant figures.

Step 4. *Convert the supplied volume units into the requested volume units.* In this case, convert the denominator, 100 mL of solution, into liters of solution. Use the metric conversion factor 1 mL = 10^{-3} L. Choose the form that has mL in the numerator, so that mL cancels with the supplied unit and L is left in the denominator.

$$\frac{1.5 \times 10^{-2} \text{ mol Na}^+}{100 \text{ mL}} \times \frac{1 \text{ mL}}{10^{-3} \text{ L}} = \frac{0.15 \text{ mol Na}^+}{\text{L}}$$

Therefore, 0.9% (m/v) NaCl solution has a molar sodium concentration of

$$\frac{0.15 \text{ mol Na}^+}{\text{L}}, \quad \text{or} \quad 0.15 \text{ M}$$

It is also convenient to know how to do the reverse process: converting from a mole-based concentration to a mass-based concentration. To see how to perform the reverse conversion, consult the flow chart again.

PRACTICE EXERCISES

E5.1 Calculate the molarity (mol/L) of Cl^- in a 0.5 g/dL solution of $CaCl_2$.

E5.2 Calculate the molarity (mol/L) of glucose in a 5% (m/v) glucose solution. The molar mass of glucose is 180.2 g/mol.

E5.3 What is the % (m/v) of sucrose in a 25 mM sucrose solution? The molar mass of sucrose is 342.3 g/mol.

E5.4 Calculate the equivalents per liter of Mg^{2+} in a solution that is 0.25% $MgCl_2$.

E5.5 How many equivalents of Na^+ are there per liter of solution in a 0.45% (m/v) NaCl solution?

E5.6 What is the % (m/v) of MgI_2 in a 0.35 eq/L Mg^{2+} solution, assuming that the only source of Mg^{2+} is the MgI_2?

E5.7 Sometimes it is necessary to convert from mol/L (M) to m/v, especially when you need to prepare a solution. How many grams of glucose should you weigh out in order to prepare 1.0 L of a 0.25 M glucose solution? The molar mass of glucose is 180.2 g/mol.

To express the concentration of a solute using equivalents, write the number of equivalents of solute per liter of solution:

$$\frac{\text{equivalents}}{L} \left(\frac{eq}{L}\right) = \frac{\text{equivalents of solute}}{L \text{ of solution}}$$

Blood test results often report calcium, sodium, potassium, and chloride ions in units of milliequivalents per liter, meq/L. Units of meq rather than eq are used because these ions are present in low concentration. Electrolyte concentrations in IV solutions are also often reported in eq/L or meq/L.

WORKED EXERCISE 5-7 Converting mmol/L to meq/L

Consider a blood test that reports a calcium level of 9.3 mmol/L. Convert this value to units of meq/L of Ca^{2+}.

SOLUTION

Step 1: Express conversions as conversion factors. In this problem, you must convert mmol to meq. Since calcium ion (Ca^{2+}) has a charge of +2, there are 2 meq of Ca^{2+} for every 1 mmol of Ca^{2+}:

$$\frac{2 \text{ meq}}{1 \text{ mmol}} Ca^{2+} \quad \text{or} \quad \frac{1 \text{ mmol}}{2 \text{ meq}} Ca^{2+}$$

Step 2: Set up the calculation so that the supplied units cancel.

$$\frac{9.3 \text{ mmol}}{1 \text{ L}} \times \frac{2 \text{ meq}}{1 \text{ mmol}} Ca^{2+} = 19 \frac{\text{meq}}{L} Ca^{2+}$$

PRACTICE EXERCISES

5.22 **a.** How many equivalents per liter of phosphate ion (PO_4^{3-}) are there in a 0.25 M solution of Na_3PO_4?
   b. How many equivalents per liter of Na^+ are there in this solution?

5.23 **a.** How many equivalents per liter of magnesium are there in 0.12 M $MgSO_4$?
   b. How many equivalents per liter of sulfate ion are there in this solution?

International Units IU, which stands for international unit, is often used in medicine. You will often see this unit on a vitamin label. Sometimes it is abbreviated "units." The IU is a unit based on mass that factors in the biological effect of the solute, and therefore varies for different substances. For example, 1 IU of vitamin C is 50 μg of vitamin C, whereas 1 IU of insulin is 45.5 μg of insulin. Thus, you can only convert from IU's of solute to mass or moles of solute if you are given information about the mass of solute per IU for a particular solute.

Dosage Calculations

Calculating dosages of drugs delivered as solutions is an important part of the health care worker's responsibility. Oral medications are commonly administered in liquid form, especially to children, and the total volume and drip rate of IV solutions often must be calculated. At times, conversion between units of measurement may be necessary. In Chapter 1 you learned how to perform metric to English conversions. The following exercises show how to calculate the proper dosage for oral suspensions and IV medications.

Dosage of Oral Suspensions A drug suspension is a mixture in which the drug is in solid form suspended as fine particles throughout the liquid. You will learn more about the properties of suspensions in the next section of this chapter.

WORKED EXERCISE 5-8 Calculating the Dose of an Oral Drug Suspension

An order is given for 500 mg of amoxicillin to be administered to a patient every 6 hr. For an oral suspension of amoxicillin that contains 250 mg of amoxicillin in every 5 mL, how many milliliters of the suspension should be administered to the patient every 6 hr?

SOLUTION

Step 1: **Express conversions as conversion factors.** The supplied unit is 500 mg of drug. The requested unit is mL of suspension. You are given the concentration of the drug as 250 mg amoxicillin per 5 mL of suspension. The concentration can be used a conversion factor between mass of drug and volume of solution:

$$\frac{250 \text{ mg}}{5 \text{ mL}} \text{ or } \frac{5 \text{ mL}}{250 \text{ mg}}$$

Step 2: **Set up the calculation so that the supplied units cancel.**

$$500 \text{ mg} \times \frac{5 \text{ mL}}{250 \text{ mg}} = 10 \text{ mL of suspension every 6 hr.}$$

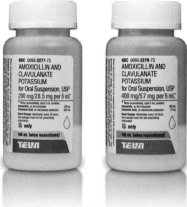

Amoxicillin is an antibiotic prepared in the form of a suspension.
[Courtesy of Teva Pharmaceuticals]

PRACTICE EXERCISES

Problem-Solving Tutorial:
Dosage Calculations

5.24 An order is given for 60 mg of Tylenol to be administered to a patient every 4 hr. You are given a suspension containing 160 mg of Tylenol in every 4.0 mL. How many milliliters of the suspension should you administer to the patient every 4 hr?

5.25 Tegretol is commonly used as an anticonvulsant and mood stabilizing drug. The order is given to administer 0.25 g as needed. The suspension contains 100.0 mg in every 5 mL. How many milliliters of the suspension should be administered to the patient as needed?

5.26 Diuretics increase the elimination of water from the body, and are frequently given to patients to control edema (swelling). Diuretics, however, often cause potassium deficiency. A patient on diuretics is prescribed 30 meq of potassium (K^+) every day. The solution supplied contains 40 meq of KCl in every 15 mL. How many milliliters of this solution should you give to the patient every day?

5.27 An order is given to administer 20 mg of Atarax, an antihistamine, every 4 hr. The suspension contains 10 mg in every 5 mL. How many milliliters should you administer to the patient every 4 hr?

Dosage of IV Solutions Certain IV medications must be administered at a specific dosage per unit of time, known as a **flow rate.** Flow rates are often given in milligrams per minute, micrograms per minute, or units per hour. Calculations can be made to determine flow rate, or when the flow rate is known, to assess the dosage being administered per unit time. It is often necessary to add medication gradually over time, because the body metabolizes drugs. The rate at which a medication must be administered depends in part on the rate at which the drug is metabolized.

Recall that the term "units" is used as an abbreviation for International Units, IUs.

Problem-Solving Tutorial:
Dosage Calculations

WORKED EXERCISE 5-9   Calculating Flow Rate of IV Medications

An order is given to infuse 500 units per hour of heparin, an anticoagulant. The IV bag supplied contains 25,000 units in 250 mL. At what flow rate in milliliters per hour should the IV solution be delivered to the patient?

SOLUTION

Step 1: Express conversions as conversion factors. In this problem, you are supplied with two values that can be used as conversion factors—500 units/hr and 25,000 units/250 mL. The requested unit is the ratio mL/hr.

$$\frac{500 \text{ units}}{\text{hr}} \quad \text{or} \quad \frac{\text{hr}}{500 \text{ units}}$$

$$\frac{25,000 \text{ units}}{250 \text{ mL}} \quad \text{or} \quad \frac{250 \text{ mL}}{25,000 \text{ units}}$$

Step 2: Set up the calculation so that certain supplied units cancel. Only the requested units of mL/hr should remain.

$$\frac{500 \text{ units}}{\text{hr}} \times \frac{250 \text{ mL}}{25,000 \text{ units}} = 5 \text{ mL/hr}$$

WORKED EXERCISE 5-10   Calculating the Administered Dose of IV Medications

Isuprel is used to treat asthma, chronic bronchitis, and emphysema. A patient is receiving 30 mL/hr of a solution that contains 2 mg of Isuprel in 250 mL D5W. How many μg/min is the patient receiving? *Note:* There are 1000 μg in 1 mg.

SOLUTION

Step 1: Express conversions as conversion factors. You are supplied with two conversion factors: 30 mL/hr and 2 mg/250 mL. You also need to convert hours to minutes and micrograms to grams. The requested unit is μg/min.

$$\frac{30 \text{ mL}}{\text{hr}} \quad \text{or} \quad \frac{\text{hr}}{30 \text{ mL}}$$

$$\frac{2 \text{ mg}}{250 \text{ mL}} \quad \text{or} \quad \frac{250 \text{ mL}}{2 \text{ mg}}$$

$$\frac{60 \text{ min}}{\text{hr}} \quad \text{or} \quad \frac{\text{hr}}{60 \text{ min}}$$

$$\frac{1 \text{ mg}}{1000 \text{ μg}} \quad \text{or} \quad \frac{1000 \text{ μg}}{1 \text{ mg}}$$

Step 2: Set up the calculation so that the supplied units cancel. Remember, your final answer should be in units of μg/min:

$$\frac{30 \text{ mL}}{\text{hr}} \times \frac{2 \text{ mg}}{250 \text{ mL}} \times \frac{1 \text{ hr}}{60 \text{ min}} \times \frac{1000 \text{ μg}}{1 \text{ mg}} = 4 \text{ μg/min}$$

Remember, it doesn't matter in what order you multiply these conversion factors because multiplication and division can always be done in any order.

PRACTICE EXERCISES

5.28 Digoxin is often used to treat a variety of heart conditions, for example, heart failure, atrial flutter, or atrial fibrillation. An order is given to administer 0.25 mg digoxin by IV, over a period of 5 min. The solution supplied contains 5 μg digoxin in 2 mL. What should the flow rate be in milliliters per minute?

5.29 Droperidol is often used to treat postoperative nausea. An order is given to administer 0.275 mg droperidol by IV over a period of 3 hr. The IV bag contains 5 mg droperidol in every 2 mL. What should the flow rate be in milliliters per minute?

5.30 Heparin is an anticoagulant, often given to patients after hip replacement surgery. A patient is receiving a heparin solution by IV at a rate of 6 mL/hr. The concentration of the heparin solution is 25,000 units in 250 mL. How many units of heparin per hour is the patient receiving?

5.31 A diabetic patient suffering from hyperglycemia is receiving an insulin solution by IV at a rate of 25 mL/hr. The concentration of the insulin solution is 200 units in every 500 mL. How many units of insulin per hour is the patient receiving?

5.3 | Colloids and Suspensions

Solutions are one type of mixture. You learned in the previous section how solutions are relevant to the medical field, and why it is impotant to know how to perform calculations based on solution concentrations for dosages of medications. Other types of mixtures, known as colloids and suspensions, also have relevance in medicine and biology, as you will see in this section.

Colloids

You may recall using finger paints as a child to make colorful pictures. Would you describe finger paint as a solution? Probably not, because by touching it, you can feel large solid particles that do not dissolve in the paint. Chemically, finger paint is considered a **colloid,** and is classified as a homogeneous mixture, but not a solution. *The distinguishing feature of a colloid is that it contains particles that are much larger than the solute particles of a solution.* Solute particles typically have a diameter less than 1 nm, on the order of the atomic scale; while colloidal particles have diameters between 1 nm and 1000 nm (1 μm), on the order of the microscale. Colloidal particles can be either large molecules or many small molecules grouped together, known as **aggregates.** As with solutes in a solution, colloidal particles are evenly dispersed throughout the mixture.

The component of the colloid present in the greatest amount, and analogous to the solvent in a solution, is known as the **medium.** In a colloid, the particles and the continuous medium can be either a gas, a liquid, or a solid; however, both cannot be gases. Gas mixtures are infinitely soluble in one another and therefore always exist as solutions.

Suspensions

While both solutions and colloids are homogeneous mixtures, suspensions are a type of heterogeneous mixture. In contrast to colloids and solutions, the particles in a **suspension** are not uniformly distributed throughout the medium and eventually settle, like sand in water. The particles in a suspension are even larger than 1 μm and can be filtered, a physical separation technique. The component of the suspension in the greatest amount, analogous to the solvent in a solution, is the **dispersion medium.** In a suspension, the particles are usually solid or liquid, while the dispersion medium is a solid, liquid, or gas. Table 5-5 compares the properties of solutions, colloids, and suspensions, and Table 5-6 lists some familiar examples of colloids and suspensions.

Finger paint is a colloid.
[David Young-Wolff/Alamy]

Table 5-5 Comparison of Solutions, Colloids, and Suspensions

Characteristic	Solutions	Colloids	Suspensions
Particle size	< 1 nm	1 nm–1000 nm	> 1000 nm (1 μm)
Composition	Single atoms, small molecules, ions, or polyatomic ions are dissolved in a solvent.	Aggregates of atoms, molecules, or ions, or large molecules, like proteins and starch, are uniformly distributed throughout the medium.	Large insoluble particles or aggregates, such as red blood cells in whole blood, are distributed nonuniformly throughout the medium.
Appearance	Transparent	Light is scattered along the path of the light beam by the colloidal particles—known as the *Tyndall effect*.	Cloudy, with visible particles. Light is scattered by these larger particles in all directions.
Motion	Solutes and solvent move by molecular motion.	*Brownian motion* occurs—colloidal particles move randomly throughout the continuous medium.	The motion of the particles is influenced by gravity.
Settling	Solutes **never** settle.	Colloidal particles may settle over time due to coagulation.	Will settle some time after mixing.
Examples	Laboratory solutions [Brand X/Corbis]	Mayonnaise [Glowimages/Getty]	Mud [Travel Ink/Getty]

Table 5-6 Examples of Common Colloids and Suspensions

Colloids	Suspensions
Blood serum	Mud
Smoke	Soot suspended in air
Cement	
Hand lotion	Flour suspended in water
Cheese	
Whipped cream	Calamine lotion
	Barium enema
	Acetaminophen suspended in cough syrup

Solutions, Colloids, and Suspensions in Your Body

Whole blood has the characteristics of a solution, a colloid, and a suspension, because it contains particles ranging in size from the nanometer to greater than a micrometer. It is a solution because it contains solutes; it is a colloid because it contains larger colloidal particles, such as proteins; and it is a suspension because it contains large particles, such as red blood cells, suspended in water.

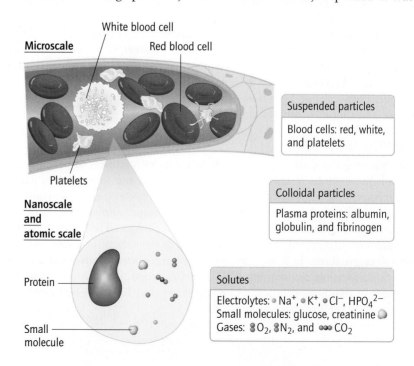

Figure 5-7 The different types of particles in whole blood allow it to be classified as a solution, a colloid, and a suspension.

Figure 5-7 shows some of the components of whole blood according to the type of particle.

You have already seen that medicines are often prepared as suspensions. Some medicines are insoluble in all acceptable media and must be administered as a suspension. Because of their fluid properties, suspensions are an ideal dosage form for patients such as children who have difficulty swallowing tablets or capsules.

Many important processes in your body must be able to separate smaller molecules and ions from larger molecules and aggregates. An example is the filtering of waste products, described in Section 5.4.

WORKED EXERCISE 5-11 Distinguishing Solutions, Colloids, and Suspensions

Do the following examples represent solutions, colloids, or suspensions? Explain your reasoning.
a. egg white (contains the protein albumin)
b. a glass of iced tea
c. the air you breathe
d. chalk dust in water

SOLUTIONS
a. Egg white is a colloid, because it is composed of proteins, which are large molecules, having a diameter greater than 1 nm.
b. Iced tea is a solution; it is a homogeneous mixture.
c. Air is a solution because gases are infinitely soluble in one another.
d. Chalk dust in water is an example of a suspension—after a while the larger chalk particles will settle to the bottom of the container and the smaller ones will float.

PRACTICE EXERCISES

5.32 Describe some differences between a solution, a colloid, and a suspension.

5.33 Do the following examples represent solutions, colloids, or suspensions?
　a. smoke
　b. mud
　c. calamine lotion
　d. 5% (m/v) glucose (aq)

5.34 Explain why humid air is considered a solution, while mist, which is also a mixture of water in air, is considered a colloid.

5.35 Explain why blood has characteristics of a solution, a colloid, and a suspension.

5.4 Membranes, Osmosis, and Dialysis

In this section, you will learn how solutions and colloids are employed in the body. Osmosis, dialysis, and diffusion are all ways that a solution can pass through a membrane. These processes determine the concentration of molecules and ions in different regions and thereby maintain concentration differences required for critical biological functions.

Semipermeable Membranes

A **membrane** is a structure that acts as a barrier between two environments. For example, the outermost layer of your skin serves as a barrier between your body and its surroundings, preventing pathogens from easily entering the body. A **semipermeable membrane** is a membrane that allows the passage of certain small molecules and ions across the membrane while preventing the passage of larger molecules and ions. Most biological membranes are semipermeable.

There are a variety of ways a molecule may pass through a membrane, sometimes requiring the assistance of other molecules. The simplest method is through **simple diffusion,** the spontaneous movement of a molecule or ion from a region of higher concentration to a region of lower concentration until the concentration is the same. Simple diffusion is illustrated in Figure 5-8.

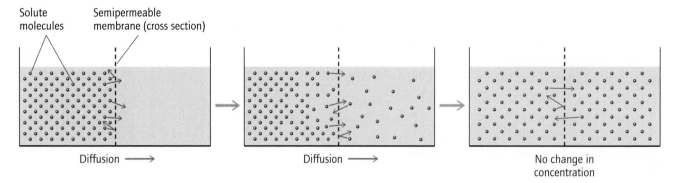

Figure 5-8 Simple diffusion is the spontaneous movement of a molecule or ion from a region of higher concentration (left) to a region of lower concentration (right).

Cell Membranes The cell membrane is a complex semipermeable membrane that separates the interior of the cell from the exterior of the cell, maintaining separate chemical environments in the solutions on either side of the membrane. For example, the concentration of various ions inside the cell is quite different from the concentration outside the cell. Table 5-7 lists the concentrations of some common electrolytes that are found in the interior and exterior of a blood cell. Notice that the units of concentration are mmol/L, a mole-based concentration unit.

Water and certain small molecules are able to cross the cell membrane by simple diffusion, but large molecules and charged species like ions cannot. Particles that cannot cross the cell membrane must be transported across it by special carrier molecules.

Osmosis and Dialysis

Osmosis and dialysis describe the movement of solutes and/or solvent across a semipermeable membrane. These are important processes that regulate the distribution of nutrients in the cell and the removal of waste products from the cell. **Osmosis** refers to the flow of *solvent* (usually water) through a semipermeable membrane. **Dialysis** refers to the movement of small *solutes* through a semipermeable membrane. The physical properties and chemical composition of the membrane determine whether osmosis, dialysis, or a combination of the two occurs. Colloidal particles and suspensions cannot cross a

Table 5-7 Typical Concentrations of Electrolytes Inside and Outside a Blood Cell

Electrolyte	Concentration Inside the Cell (mmol/L)	Concentration Outside the Cell (mmol/L)
Na^+	10	140
Cl^-	4	100
K^+	140	4
Ca^{2+}	1×10^{-4}	2.5

Process	Substance crossing membrane	How does it cross the membrane?	
			Semipermeable membrane Initial solutions
Osmosis	Water (solvent)	Simple diffusion of water from a region of lower solute concentration (higher water concentration) to a region of higher solute concentration (lower water concentration), across a semipermeable membrane.	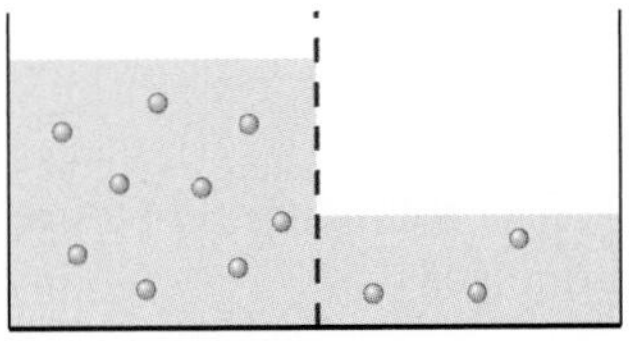 Osmosis: solvent crosses membrane
Dialysis	Solutes	Simple diffusion of solute from a higher concentration of solute to a lower concentration of solute, across a semipermeable membrane	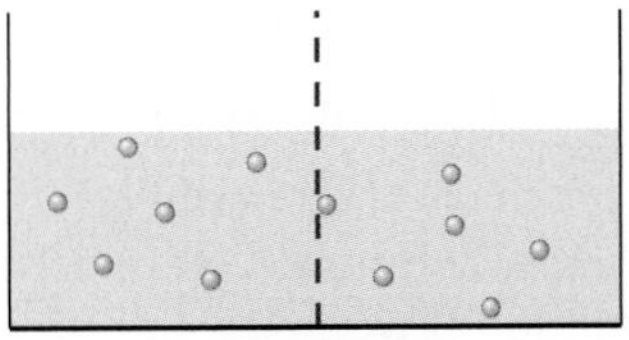 Dialysis: solute crosses membrane

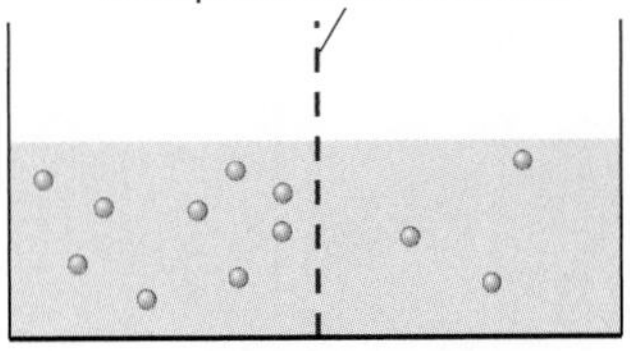

Figure 5-9 Osmosis is the movement of solvent across a membrane. Dialysis is the movement of solute across a membrane.

membrane by dialysis. Thus, dialysis effectively separates large molecules such as proteins and starch from small solute particles. The kidneys, for example, contain semipermeable membranes that allow a natural dialysis process to purify the blood. Figure 5-9 illustrates osmosis and dialysis.

Osmosis Osmosis is a form of simple diffusion. The solvent, usually water, flows from a region of lower solute concentration (higher water concentration) to a region of higher solute concentration (lower water concentration), across a semipermeable membrane. Milk of Magnesia, for example, works as a laxative by drawing water out of the tissues osmotically, thereby increasing the water content of the stool. The flow of water through a membrane is governed by the total number of solute particles, not their mass, size, or identity. For example, a solution may contain 2 mmol of glucose and 3 mmol of urea; however, the important factor is that the solution contains 5 total mmol of particles (2 mmol glucose + 3 mmol urea). Therefore, a mole-based concentration unit, such as molarity, is most practical in discussions of osmosis, since it assesses the total number of particles on either side of the membrane.

Figure 5-10a shows two solutions of equal volume separated by a semipermeable membrane. Water (the solvent) will always diffuse in the direction of the more concentrated solution. Thus, if solution A has a concentration of 0.2 mM, and solution B has a concentration of 0.4 mM, water will diffuse from solution A to solution B. Osmosis will continue until the solute concentration on both sides of the membrane is the same, in this case, 0.3 mM (Figure 5-10b). At that point, the net flow of water will stop. After osmosis, there will be a greater volume of solution on the side of the membrane that initially had the more concentrated solution.

Figure 5-10 The process of osmosis. (a) Before osmosis: Two aqueous solutions with different concentrations are separated by a semipermeable membrane. (b) After osmosis: Solvent has moved into the solution of higher solute concentration, and the two solutions now have equal concentrations. The volume of solution increases on the side that originally was more concentrated.

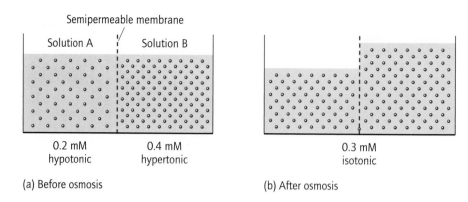

0.2 mM
hypotonic

0.4 mM
hypertonic

0.3 mM
isotonic

(a) Before osmosis

(b) After osmosis

The following terms are frequently used to describe the relative concentrations of two solutions on either side of a semipermeable membrane:

- **Hypertonic**—describes the solution with the higher solute concentration.
- **Hypotonic**—describes the solution with the lower solute concentration.
- **Isotonic**—describes solutions with equal solute concentrations.

Let's apply these terms to the example shown in Figure 5.10: Before osmosis occurs, solution A would be the hypotonic solution (0.2 mM glucose) and solution B would be the hypertonic solution (0.4 mM glucose). After osmosis occurs, solutions A and B are isotonic, both having a concentration of 0.3 mM glucose.

Osmosis can be stopped by placing external pressure on the hypertonic solution. The minimum amount of pressure that must be applied to the hypertonic solution to stop water from flowing across the membrane from the hypotonic solution is called the **osmotic pressure.** When a pressure *greater* than the osmotic pressure is applied to the hypertonic solution, **reverse osmosis** occurs, and water flows in the opposite direction, from the hypertonic solution to the hypotonic solution. In areas of the world where fresh water is scarce, reverse osmosis is used to produce fresh water from seawater (see Figure 5-11), a process known as desalination. Reverse osmosis is also used in hemodialysis, a life support treatment for individuals with kidney disease. It forces excess water out of the blood, while dialysis filters waste products out of the blood.

Osmolarity is a term that defines, in moles per liter, the number of particles in a solution that contribute to its osmotic pressure.

Figure 5-11 In reverse osmosis, pressure applied to the solution of higher solute concentration (hypertonic) forces the solvent (water in this case) into the solution of lower solute concentration (hypotonic). Reverse osmosis is commonly used to desalinate seawater in regions where fresh water is scarce.

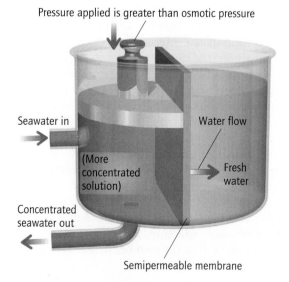

Pressure applied is greater than osmotic pressure

Seawater in

Water flow

(More
concentrated
solution)

Fresh
water

Concentrated
seawater out

Semipermeable membrane

(a) Isotonic solution   (b) Hypotonic solution   (c) Hypertonic solution

Figure 5-12 Red blood cells immersed in solutions of differing concentrations. [Dr. David M. Phillips/Visuals Unlimited]

Osmosis in Red Blood Cells The effects of osmosis can be demonstrated using red blood cells, and you will see why a person receiving IV fluids must always be given an isotonic solution. Red blood cells placed in an isotonic solution have a biconcave disk shape, as shown in Figure 5-12a. Consider what happens to a red blood cell if you immerse it in various solutions with concentrations different from those inside the red blood cell. If the solute concentration outside the cell is equal to the solute concentration inside the cell (an isotonic solution), the cell maintains its shape because there is no net flow of water between the inside and the outside of the cell, through the cell membrane.

When a red blood cell is immersed in a hypotonic solution, water will flow from the outside of the cell to the inside of the cell where solute is more concentrated. You can see from Figure 5-12b that the red blood cell swells visibly and loses its biconcave shape. Red blood cells immersed in hypotonic solutions will swell so much that they eventually burst—an event known as **hemolysis.**

When a red blood cell is immersed in a hypertonic solution, water will flow from the inside of the cell to the outside of the cell, where solute is more concentrated. The red blood cell collapses and assumes a shrunken appearance, a process known as **crenation,** as illustrated in Figure 5-12c.

WORKED EXERCISES 5-12   Understanding Osmosis

1. Consider osmosis in the solutions shown at right separated by a semipermeable membrane. Indicate the direction of water flow between solution A and solution B or indicate if no flow occurs. The density of the dots represents the concentration of the solution. Which solution is hypertonic?

2. To which solution in the example above would you need to apply pressure in order to stop osmosis?

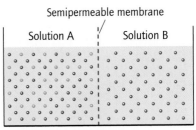

Solution A is 0.01 M glucose and 0.02 M sucrose.

Solution B is 0.02 M sucrose.

SOLUTIONS

1. Water will flow from solution B to solution A, because solution B is hypotonic, and solution A is hypertonic. Remember that the total number of particles is what counts in osmosis, not their mass, size, or identity. Solution A has a total solute concentration of 0.03 M (0.01 M glucose plus 0.02 M sucrose), while solution B has a total solute concentration of 0.02 M (0.02 M sucrose).

2. Pressure would need to be applied to the hypertonic solution A, because water is diffusing through the membrane from B to A.

Problem-Solving Tutorial:
Osmosis and Dialysis

PRACTICE EXERCISES

5.36 Consider the following solutions separated by a semipermeable membrane. Indicate the direction of water flow between solution A and solution B, or state if no flow occurs.

a.
Semipermeable membrane

| Solution A | Solution B |

Solution A is 0.05 M NaCl.
Solution B is 0.05 M NaCl.

b.

| Solution A | Solution B |

Solution A is 0.05 M Glucose.
Solution B is 0.05 M NaCl.

5.37 **Critical Thinking Question:** If you are stranded on a desert island, it is not wise to drink large quantities of seawater in place of fresh water. Considering the consequences of osmosis on red blood cells, why is this true? *Hint:* Seawater has a much higher solute concentration than 0.15 M NaCl, which is isotonic with red blood cells.

Dialysis *Dialysis is the simple diffusion of small solutes across a semipermeable membrane from the area of higher solute concentration to the area of lower solute concentration.* Dialysis can be used to separate solutes from colloidal particles, so, unlike osmosis, the identity of the particles in solution *does* matter.

Another distinguishing feature of solutions and colloids is that *solutes can cross a semipermeable membrane, whereas colloids cannot.* Figure 5-13 shows a solution of glucose and starch separated from pure water by a semipermeable membrane. Glucose is a small molecule, while starch is a colloidal particle.

Glucose molecules will readily dialyse from solution A to pure water because that is the direction of lower glucose concentration. The starch molecules, however, are too large to cross the semipermeable membrane. If you periodically replace solution B with pure water, you can remove almost all of the glucose from solution A, thereby separating glucose from starch (Figure 5-13c). Thus, dialysis is a technique for the physical separation of the components of an aqueous mixture.

Your kidneys perform a natural dialysis process when they separate metabolic waste products such as urea and creatinine from your blood, while retaining

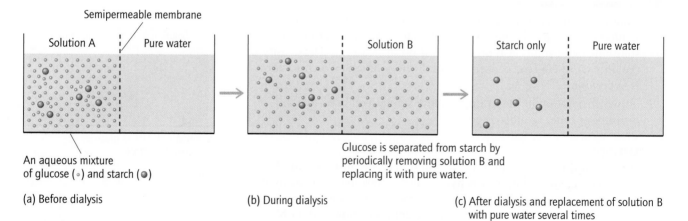

(a) Before dialysis

An aqueous mixture of glucose (○) and starch (●)

(b) During dialysis

Glucose is separated from starch by periodically removing solution B and replacing it with pure water.

(c) After dialysis and replacement of solution B with pure water several times

Figure 5-13 The process of dialysis. (a) Before dialysis: Two aqueous solutions with different concentrations are separated by a semipermeable membrane. (b) During dialysis: Small solute particles move into the solution of lower solute concentration until the two solutions have equal concentrations of the smaller solute particles. (c) After dialysis: If you periodically remove solution B and replace it with pure water, you can remove almost all the small solute from solution A, leaving only the larger starch particles.

important proteins and electrolytes within the blood. People whose kidneys have failed must have dialysis performed artificially on a routine schedule. Two forms of dialysis, hemodialysis and peritoneal dialysis, are life support treatments used to treat patients with kidney disease. See this chapter's *Chemistry in Medicine: Kidney Dialysis*, for a detailed description of kidney dialysis.

Problem-Solving Tutorial:
Osmosis and Dialysis

WORKED EXERCISE 5-13   Understanding Dialysis

Consider the following dialysis: in one compartment is a solution of urea (a small molecule) and albumin (a large protein molecule); in the other compartment is pure water. The two compartments are separated by a semipermeable dialysis membrane. Describe how you would separate urea from albumin, based on the principles of dialysis.

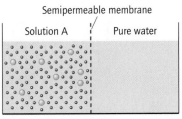

Solution A is an aqueous
mixture of urea (○) and albumin (◎).

Solution B is pure water.

SOLUTION

Since urea is a small solute molecule, it will dialyse through the membrane from solution A to the pure water side B, from higher solute concentration to lower concentration, until the concentrations are equal. Albumin will be unable to diffuse through the membrane because of its large size. By periodically replacing solution B with pure water, you can separate almost all of the urea from the albumin. At the end, solution A will contain only albumin. The multiple solutions of water will contain only urea.

PRACTICE EXERCISES

5.38 Consider the following dialysis: in one compartment is a solution of sucrose (a small molecule) and glycogen (a large starch molecule); in the other compartment is pure water. The two compartments are separated by a semipermeable dialysis membrane. Describe how you would separate sucrose from glycogen, based on the principles of dialysis.

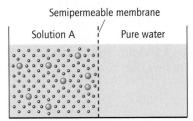

Solution A is an aqueous
mixture of sucrose (○) and glycogen (◎).

Solution B is pure water.

5.39 **Critical Thinking Question:** Consider two solutions that are separated by a semipermeable membrane that is designed for dialysis. One solution is pure water; the other is a solution (0.9% (m/v) NaCl) containing red blood cells. Based on the principles of dialysis, would you expect hemolysis, crenation, or neither to take place? Explain.

Chemistry in Medicine Kidney Dialysis

Kidney dialysis is a treatment that simulates the natural function of the kidneys, by removing waste products and excess water from the blood, while retaining colloidal particles such as proteins. There are two types of kidney dialysis—*hemodialysis* and *peritoneal* dialysis.

Hemodialysis

Hemodialysis is performed at a medical center, typically three times a week. Each treatment takes about 5 hr.

During the hemodialysis procedure, blood is allowed to flow through a dialyzer, such as the one illustrated in Figure 5-14a, in 100-mL increments.

The figure illustrates the steps involved in hemodialysis. (1) Blood is removed from the body and pumped through a tube into a dialyzer, (2). The dialyzer contains an aqueous solution—the **dialysate**—which is similar in composition to blood serum. The tubing that carries blood through the dialysate is made of cellulose, which serves as a

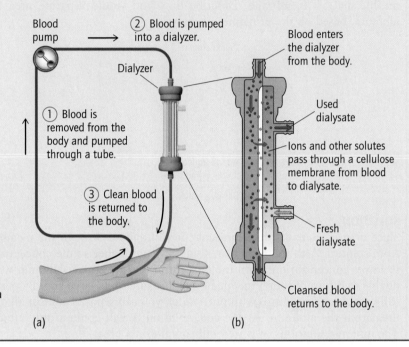

Figure 5-14 In hemodialysis, blood is removed from the body and pumped through a tube into a dialyzer. (a) Steps in hemodialysis. (b) Cross section of a dialyzer in detail.

Most fluids in the body are, like blood, aqueous mixtures containing various types of solutes and colloidal particles. Many processes in the body depend on the successful separation of solutes and colloidal particles, as well as the successful maintenance of concentrations of molecules, ions, and colloidal particles. Knowing how to interpret concentrations of solutions, and understanding how to use concentrations in calculations, is critical to key tasks in the health field such as preparing IV solutions and delivering correct dosages to patients. As you can see, knowledge of solutions, colloids, and membranes provides a foundation for a successful carreer in the health field. In the next chapter, you will learn the details about the chemical *structure* of many molecules that are involved in sustaining life as you begin to explore the topic of organic chemistry.

Chapter Summary

Mixtures and Solutions

- A mixture maintains its identity regardless of the proportions of its components.
- A mixture can be separated into its pure components through physical separation techniques.
- The components are distributed uniformly throughout a homogeneous mixture.

semipermeable membrane, separating the blood from the dialysate (Figure 5-14b). As the blood flows through the dialyzer, certain ions and other solutes such as urea and creatinine dialyse through the cellulose membrane, out of the blood and into the surrounding dialysate. Additional pressure is applied to the tubing—reverse osmosis—to force excess water from the blood into the dialysate. By removing waste and excess salt and fluids, dialysis helps control the patient's blood pressure and keeps electrolytes in proper balance. Hemodialysis also prevents urea levels from getting dangerously high. A typical hemodialysis treatment removes 2 to 10 L of dialysate containing waste products.

Peritoneal Dialysis

In contrast to hemodialysis, peritoneal dialysis can be performed at home, by patients themselves. The peritoneal membrane is a membrane lining the abdominal cavity. The membrane is naturally semipermeable, so it can act as a filter in much the same way as a kidney.

Figure 5-15 illustrates peritoneal dialysis. A specially formulated dialysis fluid is introduced into the abdominal cavity, using a catheter placed in the person's abdominal cavity, near the navel. Waste products, such as urea, diffuse from the blood, across the peritoneal membrane, and into the dialyzing fluid. The composition of the dialyzing fluid can be adjusted so that excess water will diffuse out of the blood by osmosis. After about 15 min, the fluid containing the body's wastes is drained from the peritoneal cavity through another catheter into a waste bag, which is discarded. A fresh bag of dialysate is then connected, and the process is repeated. Depending on the type of peritoneal dialysis, a patient may repeat this process four times a day, or the fluids may be drained and refilled automatically overnight while the patient is sleeping, using a special apparatus.

As you can see, dialysis is an important part of keeping failing kidneys healthy. It is based on the principles of solutions, colloids, and membranes that you learned about in this chapter.

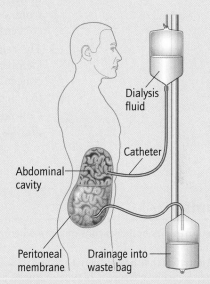

Figure 5-15 In peritoneal dialysis, which can be performed at home, a specially formulated dialysis fluid is introduced into the abdominal cavity.

- The components are *not* distributed uniformly throughout a heterogeneous mixture.
- A solution is composed of a solvent and one or more solutes. The solvent is present in the greater amount. Water is the solvent in an aqueous solution.
- Polar solvents tend to dissolve polar solutes, and nonpolar solvents tend to dissolve nonpolar solutes.
- When a molecule dissolves in solution, the chemical bonds of the molecule remain intact.
- When an ionic compound dissolves in solution, the lattice dissolves and the ions are surrounded by water molecules.
- An electrolyte is an ion dissolved in water.

Solution Concentration
- Concentration is expressed as a ratio: The amount of solute is the numerator and the total amount of solution is the denominator.
- The most commonly encountered units of concentration in the medical field are mass/volume, %mass/volume, moles/volume, and equivalents/volume.

Colloids and Suspensions
- Colloids contain particles that are much larger than typical solute particles.
- Suspensions are mixtures in which particles are unevenly distributed throughout the medium, and eventually settle.

Membranes, Osmosis, and Dialysis

- Substances can move across a semipermeable membrane through osmosis and dialysis.
- Osmosis and dialysis regulate the distribution of nutrients in the cell and the removal of waste products from the cell.
- Simple diffusion is the spontaneous movement of a molecule or ion from a region of higher concentration to a region of lower concentration.
- In osmosis, water crosses a semipermeable membrane by simple diffusion from a region of lower solute concentration to a region of higher solute concentration.
- In dialysis, solutes cross a semipermeable membrane by simple diffusion from a region of higher solute concentration to a region of lower solute concentration.
- The relative concentrations of solutions on either side of a semipermeable membrane can be described as hypertonic (having the higher concentration), hypotonic (having the lower concentration), or isotonic (having the same concentration).
- Colloidal particles cannot cross a semipermeable membrane; therefore, dialysis is a technique used to separate solutes from colloidal particles.

Key Words

Aqueous solution A homogeneous mixture where water is the solvent.

Colloid A mixture that has properties of both a solution and a heterogeneous mixture. Colloidal particles are much larger than solute particles, having diameters between 1 nm and 1000 nm.

Concentration A quantitative measure of how much solute is dissolved in a given amount of solution. Some common concentration units are mg/dL, (%)m/v, meq/L, and moles/L.

Crenation The collapse of a red blood cell when it is immersed in a hypertonic solution.

Dialysis The flow of certain solutes, from a region of higher solute concentration to a region of lower solute concentration, through a semipermeable membrane. Also, a life support treatment that performs the functions the kidneys can no longer perform.

Electrolyte A substance that produces ions when dissolved in water.

Hemolysis The bursting of a red blood cell when it is immersed in a hypotonic solution.

Heterogeneous mixture A mixture whose components are unevenly distributed.

Homogeneous mixture A mixture whose components are evenly distributed.

Hypertonic solution The solution on one side of a semipermeable membrane that has the higher solute concentration.

Hypotonic solution The solution on one side of a semipermeable membrane that has the lower solute concentration.

Isotonic solutions Solutions on either side of a semipermeable membrane that have equal solute concentrations.

Membrane A structure that acts like a barrier between two environments.

Molarity Moles of solute per liter of solution: mol/L, M.

Osmosis Simple diffusion of a solvent like water from a region of lower solute concentration (higher water concentration) to a region of higher solute concentration (lower water concentration) across a semipermeable membrane.

Osmotic pressure The minimum pressure required to stop osmosis.

Reverse osmosis The application of pressure greater than the osmotic pressure to a hypertonic solution, causing the flow of solvent in the opposite direction from osmosis.

Saturated solution A solution that contains the maximum amount of solute in a given solvent at a given temperature.

Semipermeable membrane A membrane that allows the passage of certain molecules and ions across the membrane while preventing the passage of others.

Simple diffusion The spontaneous movement of molecules or ions from a region of higher concentration to a region of lower concentration.

Solute The substance in a solution that is present in the lesser amount and is dissolved in the solvent.

Solution A homogeneous mixture that contains one or more solutes and a solvent.

Solvent The substance in a solution that is present in the greatest amount and dissolves the solute(s).

Suspension A mixture in which particles are unevenly distributed throughout the medium.

Additional Exercises

Mixtures and Solutions

5.40 What are the two components of a solution?

5.41 Which of the two components of a solution is present in the greater amount?

5.42 Identify the solute and the solvent in each solution below. State whether the solute is a molecule or electrolyte. If the solute is a molecule, state whether it is a gas, solid, or liquid.
a. An IV solution made of glucose and water.
b. A glass of carbonated water.
c. A tincture of iodine (I_2 in ethanol).
d. An IV solution of magnesium sulfate in water, used to prevent seizures in pregnant women with pre-eclampsia.

5.43 Identify the solute and solvent in each solution below. For (a)–(c), consult Table 5-1.
a. bronze
b. lemon-lime soda
c. beer
d. 3 mL isopropyl alcohol mixed with 20 mL water.

5.44 Indicate whether the following statements are true or false.
a. When a molecule dissolves in water, the covalent bonds of the molecule break.
b. Hydrogen bonding aids in the dissolution of polar molecules in water.
c. Solutions containing electrolytes conduct electricity.
d. Dissolution of electrolytes is a chemical change.

Solution Concentrations

5.45 In your own words, explain what "concentration" indicates for a solution.

5.46 A patient's blood test shows that his cholesterol level is 161 mg/dL. Convert this amount to grams per liter. Is cholesterol ($C_{27}H_{46}O$) a molecule or an electrolyte?

5.47 A patient's blood test shows that her iron level is 125 μg/dL. Convert this concentration to milligrams per liter. Is iron a molecule or an electrolyte?

5.48 A patient's glucose level is 78 mg/dL. Convert this concentration to micrograms per milliliter.

5.49 A patient's blood test shows that her hemoglobin level is 14 g/dL. Convert this concentration to milligrams per milliliter.

5.50 You are asked to prepare 1 L of 3.3% dextrose (m/v) solution for IV therapy. How much dextrose do you need to weigh out to prepare the solution?

5.51 How many grams of sodium chloride do you need to prepare 500 mL of a 0.3% normal saline (m/v) solution for IV therapy?

5.52 How much sodium chloride do you need to prepare 250 mL of a 0.9% saline (m/v) solution for IV therapy?

5.53 A patient's blood test shows that he has a bilirubin level of 12 mmol/L. Convert this concentration to units of molarity (mol/L).

5.54 A patient's blood test shows that she has elevated levels of LDL (low density lipoproteins) at 4.14 mmol/L. Convert this concentration to units of molarity (mol/L).

5.55 A patient's blood test shows that his insulin level is 120 pmol/L, a typical reading after fasting. Convert this concentration to units of molarity (mol/L). *Note:* There are 10^{12} picomoles (pmol) in 1 mol.

5.56 How many moles of calcium ions (Ca^{2+}) are there in 0.52 L of 0.45 M $CaCl_2$?

5.57 **Critical Thinking Question:** If a solution of K_2CO_3 contains 9.65×10^{25} ions of K^+ in 3 L of solution, what is the concentration in millimoles of K_2CO_3 of this solution? Recall that 1 mol = 6.02×10^{23} atoms, molecules, or ions.

5.58 A patient's blood test reports a magnesium level of 0.9 mmol/L. Convert this concentration to milliequivalents per liter of Mg^{2+}.

5.59 A patient's blood test reports a potassium level of 4.0 mmol/L. Convert this concentration to milliequivalents per liter of K^+.

5.60 How many equivalents per liter of sulfate (SO_4^{2-}) are there in a 0.15 M solution of $MgSO_4$? How many equivalents per liter of Mg^{2+} are there in this solution?

5.61 What additional information about the solute does the international unit (IU) take into account that other concentration units do not?

5.62 50 mg of Vistaril, an antihistamine, is to be administered every 3 hr. The suspension contains 25 mg/5 mL. What dose in milliliters should be administered to the patient every 3 hr?

5.63 An order is given to administer 300,000 units of Betapen-VK, an oral formulation of penicillin. 125 mg of Betapen-VK is equivalent to 200,000 units. The suspension provided contains 125 mg of Betapen-VK in every 5 mL. What dose in milliliters should be administered to the patient?

5.64 An order is given to administer 15 g of lactulose syrup, often used for treating complications of liver disease. The suspension contains 10 g/15 mL. What dose in milliliters should be given to the patient?

5.65 An order is given to administer methylprednisolone, an anti-inflammatory drug, by IV at a rate of 250 mg every 6 hr. The IV bag contains 125 mg methylprednisolone in every 2 mL. What should the flow rate be in milliliters per minute?

5.66 Morphine is a potent painkilller that acts directly on the central nervous system to relieve pain. An order is given to administer 2 mg of morphine to a patient as needed for breakthrough pain. The solution supplied contains 2 mg/mL of morphine. If the morphine dose should be administered over a period of 5 min, what should the flow rate be in milliliters per minute?

5.67 Nipride is often ordered for patients who have experienced severe heart failure. A patient is receiving a Nipride solution by IV, at a rate of 60 mL/hr. The concentration of the nipride solution is 50 mg/250 mL D5W. How many milligrams of nipride per hour is the patient receiving?

5.68 Aminophylline is used in the treatment of bronchial asthma. A patient is receiving an aminophylline solution by IV, at a rate of 22 mL/hr. The concentration of the solution is 2 g/1000 mL D5W. How many milligrams of aminophylline per hour is the patient receiving?

5.69 Erythromycin is ordered for a patient weighing 75 lb. The dose is 300 mg every 6 hr. The recommended dose is 30–50 mg/day per kilogram body weight, equally divided every 6 hr. Is the ordered dose in the safe range?

5.70 Nitroglycerine, used to treat patients with angina, is prescribed for a patient by IV at a rate of 3 μg/min per kilogram body weight. The IV bag contains 50 mg nitroglycerine/250 mL D5W, and the patient weighs 75 kg. How many milliliters per hour should the patient receive?

Extension Topic 5-1: Converting Between Different Concentration Units

5.71 Calculate the moles per liter of glucose in a 3.3% (m/v) dextrose solution.

5.72 Calculate the moles per liter of Na^+ in a 0.45% (m/v) saline solution. How many equivalents of Na^+ are there in one liter of the solution?

5.73 **Critical Thinking Question:** Calculate the moles per liter of Na^+ in a 0.6 mg/dL solution of Na_3PO_4.

Colloids and Suspensions

5.74 In a colloid, what is analogous to the solute in a solution? In a solution, what is analogous to the medium of a colloid?

5.75 In a suspension, what is analogous to the solute in a solution? To the solvent in a solution?

5.76 Identify the following as either a colloid or a solution.
  a. a glass of milk   **e.** a glass of wine
  b. a glass of lemonade   **f.** a spoonful of
  c. a carbonated soda   mayonnaise
  d. marshmallows   **g.** whipped cream

5.77 Identify the following as either a solution or a colloid.
  a. 5% dextrose   **e.** mercury amalgam used
  b. Jello   in dentistry
  c. blood   **f.** butter cookies
  d. air

5.78 Why must some medications be prepared as suspensions?

5.79 List at least one example each of a solute, a colloidal particle, and a suspended particle in whole blood.

Membranes, Osmosis, and Dialysis

5.80 What is the difference between osmosis and dialysis? What do they have in common?

5.81 What is a semipermeable membrane?

5.82 What is a primary function of the cell membrane?

5.83 What is simple diffusion?

5.84 What is the difference between hypertonic, hypotonic, and isotonic solutions?

5.85 Consider the following solutions separated by a semipermeable membrane. Indicate the direction of water flow between solution A and solution B, or state if no flow occurs.

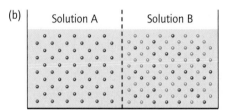

Solution A is 0.25 M NaCl.

Solution B is 0.05 M NaCl.

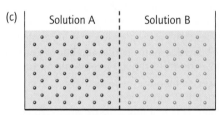

Solution A is 0.02 M sucrose.

Solution B is 0.02 M glucose and 0.01 M sucrose.

(c)

Solution A is 0.02 M sucrose.

Solution B is 0.02 M glucose.

5.86 Pre-eclampsia is a disorder that only occurs in pregnancy. The main symptoms are high blood pressure and protein in the urine; however, it can affect the brain, the kidneys, the liver, and the cardiovascular system. The only treatment for this disease is delivery of the baby. Pregnant women who have pre-eclampsia are at increased risk of hemolysis of the red blood cells. What happens to the red blood cells in this process?

5.87 In which direction does the water flow in osmosis of two solutions, from hypertonic to hypotonic or from hypotonic to hypertonic?

5.88 When dialysis is performed, why does the smaller molecule or particle move through the membrane, rather than the larger one?

5.89 If two solutions on either side of a membrane are isotonic, do you expect dialysis to occur? Explain your answer.

5.90 Consider the example of glucose dialysis discussed in Figure 5-9. Why does solution B periodically need to be replaced with water?

5.91 Consider the following dialysis: in one compartment is a solution of creatine (a small molecule) and globulin (a type of large protein molecule); in the other compartment is pure water. The two compartments are separated by a semipermeable dialysis membrane. Describe how you would separate creatine from globulin, based on the principles of dialysis.

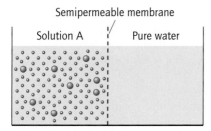

Semipermeable membrane

Solution A | Pure water

Solution A is an aqueous mixture of creatine (○) and globulin (●).

Solution B is pure water.

5.92 **Critical Thinking Question:** Two aqueous solutions are separated by a semipermeable membrane. Solution A contains 0.25 g/mL of KCl and solution B contains 0.75 g/mL of Na$_2$S. Will water flow from compartment A to B, or vice-versa? Show all calculations.

5.93 **Critical Thinking Question:** Two aqueous solutions are separated by a semipermeable membrane. Solution A is half-normal saline (0.45% NaCl) and solution B

contains 0.95 meq/L of MgCl$_2$. Will water flow from compartment A to B, or vice-versa? Show all calculations.

Chemistry in Medicine

5.94 Why do patients with renal failure require dialysis?

5.95 Briefly describe how the kidneys act as a semipermeable membrane.

5.96 What are some symptoms of kidney disease?

5.97 In hemodialysis, what substances pass through the cellular membrane out of the blood and into the surrounding dialysate? Explain why this process occurs.

5.98 Define the term dialysate.

5.99 Name and explain the difference between two types of kidney dialysis.

5.100 Explain the process of peritoneal dialysis.

Critical Thinking Questions

5.101 Sometimes KCl will be added to IV solutions. It will be added in concentrations of 10 meq/500 mL. How many moles of K$^+$ will there be in 1 L of solution?

5.102 A solution of calcium ions contains 5 meq/500 cm^3 of calcium ion. How many moles of calcium ion does this solution contain? If this were a calcium phosphate solution, how many moles of phosphate would be in the solution?

5.103 Two aqueous solutions are separated by a semipermeable membrane. Solution A contains 0.25 M KCl and solution B contains 0.25 M Na$_2$S. Will water flow from compartment A to B, or vice-versa? Show all calculations.

5.104 Two aqueous solutions are separated by a semipermeable membrane. Solution A contains 0.45 M NaCl and solution B contains 0.95 M MgCl$_2$. Will water flow from compartment A to B, or vice-versa? Show all calculations.

Answers to Practice Exercises

5.1 A heterogeneous mixture has the components unevenly distributed throughout the mixture, while a homogeneous mixture has the components evenly distributed throughout the mixture.

5.2 **a.** solute, ethanol; solvent, water
b. solute, NaCl; solvent, water
c. solute, CO$_2$; solvent, O$_2$

5.3 Polar solvents will dissolve polar solutes, and nonpolar solvents will dissolve nonpolar solutes.

5.4 Vinegar is a solution that contains water as a solvent. Water is a polar molecule, while oil is composed of mostly nonpolar molecules. The nonpolar oil molecules cannot hydrogen bond with the water molecules.

5.5 A saturated solution is a solution that cannot dissolve any more solute. When a solution is saturated, a precipitate will appear because the solute separates from the solvent.

5.6 Kidney stones form because the solution within the kidneys has become saturated in calcium salts.

5.7 **a.** molecule; **b.** electrolyte; **c.** gas

5.8 **a.** MgCl$_2$ → Mg$^+$ + 2 Cl$^-$; one magnesium ion, two chloride ions
b. K$_3$PO$_4$ → 3 K$^+$ + PO$_4$$^-$; three potassium ions, one phosphate ion
c. NaHCO$_3$ → Na$^+$ + HCO$_3$$^-$; one sodium ion, one bicarbonate ion

5.9 As glucose dissolves in water, the glucose molecules disperse uniformly throughout the aqueous solution. After dissolution, several water molecules surround each glucose molecule. Hydrogen bonding between glucose and water molecules aids the dissolution process. The molecular structure of glucose remains intact throughout the dissolution process.

5.10 Potassium iodide is an ionic compound. As it dissolves in water, the potassium (K^+) and iodide (I^-) ions separate from the solid KI lattice, and disperse uniformly throughout the aqueous solution. The partially negative oxygen atoms in the water molecules attract the positively charged potassium cations, K^+, drawing the cations into solution. Likewise, the partially positive hydrogen atoms in the water molecules attract the negatively charged iodide anions, I^-, drawing the anions into solution.

5.11 4×10^2 mg/dL

5.12 0.008 g/L

5.13 9.9×10^5 µg/L

5.14 1.22×10^6 µg/L

5.15 3.8% (m/v) I_2

5.16 8.9% (m/v) sucrose

5.17 4.5 g of NaCl

5.18 22.5 g of NaCl

5.19 1.2×10^{-2} mol

5.20 7.7×10^{-3} M

5.21 7.5 mmol

E5.1 0.1 M

E5.2 0.3 M

E5.3 0.86% (m/v) sucrose

E5.4 5.3×10^{-2} eq/L Mg^{2+}

E5.5 7.7×10^{-2} eq/L Na^+

E5.6 4.9% (m/v) MgI_2

E5.7 45 g glucose

5.22 a. 0.75 eq PO_4^{3-} in the solution;
    b. 0.75 eq Na^+ in the solution

5.23 a. 0.24 eq Mg^{2+};
    b. 0.24 eq SO_4^{2-}

5.24 1.5 mL

5.25 12.5 mL

5.26 11.25 mL

5.27 10 mL

5.28 20 mL/min

5.29 6×10^{-4} mL/min

5.30 6.0×10^2 units/hr

5.31 10 units/hr

5.32 A solution has a solute dissolved in a solvent. A colloid has the properties of a solution and a homogeneous mixture. The particles in a colloid are much bigger than the particles in a solution. A suspension has large particles and the particles are not uniformly distributed throughout the medium.

5.33 a. colloid
    b. suspension
    c. suspension
    d. solution

5.34 In humid air, the water is evenly distributed throughout the air. The solution is transparent and the water never settles out of the air. In mist, the aggregates of water are bigger and not evenly distributed throughout the air. The solution is not transparent and water will settle out of the mist.

5.35 Blood has the characteristics of a solution, a colloid, and a suspension because it contains particles ranging upward in size from the nanometer to the micrometer. It is a solution because it contains solutes dissolved in water. It is a colloid because it contains larger colloidal particles; and it is a suspension because it contains large particles suspended in water.

5.36 a. no flow occurs;
    b. flow occurs from solution A to solution B.

5.37 The concentration of solutes in seawater is much higher than the concentration inside the red blood cell, so the water inside the red blood cells would flow out, in an attempt to equalize concentration with the seawater. The red blood cell would collapse (crenation would occur).

5.38 Since sucrose is a small solute molecule, it will diffuse through the membrane from solution A to the pure water side, B, from higher concentration to lower concentration, until the concentrations are equal. Glycogen will be unable to diffuse through the membrane because of its large size. By periodically replacing solution B with water, the sucrose can be separated from the glycogen. In the end, solution A will contain only glycogen. The solutions of water will contain sucrose.

5.39 Hemolysis would take place. The NaCl would diffuse across the membrane to the side containing pure water. The concentration of NaCl in the solution containing the red blood cells would be lower, so that the red blood cells would be in a hypotonic solution. The water from the hypotonic solution would flow into the red blood cell.

Hydrocarbons and Structure

CHAPTER 6

Foods containing omega-3-fatty acids
[Tracey Kusiewicz/Foodie Photography/
Jupiterimages]

OUTLINE

 This icon indicates that a **Problem-Solving Tutorial**
is available at www.whfreeman.com/gob

179

Good Fats . . . Bad Fats . . . What Does It All Mean?

"Eat a diet low in fat and cholesterol" used to be the standard recommendation for a heart-healthy diet. As a result of studies investigating the link between the amount of fat in a woman's diet and her risk of developing heart disease, breast cancer, and colon cancer, this recommendation has since changed. Studies have shown that the *total* amount of fat in the diet—high or low—is not the culprit in heart disease and cancer; instead, it is the *type* of fat that matters. The landmark *Nurses' Health Study*, which followed over 100,000 nurses over several years through surveys and blood tests, showed that women who consumed the most saturated and trans fats were much more likely to develop heart disease and type 2 diabetes than women who consumed the least amount of these fats—thus the label "bad fats." In fact, unsaturated fats were actually found to lower the risk of developing heart disease and cancer—thus the label "good fats." The key to a healthy diet, therefore, is not necessarily to increase or decrease *total* fat, but to substitute *good fats* for *bad fats*!

What is the difference in the chemical structure of a good fat and a bad fat? Fats, like most biological molecules, are *organic* molecules; that is, they contain *carbon* atoms. A fat molecule contains mostly carbon and hydrogen atoms, making it similar to a type of molecule known as a *hydrocarbon,* which contains only carbon and hydrogen atoms. It is the hydrocarbon component of fats that makes them insoluble in water.

Saturated fats are found in both animal and vegetable sources. Animal sources high in saturated fats include whole milk, cheese, butter, ice cream, and red meat. Vegetable oils high in saturated fat include coconut, palm, and palm kernel oil. Coconut oil has the highest saturated fat content of any food—even higher than lard or beef.

Saturated fats contain long hydrocarbon chains *without* any double bonds, as you can see in the structure of the fat derived from three palmitic acid molecules in Figure 6-1. Saturated hydrocarbons have an overall linear shape, which allows them to pack together in a tight and organized manner, maximizing dispersion forces between fat molecules. The result is a fat with a higher melting point; and therefore, a waxy solid consistency at room temperature.

Unsaturated fats are found in most vegetable oils such as canola, olive, and peanut oil—but not in tropical oils. Nuts, avocados, and cold-water fish such as tuna and salmon are also high in unsaturated fats.

Chemically, an unsaturated fat contains one or more double bonds in its hydrocarbon chains. The double bond in an unsaturated fat disrupts the otherwise linear shape of the hydrocarbon chain, reducing the number of points of contact between fat molecules, as seen in the fat derived from three linolenic acid molecules shown in Figure 6-2. Consequently, there are fewer dispersion forces, resulting in a solid that melts at a lower temperature. Thus, unsaturated fats have a liquid consistency—they are oils—at room temperature. In this chapter you will also learn about *trans* fats, which are a special type of unsaturated fat that is even less healthy than a saturated fat.

Again, chemistry supplies the tools for understanding many current issues in health and nutrition, allowing us to make more informed decisions regarding nutritional choices.

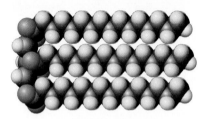

Palmitic acid (saturated)

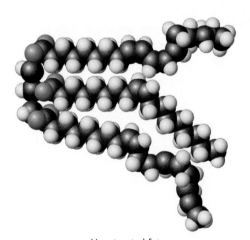

Saturated fat

Figure 6-1 A simple saturated fat molecule derived from three palmitic acid molecules. Note that there are no carbon–carbon double bonds; hence, it is classified as a saturated fat.

Linolenic acid (polyunsaturated)

Unsaturated fat

Figure 6-2 An unsaturated fat derived from three linolenic acid molecules. Linolenic acid is an omega-3 fatty acid, where the "3" indicates that a double bond occurs on the third carbon atom in the chain.

Organic chemistry is the branch of chemistry devoted to the study of carbon-containing compounds and their chemical reactions. Carbon is unique among the elements in that it is able to form many covalent bonds to other carbon atoms that result in long straight and branched chains of carbon atoms. These carbon chains also contain many C–H bonds. In addition, carbon also forms covalent bonds to oxygen, nitrogen, phosphorous, and sulfur—atoms often referred to as **heteroatoms.** Because of these properties, it's easy to see why 95% of all known compounds are organic. Indeed, most biologically important compounds are organic, with the exception of water. Therefore, to understand the chemistry of the human body, an understanding of organic chemistry is essential.

At one time it was believed that organic compounds could *not* be prepared in the laboratory, but could only be produced by living plants and animals. This belief is the origin of the term *organic*, which means "from living things." At the same time, many inorganic substances—substances that do not contain carbon—had been synthesized in the laboratory. We now know that almost any organic compound can be prepared in the laboratory, even extremely complex molecules. Nevertheless, the terms *organic* and *inorganic* remain with us today.

In our current day and age, substances produced in nature can also be prepared in the laboratory and many are used as **pharmaceuticals**—drugs used for therapeutic purposes. Plants and animals are still the original sources of many medicinally valuable organic compounds, such as taxol, the life-saving cancer drug first isolated from the yew tree. Yet, many pharmaceuticals are entirely man-made; that is, they are not produced naturally by any plant or animal. You will see examples of pharmaceuticals throughout this chapter and the next, as most are organic compounds.

Many important pharmaceuticals were first isolated from plant or animal sources. Some of the toxins produced by the poison dart frog, for example, are being considered for development as pain medicines. [Peter A. Vogel]

Organic: contains carbon.
Inorganic: does not contain carbon.

6.1 Hydrocarbons

In this chapter you will focus on a class of organic compounds known as **hydrocarbons**, which are molecules that contain exclusively *carbon* and *hydrogen* atoms.

Types of Hydrocarbons

Pure hydrocarbons are divided into four categories: alkanes, alkenes, alkynes, and aromatic hydrocarbons. **Alkanes** contain only *single bonds.* **Alkenes** contain one or more carbon–carbon *double bonds* and **Alkynes** contain one or more carbon–carbon *triple bonds.* **Aromatic** hydrocarbons have a unique ring structure containing several double bonds.

Alkanes are further classified as **saturated hydrocarbons** because they contain the maximum number of hydrogen atoms for a given number of carbon atoms; they are "saturated" with hydrogen. For example, the alkane C_2H_6, shown in Figure 6-3 has six hydrogen atoms, the maximum number of hydrogen atoms that can be present in a molecule containing two carbon

1A							8A
1 H	2A	3A	4A	5A	6A	7A	2 He
3 Li	4 Be	5 B	6 C	7 N	8 O	9 F	10 Ne

HYDROCARBONS

Alkanes	Alkenes	Alkynes	Aromatic
H–C–C–H with H H / H H	$C=C$ with H, H / H, H	$H–C≡C–H$	(ring)
C_2H_6	C_2H_4	C_2H_2	C_6H_6
Bond angle: 109.5°	120°	180°	120°
Saturated hydrocarbons	**Unsaturated hydrocarbons**		

Figure 6-3 Classification of hydrocarbons.

Gasoline is a mixture of several different hydrocarbons. [Oleksiy Maksymenko/Alamy]

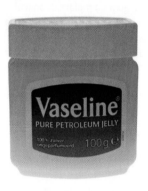

Vaseline is a hydrocarbon—it is composed of only carbon and hydrogen atoms. It is also called petroleum jelly, because, like many hydrocarbons, it is obtained from petroleum products. [Lenscap/Alamy]

atoms. The general formula for a saturated hydrocarbon is C_nH_{2n+2}, where n is equal to the number of carbon atoms in the molecule.

> Formula for a saturated hydrocarbon:
> C_nH_{2n+2}, n = number of carbon atoms.

Alkenes, alkynes, and aromatic compounds are classified as **unsaturated hydrocarbons**, because they contain less than the maximum number of hydrogen atoms per carbon atom. Unsaturated hydrocarbons contain fewer hydrogen atoms than saturated hydrocarbons because they contain one or more multiple bonds. Figure 6-3 shows that the two-carbon alkene, C_2H_4, has two fewer hydrogen atoms than the two-carbon alkane, as a result of one double bond. The two-carbon alkyne, C_2H_2, has four fewer hydrogen atoms, as a result of one triple bond.

You are probably already familiar with the terms *saturated* and *unsaturated* in the context of "bad" and "good" fats, respectively, first described in the opening vignette. Saturated fats contain only carbon–carbon single bonds, while unsaturated fats contain one or more double bonds. Strictly speaking, fats are not true hydrocarbons because they contain oxygen atoms; nevertheless, a significant portion of their structure is like a hydrocarbon. You will learn more about fats when you study lipids in Chapter 13. Lipids are a broad class of compounds that include all the biological molecules that are insoluble in water, which includes fats.

WORKED EXERCISES 6-1   Saturated and Unsaturated Hydrocarbons

1 What is the chemical formula for a saturated hydrocarbon with seven carbon atoms?

2 From the chemical formula, determine which of the hydrocarbons listed below is a saturated hydrocarbon.
   a. C_5H_8 b. C_5H_{10} c. C_5H_{12}

3 For each of the structural formulas shown below, identify whether it is an alkane, alkene, or alkyne. Explain why both compounds are classified as hydrocarbons.

a.

b.

SOLUTIONS

1 The formula for a saturated hydrocarbon is C_nH_{2n+2}. Substituting the number 7 for n gives the number of hydrogen atoms: $(2 \times 7) + 2 = 16$. Therefore, the formula for a saturated hydrocarbon with 7 carbons atoms is C_7H_{16}.

2 Compound (c) is the only saturated hydrocarbon because it contains 12 hydrogen atoms when 5 is substituted for n in the formula C_nH_{2n+2}.

3 a. The compound is an alkene because it contains a carbon–carbon double bond.

   b. The compound is an alkane because it contains carbon–carbon single bonds and carbon–hydrogen bonds; there are no double or triple bonds present.

   Both compounds are classified as hydrocarbons because they contain exclusively carbon and hydrogen atoms.

PRACTICE EXERCISES

6.1 Which of the following are unsaturated hydrocarbons (more than one choice is possible)?

    a. alkanes **b.** alkenes **c.** alkynes **d.** aromatic hydrocarbons

6.2 Which of the following are saturated hydrocarbons?

    a. C_7H_{14} **b.** C_6H_{14} **c.** C_3H_8 **d.** C_2H_2

6.3 By modern standards, what distinguishes an organic compound from an inorganic compound?

6.4 How many covalent bonds does carbon form in a molecule? Explain why.

6.5 For each compound shown below, identify whether it is an alkane, alkene, or alkyne.

a.
```
          H
          |
    H─C─H
      H  |  H
      |  |  |
  H─C─C─C─H
      |  |  |
      H  H  H
```

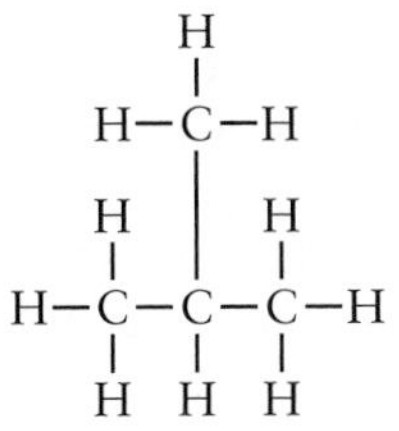

b.
```
    H           H
    |           |
H─C─C≡C─C─H
    |           |
    H           H
```

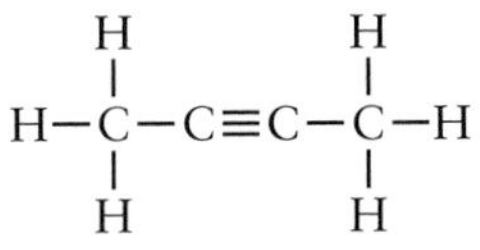

Physical Properties of Hydrocarbons

In Chapter 3 you learned that carbon–carbon and carbon–hydrogen bonds are nonpolar covalent bonds (Section 3.2). Therefore, all hydrocarbons are nonpolar *molecules*. Nonpolar molecules interact through dispersion forces (Section 3.3), the weakest of the intermolecular forces of attraction. Consequently, hydrocarbons have relatively low boiling points.

Hydrocarbons are insoluble in water because they are nonpolar and because water is polar—water molecules interact through hydrogen bonding. Consider an oil spill on the ocean. Since oil is a hydrocarbon, it is insoluble in water and does not mix with water. Oil is also less dense than water, and so it floats on top of the water. The inability of hydrocarbons to dissolve in water is often referred to as having **hydrophobic**—water fearing—properties. In contrast, polar substances are soluble in water and are said to display **hydrophilic**—water loving—properties. While hydrocarbons are hydrophobic and will not dissolve in water, they do dissolve in other hydrocarbons (like molecules).

Hydrophobic or "water fearing":	insoluble in water soluble in nonpolar substances
Hydrophilic or "water loving":	soluble in water insoluble in nonpolar substances

WORKED EXERCISE 6-2 Physical Properties of Hydrocarbons

Based on your everyday experience, which of the following household substances are hydrophobic?

a. rubbing alcohol   **b.** Chap-stick   **c.** vegetable oil   **d.** table salt

SOLUTION

(b) and (c). Chap-stick and vegetable oil do not dissolve in water and so display hydrophobic properties. Alcohol and table salt are hydrophilic, as they readily dissolve in water.

PRACTICE EXERCISES

6.6 Based on your everyday experience, which of the following household substances are hydrophobic?

    a. orange juice **b.** motor oil **c.** sugar **d.** tanning lotion

6.7 Why are hydrocarbons nonpolar? Explain in terms of the electronegativity of atoms.

Methane is the alkane present in natural gas, the fuel used by a gas stove.
[Tetra images/punchstock]

Figure 6-4 The three simplest alkanes.

6.2 | Saturated Hydrocarbons: The Alkanes

Alkanes are hydrocarbons that contain only carbon–carbon single bonds as well as carbon–hydrogen bonds; they contain no multiple bonds. The simplest alkane is methane, CH_4, shown in Figure 6-4a. You learned in Section 3.1 that methane has a tetrahedral geometry with 109.5° bond angles. These are the bond angles for all alkanes. The alkane with two carbon atoms, C_2H_6, is called ethane (Figure 6-4b), and the alkane with three carbon atoms, C_3H_6, propane (Figure 6-4c). Since every carbon atom in an alkane must have a tetrahedral geometry, the overall shape of the molecule takes on a zigzag appearance in alkanes containing three or more carbon atoms.

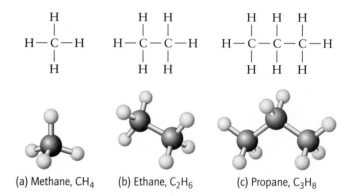

(a) Methane, CH_4 (b) Ethane, C_2H_6 (c) Propane, C_3H_8

Alkane Conformations

A molecule is not stationary and inflexible. Indeed, there is free rotation about every carbon–carbon single bond (unless the bond is part of a ring or other structural constraint). Consequently, a molecule can exist in many different rotational forms, known as **conformations** of the molecule. For example, if you rotate 180° about the carbon–carbon bond located between the second and third carbon atoms of the zigzag conformation shown in Figure 6-5a, the conformation shown in Figure 6-5b is formed.

Figure 6-5 The two conformations of C_5H_{12} shown in the ball-and-stick models are generated by a 180° bond rotation about the C–C bond that is labeled with a red ring.

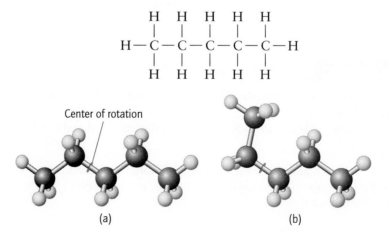

Center of rotation

(a)   (b)

The Model Tool 6-1 Alkane Conformations

I. Construction of Pentane

1. Obtain 5 black carbon atoms, 12 light-blue hydrogen atoms, and 16 straight bonds.
2. Construct a model of the structural formula shown below.

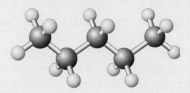

3. Rotate around each of the carbon–carbon single bonds until you arrive at the zigzag conformation shown below.

II. Observation Questions

4. **a.** What is the geometry around each carbon atom? How does the geometry around each carbon atom affect the overall shape of the molecule?
   b. What are the H—C—H bond angles? Why aren't the bond angles 90° as they appear in the structural formula?
5. Do you ever have to break a C—C or C—H bond to arrive at the zigzag conformation? Is it true that different conformations are obtained merely by rotating C—C single bonds?

Another way to envision the concept of conformations is to note that molecules are constantly moving around and rotating about each of their carbon–carbon single bonds. Remember the kinetic molecular view from Chapter 4: the higher the temperature, the more kinetic energy there is available to the molecule, and the more rapidly the molecule assumes different conformations. Drawing different conformations is equivalent to taking different snapshots of a moving object at various instants in time. Although different conformations may appear as though they are different molecules, they are, in fact, the same molecule. *Anytime a structure can be converted into another structure by merely* rotating *about one or more carbon–carbon single bonds, the two structures are the same compound, just different conformations.* Perform *The Model Tool 6-1: Alkane Conformations* to experience first-hand the free rotation around the carbon–carbon single bonds of an alkane.

WORKED EXERCISE 6-3 Alkane Conformations

Do the following pairs of molecules represent different conformations of the same molecule or different molecules? If they are different molecules, explain what makes them different.

a.

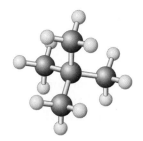 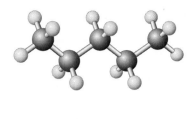

b.

SOLUTION

a. The two ball-and-stick models represent different molecules, because the connectivity of the atoms is different. Although both models contain five carbon atoms and 12 hydrogen atoms, in the first model a central atom is bonded to four carbon atoms, whereas in the second model all five carbon atoms are bonded to one or two carbon atoms in a linear continuous chain.

b. The two structures shown represent two different conformations of the same compound because the connectivity of the atoms is the same. That is, the five carbon atoms in both structures are connected in a linear chain. They differ only in rotation around a carbon–carbon bond.

PRACTICE EXERCISE

6.8 Which of the following pairs of molecules represent different conformations of the same molecule? Which of the following pairs of molecules represent different molecules? For any pairs that are different molecules, explain what makes them different.

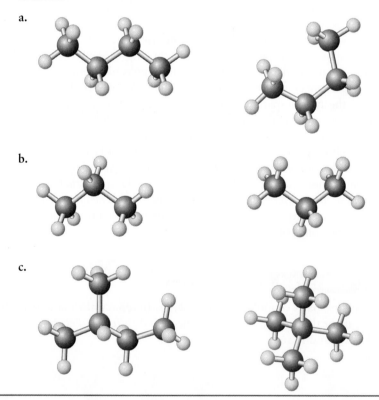

a.

b.

c.

Structural Isomers

Structural isomers are also known as constitutional isomers.

You have seen that an alkane can exist in many different conformations. What structural characteristics make alkanes different? Obviously, when two alkanes have different chemical formulas, they are unequivocally different compounds. For example, the formula of methane is CH_4 and the formula of ethane is C_2H_6, so ethane and methane are different compounds. However, two molecules with the *same* chemical formula may not be the same. It depends on whether the arrangement of the carbon atoms—the connectivity of the atoms—is different. *Molecules with the same chemical formula, but a different connectivity of atoms, are known as* **structural isomers.** Structural isomers are molecules with different physical properties, and a different chemical name. Figure 6-6 shows the three structural isomers of C_5H_{12}. Note that the connectivity of the atoms differs in each structural isomer.

Figure 6-6a shows the structure of a linear chain of five carbon atoms, referred to as the **straight-chain** isomer because of its end-to-end arrangement of carbon atoms. The structural isomers in Figure 6-6b and 6-6c both contain branch points along the main hydrocarbon chain; hence, they are called **branched-chain** isomers. In a branched-chain alkane at least one hydrogen atom in the chain has been replaced by a carbon atom or chain of carbon atoms. The branch point in the structural isomers in Figure 6-6b and 6-6c is shown in red in the structural formula. The majority of structural isomers are branched-chain isomers. Perform *The Model Tool 6-2: Structural Isomers* to gain experience with structural isomers. Keep in mind that each structural isomer exists in many different conformations.

A branched-chain alkane has carbon atoms branching off the main carbon chain, like the branches of a tree.
[Gerolf Kalt/zefa/Corbis]

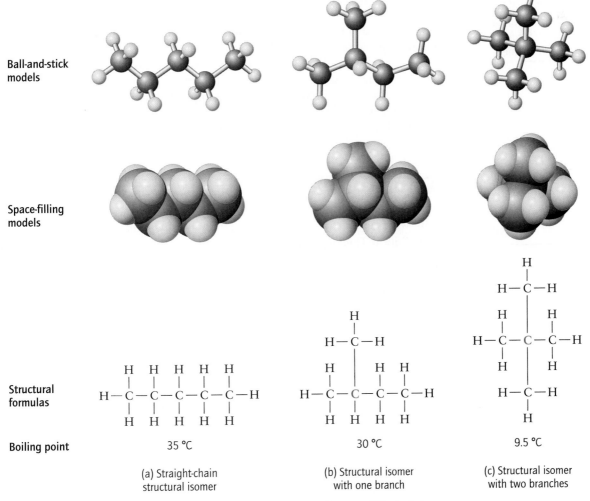

Figure 6-6 The three structural isomers of C_5H_{12}. The three structural isomers differ in their connectivity of atoms: (a) straight chain, (b) one branch, (c) two branches. The branch point is shown in red.

The Model Tool 6-2 Structural Isomers

I. Construction of the Structural Isomers of Butane, C_4H_{10}

1. Obtain 4 black carbon atoms, 10 light-blue hydrogen atoms, and 13 straight bonds.
2. Construct a model of the straight-chain alkane with formula C_4H_{10}.
   a. Draw a structural formula of this compound.
   b. Can you rotate the C—C bonds in this molecule? Why are there many different conformations of this isomer?
3. Make a model of the branched-chain isomer using only the atoms from the model you constructed in Step 2.
   a. Draw a structural formula of this branched-chain structural isomer.
   b. How is this isomer different from the one you made in Step 1?
   c. How does the connectivity of the carbon atoms differ from the straight-chain isomer?
   d. Are these two structural isomers different compounds? Explain.
   e. Would you expect these two compounds to have identical or different physical properties?
   f. Did you have to break and remake bonds to build the branched-chain isomer from the straight-chain isomer? How does this tell you that they are not simply different conformations?

Table 6-1 Number of Structural Isomers for Different Hydrocarbons

Chemical Formula	Number of Structural Isomers
CH_4	1
C_2H_6	1
C_3H_8	1
C_4H_{10}	2
C_5H_{12}	3
C_6H_{14}	5
C_7H_{16}	9
C_8H_{18}	18
C_9H_{20}	35
$C_{10}H_{22}$	75
$C_{20}H_{42}$	366,319

Since structural isomers are different chemical compounds, they display different physical and chemical properties. For example, each structural isomer with the formula C_5H_{12} has a unique boiling point (a physical property), which is listed in Figure 6-6. The boiling points are all relatively low, as expected for a hydrocarbon.

The number of different structural isomers that exist for any given chemical formula increases as the number of carbon atoms in the chemical formula increases, as shown in Table 6-1. For example, C_7H_{16} has nine structural isomers, while $C_{20}H_{42}$ has 366,319 structural isomers! The number of branching possibilities increases with each additional carbon atom in the formula, a characteristic unique to the element carbon.

PRACTICE EXERCISES

6.9 Write the structural formula for the two structural isomers of C_4H_{10}.

6.10 Write the structural formula for the three structural isomers of C_5H_{12}.

6.11 Critical Thinking Question: Write the five structural isomers of C_6H_{14}.

6.12 Why does the number of structural isomers increase with increasing number of carbon atoms?

6.13 Which of the following pairs are not structural isomers?

a.

```
            H                        H
            |                        |
         H−C−H                    H−C−H
   H  H  H  |  H            H  |  H  H  H
   |  |  |  |  |            |  |  |  |  |
 H−C−C−C−C−C−H            H−C−C−C−C−C−H
   |  |  |  |  |            |  |  |  |  |
   H  H  H  H  H            H  H  H  H  H
```

b.

```
                                     H
                                     |
                                  H−C−H
   H  H  H  H               H  |  H
   |  |  |  |               |  |  |
 H−C−C−C−C−H             H−C−C−C−H
   |  |  |  |               |  |  |
   H  H  H  H               H  H  H
```

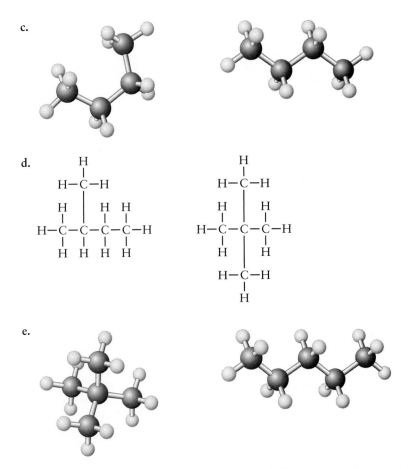

c.

d.

$$H-\overset{\overset{\displaystyle H}{|}}{\underset{}{C}}-H$$

$$H-\overset{\overset{\displaystyle H}{|}}{\underset{\overset{|}{H}}{C}}-\overset{\overset{\displaystyle H}{|}}{\underset{\overset{|}{H}}{C}}-\overset{\overset{\displaystyle H}{|}}{\underset{\overset{|}{H}}{C}}-\overset{\overset{\displaystyle H}{|}}{\underset{\overset{|}{H}}{C}}-H$$

$$H-\overset{\overset{\displaystyle H}{|}}{\underset{}{C}}-H$$

$$H-\overset{\overset{\displaystyle H}{|}}{\underset{\overset{|}{H}}{C}}-\overset{}{\underset{}{C}}-\overset{\overset{\displaystyle H}{|}}{\underset{\overset{|}{H}}{C}}-H$$

$$H-\overset{}{\underset{\overset{|}{H}}{C}}-H$$

e.

6.14 Which pairs of compounds in Exercise 6.13 have different physical properties?

Cycloalkanes

An unbranched **cycloalkane** is a molecular ring created when a carbon–carbon bond is formed between the two ends of an alkane chain, as illustrated in Figure 6-7. Two hydrogen atoms from each end of the alkane chain must be removed to create a cycloalkane. Substituted cycloalkanes have one or more branch points on the ring.

Cycloalkanes are written as polygons with three or more sides. For example, a five-membered cycloalkane looks like a pentagon, a five-sided polygon. A carbon atom is present at each corner of the polygon with the two attached hydrogen atoms not shown.

The most common ring sizes encountered in nature are five- and six-membered rings, but other ring sizes also exist. Three- and four-membered rings are less common because the rings are strained. Ring strain arises in

Pentane

$$H-\overset{\overset{\displaystyle H}{|}}{\underset{\overset{|}{H}}{C}}-\overset{\overset{\displaystyle H}{|}}{\underset{\overset{|}{H}}{C}}-\overset{\overset{\displaystyle H}{|}}{\underset{\overset{|}{H}}{C}}-\overset{\overset{\displaystyle H}{|}}{\underset{\overset{|}{H}}{C}}-\overset{\overset{\displaystyle H}{|}}{\underset{\overset{|}{H}}{C}}-H$$

Remove a hydrogen atom
from each end of the
carbon chain.

$\Longrightarrow$

Cyclopentane

Make a new C–C bond.

$=$

Figure 6-7 Generating a cycloalkane from a straight-chain alkane.

Figure 6-8 (a) Ball-and-stick and
(b) tube model of cyclohexane. Note, it
is not a flat molecule.

(a) Ball-and-stick model   (b) Tube model

such small rings because they are unable to achieve the 109.5° bond angle
that is required by a tetrahedral carbon atom. Bond angles are forced to be
60° for three-membered rings (drawn as a triangle) and approximately 90° for
four-membered rings (drawn as a square).

The polygons used to depict cycloalkanes might lead you to believe that
cycloalkanes are flat molecules. However, only a three-membered ring cycloalkane is flat. If you look at the models of the six-membered ring cycloalkane
shown in Figure 6-8, you can see that it does not look like a hexagon when
drawn in three dimensions. However, a top (or bottom) view of the ring does
look like the hexagon used to depict it. Perform *The Model Tool 6-3: Alkanes
and Cycloalkanes*, to gain more familiarity with cycloalkanes.

 The Model Tool 6-3 Alkanes and Cycloalkanes

I. Construction of hexane, C_6H_{14}

1. Obtain 6 black carbon atoms, 14 light-blue hydrogen atoms, and 19 straight bonds.

2. Construct a model of the straight-chain structural isomer of C_6H_{14}:

    a. Is free rotation possible around every C—C bond?

    b. How many C—H bonds are there on carbons 1 and 6? How many C—H bonds are
    there on all the other carbon atoms?

II. Construction of Cyclohexane, C_6H_{12}

3. Remove one hydrogen atom from carbon 1 and one hydrogen atom from carbon 6.
    Remove one of the C—H bonds as well.

4. Use the remaining bond to join carbon 1 and carbon 6. Your model should look similar to
    the tube model below.

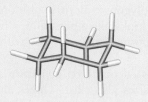

III. Observation Exercises

5. Is there still free rotation available to every C—C bond? Is there some flexibility in the
    molecule?

6. Why has the formula of the new model changed?

7. How many C—H bonds are there on each carbon atom? How does this compare to C_6H_{14}?

8. Is the ring flat? Can you force it to be flat?

    a. Rotate the model so that you have a top view of the ring. What type of polygon does it
    look like?

    b. Rotate the model so that you have a side view. Does the molecule appear as shown
    above? Why is the molecule not flat?

WORKED EXERCISE 6-4 Cycloalkanes

Write the structure used to depict a cycloalkane containing six carbon atoms in the ring. Would you expect this ring to be more or less common in nature than a three-membered ring? Explain.

SOLUTION

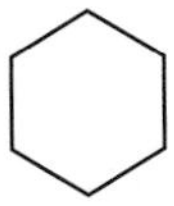

The depiction of a six-carbon ring looks like a hexagon. Each corner represents a carbon atom with the two hydrogen atoms attached not shown. This ring would be more common than a three membered ring, because the bond angles can be 109.5°.

PRACTICE EXERCISES

6.15 How many carbon atoms are there in the following cycloalkanes?

a.   b.   c.

6.16 **Critical Thinking Question:** Examine the bond angles in the ball-and-stick model of cyclopropane shown below. What is the preferred C—C—C bond angle and why? What is the actual bond angle in cyclopropane? Why might you break some pieces if you tried to make this model with the parts in your model kit? Why is this ring size less common in nature?

6.17 Referring to the tube model of the six-membered cycloalkane shown below, answer the following questions.

    a. How many hydrogen atoms are on each carbon atom?
    b. Are the C—C bonds all in the same plane? Are the C—H bonds all in the same plane? Is the model flat?

6.3 | Writing Structures

Structural formulas depict the connectivity of atoms in an organic molecule by showing all covalent bonds and atoms. Writing structural formulas of larger organic molecules, however, can become tedious, and therefore, two simpler ways of writing organic structures have been developed that are in common use: *condensed structural formulas* and *skeletal-line structures*.

Table 6-2 Condensed Structural Formulas for Straight-Chain Alkanes

Number of Carbon Atoms	Condensed Structural Formula	Alternative Condensed Structural Formula
1	CH_4	—
2	CH_3CH_3	—
3	$CH_3CH_2CH_3$	—
4	$CH_3CH_2CH_2CH_3$	$CH_3(CH_2)_2CH_3$
5	$CH_3CH_2CH_2CH_2CH_3$	$CH_3(CH_2)_3CH_3$
6	$CH_3CH_2CH_2CH_2CH_2CH_3$	$CH_3(CH_2)_4CH_3$
7	$CH_3CH_2CH_2CH_2CH_2CH_2CH_3$	$CH_3(CH_2)_5CH_3$
8	$CH_3CH_2CH_2CH_2CH_2CH_2CH_2CH_3$	$CH_3(CH_2)_6CH_3$
9	$CH_3CH_2CH_2CH_2CH_2CH_2CH_2CH_2CH_3$	$CH_3(CH_2)_7CH_3$
10	$CH_3CH_2CH_2CH_2CH_2CH_2CH_2CH_2CH_2CH_3$	$CH_3(CH_2)_8CH_3$

Condensed Structural Formulas

In a **condensed structural formula**, each carbon atom and its attached hydrogen atom or atoms is written as a group: CH, CH_2, or CH_3. Each such group is then written sequentially according to its order in the chain. Except in the case of branch points, bonds are omitted. For example, the condensed structural formula for C_3H_8 would be written as $CH_3CH_2CH_3$.

Condensed Formulas for Straight-Chain Alkanes In writing the condensed formula for a straight-chain alkane, the first and last carbons in the chain are written as a CH_3 group, while all the other carbon atoms are written as CH_2 groups. For example, the straight-chain C_5H_{12} structural isomer can be written as

Structural formula

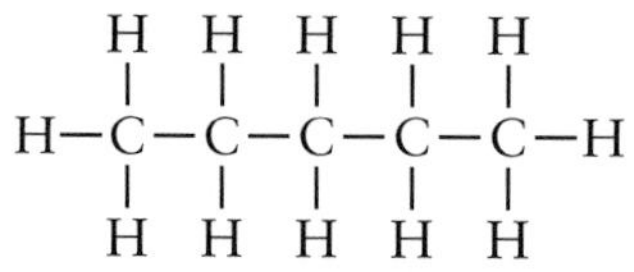

$$
\begin{array}{ccccc}
\text{H} & \text{H} & \text{H} & \text{H} & \text{H} \\
| & | & | & | & | \\
\text{H}-\text{C}-\text{C}-\text{C}-\text{C}-\text{C}-\text{H} \\
| & | & | & | & | \\
\text{H} & \text{H} & \text{H} & \text{H} & \text{H}
\end{array}
$$

Condensed formula $CH_3CH_2CH_2CH_2CH_3$

In writing the condensed formula for long straight-chain alkanes, the repeating CH_2 groups are often indicated by writing a single CH_2 enclosed in parentheses, followed by a subscript that indicates the number of times the CH_2 group repeats. Thus, an alternative way of writing the C_5H_{12} straight-chain isomer is $CH_3(CH_2)_3CH_3$, where the subscript 3 following the parentheses indicates that there are three CH_2 groups attached one after the other. Table 6-2 lists the two ways to write the condensed structural formulas for straight-chain alkanes with from 1 to 10 carbon atoms.

Condensed Formulas for Branched-Chain Alkanes To write the condensed structure for a branched-chain hydrocarbon, follow the guidelines given for writing straight-chain alkanes, then indicate any branch point along the chain by inserting a bond (pointing either up or down) from the carbon atom where the branch occurs. For example, the condensed structural formula for one of the branched-chain C_5H_{12} isomers is shown on the facing page. The carbon atom at which the branch occurs is shown in red.

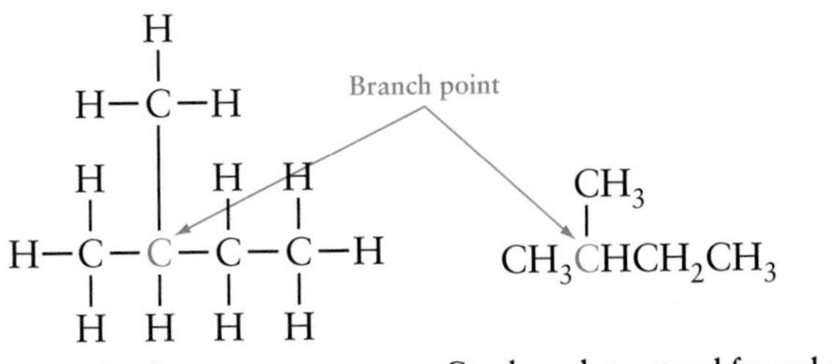

H
|
H—C—H   Branch point
|
H | H H   CH$_3$
| | | | |
H—C—C—C—C—H   CH$_3$CHCH$_2$CH$_3$
| | | |
H   H   H H

Lewis structure   Condensed structural formula

WORKED EXERCISES 6-5   Writing Condensed Structural Formulas

For each condensed structural formula listed below, determine whether it represents a *straight-chain* alkane or a *branched-chain* alkane; then write the corresponding structural formula.

1 CH$_3$(CH$_2$)$_5$CH$_3$

2   CH$_3$
       |
  CH$_3$CCH$_2$CH$_3$
       |
       CH$_3$

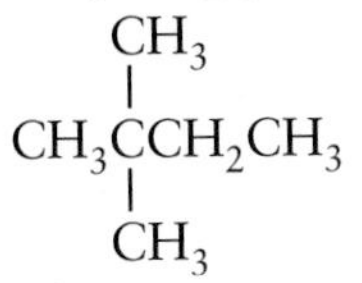

SOLUTIONS

1 Straight-chain alkane, because there are no bonds indicating branch points along the chain.

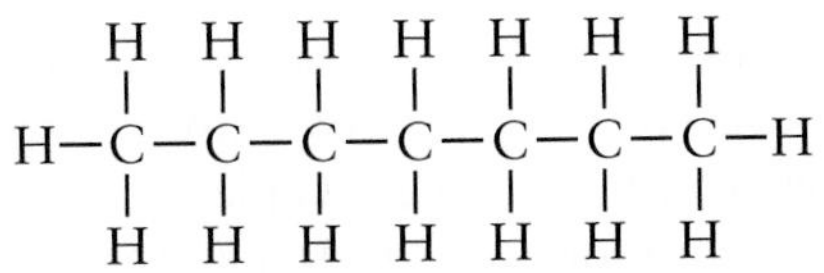

H   H   H   H   H   H   H
| | | | | | |
H—C—C—C—C—C—C—C—H
| | | | | | |
H   H   H   H   H   H   H

2 Branched-chain alkane. There are two CH$_3$ groups branching off the second carbon atom in the chain, indicated by the two bonds shown projecting from the second carbon atom.

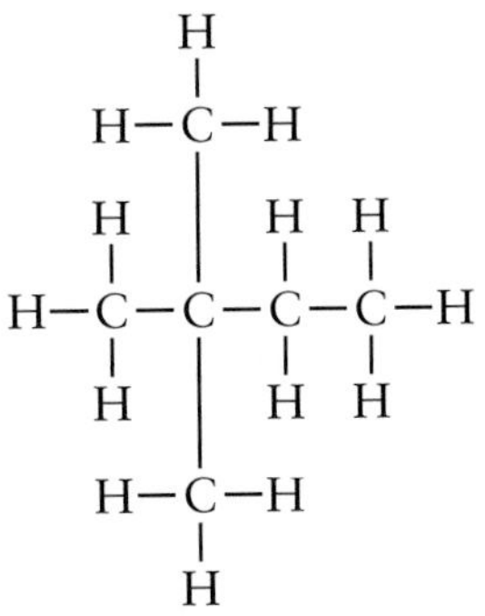

H
|
H—C—H

H | H H
| | | |
H—C—C—C—C—H
| | | |
H | H H

H—C—H
|
H

PRACTICE EXERCISES

6.18 For each condensed structural formula listed below, determine whether it represents a *straight-chain* alkane or a *branched-chain* alkane, then write the corresponding structural formula.

a. CH$_3$CH$_2$CH$_2$CH$_2$CH$_2$CH$_3$

            CH$_2$CH$_3$
            |
b. 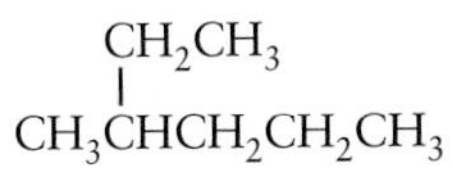 CH$_3$CHCH$_2$CH$_2$CH$_3$

                    CH$_3$
                    |
            CH$_3$CHCH$_2$CHCH$_2$CH$_3$
                         |
c. 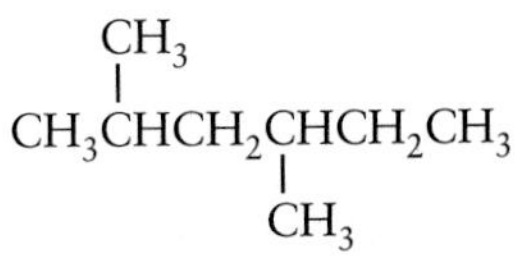   CH$_3$

6.19 Write a condensed structural formula for the following compound.

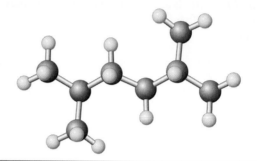

6.20 Write a condensed structural formula and a structural formula for the compound shown in the ball-and-stick model below.

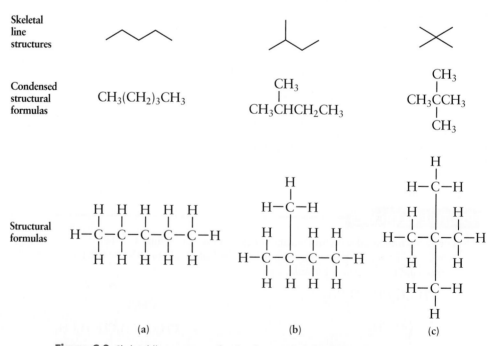

Skeletal Line Structures

You have seen that writing a condensed structural formula simplifies the task of writing complex chemical structures. **Skeletal line structures** are an even more convenient shorthand for writing complex structures, and they have a clean and uncluttered appearance. For example, the three structural isomers of C_5H_{12} shown in Figure 6-6 can be written as the skeletal line structures shown in Figure 6-9. The general guidelines for writing skeletal line structures are described in the *Guidelines* box on the facing page.

Figure 6-9 Skeletal line structures for the three structural isomers of C_5H_{12}.

Guidelines for Writing Skeletal Line Structures

1. All carbon-carbon single bonds are shown as a single line: —.

2. Double bonds are shown as two parallel lines, =, and triple bonds are shown as three parallel lines, ≡.

3. The chemical symbol for carbon, C, is omitted. The presence of a carbon atom is implied wherever two lines join, *as well as at the end of a line*. A continuous chain of carbon atoms is represented as a zigzag arrangement of lines.

Implied carbon atoms

4. Atoms other than carbon and hydrogen (O, N, S, and P) must be written in, in order to distinguish them from carbon atoms.

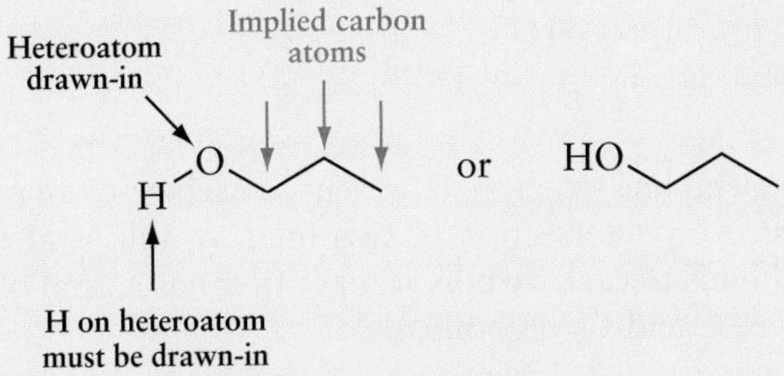

Heteroatom drawn-in   Implied carbon atoms

H on heteroatom must be drawn-in

5. Carbon–hydrogen bonds as well as the H atoms bonded to carbon are omitted from skeletal line structures altogether.

- Hydrogen atoms attached to a heteroatom (O, S, N, P), however, must be written in.

- Since a carbon atom always has *four* bonds (the octet rule, Sections 3.2 and 3.3), you can determine the number of H atoms bonded to a particular carbon atom in a skeletal line structure by counting the number of bonds and subtracting this value from *four*:

Number of C—H bonds on a C atom = 4 − (number of lines to C atom)

Since there is usually more than one way to write a skeletal line structure, remember that, as with writing structural formulas, the unique connectivity of the atoms determines the identity of a given structural isomer. For example, six equally correct ways that you could represent the isomer in Figure 6-9b are shown below. You know they all represent the same compound because they all have the same connectivity: a CH_3 group branching off the second carbon atom from one end of a four-carbon main chain.

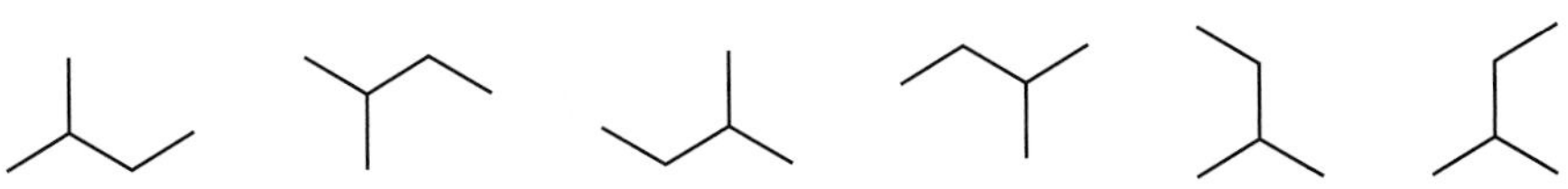

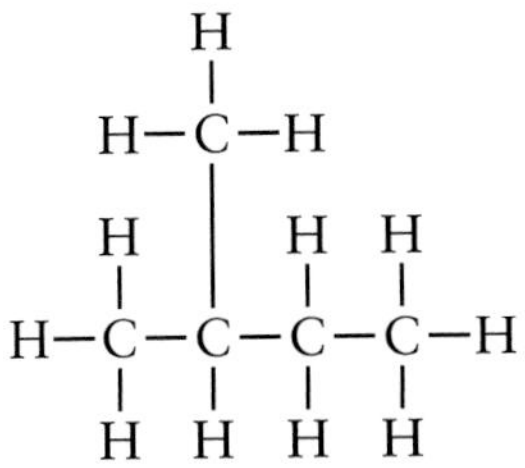

The guidelines for drawing structures are somewhat flexible, so you will often find that a combination of the skeletal line structure and condensed formula is used. For example, some CH_3 groups may be written in an otherwise pure skeletal line structure. If you are familiar with both systems, you should be able to interpret most of the structures that you encounter.

Problem-Solving Tutorial:
Representations of Molecules

WORKED EXERCISE 6-6   Writing Skeletal Line Structures

Write the structural formula for the skeletal line structure shown below.

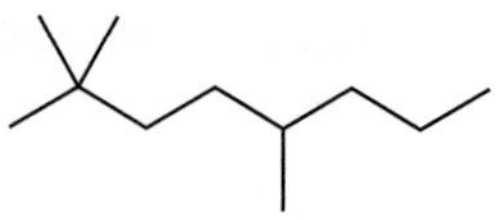

SOLUTION

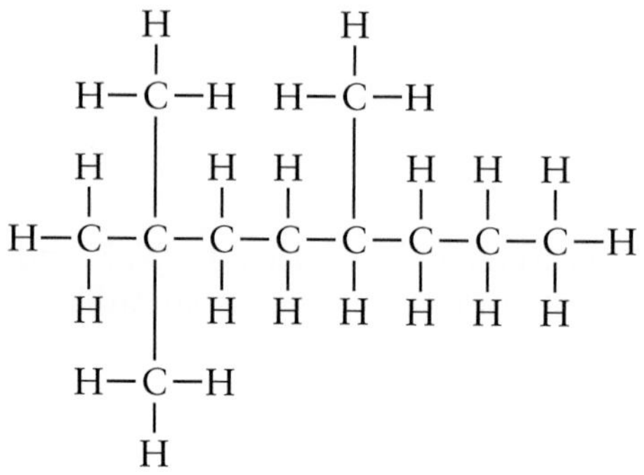

Every line in a skeletal line structure represents a carbon–carbon bond. Place a carbon atom at every intersection of two lines as well as at the end of a line. Add C—H bonds to each carbon atom until it has a total of four bonds (the sum of the C—C and C—H bonds).

PRACTICE EXERCISES

6.21 Fill in the table below for the two C_4H_{10} structural isomers.

Structural Formula	Skeletal Line Structure

6.22 Write the structural formula, the condensed structural formula, and the skeletal line structure for the molecule shown below.

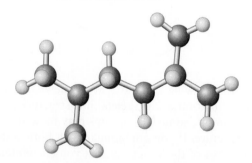

6.23 Write the structural formula and the condensed structural formula for the skeletal line structures shown below.

a.

b.

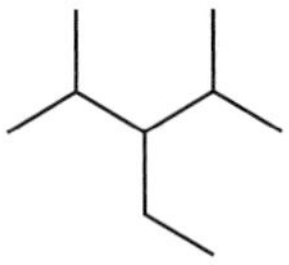

6.24 Write a skeletal line structure that corresponds to each of the condensed structural formulas shown below:

a. $CH_3(CH_2)_8CH_3$

b.
$$CH_3$$
$$|$$
$$CH_3CHCHCH_3$$
$$|$$
$$CH_3$$

6.4 | Unsaturated Hydrocarbons: Alkenes and Alkynes

Alkenes, alkynes, and aromatic hydrocarbons are unsaturated hydrocarbons. Here we will describe alkenes and alkynes, and in Section 6.6 you will learn about aromatic hydrocarbons.

Alkenes

An **alkene** is a hydrocarbon that contains one or more carbon–carbon double bonds. In Chapter 3 you were introduced to the simplest alkene, C_2H_4, which is known as ethene, or by its common name, ethylene. Recall from Chapter 3 that the bond angles around each of the carbon atoms that make up a double bond is 120°, giving this part of the molecule a trigonal planar molecular geometry.

Ethene (ethylene)
C_2H_4

> Ethylene, a gas, is a plant hormone that promotes ripening in fruits.

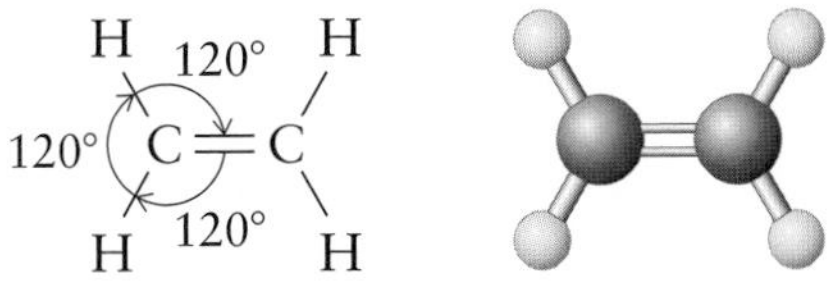

An alkene containing two carbon–carbon double bonds is referred to as a **diene**, from "di" for "two." An alkene containing several carbon–carbon double bonds is called a **polyene**, from "poly" for "many"—in this case, many double bonds. The hydrocarbon β-carotene, from which vitamin A is derived, is an example of a polyene. The skeletal line structure of β-carotene shows its 11 double bonds.

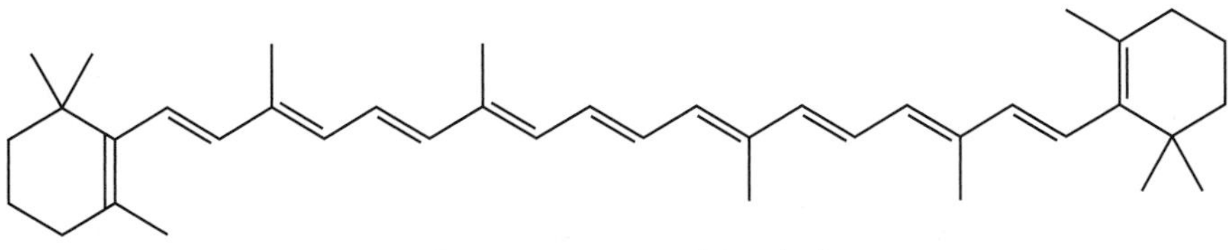

β-Carotene, a source of vitamin A

Remember when writing the skeletal line structure of an alkene to use two parallel lines to represent a double bond.

β-Carotene is the polyene found in carrots that gives them their distinctive orange color. [Sue Wilson/Alamy]

WORKED EXERCISES 6-7   Alkenes

1 Identify the skeletal line structure below as a simple alkene, a diene, or a
   polyene. Then rewrite the structure as a structural formula showing the
   correct number of hydrogen atoms on each carbon atom:

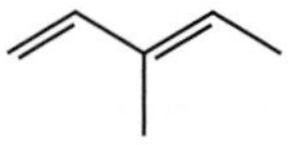

2 Fill in the table below and write the structural formula for the skeletal
   line structure shown.

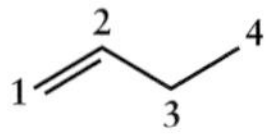

Carbon Atom Number	Number of Bonds to Carbon	Apply Formula 4 − (No. bonds)	Number of Hydrogen Atoms on This Carbon Atom
1			
2			
3			
4			

SOLUTIONS

1 The compound is a diene because there are two double bonds present in
   the molecule. Add C—H bonds so that each carbon atom has four bonds.

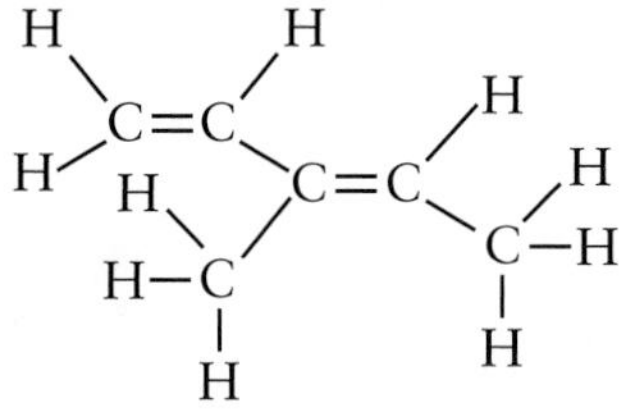

2

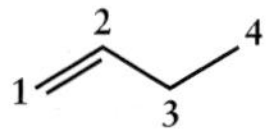

Carbon Atom Number	Number of Bonds to Carbon	Apply Formula 4 − (No. bonds)	Number of Hydrogen Atoms on This Carbon Atom	Structural Formula with Hydrogen Atoms Highlighted
1	2	4 − 2	2	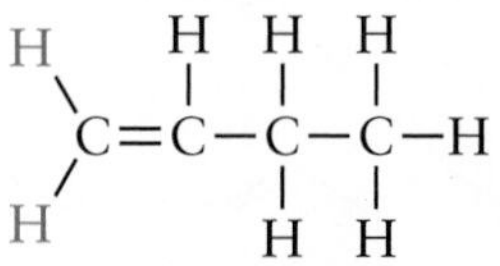
2	3	4 − 3	1	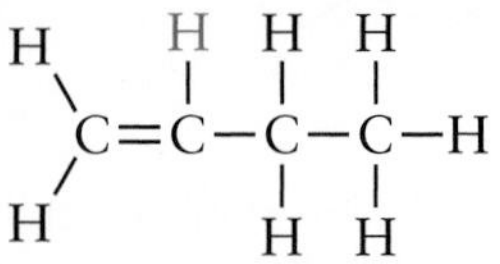

Carbon Atom Number	Number of Bonds to Carbon	Apply Formula 4 − (No. bonds)	Number of Hydrogen Atoms on This Carbon Atom	Structural Formula with Hydrogen Atoms Highlighted
3	2	4 − 2	2	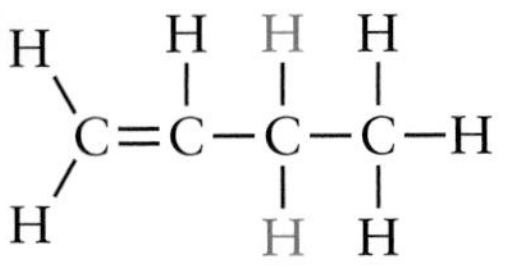 H\C=C−C−C−H with H H H on top and H, H H below
4	1	4 − 1	3	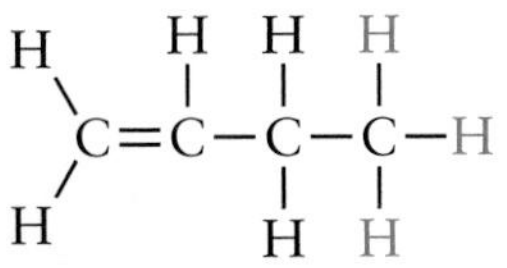 H\C=C−C−C−H with H H H on top and H, H H below

PRACTICE EXERCISES

6.25 Identify the compounds shown below as simple alkenes, dienes, or polyenes.

a.

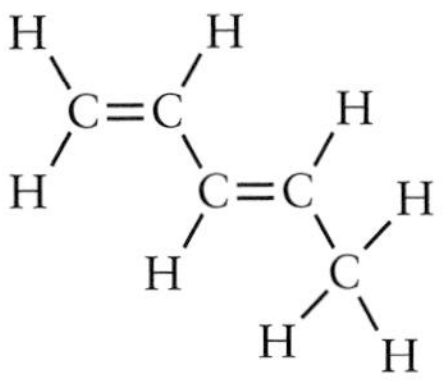

b.

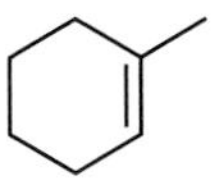

c.

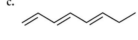

6.26 What are the bond angles around each of the carbon atoms in compound (a) of Exercise 6.25?

6.27 The structure of vitamin A is shown below.

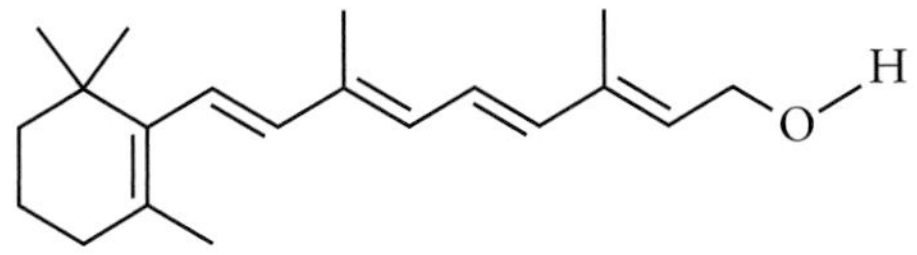

Retinol (vitamin A)

a. How many double bonds does vitamin A contain?
b. In what way does vitamin A resemble β-carotene?
c. In this molecule, how many carbon atoms have three hydrogen atoms? How many carbon atoms do not have any hydrogen atoms?
d. Vitamin A is not classified as a hydrocarbon. Explain why.
e. Would you expect vitamin A to be hydrophobic or hydrophilic? Explain.
   Hint: Vitamin A is one of the four fat-soluble vitamins.

Geometric Isomers

There is no free rotation around a carbon–carbon double bond. As a result, the groups or atoms attached to a double-bond carbon are locked into a fixed spatial orientation. The fixed three-dimensional orientation of these groups or atoms creates the possibility for **geometric isomers** in some alkenes. *Geometric isomers are compounds with the same chemical formula* and *the same connectivity of atoms, but with a different three-dimensional orientation in space, as the result of a double bond.* As with structural isomers, geometric isomers have different chemical and physical properties.

Figure 6-10 Geometric isomers of $CH_3CH=CHCH_3$ differ in their orientation in three-dimensional space, as a result of the restricted rotation around a C=C double bond.

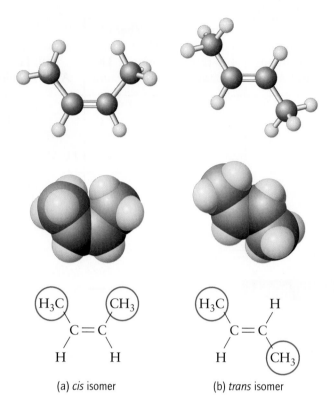

(a) *cis* isomer   (b) *trans* isomer

For example, $CH_3CH=CHCH_3$ exists as the two geometric isomers shown in Figure 6-10. Both geometric isomers have the same chemical formula: C_4H_8. Moreover, the atoms of both isomers are connected in the same way: $CH_3CH=CHCH_3$. However, the isomers are not identical compounds because the two CH_3 groups attached to the double-bond carbons (shown circled in blue) occupy different regions of three-dimensional space. The difference in shape is particularly apparent when you compare the ball-and-stick and space-filling models of these geometric isomers.

When the two CH_3 groups on the double-bond carbons are on the *same side* of the double bond, the geometric isomer is called the **cis** isomer (Figure 6-10a). When the two CH_3 groups are on the *opposite sides* of the double bond, the geometric isomer is called the **trans** isomer (Figure 6-10b).

---

cis: The groups on the double-bond carbons are on the *same side* of the double bond.

trans: The groups on the double-bond carbons are on *opposite sides* of the double bond.

If a double bond has more than two groups, that is, three or four groups, the cis/trans designation does not apply. The alternative system used to name those alkenes is more complex and will not be described.

Geometric isomers do not exist for alkanes because of the free rotation that occurs around a carbon–carbon single bond, as illustrated in the analogous four-carbon alkane in Figure 6-11a. Here the two structures shown represent two conformations of the same molecule. Contrast this to the four-carbon alkene in Figure 6-11b, where the structures represent two different molecules, cis and trans geometric isomers. To draw the geometric isomer of an alkene, simply switch the two groups attached to either double-bond carbon atom.

(a)

Conformations of C_4H_{10}

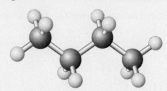

Free rotation

(b)

Geometric isomers of C_4H_8

cis   No free rotation   *trans*

Figure 6-11 Rotation about two different types of carbon–carbon bonds. (a) The alkane C_4H_{10} can rotate freely about the central C—C single bond, producing many possible conformations of the same compound. (b) No rotation is possible around the central C=C double bond of C_4H_8. As a result, this molecule has two geometric isomers.

Some alkenes do not have a geometric isomer. For example, there is only one compound with the condensed formula of CH_2=$CHCH_2CH_3$. In other words, if you switch two groups on one of the double-bond carbon atoms in this molecule, you do not create a different molecule. Geometric isomers only exist when there are two *different* atoms or groups on *both* double-bond carbon atoms. To gain familiarity with geometric isomers, perform *The Model Tool 6-4*.

 The Model Tool 6-4 Geometric Isomers

I. Construction of C_4H_{10}

1. Obtain 8 black carbon atoms, 16 light-blue hydrogen atoms, 20 straight bonds, and 4 bent bonds.
2. Construct a model of the straight-chain structural isomer of C_4H_{10}:

Is there free rotation around the middle carbon atoms?

II. Construction of cis and trans Isomers of C_4H_8

1. Remove one hydrogen atom from both carbon 2 and carbon 3 of your model. Replace the C(2)–C(3) single bond with two bent bonds to represent a double bond. You will have constructed either the cis or the trans isomer of this alkene.
2. Build a model of the other geometric isomer.

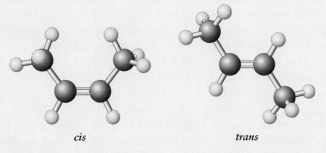

cis   *trans*

III. Observations

1. Is the connectivity of the carbon atoms the same in both models?
2. In which geometric isomer are the hydrogen atoms on carbon 2 and carbon 3 on the same side of the double bond? What is this geometric isomer called? How is the structure of the other geometric isomer different?
3. Is there free rotation about the double bond? How is the double bond responsible for the existence of cis and trans isomers?

WORKED EXERCISE 6-8 Geometric Isomers

Draw the geometric isomer of the compound below.

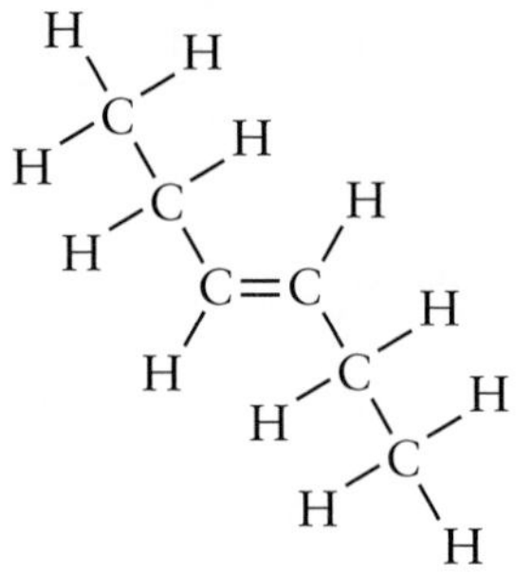

SOLUTION

Exchange the two groups on one of the double-bond carbons to generate the cis isomer. In this case, exchange a hydrogen and a CH_2CH_3 group. Remember always to draw a 120° bond angle around a double bond.

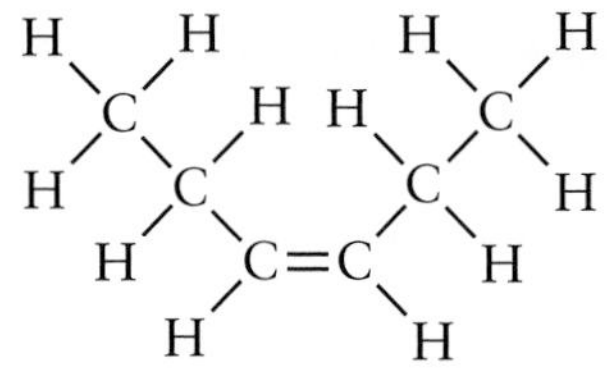

PRACTICE EXERCISE

6.28 Draw the geometric isomer for the following compounds. Indicate if the geometric isomer shown is cis or trans.

a.

b.

Cis–trans Isomers in Biochemistry You have seen that cis and trans geometric isomers are different compounds, so by definition they have different chemical and physical properties. The only way to convert one geometric isomer into the other is through a chemical reaction, the subject of Chapters 8 and 10. *A chemical reaction that converts one structural isomer or geometric isomer into another is known as an* **isomerization** *reaction.*

One classic example of a biochemical isomerization reaction occurs in the chemistry of vision. The retina of the eye contains molecules called photoreceptors that absorb light. There are two types of photoreceptors: rods and cones. Rods contain a polyene known as retinal, which is part of a protein known as rhodopsin. Light initiates a chemical reaction that converts one of the cis double bonds in retinal into the trans isomer, as shown in Figure 6-12. The change in the spatial orientation from cis to trans causes the shape of the entire molecule to change. The change in molecular shape initiates a nerve impulse that travels along the optic nerve to the brain. The nerve impulses from all the rods together are interpreted as a visual image.

Perhaps you noticed a similarity in the structures of retinal and β-carotene on page 197. The body produces retinal from β-carotene, hence their similarity. This is why your mother told you to eat your carrots, because they would be good for your eyes. She was right!

Retinal

H   O

Light energy

cis-Alkene

trans-Alkene

Protein

N — Protein

Chemical
reaction

Groups on
same side
of double bond

Groups on
opposite sides
of double bond

N

Protein

Rhodopsin

Figure 6-12 The chemistry of vision. Light induces the cis alkene in rhodopsin to undergo an isomerization reaction to the trans alkene.

PRACTICE EXERCISES

6.29 Describe the difference in overall shape between the cis and trans forms of rhodopsin.

6.30 The polyene portion of rhodopsin is a relatively flat portion of the molecule. Explain why. What is a polyene?

6.31 What type of reaction is shown in Figure 6-12?

Alkynes

An alkyne contains one or more carbon–carbon triple bonds. The simplest alkyne is ethyne, $CH{\equiv}CH$, also known by its common name, acetylene. Alkynes have a linear geometry around the carbon atoms that make up the triple bond, because the H—C—C bond angle is 180°. The zigzag convention used for writing skeletal line drawings of alkanes and alkenes is not used around the triple bond of an alkyne. Also, as a result of their linear geometry, geometric isomers do not exist for alkynes.

Acetylene, the simplest alkyne, is a gas at room temperature and is used in welding torches. [Trevor Smithers ARPS/Alamy]

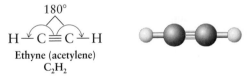

180°

$H{-}C{\equiv}C{-}H$

Ethyne (acetylene)
C_2H_2

Alkynes are not common in nature, although they do exist. For example, the poison dart frog, native to the South American rain forests, produces the toxic alkyne histrionicotoxin to ward off predators. This molecule contains two triple bonds. Notice the linear arrangement of atoms around the triple bonds when the molecule is written as a skeletal line structure.

H

N

HO —

Histrionicotoxin

Histrionicotoxin is the neurotoxin produced by the poison dart frog. The shape of the molecule is similar to the neurotransmitter acetylcholine, used to transmit nerve impulses. The similarity in structure accounts for its paralytic effect on the nervous system.

6.5 Naming Hydrocarbons

Every organic compound has a unique name based on a systematic set of rules created by an international organization known as the IUPAC. In addition to an

Table 6-3 Names for a Common Analgesic (a Painkiller)

Skeletal Line Structure	Brand Name	Generic Name	IUPAC Name	Chemical Formula	Type of Drug
	Motrin, Advil	Ibuprofen	2-[4-(2-methylpropyl)] phenylpropanoic acid	$C_{13}H_{18}O_2$	Analgesic

---

I International
U Union for
P Pure and
A Applied
C Chemistry

IUPAC name, some compounds also have a common name. In this section you will learn the basic rules for naming hydrocarbons according to the IUPAC system. An IUPAC name is composed of three parts: a prefix, a root, and a suffix (ending):

(Prefix)(Root)(Suffix)

In addition to its IUPAC name, the active pharmaceutical ingredient in every pharmaceutical is assigned both a generic name, which is a shorter name used to identify the compound, and a brand name, which is associated with the maker of the drug, as shown for the analgesic (pain killer) ibuprofen in Table 6-3. If you look at the literature supplied with most pharmaceuticals, you will see its IUPAC name included as well as its chemical structure, generic name, and brand name.

Naming Straight-Chain Hydrocarbons and Cycloalkanes

The easiest compounds to name using the IUPAC system are straight-chain hydrocarbons and cycloalkanes, because they do not contain a prefix. Straight-chain alkanes are named based on the number of carbon atoms they contain, as indicated in Table 6-4. Follow the rules in the *Guidelines* box on the facing page to assign the IUPAC name for a straight-chain hydrocarbon.

Naming Cycloalkanes A cycloalkane is assigned the same root as a straight-chain alkane, corresponding to the number of carbon atoms in the ring, with the additional term *cyclo-* inserted before the root name. Thus, a six-membered ring cycloalkane is named *cyclo*hexane. Table 6-5 shows a list of the names and skeletal line structures for the cycloalkanes with 3 to 8 carbon atoms.

Table 6-4 IUPAC Names, Condensed Formulas, and Skeletal Line Structures for the Ten Simplest Straight-Chain Alkanes

Number of Carbon Atoms	IUPAC Name Root	Condensed Formula	Skeletal Line Structure
1	Methane	CH_4	Not applicable
2	Ethane	CH_3CH_3	——
3	Propane	$CH_3CH_2CH_3$	
4	Butane	$CH_3(CH_2)_2CH_3$	
5	Pentane	$CH_3(CH_2)_3CH_3$	
6	Hexane	$CH_3(CH_2)_4CH_3$	
7	Heptane	$CH_3(CH_2)_5CH_3$	
8	Octane	$CH_3(CH_2)_6CH_3$	
9	Nonane	$CH_3(CH_2)_7CH_3$	
10	Decane	$CH_3(CH_2)_8CH_3$	

Guidelines for Naming Straight-Chain Hydrocarbons and Cycloalkanes

Rule 1: Assign the root. Count the number of carbon atoms in the chain to assign the root. The root name comes from the Greek or Latin word corresponding to that number of carbon atoms, as shown in Table 6-4 for the straight-chain alkanes containing from 1 to 10 carbon atoms. You may want to memorize these names, because they form the basis for naming most other organic compounds.

Rule 2: Assign the suffix. The suffix indicates the type of hydrocarbon: alkane, alkene, or alkyne. The suffix for an alk*ane,* including cycloalkanes, is *-ane;* for an alk*ene, -ene;* and for an alk*yne, -yne.* Aromatic hydrocarbons follow a different naming system that will be described in Section 6.6.

Rule 3: Assign a locator number to the root if a multiple bond is present. A locator number is used to indicate the location of a multiple bond in an alkene or an alkyne. Number the carbon atoms in the root chain, starting from the end closer to the multiple bond. The number corresponding to the first carbon atom of the multiple bond is the locator number. Place this number, followed by a hyphen, in front of the root name.

Remember, geometric isomers may need a cis or trans designation, which is inserted before the locator number. For example, the two structures shown below are *cis*-2-butene and *trans*-2-butene. Notice that numbers and letters are always separated by a hyphen in an IUPAC name.

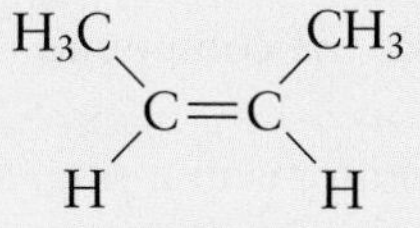

cis-2-Butene 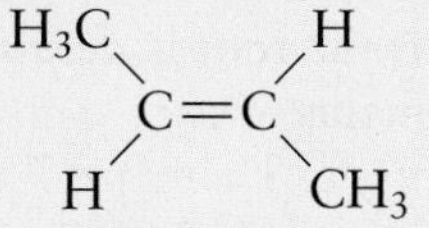   *trans*-2-Butene

Table 6-5 Skeletal Line Structure and IUPAC Name for Several Cycloalkanes

Number of Carbon Atoms in the Ring	Skeletal Line Structure	IUPAC Name
3	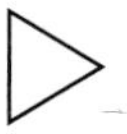	Cyclopropane
4	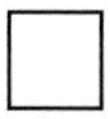	Cyclobutane
5	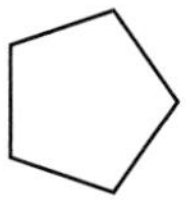	Cyclopentane
6	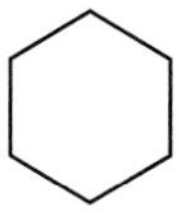	Cyclohexane
7		Cycloheptane
8	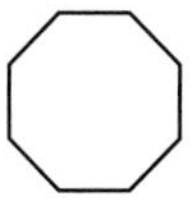	Cyclooctane

> **WORKED EXERCISES 6-9**   Assigning IUPAC Names for Straight-Chain
>   Hydrocarbons and Cycloalkanes
>
> 1 Without looking at Tables 6-4 and 6-5, write the IUPAC name for the
>   compounds below:
>   **a.** $CH_3(CH_2)_5CH_3$
>
>   **b.**
>
>   **c.** 
>
>   **d.** 
>
> 2 Draw the skeletal line structures of the following alkenes:
>   **a.** 1-octene   **b.** *trans*-4-octene   **c.** *cis*-2-octene
> 3 Draw skeletal line structures of *cis*-2-pentene and *trans*-2-pentene. What
>   do these two compounds have in common? How do they differ? Will
>   they have identical physical properties?
> 4 Write the structural formula, condensed structure, and skeletal line
>   structure of 1-butyne and 2-butyne.

SOLUTION

1 **a.** Heptane. There are 7 carbon atoms in a straight chain, so according
     to Rule 1, the root is *heptane*. Since the compound is an alkane (it
     has no multiple bonds), the suffix remains *-ane*.

  b. Decane. There are 10 carbon atoms in a straight chain, so according
     to Rule 1, the root is *decane*. Since the compound is an alkane, the
     suffix remains *-ane*.

  c. Cycloheptane. There are 7 carbon atoms in a ring, so according to
     Rule 1, the root is *heptane*. Since the 7 carbon atoms are in a ring,
     the term *cyclo-* is inserted before the root. Since the compound is an
     alkane, the suffix remains *-ane*.

  d. 1-Hexene. There are 6 carbon atoms in the chain, so the root is
     hexane. The suffix is changed to *-ene* to yield hex*ene*, because there
     is a double bond in the molecule. The chain is numbered from the
     left end because that is the end closer to the double bond. The first
     carbon atom of the double bond is carbon 1, so the locator number
     used in the name is 1.

2 **a.**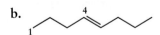

     Write a zigzag structure showing 8 points and place a double bond
     between the first two (or last two) carbon atoms.

  b.

     Write the structure as in (a) above, but instead, place the double
     bond between the fourth and fifth carbon atoms. To indicate the trans
     isomer, place the two groups on the double bond on opposite sides of
     the double bond.

  c.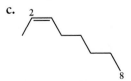

3

trans-2-Pentene   *cis*-2-Pentene

Both compounds are straight-chain alkenes with five carbon atoms and a double bond between carbons 2 and 3. They differ in the three-dimensional arrangement of atoms. In the trans isomer, the CH_2CH_3 and CH_3 groups are on opposite sides of the double bond, and in the cis isomer, they are on the same side of the double bond. Since they are geometric isomers, they are different compounds, and therefore will have different physical properties.

4 The root for these two alkynes is *butane*, which tells you that their main chain contains four carbon atoms. The *-yne* suffix indicates that a triple bond is present. The 1- in 1-butyne indicates that the triple bond is located between the first and second carbon atoms. The 2- in 2-butyne indicates that the triple bond is between the second and third carbon atoms. Remember to draw atoms attached to a triple bond in a linear, rather than a zigzag manner.

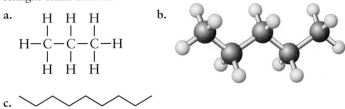

	1-Butyne	2-Butyne
Structural formula	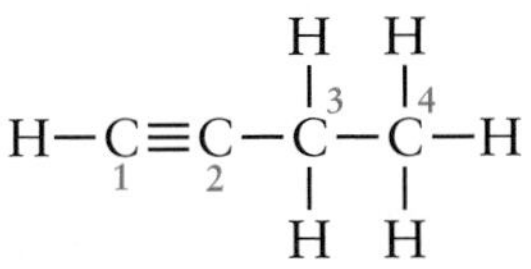 H—C≡C—C—C—H	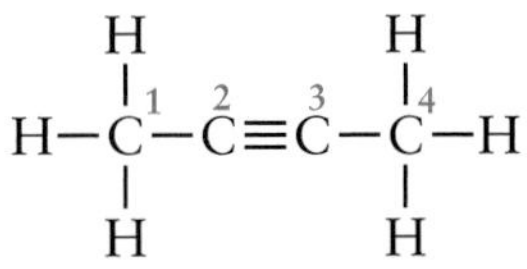 H—C—C≡C—C—H
Condensed structure	$CHCCH_2CH_3$	CH_3CCCH_3
Skeletal line drawing	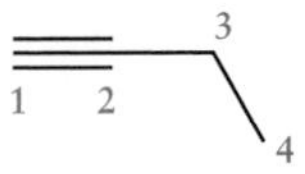	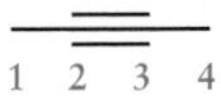

PRACTICE EXERCISES

6.32 Without looking at Table 6-4, write the IUPAC name for each of the following straight-chain alkanes:

a.

H—C—C—C—H

b.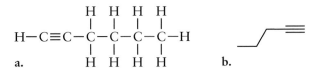

c.

6.33 Write the structural formula, condensed structure, and skeletal line structure for the following straight-chain hydrocarbons:

a. octane   b. 3-octyne   c. 1-pentene

6.34 Name the following alkynes:

H—C≡C—C—C—C—C—H

a.   b.

Naming Branched-Chain Hydrocarbons

Branched-chain hydrocarbons have one or more hydrogen atoms along the main carbon chain replaced, or "substituted," by one or more smaller carbon atom chains. Any branch replacing a hydrogen atom is called a **substituent**. A prefix is inserted in front of the hydrocarbon root name to identify the substituents *and* their location along the main chain. Rules for naming branched-chain hydrocarbons are described in the box below. Guidelines for naming prefixes are given in Rules 3 through 5.

Guidelines for Naming Branched-Chain Hydrocarbons

Rule 1: Assign the root. Determine the root name for a branched-chain hydrocarbon from the length of the *main* chain, in accordance with Table 6-4. *The main chain is the longest continuous carbon chain, containing the double or triple bond, if one exists.*

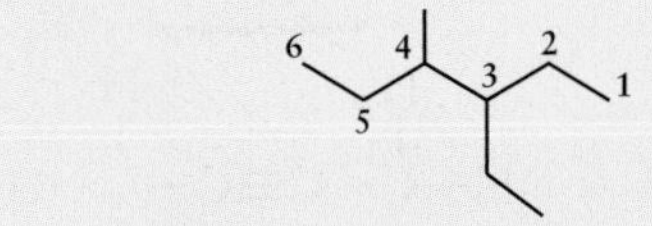

Root: hexane (6 carbon atoms in main chain)

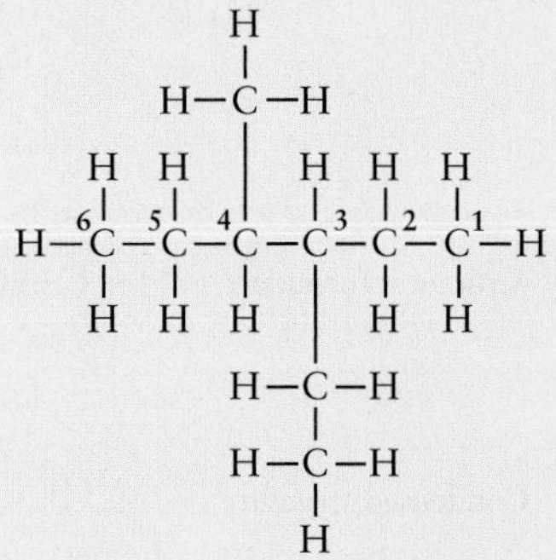

Rule 2: Assign the suffix. As with straight-chain hydrocarbons, the suffix is changed to reflect whether the molecule is an alkane, alkene, or alkyne. In addition, for a multiple bond, a locator number is placed immediately before the root name to indicate the location of the first carbon atom of the multiple bond.

Rule 3: Name each substituent. Assign a name to each substituent based on the number of carbon atoms in the substituent chain, using Table 6-4. Change the ending on a substituent from *-ane* to *-yl* to signify that it is a substituent and not the main chain. *A -yl ending always signifies a substituent.*

For example, if the substituent is a CH_3 group, the substituent name would be meth*yl*, not meth*ane*. The most common branches are methyl and ethyl.

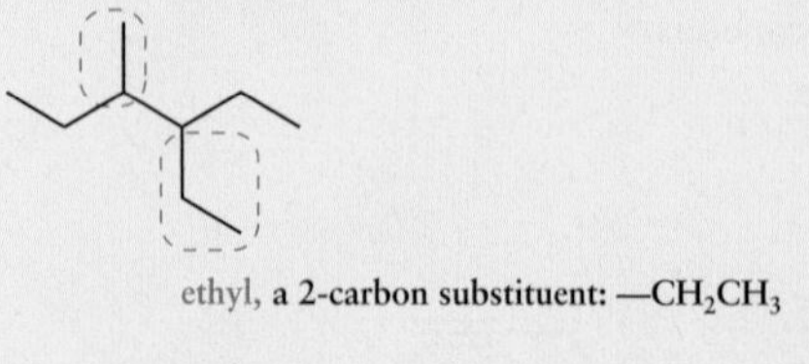

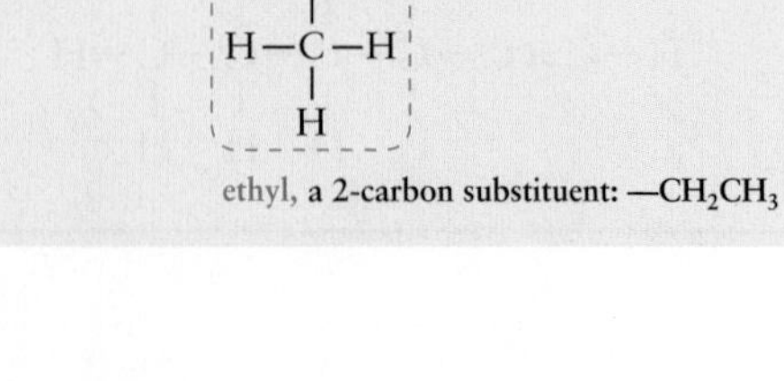

methyl, a 1-carbon substituent: —CH_3

ethyl, a 2-carbon substituent: —CH_2CH_3

methyl, a 1-carbon substituent: —CH_3

ethyl, a 2-carbon substituent: —CH_2CH_3

Rule 4: Assign a locator number to all substituents. Number each carbon atom in the main chain, beginning from the end of the chain closer to the multiple bond if one exists. Otherwise, start numbering from the end nearer a branch point. Identify by number all carbon atoms along the main chain with substituents—these are locator numbers. If there are two branch points on the same carbon atom, use the same locator number twice.

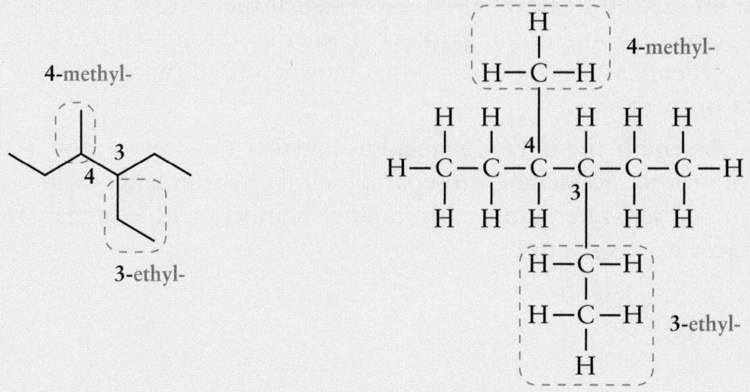

Rule 5: Assemble the prefix name. List the substituent names in alphabetical order. Place the locator number in front of each substituent name separated by a hyphen.

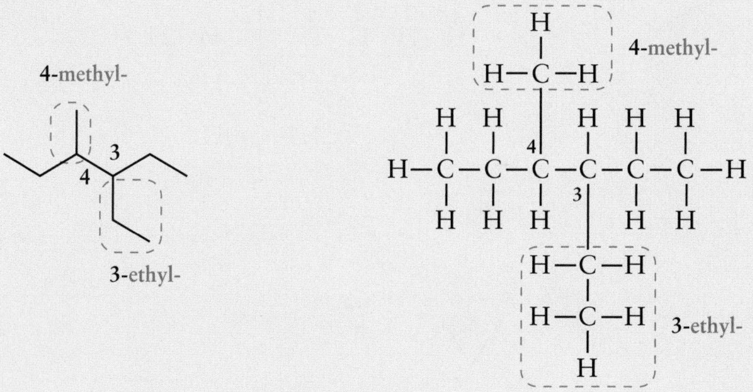

3-Ethyl-4-methyl...

Rule 5b: Assembling the prefix name. What if some substituents are repeated more than once along the main chain? In this case, insert the additional prefix *di-* (for 2), *tri-* (for 3), or *tetra-* (for 4), in front of the prefix to indicate how many times a particular substituent is repeated along the main chain. For example, the eight-carbon main chain shown below contains four methyl substituents at carbons 2, 3, and 5; therefore the prefix name would be *2,3,3,5-tetramethyl....*

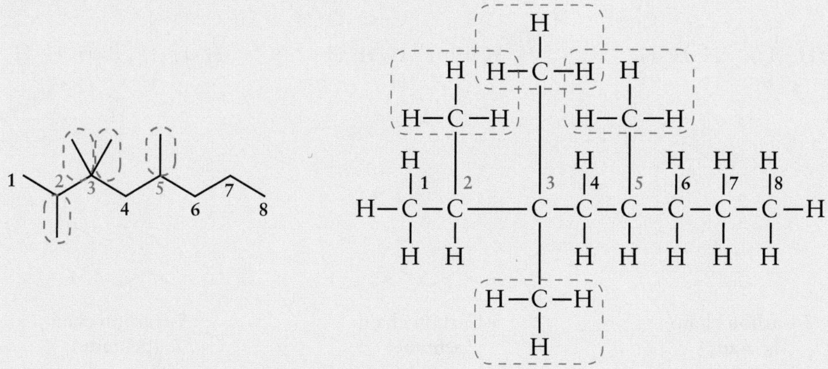

2,3,3,5-Tetramethyl...

Place the locator numbers that go with repeated substituents in front of the *di-, tri-, tetra-* designation and separate the locator numbers with a comma. For example, the prefix *2,3,3,5-tetramethyl...* indicates that there are four methyl substituents along the chain and that they are located on the second, third, and fifth carbon atoms in the chain. The fact that the number three is shown twice indicates that two of the four methyl groups branch off the third carbon atom. Note that if there are two substituents on the same carbon atom, regardless of whether the substituents are different or identical, the locator number must be listed twice, once for each substituent.

Rule 6: Assemble the IUPAC name. Put together the prefix, root, and suffix in that order. Remember to separate numbers from letters with a hyphen, and separate numbers from other numbers with commas. Do not add spaces between letters.

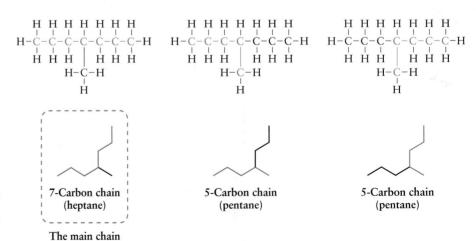

Most of the effort in naming branched-chain hydrocarbons goes into constructing the prefix. Do not let long and complicated names scare you, as the rules are very logical. Note that structural isomers will always have different IUPAC names.

Figure 6-13 Finding the main chain: The longest continuous chain in the structure shown is the chain with seven carbon atoms.

7-Carbon chain
(heptane)

The main chain

5-Carbon chain
(pentane)

5-Carbon chain
(pentane)

Structural formula

Skeletal line structure

4-Carbon chain: does contain double bond

5-Carbon chain: does not contain double bond

The main chain

Figure 6-14 Finding the main chain: In the molecule shown, the longest chain including the double bond contains four carbon atoms.

Finding the Main Chain

Note that the main chain is not always the one that lies horizontally on the page. For example, the longest continuous chain in the molecule shown in Figure 6-13 is the seven-carbon chain (printed in red) and not the five-carbon chains (printed in blue). It may help to highlight the main chain before you perform the rest of the naming steps.

For alkenes and alkynes, the main chain must contain *both* carbon atoms of the multiple bond, even if there is a longer chain. For example, in the molecule shown in Figure 6-14, the longest continuous chain containing the entire double bond is printed in red. You would select this chain as the main chain when assigning the root, even though it contains one fewer carbon atom than the five-carbon chain printed in blue.

WORKED EXERCISES 6-10   Naming Branched Hydrocarbons

1 Provide the IUPAC name for the structure shown below:

2 Provide a structural formula and a skeletal line structure for 2,2-dimethylnonane.

SOLUTIONS

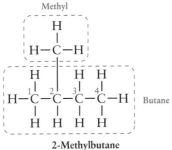

2-Methylbutane

1 The IUPAC name for the branched-chain alkane shown is 2-methylbutane.
   Rule 1: Assign the root. The longest continuous chain of carbon atoms contains four carbon atoms, so according to Table 6-4 the root name is *butane*.
   Rule 2: Assign the suffix. The suffix remains *-ane*, because the compound is an alkane.

Problem-Solving Tutorial:
IUPAC Naming of Alkanes

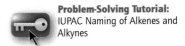

Problem-Solving Tutorial:
IUPAC Naming of Alkenes and Alkynes

Rule 3: Name each substituent. There is only one substituent, and it consists of one carbon atom: methyl.

Rule 4: Assign a locator number to the substituent. The locator number 2- is assigned to the substituent because it is located on the second carbon on the main chain when numbering from the end closer to the substituent (left end as drawn).

Rule 5: Assemble the prefix. Place the locator number in front of the substituent name to create the prefix *2-methyl*.

Rule 6: Assemble the IUPAC name. Write the prefix followed by the root and suffix: 2-methylbutane.

2 The structural formula and skeletal line structure of 2,2-dimethylnonane is shown below:

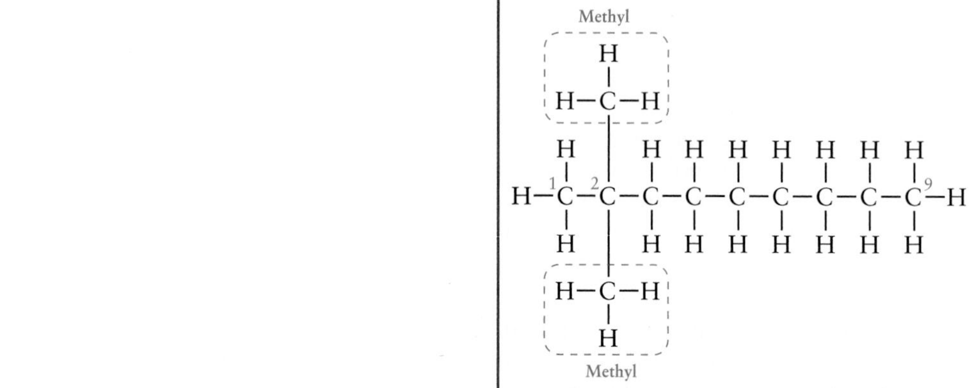

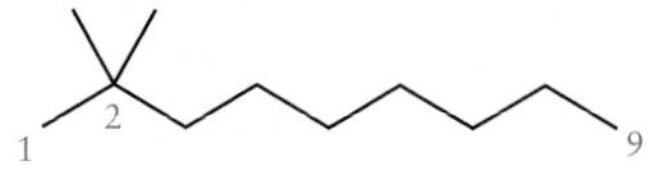

The root name *nonane* indicates a nine-carbon chain. Number the chain beginning from either end. Insert two methyl groups, both on carbon 2.

PRACTICE EXERCISES

6.35 Write the IUPAC name for the following branched-chain alkanes:

a.

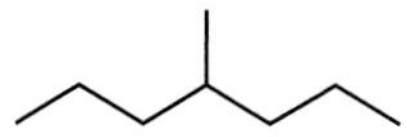

b.

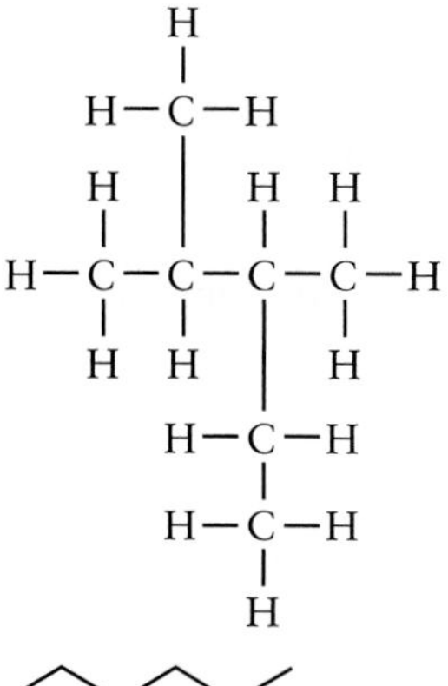

c.

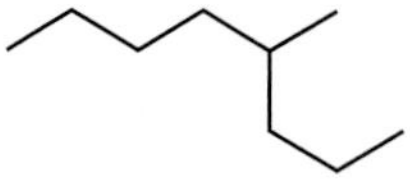

6.36 Write the structural formula and skeletal line structure for the branched-chain alkanes with the following IUPAC names:

a. 2-methyloctane   b. 3-ethylhexane

6.37 Name the following branched-chain alkenes (there are no cis–trans isomers):

a. 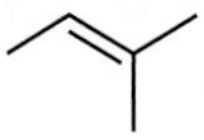 b. 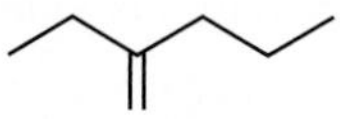 c.

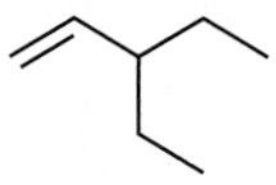

6.38 Write the skeletal line structure for the following alkenes:

    a. 2-ethyl-3-methyl-1-butene   **b.** 3-ethyl-5,5-dimethyl-3-octene

6.39 Octane ratings on gasoline fuel are an indication of how well the fuel burns compared to a mixture of isooctane (common name) and heptane—hence the term *octane rating*. Isooctane has the IUPAC name 2,2,4-trimethylpentane, which has an octane rating of 100, while heptane has an octane rating of 0. Thus, a fuel with an octane rating of "87" behaves like an 87:13 mixture of 2,2,4-trimethylpentane and heptane. A higher octane rating results in less engine knocking (premature ignition of the fuel in the engine). Provide a skeletal line structure and a condensed structure of both isooctane and heptane.

6.40 Write the skeletal line structure and provide the IUPAC name for all three structural isomers of C_5H_{12}.

6.41 **Critical Thinking Question:** Provide the IUPAC name for the following structural formula. (When two main chains of equal length exist, select the one with the most branch points.)

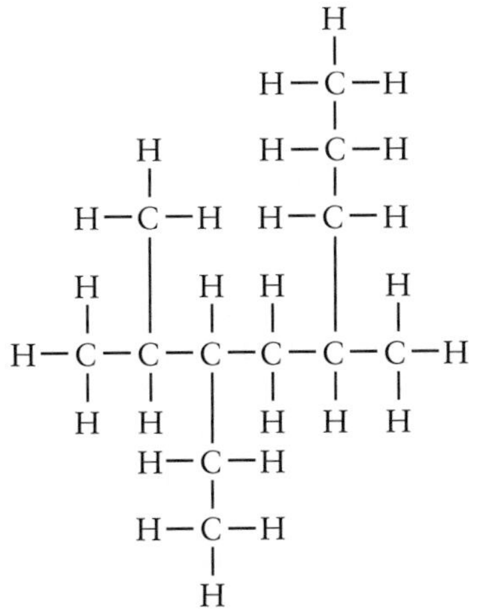

6.6 | Aromatic Hydrocarbons

In the nineteenth-century, chemists isolated a number of hydrocarbons that contained a unique unsaturated hydrocarbon ring structure. These compounds exhibited strong and mostly pleasant aromas and therefore became known as **aromatic compounds.** Vanillin is one such fragrant aromatic compound. Although the term *aromatic* remains with us today, we now define aromatic hydrocarbons by their chemical structure, not their odor.

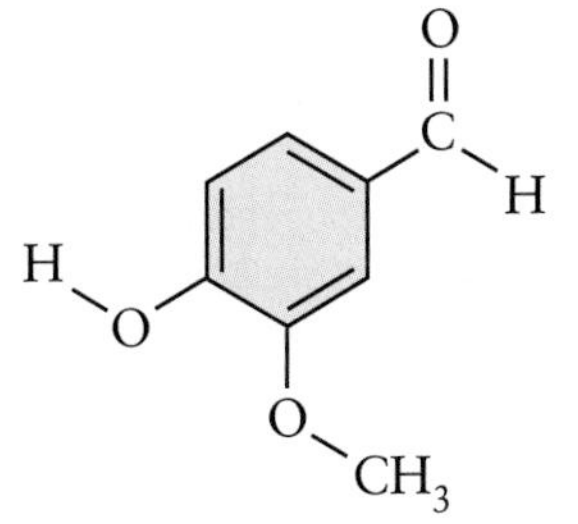

Benzene

The simplest aromatic hydrocarbon is **benzene**, shown in Figure 6-15. Benzene is an unsaturated ring of six carbon atoms in which each carbon atom is drawn so that it has one C–H bond, one C—C single bond, and one C=C double bond. As a result, each carbon atom has a trigonal planar geometry and every C—C—C bond angle is 120°, giving the overall molecule a flat two-dimensional shape. This flat shape is evident in the three views of models of benzene shown in Figure 6-15.

Benzene is not classified as an alkene despite the apparent presence of three carbon–carbon double bonds in its structure. The electrons shared between the carbon atoms (the C—C bonds) of the six-membered benzene ring are actually evenly distributed across all six carbon atoms; they are not alternating single and double bonds. When electrons are not localized between two atoms, but distributed over several atoms they are said to be **delocalized.** In the case of benzene, six electrons are actually delocalized over all six carbon atoms of the ring. Delocalization of electrons in benzene is evident from

Vanillin

Vanillin is the aromatic compound found in vanilla beans responsible for its pleasant aroma.

Benzene

Do not confuse benzene with cyclohexane. Both have six carbon atoms, but cyclohexane has 12 hydrogen atoms and the carbon atoms are tetrahedral (see Figure 6-8 on page 190), while benzene has 6 hydrogen atoms and the carbon atoms are trigonal planar. Benzene is flat, cyclohexane is not.

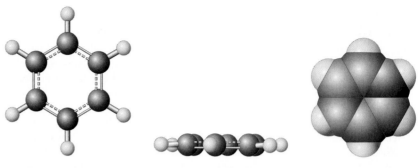

Top view, ball-and-stick   Side view, ball-and-stick   Top view, space-filling

Figure 6-15 Three molecular models illustrate the flat shape of benzene. The leftmost model shows each C—C bond as one and a half bonds. The half bonds are depicted as dashed lines.

The compound benzene is a carcinogen— a compound that causes or promotes cancer. Although benzene itself is a carcinogen, compounds containing aromatic rings are usually not carcinogenic.

experimental measurements. For example, measurement of the carbon–carbon bond length shows that each bond is the same length, 140 pm. This bond length is shorter than a single bond (153 pm) and longer than a double bond (134 pm)—about the length of 1½ bonds, supporting the fact that electrons are delocalized in benzene and all the C—C bonds are the same length.

Structural formulas are limited to showing only localized electrons and only in multiples of two (one line always equals two electrons). In order to depict delocalized electrons, resonance structures are used. When writing **resonance structures** of benzene, two skeletal line structures are drawn, separated by a double headed arrow, as shown in Figure 6-16a. The actual structure of benzene is understood to be a hybrid of its two resonance structures. As a general rule, resonance structures differ only in the placement of electrons, not atoms. If you look closely at Figure 6-16a you will see that the only difference between the two structures is the precise location of the double bonds. Usually only one resonance structure is written for benzene, and the other is implied. Alternatively, a circle in the center of a hexagon is another way to depict the six delocalized electrons in benzene, as shown in Figure 6-16b. The condensed structure for a benzene ring used as a substituent on another chain is C_6H_5— or Ph— (phenyl).

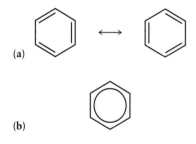

Figure 6-16 Two representations of benzene shown as skeletal line structures.

Delocalization of electrons in a molecule minimizes electron–electron repulsions, so a molecule with delocalized electrons will be more stable— lower in potential energy. *A more stable molecule is less likely to undergo chemical reactions.* It is the added stability of aromatic compounds that places them in a category separate and distinct from alkenes. The stability of aromatic hydrocarbons also accounts for their presence in many compounds found throughout nature.

Naming Substituted Benzenes

Benzene is the IUPAC name for C_6H_6, but it also serves as the root name when naming many compounds containing substituents attached to a benzene ring (see the *Guidelines* box below).

Guidelines for Naming Substituted Benzenes

Rule 1: Assign the root. The root *benzene* appears in many compounds containing an aromatic ring. Another common root is the benzene ring together with one CH_3 group. This root is called *toluene*.

Rule 2: Assign the prefix. The prefix provides the identity of the substituent(s) on the benzene ring: methyl, ethyl, etc. Substituents are named using the same system used to name the substituents on hydrocarbon chains.

• When one substituent is present on the benzene ring, no locator number is needed, because all positions are identical.

- When more than one hydrogen atom on the benzene ring has been replaced by substituents, locator numbers must be included in the prefix to indicate the relative positions of the substituents on the ring. Number in a way that gives the lower set of numbers to the substituents—for example, 1,2 rather than 1,6 and 1,3 rather than 1,5. Place the locator numbers at the beginning of the prefix, separated by commas. When repeated substituents appear, the additional prefixes *di-, tri-,* and *tetra-* are used the same as with branched chain hydrocarbons.

Three examples of a benzene ring attached to a single substituent are shown below. Notice that the root for each is *benzene* or *toluene*. The substituent is named according to the number of carbon atoms that it contains, with the usual ending -*yl*.

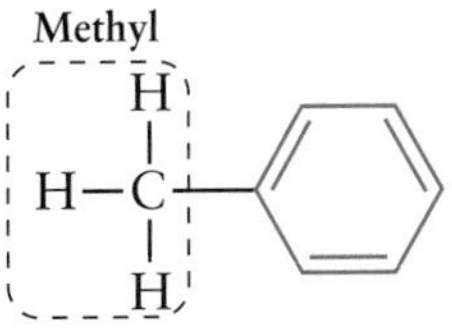

Methylbenzene (toluene)   Ethylbenzene   Propylbenzene

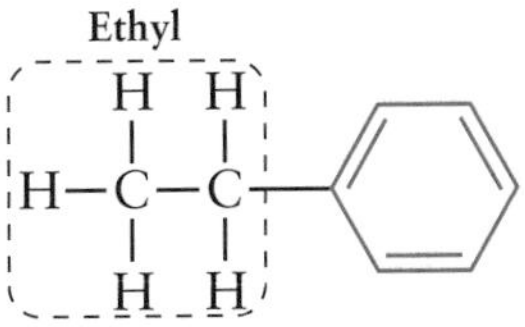

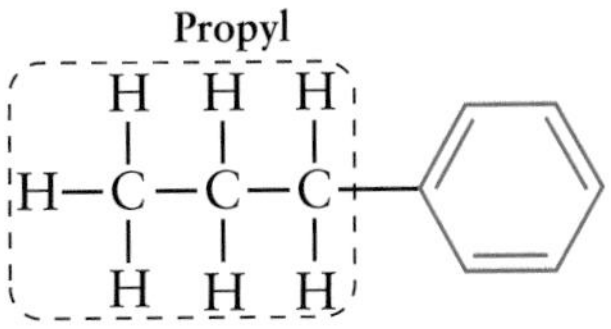

The benzene derivative 1-ethyl-3-methylbenzene contains two substituents: an ethyl group (CH_3CH_2) and a methyl group (CH_3), located at carbon 1 and carbon 3. Locator numbers are necessary when there are two or more substituents to indicate the relationship between the substituents. The locator numbers 1 and 3 are used because they are a lower set of numbers than the alternative, 1 and 5.

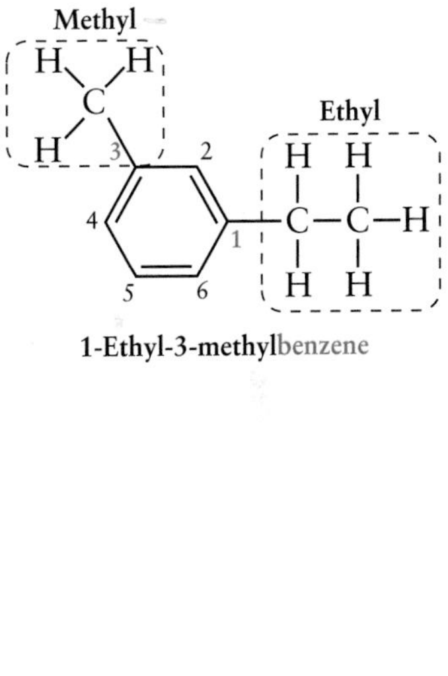

When two identical substituents are present, *di-* is inserted in front of the substituent name. *D*imethylbenzene has three structural isomers: 1,2-dimethylbenzene, 1,3-dimethylbenzene, and 1,4-dimethyl benzene, as shown in Figure 6-17. The designation 1,2- indicates that the two CH_3 groups are on adjacent carbon atoms (carbons 1 and 2); the designation 1,3 indicates that the CH_3 groups are on the first and third carbon atoms, and so forth.

Many common over-the-counter (OTC) analgesics contain aromatic rings with two substituents, including aspirin, ibuprofen (Motrin, Advil), and acetaminophen (Tylenol). These molecules are not hydrocarbons because they contain other atoms in addition to carbon and hydrogen atoms; however, they do contain an aromatic ring. To gain more familiarity with benzene and substituted benzenes, perform *The Model Tool 6-5: Benzene and Substituted Benzenes.*

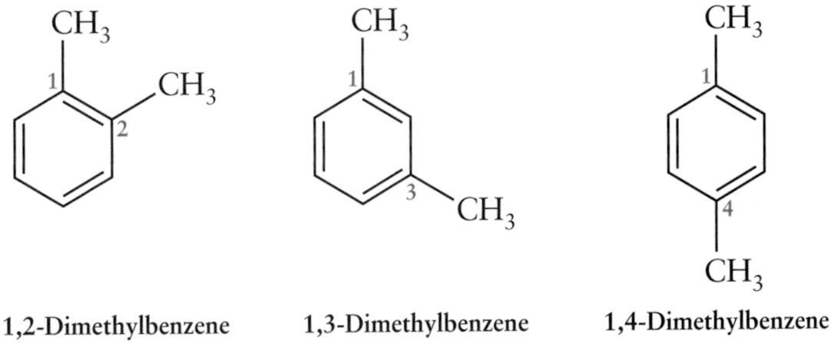

1-Ethyl-3-methylbenzene

1,2-Dimethylbenzene   1,3-Dimethylbenzene   1,4-Dimethylbenzene

Figure 6-17 Structural isomers of dimethylbenzene differ in the relative location of the two methyl substituents.

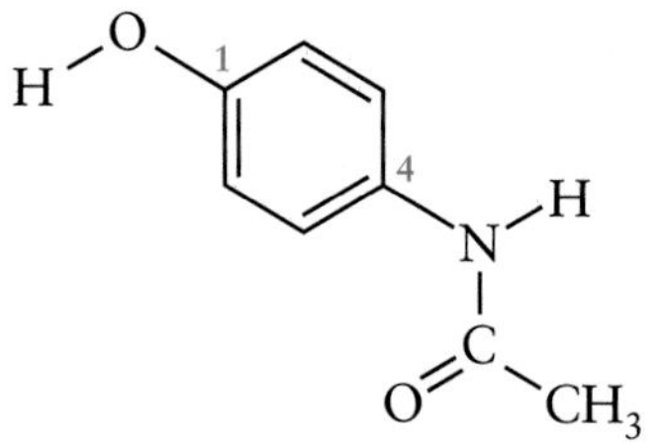

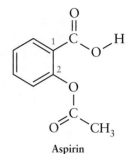

Acetaminophen 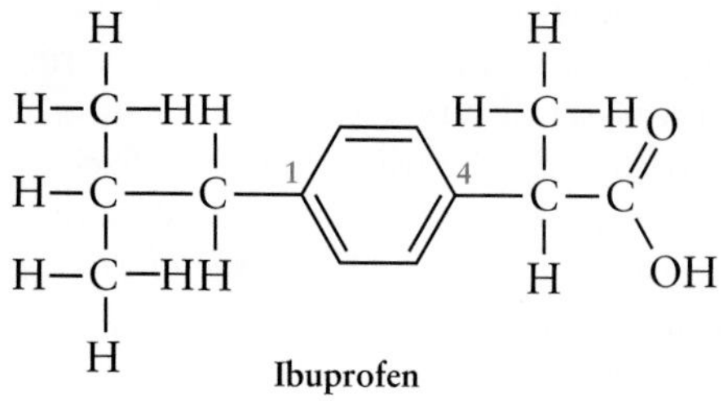   **Ibuprofen**   **Aspirin**

Problem-Solving Tutorial:
IUPAC Naming of Substituted
Benzenes

WORKED EXERCISE 6-11   Benzene and Substituted Benzenes

Provide the IUPAC names for the following substituted aromatic compounds:

a.   b.

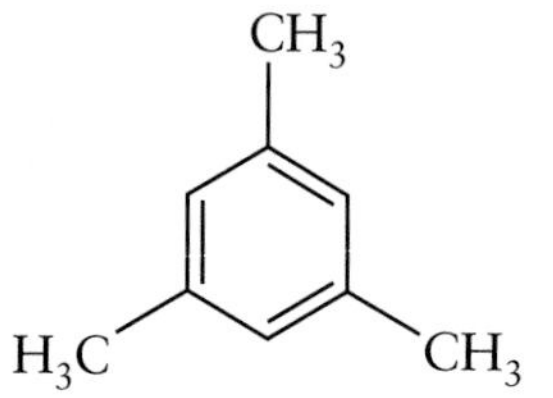

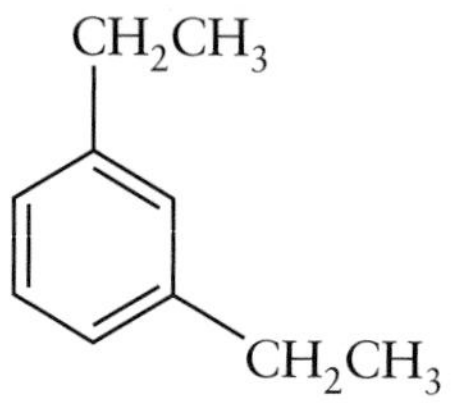

SOLUTION

a. The IUPAC name is 1,3,5-trimethylbenzene.

   Rule 1: Assign the root. The root is *benzene* because a six-membered aromatic ring is present

   Rule 2: Assign the prefix. Three of the hydrogen atoms on the aromatic ring have been replaced with a CH_3 group. Locator numbers are needed to indicate their relative locations. Starting at any of the substituted positions, numbering the carbons with substituents gives 1, 3, and 5.

b. The IUPAC name is 1,3-diethylbenzene. The root is *benzene* and there are two identical ethyl substituents on the ring, so *di-* must be inserted in front of the substituent name. Hence, the prefix is *diethyl-*. To determine the relative positions of the two ethyl substituents give one the number "1" and then count around the ring the shorter distance to the other substituent, so that you obtain the lower set of values: 1,3 (not 1,5). Insert the prefix, *1,3-diethyl-*, before the root, *benzene*.

PRACTICE EXERCISES

6.42 Provide the IUPAC names for the following substituted aromatic compounds:

a.   b. $C_6H_5CH_2CH_3$

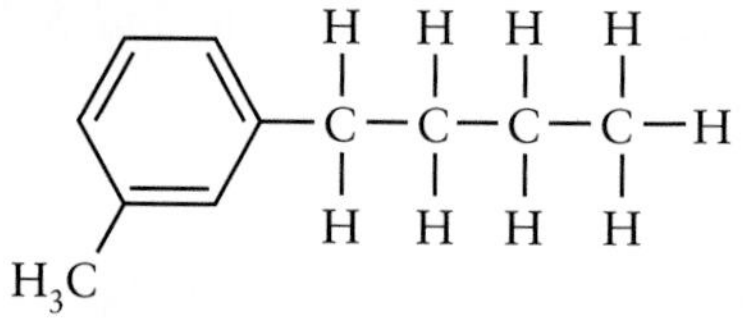

c.   d.

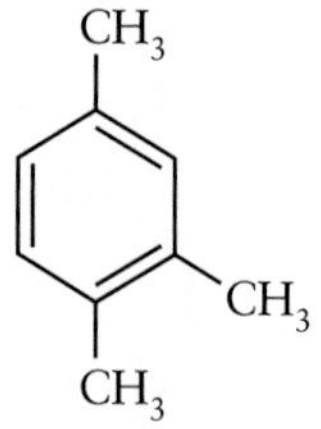

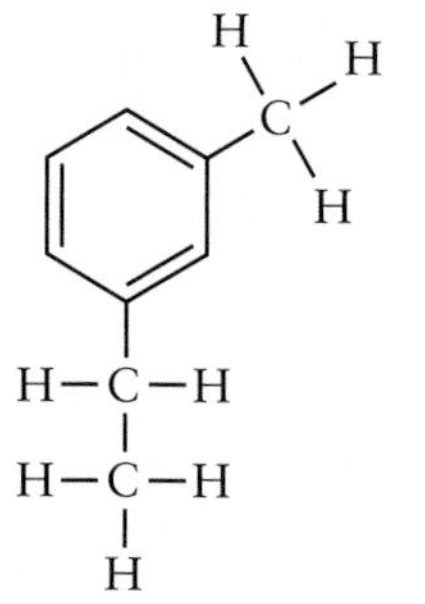

6.43 Provide the structural formula for the following compounds:
 a. 1,4-diethylbenzene
 b. 1-ethyl-4-methylbenzene (also known as 4-ethyltoluene)

6.44 Indicate whether the following statements are *true* or *false*.
 a. Aromatic hydrocarbons are unstable molecules that readily undergo chemical reactions.
 b. Benzene, C_6H_6, is a carcinogen.
 c. The molecular shape of all the carbon atoms in benzene is trigonal planar.
 d. Aromatic rings are rare in nature.
 e. Benzene has chemical properties that are distinctly different from those of an alkene.

The Model Tool 6-5 Benzene and Substituted Benzenes

I. Construction of Benzene

1. Obtain 8 black carbon atoms, 10 light-blue hydrogen atoms, 13 straight bonds, and 6 bent bonds.

2. Construct a model of benzene, C_6H_6, by joining six carbon atoms with alternating single and double bonds. Use two bent bonds for every double bond you make.

Benzene

II. Observations

3. Describe the overall shape of your benzene model?

4. Viewed from the side, what does benzene look like?

5. Viewed from above, what does benzene look like?

6. Are the C—H bonds in benzene in the same plane as the C—C bonds? That is, is the molecule flat?

7. Does benzene have the same shape as cyclohexane? Explain.

8. What are the H—C—C and C—C—C bond angles around each carbon atom in benzene?

9. Is there free rotation around any of the C—C bonds in benzene?

10. Does benzene have more or less flexibility than cyclohexane? Explain.

11. You constructed your model using alternating single and double bonds. Although this is the only way to make a ball-and-stick model of benzene, in what way is this not an accurate representation of benzene?

III. Construction of Ethylbenzene

12. Construct a model of ethylbenzene by removing one hydrogen atom from your model of benzene and replacing it with a two-carbon chain—an ethyl group.

13. Why is the model the same no matter which hydrogen atom is replaced?

IV. Construction of Dimethylbenzene Isomers

14. Remove the ethyl group from the model you constructed in Part III, and remove another H atom from the benzene ring. Replace the two hydrogen atoms with two methyl groups.
 a. Does it matter which hydrogen atoms on benzene are replaced with CH_3 groups?
 b. How does 1,3-dimethylbenzene differ from 1,2-dimethylbenzene?
 c. How does 1,4-dimethylbenzene differ from 1,2-dimethylbenzene and 1,3-dimethylbenzene?
 d. Why are there no structural isomers named 1,5- or 1,6-dimethylbenzene?
 e. Why does 1,1-dimethylbenzene not exist, whereas 1,1-dimethylcyclohexane does?
 f. How many structural isomers of dimethylbenzene are there?

Chemistry in Medicine Cholesterol and Cardiovascular Disease

Cholesterol is an important biological molecule which, except for one O—H group, is a hydrocarbon, as shown in the structures below.

Cholesterol

Although the liver has the capacity to make all the cholesterol the body needs, cholesterol is also supplied by a variety of foods that we eat such as meat, fish, and dairy products. Fruits and vegetables—and all plant-based foods—do not contain cholesterol. Our cells need cholesterol to build their cell membranes. Cholesterol is also the body's raw material for making bile, vitamin D, and many important steroid hormones. The body requires about 1 gram of cholesterol a day.

The hydrocarbon characteristics of cholesterol make it insoluble in the bloodstream, which is composed primarily of water. Thus, cholesterol is transported through the circulatory system by soluble aggregates known as **lipoproteins**, composed of a combination of *lip*ids and *proteins*. Lipids are water-insoluble biological compounds, which include cholesterol and fats. Lipoproteins are spherical in shape and contain a hydrophilic ("water loving") exterior that makes them soluble in blood. Cholesterol and fats are found nestled in the hydrophobic ("water-fearing") interior of a lipoprotein.

Lipoproteins

Lipoproteins are subdivided into five categories based on their density. Low density lipoproteins (LDLs) are high in fats and cholesterol compared to proteins. At the other end of the spectrum, the high density lipoproteins (HDLs), have the highest protein content. There are also several lipoproteins that fall in between these two extremes, but these will be considered later, when you learn about lipids in Chapter 13.

The role of the LDLs is to transport cholesterol to the tissues. However, if a cell already has enough cholesterol, LDLs are not allowed to deliver more to the cell, leaving them circulating in the bloodstream. Eventually the excess circulating LDLs leave their contents along artery walls contributing to plaque formation, resulting in a condition known as **atherosclerosis**. It is for this reason that LDLs are considered the "bad cholesterol." Over time, plaque can build up and restrict the blood flow through arteries—causing **cardiovascular disease**. One in four Americans has cardiovascular disease. The X ray and corresponding sketch in Figure 6-18 shows plaque deposits restricting blood flow through an artery.

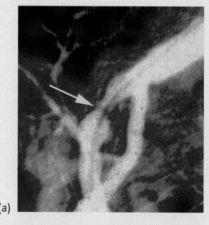

(a)

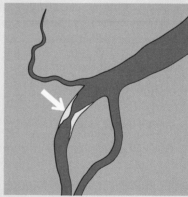

(b)

Figure 6-18 (a) An X ray shows the restricted blood flow in an artery caused by plaque. (b) The sketch highlights the location of the plaque in the X-ray image in (a).[James Cavallini/Photo Researchers, Inc.]

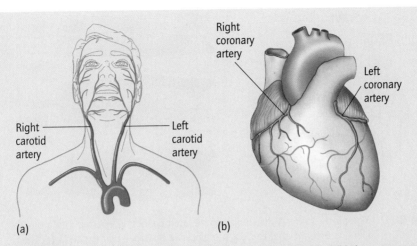

Figure 6-19 (a) The carotid arteries supply oxygen to the brain, and (b) the coronary arteries supply oxygen to the heart. Blockage of one of these arteries by plaque also blocks the oxygen supply to the organ in question.

If a complete blockage occurs, the result can be life-threatening, leading to a stroke or heart attack. If a blockage is located in the carotid artery to the brain, shown in Figure 6-19a, oxygen deprivation of the brain causes a stroke. If the blockage is in a coronary artery, shown in Figure 6-19b, oxygen deprivation of the heart muscle causes a heart attack. Cardiovascular disease is the leading cause of death in the United States in both men and women, and accounts for over 40% of all deaths.

About 60 to 75% of all blood cholesterol is found in the LDLs. The remainder of the cholesterol in blood is found primarily in HDLs. HDLs are considered the "good cholesterol," because they scavenge excess cholesterol in the bloodstream and transport it back to the liver. Higher levels of this lipoprotein circulating in the bloodstream are desirable, because it removes excess cholesterol from the blood. A healthy diet and exercise have both been shown to increase HDL levels.

Treating High Cholesterol

When a medical laboratory measures a person's cholesterol levels, they measure the concentration of the various lipoproteins as well as fats in the blood serum. The measurements are sometimes referred to as a *lipid panel*. Table 6-6 shows the normal concentration range for some lipids as recommended by the National Institutes of Health (NIH). Notice that lower levels of all but the HDLs are desirable.

When levels of cholesterol and LDLs are elevated, patients are often advised to reduce the amount of cholesterol and saturated fat in their diet as a first step

Table 6-6	NIH Recommended Lipid Levels	
Lipid Type	Desirable Level (mg/dL)*	Undesirable Level (mg/dL)*
Total cholesterol	<200	>240
HDL cholesterol	>60	<40
LDL cholesterol	<100	>160
Triglycerides (fats)	<150	>200

*(milligram per deciliter)

toward lowering LDL levels. In many cases, cholesterol-reducing drugs are prescribed.

In advanced stages of atherosclerosis, a doctor can perform a technique known as **angioplasty** to reopen the narrowed artery and restore blood flow (Figure 6-20). In this procedure, a medical balloon located at the tip of a special catheter is inserted into a blood vessel through the arm or groin, and conveyed to the narrowed section of artery. Upon reaching the constriction, the balloon is inflated, reopening the artery and restoring blood flow. An artificial reinforcement, known as a stent, can be inserted at the site to prevent the artery from narrowing again in the future.

Both diet and heredity play a role in the amount of cholesterol circulating in your blood. Although cholesterol and fats are both integral to health, too much cholesterol or LDLs circulating through the bloodstream can lead to cardiovascular disease.

A knowledge of hydrocarbons and their physical properties makes it easier to understand this link between fats, cholesterol, and heart disease.

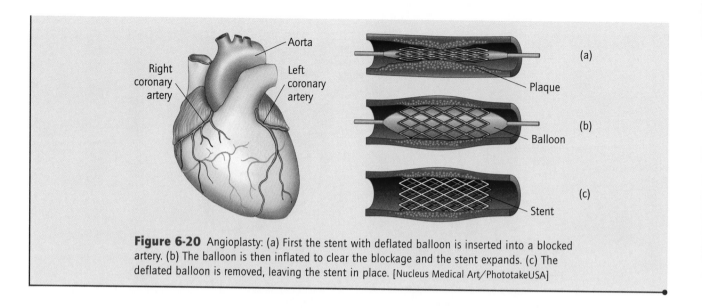

Figure 6-20 Angioplasty: (a) First the stent with deflated balloon is inserted into a blocked artery. (b) The balloon is then inflated to clear the blockage and the stent expands. (c) The deflated balloon is removed, leaving the stent in place. [Nucleus Medical Art/PhototakeUSA]

You have completed the first of two chapters that focus on the fundamentals of organic chemistry. In this chapter you were introduced to the four types of hydrocarbons: alkanes, alkenes, alkynes, and aromatic hydrocarbons, as well as some basic rules for naming them using the IUPAC system. The structural characteristics of these compounds, including the many different conformations that alkanes can adopt, were introduced. In the next chapter you will consider organic compounds containing atoms other than carbon and hydrogen. You will be introduced to the common heteroatom groupings commonly seen in organic compounds—known as functional groups.

Chapter Summary

Hydrocarbons

- Organic compounds are compounds containing one or more carbon atoms.
- Hydrocarbons are organic compounds composed of only carbon and hydrogen atoms.
- There are four types of hydrocarbons: alkanes, alkenes, alkynes, and aromatic hydrocarbons.
- Alkanes are saturated hydrocarbons.
- Alkenes, alkynes, and aromatic hydrocarbons are unsaturated hydrocarbons.
- Hydrocarbons are nonpolar and hydrophobic substances, and therefore insoluble in water. Hydrocarbons are hydrophobic—water fearing.

Saturated Hydrocarbons: The Alkanes

- Alkanes are saturated hydrocarbons containing no multiple bonds: only carbon–carbon single bonds and C—H bonds.
- Alkanes have a tetrahedral geometry around each of their carbon atoms, giving a chain of three or more carbon atoms a zigzag appearance. Bond angles are 109.5°.

- Conformations are rotational forms of the same compound. Alkanes can rotate freely around their carbon–carbon single bonds and therefore give rise to many different conformations.
- Structural isomers have the same chemical formula but a different connectivity of atoms. They are different compounds so they have different physical and chemical properties.

Drawing Structures

- Condensed structural formulas simplify the process of writing organic structures by listing each carbon atom and its attached hydrogen atom(s) in sequential order and omitting the bonds, except where a branch occurs.
- Skeletal line structures are a shorthand for writing organic structures in which the carbon and hydrogen atoms are omitted. Carbon atoms are represented at both the intersection of lines, which represent carbon–carbon bonds, and the ends of lines.
- Cycloalkanes are ring structures often depicted as skeletal line structures, which typically appear as 3- to 8-sided polygons.

Unsaturated Hydrocarbons: Alkenes and Alkynes

- Alkenes are unsaturated hydrocarbons containing one or more double bonds.
- The carbon atoms in a double bond have a trigonal planar geometry.
- Some alkenes have a geometric isomer, a different compound with the same chemical formula, same connectivity of atoms, but a different three-dimensional orientation as a result of the restricted rotation of a double bond.
- The cis geometric isomer has the carbon groups on the same side of the double bond; the trans geometric isomer has them on the opposite sides of the double bond.
- Alkynes are unsaturated hydrocarbons containing one or more triple bonds.
- The triple bond has a linear geometry.

Naming Hydrocarbons

- To create the IUPAC name, you assign a root, suffix, and prefix.
- The root name of a hydrocarbon is based on the number of carbon atoms in the main chain, according to Table 6-4.
- The suffix for an alkane is -ane; for an alkene, -ene; and for an alkyne, -yne. Locator numbers must also be supplied in alkenes and alkynes to indicate the location of the first carbon containing the multiple bond.
- Substituents are named by providing a locator number along with the name of the substituent as a prefix in the IUPAC name. Substituents are named based on the number of carbon atoms in the substituent and a change in the ending to -yl.

Aromatic Hydrocarbons

- Aromatic hydrocarbons are unsaturated hydrocarbons containing a six-membered ring with six electrons delocalized throughout the ring. The simplest aromatic hydrocarbon is benzene, C_6H_6, a flat molecule with extraordinary stability due to delocalization of electrons.
- Aromatic compounds are named using the root name benzene.
- Substituents on a benzene ring are named in the same way as substituents of alkanes. Locator numbers must be provided when there are two or more substituents, giving the relative location of the substituents.

Key Words

Alkane Hydrocarbon containing only carbon–carbon single bonds.

Alkene Hydrocarbon containing one or more carbon–carbon double bonds.

Alkyne Hydrocarbon containing one or more carbon–carbon triple bonds.

Aromatic hydrocarbon Six-membered ring written as alternating double and single bonds, although six of these electrons are actually distributed evenly over the six carbon atoms in the ring.

Benzene The simplest aromatic hydrocarbon, C_6H_6, composed of a ring of six carbon atoms, each attached to a single hydrogen atom.

Branched-chain isomer A structural isomer in which there are carbon branches along the main carbon chain.

Condensed structural formula A notation used for writing the chemical structure of a molecule such that each carbon atom and its attached hydrogen atoms are written as a group: CH, CH_2, or CH_3. Bonds are omitted except for branch points.

Conformations Different rotational forms of the same compound.

Cycloalkane An alkane chain where the first carbon is connected to another carbon atom three to eight carbon atoms later in the chain, giving the molecule the shape of a ring. Written as a polygon in skeletal line structures.

Geometric isomers Compounds with the same chemical formula and same connectivity of atoms, but a different spatial orien-tation as a result of the restricted rotation of the double bond.

Hydrocarbon An organic compound composed only of carbon and hydrogen atoms.

Hydrophilic Describing the tendency of polar organic compounds to be soluble in water.

Hydrophobic Describing the tendency of nonpolar compounds to avoid water.

Lipoproteins Spherical aggregates of lipid and protein, which are hydrophilic on the outside and hydrophobic on the inside. They carry cholesterol and other fats through the polar aqueous medium of the circulatory system.

Organic chemistry The branch of chemistry devoted to the study of carbon-containing compounds and their chemical reactions.

Pharmaceutical A drug used for therapeutic purposes.

Saturated hydrocarbon A hydrocarbon without multiple bonds in its structure: alkanes.

Skeletal line structure An abbreviated form of writing chemical structures in which carbon and hydrogen atoms are not written and C—H bonds are omitted. Carbon–carbon bonds are written as lines in a zigzag format.

Straight-chain isomer A structural isomer in which there is no branching in the carbon chain.

Structural isomers Compounds that have the same chemical formula but differ in the connectivity of the atoms. Structural isomers are different compounds that exhibit different physical properties and have different chemical names.

Substituent A carbon branch along the main chain of a molecule.

Unsaturated hydrocarbon A hydrocarbon containing one or more multiple bonds in its structure. These could be alkenes, alkynes, or aromatic hydrocarbons.

Additional Exercises

Hydrocarbons

6.45 What is the chemical difference between an organic compound and an inorganic compound?

6.46 What type of organic compound contains only carbon and hydrogen atoms?

6.47 List the four types of hydrocarbons.

6.48 What is a heteroatom?

6.49 Indicate whether each of the following statements is *true* or *false*.
  a. Organic compounds cannot be synthesized in the laboratory.
  b. Pharmaceuticals can be isolated from plants and animals.
  c. Pharmaceuticals can be prepared in the chemical laboratory.
  d. Vaseline is a hydrocarbon.

6.50 Which of the following are unsaturated hydrocarbons?
  a. C_3H_8 **c.** C_5H_{10}
  b. C_4H_{10} **d.** C_2H_2

6.51 Circle all of the terms below that apply to hydrocarbons. There is more than one correct answer.
  a. They are hydrophobic.
  b. They are hydrophilic.
  c. They are capable of hydrogen bonding.
  d. They are insoluble in water.
  e. They are soluble in other hydrocarbons.

6.52 What type of intermolecular force of attraction exists between hydrocarbon molecules? Is this force of attraction strong or weak compared to the other intermolecular forces? What are the other two main types of intermolecular forces that exist between molecules?

6.53 Does water or methane have a higher boiling point? Explain why.

6.54 Which of the following household substances are hydrophobic, based on your everyday experience?
  a. bleach   **c.** waterproof mascara
  b. butter   **d.** regular mascara

Saturated Hydrocarbons

6.55 What is an alkane? Why are alkanes classified as saturated hydrocarbons?

6.56 Indicate whether each of the following statements is *true* or *false*:
  a. Two different conformations of a compound will have a different connectivity of atoms.
  b. One conformation of a molecule can be converted into another conformation without having to break any bonds.
  c. Carbon–carbon single bonds are not free to rotate.
  d. Alkanes are stationary molecules.

6.57 Why does an alkane chain of three or more carbon atoms have an overall zigzag appearance? Can it have a different appearance?

6.58 Draw two different conformations of C_5H_{12}.

6.59 What is the geometry of a carbon atom in an alkane: *tetrahedral, trigonal planar,* or *linear*?

6.60 What is the geometry of a carbon atom in the double bond of an alkene: *tetrahedral, trigonal planar,* or *linear*?

6.61 What is the geometry of a carbon atom in the triple bond of an alkyne: *tetrahedral, trigonal planar,* or *linear*?

6.62 Write the structural formula of the two structural isomers of C_4H_{10}. Label the branched-chain and straight-chain structural isomer.

6.63 Indicate whether the following pairs of compounds are:
i. different compounds because they have different chemical formulas.
ii. different compounds because they are structural isomers.
iii. identical compounds, but different conformations.

a.
```
    H  H  H              H  H
    |  |  |              |  |
H — C — C — C — H    H — C — C — H
    |  |  |              |  |
    H  H  H              H  H
```

b.
```
    H  H  H  H              H  H  H
    |  |  |  |              |  |  |
H — C — C — C — C — H    H — C — C — C — H
    |  |  |  |              |  |  |
    H  H  H  H              H  |  H
                             H — C — H
                                 |
                                 H
```

c.
```
    H  H  H  H              H  H  H  H
    |  |  |  |              |  |  |  |
H — C — C — C — C — H    H — C — C — C — C — H
    |  |  |  |              |  |  |  |
    H  H  H  H              H  |  H  H
                             H — C — H
                                 |
                                 H
```

d.
```
    H  H  H  H                  H
    |  |  |  |                  |
H — C — C — C — C — H       H — C — H
    |  |  |  |                  |
    H  |  H  H         H  H     H
    H — C — H          |  |     |
        |          H — C — C — C — C — H
        H              |  |  |  |
                       H  H  H  H
```

6.64 Why does $C_{10}H_{22}$ have so many more structural isomers than C_4H_{10}?

6.65 Explain how the structural isomers below are *different* from one another. What similarity do structural isomers have?

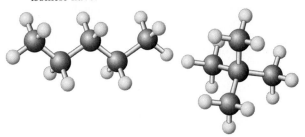

6.66 Some very large rings are found in nature. How many carbon atoms are in the ring shown below?

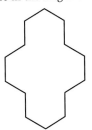

Writing Structures

6.67 Write the condensed structural formulas for the following structural formulas:

a.
```
                 H
                 |
             H — C — H
    H  H      |      H  H  H  H
    |  |      |      |  |  |  |
H — C — C — C — C — C — C — H
    |  |      |  |  |  |  |
    H       H  H  H  H  H
        H — C — H
            |
            H
```

b.
```
    H  H  H  H  H  H  H
    |  |  |  |  |  |  |
H — C — C — C — C — C — C — C — H
    |  |  |  |  |  |  |
    H  |  H  H  H  H  H
    H — C — H
        |
    H — C — H
        |
        H
```

c.
```
        H
        |
    H — C — H
    H   |    H  H
    |   |    |  |
H — C — C — C — C — H
    |   |    |  |
    H   |    H  H
    H — C — H
        |
        H
```

d.
```
    H  H  H  H
    |  |  |  |
H — C — C — C — C — H
    |  |  |  |
    H  H  H  H
```

e.
```
    H  H  H  H
    |  |  |  |
H — C — C — C — C — H
    |  |  |  |
    H  |  H  H
    H — C — H
        |
        H
```

f.
```
    H  H  H  H  H  H  H
    |  |  |  |  |  |  |
H — C — C — C — C — C — C — C — H
    |  |  |  |  |  |  |
    H  H  H  H  H  H  H
```

6.68 Write the structural formulas for the following condensed formulas:

a. $CH_3(CH_2)_6CH_3$

b. 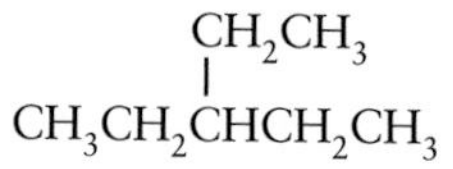
$$CH_3CH_2\overset{\overset{\displaystyle CH_2CH_3}{|}}{C}HCH_2CH_3$$

c.
$$CH_3CH_2\overset{\overset{\displaystyle CH_3}{|}}{\underset{\underset{\displaystyle CH_3}{|}}{C}}CH_2CH_3$$

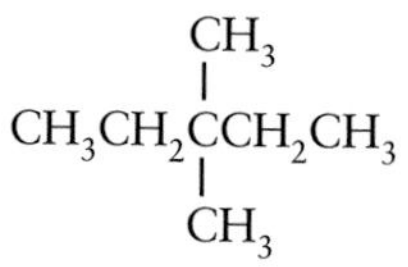

6.69 Provide a structural formula and a condensed formula for the following skeletal line structures.

a.

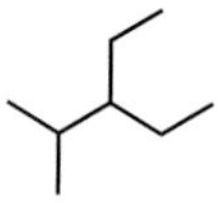

c.

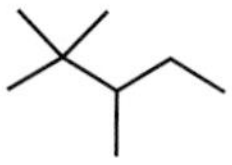

b.

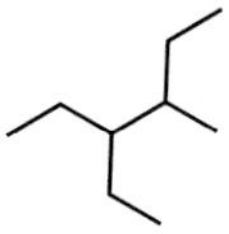

d.

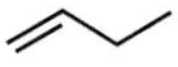

6.70 Provide skeletal line structures for the following structural formulas.

a. 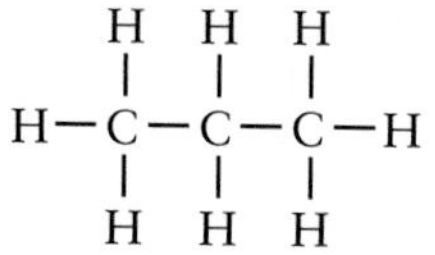
$$H-\overset{\overset{\displaystyle H}{|}}{\underset{\underset{\displaystyle H}{|}}{C}}-\overset{\overset{\displaystyle H}{|}}{\underset{\underset{\displaystyle H}{|}}{C}}-\overset{\overset{\displaystyle H}{|}}{\underset{\underset{\displaystyle H}{|}}{C}}-H$$

b.
$$H-\overset{\overset{\displaystyle H}{|}}{C}-H$$
$$H-\overset{\overset{\displaystyle H}{|}}{\underset{\underset{\displaystyle H}{|}}{C}}-\overset{\overset{\displaystyle H}{|}}{\underset{\underset{\displaystyle H}{|}}{C}}-\overset{\overset{\displaystyle H}{|}}{\underset{\underset{\displaystyle H}{|}}{C}}-\overset{\overset{\displaystyle H}{|}}{\underset{\underset{\displaystyle H}{|}}{C}}-H$$
$$H-\overset{\overset{\displaystyle H}{|}}{\underset{\underset{\displaystyle H}{|}}{C}}-H$$

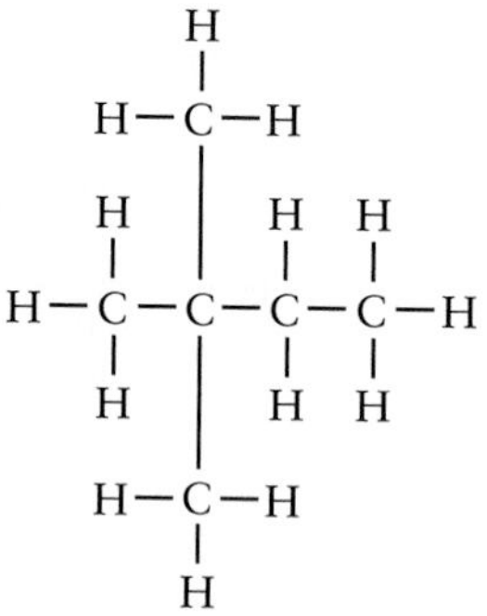

c.
$$H-\overset{\overset{\displaystyle H}{|}}{\underset{\underset{\displaystyle H}{|}}{C}}-\overset{\overset{\displaystyle H}{|}}{\underset{\underset{\displaystyle H}{|}}{C}}-\overset{\overset{\displaystyle H}{|}}{\underset{\underset{\displaystyle H}{|}}{C}}-\overset{\overset{\displaystyle H}{|}}{C}-H$$
$$H-\overset{\overset{\displaystyle H}{|}}{\underset{\underset{\displaystyle H}{|}}{C}}-H$$

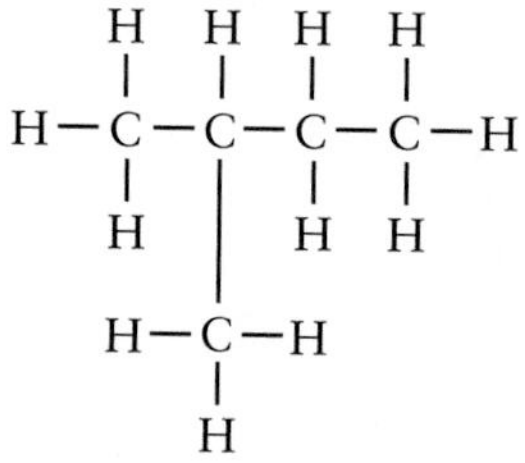

d. 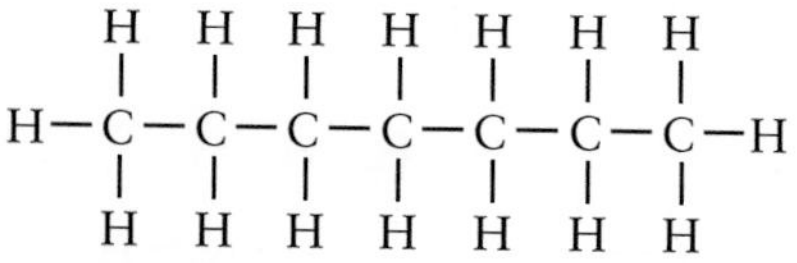
$$H-\overset{\overset{\displaystyle H}{|}}{\underset{\underset{\displaystyle H}{|}}{C}}-\overset{\overset{\displaystyle H}{|}}{\underset{\underset{\displaystyle H}{|}}{C}}-\overset{\overset{\displaystyle H}{|}}{\underset{\underset{\displaystyle H}{|}}{C}}-\overset{\overset{\displaystyle H}{|}}{\underset{\underset{\displaystyle H}{|}}{C}}-\overset{\overset{\displaystyle H}{|}}{\underset{\underset{\displaystyle H}{|}}{C}}-\overset{\overset{\displaystyle H}{|}}{\underset{\underset{\displaystyle H}{|}}{C}}-\overset{\overset{\displaystyle H}{|}}{\underset{\underset{\displaystyle H}{|}}{C}}-H$$

e. 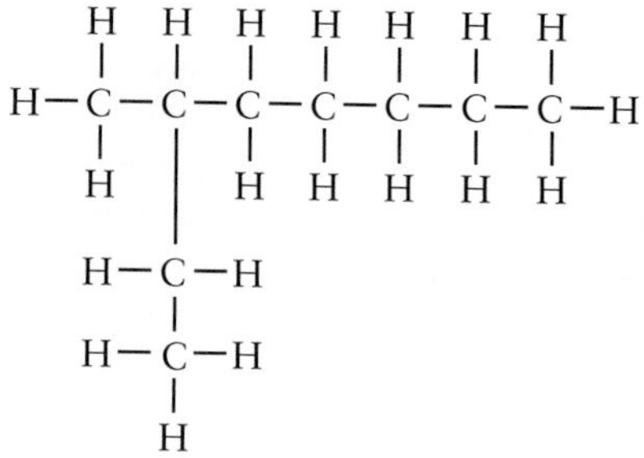
$$H-\overset{\overset{\displaystyle H}{|}}{\underset{\underset{\displaystyle H}{|}}{C}}-\overset{\overset{\displaystyle H}{|}}{\underset{\underset{\displaystyle H}{|}}{C}}-\overset{\overset{\displaystyle H}{|}}{C}-\overset{\overset{\displaystyle H}{|}}{\underset{\underset{\displaystyle H}{|}}{C}}-\overset{\overset{\displaystyle H}{|}}{\underset{\underset{\displaystyle H}{|}}{C}}-\overset{\overset{\displaystyle H}{|}}{\underset{\underset{\displaystyle H}{|}}{C}}-\overset{\overset{\displaystyle H}{|}}{\underset{\underset{\displaystyle H}{|}}{C}}-H$$
$$H-\overset{\displaystyle }{C}-H$$
$$H-\overset{\overset{\displaystyle }{|}}{\underset{\underset{\displaystyle H}{|}}{C}}-H$$

6.71 Provide skeletal line structures for the following condensed formulas.

a. 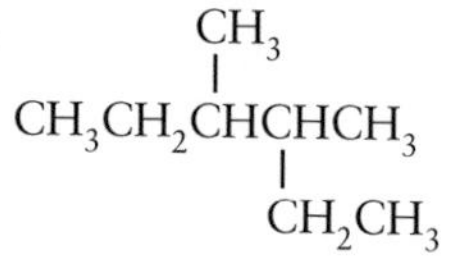
$$CH_3CH_2\overset{\overset{\displaystyle CH_3}{|}}{C}HCHCH_3$$
$$\underset{\underset{\displaystyle CH_2CH_3}{|}}{}$$

b. 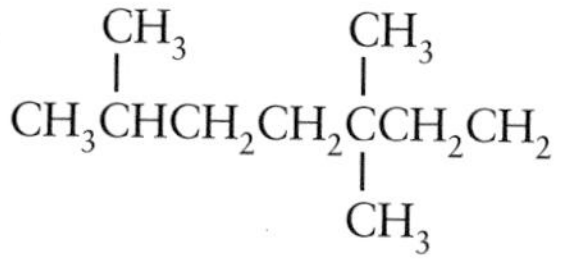
$$CH_3\overset{\overset{\displaystyle CH_3}{|}}{C}HCH_2CH_2\overset{\overset{\displaystyle CH_3}{|}}{\underset{\underset{\displaystyle CH_3}{|}}{C}}CH_2CH_2$$

c. $CH_3(CH_2)_{12}CH_3$

6.72 Indicate whether the following pairs of skeletal line structures represent a pair of structural isomers or two different conformations of the same compound:

a.

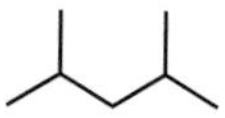

and

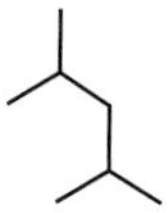

b.

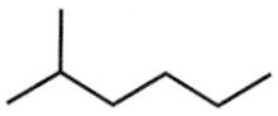

and

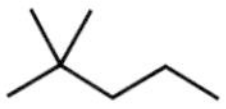

c.

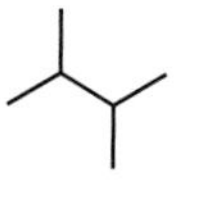

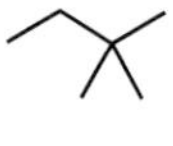

d.

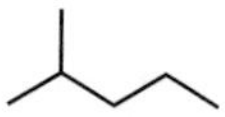

and

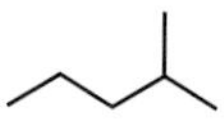

6.73 Write the skeletal line structure for a cycloalkane containing three carbon atoms in the ring and one containing five carbon atoms in the ring. Which is more commonly found in nature and why?

6.74 How many carbon atoms are there in the cycloalkane shown below? How many hydrogen atoms are there on each carbon atom? How is the number of hydrogen atoms in this cycloalkane different from the straight-chain hydrocarbon with the same number of carbon atoms?

Unsaturated Hydrocarbons: Alkenes and Alkynes

6.75 Indicate whether each of the following alkene structures should be classified as a *simple alkene*, a *diene*, or a *polyene*.

a.
b.
c.
d.

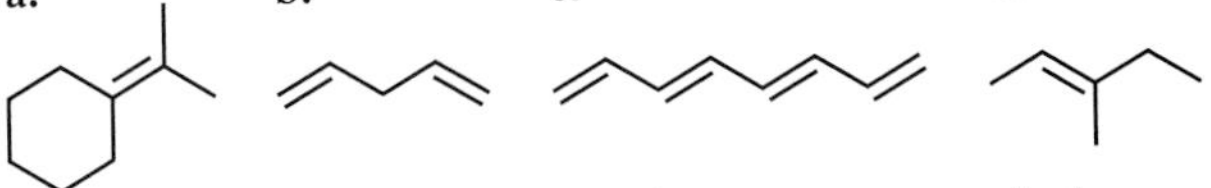

6.76 Indicate the number of hydrogen atoms attached to each of the carbon atoms in the structure shown below:

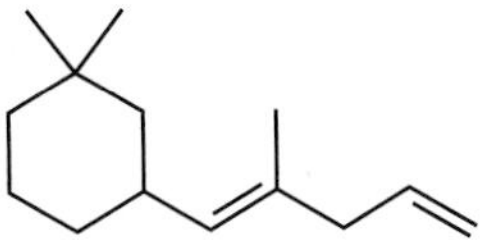

6.77 Write the geometric isomer of the compound shown below:

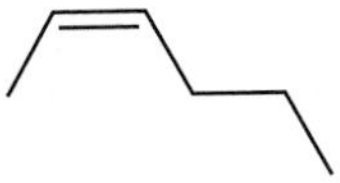

6.78 The structure of rhodopsin is shown below. Answer the questions by referring to the corresponding lettered arrow next to the structure.
a. Is this double bond cis or trans?
b. Is this double bond cis or trans?
c. If this double bond isomerized would the overall molecule have a different shape?
d. Which double bond undergoes isomerization during the chemical process of vision?

(a) (c)   (b)

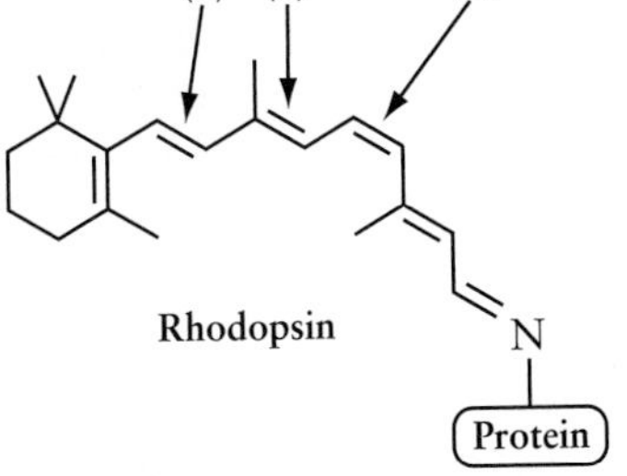

Rhodopsin

Protein

6.79 Is the rotational freedom around a carbon–carbon double bond different from that around a carbon–carbon single bond?

6.80 The structure of linoleic acid, found in plants and animals, is shown below. Are the double bonds cis or trans? How can you tell? Would this compound be classified as a trans fat or is it the natural cis form of the fatty acid?

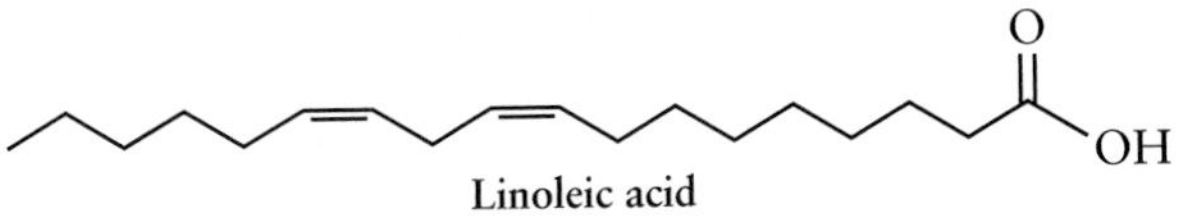

Linoleic acid

6.81 Indicate the C–C–C bond angle around the carbon atoms.

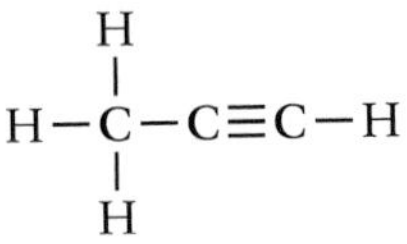

$$H-\overset{\overset{\displaystyle H}{|}}{\underset{\underset{\displaystyle H}{|}}{C}}-C\equiv C-H$$

Naming Hydrocarbons

6.82 The analgesic acetaminophen is the active pharmaceutical ingredient in Tylenol. Which name is the generic name and which is the brand name?

6.83 What are the three parts of an IUPAC name?

6.84 Write the structural formula, condensed formula, and skeletal line structure for the following alkanes:
a. pentane
b. 2-methylpentane
c. 3-ethyl-2-methyloctane.

6.85 Name the following straight-chain alkanes:
a.
b.
c.

6.86 Name the following branched-chain alkanes:
a. 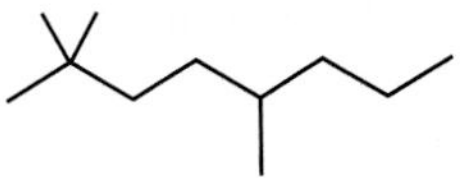
b. 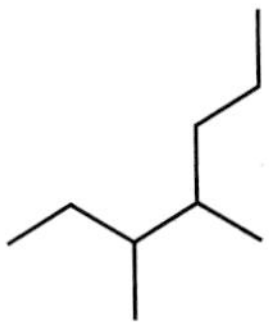
c.

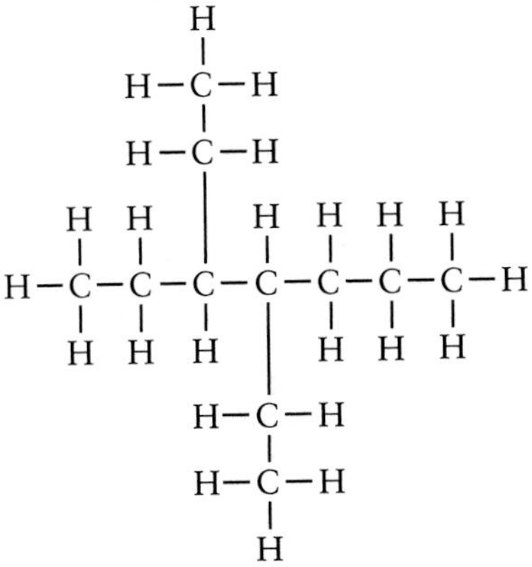

$$H-\overset{H}{\underset{H}{C}}-H$$

d.

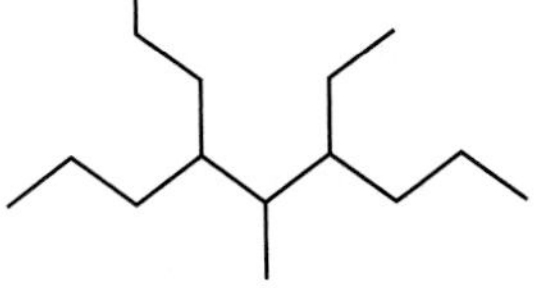

6.87 Write the skeletal line structure for all the structural isomers with the formula C_6H_{14}. Provide the IUPAC name for each structural isomer. If two compounds have the same name, what mistake did you make in identifying the structural isomers?

6.88 Provide skeletal line structures for the IUPAC names below:
a. *cis*-4-ethyl-2-hexene
b. 2,2-dimethyl-3-ethylhexane

6.89 Name the following compounds:
a.
b.
c.

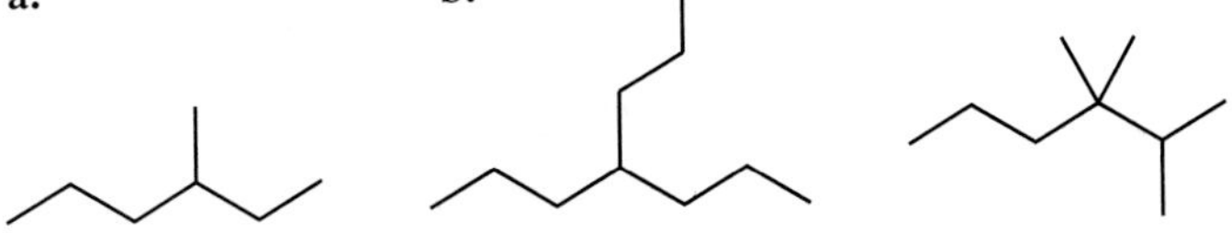

6.90 Provide skeletal line structures for the following cycloalkanes:
 a. cyclopropane
 b. cyclohexane
 c. cyclobutane
 d. methylcyclobutane.

6.91 Give the IUPAC names for the following cycloalkanes:

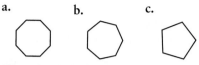

a.   b.   c.

6.92 Write the skeletal line structures for the pairs of compounds listed below, along with their chemical formulas. In what way are their structural formulas different? In what way are they the same?
 a. hexane, cyclohexane
 b. propane, cyclopropane.

6.93 Name the following alkenes:

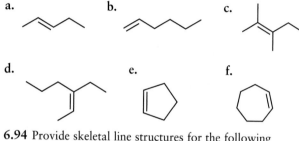

a.   b.   c.

d.   e.   f.

6.94 Provide skeletal line structures for the following alkenes:
 a. propene
 b. ethene
 c. 1-heptene
 d. *trans*-2-octene
 e. 2-methyl-2-butene
 f. 2,3,4-trimethyl-2-hexene
 g. 3-ethyl-2-pentene
 h. cyclohexene

6.95 Provide the skeletal line structure for the following compounds:
 a. *cis*-3-hexene   c. *cis*-4-methyl-2-pentene
 b. *trans*-2-pentene   d. *trans*-5-ethyl-3-heptene

6.96 Write the skeletal line structure for the following alkynes:
 a. 2-pentyne
 b. 1-pentyne
 c. 3-heptyne
 d. 3-methyl-1-pentyne

6.97 Name the following alkynes:

a.
$$H-\underset{\underset{H}{|}}{\overset{\overset{H}{|}}{C}}-\underset{\underset{H}{|}}{\overset{\overset{H}{|}}{C}}-\underset{\underset{H}{|}}{\overset{\overset{H}{|}}{C}}-C\equiv C-\underset{\underset{H}{|}}{\overset{\overset{H}{|}}{C}}-\underset{\underset{H}{|}}{\overset{\overset{H}{|}}{C}}-\underset{\underset{H}{|}}{\overset{\overset{H}{|}}{C}}-H$$

$$H-\overset{|}{\underset{|}{C}}-H$$
$$H-\overset{|}{\underset{|}{C}}-H$$
$$H$$

b.
$$H-C\equiv C-\underset{\underset{H}{|}}{\overset{\overset{H}{|}}{C}}-\underset{\underset{H}{|}}{\overset{\overset{H}{|}}{C}}-\underset{\underset{H}{|}}{\overset{\overset{H}{|}}{C}}-\underset{\underset{H}{|}}{\overset{\overset{H}{|}}{C}}-\underset{\underset{H}{|}}{\overset{\overset{H}{|}}{C}}-H$$

c.

d.
$$CH_3CH_2CH_2CH_2CCCH_2CH_2\underset{\underset{CH_3}{|}}{\overset{\overset{CH_3}{|}}{C}}HCH_3$$

e.

Aromatic Hydrocarbons

6.98 Write the structure of benzene with all carbon and hydrogen atoms included. What is the H—C—C bond angle at every carbon atom in the molecule?

6.99 Why is benzene classified as an unsaturated hydrocarbon?

6.100 How does delocalization affect the chemical reactivity of benzene? What skeletal line structures are used to depict benzene with its delocalized electrons?

6.101 In what way are benzene and cyclohexane similar? In what way are they different?

6.102 Indicate whether each of the following statements about benzene, C_6H_6, is *true* or *false*.
 a. Aromatic rings are stable and are found in many compounds throughout nature.
 b. Exposure to benzene, C_6H_6, can cause cancer.
 c. Benzene is a flat molecule.
 d. The skeletal line structure for benzene is

6.103 Write the structure of 1-ethyl-2-methylbenzene.

6.104 Which of the following are acceptable ways to represent benzene? For those selected as acceptable, explain what they are intended to show.

a.

b.

c.

d.

e. C_6H_6

6.105 Name the following substituted benzene compounds:

a.

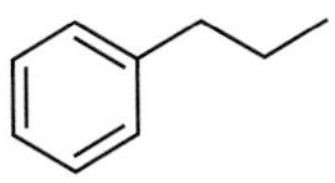

b.

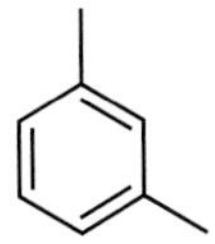

c.

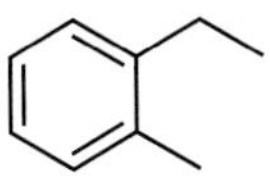

6.106 Many insect repellents contain DEET, whose chemical structure is shown below. Circle the aromatic ring and indicate whether the substitution on the ring is 1,2-, 1,3-, or 1,4-.

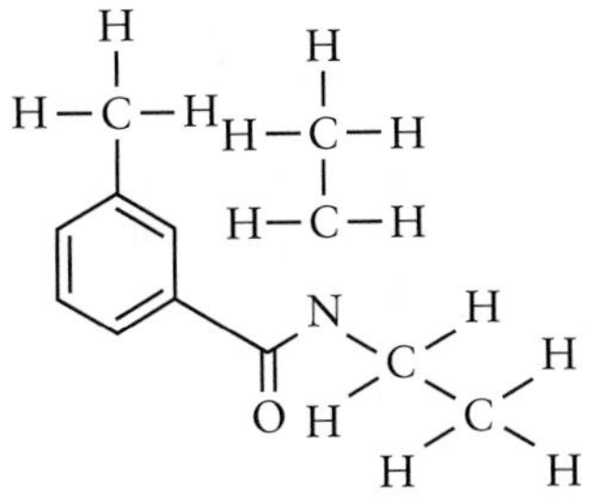

DEET

6.107 PABA, found in many sunscreens, is an abbreviation that stands for para-aminobenzoic acid. The structure of PABA is shown below.

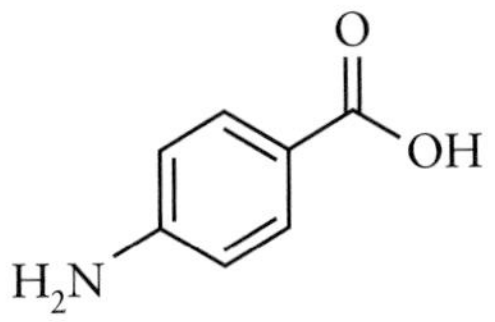

PABA

a. Circle the aromatic ring in the structure of PABA. What part of the root name *benzene* appears in the IUPAC name of PABA?
b. Is the substitution on the aromatic ring 1,2-; 1,3-; or 1,4-?

6.108 Methadone is used to treat heroin and morphine addiction. Circle the aromatic rings in both molecules. How is the structure of methadone similar to methamphetamine? How is it different?

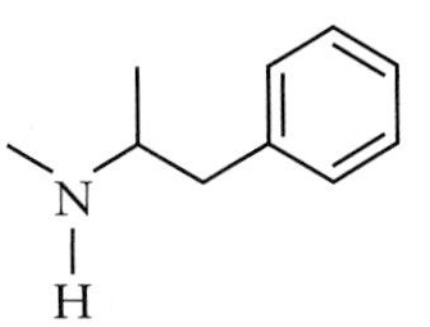

Methamphetamine

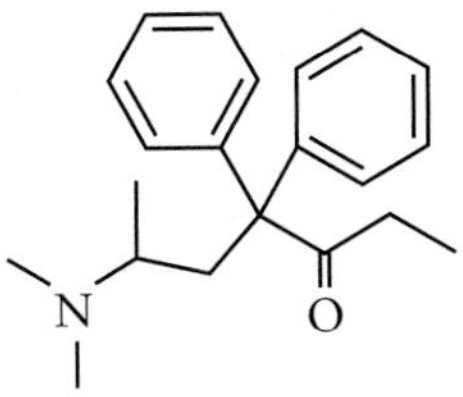

Methadone

6.109 Critical Thinking Question: The chemical structures for the four fat-soluble vitamins (vitamins D, A, E, and K) are shown below.

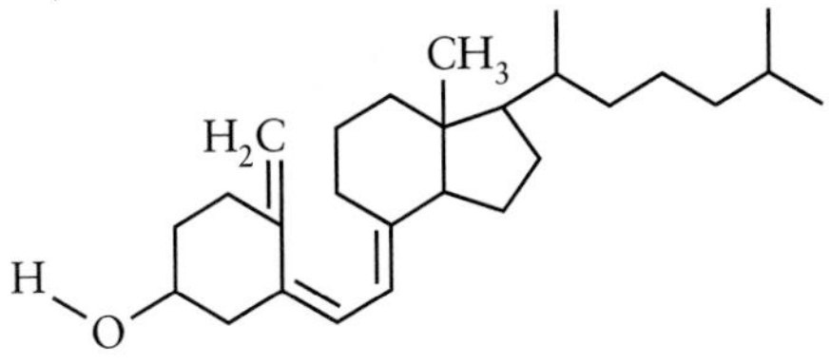

Vitamin D$_3$

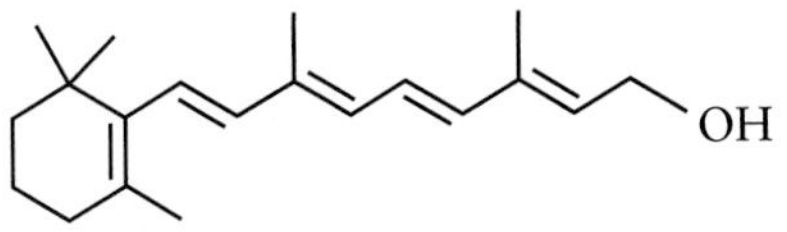

Vitamin A
(retinol)

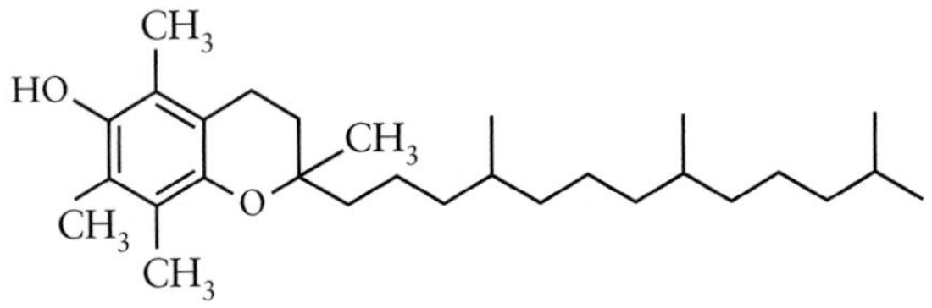

Vitamin E

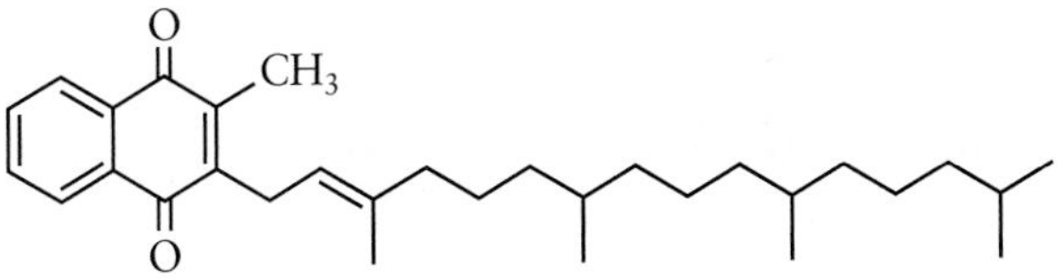

Vitamin K$_1$

a. Which vitamins contain aromatic rings? Which vitamins can be classified as polyenes?
b. How many carbon atoms are there in the long hydrocarbon chain in vitamin E? Is this a straight-chain hydrocarbon or a branched-chain hydrocarbon?
c. There are two heteroatoms in vitamin E; what are they?
d. Are these vitamins hydrophobic or hydrophilic? Explain why this makes them fat soluble.
e. These vitamins are not readily excreted through the urine like the B-vitamins. Offer an explanation.
f. Which vitamin plays a role in the chemistry of vision?
g. How many substituents are there on the aromatic ring in vitamin E?
h. How is the structure of vitamin D$_3$ different from cholesterol? Your body is able to synthesize vitamin D from cholesterol. Does this seem plausible, from the standpoint of chemical structure?

6.110 Muscone is a compound used in the perfume industry. Does this compound contain a cycloalkane? How many carbon atoms are in the ring?

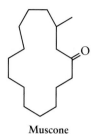

Muscone

6.111 Substituted benzenes are not the only aromatic hydrocarbons. The fused aromatic compounds naphthalene and anthracene, shown below, are also aromatic hydrocarbons. Given that these are aromatic compounds, answer the following questions.

     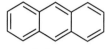

Naphthalene   Anthracene

a. Would you expect some electrons to be delocalized?
b. Would you expect naphthalene to be chemically reactive?
c. Would you expect naphthalene and anthracene to be flat molecules?
d. What is the bond angle around each carbon atom?
e. Write an alternate skeletal line structure for naphthalene.

6.112 Cyclopropane is not a very common ring size in nature. Offer an explanation.

6.113 What is a carcinogen? Give an example of a carcinogen.

6.114 The explosive TNT stands for 2,4,6-trinitrotoluene. Given that a nitro group is —NO_2, what is the structure of this tetrasubstituted benzene compound? *Hint:* Begin numbering at the methyl substituent.

Chemistry in Medicine

6.115 Draw the structure of cholesterol and circle the part of the molecule that is not a hydrocarbon.

6.116 Where does your body obtain cholesterol?

6.117 What function does cholesterol perform in the body?

6.118 Answer the following questions about lipoproteins:
a. What function do they serve?
b. Which type of lipoprotein scavenges cholesterol from the circulatory system?
c. Which type of lipoprotein carries cholesterol to the cells that need it?
d. Name a difference between an LDL and an HDL?
e. Which is more dense, an HDL or an LDL?
f. Which type of lipoprotein is considered "good cholesterol"?
g. Which is a more desirable LDL level: 90 mg/dL or 160 mg/dL?

6.119 What is plaque?

6.120 Why is cholesterol not soluble in the blood?

6.121 If a person has a stroke as a result of a blocked artery, which artery was most likely blocked?

6.122 What does the technique of angioplasty accomplish?

6.123 In what organ is cholesterol synthesized?

6.124 What is cardiovascular disease? Is it a relatively common or rare?

Answers to Practice Exercises

6.1 b, c, and d

6.2 b and c

6.3 An organic compound contains one or more carbon atoms. An inorganic compound does not contain carbon.

6.4 Carbon forms four bonds because it has four valence electrons and needs another four electrons to achieve an octet—a stable arrangement of electrons.

6.5 **a.** alkane   **b.** alkyne

6.6 b and d

6.7 Hydrocarbons are nonpolar because they contain nonpolar bonds C—C and C—H bonds. C—C bonds are nonpolar because two identical atoms have the same electronegativity. C—H bonds are nonpolar because carbon and hydrogen have very similar electronegativities. Thus, hydrocarbons do not contain any bond dipoles.

6.8 **a.** different conformations of the same molecule
b. different conformations of the same molecule
c. different compounds, because the atoms are connected in a different way; they differ in more than just bond rotations.

6.9

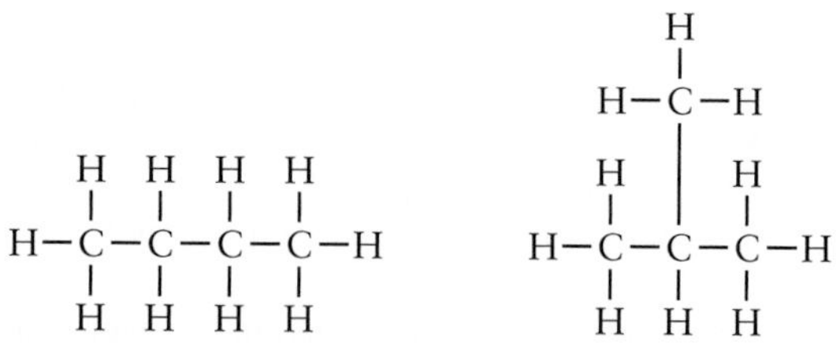

```
         H
         |
     H—C—H
 H  H  H  H            H   |   H
 |  |  |  |            |   |   |
H—C—C—C—C—H        H—C—C—C—H
 |  |  |  |            |   |   |
 H  H  H  H            H   H   H
```

6.10
```
 H  H  H  H  H
 |  |  |  |  |
H—C—C—C—C—C—H
 |  |  |  |  |
 H  H  H  H  H
```

```
         H                          H
         |                          |
     H—C—H                      H—C—H
         |  H  H                    |   H
 H       |  |  |            H       |   |
 |       |  |  |            |       |   |
H—C———C—C—C—H           H—C—C—C—H
 |       |  |  |            |   |   |
 H  H  H  H  H              H   |   H
                                |
                            H—C—H
                                |
                                H
```

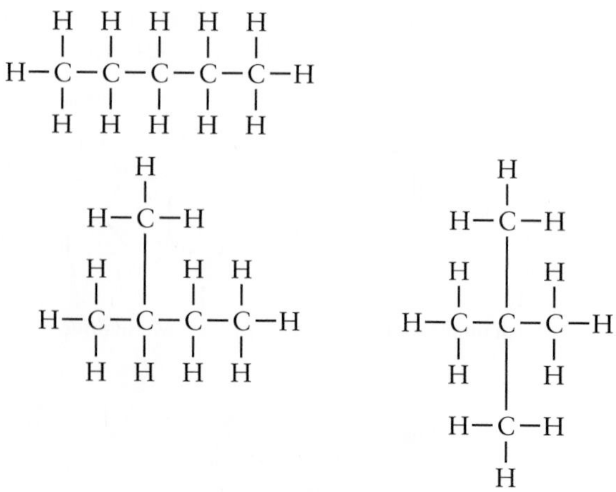

6.11

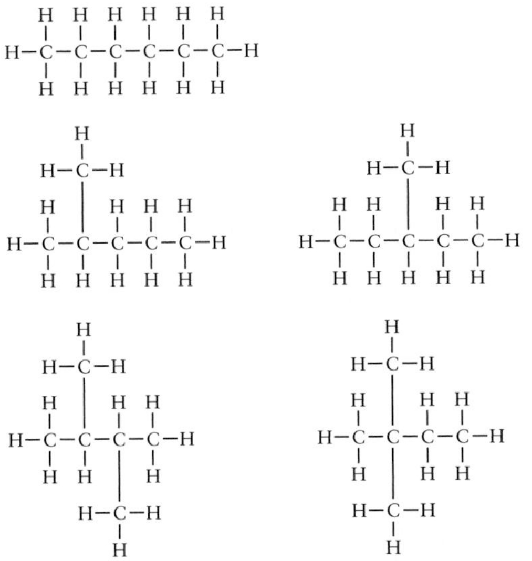

```
    H H H H H H
    | | | | | |
H - C-C-C-C-C-C - H
    | | | | | |
    H H H H H H
```

```
      H
      |
    H-C-H
   H  |  H H H
   |  |  | | |
H- C -C- C-C-C -H
   |  |  | | |
   H  H  H H H
```

```
        H
        |
      H-C-H
   H H  |  H H
   | |  |  | |
H- C-C- C- C-C -H
   | |  |  | |
   H H  H  H H
```

```
      H
      |
    H-C-H
   H  |  H H
   |  |  | |
H- C- C-C-C -H
   |  |  | |
   H  H  | H
       H-C-H
         |
         H
```

```
      H
      |
    H-C-H
   H  |  H H
   |  |  | |
H- C- C-C-C -H
   |  |  | |
   H  H  H H
       H-C-H
         |
         H
```

6.12 The more carbon atoms there are, the more possibilities for branching.

6.13 a, c

6.14 b, d, and e

6.15 a. four **b.** five **c.** eight

6.16 Preferred C—C—C bond angle is 109.5° because of the requirements of a tetrahedral carbon. The actual bond angle in an equilateral triangle is 60°. You might break pieces on your model kit as you try to bend the bonds into such an acute angle. This is known as ring strain and accounts for a three-membered ring not being a common ring size in nature.

6.17 a. Two hydrogen atoms
b. No, the C—C bonds are not in the same plane and neither are the C—H bonds. The model is not flat, but has a three-dimensional shape like a lounge chair.

6.18 a. straight chain:

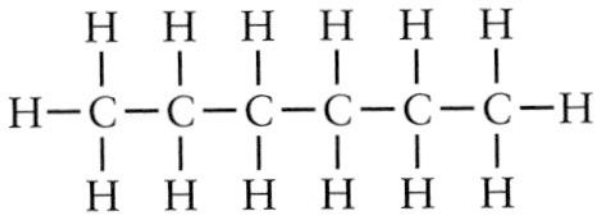

```
    H H H H H H
    | | | | | |
H - C-C-C-C-C-C - H
    | | | | | |
    H H H H H H
```

b. branched chain:

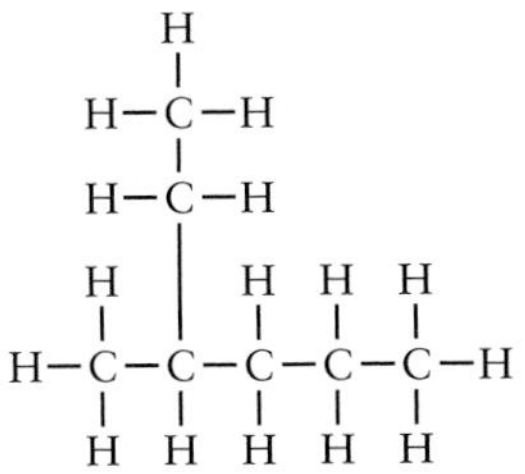

```
      H
      |
    H-C-H
      |
    H-C-H
   H  |  H H H
   |  |  | | |
H- C- C-C-C-C -H
   |  |  | | |
   H  H  H H H
```

c. branched chain:

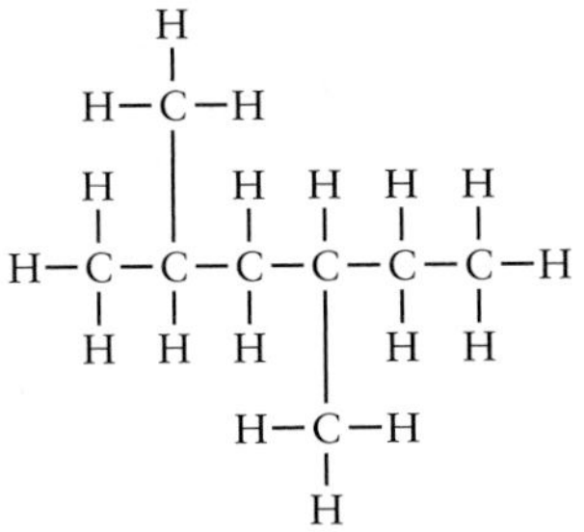

```
      H
      |
    H-C-H
   H  |  H H H H
   | |  | | | |
H- C-C- C-C-C-C -H
   | |  | | | |
   H H  H H H H
       H-C-H
         |
         H
```

6.19

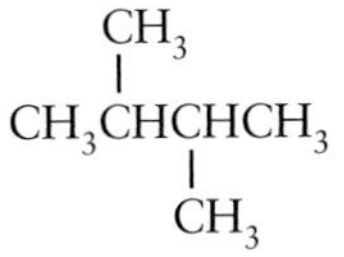

```
        CH₃
        |
CH₃CHCHCH₃
        |
        CH₃
```

6.20

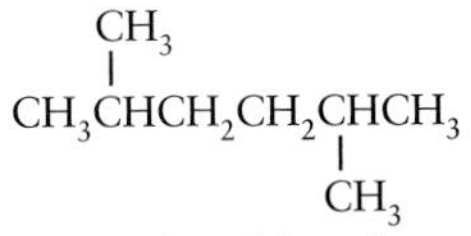

```
    CH₃
    |
CH₃CHCH₂CH₂CHCH₃
            |
            CH₃
```
Condensed formula

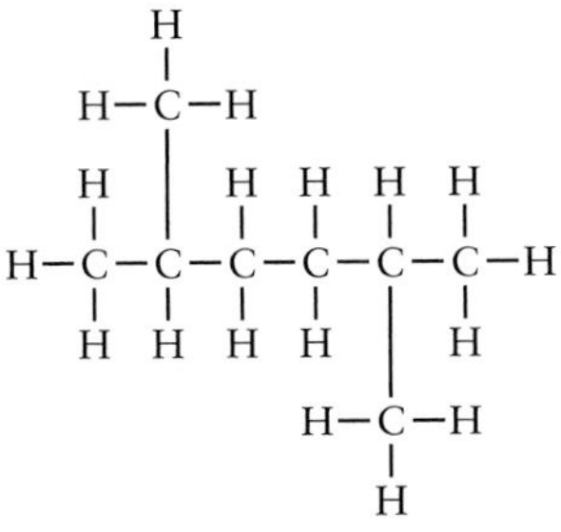

```
          H
          |
        H-C-H
   H  H  |  H H H
   |  |  |  | | |
H- C- C- C-C-C-C -H
   |  |  |  | | |
   H  H  H  H | H
           H-C-H
             |
             H
```
Structural formula

6.21

Structural Formula	Skeletal Line Structure
`H H H H` `|		|` `H-C-C-C-C-H` `|		|` `H H H H`	
`    H` `    |` `  H-C-H` ` H	H` `		|` `H-C-C-C-H` `		|` ` H H H`	

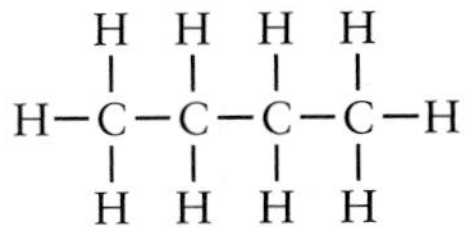

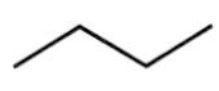

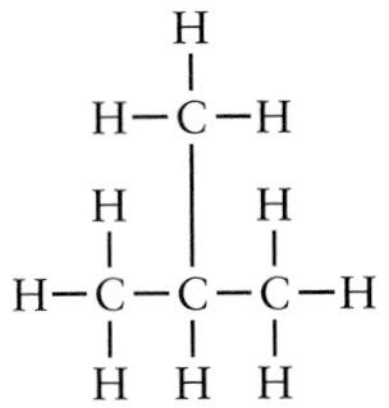

6.22

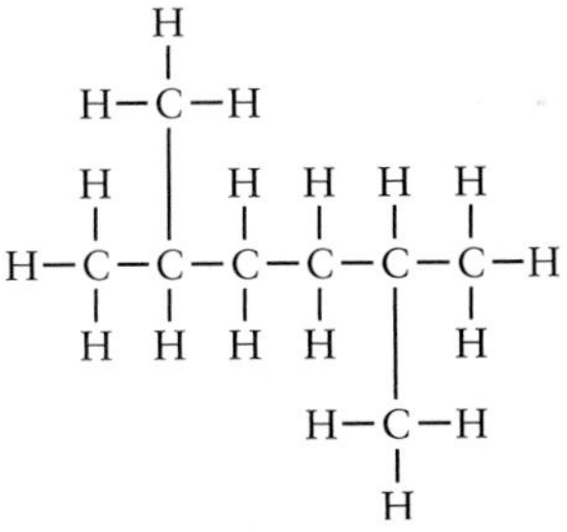

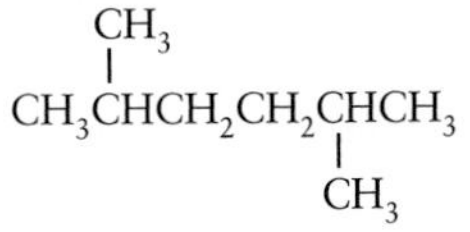

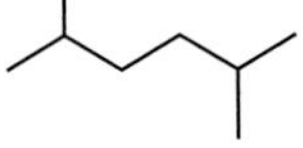

```
      H
      |
    H-C-H
   H  |  H H H H
   |  |  | | | |
H- C- C-C-C-C-C -H
   |  |  | | | |
   H  H  H H | H
           H-C-H
             |
             H
```

```
CH₃
|
CH₃CHCH₂CH₂CHCH₃
            |
            CH₃
```

6.23 a.

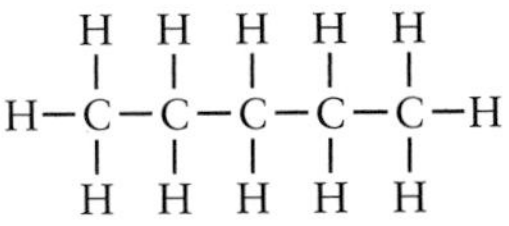

```
    H H H H H
    | | | | |
H - C-C-C-C-C - H
    | | | | |
    H H H H H
```
CH₃(CH₂)₃CH₃

b.

```
      H   H
      |  /H  |
    H-C   H-C-H
   H  |  H  |  H
   |  |  |  |  |
H- C- C- C- C-C -H
   |  |  |  |  |
   H  H  | H  H
       H-C-H
         |
       H-C-H
         |
         H
```
```
CH₃  CH₃
 |    |
CH₃CHCHCHCH₃
      |
      CH₂CH₃
```

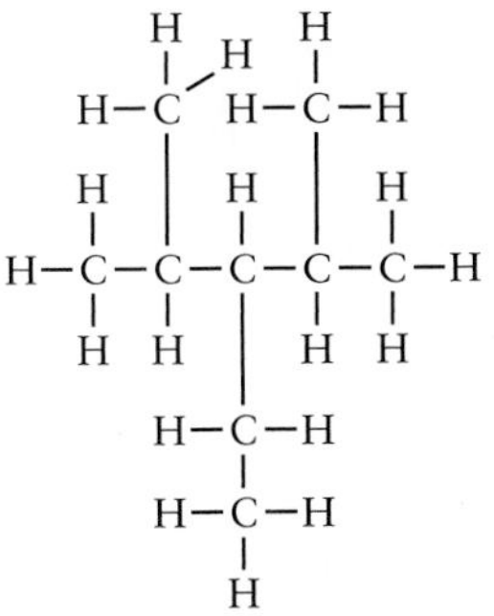

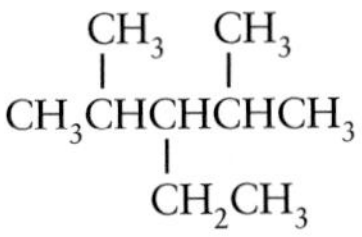

6.24

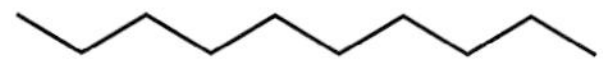

a. 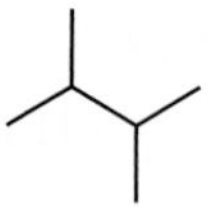   b.

6.25 a. diene   **b.** simple alkene   **c.** polyene

6.26 The bond angles are 120° for the carbon atoms with a double bond, and 109.5° for the tetrahedral carbon on the far right.

6.27 a. five

b. Vitamin A is similar to half of a β-carotene molecule with an added OH group.

c. Three H atoms

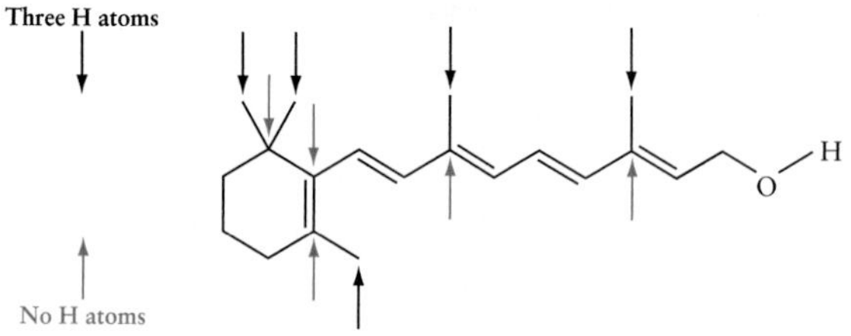

Retinol (vitamin A)

No H atoms

d. Vitamin A is not a hydrocarbon because it is not composed of only carbon and hydrogen atoms; it contains one oxygen atom.

e. Hydrophobic, because most of its structure is hydrocarbon. Thus, it is soluble in other fats and organic solvents but not in water.

6.28 a.   **b.**   cis

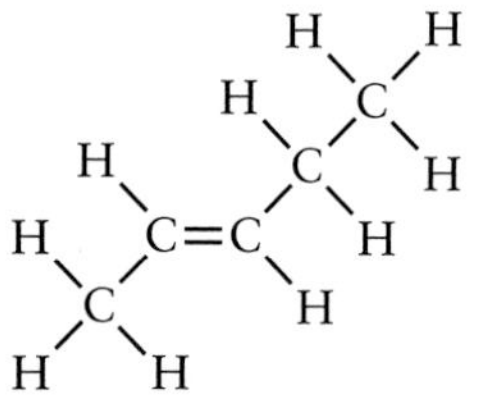

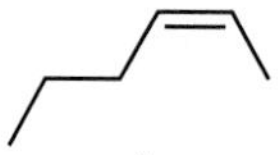

trans

6.29 The cis form of rhodopsin has a curved shape, while the trans form has a linear shape.

6.30 It is flat because there is a continuous array of double bonds, and these double-bond carbon atoms have a trigonal planar geometry—hence they are all in the same plane (flat). A polyene is an alkene with many double bonds.

6.31 An isomerization reaction

6.32 a. propane   **b.** pentane   **c.** nonane

6.33 a.

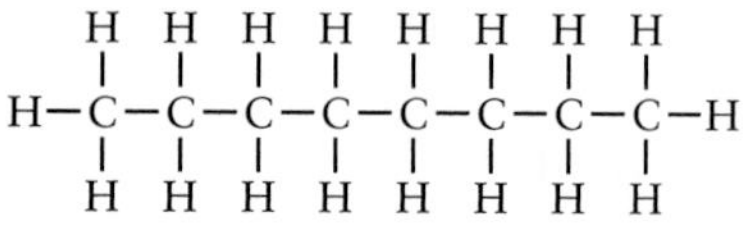

$CH_3(CH_2)_6CH_3$

b.

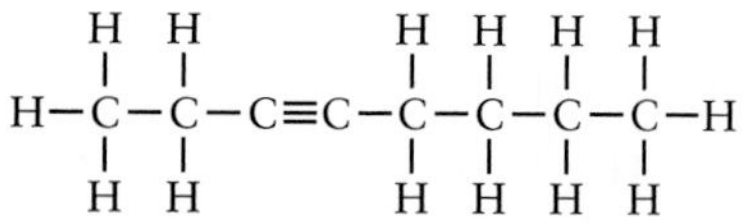

$CH_3CH_2CCCH_2CH_2CH_2CH_3$
$CH_3CH_2CC(CH_2)_3CH_3$

c.

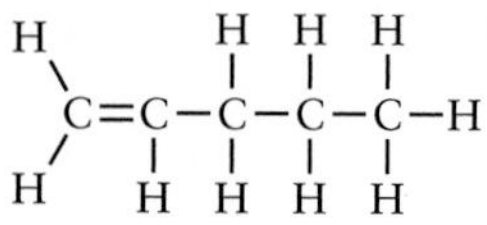

$CH_2CHCH_2CH_2CH_3$

6.34 a. 1-hexyne   **b.** 1-pentyne

6.35 a. 4-methylheptane   **b.** 2,3-dimethylpentane
    c. 4-methyloctane

6.36 a.

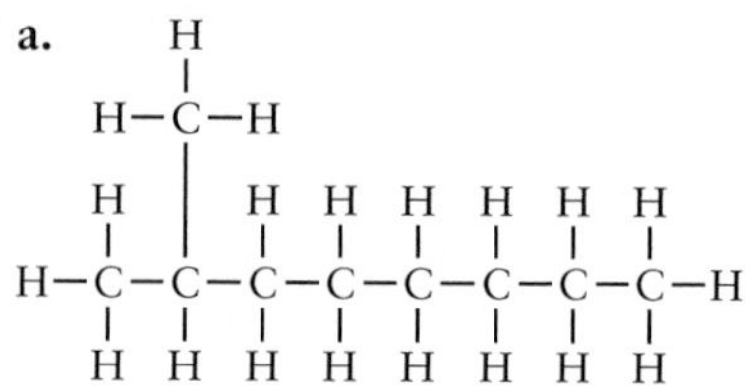

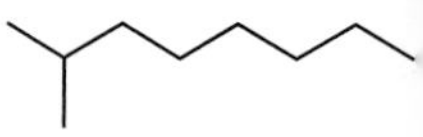

b.

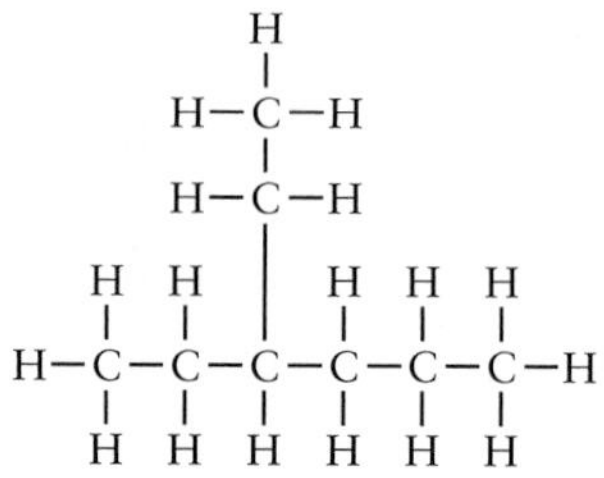

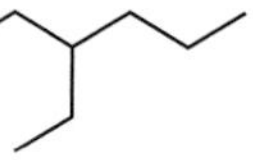

6.37 a. 2-methyl-2-butene   **b.** 2-ethyl-1-pentene
    c. 3-ethyl-1-pentene

6.38

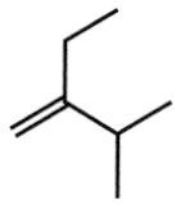

a. 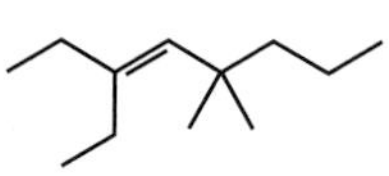   b.

6.39   Isooctane
(2,2,4-trimethylpentane)   Heptane

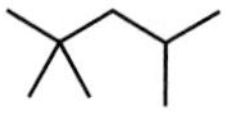

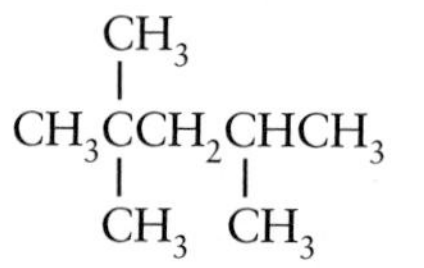

$CH_3CCH_2CHCH_3$ with CH_3 groups $CH_3(CH_2)_5CH_3$

6.40

Pentane 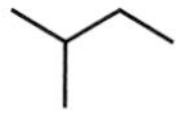   2-Methylbutane    2,2-Dimethylpropane

6.41 3-ethyl-2,5-dimethyloctane

6.42 a. 1-butyl-3-methylbenzene
    b. ethylbenzene
    c. 1,2,4-trimethylbenzene
    d. 1-ethyl-3-methylbenzene

6.43

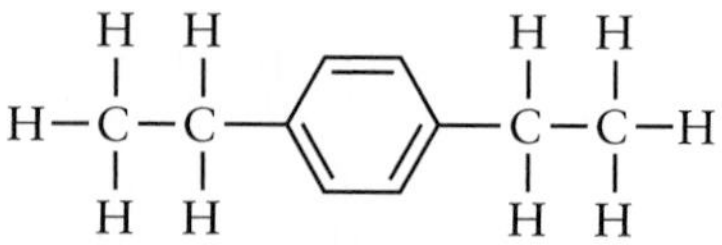

a.

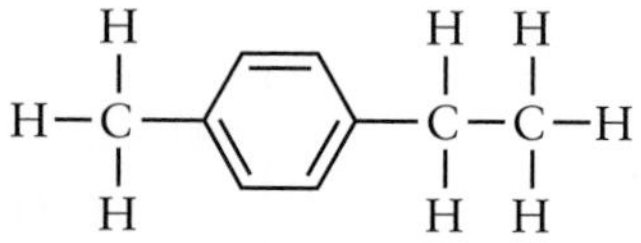

b.

6.44 a. false
    b. true
    c. true
    d. false
    e. true

Organic Functional Groups

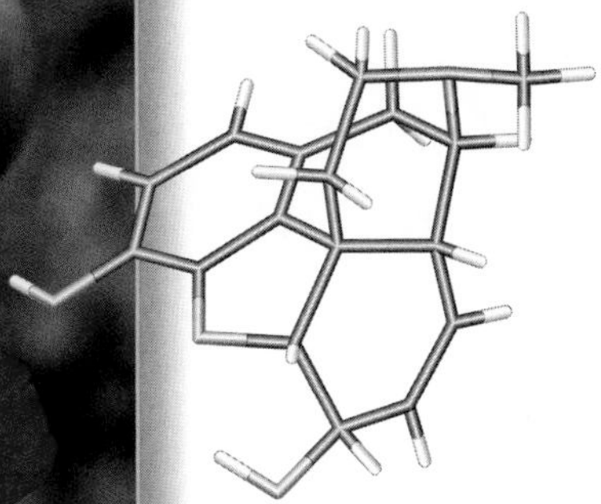

Papaver somniferum, commonly known as the poppy plant, contains morphine, a powerful analgesic. An amine (—N—) functional group in the structure shown is particularly important to its pain-killing effects. [John Glover/Alamy]

OUTLINE

 This icon indicates that a **Problem-Solving Tutorial** is available at www.whfreeman.com/gob

Pain and the Opioid Analgesics

Have you ever had a headache and reached for the medicine cabinet to take some aspirin? Perhaps you have had the misfortune of having to spend a night in the hospital, where a nurse administered an even stronger medicine such as a narcotic. You are not alone—most people suffer from some kind of pain at some time in their life, and most diseases are associated with pain. Pain medicines are known as *analgesics,* a combination of the Greek words "an" meaning without and "algesic" meaning pain.

All pain medications are organic compounds containing atoms such as nitrogen and oxygen attached by covalent bonds to carbon. Characteristic arrangements of atoms and bonds, such as C—O—H and C—N—H, are known as *functional groups.* The functional groups in a molecule define its chemical and physical properties. Therefore, it is the functional groups in an analgesic that determine how it interacts with pain receptors in the brain to stop or reduce pain.

The strongest analgesics are the opioid analgesics, originally isolated from the opium poppy, *Papaver somniferum.* The opioid analgesics include morphine (Figure 7-1a), codeine (Figure 7-1b), and several other nitrogen-containing organic compounds, known as amines. The amine functional group in these compounds is essential to their ability to reduce pain. These compounds have the ability to induce a state of euphoria, but morphine and compounds structurally related to morphine can also lead to dependence and, with repeated use, addiction; therefore, they are classified as narcotics.

Opioids reduce pain by binding to the opioid receptors on the surface of brain cells. Morphine and its analogs activate the opioid receptor, initiating a sequence of biological events that, among other things, leads to a reduction in pain.

Why do neurons have opioid receptors? Pain is the body's way of signaling the brain—through communication between neurons—that it is in danger, and needs to act to avoid further injury, or rest to recover. However, if you were attacked and injured by a wild animal and had to run away to save yourself, you could not afford to be hindered by the excruciating pain of your injury. Consequently, as humans evolved, the body adapted by developing its own internal analgesics and opioid receptors, making it possible for the injured human to get out of harm's way. The body's internal analgesics are compounds known as endorphins. The word *endorphin* comes from the combination of the words *end*ogenous (internal) and m*orphin*e.

Endorphins are the body's natural ligands for the opioid receptor. They bind to the pain receptor as does morphine and its analogs, initiating the sequence of events that leads to their analgesic and euphoric effects. You will see an example of an endorphin later in the chapter.

Modifying some of the functional groups in morphine produces compounds known as morphine *analogs.* Even slight changes to the functional groups in morphine can have a profound affect on the analgesic and addictive properties of a morphine analog. For example, codeine differs structurally from morphine at only one functional group (Figures 7-1a and 7-1b). Where codeine has an *ether* functional group, morphine has a *phenol* functional group. Due to this difference, codeine has only 15% of the potency of morphine as a painkiller, but it is much less addictive. However, changing both the alcohol and the phenol functional groups in morphine to *ester* functional groups, C—O—C, produces the analog known as heroin (Figure 7-1c), which is three times more potent than morphine, and extremely addictive. Clearly, functional groups have a significant impact on the chemical behavior of an organic compound. In this chapter you will be introduced to the major functional groups in organic chemistry.

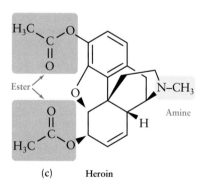

Figure 7-1 Morphine (a) and its structural analogs codeine (b) and heroin (c).

In Chapter 6 you learned about the structure and properties of hydrocarbons—molecules containing exclusively carbon and hydrogen atoms. In this chapter you will be introduced to organic compounds that contain oxygen (O), nitrogen (N), and phosphorus (P) atoms. Atoms in an organic compound that are not carbon or hydrogen are sometimes referred to as **heteroatoms.**

The addition of heteroatoms to the carbon and hydrogen skeleton of an organic compound imparts unique physical and chemical properties to the compound. Carbon–oxygen, carbon–nitrogen, and carbon–phosphorus bonds are polar, in sharp contrast to the nonpolar C—C and C—H bonds of hydrocarbons described in the previous chapter. Thus, in this chapter you will see some water soluble organic compounds. Moreover, polar bonds are chemically more reactive than nonpolar bonds, so you will see that these groupings of atoms undergo characteristic chemical reactions, which will be described in Chapters 8–10.

Certain combinations of heteroatoms and bonds are seen frequently, for example, C—O—H or C—N—H. These characteristic arrangements of atoms and bonds are known as **functional groups** *because they "function" in characteristic way in certain chemical reactions.* Therefore, functional groups determine a molecule's chemical and physical properties.

The multiple bonds in hydrocarbons are also considered functional groups. Thus, you have already seen three functional groups in Chapter 6: carbon–carbon double bonds (alkenes), carbon–carbon triple bonds (alkynes), and aromatic hydrocarbons. Note that the carbon–carbon single bonds of alkanes are not considered functional groups. Although alkanes undergo a limited number of reactions, they serve primarily as the structural scaffolding for organic molecules.

Functional groups serve as an important organizing tool for predicting chemical behavior in organic compounds, which allows us to make sense of the vast number of organic chemical reactions.

7.1 C—O-Containing Functional Groups: Alcohols and Ethers

Two functional groups are derived from water (H_2O):

- alcohols and
- ethers.

Replacement of *one* of the hydrogen atoms in a water (H_2O) molecule with a chain of one or more carbon atoms produces an **alcohol**, as shown in Figure 7-2a. Replacement of *both* hydrogen atoms of a water (H_2O) molecule with a chain of one or more carbon atoms produces an **ether**, as shown in Figure 7-2b. The geometry around the oxygen atom in water, an alcohol, and an ether is *bent*, with a bond angle of less than 109.5° (Section 3.1).

For a functional group to be classified as an alcohol, the carbon atom bearing the O—H group must be saturated—that is, it must contain four single bonds (Section 6.1). For a functional group to be classified as an ether, the carbon atoms bonded to the oxygen atom can be either saturated or part of an aromatic ring. Beyond the first carbon atom(s), the details of the carbon chain are not important to the identity of the functional group, so the carbon chain is often abbreviated "R" to indicate the generic structure of a particular functional group. Thus, the condensed structural formula is written ROH for an alcohol and ROR for an ether.

Alcohols are further subdivided into three classes: primary (1°), secondary (2°), and tertiary (3°), depending on whether they have one, two, or three

A sugar is an organic molecule containing many alcohol functional groups in its chemical structure.
[Foodcollection.com/Alamy]

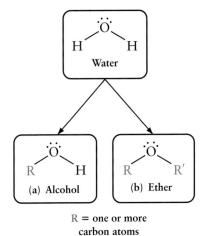

R = one or more carbon atoms

Figure 7-2 Alcohols and ethers are structurally related to water, H_2O.

Figure 7-3 Alcohols are classified as primary (1°), secondary (2°), or tertiary (3°) alcohols.

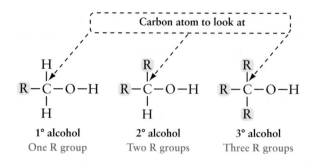

carbon atoms (or R in Figure 7-3), respectively, on the carbon atom bearing the OH group. An OH group is also sometimes called a **hydroxyl group.**

Three simple alcohols are shown in Figure 7-4. From left to right, the R groups are —CH_3, —CH_2CH_3, and —$\overset{\overset{CH_3}{|}}{CH}CH_3$. Ethanol is a primary alcohol and 2-propanol is a secondary alcohol. Methanol contains only one carbon atom, so the subclassifications do not apply to it.

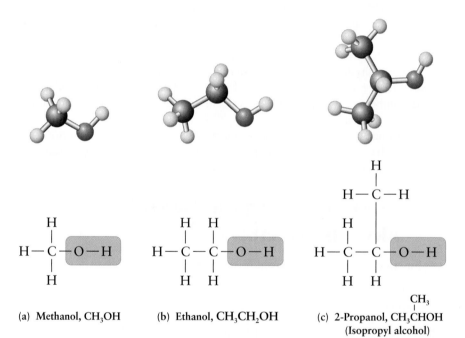

Figure 7-4 Three simple alcohols and their IUPAC and common names. (a) Methanol is poisonous and can cause blindness if ingested. (b) Ethanol is the "alcohol" in alcoholic beverages. (c) Isopropyl alcohol is the "alcohol" in rubbing alcohol.

(a) Methanol, CH_3OH (b) Ethanol, CH_3CH_2OH (c) 2-Propanol, CH_3CHOH (Isopropyl alcohol)

Diethyl ether was one of the first anesthetics. It is used infrequently today because much better anesthetics are available.

Two simple ethers are shown in Figure 7-5. Both R groups are methyl groups, —CH_3, in dimethyl ether, and both are ethyl groups, —CH_2CH_3, in diethyl ether.

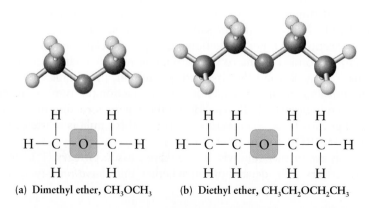

Figure 7-5 Two simple ethers and their common names.

(a) Dimethyl ether, CH_3OCH_3 (b) Diethyl ether, $CH_3CH_2OCH_2CH_3$

When writing the skeletal line structure of an alcohol or an ether, the oxygen atoms must always be written in, in order to distinguish the oxygen atom from a carbon atom. In addition, hydrogen atoms bonded to a heteroatom must always be written in; only those hydrogen atoms attached to carbon atoms may be omitted.

Skeletal line drawing of an alcohol and an ether

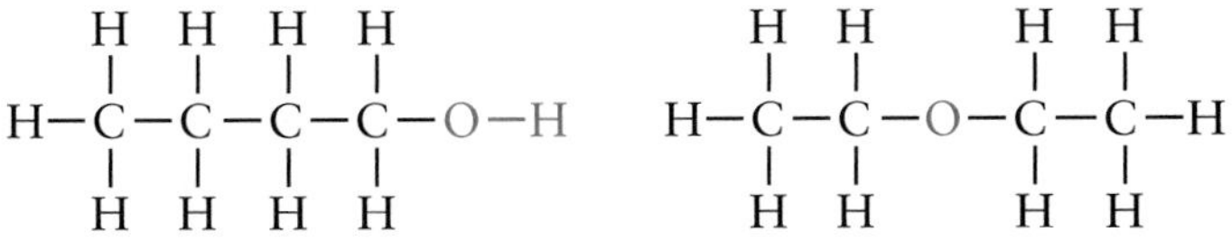

Naming Alcohols and Ethers

To name an alcohol follow the guidelines in the box below for naming alcohols.

Guidelines for Naming Alcohols

Rule 1: Assign the root. Follow the IUPAC rules established for hydrocarbons to name the root (Section 6.5). As the main chain, select the longest continuous chain of carbon atoms that contains the alcohol functional group.

Rule 2: Assign the suffix. Change the suffix of the IUPAC name from *-ane* to *-anol*, to signify that an alcohol functional group is present.

Rule 3: Assign a locator number to the root indicating the location of the alcohol. As with alkenes and alkynes, a locator number is used to indicate the carbon atom containing the hydroxyl (OH) group. Begin numbering the main chain from the end closer to the OH group. Place the locator number, followed by a hyphen, in front of the root name. If the OH group is located on an unbranched cycloalkane, it does not need a locator number.

Rule 4: Assign a prefix. If the main chain contains substituents, assign names and locator numbers as a prefix in front of the root, as described in Section 6.5.

For example, the IUPAC name for the alcohol shown below is 3-methyl-2-hexanol.

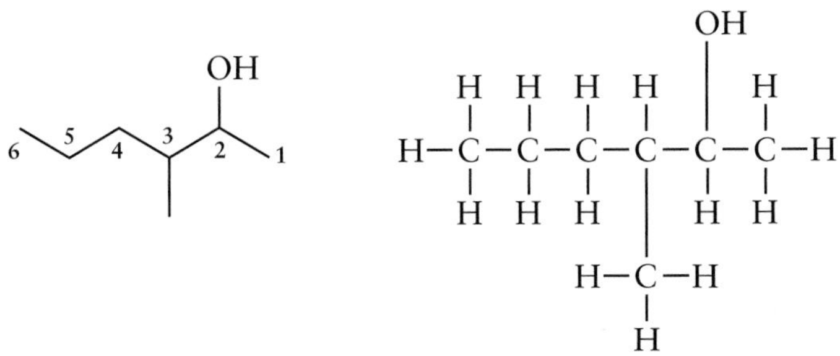

3-Methyl-2-hexanol

It contains six carbon atoms in the main chain (the main chain must always contain the functional group); therefore, it has the root *hexane*. The suffix is changed from *-ane* to *-anol* to signify that an alcohol functional group is present, yielding the term *hexanol*. The OH group is on the second carbon atom in the chain when numbering from the end closer to the OH group, so a 2- is placed before the root, yielding the name *2-hexanol*. The prefix *3-methyl-* must be added to indicate that there is a methyl substituent on the third carbon atom in the chain, so the full name is *3-methyl-2-hexanol*.

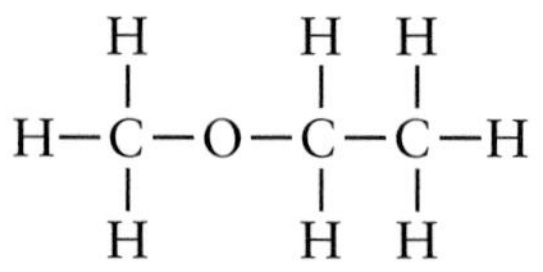

H—C—O—C—C—H
Ethyl methyl ether

OH

Phenol

OH

4-Ethylphenol

Although there are IUPAC rules for naming ethers, common names are typically used for simple ethers. The common name for an ether is constructed by naming each R group as though it were a substituent, with a *-yl* ending, and then ending the name with *ether*. If both R groups are identical, the prefix *di-* is added before the name of the R groups. For example, dimethyl ether has two —CH_3 (methyl) groups attached to the ether oxygen atom (Figure 7-5a); while diethyl ether has two —CH_2CH_3 (ethyl) groups attached to the ether oxygen (Figure 7-5b). Ethyl methyl ether has one —CH_3 group and one —CH_2CH_3 group bonded to the oxygen atom.

Phenols

When an O—H group is attached to a carbon atom that is part of an aromatic ring, the functional group is no longer considered an alcohol, but a **phenol.** For example, the female hormone estrone contains a phenol functional group, because the aromatic ring is directly attached to an OH group. The phenol functional group includes the entire benzene ring together with the OH group, highlighted in pink.

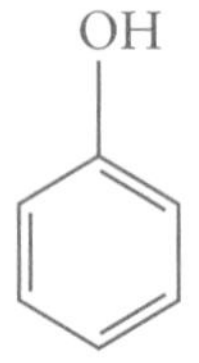

Aromatic carbon atom

Phenol

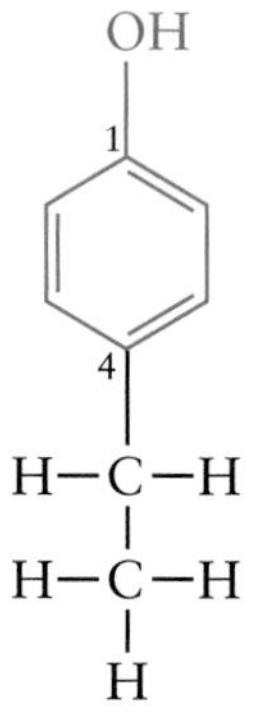

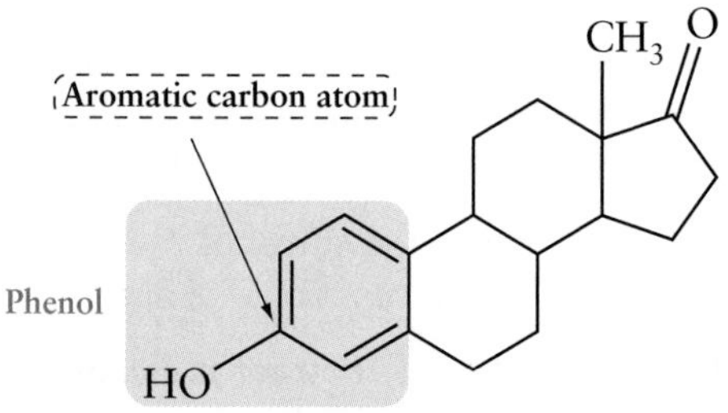

Estrone

To name a phenol, follow the rules for naming substituted benzenes (Section 6.6), but change the root name from *benzene* to *phenol*. For example, the substituted phenol shown in the margin is 4-ethylphenol. Here the root is phenol. Begin numbering the carbon atoms in a phenol at the carbon atom bonded to the OH group and then proceed clockwise or counterclockwise, in whichever direction gives the lower set of numbers. Since the ethyl group appears at carbon 4, the prefix *4-ethyl-* is added.

You may recall from the opening vignette that the structural difference between morphine and codeine was the presence of a phenol functional group in morphine where there was an ether functional group in codeine (Figure 7-1).

Hydrogen Bonding in Alcohols and Ethers

Ethers containing two or three carbon atoms are soluble in water because the oxygen atom can participate in hydrogen bonding with water molecules. However, ethers cannot hydrogen bond to one another, because they lack an O—H bond. Consequently, ethers have relatively low boiling points, because the intermolecular forces of attraction between them are the weaker dipole–dipole forces. In contrast, alcohols and phenols are capable of hydrogen bonding, as illustrated in Figure 7-6 for methanol.

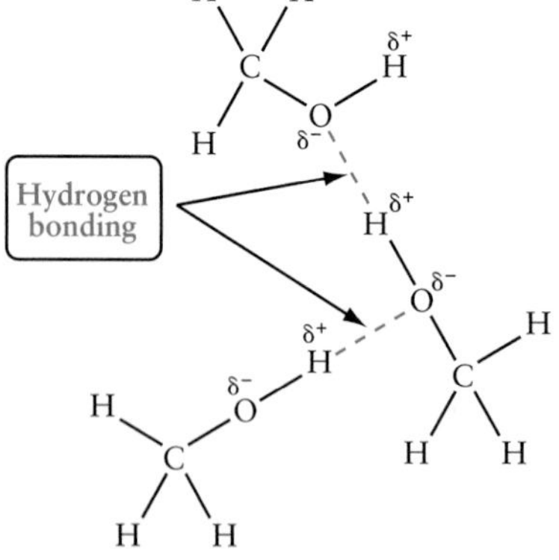

The solubility of an alcohol in water depends on the relative proportion of hydrocarbon structure to the number of OH groups in the overall molecular structure. For example, ethanol, the simple two-carbon alcohol present in alcoholic beverages, readily dissolves in water. The hydrogen bonding characteristics of the alcohol functional group dominate the small hydrocarbon portion of ethanol.

In general, the more OH groups there are in a molecule, the more water soluble it is, and the more extensive the hydrocarbon portion is, the less water soluble it is. Consider blood sugar, glucose, which is very soluble in water. Glucose contains five OH groups on a structure composed of six carbon atoms. In contrast, cyclohexanol, with six carbon atoms and one alcohol

Hydrogen bonding

Figure 7-6 Hydrogen bonding in three methanol molecules.

The Model Tool 7-1 Alcohols and Ethers

I. Construction of Water, Methanol, and Dimethyl Ether

1. Obtain 2 black carbon atoms, 6 light-blue hydrogen atoms, 1 oxygen atom, and 8 straight bonds.

2. Construct a model of water, H_2O.
   a. What is the H—O—H bond angle?
   b. What is the geometry of a water molecule?
   c. Is this molecule organic? Explain.

3. Take the water molecule you just made in Step 2 and replace one of the hydrogen atoms with a methyl group (—CH_3).
   a. What is the name of the model that you just made?
   b. What functional group does this molecule contain?
   c. What is the O—C—H bond angle? What is the geometry around the carbon atom?
   d. What is the H—O—C bond angle? What is the geometry around the oxygen atom?
   e. Is this molecule organic? Explain.

4. Take the model you just made in Step 3 and replace the other hydrogen atom attached to oxygen with another methyl group (—CH_3)
   a. What is the name of the model that you just made?
   b. What functional group does this molecule contain?
   c. What is the C—O—C bond angle? What is the geometry around the oxygen atom?
   d. What is the geometry around the carbon atoms?
   e. Is this molecule organic? Explain.

functional group, is insoluble in water. Perform *The Model Tool 7-1: Alcohols and Ethers* to gain more familiarity with these two functional groups.

Cholesterol, the important steroid introduced in Chapter 6, contains an alcohol functional group. The R group in this alcohol is a large hydrocarbon, which accounts for its insolubility in water.

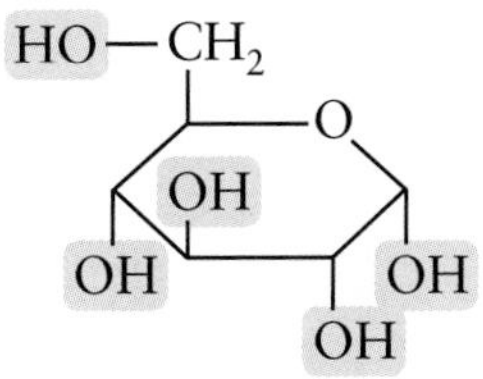

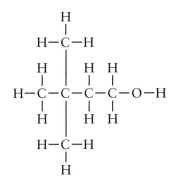

Cholesterol

HO—CH_2

Glucose

Blood sugar, glucose, is very soluble in water.

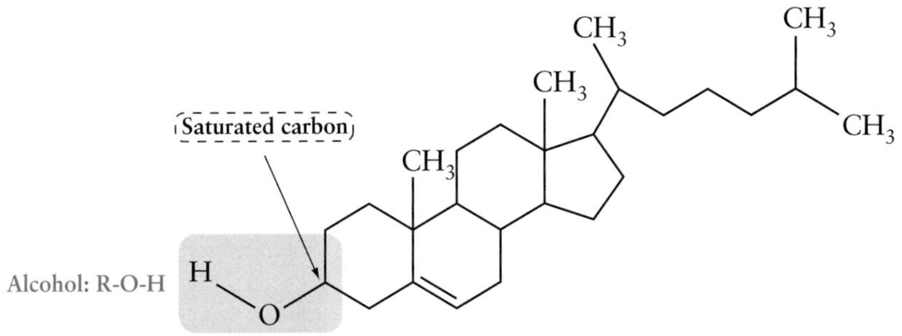

WORKED EXERCISE 7-1   Naming Alcohols

Write the IUPAC name for the alcohol shown below. Indicate whether the alcohol is a 1°, 2°, or 3° alcohol.

The following tutorials are available to help you with all the functional groups in the chapter:

 Problem-Solving Tutorial: Naming Organic Functional Groups

 Problem-Solving Tutorial: Identifying the Functional Groups in a Biomolecule

 Problem-Solving Tutorial: Properties of Functional Groups

SOLUTION

3,3-dimethyl-1-butanol

Rule 1: Assign the root. There are four carbon atoms in the main chain, so the root is *butane*.

Rule 2: Assign the suffix. There is an alcohol on the chain, so the suffix is changed to *-anol,* yielding *butanol.*

Rule 3: Assign a locator number to the root indicating the location of the alcohol. The OH group is located on the first carbon in the chain, counting from the carbon atom closer to the OH group, so the locator number *1-* is inserted before the root, yielding *1-butanol.*

Rule 4: Assign a prefix. The main chain is branched, so a prefix must be inserted before the root. Since there are two methyl groups on carbon 3, the IUPAC name is *3,3-dimethyl-1-butanol.*

The alcohol is a primary alcohol because the carbon atom to which the OH group is attached is itself also attached to one carbon atom and two hydrogen atoms.

WORKED EXERCISE 7-2 Recognizing Ethers

The antidepressant Prozac is a pharmaceutical that contains an ether functional group. Locate the ether functional group and indicate whether the R groups attached to the ether oxygen are aromatic or saturated.

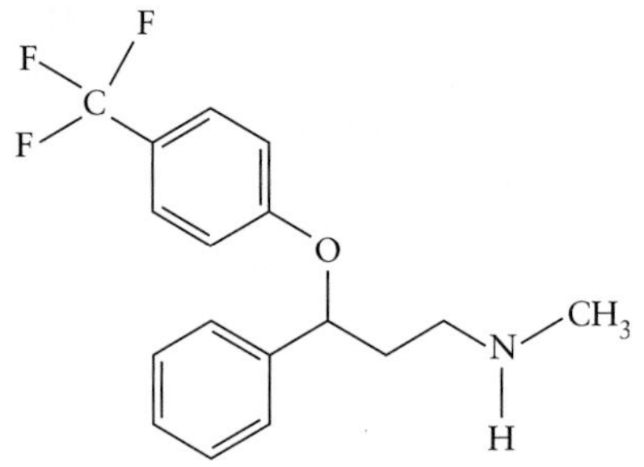

Prozac
(fluoxetine)

SOLUTION

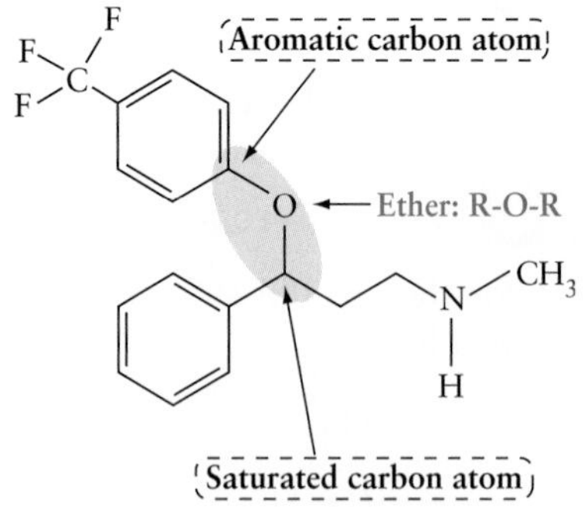

Prozac
(fluoxetine)

PRACTICE EXERCISES

7.1 What is the molecular shape around the oxygen atom of an alcohol and an ether? What are the bond angles around the oxygen atom?

7.2 Provide the IUPAC name for the alcohols listed below. Write the condensed structural formula for (a). Indicate whether each molecule is a 1°, 2°, or 3° alcohol:

a.

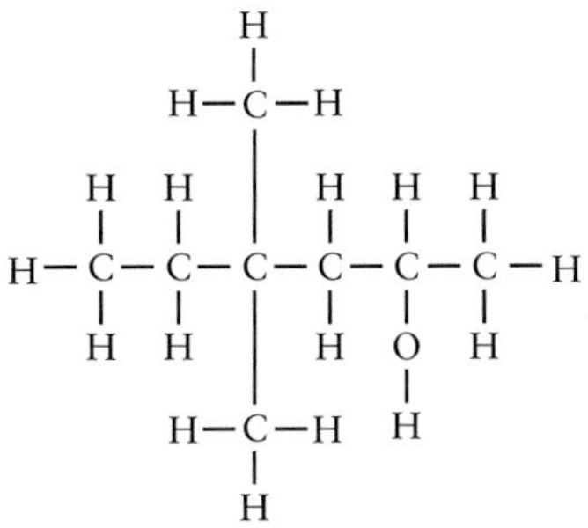

b.

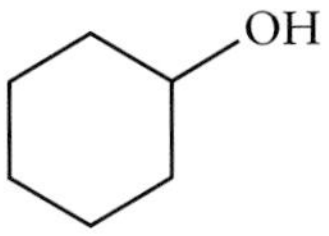

7.3 Provide a structural formula and a skeletal line drawing for the following alcohols and ethers. Indicate which of the four compounds you would expect to be the most soluble in water and explain why:

a. 2-propanol   b. 1-heptanol   c. cyclopentanol   d. ethyl cyclopentyl ether

7.4 The structures of the brain chemical dopamine and the hormone adrenaline are shown below. Circle and label the alcohol and phenol functional groups. In what way are these two molecules similar? How are they different?

a.

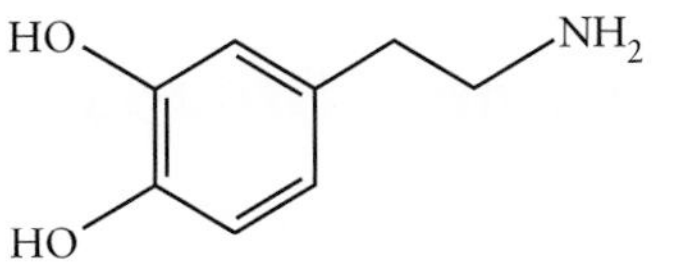

Dopamine

b.

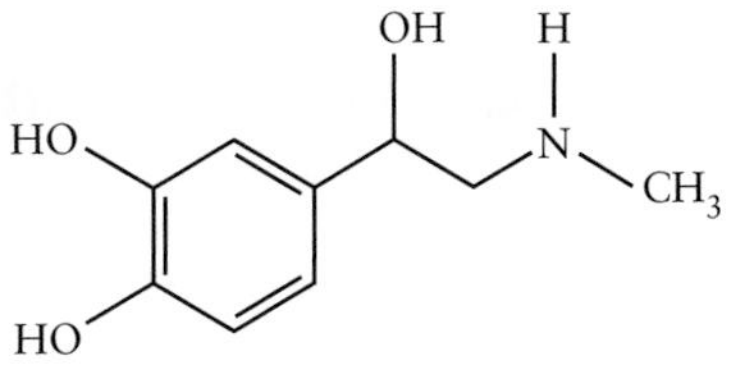

Epinephrine
(adrenaline)

7.5 Draw three ethanol molecules showing hydrogen bonding among them.

7.6 The structure of propylene glycol is $HOCH_2CHCH_3$.

a. Write a skeletal line structure and a structural formula for this compound.

b. Circle all the alcohol functional groups. Indicate whether they are 1°, 2°, or 3° alcohols.

c. Would you expect this compound to be soluble in water? Explain.

7.7 The structure of Δ^9-tetrahydrocannabinol, Δ^9-THC, is shown below. It is the psychoactive substance found in *cannabis* (marijuana), and interacts with the cannabinoid receptors in the brain.

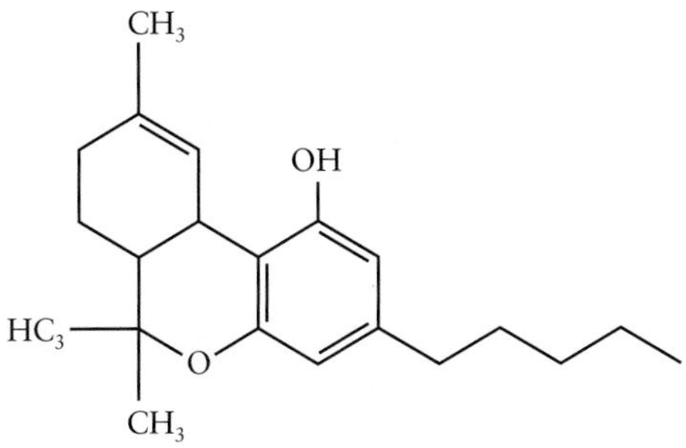

Δ^9-THC

a. Locate and identify the ether and phenol functional groups in this molecule.

b. What part of the chemical name for this compound suggests that there is an alcohol or phenol present in the molecule?

c. Would you expect this compound to be water soluble? Explain.

7.8 Critical Thinking Question: Structural isomers (Section 6.2.) also exist in organic substances with functional groups. For example, there are three structural isomers of C_3H_8O: two are alcohols and one is an ether. Each structural isomer has the same chemical formula, but the connectivity of the atoms differs. In some cases, the change in connectivity results in a different functional group. Name each of the structural isomers shown below and indicate those that are alcohols and those that are ethers.

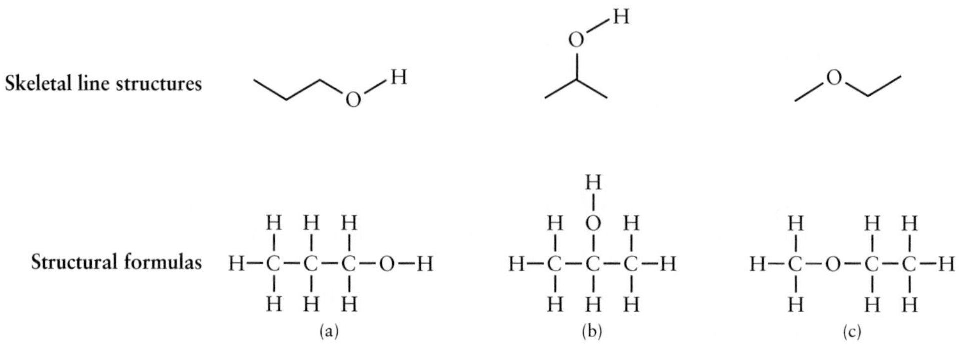

Skeletal line structures

Structural formulas

(a)   (b)   (c)

7.2 | C=O-Containing Functional Groups

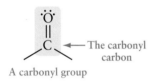

A carbonyl group

Several functional groups contain a carbon–oxygen double bond, commonly known as a **carbonyl group.** The carbonyl group itself is not a functional group, but rather a component of many functional groups. The carbon atom of a carbonyl group is always attached to two groups or atoms. One group is usually an R group.

The other group or atom is either a hydrogen atom, another R group, or a heteroatom. *The identity of the carbonyl-containing functional group depends on what this other atom or group is.* The common carbonyl-containing functional groups are shown in Figure 7-7 and listed below, followed by the identifying group attached to the carbonyl carbon:

- aldehydes, H
- ketones, R
- carboxylic acids, OH
- esters, OR
- thioesters, SR
- amides, NR_2

As expected from VSEPR theory, the geometry around a carbonyl carbon is trigonal planar with bond angles of 120°. Thus, carbonyl-containing functional groups have a two-dimensional shape.

Since oxygen is more electronegative than carbon, a carbonyl group is polar. Functional groups containing a carbonyl group, therefore, have a permanent dipole and are able to interact with one another through dipole–dipole intermolecular forces of attraction (Section 3.3). The carboxylic acid and amide functional groups exhibit the stronger hydrogen bonding forces of attraction, because they also contain the more polar O—H and N—H bonds. Thus, many carboxylic acids are soluble in water. Although aldehydes and ketones do not have an O—H or N—H bond, they are electronegative and, therefore, good hydrogen bond acceptors in aqueous solution. As such, some are soluble in water, such as acetone and formaldehyde.

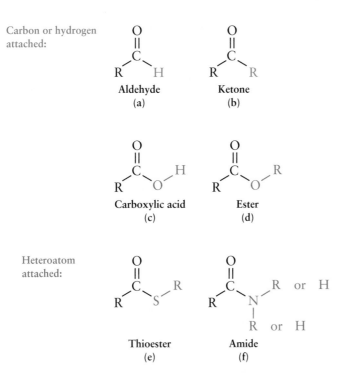

Carbon or hydrogen attached:

Aldehyde
(a)

Ketone
(b)

Carboxylic acid
(c)

Ester
(d)

Heteroatom attached:

Thioester
(e)

Amide
(f)

Figure 7-7 Functional groups that contain a carbonyl group, C=O.

Cinnamaldehyde

Aldehydes

The **aldehyde** functional group is a carbonyl group attached to a hydrogen atom and an R group, except that in the case of formaldehyde, two H atoms are connected to the carbonyl carbon. In condensed structural notation an aldehyde is written as RCHO. When writing the skeletal line structure for an aldehyde, the carbon atom of the aldehyde is not written-in, but the hydrogen attached to the carbonyl group is often shown to avoid any confusion, as shown in the skeletal line structure of cinnamaldehyde in the margin.

The three simplest aldehydes are shown in Figure 7-8. The very simplest, formaldehyde, contains a single carbon atom and no R group; it has the IUPAC name methanal. The suffix changes from *-ane* to *-anal* to signify that an **al**dehyde functional group is present. The condensed structural formula is written HCHO.

Ethanal, shown in Figure 7-8b, contains two carbon atoms: the carbonyl carbon and a methyl group attached to it. The compound is also referred to by its common name, acetaldehyde. The aldehyde containing three carbons is

Cinnamon contains cinnamaldehyde, an aldehyde responsible for its familiar fragrance and taste. [vario images GmbH & Co.KG/Alamy]

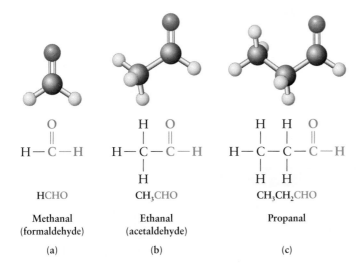

HCHO

Methanal
(formaldehyde)

(a)

CH$_3$CHO

Ethanal
(acetaldehyde)

(b)

CH$_3$CH$_2$CHO

Propanal

(c)

Figure 7-8 Three simple aldehydes and their IUPAC and common names.

Figure 7-9 Three simple ketones and their IUPAC and common names.

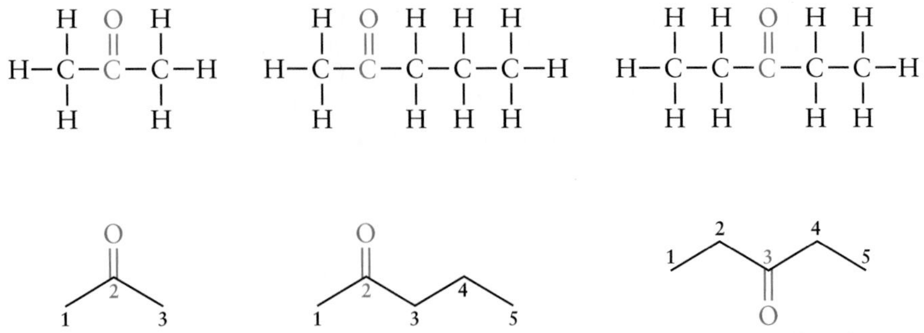

2-Propanone
(acetone)
(a)

2-Pentanone
(b)

3-Pentanone
(c)

Formaldehyde has historically been used in the embalming process, but since 2007 has been banned by the European Union due to its carcinogenic—cancer-causing—effects.

known by its IUPAC name, propanal. No locator number is needed when naming aldehydes, because the aldehyde functional group always appears at one end of a chain. When naming the substituents on branched chain aldehydes, begin numbering at the carbonyl carbon.

Ketones

A **ketone** is a carbonyl group with bonds to *two* R groups. When writing condensed structural formulas, a ketone is written as RCOR. The simplest ketone is 2-propanone, which is more commonly known as acetone, shown in Figure 7-9a.

To name a ket**one** change the suffix from -*ane* to -*anone* and insert a locator number in front of the root to indicate the location of the carbonyl carbon along the main chain. Number the main chain starting from the end closer to the carbonyl group. For example, 2-pentanone and 3-pentanone are both five-carbon chains containing a ketone functional group (Figure 7-9b and 7-9c). However, their carbonyl groups are located at different points along the chain, on the second carbon and the third carbon, respectively. In unbranched rings containing a ketone functional group, no locator number is needed.

Ketones and aldehydes are found throughout nature. For example, 11-*cis*-retinal, the molecule from which rhodopsin (Chapter 6) is produced, is an aldehyde. A ketone functional group is present in the sex hormone, testosterone. Since aldehydes and ketones have a permanent dipole, they exhibit dipole–dipole intermolecular forces of attraction.

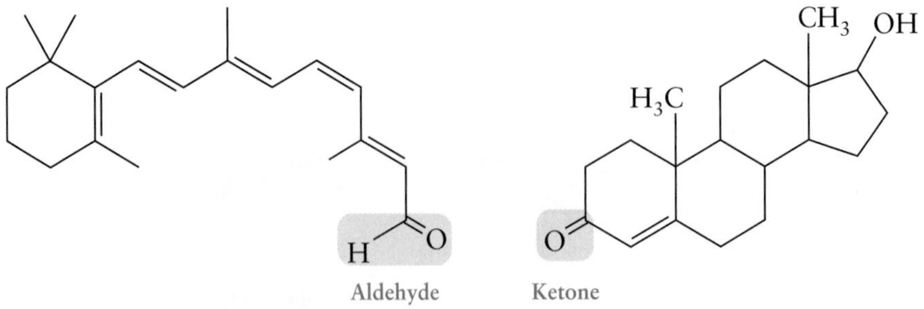

Aldehyde

11-*cis*-Retinal

Ketone

Testosterone

WORKED EXERCISES 7-3 Identifying and Naming Aldehydes and Ketones

1 Identify the following compounds as aldehydes or ketones. Write the IUPAC name for both compounds, and the common name for compound (a). Write the condensed structural formula for (a).

a.

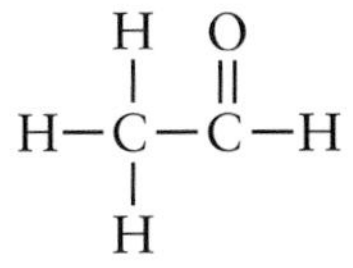

$$H-\underset{\underset{H}{|}}{\overset{\overset{H}{|}}{C}}-\overset{\overset{O}{\|}}{C}-H$$

b.

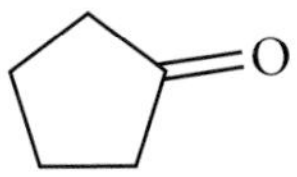

2 Zofran (ondansetron) is used to treat nausea and vomiting in cancer patients and morning sickness in pregnant women. Circle and identify the ketone functional group in this molecule.

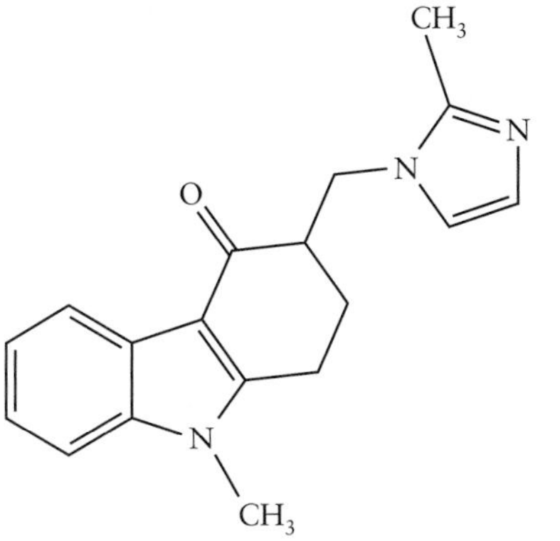

SOLUTIONS

1 a. An aldehyde. The IUPAC name is *ethanal*. Because there are two carbon atoms in the chain, the root is *ethane*. The suffix changes from -*ane* to -*anal* to indicate an aldehyde. The common name for the two-carbon aldehyde is *acetaldehyde*. The condensed structural formula is CH_3CHO.

  b. A ketone. The IUPAC name is *cyclopentanone*. The root is *cyclopentane* because there are five carbon atoms in the chain and the five-carbon chain is in the form of a ring. The suffix changes to -*anone* because there is a ketone present. No locator number is needed because all positions on an unbranched ring are the same.

2

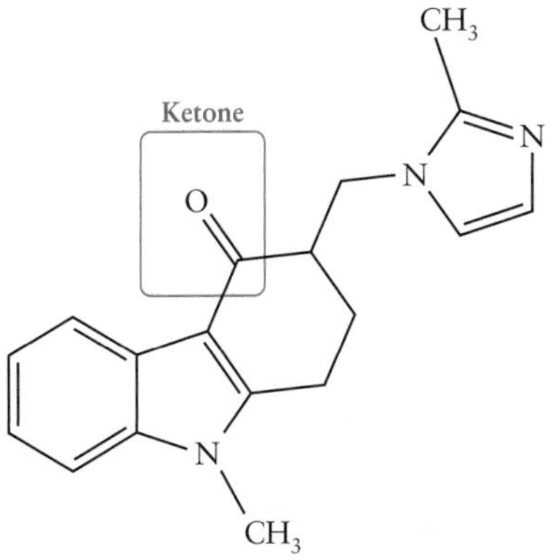

Ketone

PRACTICE EXERCISES

7.9 What is the H—C—H bond angle in formaldehyde? What is the molecular shape of formaldehyde?

7.10 Rank the compounds below from lowest to highest boiling point and explain your ranking.

  a. propanal **b.** 2-propanol **c.** propane

7.11 Name the following aldehydes and ketones. Provide the common name for (e).

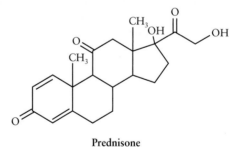

a. 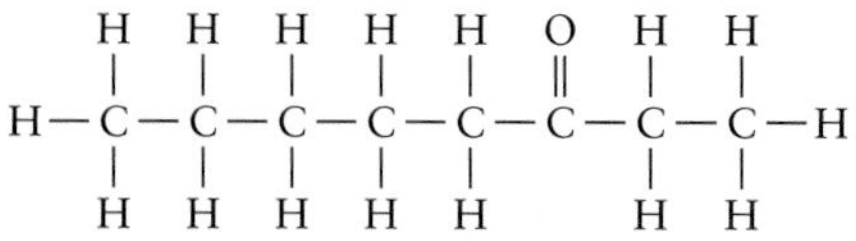   b. 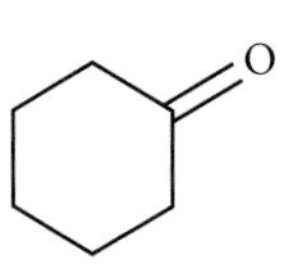   c.

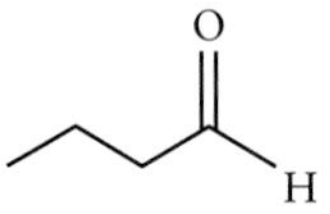

d. 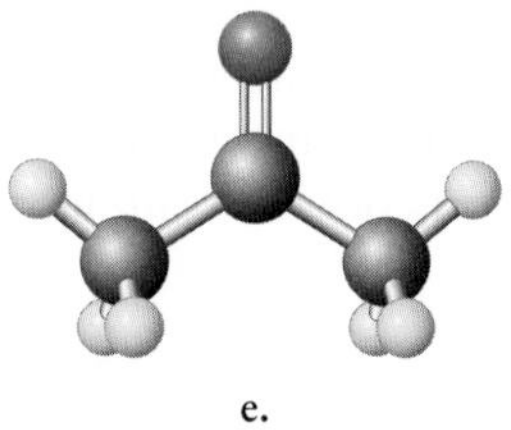   e.

7.12 One form of the sugar glucose is shown below. Circle the aldehyde functional group in this molecule. What other functional groups are present?

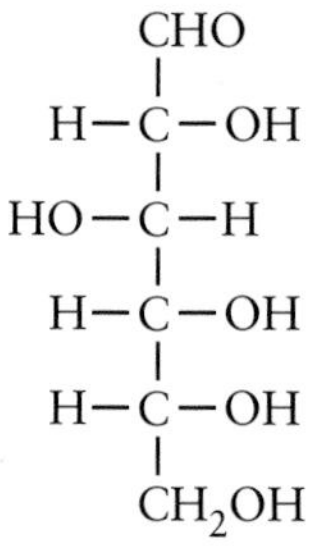

$$
\begin{array}{c}
\text{CHO} \\
|\\
\text{H}-\text{C}-\text{OH} \\
|\\
\text{HO}-\text{C}-\text{H} \\
|\\
\text{H}-\text{C}-\text{OH} \\
|\\
\text{H}-\text{C}-\text{OH} \\
|\\
\text{CH}_2\text{OH}
\end{array}
$$

D-**Glucose**

7.13 Prednisone is a synthetic steroid used as an immunosuppressant—it reduces the activity of the immune system. Prednisone is used to treat various conditions that improve when the immune system is suppressed, including asthma and rejection of an organ after an organ transplant. Circle the three ketones in Prednisone. What are the names of the other functional groups in this molecule?

Prednisone

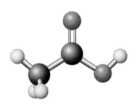

Acetic acid

Vinegar, the bottom layer in oil and vinegar, is an aqueous solution containing acetic acid, a carboxylic acid.
[George Mattei/Photo Researchers, Inc.]

Carboxylic Acids

Carboxylic acids are a common functional group in molecules found throughout nature. A **carboxylic acid** is a carbonyl group directly attached to an O—H group. The condensed structural formula for a carboxylic acid is written either as RCO_2H or $RCOOH$.

Three other common functional groups contain a carbonyl carbon attached to a heteroatom: esters, thioesters, and amides. Figure 7-10 shows how these three functional groups and the carboxylic acid are related. In Chapter 10 you will learn the important chemical reactions that transform carboxylic acids into these three functional groups. Because esters, thioesters, and amides are derived from carboxylic acids through chemical reactions, they are often called carboxylic acid *derivatives*.

Despite the presence of an O—H group in a carboxylic acid, the chemical reactivity of a carboxylic acid is notably different from that of an alcohol

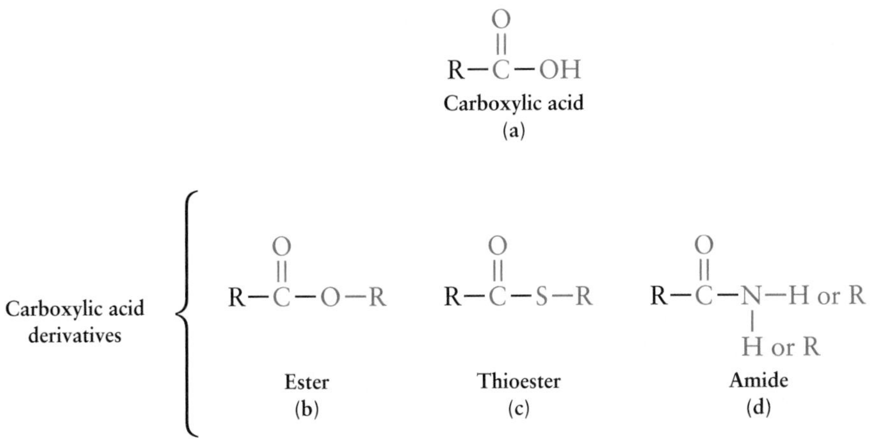

$$
\underset{\substack{\text{Carboxylic acid}\\ \text{(a)}}}{\text{R}-\overset{\displaystyle O}{\overset{\|}{\text{C}}}-\text{OH}}
$$

Figure 7-10 Esters, thioesters, and amides are carboxylic acid derivatives formed by the replacement of the OH group with other heteroatom groups.

Carboxylic acid derivatives

$$
\underset{\substack{\text{Ester}\\ \text{(b)}}}{\text{R}-\overset{\displaystyle O}{\overset{\|}{\text{C}}}-\text{O}-\text{R}}
\qquad
\underset{\substack{\text{Thioester}\\ \text{(c)}}}{\text{R}-\overset{\displaystyle O}{\overset{\|}{\text{C}}}-\text{S}-\text{R}}
\qquad
\underset{\substack{\text{Amide}\\ \text{(d)}}}{\text{R}-\overset{\displaystyle O}{\overset{\|}{\text{C}}}-\overset{\displaystyle }{\underset{\displaystyle \text{H or R}}{\text{N}-\text{H or R}}}}
$$

(Figure 7-11). Hence, alcohols and carboxylic acids are classified as entirely different functional groups. *In a carboxylic acid, the carbonyl group and the O—H together act as a single unit and thus constitute one functional group.* When you identify a carboxylic acid, remember to circle the entire COOH group as one group, as shown in pink in Figure 7-12.

Figure 7-12 shows several simple carboxylic acids. The simplest carboxylic acid is methanoic acid, also known by its common name, formic acid. It has a single carbon atom and no R group. Vinegar contains the two-carbon carboxylic acid known as ethanoic acid, or by its common name, acetic acid. The simplest aromatic carboxylic acid is benzoic acid. It consists of a carboxylic acid functional group attached to a benzene ring.

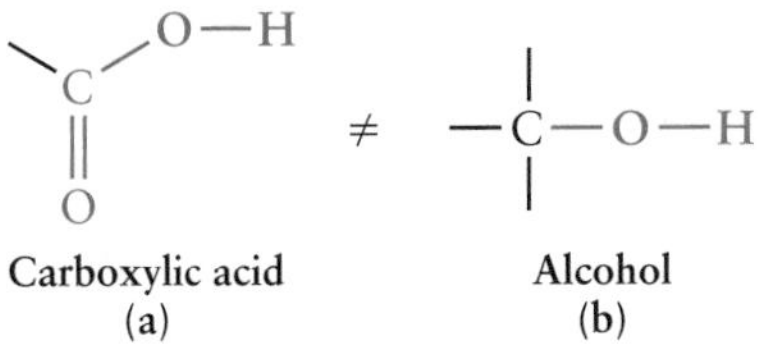

$$
\underset{\substack{\text{Carboxylic acid}\\ \text{(a)}}}{\overset{\displaystyle }{\underset{\displaystyle O}{\overset{\displaystyle \diagdown}{\underset{\|}{\text{C}}}}}\;O-H}
\qquad \neq \qquad
\underset{\substack{\text{Alcohol}\\ \text{(b)}}}{-\overset{\displaystyle |}{\underset{\displaystyle |}{\text{C}}}-O-H}
$$

Figure 7-11 The structure of a carboxylic acid compared to an alcohol. The OH group in a carboxylic acid is attached to a carbonyl group, C=O.

Naming Carboxylic Acids To name a carboxylic acid, count the number of carbon atoms in the chain, beginning from the carbonyl carbon, and assign the root based on the number of carbon atoms, in the usual fashion. Change the suffix from *-ane* to *-anoic acid*. No locator number is needed if the carboxylic acid is an unbranched chain, because carboxylic acids are always located on one end of a carbon chain.

For example, propanoic acid is composed of three carbon atoms, so the root of the IUPAC name is propane (Figure 7-12c). The suffix is changed to *-anoic acid* to reflect that the molecule contains a carboxylic acid. Common names are often used for the two simplest carboxylic acids, formic acid and acetic acid, shown in Figure 7-12a and 7-12b, respectively.

Ionic and Neutral Forms of a Carboxylic Acid The carboxylic acid functional group readily loses a proton (H^+) to form a polyatomic ion, known as a carboxylate ion $RCOO^-$ (Figure 7-13). The tendency to lose this proton is a

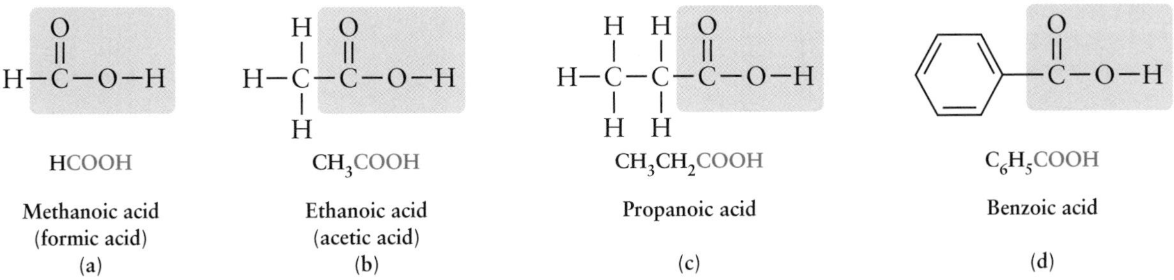

$$
\underset{\substack{\text{HCOOH}}}{\text{H}-\overset{\displaystyle O}{\overset{\|}{\text{C}}}-\text{O}-\text{H}}
\qquad
\underset{\substack{\text{CH}_3\text{COOH}}}{\text{H}-\overset{\displaystyle \overset{\text{H}}{|}}{\underset{\underset{\text{H}}{|}}{\text{C}}}-\overset{\displaystyle O}{\overset{\|}{\text{C}}}-\text{O}-\text{H}}
\qquad
\underset{\substack{\text{CH}_3\text{CH}_2\text{COOH}}}{\text{H}-\overset{\displaystyle \overset{\text{H}}{|}}{\underset{\underset{\text{H}}{|}}{\text{C}}}-\overset{\displaystyle \overset{\text{H}}{|}}{\underset{\underset{\text{H}}{|}}{\text{C}}}-\overset{\displaystyle O}{\overset{\|}{\text{C}}}-\text{O}-\text{H}}
\qquad
\underset{\substack{\text{C}_6\text{H}_5\text{COOH}}}{\underset{\displaystyle }{\bigcirc}\!-\!\overset{\displaystyle O}{\overset{\|}{\text{C}}}-\text{O}-\text{H}}
$$

Methanoic acid (formic acid) (a)   Ethanoic acid (acetic acid) (b)   Propanoic acid (c)   Benzoic acid (d)

Figure 7-12 Four simple carboxylic acids and their IUPAC and common names.

Figure 7-13 Neutral and ionic forms of a carboxylic acid. The ionic form is created upon the *loss* of a proton (H$^+$), a characteristic of acids in general.

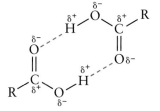

characteristic property of acids, which you will learn about in Chapter 9. Whether a carboxylic acid exists in its neutral or ionic form depends on its environment. In the body, carboxylic acids are usually in their ionic form, but you should be able to recognize both forms.

In its ionic form, RCOO$^-$, the suffix in the IUPAC name of a carboxylic acid changes from *-ic acid* to *-ate*. For example, when ace*tic* acid loses a proton, the polyatomic ion formed is called ace*tate* ion. Since the carboxylate ion is a polyatomic ion, it exists together with a cation to make an ionic compound. For example, when the cation is sodium and the anion is acetate ion, the ionic compound is called sodium acetate, according to the rules for naming polyatomic ionic compounds (see Sections 2.1 and 2.3).

Hydrogen Bonding in Carboxylic Acids Carboxylic acids are capable of hydrogen bonding, because they contain the strong O—H bond dipole and there is a partial negative charge on both oxygen atoms of the carboxylic acid functional group, as shown in Figure 7-14. Thus, carboxylic acids, like alcohols, are soluble in water provided the hydrocarbon portion of the R group is not too large to outweigh the polar characteristics of the carboxylic acid functional group. Like alcohols, they have higher boiling points than other compounds having a comparable number of atoms in their molecular formula.

The ionic form of a carboxylic acid is even more soluble in water than the neutral form. The full negative charge on a carboxylate ion allows it to form strong electrostatic interactions with water molecules.

Figure 7-14 Hydrogen bonding between two carboxylic acids.

Some Interesting Carboxylic Acids Long, unbranched hydrocarbon chains containing a carboxylic acid are known as **fatty acids.** Soap is the carboxylate ion of a fatty acid. The hydrocarbon portion of the molecule dissolves grease, which contains hydrocarbon-like substances. Meanwhile, the carboxylate end makes it possible for the entire molecule to dissolve in water, giving soap the ability to "wash away" grease and dirt. You will study the role of fatty acids when you study lipids in Chapter 13.

Soap is the carboxylate ion, RCOO$^-$, of a fatty acid. A fatty acid is a carboxylic acid containing a long, unbranched hydrocarbon chain. [Inspirestock/Punchstock]

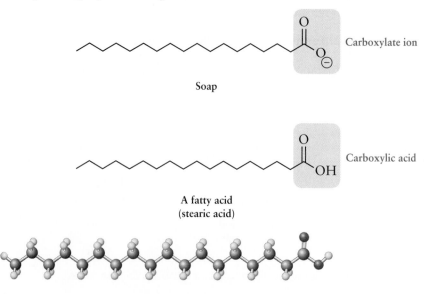

Some molecules contain two or more carboxylic acid functional groups. For example, citric acid is a *tri*carboxylic acid, a compound that contains *three* carboxylic acid functional groups. This molecule is an important molecule in biochemistry, where you will typically see it written in its carboxylate ion form:

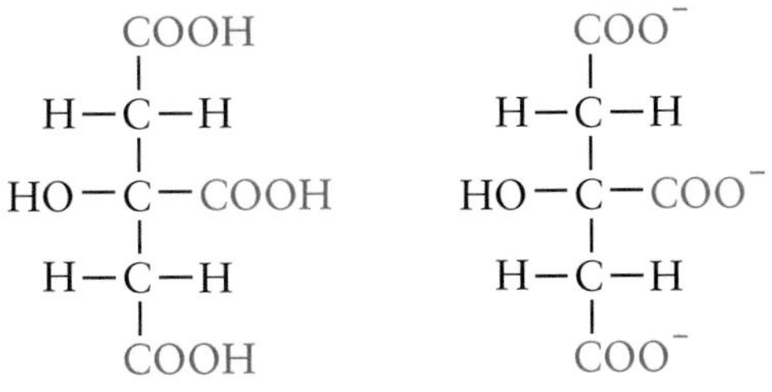

COOH
|
H—C—H
|
HO—C—COOH
|
H—C—H
|
COOH

Citric acid

COO⁻
|
H—C—H
|
HO—C—COO⁻
|
H—C—H
|
COO⁻

Citrate

Citric acid is found in citrus fruits.
[Rosemary Calvert/Punchstock]

WORKED EXERCISE 7-4   Carboxylic Acids

The structures of two common over-the-counter (OTC) analgesic drugs are shown below. Ibuprofen is the active ingredient in Motrin, and naproxen is the active ingredient in Aleve.

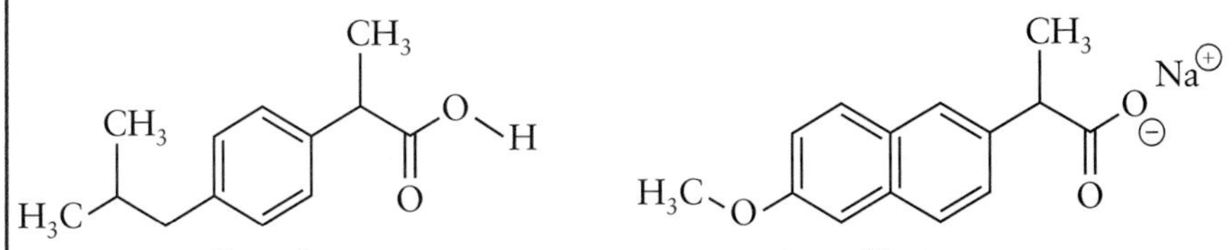

Ibuprofen

Naproxen

a. Circle and label the carboxylic acid or carboxylate functional groups.
b. Indicate which form of the carboxylic acid is present in each molecule: the carboxylic acid or the carboxylate ion.
c. Why is there a Na^+ shown next to the organic structure of naproxen?

SOLUTION

a.

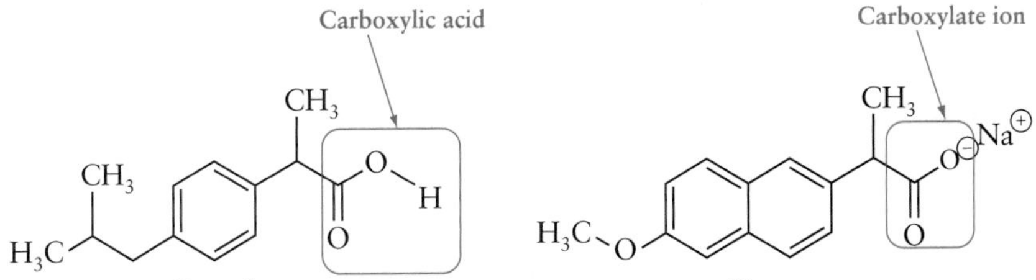

Carboxylic acid

Ibuprofen

Carboxylate ion

Naproxen

b. The carboxylic acid in ibuprofen is in its neutral form. The carboxylate ion in naproxen is in its ionic form.
c. Na^+ is the cation component of the ionic compound naproxen. The cation and anion exist in a 1:1 ratio to give an electrically neutral salt.

PRACTICE EXERCISES

7.14 What is the difference between an alcohol and a carboxylic acid? How can you readily distinguish these two functional groups?

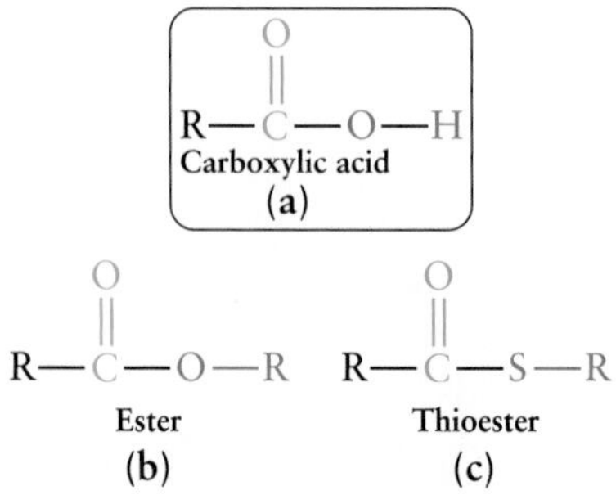

Figure 7-15 Esters and thioesters are derivatives of carboxylic acids formed by replacing the OH group of a carboxylic acid with OR or SR.

The flavor and fragrance of many fruits and flowers is due to compounds containing an ester functional group. For example, methyl jasmonate is the ester responsible for the distinctive fragrance of jasmine. [Organics image library/Alamy]

Figure 7-16 Three simple esters.

7.15 Name the following carboxylic acids:

a.

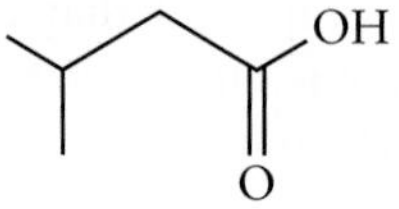

b.

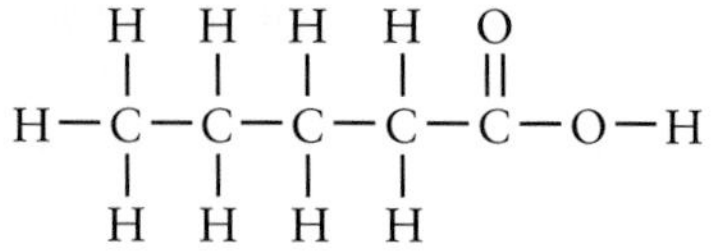

c.

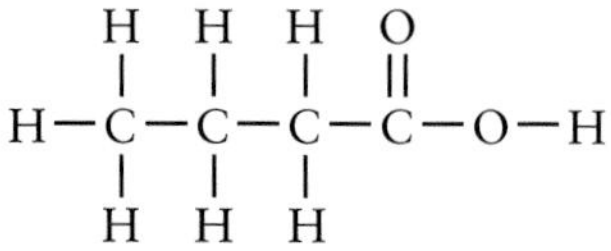

7.16 Which of the functional groups listed below are capable of hydrogen bonding to other like molecules?
a. alcohols   b. ethers   c. aldehydes   d. ketones   e. carboxylic acids

7.17 Which of the following compounds is more soluble in water? Explain why. Name these compounds.

a.

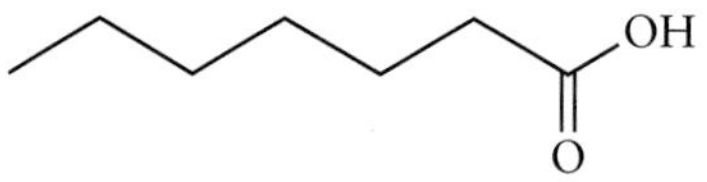

b.

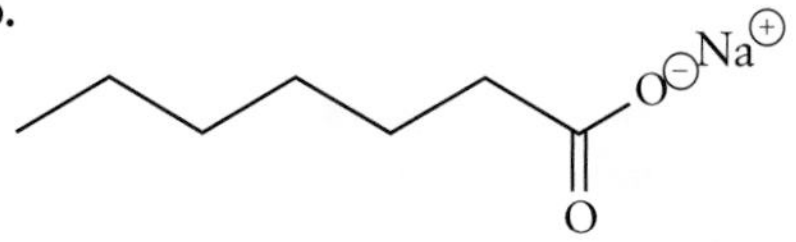

7.18 Citric acid contains three carboxylic acid functional groups. What is the other functional group in the molecule?

7.19 What does the word "fatty" in fatty acid mean?

Esters and Thioesters

Figure 7-15 shows two derivatives of a carboxylic acid. An **ester** is similar to a carboxylic acid, except that O—R instead of O—H is directly attached to the carbonyl carbon. A **thioester** is similar to an ester, except that S—R instead of O—R is attached to the carbonyl carbon.

You may recall from the opening vignette that the feature distinguishing morphine from heroin was the presence of two ester functional groups in heroin whereas morphine had an alcohol and a phenol functional group (review Figures 7-1a and 7-1c).

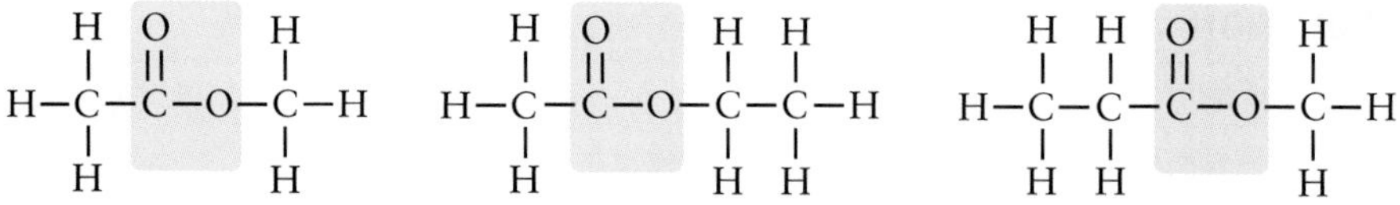

$CH_3CO_2CH_3$
(a)

$CH_3CO_2CH_2CH_3$
(b)

$CH_3CH_2CO_2CH_3$
(c)

Some simple esters are shown in Figure 7-16, and a simple thioester is shown in Figure 7-17. Esters and thioesters contain two R groups, one attached to the carbonyl carbon and the other attached to the oxygen or sulfur atom. These two R groups may be the same (R = R′) or they may be different (R ≠ R′). An ester is written as RCOOR′ or $RCO_2R′$ in condensed structural notation; whereas a thioester is written as RCOSR′.

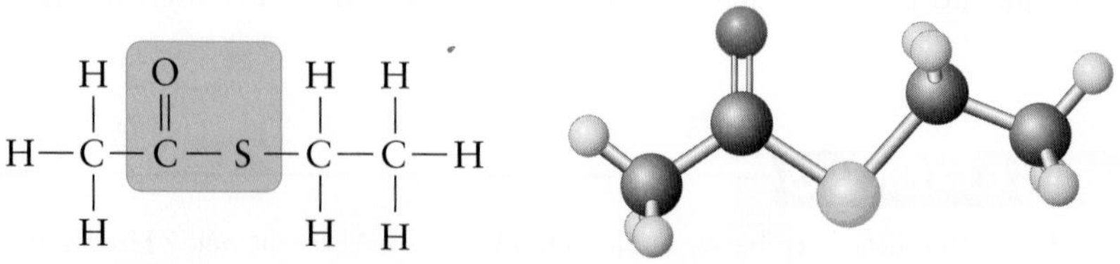

Figure 7-17 A simple thioester.

$CH_3COSCH_2CH_3$

Commonly Found Esters and Thioesters The ester functional group is present in a wide variety of biomolecules. The fat in your diet, for example, is composed of molecules known as triglycerides that contain esters. Triglycerides contain three ester functional groups derived from fatty acids, as illustrated in the representative triglyceride shown.

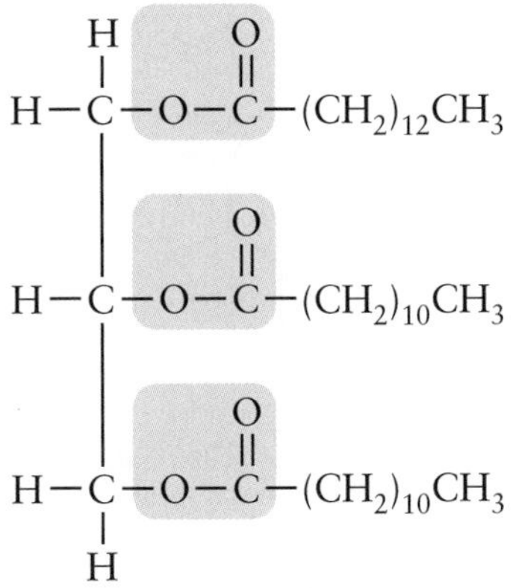

Esters

H–C–O–C–(CH$_2$)$_{12}$CH$_3$

H–C–O–C–(CH$_2$)$_{10}$CH$_3$

H–C–O–C–(CH$_2$)$_{10}$CH$_3$

Triglyceride

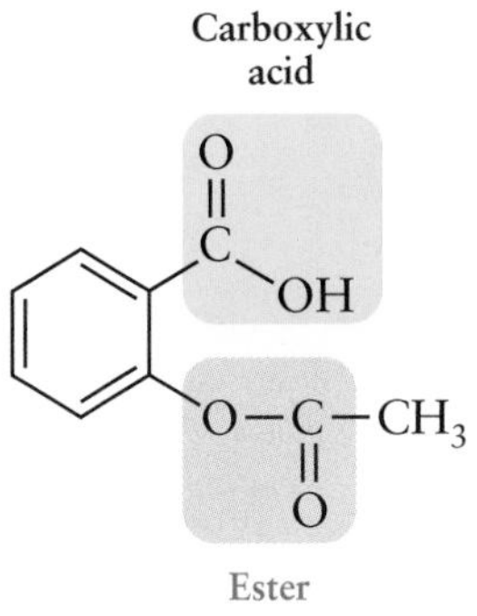

Carboxylic acid

Ester

Aspirin

A common over-the-counter medication, aspirin, contains an ester functional group as part of its structure. It also contains a carboxylic acid.

Unlike esters, thioesters are not common functional groups. However, there are a few important thioesters involved in metabolism. One of these is **acetyl coenzyme A,** abbreviated acetyl CoA. The thioester functional group is highlighted in red in the structure of acetyl CoA shown in the margin. Acetyl CoA is a complex organic molecule synthesized in the body from the vitamin pantothenic acid (B$_5$).

Thioester

$$H_3C-C-S-CoA$$

Acetyl CoA

Acetyl CoA is derived from pantothenic acid (vitamin B$_5$). Pantothenic acid is obtained in the diet from meat and whole grains. [©Don Mason/Corbis]

WORKED EXERCISE 7-5 Esters and Thioesters

Circle the ester functional groups in the structure of cocaine below.

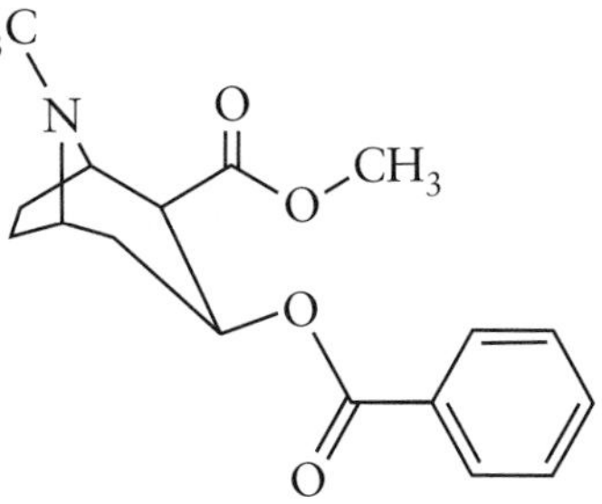

H$_3$C

Cocaine

SOLUTION

H$_3$C

Ester

Ester

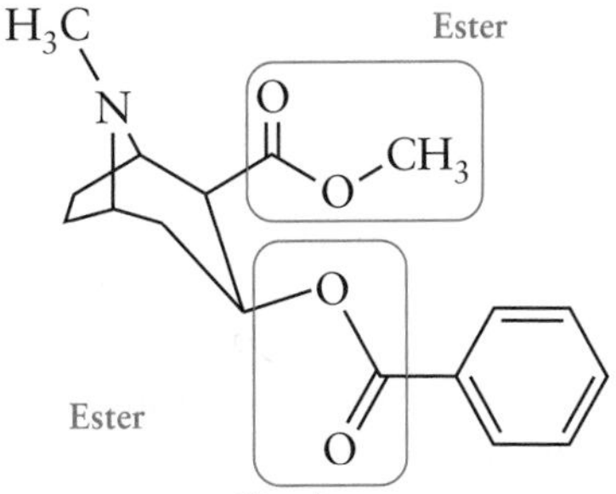

Cocaine

PRACTICE EXERCISES

7.20 Would you expect an ester to be more or less soluble in water than a carboxylic acid having the same number of carbon atoms? Explain.

7.21 Fluticasone propionate is the active ingredient in Flonase nasal spray.

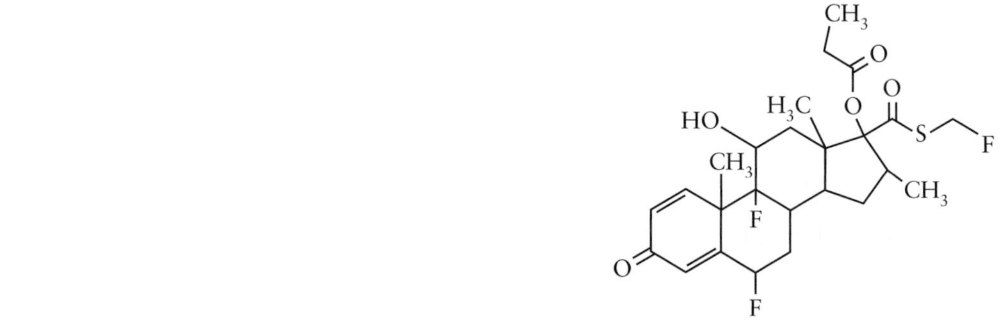

**Fluticasone propionate
(Flonase nasal spray)**

    a. Circle and label the thioester and the ester functional groups and explain how they are similar to one another.

    b. Circle and label the ketone and alcohol functional groups.

7.22 What type of molecule is shown below? Circle the ester functional groups. Would you expect this molecule to be soluble or insoluble in water? Explain.

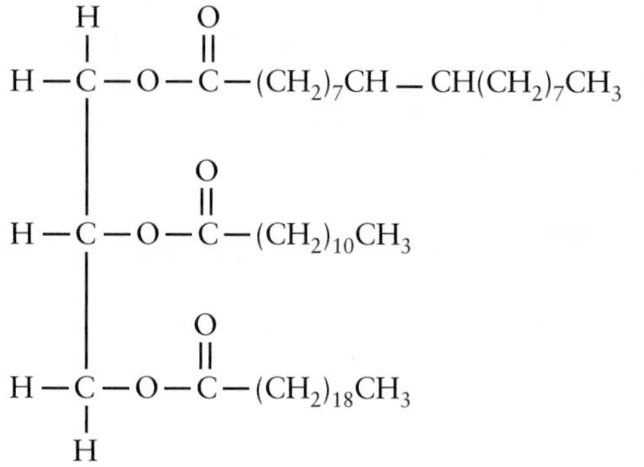

7.23 What is the difference between the two molecules shown below? Describe the difference in terms of functional groups. How are they similar?

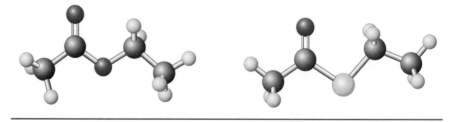

Amides

The most common nitrogen-containing derivative of a carboxylic acid is the **amide.** An amide functional group contains a nitrogen atom directly attached to the carbonyl carbon, as shown in Figure 7-18. Since nitrogen typically forms three bonds, the nitrogen atom in an amide will be bonded to two other atoms in addition to the carbonyl carbon. These other atoms may be

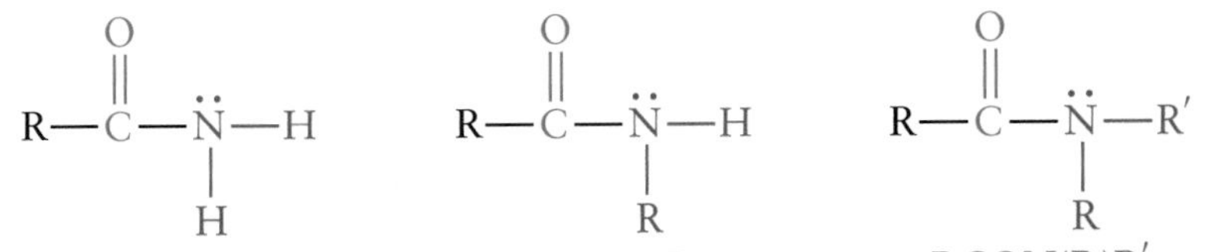

Figure 7-18 The nitrogen atom in an amide is bonded to the carbonyl group and from 0 to 2 hydrogen atoms and 2 to 0 R groups.

RCONH₂   RCONHR   RCON(R)R′

hydrogen atoms or carbon atoms (R groups). The condensed structural formula for an amide appears as $CONH_2$, CONHR, or CON(R)R′, depending on the number of R groups attached to the nitrogen atom.

As with esters, thioesters, and carboxylic acids, the carbonyl group together with the nitrogen atom behave as a single unit—a functional group.

Naming Amides To name simple amides, count the number of carbon atoms in the carbon chain containing the carbonyl carbon, as you did with esters, and assign the root according to Table 6-5. Change the suffix from *-ane* to *-anamide*.

Extension Topic 7-1 Assigning the IUPAC Name for an Ester

The process for naming an ester should sound familiar: assign the root, assign the suffix, and assign the prefix.

Rule 1: Assign the root. Count the number of carbon atoms in the carbon chain containing the carbonyl carbon—the main chain. Assign the root according to Table 6-5. For branched-chain esters, number the chain starting with the carbonyl carbon atom.

Rule 2: Assign the suffix. Change the suffix from *-ane* to *-anoate*. For one- and two-carbon roots, the common names *formate* and *acetate*, respectively, are frequently used.

Rule 3: Assign the prefix. Name the R group attached to the oxygen atom as you would any substituent, with a *-yl* ending, and insert the prefix in front of the root name.

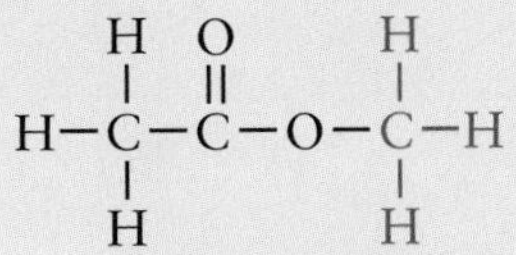

Methyl ethanoate
(methyl acetate)

For example, the ester shown above has the IUPAC name methyl ethanoate. Because the R group containing the carbonyl carbon contains two carbon atoms, the root is *ethane*. The suffix is changed from *-ane* to *-anoate*, producing *ethanoate*. The R group attached to the oxygen atom is treated as a substituent. It contains one carbon atom, so the substituent name is *methyl*, producing *methyl ethanoate*. Since the root is a two-carbon chain, this compound also has the common name *methyl acetate*. The esters shown in the next column have the

IUPAC names *ethyl ethanoate* (the common name *ethyl acetate*) and *methyl propanoate*.

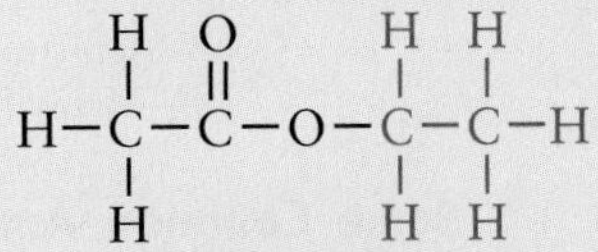

Ethyl ethanoate
(ethyl acetate)

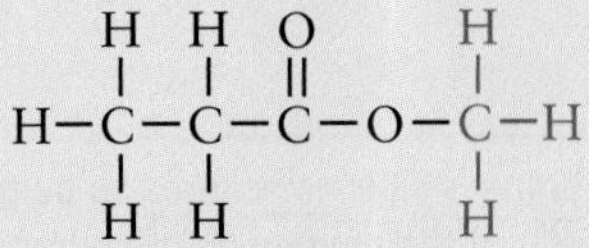

Methyl propanoate

PRACTICE EXERCISES

E7.1 Write the skeletal line structures for the following esters:

   a. ethyl propanoate **b.** propyl ethanoate **c.** methyl benzoate **d.** cyclohexyl acetate

E7.2 Write the IUPAC names for the following esters:

   a.

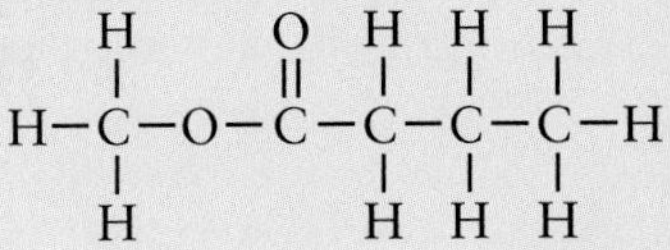

   b.

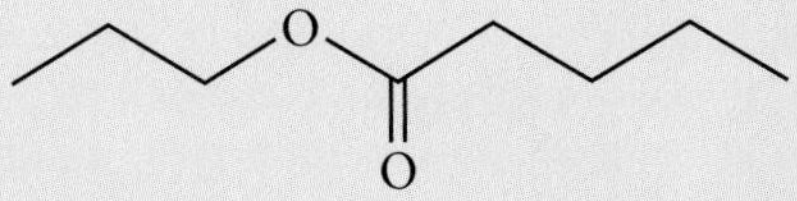

   c. $CH_3CH_2CO_2CH_3$

   d.

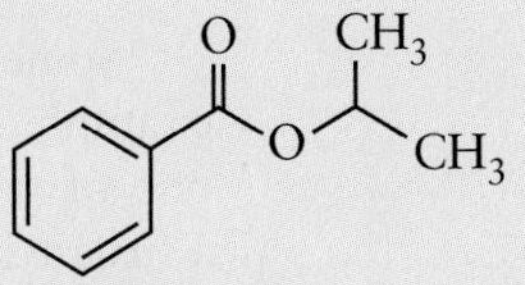

For example, the amides shown below are named methanamide, ethanamide, and propanamide, because they contain one, two, and three carbon chains. The common names for one- and two-carbon unsubstituted amides are formamide and acetamide. The simplest aromatic amide is known as benzamide.

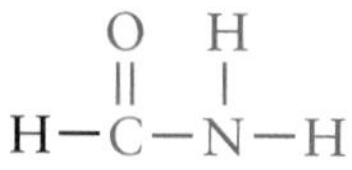

Methanamide (formamide)   Ethanamide (acetamide)   Propanamide 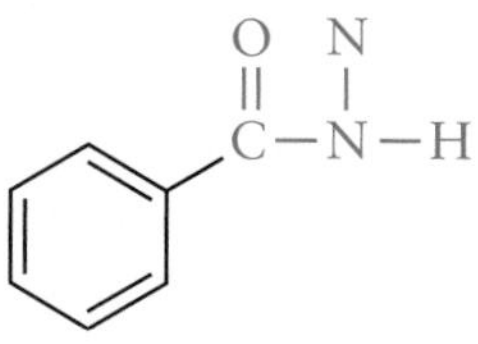   Benzamide

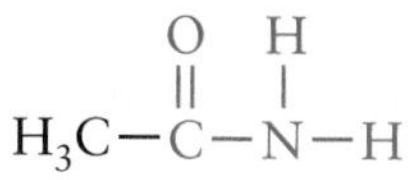

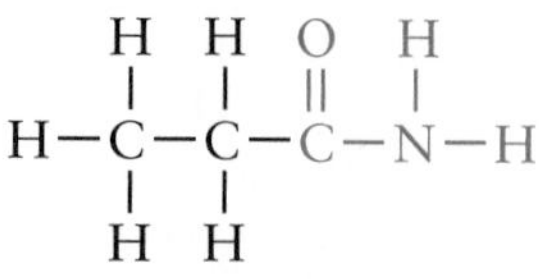

If the nitrogen atom contains one or two R groups, they are named as substituents, using the *-yl* ending, and inserted as a prefix following the letter N- to indicate that the substituent is located on the nitrogen atom rather than on the main chain. For example N-methylmethanamide and N-methylpropanamide are shown below:

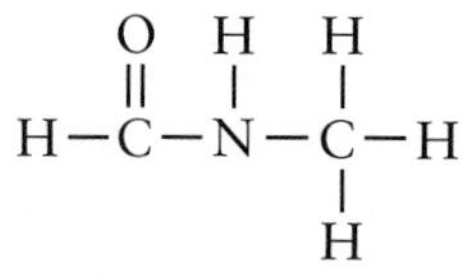

N-Methylmethanamide (*N*-Methylformamide)   *N*-Methylpropanamide

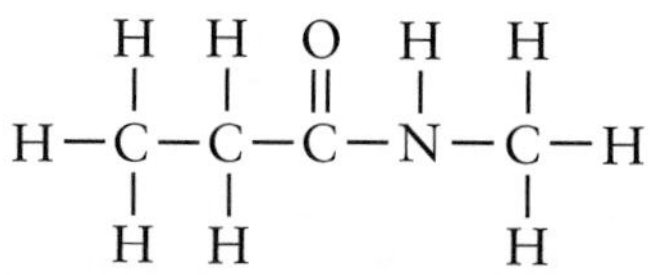

"Runner's high" is believed to be the result of the production of endorphins which bind to the brain's opioid receptors.

Some Common Amides The amide functional group is very common in nature. Molecules containing amides include proteins and the endorphins introduced in the opening vignette—the body's natural pain killers.

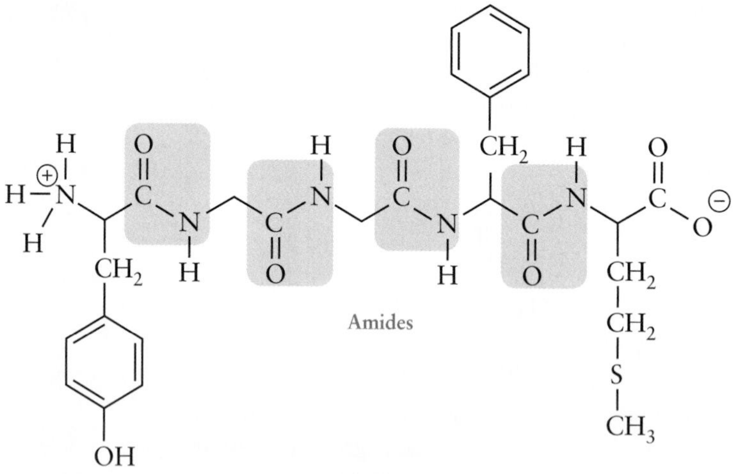

Endorphin: Met-enkephalin

Some important pharmaceuticals contain amide functional groups. For example, the penicillin antibiotics contain two amide functional groups, one of which is part of a ring, as shown in the structure of amoxicillin, a broad-spectrum antibiotic. In addition, amoxicillin contains another nitrogen-containing functional group, but since it is *not* directly attached to a carbonyl carbon, it is not an amide. It is an amine, the subject of the next section.

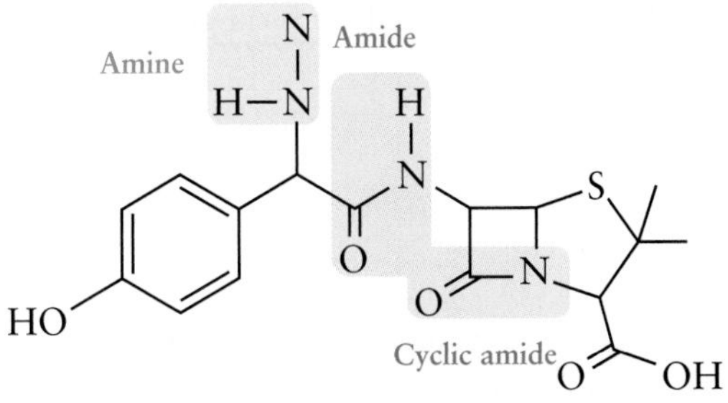

Penicillin antibiotic:
amoxicillin

WORKED EXERCISE 7-6 | Amides

Write the structural formula and the skeletal line structure for the compounds
shown below. Which compound is NOT an amide?

a. $CH_3CH_2NHCH_2CH_3$ **b.** CH_3CONH_2 **c.** $(CH_3)_2NCOCH_2CH_3$

SOLUTION

a. is not an amide because it does not contain a carbonyl group,

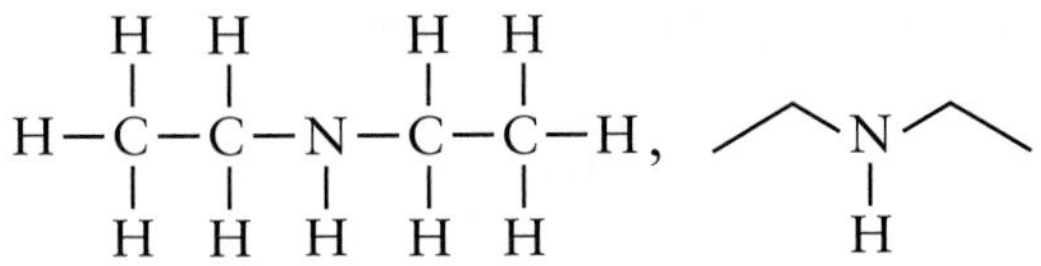

b. is an amide,

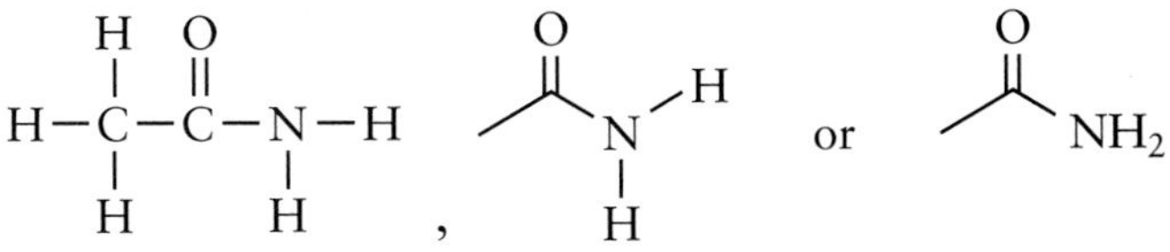

c. is an amide

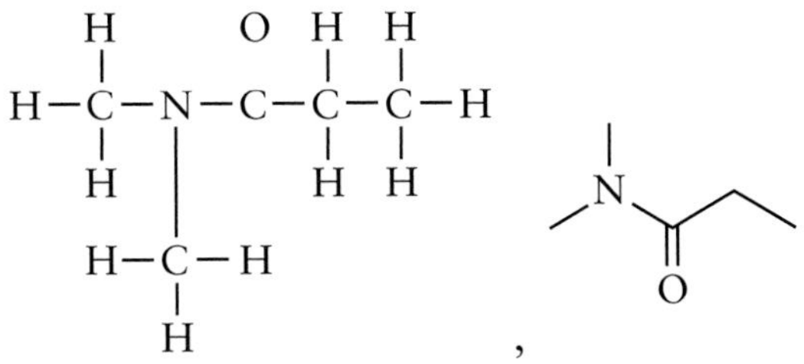

PRACTICE EXERCISES

7.24 Write the structural formula and the skeletal line structure for the compounds
shown below. Which compound is NOT an amide?

   a. $CH_3CH_2CONH_2$

   b. CH_3CONCH_3
            $|$
            CH_3

   c. $CH_3CH_2NHCH_3$

   d. CH_3CONH_2

7.25 Which of the following is an amide functional group?

a.

$$R-\overset{\overset{\displaystyle O}{\|}}{C}-S-R$$

b.

$$R-\overset{\overset{\displaystyle O}{\|}}{C}-O-H$$

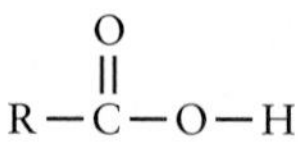

c.

$$R-\overset{\overset{\displaystyle O}{\|}}{C}-O-R$$

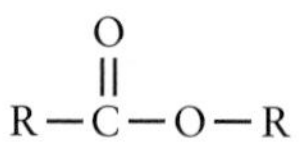

d.

$$R-\overset{\overset{\displaystyle O}{\|}}{C}-\underset{\underset{\displaystyle H}{|}}{N}-H$$

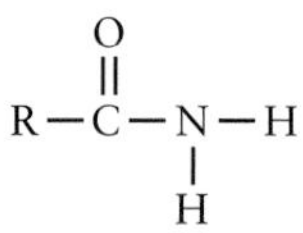

7.26 The structure of vitamin B_5, pantothenic acid, is shown below. Circle and identify the functional groups.

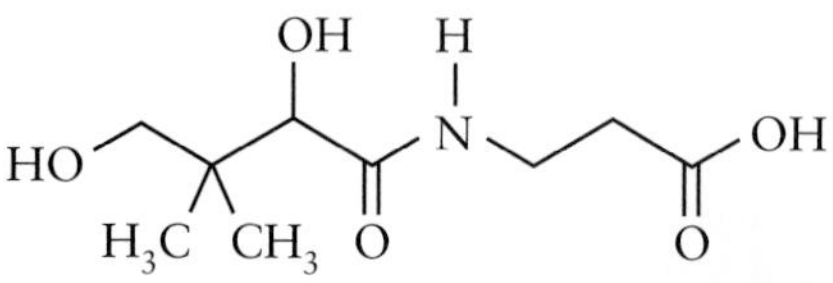

Table 7-1 provides a summary all of the carbonyl-containing functional groups described in this chapter. All but the first two are carboxylic acid derivatives. Perform *The Model Tool 7-2: Carboxylic Acids and Their Derivatives* to gain familiarity with some of these functional groups.

Table 7-1 Functional Groups Containing the Carbonyl Group

Structural Formula	Condensed Formula	Name of Functional Group
$R-\overset{\overset{O}{\|}}{C}-H$	RCHO	Aldehyde
$R-\overset{\overset{O}{\|}}{C}-R$	RCOR	Ketone
$R-\overset{\overset{O}{\|}}{C}-O-H$	RCOOH or RCO_2H	Carboxylic acid
$R-\overset{\overset{O}{\|}}{C}-O-R$	RCOOR or RCO_2R	Ester
$R-\overset{\overset{O}{\|}}{C}-S-H$	RCOSR	Thioester
$R-\overset{\overset{O}{\|}}{C}-\underset{\underset{H}{\|}}{N}-H$	$RCONH_2$	Amide
$R-\overset{\overset{O}{\|}}{C}-\underset{\underset{H}{\|}}{N}-R$	RCONHR	Amide
$R-\overset{\overset{O}{\|}}{C}-\underset{\underset{R}{\|}}{N}-R$	$RCONR$ with R below	Amide

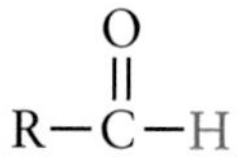

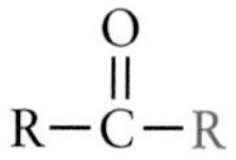

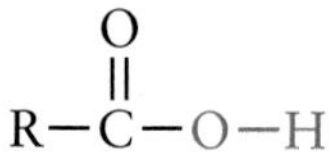

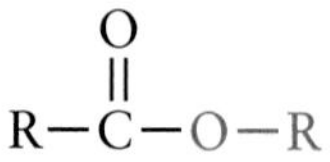

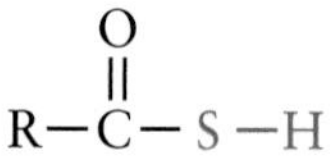

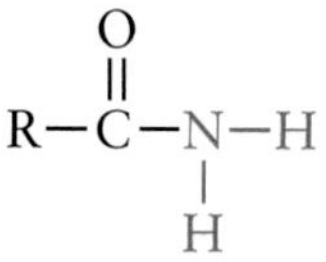

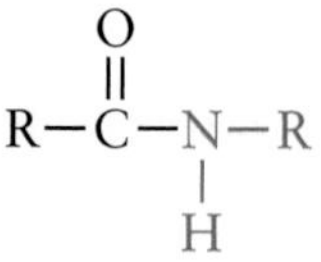

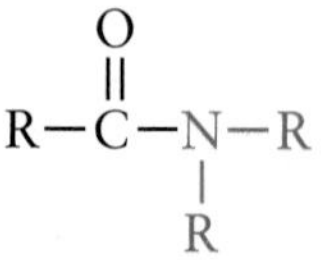

The Model Tool 7-2 Carboxylic Acids and Their Derivatives

I. Construction of Acetic Acid

1. Obtain two black carbon atoms, four hydrogen atoms, two oxygen atoms, six straight bonds, and two bent bonds.
2. Make a model of acetic acid.

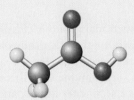

Acetic acid

a. What functional group does this molecule contain?
b. Where is the carbonyl group?
c. What is the IUPAC name of this compound? What common household substance contains this compound?
d. What is the bond angle around the carbonyl carbon? What is the bond angle around the other carbon atom?
e. What is the C—O—H bond angle?
f. Is this substance capable of hydrogen bonding with like molecules? Explain.
g. How is this functional group different from an alcohol?
h. Which hydrogen atom is lost when forming the acetate ion?

II. Construction of Methyl Acetate

3. Add a —CH$_3$ group to the oxygen atom in place of the hydrogen atom removed in Step 2(h).
a. What new functional group have you made?
b. Is this functional group capable of hydrogen bonding?
c. How is this functional group different from a carboxylic acid? How is it different from an ether?
d. If the OR oxygen atom were replaced with a sulfur atom, what functional group would the model contain?

7.3 C—N-Containing Functional Groups: Amines

Amines are functional groups containing a nitrogen atom with three single bonds, each attached to either a carbon atom or a hydrogen atom, as shown in Figure 7-19. Amines are derived from ammonia (NH_3), an inorganic substance, by the replacement of one, two, or three of the hydrogen atoms by carbon-containing groups. The carbon atoms bonded to the nitrogen atom can be either saturated or aromatic.

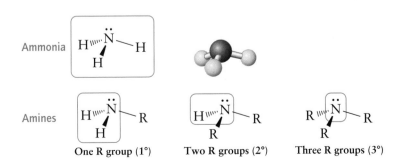

Figure 7-19 Amines are related to ammonia, NH_3.

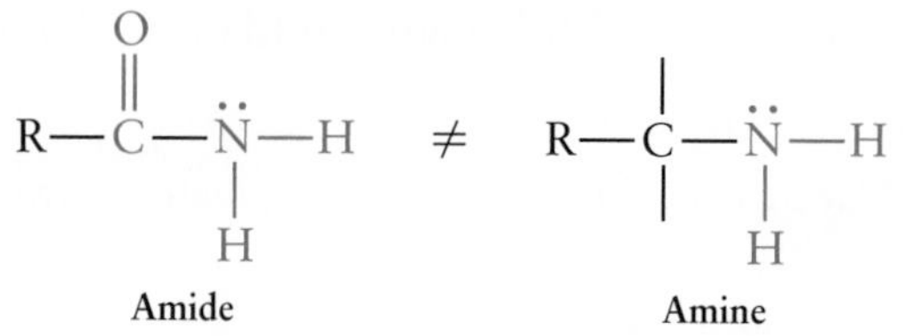

Figure 7-20 The structure of an amide compared to an amine. The nitrogen in an amide is attached to a carbonyl group.

$$R-\overset{\overset{\displaystyle O}{\|}}{C}-\overset{\overset{\displaystyle ..}{N}}{\underset{\underset{\displaystyle H}{|}}{}}-H \quad \neq \quad R-\overset{|}{\underset{\underset{\displaystyle H}{|}}{C}}-\overset{..}{\underset{\underset{\displaystyle H}{|}}{N}}-H$$

Amide   Amine

Note that a functional group is classified as an amine regardless of whether there are one, two, or three R groups attached to the nitrogen atom. These various amines are designated 1°, 2°, and 3°, respectively. The condensed structural formula for an amine will, therefore, appear as —NH₂ for a primary amine, —NHR for a secondary amine, and —NR₂ for a tertiary amine. Amines have the same geometry as ammonia around the nitrogen atom, so they are trigonal pyramidal with bond angles of approximately 109.5° (Section 3.1).

Be careful not to confuse an am*ide* with an am*ine* (Figure 7-20). They both contain a nitrogen atom, but only an amide has a carbonyl group *directly* attached to it. Amides and amines have distinctly different chemical properties, and therefore, are classified as different functional groups.

Naming Amines

Figure 7-21 shows some simple amines with one (1°), two (2°), or three (3°) methyl, —CH₃, groups. The common names for simple amines are assembled in a manner similar to ethers: The R groups are named as though they were substituents, with a -*yl* ending, and are followed by the term *amine*. For example, the amines in Figure 7-21a–c have the common names *methylamine*, *dimethylamine*, and *trimethylamine*. Aniline is the simplest aromatic amine (Figure 7-21d). The IUPAC system for naming amines is described in *Extension Topic 7-2*.

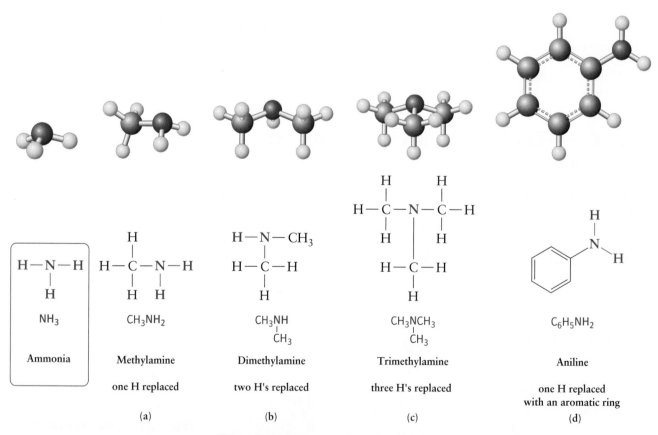

NH₃	CH₃NH₂	CH₃NH|CH₃	CH₃NCH₃|CH₃	C₆H₅NH₂
Ammonia	Methylamine	Dimethylamine	Trimethylamine	Aniline
	one H replaced	two H's replaced	three H's replaced	one H replaced with an aromatic ring
	(a)	(b)	(c)	(d)

Figure 7-21 Three simple amines and an aromatic amine, formed by replacing the Hs in ammonia with carbon-containing groups, R.

Alkaloids and Complex Amines

Amines are a common functional group in drug molecules. For example, the antidepressant Prozac contains an amine as well as the ether functional group described earlier. Recall from the opening vignette that the amine functional group is key to the physiological effects of the narcotic painkillers—a class of drugs known as opioid analgesics.

The amine functional group is also present in many natural brain chemicals, including dopamine and adrenaline. The *Chemistry in Medicine* section at the end of this chapter describes the role of dopamine in schizophrenia and Parkinson's disease. The list of physiologically active amines is extensive and includes compounds such as nicotine, the poison coniine, and cocaine, which are all derived from plants. Amines derived from plants are known as **alkaloids.**

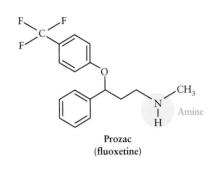

Prozac
(fluoxetine)

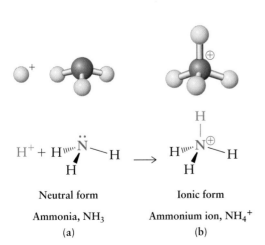

Nicotine
(tobacco)

Coniine
(hemlock)

Cocaine
(coca leaves)

Hydrogen Bonding in Amines

Amines containing one or two N—H bonds are capable of hydrogen bonding, as shown in Figure 7-22. Amines with a small hydrocarbon component are therefore soluble in water. As with alcohols, if the hydrocarbon portion of the amine is extensive, the molecule will be insoluble in water. Amines containing three R groups will not hydrogen bond to one another, but, like ethers, will be capable of hydrogen bonding with water molecules and other hydrogen-bonding substances.

Ionic and Neutral Forms of an Amine

Like ammonia, the amine nitrogen atom contains one pair of nonbonding electrons (Figure 7-23a). These nonbonding electrons readily form a bond to a proton (H^+) to make an N—H bond, giving nitrogen a fourth bond. The formation of an N—H bond converts the neutral amine into a polyatomic ion (Section 2.4) analogous to the ammonium ion, NH_4^+, shown in Figure 7-23b. You will learn more about this type of reaction in Chapter 9 when you study acid–base reactions. Whether an amine exists in its neutral or ionic form

Alkaloids are amines derived from plants. For example, nicotine is an alkaloid obtained from the tobacco plant. [WILDLIFE GmbH/Alamy]

Figure 7-22 Hydrogen bonding between two amines.

Neutral form

Ammonia, NH_3

(a)

Ionic form

Ammonium ion, NH_4^+

(b)

Figure 7-23 Neutral and ionic forms of an amine. The ionic form is created upon the addition of a proton (H^+).

The Model Tool 7-3 Amines

I. Construction of Ammonia, Methylamine, and Dimethylamine

1. Obtain two black carbon atoms, six hydrogen atoms, one nitrogen atom, and ten straight bonds.

2. Construct a model of ammonia, NH_3.
   a. Are there any nonbonding electrons on the nitrogen atom?
   b. What is the H—N—H bond angle? What is the molecular shape of the molecule?
   c. Is ammonia an organic compound? Explain.
   d. What would you need to do to turn ammonia into the ammonium ion?

3. Replace one of the hydrogen atoms in the ammonia molecule you made above with a -CH_3 group.
   a. What functional group does this molecule contain?
   b. What is the bond angle around the nitrogen atom? What is the molecular shape around the nitrogen atom?
   c. Are there any nonbonding electrons on the nitrogen atom?
   d. Is this molecule capable of hydrogen bonding?
   e. Is this molecule an organic compound? Explain.

4. Take the molecule you just made and replace one of the other hydrogen atoms attached to nitrogen with a —CH_3 group.
   a. What functional group does this molecule contain?
   b. Are there any nonbonding electrons in this molecule?
   c. Is this molecule an organic compound? Explain.

5. Add a hydrogen atom to the nitrogen atom in dimethylamine.
   a. What charge does the molecule now contain?
   b. Does this polyatomic ion contain a nonbonding pair of electrons?
   c. How is the molecular shape of this molecule different from that of ammonia?
   d. What are the bond angles around the nitrogen atom?

depends on its environment. In the cell, amines are usually in their ionic form, but you should be comfortable recognizing both forms of an amine.

The ionic form of an amine is soluble in water because there is a *full* positive charge on the polyatomic ion. Therefore, most amine-containing drugs are prepared in their ionic forms, as salts, because they are more readily absorbed by the body in this form. Perform *The Model Tool 7-3: Amines* to gain familiarity with this important functional group.

Amino acids—the building blocks of proteins—are some of the most important small molecules in living organisms. As their name suggests, they contain both a carboxylic acid and an amine. In the cell, amino acids exist with both the amine and the carboxylic acid functional group in their ionized form, as drawn in the margin. Amino acids are described in detail in Chapter 11.

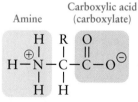

Amine — Carboxylic acid (carboxylate)

Amino acid

WORKED EXERCISE 7-7 Amines

The brain chemical dopamine is shown below.

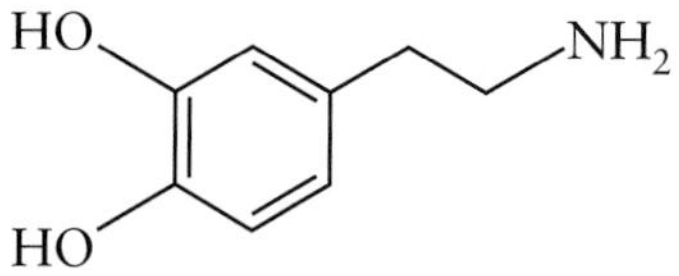

Dopamine

a. Circle the amine functional group.
b. What part of the name of this compound suggests that there is an amine functional group present?
c. What is the molecular shape around the amine nitrogen?
d. Write the ionic form of dopamine.
e. Can dopamine molecules form hydrogen bonds? Explain.

SOLUTION

a.

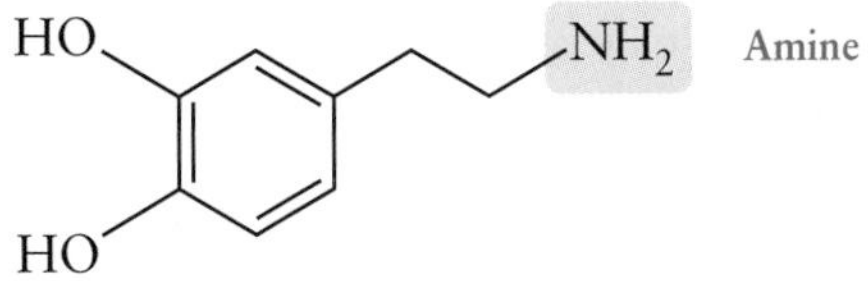

HO⟍   ⟋NH₂   Amine

HO⟋

          Dopamine

b. The *-amine* in dop*amine* suggests that an amine is present.
c. The molecular shape around the nitrogen atom is trigonal pyramidal.
d.

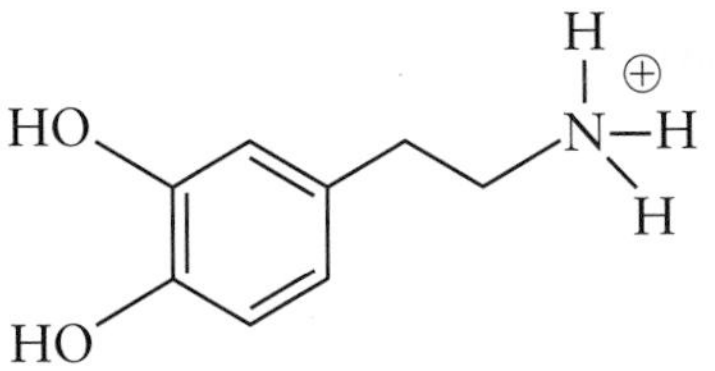

```
            H
            | ⊕
HO⟍        N—H
            |
HO⟋        H
```

e. Yes, dopamine molecules can form hydrogen bonds because they contain two N—H bonds and two OH groups.

PRACTICE EXERCISES

7.27 For the four amines with the chemical formula C₃H₉N, one has a much lower boiling point than the other three structural isomers. Write the structural formula for each and its common name, and predict which has the lowest boiling point. Explain why.

7.28 Epinephrine (adrenaline) is a hormone secreted when the body is under stress. Illicit drugs that are structurally similar to epinephrine are highly addictive stimulants. For example, the illegal and addictive drug methamphetamine is similar in structure to epinephrine. However, pseudoephedrine, which is structurally similar to epinephrine, is a decongestant and sold over the counter in pharmacies.

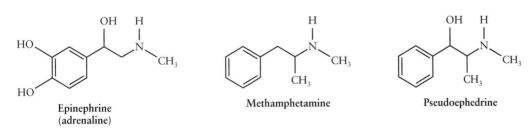

Epinephrine (adrenaline)   Methamphetamine   Pseudoephedrine

a. Circle and label all the amine functional groups in these molecules.
b. In what way is methamphetamine structurally similar to pseudoephedrine? How is it different?
c. In the body, the amine in epinephrine is in its ionic form. Write the structure of the ionic form of epinephrine.
d. In their neutral forms, which of these three compounds would you expect to be the most soluble in water? Explain why.
e. What is the C—N—C bond angle in all of these compounds?

Extension Topic 7-2 Assigning the IUPAC Name for an Amine

The IUPAC rules for naming amines are based on the rules for naming hydrocarbons, using the suffix *-anamine* to indicate an amine. Follow the guidelines below for assigning the IUPAC name for an amine.

Rule 1: Assign the root. Locate the R group attached to the nitrogen atom that has the longest carbon chain and assign the root name according to its length, using Table 6-5. Number the chain from the end closer to the nitrogen atom.

Rule 2: Assign the suffix. Change the suffix of the IUPAC name from *-ane* to *-anamine* to signify that an amine functional group is present.

Rule 3: Assign a locator number to the root indicating the location of the amine. As with alcohols, a locator number is used to indicate the carbon atom on the main chain that contains the amine. Place the locator number followed by a hyphen in front of the root name.

Rule 3: Assign a prefix. Name the other R groups attached to the nitrogen atom as you would any substituent; however, instead of a locator number, use the prefix *N-* to indicate that the substituent is directly attached to the nitrogen atom and not to a carbon atom in the main chain. If there are two R groups, write them in alphabetical order. If the R groups are the same, insert the prefix *di-*. No prefix is needed if the nitrogen atom is bonded to two hydrogen atoms (RNH_2). If there are branches along the main chain of the amine, name those in the usual fashion and list them after the *N*-substituents.

For example, the amine to the right has the IUPAC name methanamine. The only R group attached to nitrogen is a carbon chain of one carbon atom, so the root is *methane*. The suffix is changed from *-ane* to *-anamine*, giving *methanamine*. No prefix is needed, since the nitrogen atom is bonded to two hydrogen atoms.

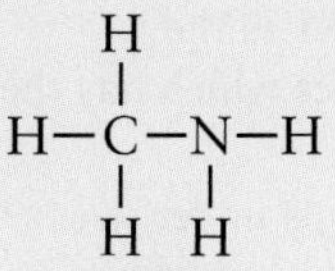

The IUPAC name for the amine at right is *N-methylmethanamine*. The root is the same as in the previous example, but the other R group is treated as a methyl substituent.

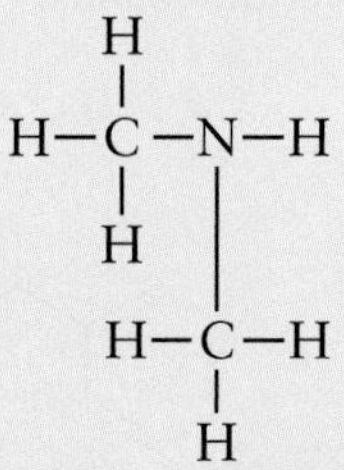

Methanamine

N-Methylmethanamine

The prefix *N-* is inserted before the substituent name to indicate that the methyl group is attached to the nitrogen atom.

The structure at right has the IUPAC name *N,N-dimethylmethanamine*. The *N,N-* indicates that there are two substituents attached to the nitrogen atom in addition to the "main chain" one-carbon chain.

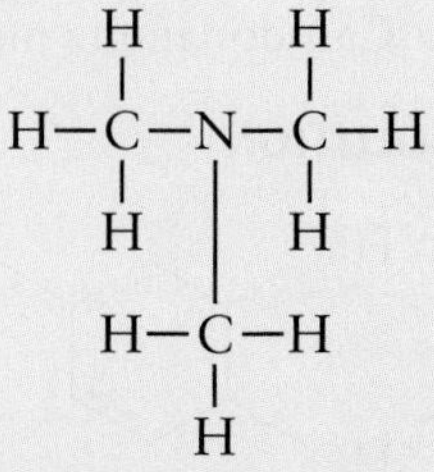

N,N-Dimethylmethanamine

WORKED EXERCISE E7-2

Write the structure of the following amines:
a. *N*-methyl-2-propanamine;
b. *N,N*-diethyl-2-methyl-1-propanamine

SOLUTION

a.

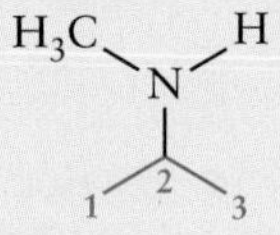

b.

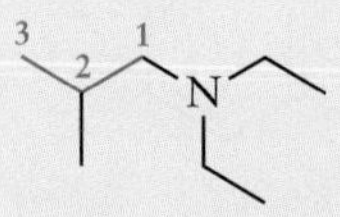

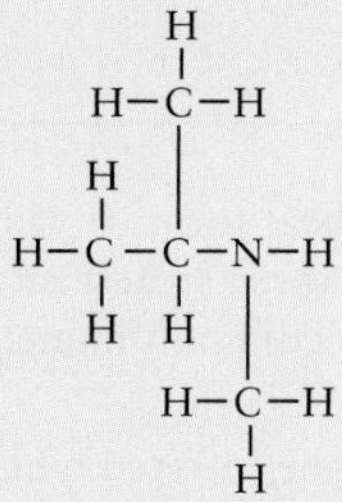

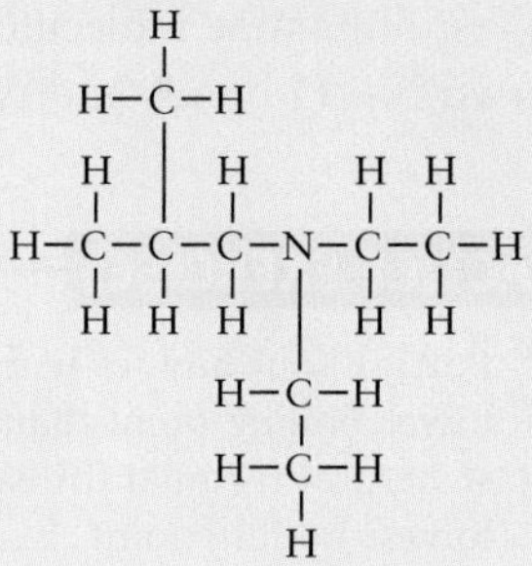

PRACTICE EXERCISE

E7.3 Write the IUPAC name for the following amines:

a.

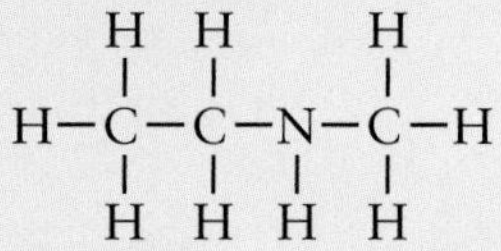

b.

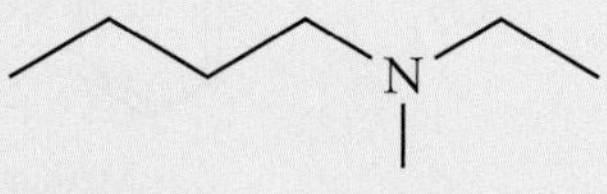

7.4 P═O–Containing Functional Groups

The last functional groups to consider are those derived from the inorganic substance phosphoric acid (H_3PO_4). These functional groups contain a phosphorus–oxygen double bond, P═O, as well as three P—O single bonds. The functional groups derived from phosphoric acid are found in important biomolecules, including DNA, RNA, ATP, and coenzyme A.

Phosphoric acid contains a central phosphorus atom surrounded by ten bonding electrons. Recall from Section 2.2 that an expanded octet is possible for phosphorus because it is a period three element. In phosphoric acid, there is a P=O bond as well as three P—O—H bonds.

Phosphoric acid can lose one, two, or all three of its hydrogen atoms to form three different polyatomic ions. The number of hydrogen atoms lost depends on the aqueous environment. In the cell the most abundant polyatomic ion is the form of phosphoric acid that has lost two hydrogen atoms. This ion is called monohydrogen phosphate because it contains a single hydrogen atom ("mono" means "one"). Monohydrogen phosphate ion, HPO_4^{2-}, is often called **inorganic phosphate** and abbreviated P_i. Recall from Chapter 2 that this polyatomic ion carries a 2− charge.

When one or more of the hydrogen atoms in phosphoric acid is replaced by one or more carbon atoms (an R group), the result is an organic functional group called a **phosphate ester.**

$$\begin{array}{c} O \\ \| \\ {}^-O-P-OH \\ | \\ O^- \end{array} \quad \begin{array}{c} \text{Monohydrogen phosphate} \\ P_i \end{array}$$

$$\begin{array}{c} O \\ \| \\ {}^-O-P-OR \\ | \\ O^- \end{array} \quad \text{Phosphate ester}$$

A unique characteristic of phosphate esters is that the phosphate group can be connected to an additional phosphate group, P_i, via a bond, called a **phosphoanhydride bond**, that forms between one of the oxygen atoms of the phosphate group and the phosphorus atom of another phosphate group. The attachment of two phosphate groups to a carbon atom creates a *di*phosphate ester (or diphosphate), while the attachment of three phosphates groups creates a *tri*phosphate ester (or triphosphate), as shown in Figure 7-24.

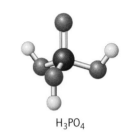

$$\begin{array}{c} O \\ \| \\ H-O-P-O-H \\ | \\ O \\ | \\ H \end{array}$$

Phosphoric acid, H_3PO_4

H_3PO_4

$$\begin{array}{cc} O & O \\ \| & \| \\ {}^-O-P-O-P-O-R \\ | & | \\ {}^-O & {}^-O \end{array}$$

Phosphoanhydride bonds

$$\boxed{\begin{array}{c} O \\ \| \\ {}^-O-P-OH \\ | \\ O^- \end{array}} \quad \begin{array}{c} \text{Monohydrogen phosphate,} \\ P_i \end{array}$$

$$\begin{array}{c} O \\ \| \\ {}^-O-P-O-R \\ | \\ O^- \end{array} \quad \text{Monophosphate ester}$$

$$\begin{array}{cc} O & O \\ \| & \| \\ {}^-O-P-O-P-O-R \\ | & | \\ O^- & O^- \end{array} \quad \text{Diphosphate ester}$$

$$\begin{array}{ccc} O & O & O \\ \| & \| & \| \\ {}^-O-P-O-P-O-P-O-R \\ | & | & | \\ O^- & O^- & O^- \end{array} \quad \text{Triphosphate ester}$$

Figure 7-24 Comparison of monohydrogen phosphate (P_i) and monophosphate, diphosphate, and triphosphate esters.

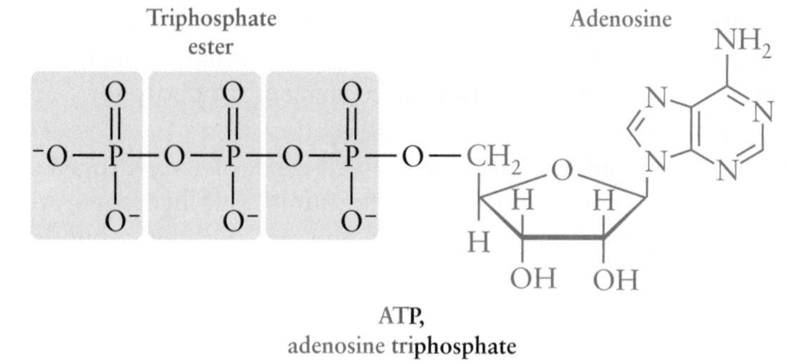

Figure 7-25 The structure of ATP, showing the triphosphate ester.

For example, adenosine triphosphate (ATP), the energy currency of the cell, is a triphosphate ester, as shown in Figure 7-25. The R group attached to the first phosphate group is known as adenosine, hence the name *adenosine* triphosphate.

The phosphoanhydride bond in these molecules is very important to their function in the cell, which usually involves storing and transferring chemical energy. Hence, you will see these functional groups and molecules again when you study biochemical reactions.

WORKED EXERCISE 7-8 Phosphate Esters

The structure of coenzyme A is shown below.

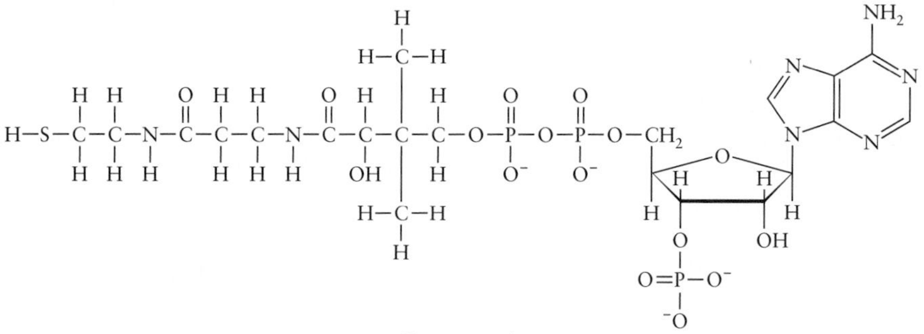

a. Circle the two phosphate esters in the molecule: One has two R groups and the other has one R group.
b. Label each phosphate ester as a monophosphate, diphosphate, or triphosphate ester, depending on how many phosphate groups are attached to the R group(s).
c. Circle and label the two amide functional groups.
d. Circle and label the alcohol functional groups. Are they primary, secondary, or tertiary alcohols?

SOLUTION
a.–d.

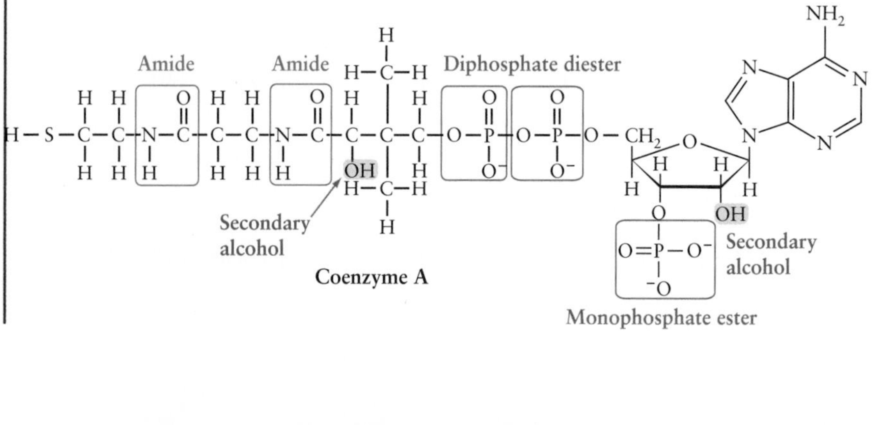

7.29 Write the structural formula for phosphoric acid. What is the most common form of phosphoric acid in the body?

7.30 The structure of **c**yclic **a**denosine **m**ono**p**hosphate, cAMP, is shown below. It is an important messenger molecule that transmits the effects of adrenaline and other hormones.

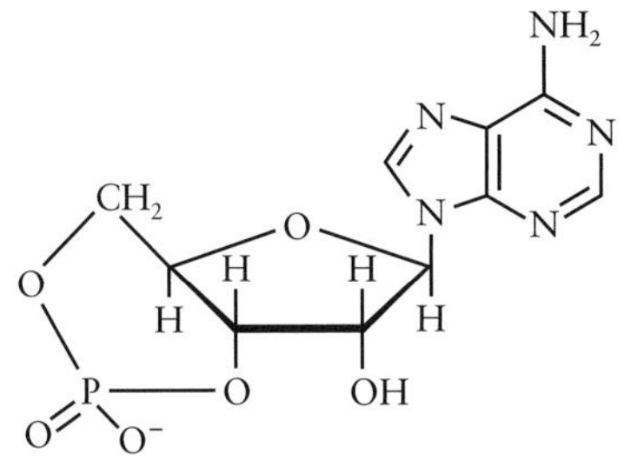

Cyclic adenosine monophosphate,
cAMP

a. How is this molecule related to phosphoric acid?
b. Why is the molecule called a monophosphate and not a di- or triphosphate?
c. This molecule is actually a monophosphate **di**ester. Explain why.
d. What is the total charge on this molecule?
e. Why is the molecule called "cyclic" AMP?

You have been introduced to all the major functional groups in organic chemistry, including many that appear in biochemistry and that are summarized in Table 7-2. You have learned to identify these functional groups within natural products as well as pharmaceuticals. In Chapter 10 you will see how these functional groups behave in chemical reactions. But first, you will learn how a chemical reaction works and consider the important role of energy in chemical reactions, the topic of the next chapter.

Table 7-2 The Common Functional Groups

Functional Group Name	Structural Formula	Functional Group Name	Structural Formula
Alkene	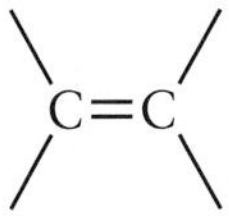 $\diagup C{=}C\diagdown$	Aldehyde	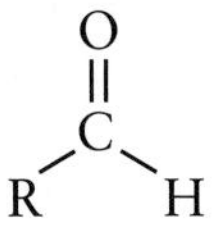 $R{-}\overset{\overset{O}{\|}}{C}{-}H$
Alkyne	$-C{\equiv}C-$	Ketone	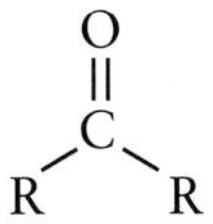 $R{-}\overset{\overset{O}{\|}}{C}{-}R$
Aromatic hydrocarbon	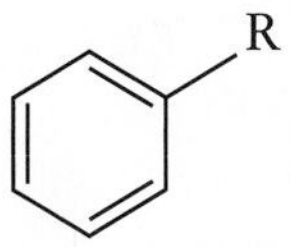 (ring)—R	Carboxylic acid	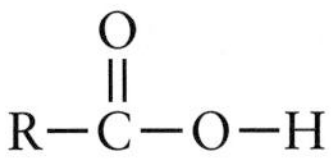 $R{-}\overset{\overset{O}{\|}}{C}{-}O{-}H$
1° Alcohol	$R{-}\overset{\overset{H}{\|}}{\underset{\underset{H}{\|}}{C}}{-}O{-}H$	Ester	$R{-}\overset{\overset{O}{\|}}{C}{-}O{-}R$

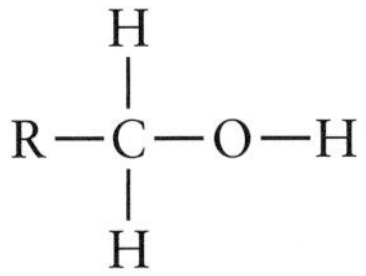

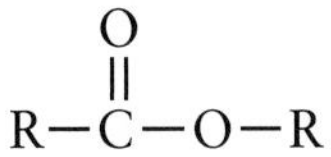

(continued)

Table 7-2 The Common Functional Groups (*continued*)

Functional Group Name	Structural Formula	Functional Group Name	Structural Formula
2° Alcohol	$R-\overset{\displaystyle R}{\underset{\displaystyle H}{C}}-O-H$	Thioester	$R-\overset{\displaystyle O}{C}-S-R$
3° Alcohol	$R-\overset{\displaystyle R}{\underset{\displaystyle R}{C}}-O-H$	Amide	$R-\overset{\displaystyle O}{C}-\overset{..}{N}-R \text{ or } H$, $R \text{ or } H$
Phenol	⬡—OH	1° amine	$R-\overset{..}{N}-H$, H
		2° amine	$R-\overset{..}{N}-H$, R
		3° amine	$R-\overset{..}{N}-R$, R
Ether	$R-O-R$	Phosphate ester	$^{-}O-\overset{\displaystyle O}{\underset{\displaystyle O^{-}}{P}}-O-R$

Chemistry in Medicine Parkinson's Disease and Schizophrenia: The Role of Dopamine

Actor Russell Crowe starred in the 2001 Academy Award winning movie, *A Beautiful Mind,* chronicling the life of Nobel Prize winning mathematician John Nash, who suffers from the debilitating mental disorder schizophrenia. Schizophrenia afflicts 1% of the population worldwide. Its symptoms can include hallucinations, delusions, disordered thinking, movement disorders, and social withdrawal. The disease usually strikes in late adolescence to early adulthood. The root cause of schizophrenia is still unknown, although the brain chemical dopamine is clearly involved. It is believed that certain symptoms of schizophrenia—hallucinations, delusions, and disordered thinking—are associated with an *excess* of dopamine in certain parts of the brain.

Parkinson's disease is somewhat better understood and is believed to arise from *decreased* dopamine activity in another part of the brain. Parkinson's disease is a chronic neurodegenerative disease characterized by a loss of coordination and movement. The disease tends to strike the elderly, usually after age 60. Symptoms of Parkinson's disease include tremor (shaking), bradykinesia (slow movement), rigidity (stiffness of limbs), and impaired balance. About 60,000 new cases of Parkinson's disease are diagnosed in the United States each year.

Russell Crowe starred in A Beautiful Mind, the Academy Award winning film chronicling the life of mathematician John Nash, who suffers from schizophrenia. [Everett Collection]

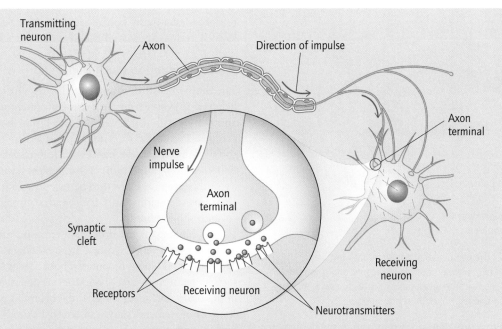

Figure 7-26 Neurons communicate through chemicals called neurotransmitters. A nerve impulse travels along the axon to the axon's terminal, where it stimulates the release of neurotransmitter into the synaptic cleft. Neurotransmitters bind to receptors on the receiving neuron, initiating a new nerve impulse.

Schizophrenia and Parkinson's disease are two very different diseases, but both involve an imbalance of dopamine in the brain. Although there is no cure for these diseases, prescription drugs are available to treat some of the more debilitating symptoms. By considering the chemistry involved in these diseases, you will gain a better understanding of the disease process, and you will also learn something about drug action as we look at the rationale used to develop drugs for the treatment of the symptoms of these two diseases. First, consider how nerve cells communicate on a molecular level.

Communication Between Neurons

Your brain has about one hundred billion (1×10^{11}) nerve cells, known as **neurons.** Most neurons communicate through chemical messengers known as **neurotransmitters:** These are organic molecules that often contain an amine functional group. The neurotransmitter dopamine, for example, has the relatively simple chemical structure shown below.

HO

HO—[ring]—C—C—N—H (H H H, Amine)

Dopamine

Communication between neurons begins with an electrical impulse that travels along the **axon,** the long shaft of a nerve cell.

The arrival of the impulse at the end of the axon causes neurotransmitters to be released into the **synaptic cleft,** the gap between two neurons, as shown in the inset in Figure 7-26. Once released, the neurotransmitters diffuse to the next neuron and bind to specific receptors on the cell membrane, setting off an impulse in the receiving neuron. Impulses travel from neuron to neuron, creating what is known as a **neuronal pathway.** The transmission of these impulses through a neuronal pathway ultimately leads to a physiological response. There are two neuronal pathways in the brain that use dopamine as the neurotransmitter, highlighted in Figure 7-27.

Parkinson's Disease: Too Little Dopamine

The impaired movement associated with Parkinson's disease is believed to result from the loss of dopamine-producing neurons along one of the two neuronal pathways in the brain that rely on dopamine, highlighted in red in Figure 7-27. Neurons differ from most other cells in the body in that they do not regenerate themselves, so the loss of these cells is permanent. Patients with Parkinson's disease show 80% less dopamine activity in this neuronal pathway.

Effective prescription drugs are available to treat many of the symptoms of Parkinson's disease. Dopamine itself cannot be administered as a drug, because it cannot pass through the blood–brain barrier. Instead, the drug L-dopa is administered. Notice from the structure of L-dopa on the next page that the only difference between the chemical structures of L-dopa and dopamine is the presence of a carboxylic acid functional group in

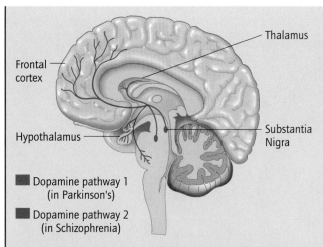

Frontal cortex

Thalamus

Hypothalamus

Substantia Nigra

■ Dopamine pathway 1 (in Parkinson's)

■ Dopamine pathway 2 (in Schizophrenia)

Figure 7-27 The two neuronal pathways that involve dopamine. The red pathway controls movement and is involved in Parkinson's disease. The green pathway affects desire, memory, and motivation and is involved in schizophrenia.

L-dopa. This functional group is removed in the neuron in a chemical reaction that converts L-dopa into dopamine:

L-Dopa

Carboxylic acid removed ⟶

Dopamine

The prolonged use of L-dopa eventually reduces its ability to activate the dopamine receptor. At that point other drugs must be employed. Unfortunately, dopamine receptors eventually become less sensitive to these drugs as well and symptoms of the disease return.

Schizophrenia: Too Much Dopamine

The thought disorders associated with schizophrenia are believed to result from an excess of dopamine activity in another neuronal pathway in the brain, highlighted in green in Figure 7-27. Stimulation of this particular neuronal pathway is associated with feelings of reward and desire, memory, and motivation. Excess dopamine activity in this area is believed to be responsible for the hallucinations, delusions, and disorganized thought associated with schizophrenia.

By binding to the dopamine receptor, **antipsychotic** medications prevent dopamine from binding to its receptor. The first generation of antipsychotic drugs, compounds such as chlorpromazine, bound to all dopamine receptors in the brain, including those in the neuronal pathway that controls movement (shown in red in Figure 7-27). Consequently, some of the side effects of these drugs were tremors and other symptoms like those from Parkinson's disease. In the early 1990s a new class of antipsychotic medications became available, known as *atypical antipsychotics*. These newer medicines also block dopamine receptors, but they specifically target a type of dopamine receptor that is found primarily in the neuronal pathway associated with schizophrenia (shown in green). Therefore, atypical antipsychotic drugs have fewer movement related side effects. The newest atypical antipsychotics have even fewer side effects.

In schizophrenia and Parkinson's disease there is too much or too little of the brain chemical dopamine, respectively, leading to distinctively different diseases. There have been significant advances in the treatment of these two diseases, as scientists have unraveled the mechanism of the disease on the molecular level. Further research should lead to even better treatment options for these patients in the future.

Chapter Summary

C—O-Containing Functional Groups: Alcohols and Ethers

- An alcohol functional group contains an OH group bonded to a saturated carbon atom, ROH.
- Alcohols are classified as 1°, 2°, and 3° alcohols, depending on whether the carbon atom to which the OH group is attached has one, two, or three R groups bonded to it.
- An ether functional group contains two R groups bonded to an oxygen atom: ROR.
- When naming an alcohol, change the suffix from *-ane* to *-anol* and indicate the location of the alcohol along the main chain by a number, placed before the root name.

- A phenol functional group contains an OH group bonded to an aromatic ring.
- Alcohols are able to form hydrogen bonds.
- The more OH groups there are in a molecule, the more water soluble it is. The more extensive the hydrocarbon portion of a molecule is, the less water soluble it is.

C=O-Containing Functional Groups

- A carbonyl group is a carbon atom and an oxygen atom connected by a double bond.
- Carbonyl-containing functional groups include aldehydes, ketones, carboxylic acids, esters, thioesters, and amides.
- A carbonyl group is polar.
- An aldehyde contains a carbonyl group attached to a hydrogen atom and an R group: RCHO. The one exception is formaldehyde, which is attached to two hydrogen atoms.
- To name an aldehyde, change the suffix from *-ane* to *-anal*.
- A ketone contains a carbonyl group attached to two R groups: RCOR. The R group must contain one or more carbon atoms.
- To name a ketone, change the suffix from *-ane* to *-anone* and insert a locator number in front of the root to indicate the location of the carbonyl group.
- Carboxylic acids contain a carbonyl group directly attached to an OH group.
- To name a carboxylic acid, change the suffix from *-ane* to *-anoic acid*.
- A carboxylic acid can exist in both a neutral form and an ionic form. The ionic form is much more water soluble than the neutral form.
- Both oxygen atoms in carboxylic acids are able to form hydrogen bonds.
- An ester contains an OR group attached to the carbonyl carbon: RCO_2R'.
- A thioester contains an SR group attached to the carbonyl carbon: RCOSR'.
- An amide functional group contains NH_2, NHR, or NR_2 directly attached to the carbonyl carbon.

C—N-Containing Functional Groups: Amines

- Amines are functional groups that contain a nitrogen atom with three single bonds, each attached to either an R group or a hydrogen atom, but not a carbon–oxygen double bond.
- Amines with one or two N—H bonds are capable of hydrogen bonding.
- An amine can exist in both a neutral and an ionic form. The ionic form is much more water soluble.
- To name an amine, identify the longest R group to define the root, and change the suffix to *-anamine*. If there are other R groups on the nitrogen atom, identify them as you would a substituent using the prefix *N-* to indicate that the substituent is located on the nitrogen atom rather than the main chain.

P=O-Containing Functional Groups

- A phosphate ester is produced when one or more carbon atoms replace the hydrogen atoms in phosphoric acid.
- The phosphate group of a phosphate ester can be connected to additional phosphate groups, P_i's, via phosphoanhydride bonds.

Key Words

Alcohol A functional group having the form R—OH.

Aldehyde A carbonyl group attached to a hydrogen and an R group. In the case of formaldehyde, the R group is also hydrogen.

Amide A carbonyl group attached to an R group and NH_2, NHR, or NR_2.

Amines Functional groups that contain a nitrogen atom with three single bonds to H or R, not attached to a carbonyl group.

Analgesic A pain medication.

Carbonyl group A carbon–oxygen double bond, C=O.

Carboxylic acid A carbonyl group attached to an OH group and an R group.

Ester A carbonyl group attached to an OR′ group and an R group (or H atom).

Ether A functional group that contains R—O—R′.

Functional group Groups of atoms containing O, N, or P covalently bonded to carbon, as well as the multiple bonds in hydrocarbons that react in a characteristic way in chemical reactions.

Heteroatom An atom in an organic compound that is not carbon or hydrogen.

Inorganic phosphate Monohydrogen phosphate ion, HPO_4^{2-}, abbreviated P_i. The most abundant ion of phosphoric acid in the cell.

Ketone A carbonyl group attached to two R groups.

Phenol An OH group attached to an aromatic ring.

Phosphate ester The structure formed by replacing one or more of the hydrogen atoms on phosphoric acid with an R group.

Thioester A carbonyl group attached to an SR′ group and an R group.

Additional Exercises

C—O-Containing Functional Groups: Alcohols and Ethers

7.31 Name the following alcohols and indicate whether they are 1°, 2°, or 3° alcohols. Rewrite (a) and (b) as condensed structural formulas:

a.

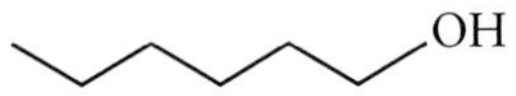

b.

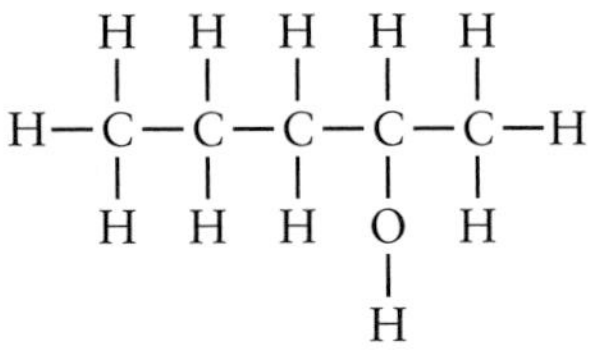

c.

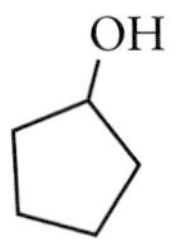

d.

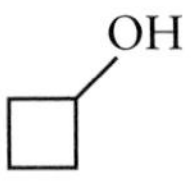

e.

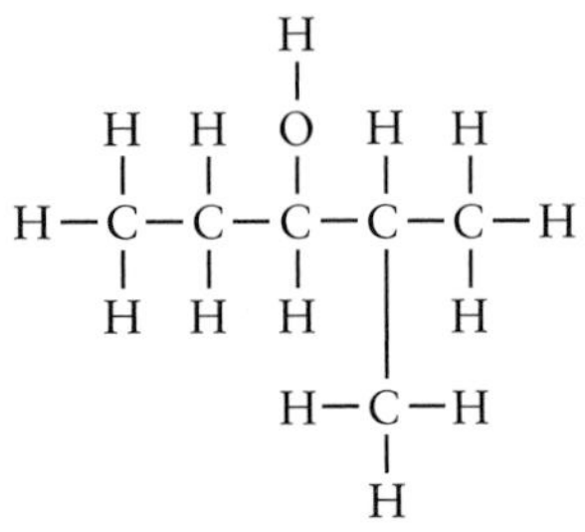

7.32 Write the skeletal line structures for the following alcohols:
   a. 1-nonanol
   b. 3-methyl-3-heptanol
   c. cyclopropanol
   d. 3-ethyl-2-methyl-2-pentanol
   e. 2-butanol

7.33 Estradiol is a sex hormone and betamethasone is a drug often given to pregnant women to help mature the lungs of a fetus in cases where the child is expected to be born prematurely, as in the case of twins and triplets. The structures for both hormones are shown below. Circle the alcohol and phenol functional groups.

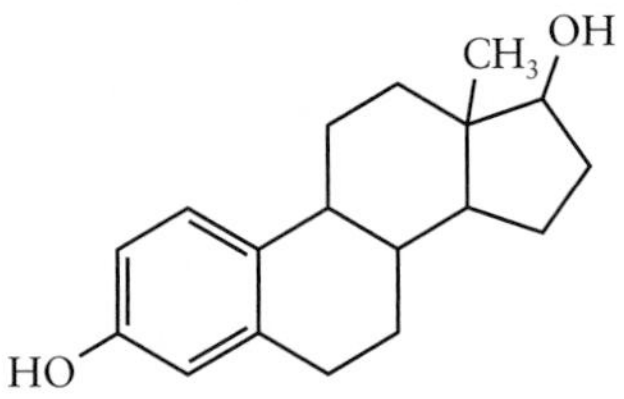

Estradiol

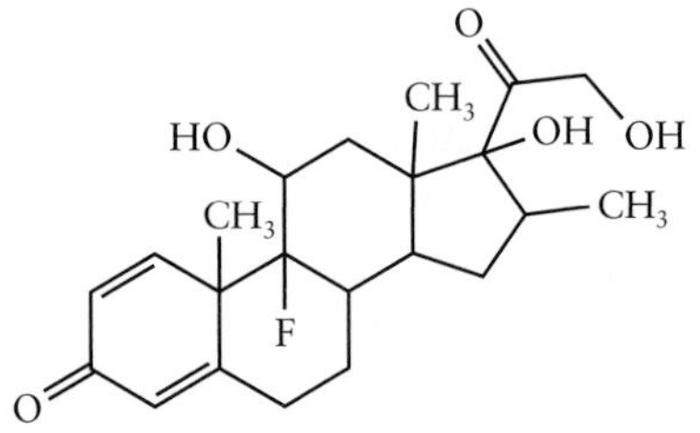

Betamethasone

7.34 How does hydrogen bonding influence whether an alcohol is soluble in water?

7.35 Which of the following compounds do you expect to be more soluble in water? Explain your answer.

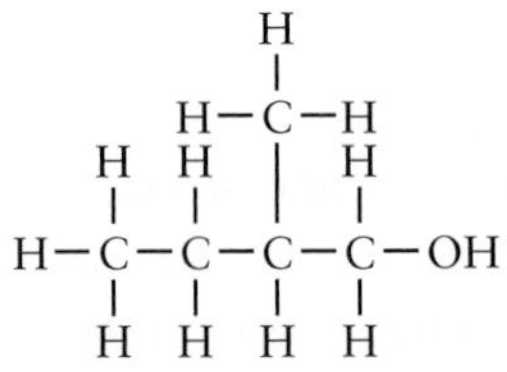

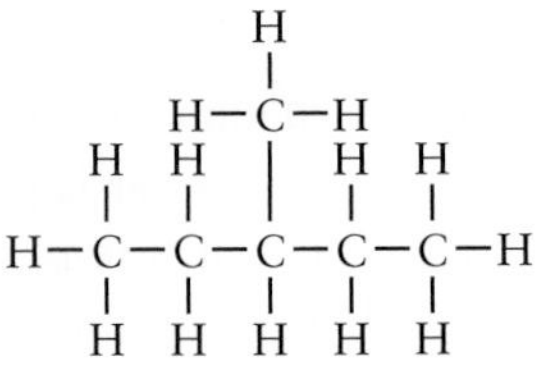

7.36 Arrange the following compounds in order of increasing boiling point.

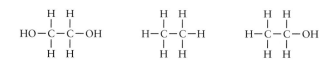

7.37 What is the structural difference between an ether and an alcohol?

7.38 Are simple ethers generally soluble in water? Explain why or why not.

7.39 Imdur (isosorbide mononitrate), shown below, is used to treat angina. It dilates the blood vessels to reduce blood pressure.

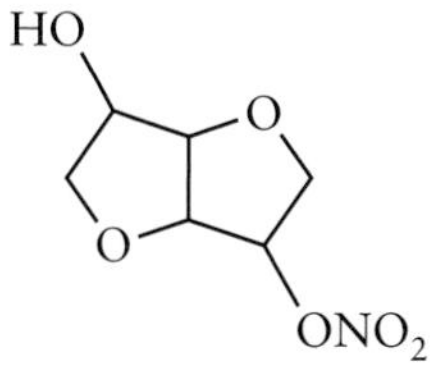

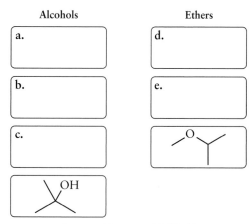

a. Circle and identify the ether functional groups.
b. Circle and identify the alcohol functional group.

7.40 Critical Thinking Question: Complete the boxes (a)–(e) to depict all the structural isomers with the formula $C_4H_{10}O$. *Hint:* Four of the isomers are alcohols and three are ethers, and none contain multiple bonds.

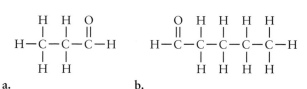

Structural isomers of $C_4H_{10}O$

Aldehydes and Ketones

7.41 Name the following aldehydes and ketones.

a. H H O
   | | ||
   H—C—C—C—H
   | |
   H H

b. O H H H H
   || | | | |
   H—C—C—C—C—C—H
   | | | |
   H H H H

c. (structure shown)

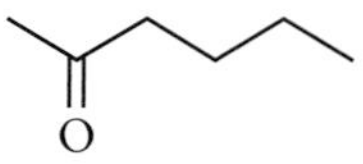

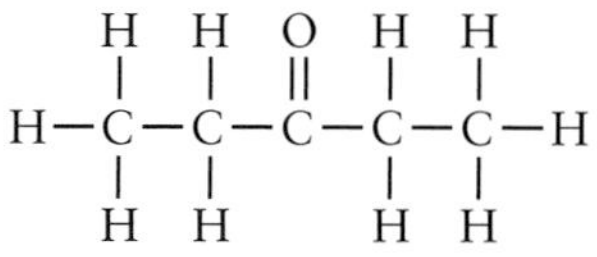

d. H H O H H
   | | || | |
   H—C—C—C—C—C—H
   | | | |
   H H H H

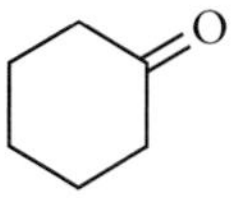

e. (structure shown)

7.42 Write the condensed structural formula for the compounds (a)–(d) in Exercise 7.41.

7.43 TriCor (fenofibrate), shown below, is used to lower cholesterol levels in patients at high risk of developing cardiovascular disease. It is a drug that lowers low-density lipoprotein (LDL) levels and very-low-density lipoprotein (VLDL) levels, while increasing high-density lipoprotein (HDL) levels.

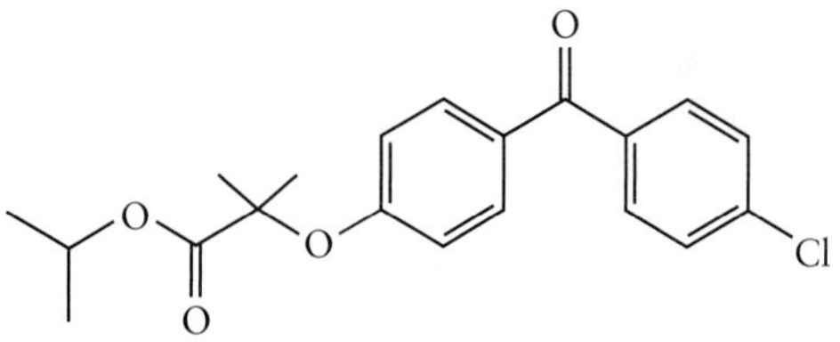

(structure shown)

a. Circle and identify the ketone functional group.
b. What are the R groups attached to the carbonyl carbon of the ketone?
c. Circle and identify the ether and the aromatic ring functional groups

7.44 Cinnamaldehyde, shown below, occurs naturally in the bark of cinnamon trees and gives cinnamon its flavoring.

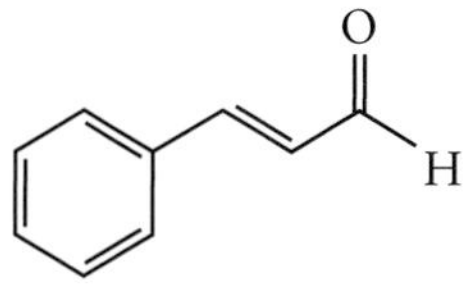

(structure shown)

a. Circle and identify the aldehyde functional group.
b. Circle and identify the double bond. Is the double bond cis or trans?

Carboxylic Acids and Their Derivatives

7.45 State the difference between the following functional groups:
a. a carboxylic acid and an ester
b. a carboxylic acid and a thioester
c. an ester and a thioester
d. a carboxylic acid and an amide
e. an amide and an amine
f. a carboxylic acid and an alcohol

7.46 Zyrtec (cetirizine) is an antihistamine used as an allergy medicine.

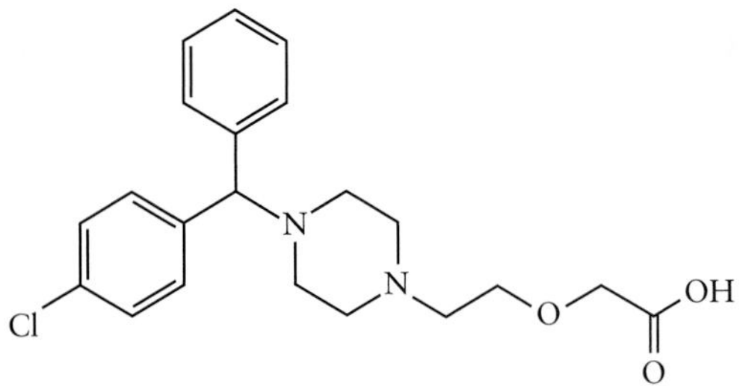

a. Circle and identify the carboxylic acid functional group.
b. Is the carboxylic acid in its neutral or ionic form?
c. Circle and identify the amine functional groups in the molecule. Are they in their neutral or ionic form?
d. What other functional groups are present in Zyrtec?

7.47 Tartaric acid, shown below, is one of the main carboxylic acids found in wine.

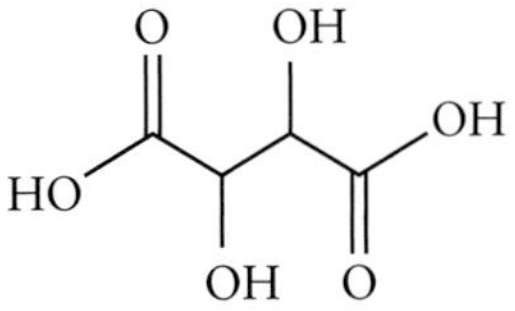

a. Circle and identify the carboxylic acid functional groups in tartaric acid.
b. What other functional group is present in tartaric acid?
c. Would you expect this molecule to be water soluble? Explain.

Esters and Thioesters

7.48 Vioxx, a nonsteroidal anti-inflammatory drug (NSAID), was recently taken off the market by the pharmaceutical company Merck because of concerns that it increases the risk of heart attacks and strokes.

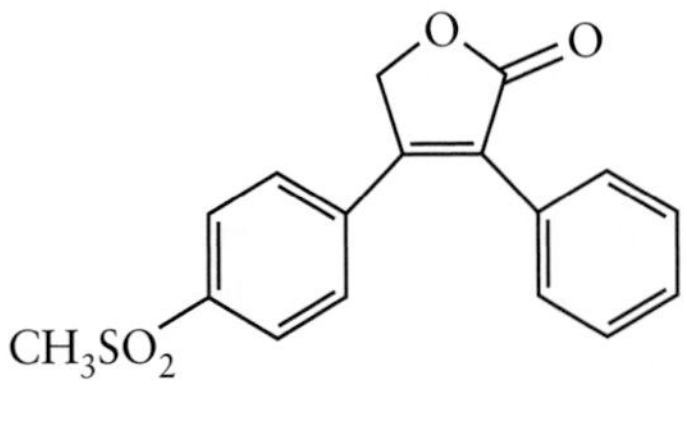

Circle and identify the ester in Vioxx. Would you expect Vioxx to be water soluble? Why or why not?

7.49 Novocain, shown below, is commonly used as a local anesthetic, particularly in dentistry.

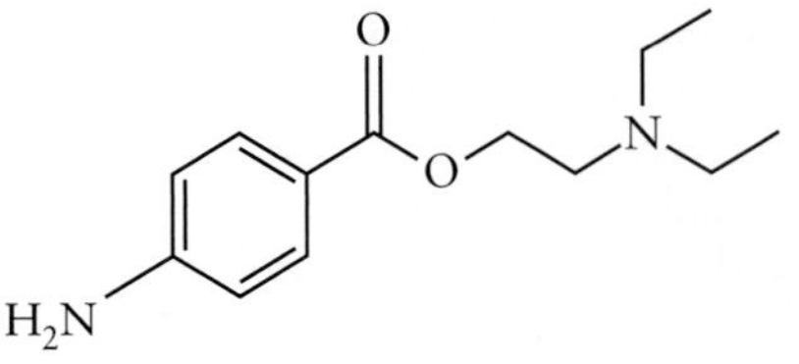

a. Circle and identify all the functional groups in Novocain.
b. Identify whether the amines are in their *neutral* or *ionic* form.
c. How many R groups are on each amine?

7.50 Fatty acids such as palmitic acid are transported into and out of mitochondria as thioesters of coenzyme A. Mitochondria are the "power plants" of cells. The thioester derived from palmitic acid is shown below.

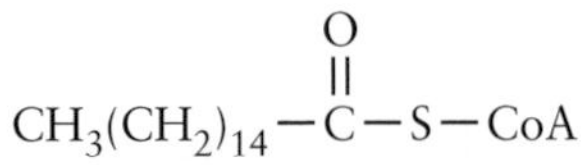

$$CH_3(CH_2)_{14}-\overset{\displaystyle O}{\overset{\|}{C}}-S-CoA$$

a. Write the skeletal line structure of palmitic acid.
b. Circle and identify the thioester.

7.51 Succinyl-CoA, shown below, is an intermediate in the citric acid cycle.

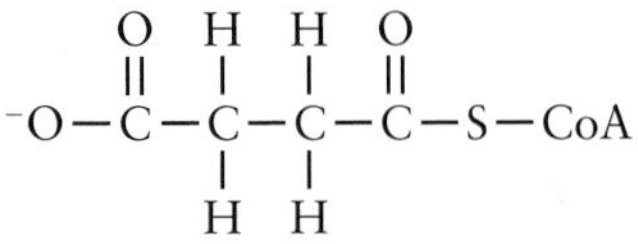

$$^-O-\overset{O}{\overset{\|}{C}}-\overset{H}{\overset{|}{\underset{H}{C}}}-\overset{H}{\overset{|}{\underset{H}{C}}}-\overset{O}{\overset{\|}{C}}-S-CoA$$

a. Circle and identify the thioester.
b. Circle and identify the other functional group in succinyl-CoA. In what form is this other functional group?

Amides

7.52 Write the condensed structural formulas for the following amides:

a.

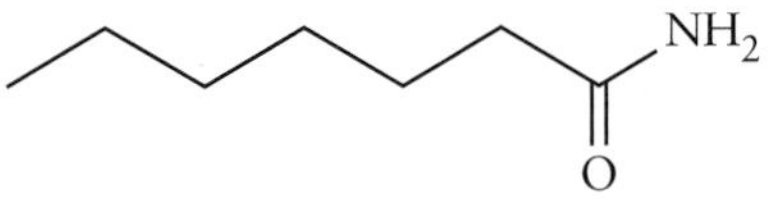

b.

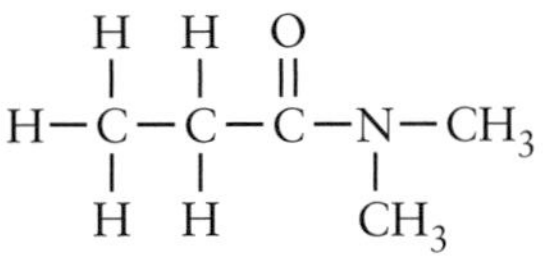

c.

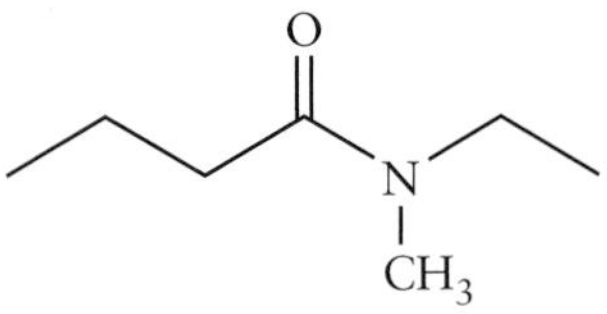

7.53 Ambien (zolpidem) is used as a sleeping pill. It is fast acting (15 minutes) and eliminated from the body quickly. The structure is shown below. Circle and identify the amide functional group. What are the R groups on the amide?

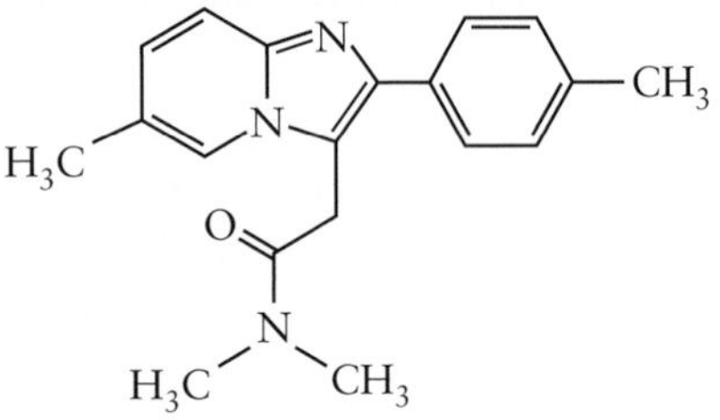

7.54 The structure of Viagra (sildenafil) is shown below. Circle and identify the following functional groups: the amide, the ether, and the aromatic ring.

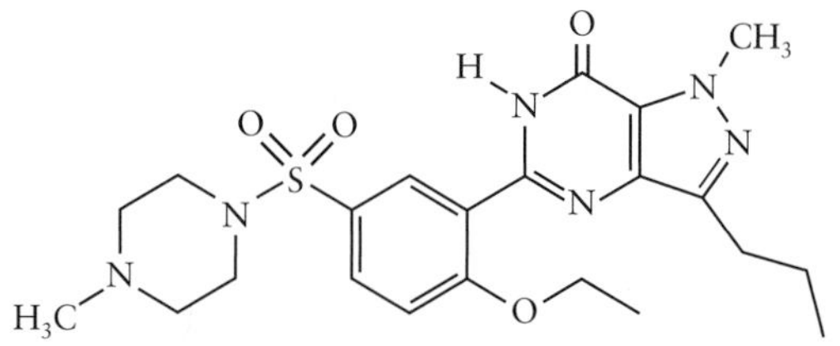

C—N-Containing Functional Groups: Amines

7.55 Benadryl, shown below, is an over-the-counter antihistamine.

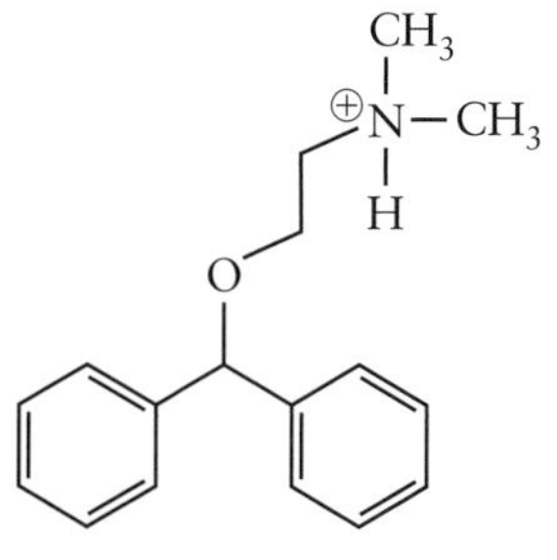

a. Circle and identify the amine functional group.
b. Is the amine in Benadryl in its *neutral* or *ionic* form?
c. What other functional groups are present in Benadryl?

7.56 Tamoxifen is a drug used to treat breast cancer.

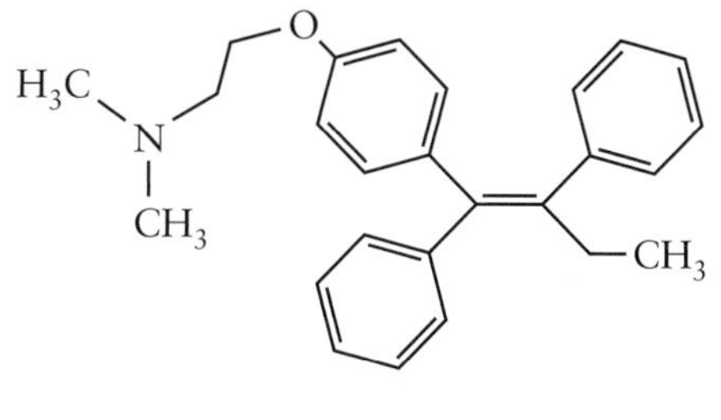

a. Circle and identify the amine functional group.
b. How many R groups are attached to the nitrogen atom?
c. Is the amine in its *neutral* or *ionic* form?
d. What other functional groups does Tamoxifen contain?
e. Would you expect Tamoxifen to be water soluble? Explain.

Phosphate Esters and ATP

7.57 Write a Lewis dot structure for the most common form of phosphoric acid in the body. What is this ion called? What is its abbreviation?

7.58 Identify the structures below as mono-, di-, or triphosphate esters. Place a box around each phosphate group. What other common functional group do you recognize in these molecules?

a.

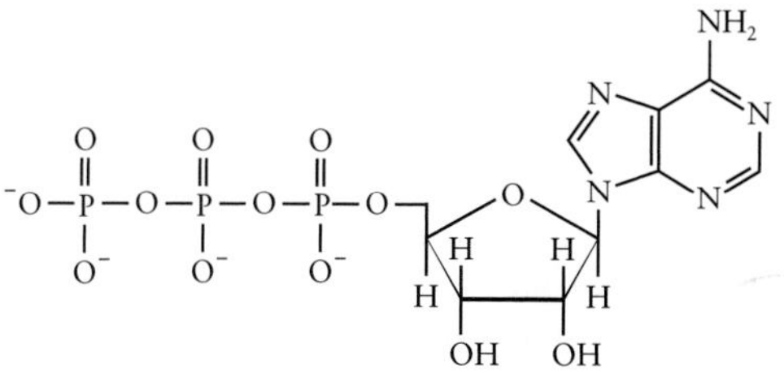

b.

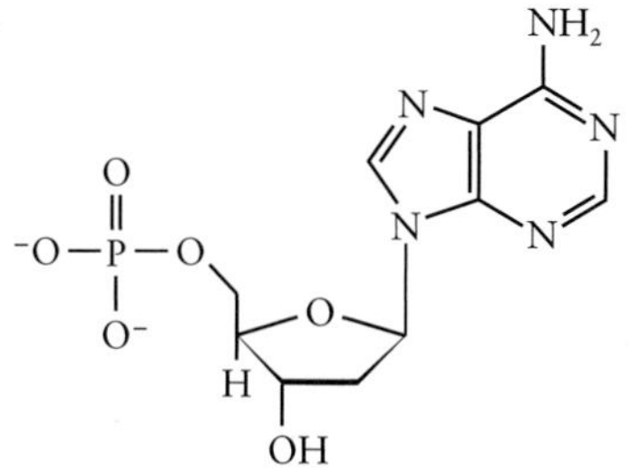

Chemistry in Medicine

7.59 How is dopamine activity different in the brain of someone who has schizophrenia compared to someone who has Parkinson's disease?

7.60 Write the structure of dopamine and circle and label all its functional groups.

7.61 Neurons use chemical messengers called _____ to communicate within the brain. The main functional group found in these chemical messengers is an _____.

7.62 What is a neuron? Briefly describe how neurons communicate with one another.

7.63 Explain why dopamine cannot be administered as a drug.

7.64 How are antipsychotic drugs able to reduce the symptoms of schizophrenia (i.e., what do they do chemically)?

7.65 What functional group is removed from L-dopa when it is converted into dopamine?

7.66 What drugs are selective for the dopamine receptors in the neuronal pathway controlling reward and desire, memory, and motivation and not in the neuronal pathway controlling movement?

7.67 Which disease involves a loss of dopamine-producing neurons along the neuronal pathway controlling movement?

7.68 Circle the functional groups in the atypical antipsychotic aripiprazole, shown below. What is "atypical" about these antipsychotics?

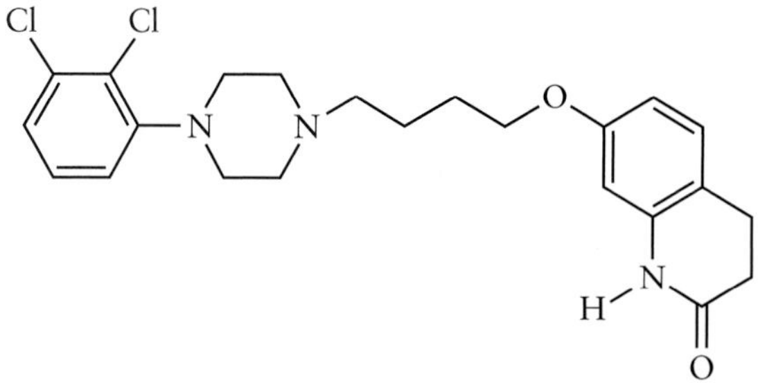

7.69 Critical Thinking Question: Taxol—also known as paclitaxel—was first isolated from the yew tree.

Taxol is used to treat different forms of cancer. This natural product has the complex structure shown below. Circle and label the following functional groups:

a. one amide
b. three alcohols
c. three aromatic hydrocarbon rings
d. one ether
e. one ketone
f. four esters
g. one double bond

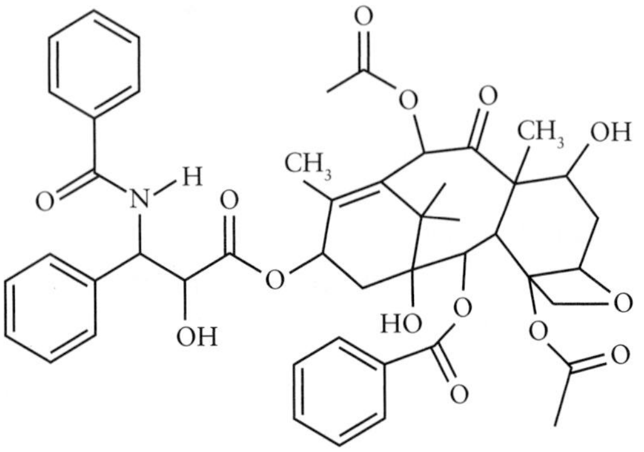

7.70 The structure of morphine is shown below.

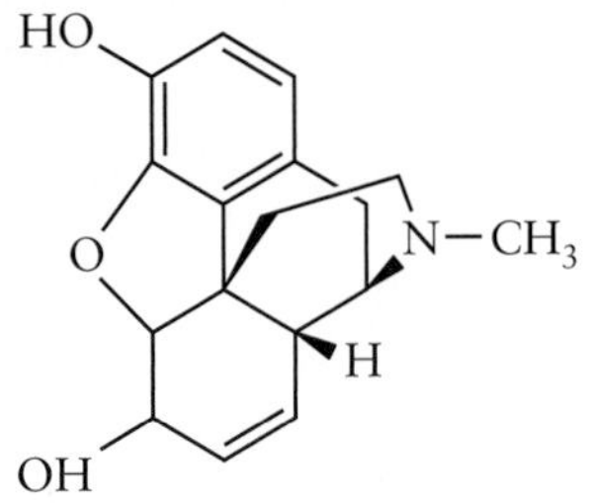

Morphine

a. Circle and label all the functional groups.
b. Codeine has the same structure as morphine, except that it contains a methyl (CH_3) ether where morphine has a phenol. Would you expect this change to impact its analgesic and addictive effects?
c. Vicodin is similar to codeine, except that it contains a ketone where codeine contains an alcohol. Write the structure of vicodin.
d. If both the alcohol and the phenol in morphine are replaced with methyl esters, heroin is the result. Which would you expect to be more soluble in water, heroin or morphine? Explain.

Answers to Practice Exercises

7.1 The molecular shape is bent around an alcohol and an ether. Therefore, the bond angles are less than 109.5°.

7.2

a. 4,4-dimethyl-2-hexanol; $CH_3CH_2CCH_2CHCH_3$; 2° alcohol
b. cyclohexanol; 2° alcohol

7.3 a.

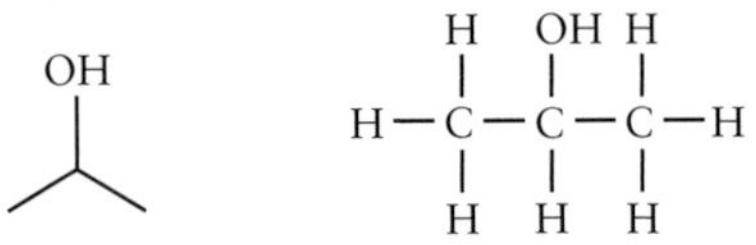

b.

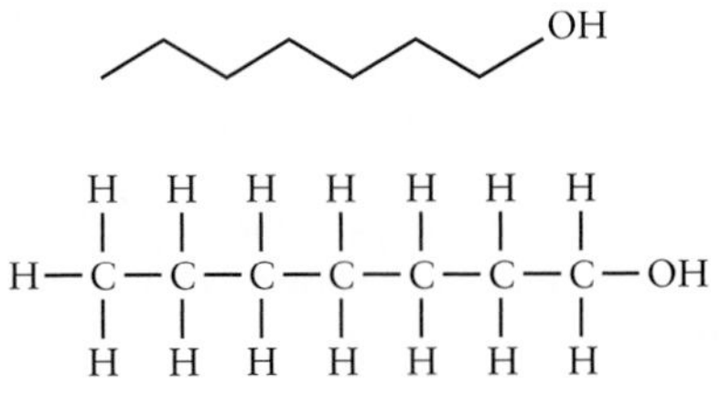

c.

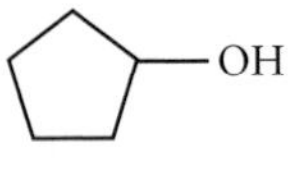

d.

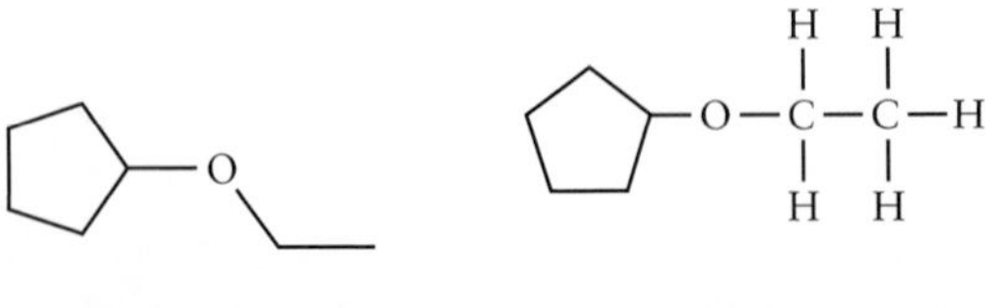

Structure (a) would be the most soluble in water, because it is the alcohol with the least amount of hydrocarbon per hydroxyl functional group.

7.4

a.

b.

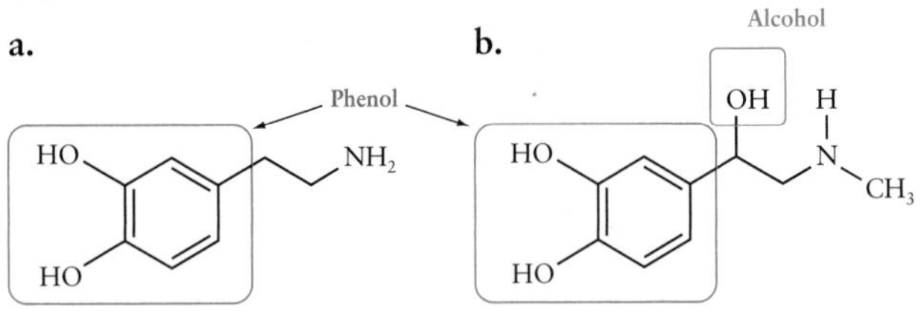

Dopamine   Epinephrine (adrenaline)

The two structures have the same phenol, and both have a nitrogen at the end of a two-carbon chain. Adrenaline also has a 2° extra alcohol and an extra CH_3 group on the nitrogen atom.

7.5

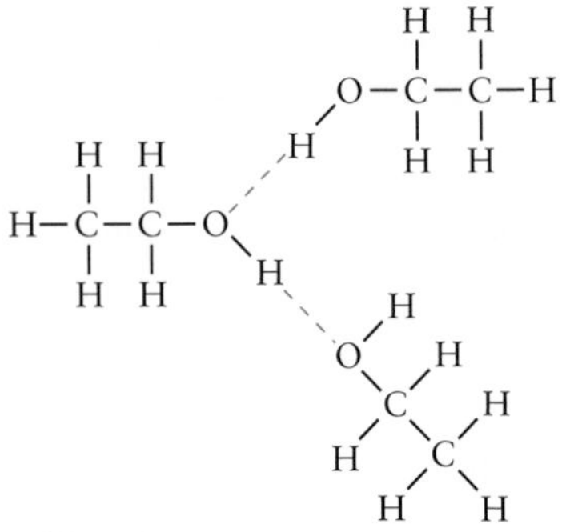

7.6 a. and b.

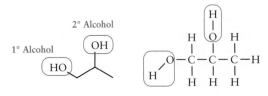

c. This compound is soluble in water because it has two polar OH groups capable of hydrogen bonding and only three carbon atoms.

7.7 a.

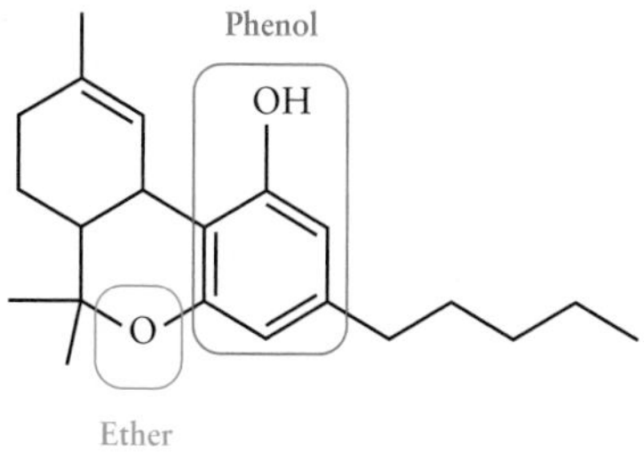

Δ⁹-THC

b. tetrahydrocannabinol, the ending *-inol* indicates an alcohol or phenol is present.

c. This compound is not water soluble because it has only one OH group compared to the much more extensive hydrocarbon structure, which is nonpolar and therefore insoluble in water.

7.8 a. 1-propanol, an alcohol **b.** 2-propanol, an alcohol
c. ethyl methyl ether, an ether

7.9 120°; trigonal planar molecular shape

7.10 propane (hydrocarbon) < propanal (aldehyde) <2-propanol (alcohol). Propane has the lowest boiling point because it is a hydrocarbon and interacts only through dispersion forces. Propanol has the highest boiling point because it is capable of hydrogen bonding interactions, which are the strongest intermolecular force of attraction. Propanal is an aldehyde with a permanent dipole, so its boiling point is higher than propane, but lower than propanol.

7.11 a. propanal **b.** 3-octanone **c.** cyclohexanone
d. butanal **e.** 2-propanone or acetone

7.12 The other functional groups are alcohols.

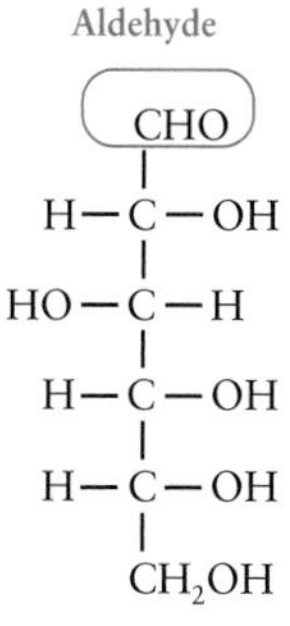

D-**Glucose**

7.13

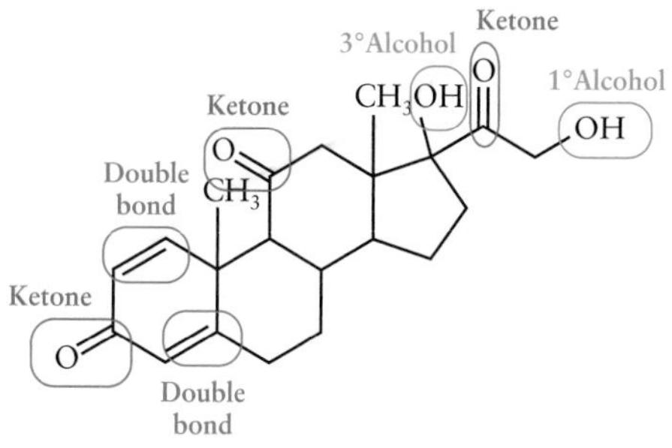

Prednisone

7.14 In a carboxylic acid, the OH group is attached to a carbonyl carbon, whereas in an alcohol it is attached to a carbon with three other single bonds.

7.15 a. 3-methylbutanoic acid **b.** pentanoic acid
c. butanoic acid

7.16 (a) and (e) because they contain O—H bonds.

7.17 a. heptanoic acid; **b.** heptanoate ion. The heptanoate ion, (b), is more water soluble because it is charged.

7.18 a tertiary alcohol

7.19 The word "fatty" refers to the extensive amount of nonpolar hydrocarbon structure—carbon and hydrogen atoms.

7.20 An ester would be less water soluble because it does not have an O—H bond, as a carboxylic acid does, and therefore cannot hydrogen bond to water molecules.

7.21

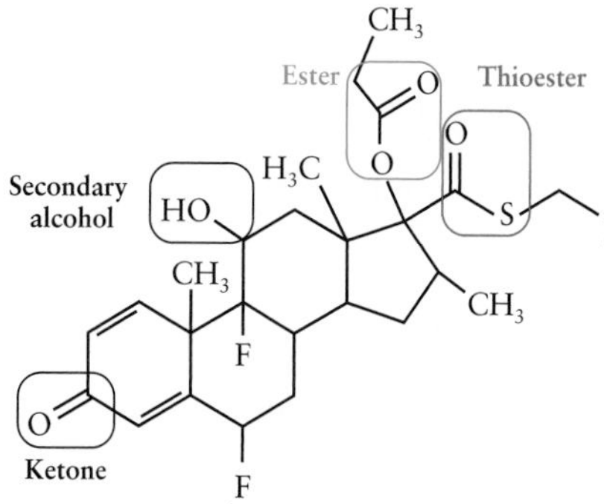

Fluticasone propionate
(Flonase nasal spray)

a.–b. A thioester has a sulfur atom where an ester has an oxygen atom. They both contain a carbonyl group directly attached to a sulfur or oxygen atom.

7.22 This is a triglyceride, or fat. This molecule is insoluble in water because of its significant amount of nonpolar hydrocarbon structures.

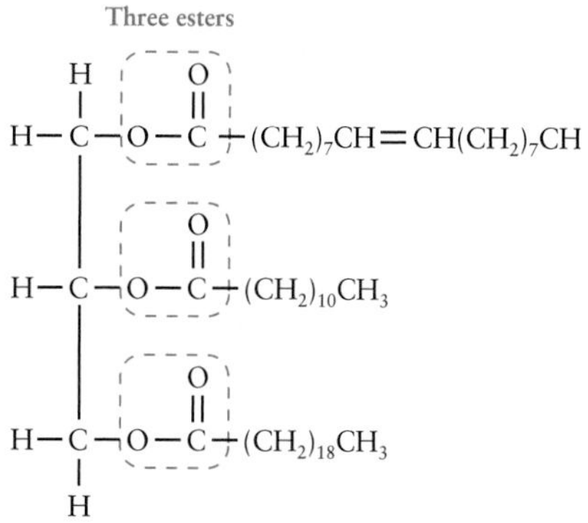

7.23 The compound on the left is an ester and the compound on the right is a thioester. There is a sulfur atom in the thioester, and an oxygen atom in the corresponding position in the ester. Both compounds have a carbonyl group and are otherwise the same.

E7.1 a.   **b.**

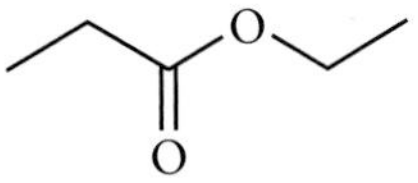

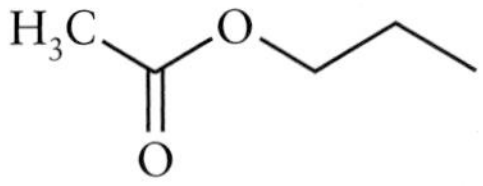

c. 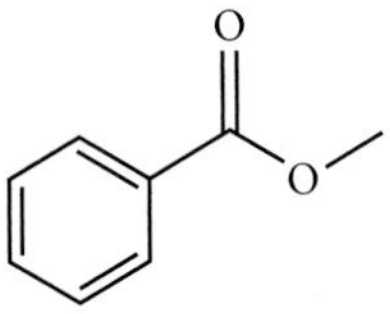   **d.**

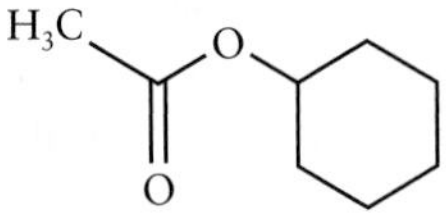

E7.2 a. methyl butanoate   **b.** propyl pentanoate
     c. methyl propanoate   **d.** isopropyl benzoate

7.24 a.

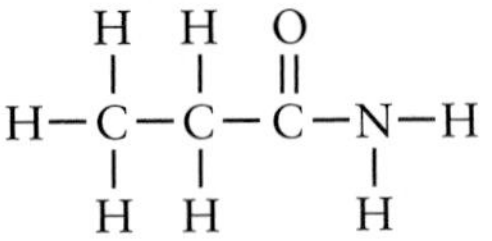

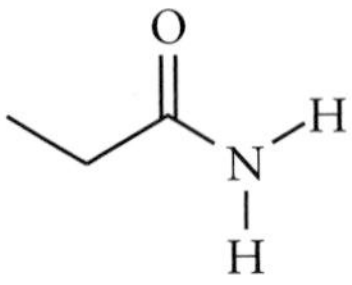

b.

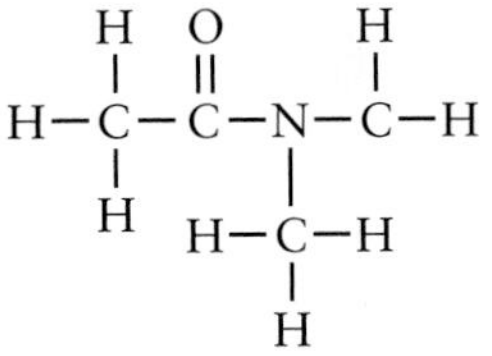

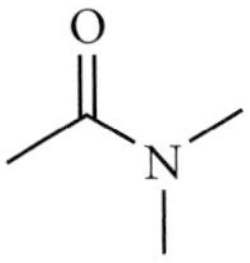

    c. This is an amine not an amide:

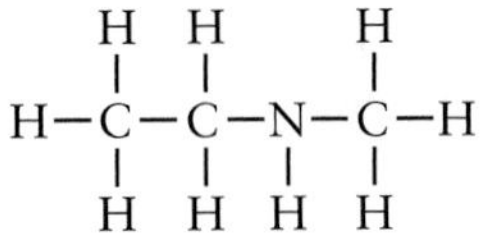

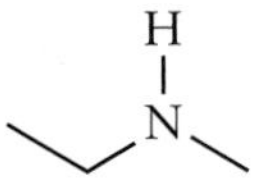

d.

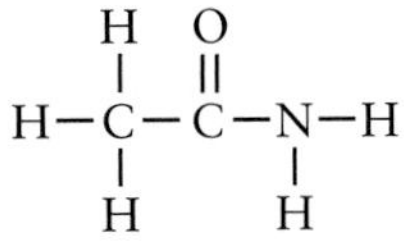

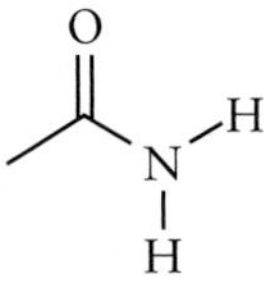

7.25 (d) is an amide.

7.26

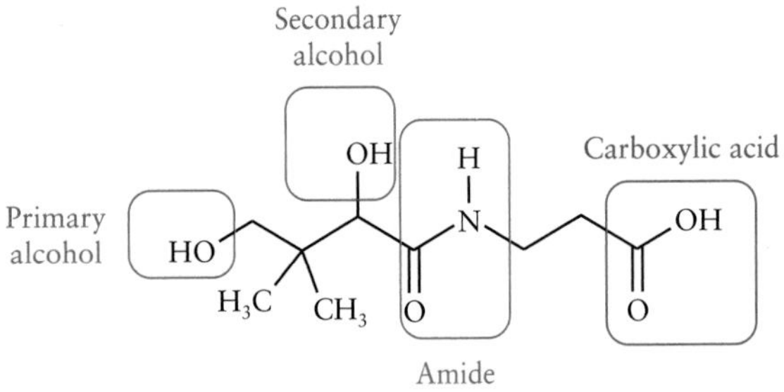

7.27

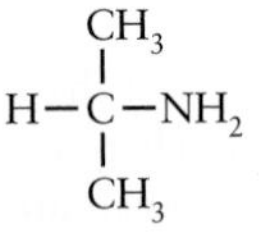

     2 - Propylamine 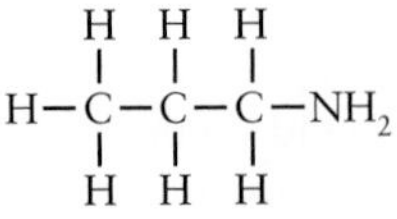   **1 - Propylamine**

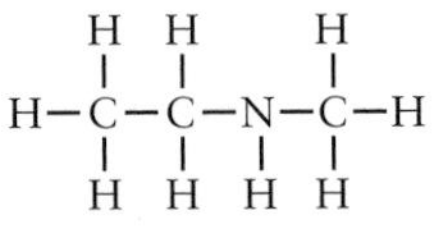

     Ethylmethylamine 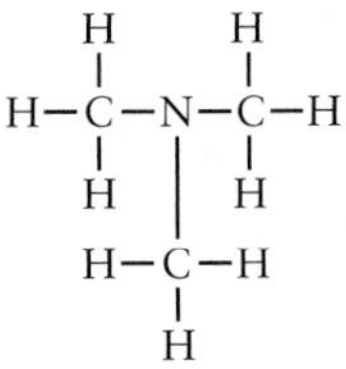   **Trimethylamine**

Trimethylamine has the lowest boiling point because there is no N—H group and thus no hydrogen bonding, which makes it easier to enter the gas phase.

7.28 a.

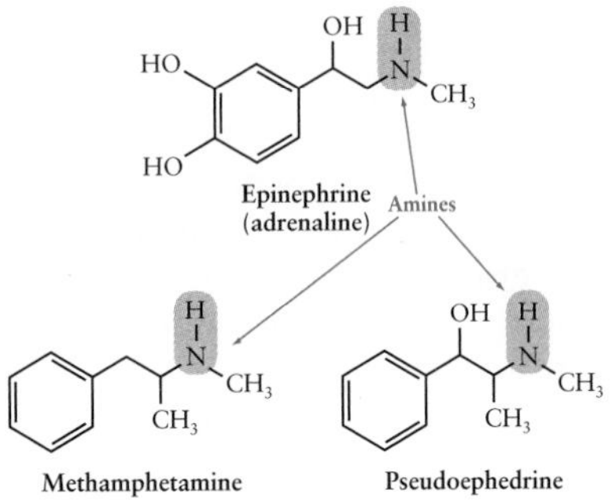

b. Methamphetamine is similar to pseudoephedrine in all ways except that it lacks an OH group on the carbon chain.

c.

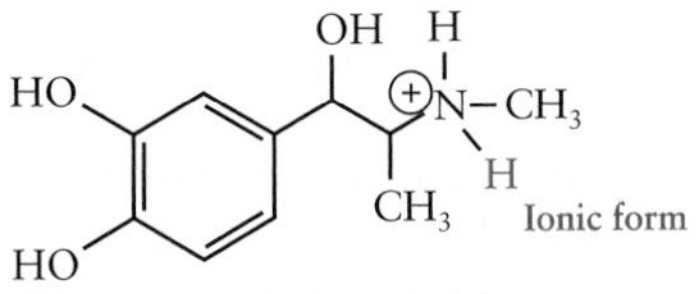

     Epinephrine
     (adrenaline)

d. Epinephrine would be the most soluble in the neutral form because it has the most OH groups on the molecule for the same number of carbon atoms.

     e. Approximately 109.5°or less

E7.3 a. *N*-methylethanamine
     b. *N*-ethyl-*N*-methyl-1-butanamine

7.29

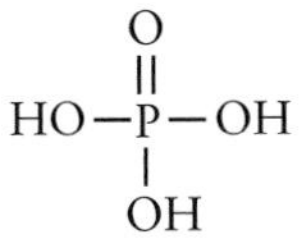

     Phosphoric acid 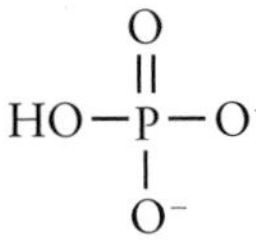   **Hydrogen phosphate ion,**
                            HPO_4^{2-} or P_i

7.30 a. It is related to phosphoric acid in that two of the hydrogen atoms have been replaced by R groups, a phosphate diester.

     b. It contains only one phosphate group

     c. Two R groups rather than one are attached to oxygen atoms of the phosphate.

     d. The total charge on this molecule is -1 due to the one O^- group on the phosphate.

     e. The two R groups of the phosphate ester are joined together, making it part of a ring—cyclic.

CHAPTER 8

Chemical Reaction Basics

Glucose

The food we eat supplies the body with molecules such as glucose—a source of energy for all cells. [Photographer's Choice/Punchstock]

OUTLINE

 This icon indicates that a **Problem-Solving Tutorial** is available at www.whfreeman.com/gob

275

Energy and Malnutrition

The world currently produces enough food to provide each person on the planet with 2720 Calories of energy per day, more than enough for the average person. Yet, the World Health Organization estimates that one-third of the world is well fed, one-third of the world is underfed, and one-third of the world is starving. From the standpoint of energy, there should be no shortage of food, but complex political, social, and cultural factors have made starvation in our modern world a very real problem.

Food, energy, and chemical reactions are all interconnected. *Energy is defined as the ability to do work or transfer heat.* Your body does work as it performs basic involuntary physiological functions, such as breathing, keeping the heart beating, repairing damaged cells, and so forth. Energy obtained from food allows this cellular work to occur, as well as outwardly visible tasks such as walking, talking, and studying.

Food is composed of three types of biological molecules: carbohydrates, proteins, and triglycerides (fats). Digestion of carbohydrates, for example, produces glucose, also known as blood sugar. When glucose enters a cell, its chemical potential energy is made available to the cell through a series of chemical reactions known as a biochemical pathway. This particular biochemical pathway, the metabolism of glucose, requires oxygen and produces carbon dioxide, water, and energy. In this chapter you will learn how to express a chemical reaction as an equation, such as the one shown below for the net reaction of glucose with oxygen.

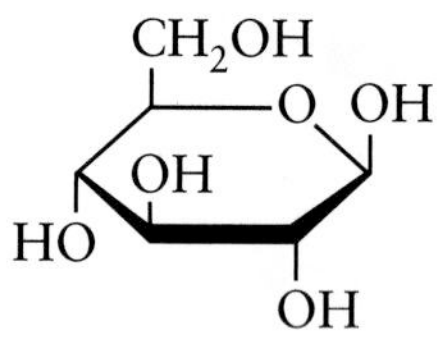

$$C_6H_{12}O_6(aq) + 6\ O_2(g) \longrightarrow 6\ CO_2(g) + 6\ H_2O(l) + \text{energy}$$

    Glucose   **oxygen**   **carbon dioxide**   **water**

In the cell, oxygen comes from the air you breathe and most of the glucose comes from the carbohydrates you eat. Most of the carbon dioxide is exhaled, and water either becomes a part of the fluid systems of your body or is eliminated.

The most important product of the reaction of glucose with oxygen is energy. Indeed, you are probably already familiar with interpreting Calories on food nutrition labels. The Calorie is a unit for measuring the amount of energy produced when a particular food reacts completely with oxygen. In order to maintain his or her weight, a person must consume enough Calories. Although fewer than 2.5% of Americans suffer from malnutrition, health care professionals can and do encounter underfeeding in the United States. The elderly, the poor, and those suffering from anorexia, bulimia, or some mental illnesses are especially susceptible to malnutrition.

Starvation occurs when the body does not take in enough Calories to maintain basic involuntary physiological functions. Because glucose is the sole source of energy for the brain, without a supply of glucose, hypoglycemic coma occurs. Therefore, when the body is starving, it begins to convert muscle and organ tissue into glucose, which eventually damages the heart muscle. Ironically, the process of converting muscle into glucose consumes as much energy as it produces. Consequently, there is no net gain in energy, and the body withers away trying to keep itself alive. Starvation for one or more months ultimately results in death.

While starvation does remain a world crisis, the number of overweight people in the world has actually overtaken the number of malnourished—there are 1 billion overweight people compared to 800 million undernourished. A 2006 study done at the University of North Carolina found that this transition from a starving world to an obese world is actually accelerating. Nutrition management is an important aspect of health care today. In this chapter, you have the opportunity to understand further, on a molecular level, how your body maintains life by converting food into energy. You will see how matter, energy, and life are all interconnected.

Glucose

One Calorie (capital C) = 1000 calories (lower-case c).

Nutritional labels list the energy content of food in units of Calories. [Tetra images/ Punchstock]

You have been introduced to the enormous variety of compounds that exist, particularly among organic compounds. Both elements and compounds have characteristic chemical properties—the tendency to react with certain substances in what is known as a **chemical reaction.** *In a chemical reaction, bonds break and new bonds are formed, yielding new chemical substances.* You witness chemical reactions every day, whether they are rust forming on a piece of metal exposed to air, food being transformed as it cooks, or burning wood seeming to disappear in a fireplace. In this chapter you will study what happens to molecules and elements as they undergo chemical reactions.

Many important chemical reactions take place in the aqueous environment of the cell. The biological molecules required to sustain life are continually undergoing chemical reactions. The chemical reactions that occur in a living cell are collectively known as metabolic reactions, or **metabolism.** These reactions make it possible for us to walk, talk, breathe, grow, and reproduce.

Intimately connected with chemical reactions is the transfer of energy. Although energy exists in many different forms, in this chapter you will focus on *chemical energy*—the potential energy of chemical bonds. You will learn that when chemical bonds are broken and formed in a chemical reaction, energy is transferred to and from the surroundings. Finally, you will consider the rates of chemical reactions and the factors that influence these rates. To understand the relationship between energy and chemical reactions, as well as reaction rates, you must first learn some of the basic principles of a chemical reaction.

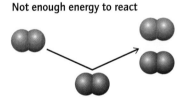

A chemical reaction: The clear, colorless liquids potassium iodide and lead nitrate react to produce the yellow solid lead iodide. [Richard Megna/Fundamental Photographs]

8.1 | Writing and Balancing a Chemical Equation

How do molecules undergo a chemical reaction? Recall from Chapter 4 that the kinetic molecular view of molecules paints a picture of atoms and molecules in constant motion, colliding with the walls of their container and occasionally colliding with one another. Recall also that temperature is a measure of the average kinetic energy of these particles, which is greatest in the gas phase. When the kinetic energy of two colliding molecules or atoms is sufficient, a chemical reaction can take place.

Kinetic Molecular View of a Chemical Reaction

Usually when two atoms or molecules collide they bounce off one another unchanged, except perhaps in the amount of kinetic energy they rebound with. Occasionally, however, a collision occurs with sufficient energy and in the proper orientation for a chemical bond to be broken or a new bond to be made or both, creating a new chemical substance. This change in chemical structure is known as a chemical reaction.

For example, consider a simple inorganic reaction: the reaction between one nitrogen molecule (N_2) and one oxygen molecule (O_2). In a mixture of these two gases at high temperature, an energetic collision between an N_2 molecule and an O_2 molecule results in a chemical reaction (Figure 8-1). By the end of the reaction, the bonds joining the two nitrogen atoms and the bonds joining the two oxygen atoms have been broken, and a multiple bond between one nitrogen atom and one oxygen atom has been formed, creating two new NO molecules.

Not enough energy to react

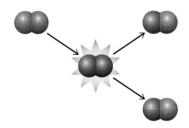

Enough energy to react

Figure 8-1 The collision of an O_2 molecule and an N_2 molecule results in a reaction only if the collision occurs with enough energy.

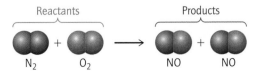

Reactants   Products

N_2 O_2 NO   NO

The reacting substance(s) are known as **reactants** and the new substance(s) formed are known as **products.** In the reaction above, nitrogen gas and oxygen gas are the reactants, and nitric oxide, NO, is the product. The reactants, nitrogen gas, N_2, and oxygen gas, O_2, are different chemical substances from the product, nitric oxide, NO.

In a chemical reaction, the same number and type of atoms appear in the product(s) as in the reactant(s), except that they are part of different substances. In the reaction above, two nitrogen atoms and two oxygen atoms are present in the reactants, and two nitrogen atoms and two oxygen atoms are likewise present in the products. The difference between reactants and products is how the atoms are joined. In the reactants, the two nitrogen atoms share electrons in the covalent triple bond of a nitrogen molecule ($N \equiv N$) and the two oxygen atoms share electrons in the covalent double bond of an oxygen molecule ($O = O$), whereas in each molecule of the product, an oxygen atom and a nitrogen atom share a covalent N—O multiple bond. The atoms involved in the reaction have been conserved: No new atoms have been added, and no atoms have been destroyed. *The conservation of atoms in a chemical reaction is a fundamental law of nature known as the **law of conservation of mass:** matter can be neither created nor destroyed.*

Consider a reaction that involves an organic molecule, methane (CH_4), and oxygen. This reaction produces the familiar blue flame seen on a gas burner. When a gas stove is turned on, methane gas (CH_4) flows out of the gas jets on the burner and reacts with the oxygen (O_2) in the air. A spark initiates the reaction. Each molecule of methane combines with two molecules of oxygen (O_2) to yield one carbon dioxide (CO_2) molecule and two water (H_2O) molecules.

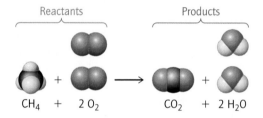

The same type and number of atoms are seen among the reactants as among the products: carbon (1), hydrogen (4), and oxygen (4), but they are joined differently. In the reactant methane, carbon has four single bonds to hydrogen, and in the products carbon has two double bonds to oxygen. In the reactants all hydrogen atoms are bonded to carbon, whereas in the products two hydrogen atoms are bonded to one oxygen atom in two water molecules. Thus, by the end of the reaction, all the C—H bonds of methane have been broken, as well as the $O = O$ bonds in the two oxygen molecules. Two $C = O$ bonds are made to create a carbon dioxide molecule and four O—H bonds are made to create two water molecules. Again, the law of conservation of mass is observed.

Although this example shows the reaction between two reactants to form two products, there is often only one reactant and/or one product. For example, one carbonic acid molecule decomposes to produce a water molecule and a carbon dioxide molecule: one reactant and two products. In theory, there are no restrictions on the number of reactants and products in a chemical reaction, although a reaction between three reactants is rare because more than two reactants seldom collide at once.

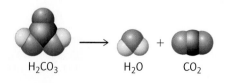

Nuclear reactions are entirely different from *chemical* reactions. In a nuclear reaction the nuclei of atoms do change. Nuclear reactions are described in Chapter 16, *Nuclear Chemistry and Nuclear Medicine.*

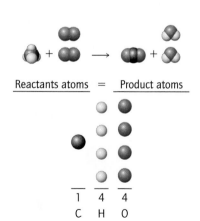

Reactants atoms	=	Product atoms
1	4	4
C	H	O

Writing a Chemical Equation

A chemical reaction is most conveniently represented as a **chemical equation.** A chemical equation conveys the identity of the reactants and products in a chemical reaction *and* their relative proportions. Other pertinent information about the reaction, such as the physical state of the reactants and products may also be conveyed.

In a chemical equation the reactants and products are represented by their chemical formulas. A reaction arrow ($\rightarrow$) is used to separate the *reactants* on the left side of the arrow from the *products* on the right side of the arrow. For example, the equation in Figure 8-2 represents the reaction between methane and oxygen to produce carbon dioxide and water, described earlier.

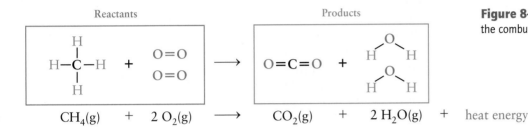

$$CH_4(g) + 2\,O_2(g) \longrightarrow CO_2(g) + 2\,H_2O(g) + \text{heat energy}$$

Figure 8-2 The chemical equation for the combustion of methane.

Whole numbers called **coefficients** are placed before each chemical formula to indicate the ratios of each reactant and each product. When no number is shown, the coefficient is assumed to be 1. Hence, the coefficients in the equation in Figure 8-2 indicate that one molecule of methane reacts with two molecules of oxygen to produce one molecule of carbon dioxide and two molecules of water. When you scale up the reaction to practical measurable quantities, the coefficients also represent molar ratios: One mole of methane reacts with two moles of oxygen to produce one mole of carbon dioxide and two moles of water. Recall from Chapter 2 that molar quantities of a substance can be calculated from the mass of that substance, using the molar mass of the substance as a conversion factor:

$$CH_4(g) + 2\,O_2(g) \longrightarrow CO_2(g) + 2\,H_2O(g) + \text{heat energy}$$

1 mol	2 mol	1 mol	2 mol
16.042 g	64 g	44.01 g	36.03 g

(8-1)

Note that the identity of the *atoms* themselves always remains the same throughout the chemical reaction. For example, the oxygen atoms in this reaction are part of molecular oxygen in the reactants, but part of carbon dioxide and water in the products. They are still oxygen atoms and have not been transformed into atoms of a different element. The parts of a chemical equation are summarized in the box below.

Parts of a Chemical Equation

- **Reactants.** The compounds or elements that are combined in a chemical reaction are shown to the left of the arrow, and referred to as the *reactants.* Bonds are broken in the reactant molecules. *For example, methane and oxygen are the reactants in the equation in Figure 8-2 and C—H bonds and O=O bonds are broken.*

- **Products.** The compounds or elements produced in a chemical reaction are shown to the right of the arrow, and referred to as *products.* Products are formed as new bonds are made between different atoms. *For example, carbon dioxide and water are the products in the reaction of methane and oxygen. New bonds are formed between carbon and oxygen to make carbon dioxide, and between oxygen and hydrogen to make water.*

- **A + sign** is used to separate the individual reactants and the individual products when more than one is present.

- **Reaction Arrow.** An arrow pointing from left to right separates the reactants from products. The arrow symbolizes the transformation of reactants into products in the chemical reaction.

- **Coefficients.** The coefficients denote the ratio of each reactant and product to the others. This ratio refers to individual atoms or molecules as well as molar quantities.

- **Physical State.** The physical state of each compound or element is sometimes included in parentheses following the formula of the compound: solid (s), liquid (l), gas (g), or aqueous solution (aq). *For example, all the reactants and products shown in Figure 8-2 are in the gas phase.*

Combustion Reactions One of the most common reactions of organic substances is **combustion.** Combustion reactions include any reaction in which an organic compound containing carbon, hydrogen, and sometimes oxygen reacts completely with oxygen (O_2). The products are carbon dioxide, CO_2, and water, H_2O. Combustion reactions are the reactions that we refer to in everyday language as burning. The previously described reaction of methane and oxygen in a gas burner is an example of a combustion reaction.

Instead of burning hydrocarbons, cells burn fats and the sugar glucose ($C_6H_{12}O_6$) as a source of energy; yet, the ultimate products are the same: carbon dioxide and water. While the combustion of glucose in air happens in a single reaction, the transformation of glucose into carbon dioxide, water, and energy in cells occurs as a sequence of chemical reactions, known as a biochemical pathway. The equation for the complete combustion of glucose is

$$\underset{\text{Reactants}}{C_6H_{12}O_6(aq) \;+\; \overset{\text{Coefficients}}{6\,O_2(g)}} \longrightarrow \underset{\text{Products}}{6\,CO_2(g) \;+\; 6\,H_2O(g)}$$

The oxygen you see as a reactant in the equation is essential for the reaction to occur, and it is the reason why you need to breathe in oxygen to live.

WORKED EXERCISE 8-1   Chemical Reactions and Equations

For the equation below,

$$4\,Fe(s) + 3\,O_2(g) \longrightarrow 2\,Fe_2O_3(s):$$

a. Indicate the reactant(s).
b. Indicate the product(s).
c. Indicate the ratio in which the reactants and products react.
d. If 4 mol of iron react with 3 mol of oxygen, how many moles of iron(III) oxide are produced?

e. If 8 mol of iron react with 6 mol of oxygen, how many moles of iron(III) oxide are produced?

f. Indicate the physical state (s, l, g, aq) of each reactant and product.

SOLUTION

a. The reactants appear to the left of the reaction arrow: iron (Fe) and oxygen (O_2).

b. The product appears to the right of the reaction arrow: iron oxide (Fe_2O_3).

c. Four iron atoms react with 3 oxygen molecules to produce 2 iron oxide formula units.

d. 2 mol of iron(III) oxide, as determined from the coefficients.

e. 4 mol of iron(III) oxide, because there is half as much iron(III) oxide formed as the number of moles of iron, based on the coefficients 2 and 4.

f. Fe and Fe_2O_3 are in the solid state and oxygen is in the gas state.

PRACTICE EXERCISE

8.1 For each of the following chemical equations,

    i. $C_3H_8(g) + 5\ O_2(g) \rightarrow 3\ CO_2(g) + 4\ H_2O(g)$,

    ii. $C_4H_6(g) + 2\ H_2(g) \rightarrow C_4H_{10}(g)$,

    iii. $Ba(OH)_2(aq) + 2\ HCl(aq) \rightarrow BaCl_2(aq) + 2\ H_2O(l)$:

a. Indicate the reactant(s).

b. Indicate the product(s).

c. Indicate the ratio in which the reactants and products react.

d. Indicate the physical state (s, l, g, or aq) of each reactant and product.

e. Which reaction above represents a combustion reaction? Explain.

Balancing a Chemical Equation

The law of conservation of mass states that matter can be neither created nor destroyed. *Applied to a chemical equation, the law of conservation of mass requires that the total number of each type of atom on the reactant side of a chemical equation must always equal the total number of each type of atom on the product side of the equation.* In other words, no new or additional atoms can be introduced; and no atoms may be lost in a chemical reaction.

To calculate the number of atoms of one type on one side of the equation, count the number of atoms of that type in each substance: Multiply the coefficient before the formula by the numerical subscript for that atom, as shown below. Then sum the results from all substances on that side of the equation. For example, the number of oxygen (O) atoms on each side of the equation below is 18.

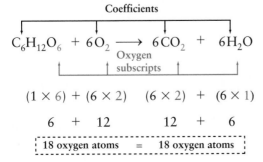

Law of conservation of mass

Coefficients

$$C_6H_{12}O_6 + 6\,O_2 \longrightarrow 6\,CO_2 + 6\,H_2O$$

Oxygen subscripts

$(1 \times 6) + (6 \times 2) \quad (6 \times 2) + (6 \times 1)$

$6 + 12 \qquad 12 + 6$

18 oxygen atoms $=$ 18 oxygen atoms

Similar calculations for the carbon atoms (C) and hydrogen atoms (H) show that there are 6 carbon atoms and 12 hydrogen atoms on each side of the equation. *A chemical equation that contains an equal number of each type of atom on both sides of the equation is known as a **balanced chemical equation.***

The process of balancing a chemical equation involves inserting the correct coefficients. To arrive at the correct coefficients, follow the guidelines for balancing an equation described in the box below.

Guidelines for Balancing a Chemical Equation

Step 1: Assess the equation. Determine the number of each type of atom on the reactant side and product side of the equation. If the two numbers do not match, the equation is not balanced.

Step 2: Balance the equation one atom type at a time by inserting coefficients. Systematically place coefficients in front of each reactant and each product as necessary, to arrive at an equal number of each type of atom on both sides of the equation.

- Never change the numerical subscripts in a chemical formula, because this changes the identity of the chemical compound. *For example, changing the subscript following oxygen from 1 to 2 turns water into hydrogen peroxide: H_2O versus H_2O_2—two entirely different molecules.*

- Remember that adding a coefficient alters the number of every type of atom in the formula it precedes. *For example, placing a 2 in front of CO_2 changes both the number of carbon atoms and oxygen atoms: 2 carbon atoms, 4 oxygen atoms.*

- The balancing process is simplified if you wait to balance last any atom type that appears in more than one compound or element on one side of the equation. *For example, since oxygen appears in both product molecules in the combustion of methane, it is easiest to balance oxygen last.*

Step 3: Check that the coefficients cannot be divided by a common factor (divisor). Convention requires that coefficients be whole numbers that cannot be divided by a common divisor. *For example, the coefficients 4, 8, and 12 can all be divided by the greatest common factor, 4, so the coefficients used should be 1, 2, and 3 instead.*

- Fractions are sometimes used temporarily in the *process* of balancing a chemical equation; however, the final balanced equation must contain only the lowest whole number coefficients. See Worked Exercise 8-2.

The law of conservation of mass is demonstrated in *The Model Tool 8-1: Balancing a Chemical Equation.* To calculate the amount (in grams or moles) of product formed in a reaction, see *Extension Topic 8-1: Reaction Stoichiometry Calculations.*

Problem-Solving Tutorial:
Chemical Equations

WORKED EXERCISE 8-2   Balancing a Chemical Equation

The hydrogen fuel cell in a revolutionary new type of car generates energy from the following unbalanced equation. Balance the equation.

$$\_\_\_ H_2(g) + \_\_\_ O_2(g) \longrightarrow \_\_\_ H_2O(g)$$

SOLUTION

Step 1: Assess the equation.

Kind of Atom	No. on Reactant Side	No. on Product Side
H	2	2
O	2	1

The equation is not balanced because the values for oxygen are not equal in the reactant and product columns.

Step 2: **Balance the equation one atom type at a time by inserting coefficients.** Since hydrogen is already balanced, balance oxygen by inserting the coefficient ½ in front of oxygen. Since this is a fraction, use this coefficient only temporarily.

$$H_2 + \tfrac{1}{2} O_2 \longrightarrow H_2O \quad \text{(hydrogen and oxygen are now balanced)}$$

Turn the fraction into a whole number by multiplying *every* coefficient in the equation by 2, because $2 \times \tfrac{1}{2} = 1$, a whole number:

$$2 H_2 + O_2 \longrightarrow 2H_2O \quad \text{(hydrogen and oxygen are still balanced,}$$
but now whole numbers are in place as coefficients, as required)

Step 3: **Check that the coefficients cannot be divided by a common factor (divisor).** There is no common divisor.

The Model Tool 8-1 Balancing a Chemical Equation

I. Construction Exercise Part I

1. Obtain one black carbon atom, four red oxygen atoms, and four light-blue hydrogen atoms. Obtain four bent double bonds and four single bonds.

2. Construct a model of methane, CH_4, and two models of oxygen, O_2. Remember, an oxygen molecule contains an oxygen–oxygen double bond requiring two bent connectors. Imagine that these are the reactants for a chemical reaction.

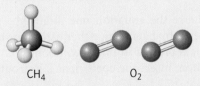

CH_4 O_2

3. How many C, O, and H atoms are present in the reactants? Fill in the table below.

Kind of Atom	No. on Reactant Side
C	
H	
O	

II. Construction Exercise Part II

4. Simulate a chemical reaction by breaking *all* the bonds in your methane and oxygen molecules.

5. Using *only* these atoms and bonds, construct as many carbon dioxide, CO_2, and water, H_2O, molecules as you can from the atoms you have available from the reactants. Use the bond connectors to construct your new models. (Remember that CO_2 contains two C=O double bonds, requiring two bent connectors per double bond).

continued

III. Inquiry Questions

6. How many CO_2 molecules were you able to build? Therefore, what coefficient should be placed before CO_2 in the balanced equation?

7. How many water molecules were you able to build? What coefficient should be placed before H_2O in the balanced equation?

8. Were any atoms left over after the exercise? Why or why not?

9. How many C, O, and H atoms are present in the product molecules? Fill in the table below. How do these numbers compare with those on the reactant side?

Kind of Atom	No. on Reactant Side	No. on Product Side
C	1	
H	4	
O	4	

10. How does your answer to Question 9 illustrate the law of conservation of matter?

11. Write a balanced equation for the reaction between methane and oxygen simulated here.

WORKED EXERCISE 8-3   Balancing a Combustion Equation

Balance the following combustion equation:

$$C_2H_6(g) + O_2(g) \longrightarrow CO_2(g) + H_2O(g)$$

SOLUTION

Step 1: Assess the equation.

Kind of Atom	No. on Reactant Side	No. on Product Side
C	2	1
H	6	2
O	2	3

The equation is not balanced because the values in the reactant column do not equal the values in the product column for all types of atoms.

Step 2: Balance the equation one atom type at a time by inserting coefficients. Begin by balancing either carbon or hydrogen. Balance oxygen last because it is present in both compounds on the product side. If you begin with carbon, insert the coefficient 2 in front of CO_2:

$$C_2H_6 + O_2 \longrightarrow 2\ CO_2 + H_2O \qquad \text{(carbon is now balanced)}$$

Balance hydrogen by inserting the coefficient 3 in front of water:

$$C_2H_6 + O_2 \longrightarrow 2\ CO_2 + 3\ H_2O \qquad \text{(carbon and hydrogen are now balanced)}$$

To balance oxygen, insert the fraction 7/2 (or 3½) in front of oxygen as a temporary coefficient, because 7 oxygen atoms are required to balance the 7 total oxygen atoms on the right side of the equation (4 from 2 CO_2 and 3 from 3 H_2O):

$$C_2H_6 + \frac{7}{2} O_2 \longrightarrow 2\ CO_2 + 3\ H_2O \qquad \text{(oxygen is now balanced)}$$

Turn the fraction into a whole number by multiplying *every* coefficient in the equation by 2:

$$2\ C_2H_6 + 7\ O_2 \longrightarrow 4\ CO_2 + 6\ H_2O$$

Step 3: Check that the coefficients cannot be divided by a common factor (divisor). There is no common factor for the coefficients: 2, 7, 4, and 6; therefore, you already have the lowest set of whole number coefficients in the balanced equation. The values in the second column now equal the values in the third column in every row; hence, the equation is balanced.

Kind of Atom	No. on Reactant Side	No. on Product Side
C	4	4
H	12	12
O	14	14

PRACTICE EXERCISES

8.2 Balance the following chemical equations by inserting the appropriate coefficients into the blanks:

a. ___ $N_2O_5(g) \rightarrow$ ___ $NO_2(g) +$ ___ $O_2(g)$
b. ___ $C_6H_{12}O_2(l) +$ ___ $O_2(g) \rightarrow$ ___ $CO_2(g) +$ ___ $H_2O(g)$
c. ___ $CaCO_3(s) +$ ___ $HCl(aq) \rightarrow$ ___ $CO_2(g) +$ ___ $H_2O(l) +$ ___ $CaCl_2(aq)$
d. ___ $Mg(s) +$ ___ $HCl(aq) \rightarrow$ ___ $MgCl_2(aq) +$ ___ $H_2(g)$
e. ___ $NH_3(g) +$ ___ $O_2(g) \rightarrow$ ___ $N_2(g) +$ ___ $H_2O(l)$

8.3 Is the reaction below balanced? Explain why or why not. *Hint:* When a subscript follows parentheses, it applies to all atoms enclosed within the parentheses. For example, there are 3 Ca atoms, 6 H atoms, and 6 O atoms represented in 3 $Ca(OH)_2$.

$$2\ Na_3PO_4(aq) + 3\ Ca(OH)_2 \longrightarrow 6\ NaOH(aq) + Ca_3(PO_4)_2(aq)$$

Extension Topic 8-1 Reaction Stoichiometry Calculations

From a balanced equation, it is possible to calculate the amount (in grams or moles) of product formed, if the mass of one of the reactants is known, provided there is an ample supply of the other reactant. This type of calculation is known as a *stoichiometry* calculation. Stoichiometry calculations are made routinely in the laboratory whenever a chemical reaction is performed.

Reaction stoichiometry calculations begin with a balanced chemical equation. *The coefficients in a chemical equation represent not only the ratio in which the individual molecules react with one another, but also the molar ratio in which they react.* For example, one mole of ethanol, CH_3CH_2OH, reacts with three moles of oxygen to form two moles of carbon dioxide and three moles of water, according to the balanced chemical equation

$$CH_3CH_2OH(l) + 3\ O_2(g) \longrightarrow 2\ CO_2(g) + 3\ H_2O(l)$$

    Ethanol   oxygen   carbon dioxide   water

It is important to recognize that coefficients represent a molar ratio, never a mass ratio. The ratio is a molar ratio because moles represent a certain number of molecules or atoms, which must react in defined proportions.

Suppose you want to calculate how many grams of CO_2 could be produced from the combustion of 12.0 g of ethanol. You would perform the stoichiometry calculation according to the following three steps.

Step 1: Convert the known mass of reactant to moles of reactant. Remember, by converting a mass into moles, you are effectively keeping track of the number of molecules—in this case, the number of ethanol molecules. Convert the mass of ethanol into moles of ethanol using one of the forms of the molar mass of ethanol as a conversion factor.

$$12.0\ \text{g ethanol} \times \frac{1\ \text{mol ethanol}}{46.07\ \text{g ethanol}} = 0.260\ \text{mol ethanol}$$

Mass of ethanol   Inverted molar mass   mol ethanol
                      of ethanol

Remember that the molar mass of a compound is the sum of the molar masses of each atom in the compound (listed in the periodic table). Hence, the molar mass of ethanol is calculated from the molar masses of the elements that make up a molecule of ethanol, C_2H_6O:

$$(2 \times 12.01\ \text{g/mol}) + (6 \times 1.008\ \text{g/mol}) + (1 \times 16.00\ \text{g/mol})$$
$$= 46.07\ \text{g/mol}$$

continued

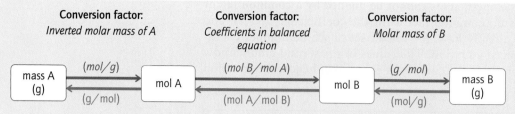

Figure E8-1 Steps in performing a reaction stoichiometry calculation. A and B can be any two reactants or products in the equation.

Step 2: convert moles of reactant to moles of product. Calculate the number of moles of carbon dioxide that can be produced from **0.260 mol** of ethanol. *Use the coefficients from the balanced equation to create a conversion factor: 2 mol CO_2 per 1 mol ethanol.* (Note: Exact ratios do not affect significant figures.)

$$0.260 \text{ mol ethanol} \times \frac{2 \text{ mol } CO_2}{1 \text{ mol ethanol}} = 0.520 \text{ mol } CO_2$$

mol ethanol   conversion factor   mol CO_2
obtained from coefficients
in balanced equation

Remember to arrange the conversion factor so that the units *mol ethanol* are in the denominator and the units *mol CO_2* are in the numerator, so that the units *mol ethanol* cancel, leaving the requested units *mol CO_2* in the answer.

Step 3: Convert moles of product to mass of product, the requested unit. Convert **0.520 mol CO_2** into mass CO_2, using the molar mass of CO_2 as a conversion factor.

$$0.520 \text{ mol } CO_2 \times \frac{44.01 \text{ g } CO_2}{1 \text{ mol } CO_2} = 22.9 \text{ g } CO_2$$

mol CO_2 molar mass of CO_2 mass of CO_2

Notice that the conversion factor, molar mass of CO_2, is arranged with *mol CO_2* in the denominator and *g CO_2* in the numerator. Placing moles of CO_2 in the denominator insures that the units *mol CO_2* cancel, leaving you with the final answer in the requested unit *g CO_2*.

Figure E8.1 summarizes the steps required to perform a reaction stoichiometry calculation.

WORKED EXERCISE E8.1 Reaction Stoichiometry Calculation

Calculate the mass of water produced if 12.0 g of ethanol undergoes the reaction described previously:

$$CH_3CH_2OH(l) + 3\ O_2(g) \longrightarrow 2\ CO_2(g) + 3\ H_2O(l)$$
Ethanol   oxygen   carbon dioxide   water

SOLUTION
Step 1: Convert grams of ethanol to moles of ethanol. Use the inverted molar mass of ethanol as the conversion factor:

$$12.0 \text{ g ethanol} \times \frac{1 \text{ mol ethanol}}{46.07 \text{ g ethanol}} = 0.260 \text{ mol ethanol}$$

Mass of ethanol   Inverted molar   mol ethanol
mass of ethanol

Step 2: Convert moles of ethanol to moles of water. Use the coefficients in the balanced equation to set up a conversion factor between mol ethanol and mol water. Remember that when no number is shown, the coefficient is 1:

$$0.260 \text{ mol ethanol} \times \frac{3 \text{ mol } H_2O}{1 \text{ mol ethanol}} = 0.780 \text{ mol } H_2O$$

mol ethanol   conversion factor   mol H_2O
generated from coefficients
in balanced equation

Note that mol ethanol is in the denominator so that it cancels with mol ethanol, leaving mol water, the desired unit.

Step 3: Convert moles of water to grams of water. Use the molar mass of water as a conversion factor.

$$0.780 \text{ mol } H_2O \times \frac{18.01 \text{ g } H_2O}{1 \text{ mol } H_2O} = 14.0 \text{ g } H_2O$$

mol H_2O molar mass H_2O mol H_2O

Thus, combustion of 12.0 g of ethanol can produce 14.0 g of water.

PRACTICE EXERCISES

E8.1 How many grams of oxygen are needed to convert 12.0 g of ethanol into products in the reaction described previously?

E8.2 Balance the chemical equation below and then determine how many grams of oxygen are needed to react with 500.0 g of propane from a barbeque grill. How many grams of CO_2 will be produced?

$$C_3H_8(g) + O_2(g) \longrightarrow CO_2(g) + H_2O(g)(unbalanced)$$
Propane

E8.3 How many grams of carbon dioxide are formed when 10.0 g of glucose undergo combustion to form carbon dioxide and water? Begin by writing the complete balanced equation.

E8.4 When 20.0 g of HCl react with excess NaOH, how many grams of NaCl can be produced?

$$HCl(aq) + NaOH(aq) \longrightarrow NaCl(aq) + H_2O(l)$$

8.2 Energy and Chemical Reactions

If you pick up a glass of milk to drink, you have done *work* and therefore expended *energy*. Recall from Chapter 4 that **energy** *is defined as the capacity to do work or to transfer heat*. **Work** is the act of moving an object over a distance against an opposing force. You obtain the energy to do work from the food in your diet, through a series of chemical reactions that take place in your cells.

Milk and other foods contain *chemical energy*, potential energy stored within the chemical bonds of the carbohydrates, proteins, and fats in food. Carbohydrates, proteins, and fats are large organic molecules that undergo chemical reactions that break them down into smaller molecules. In the process, the chemical energy stored within these molecules is used to drive other chemical reactions of the cell that require an input of energy, so that you may once again do work and pick up the glass of milk for another drink.

Bioenergetics is the field of study concerned with the transfer of energy in reactions occurring in living cells. Potential energy in the form of chemical energy exists wherever there are chemical bonds. To understand bioenergetics and the transfer of energy in living cells, we begin with the common units used to measure energy.

Units of Energy in Science and Nutrition

As with any quantity, energy can be measured. Table 8-1 lists several frequently used units of energy, along with the conversions between them. A **calorie** *(cal) is the amount of heat energy required to raise the temperature of 1 gram of water by 1°C (celsius)*. The calorie listings that you see on nutritional labels, which are spelled with a capital C, are actually kilocalories (kcal). Recall from Chapter 1 that the metric prefix "kilo" represents 10^3 (1000) times the base unit, which in this case is the calorie. Therefore, one **Calorie** is equal to one **kilocalorie,** which is equal to 10^3 cal. Remember to pay close attention to whether a capital C or a lowercase c follows the numerical value.

Another standard unit of energy used in science is the joule, J. A **joule** is defined as the amount of energy required to lift a one kilogram weight a height of 10 centimeters. The calorie is defined in terms of the joule, as shown in Table 8-1: 1 cal = 4.187 J. The value 4.187 is an exact number and therefore has an infinite number of significant figures.

Table 8-1 Common Units of Energy and Their Conversions

Unit	Conversion
calorie (cal)	1 cal = 4.184 J (exact)
Calorie (Cal) (note capital C)	1 Cal = 1 kcal
kilocalorie (kcal)	1 kcal = 10^3 cal

Remember, 1 Calorie is equal to 1 kcal or 1000 cal.

WORKED EXERCISE 8-4 Conversions Between Energy Units

How many calories (cal) of energy are represented by 1.56×10^4 J?

SOLUTION

Step 1: **Identify the conversion.** Write the expression that equates the supplied unit (joules) and the requested unit (calories), using Table 8-1:

$$1 \text{ cal} = 4.184 \text{ J (exact)}$$

Step 2: **Express the conversion as a conversion factor.** Turn the conversion into two possible conversion factors.

$$\frac{1 \text{ cal}}{4.184 \text{ J}} \quad \text{or} \quad \frac{4.184 \text{ J}}{1 \text{ cal}}$$

Step 3: **Set up the calculation so that the supplied units cancel.** Select the form of the conversion factor in Step 2 that has calories in the numerator and Joules in the denominator so that the supplied unit (J) cancels: Multiply the supplied unit (J) by the selected conversion factor, yielding an answer in the requested unit (cal).

$$1.56 \times 10^4 \, J \times \frac{1 \text{ cal}}{4.184 \, J} = 3.73 \times 10^3 \text{ cal} \text{ (3 significant figures)}$$

PRACTICE EXERCISES

8.4 Using the conversions in Table 8-1, convert the following units into calories.
   a. 5.79 kcal   **b.** 48.8 J.
8.5 How many joules are there in 2.45 cal?
8.6 How many joules are there in 2720 Cal, the amount of energy the average person consumes in a day?

Heat Energy

Recall from Chapter 4 that energy exists in several forms—chemical, mechanical, heat. Although the reactants and products of a reaction have chemical potential energy, the energy transferred in a chemical reaction can also appear in some of these other forms. For example, the energy transferred during a chemical reaction commonly appears in the form of *heat energy.*

Consider the gas grill you use for a summer barbecue. It utilizes a combustion reaction to heat your food. Once you ignite the propane gas, it reacts with the oxygen present in air, forming carbon dioxide, water, and heat energy. The reaction can be represented by the balanced chemical equation:

$$C_3H_8(g) + 5 \, O_2(g) \longrightarrow 3 \, CO_2(g) + 4 \, H_2O(g) + \text{heat energy}$$

Propane   oxygen   carbon dioxide   water

Heat is obviously a key product of this reaction; if it weren't, your hamburgers would never cook! *Indeed, most chemical reactions absorb or release heat.* In chemistry, the heat energy transferred in a chemical reaction under defined conditions is known as the **change in enthalpy, ΔH.**

It is important to note, however, that a chemical reaction, including a combustion reaction, does not necessarily transfer all of its energy in the form of heat. For example, in the engine of your car, much of the energy released in the combustion of gasoline is converted into mechanical energy that drives your car, and some is converted to heat (hence the warm engine). In the cell, glucose produced from the digestion of carbohydrates undergoes a series of chemical reactions that transfers its chemical potential energy into usable forms of energy that drive other reactions in the cell requiring energy to proceed. As in the case of the car engine, some of this energy is transferred in the form of heat, but most is not.

Exothermic and Endothermic Reactions

We have seen that combustion reactions release heat energy. On the other hand, some reactions absorb heat. For example, photosynthesis—the sequence of reactions in a plant that produce carbohydrates—absorbs energy overall. The energy is supplied in the form of sunlight. *Whether heat is released or absorbed by a chemical reaction depends on the reaction under consideration.*

Where does the heat released from a chemical reaction come from? Recall from Chapter 4 that a chemical bond represents chemical potential energy. *Breaking a bond requires energy, while making a bond releases energy.* When energy is absorbed or released in the form of heat, it can be measured as the change in enthalpy, ΔH.

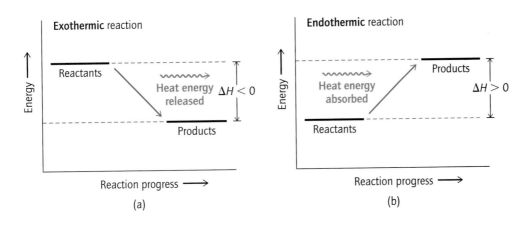

Figure 8-3 Two energy diagrams. (a) An exothermic reaction: The products have a lower energy than the reactants, so heat is released. (b) An endothermic reaction: The products have a higher energy than the reactants, so heat is absorbed.

Since not all chemical bonds are identical, different molecules represent different amounts of potential energy. Some chemical bonds are stronger than others and therefore have lower chemical potential energy. So, in a chemical reaction—where reactant bonds are broken and new product bonds are formed—heat energy is released when the covalent bonds in the products are stronger than the bonds in the reactants. Stronger bonds have lower potential energy, and the difference in energy appears as heat in the surroundings. Since carbon dioxide and water contain some of the strongest bonds, combustion reactions release energy.

When the products of a reaction have stronger bonds than the reactants, the difference in potential energy between reactants and products appears as heat energy released to the surroundings. The heat coming off your barbeque warms the surroundings, for example. This type of reaction is known as an **exothermic reaction.** The change in enthalpy in an exothermic reaction is negative ($\Delta H < 0$) because the products are lower in chemical potential energy than the reactants and ΔH is defined as the enthalpy of the products minus the enthalpy of the reactants.

Conversely, when the reactants are lower in energy than the products, heat is absorbed from the surroundings, and the surroundings become cooler. This type of reaction is known as an **endothermic reaction.** The change in enthalpy in an endothermic reaction is positive ($\Delta H > 0$) because the products are higher in chemical potential than the reactants. In an endothermic reaction, energy must be continuously supplied to sustain the reaction.

Exothermic and endothermic reactions can be illustrated in an energy diagram, such as those shown in Figure 8-3. In an **energy diagram,** energy appears on the y-axis and the progress of the reaction as you go from reactants to products is shown along the x-axis.

Note that in both exothermic and endothermic reactions energy is never lost and never created. *The conservation of energy is recognized in the first law of thermodynamics, which states that energy can be neither created nor destroyed.* Energy can only be transformed. It follows that if a reaction is exothermic, the reverse reaction is endothermic, and by the same numerical value of ΔH. Only the sign ($+$ or $-$) of ΔH is different.

$$\Delta H_{reaction} = \Delta H_{products} - \Delta H_{reactants}$$

WORKED EXERCISE 8-5 Understanding Enthalpy

Problem-Solving Tutorial:
Enthalpy and Chemical Reactions

For the combustion of methane, perform the following steps:

a. Write a balanced chemical equation.
b. Add heat to the appropriate side of the equation.
c. Is $\Delta H < 0$ or is $\Delta H > 0$? Explain.
d. Is the reaction exothermic or endothermic? Explain.
e. Construct an energy diagram for the reaction.
f. Label the axes, reactants, and products.

g. Show how the value of ΔH is represented on the energy diagram.

h. Do the surroundings become warmer or cooler as a result of this reaction? Explain.

i. Are the covalent bonds in the products stronger or weaker than the covalent bonds in the reactants overall? Explain.

SOLUTION

a. $CH_4(g) + 2\,O_2(g) \rightarrow CO_2(g) + 2\,H_2O(g)$

b. $CH_4(g) + 2\,O_2(g) \rightarrow CO_2(g) + 2\,H_2O(g)$ + heat

c. $\Delta H < 0$ (ΔH is negative), because the potential energy of the products is lower than the potential energy of the reactants.

d. The reaction is exothermic, because heat is released.

e.–g.

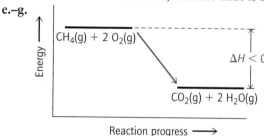

h. The surroundings become warmer because heat is released.

i. The covalent bonds in the products (CO_2 and H_2O) are stronger than the bonds in the reactants (CH_4 and O_2). Stronger bonds have lower potential energy, and the difference in energy appears as heat to the surroundings.

PRACTICE EXERCISES

8.7 For each of the following reactions, indicate whether it is an endothermic or an exothermic reaction.

    a. $8\,H_2S(g) \rightarrow 8\,H_2(g) + S_8(s)$; heat is absorbed

    b. $2\,CO_2 \rightarrow 2\,CO + O_2$; $\Delta H = +566$ kJ

8.8 The $C{=}O$ and $O{-}H$ bonds are two of the strongest covalent bonds that exist. Explain why the strength of these bonds causes combustion to be an exothermic process.

8.9 Provide an energy diagram for the reaction in Practice Exercise 8-7b.

    a. Label ΔH.

    b. Label the reactants and products.

    c. Provide a title for the x- and y-axes.

8.10 How does an endothermic reaction differ from an exothermic reaction?

Calorimetry

Calorimetry is the experimental technique used to measure enthalpy changes (ΔH) in chemical reactions, using an apparatus known as a calorimeter, as illustrated in Figure 8-4. A calorimeter is designed to measure the heat released in a combustion reaction.

To measure the caloric content of a sample of food, the food is dried and sealed in the inner chamber of a calorimeter, and then combined with oxygen. The combustion reaction is initiated by applying a spark. The heat released from the reaction in the calorimeter warms an outer chamber of water that completely surrounds the inner chamber of the calorimeter. The heat released causes the temperature of the water in the outer chamber of the calorimeter to rise. The temperature increase of the water is used to calculate the amount of heat energy (in Calories) that the water absorbed, which is equal to the heat released by the reaction. This calculation provides the caloric content of the food to a good approximation.

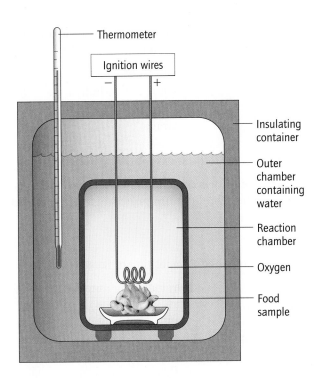

Thermometer

Ignition wires

−   +

Insulating container

Outer chamber containing water

Reaction chamber

Oxygen

Food sample

Figure 8-4 Cross-section of a calorimeter. A calorimeter is used to measure the change in enthalpy for various reactions, including the complete combustion of different types of food.

The heat released during the combustion of carbohydrates, proteins, and fats present in a sample of food can be measured in a calorimeter. This change in enthalpy is reported on the nutritional labels you see on most foods. The preferred unit of energy for nutritional applications is the Calorie. According to the first law of thermodynamics, the total energy released during the metabolism of a particular food item is equal to the enthalpy of the combustion of the same food item in a calorimeter, even though inside a living cell a series of reactions is required to release the same energy from the food. Combustion in a calorimeter occurs quickly as a single reaction, and the energy is released entirely in the form of heat, whereas inside the cell, energy is not released as heat. See *Chemistry in Medicine: Critical Needs for Human Calorimetry in Medicine* at the end of the chapter to learn how calorimetry is also used to determine human energy requirements, especially for individuals on a mechanical ventilator.

Table 8-2 compares the average caloric content per mass (Cal/g) for the three basic types of food molecules: carbohydrates, proteins, and fats. As you can see, both the type and amount of food you eat determines your total caloric intake. From Table 8-2, you can see that we obtain twice as many calories per gram of food from fats as from carbohydrates and proteins. Indeed, the high caloric content of fat is why fat molecules so effectively serve as the body's long-term source of chemical potential energy.

From the mass of each biomolecule (carbohydrate, protein, and fat) contained within a particular food item you can calculate the total caloric value of a particular food sample, using Table 8-2. After calculating the number of

Table 8-2 Caloric Content of Carbohydrates, Proteins, and Fats

Biomolecule	Foods Containing This Biomolecule	Caloric Content (Cal/g)
Carbohydrates	Rice, potatoes, bread, vegetables, fruit, milk	4
Proteins	Fish, meat, dairy products, beans, legumes, milk	4
Fats	Oils, butter, margarine, animal fats, milk	9

calories provided by each type of molecule, sum these values and round to the nearest tens place to obtain the total caloric value of the food item:

Calories provided by fat	=	Grams of fat	×	9 Cal/g
Calories provided by carbohydrate	=	Grams of carbohydrate	×	4 Cal/g
Calories provided by protein	=	Grams of protein	×	4 Cal/g

Total = Calories fat + Calories carbohydrate + Calories protein

Table 8-3 shows the caloric content and composition of some foods. The shading indicates whether the foods are primarily a source of carbohydrates, fats, or proteins.

Table 8-3 Caloric Content and Composition of Some Foods*

Food	Carbohydrate (g)	Protein (g)	Fat (g)	Calories
Potato, baked (~3.6 ounces)	27	2.5	Trace	110
Rice, brown, cooked (1 cup)	46	5	2	220
Pasta, plain, cooked (1 cup)	32	5	1	115
Apple, raw (medium)	21	0	1	90
Beef, ground (3 ounces)	0	15	18	222
Cheese, cheddar (1 ounce)	0	7	9	113
Chicken breast roasted, no skin (medium)	0	54	6	250
Flounder, baked (3.4 ounces)	0	21	5	130
Olive Oil (1 tbsp)	0	0	14	119
Avocado (1/2 cup)	9	2	18	185
Mayonnaise (1 tbsp)	0	0	11	100
Sour Cream (1 tbsp)	1	0	3	26
Cream Cheese (1 tbsp)	0	1	5	51
Butter (1 tsp)	0	0	4	36

* Values from *The Joy of Cooking*, Rombauer et al., Scribner 1997.

Nutrition Facts
Per container (175 g)

Amount	% Daily Value
Calories 170	
Fat 4.5g	7%
Saturated Fat 3.5 g + Trans Fat 0 g	18%
Cholesterol 10 mg	
Sodium 85 mg	4%
Carbohydrate 27 g	9%
Fiber 0 g	0%
Sugars 26 g	
Protein 6 g	
Vitamin A **15%** Vitamin C	0%
Calcium **20%** Iron	0%

WORKED EXERCISE 8-6 Calculating the Calories in a Sample of Food

Calculate the caloric content of 1 cup of full-fat blueberry yogurt from the nutritional label shown in the margin.

SOLUTION

The label tells you that 1 cup of blueberry yogurt contains 27 g carbohydrate, 6 g protein, and 4.5 g fat. Use the values provided in Table 8-2 to determine the total Calories supplied by each type of biomolecule:

$$\text{Calories provided by fat} = 4.5 \, \cancel{g} \times \frac{9 \, \text{Cal}}{\cancel{g}} = 41 \, \text{Cal}$$

$$\text{Calories provided by carbohydrate} = 27 \, \cancel{g} \times \frac{4 \, \text{Cal}}{\cancel{g}} = 108 \, \text{Cal}$$

$$\text{Calories provided by protein} = 6 \, \cancel{g} \times \frac{4 \, \text{Cal}}{\cancel{g}} = 24 \, \text{Cal}$$

Sum these values to obtain the total Calories supplied by the yogurt:

$$41 \, \text{Cal} + 108 \, \text{Cal} + 24 \, \text{Cal} = 173 \, \text{Cal or } \mathbf{170 \, Cal}$$

This answer agrees with the "Calories" shown at the top of the label.

PRACTICE EXERCISES

8.11 A piece of angel food cake contains 33 g carbohydrate, 3 g protein, and 0 g fat. How many total Calories does a piece of angel food cake contain?

8.12 A cup of pasta contains 32 g carbohydrates, 5 g protein, and 1 g fat. How many total Calories does a cup of pasta contain?

An Overview of Energy and Metabolism

The cells in your body perform thousands of chemical reactions every day. The chemical reactions of the cell are referred to as **biochemical reactions** and typically occur as a sequence of reactions. A **biochemical pathway** is defined by a particular sequence of reactions, many of which are common to many organisms. In a biochemical pathway the product of one reaction becomes the reactant in the next reaction, and so forth.

There are two basic types of biochemical pathways:

* catabolic and
* anabolic.

Catabolic and anabolic pathways together are known as **metabolism.** **Catabolic pathways** convert large molecules, such as carbohydrates, proteins, and fats from your diet, into smaller molecules (the left side of Figure 8-5). **Anabolic pathways** build larger molecules, such as proteins, lipids, and DNA, from smaller molecules (the right side of Figure 8-5). The most important feature distinguishing catabolic and anabolic pathways is the direction in which energy flows from these pathways. *Catabolic pathways release energy, overall, while anabolic pathways absorb energy, overall.*

The energy required to build molecules via anabolic pathways comes from the energy released in catabolic processes (illustrated by the red squiggly arrows in Figure 8-5). Instead of energy being transferred as heat, however, in biochemical reactions, energy is transferred in the form of chemical energy to and from specialized energy-carrier molecules. These specialized molecules capture the energy released in catabolic reactions in the form of chemical potential energy

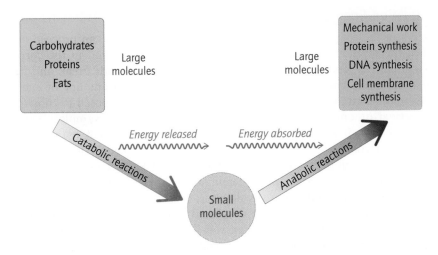

Figure 8-5 Overview of catabolic and anabolic reactions, which together constitute metabolism. The breakdown of large molecules in catabolism releases energy used to build large molecules in anabolism.

stored in covalent bonds. The most important of these energy-transfer molecules is **A**denosine **Tri**phosphate, **ATP.** You will learn more about the role of ATP and other aspects of bioenergetics in Chapters 10 and 14.

PRACTICE EXERCISES

8.13 What are the two main distinctions between catabolic and anabolic pathways?

8.14 Indicate whether each of the following processes is catabolic or anabolic. Indicate whether each releases energy or absorbs energy overall:

    a. building muscle protein from small amino acid molecules;

    b. a bear burning fat during hibernation;

    c. glucose molecules forming large polymers, known as glycogen.

8.3 | Kinetics: Reaction Rates

If a reaction does proceed, releasing or absorbing energy, how *fast* is the reaction? **Chemical kinetics** is the study of reaction rates—how fast reactants in a chemical reaction are converted into products. For example, over time, diamond reacts to form graphite, but this process is extremely slow. If you own a diamond ring, it is safe to assume that you can pass it down through hundreds of generations and it will still remain a diamond. At the other extreme, chemical reactions that are explosive occur very quickly. In an explosion, a reaction generates gases, which expand rapidly in volume. The high speed of the gas production, and the corresponding sudden volume expansion, is what creates the explosion.

The biochemical reactions that take place in your body proceed at rates that are favorable for the support of physiological processes. They neither take millions of years, nor are they explosive. Molecules called enzymes are the key to maintaining appropriate reaction rates in the body.

How fast a reaction proceeds, known as the **reaction rate,** is determined by measuring the consumption of reactants, or the formation of products, over time. The rate of reaction indicates how the concentration of reactants or products changes over time. A rate of reaction of 0.05 mol/min, for example, indicates that every minute, 0.05 mol of reactants is converted into products.

Activation Energy

Earlier in the chapter you learned that a chemical reaction can proceed only when the reactant molecules collide with enough energy and in the correct spatial orientation. When both these criteria are met, bonds break and new bonds form—a chemical reaction occurs. The minimum amount of energy that must be attained by the reactants for a reaction to proceed is known as the **activation energy, E_A.** Without this initial input of energy, molecules may collide, but they will only bounce off one another without reacting. In the example of combustion on a gas stove, the activation energy is supplied by the spark that ignites the methane on the gas burner. Once ignited, the reaction is self-sustaining and no further input of energy is necessary, provided ample methane and oxygen are available.

The activation energy can be illustrated in an energy diagram, as shown in Figures 8-6a and 8-6b. These diagrams illustrate the energy pathway leading from reactants to products for an endothermic and an exothermic reaction, respectively. As in Figures 8-3a and 8-3b, the *y*-axis represents energy and the *x*-axis represents time—or the progress of the reaction. In both exothermic and endothermic reactions, the magnitude of the activation energy,

Both diamond and graphite are composed of elemental carbon. However, they differ in the arrangement of the carbon atoms. Over time, carbon-carbon bonds in diamond break and reform as carbon-carbon bonds of graphite—a very slow chemical reaction.

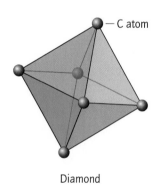

— C atom

Diamond

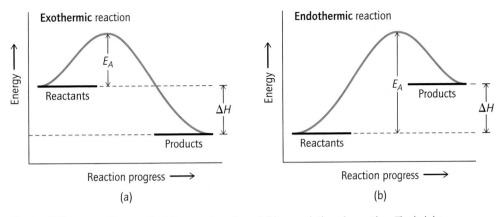

Figure 8-6 Energy diagrams for (a) an exothermic and (b) an endothermic reaction. The height of the red "hill" represents the activation energy, E_A, needed for each reaction to proceed. In this case, both reactions have the same activation energy.

E_A, is depicted by the height of the initial "hill" in the energy diagram. Consider climbing a snowy hill, so that you can slide down the other side. You need a minimum amount of energy to reach the top of the hill (your "activation energy"), but once there, you can easily slide down the other side to get to the bottom. This is true for a chemical reaction as well. Once the activation energy is attained, the reaction can proceed forward.

It is important to recognize that the activation energy, E_A, for a reaction has no effect on the potential energy of either the reactants or the products, and therefore it also has no effect on the change in enthalpy, ΔH. Thus, the activation energy, E_A, is different from the change in enthalpy, ΔH, for the reaction. Factors that influence E_A typically have no effect on ΔH.

The activation energy, E_A, has a profound effect on the rate of a reaction. The lower the activation energy, E_A, the greater the percentage of collisions that lead to a chemical reaction and vice versa. Therefore, *reactions that have low activation energies proceed faster than reactions that have high activation energies*. Thus, an energy diagram with a greater activation energy, E_A—a higher hill—represents a slower reaction. For example, Figures 8-7a and 8-7b show energy diagrams for two exothermic reactions with the same enthalpy, ΔH. Figure 8-7a represents the slower reaction, because E_A is greater.

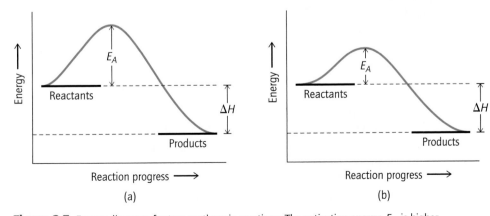

Figure 8-7 Energy diagrams for two exothermic reactions: The activation energy, E_A, is higher for reaction (a) than (b), so reaction (a) is slower than reaction (b).

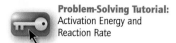
Problem-Solving Tutorial:
Activation Energy and
Reaction Rate

WORKED EXERCISES 8-7   Energy Diagrams and Activation Energy, E_A

1 Examine the energy diagrams below, representing two different reactions.
   a. Do these diagrams represent endothermic or exothermic reactions?
   b. Which reaction would you predict to proceed faster, (a) or (b)? Explain.

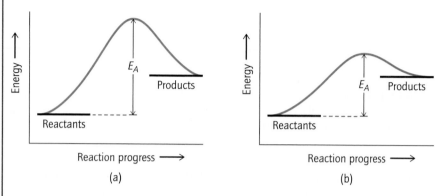

2 Suppose you mix two colorless chemicals together in aqueous solution at room temperature, and no reaction appears to occur. You find that if you increase the temperature of the reaction to 40°C by adding energy in the form of heat, your solution turns blue, indicating that a reaction has occurred. In terms of activation energy, explain why the addition of heat caused the reaction to proceed.

SOLUTIONS

1 **a.** These diagrams represent endothermic reactions, because the energy of the products is greater than the energy of the reactants.
   b. The reaction represented by diagram (b) would be the faster reaction, because it has the lower activation energy, E_A.

2 When two reactants are mixed together, there may be insufficient energy available to reach the activation energy required for the reaction to proceed: collisions may occur but with insufficient energy. When the temperature is raised, molecules have more kinetic energy, and this greater kinetic energy allows more frequent collisions and collisions with greater energy. In this example, the activation energy is apparently attained at 40°C and the reaction proceeds.

PRACTICE EXERCISES

8.15 Define what the activation energy for a reaction is.

8.16 Draw energy diagrams for two exothermic reactions with different activation energies.

   a. Label each axis.
   b. Indicate which reaction would proceed at the faster rate.
   c. Indicate the distance on the y-axis that represents the activation energy, E_A.
   d. Label ΔH, using a double-headed arrow parallel to the y-axis.
   e. Label the position on the graph that represents the reactants. Do the same for the products.

Factors Affecting the Rate of a Reaction

Whether in the lab or a cell, reactions need to proceed quickly enough to be useful, yet not so fast that an uncontrolled reaction or explosion results. The reaction rate can often be controlled by adjusting certain fac-

Reaction: ○ ⟶ ●

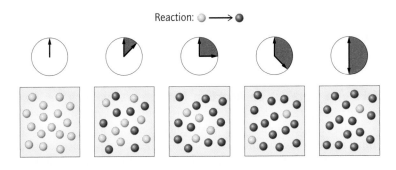

Figure 8-8 As a reaction proceeds, the concentration of reactants (white spheres) decreases and that of products (blue spheres) increases. As a result of the decreasing concentration of reactants, the reaction rate decreases throughout the course of the reaction. [From *Chemistry,* 4th ed., by Martin S. Silberberg. © The McGraw-Hill Companies, Inc.]

tors. The following factors have the most influence on the rate of a chemical reaction:

- the concentration of the reactants,
- the temperature of the reaction, and
- whether or not a catalyst is present.

The Effect of Concentration on Reaction Rate In most reactions the rate of reaction decreases as the concentrations of the reactants decrease, because as reactants are consumed there are fewer reactant molecules and therefore fewer collisions leading to products (Figure 8.8). *Therefore, the rate of a reaction decreases with time.*

Consider the decomposition reaction of N_2O_5 gas in a 1-L container:

$$2\ N_2O_5(g) \longrightarrow 4\ NO_2(g)\ +\ O_2(g)$$

As N_2O_5 molecules collide with one another at constant temperature, they decompose to four NO_2 molecules and one O_2 molecule. Notice that the concentration of the reactant, shown in the second column of Table 8-4, decreases as it is consumed in the reaction. The third column of Table 8-4 shows how the rate of this reaction changes over time. Figure 8-9 shows a graph of this data with respect to time. Notice that the curve in Figure 8-9 starts out steep, indicative of a faster rate, and over time becomes less steep, indicative of a slower rate. Although not all reactions follow this type of reaction kinetics, this reaction illustrates the importance of the relationship between concentration and the rate of a reaction.

The Effect of Temperature on Reaction Rate Recall from the kinetic molecular theory presented in Chapter 4 that, as heat is added to a substance and its temperature increases, the molecules or atoms move faster. As expected, molecules or atoms collide more frequently and with greater energy at higher temperature, increasing the likelihood that the reactants will achieve the activation energy for the reaction. *A common rule of thumb is that for every 10°C increase in temperature, the rate of reaction doubles.* Conversely, if the temperature of a reaction is lowered, the rate of reaction decreases, because the molecules have less kinetic energy, and therefore there are fewer collisions, and fewer collisions with sufficient

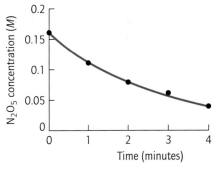

Figure 8-9 A graph of the data in columns 1 and 2 in Table 8-4 shows how the concentration of N_2O_5 decreases with time.

Table 8-4 Rate of Decomposition of N_2O_5

Time (min)	N_2O_5 Concentration (*M*)	Rate of Reaction (*M*/min)
0	0.16	0.06
1	0.11	0.04
2	0.08	0.03
3	0.06	0.02
4	0.04	0.01

Recall that the capital letter M stands for molarity (mol/L). Therefore, the M/min term in Table 8-4 stands for moles per liter per minute.

energy to reach the activation energy. Indeed, the reason you refrigerate your food is to slow the reactions that cause food to spoil. In the body, however, the temperature remains relatively constant at 37°C, so temperature is generally not a variable in biochemical reactions.

The Effect of a Catalyst on Reaction Rate A *catalyst* is a substance that *increases the rate of a chemical reaction by lowering the activation energy for the reaction.* If the activation energy is analogous to a mountain, then a catalyst is like a tunnel through the mountain. Significantly less energy is needed to cross to the other side of a mountain if a tunnel provides an alternative, lower-energy pathway. Similarly, a catalyst provides a lower-energy pathway to products for a reaction. The reaction diagram in Figure 8-10 illustrates how a catalyst lowers the activation energy (red line) compared to the reaction without a catalyst (blue line). Remember that when E_A is lower, the reaction rate is faster.

A catalyst, however, does not influence the change in enthalpy, ΔH, for a reaction. A catalyst only alters the energy pathway leading from reactants to products, as shown in Figure 8-10. The energy of the reactants and products themselves is unchanged.

Catalysts accomplish their rate enhancement in various ways, but in general they work by facilitating the collision process, making it easier for reactants to come together and react. Often they provide a surface that brings the reactants in close proximity and/or in the proper orientation for bond breaking and making to ensue.

By definition, a catalyst is not considered a reactant in the reaction, because it does not itself undergo a chemical change. This means that a single catalyst molecule or atom can be reused over and over, acting on many reactant molecules. Consequently, catalysts are required in very small molar amounts.

In the cell, chemical reactions occur at the normal body temperature of 37°C and at a relatively constant concentration. Therefore, to increase the rate of a biochemical reaction, a catalyst is necessary. Biological catalysts are molecules called **enzymes.** Most are proteins. Enzymes are capable of remarkable rate enhancements: Reaction rates are from 10^5 to 10^{17} times faster in the presence of an enzyme. Enzymes work by reducing the freedom of motion available to the reactants: They lower the activation energy by forcing reactants into a spatial orientation conducive to reaction. You will learn about these fascinating biomolecules in Chapter 11 when you learn about proteins.

Enzymes are so important that if one is missing or defective, a medical condition often results. Depending on the importance of the enzyme, the condition can be minor or debilitating. For example, lactose intolerance exists in a segment of the population that lacks the enzyme *lactase*, which is the catalyst for the biochemical reaction that converts the milk sugar lactose into the simpler sugars glucose and galactose in the digestion process. Individuals lacking the enzyme experience gastrointestinal discomfort when they consume dairy products.

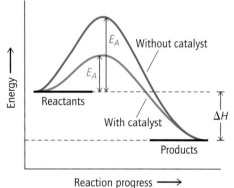

Figure 8-10 Effect of a catalyst: The activation energy is lower in the presence of a catalyst, causing the reaction rate to increase.

You can usually recognize when a substance is an enzyme by the "ase" suffix in its name. By convention, many enzyme names are also *italicized.*

Problem-Solving Tutorial:
Factors Affecting Reaction Rates

WORKED EXERCISE 8-8 The Effect of a Catalyst

A huge variety of catalysts exist and new ones are being introduced all the time. The catalytic converter in a car, for example, uses a platinum catalyst to convert the deadly gas carbon monoxide (CO) into carbon dioxide (CO_2), a nontoxic, albeit greenhouse, gas, according to the reaction

$$2\ CO(g) + O_2(g) \xrightarrow[\text{Heat}]{\text{Pt Catalyst}} 2\ CO_2(g)$$

The catalytic converter is placed near the engine of a car to ensure that it maintains a high temperature. If platinum alone is used as the catalyst, the catalytic converter needs to reach a temperature of 240°C before carbon monoxide reacts with oxygen to form carbon dioxide. If platinum is mixed with rhodium, the temperature needed to catalyze the reaction is only 150°C. Which catalyst, platinum or the platinum–rhodium mixture, lowers the activation energy of this reaction more? Explain.

SOLUTION

An increase in temperature increases the frequency with which reactant molecules collide, as well as the energy of the reactant molecules. Since the platinum–rhodium catalyst is effective at a lower temperature, it must lower the activation energy more. Remember, lower activation energy means a faster reaction.

PRACTICE EXERCISES

8.17 For the following reactions, state whether the indicated change in conditions would increase or decrease the rate of the reaction. Explain why.

    a. $2 H_2O_2(aq) \rightarrow 2 H_2O(l) + O_2(g)$: More H_2O_2 is added to the reaction.

    b. $CO(g) + NO_2(g) \rightarrow CO_2(g) + NO(g)$: The temperature is reduced from 430°C to 330°C.

    c. $2 H_2O_2(aq) \rightarrow 2 H_2O(l) + O_2(g)$: Sodium iodide, a catalyst, is added to the reaction.

8.18 Explain why the rate of a reaction usually decreases as the reaction progresses.

8.19 _____ are biological catalysts, which increase the rate of a biochemical reaction.

8.20 Draw an energy diagram showing an exothermic reaction with and without a catalyst. Label each curve, and mark the activation energy as E_A in both curves. Label the axes.

Chemistry in Medicine Critical Needs for Human Calorimetry in Medicine

On May 27, 1995, Christopher Reeve, well known for his role as Superman in the 1978 blockbuster movie, was paralyzed from the neck down after being thrown from his horse and landing on his head. For the remainder of his life, Reeve was confined to a wheelchair and unable to breathe without the assistance of a mechanical respirator. Reeve passed away on October 10, 2004. Patients like Reeve on mechanical respirators cannot eat on their own and must have their caloric and nutritional needs determined through methods based on the principles of calorimetry.

Mechanical Respiration and Critical Nutrition Decisions

Several conditions can lead to the permanent need for a mechanical respirator, including severe brain injury, spinal-cord injury, and some neurological diseases. Patients on mechanical respirators cannot eat on their own, and a medical professional must manage feedings. It is crucial not to underfeed or overfeed these patients. Malnourishment caused by underfeeding is common in patients who require permanent mechanical respirators and, in severe cases, can lead to coma and death. Then again, overfeeding increases oxygen consumption and metabolic rate. The ventilator and lungs must work harder, possibly causing respiratory muscle fatigue or even respiratory failure.

How can a health care professional manage not to overfeed or underfeed such patients? It can be difficult

Christopher Reeve (1952-2004) at a conference at MIT, March 2, 2003. Reeve sustained a spinal-cord injury in 1995, necessitating chronic use of a mechanical respirator. [Richard Ellis/Getty Images]

continued

to estimate the number of Calories that a patient on a mechanical respirator requires, especially without knowing how much energy that patient expends. Calorimetry provides a convenient method for determining a patient's energy expenditure. That information can then be used to calculate the proper number of Calories that the patient should consume per day. Recall that carbohydrates in food react with oxygen to form carbon dioxide, water, and energy.

$$Food + O_2 \longrightarrow CO_2 + H_2O + energy$$

By either measuring the amount of heat given off (**direct calorimetry**) or measuring the amount of oxygen uptake (**indirect calorimetry**), health care professionals can determine the appropriate number of Calories a person on a mechanical respirator should be consuming.

The First Human Calorimeter–Direct Calorimetry

In the 1890s, professors Wilbur Atwater and Edward Rosa of Wesleyan University in Connecticut designed and built the first human calorimeter. Figure 8-11 shows the basic layout of a human calorimeter. The energy output (expended energy) of the person inside the calorimeter was determined from the heat radiated from their body. The calorimeter consisted of an airtight room large enough for a person to live in. The device was designed to precisely measure energy expenditure in humans participating in various activities such as resting, sleeping, jogging on a treadmill, and riding a stationary bicycle. Through their experiments, Atwater and Rosa confirmed that the first law of thermodynamics—the law of conservation of energy—did indeed govern the transformation of matter and energy in the human body.

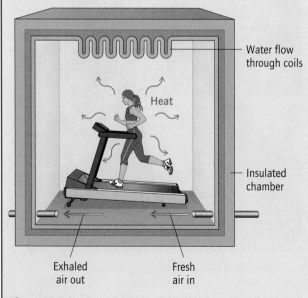

Figure 8-11 A human calorimeter.

Sites Currently Operating Human Calorimeters in the US

Site	Location
Emory Medical School	Atlanta, GA
Naval Aerospace Medical Research Laboratory	Pensacola, FL
USDA Research Center	Beltsville, MD

In a typical whole-room calorimeter, the room is insulated in such a way that any heat radiated by the individual will be absorbed by a known mass of water flowing through coils surrounding the room. The rise in the temperature of the water due to the heat absorbed corresponds directly to the energy output (heat radiated) of the individual. Adequate ventilation is ensured by continually removing moisture and exhaled carbon dioxide from the air and continually adding oxygen in order to maintain an atmosphere with the constant and natural composition of air. Research done using calorimeters continues to confirm the law of conservation of energy, as well as to yield further valuable insight on energy, human metabolism, and nutrition.

Natural Composition of Air

N_2	79.04%
O_2	20.93%
CO_2	0.03%

Direct calorimetry remains the gold standard for the accurate measurement of heat production, but it requires considerable cost, time, and engineering skills. It is impractical to use direct calorimetry to evaluate the dietary needs of patients on mechanical respirators. A more practical and common approach is to use a method known as indirect calorimetry. This technique relies on portable equipment that is more convenient for routine measurements.

Indirect Calorimetry: A Convenient Method for Measuring Human Energy Output

The catabolic reactions that produce energy require oxygen. Therefore, measuring a patient's *oxygen* uptake provides a convenient and accurate indirect way of measuring his or her energy expenditure. This is the basis for *indirect calorimetry*. The amount of air that is inhaled or exhaled is measured using a variable-volume container called a **spirometer**, shown in the photo on the facing page. In a technique known as closed-circuit spirometry, the patient breathes in 100% oxygen from a prefilled spirometer. The patient continues to re-breathe oxygen from the spirometer (usually for less than 1 min), hence the term "closed-circuit." Gases exhaled by the patient include carbon dioxide and unused oxygen. The exhaled carbon dioxide is removed by a canister of potassium hydroxide (KOH) within the breathing circuit. The amount of oxygen uptake by the patient is then determined from the decrease in volume of the spirometer, as illustrated in Figure 8-12.

Clearly, food intake and energy output are closely related. The knowledge gained from research using calorimeters has contributed to the current nutrition recommendations of the USDA (United States Department of Agriculture), as well as many advances in exercise physiology and medical care, including the vital task of managing the food intake of patients on a mechanical respirator.

A patient undergoing spirometry. [Phototake Inc./Alamy]

Figure 8-12 How closed-circuit indirect spirometry works.

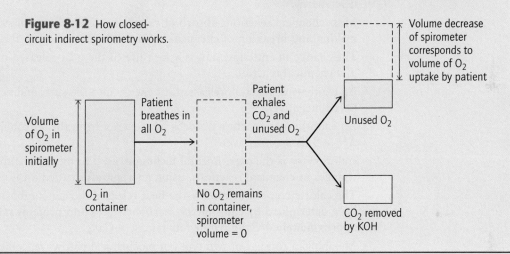

In this chapter, you have learned how energy and matter are interrelated in chemical reactions and that mass and energy are conserved in a chemical reaction. You were also introduced to the catabolic and anabolic reactions of biochemistry, which transfer energy as needed to sustain life. Finally, you learned about reaction kinetics and the role of concentration, temperature, and a catalyst on the rate of a reaction. You will build upon these concepts in the remainder of this text, as you investigate the structure and function of the biomolecules in more depth. In the next chapter you will consider one of the simplest but most important of chemical reactions: the reactions of acids and bases.

Chapter Summary

Writing and Balancing a Chemical Equation

- Chemical reactions break bonds and form new bonds, producing different substances.
- Chemical reactions either release or absorb energy.
- A chemical equation is used to represent a chemical reaction. Reactants and products represented by their chemical formulas are separated by a

reaction arrow. Coefficients placed before each chemical formula indicate the ratio of each product and reactant.

- The law of conservation of mass states that matter can be neither created nor destroyed.
- In a balanced chemical equation, the total number of each type of atom on the reactant side must always equal the total number of each type of atom on the product side.
- Balancing a chemical equation requires inserting the lowest whole-number coefficients so that the number of each type of atom is the same on the reactant side as on the product side.

Energy and Chemical Reactions

- Energy is the capacity to do work or to transfer heat.
- Energy is a measurable quantity. The units used to describe energy are the joule, calorie, and kilocalorie. The Calorie, which is equal to a kilocalorie, is used in nutritional applications.
- The energy transferred in a chemical reaction often appears in the form of heat energy.
- Most chemical reactions absorb or release heat. The heat comes from the making and breaking of chemical bonds in the molecules in the reaction.
- The change in enthalpy, ΔH, is a measure of the heat released or absorbed in a chemical reaction.
- An exothermic reaction releases heat energy into the surroundings. $\Delta H < 0$ (ΔH is negative).
- An endothermic reaction absorbs heat energy from the surroundings. $\Delta H > 0$ (ΔH is positive).
- Calorimetry is the experimental technique used to measure enthalpy changes in chemical reactions, using an apparatus called a calorimeter.
- The caloric content of foods—the heat released during combustion—has been determined by calorimetry. Carbohydrates and proteins release approximately 4 Cal/g, while fats release 9 Cal/g.
- Metabolism consists of two types of biochemical pathways, catabolic and anabolic. Catabolic pathways convert large molecules into smaller molecules, while anabolic pathways convert small molecules into larger ones.
- Catabolic pathways release energy, while anabolic pathways absorb energy.
- Adenosine triphosphate, ATP, is the molecule that carries chemical energy between the catabolic and anabolic reactions of the cell.

Kinetics: Reaction Rates

- Chemical kinetics is the study of reaction rates.
- A chemical reaction can proceed only when the reactant molecules collide with enough energy and in the correct spatial orientation.
- The minimum amount of energy that must be attained by the reactants for a chemical reaction to proceed is known as the activation energy, E_A.
- The activation energy determines the rate of a reaction; the higher the activation energy, the slower the reaction.
- The higher the concentration of reactants, the faster the rate of the reaction.
- As the temperature increases, the rate of the reaction also increases.
- A catalyst lowers the activation energy of a reaction, thus increasing the rate of the reaction.
- Biological catalysts are called enzymes.

Key Words

Activation energy, E_A The minimum amount of energy that must be attained by the reactants for a reaction to proceed.

Anabolic pathways Biochemical reactions that convert smaller molecules into larger molecules such as proteins and DNA. Anabolic reactions consume energy overall.

Bioenergetics The field of study concerned with the transfer of energy in reactions occurring in living cells.

calorie The amount of heat energy required to raise the temperature of one gram of water by 1°C.

Calorie 1000 calories, or 1 kcal.

Calorimetry The experimental technique used to measure enthalpy changes in chemical reactions.

Catabolic pathways Reactions that convert large molecules, such as carbohydrates, proteins and fats, into smaller molecules. Catabolic reactions release energy overall.

Catalyst A substance that increases the rate of a reaction by lowering the activation energy for the reaction.

Change in enthalpy, ΔH Heat energy transferred in a chemical reaction under certain defined conditions.

Chemical equation A symbolic representation of a chemical reaction that includes a reaction arrow and the formulas of the reactants and products, and coefficients before each formula.

Chemical kinetics The study of reaction rates—how fast reactants are converted into products.

Chemical reaction The transformation of one or more substances into one or more products when the chemical bonds between atoms in compounds and elements are broken and new ones are formed.

Coefficient The whole number placed in front of the formula of a reactant or product in a chemical equation. It represents the ratio of each of the reactants and products.

Combustion A reaction in which an organic substance reacts completely with oxygen (O_2) to produce carbon dioxide (CO_2) and water (H_2O).

Endothermic reaction A reaction that absorbs heat from the surroundings and has a positive change in enthalpy.

Energy The ability to do work or transfer heat.

Energy diagram A diagram that depicts the energy changes in a chemical reaction. Energy appears on the y-axis and the progress of the reaction is shown along the x-axis.

Enzyme A biological catalyst.

Exothermic reaction A reaction that releases energy to the surroundings and has a negative change in enthalpy.

Kilocalorie One thousand calories: 1 kcal = 10^3 cal.

Law of conservation of mass The universal observation that matter can be neither created nor destroyed.

Metabolism The biochemical pathways of catabolism and anabolism.

Products The compounds or elements that are produced in a chemical reaction. They are shown to the right of the arrow in a chemical equation.

Reactants The compounds or elements that are combined in a chemical reaction. They are shown to the left of the arrow in a chemical equation.

Reaction rate The change in the concentration of reactants or products over time. The speed of the reaction.

Work The act of moving an object over a distance against an opposing force.

Additional Exercises

Writing and Balancing a Chemical Equation

8.21 What do the abbreviations (s), (g), and (aq) stand for in a chemical equation?

8.22 For the following chemical equations, identify:
  i.   the reactants and products,
  ii. the coefficients for each reactant and product,
  iii. the physical state of each reactant and each product.
  a. $2\,C_4H_{10}(g) + 13\,O_2(g) \rightarrow 8\,CO_2(g) + 10\,H_2O(g)$
  b. $C_6H_{12}O_6(aq) \rightarrow 2\,CO_2(g) + 2\,CH_3CH_2OH\,(aq)$
  c. $2\,Al(s) + 3\,I_2(s) \rightarrow 2\,AlI_3(s)$
  d. $2\,H_2O_2(l) \rightarrow 2\,H_2O(l) + O_2(g)$

8.23 During a chemical reaction, why must the total number of atoms of one kind on the reactant side equal the total number of atoms of that same kind on the product side? This is an application of the law of _____.

8.24 Balance the following chemical equations:
  a. $N_2H_4(l) + N_2O_4(l) \rightarrow N_2(g) + H_2O(l)$
  b. $Li(s) + O_2(g) \rightarrow Li_2O(s)$
  c. $H_2(g) + N_2(g) \rightarrow NH_3(g)$
  d. $KClO_3(s) \rightarrow KClO_4(s) + KCl(s)$

8.25 Balance the following combustion equations:
  a. $C_2H_5OH(l) + O_2(g) \rightarrow CO_2(g) + H_2O(l)$
  b. $C_6H_{14}(l) + O_2(g) \rightarrow CO_2(g) + H_2O(l)$
  c. $CH_3OCH_3(g) + O_2(g) \rightarrow CO_2(g) + H_2O(l)$
  d. $C_3H_6(g) + O_2(g) \rightarrow CO_2(g) + H_2O(l)$

8.26 The following reactions are not balanced properly. Change the coefficients to reflect a properly balanced equation:
  a. $2\,CO(g) + 4\,H_2(g) \rightarrow 2\,CH_4O(g)$
  b. $\frac{1}{2}\,N_2(g) + \frac{1}{2}\,O_2(g) \rightarrow NO(g)$
  c. $Na(s) + Cl_2(g) \rightarrow NaCl(s)$
  d. $Al(s) + HCl(aq) \rightarrow AlCl_3(aq) + H_2(g)$

Extension Topic: Reaction Stoichiometry Calculations

8.27 In the reaction shown, determine how much carbon dioxide is produced when the body metabolizes 15.0 g of glucose.

$$C_6H_{12}O_6(aq) + 6\,O_2(g) \longrightarrow 6\,CO_2(g) + 6\,H_2O(l)$$

8.28 Balance the following equation and then determine how many grams of copper are produced when excess aluminum reacts with 10.5 g of copper oxide.

$$Al(s) + CuO(s) \longrightarrow Al_2O_3(s) + Cu(s)$$

8.29 Balance the following equation and then determine how many grams of water are produced by burning 1.05 g of butane from a butane lighter.

$$C_4H_{10}(g) + O_2(g) \longrightarrow CO_2(g) + H_2O(l)$$
Butane

Energy and Chemical Reactions

8.30 What is bioenergetics?

8.31 How many "small c" calories are found in one "capital C" Calorie? Are "small c" calories or "capital C" calories normally reported on nutritional labels for foods?

8.32 List two of the most common units of energy in science.

8.33 How many calories are there in:
  a. 0.234 Cal
  b. 0.0991 kcal
  c. 20.7 kcal
  d. 352 Cal

8.34 How many calories are there in:
  a. 4.14 J
  b. 36.2 kJ
  c. 0.0587 kJ
  d. 367 J

8.35 Define ΔH.

8.36 What is the difference between an exothermic and an endothermic reaction?

8.37 Why is the reverse of an exothermic reaction always endothermic?

8.38 Where does the heat from burning propane in your barbeque grill come from?

8.39 State whether the following chemical reactions are endothermic or exothermic:
  a. $2 H_2(g) + O_2(g) \rightarrow 2 H_2O(l) + heat$
  b. $2 C_2H_2(g) + 5 O_2 (g) \rightarrow 4 CO_2(g) + 2 H_2O(l) + heat$
  c. $2 CO_2(g) + heat \rightarrow 2 CO(g) + O_2(g)$
  d. $CH_3OH(l) + heat \rightarrow CO(g) + 2 H_2(g)$

8.40 Which process releases energy?
  a. breaking chemical bonds
  b. making chemical bonds
  c. photosynthesis

8.41 What is the name of the instrument used to measure the caloric content of various substances, including food?

8.42 Identify the parts of a calorimeter. How is a calorimeter used to measure the caloric content of a food?

8.43 Use Table 8-2 to calculate the total calories in the following foods:
  a. One cup of almonds containing 27 g protein, 71 g fat, and 28 g carbohydrate.
  b. A banana containing 1 g protein, 1 g fat, and 27 g carbohydrate.
  c. A 1-oz serving of cheddar cheese containing 7 g protein, 9 g fat, and 0 g carbohydrate.
  d. A glazed doughnut containing 2 g protein, 10 g fat, and 23 g of carbohydrate.
  e. A 3-oz serving of broiled swordfish containing 22 g protein, 4 g fat, and 0 g carbohydrate.

8.44 Explain why drinking milk supplies your body with energy, but drinking water does not.

8.45 What serves as the important energy-carrier molecule between catabolic and anabolic reactions?

8.46 What are the two types of metabolic reactions?

8.47 List the three major biomolecules that are obtained through the diet and metabolized into smaller molecules.

8.48 Which of the following apply to catabolic reactions:
  a. release energy
  b. absorb energy
  c. build larger molecules from smaller molecules
  d. break down larger molecules into smaller molecules

8.49 In the body, the synthesis of proteins requires energy. Is this an anabolic or catabolic process?

8.50 Explain why you cannot use the exhaust from your car as a fuel to drive the car.

Kinetics: Reaction Rates

8.51 Define chemical kinetics.

8.52 Which reaction is faster, one with a large or a small E_A?

8.53 Consider the energy diagram shown below.
  a. Does it illustrate an exothermic or an endothermic reaction?
  b. Redraw the diagram and label the following parts of the energy diagram:
    i. Reactants
    ii. Products
    iii. ΔH
    iv. E_A
  c. How might the energy diagram differ if a catalyst were added to the reaction? Sketch over the curve to show how the reaction would be different in the presence of a catalyst.

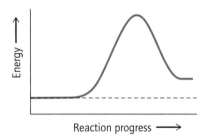

8.54 What effect would the following have on the rate of a chemical reaction?
  a. decreasing the temperature
  b. adding a catalyst
  c. increasing the concentration of a reactant

8.55 Does a catalyst affect the value of ΔH? Does a catalyst affect the value of E_A?

8.56 Indicate whether the following statements are *True* or *False*.
  _____ **a.** A catalyst is always used in high concentration.
  _____ **b.** A catalyst speeds up the rate of a reaction.
  _____ **c.** A catalyst lowers the activation energy for a reaction.
  _____ **d.** A catalyst is chemically unaltered during a chemical reaction and therefore is reused over and over again in a reaction.
  _____ **e.** Biological catalysts are known as enzymes.

8.57 Why are enzymes required for biochemical reactions? What role do enzymes play in biochemical reactions?

8.58 What enzyme is missing or not functioning properly in a person with lactose intolerance?

8.59 Reaction A is an endothermic reaction that occurs twice as fast if a catalyst is added to it. Draw side-by-side energy diagrams for reaction A occurring with and without a catalyst. Which has the higher activation energy? Explain.

8.60 The combustion of fats is a slower process than that of carbohydrates.
a. Are these reactions exothermic or endothermic?
b. Draw side-by-side energy diagrams for each, showing the difference in activation energies.
c. Which reaction has the higher activation energy, the combustion of carbohydrates or fats? Explain.

Critical Thinking Questions

8.61 Walking one mile burns about 40 Cal of energy. If you live two miles away from work, and you walk to and from work each day for five days, how many additional Calories would you burn that week, assuming that you normally drive to work?

8.62 An average size woman needs to eat about 2000 Calories per day in order to maintain her weight. If her diet consists of 30% protein, 40% carbohydrate, and 30% fat, how many grams of each should she ingest each day?

8.63 If you wanted to lose 2 lb of fat, how many Calories would you have to burn, considering that fat is 15% water? There are 454 g in 1 lb.

8.64 The invention and development of the steam engine showed that heat can be used to perform work, and therefore, heat is a form of energy. Explain, using calorimetry and the heat content of foods, why heat is a form of energy.

8.65 Both condensation (gas → liquid) and deposition (gas → solid) may occur in clouds. Condensation produces rain and deposition produces snow.
a. Is condensation an endothermic or exothermic physical process?
b. Is deposition an endothermic or exothermic physical process?
c. Just after condensation or deposition occurs, would you expect the surrounding air temperature to rise or fall?

8.66 The combustion of pentane, C_5H_{12}, produces carbon dioxide and water. Write the balanced equation for this reaction. Be sure to include heat appropriately (as a product or reactant).

8.67 If heated, zinc carbonate, $ZnCO_3$, will decompose into zinc oxide, ZnO, and carbon dioxide. Write the balanced equation for this reaction.

Chemistry in Medicine

8.68 Why is it important to know the number of calories per day to feed a person on a mechanical respirator?

8.69 Are the final products of the combustion of food always the same, whether the food is combusted in a calorimeter or the human body? Explain.

8.70 What is the difference between direct and indirect calorimetry? Which is the gold standard? Which is the more practical method to use on patients who are on mechanical respirators? Explain.

8.71 Define spirometry.

Answers to Practice Exercises

8.1. i. a. reactants: C_3H_8 and O_2
b. products: CO_2 and H_2O
c. 1 C_3H_8 to 5 O_2 yields 3 CO_2 and 4 H_2O
d. All are gases.
e. This reaction is a combustion reaction because an organic substance reacts with oxygen to produce CO_2 and water.

ii. a. reactants: C_4H_6 and H_2
b. product: C_4H_{10}
c. 1 C_4H_6 to 2 H_2 yields 1 C_4H_{10}
d. All are gases.
e. This is not a combustion reaction.

iii. a. reactants: $Ba(OH)_2$ and HCl
b. products: $BaCl_2$ and H_2O
c. 1 $Ba(OH)_2$ to 2 HCl yields 1 $BaCl_2$ and 2 H_2O
d. They are all aqueous solutions except water, which is a liquid.
e. This is not a combustion reaction.

8.2 a. $\underline{2}$ $N_2O_5(g) \rightarrow \underline{4}$ $NO_2(g) + \underline{}$ $O_2(g)$
b. $\underline{}$ $C_6H_{12}O_2(l) + \underline{8}$ $O_2(g) \rightarrow$
$\underline{6}$ $CO_2(g) + \underline{6}$ $H_2O(g)$
c. $\underline{}$ $CaCO_3(s) + \underline{2}$ $HCl(aq) \rightarrow$
$\underline{}$ $CO_2(g) + \underline{}$ $H_2O(l) + \underline{}$ $CaCl_2(aq)$

d. $\underline{}$ $Mg(s) + \underline{2}$ $HCl(aq) \rightarrow$
$\underline{}$ $MgCl_2(aq) + \underline{}$ $H_2(g)$
e. $\underline{4}$ $NH_3(g) + \underline{3}$ $O_2(g) \rightarrow$
$\underline{2}$ $N_2(g) + \underline{6}$ $H_2O(l)$

8.3 The reaction is balanced because the same number of all types of atoms are present on both sides of the equation: 6 Na, 2 P, 14 O, 6 H, and 3 Ca.

E8.1

$$12.0 \text{ g } C_2H_6O \times \frac{1 \text{ mol } C_2H_6O}{46.07 \text{ g } C_2H_6O} \times \frac{3 \text{ mol } O_2}{1 \text{ mol } C_2H_6O}$$
$$\times \frac{32 \text{ g } O_2}{1 \text{ mol } O_2} = 25.0 \text{ g } O_2$$

E8.2

$$C_3H_8(g) + 5 \text{ } O_2 \text{ } (g) \rightarrow 3 \text{ } CO_2 \text{ } (g) + 4 \text{ } H_2O(g)$$

$$500.0 \text{ g } C_3H_8 \times \frac{1 \text{ mol } C_3H_8}{44.094 \text{ g } C_3H_8} \times \frac{5 \text{ mol } O_2}{1 \text{ mol } C_3H_8} \times$$

$$\frac{32.00 \text{ g } O_2}{1 \text{ mol } O_2} = 1814 \text{ g } O_2$$

$$500.0 \text{ g } C_3H_8 \times \frac{1 \text{ mol } C_3H_8}{44.094 \text{ g } C_3H_8} \times \frac{3 \text{ mol } CO_2}{1 \text{ mol } C_3H_8} \times$$

$$\frac{44.01 \text{ g } CO_2}{1 \text{ mol } CO_2} = 1497 \text{ g } CO_2$$

E8.3

$$C_6H_{12}O_6 + 6 O_2 \rightarrow 6 CO_2 + 6 H_2O$$

$$10.0 \text{ g } C_6H_{12}O_6 \times \frac{1 \text{ mol } C_6H_{12}O_6}{180.156 \text{ g } C_6H_{12}O_6} \times$$

$$\frac{6 \text{ mol } CO_2}{1 \text{ mol } C_6H_{12}O_6} \times \frac{44.01 \text{ g } CO_2}{1 \text{ mol } CO_2} = 14.7 \text{ g } CO_2$$

E8.4

$$20.0 \text{ g } HCl \times \frac{1 \text{ mol } HCl}{36.46 \text{ g } HCl} \times \frac{1 \text{ mol } NaCl}{1 \text{ mol } HCl} \times$$

$$\frac{58.44 \text{ g } NaCl}{1 \text{ mol } NaCl} = 32.1 \text{ g } NaCl$$

8.4 **a.** $5.79 \text{ kcal} \times \frac{10^3 \text{ cal}}{1 \text{ kcal}} = 5790 \text{ cal}$

b. $48.8 \text{ J} \times \frac{1 \text{ cal}}{4.184 \text{ J}} = 11.7 \text{ cal}$

8.5 $2.45 \text{ cal} \times \frac{4.184 \text{ J}}{1 \text{ cal}} = 10.3 \text{ J}$

8.6 $2720 \text{ Calories} \times \frac{1 \text{ kcal}}{1 \text{ Calories}} \times \frac{10^3 \text{ cal}}{1 \text{ kcal}} \times \frac{4.184 \text{ J}}{1 \text{ cal}}$

$$= 1.14 \times 10^7 \text{ J}$$

8.7 **a.** endothermic **b.** endothermic

8.8 The strength of these bonds makes the products lower in energy than the reactants; therefore, the change in enthalpy is negative, releasing energy as in an exothermic process.

8.9

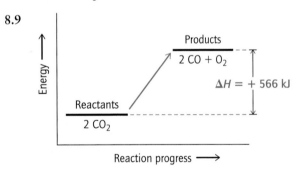

8.10 An endothermic reaction absorbs heat energy ($\Delta H > 0$) and an exothermic reaction releases heat energy ($\Delta H < 0$).

8.11 144 Cal, or 140 Cal when rounded

8.12 157 Cal, or 160 Cal when rounded

8.13 Catabolic reactions degrade larger molecules into smaller molecules and release energy in the process. Anabolic reactions build larger molecules from smaller molecules and absorb energy in the process.

8.14 **a.** anabolic, absorbs energy **b.** catabolic, releases energy **c.** anabolic, absorbs energy

8.15 The activation energy is the amount of energy that must be supplied to the reactants for a reaction to occur.

8.16 The energy diagram on the bottom with the lower "hill" is the faster reaction because it has the smaller energy of activation, E_A.

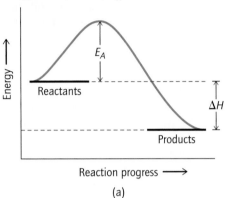

(a)

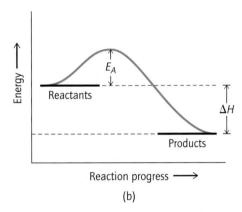

(b)

8.17 **a.** Increases the rate of the reaction because the concentration of the reactants has increased with the addition of more H_2O_2, a reactant.
b. Decreases the rate of the reaction because the temperature has been reduced, thereby reducing the kinetic energy of the reactants, and so reducing the number of collisions and their kinetic energy.
c. Increases the rate of the reaction because a catalyst speeds up a reaction by facilitating the interaction between the reactants.

8.18 The rate of a reaction decreases with time because the reactants are being consumed, which leads to fewer reactant molecules that can collide and react.

8.19 Enzymes

8.20 See Figure 8-10.

Acids, Bases, pH, and Buffers

CHAPTER 9

Acid–base imbalances in the body affect humans and animals alike. [© LWA-Dann Tardif/Corbis]

OUTLINE

 This icon indicates that a **Problem-Solving Tutorial** is available at www.whfreeman.com/gob

307

Ethylene Glycol Poisoning: A Matter of Acid Imbalance

In May of 2001, a jury found Julia Turner guilty of murdering her husband, Glen Turner, by slipping antifreeze (ethylene glycol) into his food. Ethylene glycol, $HOCH_2CH_2OH$, is a colorless, odorless liquid found in antifreeze and deicing fluids. After being ingested, ethylene glycol is broken down into toxic organic acids, such as oxalic and glyoxylic acids, causing a potentially fatal condition. Once oxalic acid reaches the kidneys, it reacts with calcium ions to form calcium oxalate crystals, which cause permanent damage to the kidneys and eventual death. Death resulting from ethylene glycol poisoning is confirmed by the presence of calcium oxalate crystals in the kidneys during an autopsy.

Acids are compounds that release protons, H^+, in aqueous solution, leading to the formation of the hydronium ion, H_3O^+. The following equation shows how oxalic acid, for example, releases a proton to produce H_3O^+:

$$H_2C_2O_4 + H_2O \longrightarrow HC_2O_4^- + H_3O^+$$
Oxalic acid   Oxalate ion Hydronium ion

The pH of the blood is a measure of H_3O^+ concentration, which must be maintained within a very narrow range. Outside this range, breathing becomes difficult, and cells are irreversibly damaged.

The symptoms of ethylene glycol poisoning include dizziness, confusion, and irregular breathing. Symptoms like these caused Glenn Turner to go to a hospital reporting "flulike" symptoms, but the hospital released him without determining the cause of his symptoms. A final dose delivered by his wife after his release killed him. In fact, no one realized that he had been murdered until his wife's boyfriend died in the same manner several years later.

Had the hospital doctors known that Turner had been poisoned with ethylene glycol, they could have treated him by administering drugs to counteract the effects of the toxic acids. Hemodialysis could also have been performed to eliminate unmetabolized ethylene glycol and toxic organic acids from the blood. See *Chemistry in Medicine: Kidney Dialysis* in Chapter 4 for more information about hemodialysis.

The effect of ethylene glycol in animals is very similar to that in humans. In fact, Julia Turner inquired about the effects of ethylene glycol on cats at a local cat shelter before poisoning her husband. Indeed, several hundred pets die each year of accidental ethylene glycol poisoning. Attracted to its sweet flavor, they may ingest antifreeze or deicing fluid that has leaked or was spilled on a garage floor. Today, the nontoxic substitute, propylene glycol, is being used in some products that formerly contained ethylene glycol. Unlike ethylene glycol, the metabolism of propylene glycol does not produce toxic substances. In this chapter, you will learn the chemistry of acids and bases, which are some of the most important and common reactants in chemical reactions. As you have seen, acids and bases play a critical role in health.

$H_2C_2O_4$
Oxalic acid

You are undoubtedly familiar with common household acids and bases. Perhaps when you were younger you performed the tried and true "volcano" science experiment: You would have added vinegar to baking soda and watched an "eruption" occur. The eruption you witnessed was actually a chemical reaction between an acid and a base—between acetic acid (vinegar) and the base sodium bicarbonate (baking soda). Acids and bases, in addition to making fun science projects, play a critical role in health and medicine. For example, the gastric juices in your stomach contain an acid, hydrochloric acid, that helps to digest food. Many drugs, such as morphine and other amines, are bases. Amino acids, the building blocks of proteins, have characteristics of both an acid and a base. In this chapter, you will learn many of the unique properties of acids and bases, as well as their role in health and medicine.

9.1 Properties of Acids and Bases

In Chapter 8, you learned the basic principles of chemical reactions. Chemical reactions of acids and bases represent the simplest type of reaction, characterized by the transfer of a single proton. Recall that a hydrogen atom has one electron and one proton. Therefore, a hydrogen ion (H^+) is a proton.

Acids and bases can be defined by how they react in aqueous solution. Sixty percent of the human body is water, and reactions in the cell occur in aqueous solution. So, all of the reactions that we will consider in this chapter will be in aqueous solution. There are several different theories that define the behavior of acids and bases in aqueous solution. One definition of an acid is that it is a substance that produces protons (H^+) when dissolved in water, while a base is a substance that produces hydroxide ions (OH^-) when dissolved in water.

A "volcano" erupts—an example of an acid–base reaction between acetic acid (vinegar) and sodium bicarbonate (baking soda). [© David Young-Wolff/Photo Edit]

Definition 1

> An acid is a substance that produces protons (H^+) when dissolved in water.
> A base is a substance that produces hydroxide ion (OH^-) when dissolved in water.

Definition 1 is commonly known as the Arrhenius definition.

Another definition classifies a molecule or ion that *donates* a proton (H^+) to another molecule or ion as an **acid,** and a molecule or ion that *accepts* a proton from another molecule or ion as a **base.** Although we will come back to the first definition, we will focus mainly on Definition 2 throughout this text.

Definition 2

> An acid is a proton donor.
> A base is a proton acceptor.

Definition 2 is commonly known as the Brønsted-Lowry definition.

Formation of Hydronium Ions

What happens to the proton released by an acid in aqueous solution? When an acid releases a proton, the proton forms a covalent bond to the oxygen atom in an H_2O molecule to make the hydronium ion, H_3O^+. The Lewis dot structure for the hydronium ion is shown in the margin.

Therefore, acids produce hydronium ions. For example, gastric acid from the stomach contains hydrochloric acid, HCl, which donates a proton to water to produce a hydronium ion and a chloride ion:

$$\begin{array}{ccccccc} HCl & + & H_2O & \longrightarrow & H_3O^+ & + & Cl^- \\ \text{Hydrochloric acid} & & \text{Water} & & \text{Hydronium ion} & & \text{Chloride ion} \end{array}$$

Table 9.1 contains the chemical structures of some common acids in biochemistry and medicine. The hydrogen atoms printed in red are released as protons in solution. Note that only certain hydrogen atoms in an acid can be released as protons from the molecule.

$$\left[\begin{array}{c} H - \overset{\displaystyle ..}{O} - H \\ | \\ H \end{array} \right]^+$$

A hydronium ion

Table 9-1 Common Acids in Biochemistry and Medicine

Molecule	Structure*	Biological Role
Hydrochloric acid (HCl)	H—Cl	HCl is the main component of gastric acid, found in the human stomach.
Phosphoric acid (H_3PO_4)	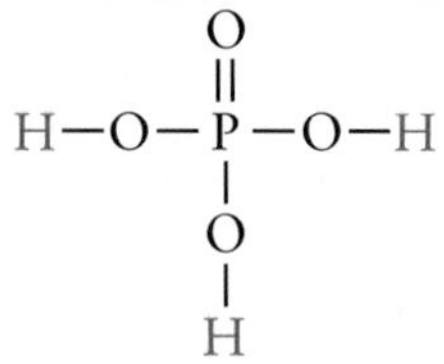 (structure: H—O—P(=O)(—O—H)—O—H)	Phosphate esters are derived from phosphoric acid. They are present in ATP, DNA, and RNA.
Carbonic acid (H_2CO_3)	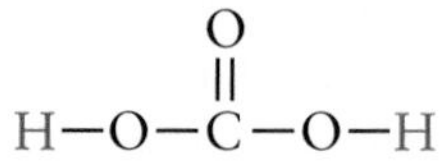 (structure: H—O—C(=O)—O—H)	When carbon dioxide (CO_2) dissolves in the blood, carbonic acid is formed.
Acetic acid ($C_2H_4O_2$)	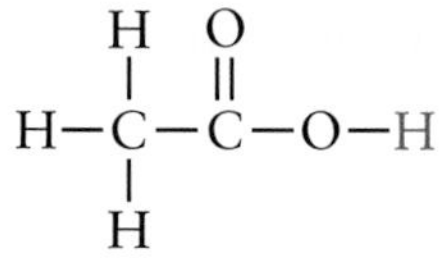 (structure: H—C(H)(H)—C(=O)—O—H)	Acetic acid is central to the metabolism of carbohydrates and fats.
Citric acid ($C_6H_8O_7$)	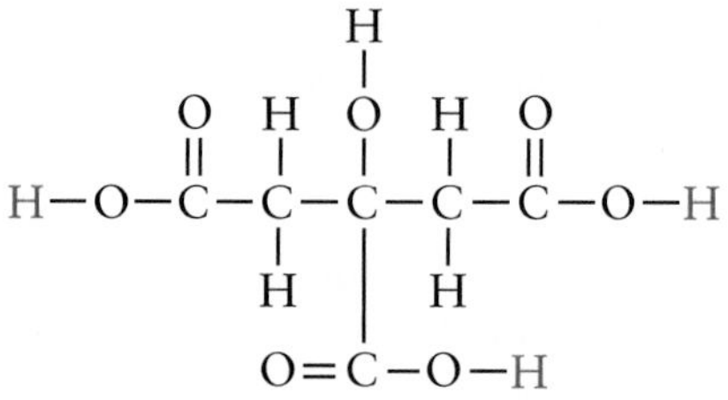 (structure: H—O—C(=O)—C(H)(H)—C(OH)(—C(=O)—O—H)—C(H)(H)—C(=O)—O—H)	Citric acid is formed in the citric acid cycle, the metabolic pathway that converts acetyl CoA into carbon dioxide, water, and energy.
Lactic acid ($C_3H_6O_3$)	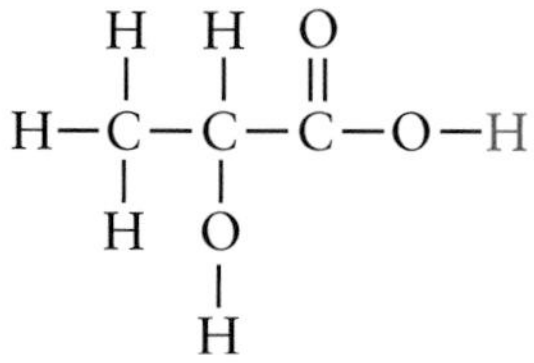 (structure: H—C(H)(H)—C(H)(O—H)—C(=O)—O—H)	Lactic acid is produced in the absence of oxygen during normal metabolism, such as during intense exercise.
Pyruvic acid ($C_3H_4O_3$)	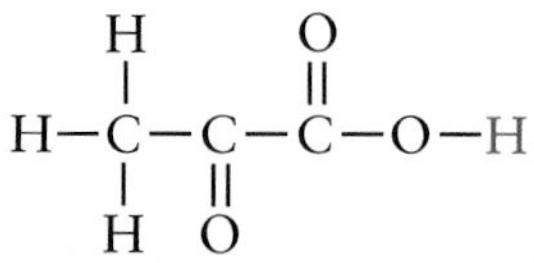 (structure: H—C(H)(H)—C(=O)—C(=O)—O—H)	Pyruvic acid is produced in the metabolism of sugar.

*The hydrogen atoms printed in red can be released as protons in aqueous solution.

Formation of Hydroxide Ions

When a base is dissolved in water, hydroxide ions (OH^-) are produced. Hydroxide ions are produced from a base in one of two ways:

- The base accepts a proton from a H_2O molecule—creating a hydroxide, OH^-, ion. The dissolution of ammonia in water is an example of this case. In aqueous solution, ammonia accepts a proton from water, producing an ammonium ion and a hydroxide ion:

$$\ddot{N}H_3 \ + \ H_2\ddot{O}: \ \rightleftharpoons \ NH_4^+ \ + \ :\ddot{O}H^-$$

Ammonia   Water   Ammonium ion   Hydroxide ion

The nonbonding pair of electrons on nitrogen allows ammonia to form a covalent bond to a proton (H⁺) donated by the water molecule. Indeed, all bases contain at least one nonbonding pair of electrons. Notice that the equation arrows point in both directions. This indicates that the reaction is reversible; the products also react to produce reactants.

In the body, about 99% of the ammonia molecules produced are converted into the ammonium ion, NH_4^+.

- The base is an ionic compound containing a hydroxide ion (OH^-) that simply dissociates in solution, releasing hydroxide ions (OH^-) into aqueous solution. In this case, a proton is not accepted, but since OH^- is released, it is a base according to Definition 1. For example, sodium hydroxide, NaOH, is a base. Known commonly as lye, NaOH is used in a variety of consumer applications ranging from cleaning drains to curing olives.

$$NaOH(s) + H_2O \longrightarrow Na^+(aq) + OH^-(aq)$$

Lye (Sodium hydroxide)   Water   Sodium ion   Hydroxide ion

All bases contain a nonbonding pair of electrons and they are either neutral like ammonia or negatively charged like the hydroxide ion (OH^-). Table 9.2 contains a list of commonly encountered organic bases in biochemistry and medicine. Notice that they all contain an amine functional group, the only basic functional group you have learned about thus far. The atom with the nonbonding pair of electrons is printed in blue in these structures.

WORKED EXERCISE 9-1   Acids and Bases

Write the equation that represents the reaction of acetic acid in water.

SOLUTION

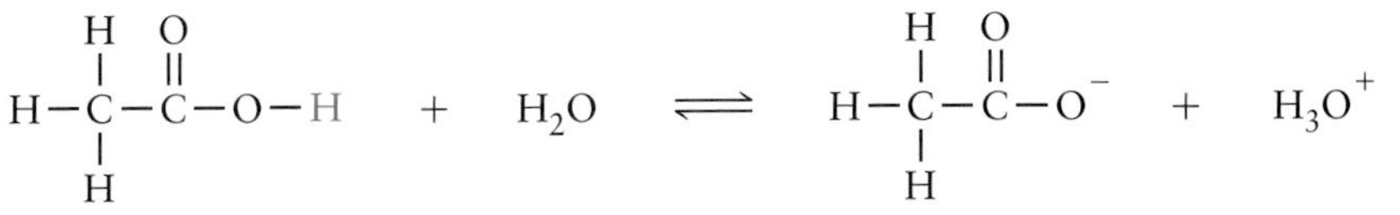

$$CH_3COOH + H_2O \rightleftharpoons CH_3COO^- + H_3O^+$$

PRACTICE EXERCISES

Fill in the blanks for the following statements.

9.1 A base is a molecule or ion that _____ a proton.

9.2 When dissolved in water, an acid will produce _____ ions.

9.3 Write the equation that represents the reaction of pyruvic acid in water.

Conjugate Acids and Bases

The properties of water itself are important for understanding how acids and bases behave in aqueous solution. In pure water, water molecules react with one another to some extent to produce a small amount of hydronium ions (H_3O^+) and hydroxide ions (OH^-):

$$H_2O + H_2O \rightleftharpoons H_3O^+ + OH^-$$

This reaction is known as the **autoionization of water.** Since water produces both H_3O^+ and OH^-, water is considered an **amphoteric molecule**—it can act as either an acid or a base. When water reacts with an acid, it

Table 9-2 Common Bases in Biochemistry and Medicine

Molecule	Structure*	Biological Role
Ammonia (NH_3)	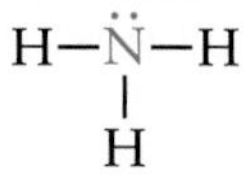	Ammonia is one of the end products of amino acid metabolism
Epinephrine (Adrenaline)	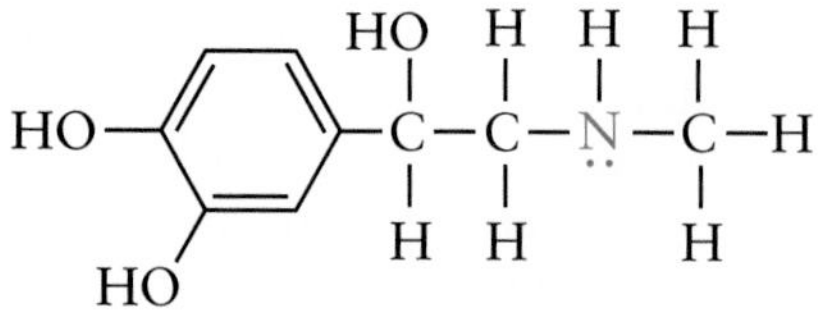	A hormone and a neurotransmitter
Dopamine	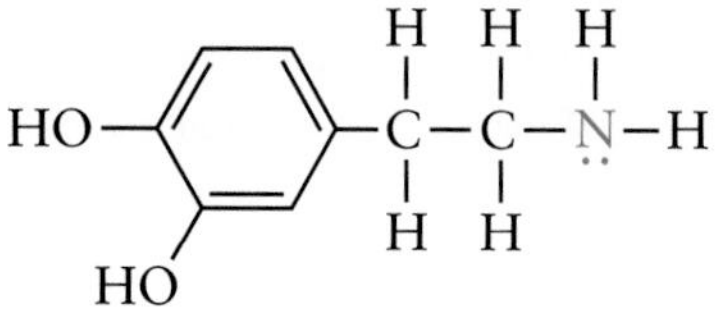	A hormone and a neurotransmitter
Adenine	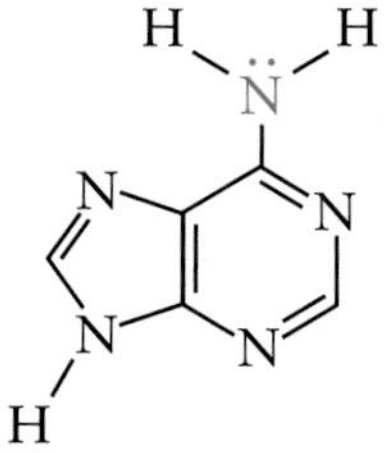	A part of many important molecules involved in energy production during metabolism
Chlorphenamine	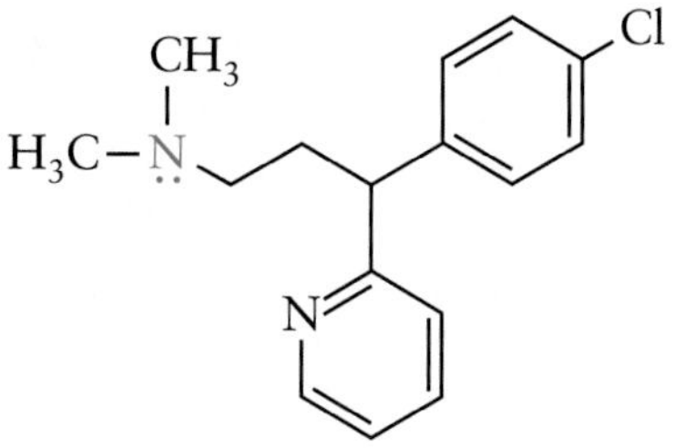	Pharmaceutical used as an antihistamine
Ephedrine	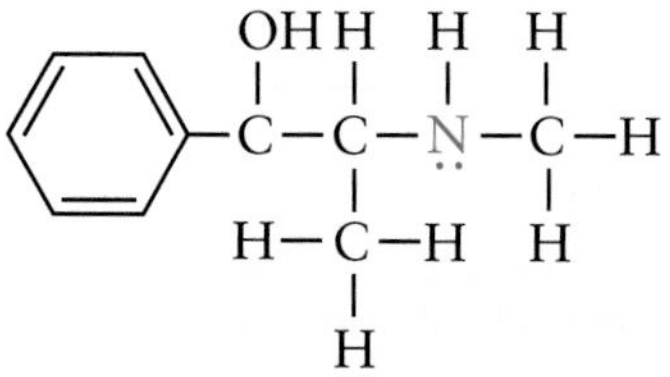	Pharmaceutical used as a decongestant

*Nonbonding pairs of electrons are shown in blue.

accepts a proton, forming the hydronium ion (H_3O^+); in this case, the water molecule acts as a base. When water reacts with a base, it donates a proton, forming hydroxide ion (OH^-); in this case, the water molecule acts as an acid.

Two molecules or ions that differ only in the presence or absence of a proton constitute a **conjugate acid–base pair.** When an acid releases a proton, the part of the acid that is left is known as the **conjugate base** of the acid. The released proton is used to produce a hydronium ion, as we have seen. For example, when pyruvic acid is added to water, it produces hydronium ions and pyruvate ions. Pyruvate is the *conjugate base* of pyruvic acid.

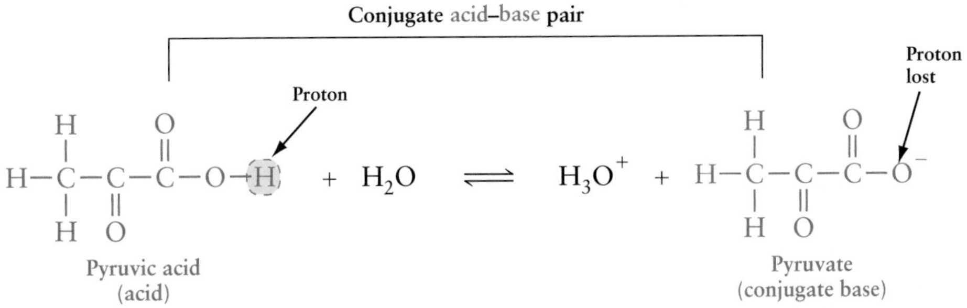

Conjugate acid–base pair

$$H-\underset{\underset{H}{|}}{\overset{\overset{H}{|}}{C}}-\underset{\underset{O}{||}}{\overset{\overset{O}{||}}{C}}-\overset{O}{\overset{||}{C}}-O-\text{H} \quad + \quad H_2O \quad \rightleftharpoons \quad H_3O^+ \quad + \quad H-\underset{\underset{H}{|}}{\overset{\overset{H}{|}}{C}}-\underset{\underset{O}{||}}{\overset{\overset{O}{||}}{C}}-\overset{O}{\overset{||}{C}}-O^-$$

Proton

Proton lost

Pyruvic acid
(acid)

Pyruvate
(conjugate base)

In this example, you can see that water acts as a base, accepting a proton from pyruvic acid. Therefore, hydronium ion is the conjugate acid of water.

When a base accepts a proton from water, the substance produced is known as the **conjugate acid** of the base. The water molecule donates a proton, thereby becoming a hydroxide ion. For example, when acetate dissolves in water, it produces hydroxide ions and acetic acid. Acetic acid is the *conjugate acid* of acetate.

Conjugate acid–base pair

$$CH_3COO^- \quad + \quad H_2O \quad \rightleftharpoons \quad OH^- \quad + \quad CH_3COOH$$

Acetate
(base)

Acetic acid
(conjugate acid)

In the above example, water acts as an acid, donating a proton to the acetate ion. Therefore, the conjugate base of water is the hydroxide ion (OH^-).

Notice from the two previous examples that acetic and pyruvic acids both contain carboxylic acid functional groups. Recall from Chapter 7 that carboxylic acids may exist in two forms, neutral and negatively charged. The negatively charged form is the carboxylate ion, which is the conjugate base of the carboxylic acid. Notice that most of the acids listed in Table 9-1 contain a carboxylic acid functional group.

WORKED EXERCISE 9-2 Conjugate Acids and Bases

Label the conjugate acid–base pairs in the reaction below. Does water act as an acid or a base in this reaction?

$$HF + H_2O \rightleftharpoons H_3O^+ + F^-$$

SOLUTION

Conjugate acid–base pair

$$HF \quad + \quad H_2O \quad \rightleftharpoons \quad H_3O^+ \quad + \quad F^-$$

Acid

Conjugate base

Water acts as a base in this reaction, accepting a proton.

PRACTICE EXERCISES

9.4 For each of the following equations, label the conjugate acid–base pairs. Indicate whether water acts as an acid or a base in the reaction.

a. $NH_4^+ + H_2O \rightleftharpoons H_3O^+ + NH_3$
b. $HCO_3^- + H_2O \rightleftharpoons OH^- + H_2CO_3$

Problem-Solving Tutorial:
Conjugate Acid-Base Pairs

9.5 Consider the autoionization of water:

$$H_2O + H_2O \rightleftharpoons H_3O^+ + OH^-$$

Is water acting as an acid, a base, or both in this reaction? Explain. Label the conjugate acid–base pairs.

9.6 Recall that carboxylic acids may exist in two forms, neutral and negatively charged. The negatively charged form is the carboxylate ion, which is the conjugate base of the carboxylic acid. For lactic acid and pyruvic acid, structures contained in Table 9-1, write the chemical equations for the dissociation of the acid in water, and label the conjugate acid–base pairs.

Strengths of Acids and Bases

An acid or a base is classified as either *strong* or *weak*, depending on the extent to which it dissociates in water. Both strong and weak acids and bases can be highly corrosive. As dilute solutions, however, most are generally safe to handle.

> Strong acids are also strong electrolytes, because they dissociate completely.

Strong Acids Every molecule of a **strong acid** donates a proton to water to form H_3O^+. Hydrochloric acid (HCl) is classified as a strong acid because all HCl molecules donate their proton to form H_3O^+ in solution; no HCl molecules remain in solution. Complete dissociation is indicated in a chemical equation by a single forward arrow:

$$\underset{\text{Hydrochloric acid}}{HCl(g)} + \underset{\text{Water}}{H_2O} \longrightarrow \underset{\text{Hydronium ion}}{H_3O^+(aq)} + \underset{\text{Chloride ion}}{Cl^-(aq)}$$

Figure 9-1 illustrates the concept of complete dissociation of a strong acid (HCl) in aqueous solution.

Hydrochloric acid exists in your stomach at concentrations around $0.1\ M$ to $0.01\ M$. This concentration is sufficiently dilute not to damage the stomach, but concentrated enough to damage the more delicate lining of the esophagus. Hence, chronic acid reflux—the upward movement of stomach acid into the esophagus—damages the esophagus. For more information on hydrochloric acid in your body, see this chapter's *Chemistry in Medicine: Why Some Drugs Cannot Be Administered Orally: The Issue of pH.*

> Chronic acid reflux disease is also known as GERD, the abbreviation for **G**astro **E**sophageal **R**eflux **D**isorder.

Hydrochloric acid is the only strong acid present in the body. In the laboratory, other strong acids include nitric acid, HNO_3, and sulfuric acid, H_2SO_4. Exposure to any of these acids in concentrated form can cause

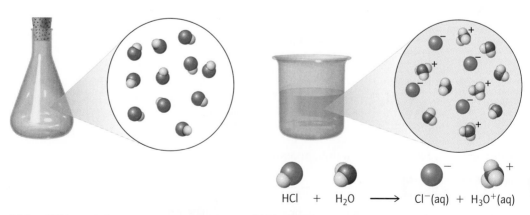

(a) Pure HCl is a colorless gas at room temperature.   (b) Dissolved in water, HCl donates a proton to an H_2O molecule.

Figure 9-1 A strong acid completely dissociates in aqueous solution. (a) Pure hydrochloric acid, HCl, has a covalent bond; it is undissociated. (b) When dissolved in water, all HCl molecules donate a proton to water molecules. The resulting solution contains only hydronium ions, chloride ions, and water molecules.

severe chemical burns. Table 9-3 lists some common strong acids and their molecular formulas.

Strong Bases **Strong bases** consist of ionic compounds containing hydroxide ions that completely dissolve in water. Sodium hydroxide, NaOH, is the most common strong base used in chemical laboratories. When added to water, the sodium hydroxide lattice completely dissociates into sodium ions and hydroxide ions; no NaOH exists in solution:

$$\underset{\text{Sodium hydroxide}}{\text{NaOH(s)}} + \underset{\text{Water}}{\text{H}_2\text{O(l)}} \longrightarrow \underset{\text{Sodium ion}}{\text{Na}^+\text{(aq)}} + \underset{\text{Hydroxide ion}}{\text{OH}^-\text{(aq)}}$$

Figure 9-2 illustrates the complete dissociation of a strong base (NaOH) in aqueous solution.

Table 9-4 lists some strong bases and their chemical formulas. Notice that the cations of these strong bases appear in group 1A and group 2A of the periodic table, the alkali and alkaline earth metals.

Weak Acids **Weak acids** are distinguished by the fact that they do not completely dissociate in water; a significant percentage of the undissociated acid is always present in solution. Indeed, only a small percentage of the molecules of a weak acid will donate a proton to water. For example, when acetic acid dissolves in water, most of it remains in the form of acetic acid molecules and only a small fraction is ionized to the conjugate base, acetate ion:

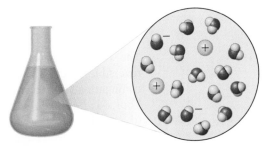

The dissociation of a weak acid in aqueous solution is a reversible reaction, as indicated by the double arrows. In a reversible reaction, both the forward and the reverse reactions are occurring simultaneously: Molecules of acetic acid are reacting with water molecules to form hydronium and acetate ions at the same time that hydronium ions and acetate ions are reacting to form acetic acid and water molecules. Hence, all species—the hydronium ion, the acetate ion, and

Table 9-3	Some Strong Acids
Formula	**Name**
HNO_3	Nitric acid
HCl	Hydrochloric acid
H_2SO_4	Sulfuric acid
$HClO_4$	Perchloric acid
HBr	Hydrobromic acid
HI	Hydroiodic acid

Table 9-4	Some Strong Bases
Formula	**Name**
KOH	Potassium hydroxide
$Ba(OH)_2$	Barium hydroxide
NaOH	Sodium hydroxide (lye)
$Sr(OH)_2$	Strontium hydroxide
$Ca(OH)_2$	Calcium hydroxide
LiOH	Lithium hydroxide

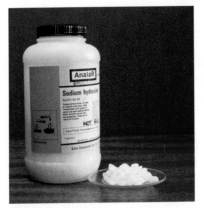

(a) Sodium hydroxide, NaOH, is a solid at room temperature.

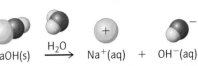

(b) Sodium hydroxide dissociates in water to form a strong base, OH⁻.

Figure 9-2 A strong base completely dissociates in aqueous solution. (a) Undissolved sodium hydroxide, NaOH, exists as a lattice containing ionic bonds. (b) When dissolved in water, NaOH completely dissociates into Na⁺ and OH⁻. [(a) Andrew Lambert Photography/Photo Researchers Inc.]

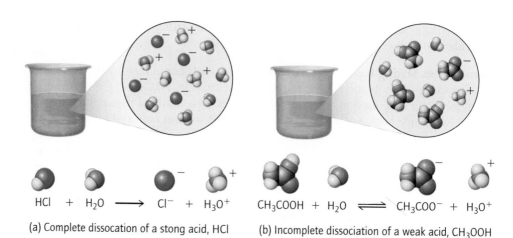

HCl + H$_2$O $\longrightarrow$ Cl$^-$ + H$_3$O$^+$ CH$_3$COOH + H$_2$O $\rightleftharpoons$ CH$_3$COO$^-$ + H$_3$O$^+$

(a) Complete dissocation of a stong acid, HCl   (b) Incomplete dissociation of a weak acid, CH$_3$OOH

Figure 9-3 The difference between a strong acid and a weak acid in aqueous solution. (a) A strong acid like HCl completely dissociates. (b) A weak acid like acetic acid does **not** completely dissociate.

Table vinegar is an aqueous solution of acetic acid. [Iconotec/Alamy]

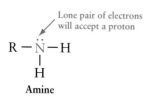

Lone pair of electrons will accept a proton

R — N̈ — H
        |
        H
Amine

acetic acid—are present together in aqueous solution. Figure 9-3 illustrates the difference between a strong acid and a weak acid dissolved in aqueous solution. Generally, if you do not see an acid listed in the table of strong acids, you may assume it is a weak acid.

Note that all organic compounds containing a carboxylic acid functional group are weak acids. The loss of a proton from a carboxylic acid results in a carboxylate ion, a weak base. Since carboxylic acids are weak acids, the concentration of the carboxylic acid will be greater than the concentration of carboxylate ion produced. Many molecules that participate in biochemical reactions in the body contain carboxylic acid functional groups. We will see later in the chapter that the ability of a carboxylic acid to exist in the carboxylate form is critical to keeping these molecules within the confines of the cell where reactions take place.

Weak Bases Weak Bases are distinguished by the fact that a significant percentage of the base is always present in solution; only a small percentage accepts a proton from water, forming the conjugate acid. Ammonia is an example of a weak base. In water, some ammonia molecules accept a proton from water, producing ammonium ions (the conjugate acid) and hydroxide ions.

$$:NH_3 \quad + \quad H_2O \quad \rightleftharpoons \quad NH_4^+ \quad + \quad OH^-$$

Ammonia   Water   Ammonium ion   Hydroxide ion

The double arrows in this equation indicate that ammonia undergoes a reversible reaction. Thus, ammonia is present in solution together with its conjugate acid, the ammonium ion, and hydroxide ion. Perform *The Model Tool 9-1: Equilibrium of a Weak Base* to gain further insight into the reaction of a weak base with water.

Weak bases have a number of common applications and biological functions. Dilute solutions of ammonia are often used as cleaning fluids, although exposure to high levels of concentrated gaseous ammonia can cause lung problems and even death. Like ammonia, amines are also weak bases. The amine functional group is found in a number of pharmaceuticals, as well as in the important neurotransmitters present in the brain. Many pharmaceutical agents containing the amine functional group are actually prepared in their conjugate acid form, because they are ions in this form, and as such are charged and therefore more soluble in aqueous solutions that can be taken orally.

The Model Tool 9-1 Equilibrium of a Weak Base

I. Construction Exercise Part I: The Forward Reaction

1. Obtain one red oxygen atom, five light-blue hydrogen atoms, and one blue nitrogen atom. Obtain 5 single bonds.
2. Construct a model of ammonia, NH_3, and a model of water, H_2O.
3. Simulate the *forward* reaction of one ammonia molecule with one water molecule, by forming a model of one ammonium ion and a model of one hydroxide ion using only the models of ammonia and water.

$$NH_3 + H_2O \longrightarrow NH_4^+ + OH^-$$

Inquiry Questions

1. Which bond did you *break* in the reactants in order to form the ammonium ion and hydroxide ion?
2. Which bond did you *make* in order to form the ammonium ion and the hydroxide ion?
3. Which is acting as the acid, NH_3 or H_2O?

II. Construction Exercise Part II: The Reverse Reaction

1. Using your ammonium ion and hydroxide ion from Part I, simulate the *reverse* reaction of one ammonium ion with one hydroxide ion, by forming one ammonia molecule and one water molecule using only your models of the ammonium ion and the hydroxide ion.

$$NH_4^+ + OH^- \longrightarrow NH_3 + H_2O$$

Inquiry Questions

1. Which bond did you *break* in order to form ammonia and water?
2. Which bond did you *make* in order to form ammonia and water?
3. Which is acting as the acid, NH_4^+ or OH^-?

III. Final Questions

1. In both the forward and reverse reactions, what type of atom was transferred?
2. Label the conjugate acid–base pairs.
3. Write a balanced equilibrium equation for the reaction between ammonia and water.
4. Write a balanced equilibrium equation for the analogous reaction between methylamine (CH_3NH_2) and water.

Amino acids, the building blocks of proteins, contain both an amine and a carboxylic acid functional group. In the cell, the amine is in its conjugate acid form and the carboxylic acid is in its conjugate base form. Therefore, amino acids are amphoteric molecules—they can react as either an acid or a base. The structures of some amino acids are shown below in the form most commonly found under physiological conditions:

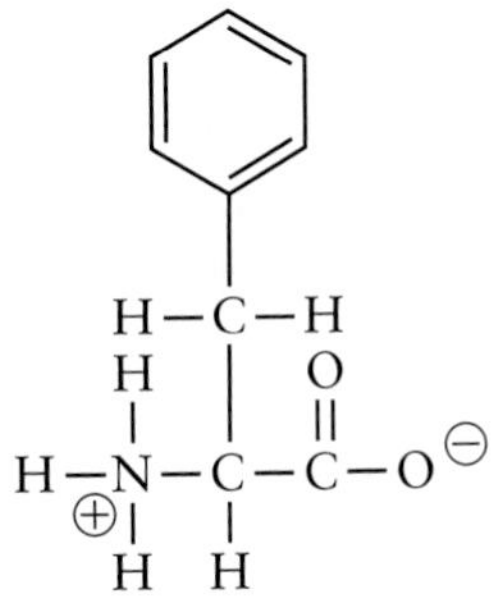

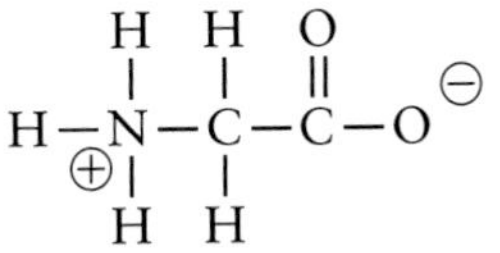

Glycine

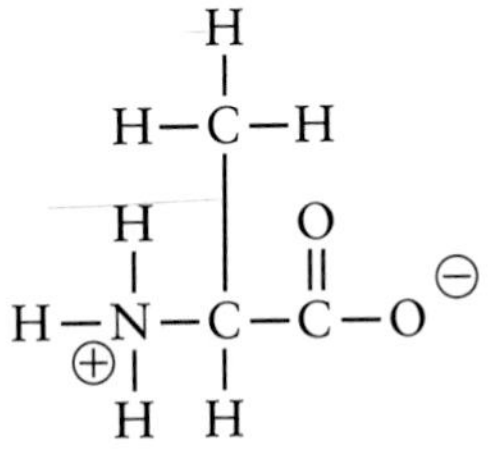

Alanine

Phenylalanine

WORKED EXERCISE 9-3   Strengths of Acids and Bases

Consult Tables 9-3 and 9-4. Does the chemical equation below represent the dissociation of a strong acid, a weak acid, a strong base, or a weak base? Explain.

$$H_2CO_3 + H_2O \rightleftharpoons H_3O^+ + HCO_3^-$$

SOLUTION

This equation represents the dissociation of a weak acid. You can recognize that H_2CO_3 is an acid because hydronium ions (H_3O^+) are produced, and you know that it is a weak acid because H_2CO_3 is not on the list of strong acids. Also, notice the double arrow, indicating a reversible reaction, characteristic of a weak acid.

PRACTICE EXERCISES

9.7 Indicate whether each of the following is a strong acid, a weak acid, a strong base, or a weak base. For each, write the chemical equations that represent their dissociation in water.

  a. LiOH   b. HNO_3 c. CH_3COO^-

9.8 Predict the products for the aqueous reactions shown below.

  a.

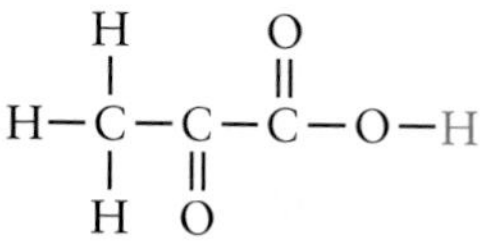

  b. $HNO_3 + H_2O$

  c.

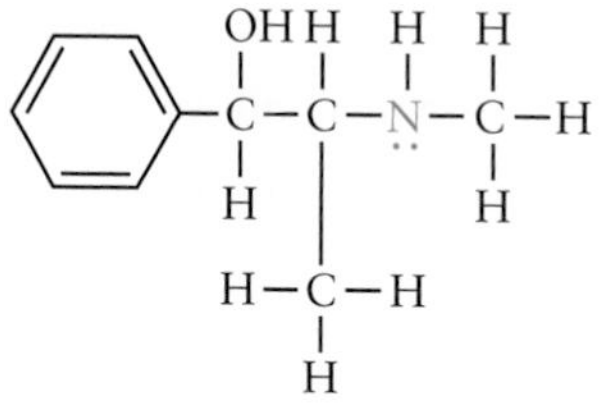

  d. $NaOH + H_2O$

  e.

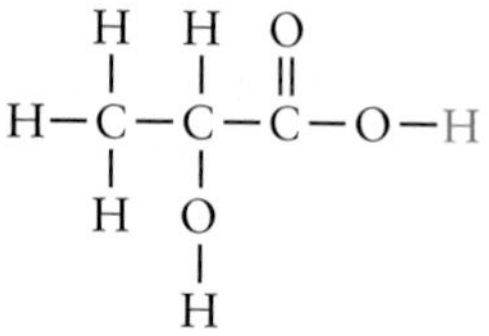

9.9 Carnosine is found in muscle and brain tissue. Is carnosine an acid, a base, or an amphoteric molecule? Explain.

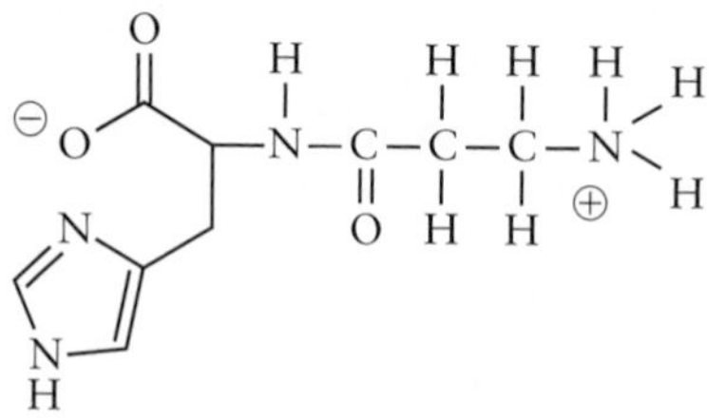

9.10 **Critical Thinking Question:** Adenine, whose structure is shown below, is a part of the structure of DNA and RNA. Would you classify adenine as an acid or a base? Explain.

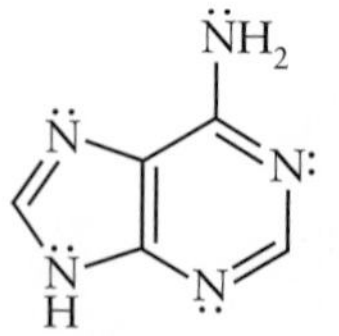

Acid–Base Equilibria

You observed in the previous section that weak acids and weak bases undergo reversible reactions. When reactants in a reversible reaction are mixed together, both the forward and the reverse reaction occur simultaneously. Therefore, both products and reactants are continuously being formed. One consequence of a reversible reaction is that there are always reactants and products present in solution. The concentration of reactants and products stops changing at a point known as **equilibrium,** which we can understand by considering the reaction rate. Equilibrium is reached in a variety of different types of reactions; here we will focus on equilibrium in acid–base reactions.

Although both forward and reverse reactions proceed simultaneously in a reversible reaction, they do not necessarily proceed at the same rate. If the forward reaction proceeds at a faster rate than the reverse reaction, the concentration of the *products* will increase. If the reverse reaction proceeds at the faster rate, the concentration of the *reactants* will increase. When the forward and reverse reactions proceed at the same rate, the concentration of reactants and products no longer changes, and the reaction has reached a state of equilibrium.

At equilibrium, the reaction has not stopped. Rather, for each set of reactants that is turned into products, another set of products is turned into reactants. The concept of equilibrium is important in understanding the reactions of weak acids and bases—and equilibrium is a factor in determining the concentration of hydronium ion in the cell.

For example, consider the following reversible reaction involving pyruvic acid, an important compound in biochemical reactions:

Conjugate acid–base pair

$$\underset{\substack{\text{Pyruvic acid}\\\text{(acid)}}}{\text{H}-\overset{\overset{\text{H}}{|}}{\underset{\underset{\text{H}}{|}}{\text{C}}}-\overset{\overset{\text{O}}{\|}}{\underset{\underset{\text{O}}{\|}}{\text{C}}}-\overset{\overset{\text{O}}{\|}}{\text{C}}-\text{O}-\text{H}} \;+\; \text{H}_2\text{O} \;\rightleftharpoons\; \text{H}_3\text{O}^+ \;+\; \underset{\substack{\text{Pyruvate}\\\text{(conjugate base)}}}{\text{H}-\overset{\overset{\text{H}}{|}}{\underset{\underset{\text{H}}{|}}{\text{C}}}-\overset{\overset{\text{O}}{\|}}{\underset{\underset{\text{O}}{\|}}{\text{C}}}-\overset{\overset{\text{O}}{\|}}{\text{C}}-\text{O}^-}$$

Initially, pyruvic acid reacts with water, forming pyruvate. As the concentration of pyruvate increases, it starts to react with hydronium ion, regenerating pyruvic acid. Initially, the forward reaction proceeds much faster than the reverse reaction because so much more pyruvic acid is present than pyruvate. As pyruvic acid reacts and pyruvate is formed, the forward reaction decreases in rate and the reverse reaction increases. This continues until the rate of the forward and reverse reactions are the same and the reaction has reached equilibrium.

Although both products and reactants are present at equilibrium, *they are not present in equal concentrations.* At equilibrium the neutral form of pyruvic acid is present in much higher concentration than the ionized form. Equilibrium favors the formation of reactants for weak acids and bases. In contrast, there are other chemical reactions that favor the products, and in those cases, the concentration of products is greater than the concentration of reactants at equilibrium. Figure 9-4 illustrates the general case of equilibrium for a weak acid (acetic acid), compared to the dissociation of a strong acid (HCl).

Le Châtelier's Principle

A reaction at equilibrium can be disturbed by a change in temperature or by a change in the concentration of a reactant or product. Recall learning in Chapter 8 that for most reactions, the higher the concentration of reactants, the faster the rate of reaction observed. In a reaction at equilibrium, increasing the concentration of a reactant will increase the rate of the reaction in the

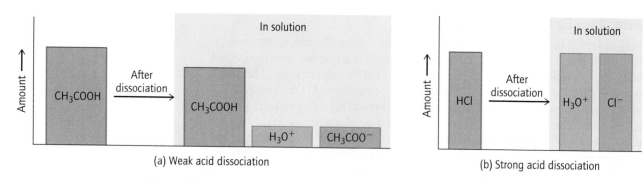

(a) Weak acid dissociation   (b) Strong acid dissociation

Figure 9-4 (a) In the reaction of a weak acid (acetic acid) with water (H_2O), the equilibrium favors the formation of reactants. (b) In the reaction of a strong acid (HCl) with water, all reactants are transformed into products.

forward direction until a new equilibrium is established, whereas increasing the concentration of a product will increase the rate of reaction in the reverse direction, again until a new equilibrium is established. ***Le Châtelier's Principle*** *states that when a reaction at equilibrium is disturbed, the reaction responds by shifting in the direction that restores equilibrium: either the forward direction (to the right, →) or the reverse direction (to the left, ←).*

For example, if a product is removed from a reversible reaction, the forward reaction will proceed at a faster rate than the reverse reaction, to increase the concentration of products until equilibrium is restored. We say that the reaction "shifts to the right." Conversely, if a product is added to a reversible reaction, the reverse reaction will proceed at a faster rate than the forward reaction, to increase the concentration of reactants until equilibrium is restored. We say that the reaction has "shifted to the left." The addition of a product (or reactant) causes the opposite shift in equilibrium from its removal. Table 9-5 summarizes the direction in which a reaction at equilibrium shifts, depending on how the concentration of one of the reactants or products is changed. Figure 9-5 illustrates Le Châtelier's principle.

Table 9-5 How Changes in Concentration Shift Equilibrium

Concentration Change	Direction Reaction Shifts	
Addition of reactant	Right	Forward →
Removal of reactant	Left	Reverse ←
Addition of product	Left	Reverse ←
Removal of product	Right	Forward →

Consider, for example, the equilibrium of the weak acid pyruvic acid in solution. An increase in hydronium ion concentration will disturb the pyruvate equilibrium. Since the concentration of one of the products, hydronium ion, has increased, Le Châtelier's principle predicts that the reaction will respond by shifting to the left: the rate of the reverse reaction increases more than the rate of the forward reaction. A shift in the equilibrium to the left consumes some of the excess H_3O^+, thereby restoring equilibrium.

$$H-\underset{\underset{H}{|}}{\overset{\overset{H}{|}}{C}}-\underset{\underset{O}{\|}}{C}-\overset{\overset{O}{\|}}{C}-O-H \ + \ H_2O \ \rightleftharpoons \ H_3O^+ \ + \ H-\underset{\underset{H}{|}}{\overset{\overset{H}{|}}{C}}-\underset{\underset{O}{\|}}{C}-\overset{\overset{O}{\|}}{C}-O^-$$

Pyruvic acid
(acid)

Increase
concentration

Pyruvate
(conjugate base)

Temporary shift to
the left eventually
restores equilibruim

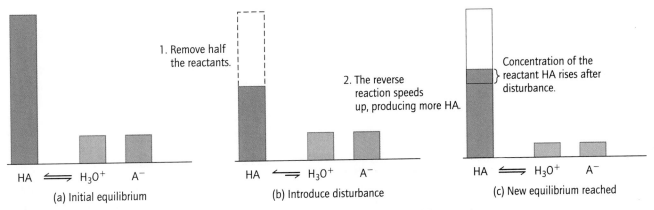

Figure 9-5 Le Châtelier's principle. Removing reactants (b) from a reaction at equilibrium (a) causes the reverse reaction to increase in rate (c): The number of reactant molecules increases and the number of product molecules decreases.

In biochemical systems equilibrium is usually never reached, but is shifting either to the right or left, according to Le Châtelier's principle.

Problem-Solving Tutorial:
Weak Acid Dissociation and Le Châtelier's Principle

WORKED EXERCISES 9-4 **Acid–Base Equilibrium and Le Châtelier's Principle**

1 Formic acid, HCOOH, is produced by bees (as a defense against predators) and in the laboratory (as a preservative and antibacterial agent in livestock feed). In aqueous solution, formic acid undergoes the following reversible reaction:

$$\underset{\text{Formic acid}}{\text{H}-\overset{\overset{\text{O}}{\|}}{\text{C}}-\text{OH}} + \underset{\text{Water}}{\text{H}_2\text{O}} \rightleftharpoons \underset{\text{Formate ion}}{\text{H}-\overset{\overset{\text{O}}{\|}}{\text{C}}-\text{O}^-} + \underset{\text{Hydronium ion}}{\text{H}_3\text{O}^+}$$

a. What substances are present at equilibrium?
b. At equilibrium, is the concentration of formic acid and formate ion constant or changing?
c. What is of the meaning of the two opposing arrows in this equation?
d. How will the equilibrium shift if additional formic acid is added to the solution?
e. How will the equilibrium shift if formate ion is removed from the solution?

2 Consider the reversible reaction shown below, which occurs in your body:

$$\underset{\text{Carbonic acid}}{\text{H}_2\text{CO}_3(\text{aq})} \rightleftharpoons \underset{\text{Carbon dioxide}}{\text{CO}_2(\text{g})} + \underset{\text{Water}}{\text{H}_2\text{O}(\text{l})}$$

This reversible reaction demonstrates how an increase or a decrease in the concentration of carbon dioxide causes a shift in the reaction to re-establish equilibrium.
a. Carbon dioxide is produced as a waste product in the tissues of the body. Does this addition of CO_2 in solution in the blood cause the reaction above to *shift to the left or to the right?* Explain.
b. In the lungs, carbon dioxide is removed as it is exhaled. Does the removal of CO_2 cause the reaction above to *shift to the left or to the right?* Explain.

Recall that CO_2 is a gas and will exist in solution according to Henry's law; see Chapter 5.

SOLUTIONS
1 a. Formic acid, water, formate ion, and hydronium ion are all present at equilibrium.
b. The concentration of formic acid and formate ion are constant at equilibrium.

c. The two opposing arrows indicate that both the forward and reverse reactions occur simultaneously.

d. The equilibrium will shift to the right, since reactant was added.

e. The equilibrium will shift to the right, since product was removed.

2 a. The addition of carbon dioxide will cause a *shift to the left*, in order to consume the excess CO_2 being produced and restore equilibrium.

b. The removal of carbon dioxide will cause a *shift to the right*, in order to form more CO_2, and restore equilibrium.

PRACTICE EXERCISES

9.11 Lactic acid is produced in muscle cells during intense exercise after all available oxygen has been consumed. Write the reaction of lactic acid in aqueous solution.
 a. What substances are present at equilibrium?
 b. Is the concentration of lactic acid and lactate ion constant or changing at equilibrium?

9.12 Write the equilibrium reaction for acetic acid in aqueous solution.
 a. If H_3O^+ is added, will the reaction shift to the left or to the right? Explain.
 b. If acetate ion is removed, will the reaction shift to the left or to the right? Explain.

9.13 **Critical Thinking Question:** Although it is not an acid–base reaction, the reaction of oxygen with hemoglobin is an important biochemical equilibrium reaction. In blood, hemoglobin reacts with oxygen to form oxyhemoglobin:

$$\text{Hemoglobin} + O_2 \rightleftharpoons \text{oxyhemoglobin}$$

Oxyhemoglobin transports oxygen to cells, where the oxygen is released and used to metabolize glucose to carbon dioxide.

 a. In the lungs, there is an excess of oxygen available. Will this shift the reaction to the left or the right? Explain.
 b. At the ends of small capillaries that deliver blood to tissues, the oxygen concentration is no longer in excess, and in fact has decreased due to consumption of oxygen by cells along the way. Will the reaction shift to the left or to the right? Explain.
 c. How does this equilibrium facilitate the transfer of oxygen from the lungs to oxygen-depleted tissues?

Acid–Base Neutralization Reactions

Recall that in Chapter 8 you learned about chemical reactions in general. Now you will be introduced to one of the most common chemical reactions—acid–base neutralization.

Have you ever noticed the wide variety of antacid tablets available in pharmacies? Some are listed in Table 9-6. How do these popular home remedies for acid indigestion work? Antacids are bases, and they react with stomach acid in what is known as an acid–base neutralization reaction. In an acid–base **neutralization reaction,** a proton is transferred from an acid to a base, producing *water* and a *salt*.

Consider the reaction of the acid, HI, with the base, KOH:

$$\underset{\text{Acid}}{HI(aq)} + \underset{\text{Base}}{KOH(aq)} \longrightarrow \underset{\text{Water}}{H_2O(l)} + \underset{\text{Salt}}{KI(aq)}$$

Table 9-6 Some Common Over-the-Counter Antacids

Brand Name	Active Ingredient(s)
Amphogel	Aluminum hydroxide, $Al(OH)_3$
Phillips' Milk of Magnesia	Magnesium hydroxide, $Mg(OH)_2$
Maalox, Mylanta	$Al(OH)_3$ and $Mg(OH)_2$
Tums, Rolaids	Calcium carbonate, $CaCO_3$

The acid, HI, donates a proton, H^+, to the base, OH^-, to produce water—a neutral substance, and hence the term "neutralization". The net reaction can be written as:

$$H^+ \; + \; OH^- \; \longrightarrow \; H_2O$$
$$\text{Acid} \qquad \text{Base} \qquad \text{Neutral water}$$

The dissolved salt, KI(aq), also a neutral substance, exists in solution as dissolved ions:

$$K^+(aq) + I^-(aq)$$

To balance a neutralization reaction that produces water, keep in mind that H^+ ions and OH^- ions react in a one-to-one ratio to form one H_2O molecule.

While KOH is highly soluble in water, several hydroxide salts, such as aluminum hydroxide ($Al(OH)_3$) and magnesium hydroxide ($Mg(OH)_2$), have limited solubility in water and consequently they are not classified as strong bases; therefore, they can be used as antacids. At high concentrations, free hydroxide ion can be quite corrosive, but these commercial preparations are safe when used as directed, because due to their limited solubility, they release only small amounts of hydroxide ion into solution.

The reaction of an antacid with an acid is only one example of how neutralization reactions are important in health care. Many pharmaceuticals are produced in the form of a neutral base, for example, the amine. As the last step in manufacturing, an acid–base reaction similar to a neutralization is performed on the amine functional group. This change allows the drug to be administered in the form of a salt, which, in many cases, increases the solubility of the drug in body fluids. For example, ephedrine, a common decongestant, is sold as the *hydrochloride salt*. To form the salt, a proton is transferred from the acid, HCl, to the base, ephedrine, in the following acid–base reaction:

Base (ephedrine)   Acid   Hydrochloride salt

WORKED EXERCISE 9-5 Acid–Base Neutralization Reactions

A base can be used to neutralize the effects of excess stomach acid. For example, magnesium hydroxide is a base commonly found in over-the-counter antacid medications, such as Phillips' Milk of Magnesia, used to treat heartburn caused by excess stomach acid. Write and balance the neutralization reaction of magnesium hydroxide, $Mg(OH)_2$ with stomach acid, HCl.

SOLUTION

$$Mg(OH)_2(s) \; + \; 2\,HCl \; \longrightarrow \; 2\,H_2O \; + \; MgCl_2(aq)$$
$$\text{Base} \qquad\quad \text{Acid} \qquad\qquad \text{Water} \qquad \text{Salt}$$

Since magnesium hydroxide yields two OH^- per formula unit, two HCl molecules are required to neutralize one formula unit of $Mg(OH)_2$. Thus, the coefficient 2 must be placed before HCl (yielding $2H^+$). Two water molecules are produced as well as the salt $MgCl_2$, which is electrically neutral.

Commercial antacids neutralize stomach acid. [© 1995 Michael Dalton, Fundamental Photographs]

PRACTICE EXERCISES

9.14 Write the balanced equation for the neutralization of hydrochloric acid with calcium hydroxide, $Ca(OH)_2$.

9.15 Write the balanced equation for the reaction of HNO_3 with the following:
    a. KOH   **b.** $Mg(OH)_2$ **c.** $Al(OH)_3$

9.16 Zoloft, primarily used to treat depression in adults, is sold as the hydrochloride salt of the amine. The structure of the neutral form of Zoloft is given below. Write the reaction of this molecule with HCl, showing the formation of the hydrochloride salt. Why is water not formed in this neutralization reaction?

9.17 **Critical Thinking Question:** Use Le Châtelier's principle to explain how certain bases having limited solubility in water, such as aluminum hydroxide ($Al(OH)_3$) and magnesium hydroxide ($Mg(OH)_2$), can still neutralize stomach acid.

9.2 pH

Have you ever tested the acidity or alkalinity (basicity) of a fish tank, a pool, or an aqueous solution in a high school laboratory experiment using pH paper? If you did, you were measuring the concentration of hydronium ions (H_3O^+) in solution. The higher the concentration of hydronium ions, the redder the pH paper turns; the lower the concentration of hydronium ions, the bluer the paper turns. *The pH of an aqueous solution is a quantitative measure of the concentration of hydronium ions.* Aqueous solutions have a pH ranging anywhere from 1 to 14. *A pH less than 7 indicates an acidic solution, a pH greater than 7 indicates a basic solution, and a pH equal to 7 indicates a neutral solution.*

The body regulates the pH within cells. Table 9-7 lists the normal pH range of important body fluids. Blood and intracellular fluids have a pH close to neutral. When the pH of any body fluid moves out of the normal range, biochemical reactions cannot proceed normally, and serious medical conditions result. For example, blood pH values lower than 7.35 indicate a potentially fatal condition known as acidosis, described on page 333. In this section you will learn how to interpret pH values and perform calculations involving pH, as well as learn how pH affects the environment inside the cell.

Ion-Product Constant, K_w

Recall that in the autoionization of water, H_2O ionizes to produce small amounts of hydronium ions (H_3O^+) and hydroxide ions (OH^-):

$$H_2O + H_2O \rightleftharpoons H_3O^+ + OH^-$$

In pure water, the concentration of the hydronium ion will be equal to the concentration of the hydroxide ion, which at room temperature (25 °C) is $1.0 \times 10^{-7}\ M$.

$$\boxed{[H_3O^+] = [OH^-] = 1.0 \times 10^{-7}\ M \text{ in pure water at 25°C}}$$

Litmus paper can be used to test the pH of a variety of solutions, including the juice in citrus fruit. The red color indicates an acid. [Andrew McClenaghan/Photo Researchers, Inc.]

pH   1   7   14
    Acidic   Neutral   Basic

pH paper colors.

Table 9-7 Body Fluid pH Ranges

Body Fluid	Normal pH Range
Gastric fluid	0.5–2
Urine	5–8
Saliva	6.5–7.5
Muscle cells	6.7–6.8
Blood	7.35–7.45

Brackets placed around a chemical formula indicate that the units of concentration are moles per liter, molarity, *M*.

The product of the hydronium ion concentration and the hydroxide ion concentration is equal to 1.0×10^{-14}, known as the **ion-product constant** for water, $\boldsymbol{K_w}$. Concentration units are omitted in the K_w value.

$$\boxed{K_w = [H_3O^+] \times [OH^-]}$$

In pure water, $K_w = (1.0 \times 10^{-7}\ M) \times (1.0 \times 10^{-7}\ M) = 1.0 \times 10^{-14}$

When an acid or a base is added to water so that the water is no longer pure, the hydronium and hydroxide ion concentrations are not equal to $1.0 \times 10^{-7}\ M$, but K_w is still equal to 1.0×10^{-14}, a constant. For example, when the hydronium ion concentration goes up by a factor of 10, the hydroxide ion concentration goes down by a factor of 10. If the concentration of either the hydronium ion or the hydroxide ion is known, the concentration of the other ion can be calculated using the equation for K_w.

WORKED EXERCISE 9-6 Ion-Product Constant, K_w

If an aqueous solution has a $[H_3O^+] = 5.2 \times 10^{-6}$, what is the $[OH^-]$?

SOLUTION
Use the equation for the ion-product constant:

$$K_w = [H_3O^+] \times [OH^-]$$

Set up the equation to solve for $[OH^-]$:

$$[OH^-] = \frac{K_w}{[H_3O^+]}$$

Substitute the values for K_w and $[OH^-]$ and solve:

$$[OH^-] = \frac{1.0 \times 10^{-14}}{5.2 \times 10^{-6}} = 1.9 \times 10^{-9}$$

PRACTICE EXERCISE

9.18 Complete the following table for aqueous solutions with the following concentrations:

$[H_3O^+]$	$[OH^-]$
1.0×10^{-11}	
	9.1×10^{-9}
3.3×10^{-6}	

The pH Scale

The pH scale provides a simple and convenient way to report the hydronium ion concentration without using scientific notation. The **pH** of a solution is defined as the negative log of the hydronium ion (H_3O^+) concentration:

$$\boxed{pH = -\log [H_3O^+]}$$

For example, if $[H_3O^+] = 1 \times 10^{-5}$,

$$pH = -\log (1 \times 10^{-5}) = 5.0$$

You may also be asked to calculate the hydronium ion concentration given the pH—the reverse process. To do this, apply the formula below:

$$[H_3O^+] = 10^{-pH}$$

For example, if the pH = 4.0, then the hydronium ion concentration is:

$$[H_3O^+] = 10^{-4.0} = 1 \times 10^{-4}$$

Because in this case the pH is a whole number, a calculator is not needed. However, when pH is a decimal value, you will need your calculator to obtain the correct value. Use the 10^x button on your calculator to raise "10" to the negative value of your pH for the calculation of hydronium ion concentration. Use the log x function to calculate pH, given the hydronium ion concentration.

Since pH represents a logarithmic scale, one pH unit represents a tenfold change in hydronium ion concentration. Thus, a solution with a pH of 4.0 is 10 times more acidic than a solution with a pH of 5.0. For example, if the pH of a lake changes by just one pH unit, from 5 to 4, most fish eggs will not hatch, and many adult fish will die. Figure 9-6 shows the relationship between hydronium ion concentration, $[H_3O^+]$, and pH, along with examples of solutions at different pH values.

> If the pH of a solution changes from 4 to 2, it becomes 100 times more acidic, not twice as acidic, because pH is a logarithmic scale.

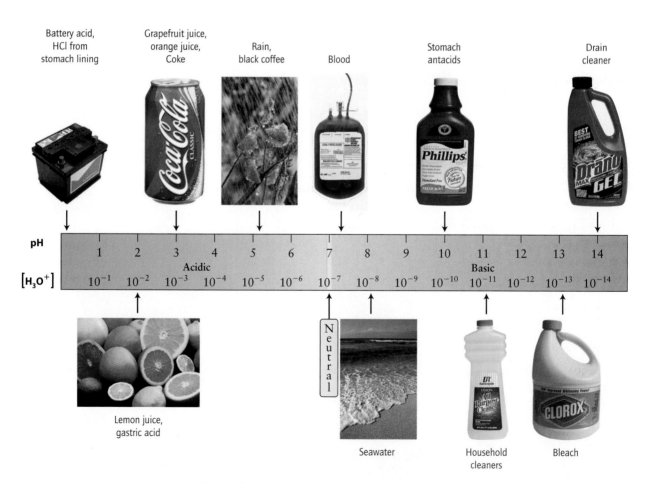

Figure 9-6 This pH scale shows the relationship between $[H_3O^+]$ and pH. [*Photos, clockwise from top left:* Ingram Publishing/Alamy; the Coca-Cola Company; Tony Cortazzi, Alamy; Davies & Starr/The Image Bank/Getty; The Photo Works (next four images); Corbis; and Photodisc Green/Getty]

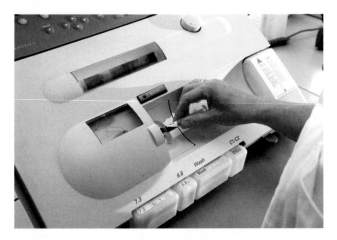

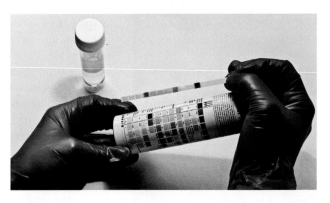

A blood gas analyzer used to test the pH of arterial blood.
[© BSIP/Phototake]

A urine dipstick. [Martin William Allen/Alamy]

change in pH considerably lowers the concentrations of the ionized forms of these key molecules within the cell. As a consequence, essential biochemical reactions within the cell do not take place. Later in this chapter, you will learn how your body responds to changes in pH and see why pH is so critical to health.

Measuring pH

When a patient presents with symptoms of an acid–base disorder, such as heavy breathing, fatigue, and confusion, the first step in assessing the patient is often to measure the pH of arterial blood. That measurement can be made with a blood gas analyzer. Although an acid–base disorder results from a change in pH within cells, arterial blood is easier to obtain and test than fluid from within the cell. Furthermore, arterial blood gives a more representative value. Often the pH of urine is tested as well.

Specialized papers incorporating pH-sensitive dyes have been designed to measure fine gradations of pH. The pH of urine is often tested using dipsticks containing these indicator dyes. In a laboratory setting, pH probes are frequently used to measure the pH of a sample. A pH probe may be connected to an electronic meter or computer. Once calibrated, the probe is placed in the sample and the meter/computer displays the pH of the solution.

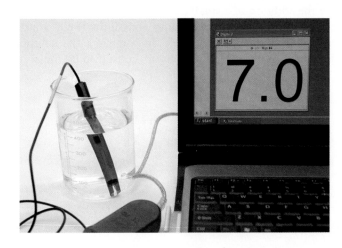

A pH probe connected to a computer.
[Martin Shields/Alamy]

The body naturally adjusts pH as it absorbs acids and bases from foods. The tomatoes and vinegar in this salad are acidic.
[Tim Hill/Alamy]

9.3 | Buffers

If the pH of blood falls below 7.35 or rises above 7.45, cells cannot function properly and severe medical conditions arise. However, the body constantly absorbs acids and bases from the food we eat, and biological processes are continually producing protons as well. So, how does the body maintain its blood pH within such a narrow range? The answer is that blood contains buffers.

Normally, when a small amount of acid or base is added to pure water, the pH will drop or rise dramatically. However, when acid or base is added to a buffer, the pH changes very little. *A **buffer** is a solution that resists changes in pH upon addition of small amounts of acid or base. A buffer is a solution containing a weak acid and its conjugate base, in relatively equal concentrations.*

Consider, for example, the acetate buffer. Acetic acid (CH_3COOH) is a weak acid, and its conjugate base is the acetate ion, CH_3COO^-. To create an acetate buffer, equal molar amounts of acetic acid and sodium acetate are combined in an aqueous solution. Sodium acetate is an ionic compound that completely dissociates into sodium ions and acetate ions in solution:

$$Na^+CH_3COO^-(s) \longrightarrow Na^+(aq) + CH_3COO^-(aq)$$

When sodium acetate is mixed with an equal amount of acetic acid, there is a substantial quantity of *both* the weak acid *and* its conjugate base in solution and they are in a state of equilibrium, creating a buffer solution.

Buffer solution

$$CH_3COOH + H_2O \rightleftharpoons H_3O^+ + CH_3COO^-$$

Weak acid   Conjugate base
Large quantity   **Large quantity**

How does a buffer solution prevent changes in pH? Adding acid to a buffer initially increases the concentration of H_3O^+, which is one of the products in the buffer equilibrium. Therefore, according to Le Châtelier's principle, the equilibrium should respond by shifting to the left, because there is a sufficient quantity of conjugate base (CH_3COO^-) available to react with the excess added H_3O^+, thus neutralizing the added H_3O^+ and maintaining a constant pH:

$$CH_3COOH + H_2O \longleftarrow H_3O^+ + CH_3COO^-$$

Weak acid   Excess H_3O^+ Conjugate base
Large quantity   added   **Large quantity**

The additional acetic acid produced from the equilibrium shift does not affect the pH because it is a weak acid and dissociates very little. The response of a buffer to the addition of a small amount of acid is illustrated in Figure 9-7. Remember, the pH of a solution will change only if the H_3O^+ concentration or OH^- concentration changes appreciably. In summary, the solution does not become acidic, because the added protons are consumed by the acetate ion.

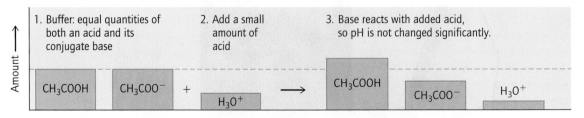

Figure 9-7 The response of a buffer to the addition of a small amount of acid.

What would happen if a small amount of strong base were added to this buffer solution? The added hydroxide ions (OH^-) would react with the weak acid (CH_3COOH) present in the buffer, forming more acetate ion and water:

$$CH_3COOH + OH^- \longrightarrow CH_3COO^- + H_2O$$

Weak acid	Small amount of	Conjugate base
Large quantity	base added	Large quantity

The reaction of hydroxide with acetic acid neutralizes the hydroxide, and again the pH will not change significantly. The concentration of acetic acid will decrease and the amount of acetate ion will increase, but these shifts in concentration will not appreciably affect the concentration of H_3O^+ or hydroxide ion. Therefore, addition of a small amount of base to a buffer will not affect the pH significantly, because the weak acid present consumes excess hydroxide. The response of a buffer to the addition of small amounts of base is illustrated in Figure 9-8.

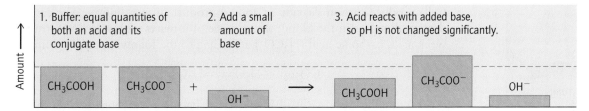

Figure 9-8 The response of a buffer to the addition of a small amount of base.

Buffer systems do have their limits, however. The greatest amount of acid or base that a buffer can accommodate while maintaining the pH is known as the **buffer capacity**—the higher the concentration of the weak acid and conjugate base in a buffer system, the greater its buffer capacity. The addition of amounts of acid or base exceeding the buffer capacity will change the pH considerably and start to disrupt key functions of the body.

WORKED EXERCISES 9-8 Buffers

1 Which of the following represents a buffer? Explain.
 a. HNO_3 (nitric acid, a strong acid) and $NaNO_3$
 b. NaCl and KCl
 c. CH_3CH_2COOH (a weak acid) and $NaCH_3CH_2COO$
 d. KOH and H_2O
2 Consider the following buffer system, consisting of the weak acid hydrofluoric acid and its conjugate base, NaF:

$$HF + H_2O \rightleftharpoons H_3O^+ + F^-$$

 a. How would this buffer react if H_3O^+ were added? Show the equation.
 b. How would this buffer react if OH^- were added? Show the equation.

SOLUTIONS
1 (c) represents a buffer, because it contains a weak acid, CH_3COOH, and its conjugate base, CH_3COO^-.
2 (a) The equilibrium would shift to the left, consuming the added acid:

$$HF + H_2O \longleftarrow H_3O^+ + F^-$$

 (b) The equilibrium would shift to the right, consuming the base:

$$HF + OH^- \longrightarrow F^- + H_2O$$

PRACTICE EXERCISE

9.23 A buffer can also be a weak base and its conjugate acid. For example ammonia, NH_3, mixed with ammonium chloride, NH_4Cl, forms a buffer.

   a. Write the equilibrium equation representing this buffer.
   b. What is the purpose of a buffer in general?
   c. What is the purpose of NH_4Cl in this buffer?
   d. How would this buffer react if H_3O^+ were added? Show the equation.
   e. How would this buffer react if OH^- were added? Show the equation.

Buffering Systems in the Body

In blood and intracellular fluids, the most important buffering systems involve bicarbonate (HCO_3^-), phosphates, and proteins. The primary buffering system in the blood consists of the weak acid, carbonic acid (H_2CO_3), and its conjugate base, the bicarbonate ion (HCO_3^-).

What are the sources of carbonic acid and carbonate ion in the blood? Cells produce carbon dioxide as an end product of metabolism. The carbon dioxide dissolves in the blood, where some of it reacts with water to form the weak acid, carbonic acid (H_2CO_3) according to the following equilibrium:

$$CO_2(g) + H_2O \rightleftharpoons H_2CO_3(aq)$$

Eventually, carbonic acid is converted back into carbon dioxide in the lungs and exhaled. However, some of the carbonic acid dissociates to form bicarbonate ion, the conjugate base of carbonic acid.

$$\underset{\text{Weak acid}}{H_2CO_3} + H_2O \rightleftharpoons \underset{\text{Conjugate base}}{HCO_3^-} + H_3O^+$$

Since the kidneys supply additional bicarbonate ion, the equal proportion of weak acid (H_2CO_3) and its conjugate base (HCO_3^-) forms a buffer. The buffering system of blood can be represented as follows.

Buffer system, Equation 9-1

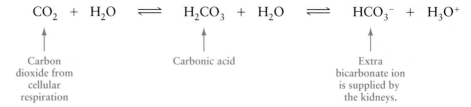

$$CO_2 + H_2O \rightleftharpoons H_2CO_3 + H_2O \rightleftharpoons HCO_3^- + H_3O^+$$

Carbon dioxide from cellular respiration   Carbonic acid   Extra bicarbonate ion is supplied by the kidneys.

Consider what happens when a small amount of acid is added to the buffer shown. The bicarbonate ion would react with the excess acid, H_3O^+, shifting the equilibrium to the left, forming water and carbonic acid. These products do not increase the concentration of H_3O^+ and therefore the pH does not change:

$$HCO_3^- + H_3O^+ \longrightarrow H_2CO_3 + H_2O \quad \text{no pH change}$$

If a small amount of base (OH^-) were added to the blood, carbonic acid would react with the OH^- ion to form bicarbonate ion and water. Again, these products do not change the concentration of H_3O^+ and therefore do not change the pH:

$$H_2CO_3 + OH^- \longrightarrow HCO_3^- + H_2O \quad \text{no pH change}$$

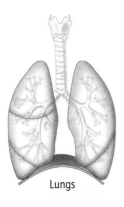

Lungs

Kidneys

The lungs and the kidneys are key organs in maintaining acid–base balance in the body.

Acid–Base Homeostasis The body's maintenance of the proper balance between acids and bases is known as **acid–base homeostasis.** The body compensates in the short term for acid–base imbalances by regulating the breathing (ventilation) rate. By changing the ventilation rate, the body can alter the concentration of carbon dioxide in the blood, which alters the pH by shifting the equilibrium in Equation 9-1. In the long term, the kidneys maintain acid–base homeostasis by altering the excretion of excess acid or base. Ventilation rate, blood pH, and carbon dioxide levels in the blood are all closely connected. Conditions causing abnormally high or low levels of any of these three will affect the other two.

Under normal circumstances, blood pH is maintained within a narrow range because of the carbonic acid/bicarbonate buffer. However, when some new condition causes excessive amounts of acid or base to enter the blood stream, the buffer cannot keep up, and the result is *acidosis* or *alkalosis*.

Acidosis When blood pH falls below its normal range, acidosis occurs. Such a fall in blood pH can occur when breathing is too weak or too slow, and carbon dioxide is not removed from the lungs rapidly enough, thereby causing excess carbon dioxide to be dissolved in the blood. Ultimately excess H_3O^+ and bicarbonate ion will form as a result of a shift to the right in Equation 9-1. Although the blood buffer will remove some of the excess H_3O^+, the ion will begin to accumulate, causing the blood pH to drop. This condition is known as **respiratory acidosis,** and can appear with a severe head injury, emphysema or asthma, or while under anesthesia.

Recall, normal physiological pH = 7.35–7.45.

Recall Henry's law—the amount of gas dissolved in a solution depends on the partial pressure of the gas above the solution.

Metabolic acidosis is another form of acidosis, caused by a metabolic disorder rather than weak respiration. Metabolic acidosis can result from kidney failure, uncontrolled diabetes, or starvation. Any of these conditions may lead to an excess of H_3O^+ entering the bloodstream, exceeding the buffer capacity, and causing blood pH to drop. Treatment of any form of acidosis usually includes IV infusion of bicarbonate ion.

Alkalosis When blood pH rises above its normal range, alkalosis occurs. Such a pH rise can occur when breathing becomes too fast, as when someone hyperventilates, and carbon dioxide is removed from the lungs faster than it is produced. Therefore, the H_3O^+ concentration in the blood drops, and the pH of the blood rises. This condition is known as **respiratory alkalosis.** Hyperventilation can result from anxiety, hysteria, altitude sickness, or intense exercise. Breathing into a paper bag often corrects the problem. The patient is forced to inhale exhaled air, which contains higher levels of carbon dioxide than normal air. The added carbon dioxide inhaled restores CO_2 levels.

Excessive vomiting, ingestion of excessive amounts of antacids, and some adrenal gland diseases can also cause **metabolic alkalosis.** Metabolic alkalosis may be treated by intravenous administration of a dilute solution of hydrochloric acid, HCl, or by hemodialysis.

In Chapters 6 and 7, you learned about the structures of molecules, and in Chapter 8 you learned the basics of chemical reactions. In this chapter you learned how the structures of many biologically important molecules are dependent on pH. You also saw how the maintenance of pH within a very narrow range allows these molecules to function appropriately in the body. Buffering systems of the body are instrumental in maintaining pH. The reactions of acids and bases are just one of many classes of chemical reactions. In the following chapter, you will study a number of other classes of chemical reactions, all important to sustaining life.

Chemistry in Medicine Why Some Drugs Cannot Be Administered Orally: The Issue of pH

Imagine as a child being told that you can no longer eat sweets, and must make regular injections into your body several times a day. This is a routine prescription given to children who have been diagnosed with diabetes. According to the National Diabetes Education Program (NDEP), over 180,000 Americans under the age of 20 have diabetes. In addition, many adults are diagnosed with diabetes each year, and must eventually give themselves injections or rely on a pump in order to keep themselves alive.

What substance do diabetics inject? The drug that must be injected at regular intervals is insulin, a hormone synthesized in the pancreas. The function of insulin is to control glucose concentration in the blood. In many children diagnosed with diabetes, the pancreas has either partially or completely ceased to produce insulin, and blood glucose levels skyrocket after a meal. The condition causes a host of medical problems ranging in severity from frequent urination to coma. In adult-onset diabetes, the problem isn't inadequate insulin production by the pancreas, but too few or ineffective insulin receptors.

Insulin is a small protein (a peptide). Peptide drugs cannot be administered orally, and must be injected under the skin. There are a number of reasons why peptide drugs cannot be ingested orally, including the following:

- Proteins degrade in the gastrointestinal tract in the presence of acids and enzymes.
- Most intact proteins cannot cross the membrane barrier in the small intestine, to proceed into the circulatory system and be delivered to the target tissues.

A diabetic child administers an insulin injection to himself.
[Horizon International Images Limited/Alamy]

- If the protein could pass through the membrane barrier in the small intestine, the protein could damage the epithelial cells lining the intestine.

Technologies are currently being developed to resolve all of these issues; here we will concentrate on why stomach acidity is a problem and what has been done to solve the problem.

General Functions and Structure of Proteins

You know that proteins are one of the main staples of the diet. We ingest proteins through a variety of food sources including meats, cheeses, and eggs. When proteins are ingested, your digestive system chemically breaks down proteins in your gastrointestinal tract by means of chemical reactions involving acid, water, and enzymes. When insulin is ingested orally, these same types of reactions break down insulin. In other words, insulin is digested in the gastrointestinal tract before it ever reaches its target tissues.

Proteins are large organic molecules consisting of amino acids linked together by amide bonds. Because of the large size of proteins, intermolecular forces such as hydrogen bonding link distant atoms within a protein molecule, causing the protein to fold into its own unique three-dimensional shape. The shape of a protein is vital to its function. Even before breaking down a protein chemically, acids in the digestive track disrupt the hydrogen bonding necessary for folding correctly, disrupting the protein's shape so that the protein can no longer perform its function.

Overview of Digestion

Figure 9-9 shows the digestive system. Digestion is a complex process, controlled by a number of factors. One important factor in the normal functioning of the digestive tract is pH. The pH of the mouth and esophagus is weakly acidic, about pH 6.8. The pH of the stomach is very acidic, in the range of pH 1–3. The mucosal tissue of the small intestine is alkaline (pH 8.5). In the small intestine, food is mixed with bile, pancreatic juices, and enzymes. As the acid level changes in the small intestine, enzymes are activated that break down the molecular structure of proteins, carbohydrates, and fats, so they may be absorbed into the circulatory system.

It takes only seconds for food to pass through the esophagus, and by the time a protein reaches your stom-

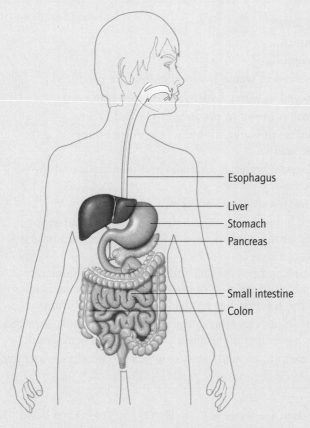

Figure 9-9 The digestive system. Digestion is a complex process in which pH plays a key role.

ach, the gastric glands in the stomach lining have begun to secrete gastric juice—a strongly acidic solution (pH 1–3) containing enzymes, hydrochloric acid, and mucous. At a pH of 1–3 in the stomach, the hydrogen ion concentration ranges from 0.1 M to 0.001 M. Hydrogen ions donated by hydrochloric acid bind to certain functional groups in a protein. By removing H_3O^+ from solution, the binding of hydrogen ions by the protein temporarily causes the pH of the stomach to rise to an alkaline level. Equilibrium is disturbed, so following Le Châtelier's principle, the stomach secretes more gastric juice, in order to restore an acidic environment:

$$H_3O^+ + \text{protein} \rightleftharpoons [\text{protein-H}^+] + H_2O$$

More H_3O^+ supplied by stomach to maintain low pH

Under acidic conditions, the shape of insulin is altered, so that it cannot carry out its normal cellular function.

Overcoming the pH Issue

As a first step in overcoming the pH issue of orally administered protein drugs like insulin, scientists have developed **enteric coatings**—barriers applied to oral medications that control where in the digestive tract the medication is absorbed. The enteric coating creates a surface on the medication that is stable in the acidic environment of the stomach, protecting the protein from exposure to the low-pH surroundings.

Large molecules that contain carboxylic acid groups are widely used as enteric coatings. In an acidic environment, the carboxylic acid groups remain in the acid form and the coating remains intact. In a basic environment, the carboxylic acid groups react to form the carboxylate ion form of the functional group, causing the coating to degrade. These coatings enable a protein to pass through the acidic environment of the stomach and into the basic environment of the small intestine. The enteric coating will then degrade in the small intestine, depositing the protein in its unaltered state, able to perform its function.

Once the intact protein is deposited in the small intestine, the challenge remains to get the intact protein across the membrane barrier within the small intestine, without damaging the epithelial cells. These are chemistry problems of another type, still under investigation. With active research continuing on this aspect of the problem, there is hope that one day diabetics will be able to simply swallow a pill, rather than injecting themselves several times every day.

The enteric coating on a drug. [© 2008 Richard Megna/Fundamental Photographs]

Chapter Summary

Properties of Acids and Bases

- Acids are proton donors.
- Bases are proton acceptors.
- When an acid is dissolved in water, hydronium ions are produced.
- When a base is dissolved in water, hydroxide ions are produced.
- Water undergoes autoionization to produce H_3O^+ and OH^-.
- When an acid dissociates in water, it produces hydronium ions and the conjugate base of the acid.
- When a base dissociates in water, it produces hydroxide ions and the conjugate acid of the base.
- Strong acids and strong bases dissociate completely in water.
- Weak acids and weak bases do not dissociate completely in water. A significant percentage of the undissociated acid or undissociated base will be present in solution.
- When both the forward and reverse reactions proceed at the same rate in a reversible reaction, the concentration of the reactants and products no longer changes, and the reaction has reached a state of equilibrium.
- Le Châtelier's principle states that when a reaction at equilibrium is disturbed, the reaction responds by shifting in the direction that restores equilibrium.
- In a neutralization reaction, an acid and a base react to produce water and a salt.

pH

- The ion-product constant for water is $K_w = [H_3O^+] \times [OH^-] = 1 \times 10^{-14}$.
- A pH less than 7 indicates an acidic solution, a pH greater than 7 indicates a basic solution, and a pH equal to 7 indicates a neutral solution.
- $pH = -\log[H_3O^+]$
- $[H_3O^+] = 10^{-pH}$
- The intracellular, or "physiological," pH is 7.35–7.45.

Buffers

- A buffer resists changes in pH upon addition of a small amount of acid or base.
- A buffered solution contains a weak acid and its conjugate base, in close to equal concentrations.
- The primary buffering system of the blood consists of carbonic acid and the hydrogen carbonate (bicarbonate) ion.
- A disturbance in the buffering system of blood can cause acidosis or alkalosis, both serious medical conditions.

Key Words

Acid A substance that produces protons (H^+) when dissolved in water; a proton donor.

Acid–base homeostasis The body's maintenance of the proper pH.

Acidosis A serious medical condition diagnosed when the pH of the blood falls below the normal range.

Alkalosis A serious medical condition diagnosed when the pH of the blood rises above the normal range.

Amphoteric molecule A molecule that can act as both an acid and a base; for example, water.

Autoionization of water The reaction of water molecules with one another to produce a small amount of

hydronium ions (H_3O^+) and hydroxide ions (OH^-) in pure water.

Base A substance that produces hydroxide ions when dissolved in water or can act as a proton acceptor.

Buffer A solution that resists changes in pH upon addition of a small amount of acid or base. A buffer is a solution composed of a weak acid and its conjugate base, in relatively equal concentrations.

Buffer capacity The amount of acid or base that can be added to a buffer before the pH will begin to change appreciably. Buffer capacity is directly related to the concentration of weak acid and conjugate base in the buffer.

Conjugate acid A base after it has accepted a proton.

Conjugate acid–base pair Molecules or ions that are related and differ only in the presence or absence of a proton.

Conjugate base An acid after it has lost a proton.

Equilibrium When both the forward and reverse reaction occur at the same rate, and the concentrations of reactants and products no longer change.

Ion-product constant, K_w The product of the hydronium ion concentration and the hydroxide ion concentration in aqueous solution at 25 °C, 1.0×10^{-14}.

Le Châtelier's principle When a reaction at equilibrium is disturbed, the reaction responds by shifting in the direction that restores equilibrium, in either the forward direction (to the right, →) or the reverse direction (to the left, ←).

Neutralization reaction A chemical reaction in which an acid reacts with a base , producing water and a salt.

pH The negative log of the hydronium ion (H_3O^+) concentration, pH = $-\log [H_3O^+]$.

Physiological pH The normal pH of a healthy cell, 7.35–7.45.

Strong acid An acid in which all molecules of the acid donate a proton to form H_3O^+ in solution; no acid molecules remain in solution.

Strong base An ionic compound that completely dissolves in water, producing hydroxide ions.

Weak acid An acid that does not dissociate completely. A significant percentage of the undissociated acid is present in solution.

Weak base A base that does not dissociate completely; only a small percentage of molecules accept a proton from water.

Additional Exercises

Properties of Acids and Bases

9.24 An acid is a molecule that _____ a proton and a base is a molecule that _____ a proton.

9.25 Write the equation that represents the reaction of lactic acid in water. (You will find the structure of lactic acid in Table 9.1.)

9.26 Write the equation that represents the reaction of carbonic acid in water. (You will find the structure of carbonic acid in Table 9.1.)

9.27 Write the equation that represents the reaction of ephedrine in water. (You will find the structure of ephedrine in Table 9.2.)

Conjugate Acids and Bases

9.28 Write the structure of the conjugate base of the following acids.
a. $CH_3CH_2CH_2COOH$
b. HCl
c. H_3O^+
d. CH_3CH_2COOH

9.29 Write the structure of the conjugate acid of the following bases.
a. H_2O
b. OH^-
c. I^-
d. $CH_3CH_2CH_2CH_2COO^-$

Strengths of Acids and Bases

9.30 Why is hydrochloric acid, HCl, classified as a strong acid?

9.31 Perchloric acid, $HClO_4$, is a strong acid. Write the equation for the reaction of perchloric acid and water.

9.32 Write the equation for the reaction of hydrobromic acid, HBr, and water.

9.33 **Critical Thought Question:** Sulfuric acid, H_2SO_4, is a strong acid. In two sequential steps it can lose two protons when it reacts with water. Write the equation for this two-step process.

9.34 How do weak acids differ from strong acids?

9.35 How do weak bases differ from strong bases?

9.36 For each of the following chemical equations, indicate whether the equation represents the dissociation of a strong acid, strong base, weak acid, or weak base.

a. $Ca(OH)_2 + H_2O \rightarrow Ca^{2+}(aq) + 2\ OH^-(aq)$
b. $HI + H_2O \rightarrow H_3O^+ (aq) + I^-(aq)$
c. $HCOO^- + H_2O \rightleftharpoons HCOOH + OH^-(aq)$

9.37 For each of the following chemical reactions indicate whether the equation represents the dissociation of a strong acid, strong base, weak acid, or a weak base.
a. $H_3PO_4 + H_2O \rightleftharpoons H_3O^+(aq) + H_2PO_4^-(aq)$
b. $NH_3 + H_2O \rightleftharpoons NH_4^+(aq) + OH^- (aq)$
c. $HClO_3 + H_2O \rightarrow H_3O^+(aq) + ClO_3^-(aq)$

9.38 Write the dissociation reaction for the following acids in water. State whether the acid is a strong acid or a weak acid. Is the reaction reversible? Identify the acid, base, conjugate acid, and conjugate base.
a. HCl
b. HNO_3
c.

$$\begin{array}{c} O \\ \| \\ H-C-O-H \end{array}$$

d. H_2CO_3

9.39 **Critical Thinking Question:** The structure of malonic acid is shown below.

$$\begin{array}{c} O\quad H\quad O \\ \|\quad\ |\quad\ \| \\ H-O-C-C-C-O-H \\ | \\ H \end{array}$$

a. Highlight the protons in malonic acid that would dissociate.
b. Write the successive dissociation reactions for each proton in malonic acid.

Acid–Base Equilibria

9.40 Butanoic acid has an acrid taste and an unpleasant odor. It is found in rancid butter and Parmesan cheese. In aqueous solution, butanoic acid undergoes the following reversible reaction:

$$\begin{array}{c} H\ H\ H\ O \\ |\ \ |\ \ |\ \ \| \\ H-C-C-C-C-OH \\ |\ \ |\ \ | \\ H\ H\ H \end{array} + H_2O \rightleftharpoons \begin{array}{c} H\ H\ H\ O \\ |\ \ |\ \ |\ \ \| \\ H-C-C-C-C-O^- \\ |\ \ |\ \ | \\ H\ H\ H \end{array} + H_3O^+$$

Butanoic acid   Water   Butanoate   Hydronium ion

a. What substances are present at equilibrium?

b. Is the concentration of butanoic acid and butanoate ion constant or changing at equilibrium?

c. What is the significance of the two opposing arrows in this equation?

9.41 Propanoic acid prevents the growth of mold and some bacteria. It is used as a preservative in animal feed and in some human food production. In aqueous solution, propanoic acid undergoes the following reversible reaction.

Propanoic acid   Water   Propanoate   Hydronium ion

a. What substances are present at equilibrium?

b. Is the concentration of propanoic acid and propanoate ion constant or changing at equilibrium?

c. What is the significance of the two opposing arrows in the equation?

9.42 The reversible reaction of acetic acid shown below occurs in the bottle of vinegar sitting on your kitchen shelf.

Acetic acid   Water   Acetate   Hydronium ion

a. What substances are present at equilibrium?

b. Are the concentrations of acetic acid and acetate constant or changing at equilibrium?

Le Châtelier's Principle

9.43 Explain Le Châtelier's principle.

9.44 The dissociation reaction of acetic acid in water is shown below.

Acetic acid   Water   Acetate   Hydronium ion

a. What happens to the equilibrium if more acetate is added to the reaction?

b. If more acetic acid is added, does the reaction shift to the left or the right?

9.45 An equilibrium reaction is shown below:

$$NH_4^+ + H_2O \rightleftharpoons NH_3 + H_3O^+$$

a. If more ammonia, NH_3, is added to the reaction, which way does the reaction shift?

b. If more NH_4^+ is added to the reaction, which way does the reaction shift?

9.46 In the blood the following reaction plays an important role:

$$H_2CO_3 + H_2O \rightleftharpoons HCO_3^- + H_3O^+$$

a. Identify two ways to shift the reaction to the right.

b. Identify two ways to shift the reaction to the left.

9.47 Critical Thinking Question: Carbon monoxide binds to hemoglobin 200 times more tightly than oxygen does. This process sets up a competing equilibrium within the body.

$$O_2\text{-hemoglobin} \rightleftharpoons O_2 + \text{hemoglobin} + CO \rightleftharpoons CO\text{-hemoglobin}$$

When a patient has carbon monoxide poisoning, there is too much of the CO-hemoglobin complex in their blood. Explain how treating the patient with more oxygen affects the equilibrium shown above.

Acid-Base Neutralization Reactions

9.48 Write the balanced equation for the reaction of HBr with the following compounds:

a. NaOH   **b.** $Al(OH)_3$ **c.** $Sr(OH)_2$

9.49 Barium hydroxide, $Ba(OH)_2$, is used as a homeopathic drug. Write the balanced equation for the neutralization of barium hydroxide with hydrochloric acid.

9.50 Write the balanced equation for the reaction of HNO_3 with the following compounds:

a. KOH   **b.** $Mg(OH)_2$ **c.** $Al(OH)_3$

9.51 Write the balanced equation for the neutralization of hydrochloric acid and aluminum hydroxide, $Al(OH)_3$.

pH

9.52 What does pH measure?

9.53 Indicate whether the following solutions are acidic, basic or neutral based on their pH.

a. 8.5   **b.** 6.3   **c.** 7.0   **d.** 1.1
e. 10.8   **f.** 11.5   **g.** 2.6

9.54 Indicate whether the following solutions are acidic, basic or neutral based on their pH.

a. 7.4   **b.** 1.9   **c.** 14.0   **d.** 3.5   **e.** 10.7

9.55 Complete the following table.

$[H_3O^+]$	$[OH^-]$	Is the Solution Acidic, Neutral or Basic?
1.0×10^{-3}		
	1.0×10^{-2}	
1.0×10^{-7}		
1.0×10^{-5}		
	1.0×10^{-5}	

9.56 Complete the following table.

$[H_3O^+]$	$[OH^-]$	Is the Solution Acidic, Neutral or Basic?
	1.0×10^{-10}	
	1.0×10^{-1}	
1.0×10^{-6}		
1.0×10^{-11}		
	1.0×10^{-4}	

9.57 What is the pH of apple juice that has an $[H_3O^+] = 3.2 \times 10^{-4}$? Calculate the $[OH^-]$ as well. Is apple juice acidic, basic, or neutral?

9.58 What is the pH of a bleach solution that has an $[H_3O^+] = 3.16 \times 10^{-13}$? Calculate the $[OH^-]$ as well. Is bleach acidic, basic, or neutral?

9.59 What is the pH of milk that has an $[H_3O^+] = 3.2 \times 10^{-7}$? Calculate the $[OH^-]$ as well. Is milk acidic, basic, or neutral?

9.60 What is the pH of a blood sample that has an $[H_3O^+] = 4.6 \times 10^{-8}$? Calculate the $[OH^-]$ as well. Is the pH of this blood sample within the normal range?

Buffers

9.61 What is the function of a buffer?

9.62 Which of the following represents a buffer? Explain.
a. HCl and HNO_3
b. CH_3CH_2COOH (propanoic acid, a weak acid) and CH_3CH_2COONa
c. NaOH and KCl
d. LiCl and H_2O

9.63 When you eat, a small amount of acid enters your bloodstream. Explain how the buffers in your blood react with the small amount of acid to maintain a constant pH.

9.64 Explain why a small amount of base added to your bloodstream does not affect the pH of your blood.

9.65 Which component of the bicarbonate (HCO_3^-)/carbonic acid (H_2CO_3) blood buffer would react with each of the following to maintain a constant pH?
a. NaOH **b.** HCl **c.** NH_3 **d.** CH_3COO^-

9.66 Hyperventilation affects the concentration of CO_2 in the blood.
a. Does the concentration of CO_2 increase or decrease in a patient who is hyperventilating?
b. How does the concentration of CO_2 in a patient who is hyperventilating affect the pH?

9.67 An emergency medical team evaluates a marathoner and determines that he has alkalosis. What component of the bicarbonate/carbonic acid buffer should be given to the marathoner to decrease the pH of the blood?

9.68 A patient has severe diarrhea, which results in an excessive loss of sodium bicarbonate from the body. How is the pH of the patient's blood affected by this loss? What component of the bicarbonate/carbonic acid buffer should be given to the patient to regulate the pH in the blood?

Chemistry in Medicine

9.69 Why do drugs that are proteins need to be injected rather than administered orally?

9.70 Which is a more acidic environment, the mouth and esophagus or the stomach?

9.71 Under the acidic conditions of the stomach, does a protein keep its original shape or cellular function?

9.72 Why does the stomach secrete more gastric acid when a protein is present in the stomach?

9.73 What is the function of the enteric coating of a pill?

9.74 Where in the body is the enteric coating degraded?

9.75 **Critical Thinking Question:** If the enteric coating of a pill is stable in an acidic environment and reacts in an alkaline environment, is that coating acidic or basic?

Answers to Practice Exercises

9.1 A base is a molecule or ion that ***accepts*** a proton.

9.2 When dissolved in water, an acid will produce ***hydronium*** ions.

9.3

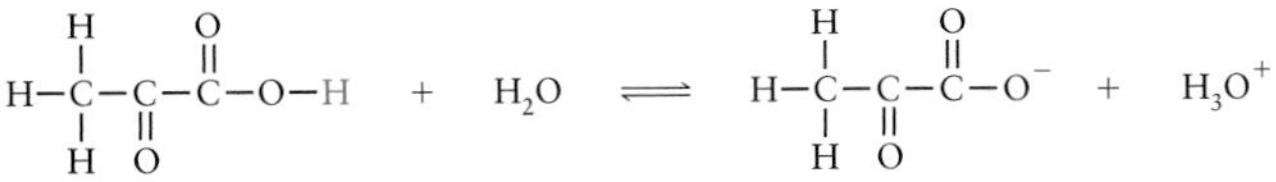

9.4 **a.**

$NH_4^+ + H_2O \rightleftharpoons H_3O^+ + NH_3$

Acid — Conjugate base

Water acts as a base in this reaction, it accepts a proton.

b.

$HCO_3^- + H_2O \rightleftharpoons OH^- + H_2CO_3$

Base — Conjugate acid

Water acts as an acid in this reaction, it donates a proton.

9.5

$H_2O + H_2O \rightleftharpoons H_3O^+ + OH^-$

Acid, Base, Conjugate acid, Conjugate base

The autoionization of water represents the dissociation of both a weak acid and weak base—water can act as either. One reactant water molecule acts as the base, accepting a proton to form H_3O^+; the other reactant water molecule acts as the acid, donating a proton to form OH^-.

9.6 lactic acid:

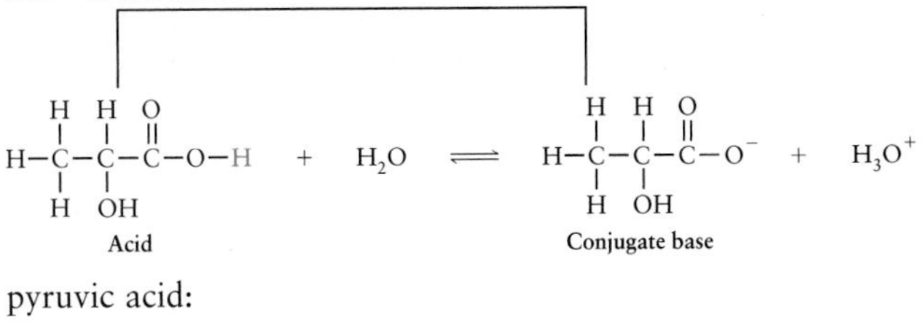

pyruvic acid:

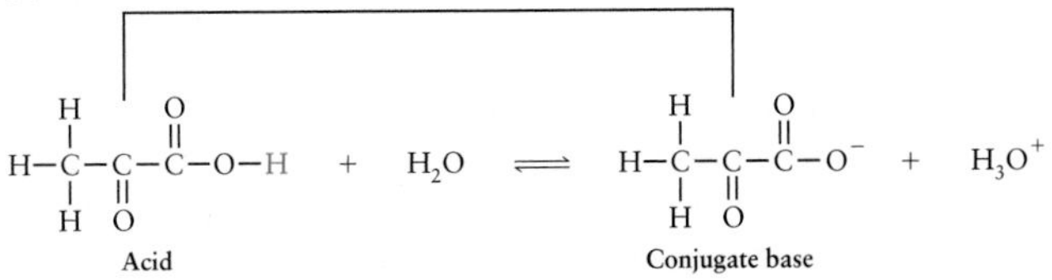

9.7 **a.** Strong base. $LiOH(s) + H_2O \rightarrow Li^+(aq) + OH^-(aq)$
b. Strong acid. $HNO_3 + H_2O \rightarrow H_3O^+(aq) + NO_3^-(aq)$
c. Weak base.
$CH_3COO^- + H_2O \rightleftharpoons CH_3COOH(aq) + OH^-(aq)$

9.8 a.

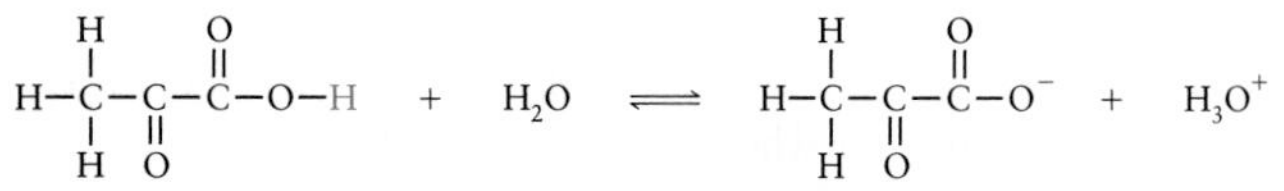

b. $HNO_3 + H_2O \rightarrow NO_3^- + H_3O^+$

c.

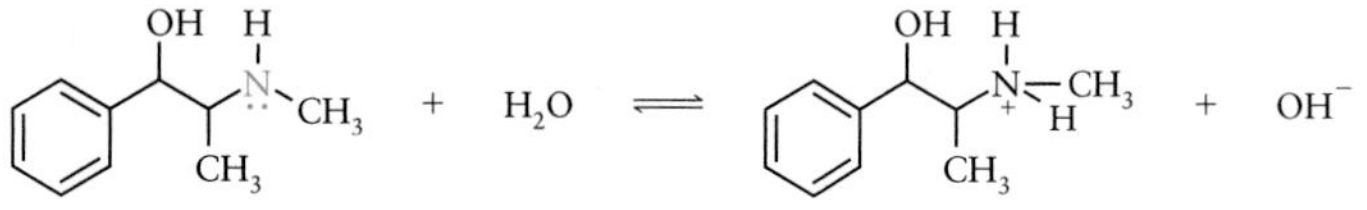

d. $NaOH(s) + H_2O \rightarrow Na^+(aq) + OH^-(aq)$

e.

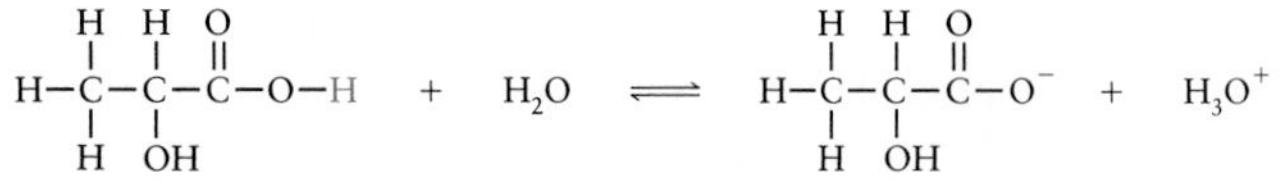

9.9 Carnosine contains a carboxylate ion and a positively charged amine group. The carboxylate ion can act as a weak base, and the protonated amine can act as a weak acid; therefore carnosine is amphoteric.

9.10 Notice the number of nitrogen atoms with lone pairs of electrons. These nitrogen atoms can accept protons; therefore adenine is considered a base.

9.11

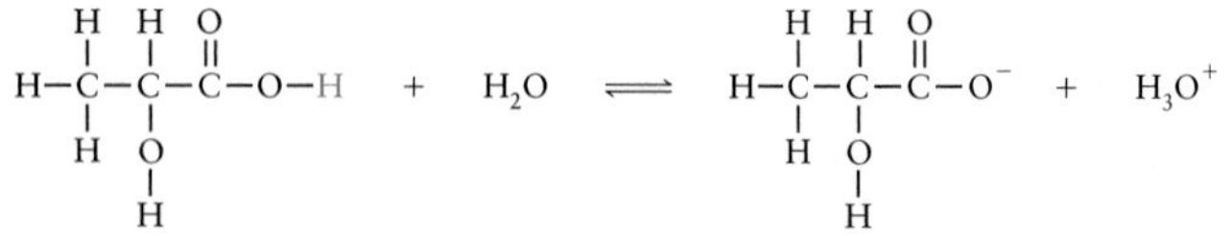

a. Lactic acid, water, lactate ion, and hydronium ion are all present at equilibrium.
b. The concentration of lactic acid and lactate ion are constant at equilibrium.

9.12

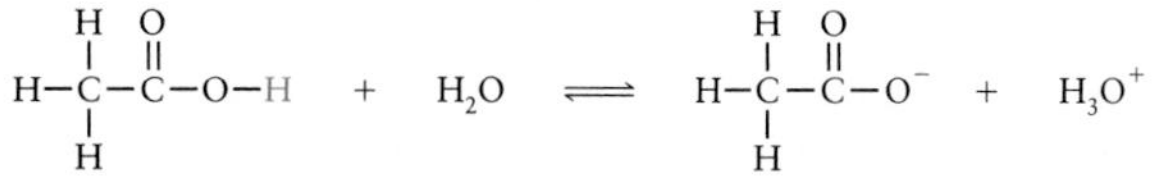

a. If H_3O^+ is added, the reaction will shift to the left in order to consume the excess H_3O^+ and restore equilibrium.
b. If acetate ion is removed, the reaction will shift to the right in order to supply more acetate ion, restoring equilibrium.

9.13 Critical Thinking Question
a. In the lungs, this reaction will shift to the right, consuming the excess oxygen and restoring equilibrium.
b. At the ends of small capillaries, this reaction will shift to the left to restore equilibrium, thus supplying oxygen to these oxygen-depleted regions.
c. The equilibrium shifts right to consume oxygen in the oxygen-rich lungs, while it shifts left, producing oxygen in oxygen-depleted tissues.

9.14 $Ca(OH)_2(s) + 2\ HCl(aq) \rightarrow 2\ H_2O + CaCl_2(aq)$

    Base   Acid   Water   Salt

9.15 a. $KOH(s) + HNO_3(aq) \rightarrow H_2O + KNO_3(aq)$
b. $Mg(OH)_2(s) + 2\ HNO_3(aq) \rightarrow 2\ H_2O +$ $Mg(NO_3)_2(aq)$
c. $Al(OH)_3(s) + 3\ HNO_3(aq) \rightarrow 3\ H_2O +$ $Al(NO_3)_3(aq)$

9.16

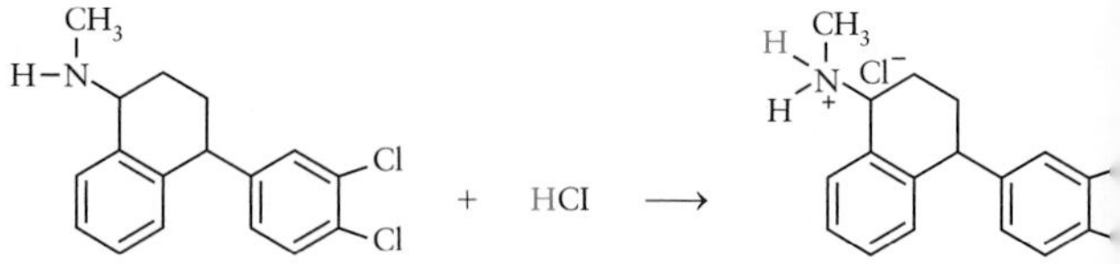

Water is not formed in this neutralization reaction because hydroxide (OH^-) is not a reactant.

9.17 Critical Thinking Question: The limited solubility of aluminum hydroxide ($Al(OH)_3$) and magnesium hydroxide ($Mg(OH)_2$) can be represented by the equilibrium equations shown below:

$$Mg(OH)_2(s) + H_2O \rightleftharpoons Mg^{2+}(aq) + 2\ OH^-(aq)$$
$$Al(OH)_3(s) + H_2O \rightleftharpoons Al^{3+}(aq) + 3\ OH^-(aq)$$

In contrast, NaOH dissolves completely in water:

$$NaOH(s) + H_2O \longrightarrow Na^+(aq) + OH^-(aq)$$

The equilibria for aluminum hydroxide ($Al(OH)_3$) and magnesium hydroxide ($Mg(OH)_2$) lie far to the left, meaning that mostly solid material is present at equilibrium. When acid is added, the equilibrium is disturbed, as the acid will react with the $OH^-(aq)$, removing OH^- and causing the equilibrium to shift to the right, dissolving more solid (Le Châtelier's principle). In this way, tissues are never exposed to excessive amounts of free hydroxide.

9.18

$[H_3O^+]$	$[OH^-]$
1.0×10^{-11}	1.0×10^{-3}
1.1×10^{-6}	9.1×10^{-9}
3.3×10^{-6}	3.0×10^{-9}

9.19 6.25
9.20 6.49
9.21 $6.0 \times 10^{-4}\ M$
9.22 $9.5 \times 10^{-3}\ M$
9.23 a. $NH_3 + H_2O \rightleftharpoons OH^- + NH_4^+$
b. The purpose of a buffer is to resist changes in pH upon addition of small amounts of acid or base.
c. NH_4^+ serves as the weak acid component of the buffer, which is present in the same amount as the base, NH_3.
d. The equilibrium would shift to the right:

$$NH_3 + H_3O^+ \longrightarrow H_2O + NH_4^+$$

e. The equilibrium would shift to the left:

$$NH_3 + H_3O^+ \longleftarrow OH^- + NH_4^+$$

Reactions of Organic Functional Groups in Biochemistry

CHAPTER 10

The ethanol in alcoholic beverages can be
produced from the fermentation of grapes.
[Walter Choroszewski/Stock Connection]

OUTLINE

10.1 The Role of Functional Groups in Biochemical Reactions

10.2 Oxidation–Reduction Reactions

10.3 Hydration–Dehydration Reactions

10.4 Acyl Group Transfer Reactions

10.5 Phosphoryl Group Transfer Reactions

 This icon indicates that a **Problem-Solving Tutorial**
is available at www.whfreeman.com/gob

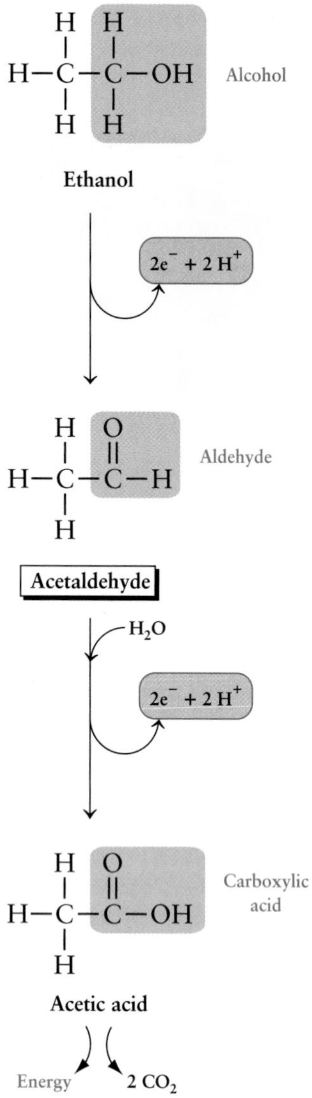

Ethanol

Alcohol

$2e^- + 2H^+$

Acetaldehyde

Aldehyde

H_2O

$2e^- + 2H^+$

Acetic acid

Carboxylic acid

Energy $2 CO_2$

Figure 10-1 Overview of the early steps in the metabolism of ethanol.

How Does the Body Handle Alcoholic Beverages?

Why does drinking too many alcoholic beverages in too short a time leave a person feeling nauseated and lightheaded, and maybe hung over into the next day? The answer lies in the little molecule ethanol, found in champagne, beer, and wine in varying concentrations. Since you have completed two chapters on organic chemistry, you already know that ethanol is an alcohol and contains the hydroxyl functional group, shown in the ball-and-stick model on the preceding page.

The human body cannot store ethanol, as it can sugars and fats, so it must *metabolize* it—chemically transform it into other compounds that are readily eliminated from the body. The body metabolizes ethanol in the liver, so it is no surprise that many alcoholics suffer from liver damage. Liver cells carry out a sequence of reactions, known as a *biochemical pathway*, that convert ethanol into carbon dioxide, which the body can exhale, and water. One of the compounds produced as part of this biochemical process, acetaldehyde, is believed to be responsible for some of the immediate negative effects of alcohol. In this chapter you will learn about many of the chemical reactions the cell employs as part of the biochemical pathways of metabolism.

The first two steps of the biochemical pathway for ethanol metabolism are shown in Figure 10-1. In the first reaction in Figure 10-1 ethanol is converted into acetaldehyde. In the second reaction acetaldehyde is converted into acetic acid. If a person consumes too much alcohol in too short a time, the enzyme catalyzing the second reaction cannot keep up with the production of acetaldehyde, and so the concentration of this compound increases. Elevated levels of acetaldehyde are believed to be responsible, in part, for the headache, nausea, dizziness, and "hangover" felt by the person consuming too much ethanol in too short a time.

Recall from Chapter 8 that energy is an important component of all chemical reactions. Indeed, the metabolism of ethanol produces energy, which also explains why a person can gain weight drinking alcohol. However, one of the problems with alcohol as an energy source is that it contains no nutrients—just calories. Indeed, alcoholics are often malnourished, because ethanol does not contain nitrogen and therefore cannot supply the building blocks for making proteins.

Some individuals have a less effective form of the enzyme that transforms acetaldehyde into acetic acid. As a result, these individuals have a low tolerance for alcohol. They exhibit symptoms of acetaldehyde poisoning after consuming only small amounts of alcohol. They are also at greater risk of developing liver diseases caused by overexposure to acetaldehyde.

As you can see from this simple example, understanding organic reactions and their role in biochemical pathways provides unique insight into how the human body works. In this chapter you will learn the major types of organic reactions carried out within all cells of the human body.

In Chapter 8 you learned the basic principles that govern a chemical reaction: the conservation of mass, the conservation of energy, and the kinetic factors that control the rate of a reaction. In Chapter 9 you were introduced to the most fundamental of all reactions, the reactions of acids and bases. In this chapter you will expand your knowledge of chemical reactions to include the chemical reactions of the important organic functional groups that you encountered in Chapter 7.

Each organic functional group is characterized by a unique chemical reactivity: a predictable set of chemical reactions that the functional group will undergo. For example, you know that one chemical reaction a *carboxylic acid* readily undergoes is losing a proton to a base to form a carboxylate ion. The main reason chemists classify organic compounds according to the functional groups they contain is that functional groups generally react in the same characteristic way under a given set of conditions, regardless of the structure of the rest of the molecule. This is true whether the reaction is performed in a flask in the laboratory or in the aqueous medium of the cell. In this chapter, you will focus primarily on chemical reactions of organic functional groups seen in biochemistry—reactions the cell can perform.

You may notice that some of the chemical structures in this chapter and subsequent ones are complex, but do not let this discourage you, because the characteristics of functional groups and their chemical reactivity are the same regardless of the complexity of the molecule. Moreover, in the cell, a chemical reaction typically affects only one functional group in the molecule. If you can learn to recognize the functional group and the type of reaction that occurs, then you will be able to predict the structure of the product regardless of its complexity.

One feature that distinguishes biochemical reactions is that they are almost always catalyzed by enzymes. In contrast, reactions that take place in the laboratory are typically facilitated by other factors such as added heat and changes in concentration. Enzymes are such a critical component of all biochemical reactions that without them these reactions would not be possible.

10.1 The Role of Functional Groups in Biochemical Reactions

Obviously, understanding biochemical reactions depends upon your recognizing the common organic functional groups. Table 10-1 summarizes the functional groups that were introduced in Chapter 7.

Remember, alkanes are not classified as a type of functional group; instead, they serve as the structural scaffolding for organic molecules. Two functional groups in Table 10-1 are relatively unreactive, and therefore will not be encountered in chemical reactions in this text: They are the ethers and tertiary alcohols.

In Chapter 8 you learned that metabolic reactions are divided into two basic biochemical pathways: anabolic and catabolic, as shown in Figure 10-2. Catabolic processes degrade larger biomolecules such as carbohydrates, proteins, and fats into smaller molecules such as carbon dioxide (CO_2), water (H_2O), and ammonia (NH_3). Catabolic processes produce energy overall. In contrast, anabolic pathways build larger biomolecules such as proteins, glycogen, lipids, and DNA from simpler organic molecules. Anabolic pathways consume energy overall. All catabolic and anabolic reactions involve organic functional groups. Thus, before

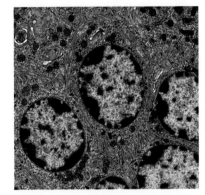

(a)

(b)

(a) Microscopic view of liver cells, where many reactions occur. (b) A chemical reaction taking place in an Erlenmeyer flask. Functional groups react in the same characteristic way whether the reaction is performed in a flask in the laboratory or in the aqueous medium of the cell.
[(a) Steve Gschmeissner/Photo Researchers, Inc.; (b) © 1994 Paul Silverman, Fundamental Photographs, NYC]

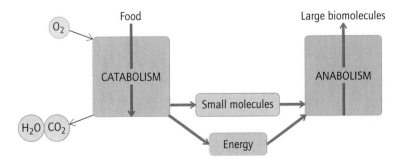

Figure 10-2 Overview of metabolism. Catabolism: Large molecules are degraded to small molecules, releasing energy. Anabolism: Small molecules are used to build larger molecules, absorbing energy.

Table 10-1 Common Organic Functional Groups*

Functional Group Name	Structure of Functional Group	Functional Group Name	Structure of Functional Group
Alkene	$\overset{\displaystyle \diagdown}{\diagup}C{=}C\overset{\displaystyle \diagup}{\diagdown}$	Amine	$-\overset{\displaystyle ..}{\underset{\displaystyle \mid}{N}}-$
Alkyne	$-C{\equiv}C-$	Aldehyde	$R-\overset{\displaystyle O}{\overset{\displaystyle \|}{C}}-H$
Aromatic ring	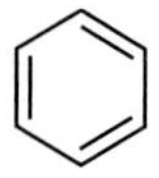 ⬡	Ketone	$R-\overset{\displaystyle O}{\overset{\displaystyle \|}{C}}-R$
1° Alcohol	$R-\overset{\displaystyle H}{\underset{\displaystyle H}{\overset{\displaystyle \|}{\underset{\displaystyle \|}{C}}}}-O-H$	Carboxylic acid	$R-\overset{\displaystyle O}{\overset{\displaystyle \|}{C}}-O-H$
2° Alcohol	$R-\overset{\displaystyle R}{\underset{\displaystyle H}{\overset{\displaystyle \|}{\underset{\displaystyle \|}{C}}}}-O-H$	Ester	$R-\overset{\displaystyle O}{\overset{\displaystyle \|}{C}}-O-R$
3° Alcohol	$R-\overset{\displaystyle R}{\underset{\displaystyle R}{\overset{\displaystyle \|}{\underset{\displaystyle \|}{C}}}}-O-H$	Amide	$R-\overset{\displaystyle O}{\overset{\displaystyle \|}{C}}-\overset{}{\underset{\displaystyle \mid}{N}}-$
Phenol	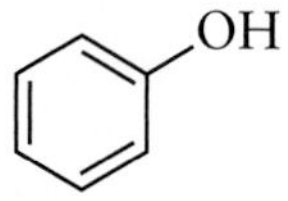 ⬡—OH	Thioester	$R-\overset{\displaystyle O}{\overset{\displaystyle \|}{C}}-S-R$
Ether	$R-O-R$	Phosphate ester	$^{-}O-\overset{\displaystyle O}{\underset{\displaystyle O^{-}}{\overset{\displaystyle \|}{\underset{\displaystyle \mid}{P}}}}-O-R$

*R is one or more carbon atoms. "—" represents a bond to hydrogen or carbon.

you can consider these biochemical pathways, you need to learn the common reactions these important functional groups undergo.

Although the cell performs a myriad of reactions, it is limited to only six basic reaction types. Therefore, by learning only a few reaction types, you will be able to recognize and understand almost all of the reactions seen in biochemistry! In this chapter we will consider the first four of these six basic reaction types, which are listed below. The last two reaction types will be covered briefly in a later chapter. *When learning these reactions, remember to focus on the change in the functional group and ignore the rest of the molecule, which remains intact.*

- oxidation–reduction reactions
- hydration–dehydration reactions

- acyl group transfer reactions (hydrolysis, esterification, and amidation)
- phosphoryl group transfer reactions
- decarboxylation reactions
- reactions that form or break carbon–carbon bonds

10.2 | Oxidation-Reduction Reactions

Oxidation–reduction reactions are some of the most important reactions in nature. You encounter oxidation–reduction reactions in everyday activities, such as burning wood, propane, gasoline, or natural gas. The burning of an organic substance is one of the most common types of oxidation–reduction reactions, known as a combustion reaction.

In Chapter 8, you learned that **combustion** of an organic compound is the reaction between an organic substance ($C_xH_yO_z$) and oxygen (O_2) to produce carbon dioxide (CO_2) and water (H_2O). For example, the combustion of natural gas (CH_4) produces the flame on a Bunsen burner, described by the chemical equation

$$CH_4(g) + 2\ O_2(g) \longrightarrow CO_2(g) + 2\ H_2O(g)$$

The most important product of a combustion reaction is energy and this is why combustion reactions are used every day to cook food, heat homes, and drive cars. Not surprisingly, the cell also uses combustion reactions to release energy from the bonds in the molecules found in food.

The primary fuel employed by the cell is glucose, also called blood sugar, because it is the form in which sugar circulates in the blood and is supplied to the cell. Glucose undergoes combustion, not in a single reaction like natural gas, but rather through a series of separate oxidation–reduction reactions that ultimately lead to the same end products: carbon dioxide, water, and energy. As with any combustion reaction, oxygen is required; hence, the term **cellular respiration** is used to describe this sequence of reactions in the cell.

Definitions of Oxidation and Reduction

Many functional groups undergo oxidation–reduction reactions, most notably carbon–carbon bonds and carbon–oxygen bonds. *An **oxidation–reduction** reaction* (sometimes abbreviated **redox**) *is characterized by the transfer of electrons from one reactant to another.* The reactant that loses electrons is said to undergo **oxidation,** and the reactant that gains electrons is said to undergo **reduction.** *Consequently, where there is oxidation, there must also be reduction.*

> **Oxidation:** loss of electrons
>
> **Reduction:** gain of electrons $\quad e^-$

The transfer of electrons in an oxidation–reduction reaction is easiest to see in simple inorganic reactions. Consider, for example, the reaction shown in the photo of the beaker in the margin on the following page. This is the reaction of magnesium metal, Mg, with aqueous hydrochloric acid, HCl, to produce aqueous magnesium chloride, $MgCl_2$, and hydrogen gas, H_2.

The flame seen on a Bunsen burner is produced from the reaction of methane with oxygen. This reaction is a combustion reaction, which is a type of oxidation-reduction reaction. [Martyn F. Chillmaid/Photo Researchers, Inc.]

"Burning" calories refers to the combustion of glucose to release the energy needed for a fast run and other activities. The products are CO_2, water, and energy. [Olivier Prevosto/TempSport/Corbis]

To Help You Remember
Loss of Electrons is Oxidation (**LEO**)
Gain of Electrons is Reduction (**GER**)
LEO the lion goes GER

Or if you prefer:

Oxidation is Loss of electrons (**OIL**)
Reduction is Gain of electrons (**RIG**)
OIL RIG

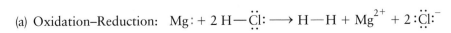

(a) Oxidation–Reduction: $\ddot{Mg}: + 2\ H-\overset{..}{\underset{..}{Cl}}: \longrightarrow H-H + Mg^{2+} + 2\ :\overset{..}{\underset{..}{Cl}}:^-$

Magnesium ribbon undergoes oxidation in an aqueous solution of HCl. The HCl is reduced to hydrogen gas (H_2), seen as bubbles emerging from the solution. The other product, magnesium chloride, is dissolved in solution. [Andrew Lambert Photography/Photo Researchers, Inc.]

In this reaction, electrons are transferred from Mg to H^+. You can show the oxidation step separately from the reduction step:

(b) Oxidation: $Mg: \longrightarrow Mg^{2+} + 2\,e^-$

(c) Reduction: $2\,e^- + 2\,H^+ \longrightarrow H—H$

Equation (b) shows the oxidation: The Mg atom loses two electrons to become a magnesium cation, Mg^{2+}. The equation is illustrated using Lewis dot structures to keep track of the electrons transferred. Equation (c) is a reduction and proceeds simultaneously with (b): Two protons (H^+) from two HCl molecules gain a net two electrons, and these electrons become the H–H covalent bond of a diatomic hydrogen molecule, H_2. Thus, magnesium has lost two electrons, and two protons (H^+) have gained these electrons, in forming an H_2 molecule. The chloride ion, Cl^-, is unchanged throughout the reaction, because it undergoes neither oxidation nor reduction. For this reason, the chloride ion in this reaction is referred to as a "spectator ion."

Generally, when a metal cation is formed from a metal, the metal has undergone oxidation; and when a nonmetal anion is formed from a nonmetal, the nonmetal has undergone reduction. Recall from Chapter 2 that metal cations and nonmetal anions form ionic compounds.

Oxidation-Reduction of Organic Molecules

The transfer of electrons between molecules is not as readily apparent in organic compounds as it is in inorganic compounds, partly because organic compounds are more complex. In organic reactions, it is easier to recognize oxidation and reduction by looking for a change in the number of hydrogen atoms and/or oxygen atoms in the reactants compared to the products. An organic molecule that gains oxygen atoms or loses hydrogen atoms or both has undergone oxidation. An organic molecule undergoing reduction experiences the reverse: It gains hydrogen atoms or loses oxygen atoms or both.

Oxidation: The number of oxygen atoms increases and/or the number of hydrogen atoms decreases.

Reduction: The number of oxygen atoms decreases and/or the number of hydrogen atoms increases.

For example, consider the first step in the metabolism of ethanol described in the opening vignette:

$$
\underset{\textbf{Ethanol}}{\underset{\text{Alcohol}}{H-\overset{\displaystyle H}{\underset{\displaystyle H}{C}}-\overset{\displaystyle H}{\underset{\displaystyle H}{C}}-OH}} \quad \xrightarrow{\text{Oxidation}} \quad \underset{\textbf{Acetaldehyde}}{\underset{\text{Aldehyde}}{H-\overset{\displaystyle H}{\underset{\displaystyle H}{C}}-\overset{\displaystyle O}{C}-H}} \; + \; 2\,e^- \; + \; 2\,H^+
$$

The functional group in this reaction changes from an alcohol to an aldehyde. Acetaldehyde, the product, has two fewer hydrogen atoms than ethanol, and there is no change in the number of oxygen atoms. Since the product has lost H atoms, printed in red, this part of the reaction represents an oxidation.

Oxidation is always accompanied by reduction, because the electrons need to be transferred to another reactant. In the cell, the reactants receiving electrons are usually **coenzymes,** special molecules that shuttle electrons and hydrogen atoms between organic molecules. Thus, enzymes that catalyze oxidation–reduction reactions typically require a coenzyme for their catalytic activity. In the reaction involving ethanol shown above, it is the coenzyme that gains the electrons and one of the hydrogen atoms lost by ethanol. You will learn more about coenzymes later in this chapter.

In oxidation–reduction reactions, the term **oxidizing agent** refers to the reactant that gets reduced, and the term **reducing agent** refers to the reactant that undergoes oxidation. For example, bleach is a well known *oxidizing agent,* which means it causes other substances to be oxidized while it is reduced in the process. Table 10-2 shows a list of some common oxidizing and reducing agents. Since there are a variety of oxidizing agents that produce the same results, the identity of the oxidizing agent is often not specified; instead, the general abbreviation [O] is placed above the reaction arrow to indicate an oxidation. Likewise, reducing agents are designated by placing the abbreviation [H] above the reaction arrow. For example, the oxidation of ethanol, described earlier, could have been written as

Table 10-2 Common Oxidizing and Reducing Agents[*]

Oxidizing Agents	
Bleach	NaOCl
Oxygen	O_2
Hydrogen peroxide	H_2O_2
Chromium(VI)	Cr^{6+}
NAD$^+$	
FAD	

Reducing Agents	
Hydrogen	H_2
Sodium borohydride	$NaBH_4$
NADH	
FADH$_2$	

[*]Shaded items are biological agents.

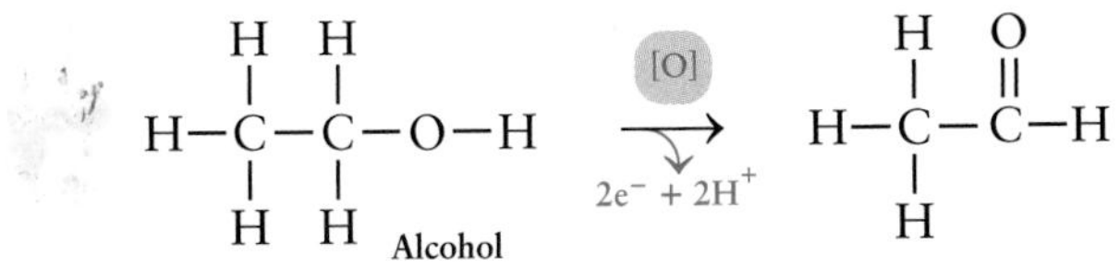

WORKED EXERCISE 10-1   Recognizing Oxidation and Reduction Reactions

Determine whether the reactant shown undergoes an oxidation or a reduction in the transformation shown. Explain your reasoning. Place an [H] or an [O] above the arrow as needed.

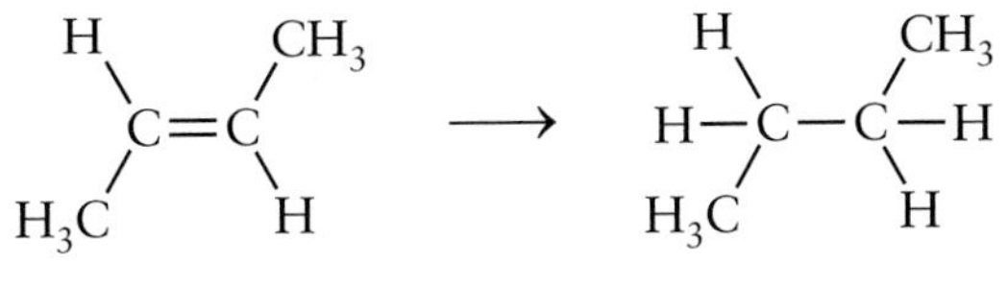

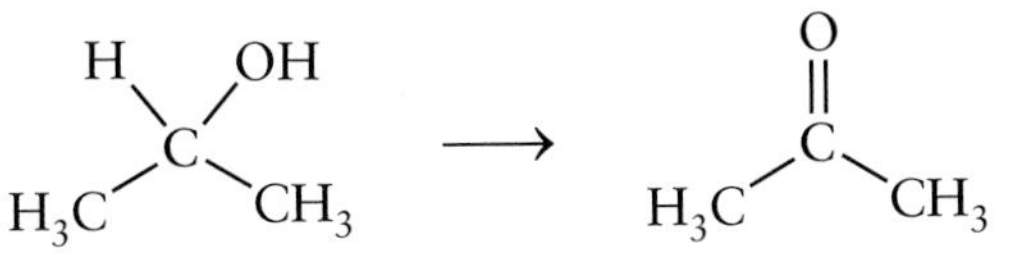

SOLUTION

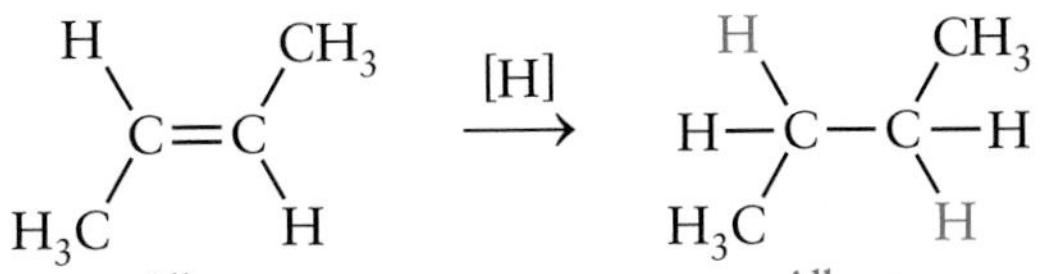

Begin by noting the change in the functional group: an alk*ene* is changed into an alk*ane*. Next, compare the number of oxygen and hydrogen atoms in the functional group of the product versus the functional group of the reactant. Since there is an increase in the number of hydrogen atoms, and no change in the number of oxygen atoms, the transformation represents a reduction.

b.

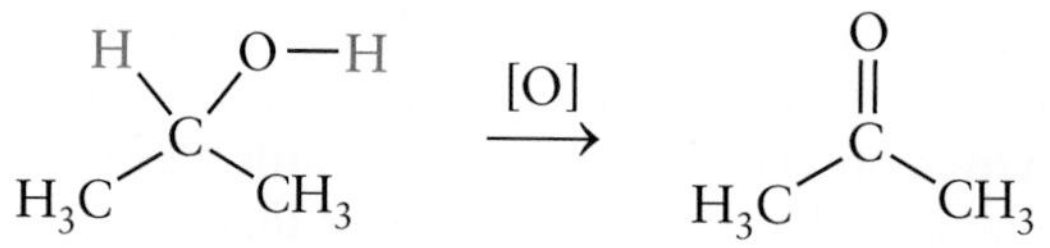

Alcohol   [O] →   Ketone

The functional group changes from an alcohol to a ketone. Since there is a decrease in the number of hydrogen atoms, and no change in the number of oxygen atoms, the reactant has undergone an oxidation.

PRACTICE EXERCISES

10.1 Determine whether the following reactions represent an oxidation or a reduction of the organic reactant shown. Explain your reasoning, by showing the hydrogen and/or oxygen atoms that have been gained or lost from the reactant. What is another name for the type of reaction shown in part (c)?

a.

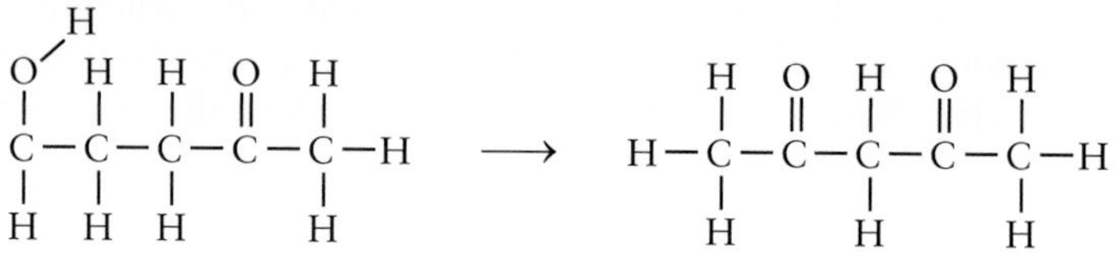

b.

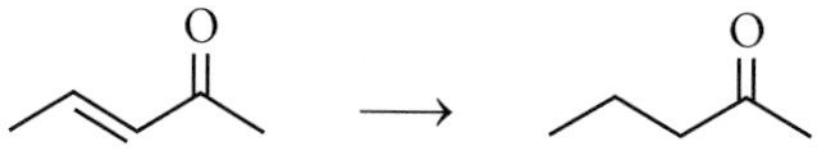

c.

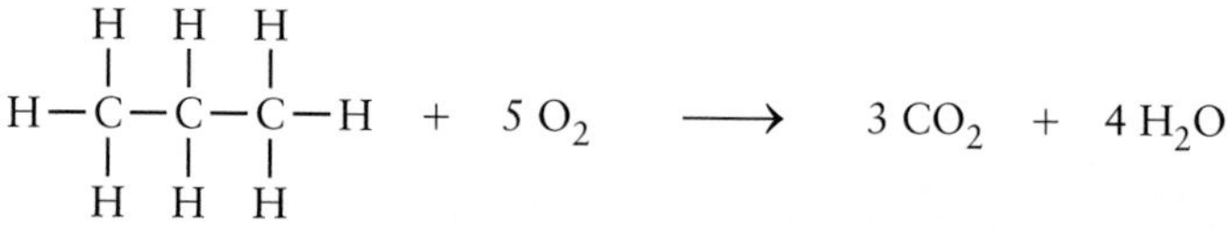

$$H-\overset{\overset{\displaystyle H}{|}}{\underset{\underset{\displaystyle H}{|}}{C}}-\overset{\overset{\displaystyle H}{|}}{\underset{\underset{\displaystyle H}{|}}{C}}-\overset{\overset{\displaystyle H}{|}}{\underset{\underset{\displaystyle H}{|}}{C}}-H \ + \ 5\,O_2 \ \longrightarrow \ 3\,CO_2 \ + \ 4\,H_2O$$

d. *Hint:* Focus on only the functional group changed.

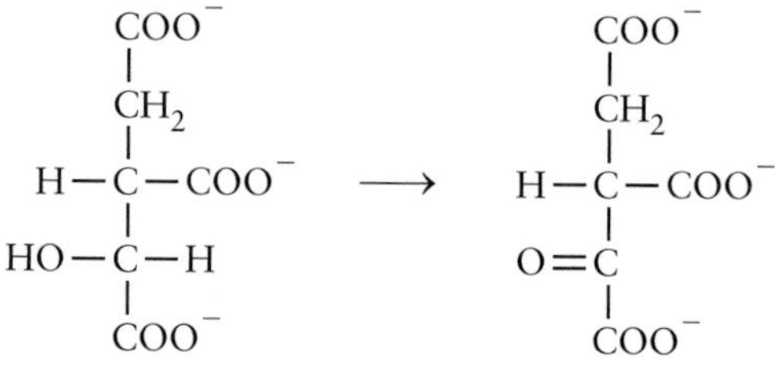

10.2 For the inorganic oxidation–reduction shown below, write the part of the reaction that undergoes oxidation and the part that undergoes reduction as two separate reactions.

$$2\,Ag^+(aq) + Cu(s) \longrightarrow Cu^{2+}(aq) + 2\,Ag(s)$$

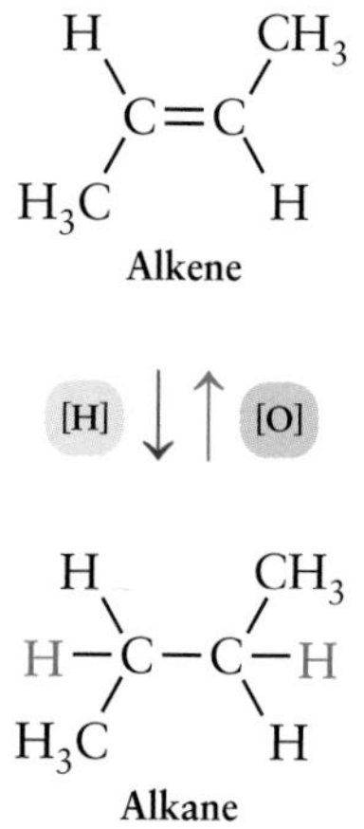

Next you will consider the oxidation and reduction of the following functional groups:

- alkenes and C—C bonds with an adjacent carbonyl group
- functional groups containing C—O and C=O

Hydrocarbon Oxidation-Reductions

Carbon–carbon double bonds undergo reduction to produce carbon–carbon single bonds, as shown in the margin. The reverse reaction is therefore an oxidation. In the laboratory, the reduction of an unsaturated hydrocarbon

---

Hydrocarbon Oxidation–Reductions

Alkene

[H] ↓ ↑ [O]

Alkane

is usually carried out with hydrogen gas (H_2) as the reducing agent in the presence of a metal catalyst (Pd, Pt, Ni, etc.), in a reaction called a **catalytic hydrogenation**:

Catalytic hydrogenation

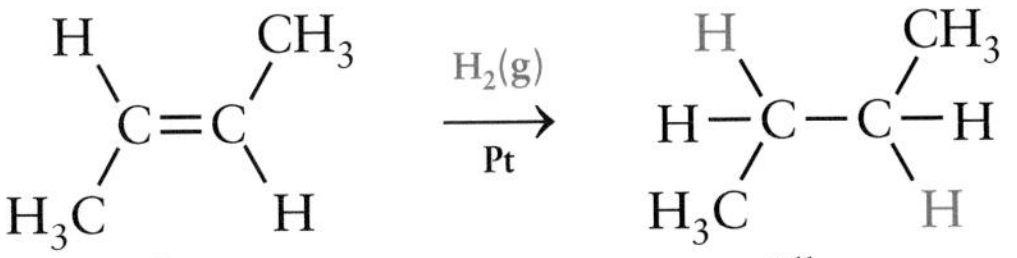

Unsaturated fats contain carbon-carbon double bonds in their chemical structure, which makes them liquids (oils) at room temperature. Saturated fats contain only carbon-carbon single bonds, giving them a solid consistency, like butter and lard. Partially hydrogenated fats have a consistency in between oils and fat. [Punchstock/Stockbyte]

Catalytic hydrogenation reactions are used by the food industry to prepare shortening from vegetable oils. Recall from Chapter 6 that unsaturated fats found in vegetable oils are healthier than saturated fats found in animal fats, but they have a shorter shelf life and tend to be liquids at room temperature. Hydrogenation of a healthy unsaturated fat to produce a saturated fat results in a fat with a more desirable solid consistency and a longer shelf life but that loses its health benefits. These foods must be labeled as "hydrogenated" or "partially hydrogenated."

One outcome of a hydrogenation reaction is that some cis double bonds undergo isomerization instead of reduction: cis double bonds are converted into trans double bonds.

Isomerization reaction

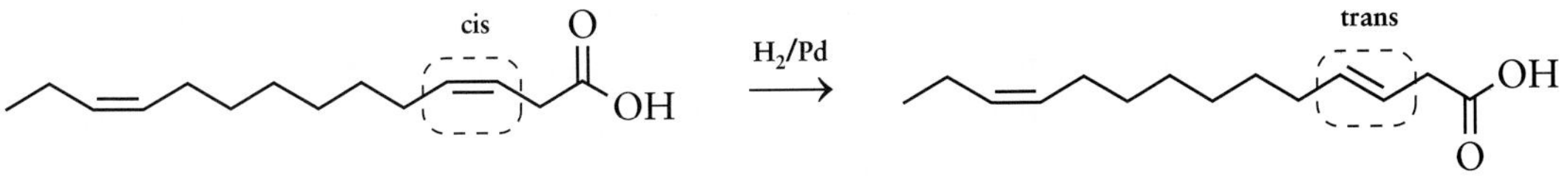

When the reactant is a fat, the product is called a *trans fat*. Unfortunately, trans fats are even less healthy than saturated fats. Indeed, trans fats have been banned from many restaurants. You have probably already noticed the trend in the food industry to produce processed foods without trans fats, often labeled "no trans fats."

Coenzyme FAD/FADH₂ Obviously a biological cell cannot use hydrogen gas as a reducing agent; instead, electrons and hydrogen atoms—the equivalent of H_2—are supplied by coenzymes. In oxidation–reduction reactions involving carbon–carbon bonds, the coenzyme **f**lavin **a**denine **d**inucleotide, abbreviated **FAD**, is typically the electron acceptor, and is reduced to **FADH₂**:

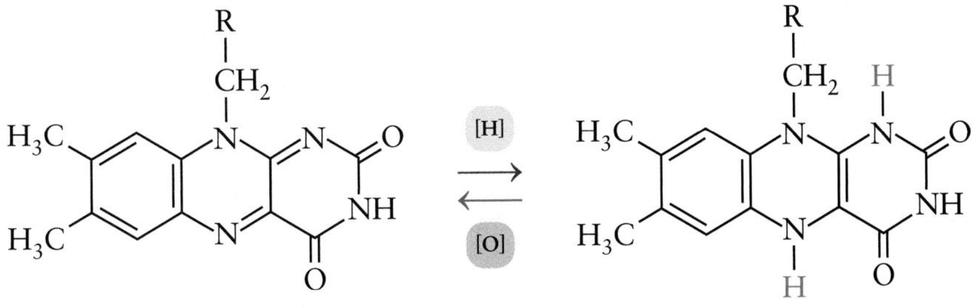

Note the presence of the two new hydrogen atoms shown in red in the structure of FADH₂, the reduced form of FAD. The reverse process is an oxidation. Thus, FADH₂ is a biological reducing agent and FAD is a biological oxidizing agent. Coenzymes are produced in the cell from vitamins, as described in *Chemistry in Medicine: Vitamins and Health* at the end of the chapter. FAD, for example, is formed in the cell from the vitamin riboflavin.

WORKED EXERCISES 10-2   Oxidation-Reduction Reactions Involving Carbon-Carbon Bonds

1 Predict the product formed in the following reactions. How has the functional group changed in (a)?

a. $\xrightarrow[\text{Pt}]{\text{H}_2(g)}$

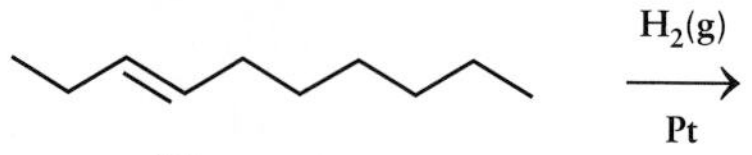

b. FAD $\xrightarrow{\text{[H]}}$

2 The following reaction occurs as part of a key metabolic pathway:

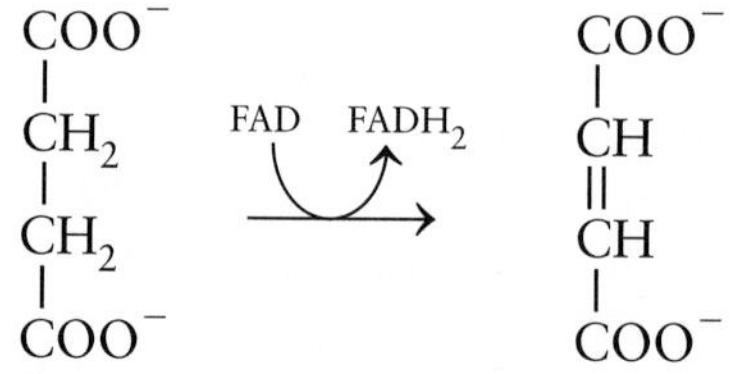

a. Is succinate oxidized or reduced in this reaction? How can you tell?
b. Is the coenzyme FAD oxidized or reduced?
c. Is FAD an oxidizing agent or a reducing agent?

SOLUTIONS

1 a. The reaction shown is a catalytic hydrogenation: It relies on hydrogen gas and a catalyst. The functional group is an alkene, which undergoes reduction in the presence of hydrogen to form the corresponding alkane, $CH_3(CH_2)_8CH_3$.
   b. Reduction of FAD produces $FADH_2$.

2 a. The loss of hydrogen atoms in going from succinate to fumarate indicates that succinate has undergone oxidation.
   b. Since succinate is undergoing oxidation, FAD must be undergoing a corresponding reduction. The gain of H atoms expected in a reduction is indicated in the formula: $FADH_2$.
   c. Since FAD is causing succinate to be oxidized, it is the oxidizing agent.

PRACTICE EXERCISE

10.3 Predict the product formed in the following reactions:

a. $\xrightarrow{\text{[H]}}$

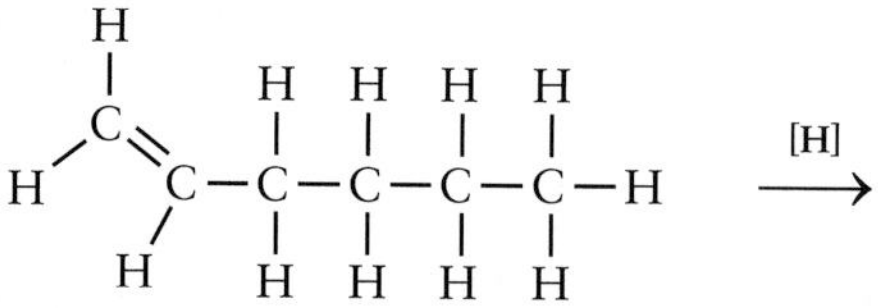

b. $\xrightarrow[\text{Pt}]{\text{H}_2(g)}$

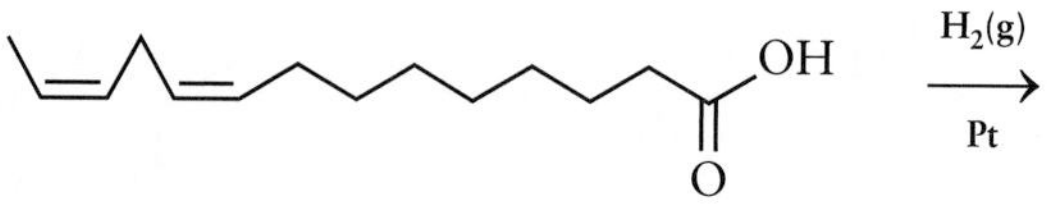

c. $FADH_2$ $\xrightarrow{\text{[O]}}$

Oxidation-Reductions Involving Carbon-Oxygen Bonds

Some of the most important oxidation–reduction reactions involve functional groups containing carbon–oxygen bonds, such as alcohols, and functional groups containing the carbonyl group, $C{=}O$.

Oxidation of Alcohols To predict the product of an alcohol oxidation you must first determine whether the alcohol is a primary, secondary, or tertiary alcohol. You saw in Chapter 7 that these designations are based on whether there are one, two, or three R groups attached to the carbon atom bearing the hydroxyl group. *Primary alcohols undergo oxidation to form aldehydes, secondary alcohols undergo oxidation to form ketones, and tertiary alcohols do not undergo oxidation. Aldehydes can undergo further oxidation in the presence of water to produce carboxylic acids.* This pattern of chemical reactivity among the different types of alcohols is summarized in Figure 10-3.

Note that in all of these oxidation reactions, there is a decrease in the number of hydrogen atoms, and in one case, an increase in the number of oxygen atoms (aldehyde to carboxylic acid); hence electrons are lost. When an alcohol undergoes oxidation, it loses the hydrogen atom on the O–H group as well as one of the hydrogen atoms on the carbon atom bearing the OH group. Indeed, it is the absence of a hydrogen atom on the carbon atom bearing the OH group that prevents a tertiary alcohol from undergoing oxidation. When an aldehyde is oxidized to a carboxylic acid, a molecule of water from solution is also required as a reactant, as shown.

Each of the reactions shown in Figure 10-3 can also proceed in the opposite direction, the corresponding *reductions*, as shown in Figure 10-4. For example, the oxidation of a 2° alcohol produces a ketone, and the reduction of a ketone produces a 2° alcohol.

Coenzyme NAD⁺/NADH A variety of oxidizing and reducing agents are available in the laboratory for the oxidation and reduction of the C–O-containing functional groups shown in Figures 10-3 and 10-4. Some examples are shown in Table 10-2. In the cell, the electron transfer agent is most often the coenzyme **n**icotinamide **a**denine **d**inucleotide, a substance derived from niacin—vitamin B₃.

The vitamin niacin is found in meats, fish, poultry, mushrooms, and whole grains. [brt FOOD/Alamy]

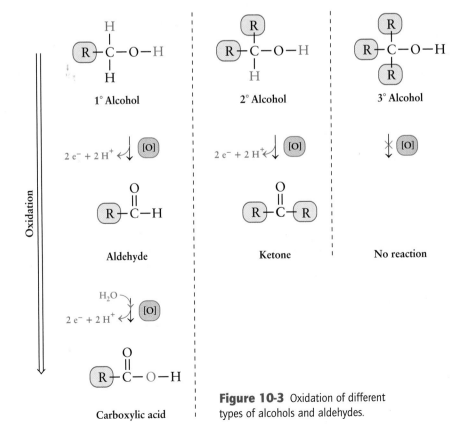

Figure 10-3 Oxidation of different types of alcohols and aldehydes.

Figure 10-4 Reduction of carboxylic acids, aldehydes, and ketones.

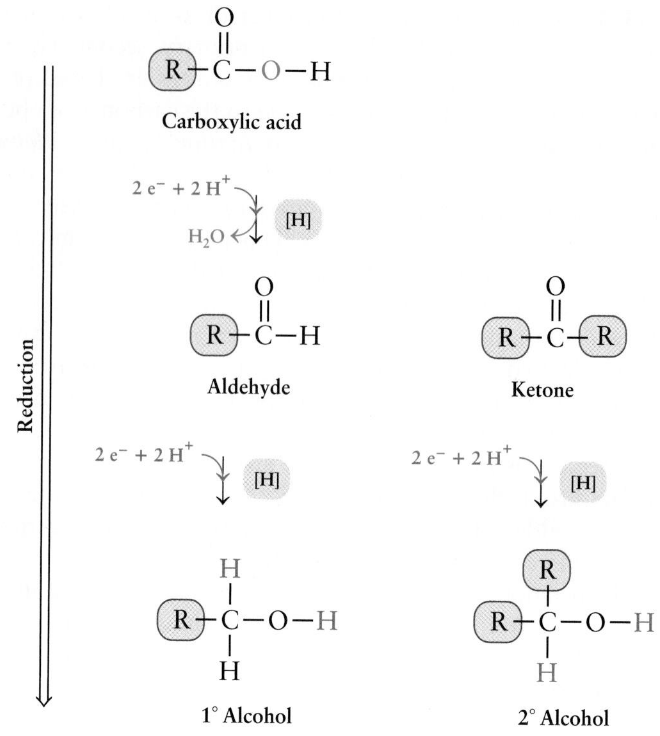

NAD$^+$ is an oxidizing agent: It causes the other reactant to undergo oxidation, or loss of electrons. One hydrogen atom and two electrons are added to the coenzyme to make NADH, and a second hydrogen atom is transferred to the solution as a proton (H$^+$), as shown in Figure 10-5. In the reverse process, NADH is a reducing agent: It causes the other reactant to undergo reduction, or a gain of electrons. A proton from the solution is also required for the reaction.

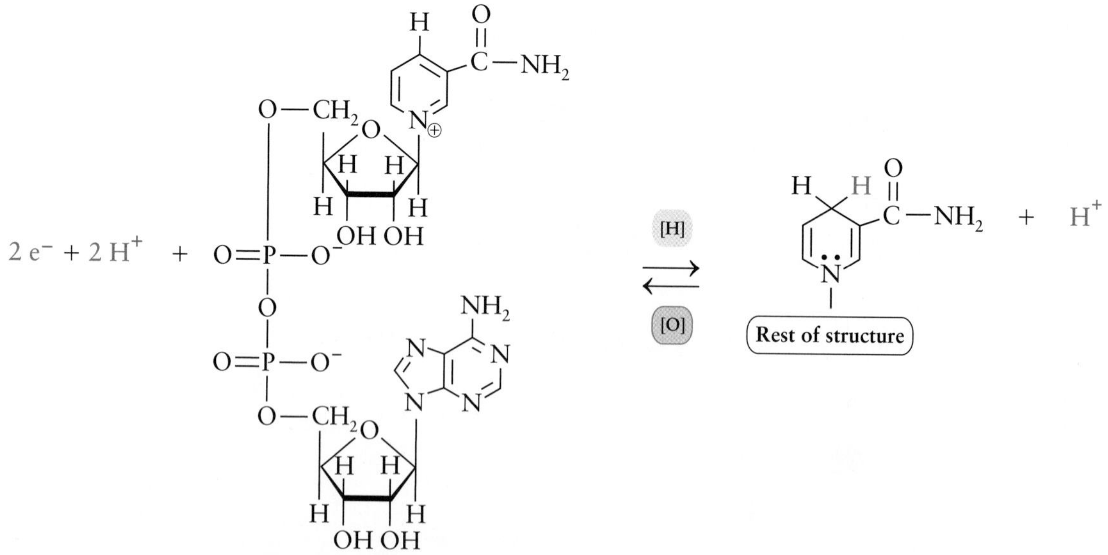

Figure 10-5 Reduction and oxidation of NAD$^+$/NADH.

NAD⁺ and NADH transfer electrons to and from carbon–oxygen bonds in a way similar to the way FAD and FADH₂ transfer electrons to carbon–carbon bonds. If you examine the first two steps in the metabolism of ethanol, introduced in the opening vignette, you will see that both reactions employ NAD⁺ as the coenzyme.

First Oxidation Steps in Ethanol Metabolism

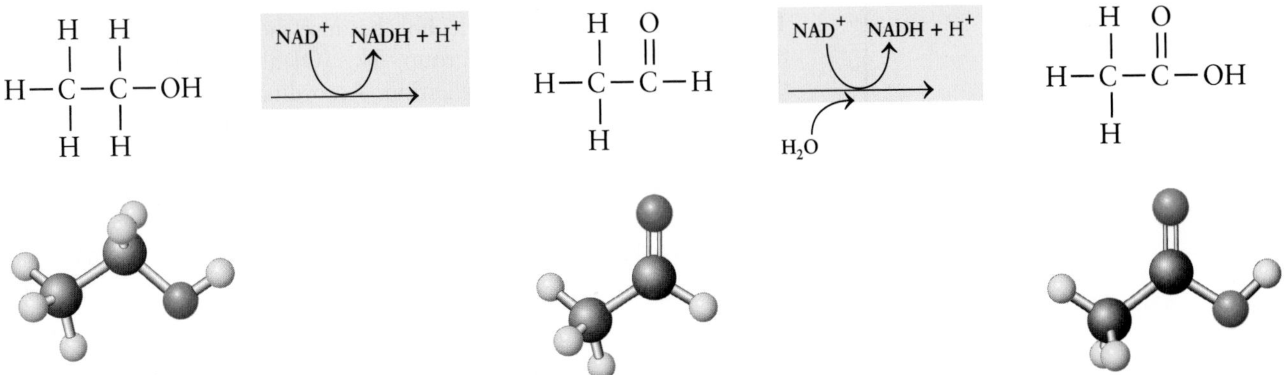

$$\underset{\text{Ethanol}}{\overset{\displaystyle\underset{|}{\overset{|}{H}}\;\underset{|}{\overset{|}{H}}}{H-\underset{|}{\overset{|}{C}}-\underset{|}{\overset{|}{C}}-OH}} \quad\xrightarrow[\text{NAD}^+\;\;\text{NADH}+\text{H}^+]{}\quad \underset{\text{Acetaldehyde}}{\overset{\displaystyle H\;\;O}{H-\underset{|}{\overset{|}{C}}-\overset{\|}{C}-H}}\quad\xrightarrow[\text{H}_2\text{O}]{\text{NAD}^+\;\;\text{NADH}+\text{H}^+}\quad \underset{\text{Acetic acid}}{\overset{\displaystyle H\;\;O}{H-\underset{|}{\overset{|}{C}}-\overset{\|}{C}-OH}}$$

The convention when writing biochemical pathways is to show the structures of the main reactant and product on either side of the main reaction arrow. The coenzyme—the oxidizing or reducing agent—is placed next to a curved arrow that intersects the main arrow. This convention makes it easier to keep track of the functional group changes in the reactant being metabolized, while at the same time following the fate of the coenzymes. Since NADH and FADH₂ are key players in biochemical pathways, especially those involved in making ATP, the convention is a convenient way to keep track of the number of NADH and FADH₂ molecules produced in a biochemical pathway.

Problem-Solving Tutorial:
Oxidation-Reduction Reactions

| **WORKED EXERCISES 10-3** | Oxidation-Reduction Reactions Involving Carbon-Oxygen Bonds |

1 Predict the structure of the product formed in each of the following reactions. Write the name of the functional groups involved. What is meant by [O] and [H] over the arrow?

a. (cyclohexanol) $\xrightarrow{\text{[O]}}$

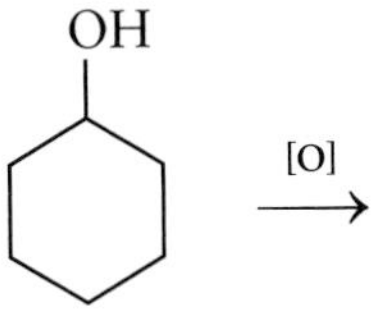

b. $\underset{}{\overset{\displaystyle H\;\;O}{H-\underset{\underset{H}{|}}{\overset{|}{C}}-\overset{\|}{C}-H}} \xrightarrow{\text{[H]}}$

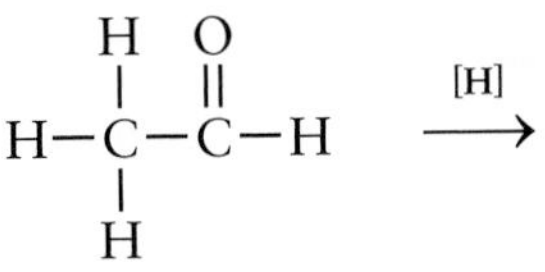

2 When muscle cells are depleted of oxygen, they form lactic acid by reducing the ketone functional group in pyruvic acid. The presence of lactic acid is responsible for the sore muscle feeling following intense exercise. Predict the structure of lactic acid:

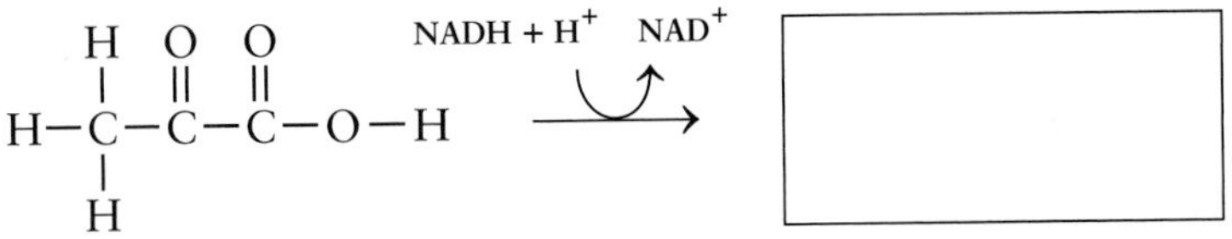

$$\underset{\text{Pyruvic acid}}{\overset{\displaystyle H\;\;O\;\;O}{H-\underset{\underset{H}{|}}{\overset{|}{C}}-\overset{\|}{C}-\overset{\|}{C}-O-H}} \quad\xrightarrow{\text{NADH}+\text{H}^+\;\;\text{NAD}^+}\quad \boxed{}$$

Lactic acid

SOLUTIONS

1 a.

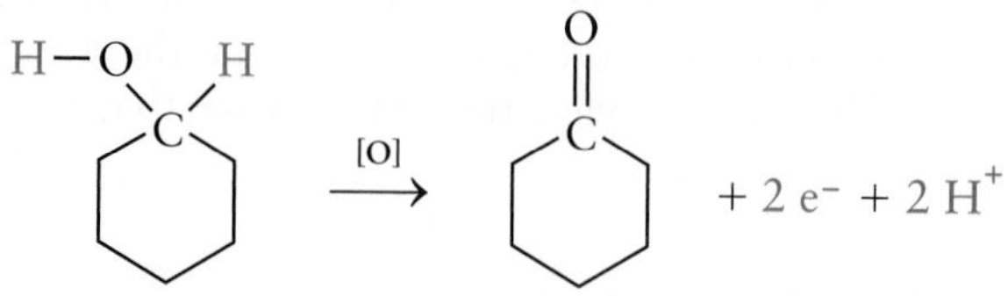

2° Alcohol   Ketone

The reactant is a secondary alcohol, because there are two R groups on the carbon bearing the OH group. The [O] above the arrow signifies that the alcohol undergoes oxidation. According to Figure 10-3, oxidation of a secondary alcohol produces a ketone. The hydrogen atoms printed in red are removed and a carbonyl group is formed: Cyclohexanol is oxidized to cyclohexanone. A word of caution when using skeletal line structures: Do not forget to count the C–H bonds when counting H atoms, since skeletal structures do not explicitly show these atoms.

b.

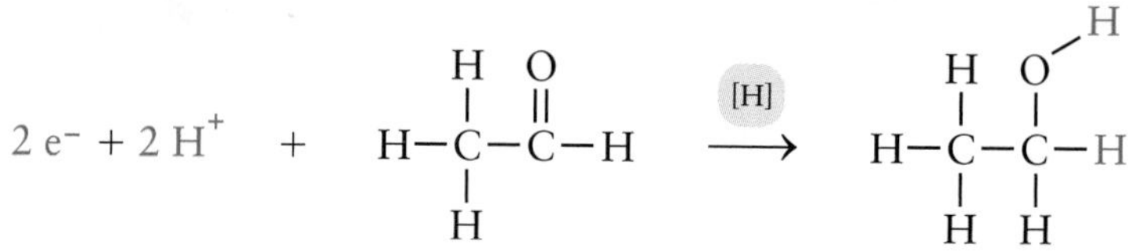

Aldehyde   1° Alcohol

The reactant is an aldehyde. The [H] above the arrow signifies that the aldehyde undergoes reduction. According to Figure 10-4, the reduction of an aldehyde produces a primary alcohol. The hydrogen atoms printed in red are inserted as shown to produce the alcohol. Thus, acetaldehyde is reduced to ethanol.

2 Since the ketone is reduced, turn the ketone into a secondary alcohol by adding two H atoms as shown.

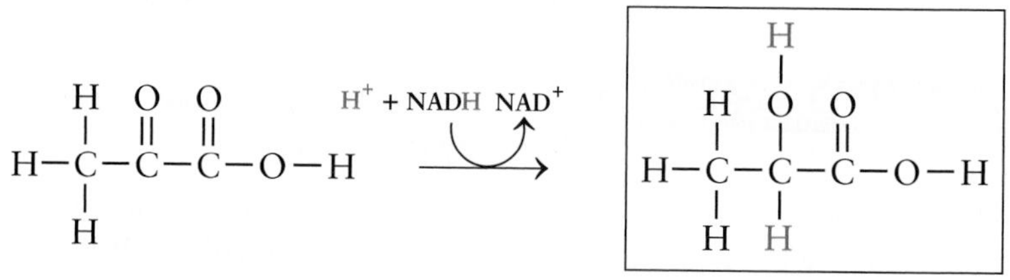

Pyruvic acid   Lactic acid

PRACTICE EXERCISES

10.4 Predict the structure of the product in the following reactions. If no reaction occurs, state so. Label the hydrogen atoms involved.

a.

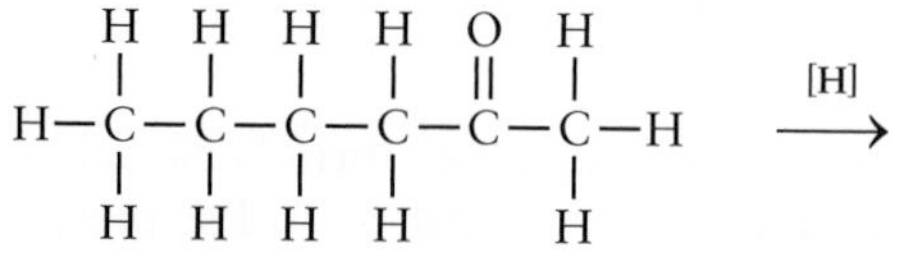

b.

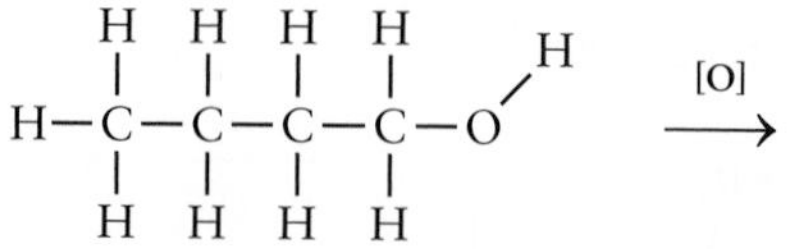

c.

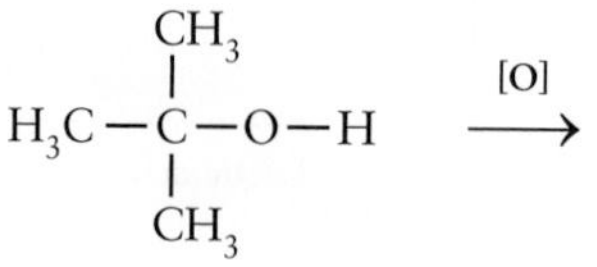

10.5 Predict the structure of the product formed in the following reactions:

a.

$$H_2O \quad + \quad H-\underset{\underset{H}{|}}{\overset{\overset{H}{|}}{C}}-\overset{\overset{O}{\|}}{C}-H \quad \xrightarrow{[O]}$$

b.

$$H-\underset{\underset{H}{|}}{\overset{\overset{H}{|}}{C}}-\underset{\underset{H}{|}}{\overset{\overset{H}{|}}{C}}-\underset{\underset{H}{|}}{\overset{\overset{H}{|}}{C}}-\overset{\overset{O}{\diagup}}{C} \quad \xrightarrow{[H]}$$

with a CH₃ branch: H−C−H on top

c.

benzoic acid structure $\xrightarrow{[H]}$

10.6 Critical Thinking Question: Law enforcement officials use the *breathalyzer* to measure blood alcohol levels. This device employs an oxidation–reduction reaction to determine the amount of ethanol (CH_3CH_2OH) in the breath. In the breathalyzer, ethanol is oxidized to acetic acid (CH_3CO_2H) by Cr^{6+}. In this reaction, chromium(VI) (Cr^{6+}) is reduced to chromium(III) (Cr^{3+}). Since Cr^{6+} is a bright orange color and Cr^{3+} is dark green, the amount of Cr^{3+} formed can be determined by the shade of green produced, which is directly related to the amount of alcohol in a person's breath (remember from Chapter 5 that color intensity is related to concentration). Based on this information, write the reaction that occurs in the breathalyzer. *Hint:* Two successive oxidations occur.

A breathalyzer determines blood alcohol level by means of an oxidation-reduction reaction. [Yellow Dog Productions/Getty Images]

Antioxidants

Antioxidants have received a great deal of attention in recent years, because of the role they are believed to play in preventing cancer and cardiovascular disease. Antioxidants have even been associated with slowing the outward signs of aging. In a nutritional context, an **antioxidant** is a substance that prevents the harmful oxidation of other substances in the cell. An antioxidant accomplishes this by reducing harmful oxidizing agents, so an antioxidant is basically a reducing agent. Fruits, vegetables, and tea are high in antioxidants such as vitamins C and E and glutathione.

If oxidation is so important in catabolic pathways, why does the body need *anti*oxidants? Oxidation reactions are known to produce free radicals, which can be damaging to cells and DNA. **Free radicals** are molecules characterized by an odd number of valence electrons. Since one of the atoms in a radical lacks an octet, radicals are extremely unstable and therefore, reactive. For example, the hydroxyl radical contains an oxygen atom with 7 rather than 8 electrons. To achieve an octet, free radicals remove electrons from other compounds—an oxidation—in a type of chain reaction that ultimately damages the cell. Antioxidants stop the chain reaction or prevent the formation of the radical species in the first place. Cells are also equipped with specific enzymes that scavenge free radicals before they can do damage.

$$H-\ddot{\overset{..}{O}}\cdot$$
Hydroxyl radical

Ascorbic acid structure (vitamin C)

10.3 Hydration-Dehydration Reactions

Hydration and dehydration reactions are common in biochemistry. As the terms suggest, water is involved. In a hydration reaction, water is a reactant, and in a dehydration reaction, water is a product.

Fruits and berries are high in antioxidants. Antioxidants prevent the damaging effects of some oxidation reactions in the cell. [Foodcollection/Punchstock]

Hydration In a **hydration reaction,** water and an alkene react with each other to produce an alcohol. In the reaction, the H and OH atoms of a water molecule each form a bond to one of the carbon atoms of the double bond. At the same time, the C=C double bond is converted into a C—C single bond:

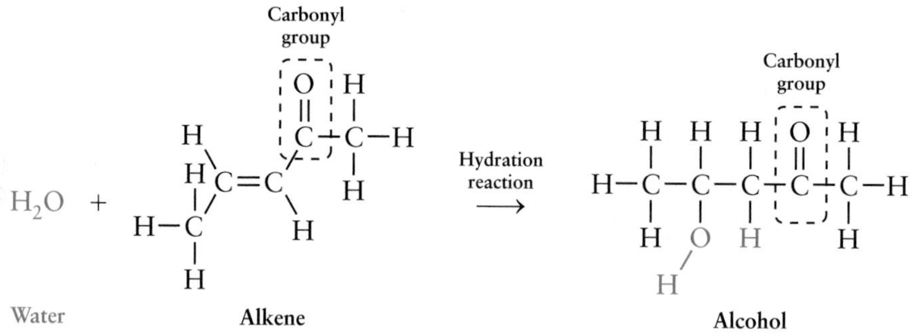

As you can see, water is one of the *reactants* in a hydration reaction, and is incorporated into the molecular structure of the product.

In biochemical applications of this reaction, a carbonyl group usually exists adjacent to the double bond, and the H and OH atoms are attached in a specific way: The H atom is attached to the double-bond carbon atom closer to the carbonyl group, and the OH group is attached to the double-bond carbon atom farther from the carbonyl group. Note that the carbonyl group itself is unchanged, but is necessary for the reaction to occur. In the laboratory, hydration reactions do not require the presence of a carbonyl group, and the OH group tends to add to the more substituted double-bond carbon atom as shown below:

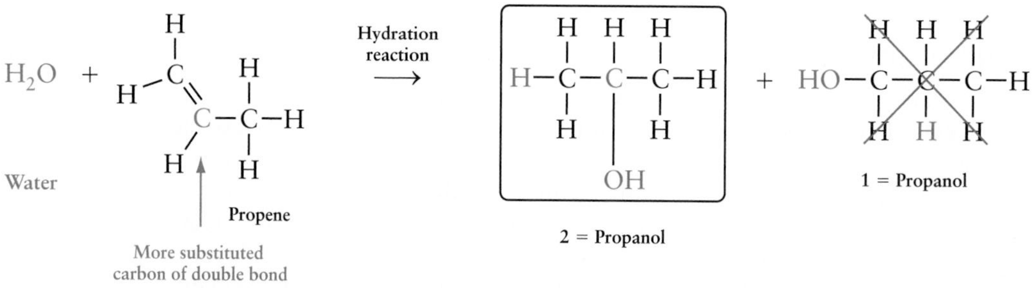

Dehydration The reverse reaction, the loss of a water molecule from an alcohol to produce a double bond, is known as a **dehydration reaction.** In a dehydration reaction OH and H are *eliminated* from the reactant to form a molecule of H_2O, and a carbon–carbon double bond is formed between the carbon atoms that had been bonded to the OH group and H atom. Remember, a carbon atom must always have four bonds.

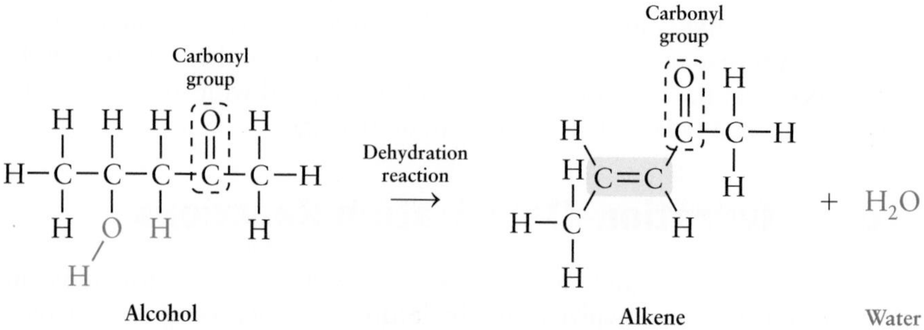

The double bond in the product appears between the carbon atom that had the OH group and the neighboring carbon atom next to the carbonyl group, as shown in the reaction above. Water is a *product* in a dehydration reaction. *The loss of OH and H in a dehydration reaction is the reverse of the addition of OH and H in a hydration reaction.*

Consider the biochemical pathway used to break down fatty acids, known as β-oxidation. This important pathway converts long-chain fatty acids into smaller two-carbon fragments. The second step of this biochemical pathway is a hydration reaction because water is incorporated into the molecule. Note that the OH group is added to the carbon atom located two carbon atoms down the chain from the carbonyl group, as expected:

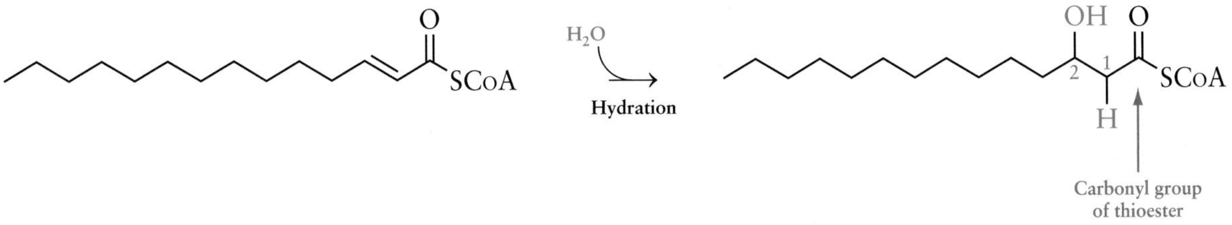

Carbonyl group
of thioester

WORKED EXERCISES 10-4   **Predicting the Product of a Hydration or Dehydration Reaction**

Problem-Solving Tutorial:
Hydration/Dehydration Reactions

1 Predict the product formed in the following reaction.

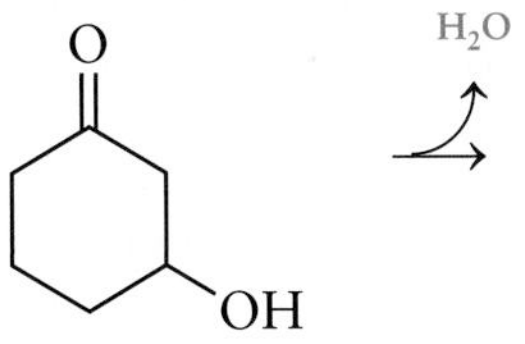

2 Why is a hydration reaction not classified as an oxidation–reduction reaction?

SOLUTIONS

1

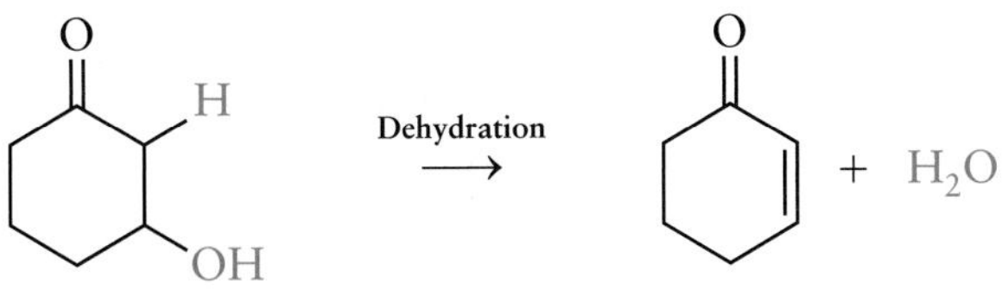

Remember to remove the OH group and the H atom closer to the carbonyl group when determining where to place the double bond in the product.

2 A hydration reaction increases both the number of oxygen atoms and the number of hydrogen atoms, so it is neither a reduction nor an oxidation.

PRACTICE EXERCISES

10.7 Predict the product formed in the following hydration reaction:

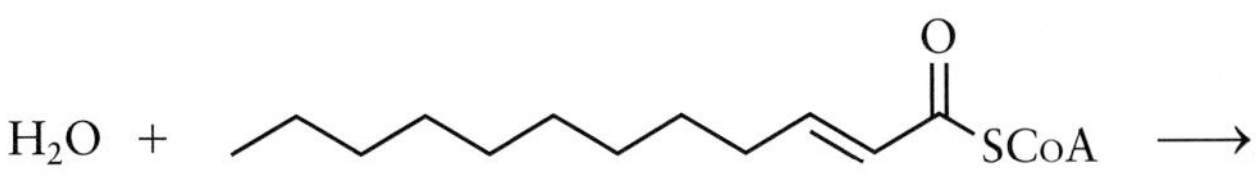

H_2O +

10.8 In Step 7 of an important biochemical pathway known as the citric acid cycle, the following reaction occurs:

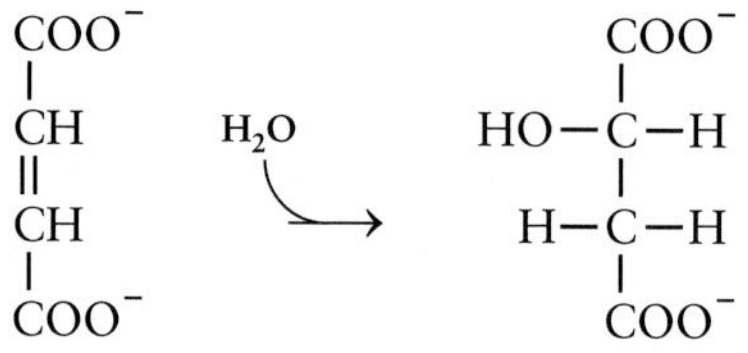

$$\begin{array}{c}
\text{COO}^- \\
| \\
\text{CH} \\
\| \\
\text{CH} \\
| \\
\text{COO}^-
\end{array}
\quad
\text{H}_2\text{O}
\quad \longrightarrow \quad
\begin{array}{c}
\text{COO}^- \\
| \\
\text{HO}-\text{C}-\text{H} \\
| \\
\text{H}-\text{C}-\text{H} \\
| \\
\text{COO}^-
\end{array}$$

Step 7

a. Is this a hydration or a dehydration reaction?

b. Indicate the OH group and H atom in the product structure that came from a molecule of water.

c. There are two carbonyl groups in the reactant. Circle them and indicate what functional group they are part of.

d. How can you quickly determine that this reaction is a hydration or dehydration reaction?

10.4 | Acyl Group Transfer Reactions

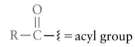

Some of the most common reactions seen in chemistry are reactions that interconvert carboxylic acids and their derivatives: esters, thioesters, and amides. Biochemists refer to this broad class of reactions as **acyl group transfer reactions** because the group transferred is an **acyl group,** which consists

of a carbonyl group and its attached R group, R—C—ξ. In this type of reaction, the acyl group appears to migrate from an O, N, or S atom to another O, N, or S atom (Figure 10-6). As with hydration and dehydration reactions, water is an essential component of these reactions, although these two types of reactions are otherwise very different.

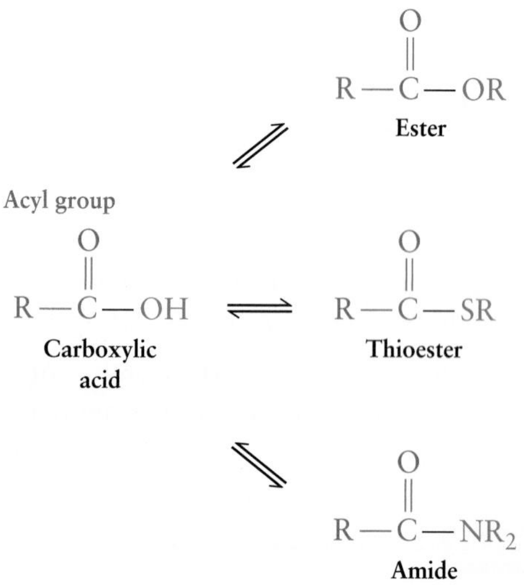

$$\underset{\substack{\text{Carboxylic} \\ \text{acid}}}{\text{R}-\overset{\overset{\text{O}}{\|}}{\text{C}}-\text{OH}}
\;\rightleftharpoons\;
\underset{\text{Thioester}}{\text{R}-\overset{\overset{\text{O}}{\|}}{\text{C}}-\text{SR}}$$

Acyl group

$$\underset{\text{Ester}}{\text{R}-\overset{\overset{\text{O}}{\|}}{\text{C}}-\text{OR}}$$

$$\underset{\text{Amide}}{\text{R}-\overset{\overset{\text{O}}{\|}}{\text{C}}-\text{NR}_2}$$

Figure 10-6 Acyl transfer reactions interconvert carboxylic acids and their derivatives: esters, thioesters, and amides. The acyl group is shown in red.

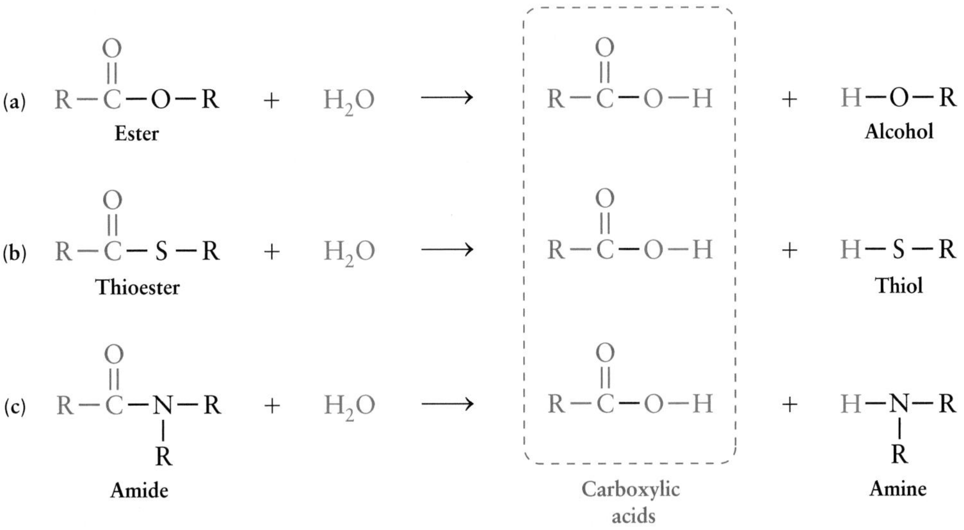

(a) $\underset{\text{Ester}}{R-\overset{\displaystyle O}{\overset{\|}{C}}-O-R}$ + H_2O ⟶ $\boxed{R-\overset{\displaystyle O}{\overset{\|}{C}}-O-H}$ + $\underset{\text{Alcohol}}{H-O-R}$

(b) $\underset{\text{Thioester}}{R-\overset{\displaystyle O}{\overset{\|}{C}}-S-R}$ + H_2O ⟶ $\boxed{R-\overset{\displaystyle O}{\overset{\|}{C}}-O-H}$ + $\underset{\text{Thiol}}{H-S-R}$

(c) $\underset{\text{Amide}}{R-\overset{\displaystyle O}{\overset{\|}{C}}-\underset{\underset{\displaystyle R}{|}}{N}-R}$ + H_2O ⟶ $\boxed{R-\overset{\displaystyle O}{\overset{\|}{C}}-O-H}$ + $\underset{\text{Amine}}{H-\underset{\underset{\displaystyle R}{|}}{N}-R}$

Carboxylic acids

Figure 10-7 Hydrolysis reactions transfer an acyl group from a carboxylic acid derivative to a water molecule. The acyl group is shown in red.

The acyl group transfer reactions that will be described here include hydrolysis reactions and esterification reactions. Amidation reactions, another important type of acyl group transfer reaction, will be described in the next chapter.

Hydrolysis Reactions

Esters, thioesters, and amides react with water in the presence of a catalyst to produce a *carboxylic acid* and either an *alcohol* (Figure 10-7a), a *thiol* (Figure 10-7b), or an *amine* (Figure 10-7c). Hydrolysis reactions are classified as acyl transfer reactions because an acyl group migrates from an oxygen, sulfur, or nitrogen atom in the carboxylic acid derivative to the oxygen atom of a water molecule, as shown in Figure 10-7.

Since *water* is the agent that breaks the bond between the carbonyl group and the O, S, or N atom, these reactions are known as **hydrolysis reactions.** The word "hydrolysis" comes from the Latin words *hydro*, for water, and *lysis*, to break. Although water is a reactant in both hydrolysis and hydration reactions, notice that hydration reactions (Section 10.3) differ from hydrolysis reactions in that there is no "lysis" in a hydration reaction; the reactant is not split into two molecules. To predict the structure of the products formed in a hydrolysis reaction, follow the guidelines below.

Guidelines for Predicting the Product of a Hydrolysis Reaction

Step 1: *Identify* the type of carboxylic acid derivative in the reactant—ester, thioester, or amide. From this information you can predict what functional groups will be produced in the product structures (right side of Figure 10-7). Remember, one of the products will always be a carboxylic acid, RCOOH (or a carboxylate ion, RCOO⁻).

Step 2: *Break* the carbonyl carbon–heteroatom bond. Break the single bond between the carbonyl carbon and the O, S, or N atom, to give two partial structures:

Note that you should not interpret these guidelines as the actual bond-breaking and -making steps that the reactants undergo to form products.

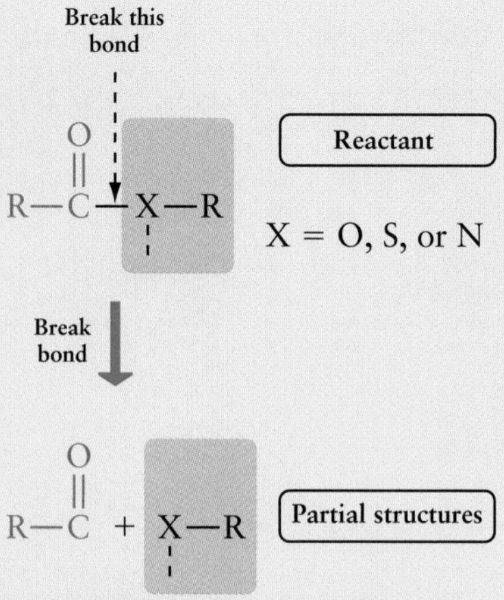

Break this bond

O
||
R—C—X—R | Reactant | X = O, S, or N

Break bond ↓

O
||
R—C + X—R | Partial structures |

Step 3: Make new bonds. Make a new bond between OH (from H_2O) and the carbonyl carbon to form a carboxylic acid *and* make a new bond between H (from H_2O) and the O, S, or N atom, to form an alcohol, —SH group, or amine.

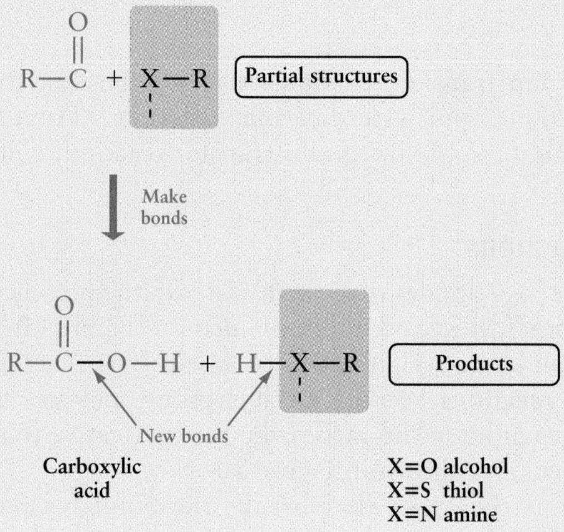

O
||
R—C + X—R | Partial structures |

Make bonds ↓

O
||
R—C—O—H + H—X—R | Products |

New bonds

Carboxylic acid

X=O alcohol
X=S thiol
X=N amine

When an amide is hydrolyzed, a carboxylic acid and an amine are produced. However, at physiological pH, the amine and the carboxylic acid immediately undergo an acid–base reaction to form the conjugate base of the acid and the conjugate acid of the amine, as expected.

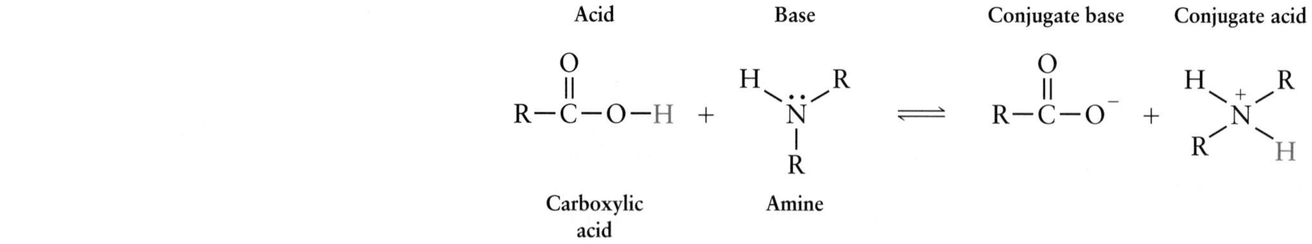

| Acid | Base | | Conjugate base | Conjugate acid |

O   H   R   O   H + R
|| \ .. / || \ N /
R—C—O—H +   N   ⇌   R—C—O⁻   +   N
                      | R   H
                      R

Carboxylic   Amine
acid

Hydrolysis of amides is an important reaction as it is the first stage in the digestion of proteins, the process of breaking a protein into its individual amino acid components.

Formation of Soaps When an ester is hydrolyzed in water under basic conditions (OH⁻), a carboxylate ion is produced instead of the expected carboxylic acid. Alkaline hydrolysis of an ester therefore yields an *alcohol* and a *carboxylate ion* in a reaction known as a **saponification,** named after the process for making soap. Saponification of fats produces glycerol and fatty acid carboxylates. After removal of the glycerol product, what remains are fatty acid carboxylate salts, the substance we know as soap.

Saponification

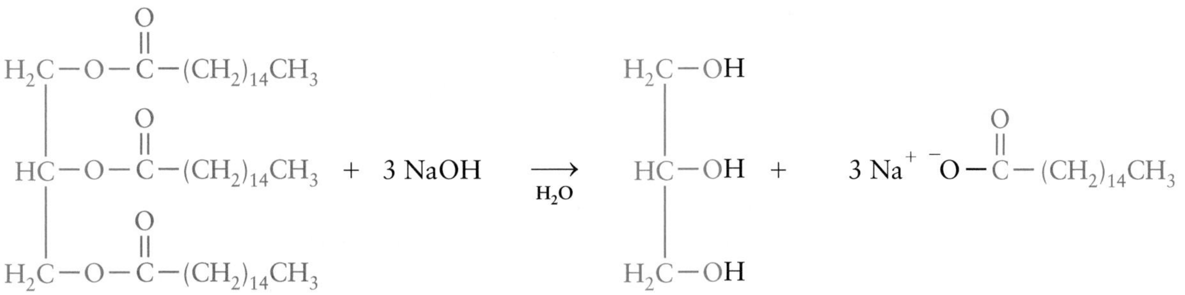

$$
\begin{array}{l}
H_2C-O-\overset{\displaystyle O}{\overset{\displaystyle \|}{C}}-(CH_2)_{14}CH_3 \\[1em]
HC-O-\overset{\displaystyle O}{\overset{\displaystyle \|}{C}}-(CH_2)_{14}CH_3 \quad + \quad 3\ NaOH \\[1em]
H_2C-O-\overset{\displaystyle O}{\overset{\displaystyle \|}{C}}-(CH_2)_{14}CH_3
\end{array}
\xrightarrow{H_2O}
\begin{array}{l}
H_2C-OH \\[1em]
HC-OH \quad + \\[1em]
H_2C-OH
\end{array}
\quad 3\ Na^+\ {}^-O-\overset{\displaystyle O}{\overset{\displaystyle \|}{C}}-(CH_2)_{14}CH_3
$$

| Fat | Glycerol | Fatty acid carboxylate salts (soap) |

WORKED EXERCISE 10-5 Hydrolysis of an Ester

> **Problem-Solving Tutorial:**
> Acyl Group Transfer Reactions

Write the structure of the products formed in the following hydrolysis reaction:

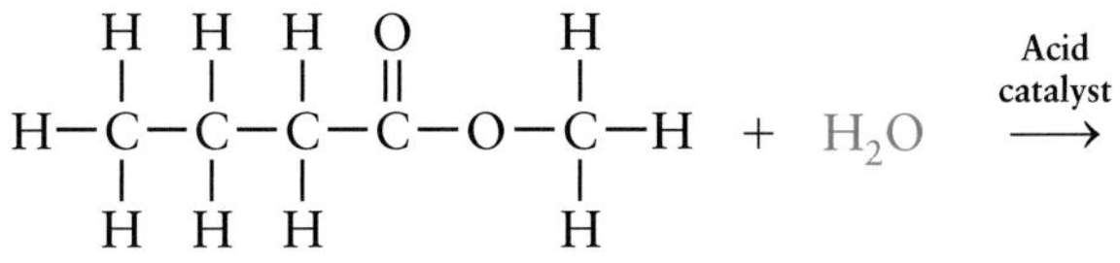

$$
\begin{array}{ccccccc}
& H & H & H & O & & H \\
& | & | & | & \| & & | \\
H- & C- & C- & C- & C- & O- & C-H \\
& | & | & | & & & | \\
& H & H & H & & & H
\end{array}
\quad +\ H_2O \xrightarrow{\ \text{Acid catalyst}\ }
$$

SOLUTION

Step 1: Identify the type of carboxylic acid derivative in the reactant. The reactant is an ester; therefore, the products should be a carboxylic acid and an alcohol.

Step 2: Break the carbonyl carbon–heteroatom bond. Break the single bond between the carbonyl carbon and the oxygen atom to yield two partial structures:

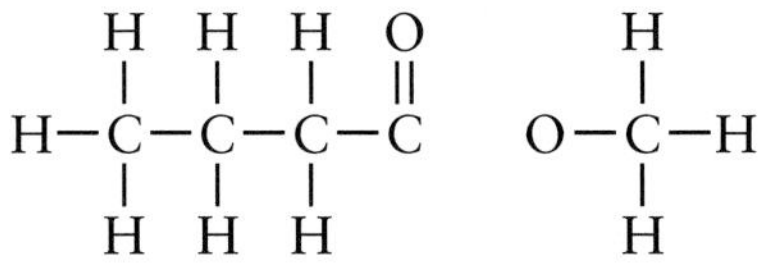

$$
\begin{array}{cccccc}
H & H & H & O & & H \\
| & | & | & \| & & | \\
H-C- & C- & C- & C & \qquad O- & C-H \\
| & | & | & & & | \\
H & H & H & & & H
\end{array}
$$

Step 3: Make new bonds. Add OH to the carbonyl carbon to form a carboxylic acid:

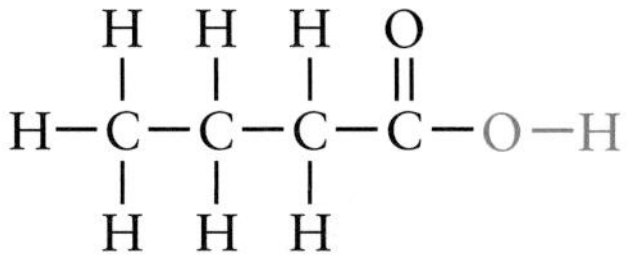

$$
\begin{array}{ccccc}
H & H & H & O & \\
| & | & | & \| & \\
H-C- & C- & C- & C- & O-H \\
| & | & | & & \\
H & H & H & &
\end{array}
$$

Carboxylic acid (butanoic acid)

Add H to the other partial structure to form an alcohol:

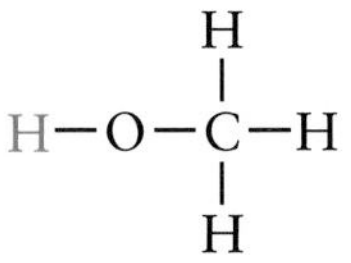

$$
\begin{array}{c}
H \\
| \\
H-O-C-H \\
| \\
H
\end{array}
$$

Alcohol (methanol)

WORKED EXERCISE 10-6   Hydrolysis of a Thioester

Write the structure of the products formed in the following hydrolysis reaction, which occurs in a biochemical pathway known as the citric acid cycle:

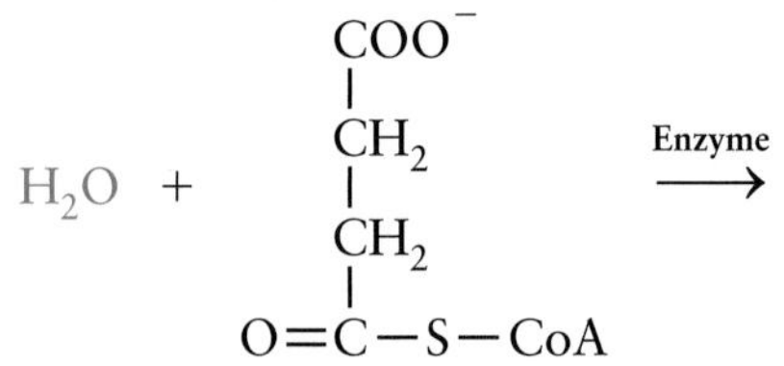

$$
\text{H}_2\text{O} \;+\;
\begin{array}{c}
\text{COO}^- \\
| \\
\text{CH}_2 \\
| \\
\text{CH}_2 \\
| \\
\text{O}=\text{C}-\text{S}-\text{CoA}
\end{array}
\;\xrightarrow{\text{Enzyme}}
$$

SOLUTION

Step 1: Identify the type of carboxylic acid derivative in the reactant. The reactant is a thioester. Therefore, the expected products are a carboxylic acid and an S–H-containing product, known as a thiol. You can ignore the other functional group in the molecule because it does not undergo a reaction.

Step 2: Break the single bond between the carbonyl carbon and the sulfur atom:

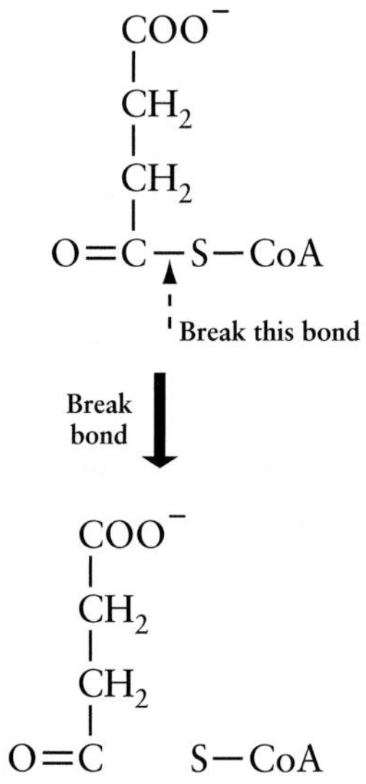

$$
\begin{array}{c}
\text{COO}^- \\
| \\
\text{CH}_2 \\
| \\
\text{CH}_2 \\
| \\
\text{O}=\text{C}-\text{S}-\text{CoA}
\end{array}
$$

Break this bond

Break bond ↓

$$
\begin{array}{c}
\text{COO}^- \\
| \\
\text{CH}_2 \\
| \\
\text{CH}_2 \\
| \\
\text{O}=\text{C} \qquad \text{S}-\text{CoA}
\end{array}
$$

Step 3: Make new bonds. Add OH to the carbonyl carbon to make a carboxylic acid:

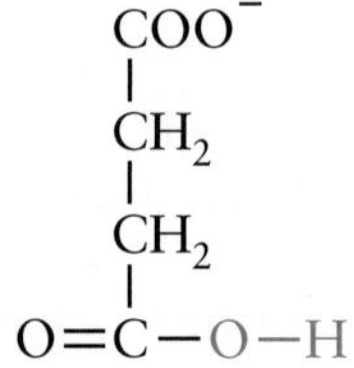

$$
\begin{array}{c}
\text{COO}^- \\
| \\
\text{CH}_2 \\
| \\
\text{CH}_2 \\
| \\
\text{O}=\text{C}-\text{O}-\text{H}
\end{array}
$$

Add H to the sulfur atom of the other partial structure to form the compound shown:

$$
\text{H}-\text{S}-\text{CoA}
$$

WORKED EXERCISE 10-7 Hydrolysis of an Amide

Write the structure of the products formed in the following hydrolysis reaction:

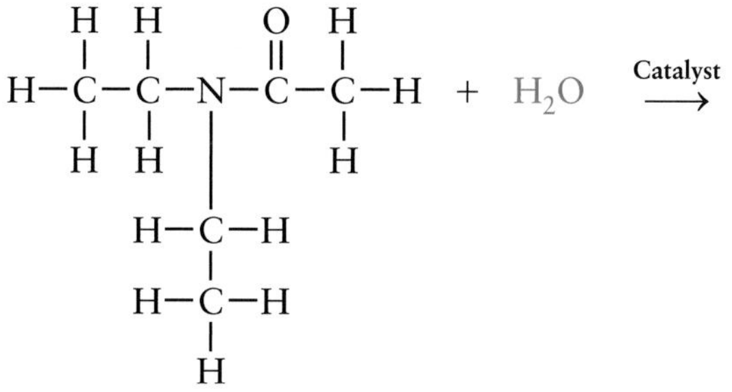

```
      H  H    O  H
      |  |    ||  |
  H—C—C—N—C—C—H  +  H₂O   --Catalyst-->
      |  |    |    |
      H  H    |    H
            H—C—H
              |
            H—C—H
              |
              H
```

SOLUTION

```
    H  H    O  H                                H  H  H                O  H
    |  |    ||  |                                |  |  |+               ||  |
H—C—C—N—C—C—H  +  H₂O  --Catalyst-->   H—C—C—N—H    +   ⁻O—C—C—H
    |  |    |    |                                |  |  |                    |
    H  H    |    H                                H  H  |                    H
          H—C—H                                       H—C—H
            |                                            |
          H—C—H                                       H—C—H
            |                                            |
            H                                            H
```

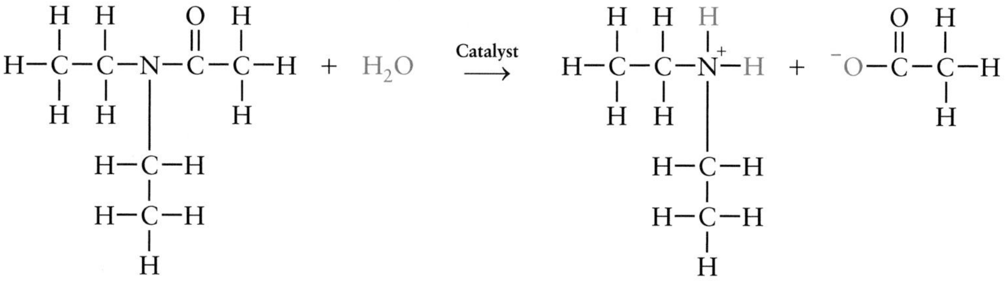

Remember that the initial amine and carboxylic acid products undergo a subsequent acid–base reaction to form the conjugate acid of the amine and the conjugate base of the carboxylic acid (a carboxylate ion).

PRACTICE EXERCISES

10.9 Write the structure of the products formed in the hydrolysis of the amide shown below.

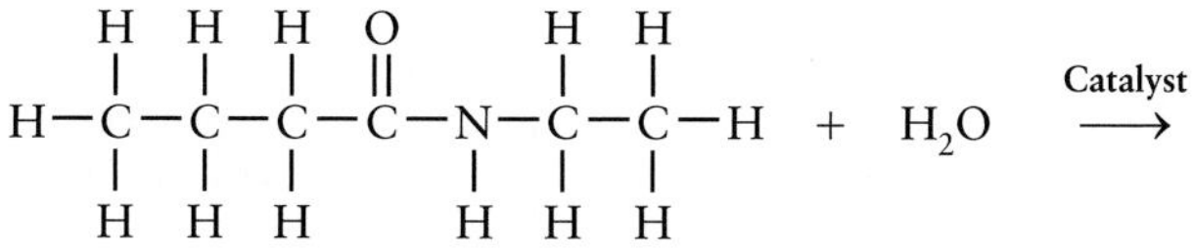

```
    H  H  H  O      H  H
    |  |  |  ||      |  |
H—C—C—C—C—N—C—C—H  +  H₂O   --Catalyst-->
    |  |  |      |  |  |
    H  H  H      H  H  H
```

10.10 Write the structure of the products formed in the following hydrolysis reactions. Identify the bond that is broken in each carboxylic acid derivative. Identify the acyl group that migrates to water. Name the products.

a.
```
    H  H  H  O      H
    |  |  |  ||      |
H—C—C—C—C—O—C—H  +  H₂O   --Catalyst-->
    |  |  |      |
    H  H  H      H
```

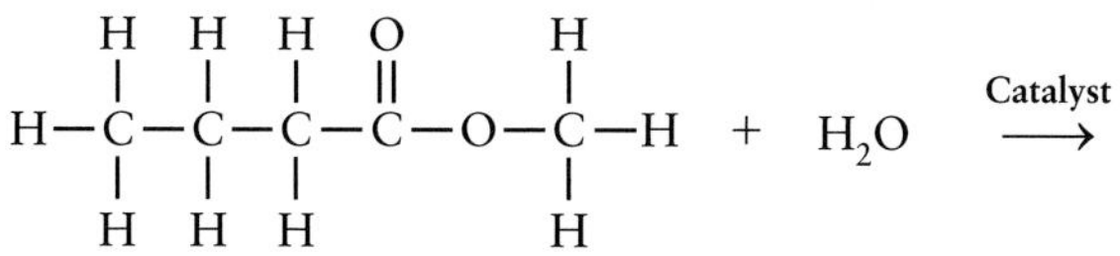

b.
```
    H  H    O  H
    |  |    ||  |
H—C—C—O—C—C—H  +  H₂O   --Catalyst-->
    |  |      |
    H  H      H
```

c.
```
    H  O
    |  ||
H—C—C—S—CoA  +  H₂O   --Catalyst-->
    |
    H
```

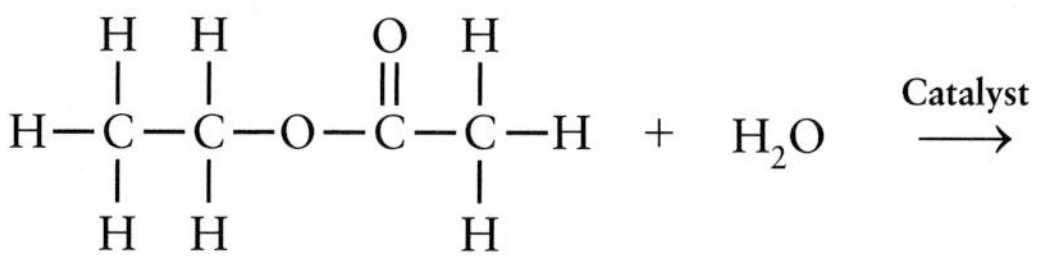

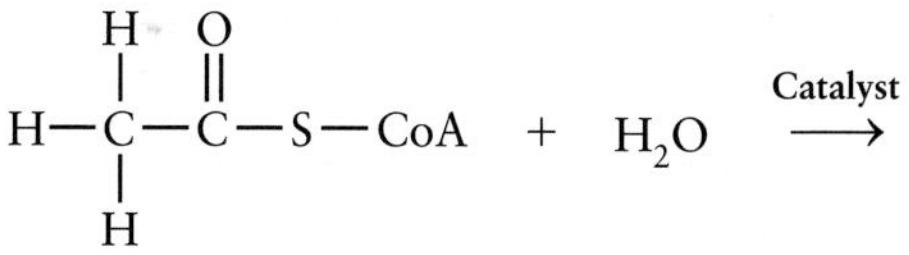

10.11 Critical Thinking Question: Hydrolysis of both ester functional groups in heroin produces morphine. What is the name of the carboxylic acid product formed along with morphine? (Two molecules of this product are formed.)

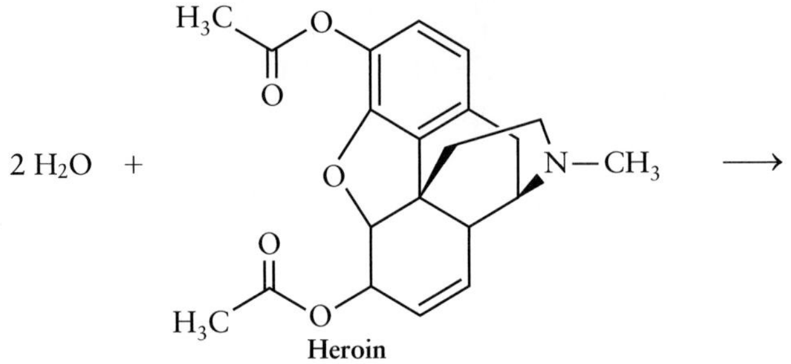

Heroin

10.12 What is the difference between the products formed in the following two hydrolysis reactions?

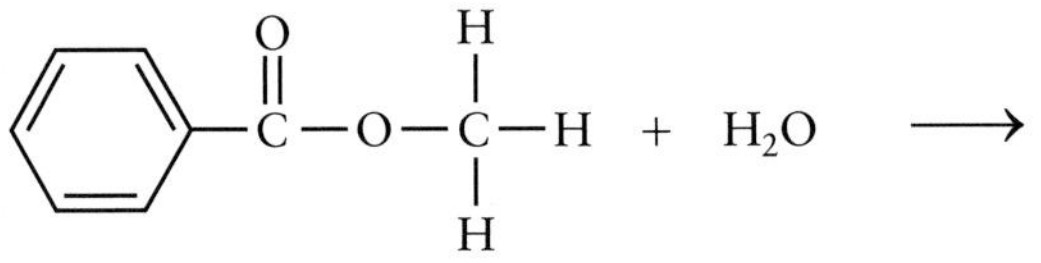

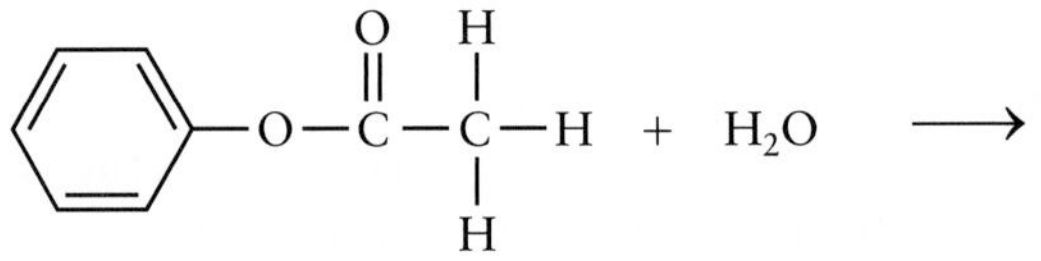

10.13 Write the structure of the products formed in the following saponification reaction:

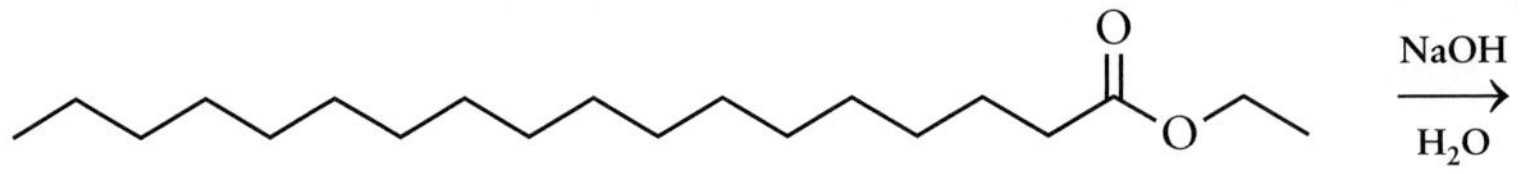

10.14 Critical Thinking Question: Write the structure of the product formed in the following hydrolysis reaction. *Hint:* Only one product is formed and it does not contain a ring.

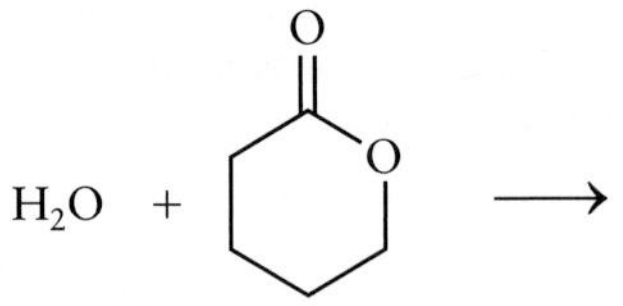

10.15 Critical Thinking Question: The products of a hydrolysis reaction performed on an unknown reactant are shown below. Write the structure of the unknown reactant.

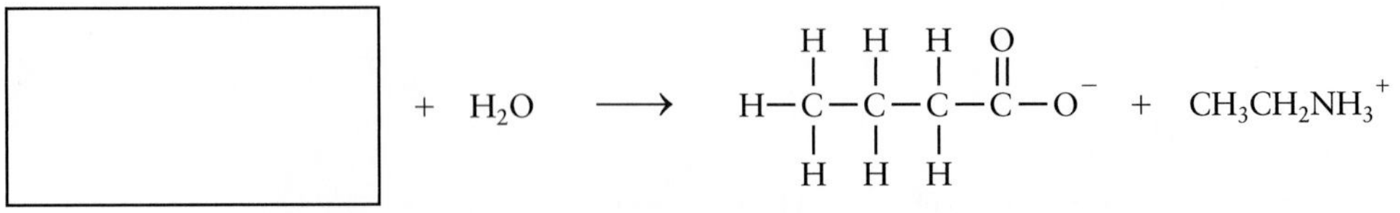

Esterification Reactions

A carboxylic acid and an alcohol can react to form an ester in an **esterification reaction,** which is the reverse of an ester hydrolysis.

Esterification

Several well-known straight-chain esters are responsible for the characteristic flavor and fragrance of fruits and perfumes. These esters can be prepared easily in the laboratory by an esterification reaction. Typically a small amount of acid (H^+) is added to catalyze this type of reaction. The number of carbon atoms in the carboxylic acid and the alcohol both influence the fragrance of the ester. For example, when acetic acid reacts with 1-pentanol, an ester with the fragrance of bananas is produced. When 1-octanol is used instead, an ester with the fragrance of oranges is produced. Not surprisingly, the perfume industry uses esterification reactions in developing the complex mixture of esters that characterize a particular brand of fragrance. Table 10-3 shows the

Table 10-3 Esterification Reactions and the Familiar Fragrance of the Ester

Carboxylic Acid	Alcohol	Ester	Fragrance
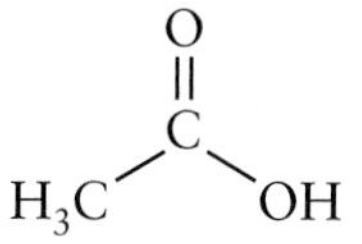 Acetic acid	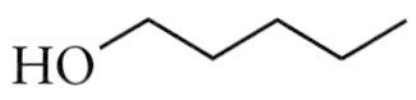 1-Pentanol	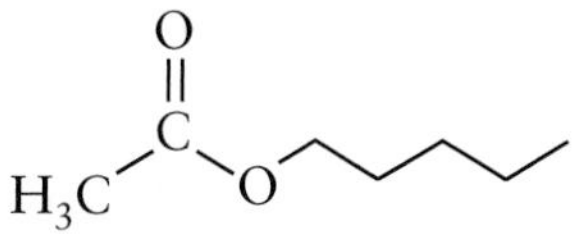	Banana
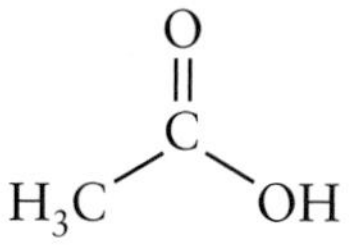 Acetic acid	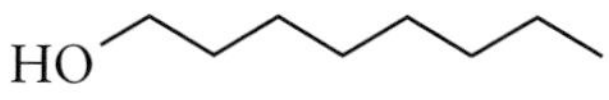 1-Octanol	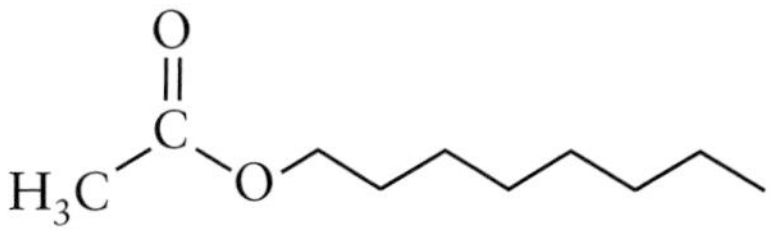	Orange
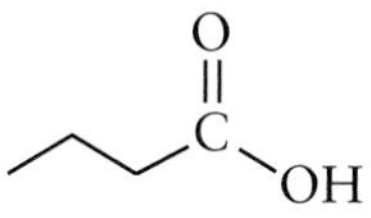 Butanoic acid	Methanol (CH_3OH)	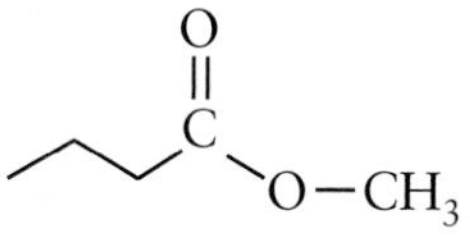	Apple
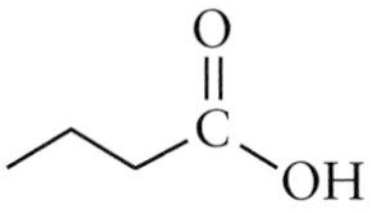 Butanoic acid	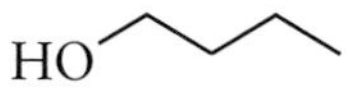 1-Butanol	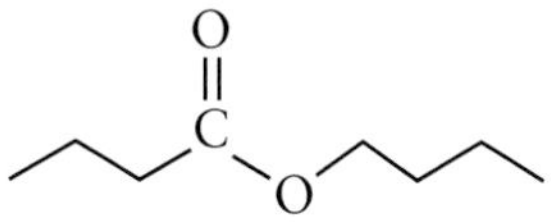	Pineapple
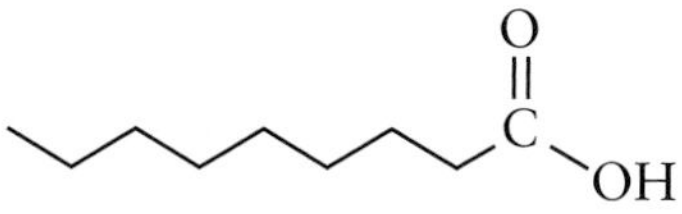 Nonanoic acid	Ethanol (CH_3CH_2OH)	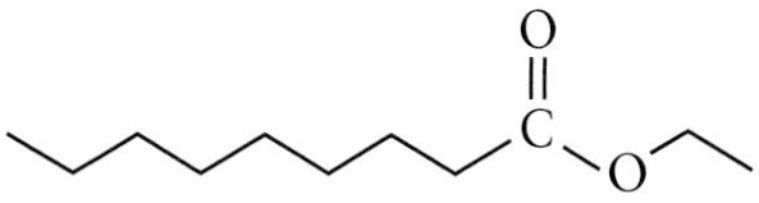	Grape
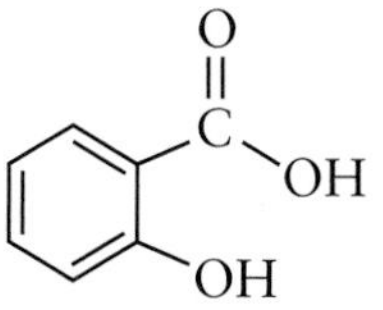 Salicylic acid	Methanol (CH_3OH)	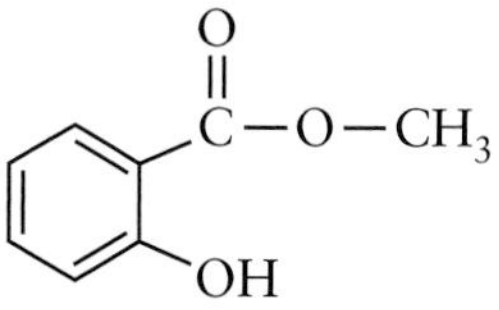	Wintergreen

alcohol and carboxylic acid components of some well-known esters and the fragrances associated with them.

Thioesters When sulfur replaces oxygen in an alcohol, the functional group is known as a **thiol.** When a thiol is used instead of an alcohol in an esterification reaction, a thioester is formed instead of an ester. The reaction is the reverse of a thioester hydrolysis.

$$\underset{\text{Carboxylic acid}}{R' - \overset{\displaystyle\overset{O}{\|}}{C} - OH} \ + \ \underset{\text{Thiol}}{H - S - R} \ \xrightarrow{\text{Catalyst}} \ \underset{\text{Thioester}}{R' - \overset{\displaystyle\overset{O}{\|}}{C} - S - R} \ + \ H_2O$$

Use the guidelines below to predict the products formed in an esterification reaction that produces either an ester or a thioester.

Guidelines for Predicting the Products Formed in Esterification Reactions

Step 1: Remove the OH group and the H atom. Remove the OH group from the carboxylic acid *and* the H atom on the oxygen atom of the alcohol or sulfur atom of a thiol to form two incomplete product structures and a water molecule:

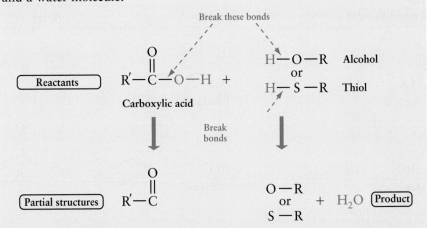

Step 2: Connect the two resulting partial structures. Form a carbon–oxygen single bond between the carbonyl carbon and the O—R group of the alcohol to produce an ester:

Form a carbon–sulfur single bond between the carbonyl carbon and the S—R group to produce a thioester:

$$
\begin{array}{ccc}
& \overset{\displaystyle O}{\overset{\displaystyle \|}{\text{R}'\text{—C}}} & \text{S—R} \qquad \boxed{\textbf{Partial structures}} \\
\\
& \big\downarrow \text{ Make bond} \\
\\
& \overset{\displaystyle O}{\overset{\displaystyle \|}{\text{R}'\text{—C}\mathbin{\underset{\displaystyle |}{\text{—}}}\text{S—R}}} & \qquad \boxed{\textbf{Product}} \\
\\
& \text{New bond} \\
& \textbf{Thioester}
\end{array}
$$

The most important thioesterification reactions occur in biochemistry when coenzyme A, a thiol, reacts with a carboxylic acid to make a thioester. Coenzyme A is produced in the body from the vitamin pantothenic acid (vitamin B_5), found in most foods. Thioesters are used in biochemistry primarily for transferring acyl groups between molecules. The most important thioester is acetyl CoA, a molecule at the center of metabolism. Acetyl CoA transfers an acetyl group, $\text{H}_3\text{C}-\overset{\displaystyle O}{\overset{\displaystyle \|}{\text{C}}}-\xi$, in acyl group transfer reactions. For example, in the opening vignette you learned about the metabolism of ethanol. The acetic acid formed from ethanol undergoes an esterification with coenzyme A to form the important molecule acetyl CoA, a thioester. Acetyl CoA then enters a common biochemical pathway for all foods, the citric acid cycle. You will learn about these important biochemical pathways in the next four chapters.

Pantothenic acid is found in most foods, but is highest in liver, yeast, egg yolk, and broccoli. Other good sources of this vitamin include fish, chicken, milk, legumes, mushrooms, avocado, and sweet potatoes.

$$
\underset{\substack{\text{Acetic acid}\\ \text{(carboxylic acid)}}}{\text{H}-\overset{\displaystyle H}{\underset{\displaystyle H}{\text{C}}}-\overset{\displaystyle O}{\overset{\displaystyle \|}{\text{C}}}-\text{OH}} \;+\; \underset{\substack{\text{Coenzyme A}\\ \text{(thiol)}}}{\text{H}-\text{S}-\text{CoA}} \;\longrightarrow\; \underset{\substack{\text{Acetyl CoA}\\ \text{(thioester)}}}{\text{H}-\overset{\displaystyle H}{\underset{\displaystyle H}{\text{C}}}-\overset{\displaystyle O}{\overset{\displaystyle \|}{\text{C}}}-\text{S}-\text{CoA}} \;+\; \text{H}_2\text{O}
$$

WORKED EXERCISE 10-8 Esterification Reactions

Write the products formed in the following esterification reaction:

$$
\underset{\textbf{Alcohol}}{\text{H}-\overset{\displaystyle H}{\underset{\displaystyle H}{\text{C}}}-\overset{\displaystyle H}{\underset{\displaystyle H}{\text{C}}}-\text{OH}} \;+\; \underset{\textbf{Carboxylic acid}}{\text{H}-\overset{\displaystyle H}{\underset{\displaystyle H}{\text{C}}}-\overset{\displaystyle H}{\underset{\displaystyle H}{\text{C}}}-\overset{\displaystyle H}{\underset{\displaystyle H}{\text{C}}}-\overset{\displaystyle O}{\overset{\displaystyle \|}{\text{C}}}-\text{OH}} \;\xrightarrow{\text{Catalyst}}
$$

SOLUTION

Step 1: Remove the OH group and the H atom. Remove the OH group on the carboxylic acid and the H atom on the oxygen atom of the alcohol to form two incomplete product structures and a water molecule:

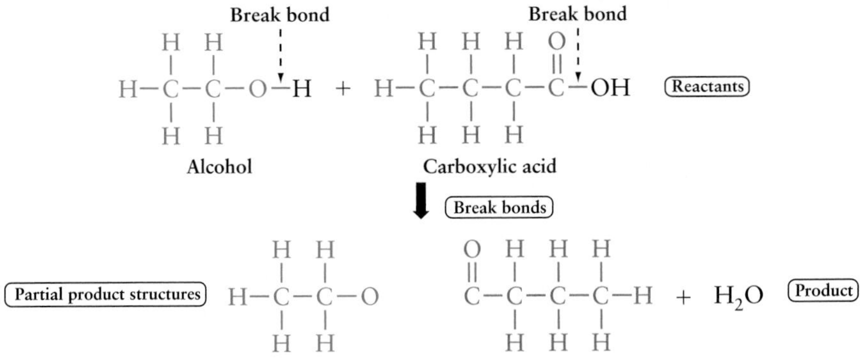

$$
\underset{\text{Alcohol}}{
\begin{array}{c}
\text{Break bond} \\
\overset{\text{H H}}{\underset{\text{H H}}{\text{H}-\overset{|}{\underset{|}{\text{C}}}-\overset{|}{\underset{|}{\text{C}}}-\text{O}\dashv\text{H}}}
\end{array}}
\;+\;
\underset{\text{Carboxylic acid}}{
\begin{array}{c}
\text{Break bond} \\
\overset{\text{H H H O}}{\underset{\text{H H H}}{\text{H}-\overset{|}{\underset{|}{\text{C}}}-\overset{|}{\underset{|}{\text{C}}}-\overset{|}{\underset{|}{\text{C}}}-\overset{\|}{\text{C}}\dashv\text{OH}}}
\end{array}}
\quad \boxed{\text{Reactants}}
$$

↓ $\boxed{\text{Break bonds}}$

$$
\boxed{\text{Partial product structures}}\;\;
\begin{array}{c}
\text{H H} \\
\text{H}-\overset{|}{\underset{|}{\text{C}}}-\overset{|}{\underset{|}{\text{C}}}-\text{O} \\
\text{H H}
\end{array}
\qquad
\begin{array}{c}
\text{O H H H} \\
\overset{\|}{\text{C}}-\overset{|}{\underset{|}{\text{C}}}-\overset{|}{\underset{|}{\text{C}}}-\overset{|}{\underset{|}{\text{C}}}-\text{H} \\
\text{H H H}
\end{array}
\;+\;\text{H}_2\text{O}\quad \boxed{\text{Product}}
$$

Step 2: Connect the two resulting partial structures. Form a carbon–oxygen single bond between the carbonyl carbon and the O—R group to produce an ester:

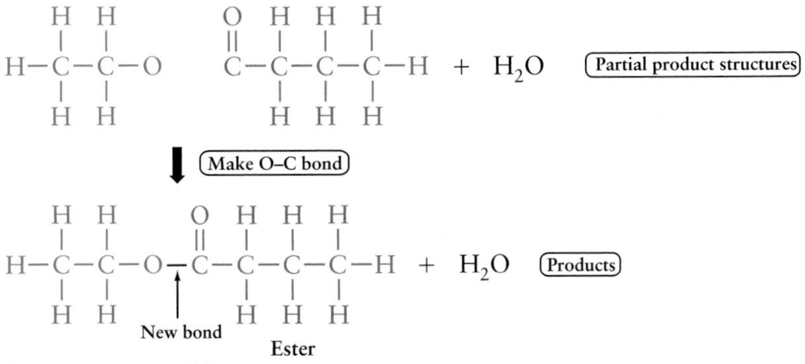

$$
\begin{array}{c}
\text{H H} \\
\text{H}-\overset{|}{\underset{|}{\text{C}}}-\overset{|}{\underset{|}{\text{C}}}-\text{O} \\
\text{H H}
\end{array}
\qquad
\begin{array}{c}
\text{O H H H} \\
\overset{\|}{\text{C}}-\overset{|}{\underset{|}{\text{C}}}-\overset{|}{\underset{|}{\text{C}}}-\overset{|}{\underset{|}{\text{C}}}-\text{H} \\
\text{H H H}
\end{array}
\;+\;\text{H}_2\text{O}\quad \boxed{\text{Partial product structures}}
$$

↓ $\boxed{\text{Make O–C bond}}$

$$
\underset{\substack{\uparrow \\ \text{New bond}}}{}
\begin{array}{c}
\text{H H} \qquad\quad \text{O H H H} \\
\text{H}-\overset{|}{\underset{|}{\text{C}}}-\overset{|}{\underset{|}{\text{C}}}-\text{O}-\overset{\|}{\text{C}}-\overset{|}{\underset{|}{\text{C}}}-\overset{|}{\underset{|}{\text{C}}}-\overset{|}{\underset{|}{\text{C}}}-\text{H} \\
\text{H H} \qquad\quad\;\; \text{H H H} \\
\text{Ester}
\end{array}
\;+\;\text{H}_2\text{O}\quad \boxed{\text{Products}}
$$

PRACTICE EXERCISES

10.16 Write the products formed in the following reactions:

a.
$$
\begin{array}{c}
\text{H H O} \\
\text{H}-\overset{|}{\underset{|}{\text{C}}}-\overset{|}{\underset{|}{\text{C}}}-\overset{\|}{\text{C}}-\text{OH} \\
\text{H H}
\end{array}
\;+\;
\begin{array}{c}
\text{H H} \\
\text{H}-\overset{|}{\underset{|}{\text{C}}}-\overset{|}{\underset{|}{\text{C}}}-\text{O}-\text{H} \\
\text{H H}
\end{array}
\;\overset{\text{Catalyst}}{\longrightarrow}
$$

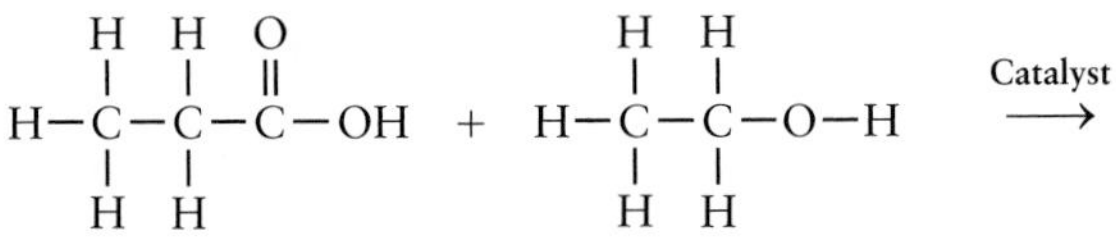

b.
$$
\underset{}{\text{C}_6\text{H}_5}\!-\!\overset{\overset{\text{O}}{\|}}{\text{C}}\!-\!\text{OH}
\;+\;
\begin{array}{c}
\text{H} \\
\text{H}-\overset{|}{\underset{|}{\text{C}}}-\text{O}-\text{H} \\
\text{H}
\end{array}
\;\overset{\text{Catalyst}}{\longrightarrow}
$$

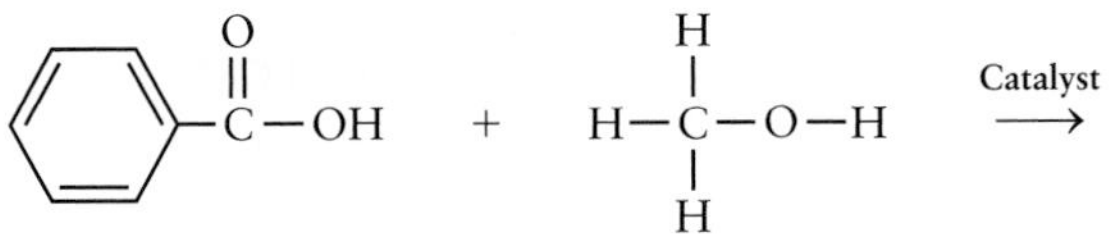

10.17 Write the structure of the product formed in the esterification reaction shown below, involving the thiol coenzyme A. This reaction is the first step in fatty acid catabolism.

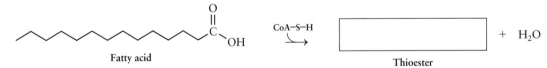

Fatty acid   Thioester

a. Circle the acyl group being transferred.
b. What are the reactants in this reaction? Why is one reactant shown next to a curved arrow above the main reaction arrow?
c. How is this reaction different from an esterification involving an alcohol?
d. Highlight the new bond formed in this reaction.

10.18 What is the structure of the ester formed between acetic acid and pentanol. What familiar smell does this ester have?

To further understand these types of reactions, see *The Model Tool 10-1: Acyl Group Transfer Reactions.*

 The Model Tool 10-1 Acyl Group Transfer Reactions

I. Construction of the Reactants

1. Obtain 4 black carbon atoms, 10 light-blue hydrogen atoms, 3 red oxygen atoms, 14 straight bonds, and 2 bent bonds.

2. Make a model of ethyl acetate, $CH_3CO_2CH_2CH_3$. Locate the carbonyl group. What functional group does ethyl acetate contain?

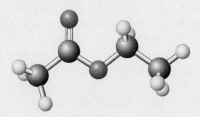

3. Make a model of a water molecule.

II. Simulation of the Hydrolysis Reaction

4. Simulate a *hydrolysis* reaction between ethyl acetate and water using only the models you have made.
   a. What bonds do you need to break?
   b. What new bonds do you need to make?
   c. What are the functional groups present in the products?
   d. What are the names of the products
   e. Write the complete chemical equation.

5. What carbon atoms of the original molecule are unchanged in the reaction?

6. Why is this reaction classified as a hydrolysis?

7. What atom is the acyl group being transferred to in this reaction?

8. What is the name of the reverse reaction?

10.5 Phosphoryl Group Transfer Reactions

Reactions that transfer phosphoryl groups, known as **phosphoryl group transfer** reactions, play a central role in the way energy is transferred in biochemical reactions. A phosphoryl group is composed of three oxygen atoms attached to a central phosphorus atom, by one P—O double bond and two P—O single bonds. The phosphorus atom is attached to the oxygen atom of another functional group, creating a phosphate ester (P—O—R). Phosphoryl group transfer reactions are similar in some respects to acyl group transfer reactions, except that a phosphoryl group migrates instead of an acyl group.

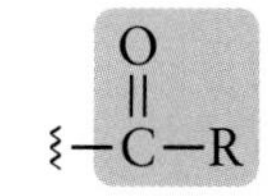

The Products of Phosphoryl Group Transfer Reactions

A phosphoryl group is unique in that it can be transferred to another phosphoryl group to produce a di- or triphosphate. Phosphoryl groups have the ability to join together via a **phosphoanhydride bond.** For example, the important triphosphate ATP contains two phosphoanhydride bonds, which join the three phosphoryl groups:

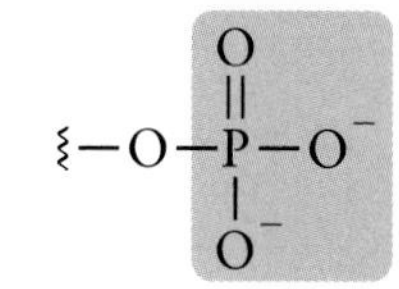

Acyl group

Phosphoryl group (in blue)

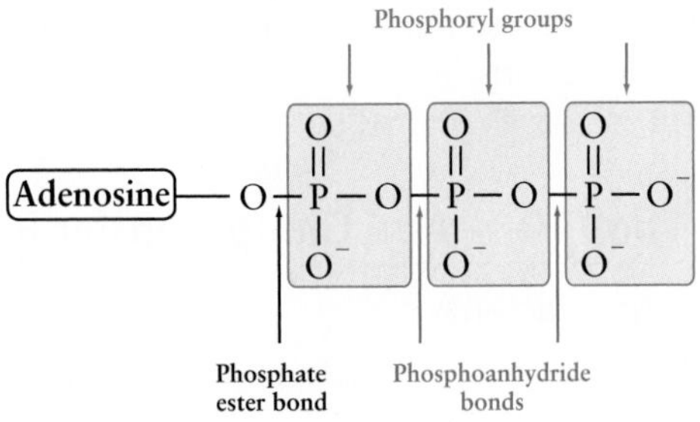

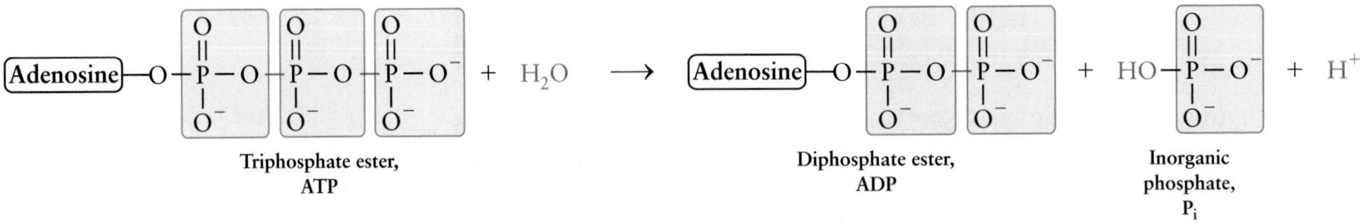

Adenosine Triphosphate
ATP

In addition, a phosphoryl group can be transferred to the oxygen atom of a water, alcohol, or carboxylic acid functional group. The transfer of a phosphoryl group to a water molecule occurs in a hydrolysis reaction. As with any hydrolysis, water is the agent that breaks a bond, in this case the phosphoanhydride bond. For example, one of the phosphoanhydride bonds in ATP is broken in the phosphoryl group transfer reaction that occurs when ATP is hydrolyzed. The phosphoryl group migrates from ATP to water, to produce ADP and an inorganic phosphate ion:

A phosphoryl group transferred to an alcohol or a carboxylic acid produces a phosphate ester. An example of phosphoryl group transfer to an alcohol is seen in the first step of glycolysis—an important catabolic pathway that breaks down glucose to produce energy. In this reaction, a phosphoryl group on ATP is transferred to the oxygen atom of a hydroxyl group on glucose.

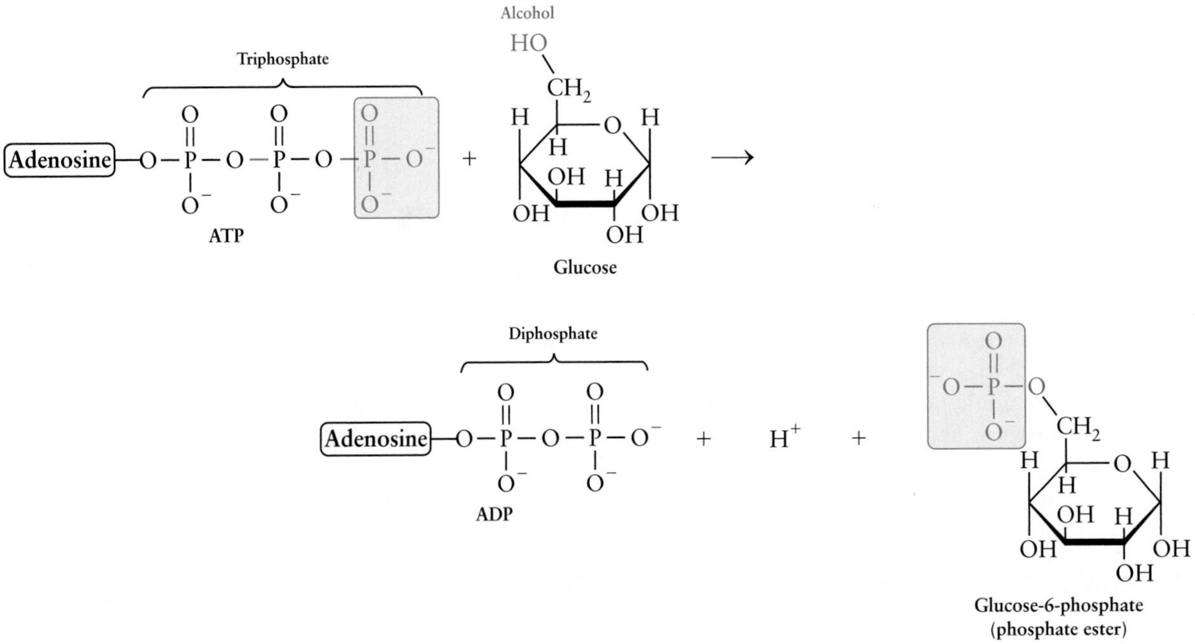

Energy Transfer in the Cell

Significant energy is transferred in a phosphoryl group transfer; therefore, reactions involving ADP and ATP are important reactions in the cell. The phosphoanhydride bond is sometimes called a "high-energy" bond, because its hydrolysis releases energy, and its formation absorbs energy; the former reactions drive the latter reactions. Indeed, ATP is called the "energy currency" of the cell because it is able to store chemical energy and transfer that energy when needed, through a phosphoryl group transfer. ATP is formed during catabolic biochemical pathways that break down carbohydrates and fats obtained from the diet. The energy released from catabolic processes can be used to transfer a phosphoryl group to ADP, thereby transferring and storing chemical energy in the phosphoanhydride bond of ATP. Generally, energy is stored as chemical energy rather than released as heat in biochemical reactions.

Similarly, when energy is needed for an anabolic reaction, the phosphoanhydride bond is hydrolyzed, producing ADP and inorganic phospate, P_i, as illustrated above, with the release of energy. You will learn the specifics of about how energy is transferred—the coupling of catabolic and anabolic reactions—in Chapter 14 in the section on bioenergetics.

WORKED EXERCISE 10-9   Phosphoryl Group Transfer Reactions

Answer the questions below for the following phosphoryl group transfer reaction seen in the third step of glycolysis.

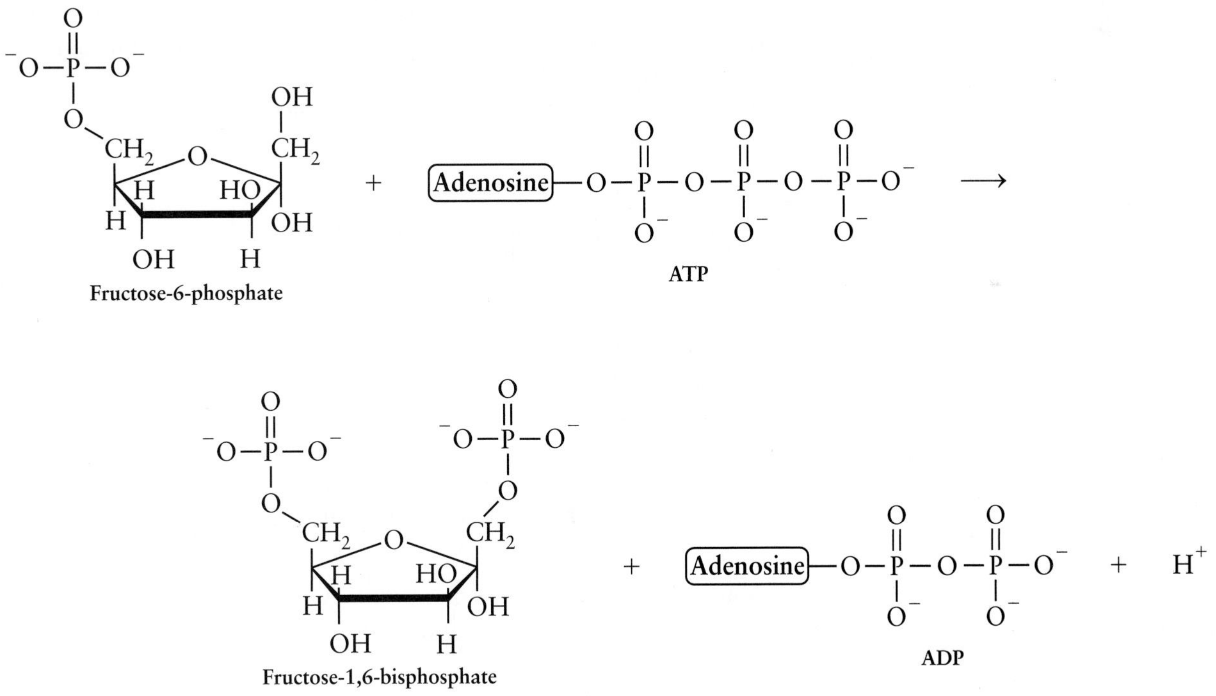

a. Circle all the phosphoryl groups and highlight the phosphoryl group that is transferred in the reaction.
b. What functional group is the phosphoryl group transferred to?
c. Why is this reaction classified as a phosphoryl group transfer reaction?
d. Identify the phosphoanhydride bond in ATP that is broken.

SOLUTION
a. Blue boxes encircle the phosphoryl groups. The shaded blue box shows the phosphoryl group that is transferred from ATP to fructose-6-phosphate to make fructose-1,6-bisphosphate.

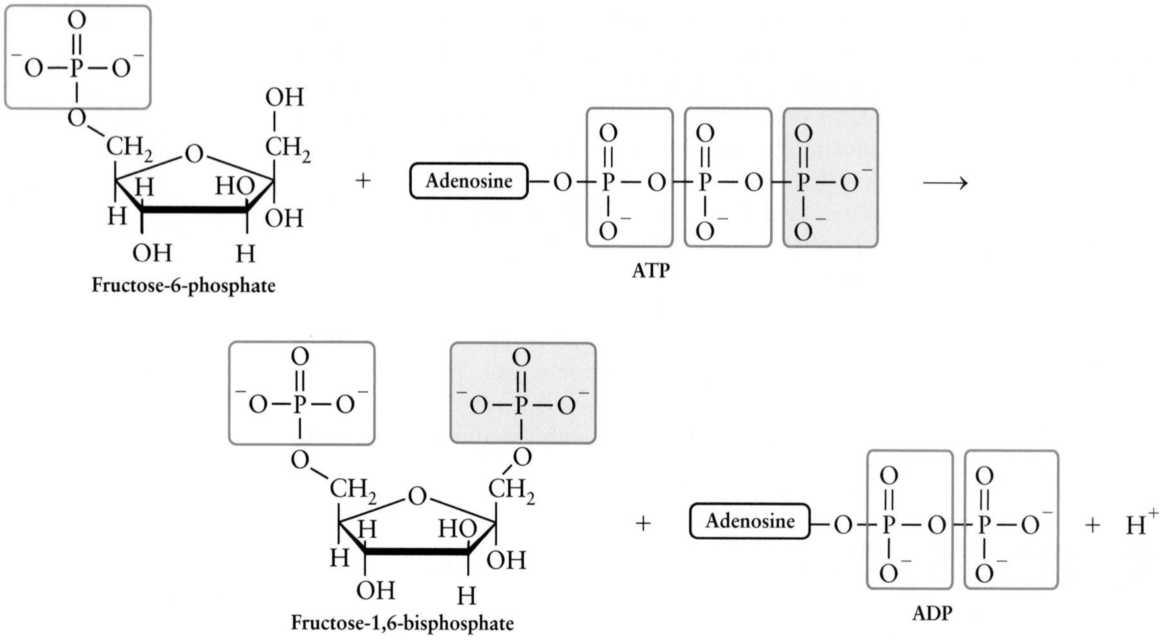

Fructose-6-phosphate + ATP $\longrightarrow$

Fructose-1,6-bisphosphate + ADP + H$^+$

b. The phosphoryl group is transferred to an alcohol functional group on fructose-6-phosphate. Note that fructose-6-phosphate already has one phosphate ester and it is gaining another (hence the "bis" in its name).

c. The reaction is a phosphoryl group transfer reaction because a phosphoryl group is transferred from one molecule (ATP) to another (fructose-6-phosphate).

d.

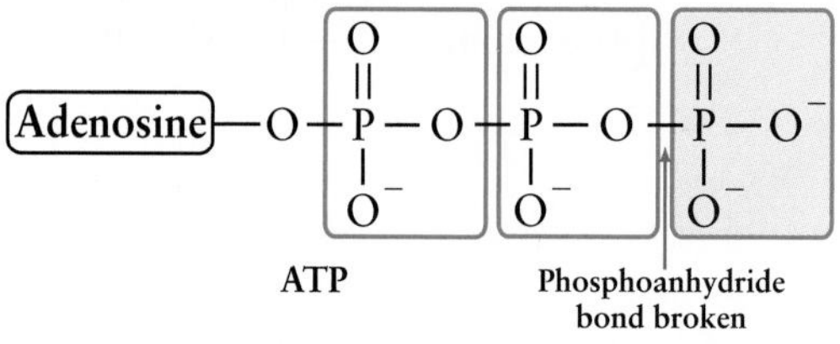

ATP   Phosphoanhydride
                bond broken

PRACTICE EXERCISES

10.19 Write the product formed in the hydrolysis of ATP. Circle the phosphoryl group transferred. Label the high-energy phosphoanhydride bond that is broken. What does the term "high energy" mean?

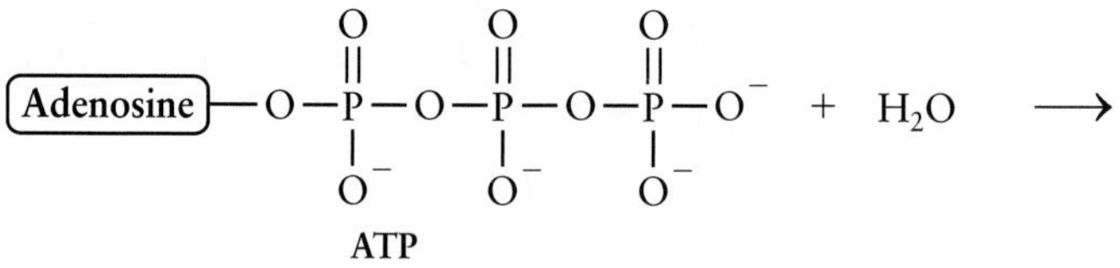

ATP

10.20 Answer the questions below regarding one of the reactions in the biochemical pathway for glucose catabolism.

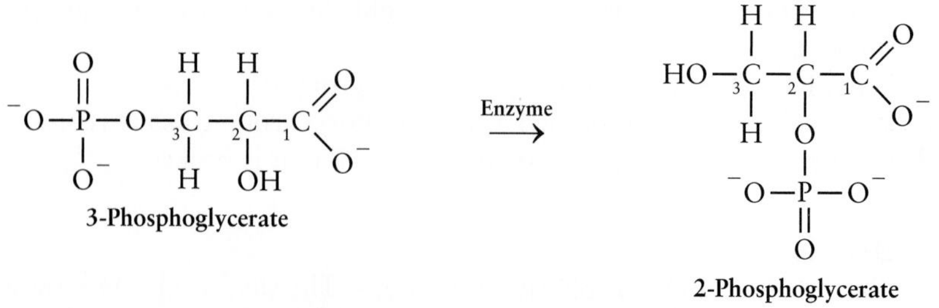

3-Phosphoglycerate   Enzyme →   2-Phosphoglycerate

a. Why is this reaction considered a phosphoryl group transfer reaction?.

b. Explain why this reaction is also considered to be an isomerization reaction.

c. How do the functional groups change in this reaction?

10.21 Explain why the reaction below is classified as a phosphoryl group transfer reaction.

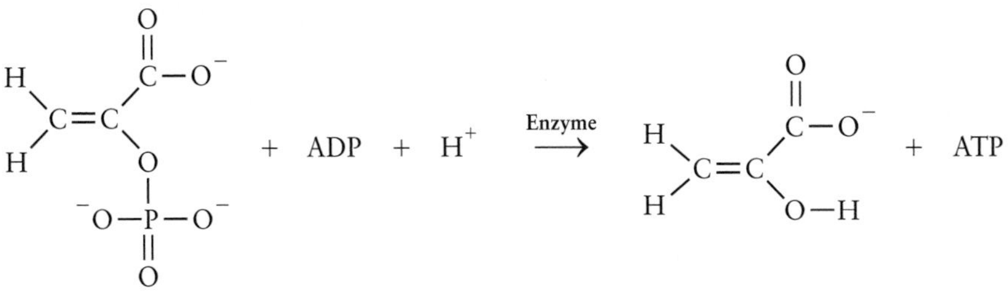

2-Phosphoglycerate   Glycerate

In this chapter you have seen that biochemistry at its core *is* organic chemistry! The reaction types described in this chapter will be seen throughout the remaining chapters of the text as you learn about the role of the biomolecules in metabolism. Table 10-4 summarizes the types of organic reactions seen in the biochemistry that were described in this chapter.

Table 10-4 Types of Organic Reactions Commonly Seen in Biochemistry

Class of Reaction	Type of Reaction	Reactant Functional Group	Product Functional Group	Change Observed
Oxidation-Reduction (Redox)	Oxidation [O] ⟶	Alkane	Alkene	Decrease in hydrogen or increase in oxygen. Loss of electrons.
		1° Alcohol	Aldehyde	
		2° Alcohol	Ketone	
		3° Alcohol	No reaction	
		Aldehyde	Carboxylic acid	
	Reduction [H] ⟶	Alkene	Alkane	Increase in hydrogen or decrease in oxygen. Gain of electrons.
		Aldehyde	1° Alcohol	
		Ketone	2° Alcohol	
		Carboxylic acid	Aldehyde	
Dehydration-Hydration	Dehydration H_2O ↗	Alcohol	Alkene	Expulsion of H and OH from molecule; form double bond.
	Hydration H_2O ↘	Alkene (adjacent to C=O)	Alcohol (two carbon atoms down from C=O)	Addition of H and OH to molecule; lose double bond.
Acyl Group Transfer	Hydrolysis	Ester	Carboxylic acid + alcohol	Acyl group migrates: H_2O absorbed
		Thioester	Carboxylic acid + thiol	
		Amide	Carboxylate + amine ion ($RCO_2^- + RNH_3^+$)	
	Esterification	Carboxylic acid + alcohol	Ester	Acyl group migrates: H_2O released
		Carboxylic acid + thiol	Thioester	
Phosphoryl Group Transfer	Hydrolysis	$ATP + H_2O$	$ADP + HOPO_3^{2-} + H^+$	Transfer of PO_3^{2-}.
	Phosphoryl group transfer to or from ATP	$ATP + ROH$	$ADP + ROPO_3^{2-} + H^+$	
		$ATP + RCO_2H$	$ADP + RCO_2PO_3^{2-} + H^+$	

Chemistry in Medicine Vitamins and Health

Vitamins are organic molecules required in trace amounts by the human body. Most vitamins cannot be synthesized in the body, so they must be supplied by the diet. Insufficient vitamin intake or poor vitamin absorption can lead to vitamin deficiency diseases.

Riboflavin (vitamin B_2), for example, is an important vitamin found in eggs, milk, cheese, meat, and fortified cereals. Symptoms of a riboflavin deficiency include inflammation of the tongue (glossitis) and skin (dermatitis), lesions in the mouth, and cataracts.

Riboflavin deficiencies are often seen in newborn babies undergoing phototherapy (light therapy), a common treatment for babies born with jaundice. Jaundice, a condition characterized by a yellow coloring of the skin and eyes, is caused by elevated levels of bilirubin, a product of the catabolism of hemoglobin. The type of light used in phototherapy causes riboflavin to undergo a chemical reaction, depleting the newborn of riboflavin.

Vitamins have been historically divided into two categories, depending on their solubility: *water-soluble* and *fat-soluble*, as shown in Table 10-5. The water-soluble vitamins include all the B complex vitamins and vitamin C. Vitamins A, D, E, and K are classified as fat-soluble vitamins. The water-soluble vitamins are readily eliminated from the body, so they must be replenished on a daily basis through the diet. The fat-soluble vitamins tend to be stored in fat tissue and therefore remain in the body longer.

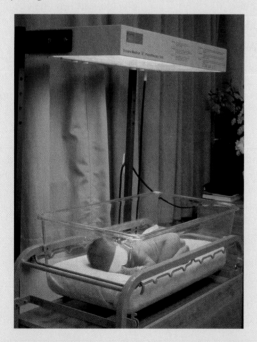

A newborn undergoing phototherapy—a treatment for jaundice. The yellow coloring of the skin and eyes seen in infants with jaundice is due to elevated levels of bilirubin in the blood. [Ron Sutherland/Photo Researchers, Inc.]

Table 10-5 Water-Soluble and Fat-Soluble Vitamins

Water-Soluble Vitamins

Thiamin	B_1
Riboflavin	B_2
Niacin	B_3
Pantothenic acid	B_5
Pyridoxamine, pyridoxal, pyridoxine	B_6
Biotin	—
Cobalamin	B_{12}
Folic acid	—
Ascorbic acid	C

Fat-Soluble Vitamins

A
D
E
K

Most of the water-soluble vitamins serve as the starting materials for the biosynthesis of the important coenzymes FAD, NAD^+, and Coenzyme A. You have already seen the role that NAD^+ and FAD play in oxidation–reduction reactions. You have also seen that Coenzyme A plays a role in acyl transfer reactions. Coenzymes are required by certain enzymes to achieve their catalytic activity. Coenzymes also serve as the electron carriers in biochemical pathways that will be described further in the next four chapters. Table 10-6 lists the vitamin precursors for each of the major coenzymes, along with the type of reaction the coenzyme is involved in.

To illustrate the role of these vitamins, consider riboflavin, whose structure is shown in Figure 10-8. In the body, riboflavin reacts with one ATP molecule to form the coenzyme flavin adenine dinucleotide (FAD), a phosphate ester formed between riboflavin and ADP. A second ATP molecule is hydrolyzed to supply the energy for the reaction.

This reaction involves two phosphoryl group transfer reactions. The phosphate groups are printed in blue, and the adenosine portion of the molecule is shown in red.

Table 10-6 Vitamin Precursors for Coenzymes

Water-Soluble Vitamins	Coenzyme	Reaction Type
Riboflavin, B_2	FAD/$FADH_2$	Oxidation-reduction
Niacin, B_3	NAD^+/$NADH_2$	Oxidation-reduction
Pantothenic acid, B_5	Coenzyme A	Acyl transfer

Phosphoryl Group Transfer Reactions

- Phosphoryl group transfer reactions are important reactions in biochemistry, because they are involved in the transfer of energy.
- The phosphoryl group is usually transferred between the oxygen atoms of water, alcohols, carboxylic acids, and other phosphates.
- Phosphoryl groups can be linked together to form diphosphates or triphosphates.
- The phosphoanhydride bond joins phosphoryl groups in molecules like ADP and ATP.
- The phosphoanydride bond is a high-energy bond that releases energy upon hydrolysis and absorbs energy when it is formed.

Key Words

Acyl group transfer reactions A type of reaction that interconverts carboxylic acids and their derivatives: esters, thioesters, and amides. $RC=O$ is an acyl group.

Antioxidant Natural substances found in fruits, vegetables, and tea that prevent the damaging effects of oxidation by reducing harmful oxidizing agents.

Catalytic hydrogenation The reduction of carbon–carbon double bonds to carbon–carbon single bonds using hydrogen gas (H_2) in the presence of a metal catalyst (Pd, Pt, etc.).

Coenzyme A biological oxidizing or reducing agent such as $NAD^+/NADH$ and $FAD/FADH_2$. They serve as electron carrier molecules in biochemical pathways. They are required by certain enzymes for catalysis to occur.

Combustion reaction The reaction between $C_xH_yO_z$ and oxygen (O_2) to produce carbon dioxide (CO_2), water, and energy.

Dehydration reaction The loss of H_2O from an alcohol to produce a double bond. This reaction is the reverse of a hydration reaction.

Esterification reaction The formation of an ester from the reaction between a carboxylic acid and an alcohol.

FAD/FADH₂ Flavin adenine dinucleotide. A coenzyme seen in oxidation–reduction reactions involving carbon–carbon bonds.

Hydration reaction The addition of H and OH from water to a double bond to form an alcohol. It is the reverse of a dehydration reaction.

Hydrolysis reaction The reaction that splits a carboxylic acid derivative (ester, thioester, or amide) into a carboxylic acid and either an alcohol, thiol, or amine, by the addition of water.

NAD⁺/NADH Nicotinamide adenine dinucleotide. A coenzyme seen in oxidation–reduction reactions involving carbon–oxygen bonds.

Oxidation The part of an oxidation–reduction reaction that loses electrons in going to product.

Oxidation–reduction reaction A reaction characterized by the transfer of electrons from one reactant to another.

Oxidizing agent The reactant that gains electrons in an oxidation–reduction reaction by oxidizing the other reactant.

Phosphoanhydride bond The high-energy P—O single bond that connects phosphoryl groups in a di- or triphosphate ester. When it is hydrolyzed, energy is released.

Phosphoryl group transfer reaction The transfer of a phosphoryl group,   , to and from water,

$$\xi-O-\overset{\overset{\textstyle O}{\|}}{\underset{\underset{\textstyle O^-}{|}}{P}}-O^-$$

an alcohol, a carboxylic acid, or another phosphoryl group.

Reducing agent The reactant that loses electrons in an oxidation–reduction reaction by reducing the other reactant.

Reduction The part of an oxidation–reduction reaction that gains electrons in going to product.

Saponification reaction The hydrolysis of an ester in the presence of hydroxide ion to produce an alcohol and a carboxylate ion. If the reactant is a fat, the carboxylate product is soap.

Additional Exercises

The Role of Functional Groups in Biochemical Reactions

10.22   What is the "function" of a functional group?

10.23   Using Table 10-1, identify the functional groups involved in acyl transfer reactions.

10.24   Describe the following functional groups:
    a. aldehyde
    b. carboxylic acid
    c. amide

10.25   Describe the following functional groups:
    a. ketone
    b. ester
    c. thioester

Oxidation–Reduction Reactions

10.26   What products are formed when an organic substance containing carbon, hydrogen, and oxygen undergoes a combustion reaction?

10.27 The reactant that undergoes loss of electrons in an oxidation–reduction reaction is said to undergo _____. [*oxidation* or *reduction*]

10.28 The reactant that gains electrons in an oxidation–reduction reaction is said to undergo _____. [*oxidation* or *reduction*]

10.29 If an organic reactant gains hydrogen atoms, has it undergone oxidation or reduction?

10.30 If an organic reactant loses hydrogen atoms, has it undergone oxidation or reduction?

10.31 If an organic reactant gains oxygen atoms, has it undergone oxidation or reduction?

10.32 If an organic reactant loses oxygen atoms, has it undergone oxidation or reduction?

10.33 For the following inorganic reactions, write the oxidation portion separate from the reduction portion of the reaction, and show electrons as Lewis dots. Label the oxidation and the reduction. If there is a spectator ion, identify it.
a. $Zn(s) + 2 HCl(aq) \rightarrow H_2(g) + ZnCl_2(aq)$
b. $2 Na(s) + Cl_2(g) \rightarrow 2 NaCl(s)$

10.34 For the oxidation–reduction reactions shown below, determine whether the organic compound shown has undergone an oxidation or a reduction by placing either an H or an O in the brackets above the arrow when indicated. Explain how you determined whether the reaction is an oxidation or a reduction.

a.

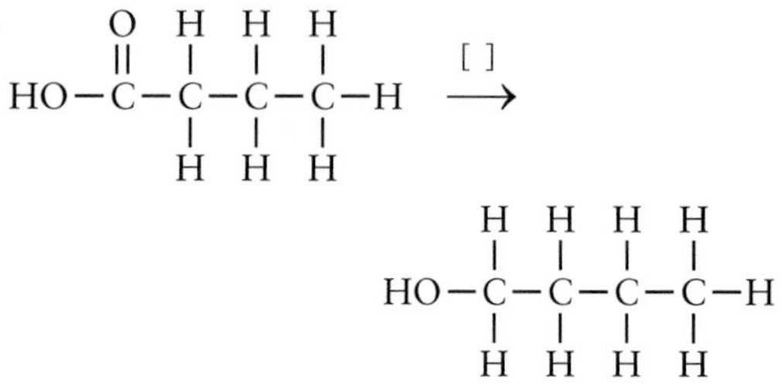

b.

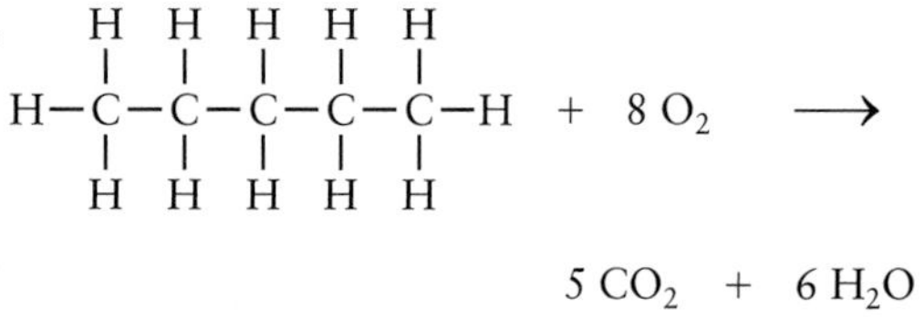

c.

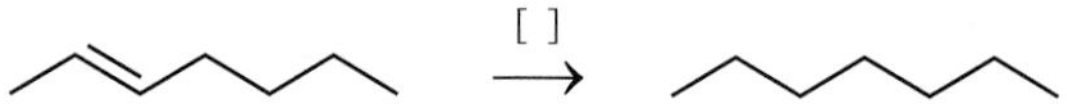

d.

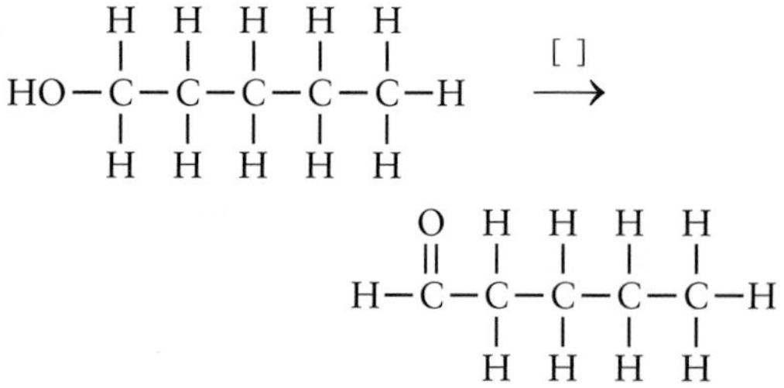

10.35 For the oxidation–reduction reactions shown below, determine whether the main reactant has undergone an oxidation or a reduction. Indicate whether the coenzyme shown has undergone an oxidation or a reduction.

a.

Acetaldehyde   Acetic acid

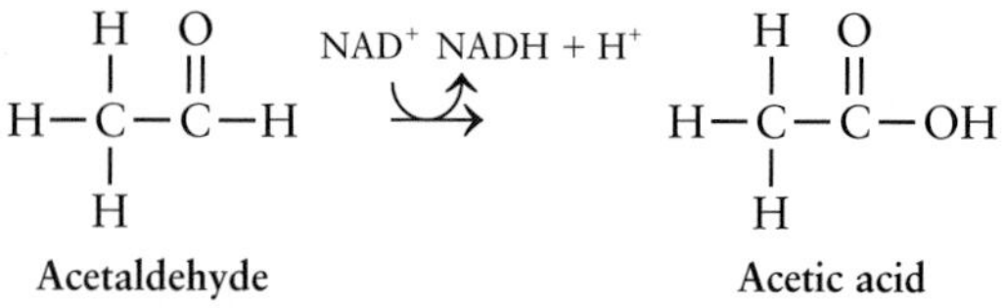

b.

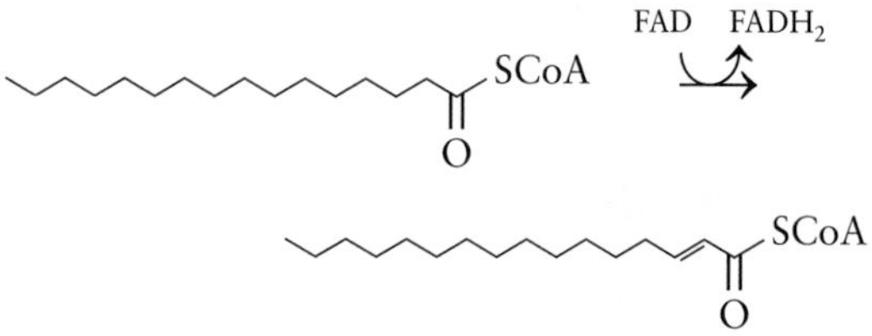

10.36 Write the structure of the product formed in the following reaction. What type of reaction is this?

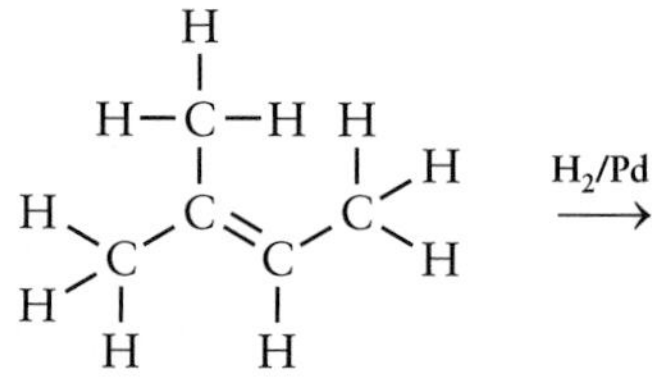

10.37 Write the structure of the product formed in the following reaction. What type of reaction is this?

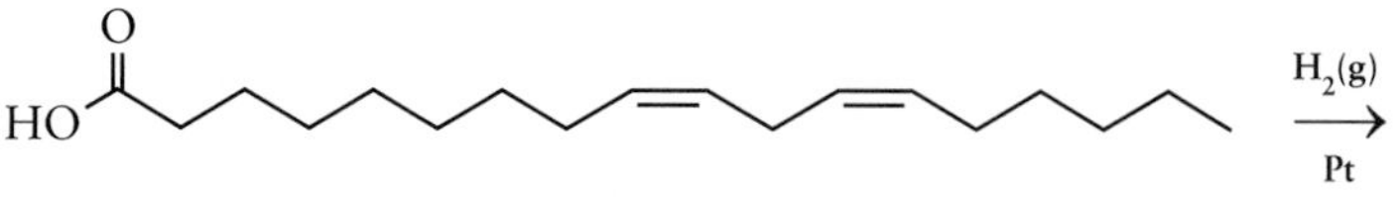

10.38 Why does the food industry use "partially hydrogenated" fats? What does it mean chemically to partially hydrogenate a fat?

10.39 Which is more likely to be a liquid?
a. a saturated fat
b. an unsaturated fat

10.40 Which coenzyme is most likely used in a biochemical oxidation of an alcohol?
a. FAD
b. $FADH_2$
c. NAD^+
d. $NADH + H^+$

10.41 Which coenzyme is most likely used in a biochemical reduction of an alkene?
a. FAD
b. $FADH_2$
c. NAD^+
d. $NADH + H^+$

10.42 Write the structure of the product formed in the following oxidation–reduction reactions. Label the functional group that changes in the reactant and identify it in the product. If the functional group is an alcohol, identify the type of alcohol.

a.

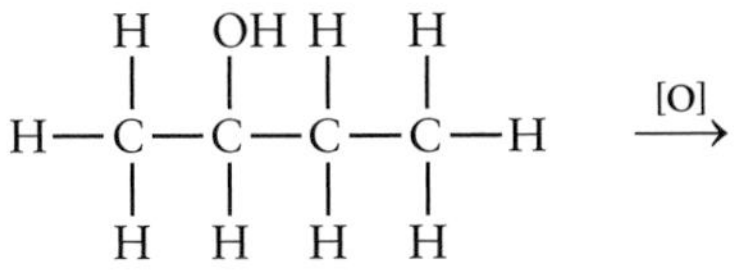

$$\xrightarrow{[O]}$$

b.

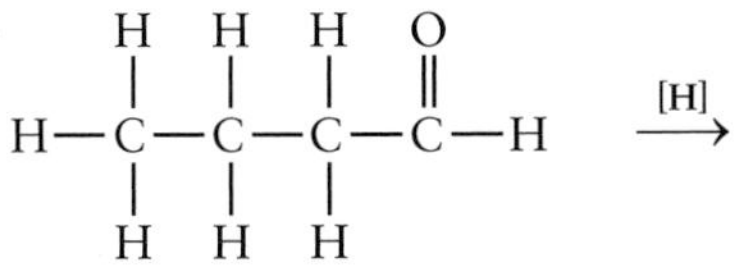

$$\xrightarrow{[H]}$$

c.

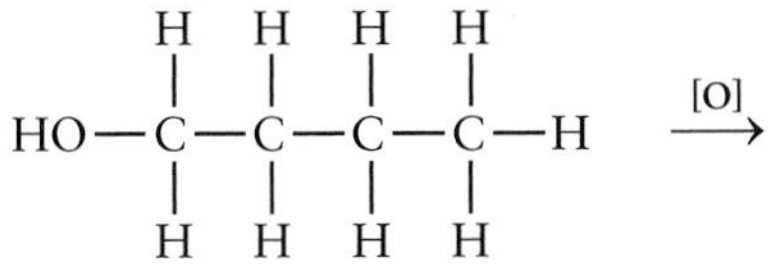

$$\xrightarrow{[O]}$$

d.

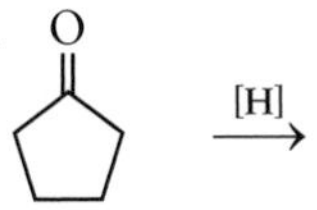

$$\xrightarrow{[H]}$$

e.

$$\xrightarrow{[O]}$$

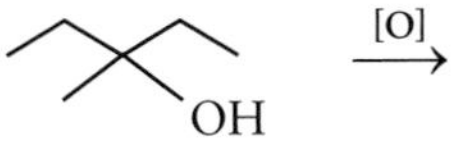

10.43 Write the structure of the organic product formed in the following reactions:

a.

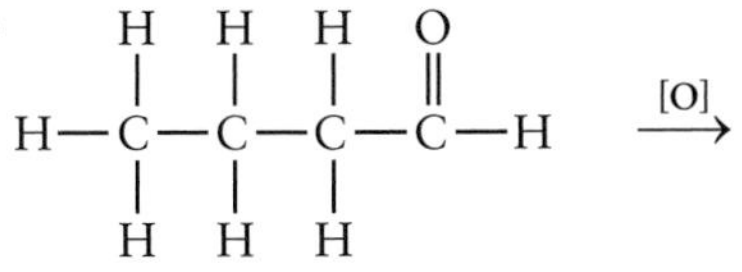

$$\xrightarrow{[O]}$$

b.

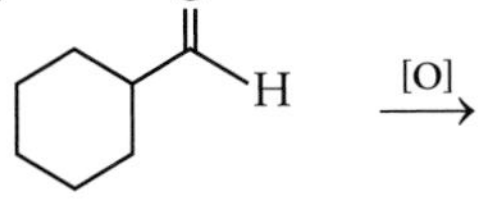

$$\xrightarrow{[O]}$$

c.

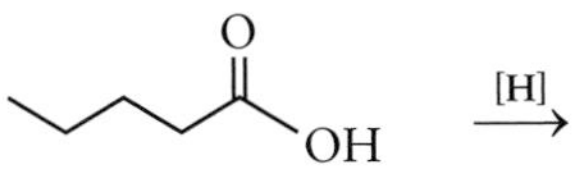

$$\xrightarrow{[H]}$$

d.

$$\xrightarrow{[H]}$$

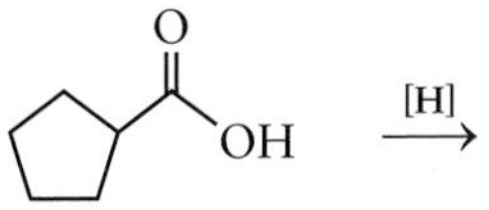

10.44 Circle the hydrogen atom removed when NADH is oxidized to NAD⁺.

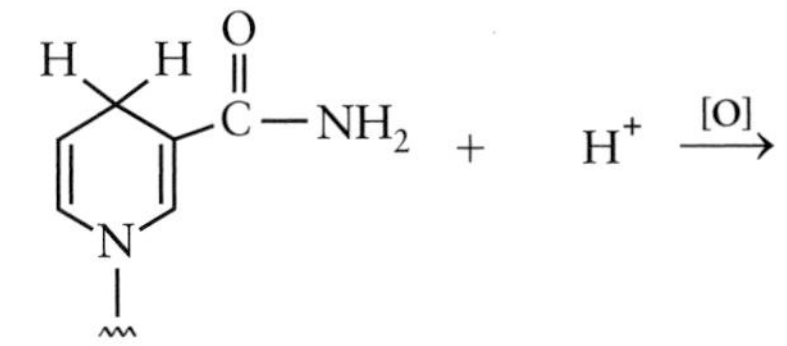

Nicotinamide adenine dinucleotide
(NADH)

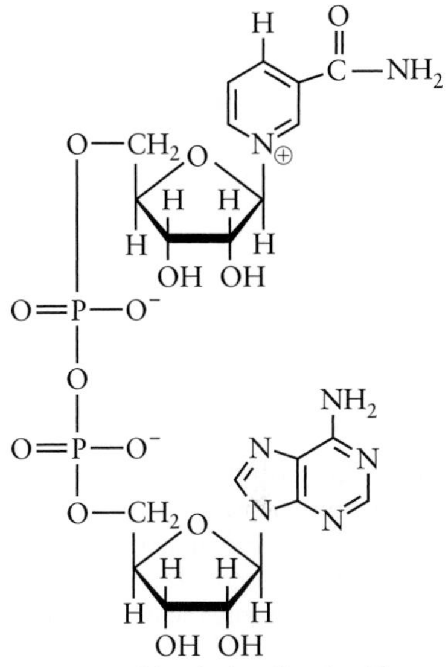

Nicotinamide adenine dinucleotide
(NAD⁺)

10.45 Write the structure of the product(s) formed in each of the reactions.

a. $C_3H_8 + 5 O_2 \rightarrow 3$ _____ $+ 4$ _____

b.

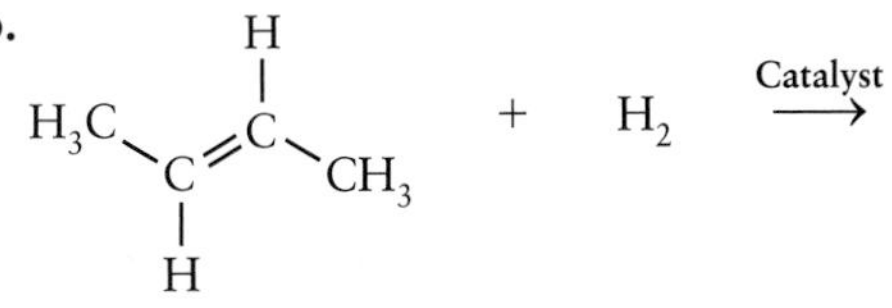

$$+ \; H_2 \xrightarrow{Catalyst}$$

c.

$$\xrightarrow{NAD^+ \quad NADH + H^+}$$

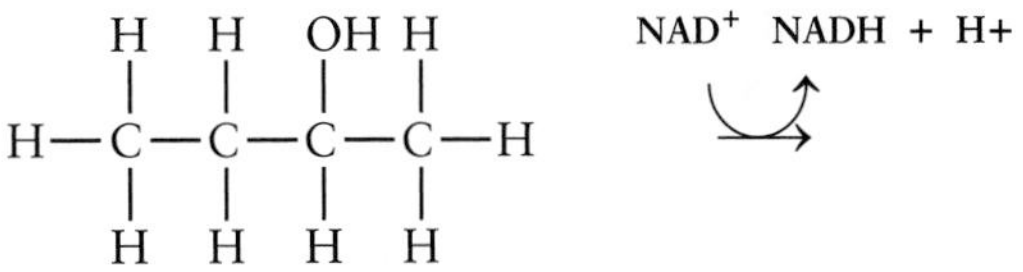

d. **Critical Thinking Question:** $FADH_2 + NAD^+ \rightarrow$

10.46 Write the structure of the *reactant* that upon reduction would yield the following product.

a.

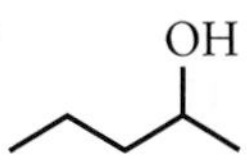

b.

$$H_3C-\underset{\underset{CH_3}{|}}{\overset{\overset{H}{|}}{C}}-\overset{\overset{O}{||}}{C}-H$$

c. 1-Butanol
d. NADH

10.47 Predict the structure of the *reactant* that upon oxidation would yield the following product.

a.

$$H_3C-\underset{\underset{CH_3}{|}}{\overset{\overset{H}{|}}{C}}-\overset{\overset{O}{||}}{C}-CH_3$$

b.

a zig-zag chain with $\overset{\overset{}{||}}{\underset{O}{C}}-S-CoA$ group

c. FAD

d.

benzene ring with $\overset{\overset{O}{||}}{C}-OH$

10.48 Which of the following is not true about an antioxidant?
a. It is a reducing agent.
b. It undergoes oxidation.
c. It prevents the oxidation of other substances in the cell.
d. It is an oxidizing agent.
e. They are believed to slow the aging process.

10.49 Which of the following is not true about a free radical?
a. It is an unstable species.
b. It contains an odd number of electrons.
c. Antioxidants prevent the formation of free radicals.
d. They contain an octet.

Hydration–Dehydration Reactions

10.50 For the reactions listed below, indicate whether they are hydration or dehydration reactions and add a curved arrow showing water in the equation.

a.

$$H-\underset{\underset{H}{|}}{\overset{\overset{H}{|}}{C}}-\underset{\underset{H}{|}}{\overset{\overset{OH}{|}}{C}}-\underset{\underset{H}{|}}{\overset{\overset{H}{|}}{C}}-\overset{\overset{O}{||}}{C}-\underset{\underset{H}{|}}{\overset{\overset{H}{|}}{C}}-H \longrightarrow$$

$$H-\underset{\underset{H}{|}}{\overset{\overset{H}{|}}{C}}-C=C-\overset{\overset{O}{||}}{C}-\underset{\underset{H}{|}}{\overset{\overset{H}{|}}{C}}-H$$

b.

ring structures with O, H_3C, OH groups $\longrightarrow$

10.51 For the reactions listed below, indicate whether they represent hydration or dehydration reactions and show water next to a curved arrow in the equation.

a.

cyclohexenone $\longrightarrow$ cyclohexanone with OH

b.

OH O structure $\longrightarrow$ O structure

c.

O structure $\longrightarrow$ OH O structure

10.52 Write the product formed in the following reactions.

a.

$$H-\underset{\underset{H}{|}}{\overset{\overset{H}{|}}{C}}-\underset{\underset{H}{|}}{\overset{\overset{H}{|}}{C}}-\overset{\overset{O}{||}}{C}-\underset{\underset{H}{|}}{\overset{\overset{H}{|}}{C}}-\underset{\underset{H}{|}}{\overset{\overset{OH}{|}}{C}}-\underset{\underset{H}{|}}{\overset{\overset{H}{|}}{C}}-H \quad \xrightarrow{\text{Dehydration}}$$

b.

$$\underset{H_3C}{\overset{H_3C}{}}C=C\underset{\underset{\underset{O}{||}}{C}-CH_3}{\overset{H}{}} \quad + \quad H_2O \quad \longrightarrow$$

10.53 Write the product formed in the following reactions.

a.

O acetyl cyclopentene $\quad + \quad H_2O \quad \longrightarrow$

b.

COO^- structure with OH $\quad \overset{H_2O}{\nearrow}$

Acyl Group Transfer Reactions

10.54 Write the structure of the products formed in the following hydrolysis reactions. Identify all functional groups in both reactants and products.

a.

$$H-\underset{\underset{H}{|}}{\overset{\overset{H}{|}}{C}}-\overset{\overset{O}{||}}{C}-O-\underset{\underset{H}{|}}{\overset{\overset{H}{|}}{C}}-\underset{\underset{H}{|}}{\overset{\overset{H}{|}}{C}}-H \quad \overset{H_2O}{\underset{\text{Catalyst}}{\searrow}}$$

b.

$$H-\underset{\underset{H}{|}}{\overset{\overset{H}{|}}{C}}-\underset{\underset{H}{|}}{\overset{\overset{CH_3}{|}}{C}}-S-\overset{\overset{O}{||}}{C}-\underset{\underset{H}{|}}{\overset{\overset{H}{|}}{C}}-\underset{\underset{H}{|}}{\overset{\overset{H}{|}}{C}}-H \quad \overset{H_2O}{\underset{\text{Catalyst}}{\searrow}}$$

c.

$$H-\underset{\underset{H}{|}}{\overset{\overset{H}{|}}{C}}-\underset{\underset{H}{|}}{\overset{\overset{H}{|}}{C}}-\underset{\underset{H}{|}}{\overset{\overset{}{}}{N}}-\overset{\overset{O}{||}}{C}-\underset{\underset{H}{|}}{\overset{\overset{H}{|}}{C}}-H \quad \overset{H_2O}{\underset{\text{Catalyst}}{\searrow}}$$

10.55 Write the structures for the following hydrolysis reactions. Identify all functional groups in both reactants and products.

a.

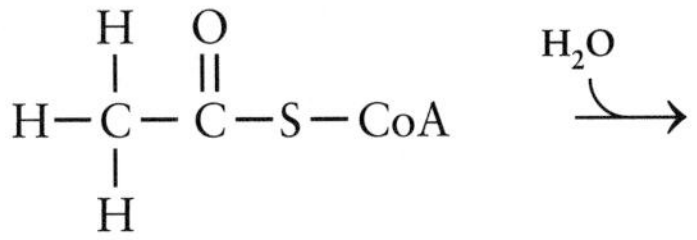

b.

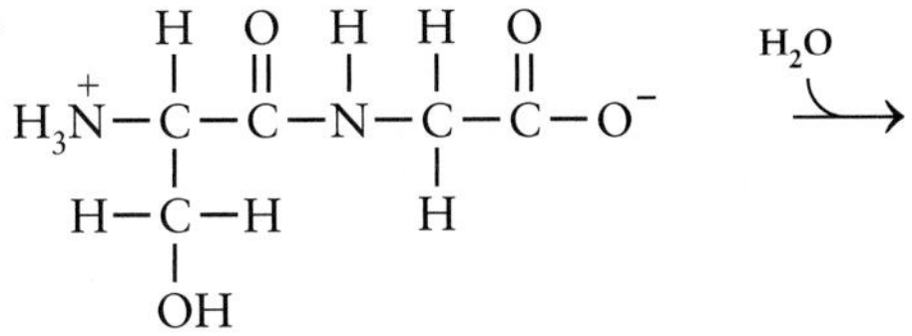

10.56 The reaction shown below occurs in the citric acid cycle.

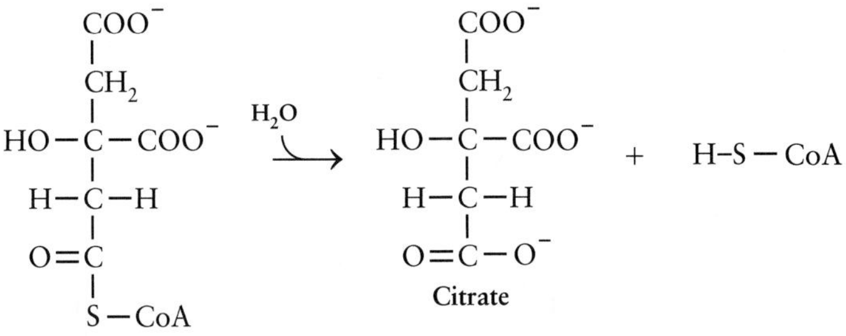

a. How many carbon atoms does the acyl group being transferred contain?

b. Between what two atoms is the acyl group being transferred?

c. Why is this reaction classified as a hydrolysis?

10.57 Write the structure of the products formed in the following hydrolysis reactions. Identify the bond broken in each carboxylic acid derivative. Identify the acyl group that migrates.

a.

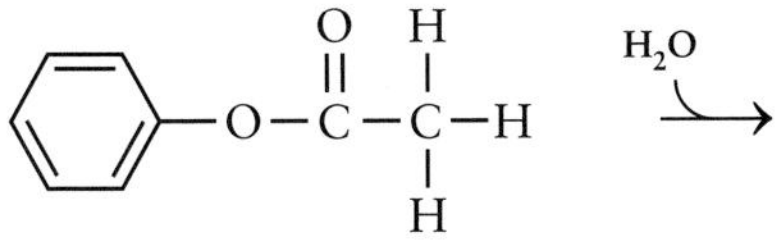

b.

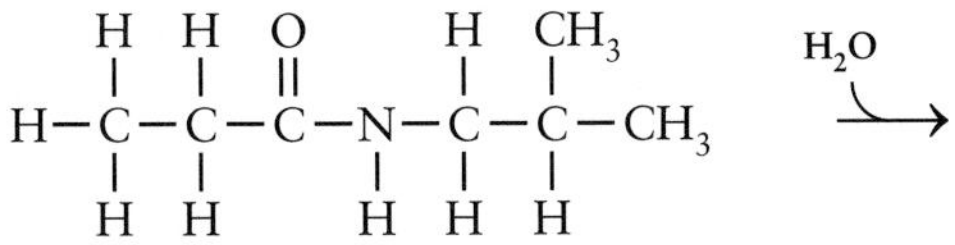

c.

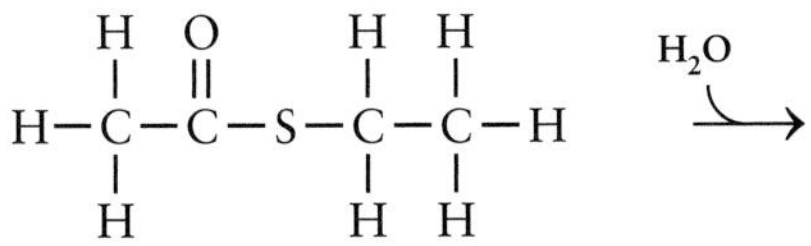

d. Critical Thinking Question: The compound shown below is known as a carboxylic acid anhydride. Predict the product formed upon hydrolysis.

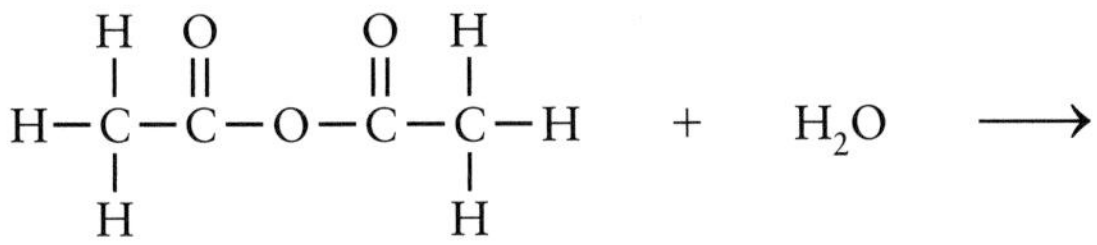

10.58 Write the products formed in the following esterification reactions.

a.

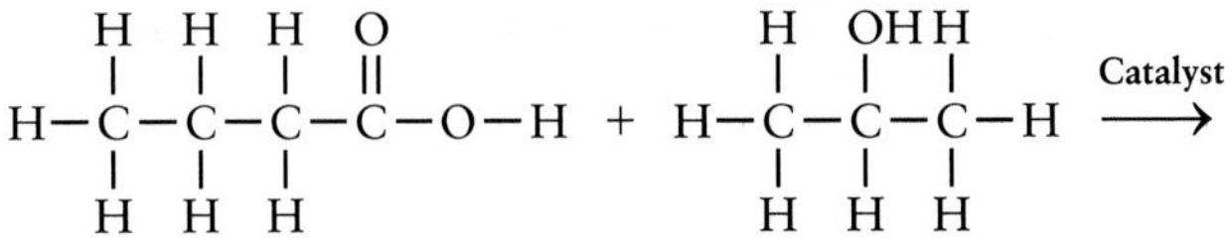

b.

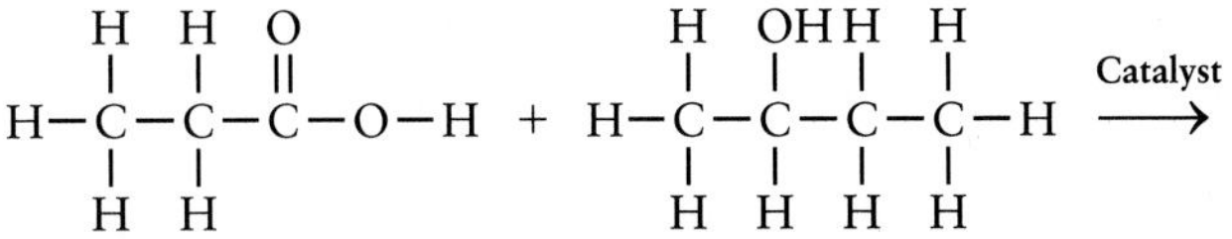

c.

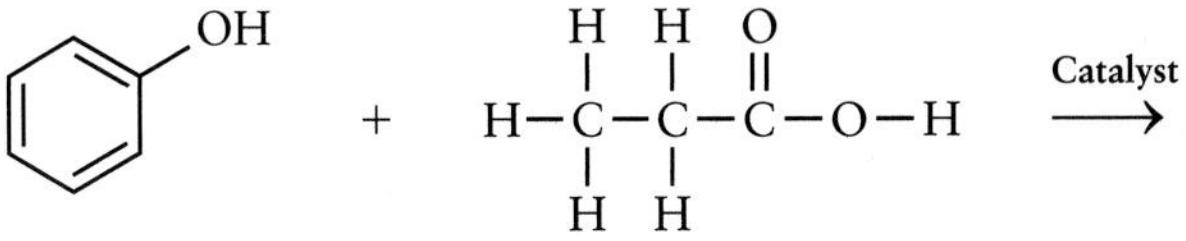

10.59 Predict the products formed in the following esterification reactions:

a.

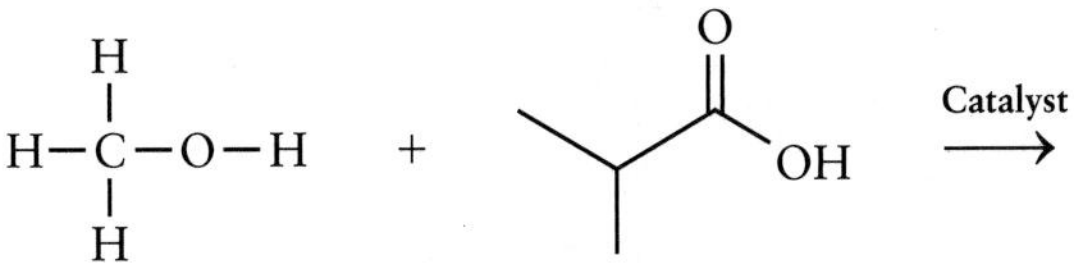

b.

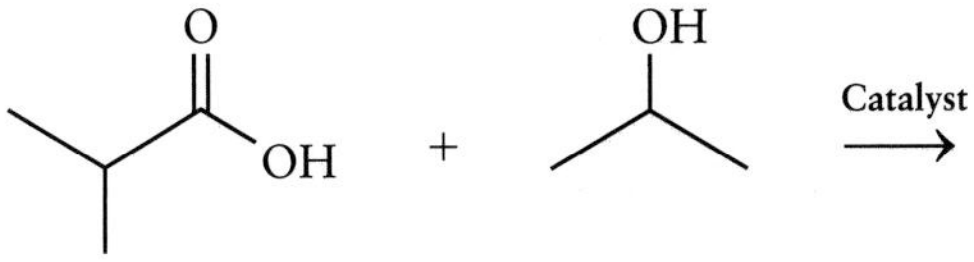

c.

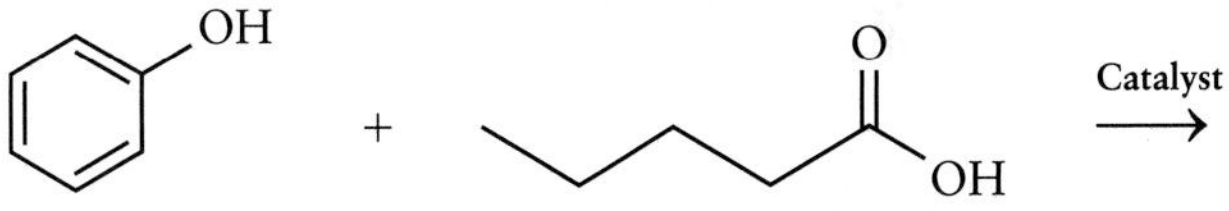

10.60 Critical Thinking Question: The structure of aspartame (Nutrasweet) is shown below.

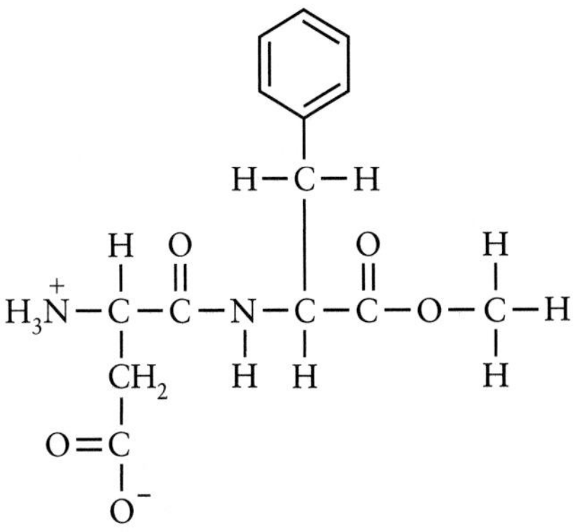

a. Circle the amide group in aspartame.

b. Write the structure of the two compounds formed when the amide in aspartame is hydrolyzed. Don't hydrolize the ester.

Phosphoryl Group Transfer Reactions

10.61 How is a phosphoryl group transfer similar to an acyl group transfer?

10.62 ADP can be hydrolyzed to AMP. Write the structure of the product and circle the phosphate group transferred.

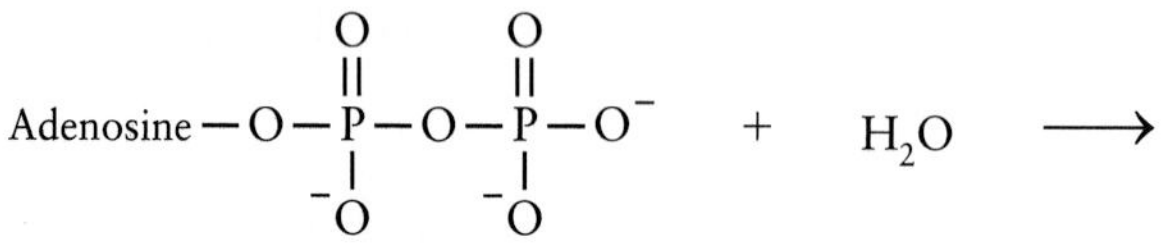

ADP

10.63 *Hexokinase* catalyzes the reaction shown below between ATP and glucose.

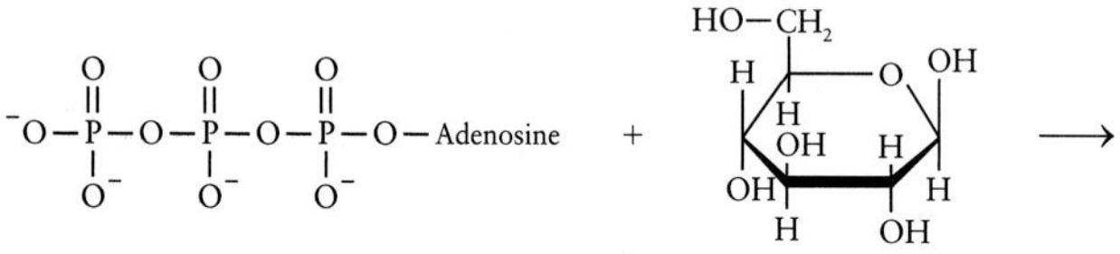

ATP   Glucose

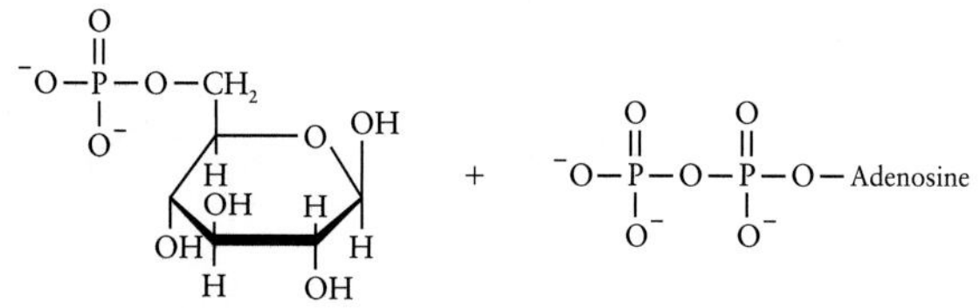

Glucose 6-phosphate   ADP

a. Highlight the phosphate group that is transferred in the reaction.
b. Is the phosphate group being transferred from a mono-, di-, or triphosphate?
c. What functional group is the phosphate group transferred to?

10.64 Explain why the reaction shown below is a phosphoryl group transfer reaction.

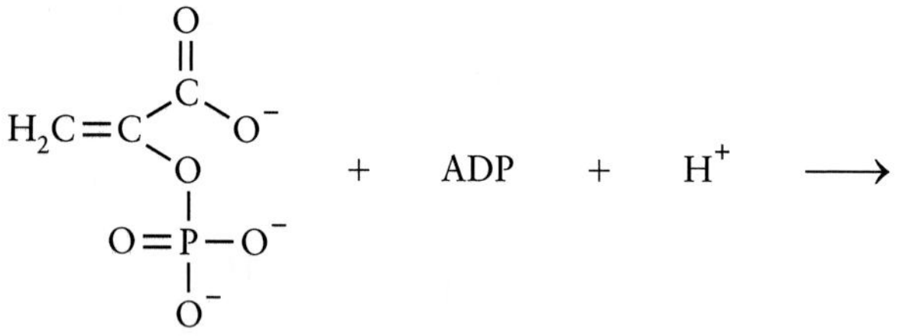

Phosphoenol pyruvate

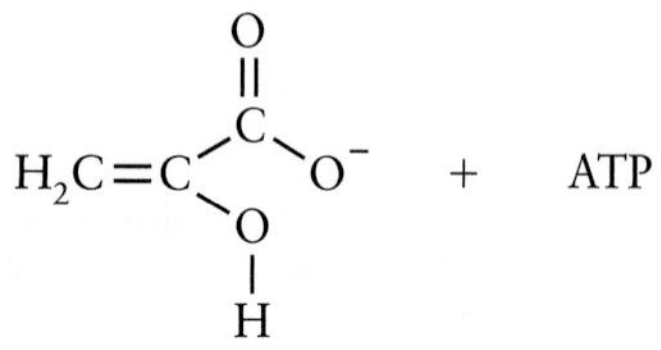

10.65 **Critical Thinking Question:** Identify the following reactions as oxidation–reduction reactions, hydration–dehydration reactions, hydrolysis reactions, acyl group transfer reactions, or phosphoryl group transfer reactions.

a.

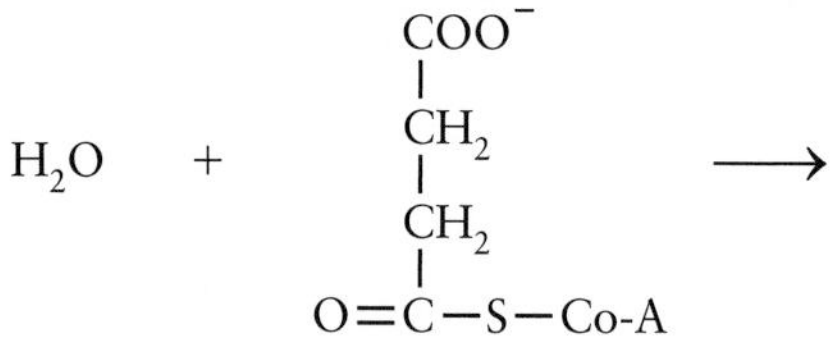

Succinyl-CoA

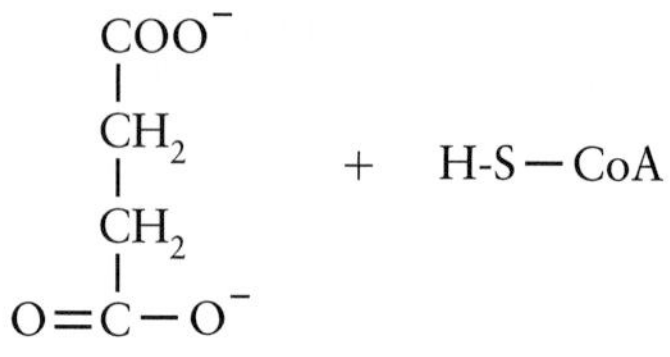

Succinate

b.

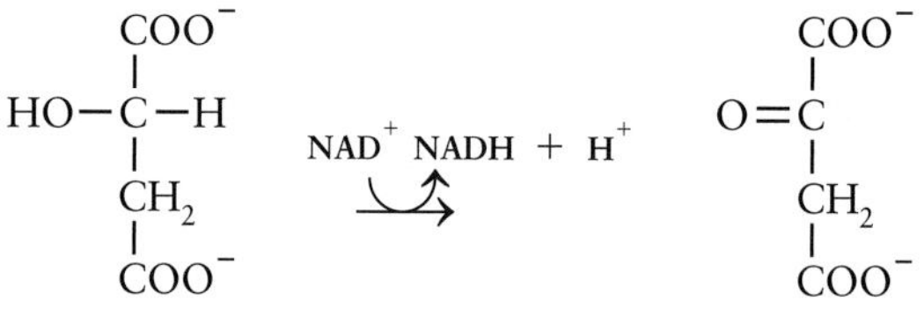

L-malate   Oxaloacetate

c.

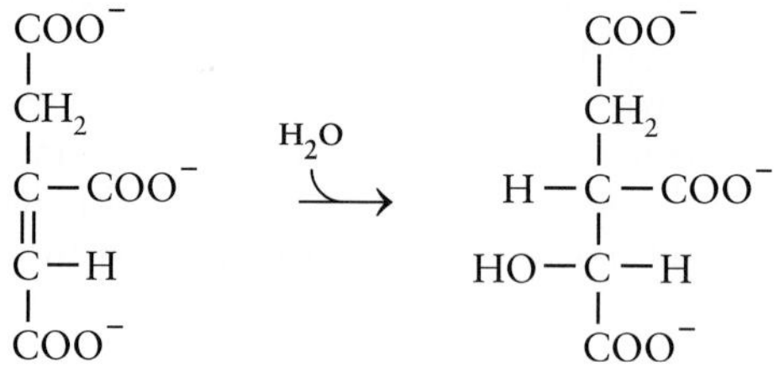

Aconitate   Isocitrate

10.66 Identify the following reactions as oxidation–reduction reactions, hydration–dehydration reactions, hydrolysis reactions, acyl group transfer reactions, or phosphoryl group transfer reactions?

a.

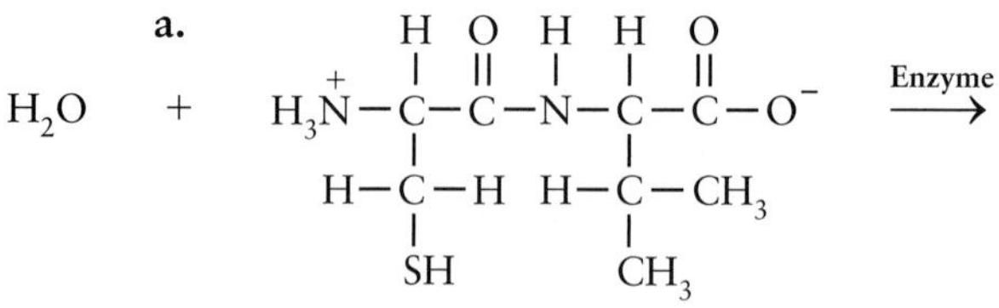

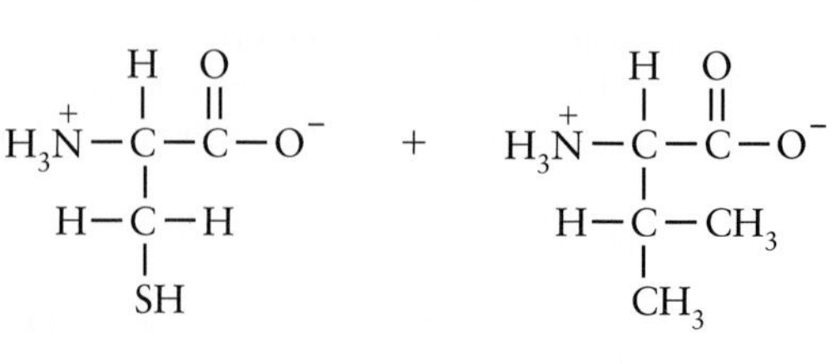

b.

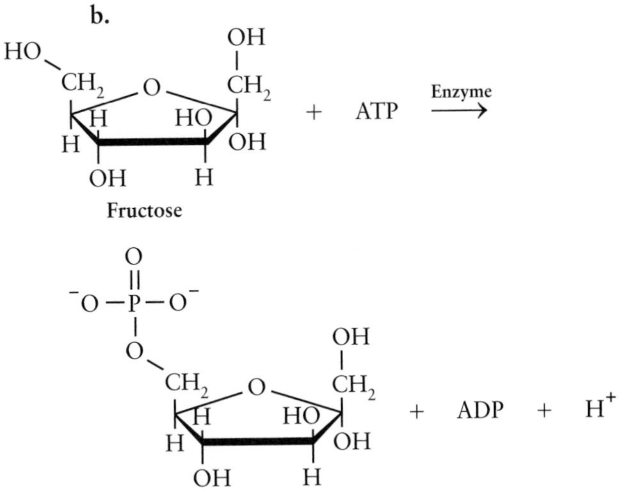

Fructose

Fructose-6-phosphate

c.

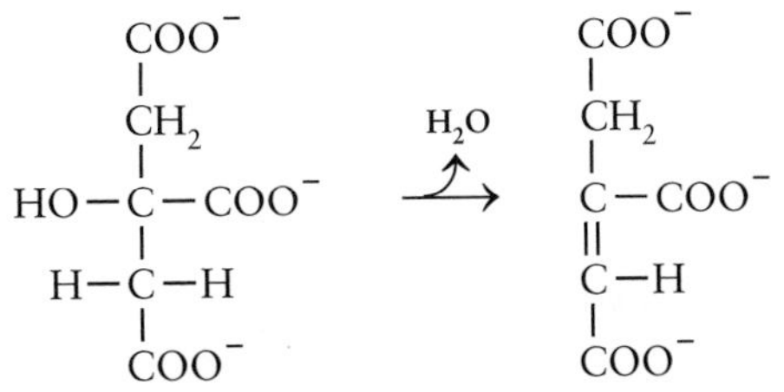

Citrate   Aconitate

10.67 Consider the reaction discussed in the opening vignette of this chapter:

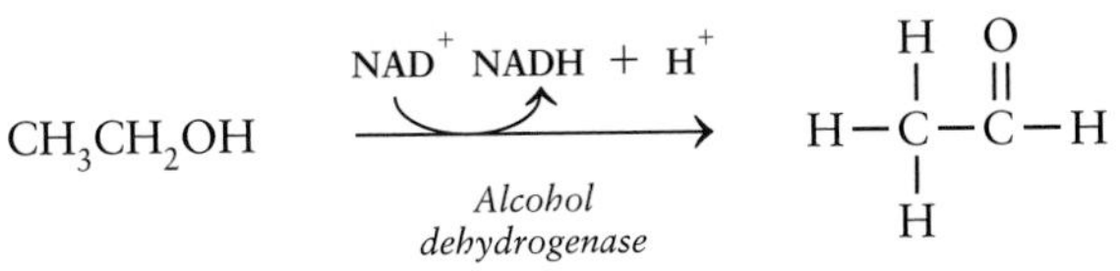

While the enzyme *alcohol dehydrogenase* catalyzes this reaction in your body, it catalyzes the reverse reaction in yeast cells during fermentation.
a. Write the reaction that occurs in yeast during fermentation.

b. In the reaction that occurs in yeast, is NADH or NAD^+ a reactant?
c. In the reaction that occurs in yeast, is acetaldehyde oxidized or reduced?

Chemistry in Medicine

10.68 Look closely at the structure of riboflavin below:

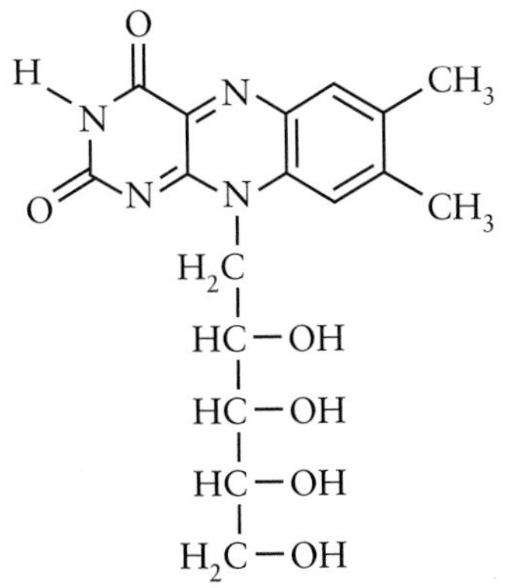

Riboflavin

a. Circle the functional groups that are responsible for the water solubility of this vitamin.
b. In the reaction between riboflavin and 2 ATP to make FAD, why are 2 ATP required?
10.69 Is the conversion of FAD into $FADH_2$ an oxidation or a reduction? Is FAD gaining or losing electrons in going to $FADH_2$? From what vitamin is FAD derived?
10.70 What vitamin deficiency in newborns is caused by phototherapy?
10.71 Name the four fat-soluble vitamins, and explain what is meant by "fat soluble."
10.72 Why does the body need to replenish water-soluble vitamins frequently?
10.73 What is a vitamin deficiency? Are vitamin deficiencies caused only by insufficient vitamins in the diet? What are the symptoms of a riboflavin deficiency?

Answers to Practice Exercises

10.1 a.

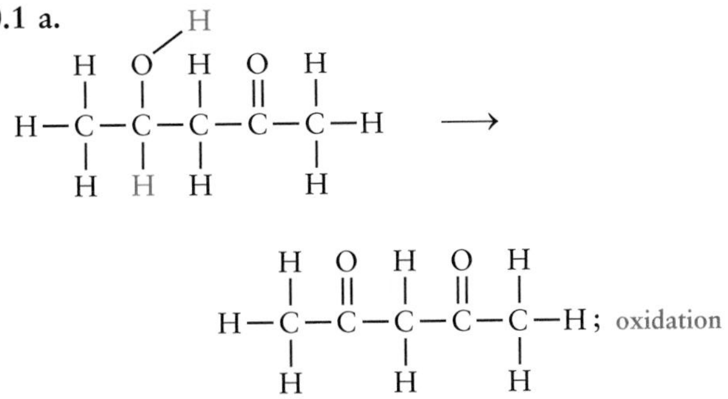

b.

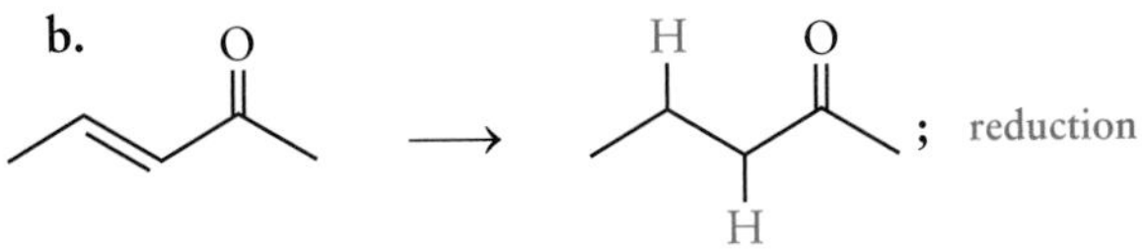

c.

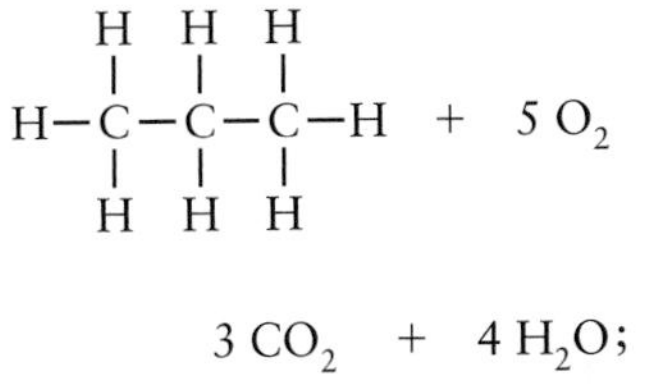

$$3\ CO_2\ +\ 4\ H_2O;$$ Combustion reaction, a type of oxidation

d.

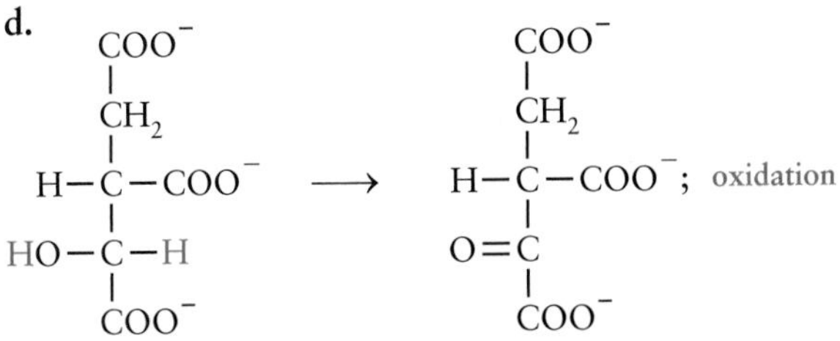

10.2 Reduction: $2\ Ag^+(aq) \rightarrow 2\ Ag(s)$; Oxidation: $Cu(s) \rightarrow Cu^{2+}(aq)$

10.3 a.

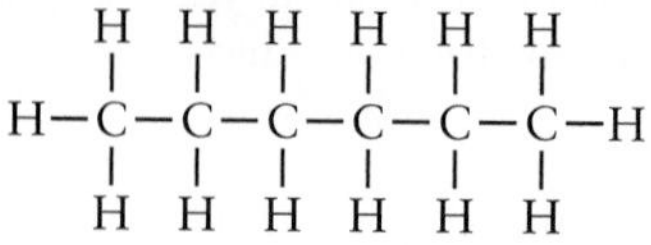

b.

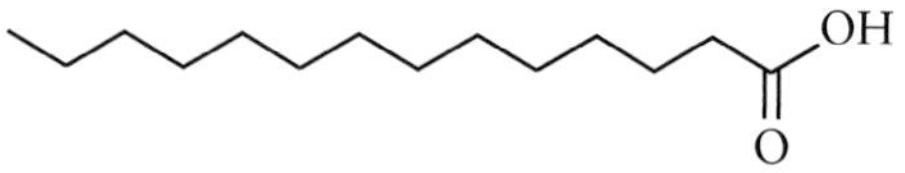

c. FAD

10.4 a.

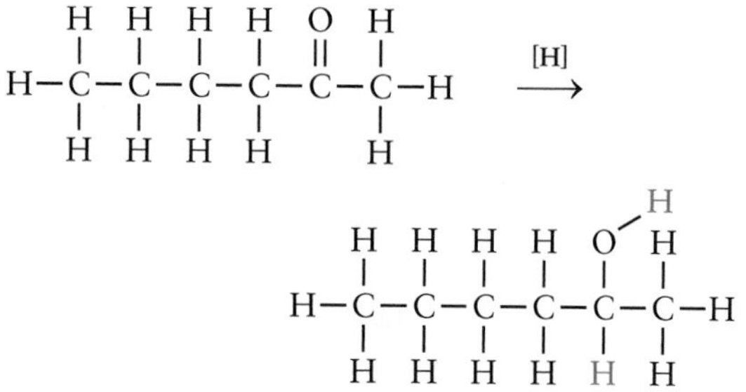

b.

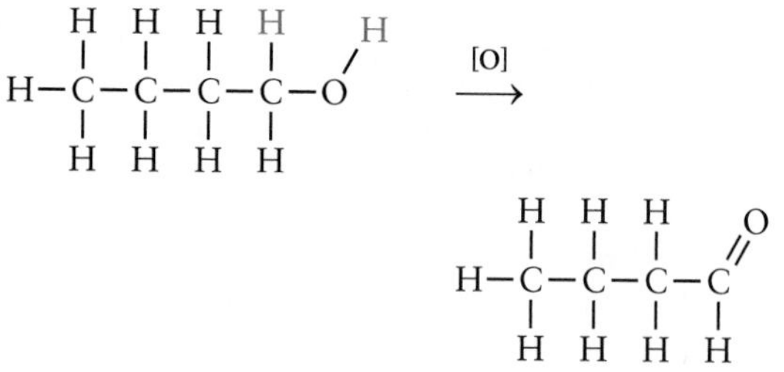

c. No oxidation occurs because the reactant is a tertiary alcohol.

10.5 a.

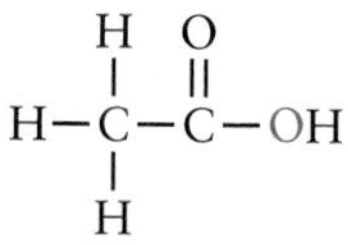

b.

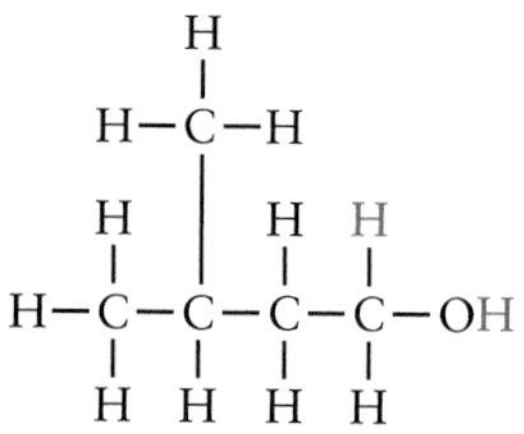

c.

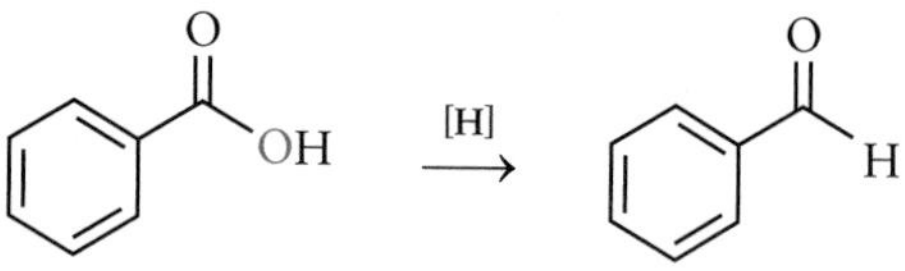

10.6

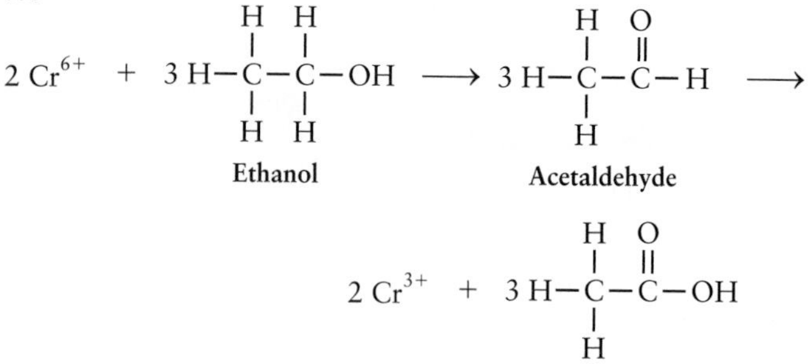

Ethanol

Acetaldehyde

$2\ Cr^{3+}\ +\ 3$

Acetic acid

10.7

$H_2O\ +$

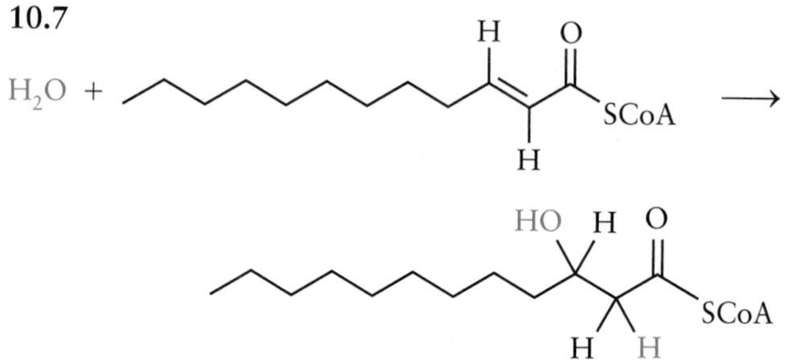

10.8 a. hydration

b.

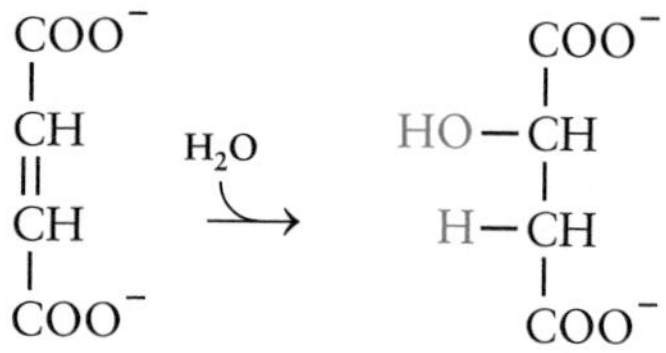

Step 7

c.

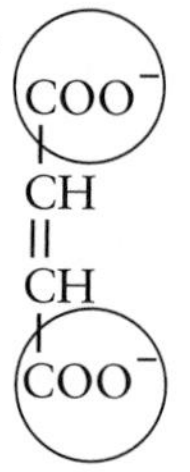

Carboxylate ions (conjugate base of a carboxylic acid)

d. It is a hydration because water is a *reactant*, and an OH group is added to form the product.

10.9.

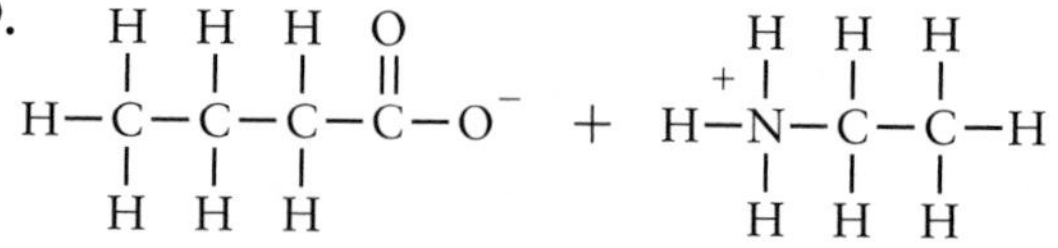

10.10 a. Acyl group

Bond broken

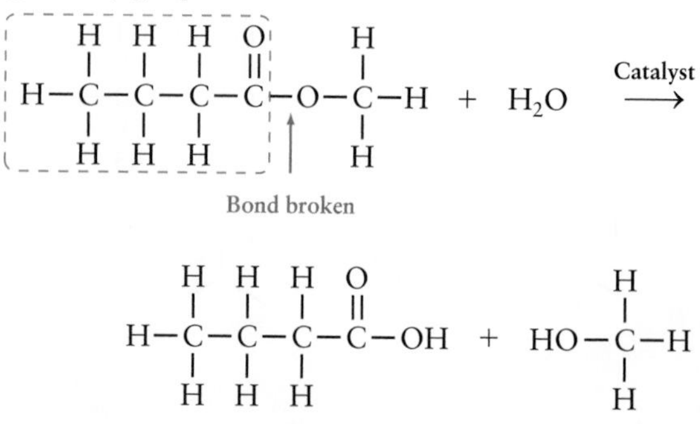

Butanoic acid

Methanol

b.

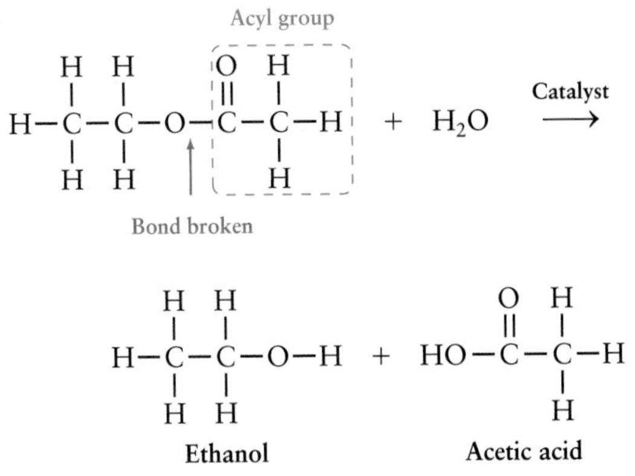

c. Acyl group

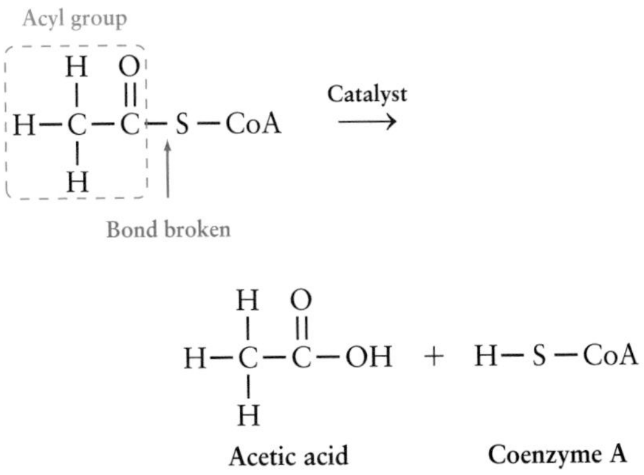

10.11

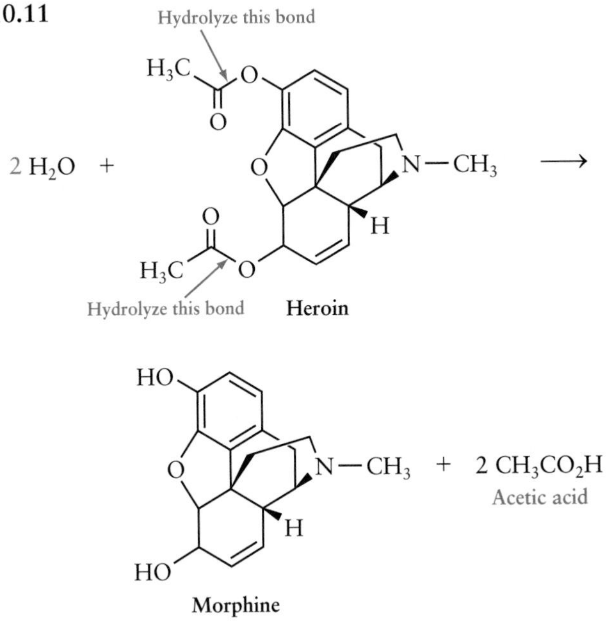

10.12 In the first reaction the aromatic ring is part of the carboxylic acid (benzoic acid) product and in the second reaction the aromatic ring is part of the phenol product.

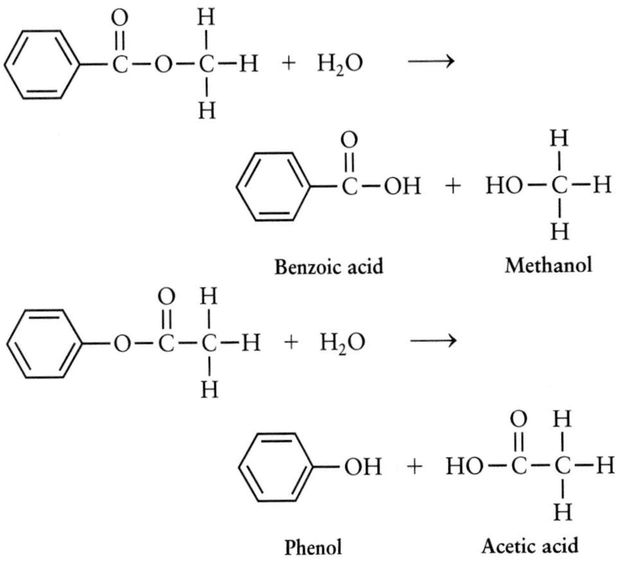

10.13

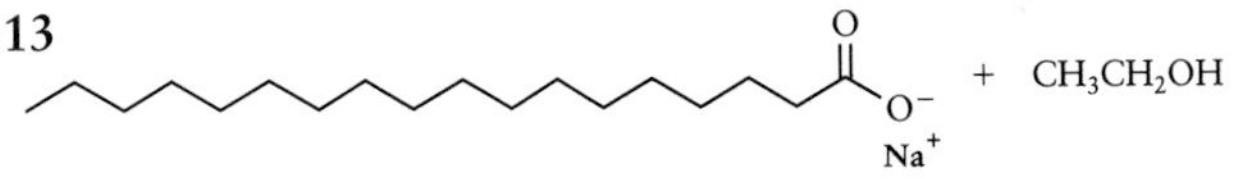

10.14

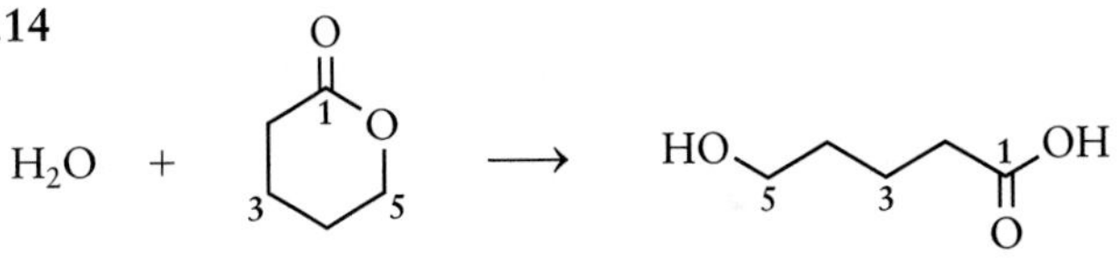

10.15

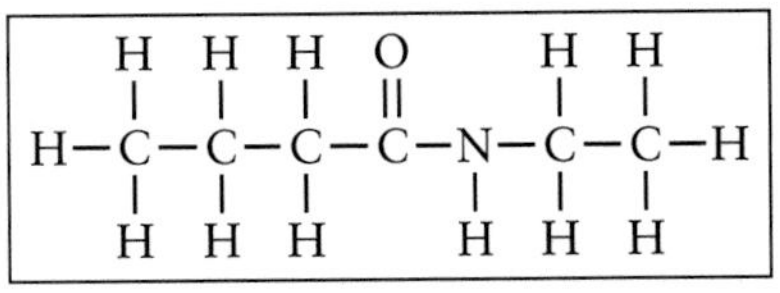

10.16 a.

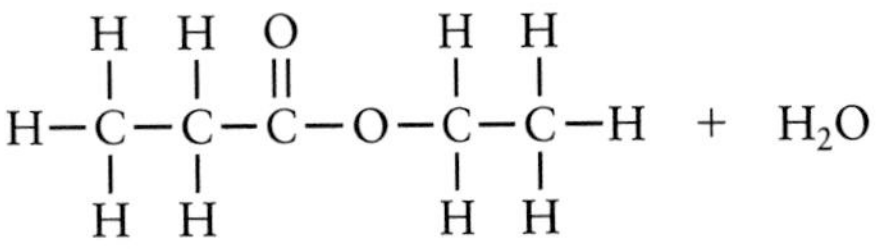

b.

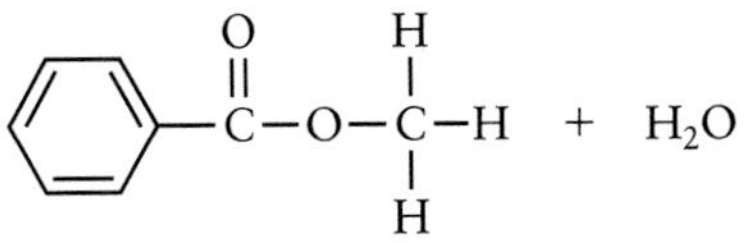

10.17 a., d.

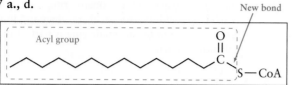

Thioester

b. The reactants are the fourteen-carbon-containing fatty acid and coenzyme A. Coenzyme A is shown next to a curved arrow because the fatty acid is the main reactant.

c. The reaction is different only in that a thiol RSH is a reactant instead of an alcohol, ROH; in other words, sulfur instead of oxygen. Note that sulfur is directly below oxygen on the periodic table, and therefore has the same Lewis dot structure.

10.18.   It would smell like bananas:

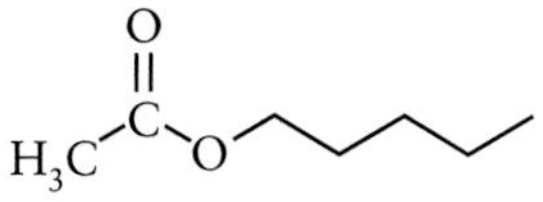

10.19.   *High-energy bond* refers to the high chemical potential energy in this covalent bond, released when hydrolyzed.

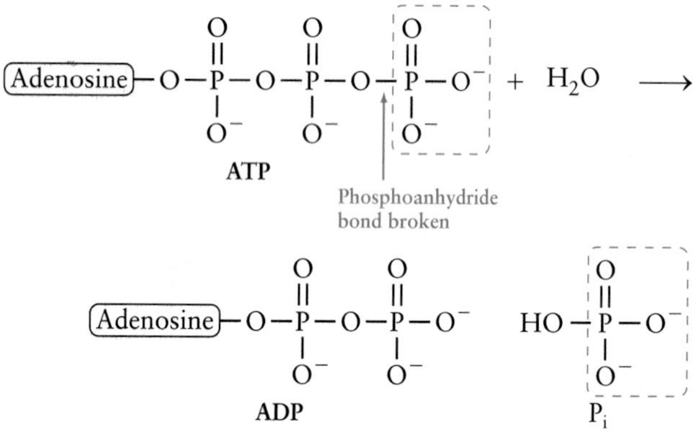

10.20. a. This reaction is considered a phosphoryl transfer reaction because a phosphoryl group migrates from one alcohol oxygen (C-3) to another alcohol oxygen (C-2) atom in glycerate.

b. It is also considered an isomerization reaction, because the reactant and product are structural isomers: same molecular formula, but different connectivity of atoms. The phosphoryl group is attached to different hydroxyl groups.

c. A phosphate ester is converted into an alcohol and a different alcohol is converted into a phosphate ester.

10.21.   A phosphoryl group is transferred from the phosphate ester in 2-phosphoglycerate to the diphosphate group of ADP.

Proteins: Structure and Function

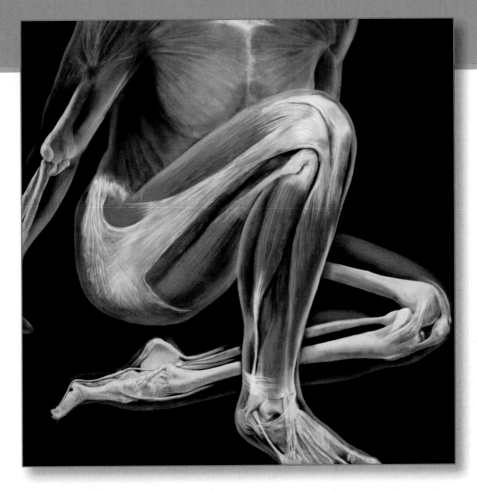

Proteins have some of the most diverse functions of all biological molecules, ranging from the hemoglobin that transports oxygen to tissues, to collagen and elastin that provide structure to ligaments, tendons, and blood vessels, to the enzymes that catalyze all biochemical reactions. [Anatomical Travelogue/Photo Researchers, Inc.]

OUTLINE

This icon indicates that a **Problem-Solving Tutorial** is available at www.whfreeman.com/gob

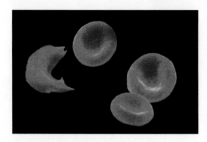

Sickle-cell anemia is characterized by crescent-shaped red blood cells, seen on the left, caused by a variant form of hemoglobin, an important protein that transports oxygen to tissues throughout the body. [© Dr. Stanley Flegler/Visuals Unlimited]

Figure 11-1 Hemoglobin, a protein composed of two identical α-chains and two identical-β chains, transports oxygen to tissues throughout the body. The α-chains are colored orange and yellow; the β-chains are colored blue and light blue.

Valine

Glutamic acid

Sickle-cell Anemia: One Wrong Amino Acid

Sickle-cell anemia is the most common inherited blood disorder in the United States. About 70,000 Americans are estimated to have the disease and 2,000,000 are believed to be carriers of the disease. One in 500 African Americans has sickle-cell anemia. The disease gets its name from the crescent (sickle) shaped red blood cells that characterize the disease, shown in the photo in the margin. Normal red blood cells have a disc shape that is concave on both sides.

Red blood cells contain hemoglobin, a type of large biological molecule known as a protein. Sickle-cell anemia is caused by an abnormal form of hemoglobin. The function of hemoglobin is to transport oxygen (O_2) to tissues throughout the body. In oxygen-depleted tissues, red blood cells carrying this abnormal form of hemoglobin become misshapen. The sickle cells aggregate and stick together, clogging delicate blood vessels and restricting blood flow, which may lead to tissue damage, severe pain, and organ loss. Individuals with sickle-cell anemia are prone to recurrent episodes of pain and tissue damage, called "sickle-cell crises." Furthermore, these abnormal cells have a much shorter life span than a normal red blood cell—15 days instead of 120 days. The number of red blood cells declines, resulting in anemia, a condition characterized by a lack of oxygen-carrying capacity that causes fatigue and lethargy.

In Chapter 1 you learned about the role of the iron (Fe) atom in a hemoglobin molecule and how a shortage of iron in the body can lead to iron deficiency anemia. In sickle-cell anemia, the problem lies with the protein portion of the hemoglobin molecule. A protein is a large molecule formed from many small molecules called amino acids. When a sequence of amino acids is connected head-to-tail they form a large molecule known as a polypeptide. When the number of amino acids in a polypeptide exceeds 50, the molecule is referred to as a protein. Hemoglobin is a protein composed of four polypeptide chains: two identical α-chains, each containing 141 amino acids, and two identical β-chains, each containing 146 amino acids, as illustrated in Figure 11-1.

In the abnormal form of hemoglobin that causes sickle-cell anemia, there is an error in the sixth amino acid of the 146-amino-acid β-chains. An amino acid known as *valine* appears where normally the amino acid *glutamic acid* should appear. This small change at the molecular level causes the protein to assume an entirely different overall three-dimensional shape, which leads to a dramatic difference in the shape of red blood cells, observable at the microscopic level, causing them to stick together when in a low-oxygen environment.

Although there is no cure for sickle-cell anemia, there are treatments for its symptoms:

- Antibiotic drugs, especially when taken in the first 5 years of life, prevent life-threatening pneumonia infections.
- Blood transfusions reduce anemia and pain by adding normal red blood cells to the circulation.
- The drug hydroxyurea is effective at reducing the incidence of "crises."

Sickle-cell anemia is just one of many inherited diseases caused by an abnormally shaped protein arising from an error in the amino acid sequence of the protein. In this chapter you will learn about amino acids and the structure and function of proteins. You will see that of all the biomolecules, proteins have the most diverse functions, from the transport of oxygen to the catalysis of chemical reactions.

Some of the most important biological molecules—biomolecules—are large complex organic molecules—macromolecules. Biomolecules are classified into four categories, according to their chemical structure and solubility characteristics:

- proteins,
- carbohydrates,
- lipids, and
- nucleic acids (DNA and RNA)

Your background in organic chemistry should provide you with the tools to recognize the structural features that characterize these biomolecules. All these biomolecules undergo the reactions of functional groups described in Chapter 10, and they obey the principles of chemical reactions in general, described in Chapter 8.

In the next five chapters you will study in detail the structure and function of each class of biomolecules listed above, including the reactions they undergo as part of metabolism. In this chapter you will focus specifically on the structure and function of **proteins**, the most abundant of all the biomolecules.

Proteins have the most diverse functions of all the biomolecules. They can act as

- enzymes, which catalyze all the reactions involved in metabolism;
- receptors, located within the cell and on the surface of cell membranes, which signal cellular events;
- structural proteins, which provide physical support to tissues such as skin, muscle, blood vessels, hair, and tendons;
- immunoglobulins, the antibodies that defend the body against infectious agents;
- transport proteins, such as hemoglobin and myoglobin that deliver oxygen to cells; and
- dietary proteins that provide important compounds used for building nitrogen-containing biomolecules and energy for the cell when all other sources are unavailable.

To understand the functions of the various proteins, you must first study their chemical structure. We begin with the basic building blocks of all proteins—amino acids.

11.1 | Amino Acids

Whether a protein is from a simple bacterium, a frog, or a human being, all proteins are constructed from the same set of 20 amino acids. As their name indicates, **amino acids** contain both an *amine* functional group ($-NH_2$) and a *carboxylic acid* functional group ($-COOH$), which are both attached to the same carbon atom, known as the **α-carbon**, as shown in Figure 11-2.

Also attached to the α-carbon is an R group, or side chain, which ranges from a simple hydrogen atom in glycine (R = H), to carbon chains of varying length and branching, and chains containing other functional groups. The structure of the R group is different for each amino acid and determines its identity. There are 20 different natural amino acids, each with a different R group.

Amino Acid Equilibria and pH

Since the amine group of an amino acid is basic and the carboxylic acid group is acidic, the charge on these groups depends on the pH of the solution. Recall from Chapter 9 that the pH of the cell, known as physiological pH, is about 7.3.

Figure 11-2 The parts of an amino acid.

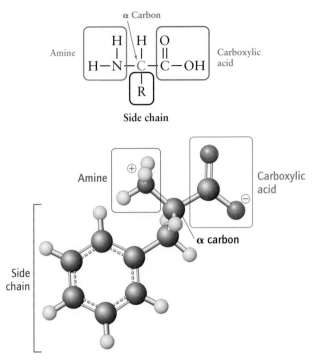

Phenylalanine (Phe)

Recall from Chapter 9 that when the pH of a solution is less than 7, the concentration of H_3O^+ exceeds the concentration of OH^-. When the pH is greater than 7, the reverse is true. At pH = 7 the concentration of OH^- is equal to the concentration of H_3O^+.

At physiological pH both these functional groups are ionized: The amine is in its conjugate acid form (RNH_3^+) and the carboxylic acid is in its conjugate base form ($RCOO^-$). A compound containing both a positive and a negative charge is known as a **zwitterion.** A zwitterion has a net neutral charge.

If the pH of solution changes, a proton is accepted by or donated to one of these two functional groups. At low pH, the carboxylate group ($RCOO^-$) gains a proton to become a neutral carboxylic acid ($RCOOH$). At high pH, the NH_3^+ group loses a proton to become a neutral amine (RNH_2). The three different pH-dependent forms of an amino acid are in equilibrium, as shown in Figure 11-3. When the pH goes up (increase in OH^-; decrease in H^+) the equilibrium shifts to the right and when the pH goes down (decrease in OH^-; increase in H^+) the equilibrium shifts to the left, as shown in Figure 11-3, in accordance with Le Chatelier's principle.

Amino Acid Side Chains

Figure 11-4 highlights the side chains, or R groups, of the 20 natural amino acids. The various side chains can be classified as either nonpolar or polar (Figure 11-4). The nonpolar side chains consist of alkanes and aromatic

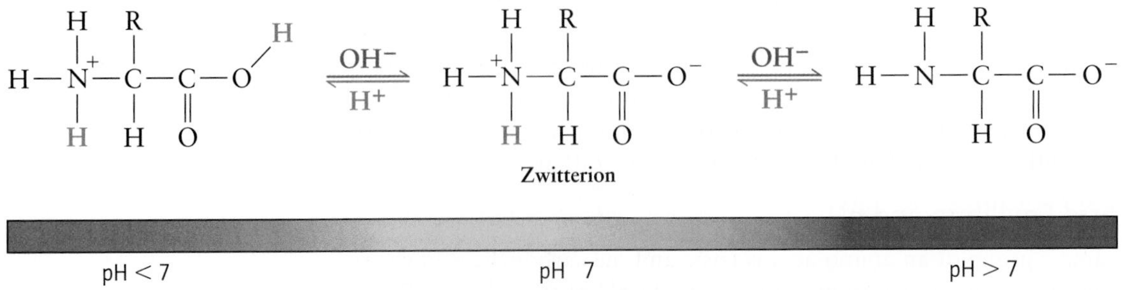

Zwitterion

pH < 7   pH 7   pH > 7

Figure 11-3 The three different pH-dependent forms of an amino acid are in equilibrium.

NONPOLAR

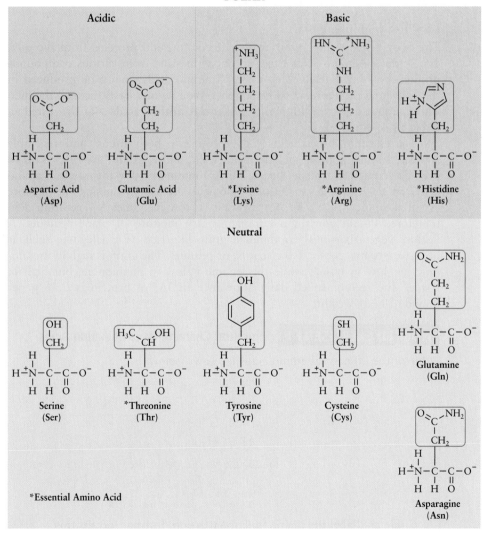

POLAR

*Essential Amino Acid

Figure 11-4 The 20 natural amino acids. The side chains are shown in red boxes.

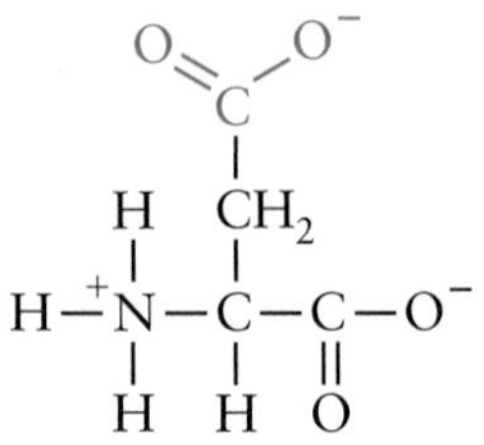

Aspartic acid
(Asp)

*Lysine
(Lys)

thiol

Cysteine

hydrocarbons, or they contain sulfur and nitrogen atoms in nonpolar covalent bonds (tryptophan, methionine).

The polar side chains can be further subdivided into acidic, basic, and neutral side chains. Two of the 20 amino acids, aspartic acid and glutamic acid, have acidic side chains because they contain a carboxylate functional group ($-COO^-$) as part of their R group. At physiological pH, aspartic acid and glutamic acid, like most carboxylic acids, exist in their conjugate base form, and therefore have a negative charge, as shown in the margin for aspartic acid.

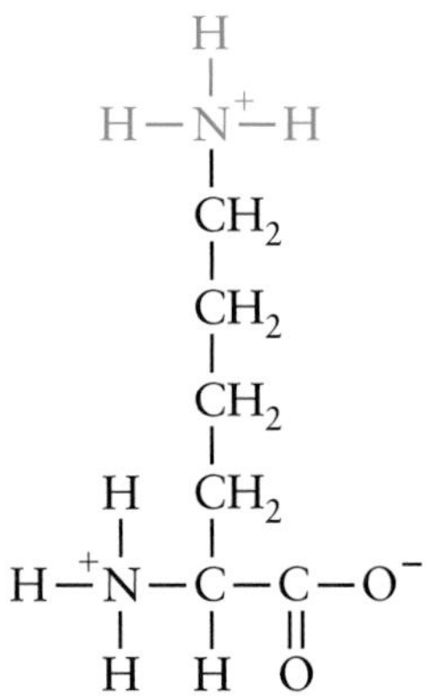

Three of the 20 amino acids have a basic side chain: These are lysine, arginine, and histidine. Lysine contains an amine, histidine contains an aromatic amine, and arginine contains another type of basic functional group not previously described. At physiological pH, basic functional groups bear a positive charge because they accept a proton and exist in their conjugate acid form, as shown in the margin for lysine.

Six of the 20 amino acids have polar side chains with no charge. These side chains are classified as neutral because in aqueous solution at physiological pH the functional group on the side chain is not ionized—it does not gain or lose a proton. Serine, threonine, and tyrosine contain an alcohol ($-OH$) on their side chain. Asparagine and glutamine contain an amide ($-CONH_2$) on their side chain. Cysteine contains an S—H group, known as a **thiol**, which is also a neutral functional group that is slightly polar.

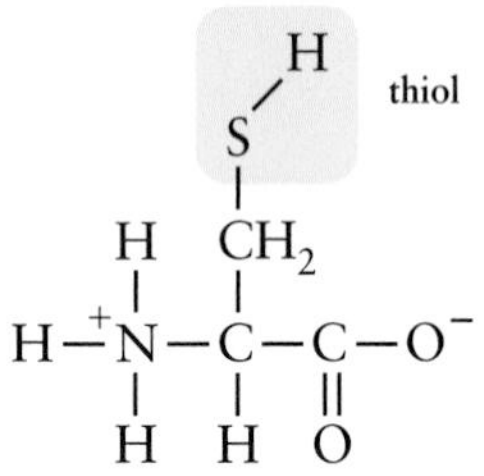

Essential Amino Acids

The body obtains the amino acids it needs to build proteins from the diet. Ten of the 20 amino acids can also be synthesized from various compounds produced in metabolism. The other 10 amino acids cannot be produced in the body, and therefore, must be supplied on a regular basis through the diet. These amino acids are known as **essential amino acids** and are listed in Table 11-1.

The essential amino acids can be found in both animal and vegetable sources. Animal sources include meat, milk, and eggs and are often said to provide "complete protein" because they contain all 10 of the essential amino acids. Single vegetable sources of amino acids are usually missing one or more essential amino acids. Rice, for example, is low in lysine. A vegetarian diet, therefore, should contain a complementary mixture of plant proteins. A healthy vegetarian meal combines a grain like rice with a legume such as beans, soybeans, peas, alfalfa, lentils, or peanuts. The grain is high in methionine but low in lysine, while the legume is low in methionine, but high in lysine. The recommended daily allowance (RDA) of protein is 0.36 g per pound of body weight.

Table 11-1 Essential Amino Acids

*Arginine
Histidine
Isoleucine
Leucine
Lysine
Methionine
Phenylalanine
Threonine
Tryptophan
Valine

*Required in children.

WORKED EXERCISE 11-1 Structural Characteristics of Amino Acids

One of the 20 natural amino acids is shown below.

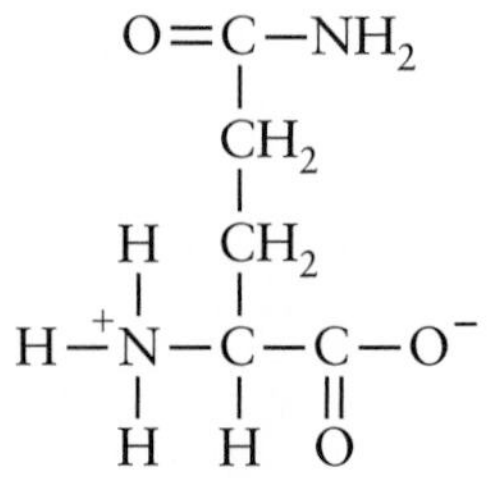

a. Circle and label the amine functional group. In what form is this functional group at physiological pH?

b. Circle and label the carboxylic acid functional group that is present in all amino acids. In what form is this functional group at physiological pH?

c. Circle and label the side chain. What is the name of this amino acid?

d. Does the side chain contain a functional group? If so, what is it?

e. Is the side chain *polar* or *nonpolar*? If it is polar, what type of polar group is it: *acidic*, *basic*, or *neutral*? Explain.

f. Write the structure of this amino acid at pH = 1. What is the net charge on the amino acid at this pH?

g. This amino acid is not an essential amino acid. What does this mean?

Essential amino acids must be supplied through the diet by eating cheese, meat, grains, or legumes because they cannot be synthesized by the body. [Rob Bartee/ Alamy]

SOLUTION

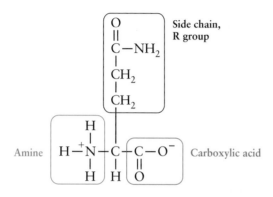

a. Normally an amine is in its conjugate acid form, —NH₃⁺, at physiological pH. See the structure above.

b. Normally a carboxylic acid is in its conjugate base form at pH 7.3, —COO⁻. See the structure above.

c. The R group is —CH₂CH₂CONH₂. See the structure above. This amino acid is glutamine.

d. Yes, the side chain contains an amide functional group.

e. The side chain is polar and neutral, because it contains an amide.

f.

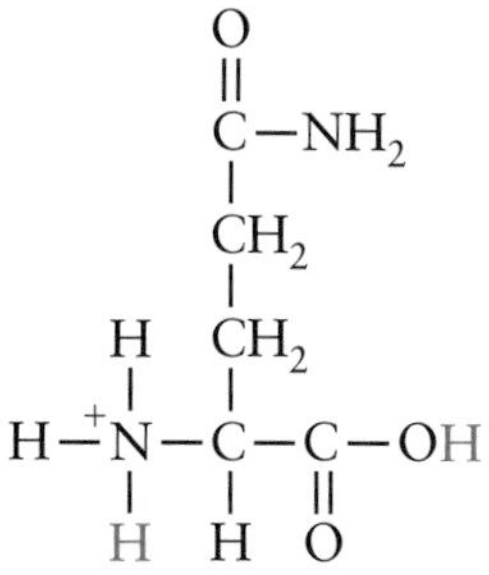

$$
\begin{array}{c}
O \\
\parallel \\
C-NH_2 \\
| \\
CH_2 \\
H \quad | \quad CH_2 \\
| \qquad | \\
H-\overset{+}{N}-C-C-OH \\
| \quad | \quad \parallel \\
H \quad H \quad O
\end{array}
$$

At low pH the amine would be in its conjugate acid form and the carboxylic acid would be in its neutral form. Therefore, the amino acid would have a net +1 charge.

g. The amino acid can be synthesized by the cell from compounds found in metabolism and does not need to be supplied through the diet.

PRACTICE EXERCISES

11.1 One of the 20 natural amino acids is shown below.

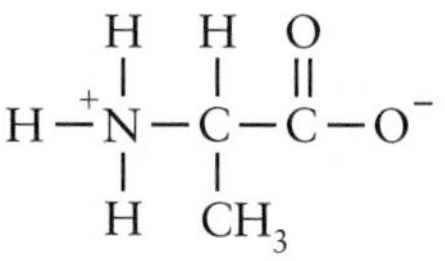

$$
\begin{array}{c}
H \quad H \quad O \\
| \quad | \quad \parallel \\
H-\overset{+}{N}-C-C-O^- \\
| \quad | \\
H \quad CH_3
\end{array}
$$

a. Circle and label the amine functional group. Is it in its neutral or ionized form?

b. Circle and label the carboxylate functional group. How does this group differ from a carboxylic acid?

c. Is this amino acid a zwitterion? At what pH does an amino acid exist as a zwitterion?

d. Circle and label the side chain. What is the name of this amino acid?

e. Is the side chain nonpolar or polar?

f. Write the structure at pH = 12. What is the net charge on the amino acid at this pH?

11.2 For each of the following amino acids place an "X" in the boxes that apply.

Amino Acid	Nonpolar	Polar	Acidic Side Chain	Basic Side Chain	Neutral Side Chain	Essential Amino Acid
Cysteine						
Lysine						
Glutamine						
Methionine						
Aspartic acid						

11.3 Which two amino acids contain a sulfur atom? Which of these two amino acids contains a thiol functional group?

11.4 One amino acid is unusual in that it contains a ring incorporating the amine functional group of the amino acid as part of its R group. What is the name of this amino acid? Write its chemical structure at physiological pH.

11.5 In the opening vignette you learned that sickle-cell anemia is a result of valine being substituted for aspartic acid as the sixth amino acid on the β-chains of hemoglobin. Write the structures of these two amino acids. Are they polar or nonpolar?

11.6 What single food products contain all the essential amino acids?

11.2 Chirality and Amino Acids

Nineteen of the twenty amino acids exhibit a property known as chirality. **Chirality** is a symmetry property present not only in molecules but in many objects. To understand chirality, you need only consider your hands. What is the difference between your left hand and your right hand? Clearly, you have the same digits on both hands, they even appear in the same order—thumb, forefinger, middle, ring, and pinky finger. Yet your left and right hand are not identical. When you lay one hand on top of the other hand you can never get all components of both hands to perfectly overlay—we say the hands are **nonsuperposable.** Either the palm and top-of-hand don't overlay or the digits don't overlay—in other words, the hands are not identical. Yet the left hand is somehow related to the right hand: They are mirror images of one another (Figure 11-5). The mirror image of an object is the reflection you see when you hold it up to a mirror. *An object is **chiral** if it is non-superposable on its mirror reflection, like the left and right hand.* An object is **achiral** if it is superposable on its mirror reflection; object and reflection are identical.

In your everyday life you have encountered objects that are chiral and objects that are achiral. A glove, a corkscrew, scissors, and shoes are all chiral objects.

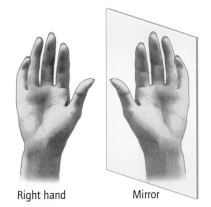

Right hand   Mirror

Figure 11-5 Chirality is a property exhibited by an object or a molecule if it is nonsuperposable on its mirror image—like your hands. Chirality is the Greek word for handedness.

Chiral objects:

Achiral objects:

Enantiomers

Molecules can also be chiral. All the amino acids except glycine, most carbohydrates, and over half of all drug molecules are chiral. A chiral molecule is nonsuperposable on its mirror image. A chiral molecule and its mirror image molecule are examples of **stereoisomers,** *molecules with the same chemical formula and connectivity of atoms but a different spatial arrangement of atoms.* Two nonsuperposable mirror image stereoisomers are known as a pair of **enantiomers,** one type of stereoisomer.

Consider, for example, the amino acid alanine shown in Figure 11-6. Since the only difference between a pair of enantiomers is the three-dimensional orientation of its atoms, we can use the dashed and wedged bonds first described in Chapter 3 to depict the different three-dimensional orientations of their atoms. In alanine, for example, you can see that when the carboxylate and CH_3 groups on the α-carbon are aligned, the NH_3^+ and H atoms do not superpose. *When evaluating whether or not a molecule is superposable on its mirror image, consider all conformations of the molecule, but do not break any bonds.* Perform *The Model Tool 11-1: Enantiomers* to get hands-on experience with chirality in molecules.

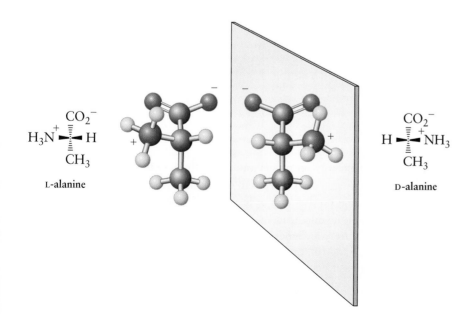

Figure 11-6 L-alanine and D-alanine are enantiomers: stereoisomers that are nonsuperposable mirror images. L-alanine is found in nature; D-alanine is not.

The Model Tool 11-1 Enantiomers

I. Construction of an Achiral Molecule

1. Obtain 6 black carbon atoms, 4 red oxygen atoms, 16 hydrogen atoms, 22 straight bonds, and 4 bent bonds.

2. Construct a model of the structure shown below, which represents 2-propanol.

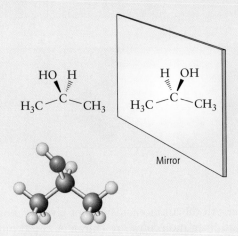

3. Imagine that there is a mirror next to your model and make a model of its mirror image.

4. See if the two models you constructed are superposable. You may rotate as many bonds as necessary, assuming any conformation, but do not break any bonds.

   a. Is your model superposable on its mirror image?
   b. Based on your answer to (a), is 2-propanol chiral?
   c. Based on your answer to (a) and (b), does 2-propanol have a stereoisomer?
   d. What is the relationship between your two models: *identical* or *enantiomers*?
   e. Do both molecules have the same molecular formula? Do both molecules have the same connectivity of atoms? Do both models have the same spatial orientation?

II. Construction of Alanine

1. Make a model of L-alanine using 1 blue nitrogen atom, 3 black carbon atoms, 7 light-blue hydrogen atoms, and 2 red oxygen atoms. Be sure to place the atoms in the three-dimensional arrangement shown.

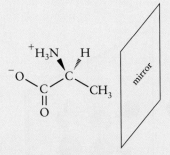

2. Imagine that there is a mirror next to your model and make a model of the mirror image.

3. See if the two models you constructed are superposable. You may rotate as many bonds as necessary, assuming any conformation, but do not break any bonds.

   a. Is your model superposable on its mirror image?
   b. Based on your answer to (a), are your models chiral?
   c. Based on your answers to (a) and (b), are these stereoisomers?
   d. What is the relationship between your two models: *identical* or *enantiomers*?
   e. Do both molecules have the same molecular formula? Do both molecules have the same connectivity of atoms? Do both models have the same spatial orientation?

In the IUPAC naming system, the prefixes D- and L- can be used to distinguish two amino acid enantiomers. Therefore, the IUPAC names of the two enantiomers of alanine are L-alanine and D-alanine. Interestingly, only L-amino acids are found in nature. Because proteins are assembled from chiral building blocks, they too are chiral.

Other prefixes in common use are listed in Table 11-2. Some of these prefixes (R and S and D and L) arise from a set of IUPAC rules that are based on chemical structure, while other prefixes, such as + and − and d and l, are assigned based on experimental measurements.

Fischer Projections

A simplified way of writing enantiomers is to show them as Fischer projections. In a **Fischer projection,** a tetrahedral carbon atom with four different atoms or groups attached to it is shown as a crosshair, ✛, and the main carbon chain is written vertically. The carboxylic acid, or most oxidized functional group, is placed at the top of the structure. Therefore, when writing an amino acid as a Fischer projection, the α-carbon appears at the cross-hair, the amine group and the hydrogen atom appear on the horizontal bonds, the carboxylate appears on the vertical bond at the top, and the side chain on the vertical bond at the bottom, as shown in Figure 11-7 for alanine. In a Fischer projection, an L-amino acid has the amine on the *left* and the H atom on the right, while a D-amino acid has these two groups reversed.

Properties of Enantiomers

Enantiomers have identical physical and chemical properties when placed in an achiral environment such as most laboratory conditions. Consequently, enantiomers are very difficult to separate, and single enantiomers are more difficult to synthesize in the laboratory. *However, when enantiomers are placed in a chiral environment, such as the body, they often exhibit profoundly different chemical properties.* For example, when a drug molecule is chiral, the two enantiomers may cause different physiological effects. Consider, for example, the two enantiomers of the morphine analog propoxyphene: Darvon and Novrad. Darvon acts as an analgesic, while Novrad acts as an antitussive (cough suppressant).

Table 11-2 Common Prefixes
Used to Identify Enantiomers

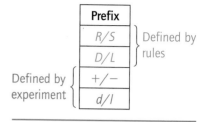

Plane polarized light, a special type of light similar to the light that comes through sunglasses, can be used to distinguish a pair of enantiomers. One enantiomer will exhibit a clockwise rotation (+) of the plane of the incoming light and the other a counterclockwise rotation (−) of the plane of the light. The magnitude of the rotation, however, will be the same for both enantiomers.

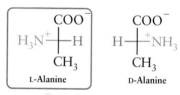

Natural amino acid

Figure 11-7 Fischer projections of L-alanine and D-alanine.

Enantiomers Darvon and Novrad

(+)-Propoxyphene (Darvon)
analgesic

(−)-Propoxyphene (Novrad)
cough suppresent

Often, particularly in man-made drugs, a chiral substance will be produced as a 50:50 mixture of enantiomers, known as a **racemic mixture.** Racemic mixtures are designated with prefixes separated by a slash: R/S, d/l, D/L, or +/−. When the medicinal value of a drug resides in only one enantiomer, the other enantiomer is either inactive or possibly even harmful. When one of the enantiomers displays adverse physiological effects, only the single enantiomer can be used as a pharmaceutical. For example, the drug L-dopa, used to treat Parkinson's disease, is administered as a single enantiomer drug.

More than half of all drugs have a single enantiomer as the active pharmaceutical ingredient.

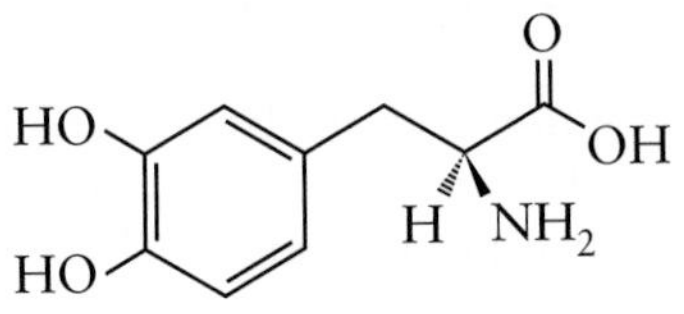

L-Dopa

WORKED EXERCISE 11-2   Chiral Molecules

The Fischer projection of one of the natural amino acids is shown below.

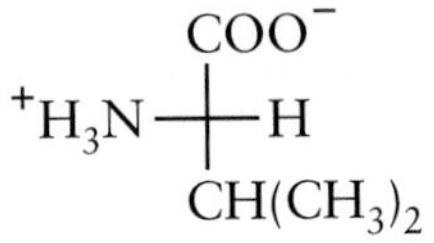

a. Which amino acid is this?
b. Is this amino acid D- or L-? Explain.
c. Place an arrow pointing to the α-carbon.
d. Is this molecule chiral? Explain.
e. Write the Fischer projection of the enantiomer of this amino acid. How would it be named?
f. A 50:50 mixture of this amino acid and its enantiomer is known as a _____ mixture.

SOLUTION
a. This is L-valine.
b. This is an L-amino acid because the amine is on the left when written as a Fischer projection.
c.

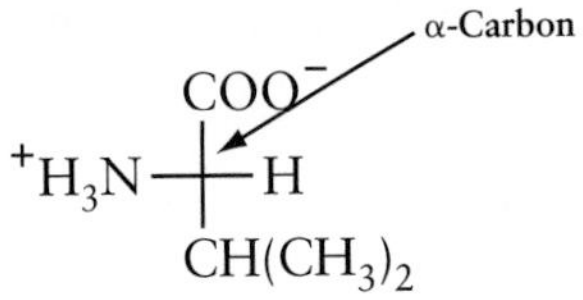

L-Valine

d. The molecule is chiral because it is nonsuperposable on its mirror image; it has an enantiomer.
e.

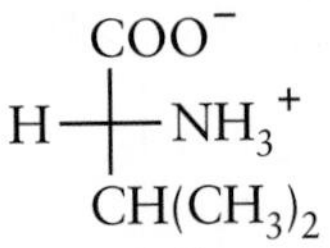

D-Valine

f. racemic

PRACTICE EXERCISES

11.7 Indicate whether the following objects are chiral or achiral:

    a. a golf club
    b. a baseball bat
    c. a grand piano
    d. a sock
    e. a glove
    f. a basketball

11.8 Indicate whether the following statements are *True* or *False*.

A chiral molecule

  a. is superposable on its mirror image.

  b. has an enantiomer.

  c. may exhibit different chemical properties from its enantiomer in the body.

  d. typically exhibits different chemical properties from its enantiomer in an achiral environment.

11.9 The molecule shown below is D-glyceraldehyde. What does the D in the name indicate about this molecule? What functional groups does this molecule contain? Write the structure of L-glyceraldehyde.

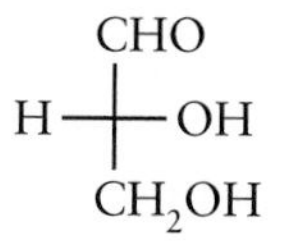

$$\begin{array}{c} \text{CHO} \\ \text{H} \!-\!\!\!\!\!\begin{array}{|c|} \hline \\ \hline \end{array}\!\!\!\!\!-\!\text{OH} \\ \text{CH}_2\text{OH} \end{array}$$

D-Glyceraldehyde

11-10 Ibuprofen, the active ingredient in Motrin and other over-the-counter analgesics, is a chiral drug that is sold as a racemic mixture. What does this mean?

11.3 | Peptides

Peptides are molecules composed of two or more amino acids. A peptide containing two amino acids is also known as a **dipeptide;** one containing three amino acids, a **tripeptide;** and so forth. Peptides containing more than about 12 amino acids are known as **polypeptides.** A polypeptide with more than about 50 amino acids folded into its active 3-D structure is referred to as a **protein.**

The Peptide Bond

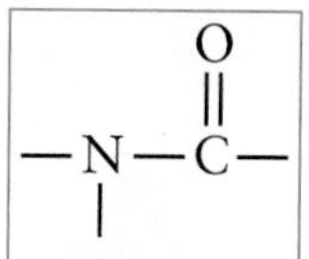

A dipeptide is formed when two amino acids are joined by a covalent bond. To form the new covalent bond, known as a **peptide bond,** the carboxylic acid of one amino acid and the amine of another amino acid react to form an amide functional group, as shown in Figure 11-8. A molecule of water is expelled as part of the reaction. This type of reaction is analogous to the reaction between a carboxylic acid and an alcohol to produce an ester, described in Chapter 10 (acyl transfer reactions).

The reverse reaction—forming two amino acids from the reaction of a dipeptide with water—is a hydrolysis reaction. In Chapter 10 you learned that esters and amides can be hydrolyzed in the presence of water to produce

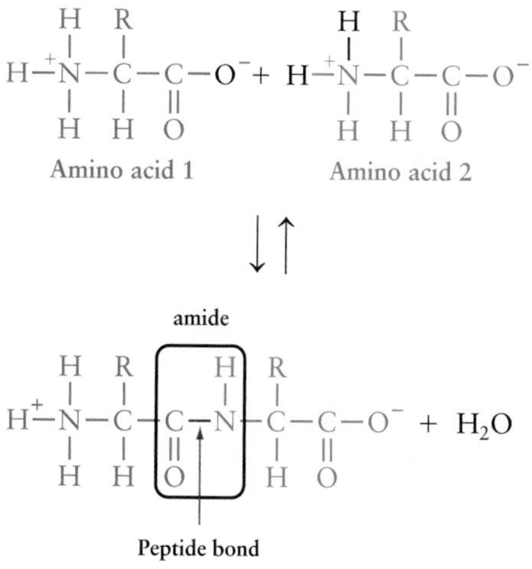

Amide

Amino acid 1   Amino acid 2

amide

Peptide bond

Figure 11-8 A dipeptide is formed when two amino acids react to form an amide functional group. The new C—N bond is known as a peptide bond.

Figure 11-9 A tripeptide is composed of three amino acids joined by two peptide bonds.

Amino acid 1   Amino acid 2   Amino acid 3

$$\text{N-terminus} \longrightarrow H{-}\overset{+}{\underset{H}{N}}{-}\overset{H}{\underset{H}{C}}{-}\overset{R}{\underset{O}{C}}{-}\overset{}{\underset{}{N}}{-}\overset{H}{\underset{H}{C}}{-}\overset{R}{\underset{O}{C}}{-}\overset{}{\underset{}{N}}{-}\overset{H}{\underset{H}{C}}{-}\overset{R}{\underset{O}{C}}{-}O^{-} \longleftarrow \text{C-terminus}$$

Peptide bonds

a carboxylic acid and either an alcohol in the case of an ester or an amine in the case of an amide. When the hydrolysis reaction involves the amide of a dipeptide, it splits the dipeptide into two free amino acids, breaking the peptide bond.

Small Peptides

The reaction of a dipeptide with a third amino acid produces a **tripeptide:** three amino acids linked together via two peptide bonds, as shown in Figure 11-9. Multiple amino acids can link together by peptide bonds to form chains of amino acids of any length.

Every amino acid is capable of forming two peptide bonds, one with its amine functional group and one with its carboxylic acid functional group. Only the first and last amino acids in a peptide have a free amine or carboxylate group. The end of a peptide containing the free amine group is known as the **N-terminus,** and the end containing the free carboxylate group is known as the **C-terminus.**

The convention for writing peptides is to list the sequence of amino acids starting from the N-terminus and continuing in order to the C-terminus. Each amino acid has a three-letter abbreviation and a one-letter abbreviation; either can be used when writing the amino acid sequence for a peptide or protein. The three-letter abbreviations for the 20 amino acids acid are given in Figure 11-4. For example, the amino acid sequence Ala-Gly-Val specifies a tripeptide that has alanine as the N-terminus and valine as the C-terminus. The amino acids are linked by peptide bonds in the order alanine, glycine, valine, as shown in Figure 11-10.

Nursing stimulates the release of oxytocin, a peptide. [Blend/Punchstock]

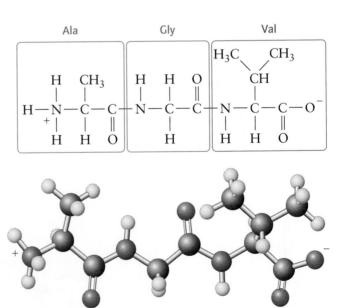

Figure 11-10 The tripeptide Ala-Gly-Val.

Many small peptides are important hormones. For example, oxytocin, which signals the start of labor at the end of pregnancy and promotes lactation, is formed from nine amino acids. The endorphins, introduced in Chapter 7, are derived from five amino acids; these hormones are released during injury or intense physical activity in order to reduce pain. To gain more familiarity with small peptides perform *The Model Tool 11-2: Building The Dipeptide Aspartame—Nutrasweet.*

The Model Tool 11-2 Building the Dipeptide Aspartame (Nutrasweet)

To perform this exercise you will need to work with a partner. Use one kit for Construction Exercise I and the other kit for Construction Exercise II. Together, perform Construction Exercises III and IV, and then answer the Inquiry Questions.

I. Construction of Aspartic Acid

1. Obtain 4 black carbon atoms, 8 light-blue hydrogen atoms, 4 oxygen atoms, 1 blue nitrogen atom, 11 straight bonds, and 4 bent bonds.

2. Construct a model of aspartic acid. Be sure to build the L-amino acid. Use Figures 11-4 and 11-6 as a reference.

3. What form should the amine be in at physiological pH? What form should the carboxylic acids be in at physiological pH? Assemble your model in this form.

II. Construction of Phenylalanine

4. Obtain 9 black carbon atoms, 11 light-blue hydrogen atoms, 1 nitrogen atom, 2 oxygen atoms, 8 bent bonds, and 19 straight bonds.

5. Construct a model of phenylalanine similar to the one shown at right. Be sure to form the L-amino acid.

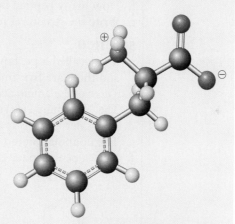

Phenylalanine (Phe)

III. Construction of the Dipeptide Asp-Phe

6. Join the two amino acids from Construction Exercises I and II to make the dipeptide Asp-Phe. You should combine the amine of one amino acid with the carboxylic acid of the other amino acid to make an amide. Think carefully about which amine and which carboxylic acid pairing to use so that you have Asp-Phe and not Phe-Asp. In forming the peptide bond, you will eliminate a molecule of water in the process: 2 hydrogen atoms and 1 oxygen atom. Write the chemical reaction represented in this exercise.

IV. Formation of the Methyl Ester, Asp-Phe-OCH$_3$

7. Obtain 1 black carbon atom, 3 light-blue hydrogen atoms, and 3 grey straight bonds.

8. Change the C-terminus from a carboxylate (COO$^-$) to a methyl ester (—COOCH$_3$). Now you have a model of aspartame, the artificial sweetener in Nutrasweet.

V. Inquiry Questions

9. Write the structural formula of aspartame.

10. Label the N-terminus of your dipeptide.

11. Label the C-terminus of your dipeptide. The C-terminus in Nutrasweet is a methyl ester; how is this different from a normal C-terminus?

12. Label the peptide bond in your dipeptide.

13. Is your dipeptide chiral? Explain.

14. Individuals with phenylketonuria (PKU) cannot metabolize phenylalanine and therefore must avoid the use of products containing aspartame. Explain why these individuals need to avoid these products. *Hint:* Consider the hydrolysis of Nutrasweet. What are the products?

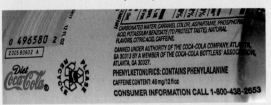

Because they contain aspartame, phenylketonurics should not drink diet soft drinks. [Bianca Moscatelli/W. H. Freeman]

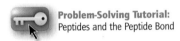

Problem-Solving Tutorial:
Peptides and the Peptide Bond

WORKED EXERCISE 11-3   Peptides

One of the endorphins, mentioned in Chapter 7, is the pentapeptide met-enkephalin, whose amino acid sequence is Tyr-Gly-Gly-Phe-Met.
a. Why is this peptide called a pentapeptide?
b. How many amino acids in this peptide are chiral?
c. What amino acid is at the N-terminus?
d. What amino acid is at the C-terminus?
e. Which amino acids in this peptide are nonpolar?
f. How many peptide bonds are there in this peptide?
g. Write the structural formula of met-enkephalin.

SOLUTION

a. It is a called a pentapeptide because the prefix "penta" indicates five, and there are five amino acids in the peptide.
b. Three of the amino acids are chiral: tyrosine, phenylalanine, and methionine.
c. The N-terminus is tyrosine, which is abbreviated Tyr.
d. The C-terminus is methionine, which is abbreviated Met.
e. The nonpolar amino acids are glycine, phenylalanine, and methionine.
f. Since a pentapeptide contains 5 amino acids, there must be 4 peptide bonds.
g.

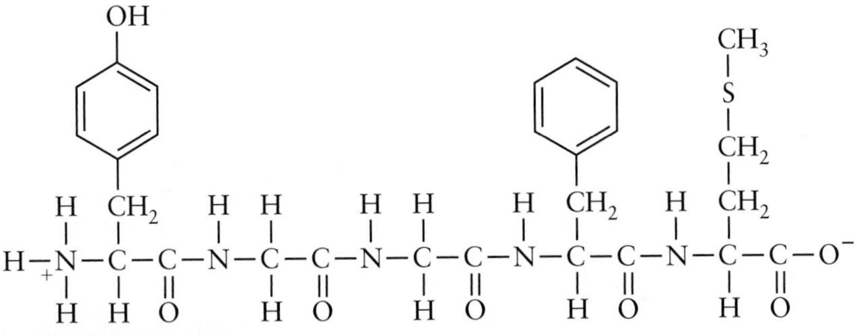

PRACTICE EXERCISES

11.11 A skeletal line structure of leu-enkephalin is shown below:

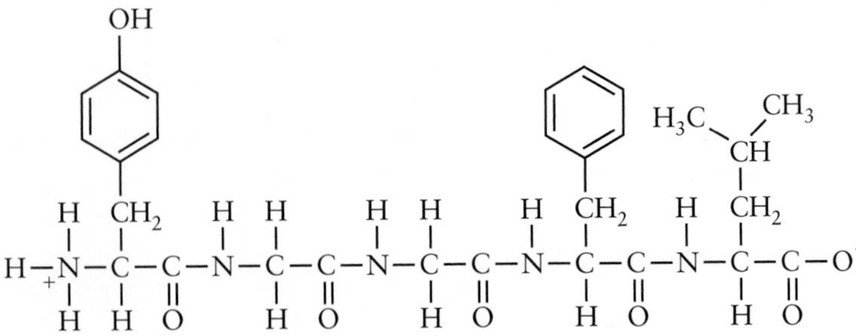

   a. Circle each amino acid in the peptide, and identify it by its three-letter
   abbreviation.
   b. Label the N-terminus and the C-terminus.
   c. What is the amino acid sequence for this peptide?
   d. How many amino acids in this peptide are chiral? Which ones are they?
   e. Identify each peptide bond.
   f. Is this peptide chiral?

11.12 Are the dipeptides Gly-Phe and Phe-Gly different compounds? Are they
   structural isomers or stereoisomers?

11.13 Write the structural formula for the tripeptide Cys-Gly-Val.

11.14 The amino acid sequence for the pregnancy hormone, oxytocin, is given
   below:

   Cys-Tyr-Ile-Gln-Asn-Cys-Pro-Leu-Gly

   a. What is the N-terminal amino acid?
   b. What is the C-terminal amino acid?
   c. How many amino acids does this peptide contain?
   d. How many peptide bonds does this peptide contain?
   e. Which amino acids in this peptide contain a thiol functional group?

11.4 Protein Architecture

Proteins are composed of anywhere from 50 to thousands of amino acids. The
largest protein, titin, is found in muscle tissue such as the heart. It contains
26,926 amino acids! Most proteins, however, contain less than 2,000 amino
acids. In addition, some proteins contain more than one polypeptide chain, such
as hemoglobin, described in the opening vignette.

All proteins are built from the 20 L-amino acids. Scientists estimate that
there could be one million different proteins in humans. How does your body
know in what order to join the amino acids necessary to produce a particular
protein? This information is encoded in your DNA. Every gene on your DNA
codes for at least one protein. The link between genes and proteins accounts
for the connection between genetics and diseases such as sickle-cell anemia,
described in the opening vignette. Since enzymes are proteins, enzyme deficiencies are also determined by your DNA. You will learn more about DNA
and how it serves as a blueprint for building proteins in Chapter 15: Nucleic
Acids: RNA and DNA.

> For a polypeptide composed of n amino acids, there are 20^n possible amino acid sequences.

The Three-Dimensional Shape of Proteins

Although the primary structure of a protein may be written as a linear sequence
of amino acids, the three-dimensional shape of a protein is not linear like
beads on a necklace. Instead, the polypeptide chain folds into a unique three-
dimensional shape, known as its **native conformation.**

A protein must fold into its specific three-dimensional shape in order
to perform its unique function. To accomplish their diverse functions, proteins come in many shapes and sizes. By analyzing the three-dimensional
shape and structure of a protein, scientists can begin to understand how a
particular protein functions, and ultimately how diseases such as sickle-cell
anemia work.

Electrostatic interactions cause proteins to fold. These interactions occur
between side chains within the protein as well as with water and other molecules in the protein's environment. Since the amine and the carboxylic acid of
each amino acid are no longer free but are part of peptide bonds, it is the amino
acid side chains that determine the polarity characteristics of the protein and,
consequently, how the protein folds. When part of a protein, amino acids are
often referred to as **residues** to emphasize that the only difference between

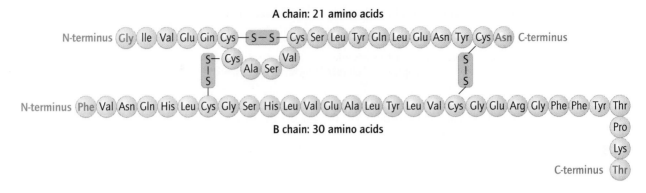

Figure 11-11 The primary structure of human insulin.

them is their side chains. For example, a valine residue contributes a —CH(CH₃)₂ group extending from the polypeptide backbone, while a glutamic acid residue contributes a —CH₂CH₂CO₂⁻ group.

Most enzymes exist in an aqueous medium; therefore, polar residues will tend to be on the exterior of the protein where they can interact with water molecules. Nonpolar residues will be on the interior.

To describe the complex three-dimensional characteristics of a protein in detail, four levels of architecture are defined:

1. primary structure
2. secondary structure
3. tertiary structure
4. quaternary structure

The Primary Structure of Proteins

*The **primary structure** of a protein is its amino acid sequence.* Errors in the primary structure of a protein can have profound effects on its overall shape, as you saw in the case of sickle-cell anemia, described in the opening vignette.

Consider the primary structure of human insulin, a relatively small protein. Insulin is composed of two polypeptides: The A chain contains 21 amino acids and the B chain contains 30 amino acids. The sequence of amino acids from the N-terminus to the C-terminus is given in Figure 11-11, where each amino acid is shown as a sphere containing its three-letter abbreviation.

Problem-Solving Tutorial:
General Protein Structure

WORKED EXERCISE 11-4   Primary Structure of Proteins

Refer to Figure 11-11 to answer the questions below about the primary structure of insulin:
a. What amino acids are at the N-termini? Why are there two N-termini in insulin?
b. What is the tenth amino acid in the A chain of insulin?
c. How many C-termini are there in insulin? What amino acids are at the C-termini?

SOLUTION
a. The N-terminus on the A chain is glycine and on the B chain is phenylalanine. There are two N-termini in insulin because two polypeptide chains make up this protein.

b. The tenth amino acid on the A chain is valine.

c. There are two C-termini in insulin, one for each of the two chains. The C-terminus in the A chain is asparagine and in the B chain it is threonine.

PRACTICE EXERCISES

11.15 If hemoglobin is composed of four polypeptide chains, how many N-termini does it have?

11.16 What are the four levels of protein architecture called?

11.17 Define the term "native conformation."

11.18 Cystic fibrosis (CF) is a disease caused by a missing phenylalanine-508 in a protein containing 1480 amino acids. The protein is necessary for chloride ions (Cl^-) to pass through the cell membrane. The abnormal CF protein does not fold correctly as a result of the missing amino acid. As a result, mucus accumulates in the lungs causing respiratory infections and difficulty breathing. Many individuals with CF die by age 30, and require a lifetime of specialized care.

    a. Write the structure of the missing amino acid.

    b. Where in the primary structure is a Phe missing?

    c. What is a gene?

    d. Would you expect cystic fibrosis to be a hereditary disease or an infectious disease? Explain.

    e. Why does the protein that causes cystic fibrosis not fold correctly?

Secondary Structure

The secondary through quaternary structures of a protein describe its folding patterns and three-dimensional structure, based on its primary structure.

Secondary protein structure refers to regular folding patterns in localized regions of the polypeptide backbone. These folds are created by *hydrogen bonds* between amino acids that are in close proximity. Hydrogen bonds are formed between N—H and carbonyl groups (C=O) in the polypeptide backbone as shown in the margin.

The most common secondary structures are the α-helix and the β-pleated sheet. Proteins differ in the amount and type of secondary structure that they contain. A protein might have exclusively α-helices throughout its structure, or it may contain some sections of α-helix and some sections of β-pleated sheets. There are also often extensive sections of the polypeptide chain with random folding in no particular pattern. For example, hair is composed of keratin, a protein made up of only α-helices. On the other hand, silk contains exclusively β-pleated sheets. Enzymes typically contain a mixture of α-helices, β-pleated sheets, and random folding.

The α-Helix An α-helix is a long or short section of polypeptide that coils into the shape of a helix, as shown in Figure 11-12. The helix is held in shape by hydrogen bonds between N—H and C=O groups on the polypeptide backbone. The carbonyl group of each amino acid forms a hydrogen bond with the N—H group of an amino acid that is situated about four amino acids farther down the sequence. The amino acid side chains extend outward from the α-helix and do not participate in the hydrogen bonding that creates the helix.

Proteins are frequently depicted using **ribbon drawings,** which trace the path of the polypeptide backbone as a ribbon. In a ribbon drawing, an α-helix appears as a coil. For example, the ribbon drawing of myoglobin

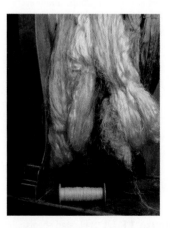

Silk fibers contain a protein known as fibroin, which contains a significant amount of β-sheet secondary structure.
[© Massimo Listri/Corbis]

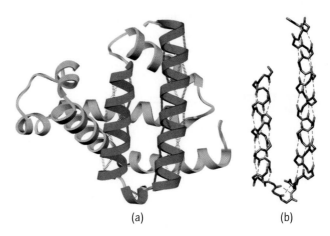

Figure 11-13 The structure of myoglobin. (a) A ribbon structure shows the seven α-helices of myoglobin. The red colored helices represent amino acids 100–148, called out as a tube model in (b). In the tube model, side chains have been omitted and hydrogen bonding is shown as dashed lines.

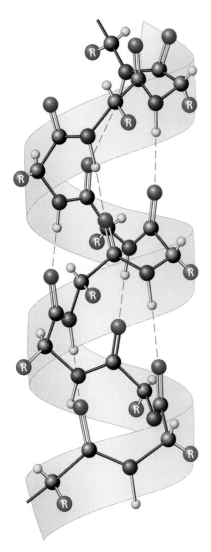

Figure 11-12 A ball-and-stick model of a section of α-helix shows hydrogen bonding (dashed red lines) between N—H and C═O groups about every four amino acids in the primary structure. Side chains, R, extend outward from the helix.

shown in Figure 11-13a shows seven α-helices as red and pink coils of differing lengths. Myoglobin is the oxygen-carrying protein in muscle tissue. A tube model of the red helices shows amino acids 100–148 in Figure 11-13b, omitting the side chains. Dashed lines show hydrogen bonding in the polypeptide backbone. As you can see, extensive hydrogen bonding exists throughout the length of the α-helix; it holds the polypeptide backbone in this folding pattern.

β-Pleated Sheets The **β-pleated sheet,** also known as a β-sheet, is formed when two or more polypeptide strands stack on top of one another in a folding pattern analogous to a pleated skirt. Hydrogen bonding between N—H and C═O groups on parallel (or antiparallel) strands stabilizes this type of secondary structure. Figure 11-14 shows hydrogen bonding between β-pleated strands. The amino acid side chains project above and below the sheets. Extensive regions of β-pleated sheet account for the strength and flexibility of silk fibers.

In a ribbon drawing, a β-pleated sheet is drawn as a wide flat arrow, with the arrow pointing in the direction of the C-terminus. For example, the ribbon drawing of a fatty-acid-binding protein containing extensive arrays of β-pleated sheets is shown in Figure 11-15.

Figure 11-14 β-pleated sheets. A ball-and-stick model of β-pleated sheets shows hydrogen bonding between two parallel or antiparallel strands of polypeptide. Notice the side chains, R, projecting above and below the pleat.

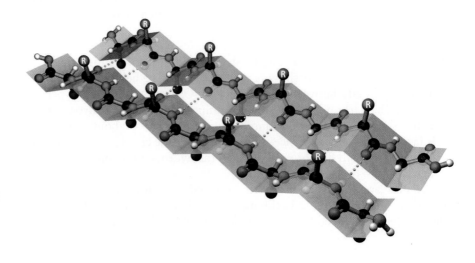

WORKED EXERCISE 11-5 Secondary Structure

A ribbon diagram of insulin is shown below.

a. What color are the α-helices? Describe their shape. What level of protein architecture do α-helices represent: primary, secondary, tertiary, or quaternary?

b. What type of electrostatic interaction holds the α-helix in its unique folding pattern? What atoms are involved?

SOLUTION

a. α-helices are shown in red and they have the shape of a coil. They represent secondary protein structure.

b. Hydrogen bonding between the N—H atom and C=O group four amino acids later stabilizes the secondary structures.

PRACTICE EXERCISES

11.19 Mad cow disease is believed to be caused by an infectious abnormal protein, known as a **prion**, an acronym for **pro**teinaceous **in**fectious agent. Prions cause normal membrane proteins in the brain to change their native conformation, thus destroying their function. The ribbon diagrams shown in the margin compare the normal protein to the expected structure of a prion. Answer the following questions based on these illustrations.

    a. Is the primary structure of the normal protein different from the primary structure of the abnormal protein?

    b. What type of secondary structure is shown in red?

    c. What type of secondary structure is shown in blue?

    d. Describe how the secondary structure of the prion is different from the secondary structure of the normal protein.

11.20 Add dashes to show hydrogen bonding between the two amide functional groups below. Label the partial positive and negative charges in the atoms involved in hydrogen bonding. What causes the partial charges on these atoms?

$$\begin{array}{ccc} \text{O} & & \text{O} \\ \| & & \| \\ -\text{C}-\text{N}- & -\text{N}-\text{C}- \\ | & | \\ \text{H} & \text{H} \end{array}$$

Figure 11-15 A fatty-acid-binding protein containing extensive β-sheet secondary structure.

Problem-Solving Tutorial: General Protein Structure

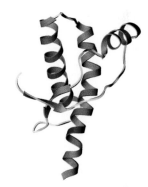

Normal protein

Prion

Tertiary Structure

The **tertiary structure** of a protein describes the complex and *irregular* folding of the polypeptide beyond its secondary structure, showing the three-dimensional picture of the protein. *Tertiary structure is determined by interactions that include distant amino acid residues as well as the surrounding environment.* For some proteins, tertiary structure also includes **prosthetic groups**—nonpeptide organic molecules or metal ions that are strongly bound to the protein and essential to its function.

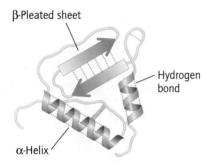

Figure 11-16 A protein's tertiary structure is a polypeptide's overall three-dimensional shape, resulting from interactions between remote residues.

Protein folding brings together amino acid residues that are not necessarily nearby in their primary structure (Figure 11-16). *The fundamental driving force behind protein folding is energy.* A folded protein and its surrounding environment are lower in potential energy than the unfolded protein in its environment. Hence, proteins fold spontaneously into their native conformation under physiological conditions. The way a protein folds may appear random, but it is strictly defined and uniquely suited to the protein's function. The three-dimensional structures of many proteins, including the examples you have seen thus far, have been determined experimentally using X-rays and other sophisticated techniques. The ribbon diagrams you have seen are based on the actual *x, y, z* coordinates of the amino acids in the native conformation of the active protein.

The main interactions between amino acid side chains, responsible for the tertiary structure of a protein, are listed below:

- Disulfide bridges (covalent bonds, hydrophobic)
- Salt bridges (ionic bonds, hydrophilic)
- Hydrogen bonding (electrostatic attraction, hydrophilic)
- Dispersion forces (electrostatic attraction, hydrophobic)

Disulfide Bridges *Disulfide bridges are covalent bonds formed between two cysteine residues (Figure 11-17c). The two thiol functional groups (—S—H) on each cysteine residue react to form a disulfide bond (—S—S—).* Disulfide bridges can form between two sections of a single polypeptide chain or between two different polypeptide chains. In the primary structure of insulin shown in Figure 11-11, there are three disulfide bonds.

Since a disulfide bridge is a covalent bond, the interaction is strong, and can only be made or broken through a chemical reaction. The reactions involved are oxidation–reduction reactions, described in Chapter 10. Thiols are readily oxidized in the presence of oxygen to disulfides; and conversely, disulfides are reduced to thiols in the presence of a reducing agent.

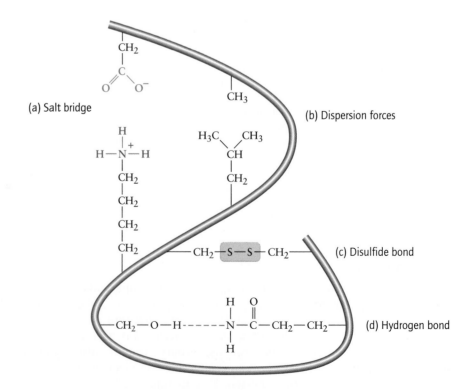

Figure 11-17 Interactions between amino acid side chains that create tertiary structure: (a) salt bridge, (b) dispersion forces, (c) disulfide bond or bridge, and (d) hydrogen bond.

If you have ever seen someone get their hair permed, you have witnessed oxidation–reduction of thiols and disulfide functional groups. The first solution that is applied to the hair in the perming process is a reducing agent, which breaks the natural disulfide bonds in the hair, a protein. Then, a second solution containing an oxidant forms new disulfide bonds, but between different thiol groups in the protein. The hair is "permanently" fixed into the shape of the rollers, at least until the hair grows out. The hair that grows in after a perm has the disulfide bridges of the natural hair, a protein formed from instructions in the individual's DNA. So you see, even hairstylists know about oxidation–reduction reactions!

Salt Bridges Ionic bonds are often called **salt bridges** in the context of proteins. *Ionic bonds form in proteins between positively and negatively charged residues: a basic and an acidic residue (Figure 11-17a).* These interactions are stronger than hydrogen bonds, because the electrostatic interaction is between a full $+1$ or -1 charge, compared to the partial charges $\delta+$ and $\delta-$ that occur in hydrogen bonding.

Hydrogen Bonding *Hydrogen bonding occurs between polar residues when one residue contains either an N—H or an O—H bond (Figure 11-17d).* Hydrogen bonding occurs within the protein itself, the same as the intermolecular forces of attraction between separate molecules that you saw in Chapter 3. Hydrogen bonding can occur between residues quite far apart in the primary structure. Although hydrogen bonds are weaker than disulfide or ionic bonds, there are often so many hydrogen bonds in a protein that they contribute significantly to the conformational integrity of the protein. It is the same type of interaction that forms the secondary structure of proteins, except that it occurs between the amino acid side chains rather than the atoms in the polypeptide backbone.

Tertiary structure is also influenced by hydrogen bonding between polar residues and the aqueous environment of most proteins. In these proteins, polar residues tend to be located on the outside of the protein, so that they can interact with the aqueous surroundings through hydrogen bonding to water molecules. These are known as hydrophilic interactions. Hydrophilic interactions like these are an application of the rule "like dissolves like." These interactions play a significant role in the way globular proteins, such as enzymes, fold.

Dispersion Forces Nonpolar residues interact with other nonpolar residues through dispersion forces, the attractive electrostatic forces exhibited by hydrocarbons, described in Chapter 6 (Figure 11-17b). Nonpolar residues avoid interactions with polar residues and water. Thus, in an aqueous environment proteins tend to fold so that nonpolar residues are on the interior, where they can avoid water. In contrast, nonpolar residues are present on the surface of membrane proteins, because their external environment—the cell membrane—is nonpolar.

Prosthetic Groups Prosthetic groups are part of the tertiary structure of some proteins, such as hemoglobin. *Prosthetic groups are nonpeptide organic molecules or metal ions, or a combination of both, that are essential to a protein's function.* Hemoglobin contains heme as a prosthetic group (Figure 11-18). There is one heme molecule bound to each polypeptide chain in hemoglobin. Each heme contains one iron (Fe) atom that binds oxygen. In proteins that function as enzymes, the prosthetic group is called a **cofactor** or a **coenzyme.**

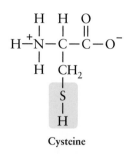

Cysteine

Perming hair involves a reduction of the disulfide bonds in hair protein, followed by an oxidation that forms new disulfide bonds, giving the hair the shape of the rollers. [Brent T. Madison, Shizuoka-Ken, Japan]

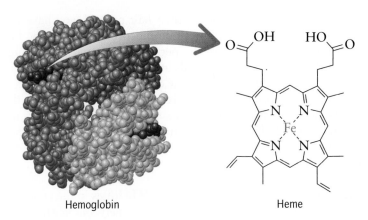

Hemoglobin   Heme

Figure 11-18 Heme, shown in purple, is a prosthetic group, a nonpeptide organic structure that is an essential part of a protein.

Problem-Solving Tutorial:
General Protein Structure

WORKED EXERCISE 11-6   Tertiary Protein Structure

Write the structure of the product formed in the oxidation reaction below. The coefficient here indicates that two identical reactants take part in the reaction.

$$2 \quad H-\overset{\overset{\displaystyle H}{|}}{\underset{\underset{\displaystyle H}{|}}{C}}-S-H \quad \xrightarrow{[O]}$$

SOLUTION

The product is the disulfide formed by removing an H atom from the —SH group in each thiol molecule and forming a disulfide bond, —S—S—, between the sulfur atoms:

$$2 \quad H-\overset{\overset{\displaystyle H}{|}}{\underset{\underset{\displaystyle H}{|}}{C}}-S-H \quad \xrightarrow{[O]} \quad H-\overset{\overset{\displaystyle H}{|}}{\underset{\underset{\displaystyle H}{|}}{C}}-S-S-\overset{\overset{\displaystyle H}{|}}{\underset{\underset{\displaystyle H}{|}}{C}}-H \quad + \quad H_2$$

Thiol   Disulfide

PRACTICE EXERCISES

11.21 Write the products formed in the following reduction. *Hint:* The reaction is the reverse of the oxidation of a thiol.

$$H-\overset{\overset{\displaystyle H}{|}}{\underset{\underset{\displaystyle H}{|}}{C}}-S-S-\overset{\overset{\displaystyle H}{|}}{\underset{\underset{\displaystyle H}{|}}{C}}-\overset{\overset{\displaystyle H}{|}}{\underset{\underset{\displaystyle H}{|}}{C}}-\overset{\overset{\displaystyle H}{|}}{\underset{\underset{\displaystyle H}{|}}{C}}-H \quad \xrightarrow{[H]}$$

11.22 To remove the odor from a dog sprayed by a skunk, the dog can be bathed in water containing a small amount of bleach (NaOCl) or hydrogen peroxide (H_2O_2). Both bleach and hydrogen peroxide are strong oxidizing agents. The chemical structure of one of the compounds that give skunks their offensive odor is shown in the margin. Write the oxidation reaction that occurs when the thiol functional group in this compound reacts with bleach.

11.23 Label the following interactions, (a) through (d), as a *disulfide bond, salt bridge, hydrogen bonding,* or *dispersion forces.*

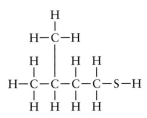

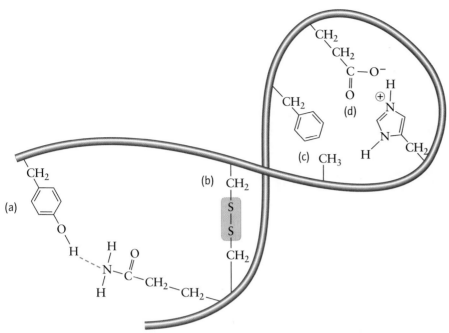

11.24 In the illustration in Exercise 11-23, assuming that the interactions are between residues on a single polypeptide, what level of architecture (*primary, secondary, or tertiary*) is created by the interactions that you labeled?

Quaternary Structure

Some proteins are comprised of more than one polypeptide chain, also called a **subunit.** *The quaternary structure of a protein describes the relative arrangement and position of the subunits within the overall three-dimensional structure of the protein.* The various subunits within a protein interact with one another and are stabilized by the same interactions that determine the tertiary structure of a protein: disulfide bridges, ionic interactions, hydrogen bonding, and dispersion forces. Hemoglobin, for example, has four subunits: two identical α-chains and two identical β-chains (note that α and β are references to the different subunits, not to secondary structure).

An analysis of the four levels of architecture in hemoglobin is summarized in Figure 11-19.

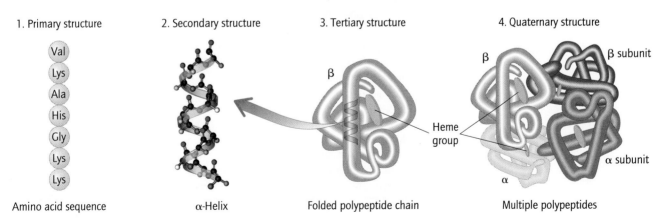

Figure 11-19 Four levels of protein architecture in hemoglobin. (1) The primary structure is the sequence of amino acids in each polypeptide. (2) Secondary structure is the localized folding patterns such as the α-helix and the β-pleated sheet, created by hydrogen bonding along the peptide backbones. (3) Tertiary structure is the result of electrostatic interactions between distant amino acid residues and prosthetic groups. (4) Quaternary structure is the result of interactions between the various polypeptides (subunits) in a protein containing more than one polypeptide.

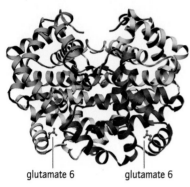

glutamate 6   glutamate 6

Normal hemoglobin

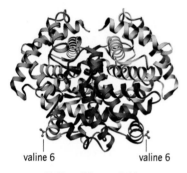

valine 6   valine 6

Sickle-cell hemoglobin

Figure 11-20 The protein structure of HbSc, the hemoglobin causing sickle-cell anemia, has a very different shape from normal (Hb) hemoglobin.

Problem-Solving Tutorial:
General Protein Structure

The way in which the four subunits of hemoglobin fold together is critical to hemoglobin's role in transporting oxygen to cells. In the hemoglobin of individuals with the disorder sickle-cell anemia, valine is substituted for glutamic acid at the sixth amino acid of the β-chains. This substitution alters protein folding, giving a slightly different tertiary and quaternary structure, as illustrated in Figure 11-20. Under conditions of low oxygen, blood cells with this variant form of hemoglobin take on a crescent shape rather than the normal biconcave shape of a red blood cell. Consequently, the cells aggregate and lose elasticity, clogging narrow capillaries.

WORKED EXERCISE 11-7 Quaternary Protein Structure

For the two proteins shown below, hemoglobin found in red blood cells and keratin found in hair, distinguish the parts of the illustration that represent secondary structure, tertiary structure, and quaternary structure.

(a) Hemoglobin molecule   (b) Keratin fiber

SOLUTION

a. Secondary structure: sections of α-helix within each of the four subunits. Tertiary structure: the folded subunits and the heme groups bound to the subunit.

Quaternary structure: the two α and two β subunits organized as shown.

b. Secondary structure: long strands of α-helix shown in brown. Tertiary structure: the linear strands represented by the yellow cylinders.

Quaternary structure: the twisting of three yellow cylinders, each representing a polypeptide, into a braid (known as a triple helix).

PRACTICE EXERCISES

11.25 For the following pairs of amino acids, indicate how they might interact to contribute to the tertiary or quaternary structure of a protein. Choose from among the following choices: *disulfide bond, salt bridge, hydrogen bonding, or dispersion forces.*

a. aspartic acid and histidine
b. serine and lysine
c. leucine and valine
d. two cysteines
e. tryptophan and isoleucine

11.26 Which level of protein architecture describes exclusively covalent bonds: *primary, secondary, tertiary, or quaternary?*

Denaturation of a Protein

Denaturation *of a protein refers to any process that disrupts the secondary, tertiary, or quaternary structure of a protein so that it no longer can perform its function, as illustrated in Figure 11-21.* Denaturing a protein does not alter its primary structure, and in some instances can be reversed.

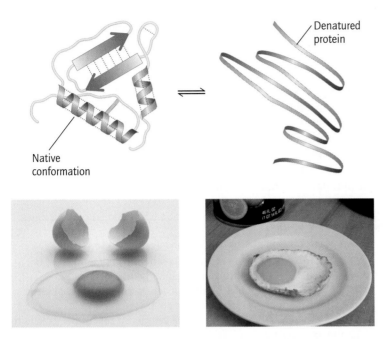

Keratin: Feathers contain keratin, a structural protein composed of parallel polypeptide chains. [Stephen G. Maka/Photex/beneluxpress/zefa/Corbis]

Figure 11-21 Denaturing a protein alters its secondary, tertiary, and quaternary structure. It leaves the covalent bonds of its primary structure intact. [Photos: left, DAJ/Punchstock; right, Ramon Rivera-Moret]

Agents that denature a protein include heat, pH changes, mechanical agitation, detergents, and some metals. A denatured protein loses its shape and hence, its biological activity. When you cook an egg, for example, you are denaturing the protein albumin. The change from colorless to white observed as the egg cooks is the result of albumin undergoing denaturation. Cooking an egg is an example of an irreversible denaturation.

Types of Proteins

There are three general classes of proteins, distinguished by their tertiary and quaternary structure as well as their solubility:

- Fibrous proteins
- Globular proteins
- Membrane proteins

Elastins: SEM (Scanning Electron Microscope) image of periodontal ligament fibers, composed of elastins, which anchor the roots of teeth to the jaw bone and provide added cushioning. [Steve Gschmeissner/Photo Researchers, Inc.]

Fibrous Proteins Fibrous proteins contain parallel polypeptide chains, resulting in long fibers or sheets. These proteins are strong and insoluble in water and are observed to have a structural role in nature. We will discuss three basic types of fibrous proteins: keratins, elastins, and collagens.

- **Keratins** are found in hair and fingernails, and in the feathers, horns, claws, and hooves of other animals.
- **Elastins** have elastic fibers that give flexibility to skin, blood vessels, the heart, lungs, intestines, tendons, and ligaments.
- **Collagens** are the main structural protein in the body. Collagens provide the rigidity of connective tissue seen in skin, tendons, bones, cartilage, and ligaments.

Elastins and collagens work together to provide the elasticity and rigidity of connective tissue. The visible affects of aging can be explained in part by the fact that the production of elastins declines with age.

Globular Proteins Globular proteins are soluble polypeptides folded into complex overall spherical shapes. Globular proteins may serve as enzymes,

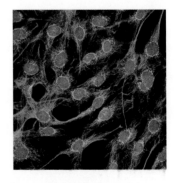

Collagen: Fibroblast cells produce collagen, the main structural protein in the body. [Dr. Gopal Murti/Photo Researchers, Inc.]

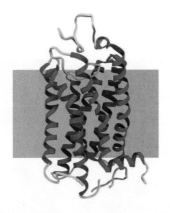

Figure 11-22 A common membrane protein containing seven α-helices (red) that span the width of the cell membrane.

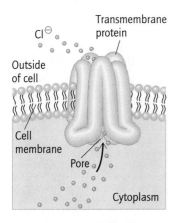

Only a handful of enzymes are not proteins. They are RNA molecules—known as ribozymes. Nevertheless, the vast majority of enzymes are proteins.

hormones, or antibodies. Hemoglobin is an example of a globular protein. The role of enzymes, which are typically globular proteins, is described in the next section.

Membrane Proteins Membrane proteins are specialized proteins that reside entirely within or in part of the cell membrane. They often serve as receptors for ligands conveying cell signals, or as ion channels that shuttle ions into and out of the cell. One common type of membrane protein contains seven α-helices, each spanning the width of the cell membrane, as illustrated in Figure 11-22. Since the cell membrane is a hydrophobic environment, these proteins have hydrophobic residues on their exterior.

PRACTICE EXERCISES

11.27 Answer the following questions about the illustration in the margin.
a. State what type of protein is shown in white: *globular, fibrous,* or *membrane*? Explain the reasoning behind your selection.
b. Would you expect residues on the outside of this protein that interact with the cell membrane to be nonpolar or polar?
c. Would you expect the residues on the cytoplasm side of the protein to be nonpolar or polar?
d. Would you expect the residues in contact with the aqueous environment on the outside of the cell to be polar or nonpolar?
e. Each membrane-spanning region of the protein is composed of an α-helix. What type of structure does this represent: *primary, secondary, tertiary,* or *quaternary*?
f. Describe the tertiary structure of this protein.

11.28 List three ways to denature a protein.

11.29 Which level of protein architecture is unaffected by denaturation: *primary, secondary, tertiary,* or *quaternary structure*?

11.30 State two differences between globular and fibrous proteins.

11.31 Keratin, collagen, and elastin all belong to what class of proteins?

11.32 Enzymes are typically _____ proteins. Choose from *fibrous, globular,* or *membrane*.

11.33 What is the difference between collagens and elastins?

11.5 | Enzymes

One of the most important roles that proteins serve is their function as enzymes, the globular proteins that catalyze most of the chemical reactions that occur in a living cell. In the absence of enzymes, most biochemical reactions are too slow to be useful; thus, enzymes are essential for life. For example, the sugar on your table can be stored indefinitely without undergoing any chemical reactions, but your body can turn sugar into CO_2, water, and the energy you need for your activities today. The difference is that enzymes in the cell greatly accelerate the reactions needed to transform sugar into useful energy.

How Do Enzymes Work?

You learned in Chapter 8 that the rate of a reaction can be increased by increasing the *concentration* of the reactants, raising the *temperature* of the reaction, or adding a *catalyst* to the reaction. Since the first two variables cannot be changed significantly in biological cells, the cell employs biological catalysts called **enzymes** to increase reaction rates. Enzymes are remarkable catalysts. A biochemical reaction catalyzed by an enzyme is 10 to 10^{16} times faster than the uncatalyzed reaction! Enzymes typically act only on one particular reactant

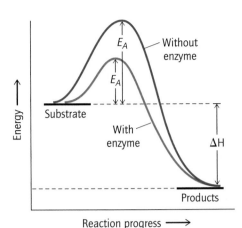

Figure 11-23 This reaction diagram shows the energy pathway for a reaction catalyzed by an enzyme (red line) and not catalyzed by an enzyme (blue line). E_A is lower in the enzyme-catalyzed reaction.

(or one with a very similar structure), referred to as the **substrate.** Moreover, if the product of an enzyme catalyzed reaction is chiral, the enzyme ensures that the correct enantiomer is produced.

As with all catalysts, enzymes work by lowering the energy of activation, E_A, for the reaction. Recall from Chapter 8 that catalysts do not alter the change in enthalphy—ΔH—of the reaction. As the energy diagram in Figure 11-23 shows, the energy pathway (red line) for the catalyzed reaction is lower than the energy pathway for the uncatalyzed reaction (blue line). Notice that the overall energy difference between reactants and products is unchanged.

The catalyst itself is unchanged at the end of a chemical reaction, and thus, the same enzyme molecule can be used over and over again to catalyze the transformation of many substrate molecules into product molecules. Therefore, in a chemical equation, enzymes are not shown as a reactant with an accompanying coefficient; instead, the enzyme name is simply written above the reaction arrow.

Enzymes are named after the substrate or the type of reaction they catalyze, with the addition of the suffix "ase." Their names are usually italicized. For example, the enzyme *lactase* catalyzes the reaction that hydrolyzes lactose (milk sugar) into two simpler sugars. Some enzymes have a common name derived from where they were first isolated. For example, *papain* is an enzyme that was first isolated from the papaya fruit.

Biochemists classify enzymes by the type of reaction they catalyze. For example, oxidoreductases catalyze oxidation–reduction reactions. Table 11-3 lists the six different classes of enzymes and the type of reaction they catalyze. The first five of these reaction types were described in Chapter 10.

The enzyme *papain* was first isolated from the papaya fruit. [ImageState/Alamy]

The Enzyme–Substrate Complex How does an enzyme lower the energy of activation for a reaction? The first step in an enzyme catalyzed reaction is the

Table 11-3 Enzyme Classes and the Types of Reactions They Catalyze

Class of Enzyme	Type of Reaction Catalyzed
1. Oxidoreductases	Oxidation-reduction (electron transfer) reactions
2. Transferases	Acyl and phosphoryl group transfer reactions ($RC{=}O$ and PO_3^{2-})
3. Hydrolases	Hydrolysis reactions
4. Lyases	Hydration and dehydration reactions
5. Isomerases	Reactions that transform one structural or geometric isomer to another
6. Ligases	Reactions that join two molecules by forming a C—C, C—O, or C—S bond; requires ATP.

binding of the substrate to the enzyme. An enzyme is a large protein molecule that is typically much greater in size than its substrate. The substrate binds to the enzyme at a pocket or cleft within the protein, known as the **active site** or **binding site.** The active site is where the chemical reaction actually takes place. *The shape of the active site is complementary to the shape of the substrate and is largely responsible for the selectivity that an enzyme has for its substrate.* The active site contains amino acid residues positioned in such a way that they optimally bind the substrate, **S**, but not other molecules. The substrate binds to the active site through ionic attractions and intermolecular forces of attraction (dispersion forces, dipole–dipole forces, and hydrogen bonding). These binding interactions, together with a complementary shape, create a good "fit" between the enzyme and its substrate, called the **enzyme–substrate complex, ES**.

The binding of a substrate to an enzyme is often described by the **lock-and-key model.** In this model the active site and the substrate have complementary shapes, which allow them to fit together like a key fits a lock (Figure 11-24). A more refined representation of the model is that the enzyme makes some minor adjustments to its conformation upon binding in order to achieve an even better fit to the substrate—an "induced fit."

In the enzyme–substrate complex (ES), the substrate is held in a position and orientation that places the reacting functional group(s) near one another, thereby facilitating the chemical reaction. *It is the placement of the substrate in an optimal geometry and proximity to the other reactant (if there is one) that lowers the energy of activation, E_A, for the reaction.* For example, the substrate molecule and a molecule of water must come together for a hydrolysis reaction to occur; two amino acids must come together for a peptide bond to be formed; and so on. When the reaction is over, the enzyme releases the product, **P**, and the original enzyme, **E**. The enzyme, **E**, is then free to bind to another substrate molecule and repeat the process. These steps, each of which is reversible, are summarized in Figure 11-24. A single enzyme molecule may catalyze a reaction at a rate as slow as 1 substrate molecule every 2 seconds up to a rate as high as 10 million substrate molecules every second.

Cofactors and Coenzymes Some enzymes require a cofactor to achieve their catalytic effect. **Cofactors** can be metal ions, such as Fe^{2+}, Mg^{2+}, and Zn^{2+}. Cofactors can also be organic compounds such as $NAD^+/NADH$, $FAD/FADH_2$, and coenzyme A, which are derived from B vitamins. Cofactors that are organic molecules are called **coenzymes.** A cofactor or coenzyme that is tightly bound or covalently bonded to the enzyme is also referred to as a prosthetic group.

Unlike the *enzyme*, a *coenzyme* undergoes a chemical change during the reaction, but it is regenerated in a subsequent reaction so that it can be

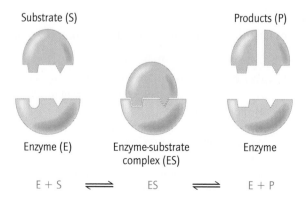

Figure 11-24 Lock-and-key model showing that substrate and active site are complementary in shape.

Substrate (S)　　　　　　　　　　　　　　　Products (P)

Enzyme (E)　　Enzyme-substrate　　Enzyme
　　　　　　　complex (ES)

E + S　⇌　ES　⇌　E + P

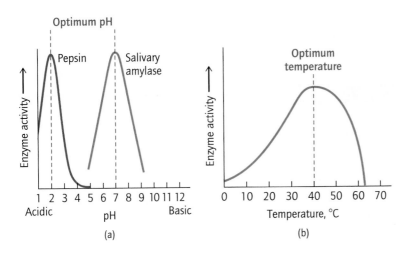

Optimum pH

Pepsin

Salivary amylase

Optimum temperature

Enzyme activity ⟶

Enzyme activity ⟶

1 2 3 4 5 6 7 8 9 10 11 12
Acidic   pH   Basic

0 10 20 30 40 50 60 70
Temperature, °C

(a)

(b)

Figure 11-25 (a) Enzyme activity as a function of pH for two representative enzymes. *Pepsin* needs to function at low pH to break down proteins in the very acid environment of the stomach. (b) Enzyme activity as a function of temperature. Optimal temperature is 40°C. Denaturation begins to occur at temperatures above 45–50°C.

employed again by the enzyme. For example, NAD^+ is converted into NADH in the enzyme catalyzed oxidation of alcohols, then NADH is converted back into NAD^+ by a different enzyme in a reduction reaction. By converting back and forth between its two forms, the coenzyme NAD^+/NADH can be used to shuttle electrons between molecules.

pH and Temperature Dependence of Enzymes

An enzyme must be in its native conformation (see Figure 11-21) to achieve its catalytic activity, which requires that both temperature and pH be in the optimal range for the enzyme. Changes in pH can alter the charge on acidic and basic residues, which in turn affects the shape of the protein. For most enzymes, physiological pH (pH = 7.3) is optimal, although some enzymes, such as those found in the stomach, like *pepsin*, function best at low pH (pH = 1−2.5) as shown in Figure 11-25a.

Acidosis, described in Chapter 9, is a serious medical condition caused by a drop in blood pH that, among other effects, causes enzymes to denature. Some poisons inflict their damage through pH changes that denature enzymes. Because a denatured enzyme can no longer catalyze reactions, important biochemical pathways will shut down.

Although increasing the temperature of a reaction in the laboratory typically increases the rate of a reaction; in the body, temperatures that are too high will cause enzymes to denature and thereby decrease reaction rates (Figure 11-25b). Most enzymes function best around normal body temperature, 37°C. Consider the case of an individual surviving a fall into extremely cold waters. The lower body temperature brought on by the cold water causes all metabolic enzymes to slow, thereby decreasing the individual's requirement for oxygen (a key reactant in metabolism), and thereby preserving brain function. The individual is thus able to survive an experience that would otherwise have resulted in certain death.

WORKED EXERCISE 11-8 Enzymes

Which of the interactions listed below are involved in binding an enzyme to its substrate?
a. disulfide bonds
b. hydrogen bonding
c. dispersion forces
d. ionic bonds

Problem-Solving Tutorial:
Protein Function: Enzymes

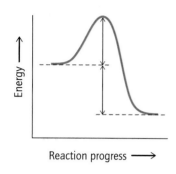

Reaction progress ⟶

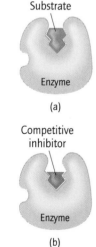

Substrate

Enzyme

(a)

Competitive inhibitor

Enzyme

(b)

Noncompetitive inhibitor

Substrate

Enzyme

(c)

Figure 11-26 (a) Normal substrate and enzyme binding. (b) A competitive inhibitor prevents the substrate from bonding by occupying the active site. (c) A noncompetitive inhibitor changes the shape of the active site so that the substrate cannot bind.

SOLUTION

The interactions involved in binding a substrate to an enzyme include (b) hydrogen bonding, (c) dispersion forces, and (d) ionic bonds.

PRACTICE EXERCISES

11.34 Explain how an "induced fit" between an enzyme and its substrate describes the model of an enzyme binding to its substrate.

11.35 For the energy diagram of an uncatalyzed reaction shown in the margin, label the following parts:
a. E_A b. ΔH c. reactants   d. products
e. Trace a second curve showing the energy path for the reaction in the presence of the appropriate enzyme.

11.36 What is the active site of an enzyme?

11.37 What is the ES complex?

11.38 Which of the following changes when a catalyst is added to a reaction:
a. E_A or ΔH?
b. The heat released from the reaction or the rate of the reaction?

11.39 To achieve its maximum catalytic power, what pH is needed for a reaction catalyzed by *pepsin*, an enzyme found in the stomach? Use the graph shown in Figure 11-25a.

11.40 FAD/FADH₂ is a coenzyme. In what way does this coenzyme change during a reaction? Why is it considered a coenzyme?

11.41 In the laboratory, what factors increase the rate of a chemical reaction?

11.42 How is the rate of an enzyme catalyzed reaction affected by the following changes?
a. an increase in pH
b. a decrease in pH
c. a decrease in temperature
d. an increase in temperature

Enzyme Inhibitors

Enzyme inhibitors are compounds that prevent an enzyme from performing its function. Many poisons and some pharmaceutical drugs work as enzyme inhibitors. pH and temperature are examples of **nonspecific enzyme inhibitors** because they can denature many different types of enzymes at the same time. **Specific enzyme inhibitors** target only one enzyme, and most references to enzyme inhibitors assume that the enzyme inhibitor is specific. There are two classes of specific enzyme inhibitors: competitive inhibitors and noncompetitive inhibitors, illustrated in Figure 11-26.

Competitive Inhibitors A **competitive inhibitor** competes with the substrate for the active site of the enzyme, because it too has a structure complementary in shape to the active site. By binding to the active site, a competitive inhibitor blocks the substrate from binding to the enzyme. As a result, the ES complex cannot form, and no catalysis occurs.

The effectiveness of a competitive inhibitor is concentration dependent: Inhibition depends on the relative concentrations of substrate and inhibitor. Like the game "musical chairs," both substrate and inhibitor compete for the same active site (the chair) but there is only room for one (someone is left standing). Thus, the more molecules of inhibitor are present, the less likely the substrate will find a free enzyme to bind to. Conversely, the more molecules of substrate are present, the more enzyme catalysis can occur.

Drugs that act as enzyme inhibitors are generally given at regular doses to maintain the concentration of inhibitor. To see an example of how a specific enzyme inhibitor is used to treat an enzyme-based disease, read *Chemistry in Medicine: ACE Inhibitors—A Treatment for Hypertension.*

Noncompetitive Inhibitors Noncompetitive inhibitors bind at a location on the enzyme other than the active site. Binding of a noncompetitive inhibitor causes the shape of the enzyme to change in such a way that the active site can no longer bind to the substrate. Therefore no ES complex is formed. *In contrast to competitive inhibitors, increasing the concentration of substrate will not restore enzyme activity when a noncompetitive inhibitor is bound to the enzyme. Enzyme activity can only be restored when the concentration of noncompetitive inhibitor is low.*

For example, cholesterol is a noncompetitive inhibitor of an enzyme required for its own synthesis, known as *HMG-CoA Reductase.* This enzyme catalyzes the second step in a multistep biochemical pathway that ultimately produces cholesterol.

$$\text{HMG-CoA} \xrightarrow[\substack{\text{HMG-CoA} \\ \text{reductase}}]{} \text{Mevalonate} \twoheadrightarrow\twoheadrightarrow\twoheadrightarrow \text{Cholesterol (noncompetitive inhibitor)}$$

Feedback inhibition

Thus, cholesterol inhibits its own production when cholesterol concentrations are high; and enzyme activity is restored when cholesterol concentrations are low, thus resuming cholesterol synthesis. In this clever way the body controls the production of cholesterol—a process known as enzyme **feedback inhibition.** As you know, too much cholesterol can have serious health effects, as can too little. This is true of many substances, so many biochemical pathways in the cell are regulated through noncompetitive inhibition of a key step in the biochemical pathway. Thus, many enzymes serve a regulatory role in addition to their role as biological catalysts.

One class of cholesterol-lowering drugs lowers cholesterol levels in the blood by acting as a competitive inhibitor of *HMG-CoA Reductase.* These are the statins, such as Lipitor.

PRACTICE EXERCISES

11.43 Which type of enzyme inhibitor binds to the active site of an enzyme: a *competitive inhibitor* or a *noncompetitive inhibitor?*

11.44 For what type of inhibition does an increase in the concentration of substrate restore enzyme activity: *competitive* or *noncompetitive* inhibition?

In this chapter you have been introduced to one of the most important biomolecules, proteins. You have seen that proteins are built from smaller building blocks known as amino acids. In Chapter 15 you will learn how proteins are assembled from amino acids using the instructions stored in your DNA. In the next chapter you will be introduced to another kind of biomolecule, carbohydrates, the primary source of energy for the cell. As with proteins, you will consider both their structure and function in the human body. When you study the metabolism of carbohydrates, you will see that enzymes play a key role in the biochemical pathways of carbohydrates.

Chemistry in Medicine ACE Inhibitors—A Treatment for Hypertension

Enzyme inhibition is the molecular basis for the action of many pharmaceuticals. Enzyme inhibitors have been used in the treatment of a wide range of diseases including cancer, AIDS, diabetes, and heart disease. Some of the earliest enzyme inhibitors were used before their mechanism of action was even understood. Aspirin, for example, was used to treat pain and inflammation long before it was discovered in the 1970s to be an inhibitor of *cyclooxygenase*, a key enzyme involved in the biochemical pathways of inflammation.

ACE inhibitors are a class of pharmaceuticals used to treat chronically elevated blood pressure, a condition known as **hypertension.** Hypertension is a major risk factor in stroke, heart attack, heart failure, and kidney failure. More than half of Americans over age 65 have hypertension.

A Biochemical Pathway for Raising Blood Pressure

The body must maintain pressure in blood vessels so that it never falls too low or rises too high. ACE inhibitors are designed to lower blood pressure in individuals who have chronically elevated blood pressure. They act by inhibiting a pathway used by the body to elevate the blood pressure.

One way the body detects low blood pressure is by monitoring levels of blood sodium, Na⁺. When the kidneys detect a drop in sodium, they release an enzyme known as *renin*. *Renin* is an enzyme that acts on a single substrate: angiotensinogen, a polypeptide produced in the liver. *Renin* cleaves the peptide bond between Leu and Val in angiotensinogen to produce the hormone angiotensin I, (see Figure 11-27).

The next step in this biochemical pathway is the conversion of angiotensin I into angiotensin II by **A**ngio-tensin **C**onverting **E**nzyme—**ACE**—an enzyme produced primarily in the lungs. ACE cleaves the Phe-His peptide bond in angiotensin I, to produce the hormone angiotensin II.

Angiotensin II initiates a number of physiological events that raise blood pressure. For example, it causes the muscles surrounding blood vessels to constrict—a process known as **vasoconstriction.** Angiotension II also triggers the release of the steroid hormone aldosterone from the adrenal cortex. That hormone stimulates the kidneys to reabsorb sodium ions rather than releasing them in the urine. Sodium ions remain in the bloodstream, holding excess water in the blood, through osmosis, which increases blood pressure. The renin-angiotensin-aldosterone system is summarized in Figure 11-28, which shows the central role of angiotensin II in the interconnected physiological events that lead to an increase in blood pressure.

Developing a New Treatment for Hypertension

Once scientists discovered the renin-angiotensin-aldosterone pathway for raising blood pressure, it was postulated that an inhibitor of ACE—a key enzyme in the pathway shown in Figure 11-28—might offer a viable treatment for high blood pressure. An ACE inhibitor is a substance that binds to ACE, preventing angiotensin I, the natural substrate, from binding to the enzyme.

The first substance that was observed to act as a competitive inhibitor of ACE was a polypeptide found in the venom of the poisonous pit viper. Snake venom in general is well known for its polypeptides that interfere with blood coagulation and other blood regulation mechanisms. Since proteolytic enzymes in the stomach hydrolyze polypeptides, these snake venom peptides are

Angiotensinogen: Asp-Arg-Val-Tyr-lle-His-Pro-Phe-His-Leu-Val-Tyr-Ser-protein

Renin ↓

Angiotensin I: Asp-Arg-Val-Tyr-lle-His-Pro-Phe-His-Leu

Angiotensin Converting Enzyme
ACE ↓

Angiotensin II: Asp-Arg-Val-Tyr-lle-His-Pro-Phe + His-Leu

Figure 11-27 Renin-angiotensin biochemical pathway.

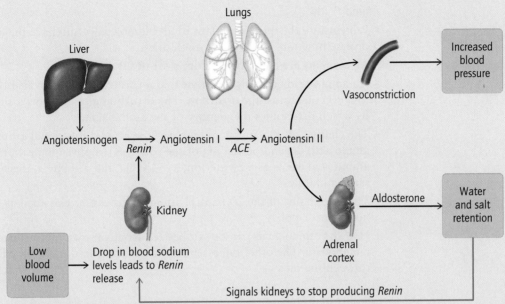

Figure 11-28 The regulation of blood pressure in the body.

not effective as orally administered drugs, so the quest for a nonpeptide ACE inhibitor began.

Captopril

Researchers experimented with structurally similar compounds that combined the characteristics of the C-terminal proline found in the snake venom and the dipeptide cleaved from angiotensin I. The first ACE inhibitor, captopril, was born. Notice the five-membered nitrogen-containing proline ring in the structure of captopril

shown here. This ACE inhibitor is extremely effective because the active site of ACE contains a zinc atom, Zn^{2+}, that binds the sulfur in captopril exceedingly well. Captopril is a competitive inhibitor of ACE, competing for the active site as well as or better than the natural substrate.

Since the introduction of captopril, many new and improved ACE inhibitors have become available for the treatment of hypertension. Most of the newer ACE inhibitors have fewer side effects than captopril and bind just as well to the enzyme. Today, ACE inhibitors are some of the most widely prescribed drugs on the market. As you can see, medicine is, at its core, chemistry.

The venom of the poisonous Brazilian pit viper contains peptides that are ACE inhibitors. [blickwinkel/Alamy]

Chapter Summary

Amino Acids

- Proteins are the most abundant of all biomolecules and have the most diverse functions of all the biomolecules.
- All proteins are constructed from a set of 20 amino acids.
- An amino acid contains an amine and a carboxylic acid covalently bonded to a carbon atom—the α-carbon. The α-carbon also contains a side chain, R, which determines the identity of the amino acid.
- The charges on the amine and the carboxylic acid functional groups in an amino acid depend on the pH of the solution. At physiological pH an amino acid exists as a zwitterion, a —COO$^-$ and a —NH$_3^+$ together in the same molecule.
- The side chains of the 20 natural amino acids can be classified as either nonpolar or polar.
- Half the amino acids can be synthesized from various intermediates of metabolism. The other ten essential amino acids need to be supplied through the diet.

Chirality and Amino Acids

- Nineteen of the amino acids are chiral. Glycerine is achiral.
- A chiral object has a nonsuperposable mirror image.
- Stereoisomers are molecules with the same chemical formula and connectivity but different spatial arrangement of atoms.
- Enantiomers are two stereoisomers that are nonsuperposable mirror images.
- In the IUPAC naming system, amino acid enantiomers can be distinguished by using the prefixes D- and L-.
- Enantiomers have identical physical and chemical properties when placed in an achiral environment. Some enantiomers exhibit profoundly different chemical properties when placed in a chiral environment, such as the body.

Peptides

- Peptides are molecules composed of two or more amino acids. The amino acids are held together by peptide bonds.
- A peptide bond is a covalent bond formed between the carboxylic acid of one amino acid and the amine of another amino acid to form an amide functional group. A molecule of water is released in the reaction.
- Every amino acid is capable of forming two peptide bonds, one with its amine functional group and one with its carboxylic acid functional group.

Protein Architecture

- Proteins are composed of anywhere from about 50 to thousands of amino acids.
- All proteins are built from the 20 L-amino acids.
- In order for a protein to perform its unique function, the chain of amino acids must fold into a specific three-dimensional shape.
- The four levels of protein architecture are defined as primary structure, secondary structure, tertiary structure, and quaternary structure.
- The primary structure of a protein is its amino acid sequence.

- Secondary protein structure refers to regular folding patterns in local regions of the polypeptide backbone. These folds are held together by hydrogen bonds between the N—H and C=O groups in the polypeptide backbone.
- The most common secondary structures are the α-helix and β-pleated sheet.
- The tertiary structure of a protein describes the complex and irregular folding of the entire polypeptide, beyond its secondary structure.
- The fundamental driving force for protein folding is energy. A folded protein and its surrounding environment are lower in potential energy than an unfolded protein and its surrounding environment.
- The main interactions between amino acid side chains that are responsible for the tertiary structure of a protein are disulfide bridges, salt bridges, hydrogen bonding, and dispersion forces.
- Prosthetic groups are nonprotein organic molecules or ions that are part of the tertiary structure of some proteins.
- The quaternary structure of a protein is the relative arrangement and position of each of the polypeptide subunits within the overall structure of a protein.
- Denaturation of a protein is the disruption of the secondary, tertiary, and quaternary structure of a protein so that it can no longer perform its function.
- There are three general classes of proteins, distinguished by their tertiary and quaternary structure as well as their solubility: fibrous proteins, globular proteins, and membrane proteins.

Enzymes

- Enzymes are proteins that act as catalysts to speed up chemical reactions. Enzymes are specific for one particular substrate.
- Enzymes work by lowering the energy of activation, E_A, for a reaction.
- The substrate binds to the enzyme at a pocket or cleft known as the active site or binding site. The shape of the active site is complementary to the shape of the substrate.
- Changes in pH or an increase in temperature can alter the shape of an enzyme so that it no longer performs its function.
- Enzyme inhibitors are compounds that prevent an enzyme from performing its function.
- A competitive inhibitor competes with the substrate for the active site of the enzyme because it also has a structure that is complementary to the active site.
- Noncompetitive inhibitors bind at a location on the enzyme other than the active site and cause the shape of the enzyme to change in such a way that the substrate can no longer bind to the enzyme.

Key Words

Achiral Superposable on its mirror image, and therefore the same as its mirror image.

Active site The pocket or cleft within a protein where the chemical reaction takes place. Also known as the binding site.

Amino acid A compound that contains both an amine functional group and a carboxylic acid functional group attached to the α-carbon. The building blocks for proteins.

β-pleated sheet A type of polypeptide secondary structure in which two or more polypeptide strands stack on top of one another in a pleat-like folding pattern.

Chiral Nonsuperposable on its mirror image.

Coenzyme A necessary component of some enzymes, it is an organic molecule that undergoes a chemical change as part of the enzyme-catalyzed reaction, but can be regenerated in a subsequent reaction.

Cofactor A metal ion or organic molecule within an enzyme that helps the enzyme achieve its catalytic effect.

Competitive inhibitor An inhibitor that competes with the substrate for the active site of the enzyme, because it also has a structure complementary to the active site.

C-terminus The end of a peptide containing a free carboxylate group.

Dipeptide A peptide created from two amino acids.

Enantiomers Two nonsuperposable mirror image stereoisomers.

Enzyme–substrate complex The complex formed when a substrate binds to an enzyme.

Essential amino acids Amino acids that must be supplied though diet because they cannot be synthesized in the body. They are arginine, histidine, isoleucine, leucine, lysine, methionine, phenylalanine, threonine, tryptophan, and valine.

Lock-and-key model A model used to describe the binding of a substrate to an enzyme. In this model the active site and the substrate have complementary shapes.

Native conformation The three-dimensional shape of a protein that allows it to perform its function.

Nonspecific enzyme inhibitors A substance or condition that can denature many different types of enzymes at the same time, such as temperature or pH.

Nonsuperposable objects Two objects that cannot be perfectly overlayed.

N-terminus The end of a peptide containing a free amino group.

Peptide A molecule composed of two or more amino acids.

Peptide bond A covalent bond in the amide formed between the carboxylic acid of one amino acid and the amine of another amino acid.

Polypeptide A peptide that contains many amino acids.

Prosthetic groups Nonpeptide organic molecules and metal ions that are strongly bound to a protein and are essential to its function.

Racemic mixture A 50:50 mixture of enantiomers.

Residue The amino acid with its side chain in the context of a polypeptide or protein.

Ribbon drawings Drawings in which the length of the peptide backbone is depicted as a ribbon.

Specific enzyme inhibitors Inhibitors that target only one enzyme.

Stereoisomers Molecules with the same chemical formula and connectivity of atoms, but a different spatial arrangement of atoms.

Substrate The reactant that binds to an enzyme in an enzyme-catalyzed reaction.

Subunit A polypeptide chain in a protein with more than one polypeptide.

Tertiary structure The complex and irregular folding of the polypeptide beyond its secondary structure, showing the three-dimensional picture of the entire protein.

Thiol A functional group that contains an —S—H group.

Tripeptide A peptide that contains three amino acids.

Zwitterion A neutral molecule that contains two opposite charges.

Additional Exercises

Amino Acids

11.45 Write the structure for each the following amino acids at physiological pH.
   a. glycine
   b. aspartic acid
   c. tyrosine
   d. cysteine
   e. glutamic acid

11.46 Write the structure for each of the following amino acids at physiological pH.
   a. alanine
   b. lysine
   c. threonine
   d. phenylalanine

11.47 Identify three amino acids with nonpolar side chains. Identify an amino acid with a basic side chain. Identify an amino acid with an acidic side chain. Identify two amino acids that have side chains that are polar, but neutral.

11.48 Identify the side chains of the following amino acids as *nonpolar*, *basic*, *acidic*, or *neutral and polar*.
   a. aspartic acid
   b. lysine
   c. cysteine
   d. proline

   e. serine
   f. tryptophan

11.49 Why are some amino acids known as essential amino acids?

11.50 List all the amino acids that have the following functional group in their side chain:
   a. an alcohol   **b.** an amine

11.51 List all the amino acids that have the following functional group in their side chain:
   a. a thiol   **b.** an amide

11.52 Which amino acid contains a thiol group? What is a thiol functional group?

11.53 Write the structure of cysteine. What functional group does cysteine contain in its side chain?

11.54 One of the 20 natural amino acids is shown below.

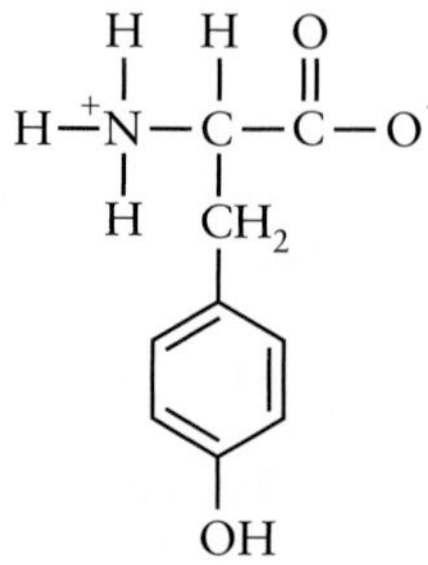

a. Circle and label the amine functional group. Is it in its neutral or ionized form?
b. Circle and label the carboxylate functional group. How does this group differ from a carboxylic acid?
c. Is this amino acid a zwitterion? At what pH does an amino acid exist as a zwitterion?
d. Circle and label the side chain. What is the name of this amino acid?
e. Is the side chain nonpolar or polar?
f. Write the structure of this amino acid at pH = 12. What is the net charge on the amino acid at this pH?

11.55 One of the 20 natural amino acids is shown below.

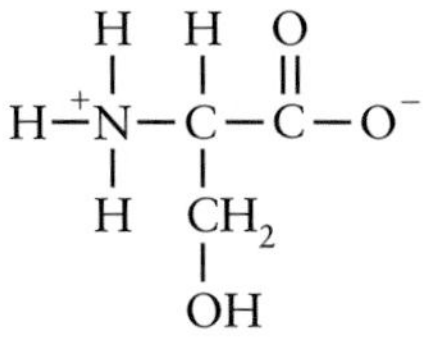

$$H-\overset{\overset{\displaystyle H}{|}}{\underset{\underset{\displaystyle H}{|}}{\overset{+}{N}}}-\overset{\overset{\displaystyle H}{|}}{\underset{\underset{\displaystyle CH_2}{|}}{C}}-\overset{\overset{\displaystyle O}{\|}}{C}-O^-$$
$$CH_2$$
$$|$$
$$OH$$

a. Circle and label the amine functional group. Is it in its neutral or ionized form?
b. Circle and label the carboxylate functional group. How does this group differ from a carboxylic acid?
c. Is this amino acid a zwitterion? At what pH does an amino acid exist as a zwitterion?
d. Circle and label the side chain. What is the name of this amino acid?
e. Is the side chain nonpolar or polar?
f. Write the structure of this amino acid at pH = 12. What is the net charge on the amino acid at this pH?

11.56 For each of the following amino acids place an X in the boxes that apply.

Amino Acid	Non-polar	Polar	Acidic Side Chain	Basic Side Chain	Neutral Side Chain	Essential Amino Acid
Threonine						
Histidine						
Isoleucine						

11.57 For each of the following amino acids place an X in the boxes that apply.

Amino Acid	Non-polar	Polar	Acidic Side Chain	Basic Side Chain	Neutral Side Chain	Essential Amino Acid
Phe						
Asn						
Met						

11.58 Does rice contain all the essential amino acids?

11.59 Does meat contain all the essential amino acids?

Chirality and Amino Acids

11.60 The Fischer projection of one of the natural amino acids is shown below.

$$^+H_3N-\overset{\overset{\displaystyle COO^-}{|}}{\underset{\underset{\displaystyle CH_2OH}{|}}{}}-H$$

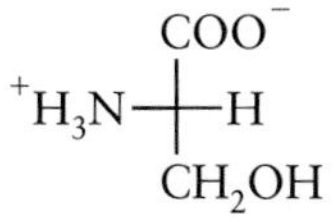

a. Which amino acid is this?
b. Is this amino acid D- or L-? Explain.
c. Place an arrow pointing to the α-carbon.
d. Is this molecule chiral? Explain.
e. Write the Fischer projection of the enantiomer of this amino acid. How would it be named?
f. A 50:50 mixture of this amino acid and its enantiomer is known as a _____ mixture.

11.61 Which of the following objects are chiral?
a. a corkscrew
b. an orange
c. a car
d. a nail

11.62 Answer the following questions for the two models of 2-butanol shown below.

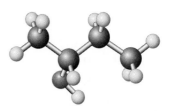

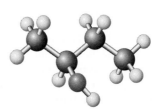

a. What functional group is present in these two models?
b. Do these two compounds have the same chemical formula and connectivity of atoms? Write the condensed structural formula for both.
c. Do these two compounds have the same three-dimensional spatial orientation?
d. As drawn, what atoms do not overlay perfectly?
e. Are these two compounds superposable? Explain.
f. Are these two compounds mirror images of one another? Explain.
g. Are these two compounds enantiomers? Explain.
h. Are these two compounds chiral? Explain.

11.63 What would a racemic mixture of alanine contain?

Peptides

11.64 What are the differences between a dipeptide, a tripeptide, and a polypeptide?

11.65 Write the structures of the two different dipeptides that can be formed from the following pairs of amino acids:
  a. alanine and methionine
  b. aspartic acid and lysine
  c. threonine and cysteine
  d. serine and glycine

11.66 Write the structures of the two different dipeptides that can be formed from the following pairs of amino acids:
  a. histidine and leucine
  b. valine and proline
  c. glutamic acid and arginine
  d. asparagine and isoleucine

11.67 Write the structure of the tripeptide Leu-Glu-His.

11.68 Write the structure of the tripeptide Tyr-Ala-Cys.

11.69 Palmitoyl-pentapeptide-3 is a component of antiaging creams. It contains the pentapeptide below.

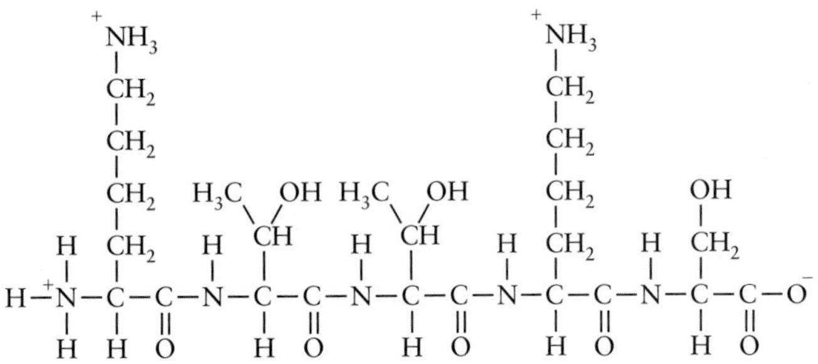

$$
\overset{+}{N}H_3 \quad\quad\quad\quad\quad\quad\quad\quad \overset{+}{N}H_3
$$

(structure of the pentapeptide with side chains CH₂-CH₂-CH₂-CH₂-NH₃⁺, CH₃/OH CH(OH), CH₃/OH CH(OH), CH₂-CH₂-CH₂-CH₂-NH₃⁺, CH₂-OH drawn across the backbone H-N-C-C-N-C-C-N-C-C-N-C-C-N-C-C-O⁻)

  a. Give the amino acid sequence for palmitoyl-pentapeptide-3.
  b. Which end is the N-terminus?
  c. Which end is the C-teminus?
  d. Identify the functional groups on the side chains.

11.70 The pentapeptide Gln-Tyr-Asn-Ala-Asp is present in human cerebrospinal fluid, and its concentration increases in demyelinating diseases. In these diseases, the myelin sheath of nerves is damaged and impairs the conduction of nerve signals.
  a. What is the N-terminal amino acid?
  b. What is the C-terminal amino acid?
  c. How many peptide bonds does this pentapeptide contain? How many amino acids does this pentapeptide contain?
  d. What functional groups are present in the side chains?
  e. Write the structural formula for this pentapeptide.

11.71 Vasopressin, also known as antidiuretic hormone (ADH), controls resorption of water by the kidneys. This hormone has a similar structure to oxytocin except that it has Phe in place of Ile and Arg in place of Leu. Based on this information, which hormone would you expect to be more polar, oxytocin or vasopressin?

11.72 The skeletal line structure of bradykinin is shown below. This hormone causes smooth muscle to contract.

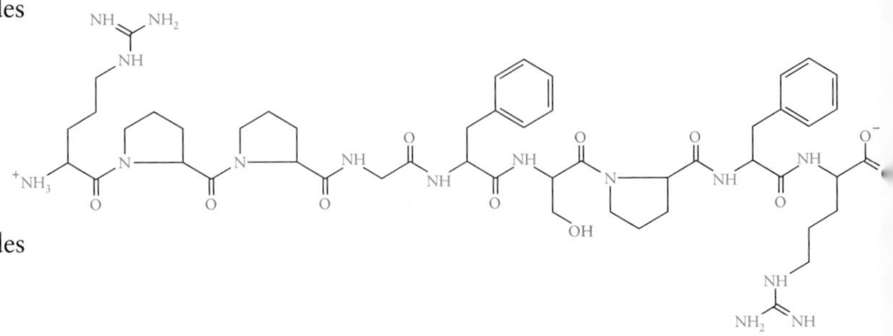

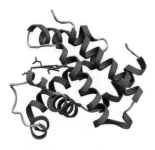

  a. How many amino acids does this peptide contain?
  b. Referring to Figure 11-4, write the amino acid sequence for bradykinin in the conventional manner using the three-letter abbreviations for the amino acids.

Protein Architecture

11.73 How does a cell know how to produce the amino acid sequence for a given protein?

11.74 What is the primary structure of a protein?

11.75 What are the two most common forms of secondary structure? How is secondary protein structure formed?

11.76 The structure of one of the subunits of hemoglobin is shown below.

  a. What is the major type of secondary structure present in this protein?
  b. Heme is part of this subunit. Is heme a protein? What is this part of the subunit called? What atom lies at the center of heme?
  c. What is a subunit? How many subunits does hemoglobin contain?

11.77 What electrostatic interactions are responsible for creating tertiary structure in a protein? Provide an example of each.

11.78 What amino acid needs to be present in a protein in order for a disulfide bridge to form? Would a disulfide bridge be part of the primary, secondary, tertiary, or quaternary structure of the protein?

11.79 Which amino acids have side chains that can form salt bridges?

11.80 Where are nonpolar regions of a protein likely to be found in a protein? Where are the polar regions of protein likely to be found?

11.81 What are the three general classes of proteins, distinguished by their quaternary structure and solubility?

11.82 Write the product formed when the thiol shown below is oxidized:

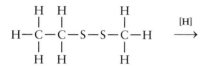

11.83 Write the products formed when the disulfide shown below is reduced. What functional group is present in the products?

H—C—C—S—S—C—H [H] →

11.84 For the following pairs of amino acids, indicate how they might interact to contribute to the tertiary or quaternary structure of a protein. Choose from among the following: *disulfide bond, salt bridge, hydrogen bonding,* or *dispersion forces.*
  a. glutamic acid and lysine
  b. tyrosine and lysine
  c. Ile and Ala
  d. two cysteines

11.85 Where in the body are most fibrous proteins found? How does their quaternary structure support the role fibrous proteins play in the body?

11.86 What shape is a globular protein? What role do globular proteins typically have in nature?

11.87 Where in the body are membrane proteins found?

11.88 Since the cell membrane is a hydrophobic environment, what types of amino acids will be found on the exterior of a membrane protein? How is this arrangement of amino acids different from proteins in an aqueous environment?

11.89 What agents can denature a protein? What happens when a protein is denatured?

Enzymes

11.90 What pH is optimal for most enzymes? What temperature is optimal for most enzymes?

11.91 Fill in the blank and indicate what each letter stands for:

$$E + S \rightleftharpoons \underline{\hspace{1.5cm}} \rightleftharpoons E + P$$

11.92 What happens to a coenzyme during an enzyme-catalyzed reaction? What are some examples of coenzymes?

11.93 What are some examples of cofactors?

11.94 An enzyme works by lowering the _____ for a reaction.

11.95 What types of electrostatic forces are responsible for forming the enzyme–substrate complex?

11.96 Describe how the conformation of the enzyme changes somewhat upon binding to its substrate. What is this model called?

11.97 Is the activation energy, E_A, higher or lower for an enzyme-catalyzed reaction compared to an uncatalyzed reaction?

11.98 What do enzyme inhibitors do to an enzyme?

11.99 How does a competitive inhibitor work? How does the shape of a competitive inhibitor compare to the shape of the substrate?

11.100 Where do noncompetitive inhibitors bind? How do noncompetitive inhibitors affect the shape of the active site?

11.101 For the energy diagram of an uncatalyzed reaction shown below, label the activation energy, E_A, and ΔH.

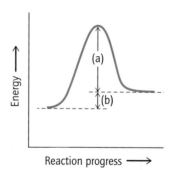

Chemistry in Medicine

11.102 What is the main symptom of hypertension? What are the health risks for someone who has hypertension?

11.103 Why does stopping the production of angiotensin II prevent hypertension?

11.104 Write the structure for the dipeptide Phe-His.

11.105 Indicate the bond broken in angiotensin I by *angiotensin converting enzyme (ACE).*

11.106 Where does captopril bind to ACE?

11.107 In the opening vignette, you learned that valine replaced glutamic acid in the hemoglobin of people who have sickle-cell anemia. Why does this small change affect the shape of the protein? What type of interaction is no longer present? What parts of the protein structure (primary, secondary, tertiary, or quaternary) have been affected?

11.108 Proteins are denatured in your stomach by exposure to a very low pH. Write the structure of the following tripeptide with all the functional groups in the correct form at pH = 1: Asp-Trp-Lys.

11.109 Why does a low pH denature most proteins? What kinds of interactions are changed within the protein?

Answers to Practice Exercises

11.1 a, b, d.

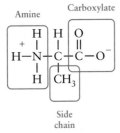

Amine and carboxylic acid are in their ionized forms. A carboxylate ion is the conjugate base of a carboxylic acid.

c. This amino acid is a zwitterion. Usually an amino acid is a zwitterion around physiological pH.

d. This amino acid is alanine.

e. The side chain is nonpolar.

f.

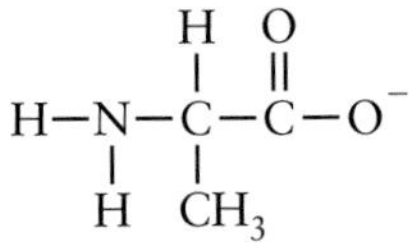

$$H-N-C-C-O^-$$

The net charge on the amino acid is -1.

11.2

Amino Acid	Non-polar	Polar	Acidic Side Chain	Basic Side Chain	Neutral Side Chain	Essential Amino Acid
Cysteine		X			X	
Lysine		X		X		X
Glutamine		X			X	
Methionine	X					X
Aspartic Acid		X	X			

11.3 Methionine and cysteine contain a sulfur atom. Cysteine contains a thiol (—S—H) functional group.

11.4 Proline.

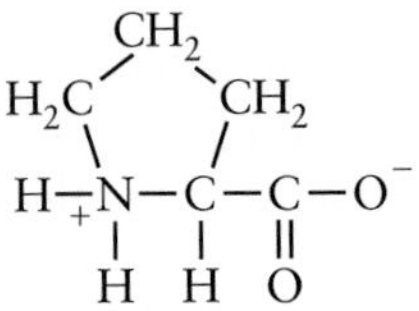

Proline (Pro)

11.5

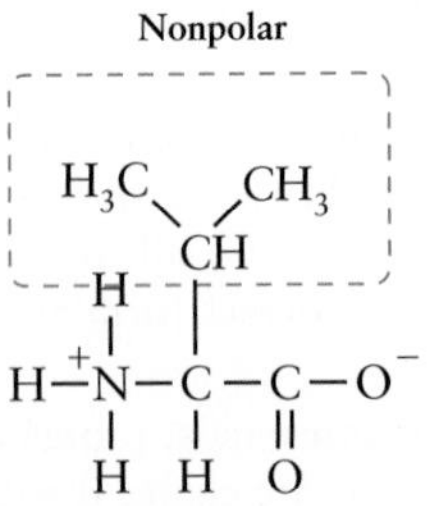

Nonpolar

Valine

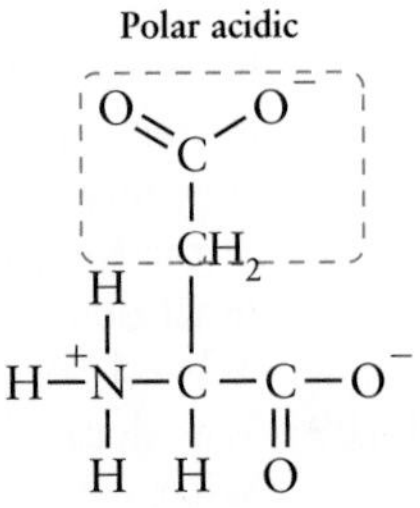

Polar acidic

Aspartic acid

11.6 Meat and fish contain all the essential amino acids.

11.7
a. A golf club is chiral.
b. A baseball hat is achiral.
c. A grand piano is chiral.
d. A sock is achiral.
e. A glove is chiral.
f. A basketball is achiral.

11.8
a. False
b. True
c. True
d. False

11.9 The D designates one of two enantiomers. The molecule contains two hydroxyl groups and an aldehyde.

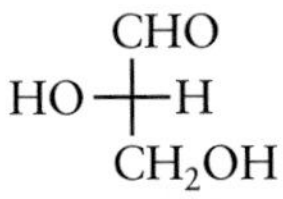

$$CHO$$
$$HO——H$$
$$CH_2OH$$
L-Glyceraldehyde

11.10 It means there is a 50:50 mixture of both ibuprofen enantiomers in the drug.

11.11 a, b.

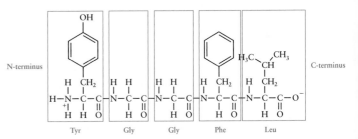

c. Tyr-Gly-Gly-Phe-Leu
d. Three amino acids are chiral: Tyr, Phe, Leu.
e.

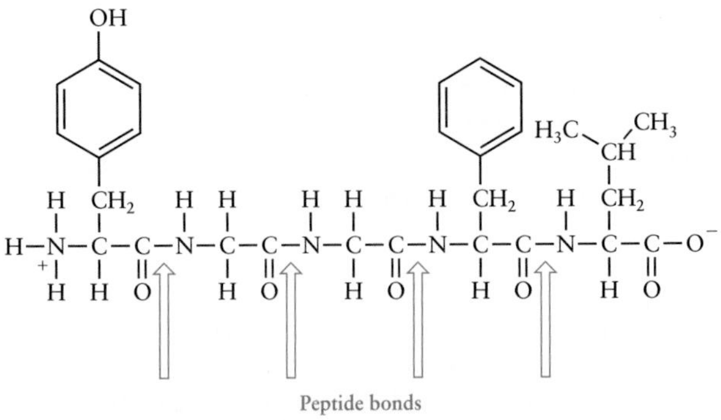

Peptide bonds

f. The peptide is chiral.

11.12 Yes, they are different compounds; they are structural isomers.

11.13

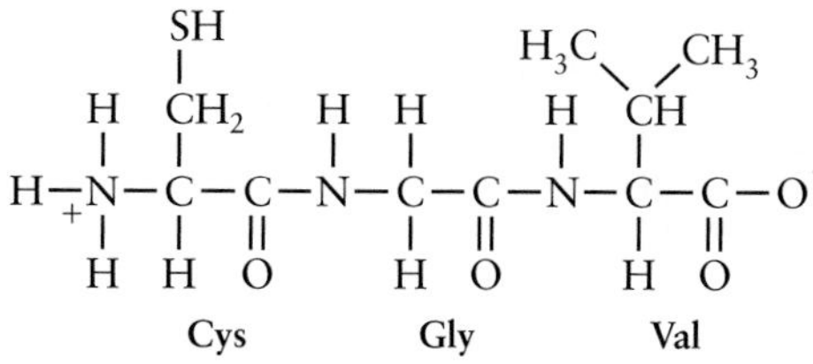

11.14 **a.** cysteine
b. glycine
c. 9
d. 8
e. the two cysteine residues contain a thiol

11.15 **a.** 4 N-termini, one for each polypeptide

11.16 Primary, secondary, tertiary, and quaternary

11.17 The native conformation of a protein is the three-dimensional structure of the protein required to perform its function.

11.18 **a.**

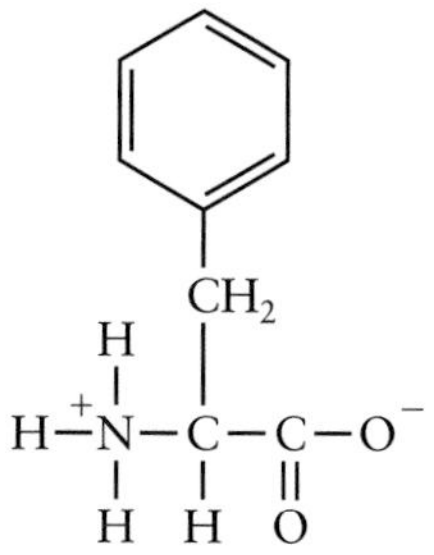

b. At amino acid 508, counting from the N-terminus.
c. A gene is a segment of DNA that codes for a particular protein.
d. You would expect cystic fibrosis to be a hereditary disease because it is a protein disorder and therefore prescribed in the DNA. DNA is the genetic material we inherit from our parents. Hence, the disease is not infectious.
e. It is missing in Phe, causing the protein to fold differently, thereby affecting its function.

11.19 **a.** No, the primary structure is not different.
b. α-helices
c. β-pleated sheets
d. In the normal protein there are more α-helices and only a small section of β-pleated sheet, whereas in the prion there is a significant amount of β-pleated sheet, and somewhat less α-helix.

11.20 Partial charges arise from polar bonds created between atoms with different electronegativities. The most polar bonds are seen for N—H and O—H, which allows for hydrogen bonding (dashed red line).

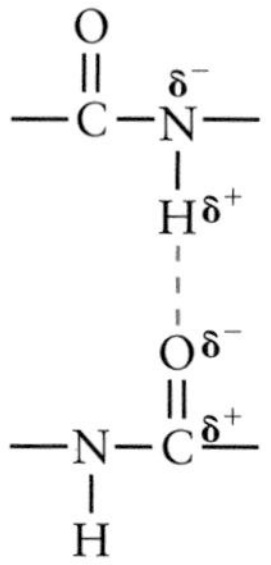

11.21

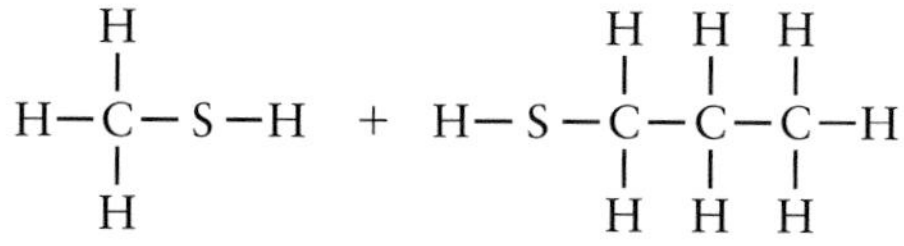

11.22

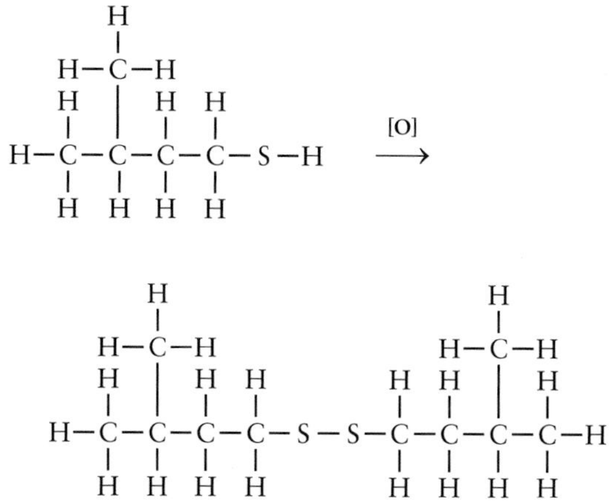

11.23 **a.** hydrogen bonding
b. disulfide bridge
c. dispersion
d. salt bridge

11.24 tertiary

11.25 **a.** salt bridge
b. hydrogen bonding
c. dispersion forces
d. disulfide bridge
e. dispersion

11.26 primary protein structure

11.27 **a.** membrane, because it appears to reside primarily within the cell membrane
b. nonpolar
c. polar
d. polar
e. secondary protein structure
f. several transmembrane-spanning regions

11.28 Detergents, change in pH, change in temperature, mechanical agitation, some metals

11.29 Primary protein structure

11.30 Fibrous proteins contain parallel peptide chains resulting in long fibers. They are insoluble in water and have a structural role. Globular proteins are folded into a complex overall spherical shape. They are soluble in water and serve as enzymes, hormones, or antibodies.

11.31 fibrous proteins

11.32 globular

11.33 Elastins are more flexible, and collagens are more rigid fibrous proteins.

11.34 The induced fit means that after the substrate binds to the enzyme (like a lock-and-key) the enzyme makes further adjustments to its conformation to create an even better fit between the substrate and the active site—the induced fit.

11.35

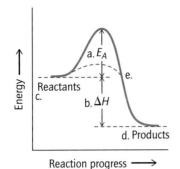

11.36 The active site is a cleft or pocket in the enzyme that binds the substrate. It is complementary in shape to the substrate.

11.37 The enzyme–substrate complex; it is the enzyme with its substrate bound to the active site.

11.38 **a.** E_A
b. the rate of the reaction

11.39 According to the graph in Figure 11-25a, *pepsin* has its maximum catalytic activity at pH = 2 (dashed line).

11.40 The coenzyme gains and loses electrons and hydrogen atoms. It is a coenzyme because it works together with an enzyme that cannot function without it, and it is chemically altered in the reaction.

11.41 In the laboratory, the rate of a reaction can be increased by raising the temperature, increasing the concentration of reactants, or adding a catalyst.

11.42 **a–d.** Rate decreases in all cases.

11.43 a competitive inhibitor

11.44 a competitive inhibitor

Carbohydrates: Structure and Function

Sugar is a carbohydrate, and sucrose is the carbohydrate in table sugar. It is a molecule composed of glucose and fructose. [Emilio Ereza/Alamy]

OUTLINE

 This icon indicates that a **Problem-Solving Tutorial** is available at www.whfreeman.com/gob

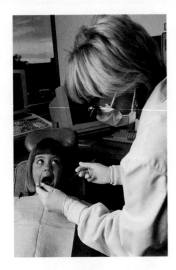

Cleanings by a dental hygienist help prevent tooth decay. [Scott T. Baxter/ Photodisc/Getty Images]

$$
\begin{array}{c}
\text{COOH} \\
| \\
\text{H} - \text{C} - \text{OH} \\
| \\
\text{CH}_3
\end{array}
$$
Lactic acid

Dental Caries and the Sweet Tooth

Do you visit the dentist to have your teeth cleaned and checked twice a year? Are you brushing your teeth at least twice a day? There seems to be no debate that these are good practices for preventing tooth decay, more formally known as *dental caries*, the most common disease in the world.

Why do teeth need to be cleaned regularly to avoid tooth decay? After all, teeth are meant for chewing food. At this point it should be no surprise to you that the explanation lies in the chemistry of the mouth.

Crucial to the formation of dental caries is a class of biomolecules known as *carbohydrates*, which are the subject of this chapter. Carbohydrates, a significant part of the diet, are large organic molecules composed of repeating simple sugars—or *monosaccharides*. For example, glucose (blood sugar) and fructose are simple sugars. Among the evidence for the role of carbohydrates in dental caries is a study showing that Eskimos who lived on a traditional diet of meat and fish that lacked carbohydrates had little or no dental caries, while Eskimos who had switched to a modern diet were found to have the same amount of dental caries as the rest of the population.

Together with fats, carbohydrates supply the body with the fuel that cells need to produce energy. Unfortunately, carbohydrates also serve as a fuel for the anaerobic bacteria that live in our mouths, the real culprit in dental caries.

Anaerobic bacteria are organisms that do not require oxygen to produce energy. An anaerobic biochemical process known as *lactic acid fermentation* is at the root of the cause of dental caries. You will learn more about lactic acid fermentation later in the chapter, but in brief, it is a biochemical process that produces energy from sugar, and in so doing, produces lactic acid as a by-product. This is the same type of fermentation used by the food industry to make sourdough bread, yogurt, sauerkraut, pickles, kimchi, and other foods. Indeed, the sour taste of sourdough bread, yogurt, and sauerkraut is due to the production of lactate by microorganisms as a by-product of fermentation.

It is lactic acid that actually causes teeth to decay. As bacteria feed on the sugar in your mouth, lactic acid is produced as a by-product. Lactic acid is a carboxylic acid and, like all acids, releases a proton, H^+, thereby increasing the acidity of local the environment to a pH of 4-5. Unfortunately, at this low pH tooth enamel literally dissolves. Although the enamel is slowly regenerating, the decay process occurs much faster than the regeneration process at this pH. Eventually a cavity is formed, and the only way to prevent further loss of tooth is to have a dentist remove the decayed area and fill the tooth with dental amalgam, an inert inorganic mixture that replaces the lost enamel.

Since anaerobic bacteria are always present in the mouth, the only way to prevent the progression of dental caries is to remove the sugar that allows these bacteria to thrive. Hence, the reason you reach for your toothbrush and schedule regular cleanings with the hygienist.

In the last chapter you learned about proteins, one of the four important biomolecules. In this chapter you will study the most abundant of the biomolecules: the carbohydrates. You will first learn about the chemical structure of carbohydrates and then study their important functions. In Chapter 8 you learned that your body derives energy from the catabolism of carbohydrates. Indeed, the oxidation of carbohydrates is the primary way for cells to obtain energy. Carbohydrates also serve an important function as cell markers, most notably in determining the four blood types A, B, AB, and O, as described at the end of the chapter.

12.1 | **Role of Carbohydrates**

Carbohydrates are found in foods such as fruits, vegetables, pasta, bread, potatoes, and rice. Half the earth's carbon exists in the form of two types of carbohydrates found in plant material: cellulose and starch. *Cellulose* provides structure and *starch* provides energy for a plant. Plants synthesize carbohydrates from carbon dioxide (CO_2) and water (H_2O) in a famous biochemical pathway known as **photosynthesis,** shown in the top half of Figure 12-1. The energy necessary to drive photosynthesis is supplied by sunlight. In the body, carbohydrates are metabolized back to carbon dioxide and water, releasing energy through another series of biochemical pathways. The result is essentially the reverse of photosynthesis.

Overview of Catabolism

Energy can be extracted from carbohydrates, fats, and when necessary, proteins, through biochemical pathways that are collectively referred to as catabolism. Catabolism is divided into three stages, as illustrated in Figure 12-2.

In the first stage of catabolism each biomolecule is hydrolyzed into its smaller building blocks: Proteins are hydrolyzed into amino acids; fats are hydrolyzed into fatty acids and glycerol; and carbohydrates are hydrolyzed into monosaccharides.

In the second stage of catabolism, each of these smaller building blocks is broken down further, by separate biochemical pathways. All these pathways converge at the same final product, acetyl CoA. Acetyl CoA is considered the central molecule of catabolism.

In the third and final stage of catabolism, acetyl CoA enters a biochemical pathway called the citric acid cycle, which extracts energy in the form of electrons from acetyl CoA. In the process, acetyl CoA is converted into two molecules of carbon dioxide and coenzyme A. The electrons harvested in the citric acid cycle

Pasta, bread, potatoes, rice, and grains are high in carbohydrates. [Charles D. Winters/Photo Researchers, Inc.]

Every year 100 billion metric tons of CO_2 and water are converted into cellulose and other carbohydrates through photosynthesis in plants. 1 ton = 1000 kg.

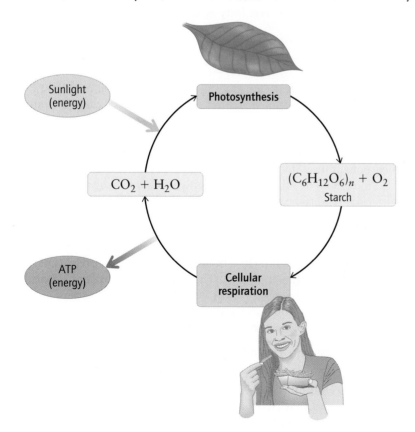

Figure 12-1 Photosynthesis and cellular respiration are two interconnected biochemical pathways: Photosynthesis produces carbohydrates and absorbs energy; while cellular respiration metabolizes carbohydrates and releases energy.

Figure 12-2 Catabolism, the biochemical pathways that produce energy from the food we eat, occurs in three stages.

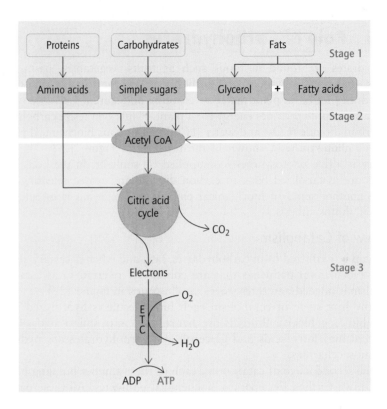

ultimately reduce oxygen to water, a process that drives the phosphorylation of ADP to ATP, the primary energy currency of the cell. You will study the details of this third stage of catabolism in Chapter 14.

In this chapter you will focus on the biochemical pathway that converts monosaccharides into acetyl CoA. In the next chapter you will learn the biochemical pathway that converts fatty acids into acetyl CoA.

The word *saccharide* comes from the Greek word sakcharon, meaning *sugar*.

The names of many carbohydrates end in "-ose" as in gluc**ose**.

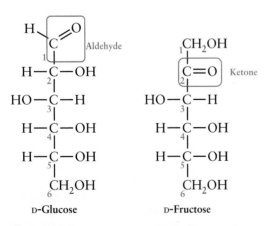

Figure 12-3 Two common sugars: D-glucose and D-fructose.

12.2 | Monosaccharides

To understand carbohydrate function and metabolism, you must first become familiar with the chemical structures of the simplest carbohydrates: the monosaccharides.

Monosaccharides *are carbohydrates that cannot be hydrolyzed into simpler carbohydrates.* Monosaccharides are the basic structural building blocks for all carbohydrates. Since monosaccharides cannot be hydrolyzed into anything *simpler*, they are also known as *simple sugars*. Monosaccharides tend to be crystalline colorless solids with a sweet taste.

Structure of Monosaccharides

Monosaccharides consist of a carbon chain containing an aldehyde (RCHO) or a ketone (RCOR) and two or more hydroxyl (OH) groups (alcohol functional groups). The most common monosaccharides are glucose and fructose, which each have five OH groups, shown in Figure 12-3. Monosaccharides are soluble in water because their OH groups are able to form hydrogen bonds with water molecules.

Chirality and the D-Sugars

Most monosaccharides are chiral. Recall from the last chapter that a chiral molecule is one that is nonsuperposable on its mirror image and that two such stereoisomers are called enantiomers. In Chapter 11 you learned that Fischer projections are often used to depict chiral molecules, because an enantiomer is readily drawn by simply exchanging the horizontal groups at every cross hair. The Fischer projection of a monosaccharide is similar, except that there are several carbon atoms represented by cross hairs, each containing a hydroxyl group. The Fischer projection of glucose is shown in Figure 12-3.

To draw the enantiomer of a monosaccharide, every group at a cross hair must be exchanged. For example, exchanging all the OH and H atoms at the cross hairs in the Fischer projection of D-glucose (Figure 12-4a) produces L-glucose (Figure 12-4b). D-glucose and L-glucose are enantiomers, nonsuperposable mirror-image stereoisomers. Similarly, D- and L-fructose are enantiomers. *In a Fischer projection, a pair of enantiomers can be readily identified because every horizontal OH and H pair at a cross hair has opposite orientations.*

The two enantiomeric forms of a carbohydrate each have the same name but are distinguished by the prefix D- or L- before the name. For example, D-glucose and L-glucose are enantiomers. D- and L-sugars are easily distinguished in a Fischer projection, because a D-sugar has the OH group at the cross hair farthest from the carbonyl group pointing to the *right*, as shown in Figure 12-4a. Conversely, in an L-sugar, the OH group at the cross hair farthest from the carbonyl group points to the *left*, as shown in Figure 12-4b.

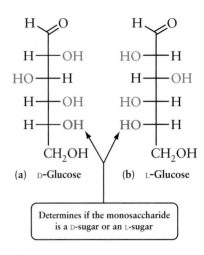

Figure 12-4 Identifying D and L sugars: In a Fischer projection, the OH at the cross hair farthest from the carbonyl group points right in a D sugar and left in an L sugar.

> **D-sugar:** The OH at the cross hair farthest from the carbonyl group points *right*.
>
> **L-sugar:** The OH at the cross hair farthest from the carbonyl group points *left*.

Nature produces only D-sugars, with only a few exceptions. Thus, you can assume that a monosaccharide is the D-sugar when it is listed without its D or L designation.

What is the stereochemical relationship between D-glucose and D-galactose, two D-sugars with the same chemical formula and the same connectivity of atoms, but that are not mirror images? D-glucose and D-galactose are diastereomers, another type of stereoisomer. ***Diastereomers** are nonsuperposable stereoisomers that are not mirror images. In a Fischer projection, a pair of diastereomers will have at least one, but not all, of the groups/atoms at cross hairs interchanged.* Thus, D-glucose and D-galactose are diastereomers, because they have the same configuration at C2, C3, and C5, but not at C4, as shown in Figure 12-5. They have different names, because they are diastereomers.

Haworth Drawings

Monosaccharides with 5 or 6 carbon atoms come in two forms: an open-chain form, which you have already seen in Figures 12-3 through 12-5, and a ring form. Most open-chain monosaccharides undergo a reaction to form a five-membered ring containing an oxygen atom, called a **furanose,** or a six-membered ring containing an oxygen atom, called a **pyranose.** The convention for drawing these rings is to place the ring oxygen at the top of the ring in a furanose, and on the top right of the ring in a pyranose, as shown in red in the margin.

The hydroxyl groups attached to the carbon atoms are oriented either above or below the ring. The ring form is written as a **Haworth projection,** which is a flattened representation of the ring that emphasizes the orientation of each hydroxyl group either *above* or *below* the ring. These hydroxyl-group orientations are important because they describe the stereoisomer and hence the identity of the monosaccharide. Recall from Figure 12-5 that the only difference between D-glucose and D-galactose is the orientation of the hydroxyl

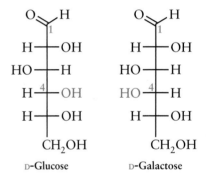

Figure 12-5 The diastereomers D-glucose and D-galactose have a different configuration at C4. They are both D-sugars because the last OH at a cross hair is oriented to the right.

group at C4. Figure 12-6 shows these two monosaccharides in their pyranose form. The C4 hydroxyl group is positioned below the ring in D-glucose and above the ring in D-galactose. An OH group that is oriented to the left in a Fischer projection appears above the ring in a Haworth projection, and an OH group that is oriented to the right in a Fischer projection appears below the ring in a Haworth projection.

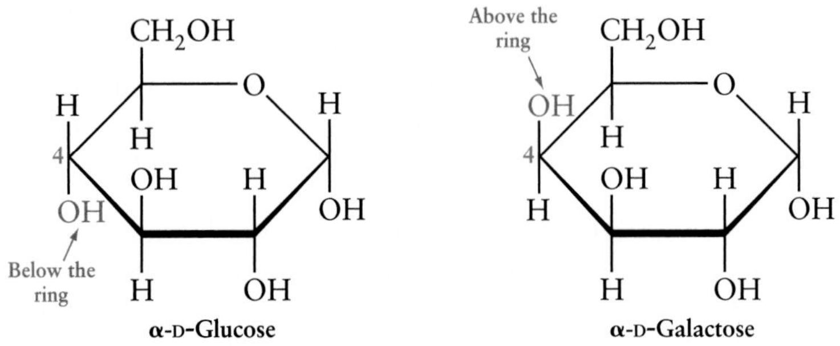

Figure 12-6 Haworth projections of the pyranose ring forms of glucose and galactose show that the two molecules differ in the orientation of the OH group at C4, either above or below the ring.

In the pyranose and furanose form of a monosaccharide, a D-sugar always has the C6 CH$_2$OH group positioned above the ring, as shown in Figure 12-7.

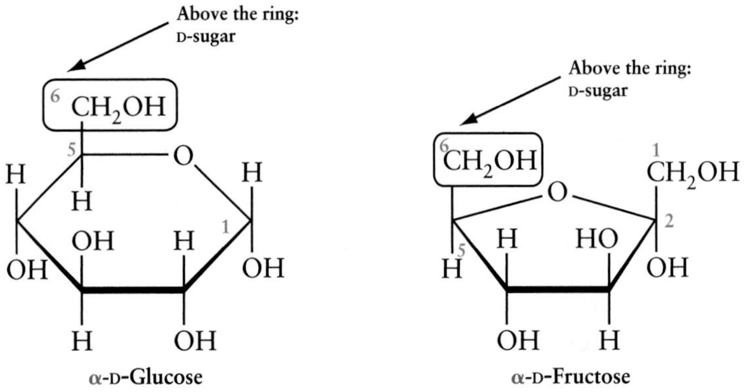

Figure 12-7 Haworth projections of D-glucose and D-fructose show that in both ring forms the CH$_2$OH group at C6 is positioned above the ring in a D-sugar.

The carbon atom bonded to both the ring oxygen atom and a hydroxyl group is known as the **anomeric carbon.** When the hydroxyl group on the anomeric carbon is below the ring, the sugar is known as the **α-anomer,** and when it is above the ring, the sugar is known as the **β-anomer.** For example, there are two pyranose ring forms of D-glucose: α-D-glucose and β-D-glucose, as shown in Figure 12-8. Both the α-anomer and the β-anomer are formed whenever a furanose or pyranose ring is formed.

> **α-anomer:** OH on the anomeric carbon is *below the ring.*
>
> **β-anomer:** OH on the anomeric carbon is *above the ring.*

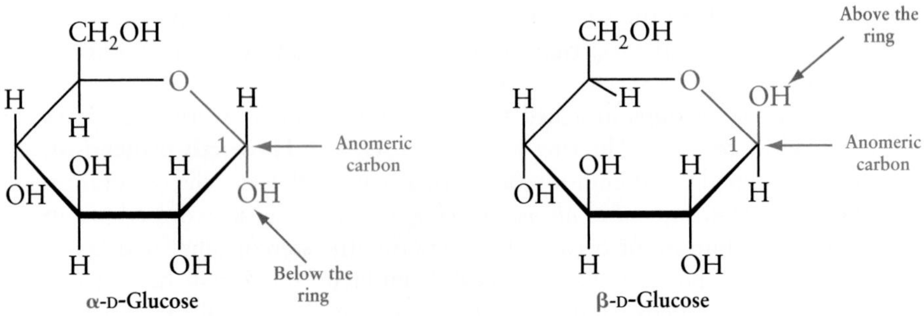

Figure 12-8 Haworth projections of α-D-glucose and β-D-glucose show the α- and β-anomeric forms of glucose.

It is the convention when writing Haworth projections of monosaccharides to place the anomeric carbon to the right of the ring oxygen. A pyranose ring is numbered beginning with C1 at the anomeric carbon.

Fructose is an example of a monosaccharide that exists as a furanose ring, as shown in Figure 12-9. The anomeric carbon appears at C2 in a furanose ring. A furanose has a CH_2OH group at the anomeric carbon, numbered C1. As with a pyranose, the OH group on the anomeric carbon can be either a or b, so the C1 CH_2OH group is also either above or below the ring. By locating the anomeric carbon, you can readily distinguish the CH_2OH group at C1 from the CH_2OH group at C6 that exists in a D-sugar. This can be a useful aid in identifying monosaccharides when the rings are rotated in the drawings of some disaccharides.

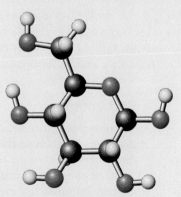

Figure 12-9 Haworth projections of α-D-fructose and β-D-fructose show the α- and β-anomeric forms of fructose.

To gain more familiarity with the ring forms of a monosaccharide, perform *The Model Tool 12-1: D-Glucose*.

The Model Tool 12-1 D-Glucose

I. Construction of β-D-Glucose

1. Obtain 6 black carbon atoms, 6 red oxygen atoms, 12 light-blue hydrogen atoms, 24 straight bonds, and two bent bonds.

2. Construct a model of β-D-glucose:

   a. Form a six-membered ring using five carbon atoms and one oxygen atom.

   b. Adjust the conformation of the ring so that it appears like the tube model shown below when viewed from the side. (Your model will not have all the OH groups attached yet.) Let the ring oxygen atom serve as a reference point by placing it in the position shown in the ball-and-stick model and tube model below.

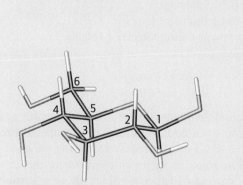

   c. To create a D-sugar, place the C6 CH_2OH group above the ring.

   d. To create the β-anomer, place the OH group on the anomeric carbon above the ring. This OH group and the CH_2OH group from part (c) should both be above the ring.

  e. At C2 and C4, place the OH groups so that they are below the ring.

  f. At C3, place the OH so that it is above the ring.

  g. Add hydrogens to all remaining open holes.

II. Inquiry Questions

3. Compare your model to the Haworth structure shown at right.

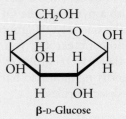

β-D-Glucose

  a. Is your molecule flat like the Haworth projection drawn? Explain. How would you describe the shape of your ring?

  b. Are the OH groups that are above the ring in the Haworth projection generally oriented above the ring in your model? You will notice that some groups are more obviously pointing above or below the ring, while other groups appear only slightly above or below the ring. This is normal.

  c. Locate the anomeric carbon in your model. How can you quickly locate this carbon atom? How is it different from all the other carbon atoms in the model?

  d. Is the OH group on the anomeric carbon in your model above or below the ring? In other words, did you build α-D-glucose or β-D-glucose?

III. Conversion to α-D-glucose

4. Convert your model of β-D-glucose into a model of α-D-glucose. What two atoms or groups do you need to exchange? Your model should look like the tube model shown, except that your model is a ball-and-stick model. Draw a Haworth structure of the model you just made using the template provided.

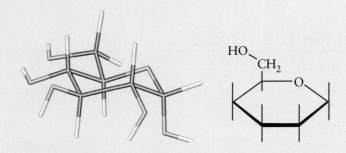

IV. Inquiry Questions

5. Examine your model and answer the questions below:

  a. Is the OH group on the anomeric carbon above or below the ring? Is it opposite C6?

  b. Are α-D-glucose and β-D-glucose superposable? Explain.

  c. Are α-D-glucose and β-D-glucose mirror image isomers? Explain.

  d. Are α-D-glucose and β-D-glucose enantiomers or diastereomers? Explain.

Mutarotation

Pyranoses and furanoses are unstable in aqueous solution and readily undergo reactions that cause the ring to open and reclose. The ring opens when the bond between the ring oxygen and C1 or C2 breaks, producing the open-chain form of the monosaccharide. In going to the open-chain form, the anomeric carbon becomes the carbonyl carbon of either an aldehyde or a ketone, and the ring oxygen becomes the hydroxyl group at C5.

When the open-chain form closes again, it can become either an α- or a β-anomer. Even when the pure crystalline form of α-D-glucose is isolated,

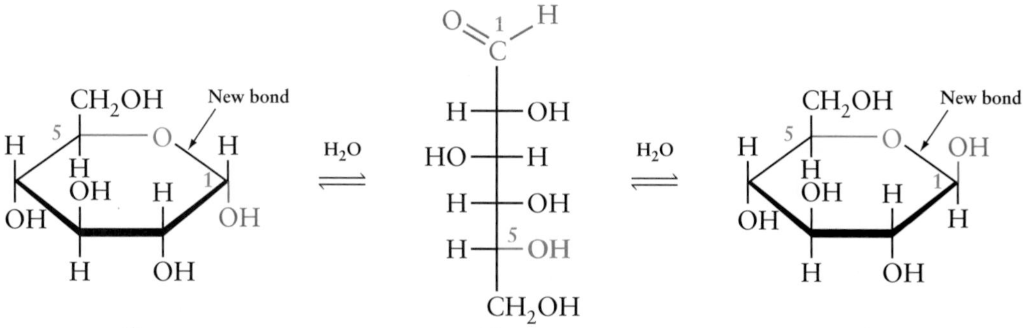

Figure 12-10 Equilibrium forms of D-glucose.

it turns into a mixture of the α- and the β- forms when placed in aqueous solution, a process known as **mutarotation.** Similarly, when the pure β-form is placed in water, the same ratio of α-and β-anomers is produced. In other words, all three forms of a monosaccharide, α-, β-, and open-chain form, are in equilibrium in aqueous solution (Figure 12-10).

Equilibrium favors both ring forms over the open-chain form, as seen by the ratios given in Figure 12-10. The aldehyde functional group in the open-chain form is capable of undergoing chemical reactions, such as oxidation or reduction. Even though only a small amount of the open-chain form is present, once it undergoes reaction, the equilibrium will shift to replenish the open-chain form until all the sugar molecules have reacted, a classic application of Le Châtelier's principle. The glucose test strip, used to test urine for the presence of glucose, is based on the reactivity of monosaccharides through their open-chain aldehyde form. In this example the aldehyde oxidizes to a carboxylic acid, which is accompanied by an obvious color change indicating the presence of glucose.

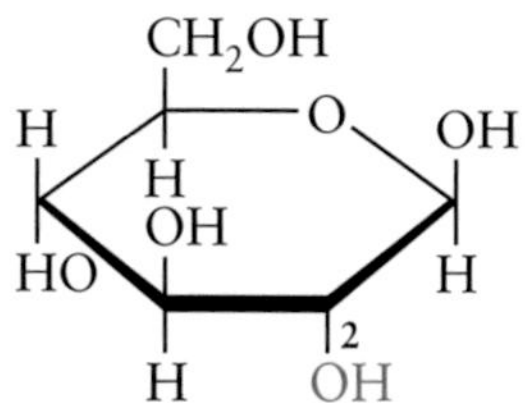

Modified Monosaccharides

Some molecules contain modified monosaccharides. These are monosaccharides with a modification to one of the hydroxyl groups on the ring:

- amino sugars have an amine functional group replacing an alcohol at one of the carbon atoms;
- phosphosugars contain a phosphate ester at one of the hydroxyl groups;
- deoxysugars lack a hydroxyl group at one position; and
- glycosides contain an OR group instead of a hydroxyl group at the anomeric carbon.

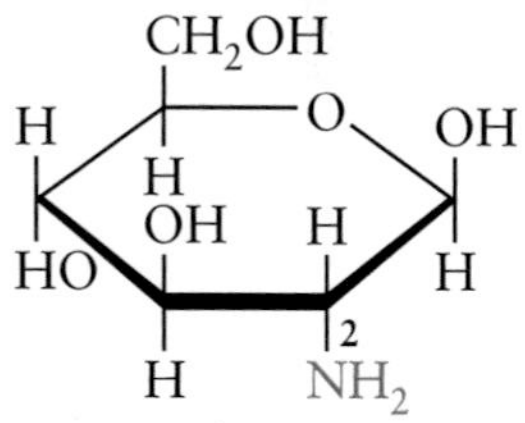

For example, D-glucosamine, shown in the margin, has an amine instead of an alcohol functional group at C2. The monosaccharide that serves as the structural backbone for DNA is 2-deoxyribose, another modified sugar. It is similar to D-ribose, except that it lacks an alcohol at C2, as shown in the margin.

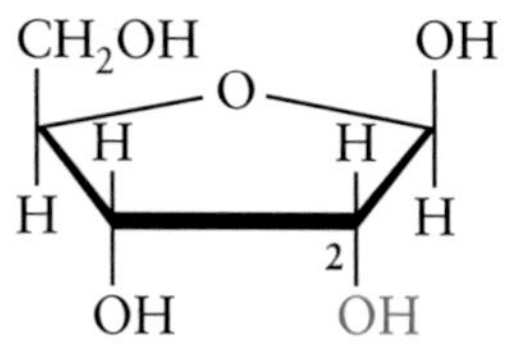

Glycosides contain an OR group at the anomeric carbon. Glycosides are formed by the reaction of a monosaccharide with an alcohol, in the presence of a catalyst. For example, methyl glycosides are formed when a monosaccharide reacts with methanol, CH_3OH, in a reaction that produces a glycoside and a molecule of water.

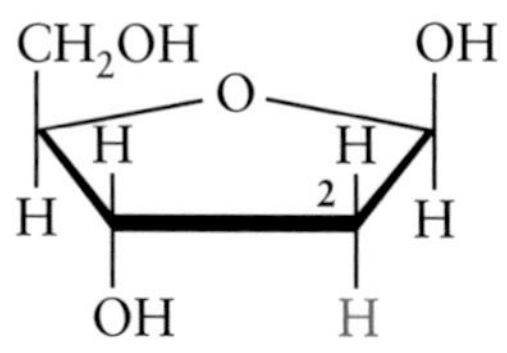

β-D-Glucose

β-D-Glucoseamine

β-D-Ribose

β-D-2-Deoxyribose

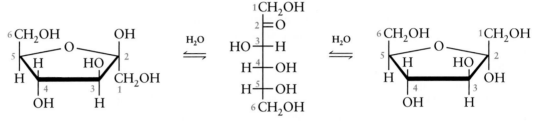

Monosaccharide   Methyl glycoside

When the alcohol is the hydroxyl group of another monosaccharide, a disaccharide consisting of two linked monosaccharides is formed. Disaccharides and polysaccharides are the subject of the next section.

Problem-Solving Tutorial:
Monosaccharides

WORKED EXERCISE 12-1   Monosaccharides

In aqueous solution D-fructose exists in three forms:

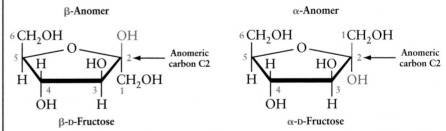

Problem-Solving Tutorial:
Ring Forms of Monosaccharides

a. Do the Haworth projections show that D-fructose is a *furanose* or a *pyranose*?
b. Label the anomeric carbon in both ring structures. Which carbon atom is the anomeric carbon in this sugar, C1 or C2?
c. Name the α-anomer and the β-anomer. How can you tell them apart?
d. Why are double arrows used in these chemical equations?
e. How can you tell that these structures represent D-fructose and not L-fructose?
f. Is α-D-fructose chiral? Is β-D-fructose chiral? Are α-D-fructose and β-D-fructose enantiomers or diastereomers?

SOLUTION
a. The Haworth projections show D-fructose as a furanose.
b and c.

β-Anomer   α-Anomer

β-D-Fructose   α-D-Fructose

Remember, you can distinguish the α- and β-anomers by whether the anomeric OH group is up (β) or down (α).
d. The double reaction arrows indicate that the reactions are reversible and that all three forms are present at equilibrium. The ring forms open and reclose so that all forms, open-chain and both ring forms, are present in aqueous solution.
e. In the furanose form, a D-sugar has the last carbon, C6 in this case, above the ring. In the Fischer projection of the open-chain form, the last OH group at a cross hair is oriented to the right in a D-sugar.
f. Yes, both are chiral. The two are diastereomers because they are nonsuperposable stereoisomers that are not mirror images.

PRACTICE EXERCISES

12.1 Two ring forms of galactose are shown below

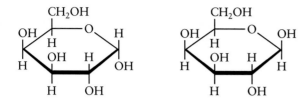

a. Are these D-sugars or L-sugars? How can you tell?
b. Name each form of galactose: ____-____galactose and ____-____galactose.
c. Do the Haworth projections represent pyranoses or furanoses? Explain.
d. Are these monosaccharides chiral?
e. What is the stereochemical relationship between these two forms of galactose: Are they enantiomers or diastereomers?
f. Number the carbon atoms in the ring. Which carbon atom is the anomeric carbon?
g. Which OH group can change its configuration: move from above the ring to below the ring as the ring opens and closes? What is this process called?
h. Would you expect galactose to be water soluble? Explain.

12.2 Fischer projections of three different monosaccharides are shown below.

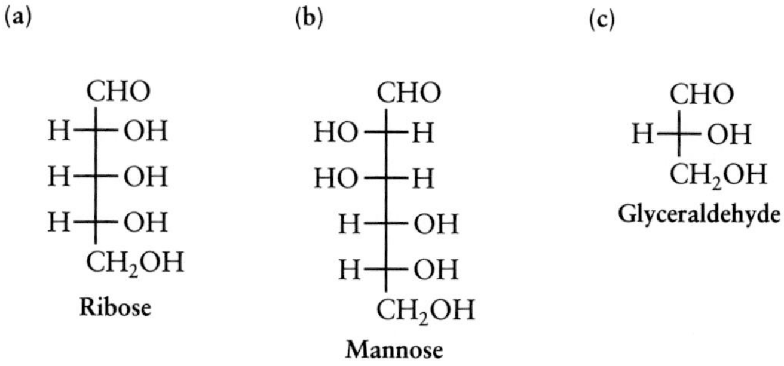

(a)

CHO
H——OH
H——OH
H——OH
CH₂OH
Ribose

(b)

CHO
HO——H
HO——H
H——OH
H——OH
CH₂OH
Mannose

(c)

CHO
H——OH
CH₂OH
Glyceraldehyde

a. Are these structures D-sugars or L-sugars? How can you tell?
b. How is D-mannose different from D-glucose? Are they enantiomers or diastereomers?
c. Write the structure of L-ribose. What is the stereochemical relationship between L-ribose and D-ribose?

12.3 Answer the questions below for the two ring forms of the monosaccharide shown.

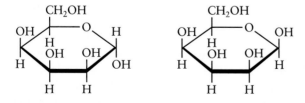

a. Label the anomeric carbons with a *.
b. Which structure is the α-anomer and which is the β-anomer?
c. What is the stereochemical relationship between these two ring forms: Are they *enantiomers* or *diastereomers*?

12.4 The exoskeleton of insects, spiders, and crustaceans is made of chitin, a carbohydrate containing a modified monosaccharide. The monosaccharide is similar to D-glucose except that the C2 hydroxyl group is replaced by an *N*-acetyl group, NHCOCH₃. Write the structure of this modified sugar.

12.3 | Complex Carbohydrates

Complex carbohydrates are composed of two or more monosaccharides linked together by a covalent bond. Starch and cellulose contain thousands of glucose units, and therefore are classified as **poly**saccharides. Carbohydrates composed of two monosaccharides are known as **di**saccharides.

Disaccharides

Disaccharides are carbohydrates that when hydrolyzed (reacted with water to break bonds) yield two monosaccharides. For example, hydrolysis of maltose yields two D-glucose molecules, as shown in Figure 12-11.

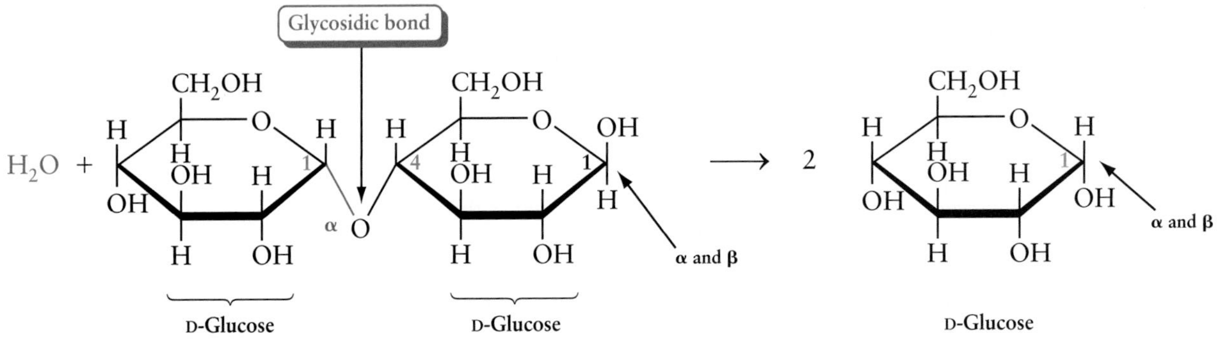

Figure 12-11 Hydrolysis of maltose produces two molecules of glucose.

The two monosaccharides that make up a disaccharide may be identical or different. For example, sucrose is composed of D-glucose and D-fructose, and therefore yields these monosaccharides upon hydrolysis:

$$\text{Sucrose} + H_2O \longrightarrow \text{glucose} + \text{fructose}$$

Three structural features characterize a disaccharide:

- the identity of the two monosaccharide components,
- which carbon atoms contain the hydroxyl group(s) that join the two monosaccharides, identified by carbon number, and
- the configuration of the anomeric carbon atom(s), α- or β-, joining the monosaccharides.

The covalent bond that joins two monosaccharides is known as a **glycosidic bond** or **glycosidic linkage.** A glycosidic bond always forms between the anomeric carbon atom of one monosaccharide (drawn on the left) and a hydroxyl group of another monosaccharide (drawn on the right). The hydroxyl group may be any involved, including the one on the anomeric carbon. The carbon numbers of the glycosidic bond are used to identify a disaccharide. For example, the two D-glucose units in maltose are linked together between the anomeric carbon (C1) of one glucose unit and the C4 alcohol of a second glucose unit, as shown in Figure 12-11. The convention is to write the sugar whose anomeric carbon is involved in the linkage on the left.

The glycosidic linkage in maltose is labeled α, because the oxygen atom on the anomeric carbon is below the ring. The glycosidic linkage in maltose is, therefore, referred to as an $\alpha(1 \rightarrow 4)$ linkage. *The α indicates the configuration of the anomeric carbon atom.* The numbers in parentheses indicate the carbon atoms on each sugar involved in the glycosidic linkage. The first number refers to the carbon atom involved in the sugar on the left and the second number refers to the carbon atom involved in the sugar on the right. The first number always refers to an anomeric carbon.

The monosaccharide component that is part of a glycosidic linkage does not open into its open-chain form in aqueous solution. This is in striking contrast to lone monosaccharides. Thus, an α-linkage remains α and a β-linkage remains β. To break a glycosidic linkage requires acid or a catalyst. Enzymes that facilitate the hydrolysis of glycosidic linkages are known as glycosidases. Humans have α-glycosidases in the stomach that hydrolyze α-linkages, but are unable to hydrolyze β-linkages.

Table 12-1 shows the structure of the common disaccharides: lactose, sucrose, cellobiose, and maltose. Sucrose is the disaccharide commonly found in table sugar and lactose is the disaccharide found in milk, sometimes referred to as milk sugar.

To gain more experience with disaccharides, perform *The Model Tool 12-2: Maltose.*

Fruits are high in carbohydrates, including the disaccharide sucrose.
[Chris Rout/Alamy]

WORKED EXERCISE 12-2 Disaccharides

Use Table 12-1 to answer the following questions about the disaccharide sucrose:
a. Identify the monosaccharide components.
b. Sucrose is unusual in that the linkage occurs between the anomeric carbons of both monosaccharides: the ____-anomer of D-glucose and the ___-anomer of D-fructose.
c. The glycosidic linkage in sucrose is classified as an α1→β2 linkage. Explain.
d. What ingredient found in most kitchens contains this disaccharide?

SOLUTION
a. D-glucose and D-fructose
b. the _α_-anomer of D-glucose and the _β_-anomer of D-fructose.
c. The linkage is an α1→β2 because it occurs between the α-anomer, C1, of glucose and the β-anomer, C2, of fructose via an oxygen atom joining the two carbon atoms. Both these carbon atoms are the anomeric carbons of their respective sugars, specifically the α-anomer of glucose (the top sugar) and the β-anomer of fructose (the bottom sugar).
d. Table sugar contains sucrose.

PRACTICE EXERCISES

12.5 Using Table 12-1, answer the following questions for cellobiose.
   a. Why is cellobiose classified as a dissacharide?
   b. What are the monosaccharide components of cellobiose?
   c. What type of linkage exists between the monosaccharides in cellobiose?
   d. How is cellobiose different from maltose?

12.6 Using Table 12-1, answer the following questions for lactose.
   a. Explain why lactose is classified as a dissacharide.
   b. What are the monosaccharide components of lactose?
   c. What type of linkage exists between the monosaccharides in lactose?
   d. How does lactose differ from maltose?
   e. What ingredient found in most kitchens contains this disaccharide?

12.7 Bees hydrolyze sucrose, obtained from the nectar of plants, to make honey. Based on this information, what monosaccharides does honey contain?

12.8 Critical Thinking Question: The blood sugar found in insects is not glucose but trehalose, a disaccharide that can withstand temperature differences better than glucose. Trehalose is a disaccharide composed of only glucose, in an α-1→ α-1 linkage.
   a. What monosaccharide(s) are produced upon hydrolysis of trehalose?
   b. Could an α-glycosidase catalyze the hydrolysis of this disaccharide?

Table 12-1 Some Common Disaccharides

Disaccharide	Structure	Monosaccharide Components	Linkage	Configuration of Anomeric Carbon in Linkage
Lactose	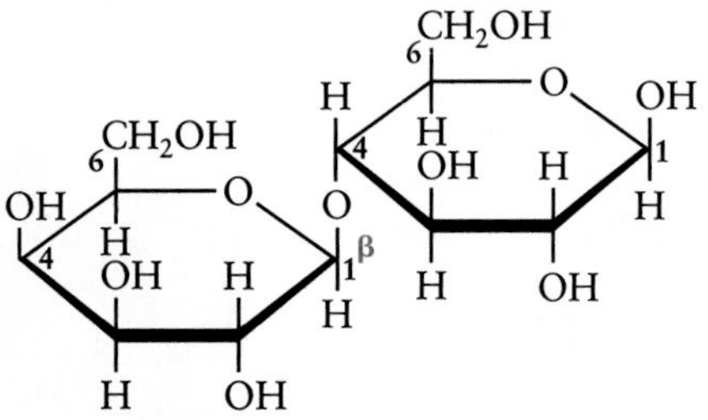 β(1→4)	Galactose and glucose	1, 4	β
Sucrose	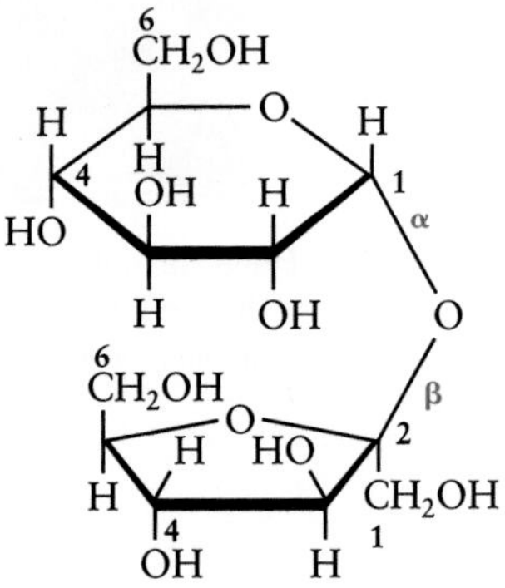 α1→β4	Glucose and fructose	1, 2	α, β
Cellobiose	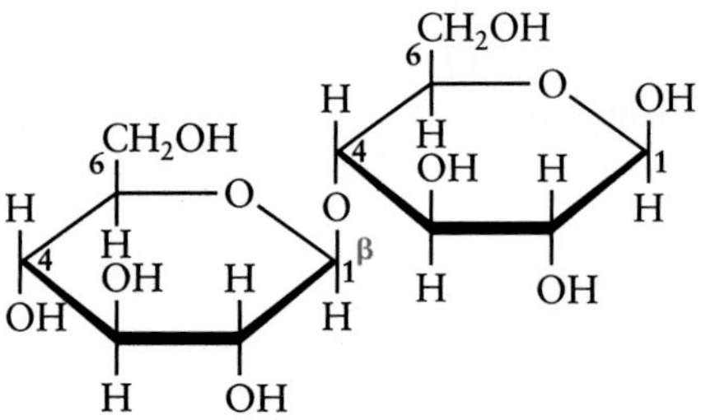 β(1→4)	Glucose	1, 4	β
Maltose	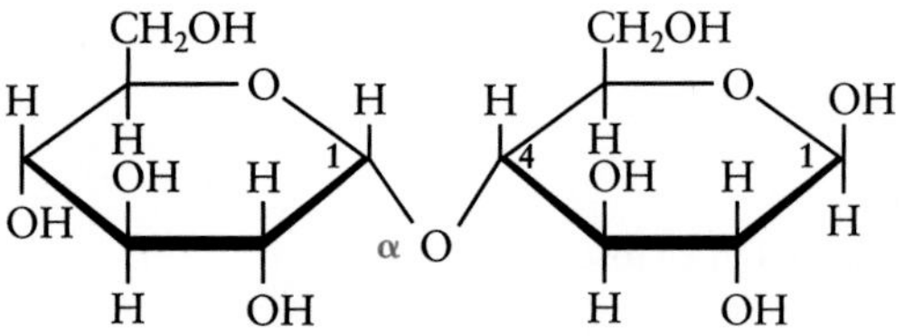 α(1→4)	Glucose	1, 4	α

The Model Tool 12-2 Maltose

This model tool exercise requires that you work with a partner or use two model kits.

I. Construction of Maltose

1. Build a model of α-D-glucose as described in *The Model Tool 12-1: Glucose.*

2. Find a partner who has also built a model of either α- or β-D-glucose.

3. To make the disaccharide maltose, you will need to link your two glucose models together. Examine the structure of maltose in Table 12-1. You must decide where to connect the sugars and how to construct the connection. Remember, the linkage in maltose is α(1→4). This means that one α-D-glucose will connect at the anomeric carbon to the other glucose molecule at its C4 hydroxyl group. One H_2O molecule in the form of OH and H will be removed from your two models in the process. The OH group on the anomeric carbon will be lost and replaced by the oxygen atom of the C4 alcohol on the other sugar, which loses only an H atom.

II. Inquiry Questions

4. Did you produce a water molecule as a by-product when you joined the two monosaccharides?

5. Find the two anomeric carbons in your model. Which anomeric carbon is still free to open to its open-chain form? What is the difference between the two anomeric carbons that accounts for this distinction?

6. Why is your model classified as a disaccharide? What bond is broken when your model is subjected to α-glycosidases?

Polysaccharides

Carbohydrates containing from 3 to 100 monosaccharides are known as **oligosaccharides** and those containing more than a hundred are known as **polysaccharides.** Most of the carbohydrates in your diet consist of polysaccharides containing thousands of glucose units linked through glycosidic bonds.

When the same monosaccharide is used repeatedly to build a polysaccharide, the polysaccharide is basically a polymer. A **polymer** is any large molecule with the same repeating structural component (poly means many). An individual unit of the repeating component is known as a monomer (mono means one). All the important polysaccharides in plants and animals are polymers composed of D-glucose. They differ primarily in whether the glycosidic linkage is α- or β- and in the amount of branching in the polysaccharide. The important polysaccharides found in plants or animals are

- starch,
- cellulose, and
- glycogen.

Plants store energy in the form of **starch,** a mixture of 20% amylose and 80% amylopectin. Humans and animals store energy in the form of glycogen. Glycogen, amylose, and amylopectin are all polysaccharides composed of glucose monomers connected mostly by α(1→4) glycosidic bonds. This is the same linkage found in maltose, but there are 250–4000 glucose monomers rather than just two!

Amylose, the minor constituent of starch, is an unbranched polysaccharide: each glucose monomer is connected to the next glucose in an α(1→4) linkage. The overall macromolecule adopts a helical shape, as shown in Figure 12-12a. Extensive hydrogen bonding between hydroxyl groups supports the helical shape

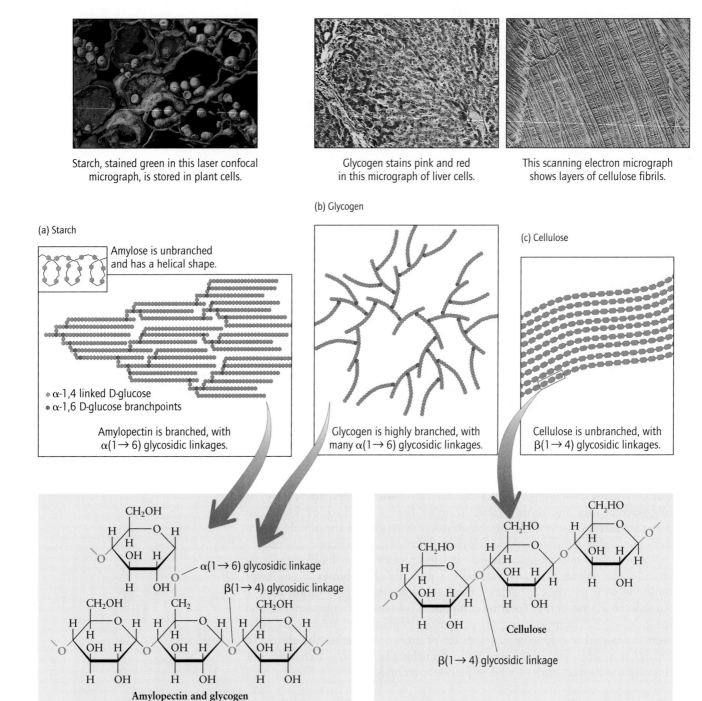

Starch, stained green in this laser confocal micrograph, is stored in plant cells.

Glycogen stains pink and red in this micrograph of liver cells.

This scanning electron micrograph shows layers of cellulose fibrils.

(b) Glycogen

(a) Starch

Amylose is unbranched and has a helical shape.

• α-1,4 linked D-glucose
• α-1,6 D-glucose branchpoints

Amylopectin is branched, with α(1 → 6) glycosidic linkages.

Glycogen is highly branched, with many α(1 → 6) glycosidic linkages.

(c) Cellulose

Cellulose is unbranched, with β(1 → 4) glycosidic linkages.

α(1 → 6) glycosidic linkage

β(1 → 4) glycosidic linkage

Amylopectin and glycogen

Cellulose

β(1 → 4) glycosidic linkage

Figure 12-12 The common complex carbohydrates—starch, glycogen, and cellulose—are polymers formed of thousands of glucose units linked together. They differ in the amount of branching and in the type of linkage, α or β. (a) Amylose and amylopectin are the components of starch, the storage form of carbohydrates in plants. (b) Glycogen is the storage form of carbohydrates in animals. (c) Cellulose plays a structural role in plants. [Top, left to right: Phototake Inc./Alamy; Educational Images Ltd./CMSP; Biophoto Associates/Photo Researchers, Inc.]

of the polymer. Amylose is not very soluble in water, and readily reacts with iodine, I_2, to form a distinctive blue color, a simple test that is often used to indicate the presence of starch.

Amylopectin, the major constituent of starch, is similar to amylose except that it is a branched polymer. The branches are created by $\alpha(1\rightarrow6)$ glycosidic linkages every 25–30 glucose units, as shown by the red hexagons in Figure 12-12a.

In humans and animals, glucose is stored as **glycogen** in liver and muscle cells. Glycogen is similar to amylopectin, except that it is even more highly branched: $\alpha(1\rightarrow6)$ glycosidic linkages are present every 8–12 glucose units (Figure 12-12b). Glycogen is synthesized by the body from glucose when glucose is in excess, following a meal, as a means of storing this essential fuel. Your body converts glycogen into glucose whenever blood glucose levels drop, in between meals. Thus, glycogen serves as a short-term reservoir of energy, supplying the body with glucose for about a day, under normal activity. Glycogen stores are depleted more quickly with vigorous activity.

Cellulose, the main component of wood, paper, and cotton, is the third polysaccharide composed of glucose. However, the glycosidic bonds in cellulose are $\beta(1\rightarrow4)$ linkages. These distinctive linkages give cellulose an entirely different overall shape, as illustrated in Figure 12-12c. Cellulose has a flat layered sheet-like appearance, which provides structural rigidity for a plant. Additional structural rigidity arises from the extensive hydrogen bonding that occurs between sections of the polysaccharide.

In the laboratory, cellulose, starch, and glycogen can all be hydrolyzed into D-glucose in the presence of an acid. In the body, starch taken in through the diet is hydrolyzed in the mouth and small intestine by enzymes that specifically hydrolyze $\alpha(1\rightarrow4)$ linkages. These enzymes are known as *α-glycosidases*. Since humans do not have enzymes that hydrolyze $\beta(1\rightarrow4)$ linkages, cellulose cannot be digested, and therefore it does not serve as a source of energy. Cellulose does, however, serve an important role as the indigestible carbohydrate we call dietary fiber. Some animals, however, can digest cellulose, because bacteria that live in their digestive tracts produce *β-glycosidases*. These animals include cows, giraffes, and other ruminants (animals with a second stomach— a rumin). This is why you see these animals grazing on grass.

WORKED EXERCISE 12-3 Polysaccharides

For each statement below, indicate the polysaccharide(s) for which the statement is true. Choose among the following (more than one may apply).

i. amylose; ii. amylopectin; iii. glycogen; iv. cellulose.

a. _____ contains $\alpha(1\rightarrow4)$ glycosidic bonds.
b. _____ is a polymer composed of glucose.
c. _____ provides structural rigidity to plants.
d. _____ is a branched polysaccharide.
e. _____ has a helical shape.
f. _____ has a layered sheet like appearance.

SOLUTION
a. amylose, amylopectin, glycogen
b. amylose, amylopectin, glycogen, cellulose
c. cellulose
d. amylopectin, glycogen
e. amylose
f. cellulose

PRACTICE EXERCISES

12.9 For each of the statements below, indicate the polysaccharide(s) for which the statement is true. (More than one selection may be chosen.)

i. amylose; **ii.** amylopectin; **iii.** glycogen; **iv.** cellulose

a. _____ provides energy for plants.
b. _____ cannot be digested by humans.
c. _____ is a type of starch.
d. _____ is produced by plants (through photosynthesis).
e. _____ serves as dietary fiber.
f. _____ serves as a glucose storage molecule in animals and humans.
g. _____ is hydrolyzed by *α-glycosidases*.

12.10 Dextrins are composed of 3–10 glucose molecules in α(1→4) linkages.
   a. Are dextrins mono-, di-, oligo-, or polysaccharides?
   b. What type of enzyme would hydrolyze dextrins into glucose?

12.11 The rumin is the second compartment in the stomach of certain animals such as giraffes, cattle, and camels. What type of enzyme is present in the bacteria living in the rumin: *α-glycosidases* or *β-glycosidases*? What reaction do these enzymes catalyze?

12.12 What is the similarity and what is the difference between cellulose and cellobiose?

12.13 What is the fundamental structural difference between starch and cellulose?

12.4 | Carbohydrate Catabolism

The main function of carbohydrates in humans is to provide energy for the cell. The catabolism of carbohydrates can be divided into three stages:

- In the first stage, starch is hydrolyzed into monosaccharides, primarily glucose.

- In the second stage, glucose is converted into pyruvate, through a 10-step biochemical pathway known as **glycolysis,** which connects ADP to ATP. Pyruvate has two possible catabolic fates depending on whether or not oxygen is available. In an aerobic environment (oxygen is available), pyruvate is converted into acetyl CoA, the central molecule of metabolism.

- In the third stage, acetyl CoA is converted into carbon dioxide in the citric acid cycle. Electrons are extracted from acetyl CoA in the citric acid cycle and carried by NADH and $FADH_2$ into the electron transport chain, a series of oxidation-reduction reactions. The reduction of oxygen to water by these electrons drives the formation of ATP from ADP and P_1 in a process called oxidative phosphorylation.

Here we will focus on the first two stages of carbohydrate catabolism (Figure 12-13). Since the catabolism of carbohydrates, proteins, and fats all converge at acetyl CoA at the end of the second stage of catabolism, the third stage of catabolism is common to all the biomolecules and will be described later, in Chapter 14.

Stage 1: Polysaccharides Are Hydrolyzed into Monosaccharides

For carbohydrates, stage 1 of catabolism, known as digestion, begins in the mouth. The enzyme *amylase,* an *α-glycosidase* present in the saliva, catalyzes hydrolysis of amylose and amylopectin, the two components of starch, into the carbohydrates

- glucose,
- maltose, and
- **dextrins,** oligosaccharides composed of 3–12 glucose molecules in α(1→4) linkages.

These reactions are represented in Figure 12-14.

COO⁻
|
C=O
|
CH₃
Pyruvate

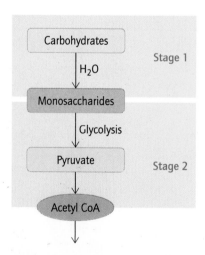

Figure 12-13 The first two stages of carbohydrate catabolism.

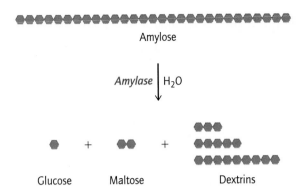

In the small intestine, maltose and dextrins are further hydrolyzed into glucose. Here, the other important disaccharides, lactose and sucrose, are also hydrolyzed into their component monosaccharides.

In Chapter 10 you learned that *hydrolysis* reactions use water to break covalent bonds. Hydrolysis of complex carbohydrates occurs at the glycosidic bonds to produce monosaccharides. In the laboratory, hydrolysis of a carbohydrate requires an acid catalyst. In the body, this type of reaction requires a *glycosidase* enzyme specific for either the α or β linkage. The enzyme that catalyzes each of the dietary disaccharides bears its name as shown in the table below.

				Maltase			
(a)	Maltose	+	H_2O	$\longrightarrow$	2 D-Glucose		
				Lactase			
(b)	Lactose	+	H_2O	$\longrightarrow$	D-Glucose	+	D-Galactose
				Sucrase			
(c)	Sucrose	+	H_2O	$\longrightarrow$	D-Glucose	+	D-Fructose

For example, *lactase* catalyzes the hydrolysis of lactose into galactose and glucose. When water breaks the glycosidic bond, H and OH are incorporated and two monosaccharides are formed, as shown in Figure 12-15. The OH from the water molecule becomes the OH on the anomeric carbon in galactose, and the H from the water molecule becomes the hydrogen atom for the hydroxyl functional group (OH) at C4 of D-glucose. The digestive systems of people with lactose intolerance are unable to perform this reaction as a result of a deficiency of the enzyme *lactase*. The undigested lactose causes intestinal discomfort.

Dietary monosaccharides diffuse through the small intestine and make their way into the bloodstream. Glucose is distributed via the blood as fuel to cells. Fructose and galactose are carried to the liver instead, where they are converted into glucose and re-enter the blood. Once glucose enters a cell, the second stage of catabolism begins.

Figure 12-15 Hydrolysis of lactose produces galactose and glucose.

Problem-Solving Tutorial:
Carbohydrate Catabolism Stage 1:
Polysaccharide Hydrolysis

WORKED EXERCISE 12-4 Hydrolysis of Polysaccharides

The structure of sucrose is shown below.

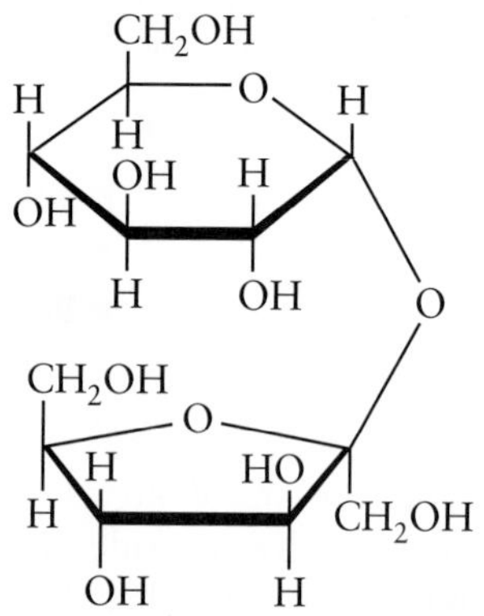

a. Is sucrose a monosaccharide, disaccharide, oligosaccharide, or polysaccharide?
b. What small molecule reactant is required to hydrolyze sucrose? What is the name of the enzyme that catalyzes the hydrolysis of sucrose?
c. Write the structures of the products formed when sucrose is hydrolyzed. Name the products.
d. Where in the body is sucrose hydrolyzed: in the mouth, liver, small intestine, or stomach?
e. What type of glycosidic linkage does sucrose contain?

SOLUTION

a. Sucrose is a disaccharide because it is composed of two monosaccharides.
b. Water, H_2O, is required to hydrolyze sucrose. The enzyme required is *sucrase*.
c.

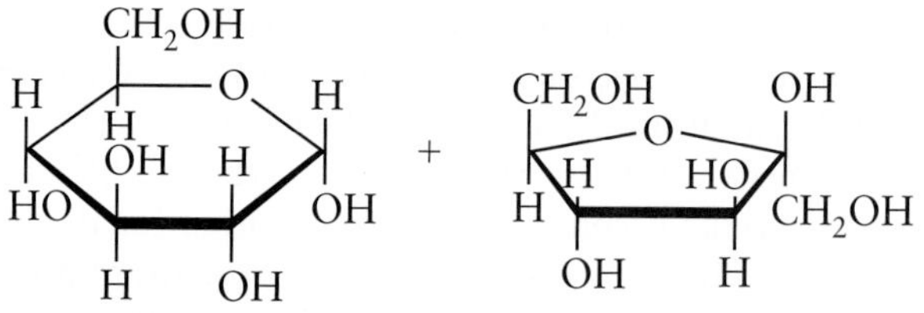

Glucose
(both α- and β-anomers formed)

Fructose
(both α- and β-anomers formed)

d. Sucrose is hydrolyzed in the small intestine.
e. Sucrose contains an α1→β2 linkage.

PRACTICE EXERCISES

12.14 Write the balanced chemical equation for the hydrolysis of maltose, showing the chemical structures for all reactants and products.

12.15 Where in the body is maltose hydrolyzed?

12.16 Where in the body is starch hydrolyzed?

12.17 Describe the chemical structure of dextrins. How are they different from maltose?

12.18 Beans contain the undigestible carbohydrate raffinose, whose structure is shown on the facing page. Bacteria in the lower intestine hydrolyze raffinose into monosaccharides and then further degrade them, producing gas as a by-product and resulting in the well known flatulence associated with eating beans. Digestive aids like Beano contain the enzyme *α-galactosidase*, which helps digest α-galactose linkages, so that no raffinose is available for bacteria to digest and no gas by-products are produced.

$$\textbf{Raffinose}$$

a. Is raffinose a monosaccharide, disaccharide, oligosaccharide, or polysaccharide? Explain.
b. Label all the anomeric carbons in raffinose with a *.
c. Label and identify the two glycosidic bonds in raffinose.
d. What monosaccharides are produced from the complete hydrolysis of raffinose into its individual monosaccharide components?
e. Which glycosidic linkage is hydrolyzed by the digestive enzymes found in Beano?

Stage 2 : Glycolysis

After passing through the cell membrane of a cell, glucose undergoes further degradation in the cytoplasm (inside of the cell). Here, glycolysis converts one molecule of glucose—containing six carbon atoms—into two molecules of pyruvate, containing three carbon atoms. The main purpose of glycolysis is to harvest energy from glucose. In the presence of oxygen, pyruvate can be oxidized further to acetyl CoA, the central molecule of catabolism.

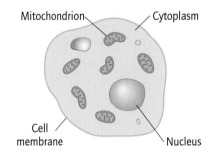

Glycolysis takes place in the cytoplasm.

Glycolysis

D-Glucose → 10 steps → 2 Pyruvate + energy

Glycolysis is the sole source of metabolic energy for some cells, including red blood cells, brain cells, and sperm cells. It is also the primary source of energy for muscle cells. Glycolysis is an *anaerobic* pathway, which means that it does *not* require oxygen. For this reason many microorganisms use glycolysis as their primary metabolic source of energy, including the microorganisms responsible for dental caries described in the opening vignette. Microorganisms causing well-known diseases such as tetanus, botulism, and gas gangrene also use glycolysis as a metabolic source of energy. Indeed, the earliest living organisms on the earth, before there was any molecular oxygen on the planet, are believed to have derived their energy from glycolysis.

The complete 10-step sequence of reactions in glycolysis were first identified in yeast cells.

The Steps of Glycolysis Figure 12-16 shows the entire 10-step biochemical pathway of glycolysis. Notice that each step requires an enzyme, which is

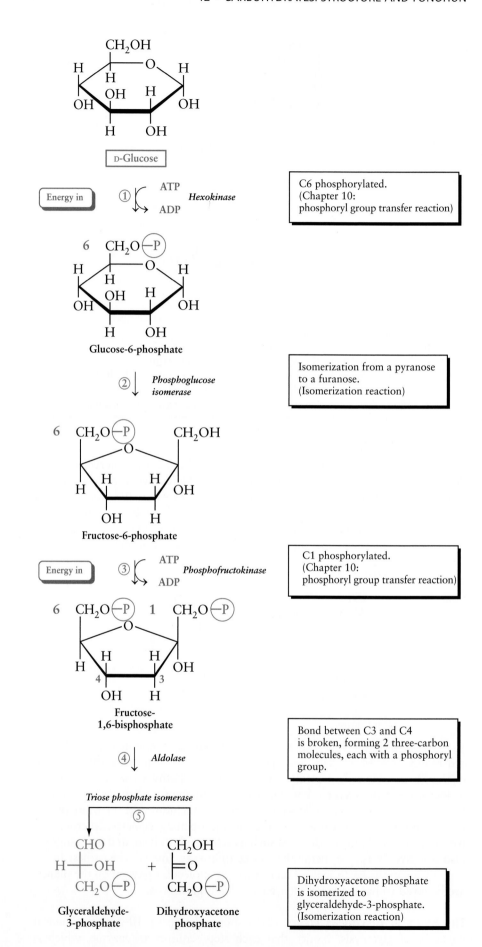

Figure 12-16 The 10 steps of glycolysis.

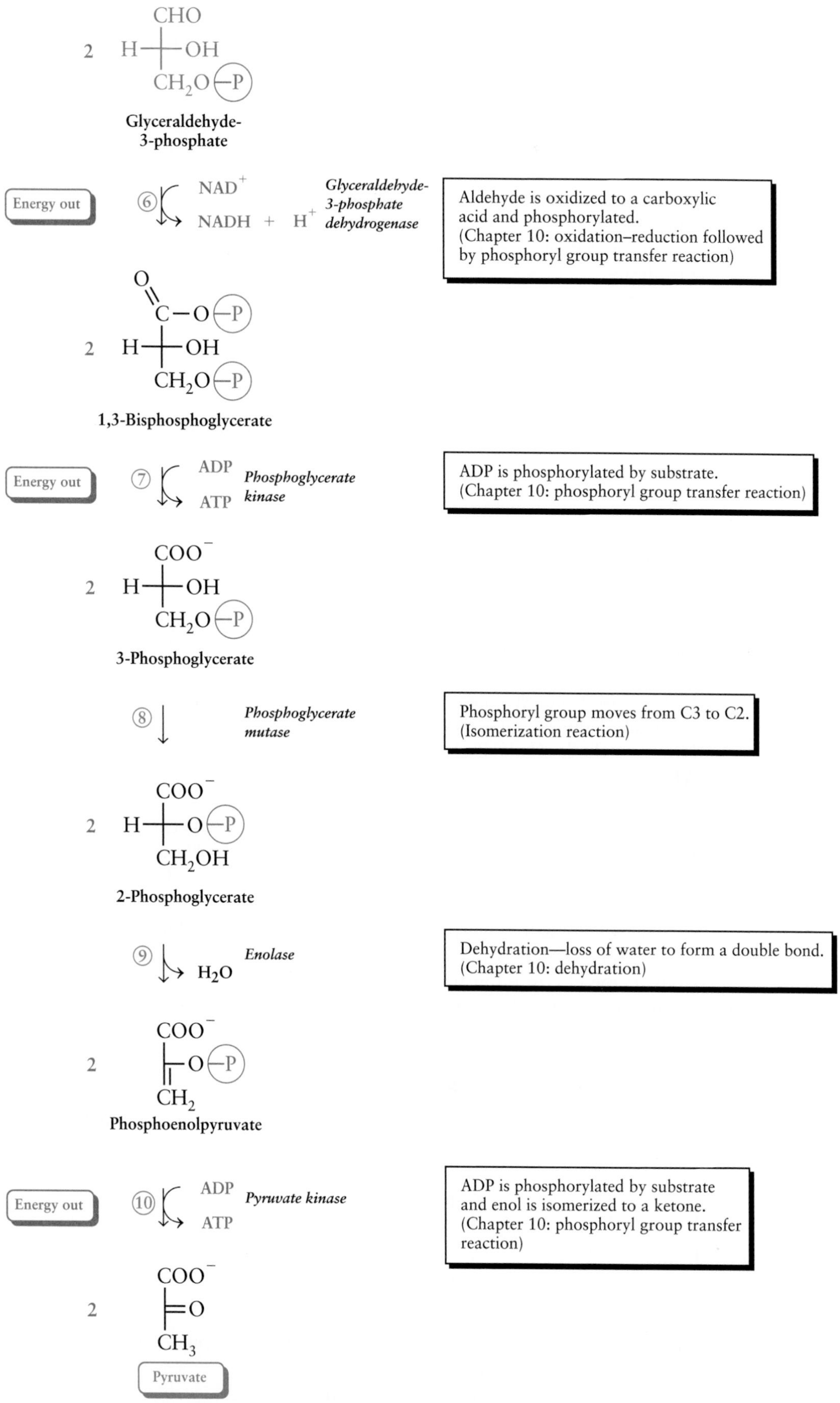

CHO

2   H—OH

CH$_2$O—Ⓟ

**Glyceraldehyde-
3-phosphate**

Energy out	⑥	NAD$^+$	*Glyceraldehyde-*	Aldehyde is oxidized to a carboxylic

NAD$^+$

NADH + H$^+$

*Glyceraldehyde-
3-phosphate
dehydrogenase*

Aldehyde is oxidized to a carboxylic
acid and phosphorylated.
(Chapter 10: oxidation–reduction followed
by phosphoryl group transfer reaction)

O
‖
C—O—Ⓟ

2   H—OH

CH$_2$O—Ⓟ

1,3-Bisphosphoglycerate

Energy out   ⑦   ADP   *Phosphoglycerate
kinase*

ATP

ADP is phosphorylated by substrate.
(Chapter 10: phosphoryl group transfer reaction)

COO$^-$

2   H—OH

CH$_2$O—Ⓟ

3-Phosphoglycerate

⑧   *Phosphoglycerate
mutase*

Phosphoryl group moves from C3 to C2.
(Isomerization reaction)

COO$^-$

2   H—O—Ⓟ

CH$_2$OH

2-Phosphoglycerate

⑨   *Enolase*

H$_2$O

Dehydration—loss of water to form a double bond.
(Chapter 10: dehydration)

COO$^-$

2   ‖—O—Ⓟ

CH$_2$

Phosphoenolpyruvate

Energy out   ⑩   ADP   *Pyruvate kinase*

ATP

ADP is phosphorylated by substrate
and enol is isomerized to a ketone.
(Chapter 10: phosphoryl group transfer
reaction)

COO$^-$

2   ‖—O

CH$_3$

Pyruvate

typically listed above the reaction arrow. The type of reaction is described in a box to the right of each reaction, including references to Chapter 10 where these reaction types were described. The organic compounds in the sequence of reactions that define a particular biochemical pathway like glycolysis are called *intermediates*. For example, in a biochemical pathway A→B→C→D→E→F, the compounds B, C, D, and E are considered intermediates of metabolism.

Notice that every intermediate in glycolysis contains one or more phosphoryl groups in its structure—indicated with a $\textcircled{P}$. The -2 charge on the phosphoryl group prevents these intermediates of glycolysis from diffusing out of the cell, since polyatomic ions cannot freely pass through a biological membrane.

In the fourth step of glycolysis a six-carbon intermediate is split into two three-carbon intermediates, which are structural isomers of each other. One of these isomers is converted into the other in Step 5, so the first half of glycolysis produces two molecules of glyceraldehyde-3-phosphate, an aldehyde containing a phosphoryl group at C3. The second half of glycolysis converts these two glyceraldehyde-3-phosphate molecules into two pyruvate molecules.

$$\begin{array}{c} CHO \\ H\!-\!\!\!\!\!\!\!\!-\!OH \\ CH_2O\!-\!\textcircled{P} \end{array}$$

Glyceraldehyde-3-
phosphate

Energy Production in Glycolysis Red boxes in Figure 12-16 indicate the key steps in glycolysis in which there is either an input or output of energy. The first half of glycolysis (Steps 1–5) requires an input of energy: Two ATP molecules are required for every one glucose molecule that enters the pathway. However, the last half of glycolysis (Steps 6–10) has an energy output of four ATP molecules: two ATP for each glyceraldehyde-3-phosphate that is converted into pyruvate. *Note that since glucose is converted into two 3-carbon molecules, and each of these three-carbon molecules produces two ATP, there is a net output of four ATP for every one glucose that enters the pathway.* If you subtract the two ATP that were input into the cycle in the first half of glycolysis, the net output of energy in glycolysis is two ATP $(4 - 2 = 2)$.

The second half of glycolysis also produces two NADH molecules for every one glucose. NADH is an energy carrier molecule that eventually transfers electrons to the electron-transport chain. *Hence, the net result of glycolysis is an output of energy in the form of two ATP molecules and two NADH molecules.*

To the left of the main text:

$$R\!-\!O\!-\!\textcircled{P} = R\!-\!O\!-\!\!\!\!\!\overset{\displaystyle O}{\underset{\displaystyle O^-}{\overset{\displaystyle \|}{P}}}\!\!\!\!\!-\!O^-$$

Phosphoryl group

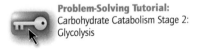

Problem-Solving Tutorial:
Carbohydrate Catabolism Stage 2:
Glycolysis

WORKED EXERCISES 12-5 Glycolysis

Step 4 of glycolysis produces glyceraldehyde-3-phosphate and dihydroxyacetone phosphate.

$$\begin{array}{ccc} CHO & & CH_2OH \\ | & & | \\ H\!-\!C\!-\!OH & & C\!=\!O \\ | & & | \\ CH_2OPO_3^{2-} & & CH_2OPO_3^{2-} \end{array}$$

1 Which of the compounds is glyceraldehyde-3-phosphate? Which is dihydroxyacetone phosphate?

2 Explain why these two compounds are classified as structural isomers.

3 What chemical change occurs in the isomerization reaction that converts dihydroxyacetone phosphate to glyceraldehyde-3-phosphate (Step 5)?

SOLUTIONS

1

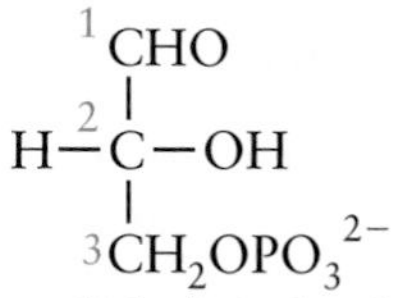

$$^1CHO$$
$$H-\overset{2}{C}-OH$$
$$^3CH_2OPO_3^{2-}$$
Glyceraldehyde-3-phosphate

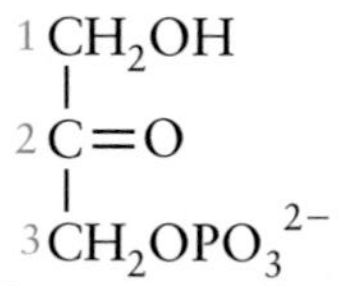

$$^1CH_2OH$$
$$^2C=O$$
$$^3CH_2OPO_3^{2-}$$
Dihydroxyacetone phosphate

2 They have the same chemical formula ($C_3H_5O_6P$), but a different connectivity of atoms: aldehyde at C1 and alcohol at C2 in glyceraldehyde-3-phosphate versus alcohol at C1 and ketone at C2 in dihydroxyacetone phosphate.

3 The alcohol functional group at C1 in dihydroxyacetone phosphate moves to C2, and the ketone carbonyl group at C2 moves to C1, becoming an aldehyde functional group.

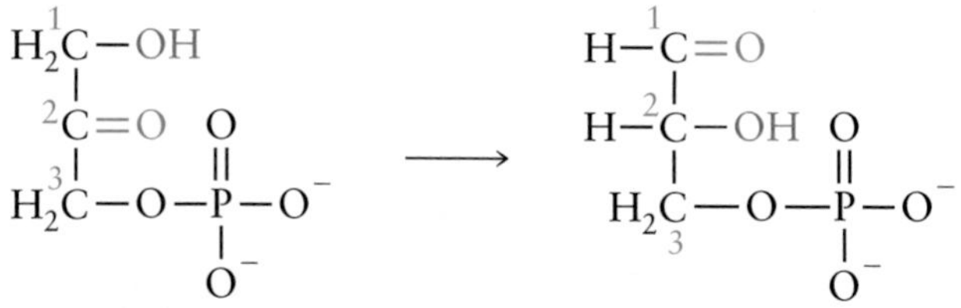

$$\overset{1}{H_2C}-OH$$
$$\overset{2}{C}=O \quad O$$
$$\overset{3}{H_2C}-O-\overset{||}{P}-O^-$$
$$\quad\quad\quad O^-$$

$\longrightarrow$

$$\overset{1}{H}-C=O$$
$$H-\overset{2}{C}-OH \quad O$$
$$H_2\underset{3}{C}-O-\overset{||}{P}-O^-$$
$$\quad\quad\quad O^-$$

PRACTICE EXERCISES

12.19 Glycolysis is a ____-step biochemical pathway that converts one molecule of _____ into two molecules of _____.

12.20 What functional group is present in every intermediate of glycolysis? How does it ensure that the intermediate stays inside the cell?

12.21 Is glycolysis an *anaerobic* or an *aerobic* process? Define these terms.

12.22 Write the chemical structure of glucose-6-phosphate. In which step of glycolysis is this intermediate produced? Is this a modified monosaccharide?

12.23 Which two steps of glycolysis require an input of energy in the form of ATP? Which two steps produce ATP? Does glycolysis have a net input or output of energy?

12.24 There are two isomerization steps in the 10-step sequence of glycolysis. In an isomerization reaction one structural or geometric isomer is converted into another. Which steps in glycolysis are isomerization reactions? What structural change occurs in each case?

12.25 Match the following compounds to its name below.

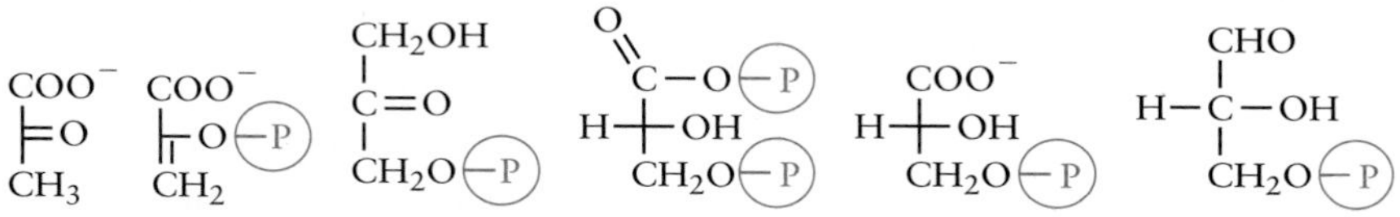

$$COO^-$$
$$\overset{||}{C}=O$$
$$CH_3$$
(a)

$$COO^-$$
$$\overset{||}{C}-O\!-\!\textcircled{P}$$
$$CH_2$$
(b)

$$CH_2OH$$
$$C=O$$
$$CH_2O\!-\!\textcircled{P}$$
(c)

$$\overset{O}{\underset{||}{C}}-O\!-\!\textcircled{P}$$
$$H\!-\!\!-\!\!OH$$
$$CH_2O\!-\!\textcircled{P}$$
(d)

$$COO^-$$
$$H\!-\!\!-\!\!OH$$
$$CH_2O\!-\!\textcircled{P}$$
(e)

$$CHO$$
$$H-C-OH$$
$$CH_2O\!-\!\textcircled{P}$$
(f)

_____ **1.** Glyceraldehyde-3-phosphate
_____ **2.** Dihydroxyacetone phosphate
_____ **3.** 1,3-Bisphosphoglycerate
_____ **4.** Phosphoenolpyruvate
_____ **5.** 3-Phosphoglycerate
_____ **6.** Pyruvate

The Fate of Pyruvate

What happens to the pyruvate produced in glycolysis? In human cells, pyruvate can follow one of two pathways, depending on whether or not oxygen is present. Conditions in which oxygen is present are termed **aerobic,** and conditions in which oxygen is absent are termed **anaerobic.**

- Aerobic conditions: Pyruvate is oxidized, producing acetyl CoA and CO_2. Acetyl CoA is the entry point of the citric acid cycle and the beginning of the third stage of catabolism.
- Anaerobic conditions: Pyruvate is reduced to lactic acid, regenerating NAD^+.

Oxidation of Pyruvate to Acetyl CoA The reaction that converts pyruvate into acetyl CoA requires coenzyme A as a reactant, and NAD^+ serves as an electron acceptor. Pyruvate is oxidized with the simultaneous loss of CO_2, producing the thioester acetyl CoA, CO_2, and NADH.

$$O_2$$

$$H_3C-\overset{\overset{O}{\|}}{C}-\overset{\overset{O}{\|}}{C}-O^- + H-S\text{-}CoA + NAD^+ \longrightarrow H_3C-\overset{\overset{O}{\|}}{C}-S\text{-}CoA + CO_2 + NADH + H^+$$

Pyruvate   **Acetyl CoA**

Although oxygen does not appear as a reactant per se, it is necessary to regenerate NAD^+ from NADH, which occurs later in catabolism.

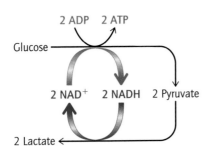

Figure 12-17 Lactic acid fermentation regenerates NAD^+ needed for glycolysis.

Reduction of Pyruvate to Lactic Acid When oxygen in cells has been depleted, as in the case of muscle cells during strenuous exercise, anaerobic conditions exist. Under these conditions, after glucose is converted to pyruvate, pyruvate is reduced to L-lactate (the conjugate base of lactic acid) by NADH, in a process known as **lactic acid fermentation** (Figure 12-17).

$$H_3C-\overset{\overset{O}{\|}}{C}-\overset{\overset{O}{\|}}{C}-O^- + NADH \longrightarrow H_3C-\overset{\overset{HO}{|}}{\underset{\underset{H}{|}}{C}}-\overset{\overset{O}{\|}}{C}-O^- + NAD^+$$

Pyruvate   L-**Lactate**

The build-up of lactic acid is believed to be responsible for the "burn" experienced during strenuous exercise such as sprinting. Once the strenuous activity is over, excess lactic acid is transported to the liver where it is converted back into pyruvate as oxygen becomes available again.

Reduction of pyruvate to lactate ensures that NAD^+ is regenerated so that glycolysis can continue. Recall that Step 6 of glycolysis requires NAD^+. Indeed, regeneration of NAD^+ by this means explains how microorganisms can continue to produce ATP in the absence of oxygen. Many anaerobic bacteria use lactic acid fermentation to produce energy in an environment where oxygen is not present. For example, the bacteria that cause dental caries—tooth decay—described in the opening vignette produce lactic acid, creating an environment acidic enough to dissolve tooth enamel.

PRACTICE EXERCISE

12.26 Answer the following question for the reaction that converts pyruvate into L-lactate:

    a. What compound is oxidized? What compound is reduced? Explain how you can tell. Refer to Chapter 10 if necessary.

b. Does this reaction occur in the presence or absence of oxygen? Are these aerobic or anaerobic conditions?

c. How does this reaction ensure that glycolysis continues even if oxygen is absent?

d. How is lactic acid related to lactate?

e. What is lactic acid fermentation? What is its role in dental caries?

Other Metabolic Roles of Glucose

You have seen that glycolysis is the biochemical pathway that converts glucose into pyruvate. Other biochemical pathways that begin or end with glucose include:

- glycogenesis—formation of glycogen from glucose,
- glycogenolysis—degradation of glycogen to form glucose,
- gluconeogenesis—formation of glucose from pyruvate, and
- the pentose phosphate pathway—formation of pentose phosphate from glucose.

Clearly, glucose plays a central role in metabolism, as summarized in Figure 12-18. In the **pentose phosphate pathway,** glucose serves as the starting material for the synthesis of ribose, an important sugar that provides the backbone for RNA—the nucleic acids involved in relaying information from DNA during the construction of proteins.

The synthesis of the polysaccharide glycogen begins with glucose in a process known as **glycogenesis,** which requires that glucose and the hormone insulin both be present. The reverse processs, degradation of glycogen into glucose, follows a similar but not identical pathway known as **glycogenolysis.** The process of glycogenolysis occurs when glucose is absent and the hormone glucagon is present or when muscles require energy.

When all glycogen stores are depleted and no glucose is available, the body can synthesize glucose in a biochemical pathway known as **gluconeogenesis.** Gluconeogenesis is a biochemical pathway similar to but not identical to the reverse of glycolysis. During starvation, the body uses gluconeogenesis to generate glucose from proteins, a process that requires the breakdown of muscle tissue. Most of the biochemical pathways involving glucose play a role in energy storage and use.

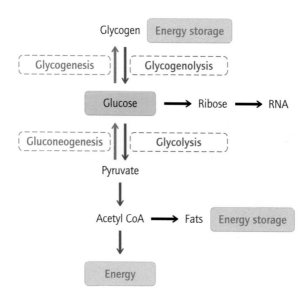

Figure 12-18 The central role of glucose in metabolism.

You have seen on a molecular level how carbohydrates produce energy. *Supplying energy to individual cells throughout the body is the primary role of carbohydrates.*

PRACTICE EXERCISES

12.27 What is the difference between gluconeogenesis and glycolysis? Which pathway is anabolic and which is catabolic?

12.28 When does the body form glycogen? Why does it form glycogen?

12.29 Which hormone promotes glycogenesis? Which hormone promotes glycogenolysis?

12.30 Indicate which of the following processes are catabolic:

  a. gluconeogenesis   **b.** glycogenesis
  c. glycogenolysis   **d.** glycolysis

12.5 | Oligosaccharides as Cell Markers

One reason why organs are rejected after an organ transplant is that the body's own cells recognize the transplanted organ as foreign and mount an immune response against it that ultimately destroys the organ. How did these cells know that the organ was foreign? The answer lies in differences in carbohydrate markers on the surface of cells from the donated organ tissue and the cells of the host.

Cell markers are chemical tags that identify a cell—like a fingerprint. They allow cells to recognize one another and to distinguish host from foreigner. Carbohydrate markers are typically oligosaccharides covalently bonded to the cell membrane or a protein within the cell membrane. They project out into the extracellular fluid, as shown in Figure 12-19.

A person's blood type, A, B, AB, or O, is determined by the oligosaccharide markers on their red blood cells. The markers of all four blood groups contain the same core trisaccharide, but differ in whether they have a fourth monosaccharide and what that monosaccharide is, as shown in Table 12-2.

Type O blood has no additional monosaccharides attached to the core trisaccharide (first row). Type A blood has an additional monosaccharide, N-acetyl-D-galactosamine, covalently bonded to the core trisaccharide (second row). Type B blood has an additional D-galactose covalently bonded to

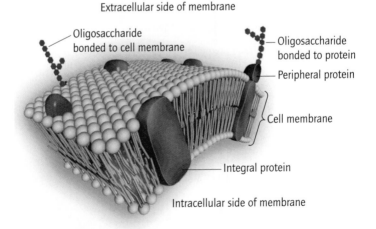

Figure 12-19 Oligosaccharide markers on the surface of the cell, attached to either the cell membrane or a membrane protein, allow cells to recognize other cells. Hexagons represent the monosaccharide components of the oligosaccharide marker.

Table 12-2 Blood Type and Oligosaccharide Marker

Blood Type (% of Population)	Oligosaccharide Marker on Red Blood Cell (RBC)	Can Receive blood from
O (43%)	RBC — N-acetyl-D-glucosamine — D-galactose — L-fucose	O (Universal donor)
A (40%)	N-acetyl-D-galactosamine (attached above) RBC — N-acetyl-D-glucosamine — D-galactose — L-fucose	A, O
B (12%)	D-galactose (attached above) RBC — N-acetyl-D-glucosamine — D-galactose — L-fucose	B, O
AB (5%)	Contains both the A and B oligosaccharides shown above.	A, B, AB, O (Universal recipient)

the core trisaccharide (third row). Type AB blood contains both type A and type B markers (last row).

Type O blood is often called the **universal donor** because people with type O blood can donate to any recipient, although they can only receive blood from type O donors. The chemical structure of the oligosaccharide marker explains why. People with type O blood have the trisaccharide component of all four blood types, so their blood is recognized as part of the recipient's blood by all recipients. However, their body rejects type A, B, or AB blood because the fourth monosaccharide is unfamiliar. An immune response occurs, which can lead to life-threatening blood clotting. Blood compatibility is always determined before any medical procedure that may require a blood transfusion.

A similar reasoning accounts for why individuals with type AB blood can receive A, B, or O type blood but cannot donate their blood. Type AB blood is called the **universal recipient.**

PRACTICE EXERCISES

12.31 At the molecular level, how is type A blood different from type O blood?

12.32 Why is type AB blood considered the universal recipient?

12.33 How is type B blood different from type O? Can people with type B blood accept a blood donation from someone with type O blood? Can people with type O blood accept a blood donation from someone with type B blood? Explain.

You have seen that the catabolism of carbohydrates yields glucose, the most important source of energy for all cells. The biochemical pathway of glycolysis converts glucose into two pyruvate molecules, while producing two molecules of ATP and reducing two NAD^+ to NADH. Pyruvate can undergo further oxidation to acetyl CoA, which produces more energy in the third stage of catabolism. In the next chapter, you will consider another form of energy storage as you learn about fats, the primary means of *long-term* energy storage. Then finally in Chapter 14 you will see how acetyl CoA enters the citric acid cycle and how electrons are used in the electron-transport chain to produce ATP, the energy currency of the cell.

Chemistry in Medicine What Happens When Blood Sugar Gets Too High or Too Low?

Marcia woke up one morning and walked into the kitchen to make herself breakfast. As she began assembling the ingredients for her meal, her speech became slurred and she appeared disoriented. When she reached for the counter for support, her husband grabbed some orange juice and forced her to drink it. Within seconds, she sprang up as though nothing at all unusual had happened, asking "What's going on? Why are you standing over me?"

Marcia's experience is not uncommon; in fact, it is something paramedics see routinely. What happened? Marcia is diabetic, a condition that prevents her cells from utilizing glucose properly. Her blood glucose concentration had fallen to a dangerously low level, making her body unable to perform some basic functions. Had someone not provided her with a quick source of glucose, Marcia could have lapsed into a coma, and even died. How is it possible that a few sips of orange juice could bring Marcia back to her normal self in only a matter of seconds?

Every cell in the body—especially brain cells—need a constant supply of glucose, the monosaccharide that we call blood sugar. Marcia's case illustrates the dangerous effects of low glucose levels. On the other hand, glucose levels that are consistently high are also unhealthy. In a normal person hormones control the concentration of glucose in the blood within a very narrow range. Normal fasting glucose levels are 70–110 mg per deciliter (dL), as determined by a **F**asting **P**lasma **G**lucose (FPG) test, a blood test often used to measure blood glucose concentration after a period of fasting. When the concentration of glucose is *below* the normal range, the person is said to be *hypoglycemic*; when it is *above* the normal range, the person is said to be *hyperglycemic*. Your body regulates glucose levels using two important hormones: *insulin* and *glucagon* (Figure 12-20). The β cells of the pancreas release insulin when blood sugar levels rise, such as after a meal. As you know from Chapter 11, the hormone insulin is a protein. Insulin binds to receptors on the cell membrane, which signal the cell to allow glucose to enter. Moreover, insulin directs the liver to store excess glucose by converting it into glycogen, the storage form of glucose.

When glucose levels drop, as for example in between meals and during exercise, the pancreas stops producing insulin and releases glucagon, instead. Glucagon stimulates liver cells to convert glycogen back into glucose—the opposite effect of insulin. Insulin and glucagon work together to restore blood sugar levels to within the normal range, as shown in Figure 12-21.

Disorders of metabolism result when the production or utilization of these hormones is compromised. Diabetes is the most common metabolic disorder. *Diabetes* is a condition characterized by hyperglycemia, caused by insufficient insulin production or an inability to utilize insulin, known as **insulin resistance.** It is estimated that 7% of the United States population has diabetes— 20.8 million children and adults.

There are two types of diabetes: Type I and Type II. Type I diabetes, also known as insulin-dependent diabetes mellitus, usually begins in childhood. It is an autoimmune disease, in which the body destroys its own β cells—the insulin-producing cells of the pancreas. Since these individuals do not produce enough insulin, too much glucagon is also present. Consequently, glucose does not enter fat cells and muscle cells, and the absence of insulin in these cells causes the liver to unnecessarily synthesize glucose through gluconeogenesis, the same biochemical pathway that is triggered during starvation. Glucose is left circulating in the blood, unable to be taken up by cells that need it. Glucose filtered into the kidneys cannot be reabsorbed. Therefore, osmosis causes large volumes of urine, leading to common symptoms of diabetes, excessive thirst and frequent urination.

People with Type I diabetes must manage their diets carefully and, in particular, control the amount of

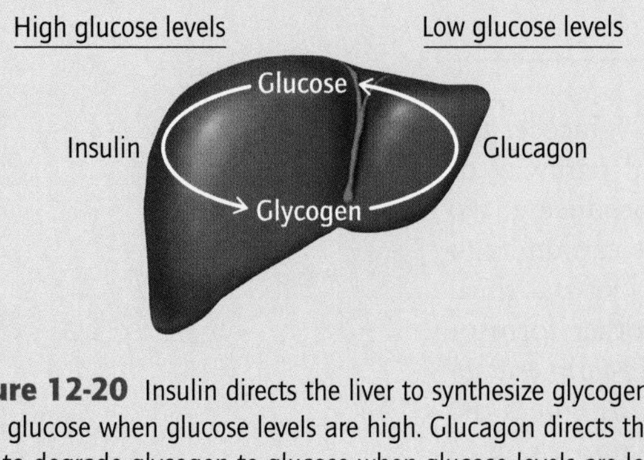

Figure 12-20 Insulin directs the liver to synthesize glycogen from glucose when glucose levels are high. Glucagon directs the liver to degrade glycogen to glucose when glucose levels are low. Glycogen is stored in liver and muscle cells.

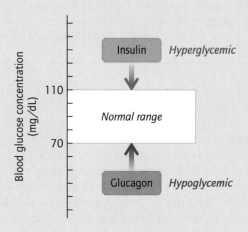

Figure 12-21 The body's response to blood glucose levels.

carbohydrates that they consume. They must also receive regular injections of insulin, since their bodies cannot produce the hormone. Patients can inject themselves or they can wear an insulin pump. Since insulin is a protein, it cannot be taken orally (see Chapter 11 *Chemistry in Medicine* box). Thus, subcutaneous injections of insulin must be made regularly throughout the day, based on the patient's blood glucose levels.

Type II diabetes, or insulin-resistant diabetes, generally appears in adulthood, and even though insulin is produced, cells do not respond properly to it. Normally when insulin is released, it mobilizes the "glucose transporters" to the cell surface. These transporters are proteins that allow the cell to take up glucose that is circulating in the bloodstream following a meal. For some reason, people with Type II diabetes are unable to mobilize the glucose transporters, so that even though there is glucose available, it cannot enter the cell. Cells are thereby starved of the fuel they need.

Once again, biochemical events at the molecular level provide insight into diseases with symptoms we see on a macroscopic level.

Chapter Summary

Role of Carbohydrates

- Carbohydrates are found in starch and cellulose, which are produced by plants from carbon dioxide and water through a biochemical process known as photosynthesis.
- Cells extract energy from carbohydrates, in the form of ATP, via multiple biochemical pathways collectively known as cellular respiration.
- Catabolism of carbohydrates occurs in three stages: (1) the hydrolysis of polysaccharides into monosaccharides (digestion), (2) the degradation of glucose to pyruvate via glycolysis and the oxidation of pyruvate to acetyl CoA, and (3) the production of ATP from ADP and P_1 through the citric acid cycle and oxidative phosphorylation.

Monosaccharides

- Monosaccharides are carbohydrates that cannot be hydrolyzed into simpler sugars.
- Common monosaccharides are glucose, fructose, and galactose.
- Monosaccharides contain a furanose or pyranose ring with many hydroxyl groups.
- Monosaccharides are chiral and can be drawn as Haworth structures, which show the hydroxyl groups above or below the ring. The orientation of the hydroxyl groups determines the identity of the monosaccharide.
- The anomeric carbon atom is the carbon atom in a monosaccharide that is bonded to both the ring oxygen and a hydroxyl group. The hydroxyl group lies above the ring in a β-anomer and below the ring in an α-anomer.
- The α- and β-forms of a monosaccharide are in equilibrium with the open-chain form of the monosaccharide, favoring the ring form.
- Most monosaccharides are chiral. Nature produces the D-sugars and not the L-sugars. In a Fischer projection, a D-sugar has the hydroxyl group on the cross hair farthest from the anomeric center pointing to the right.
- Diasteromers are sugars with the same formula and connectivity, but a different orientation of one or more, but not all, hydroxyl groups.
- Diastereomers are nonsuperposable stereoisomers that are not mirror images.

Complex Carbohydrates

- The common disaccharides are maltose, sucrose, lactose, and cellobiose.
- Disaccharides are joined between the anomeric carbon of one monosaccharide and the hydroxyl group of a second monosaccharide, in a bond known as a glycosidic bond.

- Disaccharides are defined by three structural characteristics: (1) the identity of the two monosaccharide components, (2) the carbon atoms are involved in the linkage, and (3) whether the α or β anomer is involved.
- Starch and glycogen are polysaccharides with over 1000 repeating glucose monomers connected by α(1→4) linkages. Starch is stored in plants and used for energy by humans and animals. Glycogen is the storage form of glucose in animals.
- Cellulose is a polysaccharide similar to starch but having β(1→4) linkages.

Carbohydrate Catabolism

- Carbohydrates are hydrolyzed in the presence of enzymes at the glycosidic bond into their component monosaccharides.
- The first stage of catabolism begins with the hydrolysis of starch into glucose.
- Glycolysis is a 10-step biochemical pathway that converts a molecule of glucose into two molecules of pyruvate, phosphorylates two ADP to ATP, and reduces two NAD^+ to two NADH.
- Glycolysis is an anaerobic process that produces energy.
- Pyruvate is oxidized to acetyl CoA and CO_2 when oxygen is available.
- Pyruvate is reduced to lactate in the absence of oxygen, a reaction that regenerates NAD^+ and allows glycolysis to proceed.
- Glucose is the starting material or product in several biochemical pathways, including glycogenesis, glycogenolysis, gluconeogenesis, and the pentose phosphate pathway.

Oligosaccharides as Cell Markers

- Oligosaccharides projecting from the cell membrane into the extracellular fluid serve as cell markers that allow other cells to recognize the cell.
- The ABO blood groups are defined by the monosaccharides that make up the oligosaccharide markers on red blood cells.

Key Words

Aerobic conditions Conditions in which oxygen is present.

Amylopectin The major constituent of starch; it is a branched polysaccharide with α(1→4) linkages.

Amylose The minor constituent of starch; it is an unbranched polysaccharide with α(1→4) linkages.

Anaerobic conditions Conditions in which oxygen is not present.

α-anomer A monosaccharide that has the OH group on the anomeric carbon below the ring.

Anomeric carbon The carbon atom in the ring form of a monosaccharide that is bonded to the ring oxygen atom and bears a hydroxyl group.

β-anomer A monosaccharide that has the OH group on the anomeric carbon above the ring.

Cellulose A polysaccharide that provides structure for a plant, formed of glucose units in β(1→4) linkages.

Dextrin An oligosaccharide composed of 3 to 12 D-glucose monomers in α(1→4) linkages.

Diastereomers Nonsuperposable stereoisomers that are not mirror images.

Disaccharides Carbohydrates that, when hydrolyzed, yield two monosaccharides.

D-Sugar A monosaccharide that has the OH group at the crosshair farthest from the carbonyl group pointing to the right in a Fischer projection and the C6 CH_2OH group above the ring in a Haworth projection. These are the natural monosaccharides produced in nature.

Furanose A monosaccharide in a five-membered ring that contains one oxygen atom in the ring.

Gluconeogenesis A biochemical pathway that produces glucose from pyruvate, used when glycogen stores are depleted and glucose is not available.

Glycogenesis The biochemical pathway that converts glucose into glycogen.

Glycogenolysis The biochemical pathway that converts glycogen into glucose.

Glycolysis The ten-step biochemical pathway that converts glucose into two molecules of pyruvate, two ADP to two ATP, and one NAD^+ to NADH, net.

Glycosidic bond The covalent bond joining two monosaccharides formed between the anomeric carbon atom of one monosaccharide and a hydroxyl group on a carbon atom of another monosaccharide. It also known as a glycosidic linkage.

Haworth projection A flattened representation of the ring form of a monosaccharide that conveys the three-dimensional orientation of each hydroxyl group with respect to the ring—either above or below the ring.

Lactic acid fermentation An anaerobic process in which glucose is converted to pyruvate and pyruvate is converted to L-lactate.

L-Sugar A monosaccharide that has the OH group at the cross hair farthest from the carbonyl group pointing to the left in a Fischer projection. An L-sugar is the enantiomer of the corresponding D-sugar with the same name.

Monosaccharide A carbohydrate that cannot be hydrolyzed into simpler carbohydrates.

Mutarotation The reaction that cause the α-anomer of a monosaccharide to open and reclose to form a mixture of both α and β anomers. Similarly, the β-anomer opens and recloses to the same ratio of α- and β-anomers.

Oligosaccharide A carbohydrate containing from 3 to 100 monosaccharides.

Pentose phosphate pathway The biochemical pathway that converts glucose to ribose.

Photosynthesis The biochemical pathway used by plants to synthesize carbohydrates from carbon dioxide and water using sunlight as the source of energy.

Polysaccharide A carbohydrate containing more than 100 monosaccharides.

Pyranose A monosaccharide in the form of a six-membered ring that contains one oxygen atom in the ring.

Starch A carbohydrate that is a mixture of 20% amylose and 80% amylopectin. It serves as stored energy for a plant and provides humans and animals with a major source of energy.

Universal donor A person with blood type O, who can donate blood to an individual of any blood type.

Universal recipient A person with blood type AB, who can receive any type of blood.

Additional Exercises

Carbohydrate Structure

12.34 What two forms of carbohydrates are found in plant material?

12.35 What role do starch and cellulose play in a plant?

12.36 What is the structural difference between a D-sugar and an L-sugar?

12.37 Fischer projections of two sugars are shown below.

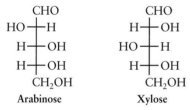

Arabinose   Xylose

    a. Are the Fischer projections shown D- or L-sugars?
    b. Write the Fischer projection of L-arabinose.
    c. How is D-arabinose different from D-xylose?

12.38 The structure of sorbose is shown below. Is the Fischer projection shown a D-sugar or an L-sugar? Explain.

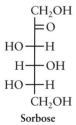

Sorbose

12.39 What is the structural difference between an α- and a β-anomer?

12.40 Indicate whether each structure below is a furanose or a pyranose. Identify the anomeric carbon in each structure. Indicate whether the α- or β-anomer is shown.

    a.

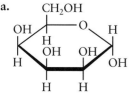

    b.

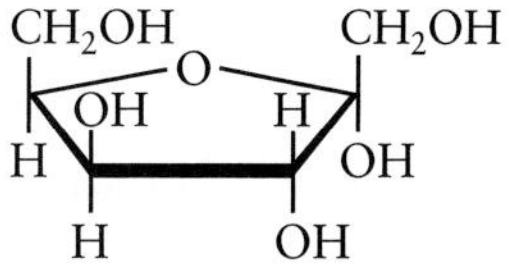

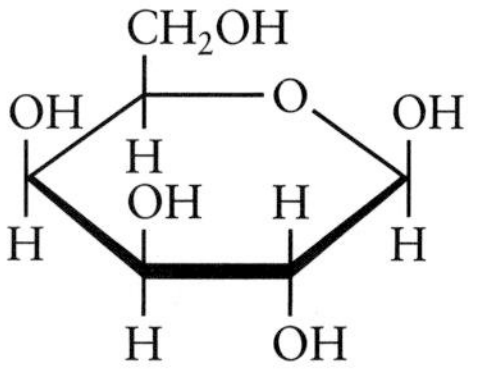

CH₂OH   CH₂OH

12.41 Indicate whether each structure below is a furanose or a pyranose. Identify the anomeric carbon in each structure. Indicate whether the α- or β-anomer is shown.

    a. CH₂OH   OH

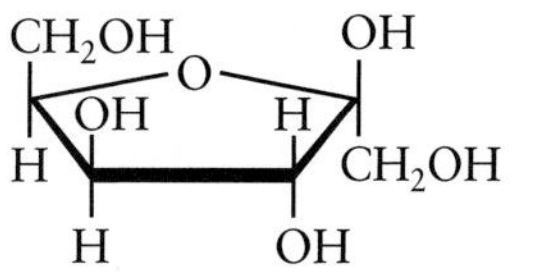

    b.   CH₂OH

12.42 The Fischer projection of D-gulose is shown below.

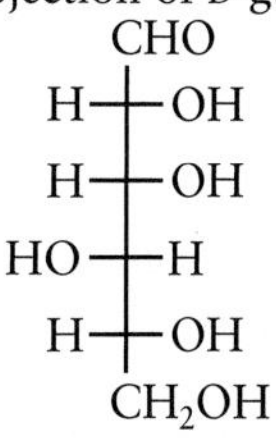

CHO
H——OH
H——OH
HO——H
H——OH
CH₂OH

D-Gulose

    a. Number the carbon atoms from 1 to 6.
    b. Which OH group in this sugar determines that it is a D-sugar?
    c. Write the structure of L-gulose. What is the stereochemical relationship between D- and L-gulose?
    d. Compare the open-chain structure of D-gulose to D-glucose. What is the stereochemical relationship between the two sugars?

12.43 The Fischer projection of L-idose is shown below.

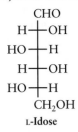

L-**Idose**

a. Number the carbon atoms from 1 to 6.
b. Which OH in this sugar determines that it is a L-sugar?
c. Write the structure of L-idose. What is the stereochemical relationship between D- and L-idose?
d. Compare the structure of D-idose to D-glucose. What is the stereochemical relationship?

12.44 Answer the questions below for the following two ring forms of a monosaccharide.

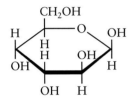 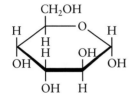

a. Label the anomeric carbon with an *.
b. Which structure is the α-anomer and which is the β-anomer?
c. What is the stereochemical relationship between these two structures: Are they *enantiomers* or *diastereomers*?
d. How do these sugars interconvert? What is this process called?

12.45 Answer the questions below for the following two ring forms of a monosaccharide.

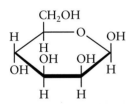 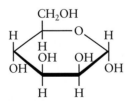

a. Label the anomeric carbon with an *.
b. Which structure is the α-anomer and which is the β-anomer?
c. What is the stereochemical relationship between these two structures: Are they *enantiomers* or *diastereomers*?
d. How do these sugars interconvert? What is this process called?

12.46 What is a disaccharide? How are disaccharides different from monosaccharides?

12.47 People with lactose intolerance do not produce enough of the enzyme *lactase*. Using Table 12-1, identify the bond that cannot be broken without the enzyme *lactase*.

12.48 What is the difference between maltose and cellobiose? How are these two molecules similar?

12.49 Isomaltose is similar to maltose except that it contains an α(1→6) linkage. Write the structure of isomaltose

12.50 What is the difference between an oligosaccharide and a polysaccharide?

12.51 Explain why giraffes can use cellulose as an energy source, while humans can't.

12.52 What is the difference in the glycosidic bonds in starch, cellulose, and glycogen?

12.53 How is the structure of D-glucosamine different from D-glucose?

Carbohydrate Catabolism

12.54 Write the balanced chemical equation for the hydrolysis of lactose showing the chemical structure for all reactants and products. What enzyme catalyzes the hydrolysis of lactose?

12.55 Maltulose, shown below, is found in honey and beer.

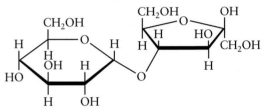

a. Is maltulose a mono-, di-, oligo-, or polysaccharide? Explain.
b. Label all the anomeric carbons in maltulose with a *.
c. Label and identify the glycosidic bond in maltulose.
d. What monosaccharides are produced from the hydrolysis of maltulose into its individual monosaccharide components?

12.56 Melibiose, shown below, is found in cacao beans.

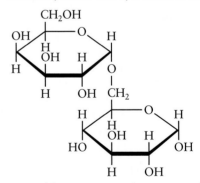

a. Is melibiose a mono-, di-, tri-, or polysaccharide? Explain.
b. Label all the anomeric carbons in melibiose with a *.
c. Label and identify the glycosidic bond in melibiose.
d. What monosaccharides are produced from the hydrolysis of melibiose into its individual monosaccharide components?

12.57 Glycolysis is a process that converts _____ into _____.

12.58 What kinds of cells use glycolysis as their sole source of metabolic energy?

12.59 Every intermediate in glycolysis contains a _____ functional group. State one purpose of this functional group.

12.60 What is the fate of pyruvate under aerobic conditions? What is the fate of pyruvate under anaerobic conditions?

12.61 What is the primary role of carbohydrates in the body?

12.62 Which steps in glycolysis produce a molecule that contains two phosphoryl groups?

12.63 Which step of glycolysis converts a larger molecule into two smaller molecules? This step is the origin of the *lysis* in gly*colysis*; "*lysis*" means "*to break.*"

12.64 In Step 5 of glycolysis, dihydroxyacetone phosphate is converted into glyceraldehyde-3-phosphate. What specific structural changes are made to dihydroxyacetone phosphate in this reaction?

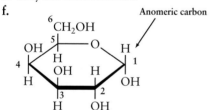

12.65 What structural changes occur in Step 2 of glycolysis, where glucose-6-phosphate is converted into fructose-6-phosphate?

Oligosaccharides as Cell Markers

12.66 Why can people with type O blood only receive blood from other type O donors? Why is type O blood the universal donor?

12.67 Why are individuals with type AB blood universal recipients? What blood types can they receive?

12.68 In 2003, a 17-year-old girl died from complications of a heart-lung transplant. She was mistakenly given organs from a person with type A blood, while she had type O blood. Explain why the difference in blood type caused a problem.

Chemistry in Medicine

12.69 When an individual's glucose levels fall:
a. Is he or she hyperglycemic or hypoglycemic?
b. What biochemical pathway is activated first to increase glucose levels?
c. When do glucose levels typically fall?
d. What hormone is secreted when glucose levels fall?

12.70 What is the difference between Type I and Type II diabetes?

12.71 What are normal fasting blood glucose levels?

12.72 Which cells produce insulin? In which organ are these cells found?

12.73 Why does someone with diabetes have to watch their diet carefully?

12.74 Why can insulin not be administered orally?

12.75 What hormone has a complementary role to insulin?

Answers to Practice Exercises

12.1 a. These are D-sugars because the C6 CH$_2$OH group lies above the pyranose ring.
b. α-D-galactose is on the left and β-D-galactose is on the right.
c. These are pyranoses, because they have a six-membered ring containing an oxygen atom.
d. Yes.
e. They are diastereomers.
f.

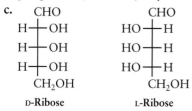

g. The OH group at C1 (the anomeric carbon); this process is called mutarotation.
h. Yes. It has many OH groups, so it is capable of hydrogen bonding with water molecules.

12.2 a. They are all D-sugars, because the OH group farthest from the aldehyde points right in each.
b. They have a different configuration of the OH and H groups at C2; therefore, they are diastereomers.
c.

```
      CHO              CHO
   H──┼──OH        HO──┼──H
   H──┼──OH        HO──┼──H
   H──┼──OH        HO──┼──H
      CH₂OH            CH₂OH
   D-Ribose         L-Ribose
```

Enantiomers

12.3 a–b.

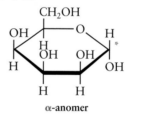

α-anomer   β-anomer

c. diastereomers

12.4

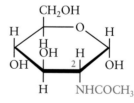

12.5 a. Cellobiose is a disaccharide because when hydrolyzed, it produces *two* monosaccharides.
b. D-glucose
c. A β(1→4) linkage
d. Cellobiose has a β(1→4) linkage while maltose has an α(1→4) linkage.

12.6 a. Lactose is a dissacharide because when hydrolyzed, it produces two monosaccharides.
b. D-galactose and D-glucose
c. A β(1→4) linkage
d. Lactose has a β linkage instead of an α linkage and it is composed of different monosaccharides, galactose and glucose, rather than just glucose.
e. Milk contains lactose.

12.7   D-glucose and D-fructose

12.8   **a.** D-glucose
b. Yes.

12.9   **a.** i, ii   **b.** iv   **c.** i–ii   **d.** iv   **e.** i, ii, iv   **f.** iii   **g.** i–iii

12.10 **a.** oligosaccharides
b. *α-glycosidases*

12.11 *β-Glycosidases*; they catalyze the hydrolysis of β-glycosidic linkages.

12.12 Cellulose is a polysaccharide containing hundreds of glucose units, while cellobiose is a disaccharide and only contains two glucose units.

12.13 The fundamental difference is the glycosidic linkage, which is α in starch and β in cellulose.

12.14

Maltose
α-(1-4)

α- and β-D-Glucose

12.15 Maltose is hydrolyzed in the small intestine.

12.16 Starch is hydrolyzed in the mouth and in the stomach.

12.17 Dextrins are oligosaccharides containing 3–10 glucose units. They contain more glucose units than maltose, which is a dissaccharide with only two glucose units.

12.18 **a.** Raffinose is an oligosaccharide or a trisaccharide, because it contains three monosaccharides.
b.–e.

Galactose

Glycosidic linkage hydrolyzed by *Beano*

Glucose

Glycosidic linkage

Fructose

Raffinose

12.19 Glycolysis is a __10__-step biochemical pathway that converts one molecule of _glucose_ into two molecules of _pyruvate_.

12.20 A phosphoryl group; its −2 charge prevents the intermediate from diffusing through the cell membrane, keeping the molecule inside the cell.

12.21 Glycolysis is an anaerobic process, which means it occurs in the absence of oxygen.

12.22

Glucose-6-phosphate

Glucose-6-phosphate is produced in the first step of glycolysis. This is a modified monosaccharide.

12.23 Steps 1 and 3 require an input of ATP. Steps 7 and 10 yield an output of ATP and Step 6 yields an output of NADH which is ultimately converted into ATP. Glycolysis has a net output of two ATP and one NADH.

12.24 Step 2: Glucose-6-phosphate is turned into fructose-6-phosphate; a change from a pyranose to a furanose ring.
Step 5: Dihydroxyacetone phosphate is turned into glyceraldehyde-3-phosphate; an exchange of the location of a carbonyl and an OH group.

12.25 1: (f); 2: (c); 3: (d); 4: (b); 5: (e); 6: (a).

12.26 **a.** NADH is oxidized to NAD$^+$ because it loses hydrogen. Pyruvate is reduced, because it gains hydrogen atoms.
b. This reaction occurs in the absence of oxygen. The conditions are anaerobic.
c. This reaction ensures that glycolysis can continue, because it reoxidizes NADH to NAD$^+$, which is essential for Step 6 of glycolysis.
d. Lactate is the conjugate base of lactic acid.
e. Lactic acid fermentation is the process of glucose being converted to pyruvate and then reduced to b-lactate under anaerobic conditions. Lactic acid fermentation creates an acidic environment that causes dental caries.

12.27 Gluconeogenesis produces glucose from pyruvate, and glycolysis accomplishes the reverse. Gluconeogenesis is anabolic and glycolysis is catabolic.

12.28 The body forms glycogen when glucose is in excess. It does this to store glucose in its storage form, glycogen.

12.29 Glycogenesis is promoted by insulin and glycogenolysis is promoted by glucagon.

12.30 **c.** and **d.**

12.31 Type A has an extra monosaccharide (N-acetyl-D-galactosamine) attached to the core trisaccharide.

12.32 Because it contains the monosaccharides markers present in all the other blood types.

12.33 B has an additional monosaccharide (D-galactose) attached to the core trisaccharide. Yes, they can accept type O blood because it only contains the core trisaccharide also present in type B. The reverse is not true because people with type O blood will not recognize the added monosaccharide in type B blood.

Lipids: Structure and Function

CHAPTER 13

Testosterone

Anabolic steroid use is banned from all athletic competitions. These steroids build muscle tissue, much like the natural hormone testosterone, but have serious health risks. [Tim de Waele/Corbis]

OUTLINE

 This icon indicates that a **Problem-Solving Tutorial** is available at www.whfreeman.com/gob

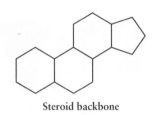

Steroid backbone

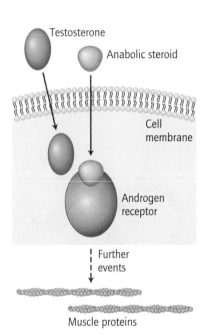

Figure 13-1 A signaling cascade occurs in muscle cells when testosterone or an anabolic steroid binds to the androgen receptor. The end result is protein production for muscle growth.

The Effects of Anabolic Steroids on the Body

American cyclist Floyd Landis forfeited his 2006 Tour de France title after testing positive for steroids. Although possession of anabolic steroids without a prescription is a federal crime, they have been widely used in sports. Athletes who take anabolic steroids do so to build muscle mass and increase physical strength. However, steroid use not only creates an unfair competitive advantage, it also poses a health risk to the athletes using them. What are anabolic steroids and how are they able to build muscle mass?

From a chemical perspective, a steroid is an organic compound that contains the characteristic four fused rings shown in the margin. Steroids contain a significant amount of hydrocarbon character, which makes them insoluble in aqueous solution and soluble in organic solvents. Biological molecules like steroids that do not dissolve in water fall into a class of biomolecules known as lipids, and are the subject of this chapter. Lipids serve many roles, ranging from long-term fat storage to the main structural component of cell membranes.

Steroids are lipids with a broad range of functions. In humans, they include all the sex hormones (androgens and estrogens), cholesterol, and the corticosteroids, a class of hormones that regulate metabolism and electrolyte balance. In general, a hormone serves as a chemical messenger, traveling through the bloodstream to hormone receptors located on or in cells in various tissues throughout the body.

Testosterone is a male sex hormone, or androgen. During puberty in males, testosterone signals the body to make the changes observed in male adolescence: growth of muscle tissue, appearance of body and facial hair, and deepening of the voice. The term *anabolic* means growth of tissue, and in the case of anabolic steroids, the tissue is muscle tissue. Testosterone acts by binding to the androgen receptor, which initiates a cascade of biochemical events that ultimately leads to protein production and muscle growth (Figure 13-1). Testosterone's other effects, the production of male sexual characteristics, are termed *androgenic* effects.

Anabolic steroids are made in the laboratory. Anabolic steroids bind to the androgen receptor as well as or even better than testosterone, and thus initiate a similar cascade of events. The chemical structure of an anabolic steroid, such as stanazolol or nandrolone, is similar to that of testosterone, but it has been chemically altered in a way that maximizes the anabolic effects while minimizing the androgenic effects.

Testosterone

Stanazolol

Nandrolone

Anabolic steroids have undesirable side effects. They disrupt the normal production of hormones, so in males their use can lead to infertility, gynecomastia (development of breasts), testicular atrophy (shrinkage of the testes), and premature baldness. Prolonged use of steroids can lead to liver damage, prostate cancer, and cardiovascular disease. In adolescents, prolonged use may stunt growth, because high levels of anabolic steroids signal the body that puberty is over. As a result, bones stop growing.

Given all the negative health effects of anabolic steroid use and its illegality, why are anabolic steroids available in the first place? Under the supervision of a doctor, anabolic steroids are an effective treatment for a number of serious disorders. For example, anabolic steroids are used to treat muscle wasting diseases, such as those experienced with advanced-stage HIV infection (AIDS). The tissue-building properties of anabolic steroids have also been shown to aid the recovery from severe burns and the treatment of hypogonadism, a condition in which the male body does not produce sufficient testosterone. In all of these cases, the dosage of steroid is low and usage is closely monitored by a doctor.

In Chapters 11 and 12 you learned about two important classes of biomolecules: proteins and carbohydrates. In this chapter you will study the structure and function of another class of biomolecules: lipids. In contrast to proteins and carbohydrates, lipids are defined primarily by their physical properties rather than their chemical structure. ***Lipids include all the biological molecules that are insoluble in water; they are soluble in nonpolar organic solvents.*** The solubility characteristics of lipids are a result of their extensive hydrocarbon structure.

Lipids serve a variety of functions, which depend on their structure. The different types of lipids, listed in order of abundance in the body, include:

- triglycerides, the fats and oils that provide the body with a long-term supply of energy;
- phospholipids and glycolipids, the main components of cell membranes;
- steroids, including cholesterol and the compounds derived from cholesterol, such as bile acids, vitamin D, and the sex hormones; and
- eicosanoids, lipids that are involved in the body's inflammatory response to trauma and infection.

13.1 | Energy Storage Lipids: Triglycerides

The most abundant lipids in the body are **triacylglycerols,** also known as **triglycerides** or, even more simply, as "fats." Triglycerides are obtained through the diet from both animal and vegetable sources. In the body, triglycerides are metabolized in muscle cells or stored in fat cells, more formally known as **adipocytes.** Triglycerides stored in fat cells serve as a reservoir of energy for the body's long-term energy needs. In contrast to glycogen, which can meet the body's energy needs for about a day, stored fat can provide energy for months, depending on the amount of stored fat.

Triglycerides are derived from fatty acids. A **fatty acid** is a long unbranched hydrocarbon chain with a carboxylic acid (RCOOH) at one end. Fatty acids contain an even number of anywhere from 12 to 24 carbon atoms. Table 13-1 shows the chemical structures of some of the most common fatty acids.

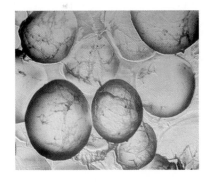

Fat cells, also known as adipocytes, store triglycerides. [© Collection CNRI/Phototake. All rights reserved.]

Saturated and Unsaturated Fatty Acids

In Chapter 6 you learned that a fatty acid may be either saturated or unsaturated. A saturated fatty acid contains no carbon-carbon double bonds, while an *un*saturated fatty acid contains one or more carbon-carbon double bonds. For example, stearic acid, shown in Figure 13-2a, is a carboxylic acid containing 18 carbon atoms. It contains no carbon-carbon double bonds, so it is a saturated fatty acid.

Fatty acids with only one double bond are also referred to as **monounsaturated** fatty acids. Oleic acid is the most abundant unsaturated fatty acid and it is a monounsaturated fatty acid, shown in Figure 13-2b.

Table 13-1 Common Fatty Acids

Fatty Acid (Common Name)	Number of Carbon Atoms: Number of Double Bonds	Condensed Structural Formula	Common Sources	Melting Point (°C)
Myristic	14:0	$CH_3(CH_2)_{12}COOH$	Nutmeg	53
Palmitic	16:0	$CH_3(CH_2)_{14}COOH$	Palm	63
Stearic	18:0	$CH_3(CH_2)_{16}COOH$	Lard (animal fat)	70
Oleic	18:1	$CH_3(CH_2)_7CH{=}CH(CH_2)_7COOH$	Olive oil, corn oil	13
*Linoleic	18:2	$CH_3(CH_2)_4(CH{=}CHCH_2)_2(CH_2)_6COOH$	Grains, eggs, plant oils	−9
*Linolenic	18:3	$CH_3CH_2(CH{=}CHCH_2)_3(CH_2)_6COOH$	Tuna, salmon, herring	−11
Arachidonic	20:4	$CH_3(CH_2)_4(CH{=}CHCH_2)_4(CH_2)_2COOH$	–	−49.5

*Essential fatty acids.

Polyunsaturated fatty acids contain more than one double bond. Examples are the unsaturated fatty acids linoleic acid and linolenic acid. Both contain 18 carbon atoms, but linoleic acid has two double bonds and linolenic acid has three.

Oleic acid and other naturally occurring unsaturated fatty acids have a *cis* double bond. The shape of a cis double bond gives the fatty acid a bend or kink in its otherwise linear shape. You can see this bend in the space-filling model of oleic acid in Figure 13-2b. In contrast, fatty acids with trans double bonds have a shape similar to saturated fatty acids. Notice the similarity in shape between the trans fatty acid in Figure 13-2c and the saturated fatty acid in Figure 13-2a.

The unsaturated fatty acids linoleic acid and linolenic acid are **essential fatty acids**, named so because they are essential nutrients that the body cannot synthesize from other metabolic intermediates. Therefore, they must be obtained from the diet. These two essential fatty acids are the well known omega-6 (ω-6) and omega-3 (ω-3) fatty acids, respectively. Arachidonic acid, an important fatty acid in the immune system described in the *Chemistry in Medicine* section, is synthesized from linoleic acid.

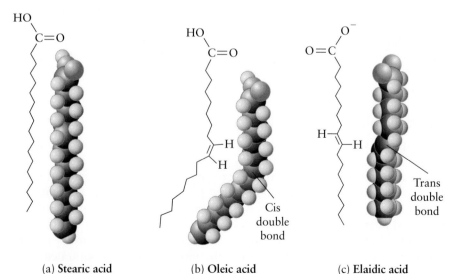

Figure 13-2 Three fatty acids that contain 18 carbon atoms: (a) stearic acid is saturated; (b) oleic acid is cis unsaturated; and (c) elaidic acid is trans unsaturated. Trans unsaturated and saturated fatty acids have similar shapes; they are both linear. Cis unsaturated fats contain a bend in their shape.

(a) **Stearic acid** Melting point 70°C

(b) **Oleic acid** Melting point 13°C

(c) **Elaidic acid** Melting point 46.5°C

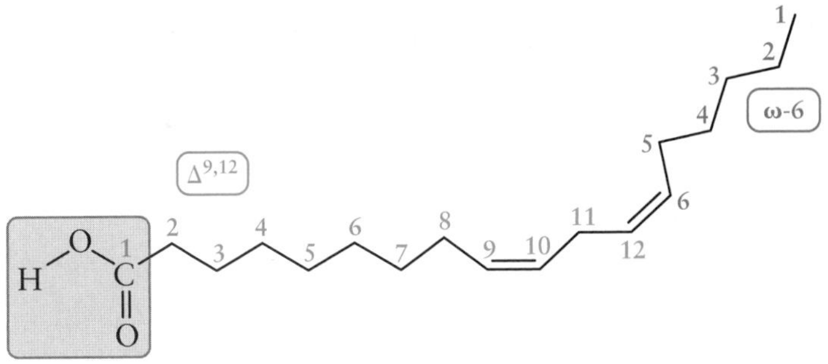

Carboxylic acid

Linoleic acid

Two different numbering systems are used to indicate the location of a double bond in a fatty acid: the omega system and the delta system. In the delta system, numbering begins with the carbonyl carbon (C1) and the location of each double bond is indicated by a number written as a superscript following the Greek letter delta (Δ). In the omega system, numbering begins at the end of the hydrocarbon chain farthest from the carbonyl group, and the location of only the first double bond is given. That location is indicated by a number following the Greek letter omega (ω). For example, linoleic acid is a $\Delta^{9,12}$ fatty acid in the delta system and an omega-6 (ω-6) fatty acid in the omega system (Figure 13-3). Nutritional scientists typically use the omega system, whereas biochemists prefer the delta system.

Melting Points of Fatty Acids

The melting point of a fatty acid increases as the number of carbon atoms in its formula increases. You can see this trend in the first three entries in Table 13-1: Myristic acid (C_{14}) has a melting point of 53°C; palmitic acid (C_{16}) a melting point of 63°C; and stearic acid (C_{18}) a melting point of 70°C. Since dispersion forces increase with surface area, this is the expected trend. More energy is needed to separate the longer fatty acid chains in the phase change from solid to liquid.

The melting point of a fatty acid drops significantly with the addition of a cis double bond. This trend is evident in the next three entries in Table 13-1: stearic acid, with no double bond, has a melting point of 70°C; oleic acid, with one double bond, has a melting point of 13°C; and linoleic acid, with two double bonds, has a melting point of −9°C. Basically, the bend created by the cis double bond(s) reduces the number of points of contact between fatty acid molecules, leading to fewer dispersion forces and, therefore, a lower melting point. In other words, it is easier to disrupt the intermolecular forces of attraction between unsaturated (cis) fatty acids than between saturated fatty acids, because they are less tightly packed.

Problem-Solving Tutorial:
Fatty Acid Structure and Physical Properties

WORKED EXERCISE 13-1 Fatty Acids

Palmitoleic acid is similar to palmitic acid except that it is a cis Δ^9 fatty acid and not a saturated fatty acid. Write the structure of palmitoleic acid and indicate which of the following statements are true:

a. It is an unsaturated fatty acid.
b. It is a polyunsaturated fatty acid.
c. It has an overall linear shape.
d. It has a higher melting point than palmitic acid.
e. It is polar.
f. It is an omega-6 fatty acid.

SOLUTION

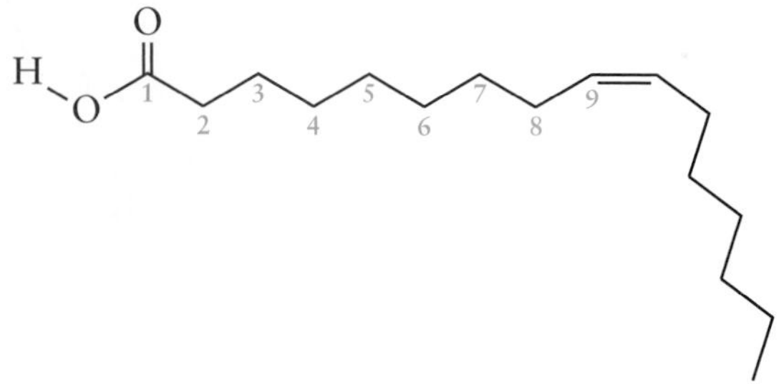

a. True.
b. False. It is a monounsaturated fatty acid.
c. False. The cis double bond creates a bend in the hydrocarbon chain.
d. False. The cis double bond lowers the melting point, because the bend created by the double bond leads to fewer dispersion forces, so less energy is required to go from the solid to the liquid state and the substance melts at a lower temperature.
e. False. The carboxylic acid is polar, but the rest of the molecule is nonpolar, therefore the molecule is an amphipathic molecule.
f. False. It is not an omega-6 fatty acid, because the double bond does not appear on the sixth carbon atom, counting from the end of the chain farthest from the carbonyl group.

PRACTICE EXERCISES

13.1 Explain the following:

   a. Palmitic acid has a higher melting point than myristic acid.
   b. Oleic acid has a lower melting point than stearic acid.

13.2 Which of the following are omega-6 fatty acids?

   a. arachidonic acid
   b. linolenic acid
   c. linoleic acid
   d. oleic acid

13.3 Fill in the blanks: In the delta numbering system, arachidonic acid is a $\Delta^{-,-,-,-}$ fatty acid. In the omega system, it is an omega-_____ fatty acid.

13.4 Which is more likely to be a liquid at room temperature: linolenic or stearic acid? Which is saturated?

13.5 In what way is the shape of a cis unsaturated fatty acid different from the shape of a saturated fatty acid? How does this difference impact its physical properties? Is a trans unsaturated fatty acid more similar to a cis unsaturated fatty acid or to a saturated fatty acid?

Triglycerides

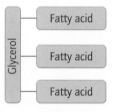

Triacylglycerols, also known as triglycerides, are assembled from one *glycerol* molecule and three *fatty acid* molecules. Glycerol is a three-carbon chain with a hydroxyl group (OH) on each carbon atom, as shown in Figure 13-4a. A simple triglyceride is shown in Figure 13-4b. This triglyceride is constructed from glycerol and three palmitic acid molecules, the fatty acid component (Figure 13-4c). Fatty acids are joined to the glycerol backbone at the hydroxyl groups by three ester linkages.

Although glycerol and fatty acid carboxylates (RCOO⁻) are water soluble, triglycerides are not, because ester linkages replace these polar functional

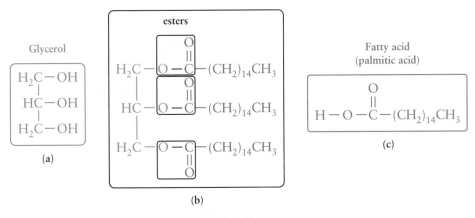

Figure 13-4 The components of a simple triglyceride.

groups. Thus, triglycerides are nonpolar molecules that exhibit hydrophobic properties (they avoid water). Indeed, this is why they are classified as lipids

Simple triglycerides contain only one kind of fatty acid, like the triglyceride formed from palmitic acid shown in Figure 13-4b. **Mixed triglycerides** are derived from two or three different types of fatty acids. Triglycerides can be characterized as saturated or unsaturated, depending on whether the fatty acids contain double bonds.

Physical Properties of Triglycerides

Triglycerides derived from saturated fatty acids tend to be solids at room temperature (25°C), and are commonly referred to as **fats.** These triglycerides are found in meat, whole milk, butter, and cheese—all animal sources. Tropical oils from plant sources such as coconut and palm oil actually have the highest concentration of saturated fats. Most other plant sources contain mainly unsaturated fats.

Triglycerides derived from unsaturated fatty acids tend to be liquids at room temperature, and are commonly referred to as **oils.** They are found primarily in plant sources.

Saturated fats are more tightly packed than unsaturated fats, because the bend created by the cis double bond(s) limits the number of points of contact in an unsaturated fat. Hence, unsaturated triglycerides have lower melting points, which explains why they are oils at room temperature; while saturated triglycerides have higher melting points, which explains why they are solids at room temperature.

Trans fats are generally not found in nature; they are produced artificially during the catalytic hydrogenation of unsaturated triglycerides (see Section 10.2, catalytic hydrogenation). While some double bonds are reduced to single bonds in the reaction, others are isomerized from the cis to the trans geometric isomer. A trans double bond gives these triglycerides an overall shape and physical properties resembling a saturated fat.

You learned that diets high in saturated and trans fats have been linked to cardiovascular disease, diabetes, and cancer. The American Heart Association recommends a diet low in saturated and trans fats. Compare the amount of saturated and unsaturated fats in various foods using the SatFat Graph in Figure 13-5, published by the National Institutes of Health. Canola oil, for example, has the least amount of saturated fat, while animal fats are at the other end of the spectrum. Palm and coconut oil have the highest saturated fat content.

Figure 13-5 The SatFat Graph shows the relative amount of saturated and unsaturated fats in various commercial oils and fats, reported here as a percentage of saturated and unsaturated fats.

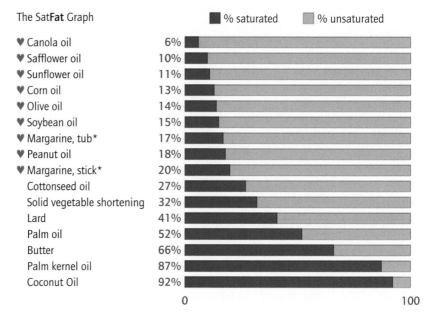

The Sat**Fat** Graph   ■ % saturated   ▨ % unsaturated

	%
♥ Canola oil	6%
♥ Safflower oil	10%
♥ Sunflower oil	11%
♥ Corn oil	13%
♥ Olive oil	14%
♥ Soybean oil	15%
♥ Margarine, tub*	17%
♥ Peanut oil	18%
♥ Margarine, stick*	20%
Cottonseed oil	27%
Solid vegetable shortening	32%
Lard	41%
Palm oil	52%
Butter	66%
Palm kernel oil	87%
Coconut Oil	92%

0   100

*An average of margarines listing liquid oil as the first ingredient.

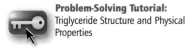

Problem-Solving Tutorial: Triglyceride Structure and Physical Properties

---

WORKED EXERCISE 13-2   Triglycerides

Answer the questions below for the following molecule:

a. What type of biomolecule is this: protein, carbohydrate, or lipid? What function does it serve in the body?

b. Is this a saturated or an unsaturated fat?

c. Circle the ester functional groups.

d. Outline the glycerol backbone in the molecule.

e. What fatty acids were used to make this molecule: palmitic acid, oleic acid, linoleic acid, or linolenic acid? Indicate all that apply.

f. Is this a simple triglyceride or a mixed triglyceride?

g. Are the double bonds cis or trans? Which is healthier?

h. In what type of cell would your body store this type of molecule?

SOLUTION

a. A lipid; it serves as a long-term source of energy for the body.

b. This is an unsaturated fat, because it is derived from fatty acids containing double bonds.

(c), (d), (e)

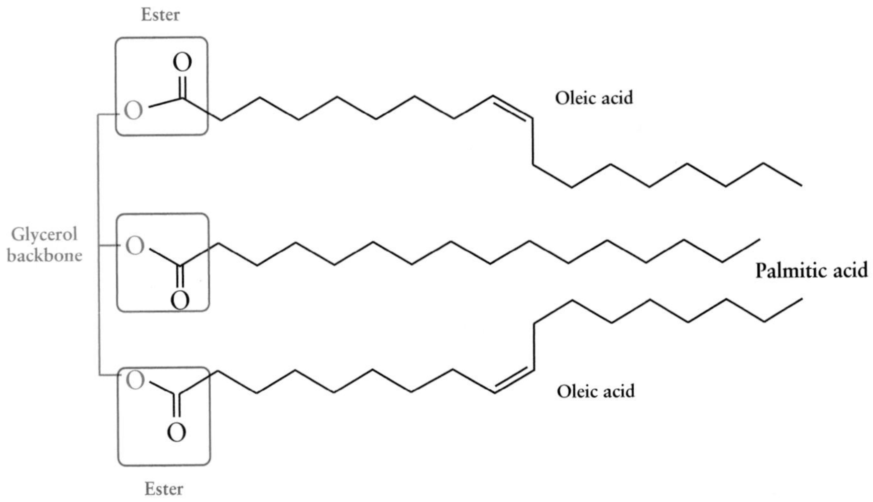

Ester

Glycerol backbone

Oleic acid

Palmitic acid

Oleic acid

Ester

f. It is a mixed triglyceride, because more than one type of fatty acid component is present.

g. The double bonds are all cis. Unsaturated fats containing cis double bonds are healthier than those containing trans double bonds, and are the natural form of a double bond found in fatty acids.

h. Triglycerides are stored in adipocytes (fat cells).

PRACTICE EXERCISES

13.6 Write the structure of a triglyceride composed of linoleic acid, linolenic acid, and oleic acid.

    a. Would this triglyceride be considered a saturated or an unsaturated fat?

    b. Would this triglyceride be considered a *simple* or a *mixed* triacylglycerol?

    c. Are any of the fatty acid components omega-3 or omega-6 fatty acids? If so, which ones?

13.7 Which vegetable oils are high in saturated fat and should, therefore, be included in limited amounts in the diet?

13.8 Oleic acid is an omega-___ fatty acid in the omega system and a Δ— in the delta system.

13.9 Which product is higher in saturated fats: peanut oil or corn oil?

13.10 Which product is healthier: coconut oil or canola oil? Explain.

13.11 Are unsaturated fats primarily found in vegetable or animal sources?

13.12 Why does arachidonic acid have a lower melting point than linolenic acid, even though it has two more carbon atoms in its chemical formula?

The Model Tool 13-1 Fatty Acids and Triglycerides

This exercise requires that you work as a group of three or obtain three model kits.

I. Construction of Glycerol

1. Using one model kit, have one person obtain 3 black carbon atoms, 3 red oxygen atoms, 8 light-blue hydrogen atoms, and 13 single bonds.

2. Construct a model of glycerol. Is glycerol polar? Explain.

II. Construction of Palmitoleic Acid

3. Using two model kits, obtain 16 black carbon atoms, 2 red oxygen atoms, and 28 light-blue hydrogen atoms, several single bonds, and 2 bent bonds.

4. Construct a model of palmitoleic acid, a cis 16:1 Δ^9 fatty acid. Arrange the model so it is in a zigzag conformation.
   a. What is the name of the acid on the omega system?
   b. Is this molecule polar, nonpolar, or amphipathic? Explain.
   c. Is this a saturated or an unsaturated fatty acid?
   d. Describe the overall shape of the fatty acid. Is it linear or is there a kink or bend in the chain? If so, where is it, and what causes it?

5. Convert the fatty acid into a trans fatty acid. How did you do this? How is the shape of the molecule different from the cis fatty acid?

6. Convert the trans fat into a saturated fat. What two atoms must be added to make this change? What is the name of this reaction? How does the shape of the saturated fatty acid compare to the trans fat?

III. Construction of Glycerol Attached to a Single Fatty Acid (a Monoacylglycerol)

7. Using the models of glycerol and palmitoleic acid that you made above, construct a monoacylglycerol by forming an ester at C1 between glycerol and your fatty acid.
   a. In forming the ester, what small molecule is lost? What type of reaction is this (see Chapter 10)?
   b. How would you make a simple triglyceride from the monoacylglycerol that you just built? Would the triacylglycerol be polar or nonpolar? Saturated or unsaturated? Solid or liquid?

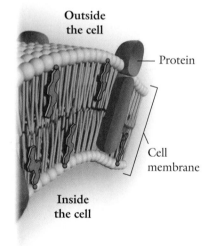

Outside the cell

— Protein

Cell membrane

Inside the cell

Figure 13-6 The cell membrane is a semipermeable barrier separating the inside and outside of the cell.

$$\begin{array}{c} O \\ \| \\ RO-P-OR \\ | \\ O^- \end{array}$$

Phosphate group

13.2 | Membrane Lipids: Phospholipids and Glycolipids

The **cell membrane** is a part of all living cells. This semipermeable barrier separates two environments: the inside and the outside of the cell (Figure 13-6). The cell membrane, also called a **plasma membrane**, controls the flow of ions and molecules in and out of the cell, allowing the cell to receive nutrients and dispose of waste. The cell membrane also plays a critical function in cell recognition and communication between cells (see Section 12.5). Although proteins are interspersed throughout the cell membrane, and carbohydrates protrude from it, the underlying components of these remarkable structures are lipids.

Phospholipids and Glycolipids

Two types of lipids are found in human cell membranes:

- phospholipids and
- glycolipids.

Phospholipids are the most common type of membrane lipids. They contain a phosphate group as the distinguishing feature of their structure, whereas **glycolipids** contain a carbohydrate instead. Cholesterol is another lipid component of the cell membrane; it will be described in Section 13.4.

The carbon backbone of membrane lipids is derived from either glycerol (Figure 13-7a) or sphingosine (Figure 13-7b). Sphingosine resembles glycerol except that it contains a long hydrocarbon chain as part of the backbone and an amine functional group in place of the middle hydroxyl group in glycerol.

A phospholipid with a sphingosine backbone is called a **sphingomyelin**, while a phospholipid with a glycerol backbone is called a **glycerophospholipid.** Glycolipids found in human cells all contain a sphingosine backbone, and therefore, they are called **sphingolipids.** These classes of membrane lipids are summarized in Figure 13-8.

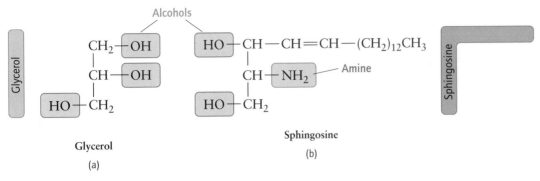

Figure 13-7 Glycerol (a) and sphingosine (b) serve as the carbon backbone of membrane lipids.

Regardless of the backbone, membrane lipids have two long hydrocarbon chains as a distinguishing feature of their chemical composition. When the backbone of a membrane lipid is glycerol, two long-chain fatty acids are linked to the first two alcohol functional groups by ester linkages in the same way that three fatty acids were linked to all three alcohols in glycerol to make a triglyceride. When the backbone is sphingosine, one fatty acid is attached to the amine by an amide linkage and one long hydrocarbon chain comes from the sphingosine backbone itself. Since hydrocarbon chains are nonpolar, this part of a membrane lipid is often referred to as its nonpolar "tail."

The remaining alcohol on each backbone is attached to either a carbohydrate, in the case of a glycolipid, or a phosphate group, in the case of a phospholipid.

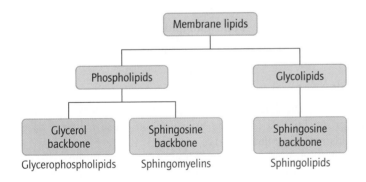

Figure 13-8 Types of membrane lipids.

Phospholipids

Phospholipids contain a phosphate group at the terminal alcohol of either sphingosine or glycerol. The phosphate group is also attached to an amino alcohol as illustrated schematically in Figure 13-9.

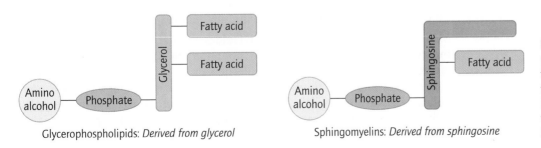

Figure 13-9 Parts of a phospholipid. There are two types of phospholipids: those with a glycerol backbone (left) and those with a sphingosine backbone (right).

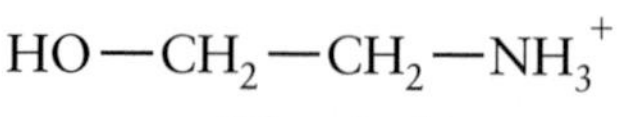

HO—CH$_2$—CH$_2$—NH$_3^+$

Ethanolamine

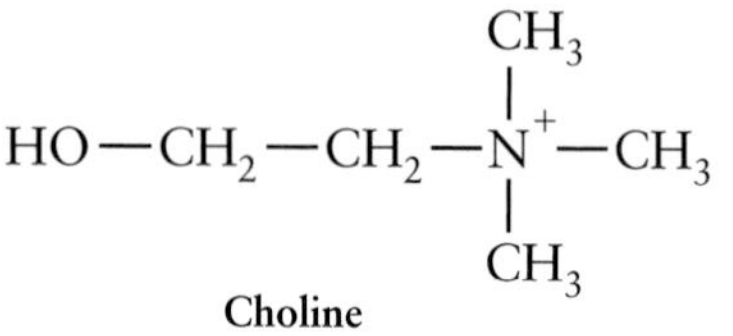

$$HO—CH_2—CH_2—\overset{\overset{\displaystyle CH_3}{|}}{\underset{\underset{\displaystyle CH_3}{|}}{N^+}}—CH_3$$

Choline

$$HO—CH_2—\overset{\overset{\displaystyle H}{|}}{\underset{\underset{\displaystyle COO^-}{|}}{C}}—NH_3^+$$

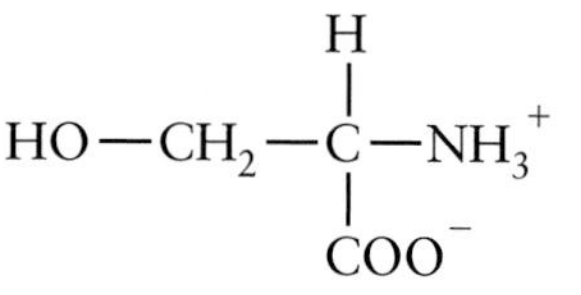

Serine
(only in glycerophospholipids)

Figure 13-10 Amino alcohols commonly found in phospholipids.

The three amino alcohols most commonly attached to the phosphate group of a phospholipid are:

- ethanolamine,
- choline, and
- serine, in the case of glycerophospholipids.

These amino alcohols are shown in Figure 13-10. Note that the amines in ethanolamine and serine are positively charged because they are in their conjugate acid form at physiological pH. Choline, on the other hand, is always positively charged because it contains four R groups covalently bonded to the nitrogen atom (a special type of amine).

The phosphate group is linked to both the terminal hydroxyl group on glycerol or sphingosine *and* the hydroxyl group of the amino alcohol, thereby creating a phosphodiester functional group. For example, the phospholipid shown in Figure 13-11 contains choline as the amino alcohol, glycerol as the backbone, and two stearic fatty acids as the nonpolar tails.

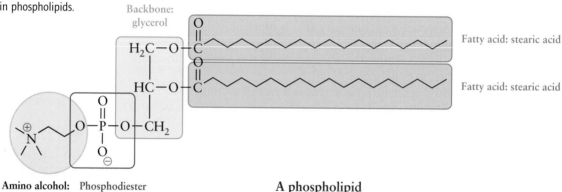

A phospholipid

Figure 13-11 A glycerophospholipid derived from two stearic acid molecules and the amino alcohol choline.

Glycolipids

In human cells, the backbone of a glycolipid is always sphingosine. The hydroxyl group farthest from the hydrocarbon chain is linked to a carbohydrate as illustrated in Figure 13-12.

The carbohydrate component of a glycolipid is a mono-, di-, or oligosaccharide (3 to 10 monosaccharides). Glycolipids containing a monosaccharide are known as **cerebrosides.** Cerebrosides containing galactose are found primarily in cells of the central nervous system and brain, while those containing glucose are found in most other tissues. Glycosides containing an oligosaccharide are known as **gangliosides.**

Consider for example, the cerebroside shown in Figure 13-13. The carbohydrate is galactose; and one of the hydrocarbon tails is derived from stearic acid. The other hydrocarbon tail is from the sphingosine backbone.

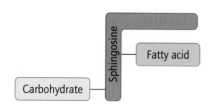

Figure 13-12 Parts of a glycolipid.

Membrane Lipids Are Amphipathic Molecules

Membrane lipids differ from triglycerides in that they are **amphipathic molecules:** polar on one end and nonpolar on the other end. You were introduced to two amphipathic molecules in Chapters 6 and 7: cholesterol and soap. Membrane lipids are amphipathic because they have two nonpolar hydrocarbon "tails" and a polar "head." In a glycolipid, the polar group is the carbohydrate. Recall that carbohydrates are polar because they contain numerous hydroxyl groups. In a phospholipid, the polar

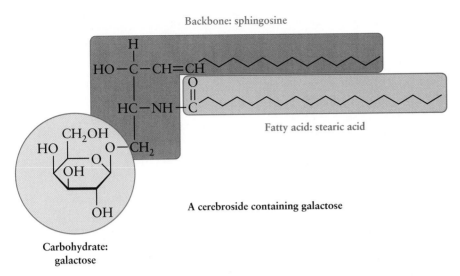

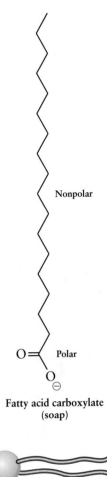

Backbone: sphingosine

Fatty acid: stearic acid

A cerebroside containing galactose

Carbohydrate:
galactose

Figure 13-13 A cerebroside derived from stearic acid and the carbohydrate galactose.

group is the negatively charged phosphate group together with the positively charged amine.

Consider the glycerophospholipid shown in Figure 13-14 as an example. The two nonpolar tails are both derived from the fatty acid stearic acid (green). Keep in mind that a variety of different fatty acids can be employed for the two "tails." The polar head is created by the phosphate group (purple) and the amino alcohol choline (yellow). The backbone is glycerol (blue).

Since membrane lipids contain a polar head and two nonpolar tails, they are often depicted as a sphere—the polar head—attached to two wavy lines—the nonpolar tails, as shown in the margin.

Membrane Lipids and the Structure of the Cell Membrane

The cell membrane is assembled from two layers of phospholipids and glycolipids called a **bilayer.** The two layers in a bilayer are aligned in a way that brings the polar heads in contact with water, and places the nonpolar tails

Nonpolar

Polar

Fatty acid carboxylate
(soap)

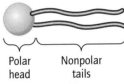

Polar
head

Nonpolar
tails

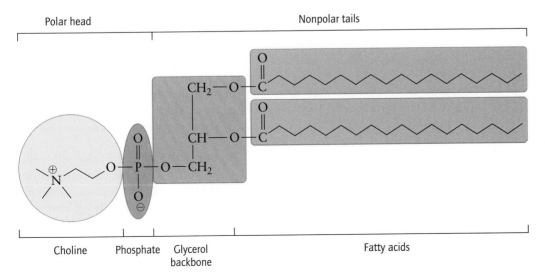

Polar head   Nonpolar tails

Choline   Phosphate   Glycerol
backbone

Fatty acids

Figure 13-14 The polar head and nonpolar tails of a glycerophospholipid.

Figure 13-15 The cell membrane is a lipid bilayer: two rows of lipids.

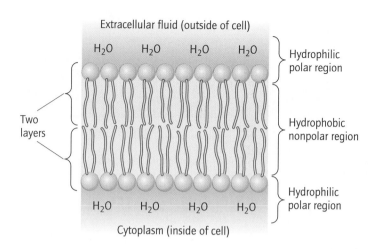

Extracellular fluid (outside of cell)

H_2O H_2O H_2O H_2O — Hydrophilic polar region

Two layers — Hydrophobic nonpolar region

H_2O H_2O H_2O H_2O — Hydrophilic polar region

Cytoplasm (inside of cell)

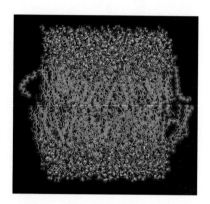

Molecular simulation showing a segment of the lipid bilayer. The polar heads shown in pink interact with water molecules (red and white). Nonpolar tails, shown in green, line up tail to tail. [Laguna Design/Photo Researchers, Inc.]

in a nonpolar environment (Figure 13-15). The lipids are organized tail to tail, creating a hydrophobic environment in the interior of the membrane. The hydrophilic polar heads lie on the outer surfaces of the membrane, forming a hydrophilic interface with the aqueous environment inside and outside of the cell. The distance between the polar heads on opposite sides of the bilayer constitutes the width of the cell membrane, which is approximately 7 nm.

The lipid bilayer is often described as a "fluid mosaic," because the phospholipid and glycolipid molecules are not covalently bonded to one another, but interact through weaker, noncovalent intermolecular forces of attraction, creating a flexible fluid-like assembly. Embedded within the cell membrane are cholesterol molecules, amphipathic molecules that provide added rigidity to the cell membrane. The polar alcohol functional group of cholesterol lies among the polar heads and the hydrocarbon portion of the molecule is embedded within the nonpolar tails. Various proteins are also found throughout the cell membrane. Transmembrane proteins, described in Chapter 11, span the entire membrane, while peripheral proteins lie on the outer and inner portions of the membrane. If you were to view a larger section of the cell membrane, you would see something like Figure 13-16.

Membrane Transport

Small nonpolar molecules, such as oxygen and carbon dioxide, pass freely through the cell membrane by simple diffusion. Water passes slowly through the cell membrane by the process of osmosis, described in Chapter 5. Since the interior of the cell membrane is nonpolar, ions and polar organic molecules cannot readily pass through the bilayer. For example, glucose must be transported through the cell membrane by specialized proteins, a process that first requires that insulin bind to receptors on the cell. People with Type II diabetes are "resistant to insulin," which means that the cell never receives the signal to transport glucose across the membrane.

In Chapter 5 you learned that one important function of the cell membrane is to ensure that different concentrations of ions inside and outside of the cell are maintained. In many cases, special transport proteins are necessary to move ions from one side of the membrane to the other. The energy needed to operate these "ion pumps" comes from ATP. Which ions move in and out of the cell is regulated by hormones and other chemical messengers, which interact with proteins on the cell surface. The ability to transport ions is crucial to cell survival. See Chapter 5 for a review of osmosis and dialysis.

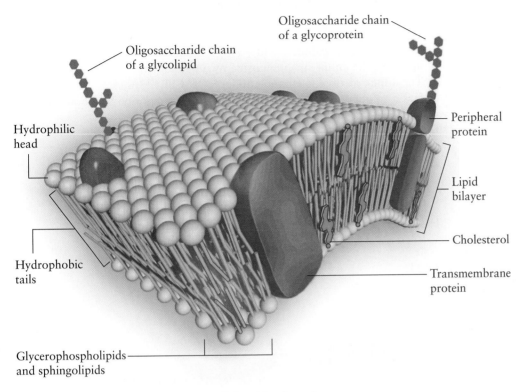

Extracellular side of membrane

Oligosaccharide chain of a glycoprotein

Oligosaccharide chain of a glycolipid

Hydrophilic head

Peripheral protein

Lipid bilayer

Cholesterol

Hydrophobic tails

Transmembrane protein

Glycerophospholipids and sphingolipids

Intracellular side of membrane

Figure 13-16 The cell membrane is a lipid bilayer composed of glycerophospholipids and glycolipids (in beige) and cholesterol (in red). Proteins (in brown) are interspersed throughout. Oligosaccharides protrude from the membrane (Chapter 12).

WORKED EXERCISE 13-3 Membrane Lipids

Refer to the membrane lipid shown below to answer the following questions.

a. Is it a phospholipid or a glycolipid? How can you tell?
b. Is the backbone derived from glycerol or sphingosine?
c. What is the fatty acid component and what functional group connects it to the backbone?
d. If it is a phospholipid, what is the amino alcohol: serine, choline, or ethanolamine?
e. Is it a sphingomyelin, cerebroside, or ganglioside?
f. Circle the part of the molecule that defines the "polar head."
g. Circle the part of the molecule that defines the two tails.

SOLUTION

a. It is a phospholipid because it contains a phosphate group and not a carbohydrate.

b. Sphingosine

c. The fatty acid is palmitic acid—a 16-carbon saturated fatty acid. It is joined to the amine in sphingosine by an amide linkage.

d. The amino alcohol is ethanolamine: $H_3N^+CH_2CH_2OH$.

e. It is a sphingomyelin because it contains a sphingosine backbone and a phosphate group.

f. and **g.**

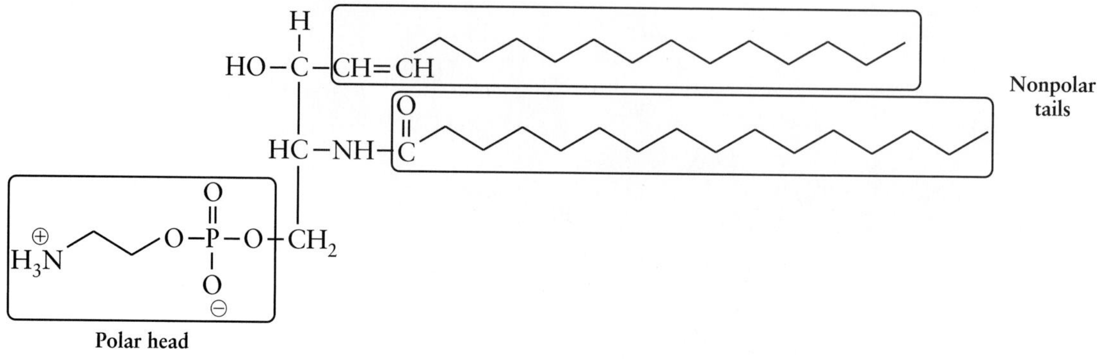

Nonpolar tails

Polar head

PRACTICE EXERCISES

13.13 For the membrane lipid below, answer the following questions:

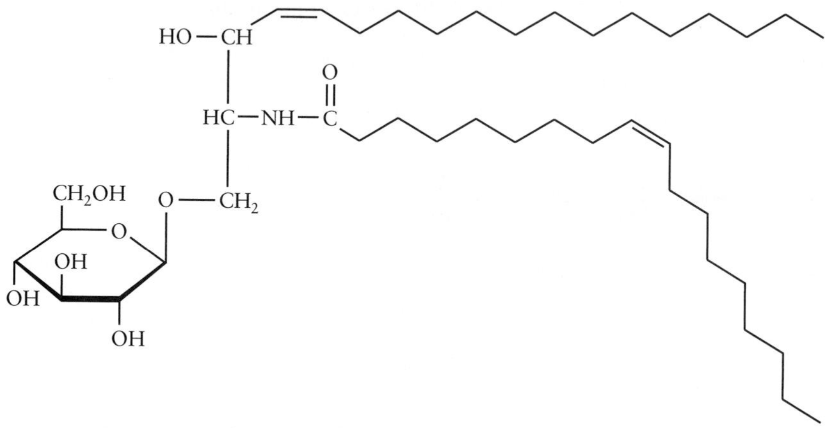

a. Is it a phospholipid or a glycolipid? How can you tell?

b. Is the backbone derived from sphingosine or glycerol? Circle the backbone.

c. What is the fatty acid from which the lipid is derived? Is it saturated or unsaturated? What functional group connects it to the backbone?

d. Is it a cerebroside or a ganglioside? Explain how you can tell the difference.

e. If it is a glycolipid, is it a glucose or galactose derivative? Is the glycosidic linkage α or β?

f. Circle and label all functional groups.

13.14 Explain what an amphipathic molecule is. How is it different from a nonpolar molecule? How is it different from a polar molecule?

13.15 How are triglycerides different from glycerophospholipids?

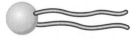

13.16 Explain what the illustration in the margin represents in molecular terms. If two of these molecules were to assemble, draw two ways in which they might align with one another.

13.17 Draw an illustration of 10 membrane lipids arranged to form a bilayer such as the cell membrane.

13.3 Fatty Acid Catabolism

We turn now to triglycerides, the most abundant type of lipid. Triglycerides are stored in fat cells and serve as the body's fuel during periods when other sources of energy, such as carbohydrates, are not available. Indeed, it is triglycerides that provide hibernating animals with energy while they are hibernating. The body uses glycogen as a short-term source of fuel and triglycerides for its longer-term energy needs. An adult man, for example, stores 300 g of glycogen, but about 15,000 g of triglycerides. The former lasts about a day, while the latter lasts 12 weeks.

The catabolism of triglycerides to provide energy in the form of ATP proceeds in three stages:

- Stage 1 (digestion): A triglyceride is hydrolyzed into glycerol and three fatty acids.
- Stage 2: Fatty acids are oxidized to acetyl CoA.
- Stage 3: Acetyl CoA is converted into carbon dioxide and energy in the citric acid cycle and oxidative phosphorylation.

Here we describe the first two stages of lipid catabolism to produce acetyl CoA (Figure 13-17). In Chapter 14 you will consider the fate of acetyl CoA in the third stage of catabolism.

Fatty acid oxidation is the central energy producing pathway for many cells. For example, it provides 80% of the energy needs for heart and liver cells. The highly reduced (saturated or close to saturated) form of the hydrocarbon chain of a fatty acid makes it an excellent fuel—much as the hydrocarbon gasoline is a good fuel for your car engine. In fact, you saw in Chapter 8 that triglycerides provide more than double the energy of carbohydrates and proteins of comparable mass (9 Cal/g rather than 4 Cal/g).

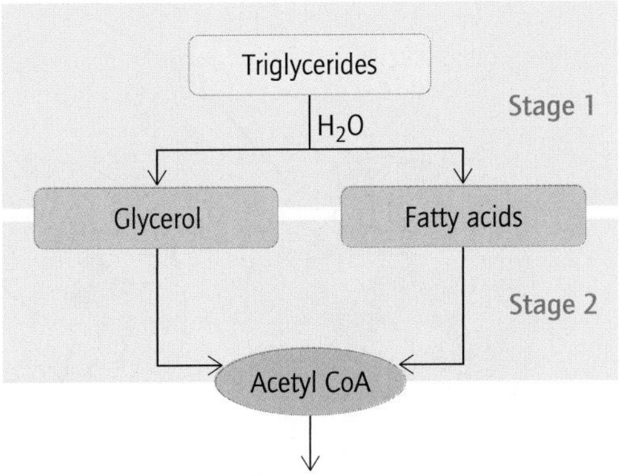

Figure 13-17 The first two stages of triglyceride catabolism.

Stage 1: Triglyceride Transport and Hydrolysis

Dietary fat must first be degraded into fatty acids and glycerol in order to deliver these molecules to the cells that store them (adiposytes) or use them for energy. However, since triglycerides are not soluble in the aqueous environment of the blood, they require specialized aggregates to transport them: Lipoproteins transport triglycerides to their target cells.

The degradation of dietary fat begins in the small intestine. Water-soluble enzymes called lipases hydrolyze triglycerides into smaller components that can cross the cell membrane of cells lining the intestine (the intestinal mucosa). Triglycerides from the diet enter the small intestine as large globules of insoluble fats. Before degradation can begin, these globules of fat must be turned into a colloid of finely dispersed microscopic fat droplets suspended in water, a process called **emulsification.** This process is aided by **bile salts**—nature's detergents—which are released into the small intestine from the gall bladder after a fatty meal. Emulsification is necessary to bring the water-insoluble triglycerides in contact with the water-soluble lipases. Hydrolysis of triglycerides produces a combination of fatty acids, glycerol, and partially hydrolyzed *tria*cylglycerols with only one or two fatty acids attached (*mono*acylglycerols and *di*acylglycerols).

Once these hydrolyzed molecules have entered intestinal mucosa cells, they undergo esterification reactions to reform triglycerides. Triglycerides formed in intestinal mucosa cells are then packaged along with cholesterol into chylomicrons, a type of lipoprotein, for transport through the lymph system and bloodstream to cells that need energy or store triglycerides. See Steps 1 to 5 in Figure 13-18.

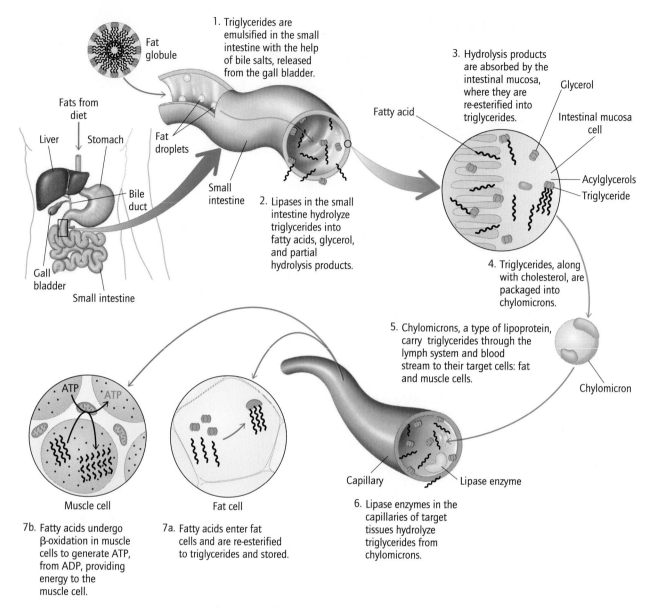

1. Triglycerides are emulsified in the small intestine with the help of bile salts, released from the gall bladder.

2. Lipases in the small intestine hydrolyze triglycerides into fatty acids, glycerol, and partial hydrolysis products.

3. Hydrolysis products are absorbed by the intestinal mucosa, where they are re-esterified into triglycerides.

4. Triglycerides, along with cholesterol, are packaged into chylomicrons.

5. Chylomicrons, a type of lipoprotein, carry triglycerides through the lymph system and blood stream to their target cells: fat and muscle cells.

6. Lipase enzymes in the capillaries of target tissues hydrolyze triglycerides from chylomicrons.

7a. Fatty acids enter fat cells and are re-esterified to triglycerides and stored.

7b. Fatty acids undergo β-oxidation in muscle cells to generate ATP, from ADP, providing energy to the muscle cell.

Figure 13-18 Triglyceride digestion: Triglycerides are transported from the small intestine through the bloodstream to fat and muscle cells.

Chylomicrons are spheres formed of a single layer of phospholipids, 1 μm (micrometer) in diameter. Since proteins are embedded throughout the lipid layer, chylomicrons are classified as a type of *lipoprotein* (lipid + protein). The lipids forming a chylomicron have a polar head and a nonpolar tail. The polar head is positioned on the exterior of the chylomicron, where it interacts with the aqueous external environment. The nonpolar tails form a hydrophobic interior, which encapsulates the triglycerides, as illustrated in Figure 13-19. Since lipoproteins are soluble in aqueous solution, chylomicrons serve as a vehicle for carrying triglycerides, transporting them through the lymph system and the bloodstream to muscle and adipose tissue. See Steps 4 and 5 in Figure 13-18.

Upon arrival at the capillaries of their target tissues, enzymes located outside the cell hydrolyze the triglycerides delivered by the chylomicrons back into fatty acids and glycerol, substances that are small enough to pass through the cell membrane. See Step 6 in Figure 13-18.

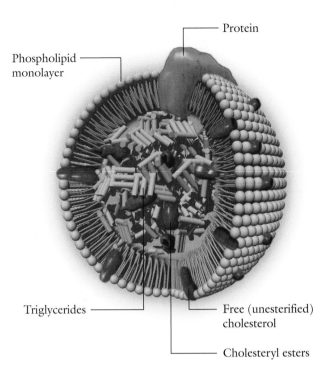

Protein

Phospholipid
monolayer

Triglycerides

Free (unesterified)
cholesterol

Cholesteryl esters

Figure 13-19 The structure of a lipoprotein: a single spherical layer of phospholids with a polar exterior and a nonpolar interior, embedded with proteins. Lipoproteins transport triglycerides and cholesterol esters, in the nonpolar interior of the lipoprotein, through the aqueous medium of the lymph system and bloodstream.

Inside an adipocyte, fatty acids and glycerol are converted back into triglycerides and stored (Figure 13-18, Step 7a). In muscle cells, fatty acids enter the second stage of triglyceride catabolism (Figure 13-18, Step 7b).

Lipoproteins

Chylomicrons are one of several types of serum **lipoproteins**, all of which transport lipids through the blood. They differ in their density, as shown in Table 13-2. Chylomicrons are the largest of the lipoproteins (1 μm (1 micron) in diameter). The various types differ in their specific transport functions.

The various types of lipoproteins are categorized by their protein to lipid ratio. The higher the protein/lipid ratio, the more dense the lipoprotein. Chylomicrons are the least dense because they have the highest lipid content—mainly triglycerides. Low-density lipoproteins (LDLs) are also high in triglycerides and cholesterol compared to protein. At the other end of the spectrum, high-density lipoproteins (HDLs) have the greatest proportion of protein.

After chylomicrons deliver their contents to cells, they return to the liver as chylomicron remnants (leftovers), where they are either degraded or, when there is an excess of fatty acids, converted into triglycerides and packaged into VLDLs. VLDLs also travel through the bloodstream to deliver triglycerides to adipose and muscle cells. Eventually, as they unload cholesterol and triglycerides to cells, they become IDLs and LDLs. In contrast, the role of HDLs is to scavenge cholesterol from the blood and return it to the liver.

You may have heard the terms "bad cholesterol" and "good cholesterol"; in reality, there is only one cholesterol molecule. More accurately, these terms refer to the lipoproteins and the cholesterol contained therein. The cholesterol in a lipoprotein is actually a cholesterol ester formed between a fatty acid and cholesterol. HDLs are considered the "good cholesterol" because they scavenge excess cholesterol in the bloodstream and transport it back to the liver, where it is converted into bile salts and excreted. Higher levels of this lipoprotein circulating in the bloodstream is desirable because its removal of cholesterol from the blood prevents plaque formation. A healthy diet and exercise have both been shown to increase HDL levels.

Table 13-2 Properties of Lipoproteins

Lipoprotein	Size (drawn to scale)		Density (g/mL)	Lipid/ Protein Ratio	Triglyceride/ Cholesterol Ester Ratio
Chylomicrons	(10× larger than VLDL)	1 μm (1000 nm) = 1 micron	0.95	66	29
VLDL (very-low-density)		70 nm	0.98	11	3.9
IDL (intermediate-density)		40 nm	1.01	8	0.82
LDL (low-density)		20 nm	1.04	3.8	0.18
HDL (high-density)		10 nm	1.13	1.2	0.16

Problem-Solving Tutorial:
Fatty Acid Catabolism Stage 1:
Triglyceride Hydrolysis

WORKED EXERCISE 13-4 Triglyceride Transport and Hydrolysis

The following events occur in the digestion of triglycerides. Place them in the correct chronological order.

a. Lipases hydrolyze triglycerides into fatty acids, glycerol, and partial hydrolysis products.
b. Triglycerides are emulsified in the small intestine.
c. Chylomicrons move through the lymph system and bloodstream carrying triglycerides.
d. Enzymes in target tissue capillaries hydrolyze triglycerides so that they can diffuse through the cell membranes of muscle and fat cells.
e. Fatty acids enter fat cells and are converted into triglycerides and stored or enter muscle cells and are used to generate energy.
f. Fatty acids are absorbed by intestinal mucosa cells where they are converted into triglycerides and packaged into chylomicrons.

SOLUTION
(b), (a), (f), (c), (d), (e).

PRACTICE EXERCISES

13.18 Which biomolecules produce the most energy per gram: proteins, carbohydrates, or triglycerides?

13.19 The egg in mayonnaise serves as an emulsifying agent for the other two main ingredients: oil and vinegar. What is an emulsifying agent and what type of molecule in eggs serves this function in mayonnaise?

13.20 What role does bile serve in the digestion of lipids?

13.21 What type of reactions do *lipases* catalyze and in which organ of the body are they found?

13.22 What function do chylomicrons serve? What is the composition of a chylomicron?

13.23 Extracellular enzymes hydrolyze triglycerides before they enter fat and muscle cells. What happens to the hydrolysis products after they enter adipocytes? What happens to the hydrolysis products after they enter muscle cells?

13.24 In which organ in the body is bile
  a. stored?
  b. synthesized from cholesterol?
  c. used to emulsify fats after consumption of a high-fat meal?

13.25 The structure of one of the bile salts, from cholic acid, is shown below. Circle and label the functional groups in bile. Is this molecule polar, nonpolar, or amphipathic?

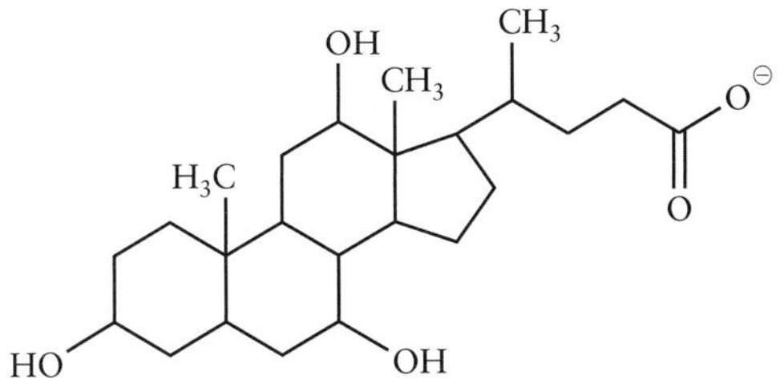

Bile salt
(cholate)

Stage 2: Fatty Acid Oxidation

Oxidation of a single fatty acid produces several acetyl CoA molecules that can proceed on to the third stage of catabolism. In addition, NADH and $FADH_2$ are produced. Both acetyl CoA and the electrons carried by NADH and $FADH_2$ supply the energy for the phosphorylation of ADP to ATP.

After an initial activation step, β-oxidation takes place in the mitochondria, organelles often described as the "power plants" of the cell, because they are where most ATP producing pathways occur, including the citric acid cycle, the electron-transport chain, and β-oxidation.

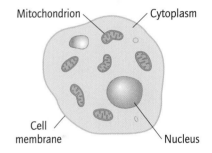

The power plants of the cell, mitochondria, are found in all cells containing a nucleus. The citric acid cycle, β-oxidation, and oxidative phosphorylation occur in the mitochondria.

Initial Activation Step Fatty acid catabolism begins in the cytoplasm of muscle and liver cells when a fatty acid reacts with coenzyme A to produce fatty acyl CoA, a thioester similar to acetyl CoA but having a much longer hydrocarbon chain. This reaction activates the fatty acid molecule, making it ready for degradation by the biochemical process known as β-oxidation. Activation of a fatty acid requires ATP, which is converted into AMP and 2 P_i (inorganic phosphate).

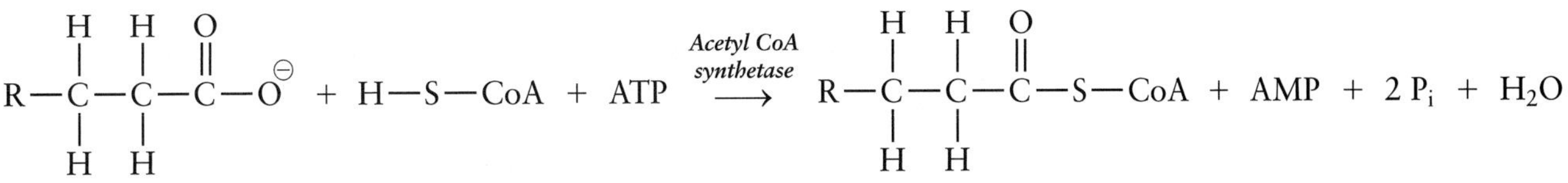

Fatty acid + Coenzyme A + ATP $\xrightarrow{\text{Acetyl CoA synthetase}}$ Fatty acyl CoA + AMP + 2 P_i + H_2O

Note that the conversion of ATP to AMP and 2 P_i is the equivalent of converting two ATP to ADP. The activated fatty acid, fatty acyl CoA, is then transported into the mitochondrion for β-oxidation.

β-Oxidation The **β-oxidation** of a fatty acid is a sequence of four reactions that are repeated over and over again. With each pass through β-oxidation, a two-carbon fragment is removed from the fatty acid, shortening its length by two carbon atoms. By repeated removal of two-carbon fragments, a long-chain fatty acid is degraded into many two-carbon units of acetyl CoA. Thus, a 14-carbon fatty acid is converted into seven acetyl CoA molecules.

The four steps of β-oxidation are shown in Figure 13-20. Two of the reactions are oxidation–reduction reactions and one is a hydration reaction (see Chapter 10). Since half the reactions involve oxidation of the β-carbon, the pathway is termed β-oxidation.

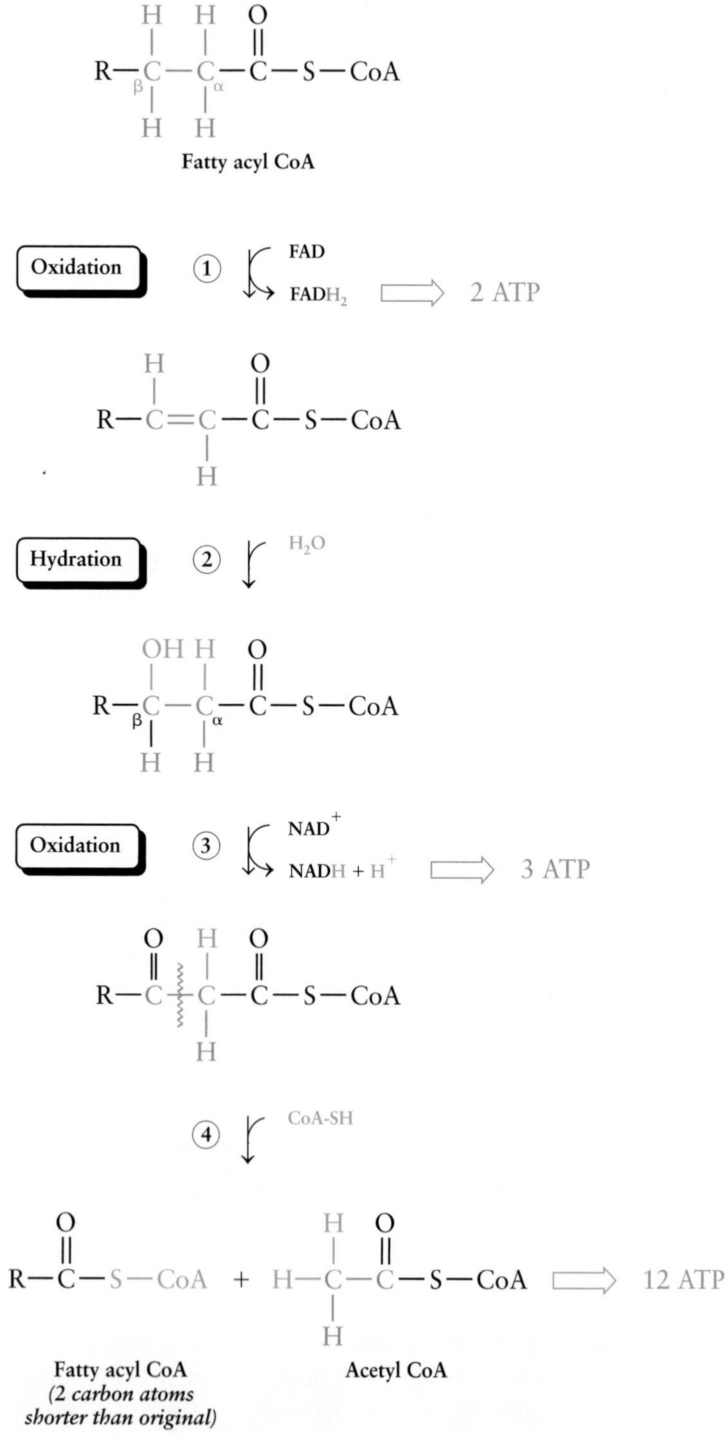

Figure 13-20 One round of β-oxidation.

Step 1: Oxidation. The carbon–carbon single bond, between the α- and β-carbon atoms, is oxidized to a carbon–carbon double bond. The loss of two hydrogen atoms indicates that this reaction is an oxidation. In Chapter 10 you learned that this type of reaction uses FAD as a coenzyme, which is reduced to $FADH_2$ in the process.

Step 2: Hydration. An OH group and an H atom from water are added to the double bond in the normal fashion: the OH group to the β-carbon and the H atom to the α-carbon (see Chapter 10). Therefore, the product contains a hydroxyl group at the β-carbon.

Step 3: Oxidation. The newly installed hydroxyl group at the β-carbon is oxidized to a ketone. The loss of two hydrogen atoms indicates that this reaction is also an oxidation. In Chapter 10, you learned that this type of oxidation requires NAD^+ as a coenzyme, which is reduced to NADH in the process.

Step 4: Carbon-carbon bond cleavage. The single bond between the α- and the β-carbon atoms breaks, and the acyl group, R—C=O, is transferred to a new molecule of coenzyme A. Two products are formed: acetyl CoA and a new fatty acyl CoA molecule, with a carbon chain that is two carbon atoms shorter than the fatty acyl CoA molecule in the first step.

The cycle then repeats until in the final step a four-carbon fatty acyl CoA molecule is converted into two acetyl CoA molecules. The repetitive nature of the β-oxidation process is demonstrated with palmitic acid in Figure 13-21. Note that the systematic removal of acetyl CoA molecules occurs from the thioester end of the molecule.

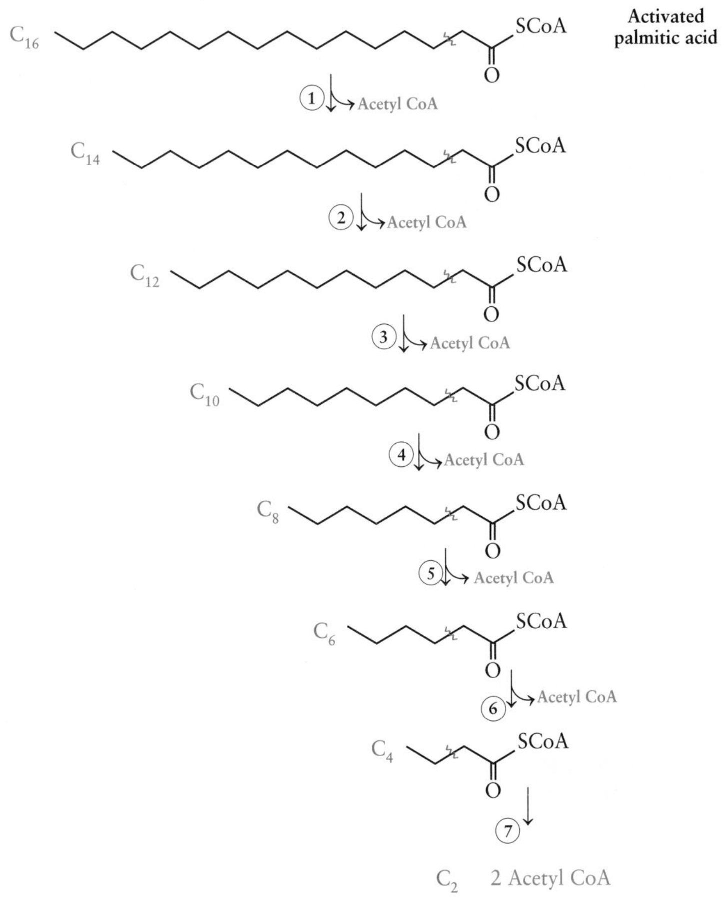

Figure 13-21 Palmitic acid, a C_{16} fatty acid, undergoing repeating cycles of β-oxidation. ⁴⌐ indicates C—C bond broken.

Energy from β-Oxidation

All the acetyl CoA molecules produced through β-oxidation can enter the citric acid cycle, where one acetyl CoA molecule produces two carbon dioxide molecules and ultimately 12 ATP. Later, in the electron-transport chain, $FADH_2$ formed in Step 1 will be converted back into FAD, while converting two ADP into two ATP in the process—thus yielding energy. Also in the electron-transport chain, the NADH formed in Step 3 will be converted back into NAD^+, producing 3 ATP in the process—yielding still more energy.

To calculate how many ATP molecules will be produced from complete catabolism of a particular fatty acid, we will consider palmitic acid. Palmitic acid is composed of 16 carbon atoms, and will undergo seven β-oxidation cycles, yielding eight acetyl CoA molecules (see Figure 13-21). Every cycle produces one NADH and one $FADH_2$ molecule. In the electron-transport chain every $FADH_2$ molecule yields 2 ATP molecules, and every NADH molecule yields 3 ATP molecules, thus 5 ATP are produced in each β-oxidation cycle from reduced coenzymes. Add 12 ATP for each acetyl CoA molecule that is formed. Subtract the 2 ATPs used to activate the initial fatty acid, a one-time expenditure of ATP. So, in a process similar to balancing your check-book, you can calculate the amount of energy produced from palmitic acid:

> **Palmitic acid: C_{16}**
>
> | | | |
> |---|---|---|
> | 7 cycles of β-oxidation | × 5 = | 35 ATP |
> | 8 acetyl CoA molecules | × 12 = | 96 ATP |
> | One time activation of palmitic acid | | −2 ATP |
> | **Total** | | **129 ATP** |

Note, that the number of β-oxidation cycles is half the total number of carbon atoms in the fatty acid *minus one*, because the last cycle produces two acetyl CoA molecules (a four-carbon chain split into two).

Problem-Solving Tutorial:
Fatty Acid Catabolism Stage 2:
Fatty Acid Oxidation

WORKED EXERCISES 13-5   β-Oxidation

1   Calculate the number of ATP molecules produced from stearic acid after β-oxidation and complete oxidation to CO_2.

2   In which step of β-oxidation is FAD the coenzyme and in which is NAD^+ the coenzyme?

SOLUTIONS

1   **Stearic Acid: C-18**

8 β-oxidation steps	× 5 =	40 ATP
9 acetyl CoA molecules	× 12 =	108 ATP
One time activation of fatty acid	−	2 ATP
Total		**146 ATP**

2   NAD^+/NADH is typically the coenzyme when an oxidation involves a C—O bond, as in the third step, while FAD/$FADH_2$ is the coenzyme when an oxidation involves a C—C bond, as in the first step. See Chapter 10 to review oxidation–reduction reactions.

PRACTICE EXERCISE

13.26   Calculate the number of ATP molecules produced from the complete catabolism of myristic acid to CO_2.

13.27   Explain why the first step in β-oxidation is considered an oxidation.

13.28   What is the function of the mitochondria? What part of lipid metabolism occurs in the mitochondria?

13.29   Before entering a mitochondrion, what chemical reaction must a fatty acid undergo? Does this reaction release or absorb energy?

13.4 Cholesterol and Other Steroid Hormones

Cholesterol and the steroid hormones derived from cholesterol are characterized by a common core ring system: three six-membered rings and one five membered ring, fused together as shown in the margin. Steroids are classified as lipids, because they are biomolecules insoluble in water and soluble in nonpolar organic solvents.

Cholesterol is an amphipathic molecule with a polar head (OH) and a nonpolar tail. Its presence in the cell membrane adds rigidity to that structure. Cholesterol also serves as the biochemical precursor (starting material) from which the body synthesizes bile salts, vitamin D, and many important hormones, as summarized in Figure 13-22.

Bile salts, such as cholate, are a group of amphipathic molecules that serve as emulsifying agents: They break up globules of dietary fat so that water soluble lipase enzymes can hydrolyze triglycerides into their individual fatty acids and glycerol. Bile is produced in the liver and stored in the gall bladder, from where it is released into the small intestine in response to a high-fat meal.

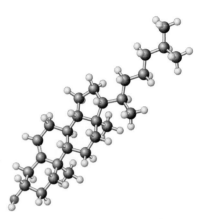

Steroid core

Ball-and-stick model of cholesterol.

Bile salt
(cholate)

Five classes of important steroid hormones are produced from cholesterol, as summarized in Figure 13-23.

The **glucocorticoids** regulate carbohydrate, protein, and lipid metabolism. They also have powerful anti-inflammatory and immunosuppressent activity. For example, an ointment of 1% hydrocortisone is an over-the-counter steroid medication used to treat dermatitis and other skin conditions. Prednisone is a synthetic glucocorticoid used as an immunosuppressent in the treatment of asthma and arthritis.

The **mineralocorticoids** are produced in the adrenal gland and regulate ion (Na^+, K^+, Cl^-) balance in tissues.

Cholesterol

Bile acids | Vitamin D | Hormones

Figure 13-22 Cholesterol is the starting material for bile acids, vitamin D, and a variety of hormones.

Figure 13-23 The five classes of steroid hormones produced from cholesterol.

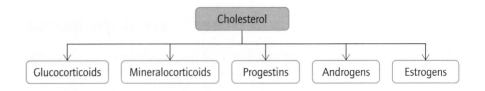

Progestins, estrogens, and androgens collectively represent the major sex hormones. Progestins, such as progesterone, control the menstrual cycle and pregnancy; androgens (testosterone) and estrogens (estradiol) control the development of sexual characteristics in the fetus and during puberty.

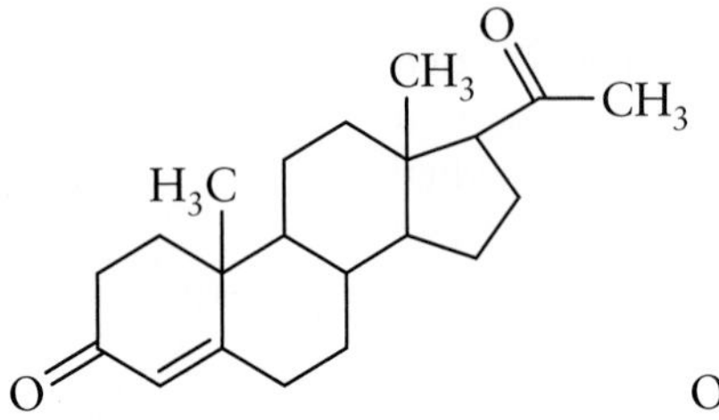

Progesterone
(progestin)

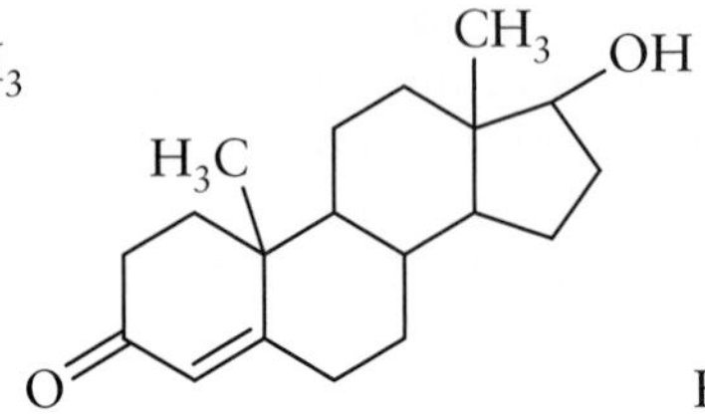

Testosterone
(androgen)

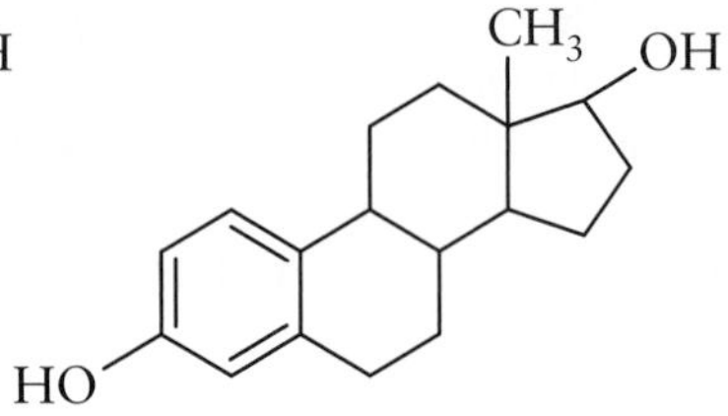

Estradiol
(estrogen)

The anabolic steroids described in the opening vignette are man-made steroids. They contain the characteristic four fused rings of a steroid and mimic androgens like testosterone. Their structures have been altered chemically to improve their ability to build muscle while lessening their other effects.

PRACTICE EXERCISES

13.30 What structural feature characterizes a steroid?

13.31 Prednisone reduces itching, swelling, redness, and allergic reactions. Is prednisone a steroid? Naproxen also reduces pain, fever, and swelling. Is naproxen a steroid?

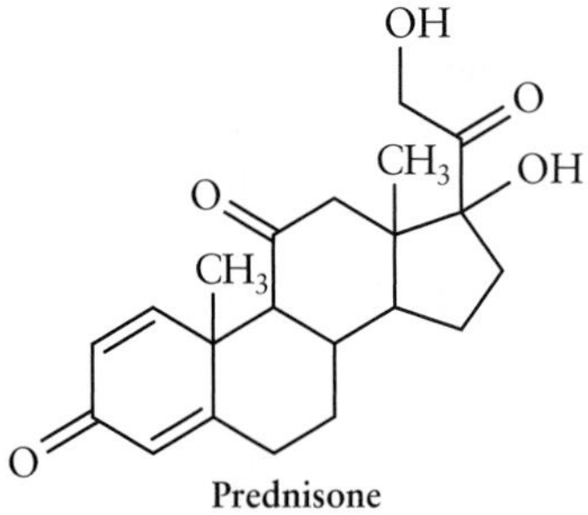

Prednisone

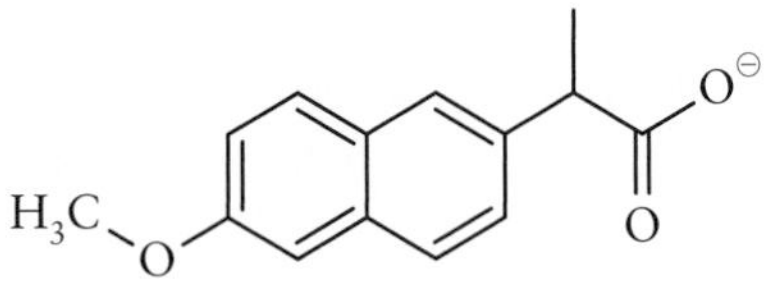

Naproxen

13.32 What do all of the following classes of compounds found in the body have in common? From what molecule are they synthesized?
a. mineralocorticoids
b. androgens
c. bile salts
d. vitamin D
e. estrogens
f. glucocorticoids

Lipids as a class are biological molecules that are insoluble in water but soluble in organic solvents. Lipid structures vary depending on their function. Lipids make up the cell membrane and serve as important hormones, but the most abundant lipids are the triglycerides (triacylglycerols) that serve as a supply of long-term energy for cells. You have learned the biochemical pathways involved in the first two stages of triglyceride catabolism, as well as the biochemical pathways involved in the catabolism of carbohydrates. All these pathways converge at the same central molecule of catabolism, acetyl CoA. In the next chapter you will see how acetyl CoA is converted into energy through the citric acid cycle and oxidative phosphorylation.

Chemistry in Medicine Inflammation and the Role of Eicosanoids

What do a broken leg, a bee sting, and acute bronchitis have in common? They all invoke the body's natural inflammatory response. Inflammation is part of the body's immune response to trauma (for example, a broken leg), invasion by a foreign substance (for example, a bee stinger), or infection (for example, *Streptococcus pneumoniae* infection). Inflammation is part of the body's mechanism for repairing the injury or removing the foreign substance. The fundamental signs of inflammation are redness, swelling, pain, and warmth. The eicosanoids, a class of lipids, serve as signaling molecules that activate these signs of inflammation. Drugs used to reduce inflammation are targeted at specific reactions within the biochemical pathways that produce these eicosanoids.

Eicosanoids are derived from arachidonic acid, a fatty acid containing 20 (eicosa) carbon atoms and four double bonds. (It is an omega-6 fatty acid, or a $\Delta^{5,8,11,14}$ fatty acid.) When the body experiences a trauma,

A bee sting causes inflammation. [Dr. Jeremy Burgess/Science Photo Library]

invasion by a foreign substance, or infection, arachidonic acid is hydrolyzed from glycerophospholipids in cell membranes by the hydrolysis reaction shown in Figure 13-24. Arachidonic acid is then converted by several different pathways, known as the arachidonic acid cascade, into the various eicosanoids (Figure 13-25).

The three major classes of eicosanoids are

- prostaglandins,
- thromboxanes, and
- leukotrienes.

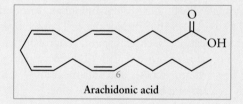

Arachidonic acid

Figure 13-24 Hydrolysis of a membrane lipid produces arachidonic acid.

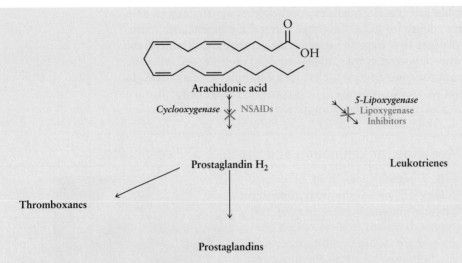

Figure 13-25 NSAIDs are enzyme inhibitors of *cyclooxygenase*, and lipoxygenase inhibitors inhibit *5-lipoxygenase*.

What role do these eicosanoids play in inflammation? Let's consider the bee sting. Someone stung by a bee experiences several signs of inflammation. In most cases, the inflammatory process assists immune system cells in their task of destroying the stinger that entered the tissue.

Redness—If you've ever been stung by a bee, you know that the site of the sting is first pale, and then eventually turns red. Thromboxanes serve as short-acting vasoconstrictors (they constrict blood vessels) that are released quickly after the bee sting occurs. They cause the site to momentarily turn pale. Thromboxanes then signal local cells to release prostaglandins and leukotrienes, which are vasodilators that cause the blood vessels to engorge and the injury to redden. The increased blood flow brings immune system cells to the damaged tissue.

Swelling—Leukotrienes make blood vessels more permeable. As a result, plasma leaks out into the connective tissues, and the tissues swell. Permeability allows fluid containing immune system cells to enter the damaged tissues from the bloodstream.

Pain—Elevated levels of prostaglandins sensitize pain neurons. This alerts the body to get out of harm's way. In other words, run away from those bees.

Warmth—Release of prostaglandins also causes fever, or in the case of a localized invasion such as a bee sting, warmth at the site of the sting. Elevated temperatures can decrease the ability of some pathogens to replicate.

Drugs that inhibit the two main enzymes in the arachidonic acid cascade are used in the treatment of many inflammatory diseases. Lipoxygenase inhibitors are a class of drugs that reduce the production of leukotrienes. They include some anti-asthma medications, such as Singulair and Accolate. These medications are often used to treat inflammation of the bronchial passages in the case of acute bronchitis.

Singulair

Inhibition of *cyclooxygenase*, the key enzyme involved in the formation of prostaglandin H_2, is the biochemical basis through which nonsteroidal anti-inflammatory drugs (NSAIDs) work. NSAIDS include the well known over-the-counter medications such as aspirin, ibuprofen (Motrin, Advil), and naproxen (Aleve). NSAIDs reduce pain (analgesics), fever (antipyretics), and inflammation—welcome relief in the case of a broken leg or bee sting.

Aspirin

Other medications act to relieve inflammation by inhibiting other key reactions in the arachidonic acid cascade. Some of these medications are used to treat conditions such as glaucoma, asthma, and pulmonary and ocular hypertension. In this chapter, you learned that lipids function to store energy and act as structural components of cell membranes. As you can see, lipids are also important signaling molecules. Understanding the role of eicosanoids in inflammation at the molecular level has made possible the development of effective drugs for the treatment of inflammation.

Chapter Summary

Energy Storage Lipids: Triglycerides

- The major categories of lipids are triglycerides, phospholipids, glycolipids, steroids, and eicosanoids.
- The most abundant lipids in the body are triglycerides, also known as triacylglycerols, which are stored in fat cells as a long-term supply of energy.
- Triglycerides are assembled from one glycerol molecule and three fatty acid molecules, which are joined by three ester linkages.
- Fatty acids may be either saturated (have no double bonds) or unsaturated (have one or more double bonds).
- The double bond in a natural unsaturated fatty acid is a cis double bond, which gives the fatty acid a bend in its otherwise overall linear shape.
- The melting point of fatty acids increases as the number of carbon atoms in their formulas increases. The melting point of a fatty acid drops significantly with the addition of each double bond.
- Triglycerides are nonpolar molecules that exhibit hydrophobic properties.
- Triglycerides derived from saturated fatty acids tend to be solids at room temperature and are commonly referred to as fats.
- Triglycerides derived from unsaturated fatty acids tend to be liquids at room temperature and are commonly referred to as oils.
- Trans fats are generally not found in nature; they are produced artificially by the catalytic hydrogenation of unsaturated cis triglycerides, when isomerization instead of reduction occurs.

Membrane Lipids: Phospholipids and Glycolipids

- The cell membrane is a selectively permeable membrane that separates the inside of the cell from the outside of the cell.
- Phospholipids and glycolipids are the lipids found in human cell membranes.
- A phospholipid with a sphingosine backbone is called a sphingomyelin, while a phospholipid with a glycerol backbone is called a glycerophospholipid.
- Glycolipids found in human cells contain a sphingosine backbone and are called sphingolipids.
- Membrane lipids have two long hydrocarbon chains as a key part of their chemical composition. These long chains are the nonpolar "tail" of the molecule.
- Membrane lipids have a polar "head." In phospholipids the polar head is a phosphate group attached to an amino alcohol. In glycolipids the polar head is a carbohydrate.
- The cell membrane is assembled from two layers, a bilayer, of phospholipids and glycolipids, aligned in a way that brings the polar heads in contact with water and places the nonpolar tails next to each other, creating a nonpolar environment.
- Small nonpolar molecules pass freely through the cell membrane by simple diffusion. Water passes slowly through the membrane by osmosis.
- Special transport proteins are needed to move ions and polar molecules across the nonpolar interior of the cell membrane.

Fatty Acid Catabolism

- Fatty acid oxidation is the central energy producing pathway for heart and liver cells.

- The catabolism of triglycerides proceeds in three stages: Stage 1, a triglyceride is hydrolyzed into glycerol and three fatty acids. Stage 2, fatty acids are oxidized to acetyl CoA. Stage 3, acetyl CoA is converted into carbon dioxide and energy via the citric acid cycle and oxidative phosphorylaion.
- Serum lipoproteins are spherical aggregates formed of a single layer of lipids, with proteins embedded throughout the lipid layer. They transport triglycerides and cholesterol.
- β-oxidation of a fatty acid is a biochemical sequence of four reactions which are repeated over and over until a long-chain fatty acid is degraded completely into several two-carbon units of acetyl CoA.
- β-oxidation occurs in the mitochondria.
- A single cycle of β-oxidation produces one $FADH_2$ molecule, which eventually generates two ATP molecules, and one NADH molecule, which eventually generates 3 ATP molecules.
- The acetyl CoA molecules produced through repetitive β-oxidation cycles can enter the citric acid cycle. Each acetyl CoA molecule can produce 12 ATP molecules.

Cholesterol and Other Steroid Hormones

- Cholesterol and the steroid hormones derived from cholesterol are characterized by a common core ring system: three six-membered rings and one five membered ring fused together.
- Cholesterol is amphipathic with a polar head (OH) and a nonpolar tail that adds rigidity to the cell membrane.
- Cholesterol is the biological precursor (starting material) for bile acids, vitamin D_3, and many important hormones, including the major sex hormones: the progestins, androgens, and estrogens.

Key Words

Adipocytes Fat cells

Amphipathic molecules Molecules that are polar on one end and nonpolar on the other. They are effective emulsifying agents.

β-Oxidation A biochemical sequence of reactions that breaks down a fatty acyl CoA molecule into acetyl CoA and a fatty acyl CoA that is two carbons shorter in the mitochondria. The sequence of reactions can be repeated over and over again until the fatty acid is completely degraded into acetyl CoA molecules.

Bile salts Nature's detergent; bile salts facilitate emulsification of fats in the small intestine.

Cell membrane A selectively permeable barrier that separates the environment of the inside of the cell from the environment of the outside of the cell. Also known as a plasma membrane. Consists of a lipid bilayer.

Cerebroside A glycolipid containing a monosaccharide.

Chylomicrons Serum lipoproteins formed from a single layer of lipids, one micrometer in diameter. They carry triglycerides from the cells lining the small intestine to their destination through the lymph system and bloodstream.

Essential fatty acids Fatty acids that are essential nutrients that the body cannot synthesize and therefore must be supplied by the diet. Examples are linoleic acid and linolenic acid.

Fats Triglycerides derived from saturated fatty acids; they tend to be solids at room temperature (25°C).

Fatty acid A long unbranched hydrocarbon with a carboxylic acid group (RCO_2H) at one end.

Ganglioside A glycolipid containing an oligosaccharide.

Glucocorticoids A class of steroid hormones derived from cholesterol that regulate carbohydrate, protein, and lipid metabolism.

Glycerophospholipid A lipid that contains a phosphate group and a glycerol backbone.

Glycolipid One of the two main classes of membrane lipids. They contain a carbohydrate linked to a sphingosine backbone.

Lipids The biomolecules that are insoluble in water.

Mineralocorticoids A class of steroid hormones derived from cholesterol that regulate ion balance in tissues.

Mixed triglyceride A triglyceride that contains two or three different types of fatty acids.

Monounsaturated fatty acid A fatty acid with only one double bond.

Oils Triglycerides derived from unsaturated fatty acids. They tend to be liquids at room temperature (25°C).

Phospholipid The most common type of membrane lipid. They contain a phosphate group at the terminal end of either sphingosine or glycerol.

Simple triglyceride A triglyceride derived from only one kind of fatty acid.

Sphingolipid A lipid that contains a sphingosine backbone.

Sphingomyelin A phospholipid with a sphingosine backbone.

Triglycerides The most abundant type of lipid in the body, also known as triacylglycerols or fats. They serve as the body's long-term energy supply.

Additional Exercises

Energy Storage Lipids: Triglycerides

13.33 Name two physical properties of lipids.

13.34 What are the four major classes of biomolecules?

13.35 What structural feature makes lipids insoluble in water?

13.36 What is the difference between a saturated fat and an unsaturated fat?

13.37 What type of fatty acid is oleic acid? Answer using the delta system and the omega system.

13.38 What type of fatty acid is linoleic acid? Answer using the delta system and the omega system.

13.39 Flaxseed oil is rich in linolenic acid. Linolenic acid is an omega- _____ fatty acid.

13.40 List the following fatty acids in order of increasing melting point (lowest to highest). Explain why you chose that order.
a. stearic acid
b. linolenic acid
c. oleic acid
d. linoleic acid
e. palmitic acid

13.41 Why does the presence of a double bond in a fatty acid affect its melting point?

13.42 What functional group is present in all triglycerides?

13.43 Why are fats generally solids at room temperature?

13.44 Why are oils liquids at room temperature?

13.45 Coconut oil contains 92% saturated fat. Do you expect it to be a solid or a liquid at room temperature?

13.46 Canola oil contains 6% saturated fat. Do you expect it to be a solid or a liquid at room temperature?

13.47 Olive oil contains oleic acid. Draw the simple triglyceride that would be derived from just this fatty acid. Why is olive oil considered a healthy fat?

13.48 Draw a triglyceride derived from one molecule of oleic acid, one molecule of stearic acid, and one molecule of palmitic acid.

13.49 Would you expect a triglyceride to be soluble in water? Why or why not?

13.50 Which product is higher in saturated fats, lard or coconut oil?

13.51 Which product is lower in saturated fats, safflower oil or cottonseed oil?

13.52 Which is a healthier food, soybean oil or sunflower oil?

13.53 Which is a healthier food, olive oil or butter?

13.54 Which lipids pack more closely, saturated fats or unsaturated fats? Which lipids have stronger intermolecular forces of attraction, saturated or unsaturated fats?

13.55 **Critical Thinking Question:** Which fat molecules would you expect to pack more closely, unsaturated trans fats or unsaturated cis fats? Which type of fat would have a higher melting point? Explain your answer.

Membrane Lipids: Phospholipids and Glycolipids

13.56 What part of the cell controls the flow of ions and molecules in and out of the cell?

13.57 What type of biomolecule constitutes the main component of the cell membrane?

13.58 How are the structures of glycerol and sphingosine different?

13.59 Which of the following membrane lipids contain a phosphate?
a. glycolipids
b. sphingomyelins
c. glycerophospholipids

13.60 Which membrane lipids are derived from sphingosine?

13.61 In a phospholipid, what functional groups form the polar head of the molecule?

13.62 How does the structure of a glycolipid differ from the structure of a phospholipid?

13.63 For the membrane lipid shown below, answer the following questions.

a. Is it a phospholipid or glycolipid?
b. Is it a sphingolipid or a glycerophospholipid? Outline the backbone.
c. What is the fatty acid from which this lipid is derived? Is the fatty acid component saturated or unsaturated? What functional group connects the fatty acid to the backbone?
d. Is this lipid a cerebroside or a ganglioside? How can you tell the difference?
e. If it is a glycolipid, does it contain a glucose or a galactose derivative? Is the glycosidic linkage α or β?
f. Circle and label all functional groups.

13.64 For the membrane lipid shown below, answer the following questions.

a. Is it a phospholipid or a glycolipid?

b. Is it a sphingomyelin or a glycerophospholipid? Outline the backbone.

c. What is (are) the fatty acid(s) from which this lipid is derived? Are the fatty acid components saturated or unsaturated? What functional group connects the fatty acids to the backbone?

d. Does the lipid contain a phosphate or a carbohydrate?

e. If the lipid contains a phosphate, what is the amino alcohol: serine, choline, or ethanolamine?

f. Is the lipid a sphingomyelin, cerebroside, or ganglioside?

g. What part of the lipid constitutes the polar head?

h. Circle and label all functional groups.

13.65 What type of intermolecular forces are responsible for the way membrane lipids organize to form a cell membrane?

13.66 Draw 12 membrane lipids arranged to form a bilayer, such as the cell membrane.

13.67 What role does cholesterol play in the cell membrane?

13.68 How are ions transported from one side of the cell membrane to the other side? What molecule provides the energy for this process?

13.69 What type of molecules can pass freely through the cell membrane?

Fatty Acid Catabolism

13.70 What molecules are produced after the first stage of triglyceride catabolism? What molecules are produced after the second stage of triglyceride catabolism?

13.71 From where does the body obtain fatty acids?

13.72 How does the body make triglycerides from a meal more susceptible to hydrolysis by the water soluble lipases during digestion?

13.73 How are triglycerides transported through the lymph system and the bloodstream?

13.74 Is the interior of a lipoprotein hydrophobic or hydrophilic?

13.75 Are triglycerides found on the interior or exterior of a chylomicron? Explain your answer.

13.76 Where in the cell does β-oxidation occur?

13.77 Calculate the number of ATP molecules produced when oleic acid undergoes β-oxidation and the products go through the citric acid cycle. Assume that unsaturation does not affect the result.

13.78 Calculate the number of ATP molecules produced when linoleic acid undergoes β-oxidation and the products go through the citric acid cycle. Assume that unsaturation does not affect the result.

13.79 Which steps in the β-oxidation biochemical pathway are oxidation steps?

Cholesterol and Other Steroid Hormones

13.80 Why are steroids classified as lipids?

13.81 **Critical Thinking Question:** How is the structure of cholesterol similar to the structure of a fatty acid?

13.82 What compounds emulsify fat globules during lipid digestion?

13.83 What do glucocorticoids do in the body?

13.84 What do mineralcorticoids do in the body?

13.85 Identify the functional groups that have been modified or added to a cholesterol molecule to form a molecule of estradiol.

13.86 Identify the functional groups that have been modified or added to a cholesterol molecule to form a molecule of bile acid (cholic acid).

Chemistry in Medicine

13.87 What is an eicosanoid?

13.88 What are the three major classes of eicosanoids?

13.89 What are the symptoms of inflammation resulting from the production of eicosanoids?

13.90 Is arachidonic acid a saturated or an unsaturated fatty acid? Are the double bonds cis or trans or a mixture?

13.91 What important eicosanoids(s) will not be produced under the following conditions?

a. A *lipoxygenase* inhibitor is present.

b. A *cyclooxygenase* inhibitor is present.

13.92 What eicosanoid is formed by the action of cyclooxygenase on arachadonic acid?

13.93 On a molecular level, how does aspirin work?

13.94 What is the name of the enzyme that catalyzes the release of arachidonic acid from glycerophospholipids? What type of reaction does this enzyme catalyze?

13.95 Define an NSAID and provide two examples.

13.96 What process does a NSAID inhibit?

Answers to Practice Exercises

13.1 **a.** Both fatty acids are saturated, but myristic acid contains 14 carbons while palmitic acid contains 16. The more carbon atoms, the more dispersion forces and, therefore, the higher the melting point for palmitic acid.
b. Both fatty acids contain 18 carbon atoms, but oleic acid is unsaturated, containing one double bond. The double bond prevents the molecules from packing closely, and thereby reduces the number of dispersion forces, causing oleic acid to have a lower melting point.

13.2 **a.** and **c.** are omega-6 fatty acids.

13.3 Δ 5, 8, 11, 14 fatty acid; omega-6 fatty acid.

13.4 Linolenic would more likely be a liquid since, although both fatty acids contain 18 carbon atoms, linolenic acid is unsaturated and stearic is saturated.

13.5 A saturated fatty acid has a linear shape and an unsaturated cis fatty acid has a bend or kink in the chain that prevents these fatty acids from packing closely. The loose packing makes the melting point of unsaturated fatty acids lower, giving them the consistency of a liquid (an oil) rather than a solid. A trans unsaturated fat is more similar to a saturated fat.

13.7 palm and coconut oils

13.8 Oleic acid is an omega-9 fatty acid in the omega system and a Δ^9 in the delta system.

13.9 Peanut oil (18%) is higher in saturated fats than corn oil (13%).

13.10 Canola oil is healthier than coconut oil, because it contains less saturated fat.

13.11 vegetable sources

13.12 Arachidonic acid (−49.5°C) has 4 double bonds, while linolenic (−11°C) has 3, and the number of double bonds impacts the melting point more than the number of carbon atoms.

13.13 **a.** a glycolipid, because it contains a carbohydrate
b. sphingosine

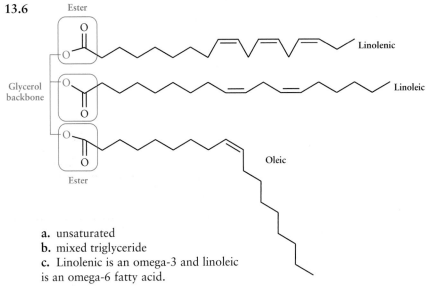

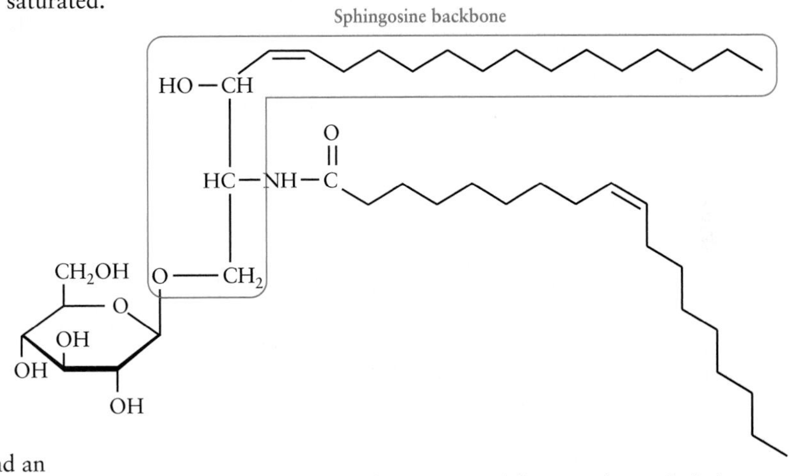

c. oleic acid; unsaturated fatty acid; amide linkage between backbone and fatty acid
d. a cerebroside because it contains a monosaccharide rather than an oligosaccharide
e. a glycolipid derived from glucose. It has a β-glycosidic linkage to the backbone.

13.6

a. unsaturated
b. mixed triglyceride
c. Linolenic is an omega-3 and linoleic is an omega-6 fatty acid.

f.

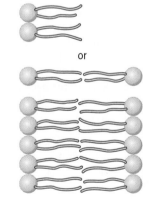

13.14 An amphipathic molecule is polar on one end and nonpolar on the other, so it has properties of both, and is soluble in both water and nonpolar solvents. A nonpolar substance is only soluble in nonpolar solvents, and a polar substance is only soluble in polar substances.

13.15 Triglycerides contain three fatty acids, linked as esters to each hydroxyl group in glycerol. A glycerophospholipid has only two fatty acids, linked as esters to the first two hydroxyl groups in glycerol. The third hydroxyl group is linked to a phosphate, which is also linked to an amino alcohol. Triglycerides are nonpolar, while glycerophospholipids are amphipathic.

13.16 The sphere represents the polar head of a membrane lipid and the two wavy lines represent the two nonpolar tails.

13.17

13.18 triglycerides

13.19 An emulsifying agent is an amphipathic molecule that breaks up and evenly distributes a fat, or nonpolar substance, in an aqueous medium to form a colloid. Egg serves as the emulsifying agent in mayonnaise, because the proteins in egg are amphipathic molecules.

13.20 Bile is an amphipathic molecule that serves as an emulsifying agent for dietary fat.

13.21 *Lipases* catalyze the hydrolysis of esters in triglycerides to form fatty acids and glycerol. *Lipases* are found in the small intestine.

13.22 Chylomicrons transport triglycerides to muscle and fat cells. Chylomicrons are composed of membrane lipids arranged in a spherical shape with triglycerides in the interior near the nonpolar tails of the membrane lipids.

13.23 Upon entering an adipocyte, fatty acids and glycerol react to reform triglycerides, and these are stored until needed. In a muscle cell fatty acids undergo β-oxidation to produce energy.

13.24 a. the gall bladder
   b. the liver
   c. the small intestine

13.25 amphipathic

13.26 Myristic Acid: C_{14}

$$
\begin{array}{lr}
\text{6 β-oxidation steps} \times 5 = & \text{30 ATP} \\
\text{7 acetyl CoA molecules} \times 12 = & \text{84 ATP} \\
\text{One time activation of myristic acid} & -2\ \text{ATP} \\
\hline
\text{Total} & \text{112 ATP}
\end{array}
$$

13.27 The first step is an oxidation because two hydrogen atoms are lost when going from a single bond to a double bond.

13.28 The mitochondria are the energy-producing organelles of the cell. β-Oxidation, the citric acid cycle, and oxidative phosphorylation occur in the mitochondria.

13.29 A fatty acid must first be activated to a fatty acyl CoA molecule, which requires ATP, thus absorbing energy.

13.30 A steroid is characterized by four fused hydrocarbon rings: three six-membered rings and one five-membered ring.

13.31 Prednisone is a steroid. Naproxen is not a steroid.

13.32 They are all steroids synthesized from cholesterol.

Metabolism and Bioenergetics

CHAPTER 14

Physical activity is possible because of catabolism, which converts ADP to ATP.
[Toshio Hoshi/Jupiterimages]

OUTLINE

 This icon indicates that a **Problem-Solving Tutorial** is available at www.whfreeman.com/gob

The Role of Physical Activity in Weight Loss

In Chapter 8 you learned about the problems of starvation in the world. Now consider the millions of people who wrestle with the opposite problem—excess weight. Why does it seem so much harder to lose weight than it is to gain weight? Isn't it simply a matter of balancing the number of calories consumed (energy in) with the number of calories expended (energy out)? The biochemistry of metabolism—the subject of this chapter—can help explain some of the complexities involved in weight loss.

Why do diets that call for a severe reduction in calories usually not work? The problem with a drastic reduction in calories is that the brain needs a constant supply of glucose, in the amount of 70–110 mg per dL of blood. However, there are no metabolic pathways that produce glucose from stored fat. Once the body's limited stores of glycogen have been converted to glucose in response to a severe reduction in calories, the body begins to degrade proteins in muscle tissue in order to supply glucose to the brain. Furthermore, plummeting glucose levels make it difficult to stick to a weight loss plan that involves a severe reduction in calories. Low blood glucose levels signal the brain that it needs sugar, which initiates the desire to consume food.

Effective diets usually recommend a moderate intake of calories combined with exercise. Indeed, the U.S. government's updated food pyramid shows "daily physical activity" as one of the seven components of a healthy diet (Figure 14-1). Why do most weight loss programs recommend exercise? A person trying to lose weight needs to metabolize the triglycerides in fat cells through β-oxidation of fatty acids, as we saw in Chapter 13. β-Oxidation produces acetyl CoA, which feeds into the citric acid cycle and ultimately converts ADP to ATP—energy. Therefore, the best way to facilitate the conversion of stored fat into energy is to stimulate the citric acid cycle. This can be accomplished by placing higher energy demands on the body, and one way to do that is through exercise. Consumption of fewer calories, combined with regular exercise, while maintaining steady blood glucose levels, causes the body to convert fatty acids into acetyl CoA, literally "burning fat." These catabolic processes occur in organelles of the cell known as mitochondria. Studies have shown that well trained and extremely fit endurance athletes actually have more mitochondria in their muscle cells than the average person.

In addition to the caloric content of food, the type of food a person consumes when dieting is important. Nutritious foods that are high in B vitamins provide the body with a steady supply of the starting materials for the coenzymes NAD^+, FAD, and coenzyme A, which are all required in the citric acid cycle.

Although weight loss involves hormones, brain chemicals, and many other complex processes and interwoven factors not described here, the biochemistry of metabolism explains part of the weight loss equation. In this chapter you will study the final stage of catabolism, which contains the major energy producing pathways: the citric acid cycle and oxidative phosphorylation.

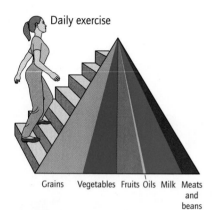

Daily exercise

Grains Vegetables Fruits Oils Milk Meats and beans

Figure 14-1 Current United States Government food pyramid.

In Chapter 8 you learned that the cells in your body are constantly performing chemical reactions to produce both energy and important cellular compounds. Most of these reactions are part of biochemical pathways that are either *catabolic*—they break down larger molecules into smaller molecules—or *anabolic*—they build larger molecules from smaller molecules. Equally important is that catabolism produces energy, while anabolism consumes energy.

In the last three chapters you learned about the first two stages of catabolism for two of the three major types of biomolecules: carbohydrates and triglycerides. In these first two stages of catabolism, these biomolecules are each broken down by separate biochemical pathways. However, both pathways produce acetyl coenzyme A, as does the breakdown of proteins. Proteins are hydrolyzed to amino acids, which can be further metabolized to acetyl CoA or other

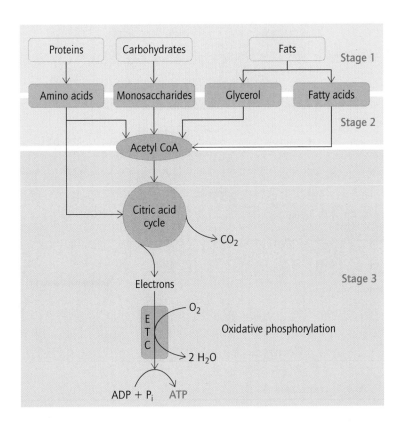

Figure 14-2 Overview of catabolism showing the central role of acetyl CoA.

citric acid intermediates. Thus, acetyl coenzyme A is the point of convergence for all three pathways, summarized in Figure 14-2. In this chapter you will focus on what happens to acetyl coenzyme A in the final stage of catabolism.

Acetyl coenzyme A can enter an eight-step biochemical pathway called the **citric acid cycle,** where it is converted into two molecules of carbon dioxide (CO_2) and coenzyme A. In the process, three molecules of NAD^+ are reduced to NADH and one molecule of FAD is reduced to $FADH_2$. The electrons carried by these reduced coenzymes enter the electron-transport chain, where a series of oxidation–reduction reactions occurs. The oxidation–reduction reactions of the electron-transport chain drive the reaction of ADP with inorganic phosphate (P_i) to produce the energy currency of the cell—ATP. This process is known as *oxidative phosphorylation.* Oxygen is an essential component of oxidative phosphorylation and is the reason why catabolism is called **cellular respiration.**

The generation of energy is the main purpose of cellular respiration. The extraction of energy from the food we eat (catabolism) drives the chemical reactions that build molecules (anabolism) and runs the pumps that transport ions and molecules across cell membranes. The study of the role of energy in metabolism is known as bioenergetics, and will be described in further detail at the end of this chapter. There you will revisit the concept of enthalpy, the heat energy of chemical bonds that was first introduced in Chapter 8. And you will be introduced to a new energy concept called entropy, the tendency toward disorder.

14.1 Acetyl Coenzyme A and the Citric Acid Cycle

Acetyl CoA

At the center of metabolism is a molecule known as acetyl coenzyme A, abbreviated acetyl CoA, which plays a central role in both anabolic and catabolic pathways. In this chapter we describe its role in catabolic pathways, as

Acetyl group

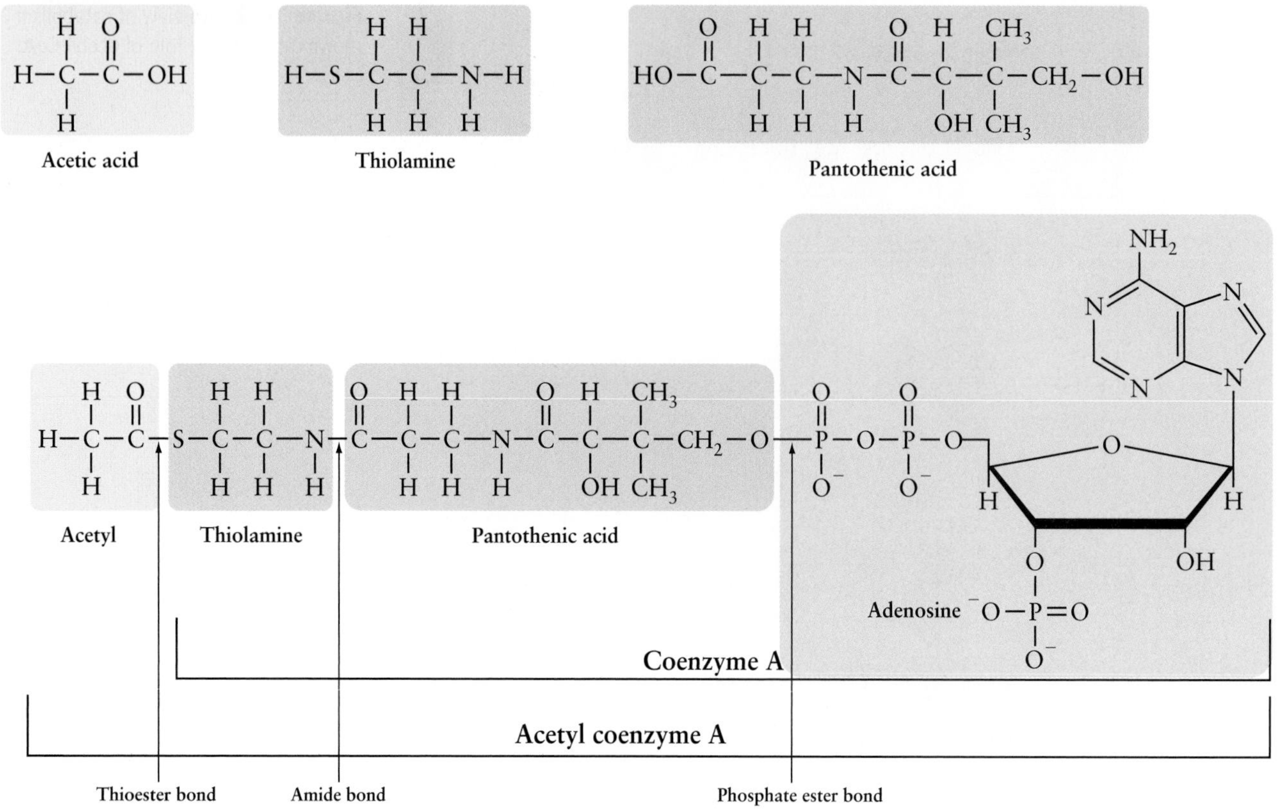

Figure 14-3 The chemical structure of coenzyme A and acetyl CoA.

illustrated in Figure 14-2. In Chapters 12 and 13, you learned that acetyl CoA is produced from:

- glucose, through glycolysis and the oxidation of pyruvate to acetyl CoA under aerobic conditions (Chapter 12) and
- fatty acids, through β-oxidation (Chapter 13).

The chemical structure of acetyl CoA, shown in Figure 14-3, is derived from four recognizable compounds: adenosine, pantothenic acid (vitamin B_5), the thiol amine $HSCH_2CH_2NH_2$, and acetic acid (CH_3CO_2H). These four molecules are joined by ester, amide, and phosphate ester functional groups to form a molecule of acetyl CoA.

Hydrolysis of acetyl CoA produces acetic acid and the thiol coenzyme A. Coenzyme A can be attached to other acyl groups via a thioester linkage to carry out the biochemical function of transferring acyl groups. Indeed, it delivers an acetyl group to the citric acid cycle and a fatty acyl group in β-oxidation. These are all examples of acyl group transfer reactions (Chapter 10).

The Citric Acid Cycle

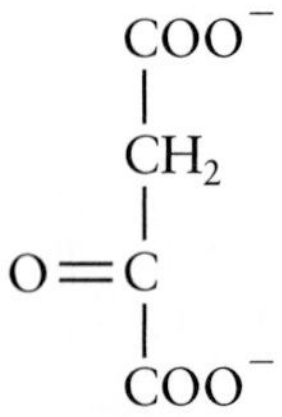

The third stage of catabolism begins with the citric acid cycle. The citric acid cycle is an eight-step biochemical pathway that begins with the reaction of acetyl CoA with oxaloacetate to produce citric acid. Since the last step of the citric acid cycle regenerates oxaloacetate, the citric acid cycle has a unique and distinguishing circular appearance when written.

In the citric acid cycle, acetyl CoA is degraded into two molecules of carbon dioxide (CO_2) and a molecule of coenzyme A. Over the course of the cycle, three molecules of NAD^+ are reduced to NADH, one molecule of FAD is reduced to FADH$_2$, and one molecule of GDP is phosphorylated to GTP. Reduction of NAD$^+$ and FAD is the most important outcome of the citric acid cycle because most of the energy harvested comes from the electrons delivered to the electron-transport chain by these reduced coenzymes.

> The citric acid cycle is also known as the Krebs cycle or the **tricarboxylic a**cid (TCA) cycle.

> Remember that carboxylic acid functional groups appear in their ionized (carboxylate) form at physiological pH. For example, citrate is the conjugate base of citric acid.

$$
\begin{array}{cc}
\text{COO}^- & \text{COOH} \\
| & | \\
\text{CH}_2 & \text{CH}_2 \\
| & | \\
\text{HO}-\text{C}-\text{COO}^- & \text{HO}-\text{C}-\text{COOH} \\
| & | \\
\text{CH}_2 & \text{CH}_2 \\
| & | \\
\text{COO}^- & \text{COOH} \\
\text{Citrate} & \text{Citric acid} \\
\text{(physiological pH)} & \text{(low pH)}
\end{array}
$$

In the **electron-transport chain,** the electrons carried by one NADH molecule supply the energy to phosphorylate three ADP molecules to form three ATP, and those from one FADH$_2$ supply the energy to phosphorylate two ADP molecules to form two ATP. In addition, the single molecule of GTP that is produced in the citric acid cycle is ultimately converted into one molecule of ATP. *Thus, one round of the citric acid cycle can phosphorylate 12 ADP molecules to 12 ATP during the biochemical process of oxidative phosphorylation—the last stage of catabolism:*

$$
\underbrace{(1\ \cancel{\text{GTP}} \times \frac{1\ \text{ATP}}{\cancel{\text{GTP}}})}_{\text{Step 5}} + \underbrace{(3\ \cancel{\text{NADH}} \times \frac{3\ \text{ATP}}{\cancel{\text{NADH}}})}_{\text{Steps 3, 4, \& 8}} + \underbrace{(1\ \cancel{\text{FADH}_2} \times \frac{2\ \text{ATP}}{\cancel{\text{FADH}_2}})}_{\text{Step 6}} = 12\ \text{ATP}
$$

The overall carbon balance for the citric acid cycle is illustrated in Figure 14-4. In the first step, the two carbon atoms (shown in blue) of the acetyl group, carried by coenzyme A, form a carbon–carbon bond to the four-carbon oxaloacetate molecule (shown in red), to produce the six-carbon citrate molecule, releasing CoA. The biochemical pathway is named after citrate, the product of this reaction, which is the ionized form of citric acid.

The first step of the citric acid cycle generates citrate (six carbon atoms) and then the next three steps break citrate down into the four-carbon molecule succinyl CoA and two molecules of carbon dioxide, CO$_2$. Then, in the second half

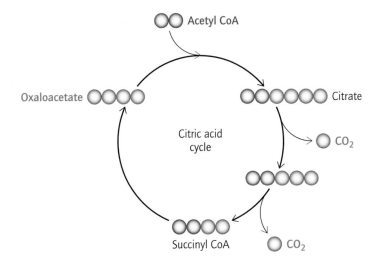

Figure 14-4 Carbon balance in the citric acid cycle.

of the citric acid cycle, oxaloacetate is regenerated from succinyl CoA in four steps. Oxaloacetate is then reused in the next pass through the citric acid cycle. Interestingly, it is not the two carbon atoms of the original acetyl CoA molecule that are released as CO_2, but two of the four carbon atoms from oxaloacetate.

The "huffing and puffing" you experience when you exert yourself is a direct result of the increased energy demands placed on your body, which cause more CO_2 to be produced via the citric acid cycle, as well as more oxygen to be consumed during oxidative phosphorylation. The eight steps of the citric acid cycle are shown in Figure 14-5. Each step involves one to three reactions, and each step is catalyzed by one specific enzyme. Four of these enzymes require either FAD or NAD^+ as a coenzyme. Extension Topic 14-1 describes each of the reactions in the citric acid cycle in more detail.

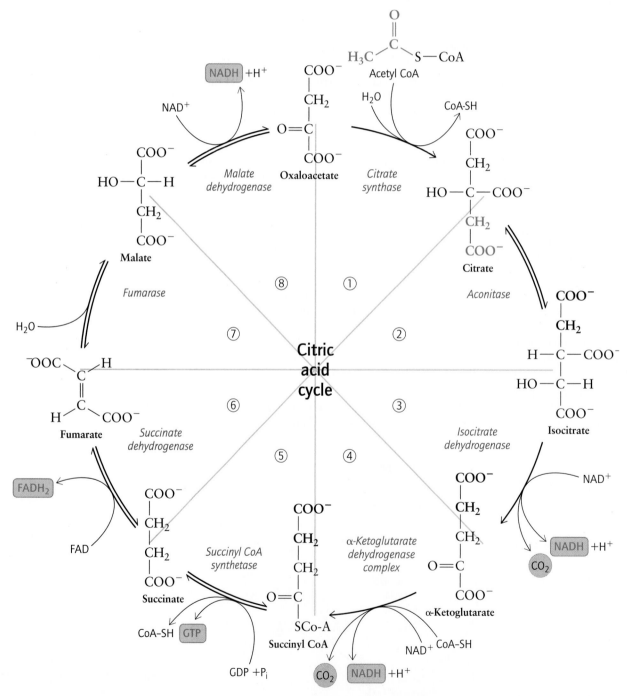

Figure 14-5 The eight steps of the citric acid cycle and the enzymes involved.

WORKED EXERCISE 14-1   Acetyl CoA and the Citric Acid Cycle

Problem-Solving Tutorial:
Steps in the Citric Acid Cycle

Oxaloacetate is produced in the last step of the citric acid cycle.

$$
\begin{array}{c}
COO^- \\
| \\
O = C \\
| \\
CH_2 \\
| \\
COO^-
\end{array}
$$

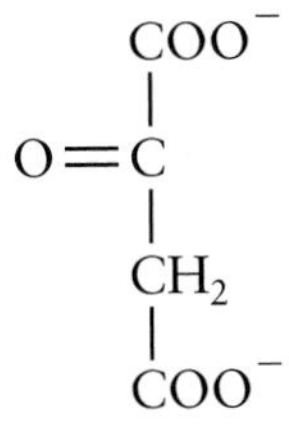

Oxaloacetate

a. What structure would this compound have at low pH? Write the carboxylic acid functional groups in their un-ionized form. What other functional group is present in the molecule?
b. How many carbon atoms does oxaloacetate contain?
c. Name the compound that oxaloacetate reacts with in the first step of the citric acid cycle. How many carbon atoms do each of the reactants and the product contain?

SOLUTION

a.

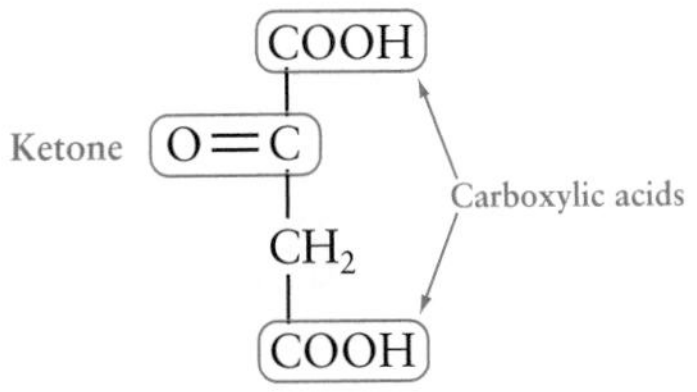

Ketone

$$
\begin{array}{c}
\boxed{COOH} \\
| \\
\boxed{O = C} \\
| \\
CH_2 \\
| \\
\boxed{COOH}
\end{array}
$$

Carboxylic acids

Oxaloacetate

b. Oxaloacetate contains four carbon atoms.
c. Oxaloacetate reacts with acetyl CoA in the first step of the citric acid cycle to form citrate. Oxaloacetate contains four carbon atoms, and acetyl CoA carries two carbon atoms in the acetyl group. Therefore, citrate contains 6 (4 + 2) carbon atoms.

PRACTICE EXERCISES

14.1   Answer the following questions about carbon dioxide in the citric acid cycle:
 a. Are carbon dioxide molecules lost during the first half or the second half of the citric acid cycle?
 b. Write a Lewis dot structure for carbon dioxide.
 c. How many carbon dioxide molecules are released in one pass through the citric acid cycle?
 d. Do the carbon atoms released as carbon dioxide originate from oxaloacetate or acetyl CoA?

14.2   How many carbon atoms are there in citrate? In which step of the citric acid cycle is citrate formed?

14.3   Based on the structure of citrate, why is an alternative name for the citric acid cycle the tricarboxylic acid (TCA) cycle?

14.4   How many molecules of NAD^+ are reduced to NADH in the citric acid cycle?

14.5   How many molecules of FAD are reduced to $FADH_2$ in the citric acid cycle?

14.6   How many molecules of ADP are converted into ATP as a result of one pass through the citric acid cycle, assuming that all the NADH and $FADH_2$ are used to phosphorylate ADP to ATP?

14.7   Name two biochemical pathways that produce acetyl CoA.

14.8   What acyl group is carried by acetyl CoA? What acyl group is carried by succinyl CoA?

Extension Topic 14-1 The Chemical Reactions of the Citric Acid Cycle

Information on the different reaction types can be found in Chapter 10, which describes the details of oxidation–reduction reactions, hydration and dehydration reactions, acyl transfer reactions, and phosphoryl group transfer reactions.

Step 1 The citric acid cycle begins when acetyl CoA reacts with oxaloacetate to produce citrate. The thioester functional group is also hydrolyzed in this step, releasing coenzyme A in an acyl transfer reaction. Thus, a two-carbon molecule combines with a four-carbon molecule to produce the six-carbon molecule citrate. The new carbon–carbon bond in citrate is highlighted below.

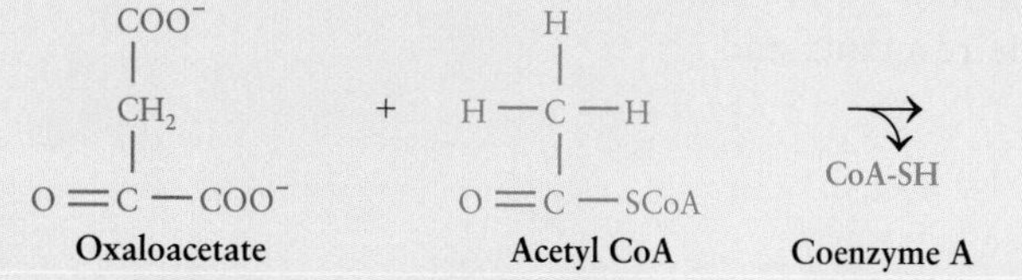

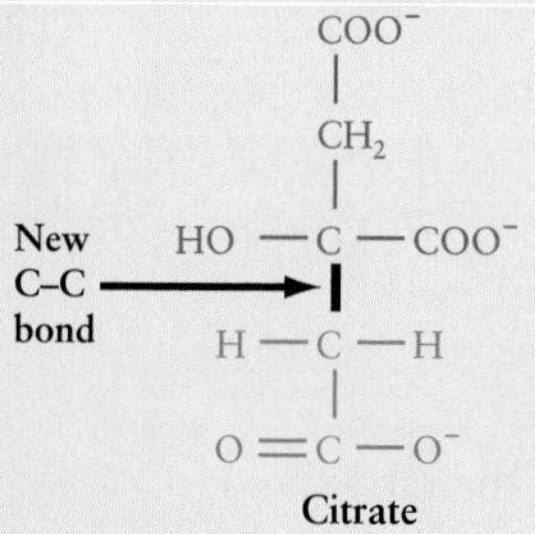

Step 2 An isomerization reaction converts citrate into its structural isomer, isocitrate. The isomerization is accomplished by a dehydration reaction followed by a hydration reaction (the sequential removal and addition of water):

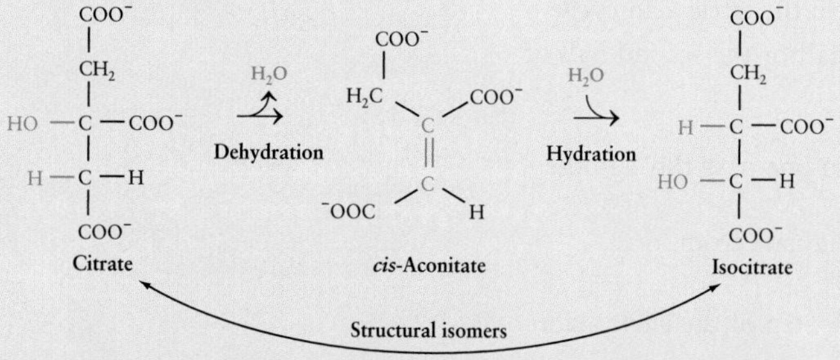

Dehydration of citrate removes an —OH group and a hydrogen atom from adjacent carbon atoms, expelling a water molecule. Then, hydration reintroduces the —OH group and hydrogen atom, but at different carbon atoms.

Step 3 Step 3 produces the first molecule of carbon dioxide (CO_2) and reduces the first of three NAD^+ molecules to NADH. Two chemical reactions are involved in this step:

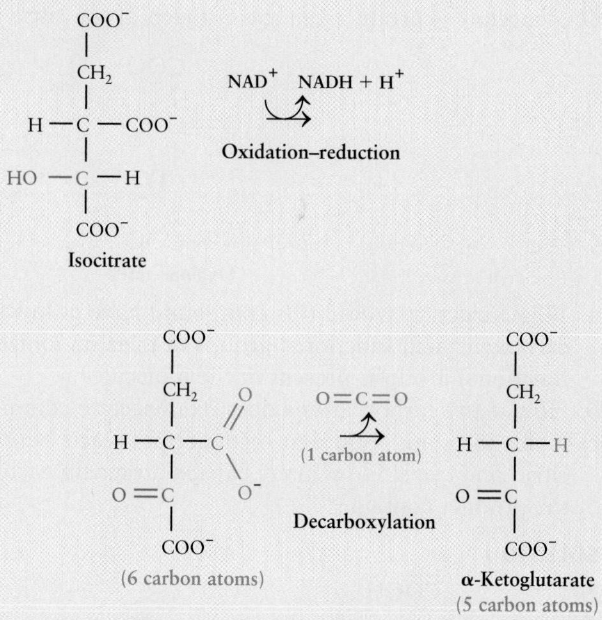

In the first step an oxidation–reduction reaction transfers electrons to NAD^+ as the secondary alcohol is oxidized to a ketone. The second reaction is a **decarboxylation reaction**, a type of reaction in which carbon dioxide (CO_2) is lost from the substrate and replaced with a proton (H^+) from solution, thus removing one carbon atom from the molecule in the form of carbon dioxide (CO_2). The product, α-ketoglutarate, has one less carbon atom than isocitrate.

Step 4 Step 4 produces the second and final molecule of CO_2 and reduces a second molecule of NAD^+ to NADH. This step involves three reactions all performed by one enzyme:

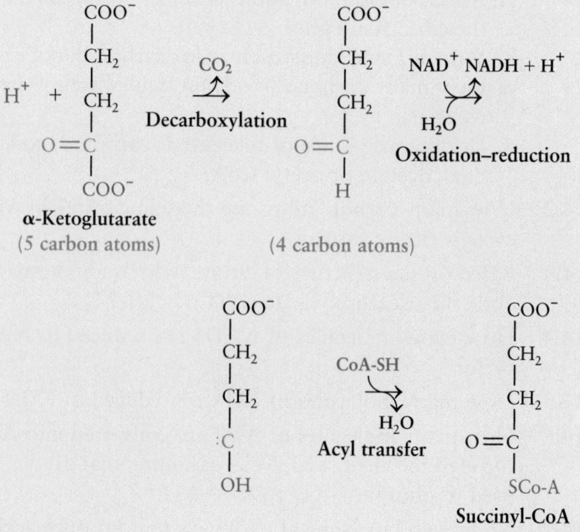

The first reaction is a decarboxylation reaction that expels CO_2, thereby shortening the five-carbon α-ketoglutarate molecule by one carbon atom, similar to the decarboxylation reaction in Step 3. The intermediate aldehyde is then oxidized to a carboxylic acid by NAD^+. In the final reaction, the acyl group is transferred to the thiol group on coenzyme A in a thioesterification reaction to produce the four-carbon thioester succinyl-CoA.

Step 5 Succinyl CoA is hydrolyzed into succinate and coenzyme A. This reaction releases enough energy to simultaneously phosphorylate guanosine diphosphate (GDP) to guanosine triphosphate (GTP):

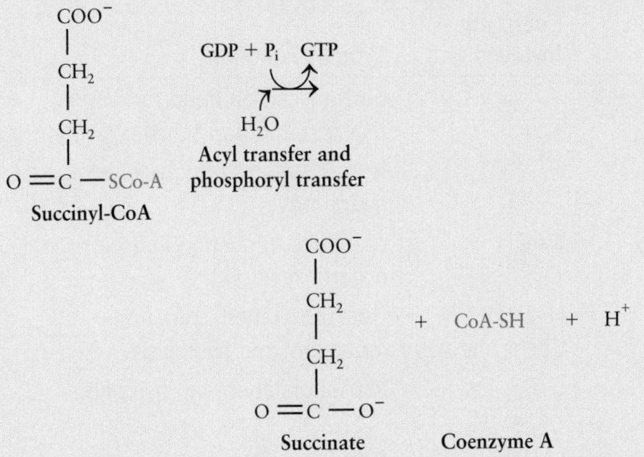

GTP is similar to ATP, except that it contains guanosine instead of adenosine as part of its structure. Eventually GTP converts one molecule of ADP into ATP, so in terms of energy it is equivalent to one ATP molecule.

Step 6 Step 6 reduces one molecule of FAD to $FADH_2$ by oxidizing the carbon–carbon single bond in succinate to a carbon–carbon double bond, forming fumarate:

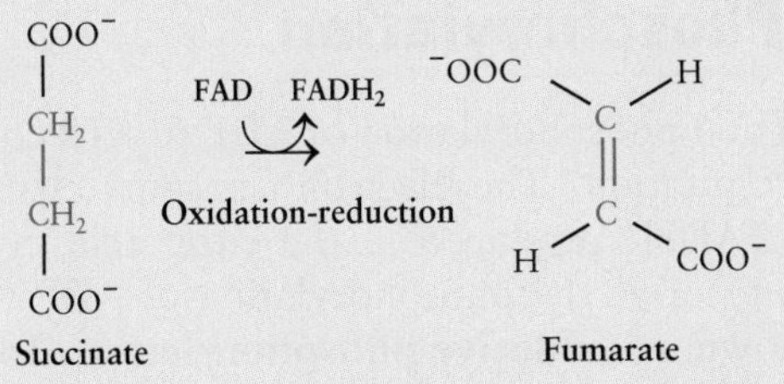

Step 7 A hydration reaction produces the chiral alcohol L-malate from fumarate. In this step, water adds to the double bond as shown to form L-malate.

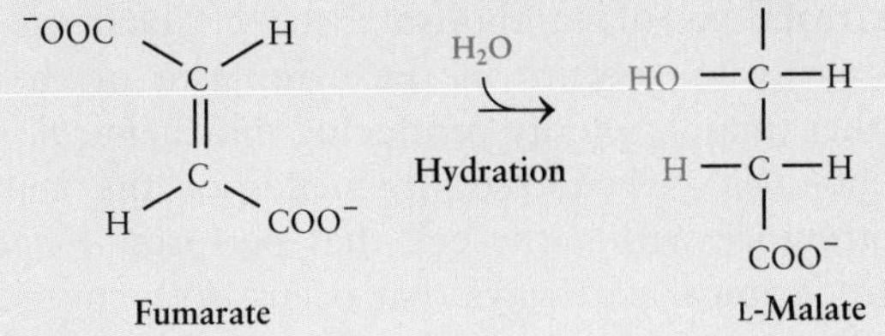

Step 8 The final step of the citric acid cycle is an oxidation–reduction reaction that produces the third molecule of NADH from NAD^+. In this reaction, the secondary alcohol in L-malate is oxidized to the ketone oxaloacetate, an achiral molecule. Oxaloacetate is then reused in the next pass through the citric acid cycle.

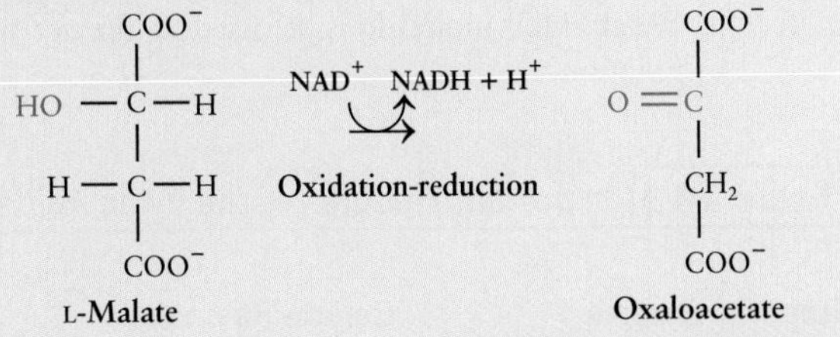

Table 14-1 summarizes the important features of each step in the citric acid cycle as well as the enzymes responsible for catalyzing each step.

WORKED EXERCISE E14-1 Reactions in the Citric Acid Cycle

Refer back to Step 4 of the citric acid cycle (p. 508).
a. What type of reaction is shown in the first reaction in this step? Explain what changes to the structure of the substrate occur in this reaction.
b. In the second reaction, what substance is oxidized and what substance is reduced?
c. What is the significance of the NADH produced in the second reaction in terms of cellular energy?
d. What is the significance of the CO_2 produced in the first reaction?
e. What type of reaction is the third reaction? What does CoA-SH stand for?
f. Which substance in Step 4 contains a thioester?

SOLUTION
a. The first reaction in the sequence is a decarboxylation reaction. In a decarboxylation reaction CO_2 is lost from the substrate, decreasing the number of carbon atoms in the molecule by one.
b. NAD^+ is reduced to NADH and the aldehyde is oxidized to a carboxylic acid.
c. NADH provides the electrons necessary to phosphorylate three ADP to ATP during oxidative phosphorylation, producing energy.
d. The CO_2 produced in the first reaction is one of two CO_2 molecules produced in the citric acid cycle, which degrades acetyl CoA into two carbon dioxide molecules and coenzyme A.
e. The reaction is a thioesterification reaction; coenzyme A.
f. Succinyl-CoA contains a thioester.

PRACTICE EXERCISES

E14.1 Refer to Step 3 of the citric acid cycle (p. 508).
  a. What type of reaction is the first reaction in this step?
  b. What functional groups are unchanged in this reaction? What functional group changes?
  c. What is the significance of the NADH produced in the first reaction in terms of energy?
  d. What small molecule is released in the second reaction?

E14.2 Refer to Step 2 in the citric acid cycle (p. 508).
  a. How many reactions are involved in Step 2? What type of reactions are they?
  b. What are the chemical formulas for citrate and isocitrate?
  c. What is the structural relationship between citrate and isocitrate?
  d. Why is a water molecule a product in the first reaction and a reactant in the second reaction?
  e. Why does the second reaction not reform citrate?

Table 14-1 Important Features of the Citric Acid Cycle

Step	Enzyme	Carbon Balance	Coenzymes Involved	Type of Reaction
1	Citrate synthase	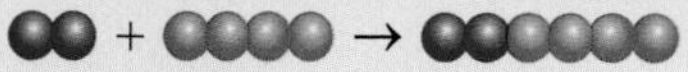 ●● + ●●●● → ●●●●●●	–	Carbon–carbon bond forming reaction and acyl transfer reaction
2	Aconitase	 ●●●●●●	–	Isomerization through dehydration-hydration
3	Isocitrate dehydrogenase	 ●●●●●● → ●●●●● + CO_2	NADH	Oxidation-reduction and decarboxylation
4	α-Ketoglutarate dehydrogenase	 ●●●●● → ●●●● + CO_2	NADH Coenzyme A	Decarboxylation, oxidation-reduction, and hydrolysis
5	Succinyl-CoA synthetase	 ●●●●	GTP	Phosphorylation and hydrolysis
6	Succinate dehydrogenase	 ●●●●	$FADH_2$	Oxidation-reduction
7	Fumarase	 ●●●●	–	Hydration
8	Malate dehydrogenase	 ●●●●	NADH	Oxidation-reduction

14.2 Oxidative Phosphorylation

The electron-transport chain and phosphorylation of ADP to ATP constitute the final stage of cellular respiration. The electron-transport chain takes electrons from NADH and $FADH_2$ produced in the citric acid cycle and other biochemical pathways to drive the phosphorylation of ADP to make ATP. These processes are known as **oxidative phosphorylation.** To understand the last stage of catabolism you need to be familiar with the structure of a mitochondrion, the site of most of the energy-producing pathways in the cell.

Where It All Happens: The Mitochondria

Where in the cell do the reactions of catabolism occur? Glycolysis—the conversion of glucose to pyruvate—occurs in the cytoplasm of the cell. However, most of the other major energy-producing biochemical pathways occur in the mitochondria, specialized organelles of the cell. An organelle is an enclosed structure within the cell that performs a specific function. The important biochemical pathways that occur in the mitochondria include β-oxidation of fatty acids, the citric acid cycle, and oxidative phosphorylation.

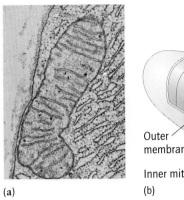
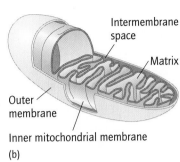

(a)

(b)

Intermembrane space

Matrix

Outer membrane

Inner mitochondrial membrane

Figure 14-6 (a) Microscopic image of a mitochondrion; (b) cross section of a mitochondrion.

The key parts of a mitochondrion are illustrated in Figure 14-6 and include

- the outer membrane,
- the intermembrane space,
- the inner membrane, and
- the matrix.

The **outer membrane** encloses and defines a mitochondrion and is permeable to many ions and small molecules. Inside the mitochondrion is a second selectively permeable membrane, called the **inner membrane.** This highly folded inner membrane separates the interior of a mitochondrion into two separate aqueous solutions: (1) the **intermembrane space** located between the outer membrane and the inner membrane, and (2) the **matrix,** located in the space enclosed by the inner membrane.

You learned in Chapter 5 that particles dialyse through a membrane in the direction that equalizes the concentration of ions. But ions, including protons (H^+), are unable to diffuse through the inner membrane by simple diffusion. Since the inner membrane is not permeable to protons, proton concentrations on either side of the membrane are not able to equalize. The analogy could be made to a compressed spring, which has a natural tendency to stretch out, unless something prevents it from doing so. Therefore, the concentration of protons—the pH—is different on either side of this membrane: the matrix has a lower concentration of protons (higher pH) than the intermembrane space (lower pH). A difference in proton concentration between adjacent regions is known as a **proton gradient** (Figure 14-7).

Most of the enzymes involved in the citric acid cycle reside in the matrix. The proteins involved in oxidative phosphorylation are located along or embedded within the inner membrane. The important enzyme *ATP synthase,* which catalyzes the phosphorylation of ADP to produce ATP, is found in the inner membrane with a significant portion of this large enzyme extending into the matrix.

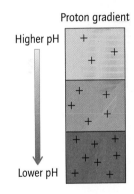

Proton gradient

Higher pH

Lower pH

Figure 14-7 The proton concentration increases from top to bottom in this sketch of a proton gradient.

PRACTICE EXERCISE

14.9 Fill in the blank with the appropriate part of the mitochondrion described: *inner membrane, outer membrane, matrix,* or *intermembrane space.*

a. The _____ is a selectively permeable membrane through which protons cannot pass.

b. The _____ and _____ contain the proteins and molecules involved in oxidative phosphorylation.

c. The _____ contains the enzymes of the citric acid cycle.

d. The _____ is the region with the lowest proton concentration.

e. The _____ and _____ consist of an aqueous medium.

f. The _____ and _____ consist of a lipid medium.

Figure 14-8 Components of the electron-transport chain. The electron-transport chain is composed of the protein complexes I–IV and the mobile proteins coenzyme Q and cytochrome *c*.

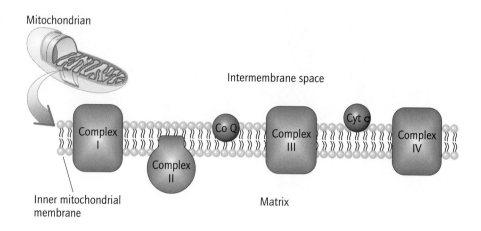

The Electron-Transport Chain

Electrons carried by the coenzymes NADH and FADH$_2$ produced in the citric acid cycle and other biochemical pathways are relayed through a series of protein complexes to molecular oxygen, O$_2$, whereupon the oxygen is reduced to water. Oxygen is critical to the electron-transport process, the reason why we need to breathe oxygen and the reason why the overall process is called cellular respiration.

These protein complexes are part of what is known as the **electron-transport chain.** The electron-transport chain contains four large multienzyme protein complexes, identified as **Complexes I, II, III,** and **IV,** along with two mobile proteins, **coenzyme Q** and **cytochrome *c*.** The protein complexes are shown in green in Figure 14-8 and the mobile carriers are shown in blue.

Each protein complex is a membrane protein, and most span the inner mitochondrial membrane, extending from the matrix to the intermembrane space. Sites within each protein complex accept electrons, then transmit them to the next protein like the baton in a relay race. The mobile proteins carry electrons within the membrane between the protein complexes.

The Proton-Motive Force

The flow of electrons through protein Complexes I, III, and IV creates the energy necessary to pump protons from the matrix across the inner membrane and into the intermembrane space, generating a proton gradient. Since the movement of protons is against the proton gradient, the process is referred to as a **proton pump.**

A proton gradient represents potential energy, which ultimately provides the energy for oxidative phosphorylation, the final step in cellular respiration. The accumulation of protons in the intermembrane space represents both chemical and electrical potential energy. Since protons are charged ($+1$), there is a difference in charge between the two spaces, and so the proton gradient also has electrical potential much like a battery. The stored potential energy in a mitochondrion is called the **proton-motive force.**

Oxidation–Reduction in the Electron-Transport Chain

How are electrons transferred through the electron-transport chain? Electrons are transferred between atoms and molecules through a series of oxidation–reduction reactions that occur at the various metal atom centers (copper and iron) as well as at organic cofactors such as coenzyme Q.

$$2 \; H^+ + 2e^- +$$

Coenzyme Q
(oxidized form)

Reduction ⇌ Oxidation

Coenzyme Q
(reduced form)

The flow of electrons through the electron-transport chain occurs by a sequence of oxidation–reduction reactions. Recall from Chapter 10 that oxidation is defined as a loss of electrons, and reduction as a gain of electrons. For example, at an iron center in the chain, Fe^{3+} is reduced to Fe^{2+} when accepting electrons, and oxidized back to Fe^{3+} when transmitting electrons to the next electron acceptor.

$$Fe^{3+} + e^- \longrightarrow Fe^{2+} \qquad \text{reduction}$$
$$Fe^{2+} \longrightarrow Fe^{3+} + e^- \qquad \text{oxidation}$$

The terms *oxidation* and *reduction* allow you to keep track of which species are on the donating and receiving ends of the electron transfer.

A metal or molecule's intrinsic affinity for electrons determines the direction of electron flow in an oxidation–reduction reaction. For example, in the laboratory you can demonstrate that copper metal (Cu) will transfer electrons to silver ion (Ag^+), but silver metal will not transfer electrons to copper ions; electrons will always flow from copper to silver but not the reverse. Similarly in the electron-transport chain, electrons are passed to centers with increasingly greater affinity for electrons, as shown in the energy diagram in Figure 14-9. Thus electron flow begins with NADH or $FADH_2$, which have the least electron affinity, and ends with the acceptance of electrons by molecular oxygen, which has the greatest electron affinity.

Silver (Ag) crystals have formed on a strip of copper metal. Silver ions (Ag^+) in solution have been reduced (accepted electrons) by copper metal (Cu), which has been oxidized (lost electrons) to copper ion (Cu^+). [Peticolas/Megna/Fundamental Photographs]

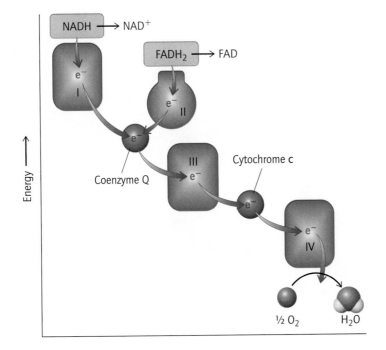

Figure 14-9 Electrons are transferred from NADH and $FADH_2$ to the various protein complexes in the electron-transfer chain in the direction that takes them from species higher in potential energy to species lower in potential energy. The lowest energy species, and final resting place of the electrons, is oxygen, which is reduced to water.

Atom centers with greater electron affinity represent species with lower potential energy. Indeed, electrons cannot flow in the opposite direction without an input of energy. The energy needed to create the proton gradient comes from the energy released by the stepwise oxidation–reduction reactions that occur as electrons are transferred to centers of increasingly lower potential energy in the chain. The energy released is used to pump protons against the proton gradient from the matrix to the intermembrane space through the inner membrane. Protons are pumped through Complexes I, III, and IV as electrons pass through these complexes.

Complex I accepts electrons from NADH and Complex II accepts electrons from $FADH_2$. Upon receipt of electrons, a metal ion (copper or iron) or disulfide group within the protein complex is reduced and NAD^+ and FAD, the oxidized forms of the coenzymes, are regenerated. Regenerating NAD^+ and FAD is important because it allows processes like the citric acid cycle, which requires NAD^+ and FAD, to continue. You may recall from Chapter 12 that glycolysis can continue through an alternative anaerobic process to meet the energy needs of muscle cells, by regenerating NAD^+ through lactic acid fermentation, an alternate route for regenerating NAD^+.

Complex I Consider what happens to the electrons from NADH when they are transferred to Complex I (Figure 14-10). Complex I is an enormous protein consisting of 46 polypeptides. Part of the protein lies within the inner membrane and another part protrudes into the matrix. Two electrons from NADH enter Complex I in the matrix and pass through a series of oxidation–reduction sequences to coenzyme Q, a mobile electron carrier within the inner membrane. In the process, two electrons (and two protons) are received by coenzyme Q and four protons are pumped from the matrix to the intermembrane space.

Complex II $FADH_2$ enters the electron-transport chain at Complex II, which is lower in energy than NADH and does *not* pump protons into the inter membrane space. Therefore, $FADH_2$ isn't capable of producing as much ATP as NADH. Complex II is another very large protein, which contains the *succinate dehydrogenase* enzyme that generates $FADH_2$ in the citric acid cycle. From Complex II, electrons are passed to the mobile carrier coenzyme Q, which is also the recipient of electrons from Complex I.

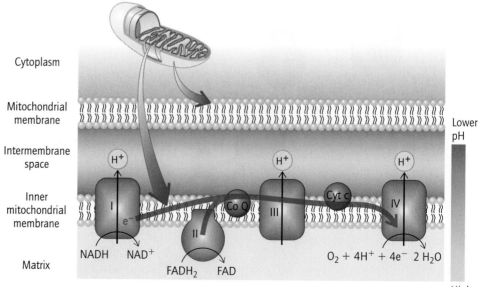

Figure 14-10 As electrons move through the electron-transport chain, protons are pumped across the inner mitochondrial membrane at Complexes I, III, and IV, creating a proton gradient.

Complex III Coenzyme Q transfers electrons to Complex III, whereupon protons are pumped from the matrix to the intermembrane space. Electrons from Complex III are transferred to the water-soluble mobile electron-carrier protein cytochrome *c*.

Complex IV Cytochrome *c* delivers electrons to the last protein complex, Complex IV, which transfers electrons from cytochrome *c* to molecular oxygen, the final repository of electrons. Reduction of oxygen generates water. Protein Complex IV also pumps protons across the inner membrane.

$$O_2 + 4\,H^+ + 4\,e^- \longrightarrow 2\,H_2O$$

Oxygen is the final destination for the electrons we extract from our food, and the energy released in the process is used to create a protein gradient that drives phosphorylation of ADP to ATP, the energy currency of the cell. Without oxygen, the phosphorylation of ADP described in the next section could not occur, and cells would be deprived of the energy they need to run the anabolic pathways necessary for life.

Cyanide Poisoning The last oxidation–reduction step that occurs in Complex IV is the transfer of electrons from an iron atom in the enzyme *cytochrome c oxidase* to oxygen. Cyanide, CN^-, is a fatal poison because it is an irreversible inhibitor of this enzyme. In Chapter 11 you learned that competitive enzyme inhibitors bind to the active site of an enzyme as well as or better than the natural ligand, preventing the enzyme from performing its function. Here the function inhibited is the crucial and final oxidation–reduction step in the electron-transport chain. Cyanide poisoning brings the electron-transport chain to a halt, thereby preventing phosphorylation of ADP and depriving the cell of energy.

One treatment for cyanide poisoning is hyperbaric oxygen treatment, a topic covered in Chapter 5. This treatment floods the mitochondria with the natural ligand, oxygen, allowing the electron-transport chain to resume.

WORKED EXERCISES 14-2   The Electron-Transport Chain

1   What is meant by "mobile" in the term "mobile electron-carriers"? What molecules are mobile carriers of electrons in the electron-transport chain?

2   Fill in the blanks for the following chain of events that describes the path electrons take through the electron-transport chain when introduced from NADH:

   NADH $\longrightarrow$ Complex ____ $\longrightarrow$ coenzyme ____ $\longrightarrow$ Complex ____
   $\longrightarrow$ cytochrome ____ $\longrightarrow$ Complex ____ $\longrightarrow O_2$.

3   Which of the four protein complexes does not pump protons from the matrix into the intermembrane space?

4   Where is the proton concentration greater, in the intermembrane space or the matrix? Where is the pH lower, in the intermembrane space or the matrix? Why do protons in the intermembrane space not diffuse into the matrix, equalizing the concentration?

SOLUTIONS

1   Mobile carriers of electrons in the electron-transport chain are proteins that are not bound to the membrane, but can move between complexes. Coenzyme Q and cytochrome *c* are mobile carriers of electrons in the electron-transport chain.

2 NADH → Complex <u>I</u> → Coenzyme <u>Q</u> → Complex <u>III</u> → cytochrome <u>c</u> → Complex <u>IV</u> → O_2.

3 Complex II does not pump protons from the matrix into the intermembrane space.

4 The proton concentration is greater in the intermembrane space, thus the pH is lower in the intermembrane space. Protons do not diffuse into the matrix because the inner mitochondrial membrane is a selectively permeable membrane that does not allow protons to pass freely between the intermembrane space and the matrix.

PRACTICE EXERCISES

14.10 How many multienzyme complexes make up the electron-transport chain? Which complex contains the enzyme *succinate dehydrogenase*? Which step in the citric acid cycle is catalyzed by *succinate dehydrogenase*? What are the products of this step of the citric acid cycle?

14.11 Fill in the blanks for the following chain of events that describes the path electrons take through the electron-transport chain when delivered by $FADH_2$:

$FADH_2 \longrightarrow$ Complex ____ $\longrightarrow$ coenzyme ____ $\longrightarrow$ Complex ____

$\longrightarrow$ cytochrome ____ $\longrightarrow$ Complex ____ $\longrightarrow O_2$.

14.12 What important role does oxygen play in the electron-transport chain?

14.13 Explain how cyanide shuts down the electron-transport chain.

ATP Synthesis

In Chapter 10 you learned that the phosphorylation of ADP to ATP requires energy:

$$\text{Energy} + \text{ADP} + \text{P}_i \longrightarrow \text{ATP} + \text{H}_2\text{O}$$

Where does the energy needed for this important reaction come from? Within the inner mitochondrial membrane is a remarkable enzyme complex known as *ATP synthase*, sometimes referred to as *Complex V*, which catalyzes this reaction. *ATP synthase uses the proton-motive force to drive phosphorylation of ADP to ATP.*

ATP synthase is a multienzyme protein complex that looks like a cylindrical motor on a stick. The stick portion of the enzyme is embedded in the inner membrane and the cylinder, with a diameter of 8.5 nm, extends into the matrix, as illustrated in Figure 14-11.

When the concentration of protons in the intermembrane space reaches a certain level, and ADP and P_i are both bound to *ATP synthase,* an ion channel within the "stick" portion of *ATP synthase* opens, allowing protons in the intermembrane space to flow freely back into the matrix, where the proton concentration is lower (Figure 14-12). Since protons flow through *ATP synthase* from a region of higher proton concentration to a region of lower proton concentration, energy is released—much like a hydroelectric dam generates electricity when water flows *down* over a dam. As protons flow in the direction of the proton gradient through a proton channel created by *ATP synthase,* the energy released causes a part of the enzyme to rotate, like a turnstile. Rotation brings the reactants ADP and P_i together so that phosphorylation can occur, and another turn of the turnstile releases ATP.

The discovery of how *ATP synthase* works was such a significant contribution toward our understanding of biochemistry that Paul Boyer, of UCLA, and John Walker, an English scientist, were awarded the 1997 Nobel Prize in chemistry for their discovery.

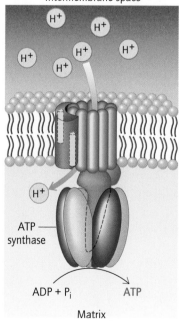

Intermembrane space

ATP synthase

ADP + P_i　　　ATP

Matrix

Figure 14-11 *ATP synthase,* the enzyme that catalyzes phosphorylation of ADP, uses energy created by the proton-motive force.

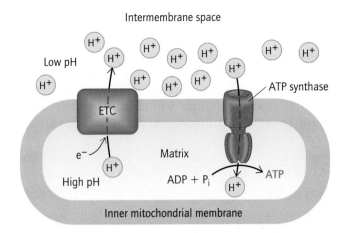

Figure 14-12 Overview of oxidative phosphorylation. The movement of electrons through the electron-transport chain (ETC) generates a proton gradient, which drives phosphorylation of ADP to make ATP.

Summary of Metabolism

You have seen that energy is extracted from the biomolecules in our diet: proteins, carbohydrates, and triglycerides. The combustion of these molecules in cellular respiration produces more than 90% of the ATP required by humans.

Catabolism of these molecules to produce energy is divided into three stages (Figure 14-13). In the first stage, proteins are hydrolyzed into amino acids, as described in Chapter 11; carbohydrates are hydrolyzed into monosaccharides, as described in Chapter 12; and triglycerides are hydrolyzed into glycerol and fatty acids, as described in Chapter 13. Each

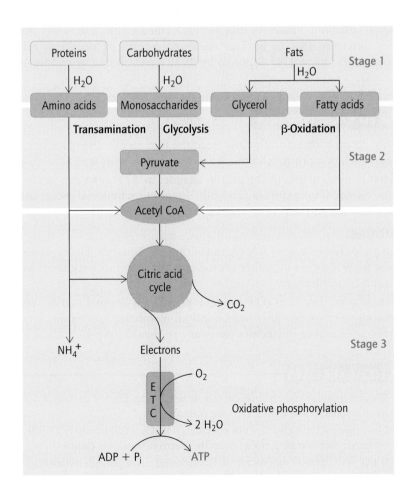

Figure 14-13 Overview of catabolism.

of these smaller biomolecular building blocks are further degraded by well-known biochemical pathways to yield acetyl CoA:

- Amino acids undergo transamination, whereupon they are converted to acetyl CoA or other intermediates of the citric acid cycle.
- Glucose undergoes the 10-step biochemical pathway known as glycolysis to produce pyruvate, a process that takes place in the cytoplasm (Chapter 12). Pyruvate under aerobic conditions is converted to acetyl CoA.
- Fatty acids undergo β-oxidation in the mitochondria to produce acetyl CoA molecules.

In this chapter you have learned about the third stage of catabolism: the citric acid cycle and oxidative phosphorylation, both processes that occur in the mitochondria. The acetyl CoA produced from all three biomolecules enters the citric acid cycle, where electrons are harvested from acetyl CoA and carried by the important coenzymes NADH and $FADH_2$ to the electron-transport chain. Electrons pass through the electron-transport chain by a series of oxidation–reduction reactions that simultaneously cause protons to be pumped from the matrix into the intermembrane space, creating a proton gradient. When a channel opens within ATP synthase, protons return to the matrix, releasing energy that drives the phosphorylation of ADP to ATP. *Thus, ultimately, one acetyl CoA molecule can phosphorylate 12 ADP to 12 ATP.*

ATP is used to perform mechanical work such as muscle contraction, active transport of molecules and ions through membranes, and the anabolic processes of building proteins, DNA, and cell membranes. In this text you have focused on the intricacies of catabolic process, but equally important are the various anabolic processes that are performed by the cell, using many of the same intermediates, to produce the biomolecules of life. These anabolic processes utilize the energy supplied by ATP, which is hydrolyzed to ADP in the process.

WORKED EXERCISES 14-3 ATP Synthesis

1 Why do protons naturally flow from the intermembrane space to the matrix when a proton channel is opened in *ATP synthase*?
2 What is the function of the proton channel in *ATP synthase,* and why do protons not simply diffuse across the inner mitochondrial membrane in the direction of the proton gradient?

SOLUTIONS

1 Simple diffusion causes ion concentrations to equalize. Here, protons flow from a region of higher proton concentration to a region of lower proton concentration.
2 The proton channel in *ATP synthase* provides a channel that allows protons to cross the inner mitochondrial membrane. A channel is necessary because the membrane is not permeable to protons.

PRACTICE EXERCISES

14.14 What important chemical reaction is driven by the flow of protons from the intermembrane space back into the matrix?

14.15 How many ADP molecules are phosphorylated to ATP for every $FADH_2$ molecule that transfers electrons to the electron-transport chain?

14.16 What is the proton-motive force? How does it drive phosphorylation of ADP?

14.3 | Entropy and Bioenergetics

Bioenergetics is the study of energy transfer in reactions occurring in living cells. Energy is central to all biochemical processes. In Chapter 8 you learned that heat energy is transferred in chemical reactions, and that the heat energy transferred is known as the change in enthalpy, ΔH. Recall that heat is released to the surroundings in an exothermic reaction and absorbed by the surroundings in an endothermic reaction. In this section you will gain a more complete picture of energy. To understand energy transfer in living cells, you must consider two additional energy concepts: entropy and free energy. The change in free energy, ΔG, for a chemical reaction allows you to predict whether or not the reaction will occur under a given set of conditions, that is, to predict whether a reaction is spontaneous or nonspontaneous.

Spontaneous and Nonspontaneous Reactions

A reaction is said to be **spontaneous** if it continues on its own once started. Conversely, a reaction is said to be **nonspontaneous** if energy must be continuously supplied to the reaction for it to proceed. Consider the analogy of movers unloading and loading boxes from a truck with a ramp (Figure 14-14). To *unload* boxes, a mover needs only to give the box a single push (the activation energy) at the top of the ramp, and the box will slide down the ramp on its own. Hence, unloading a box is analogous to a spontaneous process. To *load* boxes onto the truck up the ramp, however, requires a continuous supply of energy, not just an initial push. Hence, loading a box is analogous to a nonspontaneous process.

    Consider another example, relating to metabolism. Wood lying around in the forest will rot in a chemical reaction that yields the same end products as if you had burned the wood. No input in energy is required, so rotting is a

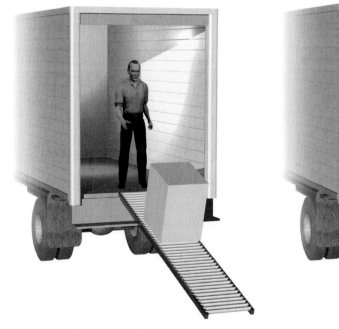

(a) A single push sends a box down a ramp.   (b) A box must be pushed continuously up a ramp.

Figure 14-14 (a) A spontaneous reaction is like pushing a box down a ramp: Only a single push is needed at the start, then the box proceeds on its own. (b) A nonspontaneous reaction is like pushing a box up a ramp: Energy must be supplied continuously.

spontaneous, albeit slow, reaction. On the other hand, a leaf on a plant requires an input of energy to grow. The energy is supplied by the sun. Indeed, if you remove the source of energy—sunlight—the leaf will not grow, because the reactions required to make the leaf need a constant input of energy. Thus, the reactions that make a leaf are nonspontaneous. *Note:* A spontaneous chemical reaction is not necessarily a fast one, but one that can proceed *on its own* in the direction written.

By definition, if a reaction is spontaneous, the reverse reaction is nonspontaneous. At normal body temperature, the hydrolysis of ATP is a spontaneous reaction:

$$\text{ATP} + \text{H}_2\text{O} \longrightarrow \text{ADP} + \text{P}_i + \text{energy} \qquad \textit{spontaneous}$$

Once initiated, this reaction will continue without further input of energy. The reverse reaction, the phosphorylation of ADP, is therefore nonspontaneous:

$$\text{Energy} + \text{ADP} + \text{P}_i \longrightarrow \text{ATP} + \text{H}_2\text{O} \qquad \textit{nonspontaneous}$$

In the last section, you saw that phosphorylation of ADP requires energy to proceed. Indeed, the electron-motive force is the main source of energy for this nonspontaneous reaction.

Biochemical reactions in a living cell occur in an aqueous environment and under conditions of constant temperature and pressure. Under these conditions, whether a reaction is spontaneous or nonspontaneous can be determined from a quantity known as the change in free energy (ΔG) for the reaction. The change in free energy depends on several factors:

- the change in enthalpy, ΔH,
- the change in entropy, ΔS, and
- the temperature, T.

Entropy

The **change in entropy**, ΔS, is a measure of the change in the degree of randomness or disorder in a reaction or physical change. Entropy increases as disorder increases. Consider the process of cleaning your room. You may make your bed, throw away papers, organize books on shelves, etc. Cleaning is a process that takes your room from a more disordered place to a less disordered place; thus, cleaning is associated with a decrease in entropy, $\Delta S < 0$. The opposite process, creating more disorder in your clean room, would be associated with an increase in entropy, $\Delta S > 0$. Similarly, entropy may increase or decrease in a chemical reaction. If the product molecules exhibit greater disorder than the reactant molecules, the entropy of the reaction has increased, $\Delta S_{\text{rxn}} > 0$. The reverse is also true: More-ordered product molecules mean the entropy of the reaction has decreased, $\Delta S < 0$.

The change in entropy, ΔS, in a chemical reaction is defined as the difference in the entropy of the products and the reactants.

$$\Delta S = S_{\text{products}} - S_{\text{reactants}}$$

The debris strewn by Hurricane Ike in Crystal Beach, Texas, is an example of an increase in entropy. [Reuters/David J. Phillip/Pool (United States)]

Thus, when the products of a reaction are *more* disordered than the reactants, the reaction *increases* in entropy, and ΔS is positive ($\Delta S > 0$). When the products of a reaction are *less* disordered than the reactants, the reaction *decreases* in entropy and ΔS is negative ($\Delta S < 0$). Table 14-2 summarizes some criteria for determining whether a reaction increases or decreases in entropy.

Table 14-2 Summary of Changes in Entropy, ΔS, for a Reaction

Disorder of Products Versus Reactants	Change in Entropy	Sign of ΔS
Products are more disordered than reactants	Entropy increases	Positive $(+)$ $\Delta S > 0$
Products are less disordered than reactants	Entropy decreases	Negative $(-)$ $\Delta S < 0$

Another, more accurate way to look at entropy is to view it as a measure of the freedom of motion in reactants and products. There are many more ways for a room to be messy than clean, so there is more freedom of motion in a messy room than a clean room. Viewed from this perspective, the change in entropy, ΔS, in a chemical reaction depends on three factors, illustrated in Figure 14-15.

- **Changes in state of reactants versus products (s, l, g).** In a physical change, entropy changes with a change of state. When a gas condenses to a liquid, entropy *decreases* ($\Delta S < 0$), because molecules in the gas state have *more* freedom of motion than those in the liquid state. Similarly, a change from the liquid to solid phase is accompanied by a decrease in entropy.

- **The difference in complexity of reactants versus products.** In a chemical reaction, entropy decreases when the products are more complex than the reactants, because there are fewer degrees of translational and rotational freedom in complex molecules ($\Delta S < 0$).

- **The difference in the number of reactant versus product atoms or molecules.** In a *chemical reaction*, a decrease in the number of products decreases entropy ($\Delta S < 0$). When one molecule breaks apart into two or more smaller molecules, and all other factors are equal, entropy *increases* because two or more individual molecules can move in more ways than one more complex molecule can; thus, the products have *more* freedom of motion than the reactants and $\Delta S > 0$.

Entropy of the Universe The second law of thermodynamics states that the entropy of the universe is always increasing, $\Delta S_{\text{universe}} > 0$. The entropy of the

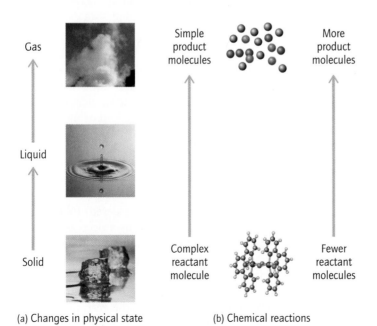

(a) Changes in physical state

(b) Chemical reactions

Figure 14-15 Factors representing an increase in entropy. (a) Entropy increases as a substance changes state from solid to liquid and liquid to gas. (b) Entropy increases in a catabolic pathway because the products are usually simpler and more numerous than the reactants. [David Arky/Corbis; SGM/Stock Connection; Zephyr/Photo Researchers, Inc.]

$$\Delta S_{universe} = \Delta S_{system} + \Delta S_{surroundings}$$

universe ($\Delta S_{universe}$) is the sum of the change in entropy of a reaction ($\Delta S_{reaction}$) (or physical change) and the change in entropy of the surroundings ($\Delta S_{surroundings}$). The second law of thermodynamics does not imply that the entropy change in a reaction or physical change is always increasing. When the products of a reaction have less freedom of motion than the reactants, there is a *decrease* in the entropy of the chemical reaction. However, the chemical reaction has affected its *surroundings*, which must have experienced an *increase* in entropy. The second law of thermodynamics predicts that the increase in entropy of the surroundings will be greater than the decrease in entropy of the reaction, so that there is always an *overall* increase in entropy in the universe.

For example, when a snowflake forms, the liquid water molecules experience a decrease in entropy ($\Delta S < 0$), because there is less disorder in a solid than a liquid. However, heat has been released into the surroundings in the process, a requirement of freezing because a change of state from liquid to solid is exothermic (Chapter 4). Therefore, the surroundings increase in kinetic energy, and thus increase in disorder. Thus, the entropy of the surroundings increases ($\Delta S > 0$). The increase in entropy of the surroundings exceeds the decrease in entropy of snowflake formation, hence the second law of thermodynamics is satisfied.

Recall learning in Chapter 11 that when an enzyme binds to its substrate, the conformation of the substrate changes and the rate of the reaction increases. Consider an enzyme (E) and a substrate (S) in solution.

$$E + S \longrightarrow E\text{–}S \text{ complex}$$

More disordered   Less disordered

Unbound, the substrate and the enzyme each possess the freedom to move in three dimensions as well as the freedom to rotate. When the enzyme and substrate interact to form a complex, this rotational and translational freedom is lost because the enzyme–substrate complex now acts as one unit. Thus, the enzyme–substrate complex is lower in entropy. Nonetheless, the entropy of the universe has increased overall, because in forming the ES complex, the surrounding atoms and molecules have become more disordered, causing an increase in entropy to the surroundings.

WORKED EXERCISES 14-4   Entropy

1   Does the following physical change represent an increase or a decrease in entropy?

$$H_2O(l) \longrightarrow H_2O(s)$$

2   Consider a polypeptide that has folded into a functional protein. In the process, hydrogen bonds, disulfide bonds, and salt bridges have been formed. Does the process of folding represent an increase or a decrease in entropy? Explain why with respect to freedom of motion.

SOLUTIONS

1   Since water is going from a liquid (*more* disordered state) to a solid (*less* disordered state), this change of state represents a *decrease in* disorder and thus a decrease in **entropy**; $\Delta S < 0$.

2   The process of a free polypeptide (*more* disordered) folding into a functional protein (*less* disordered) represents a decrease in entropy, $\Delta S < 0$. The hydrogen bonds, disulfide bonds, and ionic interactions maintain a degree of order that restricts motion in the molecule. The unfolded polypeptide is like a piece of string, with many more degrees of freedom. However, as a protein folds, the surrounding water molecules become more disordered (entropy rises), so the entropy of the universe increases in accord with the second law of thermodynamics.

14.17 Does the following physical change represent an increase or a decrease in entropy?

$$CO_2(g) \longrightarrow CO_2(s)$$

14.18 Would you expect the fourth step in glycolysis to have $\Delta S > 0$ or $\Delta S < 0$, when considering that two simpler products are formed from a more complex reactant?

14.19 Consider the condensation of steam on a cold surface. Is the entropy of the physical change greater than zero or less than zero? Is the entropy of the surroundings greater than zero or less than zero? Explain how the second law of thermodynamics is satisfied.

Free Energy, ΔG

All chemical reactions tend to attain the most stable bonding state, represented by enthalpy, and the highest state of disorder, represented by entropy. The change in free energy, ΔG, is determined by both enthalpy and entropy changes and can be used to predict whether a reaction is spontaneous or nonspontaneous. Just as ΔH and ΔS are defined in terms of the difference between products and reactants, ΔG is defined as the difference between the free energy of the products and reactants:

$$\Delta G = G_{products} - G_{reactants}$$

With regard to metabolism, **free energy** is the energy available to do cellular work. The change in free energy, ΔG, can be determined from the change in enthalpy, ΔH, the change in entropy, ΔS, and the absolute temperature (Kelvin scale), T, of a reaction or physical change. The mathematical relationship between these energy terms is shown below:

$$\underset{\substack{\text{Change} \\ \text{in free} \\ \text{energy}}}{\Delta G} = \underset{\substack{\text{Change in} \\ \text{enthalpy}}}{\Delta H} - \underset{\substack{\text{Temperature} \\ \text{multiplied by} \\ \text{change in} \\ \text{entropy}}}{(T \times \Delta S)}$$

The sign of the change in free energy of a reaction allows one to predict whether a reaction will occur, because the sign of ΔG correlates to whether a chemical reaction is spontaneous or nonspontaneous. If the change in free energy (ΔG) is negative, the reaction is spontaneous, and classified as an **exergonic reaction.** If the change in free energy (ΔG) is positive, the reaction is nonspontaneous, and classified as an **endergonic reaction.** Table 14-3 summarizes the relationship between the sign of ΔG and whether a reaction is spontaneous or nonspontaneous.

Table 14-3 Sign of ΔG and Reaction Spontaneity

Sign of ΔG	Reaction Type*	Spontaneity
Negative (−) $\Delta G < 0$	Exergonic	Spontaneous
Positive (+) $\Delta G > 0$	Endergonic	Nonspontaneous

*Note that analogous prefixes, "exo" and "endo," denote an **exo**thermic reaction ($\Delta H < 0$) and an **endo**thermic reaction ($\Delta H > 0$).

Table 14-4 The Effect of ΔH and ΔS on the Sign of ΔG

Case	Sign of ΔH	Sign of ΔS	Sign of ΔG	
1	− (exothermic)	+ (increase in entropy)	− exergonic	Spontaneous
2	+ (endothermic)	− (decrease in entropy)	+ endergonic	Nonspontaneous
3	+ (endothermic)	+ (increase in entropy)	More information is needed*	More information is needed*
4	− (exothermic)	− (decrease in entropy)	More information is needed*	More information is needed*

*The result depends on the relative values of ΔH and $T \times \Delta S$.

The sign of ΔH and ΔS together determine the sign of ΔG. Table 14-4 summarizes the four possible combinations for the sign of ΔH and ΔS and shows how each affects the overall sign of ΔG, and thus whether a chemical reaction is spontaneous or nonspontaneous.

The first two cases in Table 14-4 have obvious outcomes. Case 1 represents typical conditions for a spontaneous reaction ($\Delta H < 0$ and $\Delta S > 0$). When a chemical reaction releases heat, *and* increases in entropy, the change in free energy will be negative, and the reaction will be spontaneous. Case 2, the opposite situation, represents typical conditions for a nonspontaneous reaction ($\Delta H > 0$ and $\Delta S < 0$). Cases 3 and 4 depend on the relative values of ΔH, ΔS, and T. In these cases, temperature often plays a key role in determining whether a chemical reaction or physical change is spontaneous. Since temperature is multiplied by ΔS in the equation for free energy, higher temperatures cause entropy to have a greater influence. Thus, higher temperatures will push Case 3 to be more exergonic, because now the entropy term will dominate the enthalpy term.

For example, the free energy for the hydrolysis of ATP is −7.3 kcal/mol:

$$\text{ATP} + \text{H}_2\text{O} \longrightarrow \text{ADP} + \text{P}_\text{i} \qquad \Delta G = -7.3 \text{ kcal/mol}$$

For this reaction, the sign of ΔH is negative and the sign of ΔS is positive; therefore it is an example of Case 1 above. The reaction is spontaneous and $\Delta G < 0$. The reverse reaction is nonspontaneous, and has the same value for ΔG, but that value has a positive sign:

$$\text{ADP} + \text{P}_\text{i} \longrightarrow \text{ATP} + \text{H}_2\text{O} \qquad \Delta G = +7.3 \text{ kcal/mol.}$$

 **Problem-Solving Tutorial:** Entropy and Free Energy

---

WORKED EXERCISE 14-5 Free Energy

Using Table 14-4, determine whether the following reaction, conducted at a constant temperature and pressure, is spontaneous or nonspontaneous.

$$2 \text{ CO(g)} + \text{heat} \rightarrow 2 \text{ C(s)} + \text{O}_2\text{(g)}; \Delta S \text{ is negative}$$

SOLUTION

Heat appears on the reactant side of the chemical equation, so this is an endothermic reaction and ΔH is positive ($\Delta H > 0$). There is a decrease in entropy since ΔS is negative ($\Delta S < 0$). A positive ΔH and a negative ΔS correspond to Case 2 in Table 14-4, which results in a positive ΔG, $\Delta G > 0$. Consequently, this reaction is nonspontaneous and will not occur on its own without the continuous input of energy.

PRACTICE EXERCISES

14.20 Using Table 14-4, determine whether the following reaction, conducted at a constant temperature and pressure, is spontaneous or nonspontaneous.

$$C_{12}H_{22}O_{11}(s) + 12\ O_2(g) \longrightarrow 12\ CO_2(g) + 11\ H_2O(l) + heat \qquad \Delta S > 0$$

14.21 The following table contains the value of ΔG for some common biochemical reactions. Which of the reactions in the table are spontaneous?

Reaction*	ΔG (kcal/mol)
(a) $C_6H_{12}O_6 + 6\ O_2 \longrightarrow 6\ CO_2 + 6\ H_2O$ Glucose	−686
(b) $C_{12}H_{22}O_{11} + H_2O \longrightarrow 2\ C_6H_{12}O_6$ Maltose Glucose	−3.7
(c) $C_5H_{10}N_2O_3 + H_2O \longrightarrow C_5H_8NO_4 + NH_4^+$ Glutamine Glutamate + Ammonium ion	−3.4
(d) $C_4H_4O_5 \longrightarrow C_4H_2O_4 + H_2O$ Malate Fumarate	0.8

*The values for ΔG reported in this table are for a temperature of 25°C, a pressure of 1 atm, an initial concentration of reactants of 1 M, and a pH of 7.0. The ΔG values given for all biochemical reactions in this text are assumed to be under these conditions, and are conventionally represented as $\Delta G°'$ values.

Energy Management in Metabolism: Coupled Reactions

Many of the chemical reactions involved in metabolism are endergonic, and therefore require energy to proceed. Endergonic reactions generally use the energy from exergonic reactions to go forward. The hydrolysis of ATP to ADP is the most common exergonic reaction available to cells.

Energy is transferred from an exergonic reaction to an endergonic reaction by **coupling** the two reactions. When an exergonic reaction and an endergonic reaction occur together linked by a common intermediate, the reactions are said to be coupled. For example, the conversion of glycerol to glycerol-3-phosphate, an endergonic reaction and a step in the construction of cell membranes, is coupled to the exergonic hydrolysis of ATP:

$$\underset{\text{Glycerol}}{C_3H_8O_3} \overset{\text{ATP ADP}}{\underset{}{\rightsquigarrow}} \underset{\text{Glycerol-3-phosphate}}{C_3H_7O_3PO_3} \qquad \Delta G_{\text{overall}} = -5.1\ \text{kcal/mol}$$

The convention of placing a curved arrow above the main reaction arrow indicates that two reactions are coupled. The overall change in free energy, $\Delta G_{\text{overall}}$, for the coupled reaction in this example is negative (-5.1 kcal/mol), so the overall reaction is spontaneous and will proceed as written on its own.

Consider the free energy change of the coupled reactions separately. (1) The reaction of glycerol to glycerol-3-phosphate:

$$\underset{\text{Glycerol}}{C_3H_8O_3} + P_i \longrightarrow \underset{\text{Glycerol-3-phosphate}}{C_3H_7O_3PO_3} \qquad \Delta G_{\text{glycerol}} = +2.2\ \text{kcal/mol}$$

The change in free energy, $\Delta G_{glycerol}$, is positive and therefore the reaction is nonspontaneous. (2) The hydrolysis of ATP has a negative change in free energy, $\Delta G_{ATP} < 0$, and therefore the reaction is exergonic and spontaneous.

$$\underset{\substack{\text{Adenosine} \\ \text{triphosphate}}}{\text{ATP}} \longrightarrow \underset{\substack{\text{Adenosine} \\ \text{diphosphate}}}{\text{ADP}} + \underset{\substack{\text{Inorganic} \\ \text{phospate ion}}}{P_i} \qquad \Delta G_{ATP} = -7.3 \text{ kcal/mol}$$

The free energy change for a coupled reaction is the sum of the free energy changes of the individual reactions:

Reaction	ΔG
$P_i + C_3H_8O_3 \longrightarrow C_3H_7O_3PO_3$	$+\ 2.2$ kcal/mol
$H_2O + ATP \longrightarrow ADP + P_i$	$-\ 7.3$ kcal/mol

Net:
$$C_3H_8O_3 \overset{\text{ATP} \quad \text{ADP}}{\underset{}{\rightsquigarrow}} C_3H_7O_3PO_3 \qquad -\ 5.1 \text{ kcal/mol}$$

The coupled reaction is spontaneous because the sum of the ΔG values for the separate reactions equals a negative change in free energy, $\Delta G_{overall}$. Note, P_i is a reactant in the first reaction and a product in the second; therefore, it cancels and does not appear in the net reaction.

Once ATP is hydrolyzed to ADP, energy is required to convert ADP back into ATP. In the broader sense, catabolic pathways are exergonic overall and anabolic pathways are endergonic. Catabolic processes provide the energy for anabolic processes.

WORKED EXERCISE 14-6   Coupled Reactions

Consider the following reaction:

$$\underset{\substack{\text{Inorganic} \\ \text{phosphate}}}{P_i} + \underset{\text{Glucose}}{C_6H_{12}O_6} \longrightarrow \underset{\substack{\text{Glucose-6-} \\ \text{phosphate}}}{C_6H_{12}PO_9} + H_2O \qquad \Delta G = +3.3 \text{ kcal/mol}$$

The change in free energy, ΔG, is positive, and hence the reaction is nonspontaneous. If it were possible to couple this reaction to the conversion of ATP to ADP, would this reaction be spontaneous? Predict the value of $\Delta G_{overall}$.

SOLUTION

Yes. If coupled to the conversion of ATP to ADP, this reaction would be spontaneous because the sum of the ΔG values for each separate reaction equals a change in free energy, ΔG, less than zero ($\Delta G < 0$).

Reaction	ΔG, kcal/mol
$C_6H_{12}O_6 + P_i \longrightarrow C_6H_{12}PO_9 + H_2O$	$+3.3$
$ATP + H_2O \longrightarrow ADP + P_i$	-7.3
$ATP + C_6H_{12}O_6 \longrightarrow C_6H_{12}PO_9 + ADP$	-4.0

PRACTICE EXERCISE

14.22 Recall from Chapter 12 that the third step of glycolysis is the phosphorylation of fructose-6-phosphate to fructose 1,6-bisphosphate. This reaction is coupled to the hydrolysis of ATP, as indicated by the curved arrow.

Fructose-6-phosphate   Fructose-1,6-bisphosphate

a. If the reaction indicated by the black arrow alone has a $\Delta G = +3.9$ kcal/mol, is this reaction endergonic or exergonic?
b. Is the coupled reaction shown endergonic or exergonic? Show your calculation.
c. Is the reaction shown spontaneous or nonspontaneous?

In Chapter 12 you learned the catabolic reactions that convert polysaccharides into acetyl CoA, and in Chapter 13 you learned how fatty acids undergo β-oxidation to produce acetyl CoA. In this chapter you focused on the catabolic fate of acetyl CoA. The citric acid cycle turns the acetyl group into two CO_2 molecules, and in the process, three NAD^+ and one FAD are converted into NADH and $FADH_2$. These coenzymes transfer their electrons to O_2 through the electron-transport chain, providing the energy for the phosphorylation of ADP to ATP. The hydrolysis of ATP is a spontaneous (exergonic) reaction that can be coupled to a nonspontaneous (endergonic) reaction so that the change in free energy is overall exergonic and the nonspontaneous reaction can proceed. Enthalpy and entropy play an important role in determining the spontaneity of a reaction.

In Chapters 11, 12, and 13 you were introduced to three of the four important types of biomolecule: proteins, carbohydrates, and lipids. In the next chapter you will study the structure and function of the fourth major type of biomolecule: the nucleic acids RNA and DNA. These nucleic acids are involved in replicating and transferring the genetic information necessary to build proteins.

Chemistry in Medicine Phenylketonuria: A Metabolic Disorder

Have you ever wondered why there are warning labels on the back of some artificially sweetened products stating "Phenylketonurics: Contains phenylalanine?" These warning labels are directed at people with phenylketonuria (PKU), but what is PKU and why are the warnings necessary? PKU is a rare condition that affects about 1 in every 10,000 to 20,000 people. It is an inborn error in metabolism—which means that it is an inherited disease, and therefore, a person is born with it. In the United States, babies are screened for phenylketonuria within three days of birth. Symptoms of PKU include vomiting, irritability, a sweet odor to the urine, and a skin rash. If left untreated, PKU leads to mental retardation and regular seizures. By following a diet low in the amino acid phenylalanine, however, PKU can be managed. Until the 1930s, PKU had not been identified, and those afflicted with the condition were destined to become mentally incapacitated and remain so for the rest of their lives.

PKU is caused by a deficiency in the enzyme *phenylalanine hydroxylase (PAH)*. The disease results when a child inherits the recessive PKU gene from both parents. Thus, both parents must be carriers of the gene or have PKU themselves.

Many common diet products contain aspartame, an ingredient that phenylketonurics should not ingest. [© 1995 Michael Dalton/ Fundamental Photographs]

The Metabolic Consequences of PKU

The enzyme *phenylalanine hydroxylase (PAH)* catalyzes the conversion of phenylalanine (Phe) into the related amino acid, tyrosine (Tyr). When a person is deficient in *PAH*, tyrosine can no longer be synthesized and phenylalanine accumulates in the blood. Thus, high levels of phenylalanine are an indicator of PKU. A normal blood phenylalanine level is about 1 mg/dL. In people with PKU, phenylalanine levels range from 6 mg/dL to 80 mg/dL.

Phenylalanine normally has only two anabolic fates: incorporation into polypeptide chains, and conversion into tyrosine. However, when phenylalanine is in excess it is converted into a number of other products.

Recall from Chapter 11 that phenylalanine is a large neutral amino acid. Large neutral amino acids compete with one another for transport across the blood–brain barrier via a membrane transport protein. Excessive levels of phenylalanine in the blood saturate this transport protein, preventing other large neutral amino acids—such as valine, leucine, isoleucine, tryptophan, and tyrosine—from entering the brain. These other large neutral amino acids are vital for protein and neurotransmitter synthesis in the brain. This is the reason why phenylalanine accumulation disrupts brain development in the infant.

The label on a can of diet soda warns that it contains aspartame (Nutrasweet), which should not be consumed by phenylketonurics. Why is aspartame a problem for people with PKU? Aspartame is a dipeptide made from the two amino acids phenylalanine and aspartic acid. When aspartame is hydrolyzed in the stomach, aspartic acid, phenylalanine, and methanol are produced (Figure 14-16). It is the phenylalanine produced that causes a problem for phenylketonurics.

Treating PKU

If diagnosed at birth, phenylketonuria can be managed through a strict diet regimen, which must be followed for life. The goal is to maintain blood phenylalanine levels between 2 and 10 mg/dL. This is accomplished by severely restricting, or altogether eliminating, foods that contain phenylalanine. Such foods include beef, chicken, fish, and dairy products. In addition, any diet foods that contain aspartame must be avoided. People with PKU must also take tyrosine supplements, because they cannot synthesize this amino acid.

Figure 14-6 Hydrolysis of aspartame

Chapter Summary

Acetyl Coenzyme A and the Citric Acid Cycle

- Cellular respiration generates energy by phosphorylating ADP to make ATP through combustion of the biomolecules obtained from the food we eat.
- Acetyl coenzyme A is a central molecule in both anabolic and catabolic pathways.
- One of the main functions of coenzyme A is to transfer acyl groups in acyl transfer reactions.
- The citric acid cycle is an eight-step biochemical pathway that begins with the reaction between acetyl CoA and oxaloacetate to make citrate.
- Most of the energy released in the citric acid cycle is in the form of electrons carried by three NADH molecules and one $FADH_2$ molecule. The electrons eventually enter the electron-transport chain, where they are used to drive the phosphorylation of ADP to ATP.
- One round of the citric acid cycle eventually phosphorylates 12 ADP to 12 ATP.
- The first half of the citric acid cycle generates citric acid and then expels two CO_2 molecules, producing succinyl CoA. The second half of the citric acid cycle regenerates oxaloacetate from succinyl CoA.

Oxidative Phosphorylation

- Oxidative phosphorylation ulitizes oxygen and the electrons from NADH and $FADH_2$ to phosphorylate ADP, making ATP.
- Important biochemical pathways such as β-oxidation of fatty acids, the citric acid cycle, and oxidative phosphorylation occur in the mitochondria.
- The key parts of a mitochondrion are the outer membrane, the intermembrane space, the inner membrane, and the matrix.
- The electrons carried by the coenzymes NADH and $FADH_2$ are relayed to oxygen through the electron-transport chain.
- The electron-transport chain consists of four large membrane-bound proteins (Complexes I through IV) and two mobile electron carriers, coenzyme Q and cytochrome *c*.
- Electrons pass through the electron-transport chain through a series of oxidation–reduction reactions. Electrons are transferred from atom centers and molecules with lower electron affinity to those with higher electron affinity, with oxygen serving as the final repository of electrons.
- As electrons pass through Complexes I, III, and IV, protons are simultaneously pumped from the matrix to the intermembrane space, creating a proton gradient.
- *ATP synthase* is a multienzyme protein complex that catalyzes the phosphorylation of ADP, using the energy of the proton-motive force.

Entropy and Bioenergetics

- Bioenergetics is the study of energy transfer in reactions occurring in living cells.
- A reaction is spontaneous if it continues on its own once started. A reaction is nonspontaneous if energy must be continuously supplied for the reaction to proceed.
- The spontaneity of a reaction can be determined from the change in free energy for the reaction (ΔG). The change in free energy for the reaction depends on the change in enthalpy (ΔH), the change in entropy (ΔS), and the temperature.
- The change in entropy, ΔS, is a measure of the change in the degree of randomness or disorder in a reaction or physical change.

- Entropy increases if a process becomes more disordered; $\Delta S > 0$.
- The second law of thermodynamics states that the entropy of the universe is always increasing.
- All chemical reactions are inclined to attain the most stable bonding state, represented by enthalpy, and the highest state of disorder, represented by entropy.
- An exergonic reaction is spontaneous and has a negative ΔG. An endergonic reaction is nonspontaneous and has a positive ΔG.
- The sign of ΔH and ΔS together along with temperature determine the sign of ΔG.
- Energy transferred from an exergonic reaction to an endergonic reaction in the coupling of the two reactions allows the endergonic reaction to proceed.

Key Words

Acetyl CoA A central molecule in metabolism used to transfer acyl groups in acyl transfer reactions. Contains a thioester functional group.

Bioenergetics The study of energy transfer in reactions occurring in living cells.

Cellular respiration The combustion of carbohydrates, proteins, and triglycerides to produce energy in the form of ATP. Requires oxygen.

Change in entropy, ΔS A measure of the change in the degree of randomness or disorder in a reaction or physical change.

Citric acid cycle An important catabolic cycle that occurs in the matrix of the mitochondria. Acetyl CoA is converted into two molecules of CO^2 and coenzyme A. Three molecules of NADH and one $FADH_2$ are formed from three NAD^+ and one FAD.

Complexes I–IV Large enzyme complexes in the inner membrane of the mitochondrion where the oxidation–reduction reactions of the electron-transport chain occur.

Coupling reactions When an endergonic reaction occurs together with an exergonic reaction so that both reactions will be spontaneous.

Decarboxylation reaction A reaction in which carbon dioxide is expelled from the reactant.

Electron-transport chain A series of protein complexes that successively carry electrons from NADH and $FADH_2$ to oxygen in the last stage of catabolism. As electrons move through the electron-transport chain, protons are pumped from the matrix to the intermembrane space, creating a proton gradient.

Endergonic reaction A nonspontaneous reaction with a positive value for the change in free energy, $\Delta G > 0$.

Exergonic reaction A spontaneous reaction with a negative value for the change in free energy, $\Delta G < 0$.

Free energy ΔG, the energy available to do work. $\Delta G = \Delta H - (T\Delta S)$. A positive free energy is indicative of a nonspontaneous reaction, and a negative free energy is indicative of a spontaneous reaction.

Inner mitochondrial membrane A selectively permeable membrane that separates the matrix from the intermembrane space in a mitochondrion.

Intermembrane space The aqueous region of a mitochondrion located between the outer membrane and the inner mitochondrial membrane. It has a higher proton concentration than the matrix.

Matrix The aqueous region in the center of a mitochondrion, bounded by the inner mitochondrial membrane. It is the region where the citric acid cycle, β-oxidation, and oxidative phosphorylation occur.

Mitochondria Organelles within the cell where energy is produced: β-oxidation, the citric acid cycle, and oxidative phosphorylation.

Nonspontaneous reaction A reaction or physical change that requires the continual supply of energy in order to proceed; $\Delta G > 0$.

Outer membrane The outer semipermeable membrane of a mitochondrion—it encloses and defines the mitochondrion.

Oxidative phosphorylation The final stage of catabolism wherein the electrons from NADH and $FADH_2$ move through the electron-transport chain, while simultaneously pumping protons across the inner membrane to the intermembrane space, creating a proton gradient. *ATP synthase* allows protons to flow back into the matrix, releasing the energy to drive phosphorylation of ADP to ATP.

Proton gradient A difference in proton concentration between regions. In a mitochondrion there is a proton gradient between the intermembrane space and the matrix.

Proton-motive force The potential energy produced as a result of a proton gradient established between the matrix and the intermembrane space that is used to drive the phosphorylation of ADP to ATP.

Spontaneous reaction A reaction that continues on its own once started, without the continuous supply of energy; $\Delta G < 0$.

Additional Exercises

Acetyl Coenzyme A and the Citric Acid Cycle

14.23 What is the difference between a catabolic reaction and an anabolic reaction?

14.24 What is cellular respiration?

14.25 What catabolic biochemical pathways produce acetyl CoA?

14.26 The structure of acetyl coenzyme A is shown below. Identify the thioester, amides, and phosphate ester groups within the molecule. Which part of the molecule represents the "acetyl" group?

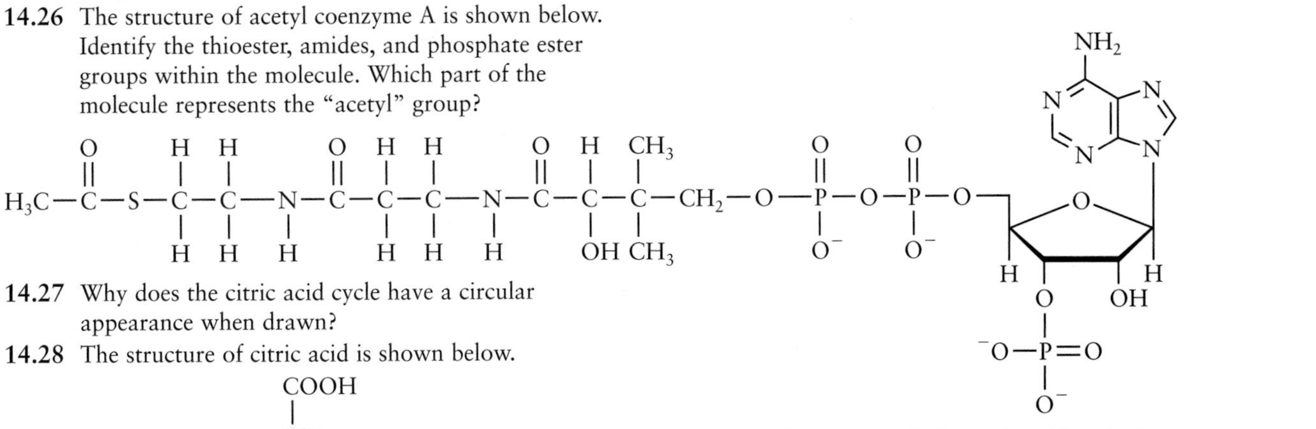

14.27 Why does the citric acid cycle have a circular appearance when drawn?

14.28 The structure of citric acid is shown below.

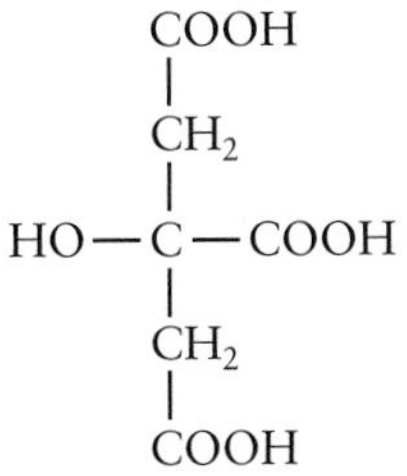

a. What structure would this compound have at physiological pH?
b. How many carbon atoms does citric acid contain?

14.29 The structure of isocitrate is shown below.

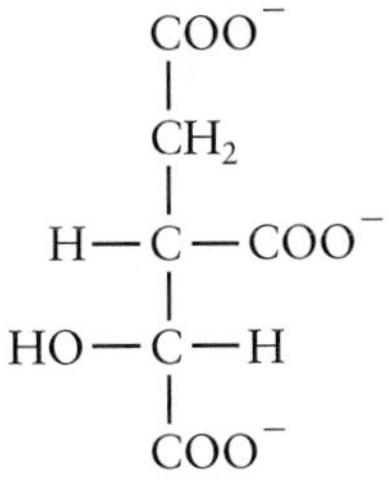

a. What structure would this compound have at a low pH? That is, show this molecule in its un-ionized form.
b. How is isocitrate similar to citrate? How is it different?
c. What type of isomers are represented by citrate and isocitrate?

14.30 The structure of α-ketoglutarate is shown below.

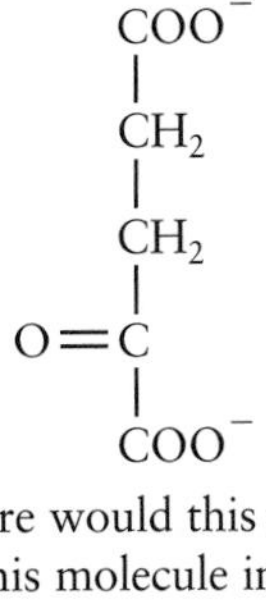

a. What structure would this compound have at low pH (i.e., show this molecule in its un-ionized form)?
b. What other functional group is present in this molecule?

14.31 Halfway through the citric acid cycle, how many carbon atoms have been removed from citrate? What molecules are expelled during the first half of the citric acid cycle in order to remove two carbon atoms from citrate? How many molecules of NADH are produced in the first half of the citric acid cycle?

14.32 How many ATP molecules are produced in one round of the citric acid cycle?

14.33 The citric acid cycle produces ____ NADH molecules. How many ATP molecules are produced for every NADH molecule?

14.34 How many ATP molecules are produced from one molecule of $FADH_2$?

14.35 What two steps of the citric acid cycle produce the CO_2 that we exhale? Where do the carbon atoms come from that are exhaled as CO_2?

14.36 The citric acid cycle produces energy in the form of NADH and $FADH_2$. Explain how this energy is used.

Extension Exercises

14.37 In Step 1 of the citric acid cycle, what kind of bond is formed? What two molecules react to form this bond?

14.38 Step 2 of the citric acid cycle is shown below.

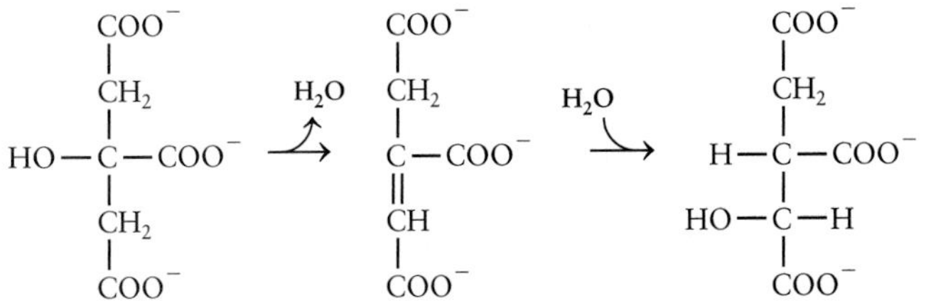

a. What type of reactions are occurring? How are these two reactions related?
b. The net reaction from citrate to isocitrate is classified as an isomerization reaction. Explain why this step is called an isomerization.

14.39 Step 6 of the citric acid cycle is shown below.

$$
\begin{array}{ccc}
\text{COO}^- & \xrightarrow{\text{FAD} \quad \text{FADH}_2} & \text{COO}^- \\
| & & | \\
\text{CH}_2 & & \text{CH} \\
| & & \| \\
\text{CH}_2 & & \text{CH} \\
| & & | \\
\text{COO}^- & & \text{COO}^- \\
\text{Succinate} & & \text{Fumarate}
\end{array}
$$

   a. What type of reaction occurs in this step?
   b. Is succinate oxidized or reduced?
   c. Is FAD oxidized or reduced?
   d. What functional group change occurs?

14.40 Step 7 of the citric acid cycle is shown below.

$$
\begin{array}{ccc}
\text{COO}^- & \xrightarrow{\text{H}_2\text{O}} & \text{COO}^- \\
| & & | \\
\text{CH} & & \text{HO}-\text{C}-\text{H} \\
\| & & | \\
\text{CH} & & \text{CH}_2 \\
| & & | \\
\text{COO}^- & & \text{COO}^- \\
\text{Fumarate} & & \text{L-Malate}
\end{array}
$$

   a. What type of reaction occurs in this step?
   b. **Critical Thinking Question:** When more than one possible product can be formed, enzymes control the outcome. What product did the enzyme prevent from being formed, that could have been produced?

14.41 Step 8, the final step of the citric acid cycle is shown below.

$$
\begin{array}{ccc}
\text{COO}^- & \xrightarrow{\text{NAD}^+ \quad \text{NADH} + \text{H}^+} & \text{COO}^- \\
| & & | \\
\text{HO}-\text{C}-\text{H} & & \text{O}=\text{C} \\
| & & | \\
\text{CH}_2 & & \text{CH}_2 \\
| & & | \\
\text{COO}^- & & \text{COO}^- \\
\text{L-Malate} & & \text{Oxaloacetate}
\end{array}
$$

   a. What type of reaction occurs in this step?
   b. What functional group changes occur?
   c. Is L-malate oxidized or reduced?
   d. What is one fate of the oxaloacetate produced in this step?

Oxidative Phosphorylation

14.42 Where in the cell do most of the energy-producing biochemical pathways occur?

14.43 What catabolic pathway occurs in the cytoplasm?

14.44 The inner membrane of the mitochondria is a selectively permeable membrane. Can ions easily pass through this membrane?

14.45 The outer membrane of the mitochondria is a semipermeable membrane. Can ions easily pass through this membrane?

14.46 Where in the mitochondrion is there a proton gradient? Provide an illustration labeling the relevant parts of the mitochondrion. Where is the concentration of protons lower?

14.47 Which has a higher pH, the intermembrane space or the matrix of a mitochondrion?

14.48 In protein Complex I of the electron-transport chain, protons are pumped through the inner mitochondrial membrane via an ion channel. Explain why this process needs energy to occur. What is the source of the energy for this process?

14.49 What is the proton-motive force?

14.50 Is there a charge difference between the matrix and the intermembrane space?

14.51 Why is cyanide lethal?

14.52 Which of the following proteins in the electron-transport chain are mobile electron carriers: coenzyme Q, Complex I, Complex II, cytochrome *c*, Complex IV?

14.53 The phosphorylation of ADP to form ATP requires energy. Where does *ATP synthase* obtain the energy needed to phosphorylate ADP?

14.54 Explain why energy is transferred when protons flow from the intermembrane space back into the matrix. What is this energy used for?

Entropy and Bioenergetics

14.55 Explain the difference between a spontaneous reaction and a nonspontaneous reaction.

14.56 Which of the following processes are spontaneous?
   a. pushing a ball uphill
   b. rolling a ball downhill
   c. building a house of cards
   d. knocking down a house of cards

14.57 Which of the following processes are nonspontaneous?
   a. propane burning in air
   b. a plant growing in sunlight
   c. swinging on a swing
   d. water evaporating from a puddle in summer

14.58 What is entropy?

14.59 Which of the following processes represent an increase in entropy?
   a. cleaning up a messy room
   b. a reaction that has more product molecules than reactant molecules (all other things being equal)
   c. a reaction that goes from simple reactant molecules to a complex product molecule
   d. snow melting
   e. liquid water becoming steam

14.60 Which of the following processes or reactions represent a decrease in entropy?
   a. breath condensing in cold air
   b. $2 \text{ SO}_2(g) + \text{O}_2(g) \rightarrow 2 \text{ SO}_3(g)$
   c. $\text{H}_2\text{O}(l) \rightarrow \text{H}_2\text{O}(g)$
   d. dropping a box of match sticks
   e. $\text{C}_6\text{H}_{12}\text{O}_6 \text{ (glucose)} + 6 \text{ O}_2 \rightarrow 6 \text{ CO}_2 + 6 \text{ H}_2\text{O}$

14.61 Consider the process of a substrate binding to an enzyme to form an enzyme–substrate complex. Do you think the process of a substrate binding to an enzyme represents an increase or decrease in entropy? Explain from the perspective of the enzyme and the substrate. What is the effect on the surroundings? How has the entropy of the universe changed?

14.62 When a reaction is spontaneous, is the change in free energy, ΔG, positive or negative? When a reaction is nonspontaneous, is the change in free energy, ΔG, positive or negative?

14.63 Are the following reactions spontaneous or nonspontaneous?
a. Lactose + H_2O → glucose + galactose
$\Delta G = -3.8$ kcal/mol
b. ADP + P_i → ATP + H_2O $\Delta G = +7.3$ kcal/mol
c. H_2CO_3 → CO_2 + H_2O $\Delta G = -3.8$ kcal/mol
d. Glucose 1-phosphate → glucose 6-phosphate
$\Delta G = -1.74$ kcal/mol

14.64 Are the following reactions spontaneous or nonspontaneous?
a. Glucose 6-phosphate + H_2O → glucose + P_i
$\Delta G = -3.3$ kcal/mol
b. Palmitic acid + 23 O_2 → 16 CO_2 + 16 H_2O
$\Delta G = -2338$ kcal/mol
c. Glucose 6-phosphate → fructose 6-phosphate
$\Delta G = 0.4$ kcal/mol

14.65 In Step 3 of the citric acid cycle, isocitrate is converted to α-ketoglutarate. The free energy, ΔG, for the reaction is -5.0 kcal/mol. Is this step spontaneous or nonspontaneous?

14.66 In Step 1 of the citric acid cycle, acetyl CoA reacts with oxaloacetate to form citrate. The free energy for the reaction, ΔG, is -7.7 kcal/mol. Is this step spontaneous or nonspontaneous?

14.67 Are the following reactions spontaneous or nonspontaneous?
a. 2 $CH_3OH(g)$ + 3 $O_2(g)$ → 2 $CO_2(g)$ + 4 $H_2O(g)$
$\Delta H < 0$
b. 3 O_2 → 2 O_3 $\Delta H > 0$; $\Delta S < 0$
c. 2 $H_2O_2(l)$ → 2 $H_2O(l)$ + $O_2(g)$
$\Delta H < 0$; $\Delta S > 0$

14.68 Fill in the blanks in the following table.

Sign of ΔH	Sign of ΔS	Sign of ΔG	Comment
−	+		Spontaneous
+	−	+	
	−	?	More information needed to determine ΔG
+		?	More information needed to determine ΔG

14.69 How does temperature affect ΔG?

14.70 Some of the reactions in catabolism are endergonic. Explain how these reactions get the energy they need to proceed.

14.71 Consider the following reaction:

CH$_2$OH
=O
HO—H
H—OH
H—OH
CH$_2$OH

ATP ADP

CH$_2$OH
=O
HO—H
H—OH
H—OH
CH$_2$O—(P)

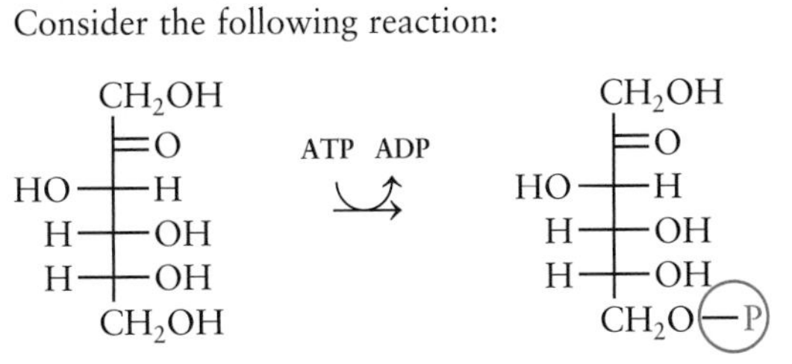

Fructose   Fructose-6-phosphate

a. If $\Delta G = +3.8$ kcal/mol for the phosphorylation of fructose (→), is the reaction endergonic or exergonic?
b. Is the overall coupled reaction endergonic or exergonic?
c. What is $\Delta G_{overall}$?

14.72 In muscle tissue, two reactions are coupled together to produce ATP. The coupled reaction is shown below.

ADP   ATP

Phosphocreatine ⤳→ creatine

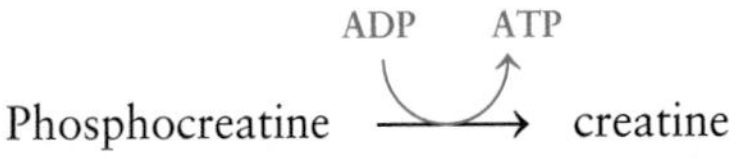

a. If $\Delta G = -10.3$ kcal/mol for the formation of creatine (black arrow), is this reaction endergonic or exergonic?
b. Is the overall coupled reaction endergonic or exergonic?
c. What is $\Delta G_{overall}$?
d. Which reaction provided the energy to drive the coupled reaction?

14.73 In the last step of glycolysis, phosphoenolpyruvate is converted to pyruvate. This reaction is coupled to the phosphorylation of ADP to form ATP. The coupled reaction is shown below.

ADP   ATP

Phosphoenolpyruvate ⤳→ pyruvate

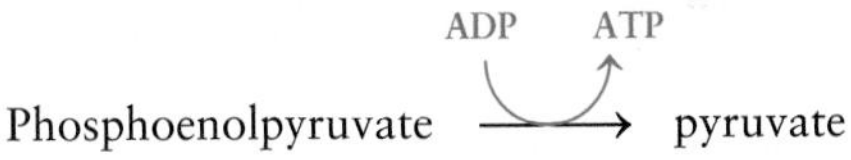

a. If $\Delta G = -14.8$ kcal/mol for the formation of pyruvate (black arrow), is this reaction endergonic or exergonic?
b. Is the overall coupled reaction endergonic or exergonic?
c. What is $\Delta G_{overall}$?
d. Which reaction provided the energy to drive the coupled reaction?

Chemistry in Medicine

14.74 What are the symptoms of PKU?

14.75 What happens if PKU is left untreated?

14.76 What is the function of *phenylalanine hyrdoxylase* in the body?

14.77 What is the normal fate of phenylalanine in a person who does not have PKU?

14.78 How does a high concentration of phenylalanine in the blood affect brain development in an infant?

14.79 **Critical Thinking Question:** One of the possible treatments of PKU uses an injectable form of *PAH*. Why does this enzyme need to be injected instead of taken in a pill form?

14.80 **Critical Thinking Question:** Examine the structure of aspartame. You probably already know that aspartame has a sweet taste, but what makes it a *zero*-calorie sweetener compared to sugar?

Answers to Practice Exercises

14.1 **a.** the first half of the citric acid cycle (Steps 3 and 4)
b. $:\overset{..}{O}=C=\overset{..}{O}:$
c. two carbon dioxide molecules
d. oxaloacetate

14.2 There are six carbon atoms in citrate. Citrate is formed in the first step of the citric acid cycle.

14.3 There are three (tri) carboxylic acid functional groups in citric acid.

14.4 Three molecules of NAD^+ are reduced to NADH.

14.5 One molecule of FAD is reduced to $FADH_2$.

14.6 Twelve molecules of ADP are converted to ATP as a result of one pass through the citric acid cycle.

14.7 β-oxidation and glycolysis followed by oxidation of pyruvate.

14.8 Acetyl CoA carries an acyl group: $CH_3C=O$. Succinyl CoA carries a succinyl group:

$$-OOCCH_2CH_2\overset{\overset{\displaystyle O}{\|}}{C}-.$$

E14.1 **a.** Oxidation–reduction: Isocitrate is oxidized; NAD^+ is reduced.
b. The carboxylate groups on both ends are unchanged. The alcohol is oxidized to a ketone.
c. The electrons from NADH ultimately phosporylate three ADP to three ATP.
d. Carbon dioxide is released.

E14.2 **a.** Two reactions: a dehydration followed by a hydration reaction.
b. $C_6H_5O_7$
c. They are structural isomers (same formula, different connectivity of atoms).
d. In a dehydration reaction, water is released from the substrate; in a hydration reaction water is added to the substrate.
e. Because the OH and H from water add to the double bond in the reverse order from the way they are positioned on citrate.

14.9 **a.** inner membrane
b. the matrix and inner membrane
c. the matrix
d. the matrix
e. the matrix and the intermembrane space
f. outer membrane and inner membrane

14.10 Four multienzyme complexes make up the electron-transport chain. Complex II contains *succinate dehydrogenase*. Succinate dehydrogenase catalyzes the sixth step of the citric acid cycle. The products of this step are fumarate and $FADH_2$.

14.11 $FADH_2 \rightarrow$ Complex $\underline{II} \rightarrow$ coenzyme $\underline{Q} \rightarrow$ Complex $\underline{III} \rightarrow$ cytochrome $\underline{c} \rightarrow$ Complex $\underline{IV} \rightarrow O_2$

14.12 Oxygen is the final resting place of the electrons that go through the electron-transport chain. Without oxygen, NADH and $FADH_2$ would not get oxidized back to NAD^+ and FAD, which would prevent further extraction of energy.

14.13 Cyanide irreversibly binds to and inhibits *cytrochrome c oxidase*, the last enzyme in the electron-transport chain.

14.14 $ADP + P_i \rightarrow ATP$

14.15 Two ADP are phosphorylated to 2 ATP.

14.16 The proton-motive force is the potential energy created by the proton gradient established between the intermembrane space and the matrix as electrons move through the electron-transport chain. When protons flow through a channel created within *ATP synthase*, the flow of protons with the gradient provides the energy to phosphorylate ADP.

14.17 a decrease in entropy, $\Delta S < 0$

14.18 an increase in entropy, $\Delta S > 0$

14.19 Condensation represents a decrease in entropy—the water molecules become more ordered as they go from gas to liquid, therefore, it represents a decrease in entropy. The entropy of the surroundings increases, because heat is released during this change of state—it is exothermic. The increase in entropy of the surroundings must be greater than the decrease in entropy of the physical change, so that the second law of thermodynamics is satisfied.

14.20 Heat is released, therefore $\Delta H < 0$. Since $\Delta S > 0$ and $H < 0$, the reaction is an example of Case I (Table 14-4). Thus, it is spontaneous, $\Delta G < 0$.

14.21 The first three examples (a.–c.) represent spontaneous reactions because $\Delta G < 0$.

14.22 **a.** This reaction is endergonic, because $\Delta G > 0$.
b. Exergonic

Fructose-6-phosphate → fructose-1,6-bisphosphate
ATP → ADP
_____
Fructose-6-phosphate + ATP →
                    fructose-1,6-bisphosphate + ADP
$\Delta G = +3.9$ kcal
$\Delta G = -7.3$ kcal
_____
$\Delta G_{overall} = -3.4$ kcal

(c) The coupled reaction is spontaneous, because $\Delta G_{overall} < 0$.

Nucleic Acids: DNA and RNA

Human cervical cancer cells during cell division. Cancer cells are abnormal, rapidly dividing cells. DNA is fluorescently labeled in blue. [Jennifer C. Waters/Photo Researchers, Inc.]

OUTLINE

This icon indicates that a **Problem-Solving Tutorial** is available at www.whfreeman.com/gob

Gene Mutations and Disease

Most diseases are caused by or are impacted in some way by our genes. Some of these diseases—about 4,000 that we know of—are hereditary diseases caused by the effects of specific gene forms. Others, such as some types of cancer, are caused by environmental factors that alter the chemical structure of genes in a particular cell type—certain compounds (carcinogens), radiation, and viruses, for example, can have that effect. In most cases, however, diseases are caused by a combination of the effects of several genes together with various environmental factors.

Modern medicine has the technology to test for the presence of altered genes that are known to predispose an individual to certain diseases. For example, the **br**east **ca**ncer genes BRCA 1 and 2 discovered in 1994 are believed to be indicators for a predisposition to developing breast cancer. A woman testing positive for these genes increases her odds to 85% that she will develop breast cancer before the age of 65. What are *genes* and why do they play such a prominent role in disease?

A gene is a section of DNA that contains the instructions for making a protein. DNA is a huge biomolecule known as a **nucleic acid.** The term *nucleic* is used because DNA is found in the *nucleus* of cells, and *acid* because it is derived from phosphoric *acid* (H_3PO_4). Genes are sections of DNA that specify the amino acid sequence for a particular protein. Your DNA contains the genes for making all the proteins you will ever require in a lifetime. Thus, your DNA determines every one of your physical traits from the color of your eyes to your basal metabolism. And now we are learning that DNA can even determine whether you have a predisposition toward developing certain diseases.

Every cell in your body contains the same DNA, subtly different from the DNA of every other individual. Indeed, forensic scientists can determine who was at the scene of a crime by analyzing any bodily fluid, hair, nail, or tissue left behind, since all are composed of cells containing the criminal's DNA fingerprint.

What is the connection between genes and disease? Genes that predispose a person to disease often contain *mutations. A mutation is a permanent alteration in the chemical structure of a gene, which in turn may produce an altered protein.* Mutations may be either acquired or inherited. *Inherited mutations* are genes that you are born with that you inherited from your mother or your father. *Acquired mutations* are alterations made to your genes sometime during your lifetime. Acquired mutations affect only one type of tissue and these mutations will not exist in the DNA of other cell types. Cervical cancer, for example, often arises from an acquired mutation in a single cell.

Various environmental triggers can cause acquired mutations. For example, some types of cervical cancer are believed to be associated with a persistent infection by the *human papillomavirus (HPV)*, one of the most common sexually transmitted diseases. Some strains of this virus can cause genes in cervical cells to mutate. The virus triggers mutations in genes that code for proteins involved in controlling cell replication, causing unrestrained growth of cervical cells. These abnormal cells (see the chapter opening photo) may develop into cervical cancer, if subsequent changes, influenced by yet other genes, also occur.

As part of an international effort known as the *Human Genome Project*, hundreds of disease-linked genes have been identified and associated with a specific chromosome. In this chapter you will learn how the cell uses the chemical information encoded within the genes of your DNA to build proteins. The process requires RNA, the other important nucleic acid in the cell. You will then be able to better understand how mutations interfere with the normal flow of information from DNA to RNA to protein synthesis.

More than 24 million women in the United States and 600 million worldwide are infected with the HPV virus.

"The ability to store and transmit genetic information from one generation to the next is a fundamental condition for life." —Albert Lehninger

The molecules responsible for storing and transmitting genetic information are the nucleic acids **d**eoxyribo**n**ucleic **a**cid (DNA) and **r**ibo**n**ucleic **a**cid (RNA). All cells, except for mature red blood cells, contain DNA and RNA in their nuclei (hence the term *nucleic* in *nucleic* acid). Within the chemical structure of DNA are all the instructions the cell needs to produce all of the proteins your body will ever require in a lifetime. These proteins include all the enzymes required for the myriad of biochemical reactions that take place in the cell. Indeed, it is ultimately chemistry that defines who you are—at least in terms of your biological traits!

The function of DNA is to store information; while the role of RNA is to transfer information from DNA to the ribosomes, the protein-making factories of the cell. Within the ribosomes, RNA directs the assembly of the sequences of amino acids that define the primary structure of proteins, according to instructions encoded within the chemical structure of DNA.

As with the study of the other biomolecules, you will begin your study of nucleic acids by learning the chemical structure of the smaller monomer units from which these biomolecules are built, molecules called **nucleotides.** You will learn how DNA replicates itself so that the genetic information it contains can be passed on to daughter cells. Then, you will learn how information flows from DNA to RNA to construct a protein from its individual amino acids.

Red blood cells are produced from stem cells in the bone marrow. When red blood cells mature, they discard their DNA to make room for more hemoglobin. Thus, they have no nucleus, no DNA, and do not divide. They produce energy through glycolysis, and therefore, do not have mitochondria.

15.1 | Nucleotides and Nucleic Acids

Nucleic acids are constructed from nucleotides, much as proteins are constructed from amino acids. However, there are 20 different amino acids from which to build a protein, and only four different nucleotides from which to build a nucleic acid.

A nucleotide contains three basic parts:

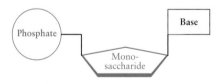

- a nitrogen-containing ring called a base,
- a monosaccharide, and
- a phosphate group.

The nitrogen base and the monosaccharide together are known as a **nucleo*side.*** A nucleoside together with the phosphate group is called a **nucleo*tide.*** The difference is in the suffix: *-side* versus *-tide.*

> **Nucleo**side = base + monosaccharide
>
> **Nucleo**tide = base + monosaccharide + phosphate

Figure 15-1 Chemical structure of D-ribose (RNA) and 2-deoxy-D-ribose (DNA).

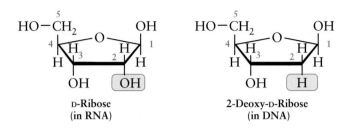

The Monosaccharide Component

The monosaccharide component of a nucleotide is D-ribose when it is part of RNA, and the related sugar, 2-deoxy-D-ribose, when it is part of DNA. These two monosaccharides differ only at carbon-2, where 2-deoxyribose lacks an OH group, as shown in Figure 15-1.

D-Ribose is a monosaccharide with five carbon atoms. In a nucleoside, ribose exists in the furanose form—a five-membered ring. As expected, it is a D-sugar, the enantiomeric series produced in nature. In nucleosides, the carbons in the monosaccharide ring are numbered, beginning at the anomeric carbon, from 1′ (pronounced "one prime") to 5′ proceeding clockwise around the ring. The prime symbol following the number is used to distinguish the carbon atoms in the monosaccharide ring (1′, 2′, 3′, …) from the carbon atoms on the nitrogen-containing ring (1, 2, 3, …).

The Nitrogen Base

The nitrogen bases found in the nucleotides that make up DNA and RNA are derived from two basic types of heterocyclic aromatic rings, known as **purines** and **pyrimidines,** as shown in Figure 15-2. DNA contains the four bases adenine, guanine, cytosine, and thymine, which are abbreviated A, G, C, and T, respectively.

RNA contains three of the same nitrogen bases as DNA, but thymine is replaced by uracil (U), as summarized in Figure 15-3. Thymine and uracil have very similar structures, differing only in the presence of a CH₃ group on thymine where there is an H atom in uracil.

In a nucleoside, the anomeric carbon atom—the 1′ carbon atom—on the monosaccharide is covalently bonded to one of these four possible nitrogen

DNA Bases

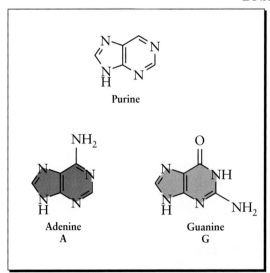

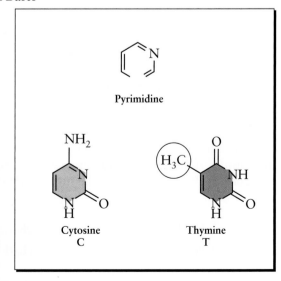

Figure 15-2 The nitrogen bases in DNA are derivatives of purine and pyrimidine.

RNA Bases

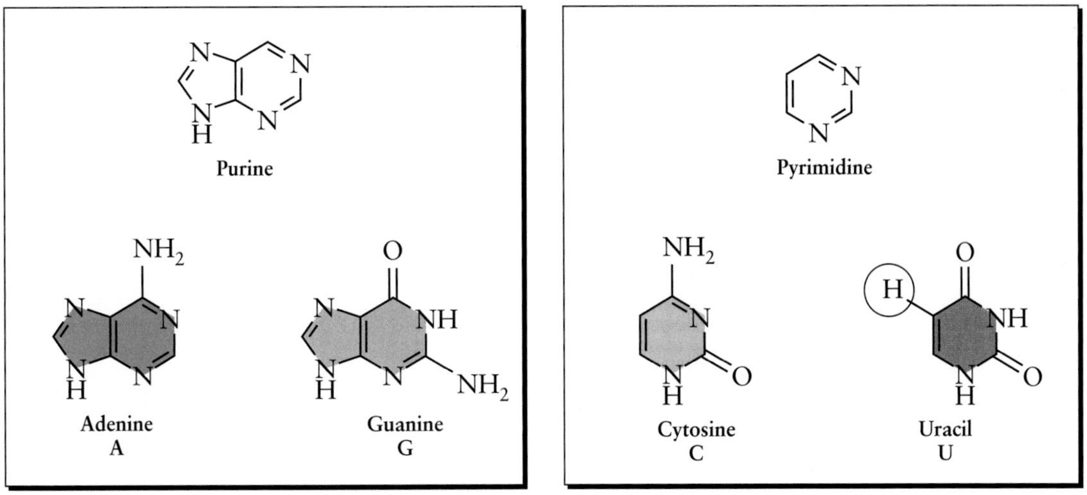

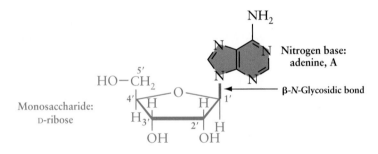

Purine

Adenine
A

Guanine
G

Pyrimidine

Cytosine
C

Uracil
U

Figure 15-3 The nitrogen bases in RNA are the same as those in DNA, except that thymine is replaced by uracil.

bases at one of the nitrogen atoms. The covalent bond between the monosaccharide and the base is called a β-*N*-glycosidic bond, because it lies above the monosaccharide ring (β), and because it is bonded to a nitrogen atom (*N*-) rather than an oxygen atom as shown in Figure 15-4.

Nitrogen base:
adenine, A

β-*N*-Glycosidic bond

Monosaccharide:
D-ribose

Figure 15-4 Chemical structure of the RNA nucleoside adenosine.

The Phosphate Group

To create a nucleo**tide** from a nucleoside, a phosphate group is attached to the 5′ alcohol of the monosaccharide (a phosphate ester functional group). Nucleotides are therefore derivatives of phosphoric acid. Figure 15-5 shows the structure of one of the four RNA nucleotides: adenosine 5′-monophosphate (AMP).

You have already seen some nucleotides in previous chapters that are not part of nucleic acids: ATP, FAD, and NAD⁺. These important molecules, described in Chapters 10 and 14, all contain the nucleotide adenosine 5′-monophosphate (AMP) at the core of their structure.

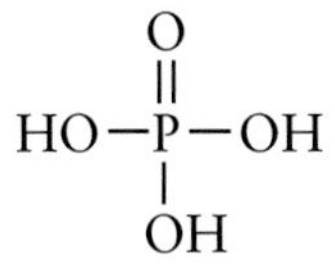

Phosphoric acid

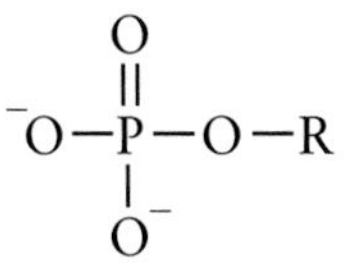

Phosphate ester
(at physiological pH)

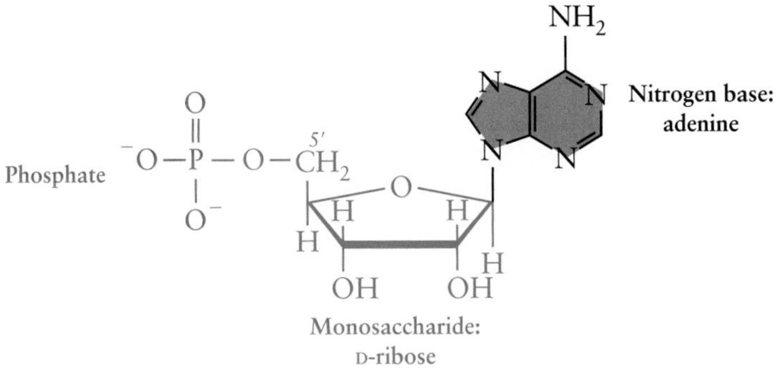

Phosphate

Nitrogen base:
adenine

Monosaccharide:
D-ribose

Figure 15-5 Chemical structure of an RNA nucleotide, adenosine 5′-monophosphate (AMP).

 **Problem-Solving Tutorial:**
Nucleotide Structure

WORKED EXERCISES 15-1 Nucleotides

1 An amino acid is to a peptide as _____ is to a nucleic acid.

2 Which of the following monosaccharides is found in RNA?

a. 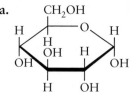 b. 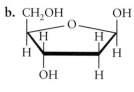 c.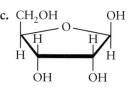

3 For the nucleotide shown below, circle and label the following:

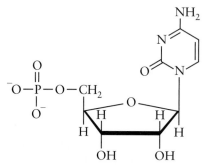

a. The nitrogen base; what is the name of this base?
b. The monosaccharide; what is the name of this monosaccharide?
c. The phosphate group.
d. The 5′ carbon.
e. The anomeric carbon. What is the carbon number for the anomeric carbon?
f. The 3′ carbon.
g. Would this nucleotide be found in DNA or RNA? Explain.

4 Name the four nucleotide bases found in DNA and give their one-letter abbreviations.

5 Which nitrogen base is not found in RNA?

SOLUTIONS

1 An amino acid is to a peptide as <u>a nucleotide</u> is to a nucleic acid.

2 (c)

3 a.–f.

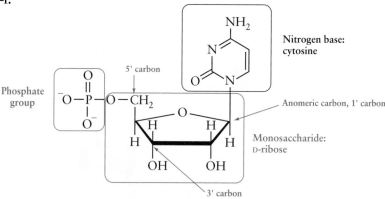

g. This nucleotide would be found in RNA, because the monosaccharide is D-ribose.

4 Adenine, A; guanine, G; cytosine, C; thymine, T.

5 Thymine is not found in RNA.

PRACTICE EXERCISES

15.1 Which of the following monosaccharides is found in DNA?

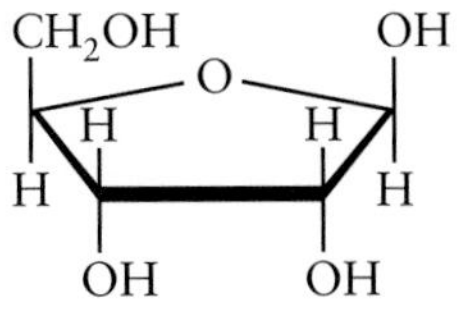

15.2 For the nucleotide shown below, circle and label the following:

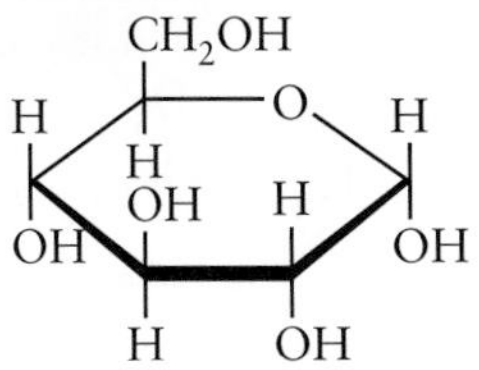

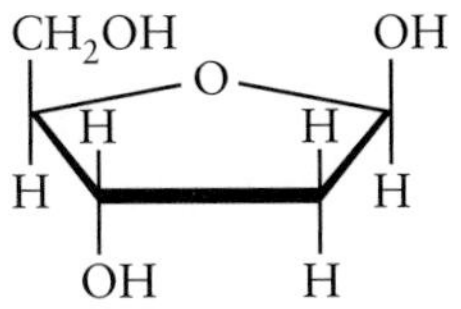

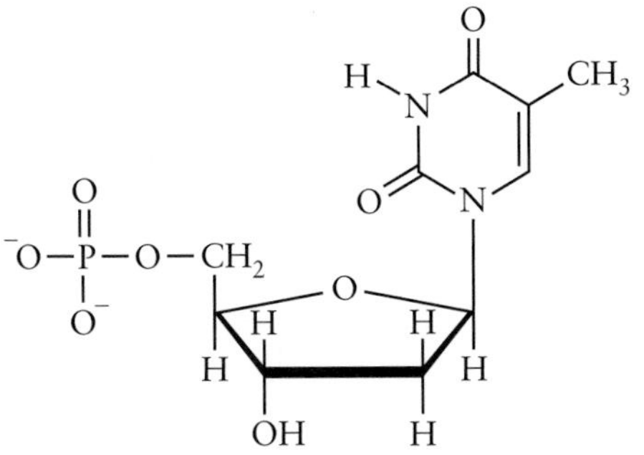

    a. The nitrogen base; what is the name of this base?
    b. The monosaccharide; what is the name of the monosaccharide?
    c. The phosphate group.
    d. The 5′ carbon.
    e. The anomeric carbon.
    f. The 3′ carbon.
    g. Would this nucleotide be found in DNA or RNA?

15.3 Name the four nucleotide bases found in RNA and give their one-letter abbreviations.

15.4 Which nitrogen base is not found in DNA?

15.5 The three basic parts of a nucleotide are given below.

    a. What part of a nucleotide is each of these molecules called?
    b. Construct a nucleotide from these three parts.
    c. Label the 3′ and the 5′ carbon atoms in your nucleotide.
    d. Would this nucleotide be found in RNA or DNA? How can you tell?

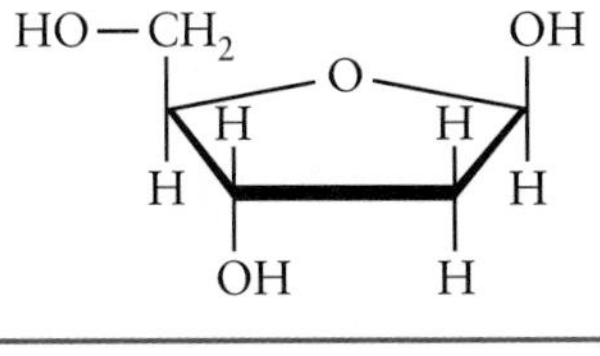

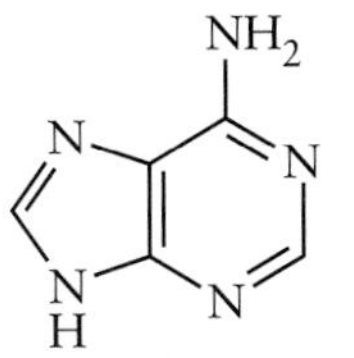

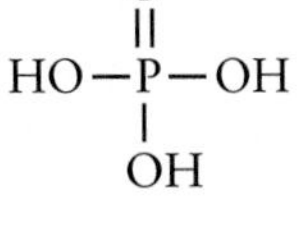

Nucleic Acids

The nucleic acids 2-deoxyribonucleic acid (DNA) and ribonucleic acid (RNA) are formed when nucleotides are joined by covalent bonds to produce a linear sequence of nucleotides. There are millions of nucleotides in the linear sequence of a DNA molecule. The bond joining two nucleotides creates a phosphate ester functional group formed between the 3′ alcohol (ROH) of one nucleotide and the 5′-phosphate group of the next nucleotide in the sequence. The bond linking two nucleotides is illustrated in Figure 15-6 for a simple RNA **di**nucleotide, a nucleic acid containing **two** nucleotides.

Figure 15-6 The chemical structure of an RNA dinucleotide, showing the phosphate ester bond joining the two nucleotides A and C.

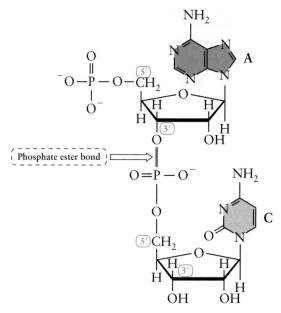

The dinucleotide contains adenine and cytosine as the nitrogen bases and therefore is abbreviated AC.

To link another nucleotide to the cytosine-containing end of the dinucleotide, known as the "three-prime end" (3′), a second phosphate ester is formed at the 3′ hydroxyl group of cytosine and the phosphate group of the next nucleotide to form a **tri**nucleotide. Figure 15-7 shows the chemical structure of the resulting trinucleotide, formed by joining a guanine-containing nucleotide to the dinucleotide above, producing the nucleotide sequence AC**G**.

Except for the nucleotides at each end of the sequence, the 5′ OH and 3′ OH groups of each nucleotide in a nucleic acid are joined to two different phosphate groups. At one end of the nucleic acid is a free phosphate group on the 5′ OH group, known as the "five-prime end." On the other end is a free OH group on the 3′ carbon, known as the "three-prime end." Thus, the backbone of a nucleic acid consists of repeating sugar–phosphate units, with different nitrogen bases attached to the sugar units. Therefore, a nucleic acid need only be identified by the sequence of its nitrogen bases; the repeating sugar–phosphate backbone of the nucleic acid is the same throughout.

The convention is to list the sequence of nitrogen bases starting from the 5′ end and working sequentially toward the 3′ end (5′ → 3′). For example, the nucleotide sequence in the trinucleotide shown in Figure 15-7 would be written ACG, because adenine is the base at the 5′ end and guanine is the base on the nucleotide at the 3′ end. If the monosaccharide is 2-deoxy-D-ribose, as in the case of DNA, a "d" is inserted as a prefix before the sequence. Thus dACG, would signify the trinucleotide in Figure 15-7 with each of the monosaccharides replaced by 2-deoxy-D-ribose.

Statistically, the number of different nucleic acids that can be formed from a sequence of n nucleotides is 4^n. For example, there are $4^3 = 4 \times 4 \times 4 = 64$ different possible nucleic acids composed of three nucleotides, of which one is the trinucleotide shown in Figure 15-7. The number of possibilities increases significantly with each added nucleotide. Given that DNA is composed of millions of nucleotides, the number of possible DNA sequences is astronomical.

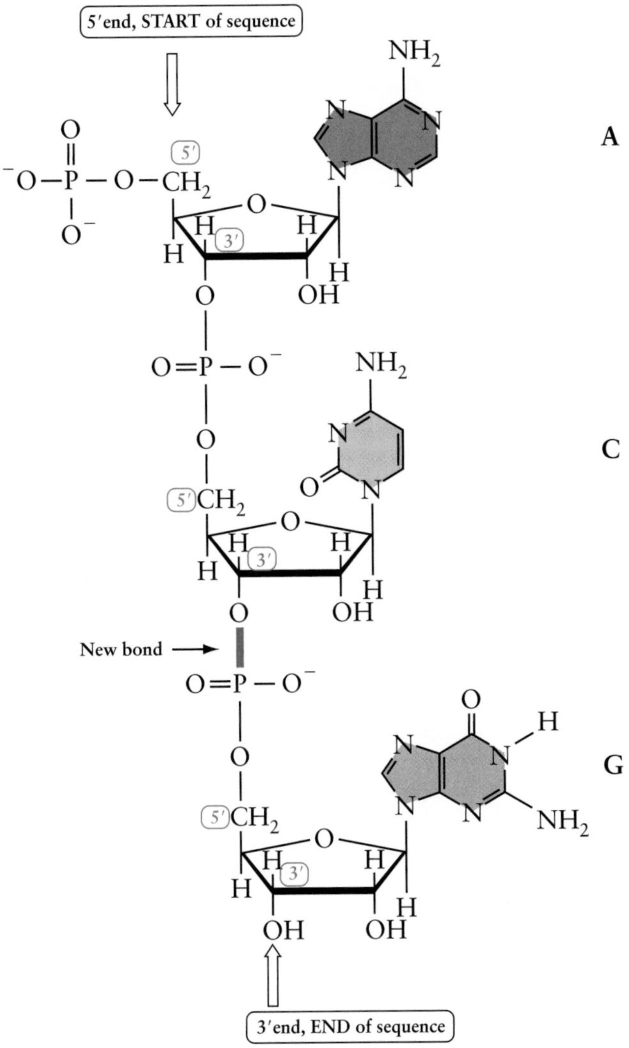

Figure 15-7 The chemical structure of the trinucleotide ACG, showing the nucleotide sequence from the 5′ end at the top of the figure to the 3′ end at the bottom of the figure.

WORKED EXERCISES 15-2 Nucleic Acids

1 Statistically, how many different **tetra**nucleotides—a sequence of four nucleotides—can be constructed? Show your calculation.

2 In the nucleoside shown below, label the 3′ hydroxyl group and the 5′ hydroxyl group. At which position is the phosphate group attached to form a nucleotide? Write the structure of the nucleotide formed from the nucleoside shown below.

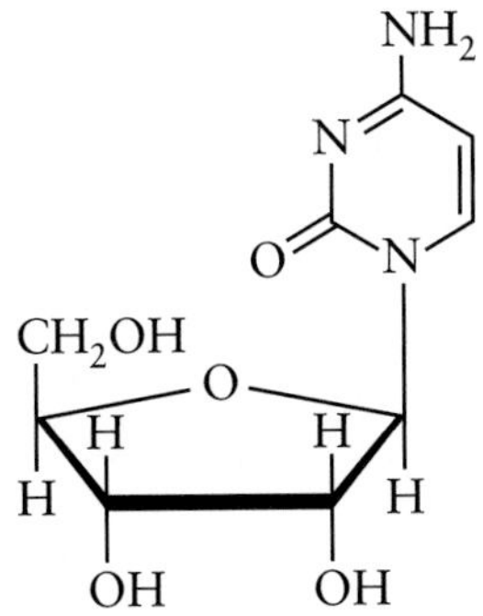

3 Two nucleotides are shown below: C and A. Write the structure of the dinucleotide CA and circle the bond joining the nucleotides. Label the 5′ and the 3′ ends of the dinucleotide. Is the dinucleotide AC the same as or different from the dinucleotide CA?

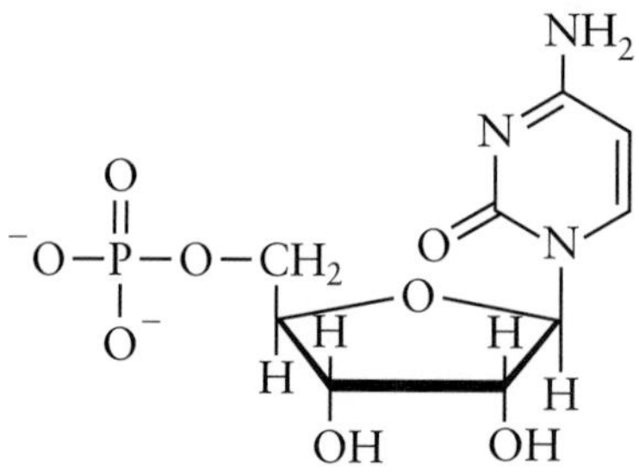

C

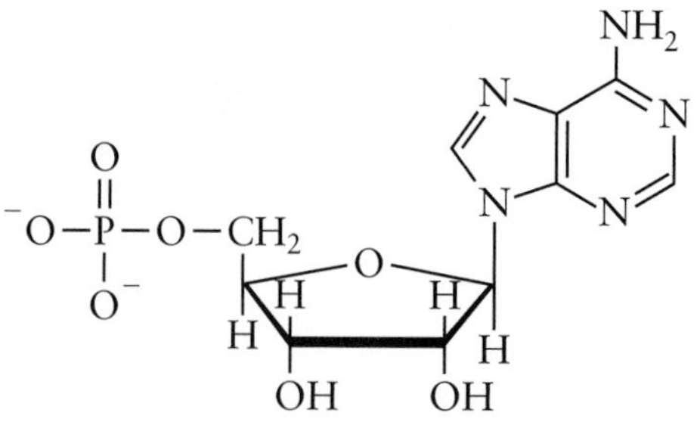

A

SOLUTIONS

1 The number of possible nucleic acid structures that can be formed from four different nucleotides is $4^4 = 4 \times 4 \times 4 \times 4 = 256$.

2 The phosphate group is attached to the 5′ hydroxyl group.

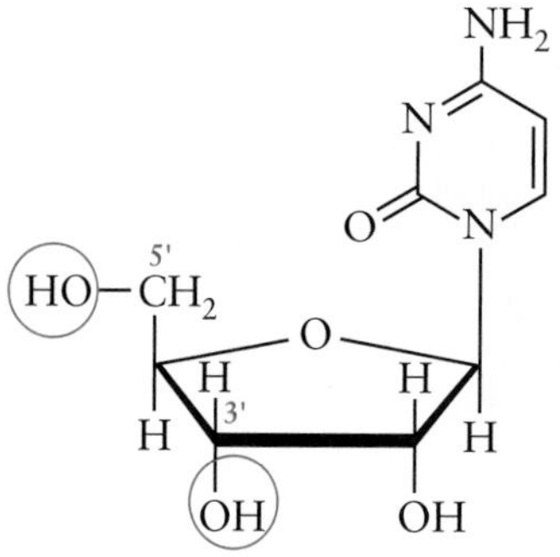

Nucleoside

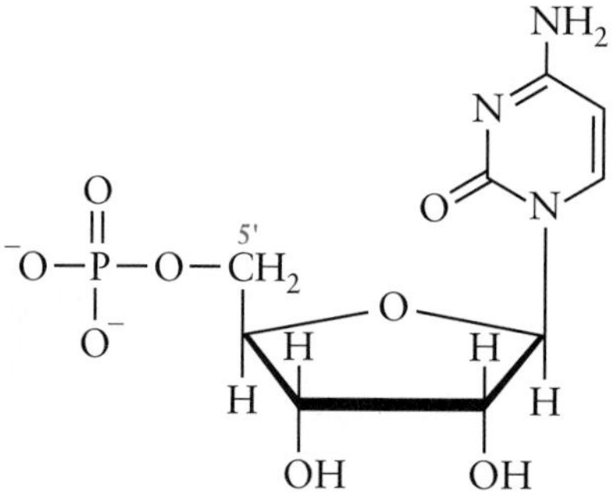

Nucleotide

3

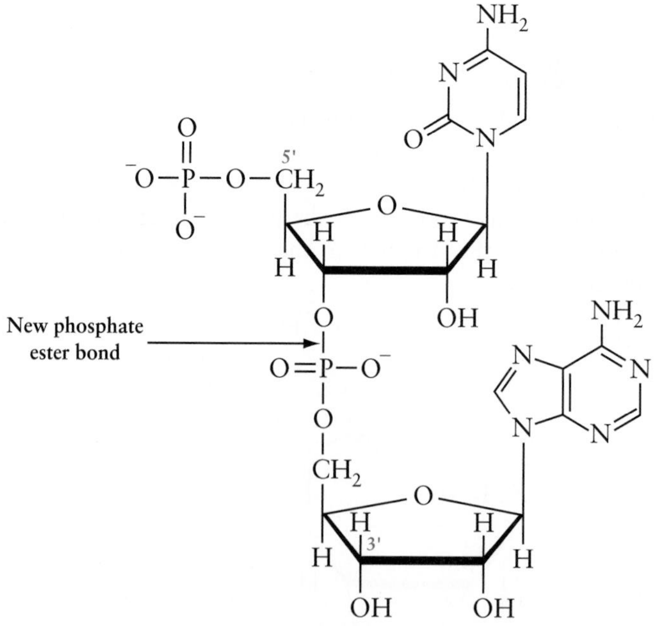

New phosphate ester bond

C

A

CA is different from AC: The C and A are at different ends.

15.6 How many different ribose-containing **di**nucleotides—a sequence of two nucleotides—can be constructed? List each dinucleotide using its two-letter abbreviation.

15.7 Draw the structure of CAC. How would the structure of dCAC be different from CAC?

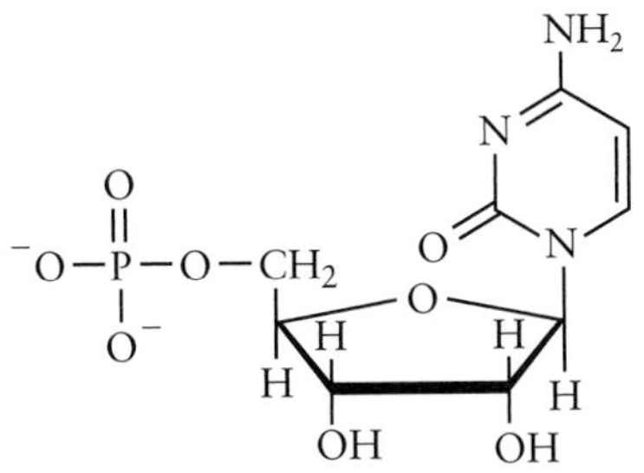

C 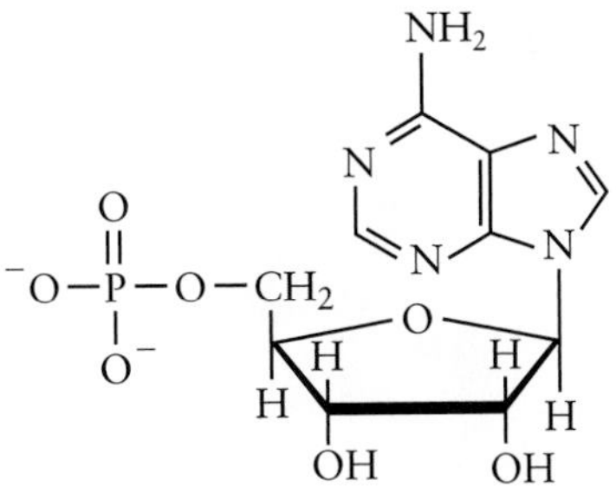   A

15.8 The structure of ATP is shown below.

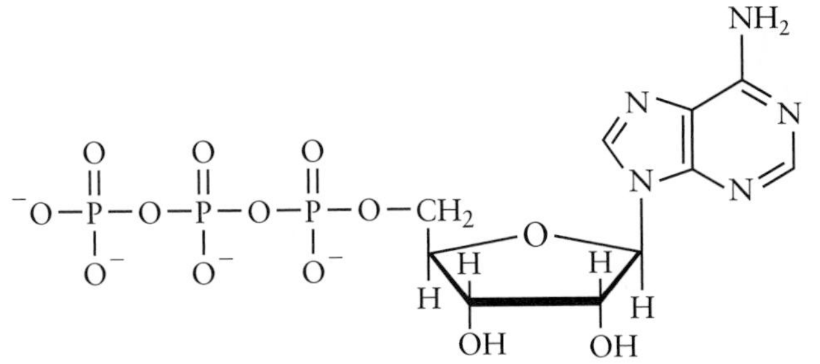

ATP
adenosine triphosphate

a. Why is this molecule classified as a nucleotide?
b. In what way is ATP different from the A-containing nucleotides in RNA?
c. What is the name of the nucleotide base present in ATP? Circle this base.

15.2 DNA

"A structure this pretty just had to exist" —James Watson

In 1953, James Watson, Francis Crick, and Maurice Wilkins determined the three-dimensional structure of DNA. They were later awarded the Nobel Prize for their discovery of the chemical structure of DNA. Using molecular models that they had built, Watson and Crick were able to determine a structure for DNA that was consistent with the experimental data available at the time, much of it collected by Rosalind Franklin. For example, experimental evidence showed that there were an equal number of adenines and thymines, and similarly an equal number of cytosines and guanines. According to Watson's own account, many of their peers frowned upon the amount of time they spent working with models, believing their time would be more wisely spent doing experiments in the laboratory. Never underestimate the value of models!

James Watson and Francis Crick standing next to their model of DNA. [A. Barrington Brown/Photo Researchers, Inc.]

Double Helix Structure of DNA

DNA consists of two nucleic acid molecules, each a linear sequence of millions of nucleotides. The two linear molecules are often referred to as two "strands" of DNA. These two strands are twisted around each other into

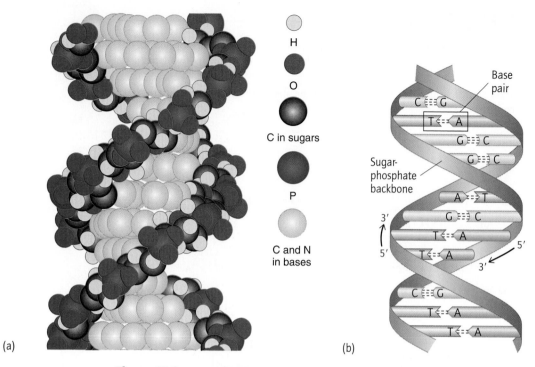

Figure 15-8 The double helix structure of DNA. (a) A space-filling model showing the base pairs in gold. (b) The stacking of base pairs is shown in this schematic drawing of DNA. It highlights the hydrogen bonding of the complementary base pairs A–T and G–C.

a double helix, as shown in Figure 15-8. The sugar–phosphate backbone of each nucleic acid forms the outer vertical portion of the DNA double helix and is represented as a brown ribbon in Figure 15-8b. The monosaccharide and phosphate groups of the backbone are hydrophilic, so they are positioned on the outside of this three-dimensional structure, which allows their polar functional groups to interact with the surrounding polar aqueous environment. Remember, "like dissolves like." The two strands of the DNA double helix are arranged antiparallel to one another, which means their 5′ to 3′ ends are oriented in the opposite direction.

The relatively nonpolar nitrogen bases project toward the interior of the DNA double helix, perpendicular to the sugar–phosphate backbone, as shown in gold in Figure 15-8a and also in Figure 15-8b. Each base forms hydrogen bonds to a complementary base on the adjacent strand, creating **base pairs,** which are stacked like rungs on a ladder. The complementary base pairs in DNA are

- A–T and
- G–C.

Thus, where there is an adenine (A) on one strand, the complementary base on the other strand is thymine (T); and where there is a guanine (G) on one strand, the complementary base on the other strand is cytosine (C). These base pairings arise from two hydrogen bonds that form between A and T and three hydrogen bonds that form between G and C, as shown in Figure 15-9. Since the nitrogen bases are hydrophobic, they are located on the inside of the DNA double helix structure, removed from the surrounding aqueous environment. The base pairs are planar, so stacking further optimizes dispersion forces between them.

The hydrogen bonds between base pairs hold the two strands of DNA together in a double helix. These bonds are significantly weaker than covalent bonds, but their cumulative strength holds the two strands together. Yet these

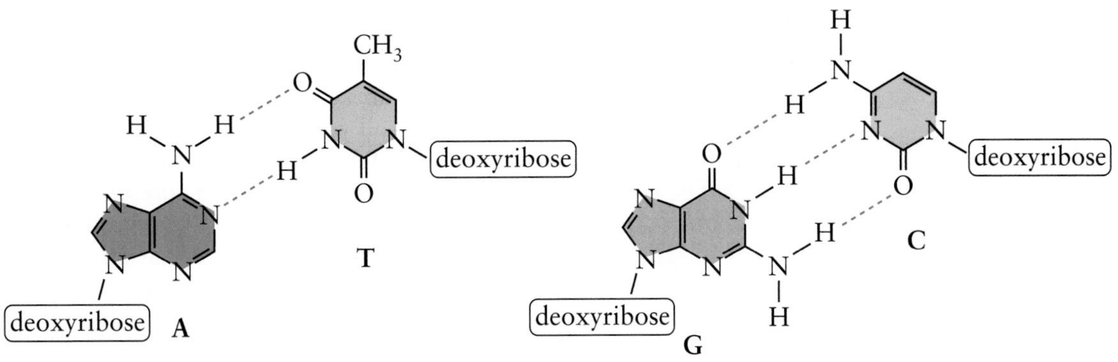

Figure 15-9 Two hydrogen bonds join the base pairs A and T in DNA, whereas three hydrogen bonds join the base pairs G and C. Recall from Chapter 3 that hydrogen bonding is represented by a dashed line.

noncovalent forces are weak enough that the two strands of DNA are able to separate during the important processes of replication and transcription, when the chemical information in a DNA strand must be read.

Problem-Solving Tutorial:
Nucleic Acid Structure

WORKED EXERCISES 15-3 DNA Double Helix

1 Below are sequences of nucleotides located on a segment of one DNA strand. Indicate the complementary sequence of base pairs that would appear on the adjacent DNA strand from the 3′ → 5′ end.
 a. dAGTCCG   **b.** dCCTTGA

2 What part of the DNA double helix has the following properties?
 a. is hydrophobic
 b. forms the "ladder rungs" on the DNA double helix
 c. is hydrophilic and interacts with the surrounding aqueous environment

3 Are the two strands of a DNA double helix *parallel* or *antiparallel*? What does this mean?

SOLUTIONS

1 The complementary sequences of base pairings would be
 a. dAGTCCG (5′ → 3′)
   dTCAGGC (3′ → 5′)
 b. dCCTTGA (5′ → 3′)
   dGGAACT (3′ → 5′)

2 **a.** the nitrogen base pairs
 b. the base pairings A–T and G–C
 c. the sugar and phosphate groups on the nucleic acid backbone

3 The two strands of DNA are arranged antiparallel to one another: the 5′ end of one strand is positioned opposite the 3′ end of the other strand.

PRACTICE EXERCISES

15.9 Below are sequences of nucleotides located on a segment of one DNA strand. Indicate the complementary sequence of base pairs on the adjacent DNA strand from the 3′ → 5′ end.
 a. dTTGGCA   **b.** dATGCCA

15.10 For each property described below, name the part of the DNA double helix that exhibits that property.
 a. is on the inside of the double helix
 b. is on the outside of the double helix
 c. interacts with the other strand through hydrogen bonding
 d. is held together by dispersion forces resulting from stacking

Higher-Order DNA Structure: Chromosomes

DNA is packaged into chromosomes found in the nucleus of cells. **Chromosomes** are highly organized compact structures containing DNA and proteins. If stretched out end to end, the DNA from all your chromosomes would span a distance greater than a meter! However, DNA is super-coiled, as illustrated in Figure 15-10. The DNA double helix is first coiled around a core of proteins, known as **histones,** creating a compact structure known as a **nucleosome** (Figure 15-10b). The histone proteins have a positive charge that attracts the negative charge of the phosphate groups on DNA. This electrostatic attraction facilitates the packing of DNA into compact nucleosomes.

Nucleosomes coil upon themselves to create a larger fibrous structure, known as a **chromatin fiber** (Figure 15-10c), which coils even further to form the familiar X-shaped structure of a chromosome (Figure 15-10d). The DNA double helix is about 2 nm wide, while a chromosome is about 1400 nm wide, wide enough to be visible under a light microscope, as seen in the image in the margin. Most human cells contain 46 chromosomes.

Genes and the Human Genome

The complete sequence of bases in your DNA, distributed over 46 chromosomes, is known as your **genome.** Every organism has a genome. These vary in size from the genome of a simple bacterium that might contain 600,000 base pairs, to the human genome, which contains about 3 billion base pairs.

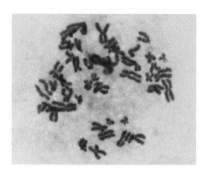

A cell nucleus showing chromosomes. Chromosomes are large enough to be visible under a microscope. [Michael Abbey/Visuals Unlimited]

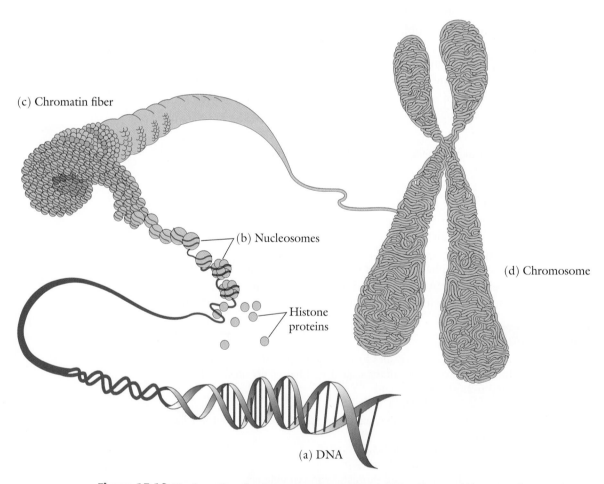

(c) Chromatin fiber

(b) Nucleosomes

Histone proteins

(d) Chromosome

(a) DNA

Figure 15-10 The formation of a chromosome from DNA. (a) DNA coils around histones to form nucleosomes (b). The nucleosomes coil upon themselves to create a larger chromatin fiber (c). Chromatin fibers coil even further to form the familiar X-shaped structure of a chromosome (d).

Surprisingly only about 2% of the human genome contains sequences of nucleotides that code for proteins. The segments of DNA containing these nucleotide sequences are known as genes (Figure 15-11). *A **gene** is a segment of DNA that contains the instructions for making a protein.* In Section 15.3 you will learn how the cell produces a protein from the instructions encoded in the nucleotide sequence of a particular gene. The average human gene contains 3,000 base pairs. There are approximately 25,000 genes, although the role of many of these genes is still unknown.

The year 2003 marked the conclusion of a 13-year international effort to map the entire human genome, known as the **Human Genome Project.** During the course of this project, the entire nucleotide sequence of the human genome was determined, and the genes associated with many proteins were identified. For example, a map of chromosome 6 is shown in Figure 15-12. Chromosome 6 is composed of 170,000,000 base pairs, and a few of the traits and disorders associated with genes on this chromosome are listed at their approximate locations on the chromosome. For example, a gene linked to breast cancer, another linked to Parkinson's disease, as well as the genes for the opioid and estrogen receptors, are located on this chromosome. The other chromosomes have been similarly mapped.

In the opening vignette, you learned that a genetic disorder results when a gene is altered and creates a dysfunctional protein. Many diseases have some genetic component, and knowing where a particular gene or set of genes lies on the genome brings us closer to understanding the mechanism of the disease, and finding treatment options.

| **WORKED EXERCISES 15-4** | DNA |

1 Which is positively charged and which is negatively charged?
   a. histone proteins   **b.** DNA

2 Approximately how many base pairs are there in the human genome? What percentage of the human genome is composed of genes?

3 What is a gene?

SOLUTIONS

1 **a.** Histones are positively charged;
   b. DNA is negatively charged.

2 There are about 3 billion base pairs in the human genome, of which 2% are genes.

3 A gene is a segment of DNA whose nucleotide sequence codes for a protein.

| **PRACTICE EXERCISES** |

15.11 Using Figure 15-12, name two diseases associated with a protein whose gene is located on chromosome 6.

15.12 How do histones facilitate the formation of nucleosomes?

15.13 What is a genome?

15.14 How many base pairs are there in the average human gene?

15.15 What does it mean that the gene for the opioid receptor is located on chromosome 6?

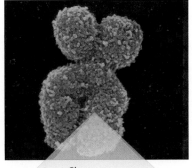

Chromosome

DNA double helix

Gene on a single strand of DNA

GGATATCCAAGC
Nucleotide sequence

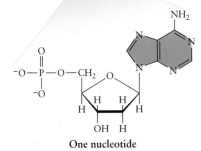

One nucleotide

Figure 15-11 Chromosomes contain DNA. Genes are segments of the DNA that contain the nucleotide sequence that specifies the amino acid sequence of a particular protein. [Photo: Biophoto Associates/Photo Researchers]

Mitochondria have their own DNA, which codes for 37 genes, essential for mitochondrial function. Your mitochondrial DNA is inherited from your mother.

DNA Replication

Every human being starts life as a fertilized egg, or *zygote*, which is a single cell containing the entire human genome. This single cell grows and divides to become a human being with one trillion cells, each with the same genome

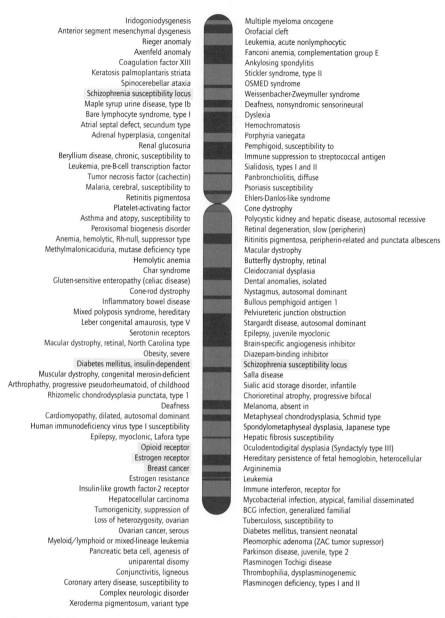

Figure 15-12 Genes on chromosome 6 that code for known proteins. The purple and green regions reflect the unique patterns of light and dark bands seen on human chromosomes that have been stained to allow viewing through a light microscope. [From the Human Genome Project Information Web site, U.S. Department of Energy]

as the zygote from which it originated. In order for the information encoded within DNA to be passed on to new cells when cells divide, DNA must be able to replicate itself before the actual cell division takes place.

Replication of DNA is an anabolic biochemical process. DNA replication begins with the unraveling of super-coiled DNA to expose the double helix. Then, the double helix itself must be unwound, a process that requires ATP and is catalyzed by the enzyme *helicase*. Next, each of the two strands of DNA is copied. Each strand of DNA serves as the template for a new **daughter** strand that is complementary to the template strand, as illustrated in Figure 15-13. At the end of the process, two new double-stranded DNA molecules, identical to the parent DNA molecule, are produced. Each new double

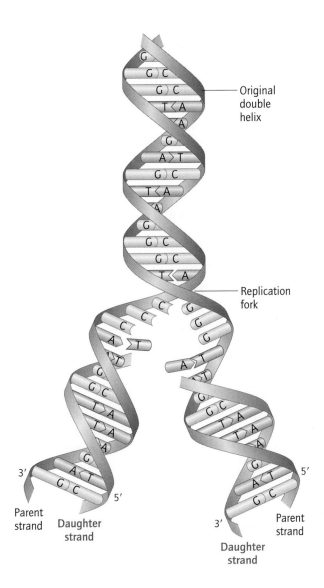

Original
double
helix

Replication
fork

3'

5'

5'

3'

Parent
strand

Daughter
strand

Parent
strand

Daughter
strand

Figure 15-13 DNA replication. Each strand of the parent molecule becomes a template for the synthesis of a new strand. Replication occurs from the 5′ to the 3′ end of the new strands; thus, in opposite directions on the parent strand.

helix contains one strand from the original double helix and one new daughter strand.

Replication begins along the DNA strand where the two strands of the double helix come apart, a site called the **replication fork,** illustrated in Figure 15-13. New nucleotides containing complementary bases then assemble along each parent strand, forming new base pairs through hydrogen bonding with the complementary base on the parent strand. The important enzyme *DNA polymerase* catalyzes the formation of phosphate ester bonds between the assembled nucleotides, forming a new nucleic acid, complementary to the parent strand.

This process occurs at an astonishing rate—about 100 nucleotides per second! *DNA polymerase* is not only fast, but it is strikingly accurate, inserting the wrong nucleotide less than once every 10,000 nucleotides. *DNA polymerase* also proofreads the daughter strands for mistakes. When errors are detected, it signals other enzymes to replace and repair incorrectly placed nucleotides. After proofreading and repair, the error rate drops to less than one in one billion nucleotides. Natural mistakes in replication do occur, however, and these mistakes account for some of the mutations that exist in the gene pool.

WORKED EXERCISES 15-5 DNA Replication

1 Consider a portion of double stranded DNA with the following complementary sequence of base pairs:

dAACCTTGG
dTTGGAACC

Write the sequence of nucleotides found in each daughter strand after replication and label the original parent strands and the daughter strands.

2 What role does *DNA polymerase* play in DNA replication?

SOLUTIONS

1   New DNA 1   New DNA 2
parent strands: dAACCTTGG   dTTGGAACC
daughter strands: dTTGGAACC   dAACCTTGG

2 *DNA polymerase* joins nucleotides by forming phosphate ester bonds and proofreads for errors in the nucleotide sequence of the daughter strands.

PRACTICE EXERCISES

15.16 Consider a portion of double-stranded DNA with the following nucleotide sequence:

dGACCTAGCCC
dCTGGATCGGG

Write the sequence of nucleotides found in each new DNA daughter strand after replication and label the parent and daughter strands.

15.17 What type of reaction does *DNA polymerase* catalyze? What type of biomolecule is *DNA polymerase*: carbohydrate, protein, lipid, or nucleic acid? How often does *DNA polymerase* make a mistake?

15.3 RNA and Protein Synthesis

Now that you are familiar with the structure of DNA, consider how a particular nucleotide sequence in DNA—a gene—is used to direct the synthesis of a protein: the assembly of amino acids in their correct order. This process requires RNA, the other important nucleic acid in the cell, found in structures outside the nucleus called **ribosomes.** An RNA molecule is much smaller than a DNA molecule, containing fewer nucleotides. There are three major forms of RNA, each with a different role in protein synthesis:

- ribosomal RNA (rRNA),
- messenger RNA, (mRNA), and
- transfer RNA (tRNA).

RNA is a single-stranded nucleic acid and contains D-ribose instead of 2-deoxy-D-ribose. Remember that the four nitrogen bases in RNA are A, C, G, and U; uracil (U) replaces thymine (T) in RNA.

Ribosomes, located in the cytoplasm, are the protein-making factories of the cell, where amino acids are assembled into proteins. Ribosomes are composed of ribosomal RNA (rRNA) and about 50 different proteins. Each rRNA strand is composed of 100–1000 nucleotides. Cells contain millions of ribosomes. Although ribosomes are located in the cytoplasm outside the nucleus, DNA never leaves the nucleus. Therefore, some way is needed to deliver the instructions encoded in the DNA to the ribosomes. That is the role of messenger RNA (mRNA).

The process of protein synthesis involves two basic steps, *transcription* and *translation* (Figure 15-14):

- **Transcription:** The section of DNA carrying the instructions for a particular protein—a gene—is copied to form a complementary strand of messenger RNA (mRNA).

- **Translation:** At the ribosome, amino acids are assembled with the help of tRNA into a polypeptide whose sequence of amino acids is determined by the instructions provided by the mRNA.

Amino acids are delivered to the ribosomes by transfer RNA (tRNA).

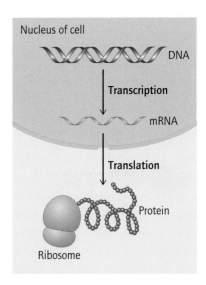

Figure 15-14 Overview of the flow of genetic information from DNA to mRNA to the ribosome.

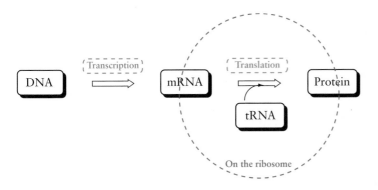

Transcription: DNA to mRNA

When a particular protein is needed by the cell, the gene that codes for it on the DNA is read—we say that the gene is "expressed." Gene expression is a complex process that is carefully regulated by the cell. Indeed, the expression of different proteins is one of the ways cells are differentiated. For example, a brain cell will produce different proteins than a muscle cell.

Gene expression begins with transcription. In this process, a nucleotide sequence on the DNA—the gene—is copied as a complementary single-stranded mRNA molecule, as illustrated in Figure 15-15. Copying begins at nucleotides known as *start codons*. Only one of the two strands forming the DNA helix is copied. The strand that is copied is known as the **template strand** or (+) strand.

The double helix must first be unraveled to expose the nucleotide sequence of the gene on the template strand. Then, *RNA polymerase* catalyzes the synthesis of the new mRNA molecule from individual nucleotides. The enzyme copies the sequence of nucleotides on the template strand of DNA from the 3′ to the 5′ end as a complementary sequence of mRNA bases. *RNA polymerase* catalyzes the formation of phosphate ester bonds between nucleotides much as *DNA polymerase* joins nucleotides during DNA replication.

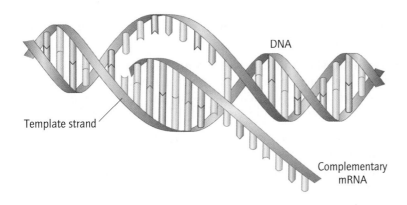

Figure 15-15 Transcription. A segment of DNA unravels to expose the nucleotide sequence on the template strand of DNA so that *RNA polymerase* can build a complementary nucleotide sequence of mRNA. DNA returns to its double helix structure and mRNA is exported out of the nucleus and to the ribosomes in the cytoplasm.

As already noted, the base sequence of the newly created mRNA is *complementary* to the DNA sequence on the template strand. For example, if the nucleotide sequence on the template strand is dGATCAT, the transcribed mRNA molecule has the sequence CUAGUA. Note that A on DNA is copied as U on RNA, because RNA employs uracil (U) instead of thymine (T).

Once transcription is complete, DNA refolds back into its double helix structure and mRNA is exported out of the nucleus and into the cytoplasm, where the ribosomes are located.

 **Problem-Solving Tutorial:** Steps in Protein Synthesis

---

WORKED EXERCISES 15-6   Transcription

1   For a DNA template strand containing the sequence dAATTGGCC, what is the sequence of nucleotides in the mRNA after transcription?
2   What is the role of *RNA polymerase*?

SOLUTIONS

1   A complementary set of bases would be found on mRNA: UUAACCGG.
2   *RNA polymerase* is the enzyme that directs the synthesis of mRNA from the template strand of DNA by joining nucleotides through phosphate ester bonds.

PRACTICE EXERCISES

15.18   What are the three types of RNA? Which type of RNA makes up part of the structure of a ribosome? What two types of biomolecules are present in a ribosome?

15.19   Describe the process of transcription.

15.20   What is the function of a ribosome?

15.21   Which contains more nucleotides: RNA or DNA?

15.22   Where in the cell is DNA found? Where in the cell is RNA found?

Translation: mRNA, tRNA, and Protein Synthesis

Ribosomes are the cellular structures where the nucleotide sequence of an mRNA molecule is read and a polypeptide—the primary structure of a protein—is built.

The Genetic Code Every grouping of *three* nucleotides on an mRNA molecule is known as a **codon** because it codes for one of the 20 natural amino acids (Figure 15-16). Since there are 64 possible codons (4^3), but only 20 amino acids, each amino acid has more than one codon. For example, UAU and UAC both code for the amino acid tyrosine. The key for the codons and the amino acids each specifies is known as the **genetic code.** A few codons do not specify an amino acid, but instead, mark the start or end of a gene. These codons are known as *start* and *stop* codons. The genetic code is shown in Table 15-1. Interestingly, all living organisms have the same genetic code.

Transfer RNA Matching a codon on an mRNA with the amino acid it encodes is the function of another type of RNA, transfer RNA (tRNA). tRNA is a smaller nucleic acid of 70–100 nucleotides characterized by a cloverleaf shape, as shown in Figure 15-17. tRNA contains two important regions:

- the anticodon loop and
- the 3′ end covalently bonded to a particular amino acid.

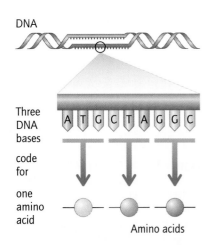

Figure 15-16 Each three-base codon on a gene codes for one amino acid.

Table 15-1 The Genetic Code, Showing mRNA Codons

		Second letter				
		U	C	A	G	
First letter	**U**	UUU } UUC } Phe UUA } UUG } Leu	UCU } UCC } UCA } Ser UCG }	UAU } UAC } Tyr UAA Stop UAG Stop	UGU } UGC } Cys UGA Stop UGG Trp	U C A G
	C	CUU } CUC } CUA } Leu CUG }	CCU } CCC } CCA } Pro CCG }	CAU } CAC } His CAA } CAG } Gln	CGU } CGC } CGA } Arg CGG }	U C A G
	A	AUU } AUC } Ile AUA } AUG Met/ Start	ACU } ACC } ACA } Thr ACG }	AAU } AAC } Asn AAA } AAG } Lys	AGU } AGC } Ser AGA } AGG } Arg	U C A G
	G	GUU } GUC } GUA } Val GUG }	GCU } GCC } GCA } Ala GCG }	GAU } GAC } Asp GAA } GAG } Glu	GGU } GGC } GGA } Gly GGG }	U C A G

(Third letter)

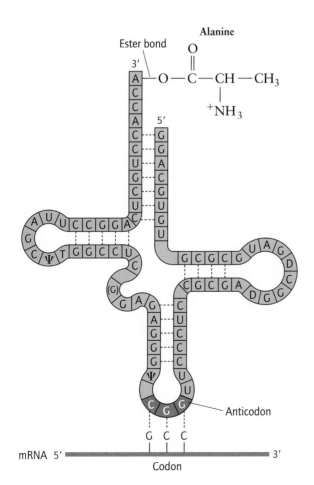

Figure 15-17 General structure of a tRNA molecule. The anticodon loop (in red) is complementary to the codon on the mRNA. The amino acid is attached via an ester linkage at the 3' end, as specified by the genetic code. Here you see that GCC codes for alanine.

The **anticodon** loop on a tRNA contains three nucleotides that hydrogen bond only to the complementary codon on mRNA. For example, a GCC codon on mRNA is complementary to a CGG anticodon on a tRNA molecule.

On its 3′ end each tRNA molecule is covalently bonded to a single amino acid. For example, a tRNA molecule with a CGG anticodon will be carrying the amino acid alanine. Thus, a GCC codon on mRNA will base pair to a tRNA molecule with the anticodon GCC that is carrying alanine, as specified by the genetic code.

Translation At the ribosome, an mRNA codon is read and a matching tRNA molecule arrives at the ribosome. Base pairs form temporarily between the anticodon on tRNA and the codon on mRNA, as illustrated in Figure 15-18a. The next matching tRNA molecule arrives at the ribosome and base pairs to the adjacent codon, and an amide bond is formed between the two amino acids on adjacent tRNA molecules, as shown in Figure 15-18b. The first tRNA molecule is released in the process, and diffuses away from the ribosome with its amino acid no longer attached. The second tRNA molecule, still bound to the ribosome, contains the growing polypeptide chain, now one amino acid longer, at its 3′ end, as shown in Figure 15-18c. The ribosome then shifts in the 5′ to 3′ direction to read the next codon on the mRNA, a process known as **translocation** (note the location of the ribosome relative to mRNA in (c) compared to (a) and (b)). The next codon on the mRNA is read, and another tRNA molecule is recruited.

These steps repeat in an overall process known as **translation,** which continues until a stop codon is reached. Each departing tRNA molecule no longer contains an amino acid, and diffuses away from the ribosome to find a replacement amino acid. A polypeptide must still undergo additional modifications to become a functional protein, folding, forming disulfide bonds, and joining with other subunits to adopt the tertiary and quaternary structures described in Chapter 11.

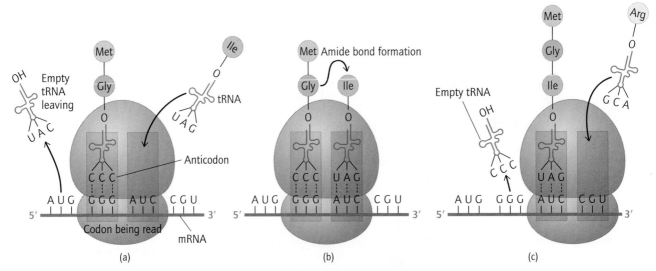

Figure 15-18 Translation. The process of building a protein from an mRNA template occurs on the ribosome (rRNA). Amino acids are delivered by tRNAs whose anticodons match the mRNA codons. (a) Codons on mRNA hydrogen bond to anticodons on tRNA. (b) A peptide bond is formed joining the growing polypeptide chain to the new amino acid. (c) Translocation: The empty tRNA leaves the ribosome and the ribosome shifts to the next codon on the mRNA. The process is then repeated over and over until a stop codon is reached.

Genetic Mutations

In the opening vignette you learned about the effects of genetic mutations. Now you are in a better position to understand how a genetic mutation affects transcription and translation to produce an abnormal protein. *A **genetic mutation** is any permanent chemical change that occurs at one or more nucleotides in the DNA sequence and affects the primary structure of a protein.* A mutation may involve a **substitution** of one nucleotide, as in the case of sickle cell anemia; or a deletion of a nucleotide. Consider the normal segment of a gene shown in Figure 15-19a, which transcribes as the mRNA shown and translates to the amino acid sequence Tyr-Leu-Ala-Leu. If there is a substitution of **T** for **G** in the gene as indicated in Figure 15-19b, transcription and translation produces the amino acid sequence Tyr-Leu-ASP-Leu instead. If the nucleotide G is deleted from the DNA, as shown in Figure 15-19c, then a **frameshift** occurs producing the sequence Tyr-Leu-Val-Tyr instead.

The effects of DNA mutations are minimized by the fact that there is more than one codon for most amino acids. For example, if a DNA mutation produces the codon CCA instead of CCC, the mutation would have no effect on the protein produced, because both codons code for the same amino acid, proline.

If a mutation changes one or more amino acids in a metabolic enzyme, the enzyme may no longer be able to perform its function. This is the case in the missing or defective enzyme *phenylalanine hydroxylase* seen in phenylketonurics, as described in *Chemistry in Medicine: Phenylketonurics*, in Chapter 14. Defective enzymes are sometimes less effective, as in the case of the hemoglobin produced in individuals with sickle-cell anemia, described in the opening vignette of Chapter 11. Similarly, an ineffective form of the enzyme *lactase* is responsible for lactose intolerance—an inability to digest lactose. When a mutation occurs in a gene that codes for an essential protein, the defect can be lethal. Babies born with a life-threatening genetic defect may die at birth or at an early age. In some instances, mutations can be beneficial.

Mutations can be caused by a number of factors including ultraviolet radiation from the sun; certain chemicals, such as those found in cigarettes; and chemotherapy. High-energy radiation such as X rays and γ rays are particularly damaging to DNA. Indeed, exposure to radiation increases the risk of cancer. You will learn about how radiation damages nucleic acids and other biomolecules when you study radiation and nuclear medicine in Chapter 16.

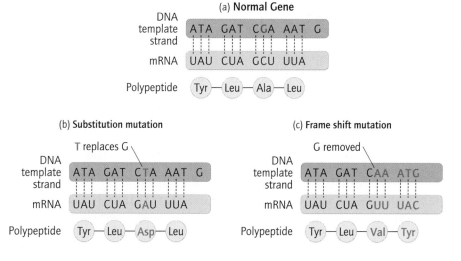

Figure 15-19 Mutations arise when there is a change in the nucleotide sequence of the DNA that leads to changes in the amino acid sequence. (a) Normal DNA segment. (b) Substitution mutation: One nucleotide is replaced by a different nucleotide. (c) Frame shift mutation: One nucleotide is deleted or added. Here a nucleotide has been removed. Note that the resulting polypeptide is different from the normal polypeptide in both types of mutations.

Problem-Solving Tutorial:
Steps in Protein Synthesis

WORKED EXERCISES 15-7 Transcription and Translation

1 Using the genetic code, determine the amino acid specified by the following codons on mRNA:
   a. AAG   **b.** UGA

2 More than one codon encodes most amino acids, and this redundancy minimizes the effects of mutations. List all the codons that code for threonine.

3 Why are there 64 different codons?

4 For the following mRNA sequences, indicate the anticodons on the three tRNA molecules recruited. Indicate the correct order in which the tRNA molecules are recruited by the ribosome and the amino acid each will be carrying. What is the amino acid sequence of the tripeptide formed? You may consult Table 15-1.
   a. AGUCCGUAC   **b.** AAUUGCUUC

5 If the sequence in Exercise 4a has a mutation where the first A is substituted by G, how will the peptide be different? If the first A is deleted, how will the peptide be different? Assume the next nucleotide in the sequences is U.

SOLUTIONS

1 **a.** AAG on mRNA codes for the amino acid lysine.
   b. UGA on mRNA is a stop codon, so it does not code for an amino acid, but signals that the polypeptide is complete.

2 The codons that code for threonine are ACU, ACC, ACA, ACG.

3 There are 64 codons because there are 64 different ways that 4 nucleotides can be arranged as a three nucleotide sequence. This result is readily calculated using the formula $4^3 = 4 \times 4 \times 4 = 64$.

4 **a.** Anticodon UCA carrying serine, followed by anticodon GGC carrying proline, followed by anticodon AUG carrying tyrosine; the tripeptide would be Ser-Pro-Tyr.
   b. Anticodon UUA carrying asparagine, followed by anticodon ACG carrying cysteine, followed by anticodon AAG carrying phenylalanine; the tripeptide would be Asn-Cys-Phe.

5 The peptide would be Gly-Pro-Tyr instead of Ser-Pro-Tyr (substitution mutation). The peptide would be Val-Arg-Thr (frameshift mutation).

PRACTICE EXERCISES

15.23 Using Table 15-1, determine the amino acid specified by the following mRNA codons:
   a. GGG   **b.** AAA

15.24 Specify all the codons that code for the following amino acids:
   a. histidine   **b.** leucine

15.25 For the following mRNA sequences, indicate the anticodons on the three tRNA molecules involved in building the protein. Indicate the order in which the tRNA molecules are recruited and the amino acid each will be carrying. What is the amino acid sequence of the tetrapeptide formed? You may consult Table 15-1.
   a. GGUACUAUUUAA   **b.** GGUGACCGACAU

15.26 For a segment of DNA that has the sequence dTTTAATCTCTGT, what would the sequence on mRNA be? What four tRNA molecules would be involved in building the protein? What is the amino acid sequence of the tetrapeptide?

15.27 What are the three stop codons? What is a stop codon?

15.28 Indicate whether the following normal mRNA sequence would produce the same or a different dipeptide if the mutation shown below occurred:

DNA: dTAATGA

mRNA: AUUACU

amino acids: Ile-Thr

a. First T on DNA is replaced by G.
b. Third nucleotide on DNA is replaced by G.
c. Third nucleotide on DNA is deleted.
d. First nucleotide on DNA is deleted.

In this chapter you have seen how proteins, the biomolecules described in Chapter 11, are synthesized using the instructions supplied by DNA. The year 2010 marked the 57th anniversary of the discovery of the structure of DNA, the molecule containing the instructions, in the form of nucleotide sequences, for assembling amino acids into proteins. In the decades since the elucidation of the structure of DNA, the entire human genome has been mapped and DNA technologies have emerged at a remarkable rate, transforming the landscape of medicine and fields such as forensics, where DNA testing has had a major impact. The twenty-first century is an exciting time to be working in the healthcare industry, where you will be supremely positioned to observe advances unfold in the field of molecular biology—the intersection of biology and chemistry.

Chemistry in Medicine Viral Nucleic Acids: The Case of HIV

HIV—the virus that causes AIDS—has infected 36 million people in the world, and 15,000 new infections are reported every day. In the United States alone, 1.2 million people are living with HIV, and 40,000 new HIV infections are reported every year. Infection by the **h**uman **i**mmunodeficiency **v**irus (HIV) ultimately leads to **a**cquired **i**mmuno**d**eficiency **s**yndrome (AIDS), a disease in which the immune system can no longer fight off life-threatening opportunistic infections, such as pneumonia.

All viruses contain nucleic acids—RNA or DNA—housed inside a protective protein coat. Viruses differ in the type of nucleic acid they contain: RNA or DNA; single-stranded or double-stranded; template strand (+) or nontemplate strand (−). These differences characterize the seven major types of viruses.

The HIV virus contains two single-stranded (+)-RNAs enclosed in a conical shaped protein coat known as a *capsid*, as shown in Figure 15-20. The HIV viral RNA contains nine genes, including genes that code for four important viral enzymes needed in order for the virus to replicate: *reverse transcriptase, protease, integrase,* and *ribonuclease.*

A virus lacks the full set of enzymes and molecular structures to reproduce its genomic material or to synthesize viral proteins on its own. Therefore, it injects its genomic material into a host cell, and hijacks the host's enzymes and nucleic acids to perform this function. The process eventually kills the host cell, but not until many new viruses have been produced that can infect other healthy cells. HIV infects T-lymphocytes, a type of white blood cell that serves a central role in the immune system.

The life cycle of the HIV virus is illustrated in Figure 15-21. The HIV virus attaches to T-lymphocytes at receptors on the cell membrane, known as CD4 receptors, in a

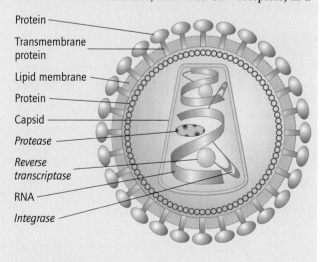

Figure 15-20 Parts of the HIV virus.

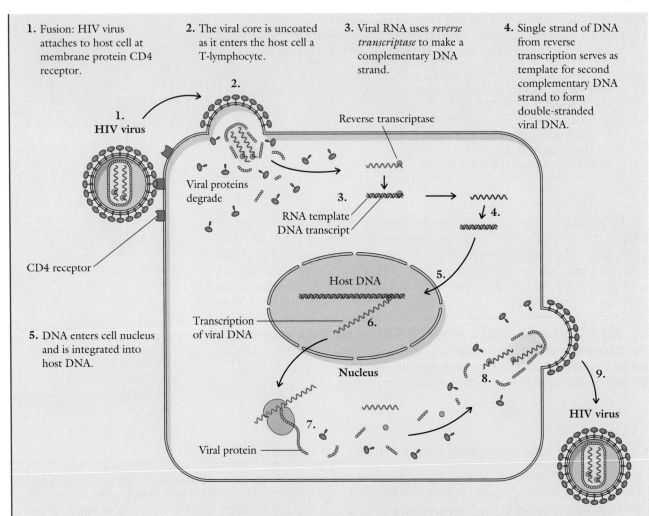

1. Fusion: HIV virus attaches to host cell at membrane protein CD4 receptor.

2. The viral core is uncoated as it enters the host cell a T-lymphocyte.

3. Viral RNA uses *reverse transcriptase* to make a complementary DNA strand.

4. Single strand of DNA from reverse transcription serves as template for second complementary DNA strand to form double-stranded viral DNA.

5. DNA enters cell nucleus and is integrated into host DNA.

6. Viral DNA is transcribed to viral mRNA, along with the cell's DNA, and then exported to the cytoplasm.

7. Viral mRNA is translated, along with the cell's mRNA. *Protease* cuts the viral protein into smaller proteins for incorporation into new viruses.

8. Viral proteins, capsids, and envelopes are assembled.

9. Assembled virus emerges from the cell and infects other T-cells.

Figure 15-21 Life cycle of the HIV virus.

process known as *fusion* (Step 1 in Figure 15-21). The virus then injects its RNA into the T-lymphocyte (Step 2).

The genomic RNA of HIV must be converted into viral DNA before transcription can occur (Steps 3 and 4). Since the conversion of RNA to DNA is the reverse of normal transcription, described in Section 15.3, HIV viruses are appropriately called **retroviruses**. The enzyme *reverse transcriptase,* coded for by one of the 9 genes of the virus, produces viral DNA by joining the complementary base pairs matched to the viral RNA template (reverse transcription).

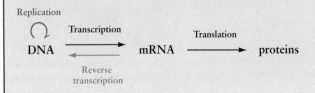

Reverse transcriptase is an error-prone enzyme that makes many mistakes during reverse transcription. Consequently, the viral proteins produced are somewhat different from one another, making the virus difficult for the host's immune system to recognize and eradicate. In essence, the virus is constantly mutating. Efforts to develop a vaccine for HIV have been unsuccessful to date in large part because of the rapidly mutating viral DNA.

Viral DNA formed in the host's cytoplasm moves into the nucleus, where it is incorporated into the host's DNA (Step 5). Once it is part of the host's DNA, viral DNA lies dormant until the T-cell's DNA is expressed, and protein synthesis begins. Ironically, activation occurs when the T-cell is called upon to fight off an infection. When the host transcribes its DNA, which contains viral DNA, viral mRNA is also produced (Step 6). Ribosomes translate viral mRNA, producing a long polypeptide that is cut by *protease* enzymes into the individual viral enzymes and other viral proteins needed by the virus (Step 7). The mRNA produced from transcription is also used as the genomic material for new viruses. This mRNA is packaged within protein layers to form additional HIV viruses (Step 8). Each new virus then leaves the host cell and infects another healthy T-cell (Step 9).

Without any drug intervention, the T-cell count of an HIV infected person eventually drops so low that the immune system cannot perform its normal function: to fight off infections. The infected individual eventually succumbs to common infections that a healthy person would be able to fight. This is the stage at which an HIV infection is elevated to AIDS. The time period between HIV infection and AIDS ranges from several months to many years, and depends on many factors.

HIV drug therapies to date have focused on developing inhibitors of the key enzymes involved in the life cycle of the virus, *reverse transcriptase* (*RT*) and *protease*, as well as the development of fusion inhibitors. Since the virus rapidly mutates and becomes resistant to any one type of drug therapy, the best treatment plans are usually combination drug therapies. *RT* inhibitors and *protease* inhibitors work according to the principles of enzyme inhibition described in Chapter 11.

These antiretroviral drugs are organic molecules that bind to the active site of a viral enzyme. *RT* inhibitors such as AZT, prevent viral RNA from being converted into DNA early in the life cycle of the virus. Note that AZT, one of the first drugs used to treat AIDS, is a nucleoside. *Protease* inhibitors like saquinavir and darunavir prevent the release of the enzymes produced late in the life cycle of the virus. The latest generation of HIV drugs are fusion inhibitors, large peptide drugs that must be administered subcutaneously, and prevent initial fusion of the virus with CD4 receptors on the T-cell. These drugs are a last resort when antiretroviral drugs are no longer an effective treatment.

As with other diseases you have learned about, understanding the biochemistry of the disease process has made it possible for scientists to develop treatments. Although there is no cure for HIV/AIDS, understanding nucleic acids and how they replicate and transcribe and translate information to make proteins has allowed scientists to understand the HIV virus and to develop targeted drug treatments that have allowed HIV-positive individuals to live much longer and healthier lives.

Chapter Summary

Nucleotides and Nucleic Acids

- Nucleotides contain a nitrogen base, a monosaccharide, and a phosphate group.
- RNA has D-ribose as its monosaccharide, and DNA has 2-deoxy-D-ribose. 2-Deoxy-D-ribose lacks an OH group on the 2-carbon of the sugar.
- The four possible nitrogen bases are covalently bonded to the monosaccharide at the 1-carbon, the anomeric carbon.
- The four nitrogen bases found in DNA are adenine, guanine, cytosine, and thymine. The four nitrogen bases found in RNA are adenine, guanine, cytosine, and uracil.
- The phosphate group of a nucleotide is attached to the 5′ alcohol of the monosaccharide, creating a phosphate ester.
- Nucleic acids are formed when nucleotides join together through phosphate ester bonds. The phosphate ester bond forms between the 3′ alcohol of one nucleotide and the phosphate group on another nucleotide.
- In RNA and DNA, the only group that changes along the sugar–phosphate backbone is the nitrogen base. Thus, nucleic acids are identified by their sequence of the nitrogen bases. The sequence is listed from the 5′ end to the 3′ end.

DNA

- DNA consists of two strands of nucleic acids, which are twisted around each other into a double helix.
- The monosaccharide and phosphate groups of the nucleic acid backbone are hydrophilic and are located on the outside of the three-dimensional structure. The base pairs are located on the hydrophobic inside.

- The two strands of DNA are arranged antiparallel to each other: $5'\rightarrow 3'$ and $3'\rightarrow 5'$.
- Base pairs on one DNA strand hydrogen bond with complementary base pairs on the other strand, thereby connecting the two nucleic acid strands in DNA. The complementary base pairs in DNA are A–T and C–G.
- Chromosomes are highly ordered compact structures containing DNA and proteins, found in the nucleus of cells.
- DNA is coiled around histones, forming nucleosomes.
- Nucleosomes coil upon themselves to form chromatin fibers, which supercoil upon themselves to form chromosomes.
- Your genome is the complete sequence of bases in your DNA. It is distributed over 46 chromosomes. The human genome contains 3 billion base pairs.
- A gene is a segment of DNA that contains the instructions for making a protein.
- DNA replication starts with the unraveling of DNA. Each strand serves as the template for the formation of a new, daughter DNA strand.
- *DNA polymerase* catalyzes the formation of phosphate ester bonds between nucleotides on the two new daughter strands of DNA during the replication process.

RNA and Protein Synthesis

- The three major forms of RNA are ribosomal RNA (rRNA), messenger RNA (mRNA), and transfer RNA (tRNA).
- Transcription is the process of copying the template strand of DNA into messenger RNA (mRNA). *RNA polymerase* controls the synthesis of mRNA.
- Codons, groups of three nucleotides on an mRNA, code for one of 20 amino acids. The amino acid specified by the codon is given by the genetic code.
- Translation is the process of building a protein from an mRNA template on the ribosome.
- tRNA delivers the appropriate amino acid when an anticodon matches an mRNA codon being read on the ribosome.
- A genetic mutation occurs when one or more nucleotides in the DNA sequence is permanently changed in a way that alters the primary structure of the protein.

Key Words

Anticodon loop The three nucleotides on a tRNA that correspond to the complementary codon on mRNA.

Base pair Complementary bases that hydrogen bond. The complementary base pairs are A–T and C–G.

Chromatin fibers The large fibrous structures formed when nucleosomes coil up on themselves.

Codon A grouping of three nucleotides on an mRNA molecule that codes for one of 20 amino acids.

Gene A segment of DNA that contains the instructions for making a protein.

Genetic code The amino acid or stop and start instruction specified by each of the 64 codons.

Genetic mutation Any permanent chemical change that occurs at one or more nucleotides in the DNA sequence, causing a change in the primary structure of a protein.

Genome The complete DNA sequence of an organism.

Histone Protein containing a positive charge that facilitates the tight packing of DNA into nucleosomes.

Nucleosome DNA coils around histones to form nucleosomes, compact structures containing DNA and proteins.

Nucleotide A molecule that contains a nitrogen base, a monosaccharide, and a phosphate group. Nucleotides are the building blocks of nucleic acids.

Ribosomes Cellular structures where ribosomal RNA (rRNA), together with many proteins, synthesize polypeptides. Ribosomes are located in the cytoplasm.

Transcription The process of synthesizing a single-stranded mRNA molecule from a DNA template.

Translation The process of building a polypeptide from the nucleotide sequence of an mRNA.

Additional Exercises

Nucleotides and Nucleic Acids

15.29 What is a gene? What is a mutation?

15.30 Which cells in the human body do not contain DNA or RNA?

15.31 What are the three basic parts of a nucleotide? Explain the difference between a nucleoside and a nucleotide.

15.32 Which of the monosaccharides shown below belongs to RNA and which belongs to DNA? What is the difference between the two monosaccharides?

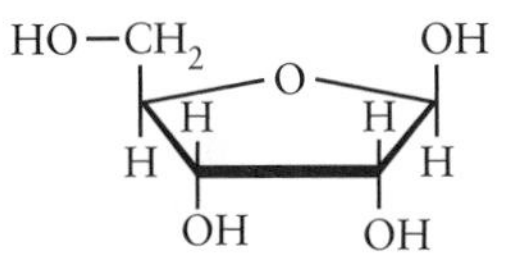

15.33 For the nucleotide shown, circle and label the following.

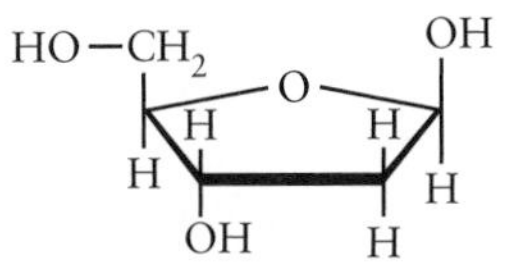

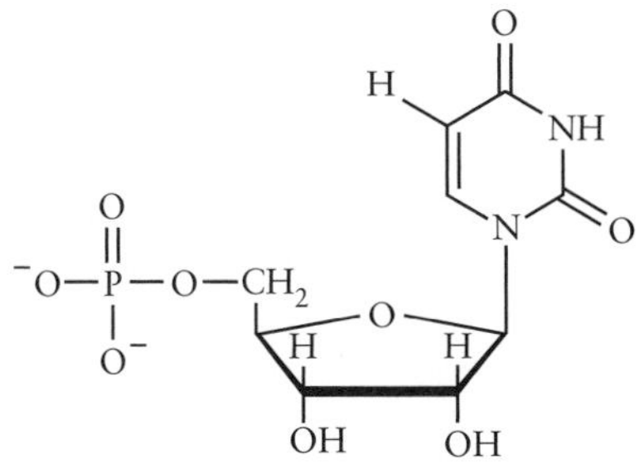

a. The nitrogen base. What is the name of the nitrogen base?
b. The monosaccharide. What is the name of the monosaccharide?
c. The phosphate group.
d. The 5′ carbon.
e. The anomeric carbon.
f. The 3′ carbon.
g. Would this nucleotide be found in RNA or DNA? Explain.

15.34 For the nucleotide shown, circle and label the following.

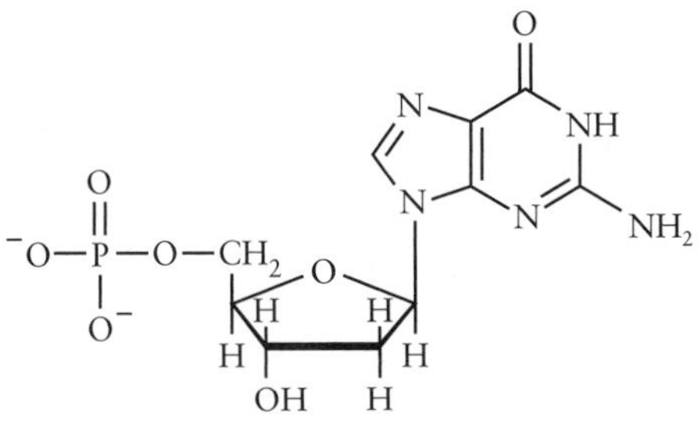

a. The nitrogen base. What is the name of the nitrogen base?
b. The monosaccharide. What is the name of the monosaccharide?
c. The phosphate group.
d. The 5′ carbon.
e. The anomeric carbon.
f. The 2′ carbon.
g. Would this nucleotide be found in RNA or DNA? Explain.

15.35 Write the structure of the nucleotide you would find in DNA containing adenine as the base.

15.36 Write the structure of the nucleotide you would find in RNA containing guanine as the base.

15.37 When two nucleotides are joined together to form a dinucleotide, what two functional groups are involved in making the new bond? What new functional group is produced?

15.38 How many different RNA pentanucleotides—a sequence of 5 nucleotides—can be constructed?

15.39 How many different RNA hexanucleotides—a sequence of 6 nucleotides—can be constructed?

15.40 Construct a nucleotide from the three basic parts shown below. Label the 3′ alcohol and the 5′ alcohol. Would this nucleotide be found in RNA or DNA? Explain.

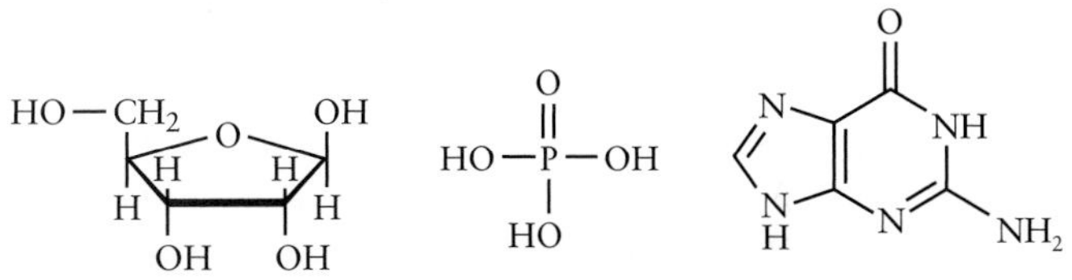

15.41 Construct a nucleotide from the three basic parts shown below. Label the 3′ alcohol and the 5′ alcohol. Would this nucleotide be found in RNA or DNA? Explain.

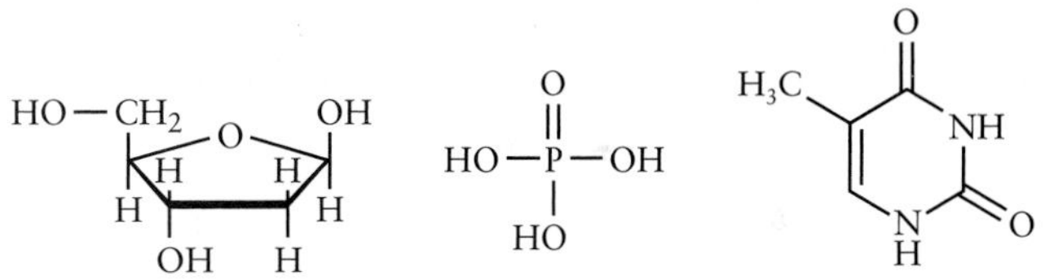

15.42 Draw the structure of dTAC.

15.43 Draw the structure of GUC.

DNA

15.44 What is the three-dimensional shape of DNA?

15.45 What parts of a nucleotide are located on the outside of the three-dimensional DNA structure? Why?

15.46 What part of a nucleotide is located on the inside of the three-dimensional DNA structure? Why?

15.47 What type of intermolecular force of attraction exists between base pairs?

15.48 How many hydrogen bonds link an A–T base pair?

15.49 How many hydrogen bonds link a C–G base pair?

15.50 For the sequence of nucleotides dTATCGC on a segment of one strand of DNA, indicate the complimentary sequence of bases that would be on the other DNA strand. How many hydrogen bonds in total hold this segment of DNA together?

15.51 For the sequence of nucleotides dCGATAG on a segment of one strand of DNA, indicate the complimentary sequence of bases that would be on the other DNA strand. How many hydrogen bonds hold this segment of DNA together?

15.52 For each sequence of nucleotides on a segment of one strand of DNA shown below, indicate the complementary sequence of bases that would be on the other DNA strand.
    a. dCTAGGC   **c.** dTTGGAA
    b. dACTGAA   **d.** dGGTACT

15.53 For each sequence of nucleotides on a segment of one strand of DNA shown below, indicate the complementary sequence of bases that would be on the other DNA strand.
    a. dTATGCC   **c.** dCCTATT
    b. dAACCTG   **d.** dGTATCC

15.54 Explain how a large molecule like DNA can fit inside the nucleus of a cell.

15.55 How many chromosomes are found in most human cells?

15.56 What is your genome? Approximately how many base pairs does your genome contain?

15.57 Name two diseases associated with a protein whose gene is located on chromosome 6.

15.58 Why is DNA replication a critical function for life?

15.59 What is the first step of DNA replication?

15.60 What is the function of the enzyme *helicase*?

15.61 Consider a portion of double-stranded DNA with the following base pairing sequence:
    dGGTACGCTT
    dCCATGCGAA
    Write the sequence of nucleotides found in each product after replication and label the original parent strands and the new daughter strands.

15.62 Consider a portion of double-stranded DNA with the following base pairing sequence:
    dCATTAAGCCG
    dGTAATTCGGC
    Write the sequence of nucleotides found in each product after replication and label the original parent strands and the new daughter strands.

15.63 What kinds of bonds are formed between the base pairs of the parent and daughter strands of DNA?

15.64 What kind of reaction does the enzyme *DNA polymerase* catalyze?

15.65 What happens when *DNA polymerase* detects a mistake in a daughter strand? What is the mistake rate after proofreading by *DNA polymerase*?

15.66 Find the errors in the following sequences of parent–daughter strands and replace the incorrect nucleotide with the correct one.
    a. dATTCCGTA parent strand
       dCAAGGTAT daughter strand
    b. dGGGCCCTTTAA parent strand
       dCCCAGGAAGTT daughter strand

15.67 Find the errors in the following sequences of parent–daughter strands and replace the incorrect nucleotide with the correct one.
    a. dCGTACTGGA parent strand
       dCCATGAACT daughter strand
    b. dGAGTATCT parent strand
       dCTCCTCGA daughter strand

RNA and Protein Synthesis

15.68 What are the three major forms of RNA?

15.69 Is RNA single or double stranded? Does it contain D-ribose or 2-deoxy-D-ribose? Does it contain the nitrogen base T or U?

15.70 What two processes are involved in synthesizing a protein from a nucleotide sequence on DNA?

15.71 What kind of reaction does *RNA polymerase* catalyze?

15.72 For the following sequences found on a DNA template strand, write the sequence of nucleotides produced on mRNA.
    a. dATAGGCCTTA   **c.** dCGATCGATCG
    b. dTTAACCGGAA

15.73 For the following sequences found on a DNA template strand, write the sequence of nucleotides produced on mRNA.
    a. dCCGGAATATA   **c.** dGTACACGTCG
    b. dAAGGCCAATT

15.74 What is a codon? What is the genetic code?

15.75 What are the two important regions of a tRNA molecule?

15.76 What are the steps involved in translation, the process of building a protein from mRNA?

15.77 What amino acid will a tRNA molecule containing the anticodon AAA be carrying?

15.78 What amino acid will a tRNA molecule containing the anticodon GAC be carrying?

15.79 What happens when the ribosome reaches a stop codon? What are the three stop codons?

15.80 Specify the codons for threonine. Why are there so many codons for threonine?

15.81 Specify the codons for alanine.

15.82 For the following nucleotide sequences on mRNA, indicate the anticodons on the three tRNA molecules recruited. What is the amino acid sequence of the tripeptide formed?
    a. AAUAGUGUG   **c.** CACCGGUGG
    b. CCCUUUGGG   **d.** UCCUUAGCA

15.83 For the following nucleotide sequences on mRNA, indicate the anticodons on the three tRNA molecules recruited. What is the amino acid sequence of the tripeptide formed?
    a. GUUGCUCGU   **c.** AGUAACUCG
    b. CUACGCGGU   **d.** UAUGAUACC

15.84 For the following sequences on DNA, write the corresponding mRNA sequence. What tRNA molecules would be involved in building the protein? What is the amino acid sequence?
    a. dTTTTCATATTAA   **c.** dGTTGCTCTGATA
    b. dCGTCAAAAAGGG   **d.** dTCATGTTGGCGA

15.85 For the following sequences on DNA, write the corresponding mRNA sequence. What tRNA molecules would be involved in building the protein? What is the amino acid sequence?
    a. dTTACCCGACGGC   **c.** dAUGACCACAGAA
    b. dGAAGCAACCATA   **d.** dCGACATCCTCTA

15.86 Indicate whether the following normal mRNA sequence would produce the same or a different dipeptide if the mutation shown occurred (assume all reading occurs from left to right):

DNA dGGTGCT

mRNA CCACGA

a. The third nucleotide on DNA is replaced by C.
b. The sixth nucleotide on DNA is replaced by A.
c. The second nucleotide on DNA is deleted.
d. The fourth nucleotide on DNA is deleted.

15.87 Indicate whether the following normal mRNA sequence would produce the same or a different dipeptide if the mutation shown occurred (assume all reading occurs from left to right):

DNA dAGTAAA

mRNA UCAUUU

a. The third nucleotide on DNA is replaced by C.
b. The sixth nucleotide on DNA is replaced by T.
c. The second nucleotide on DNA is deleted.
d. The fourth nucleotide on DNA is deleted.

15.88 **Critical Thinking Question:** A sequence of nucleotides on mRNA will code for only one sequence of amino acids. If you were given a sequence of amino acids in a protein, could you predict the unique sequence of nucleotides on mRNA that coded for that protein? Explain your answer.

Chemistry in Medicine

15.89 Does the HIV virus contain DNA or RNA as its genome?

15.90 How is the HIV virus encased?

15.91 What are the three enzymes that HIV uses for replication?

15.92 Why do viruses use the host cell to reproduce their genetic material and synthesize viral proteins?

15.93 What type of cell does HIV infect?

15.94 How is the genomic RNA of an HIV virus converted to DNA? What viral enzyme is used in this process?

15.95 What is a retrovirus? Why is HIV considered a retrovirus?

15.96 What is the function of HIV *protease*?

15.97 Why is a combination of drug therapies needed to treat HIV?

15.98 How does the drug AZT work?

15.99 How do *protease* inhibitors work?

15.100 **Critical Thinking Question:** Suppose that the sequence AUGCUGGAC is found on an HIV RNA. Write the viral DNA sequence that would be produced by *reverse transcriptase* and incorporated into the host's DNA.

Answers to Practice Exercises

15.1 b.

15.2

Phosphate
Nitrogen base: thymine
Monosaccharide: 2-deoxyribose
Anomeric carbon

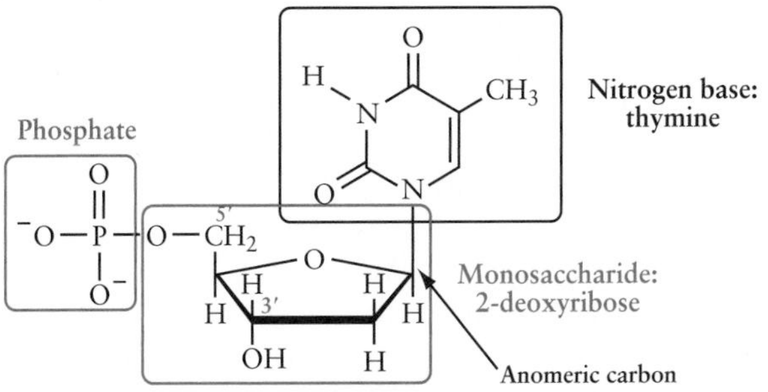

g. found in DNA

15.3 Cytosine, C; guanine, G; adenine, A; uracil, U.

15.4 Uracil, U, is not found in DNA.

15.5

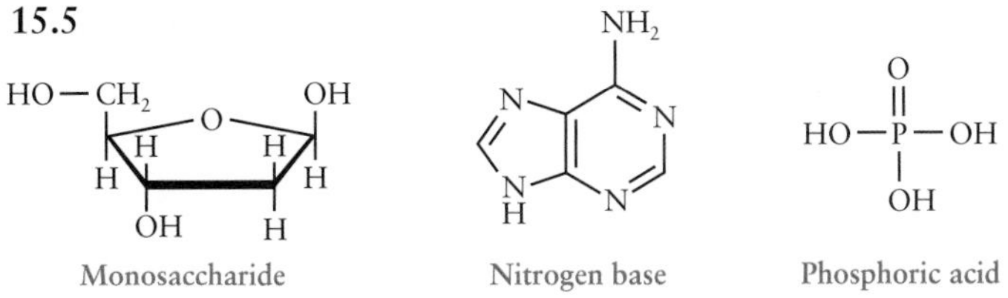

Monosaccharide
Nitrogen base
Phosphoric acid

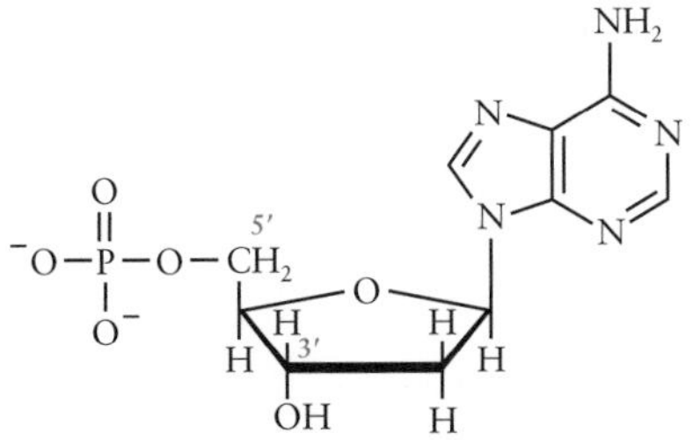

Nucleotide found in DNA, because monosaccharide is 2-deoxy-D-ribose

15.6 $4^2 = 16$

AA	UA	CA	GA
AU	UU	CU	GU
AC	UC	CC	GC
AG	UG	CG	GG

15.7

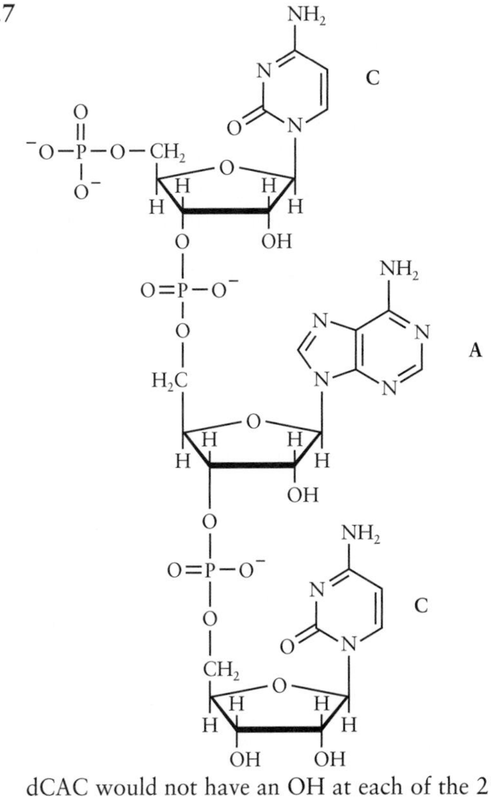

dCAC would not have an OH at each of the 2′ carbon atoms in each nucleotide.

15.8 a. ATP is classified as a nucleotide because it contains a nitrogen base, D-ribose, and a phosphate group.
b. ATP contains three phosphate groups rather than one, which makes it different from an RNA nucleotide.
c. adenine

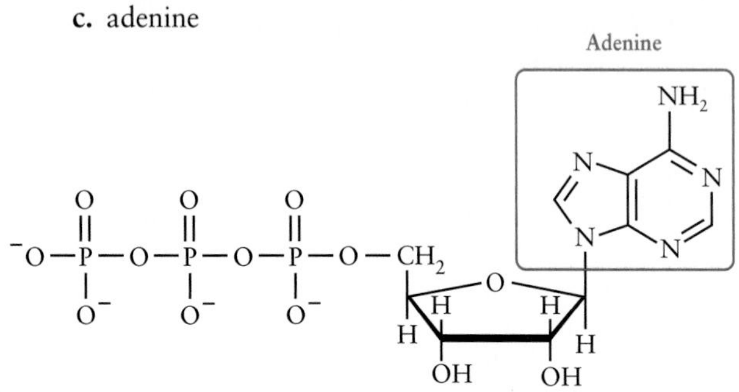

ATP
adenosine triphosphate

15.9 a. $3' \rightarrow 5'$: AACCGT
b. $3' \rightarrow 5'$: TACGGT

15.10 a. bases
b. sugar phosphate backbone
c. bases
d. bases

15.11 See Figure 15.11.

15.12 Histones are positively charged proteins that are attracted to the negatively charged DNA, neutralizing the charge, and thus allowing DNA to supercoil into a compact structure.

15.13 A genome is the entire DNA sequence of an organism.

15.14 There are 3,000 base pairs in the average human gene.

15.15 It means that the DNA in chromosome 6 contains the nucleotide sequence that codes for the amino acids that make up the primary structure of the opioid receptor, a protein.

15.16

parent 1:	dGACCTAGCCC
daughter 1:	dCTGGATCGGG
parent 2:	dCTGGATCGGG
daughter 2:	dGACCTAGCCC

15.17 *DNA polymerase* catalyzes phosphoryl transfer reactions that join nucleotides. *DNA polymerase* is an enzyme; therefore it is a protein. This enzyme makes a mistake once in every 10,000 nucleotides, but after proofreading, the error rate drops to once in every billion nucleotides.

15.18 rRNA, mRNA, and tRNA. rRNA makes up part of the ribosome. A ribosome consists of nucleic acids and proteins.

15.19 Transcription is the process in which a gene, a segment of the template strand of DNA, is used as a template for synthesizing a single strand of mRNA with a complementary sequence of nucleotides.

15.20 Ribosomes are the protein-building organelles of the cell. They produce polypeptides from the information relayed by mRNA.

15.21 DNA is a much larger molecule than RNA, containing many more nucleotides.

15.22 DNA is found in the cell nucleus (mitochondria also have their own DNA). RNA is found in ribosomes in the cytoplasm.

15.23 GGG: glycine; AAA: lysine.

15.24 histidine: CAU and CAC; leucine: UUA, UUG, CUU, CUC, CUA, CUG

15.25 a. mRNA codons: GGU ACU AUU UAA
tRNA anticodons: CCA UGA UAA AUU
amino acids: Gly - Thr - Ile - Stop
b. mRNA codons: GGU GAC CGA CAU
tRNA anticodons: CCA CUG GCU GUA
amino acids: Gly - Asp - Arg - His

15.26 mRNA codons: AAA UUA GAG ACA
tRNA anticodons: UUU AAU CUC UGU
tetrapeptides: Lys - Leu - Gly - Thr

15.27 Stop Codons: UGA, UAA, UAG. A stop codon signifies the end of a gene.

15.28 a. DNA: dGAATGA
mRNA: CUUACU
amino acids: Leu-Thr
b. DNA: dTAGTGA
mRNA: AUCACU
amino acids: Ile-Thr   no effect
c. DNA: dTA/TGA
mRNA: AUACU
amino acids: Ile-?   no effect on first amino acid
d. DNA: d/AATGA
mRNA: UUACU
amino acids: Leu-?   changes first amino acid

Nuclear Chemistry and Medicine

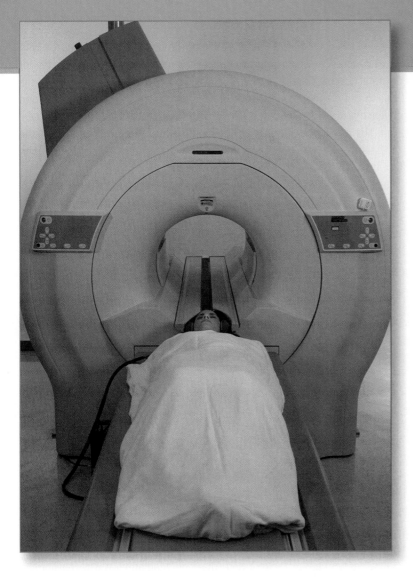

Nuclear chemistry plays an important role in therapeutic and diagnostic medicine. Here a patient is undergoing MRI, used to find injury to soft tissues.
[© Royalty-Free/Corbis]

OUTLINE

This icon indicates that a **Problem-Solving Tutorial** is available at www.whfreeman.com/gob

Diagnosing and Treating Cancer: The Powerful Role of Nuclear Chemistry

"Radioactive Man," a character on *The Simpsons* TV show, became a superhero after a radioactive shard of metal became lodged in his head from a nuclear accident. In reality, radioactive pieces of metal are actually inserted into people routinely, and not by accident. Although these people do not become superheroes, cancerous tissues are destroyed!

Each year in the United States about 0.5% of the population is diagnosed with some form of cancer. The most common cancer diagnosis in the United States for men is prostate cancer, and for women, breast cancer. Nuclear chemistry plays a powerful role in the diagnosis and treatment of many types of cancer. Consider John Doe, a 52-year-old male who visits his physician complaining of frequent and painful urination. His physician orders some screening tests, followed by a *biopsy*—the removal of tissue for examination under the microscope. The results show that John has prostate cancer.

The prostate is a gland in the male reproductive system that sits below the bladder. Cancerous tumors cause the prostate to grow, which can cause difficult and painful urination. To tell how far the prostate cancer has progressed, medical professionals use imaging techniques that allow them to "see" what's going on inside the body without performing surgery. Such techniques use the radiation from nuclear reactions to provide an image of the tissue of interest. In this chapter, you will learn about nuclear reactions as well as some important imaging techniques such as X ray, MRI, and CT scans.

Two of these imaging techniques were used to determine the progression of John Doe's prostate cancer: MRI was used to image his seminal vesicles (see Figure 16-1), and CT was used to image his pelvis. The results of these diagnostic tests allowed John Doe and his physician to determine the best course of treatment for him.

Treatments for prostate cancer include surgery and/or radiation therapy. Watchful waiting is often the choice when a slow-growing, early stage of cancer is found in an older man. Surgical removal of the prostate is usually reserved for cancer that has failed to respond to radiation therapy. Radiation therapy uses the radiation produced in a nuclear reaction to destroy cancer cells. The radiation is targeted at the cancerous area: The rapidly dividing cancer cells are more susceptible to radiation damage than the surrounding healthy tissue. John Doe and his physician decided that he would undergo a type of radiation treatment called *sealed source radiotherapy*, the most widely used treatment for prostate cancer.

Sealed source radiotherapy involves placing small rods or "seeds" containing radioactive isotopes directly into the cancerous tumors. A radioactive isotope, or *radioisotope*, is an atom with an unstable nucleus that emits radiation as it is transformed into a more stable nucleus—a process known as *nuclear decay*.

For John Doe's therapy, radioactive seeds made of the radioisotope iodine-125 were inserted directly into his tumor. Iodine-125 emits radiation in the form of high-energy electrons, called beta particles (β). These β particles collide with the tissue in the surrounding area, destroying the cancer cells. Because radioactive seeds were placed directly in the cancerous tissue, they targeted those cells preferentially while harming fewer healthy cells. After several weeks or months, the level of radioactivity within the implants eventually diminishes to undetectable levels. The seeds will remain in John Doe's body, with no lasting harmful effects.

Sealed source radiotherapy has become a common treatment for many cancers. Although the radioactive implants did not transform Mr. Doe into a superhero, they did destroy his prostate cancer.

Figure 16-1 The prostate gland is part of the male reproductive system.

Bladder

Prostate

Urine

Urethra

Seminal vesicle

Radioactive seeds about the size of a grain of rice may be implanted directly into malignant tumors. [Best Medical International, Inc.]

In the first chapter of this text, you learned about the three fundamental particles that make up the atom: protons, neutrons, and electrons. Although the majority of this text has focused on the role of the electron, in this chapter you will study changes in the *nucleus* of an atom—**nuclear chemistry.** Changes to the composition of the nucleus of an atom are known as **nuclear reactions.** Protons and neutrons are held together by the strongest forces in the universe, and harnessing or unleashing the power of the nucleus as atomic energy is the basis for nuclear power and atomic bombs.

While too much radiation can damage DNA, causing mutations and cancer, the radiation produced in a nuclear reaction can also be used in the diagnosis and treatment of disease, an area of medicine known as **nuclear medicine,** to which you were introduced in the opening vignette. One-third of all patients in US hospitals receive diagnostic or therapeutic treatments that involve nuclear medicine. This chapter gives you an overview of the fundamental principles of nuclear chemistry and applications of nuclear chemistry in the diagnosis of disease.

16.1 Radioisotopes and Nuclear Reactions

Recall from Chapter 1 that almost all elements exist in different isotopic forms. All of the isotopes of a given element have the same number of protons, which is equal to its atomic number. Each isotope of a particular element, however, will have a different number of neutrons, reflected in its mass number, the sum of the protons and neutrons in the nucleus, as shown for I-125 below:

$$\text{Mass number:} \quad {}^{125}_{53}\text{I}$$
$$\text{Atomic number:}$$

Radioisotopes

Some isotopes are unstable, due to an *imbalance* in the number of neutrons and protons in the nucleus. For most isotopes, the optimal ratio of neutrons to protons ranges from 1.0 to 1.5, depending on its mass number. If this ratio is either too high or too low, the nucleus will be unstable. All isotopes with an atomic number greater than 82 are unstable because they contain *too many* protons and neutrons. Unstable isotopes are known as **radioactive isotopes** or **radioisotopes.**

The nucleus of a radioactive isotope undergoes a natural process known as **radioactive decay** to become a more stable nucleus. Radioactive decay typically produces an isotope with a different atomic number; and hence the product isotope is a different element. Radioactive decay is always accompanied by the release of a form of energy called **radiation.**

There are more than 300 naturally occurring isotopes, and 36 of these are radioactive. About 80% of the radiation you are exposed to comes from natural sources, known as **background radiation.** These natural sources include cosmic rays from the sun and radon gas filtering up from the ground. Your body also contains some naturally radioactive substances, primarily K-40 and organic molecules containing trace amounts of C-14.

In addition to naturally occurring radioisotopes, there are the more than 2000 man-made isotopes called **artificial radioisotopes.** All man-made isotopes are radioactive. Common devices containing artificial radioisotopes include smoke detectors, computer monitors, and the instruments used for some medical imaging techniques.

Air travel increases your exposure to radiation, because cosmic radiation is 100 times greater at cruising altitudes than it is on the ground.

WORKED EXERCISE 16-1 Radioisotopes

Referring to the periodic table on the inside front cover, fill in the empty cells in the table below for the given radioisotopes. This table shows the relationship between atomic number, mass number, element symbol, number of neutrons, and number of protons.

Radioisotope	Atomic Number	Mass Number	Number of Neutrons	Number of Protons
Ba-131				
	36	79		

SOLUTION

Radioisotope	Atomic Number	Mass Number	Number of Neutrons	Number of Protons
Ba-131	56	131	75	56
Kr-79	36	79	43	36

PRACTICE EXERCISE

16.1 Referring to the periodic table on the inside front cover, fill in the empty cells in the table below for the given radioisotopes.

Radioisotope	Atomic number	Mass Number	Number of Neutrons	Number of Protons
I-131				53
	86	205		
		40		19
	79		119	
Co-60				

Radiation

Radiation from the decay of radioisotopes appears in either of two forms: high-energy electromagnetic radiation, such as X rays and gamma rays, or high-energy particles, such as α particles, β particles, and positrons (see Table 16-1).

Electromagnetic radiation is a form of energy that travels through space as a wave, at the speed of light. The **wavelength** of electromagnetic radiation is defined as the distance between wave troughs (or crests) as shown in Figure 16-2.

Electromagnetic radiation is classified into regions according to its wavelength. Electromagnetic radiation ranges from long-wavelength radio waves to short-wavelength X rays and gamma rays. The entire range is called the *electromagnetic spectrum*. Visible light, the region of the spectrum that you can actually see, is in the middle of the electromagnetic spectrum, with wavelengths in the range of 400–700 nm. The electromagnetic spectrum is shown in Figure 16-3; it includes the classifications X ray, ultraviolet, visible, infrared, microwave, and radio wave.

The energy associated with a particular type of electromagnetic radiation depends on its wavelength. The wavelength of electromagnetic radiation is inversely proportional to its energy: the shorter the wavelength, the higher the energy. Thus, at one end of the electromagnetic spectrum are low-energy radio waves and at the other end of the spectrum are high-energy gamma rays.

Table 16-1 Types of Radiation

Electromagnetic radiation	X rays
	Gamma rays
High-energy particles	α particle
	β particle
	Positron

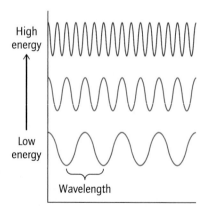

Figure 16-2 Electromagnetic radiation is defined by its wavelength—the distance between wave troughs, and its energy—the shorter the wavelength, the higher the energy.

The speed of light is 3×10^8 m/second. All forms of electromagnetic radiation travel at this speed (in a vacuum).

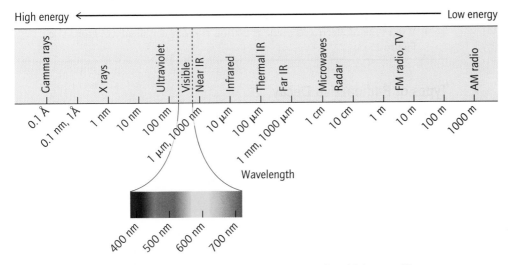

Figure 16-3 The electromagnetic spectrum is divided into regions from high energy/shorter wavelengths (left of illustration) to low energy/long wavelengths (right of illustration). The narrow region of the electromagnetic spectrum known as visible light has been enlarged.

High-energy electromagnetic radiation can damage biological tissue. The most damaging forms of electromagnetic radiation are gamma rays and X rays, both of which are emitted during certain radioactive decay processes. The biological effects of radioactive decay will be described in Section 16.2.

> If you have ever had a sunburn, you know from personal experience that ultraviolet radiation (UV) is damaging to tissue.

WORKED EXERCISES 16-2 Electromagnetic Radiation

1 Referring to Figure 16-3, answer the questions below by choosing from among the following forms of electromagnetic radiation:
i. X ray   ii. visible   iii. ultraviolet

a. Which form causes the greatest damage to biological tissue?
b. Which form is the lowest in energy?
c. Which has the shortest wavelength?
d. Which travels at the fastest speed?

2 Mice cannot see red light, but they do see blue and green light. Red light has a wavelength of 700 nm and blue light has a wavelength of 450 nm. Which type of light has more energy, red light or blue light?

SOLUTIONS

1 a. Of the three forms of electromagnetic radiation listed, X rays are the most damaging to biological tissue.
 b. Visible light is the lowest in energy of the three listed.
 c. X rays have the shortest wavelength of the three listed.
 d. All forms of electromagnetic radiation travel at the *same* speed: the speed of light—3×10^8 m/sec.

2 Blue light has a shorter wavelength than red light because 450 nm < 700 nm, therefore it has more energy.

PRACTICE EXERCISES

16.2 Referring to Figure 16-3, answer the questions below by choosing from among the following forms of electromagnetic radiation:

i. gamma ray   ii. microwave   iii. radio wave

a. Which form causes the greatest damage to biological tissue?
b. Which form is the lowest in energy?
c. Which contains less energy than visible light but more energy than radio waves?

16.3 Bumble bees can "see" UV light. How is UV light different from visible light?

16.4 What type of electromagnetic radiation is radar classified as? *Hint:* Use Figure 16-3.

Types of Radioactive Decay

To achieve stability, radioisotopes undergo a process of radioactive decay in which they emit radiation in the form of high-energy particles and/or electromagnetic radiation. The five types of radiation that can be emitted during radioactive decay are given in Table 16-2. High-energy particles include α particles, β particles, and positrons; high-energy electromagnetic radiation includes X rays and gamma rays.

α Decay and the Balanced Nuclear Equation

Certain radioisotopes undergo α decay and emit **α particles.** An α particle is a slow moving, high-energy particle consisting of 2 protons and 2 neutrons. Its nuclear symbol is $_2^4\alpha$. The composition of an alpha particle is the same as a helium nucleus; hence, its nuclear symbol is also written as $_2^4$He. Keep in mind, however, that there are no corresponding electrons, so an α particle is not the same as a helium atom. α particles are extremely dense and carry a 2+ charge. Many elements with an atomic number greater than 82, and some with a lower atomic number, have isotopes that undergo α decay.

After emission of an α particle, the mass number of the isotope decreases by four and the atomic number decreases by two. Radioactive decay can be depicted by writing a **nuclear equation,** as shown below for the α decay of Th-232:

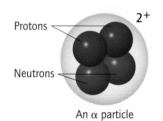

Protons

Neutrons

An α particle

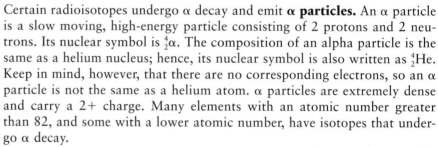

$$_{90}^{232}\text{Th} \longrightarrow \,_{88}^{228}\text{Ra} + \boxed{_2^4\alpha}$$

Parent nuclide   Daughter nuclide   α particle

The convention is to write the radioisotope, also known as the **parent nuclide,** to the left of the arrow, and the new isotope as well as the emitted particle to the right of the arrow. The atomic number and mass number for all species in the equation are included. When a different element is formed, it is referred to as a **daughter nuclide.** In the case of α decay, most of the energy released in the nuclear reaction is in the form of an α particle.

*In a **balanced nuclear equation** the sum of the atomic numbers (subscripts) of all species on the left side of the arrow must equal the sum of the*

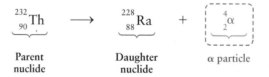

Table 16-2 Types of Radioactive Decay

Radiation Type	Nuclear Symbol	Energy Type
α particle	$_2^4\alpha$ or $_2^4$He	High-energy particles
β particle	$_{-1}^0\beta$ or $_{-1}^0$e	
Positron	$_{+1}^0\beta$	
X ray	—	Electromagnetic radiation
Gamma ray	γ	

atomic numbers of all species on the right side of the arrow; likewise for the mass numbers (superscripts). The nuclear equation below shows that the equation on the facing page is balanced.

$$(\text{Superscripts} \quad 232 \quad = \quad 228 \quad + \quad 4)$$

$$^{232}_{90}\text{Th} \longrightarrow {}^{228}_{88}\text{Ra} + {}^{4}_{2}\alpha$$

$$(\text{Subscripts} \quad 90 \quad = \quad 88 \quad + \quad 2)$$

WORKED EXERCISE 16-3 Alpha Decay

Problem-Solving Tutorial:
Alpha Decay

Predict the daughter nuclide for the α decay of U-238.

SOLUTION

Write the radioisotope with its atomic and mass numbers on the left of the reaction arrow and the α particle and the daughter nuclide on the right of the reaction arrow. Determine the identity of the daughter nuclide by subtracting 4 from the mass number and subtracting 2 from the atomic number of the parent nuclide. In this case:

$$\text{Mass number} = 238 - 4 = 234$$
$$\text{Atomic number} = 92 - 2 = 90$$

Thus, in this example, the daughter nuclide is $^{234}_{90}\text{X}$. Then, using the periodic table, replace X with the element symbol that corresponds to the atomic number. In this case, atomic number 90 corresponds to thallium, hence $^{234}_{90}\text{Th}$ is the daughter nuclide. The balanced nuclear equation is:

$$\underbrace{^{238}_{92}\text{U}}_{\substack{\text{Parent} \\ \text{nuclide}}} \longrightarrow \underbrace{^{234}_{90}\text{Th}}_{\substack{\text{Daughter} \\ \text{nuclide}}} + \underbrace{^{4}_{2}\alpha}_{\alpha \text{ particle}}$$

PRACTICE EXERCISES

16.5 List the five types of radioactive decay. Which of the five types of decay are forms of electromagnetic radiation?

16.6 Predict the daughter nuclide formed when polonium-210 undergoes α decay. Write the balanced nuclear equation.

16.7 A radioisotope produces radon-222 by α decay. Radon-222 may build up in the basements of homes located in areas with a high concentration of the parent nuclide in the surrounding earth. What is the parent nuclide? Write the balanced nuclear equation for this radioactive decay process.

β Decay

Beta decay occurs in an unstable nucleus that has *too many neutrons* compared to protons. *In the β-decay process, a neutron is converted into a proton and a high-energy electron.* The daughter nuclide will have the same mass number as the parent nuclide, but its atomic number is increased by one. Hence, a radioisotope that undergoes β decay becomes the next higher element on the periodic table.

More importantly, a high-energy electron traveling at about 90% of the speed of light is emitted from the nucleus into the surrounding area. This

Table 16-3 Nuclear Symbols for Subatomic Particles

Subatomic Particle	Symbol
Proton	$_1^1p$
Neutron	$_0^1n$
Electron	$_{-1}^0e$
β particle (high-energy electron)	$_{-1}^0\beta$

high-energy electron is known as a **β particle**, and is written as either $_{-1}^0e$ or $_{-1}^0\beta$. In this text $_{-1}^0\beta$ will be used to indicate a high-energy electron, while $_{-1}^0e$ will be reserved for orbital electrons. When subatomic particles appear in nuclear equations, the symbols shown in Table 16-3 are used to represent them.

Consider, for example, the radioactive decay of P-32, used in the detection of breast cancer and eye tumors. P-32 undergoes β decay according to the balanced nuclear equation below.

$$\text{(Superscripts: } 32 = 32 + 0\text{)}$$

$$_{15}^{32}P \rightarrow _{16}^{32}S + _{-1}^0\beta$$

$$\text{(Subscripts: } 15 = 16 + (-1)\text{)}$$

The nuclear equation shows that the radioactive decay process converts P-32, an unstable isotope of phosphorus, into S-32, a stable isotope of sulfur and the next element on the periodic table. A β particle is emitted in the process. The nuclear equation is balanced, because the subscripts on both sides of the equation are equal: $15 = 16 + (-1)$; as are the superscripts: $32 = 32 + 0$.

 **Problem-Solving Tutorial:** Beta Decay

WORKED EXERCISE 16-4 Beta Decay

The radioisotope cerium-141 is known to undergo β decay. It is often used to assess blood flow through the heart.

a. How many protons and neutrons does this radioisotope contain?
b. Write the balanced nuclear reaction.
c. What is the daughter nuclide produced from β decay of Ce-141?
d. What form of radiation is produced in this reaction? Is it electromagnetic radiation or a high-energy particle?

SOLUTION
a. There are 58 protons and 83 neutrons in a Ce-141 nucleus. Use the periodic table to obtain the atomic number, and subtract this value from the mass number to obtain the number of neutrons.
b. To write the balanced nuclear reaction, recall that the daughter nuclide is the next element on the periodic table. Alternatively, you can write the daughter nuclide as X, then insert the values for mass number and atomic number that you obtain when you balance the nuclear equation, and then look up the element symbol on the periodic table that corresponds to the atomic number calculated:

$$\text{(Superscripts: } 141 = 141 + 0\text{)}$$

$$_{58}^{141}Ce \rightarrow _{59}^{141}Pr + _{-1}^0\beta$$

$$\text{(Subscripts: } 58 = 59 + (-1)\text{)}$$

c. Pr = 141
d. A β particle is emitted, which is a type of high-energy particle.

PRACTICE EXERCISES

16.8 The radioisotope I-131 is known to undergo β decay. It is often used to diagnose thyroid conditions.

a. How many protons and neutrons does this radioisotope contain?

   b. Write the balanced nuclear equation.
   c. What is the daughter nuclide produced from the β decay of I-131?
   d. What form of radiation is produced in this reaction? Is it electromagnetic radiation or a high-energy particle?

16.9 Gold-198 is used in the diagnosis of kidney disease. Gold-198 undergoes β decay. Write the balanced nuclear equation and label the parent and daughter nuclides.

16.10 A radioisotope that undergoes β decay contains too many _____ in its nucleus compared to _____. Emission of a β particle results in a daughter nuclide that is the _____ element on the periodic table relative to the parent nuclide.

16.11 **Critical Thinking Question:** Complete the balanced nuclear equation below, showing the conversion of a neutron into a proton and a β particle, using the symbols given in Table 16-3.

$$_{0}^{1}\text{n} \longrightarrow \boxed{} + \ _{-1}^{0}\beta$$

Positron Decay

A radioisotope decays by positron emission when its nucleus contains *too many protons* compared to neutrons. A **positron** is similar to a β particle in mass and energy, but it is positively charged. A positron has the symbol $_{+1}^{0}\beta$, and is formed when a proton is converted into a neutron and a positron.

With positron emission, the daughter nuclide has the same mass number as the parent, but its atomic number is one less than the parent. Hence, a radioisotope that undergoes positron emission becomes the previous element on the periodic table.

An example of a positron emitter is N-13, shown in the equation below.

Superscripts: 13 = 13 + 0

$$_{7}^{13}\text{N} \rightarrow \ _{6}^{13}\text{C} + \underbrace{_{+1}^{0}\beta}_{\text{Positron}}$$

Subscripts: 7 = 6 + (1)

The nuclear equation shows that the radioactive decay process converts N-13, an unstable isotope of nitrogen, into carbon-13, a stable isotope of carbon and the previous element on the periodic table. A positron is emitted as part of the decay process. The nuclear equation is balanced, because the subscripts on both sides of the equation are equal: 7 = 6 + 1; as are the superscripts: 13 = 13 + 0.

Problem-Solving Tutorial:
Positron Decay

WORKED EXERCISE 16-5   Positron Decay

In brain scans both C-11 and F-18 are used. Both of these isotopes are positron emitters. Write the balanced nuclear reaction for the radioactive decay of C-11. What form of radiation is produced in these reactions? Is it electromagnetic radiation or a high-energy particle?

SOLUTION

Superscripts: 11 = 11 + 0

$$_{6}^{11}\text{C} \rightarrow \ _{5}^{11}\text{B} + \underbrace{_{+1}^{0}\beta}_{\text{Positron}}$$

Subscripts: 6 = 5 + (1)

A positron, which is a high-energy particle, is emitted.

PRACTICE EXERCISES

16.12 Write the balanced nuclear reaction for the radioactive decay of F-18, a positron emitter. What form of radiation is produced in this nuclear decay process? Is it electromagnetic radiation or a high-energy particle?

16.13 What high-energy particle has the opposite charge of a positron?

16.14 In positron emission how does the daughter nuclide differ from the parent nuclide in atomic number? In mass number?

16.15 Positron emitters contain too many _____ in their nucleus compared to _____. Positron emission turns a _____ into a _____ and a _____.

16.16 **Critical Thinking Question:** Complete the balanced nuclear equation below showing the conversion of a proton into a neutron and a positron.

$$^1_1\text{p} \longrightarrow \boxed{} + ^{\ 0}_{+1}\beta$$

X Rays and Gamma Radiation

X rays and **gamma radiation** are both short-wavelength, high-energy forms of electromagnetic radiation. X rays accompany some forms of radioactive decay, and when focused on bone or other tissue can be used to create valuable diagnostic images such as X ray and CT scans. To learn more about X rays as a diagnostic tool, see this chapter's Chemistry in Medicine: The Role of Nuclear Radiation in Imaging the Body.

Gamma radiation is of even higher energy than X rays, and can easily pass through a person. Gamma-ray emission accompanies almost all forms of radioactive decay described thus far. However, gamma emission is not usually shown in a nuclear equation because it does not affect the atomic number or mass number of the isotopes involved. Few radioisotopes are pure gamma emitters; they emit some other type of radiation as well.

After radioactive decay has occurred, the daughter nuclide is often in an **excited state,** a condition in which the nucleus still contains excess energy. An isotope in its excited state is referred to as a **metastable** isotope, and is notated by the abbreviation "m" following the mass number of the isotope. The daughter nuclide releases its excess energy in returning to the ground state by releasing a pulse of gamma radiation.

The most common gamma emitter used in medicine today is *m*etastable technetium, Tc-99m. When Tc-99m decays to Tc-99, gamma radiation is emitted:

$$^{99m}_{43}\text{Tc} \longrightarrow ^{99}_{43}\text{Tc} + \gamma$$

Since no particle is emitted, the atomic number, mass number, and element symbol for the daughter nuclide are the same as for the parent nuclide; the nuclides differ only in the amount of energy they possess.

WORKED EXERCISE 16-6 Gamma Radiation

Cesium-137 undergoes pure β decay to barium-137m, which then undergoes gamma emission to a stable nuclide.

a. Write the nuclear equations for this two-step nuclear reaction sequence.
b. What is the stable nuclide formed after this two-step process?
c. What does the "m" in barium-137m stand for?

SOLUTION

a. $^{137}_{55}\text{Cs} \longrightarrow ^{137m}_{56}\text{Ba} + ^{\ 0}_{-1}\beta \qquad ^{137m}_{56}\text{Ba} \longrightarrow ^{137}_{56}\text{Ba} + \gamma$

The food irradiation process uses gamma rays to eliminate disease-causing microorganisms such as *E. coli* and *Salmonella* from fruits and vegetables. It also reduces spoilage caused by bacteria, giving food a longer shelf-life. The food itself does *not* become radioactive, because only the electromagnetic radiation from the radioisotope, not the radioisotope itself, comes in contact with the food.

b. Ba-137.

c. The "m" in Ba-137m indicates that the isotope is a metastable isotope, an atom with its nucleus in an excited state.

PRACTICE EXERCISE

16.17 The element cobalt has one stable isotope and 22 radioisotopes, four of which are metastable states.

  a. How do the various isotopes of cobalt differ?

  b. What is a metastable state? What is the abbreviation for this state?

  c. Show the nuclear reaction that Co-60m undergoes to become Co-60.

Half-Life

You have seen that radioisotopes decay to produce radiation and often a new daughter nuclide. As a sample of a radioisotope decays, less and less of the parent nuclide remains over time. How quickly do radioactive isotopes decay? The rate of radioactive decay is unique for each type of radioisotope: Complete decay may take mere seconds or billions of years.

The time that it takes a macroscopic sample of the radioisotope to decay to one-half of its original mass is known as its **half-life.** For example, iodine-131 has a half-life of 8 days. This means that a 100.-g sample of I-131 will decay to 50.0 g of radioactive I-131 after 8 days; 25.0 g after 16 days (2 half-lives); 12.5 g after 24 days (3 half-lives); 6.25 g after 32 days (4 half-lives); and 3.13 g after 40 days (5 half-lives); and so forth. After 5 half-lives the sample is practically gone (3% of the original amount remains). This type of mathematical decrease is known as **exponential decay.** A graph showing that radioactive decay follows an exponential decrease is illustrated in Figure 16-4.

Krypton-81, which is used to diagnose lung function, has an extremely short half-life of 13 seconds. At the other extreme, carbon-14 has a half-life of 5,730 years. Carbon-14 is commonly used in carbon dating to determine the age of archeological objects. For medical applications, short half-lives are usually preferred, in order to minimize a patient's exposure to radiation. Table 16-4 lists some radioisotopes along with their half-lives and the type of radioactive emission produced.

For radioactive decay, the fraction of material remaining can be related to half-life using the equation

$$N = \left(\frac{1}{2}\right)^n$$

where N is the fraction remaining and n is the number of half-lives that have passed. Notice in the equation that n, the number of half-lives, is an exponent, thus the term "exponential decay."

Disposal of some nuclear waste is problematic because it has such a long half-life. For example, Pu-239 has a half-life of 24,100 years.

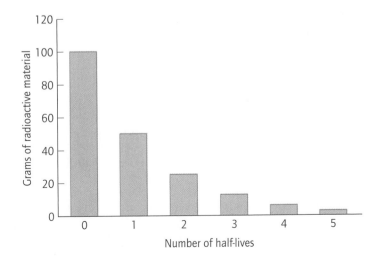

Figure 16-4 Radioactive decay is exponential. One half-life is the time it takes a radioisotope to decay to half its original amount. Two half-lives is the time it takes a radioisotope to decay to one-half of one-half, or one-fourth, of the original amount, and so forth.

Table 16-4 Radioisotopes, Radiation Type, and Half-Life

Radioisotope	Type of Emission	Half-Life
*Barium-131	γ	11.6 days
*Carbon-11	Positron, γ	20.3 minutes
Carbon-14	β	5,730 years
*Cerium-141	β	32.5 days
*Chromium-51	γ	27.8 days
*Fluorine-18	Positron, γ	109 minutes
*Gallium-67	γ	78 hours
*Gold-198	β	64.8 hours
*Iodine-131	β, γ	8 days
*Iron-59	β, γ	44.5 days
*Krypton-79	γ	34.5 hours
*Krypton-81m	γ	13.3 seconds
*Phosphorus-32	β	14.3 days
**Technetium-99m	γ	6 hours
Uranium-238	α	4×10^9 years

*Isotope used in medicine.

**Most widely used radioisotope in medicine.

Problem-Solving Tutorial: Radioactive Half-Life

WORKED EXERCISES 16-7   Half-Life

1 Iron-59 has a half-life of about 44 days. How much iron-59 would be left from a 100.-g sample after 132 days?

2 Radon-222 has a half-life of 3.8 days. How long would it take for a 20.-g sample of Ra-222 to decay to 1.3 g?

SOLUTIONS

1 First, determine how many half-lives have elapsed in the course of 132 days, given that one half-life is 44 days:

$$\frac{1 \text{ Half-life}}{44 \text{ days}} \times 132 \text{ days} = 3 \text{ half-lives}$$

Next, divide the original amount of material, 100. g, in half three consecutive times (3 half-lives):

$$100. \xrightarrow{1} 50.0 \text{ g} \xrightarrow{2} 25.0 \text{ g} \xrightarrow{3} 12.5 \text{ g}$$

Thus, 12.5 g of the original sample of Fe-59 would be left after 132 days.

2 Count the number of times that you have to divide the sample in half before arriving at 1.3 g.

$$\frac{1}{2} \times 20. \text{ g} = 10. \text{ g} \quad (1 \text{ half-life})$$

$$\frac{1}{2} \times 10. \text{ g} = 5.0 \text{ g} \quad (2 \text{ half-lives})$$

$$\frac{1}{2} \times 5.0 \text{ g} = 2.5 \text{ g} \quad (3 \text{ half-lives})$$

$$\frac{1}{2} \times 2.5 \text{ g} = 1.25 \text{ g} = 1.3 \text{ g} \quad (4 \text{ half-lives})$$

After you determine that 4 half-lives have elapsed, determine how many days 4 half-lives corresponds to, given that one half-life is 3.8 days:

$$4 \text{ Half-lives} \times \frac{3.8 \text{ days}}{\text{half-life}} = 15.2 \text{ days}$$

PRACTICE EXERCISES

16.18 Using Table 16-4, determine how many hours it would take a 1.0-g sample of gold-198 to decay to 0.5 g. What has happened to the other 0.5 g of Au-198?

16.19 For a 10.0-g sample of Cerium-141,
   a. How much Ce-141 is left after 3 half-lives?
   b. How much time has elapsed after 3 half-lives?

16.20 A 50.0-g sample of C-14 has decayed leaving only 3.1 g. How much time has elapsed since the original sample was weighed?

Artificial Radioisotopes

None of the elements with atomic numbers greater than 92 are found naturally on Earth. Elements 93–118 on the periodic table are all artificial radioisotopes. Every few years a new element is created artificially, in a particle accelerator (or "accelerator" for short). An **accelerator** is a sophisticated instrument that creates high-energy, fast moving particles. An artificial isotope is created by bombarding nuclei with a high-energy particle generated in an accelerator. Many of the radioisotopes created in an accelerator exist for only seconds, but that is long enough to detect their existence and thus, the birth of a new element.

Aerial photo of Fermilab in Batavia, Illinois. You can see the circular path of the Tevatron, one of the world's largest particle accelerators, which is 4 miles long. [Courtesy of Fermi National Accelerator Laboratory/Fred Ullrich]

The most widely used radioisotope in medicine, Tc-99m, is prepared using a small accelerator that is available in hospitals. The generation of Tc-99m proceeds according to the first two nuclear reactions shown in Figure 16-5. In the first reaction, a stable isotope Mo-98 is bombarded with high-energy neutrons ($_0^1$n) to form the radioisotope Mo-99. Absorption of a neutron causes

Step 1 Absorption of a neutron	$_{42}^{98}\text{Mo} + _0^1\text{n} \longrightarrow _{42}^{99}\text{Mo}$
Step 2 β decay	$_{42}^{99}\text{Mo} \longrightarrow _{43}^{99m}\text{Tc} + _{-1}^{0}\beta$
Step 3 Gamma emission	$_{43}^{99m}\text{Tc} \longrightarrow _{43}^{99}\text{Tc} + \gamma$

Figure 16-5 Steps in the preparation of Tc-99m.

the mass number to increase by one, creating a radioisotope with one more neutron than Mo-98. In a subsequent nuclear reaction, Mo-99 undergoes β decay ($_{-1}^{0}\beta$) naturally to produce the daughter nuclide Tc-99m. As a metastable isotope, Tc-99m emits gamma radiation as it relaxes to Tc-99, shown in the third nuclear reaction in Figure 16-5.

WORKED EXERCISE 16-8 Artificial Radioisotopes

Irene and Frederic Joliot-Curie, the daughter and son-in-law of Marie Curie, the famous woman scientist who developed the theory of radioactivity, were the first to create an artificial radioisotope. In 1934 they bombarded Al-27 with α particles. The Al-27 absorbed an α particle, producing a radioisotope that then underwent β decay to S-31. What radioactive isotope did Irene and Frederic create? Write the nuclear reactions involved.

SOLUTION

$$_{13}^{27}\text{Al} + _{2}^{4}\alpha \longrightarrow _{15}^{31}\text{P} \quad \text{New radioisotope}$$

$$_{15}^{31}\text{P} \longrightarrow _{16}^{31}\text{S} + _{-1}^{0}\beta$$

PRACTICE EXERCISES

16.21 When californium-249, named after the state of California, is bombarded with oxygen-18, a radioisotope is formed that undergoes α emission to rutherfordium-263. What is the name of the isotope formed? Write the two nuclear equations that create Rf-263 from Cf-249.

16.22 **Critical Thinking Question:** When U-235 is bombarded with a high-energy neutron, an unstable isotope of uranium, U-236, is formed. The U-236 can split in many different ways. The products include two new isotopes, three neutrons, and lots of energy. This energy is the energy of fission used in nuclear power plants and nuclear weapons. Only 1 kg of U-235 was used in the atom bomb dropped on Hiroshima, Japan. Balance the nuclear equation below, showing one way in which U-236 can undergo fission. Don't forget to consider the three neutrons when balancing the equation.

$$_{92}^{235}\text{U} + _{0}^{1}\text{n} \xrightarrow{\text{Artificial transmutation}} _{92}^{236}\text{U} \xrightarrow{\text{Fission}} 3\,_{0}^{1}\text{n} + _{56}^{141}\text{Ba} + \boxed{} + \text{energy}$$

16.2 | Biological Effects of Nuclear Radiation

The radiation emitted from a nuclear reaction is classified as **ionizing radiation,** because it has sufficient energy to dislodge an orbital electron from an atom, creating an ion. Recall from Chapter 2 that when there are fewer electrons than protons, the charge on the ion is positive (X^+) and the ion is called a *cation*.

$$\text{Radiation} + X \longrightarrow X^+ + e^-$$

Neutral atom or molecule — Cation — Orbital electron

Your body is composed of an enormous number (10^{27}) of molecules. When a molecule is ionized, it changes in a significant way. In living organisms, this change can be quite destructive—causing cell death or gene mutations. Consider the structure of DNA. The two ribose-phosphate strands are created by relatively strong phosphate ester bonds. Radiation can produce a break in a strand by destroying a phosphate ester bond. A mutation results when radiation breaks bonds sufficiently close to each other on opposite strands. These mutations are passed on when the cell reproduces, and can become the beginning of a cancer. The effect of ionizing radiation is greatest on rapidly reproducing cells such as lymphocytes (white blood cells), blood producing cells, and cancer cells. In contrast, nerve and muscle cells reproduce slowly, and therefore are the least sensitive to radiation.

The biological effects of nuclear radiation depend in large part on the type of radiation. Two characteristics of radiation determine its biological effects:

- the energy of the radiation, and
- the penetrating power of the radiation.

Penetrating power varies with the type of radiation. An α particle, for example, is relatively large and slow moving, but high in energy; therefore it is very destructive to human tissue. However, due to its size and slow speed, an α particle has little penetrating power. Light clothing or even a piece of paper is sufficient protection against α particles (see Figure 16-6). On the other hand, ingestion or inhalation of an α emitter can cause major damage to delicate internal organs due to the high energy of α particles. Few α emitters are used in medicine for this reason.

β particles and positrons have significantly less energy than α particles, but much more penetrating power because they are substantially lighter (8000 times lighter) than α particles. Specialized heavy clothing or a thick piece of aluminum is required for protection against β particles and positrons. In many respects, β particles and positrons are more dangerous than α particles because of their penetrating power.

The energy of gamma rays and X rays is less than or equal to the energy of β particles. However, gamma rays have the greatest penetrating power of all forms of radiation. Several inches of lead are required to adequately protect against gamma radiation. A thin sheet of lead is sufficient protection against X rays, as you are probably aware of from the times you have had to wear a lead apron during dental X rays.

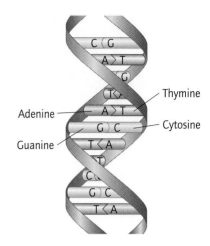

The structure of DNA. The molecule contains strong phosphate ester bonds linking nucleotides.

Gamma rays kill bacteria and other microorganisms, so they are often used to sterilize hospital instruments.

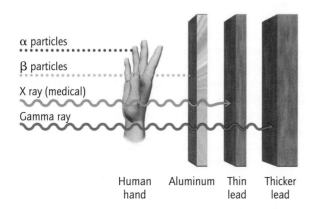

Figure 16-6 The penetrating power of different types of radiation. An α particle has the least penetrating power while a gamma ray has the greatest penetrating power.

The Geiger counter is a hand-held instrument used to measure radiation. More frequent audible clicks indicate a greater amount of radiation. [Edmund Scientific, A Division of Science Kit, LLC. www.scientificsonline.com]

A radiation badge is used to monitor an individual's exposure to radiation. It is worn by personnel, including many medical professionals, who routinely work around radioisotopes. [Cliff Moore/Photo Researchers, Inc.]

Table 16-5 Radiation Units and Their Abbreviations	
Unit	**Symbol**
(a) Amount of Radioactive Decay	
Bequerel	Bq
Curie	Ci
(b) Absorbed Dose	
Gray	Gy
Radiation absorbed dose	Rad
(c) Effective Dose	
Sievert	Sv
Roentgen equivalent man	rem

PRACTICE EXERCISES

16.23 What type(s) of radiation would be stopped by
    a. a piece of paper?   **b.** a sheet of aluminum?   **c.** a thin sheet of lead?
    d. a thick slab of lead?   **e.** skin?

16.24 Alpha particles have the least penetrating power, so why would it be dangerous to swallow a radioisotope that is an α emitter?

16.25 What subatomic particle is lost from an atom that has been subjected to ionizing radiation? How has the atom changed?

16.26 Does electromagnetic radiation have *more* or *less penetrating power* than high-energy particles?

16.27 Which is more damaging to biological tissue, an X ray or a gamma ray? Explain in terms of energy and penetrating power. What type(s) of radioactive decay produces gamma rays?

Measurement of Radiation

There are different ways to measure radiation. The **Geiger counter**, shown in the margin, is an inexpensive instrument used in the field that can detect all forms of radiation. A Geiger counter produces an audible series of clicks that increase as you move to areas with increased radiation.

A radiation badge, such as the one shown in the margin, is used to monitor radiation exposure of personnel who work in areas or use instruments that produce ionizing radiation. These badges consist of a special type of photographic film. Badges are worn throughout the day and the film is regularly processed to determine the amount of radiation exposure during a given period of time. The badge is replaced with unexposed film on a regular basis so the individual's exposure levels can be monitored. This prevents personnel who regularly work around radioactive materials from being exposed unknowingly to unsafe levels of radiation.

Units of Radiation Scientists must be able to measure the amount of radiation being produced by a radioactive source, as well as the amount of radiation absorbed by an individual. Several units of measurement are used to measure radiation. They differ in the type of information measured. These units may indicate

- the number of radioactive emissions,
- the amount of energy absorbed (absorbed dose), or
- the biological effectiveness of the energy absorbed (effective dose).

The number of radioactive emissions is the most basic unit of radiation and indicates the number of emissions per second. A Geiger counter can be used to obtain this measurement. Two common units are the **becquerel** (Bq) and the **curie** (Ci), as shown in Table 16-5a. These units of measurement indicate the rate of *all* radioactive emissions from a sample, so they do not distinguish between the different types of radioactive decay. They do not distinguish between α particles and β particles, for example.

Units of absorbed dose more adequately convey the amount of energy exposure as a result of exposure to different forms of radiation (α particle, β particle, gamma ray, etc.). An **absorbed dose** measurement indicates the energy of the radiation absorbed per mass of tissue. There are several units of absorbed dose in use (Table 16-5b), but the **gray** (Gy) is the most common in medical applications. The **Rad** is another common unit of absorbed dose.

Absorbed dose measurements still do not account for the differences in *penetrating power* of different forms of radiation. The unit that encompasses both penetrating power and amount of energy to give the actual biological effect of a particular type of radiation is the human **effective dose**, shown in

Table 16-5. This unit is calculated by multiplying the absorbed dose by a quality factor, Q, which varies for the different types of radiation. For example, $Q = 20$ for an α particle, and $Q = 1$ for gamma and X rays.

$$\text{Effective dose} = \text{absorbed dose} \times Q$$

When the unit of absorbed dose is the gray, the unit of effective dose is the **sievert** (Sv). For example, natural background radiation is approximately 0.0024 Sv per year or 2.4 mSv (millisieverts per year), although it varies with geographical location. Another common unit of effective dose is the **rem.** There are 100 rem in 1 Sv. For example, a CT scan of the head and body has an effective dose of 110 mrem per procedure.

WORKED EXERCISES 16-9 Measurement of Radiation

1 If you live within 50 miles of a nuclear power plant, you receive 0.01 mrem of additional radiation per year. Dental X rays produce 1 mrem per procedure. What type of information does the rem convey? How many years living next to a nuclear power plant is equivalent to one dental X ray?

2 International health and safety authorities endorse the safety of irradiated foods up to a dose of 10,000 Gy. What type of information does this unit convey?

SOLUTIONS

1 A rem is a unit for measuring the effective dose of radiation exposure. The rem takes into account both the energy of the radiation and its penetrating power. Living next to a nuclear power plant for 100 years is the equivalent of one dental X ray.

2 The gray is a unit of measurement that conveys the absorbed dose, a measure of energy exposure. This unit takes into account the different energies contributed by α particles, β particles, X rays, and gamma rays, but it does not take into account the penetrating power of the radiation.

PRACTICE EXERCISES

16.28 Explain the difference between the sievert and the gray.

16.29 Match the unit in the column on the left with the type of information it conveys in the column on the right.

Bq	absorbed dose (energy exposure)
Gy	effective dose
Sv	amount of radioactive decay
Ci	
rem	
Rad	

16.30 Would an effective dose of a 5-Gy α emitter be greater or less than that of a 5-Gy β emitter?

16.31 The amount of exposure to cosmic radiation that an individual receives per year depends on the altitude at which he or she lives. The value is 26 mrem at sea level, 35 mrem at 3500 ft, 52 mrem at 6000 ft, and 96 mrem at 9000 ft. Approximately, how much radiation do you receive per year where you live?

Radiation Sickness

Exposure to radiation can occur either as a single large dose (acute exposure) or as smaller doses over a longer period of time (chronic exposure).

Moreover, exposure to radiation can occur either by accident or intentionally, for example, as part of radiation therapy. **Radiation sickness** results from acute exposure to radiation. The severity of the symptoms of radiation sickness is directly proportional to the effective dose received, as shown in Table 16-6. The LD_x values shown in this table refer to the **L**ethal **D**ose of the radiation in x % of the population after 30 days. An LD_{50}, for example, is a level of exposure that would result in death in 50% of the population exposed after 30 days. Symptoms of radiation sickness often do not occur immediately after exposure, but follow a latent phase, during which time the individual shows no symptoms.

Although excessive exposure to radiation can have serious health consequences, exposure to specific forms of radiation in a controlled manner can have benefits that outweigh the risks. A case in point is the use of radiation in the diagnosis and treatment of disease.

Table 16-6 Dose Related Symptoms of Acute Radiation Sickness.

Effective Dose (Sv)	Symptoms
0.05–0.2	None
0.2–0.5	Temporary decrease in white blood cell count.
0.5–1.0	Headache and increased risk of infection. Possible temporary male sterility.
1.0–2.0	LD_{10}; nausea, hair loss, fatigue. Loss of white blood cells; temporary male sterility.
2–3	LD_{35}; loss of hair all over the body, fatigue and general illness. High risk of infection.
3–4	LD_{50}; uncontrollable bleeding in the mouth. Permanent sterility in women.
4–6	LD_{60}; death resulting from internal bleeding and infection. Permanent female sterility.
6–10	LD_{100}; death after 14 days.

PRACTICE EXERCISES

16.32 Using Table 16-6, what are the symptoms of acute exposure to 1.2 Sv of radiation?

16.33 What does it mean if the LD_{35} is 2–3 Sv?

16.34 What is the difference between acute and chronic exposure to radiation?

16.35 What level of radiation causes no overt symptoms of radiation sickness, but shows a temporary decrease in white blood cell count?

In this chapter, you were introduced in more detail to the nucleus of the atom, including the different types of nuclear decay and their effects on biological tissue. While nuclear chemistry focuses exclusively on the nuclei of atoms, the other medical and biological applications of chemistry described in this book depend on the electrons within atoms and molecules. It is the electrons that are involved in how atoms and molecules as a whole interact with one another, forming the basis for the study of biochemistry and pharmacology. Now you also know the significance of the nucleus of the atom, and how nuclear decay plays an important role in therapeutic and diagnostic medicine.

Chemistry in Medicine The Role of Nuclear Radiation in Imaging the Body

You know that visible light allows us to see our surroundings. Can other forms of electromagnetic radiation allow us to "see" images as well? Many new imaging techniques have emerged in the last 30 years that provide a view of the internal organs and systems in the body. These imaging techniques are based on nuclear reactions—radioisotopes emitting radiation as they decay to a more stable isotope. For example, a hand can be imaged using radiation from various regions of the electromagnetic spectrum, as shown in Figure 16-7.

Traditional diagnostic techniques in medicine include symptom appraisal, blood tests, and even exploratory surgery. Many times the simpler of these—symptom appraisal and blood tests—do not lead to a definitive diagnosis. Radiation-based diagnostic techniques provide a more definitive way to diagnose certain medical conditions that many times eliminates the need for exploratory surgery. These techniques let a medical professional "see" what is happening inside the body.

X rays and radio waves are the basis for different techniques used to image anatomical structures within the body. These techniques include X rays, computed tomography (CT) scans, and magnetic resonance imaging (MRI). These forms of imaging are noninvasive, relatively comfortable procedures, and are often used as complementary techniques. The instruments used for these techniques often look the same from the outside (see the photo on the chapter-opening page).

X Rays

Chances are you have had an X ray taken at the dentist's office, at the emergency room for a broken bone, or at a medical laboratory for some other health-related issue. X-ray imaging is by far the most widely used imaging technique in medicine. X rays are a form of electromagnetic radiation focused to pass through a specific area of the body.

While passing through the body, X rays will encounter structures of different densities such as bone

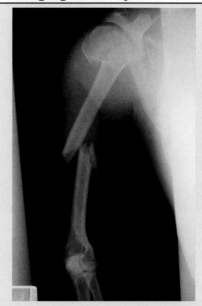

An X-ray image of a broken bone. A break in a bone will show up as a darker, discontinuous area (low-density space) in the middle of a lighter, continuous area (the high-density bone). [© Royalty-Free/Corbis]

and muscle that will absorb the radiation. High-density body structures such as bone will absorb more X rays, while lower-density structures such as muscle will absorb less. Some X rays will pass completely through the body and eventually collide with a **detector**—the part of the X-ray instrument that measures the amount of X-ray radiation that has passed through the tissue. The detector provides an image showing lighter and darker areas: Lighter areas indicate higher-density body tissues that have absorbed more X rays, and darker areas indicate lower-density tissues that have absorbed fewer X rays. Therefore, a break in a bone will show up as a darker, discontinuous area (low-density space) in the middle of a lighter, continuous area (the high-density bone).

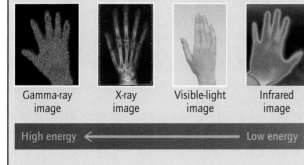

| Gamma-ray image | X-ray image | Visible-light image | Infrared image |

High energy ← Low energy

Figure 16-7 Images of a hand formed using electromagnetic radiation of various energies. Each type of radiation reveals a different feature of the hand. X rays and gamma rays reveal bones, visible light the surface of the skin, and infrared the temperature of the surface. The orange areas in the gamma-ray image are bone. [From left to right: Alfred Pasieka/Photo Researchers, Inc.; BSIP/Photo Researchers, Inc.; Yashuhide Fumoto/Getty Images; Ted Kinsman/Photo Researchers, Inc.; James Cavallini/Photo Researchers, Inc.]

continued

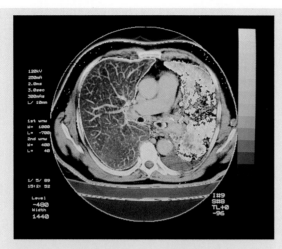

A Computed X-ray Tomography (CT) scan of the chest of a man shows cancer of the left lung. Notice the dramatic contrast in appearance between the healthy right lung (left on image) and the cancerous left lung (right on image). The heart appears in the center. [Simon Fraser/Photo Researchers, Inc.]

Exposure to X rays is kept as brief as possible to avoid damaging tissue. The typical effective dose of radiation from a chest X ray is approximately 0.1 mSv (millisieverts).

CT Scans

Like conventional X-ray instruments, **C**omputed **T**omography (**CT**) scanning instruments use X rays to create images. The difference is that the X rays are detected by a circular array of detectors surrounding the body, and the information from these detectors is downloaded to a computer for processing. CT scanning produces more detailed, three-dimensional images of body structures. Another advantage of CT scanning over X rays is that it can distinguish between two separate very similar structures that are adjacent to one another. CT is used as the definitive diagnostic technique for a variety of ailments including brain hemorrhages, pneumonia, appendicitis, and complex fractures.

Computed Tomography (CT) scans were originally termed Computed Axial Tomography scans, referred to as CAT scans.

MRI

Like CT, **M**agnetic **R**esonance **I**maging (**MRI**) produces three-dimensional images of organs and tissues. MRI is valuable for diagnosing conditions such as tumors, edema (swelling of organs and tissues), and multiple sclerosis. *MRI is best for imaging soft-tissue areas of the body such as the brain and liver, while X-ray techniques (including CT) are best for imaging denser tissue areas of the body such as bones and joints.*

MRI is a technique that employs pulses (short bursts) of electromagnetic radiation in the radio-frequency region of the electromagnetic spectrum, together with a magnetic field, to image tissues. Your body is composed mostly of water, and water contains hydrogen atoms. The protons in hydrogen atoms act like tiny magnets. When they are exposed to the strong magnetic field of the MRI instrument, the protons will align themselves with this field, similar to a compass needle aligning itself with magnetic north. As the MRI scan progresses, radio-wave pulses of various energies are applied. Protons in different environments of the body (muscle, nerve tissue, tumor tissue, etc.) will resonate with a different radio frequency, allowing the computer to distinguish body tissues using color images.

A major advantage of MRI is that it is harmless to the patient. The radio-frequency region of the electromagnetic spectrum is of low energy, and does not pose any risk of biological damage. The strong magnetic field also does not pose a risk to the patient, unless the patient is wearing a pacemaker or other similar electromagnetic device.

Imaging techniques based on nuclear radiation greatly enhance the power of diagnosis in the medical field. Medical professionals can "see" and accurately interpret conditions such as broken bones, brain hemorrhages, and even tumors. As you now have learned, medical applications based on nuclear radiation can range from the treatment of cancer (see this chapter's opening vignette) to the imaging of anatomical structures and abnormalities.

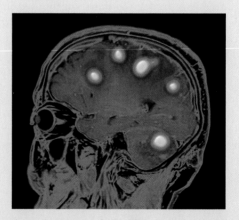

A color-enhanced MRI scan of a brain shows multiple malignant tumors (yellow-red). MRI is a cross-sectional imaging technique that uses strong magnetic fields and radiofrequency pulses to generate images. [Neil Borden/Photo Researchers, Inc.]

Chapter Summary

Radioisotopes and Nuclear Reactions

- Radioactive isotopes—radioisotopes—are unstable isotopes that achieve stability by radioactive decay, a process in which electromagnetic radiation and/or high-energy particles are emitted.

- Electromagnetic radiation is a form of energy that travels through space as a wave at the speed of light.

- The radioactive decay process can be described by a balanced nuclear equation. The subscripts and superscripts on both sides of the equation must be equal.

- There are five major types of radioactive decay: α-particle emission, β-particle emission, positron emission, X rays, and gamma-ray emission.

- An α particle ($_2^4\alpha$) is a helium nucleus: two protons and two neutrons. It carries a 2+ charge and is a high-energy, relatively slow moving particle.

- A β particle is a high-energy electron ($_{-1}^0\beta$) emitted from a nucleus that has too many neutrons.

- β decay yields a daughter nuclide with atomic number one greater than the parent nuclide and the same mass number.

- A positron is similar to a β particle, except that it carries a positive charge. It is emitted from a nucleus that has too many protons.

- Positron emission yields a daughter nuclide with atomic number one less than the parent nuclide and the same mass number.

- Gamma rays are high-energy electromagnetic radiation that accompanies most radioactive decay processes.

- The time it takes a radioisotope to decay to one-half of its original mass is known as its half-life.

- Man-made radioisotopes can be formed by bombarding a nucleus with a high-energy fast moving particle produced in a particle accelerator.

Biological Effects of Nuclear Radiation

- Nuclear reactions produce ionizing radiation, which has the ability to convert molecules into ions. Ionizing radiation destroys cells.

- The biological effects of nuclear radiation depend on the energy and the penetrating power of the radiation.

- α particles have the highest energy but least penetrating power.

- β particles, positrons, and X rays have less energy than α particles but greater penetrating power.

- Gamma rays have more energy than X rays and greater penetrating power.

- The units of measurement for the number of radioactive emissions produced by a radioisotope are the bequerel (Bq) and the curie (Ci).

- The absorbed dose is a unit of radiation measurement that takes into account the energy of the radiation. Common units are the gray (Gy) and the Rad.

- The effective dose is a unit of radiation that takes into account energy together with the penetrating power of the radiation. Common units of effective dose are the sievert (Sv) and the rem.

- Radiation sickness occurs with acute—a single large dose—exposure to radiation.

- The symptoms of radiation sickness worsen with exposure, ranging from a temporary decrease in white blood cell count for mild exposure, to nausea, hair loss, and fatigue, and, in the case of severe exposure, death.

Key Words

α particle A slow moving, high-energy particle consisting of 2 protons and 2 neutrons emitted as a result of nuclear decay. Its nuclear symbol is $^4_2\alpha$ or ^4_2He.

β particle A high-energy electron emitted as a result of nuclear decay. Its nuclear symbol is $^0_{-1}\beta$ or $^0_{-1}\text{e}$.

Absorbed dose The energy of radiation absorbed per mass of tissue.

Accelerator A sophisticated instrument that creates high-energy, fast moving particles.

Artificial radioisotopes Man-made radioisotopes.

Background radiation Environmental radiation from natural sources.

Becquerel (Bq) A unit of measurement indicating the rate of all radioactive emissions from a sample.

Curie (Ci) A unit of measurement indicating the rate of all radioactive emissions from a sample.

Daughter nuclide An isotope that results from radioactive decay.

Effective dose A quantity that encompasses both penetrating power and amount of energy to give the actual biological effect of a particular type of radiation.

Electromagnetic radiation A form of energy that travels through space as a wave at the speed of light.

Excited state A state in which the nucleus of the atom, after undergoing radioactive decay, contains excess energy.

Gamma radiation A form of electromagnetic radiation having a shorter wavelength and greater energy than X rays.

Geiger counter An inexpensive instrument used in the field that can detect all forms of radiation.

Gray (Gy) A unit of measurement indicating the energy of the radiation absorbed per mass of tissue.

Half-life The time that it takes a macroscopic sample of a radioisotope to decay to one-half of its original mass.

Ionizing radiation The radiation emitted from a nuclear reaction that has the energy to dislodge an orbital electron from a molecule, creating an ion.

Lethal dose (LD$_x$) A level of radiation exposure that would result in death in x% of the population exposed.

Metastable An isotope in an excited state.

Nuclear chemistry The study of the changes of the nucleus of an atom.

Nuclear equation The representation of decay occurring in a radioisotope and any nuclear reaction.

Nuclear medicine The use of nuclear chemistry in the diagnosis and treatment of disease.

Nuclear reactions Processes involving changes in the nucleus of an atom.

Parent nuclide A radioisotope undergoing decay.

Positron A particle having the same mass and energy as an electron, except with a positive charge, emitted as a result of nuclear decay. Its nuclear symbol is $^0_{+1}\beta$.

Radiation The energy released by an unstable nucleus.

Radiation sickness Illness resulting from acute exposure to radiation.

Radioactive decay The release of radiation by an unstable nucleus in order to achieve a more stable nucleus.

Radioisotope An isotope that is unstable due to an imbalance in the number of neutrons and protons in the nucleus.

Wavelength The distance between wave crests in electromagnetic radiation.

X rays A form of electromagnetic radiation with a short wavelength and high energy.

Additional Exercises

Radioisotopes

16.36 What part of the atom is involved in nuclear reactions? How are nuclear reactions different from chemical reactions?

16.37 What does the atomic number represent in an isotope? What does the mass number represent?

16.38 Fill in the table below showing the relationship between the radioisotope, atomic number, mass number, number of neutrons, and number of protons.

Radioisotope	Atomic Number	Mass Number	Number of Neutrons	Number of Protons
Phosphorus-32	15			
		198		79
^{169}Yb				70
		222	136	
F-18				

16.39 Fill in the table below showing the relationship between the isotope, atomic number, mass number, number of neutrons, and number of protons.

Radioisotope	Atomic Number	Mass Number	Number of Neutrons	Number of Protons
Thallium-201				
		75		34
	27	60		70
Ba-131			75	

16.40 What causes radioisotopes to be unstable? How does a radioisotope become more stable?

Electromagnetic Radiation

16.41 Define electromagnetic radiation.

16.42 Why is electromagnetic radiation often referred to as "light?"

16.43 You are continually exposed to radio waves in the environment, and need no protection. However, you need to wear a protective lead apron when you have your teeth X-rayed at the dentist. Explain.

16.44 List in order of decreasing energy the following types of electromagnetic radiation: microwave, ultraviolet, radio waves, visible, gamma rays, X rays, infrared.

16.45 Which form of electromagnetic radiation has the longer wavelength?
a. radio wave or microwave
b. X ray or gamma ray
c. visible or ultraviolet

16.46 Which form of electromagnetic radiation is higher in energy?
a. visible or infrared
b. visible or ultraviolet
c. X ray or gamma ray

16.47 Which form of electromagnetic radiation is more damaging to biological tissue? Explain why.
a. X ray or gamma ray
b. ultraviolet or visible

16.48 Which of the following forms of radiation are not considered electromagnetic radiation?
a. X ray
b. α particle
c. neutron
d. microwave
e. β particle
f. gamma ray
g. positron

Types of Radioactive Decay

16.49 Smoke detectors use a small amount of Americium-241 to detect the presence of smoke or heat sources. Americium-241 undergoes α decay. Write a balanced nuclear equation for the decay.

16.50 Lutetium-177 is an ideal therapeutic radioisotope because it is a strong β emitter and has just enough gamma radiation to enable imaging.
a. Write the balanced nuclear equation for the β decay.
b. What is the daughter nuclide produced from the β decay?
c. What types of radiation (electromagnetic or high-energy particle) are produced in the radioactive decay of lutetium-177?

16.51 What is the difference between positron decay and β decay?

16.52 Oxygen-15 undergoes positron decay. Write the nuclear reaction for this decay and identify the daughter nuclide.

16.53 Boron-12 undergoes β decay. What is the daughter nuclide?

16.54 Which type of radioactive decay produces a daughter nuclide that is a different element from the parent nuclide? More than one correct answer is possible.
a. α   **b.** β   **c.** positron   **d.** gamma

16.55 Write balanced nuclear equations for the following
a. β decay of Na-26
b. formation of Po-206 through α decay
c. formation of Mn-52 through positron emission

Half-Life

(Refer to Table 16-4 for the half-lives of the radioisotopes.)

16.56 How much of a 16-g sample of fluorine-18 is left after 2 half-lives? How much time has elapsed after 2 half-lives?

16.57 How much of an 18.0-g sample of iodine-131 is left after 32 days? How many half-lives does 32 days represent?

16.58 How long would it take a 28-g sample of iron-59 to decay to 1.75 g?

16.59 Why is Tc-99m an ideal radioisotope for use in medicine? (*Hint:* Consider only the half-life.)

16.60 If 25.0 mg of I-131 is given to a patient, how much I-131 remains in the patient after 24 days?

16.61 Tc-99m is produced from the radioactive decay of Mo-99. Mo-99 has a half-life of 66 hours. If you start with 100. g of Mo-99, how long would it take to form 75.0 g of Tc-99m?

Artificial Radioisotopes

16.62 What is a particle accelerator?

16.63 Describe how artificial radioisotopes are formed.

16.64 When plutonium-239 is bombarded with 2 neutrons, americium-241 and a β particle are produced. Write the nuclear reaction for the formation of americium-241.

16.65 Cobalt-60 is used in radiation therapy. It is formed by bombarding a cobalt isotope with a neutron. Identify the isotope of cobalt that is used to make cobalt-60. Write the nuclear reaction for the process.

16.66 Ernest Rutherford carried out one of the first reactions to produce an artificial radioisotope. He bombarded nitrogen-14 with α particles. The reaction produced a new radioisotope and a proton ($_1^1$p). What radioisotope was formed? Write the nuclear equation for the process.

16.67 Bombarding U-238 with a specific particle yields three neutrons and Pu-239. What is the particle? Write the balanced nuclear equation. *Hint:* Take into account that three neutrons are formed.

16.68 Elements 104–106 are made artificially from californium-249 by bombarding it with C-12, N-15, and O-18 nuclei, respectively. If four neutrons are formed in each reaction, write the balanced equation for these reactions.

16.69 Uranium is the heaviest naturally occurring element. Elements with atomic numbers higher than uranium are known as transuranium elements, meaning "beyond uranium." Transuranium elements are made artificially by a process using a device known as a _____ _____.
The transuranium elements belong to period _____ on the periodic table and are known as the _____.

Energy and Penetrating Power of Nuclear Radiation

16.70 Why does ionizing radiation damage biological tissue?

16.71 If an ion contains 11 protons and 10 electrons, its symbol is Na^+. What is the symbol for an ion containing 20 protons and 18 electrons?

16.72 What does ionizing radiation do to a molecule to cause it to become an ion?

16.73 How can ionizing radiation cause genetic mutations in DNA and possibly cause cancer?

16.74 It is extremely dangerous to swallow an α emitter; however, you can stop an α emitter with just a piece of paper. Explain the apparent contradiction.

16.75 Why does a lead apron give your neck and chest sufficient protection against X rays when you have a dental X ray taken?

16.76 Indicate which form of radiation has greater energy:
  a. X ray or gamma ray
  b. α particle or β particle
  c. gamma ray or β particle

16.77 Indicate which form of radiation has greater penetrating power
  a. α particle or β particle
  b. β particle or gamma ray
  c. gamma ray or X ray

16.78 What type of radiation simultaneously has the highest energy but the least penetrating power?

16.79 What two forms of radiation have the same energy and the same penetrating power?

Measurement of Radiation

16.80 How many sieverts are there in 1 rem?

16.81 Do the bequerel and the curie measure the same or different properties of radiation? Explain. What are the abbreviations for these radiation units? How would you abbreviate a millicurie?

16.82 Do the bequerel and the Rad measure the same or different properties of radiation? Explain.

16.83 What is the difference between an absorbed dose and an effective dose measurement of radiation? What unit(s) are used for absorbed dose? What unit(s) are used for effective dose?

16.84 What symptoms can you expect from acute exposure to 2.4 Sv of radiation?

16.85 What does LD_{50} mean?

16.86 Identify the following situations as either an acute or a chronic exposure to radiation.
  a. A hospital worker carries out CT scans regularly on patients.
  b. An uninformed person picks up an α emitter with his unprotected hand.
  c. A patient receives radiation therapy.
  d. A farmer is standing in his field near Chernobyl at the time of the Chernobyl nuclear reactor accident.

16.87 On a one-hour-long flight a man was exposed to 0.000003 Sv of radiation from cosmic rays. A CT scan of the head of the same man exposed him to 0.0015 Sv of radiation. Did the flight or CT scan expose him to a higher dose of radiation?

16.88 Thyroid cancer is the most common form of cancer among survivors of the Chernobyl nuclear power plant accident in 1986. What radioactive isotope do you expect is responsible for this type of cancer?

Chemistry in Medicine: The Role of Nuclear Radiation In Imaging the Body

16.89 A small child swallowed a quarter. An X ray showed the quarter clearly as a light spot in a dark area of the esophagus. Explain how the radiologist was able to see where the quarter was located in the body.

16.90 A patient came into the ER with a possible head injury. He was given a CT scan and MRI. Which one of these techniques was most likely used to determine if he had a skull fracture? Which one of these techniques was most likely used to determine if he had any brain trauma?

16.91 What are the advantages of performing an MRI instead of a CT scan? What are the disadvantages?

16.92 What are the advantages of performing an MRI or CT scan instead of taking an X ray? What are the disadvantages?

Critical Thinking Questions

16.93 On average, how long does it take a radio wave to travel from the moon to the Earth? The average distance between the moon and the Earth is 384,400 km.

16.94 Tc-99m has been used to detect infection in knee prostheses. A patient with a knee prothesis was injected with 20 mCi of Tc-99m, then monitored with a radiation-based technique. The detection technique monitored the blood flow through the prosthesis.
  a. Medical personnel will often use the curie to describe the amount of Tc-99m used rather than the mass in grams. In the example above, how much Tc-99m is left after one half-life?
  b. What type of radiation is emitted from Tc-99m?

16.95 High-dose-rate temporary brachytherapy (another name for sealed source radiotherapy) is used to treat some head and neck cancers. In temporary brachytherapy, the radioactive source is removed from the patient after a short period of time. Suppose that iridium-192 is applied to the tumor for about 6.5 minutes. The half-life of iridium-192 is 74 days.
  a. Iridium-192 undergoes β decay and release of γ radiation. Identify the daughter nuclide formed in this process. Write the nuclear equation for the decay of iridium-192.
  b. What kind of material should be used to protect the healthy tissues (i.e., the torso, the arms, and legs) of the patient from the β and γ radiation?

Answers to Practice Exercises

16.1

Radioisotope	Atomic Number	Mass Number	Number of Neutrons	Number of Protons
I-131	53	131	78	53
Rn-205	86	205	119	86
K-40	19	40	21	19
Au-198	79	198	119	79
Co-60	27	60	33	27

16.2 **a. i.** gamma rays
b. iii. radio waves
c. ii. microwave radiation

16.3 UV light is radiation of shorter wavelength and higher energy than visible light.

16.4 Radar uses electromagnetic radiation in the microwave region of the electromagnetic spectrum.

16.5 α particles, β particles, positrons, X rays, and gamma rays. Gamma rays and X rays are forms of electromagnetic radiation.

16.6 (Superscripts: $210 = 206 + 4$)

$$^{210}_{84}\text{Po} \rightarrow {}^{206}_{82}\text{Pb} + {}^{4}_{2}\alpha$$

(Subscripts: $84 = 82 + 2$)

The daughter nuclide is lead-206.

16.7 (Superscripts: $226 = 222 + 4$)

$$^{226}_{88}\text{Ra} \rightarrow {}^{222}_{86}\text{Rn} + {}^{4}_{2}\alpha$$

(Subscripts: $88 = 86 + 2$)

The parent nuclide is radium-226.

16.8 **a.** I-131 contains 53 protons and 78 neutrons ($131–53 = 78$).
b. (Superscripts: $131 = 131 + 0$)

$$^{131}_{53}\text{I} \rightarrow {}^{131}_{54}\text{Xe} + {}^{0}_{-1}\beta$$

(Subscripts: $53 = 54 + (-1)$)

c. Xe-131
d. a β particle, which is a high-energy particle.

16.9 (Superscripts: $198 = 198 + 0$)

Parent nuclide $\boxed{^{198}_{79}\text{Au}} \rightarrow \boxed{^{198}_{80}\text{Hg}} + {}^{0}_{-1}\beta$ Daughter nuclide

(Subscripts: $79 = 80 + (-1)$)

16.10 A radioisotope that undergoes β decay contains too many *neutrons* in its nucleus compared to *protons*. Emission of a β particle results in a daughter nuclide that is the *next* higher element on the periodic table relative to the parent nuclide.

16.11 $^{1}_{0}\text{n} \longrightarrow \boxed{^{1}_{1}\text{p}} + {}^{0}_{-1}\beta$

16.12 (Superscripts: $18 = 18 + 0$)

$$^{18}_{9}\text{F} \rightarrow {}^{18}_{8}\text{O} + {}^{0}_{+1}\beta$$

(Subscripts: $9 = 8 + 1$)

F-18 decays by positron ($^{0}_{+1}\beta$) emission. Positrons are high-energy particles.

16.13 A β particle, $^{0}_{-1}\beta$

16.14 In positron emission the daughter nuclide has an atomic number one less than the parent nuclide and the same mass number as the parent nuclide.

16.15 Positron emitters contain too many *protons* in their nucleus compared to *neutrons*. Positron emission turns a *proton* into a *positron* and a *neutron*.

16.16 Complete the balanced nuclear equation below showing the conversion of a proton into a neutron and a positron.

$$^{1}_{1}\text{p} \longrightarrow \boxed{^{1}_{0}\text{n}} + {}^{0}_{+1}\beta$$

16.17 **a.** The isotopes differ in the number of neutrons and whether or not they are in an excited state—the metastable state.
b. An isotope in a metastable state is one whose nucleus is in an excited state; the abbreviation for an isotope in the metastable state is "m" following the mass number.

c. $^{60\text{m}}_{27}\text{Co} \longrightarrow {}^{60}_{27}\text{Co} + \gamma$

16.18 Since 0.5 g is half of 1.0 g, the Au-198 sample would decay to this amount after 1 half-life, which is 65 hours, according to Table 16-4. The other 0.5 g of Au-198 is converted into its daughter nuclide, Hg-198.

16.19 **a.** 1.25 g of Ce-141 remain after 3 half-lives.

$$10.0\ \text{g} \xrightarrow{1} 5.00\ \text{g} \xrightarrow{2} 2.50\ \text{g} \xrightarrow{3} 1.25\ \text{g}$$

b. 97.5 days. The half-life of Ce-141, according to Table 16-4, is 32.5 days; therefore:

$$3\ \text{half-lives} \times \frac{32.5\ \text{days}}{\text{half-life}} = 97.5\ \text{days}$$

16.20 To determine how much time has elapsed since the original sample was weighed, divide the 50.-g sample in half until the amount is 3.1 g and note how many times you divided the sample in half. Then multiply the number of half-lives by the half-life of C-14, from Table 16-4:

$$50.\ \text{g} \xrightarrow{1} 25.\ \text{g} \xrightarrow{2} 12.5\ \text{g} \xrightarrow{3} 6.25\ \text{g} \xrightarrow{4} 3.125 = 3.1\ \text{g}$$

It takes 4 half-lives to diminish the sample to 3.1 g, therefore:

$$4 \; \cancel{\text{half-lives}} \times \frac{5{,}730 \text{ years}}{1 \; \cancel{\text{Half-life}}} = 22{,}920 \text{ years}$$

16.21 The new isotope is Sg-267.

$$^{249}_{98}\text{Cf} + {}^{18}_{8}\text{O} \longrightarrow {}^{267}_{106}\text{Sg}$$

$$^{267}_{106}\text{Sg} \longrightarrow {}^{263}_{104}\text{Rf} + {}^{4}_{2}\alpha$$

16.22

	Artificial transmutation		Fission

$$^{235}_{92}\text{U} + {}^{1}_{0}\text{n} \longrightarrow {}^{236}_{92}\text{U} \longrightarrow$$

$$3\,{}^{1}_{0}\text{n} + {}^{141}_{56}\text{Ba} + \boxed{{}^{92}_{36}\text{Kr}} + \text{energy}$$

16.23 **a.** α particles

b. α particles and β particles

c. α particles, β particles, and X rays

d. α particles, β particles, X rays, and gamma rays

e. α particles

16.24 Although α particles have the least penetrating power they have the most energy, so if there is no barrier preventing them from penetrating, they cause the most damage. Once an α emitter is ingested, the α particles come in direct contact with human tissue, causing significant damage.

16.25 An electron is lost. The atom now carries a positive charge, turning it into an ion, specifically a cation.

16.26 Electromagnetic radiation has *more penetrating power* than high-energy particles.

16.27 Gamma rays are more damaging to biological tissue than X rays, because they have a greater penetrating power and higher energy. Most forms of radioactive decay produce some gamma radiation.

16.28 The sievert and the gray convey different information. The gray is a unit of absorbed dose, which is information about the amount of energy exposure from radiation. The sievert is a measure of effective dose, which takes into account the energy as well as the type of radiation involved.

16.29 Bq number of radioactive emissions

Gy absorbed dose (energy exposure)

Sv effective dose

Ci number of radioactive emissions

rem effective dose

Rad absorbed dose (energy exposure)

16.30 The effective dose of a 5-Gy α emitter would be greater than that of a 5-Gy β emitter, because the quality factor for an α particle is higher, so its effective dose would be greater.

16.31 If you live at sea-level, you receive 26 mrem per year; if you live in Denver (the mile-high city), you receive close to 52 mrem per year.

16.32 nausea, hair loss, and fatigue

16.33 L stands for lethal, D stands for dose, and 35 stands for 35%. LD_{35} indicates a dose that would result in death in 35% of those exposed to 2–3 Sv.

16.34 Acute exposure is a one-time exposure, while chronic exposure is repeated exposure over a period of time.

16.35 an effective dose of 0.2–0.5 Sv

Appendix A

Basic Math Review with Tips for Using a Calculator

Students in chemistry are expected to have learned fundamental mathematical operations in earlier classes. Chemistry gives the student an opportunity to use these operations to solve practical problems. A review of these operations is provided in this Appendix to aid the student who has not used mathematical operations consistently and to offer a resource for those who need a refresher for just a few concepts.

Arithmetic

The Four Operations

There are four basic arithmetic functions: addition, subtraction, multiplication, and division.

1. Addition. Example: $2 + 5 = 7$
   - The result (7 in this case) is known as the **sum.**
2. Subtraction. Example: $5 - 2 = 3$
   - The result (3 in this case) is known as the **difference.**
3. Multiplication. Example: $4 \times 2 = 8$
   - The result (8 in this case) is known as the **product.**
   - Multiplication can be written in many different ways:
     a. $4 \times 2 = (4)(2) = 4(2) = 4 \cdot 2$
     b. 4 and 2 are known as **factors.**
4. Division. Example: $8 \div 2 = 4$
   - The result (4 in this case) is known as the **quotient.**
   - Division can be written many different ways:

$$8 \div 2 = 8/2 = \frac{8}{2}$$

Using Your Calculator

To use your calculator with just one of the four basic arithmetic functions, read the problem from left to right. Type the first number on the left followed by the indicated operation. Then, type the next number and the **ANS** (answer) or **ENTER** or = key to obtain your result.

Squaring

The superscript "2" after a number indicates that the number is "squared," or multiplied by itself. Example: $3^2 = 3 \times 3 = 9$

Using Your Calculator

To use the square function on the calculator, locate the $\mathbf{x^2}$ key. Type in the number represented by **x** and then press the $\mathbf{x^2}$ key. On some calculators you may have to press a **second** or **shift** key before pressing the $\mathbf{x^2}$ key. Try the following examples to be sure you obtain the same results.

$$4^2 = 16$$
$$8^2 = 64$$

> Most calculator keys have two functions; pressing the second or shift key lets you perform the second function, often shown in yellow next to the key.

Exponents

The "square" function above is an example of an **exponent** function. The **exponent** is a number written as a superscript after another number. For a square function, the exponent is 2. Numbers can be "raised" to any exponent, also known as "raising to a power." To solve an exponent function, multiply the number by itself the number of times indicated by the exponent. For example:

$$2^3 = 2 \times 2 \times 2 = 8$$
$$6^5 = 6 \times 6 \times 6 \times 6 \times 6 = 7,776$$

Using Your Calculator

To solve an exponent function using your calculator, find the **x^y** or **$x^{\wedge}y$** or **y^x** or **$\wedge$** key. Type in the value for the number being multiplied, press the exponent key, then type in the value for the exponent. Press the **ANS** or **ENTER** or **=** key to obtain the answer. Try the following to be sure you obtain the same results.

$$10^6 = 1,000,000$$
$$5^4 = 625$$

Order of Operations

When solving an arithmetic problem that contains several different operations, you need to apply the operations in a particular order. Follow the steps below to solve the following problem.

$$(2^2 + \frac{6}{2} \times 5) \times (4 - 2)$$

1. When there are parentheses around any part of the problem, the operations inside the parentheses must be solved first.

$$(\underline{2^2 + \frac{6}{2} \times 5}) \times (\underline{4 - 2})$$

2. Numbers that are squared are calculated first.

$$(\underline{4} + \frac{6}{2} \times 5) \times (4 - 2)$$

3. Multiplication and division are calculated next and should be performed in order from left to right.

$$(4 + \frac{6}{2} \times 5) \times (4 - 2) = (4 + \underline{3 \times 5}) \times (4 - 2) = (4 + 15) \times (4 - 2)$$

4. Addition and subtraction are performed last from left to right.

$$(\underline{4 + 15}) \times (\underline{4 - 2}) = (19) \times (2)$$

5. Once the operations inside the parentheses are solved, the rules start over again with the numbers that were obtained.

$$19 \times 2 = 38$$

Using Your Calculator

Most scientific calculators today have parentheses keys and will calculate the result of a lengthy problem using the rules listed above. If your calculator has parentheses keys, "(" and ")", try to enter the above problem as it is written. If you did not get the same result, enter each part of the problem in the order that it needs to be solved (following the underlined portions above). Try the following problem:

$$(8 + 3 \times 2 + 2) - (3 + 10 \times 3 - 37) = 20$$

Fractions and Decimals

Fractions

The numbers we have used thus far have been **integers** or **whole numbers.** In many situations, parts of a whole need to be expressed. These parts are expressed as **fractions** or **decimals.**

Example of a generic fraction: $\dfrac{\text{numerator}}{\text{denominator}}$

Decimals

Fractions can also be considered unsolved division problems. The line between the numerator and the denominator is the **dividing line.** The unsolved division problem represented by a fraction may be *solved* by dividing the numerator by the denominator. The result is a **decimal** number. A decimal number is one in which a **decimal point (.)** is used to separate whole numbers on the left from numbers representing parts of a whole number on the right.

Decimals can be calculated using long division or a calculator. The result may have a zero (0) to the left of or *in front* of the decimal point, indicating that the answer is larger than zero (0) but smaller than 1.

Using Your Calculator

To convert a fraction to a decimal, type the number for the numerator, the divide key ($\div$), the number for the denominator, and the **ANS** or **ENT** or **=** key. Try the following examples to make sure you obtain the same result.

$$\frac{3}{4} = 0.75 \qquad\qquad \frac{1}{5} = 0.2$$

Order of Operations with Fractions

The numerator and denominator of a fraction may each be a mathematical expression involving addition, subtraction, multiplication, etc.:

$$\frac{3 \times 4}{5 \times 6}$$

To solve the fraction, you should first solve the numerator and the denominator separately. Putting them in parentheses may help you see the expression:

$$\frac{(3 \times 4)}{(5 \times 6)}$$

Follow the standard order of operations within each set of parentheses. In this case there is only multiplication within the parentheses:

$$\frac{(3 \times 4)}{(5 \times 6)} = \frac{12}{30} = 0.4$$

Using Your Calculator

When using your calculator to solve the expression above, you must be sure to enter both the opening and closing parentheses in the denominator.

$$3 \times 4 \div (5 \times 6) = 0.4$$

If you do not, the answer will be incorrect: $3 \times 4 \div 5 \times 6 = 14.4 \neq 0.4$.

Alternatively, you can solve the equation without using the parentheses if you *divide* by 6 rather than multiplying. You divide by 6 because the number 6 is *under* the dividing line:

$$3 \times 4 \div 5 \div 6 = 0.4$$

You will need to remember this rule on many occasions in chemistry.

Fractions and Dimensional Analysis

Fractions are used to express relationships when setting up a **dimensional analysis** problem. **Dimensional analysis** is used to convert units, as explained on page 13 of Chapter 1. It uses the algebraic principle that when units and variables are the same in both the numerator and denominator of a fraction, they will cancel each other out.

Dimensional analysis leads to expressions such as the following that need to be solved:

$$425 \text{ mL} \times \frac{1 \text{ L}}{1000 \text{ mL}}$$

Since 425 is a **whole** number, it can be thought of as a fraction with 1 in the denominator. To solve the expression, you should solve the denominator using parentheses.

$$425 \text{ mL} \times \frac{1 \text{ L}}{1000 \text{ mL}} = \frac{(425 \text{ mL} \times 1 \text{ L})}{(1 \times 1000 \text{ mL})}$$

$$= \frac{(425 \text{ mL} \times 1 \text{ L})}{(1000 \text{ mL})} = \frac{425 \text{ mL} \cdot \text{L}}{1000 \text{ mL}}$$

The mL units cancel out and the fraction can be solved for its decimal equivalent.

$$\frac{425 \text{ mL L}}{1000 \text{ mL}} = \frac{425 \text{ L}}{1000} = 0.425 \text{ L}$$

Using Your Calculator

Enter the open parenthesis sign (, enter the expression in the numerator, then enter the close parenthesis sign). Then enter the division sign and, again within parentheses, enter the expression in the denominator. For example, the conversion of 425 mL to L above would be entered:

$$(425 \times 1\,) \div (1 \times 1000)\ \text{ANS}$$

The correct answer, 0.425, should be displayed.
Try the following example:

$$35.8 \text{ mL} \times \frac{0.889 \text{ g}}{1.00 \text{ mL}} \times \frac{1 \text{ mole}}{18.01 \text{ g}} = 1.77 \text{ mole}$$

Scientific (Exponential) Notation

In chemistry scientific notation is used to express very large and very small numbers. It is convenient to express them using powers of 10. Scientific notation has two parts: the **coefficient** and the **exponent.** The number 1.24×10^7 is in scientific notation. The coefficient is 1.24 and the exponent is 7. The expression tells you to multiply the **coefficient,** 1.24, by 10 seven times, or

$$1.24 \times 10 \times 10 \times 10 \times 10 \times 10 \times 10 \times 10 = 12,400,000$$

Each multiplication by 10 moves the decimal point one place to the right. The exponent tells you the number of multiplications and thus the number of places the decimal point is moved to the right.

$$1. \rightarrow 2 \rightarrow 4 \rightarrow 0 \rightarrow 0 \rightarrow 0 \rightarrow 0 \rightarrow 0$$
$$\quad\ 1 \quad\ 2 \quad\ 3 \quad\ 4 \quad\ 5 \quad\ 6 \quad\ 7 \text{ places}$$

A negative exponent tells you to divide the coefficient by 10 and you would divide the number of times the exponent indicates. If the exponent is 5, you divide by 10 five times:

$$5.92 \times 10^{-5} = 5.92 \div 10 \div 10 \div 10 \div 10 \div 10 = 0.0000592$$

For a negative exponent, the decimal moves to the left.

$$0.0 \leftarrow 0 \leftarrow 0 \leftarrow 0 \leftarrow 5.92$$
$$\quad 5 \quad\ 4 \quad\ 3 \quad\ 2 \quad\ 1 \text{ places}$$

An important number in chemistry is Avogadro's number: 6.02×10^{23}. This is a very large number obtained by multiplying 6.02 by 10, 23 times.

Using Your Calculator

To enter a number in scientific notation in your calculator, find the **EE** or **EXP** key. Type in the number of the coefficient, as it is written. Press the **EE** or **EXP** key, which may require pressing a shift key first, and then type in the **exponent.** For example, enter the number 6.02×10^{23} by pressing the following:

$$6.02 \text{ EE } 23$$

To enter the number 3.87×10^{-12} press the following:

$$3.87 \text{ EE } +/- 12$$

To use numbers in scientific notation in calculations, you must be able to enter them correctly in your calculator. Try the following to check if you obtain the same results:

$$4.82 \times 10^{12} - 3.33 \times 10^{10} = 4.79 \times 10^{12}$$
$$6.02 \times 10^{23} / 3.87 \times 10^{-12} = 1.56 \times 10^{35}$$

Solving Equations with Unknown Variables

Many standard equations are written with **variables,** letters used to represent numbers that vary (change). Two common equations in chemistry written with variables are:

$$P_1 V_1 = P_2 V_2$$
$$C = \frac{(F - 32)}{1.8}$$

All of the variables in these equations represent quantities. For example, F stands for the temperature in Fahrenheit and C stands for the temperature in Celsius. It is likely that the value of one of the variables will not be known and will require solving. To solve for the unknown variable, you must isolate that variable on one side of the equation, using the standard operations listed above to manipulate the equation. To maintain equality, the same operation must be performed on both sides of the equation at the same time.

In general, if the equation has a fraction, remove the denominator by multiplying both sides by the number or symbol in the denominator. To remove the numerator, divide both sides by the number or symbol in the numerator. To remove a number or symbol that is added, subtract the same number or symbol from both sides. To remove a number or symbol that is subtracted, add the same number or symbol to both sides. Regardless of the operation to be performed, *always perform the same operation on both sides to maintain equality.*

To solve the first equation, $P_1 V_1 = P_2 V_2$ for P_1, you must isolate P_1 on one side of the equation. Divide both sides of the equation by V_1.

$$\frac{P_1 \cancel{V_1}}{\cancel{V_1}} = \frac{P_2 V_2}{V_1}$$
$$P_1 = \frac{P_2 V_2}{V_1}$$

To convert the temperature from Celsius (C) to Fahrenheit (F), first eliminate the denominator by multiplying both sides of the equation by 1.8.

$$1.8 \times C = \frac{(F - 32)}{\cancel{1.8}} \times \cancel{1.8}$$
$$1.8 \, C = (F - 32)$$

To isolate F, add 32 to both sides.

$$32 + 1.8 \, C = F - 32 + 32$$
$$32 + 1.8 \, C = F$$
$$F = 1.8 \, C + 32$$

Do NOT type "$\times 10$" or your answer will be incorrect.

The [$+/-$] key or the [$-$] key is used to change the sign of a number in your calculator. Do NOT use the subtraction key.

Rounding Numbers

When performing calculations on measured values in chemistry, it is necessary to report the answer with the correct number of significant figures, as explained on pages 8–12 of Chapter 1. In order to report the correct number of significant figures, you will often need to round, or truncate, the answer. Examine the digit that is one digit beyond the correct number of significant figures.

1. Case 1: If that digit is less than five (5), the answer is truncated. For example, if the answer on the calculator reads 3.12072 and the problem dictates that there should be three significant figures, the fourth digit is examined.

$$3.12\underline{0}72$$

Since this digit is less than **5,** the answer is truncated (or cut off). The reported answer becomes

$$3.12$$

2. Case 2: If the digit that needs to be examined is equal to or greater than five (5), the last reported digit is **rounded up,** increased by one, and the answer is truncated. If the above answer, 3.12072, needs to be reported to four significant figures, the fifth digit is examined.

$$3.120\underline{7}2$$

The fifth digit is greater than **5;** therefore, the 0 would be **rounded up** or increased by one to report

$$3.121$$

A calculator will not round. Rounding is a technique that requires human examination.

Logarithms and Inverse Logarithms

The **logarithm (log)** function is explained in Chapter 9 in a Mathematical Review box on page 336. Here we give some hints on using your calculator with log functions.

Using Your Calculator
When the coefficient of a number expressed in scientific notation is 1, you can solve for the log by inspection without a calculator. In that case, the log is simply the exponent of 10: for example, the log of 1×10^5 is 5.

Most logarithms (**logs**) of numbers will be calculated with your calculator using the **log** key. To calculate the log of 3.42×10^6, you may need to consult your calculator's instruction manual. For most scientific calculators enter the following:

LOG 3.42 EE 6 ANS

You should obtain the result 6.534. Some calculators require that you enter the number first, and then press the **log** key:

3.42 EE 6 LOG

Work with your calculator or consult your instruction manual to determine which method to use. Try the following to ensure you obtain the same results:

$$\log (3.17 \times 10^{11}) = 11.501$$
$$-\log (3.17 \times 10^{-11}) = 10.499$$
$$\log (13.2) = 1.121$$

Inverse Logarithm

The **10^x** or **inverse logarithm (10^x or inv log or invlog)** function is the opposite of the logarithm function. The inverse log of 3 is 10^3 or 1000 (as above, the log of 1000 is 3).

Using Your Calculator

Locate the **10^x** or **inverse log** (**10x** or **inv log**) key. On many calculators, you must press a **second** or **shift** key followed by the log key to use the inverse log function. To solve $10^{-8.400}$ on some calculators, enter the following:

$$\text{shift } 10^x +/- \ 8.4 = 3.98 \times 10^{-9}$$

On other calculators, you enter the following instead:

$$+/- \ 8.4 \text{ second or shift } 10^x$$

Consult your instruction manual to determine the correct method for your calculation. Perform the following to make certain you can calculate 10^x correctly:

$$10^{2.68} = 478.63$$
$$10^{-2.7} = 0.001995$$
$$10^{-0.036} = 0.9204$$

Appendix B
Answers to Odd-Numbered Additional Exercises

Chapter 1

1.57 a. A skyscraper is on a macroscopic scale (you can see it).
b. A skin cell is on the microscopic scale (you need a microscope to see it).
c. DNA is on the nanoscale (it is too small to be seen with a microscope).
d. A red blood cell is on the microscopic scale (you can see it with a microscope).

1.59 pico (10^{-12}), nano (10^{-9}), micro (10^{-6}), kilo (10^3)

1.61 a. 10 mm is shorter than 1 m.
b. 10 mm is the same length as 1 cm.
c. 1 cm is shorter than one 1 dm.
d. 1 dm is shorter than 15 cm.

1.63 The volume of the lead ball is 1.5 mL.

1.65 a. 10^4 is smaller than 10^8.
b. 10^{-6} is smaller than 10^{-3}.
c. 3.7×10^{-4} is smaller than 3.7×10^4.

1.67 a. 1 ng is smaller than 1 mg.
b. 100 mg is smaller than 1 g.
c. 1000 mg is the same as 1g.

1.69 a. 1×10^{18} **b.** 2.305×10^9 **c.** 1.5×10^{-12}
d. 2.08×10^{-2}

1.71 a. 100,000 **b.** 0.0024 **c.** 165

1.73 a. false **b.** true **c.** true **d.** false

1.75 a. three **b.** one **c.** three **d.** three

1.77 a. 34,000 **b.** 0.073 g **c.** 3390 mL

1.79 a. exact number **b.** not exact **c.** not exact

1.81 a. 6000. mL **b.** 1.9 g/cm^3

1.83 $\dfrac{1 \text{ km}}{1 \times 10^3 \text{ m}}$

1.85 0.200 g

1.87 120.0 seconds

1.89 331. lb

1.91 330 L

1.93 36 mg/dose, rounded to 40 mg/dose

1.95 a. A brick has the greater density. It has more mass than a loaf of bread for the same volume.
b. A bowling ball has the greater density. It has more mass than a soccer ball and they both have the same volume.
c. The bucket of concrete is denser. It has more mass than the bucket of water, and they both have the same volume (one bucket).

1.97 44 g

1.99 206 g

1.101 Largest to smallest: 0.5000 L, 50.00 mL, 8.000 cm^3, 5,000 µL.

1.103 The diameter of the sphere with a mass of 15 g would be larger.

1.105 The electron is the lightest of the three subatomic particles.

1.107 A helium atom has the smaller diameter because it has fewer electrons. As the number of electrons increases, the outermost electrons spend more of their time in larger orbitals that extend farther from the nucleus, thereby increasing the diameter of the atom.

1.109 a. carbon, C **b.** aluminum, Al **c.** americium, Am **d.** platinum, Pt **e.** cobalt, Co.

1.111 a. cesium: 55 protons, 55 electrons **b.** rhenium: 75 protons, 75 electrons **c.** manganese: 25 protons, 25 electrons

1.113 arsenic, As

1.115 radium, Ra

1.117 Selenium is bigger. It has more electrons, which take up more space around the nucleus.

1.119 The mass number is the sum of the number of protons plus the number of neutrons.

1.121

Isotope	Mass Number	Atomic Number	Number of Protons	Number of Neutrons
Sulfur-32	32	16	16	16
Sulfur-33	33	16	16	17
Sulfur-34	34	16	16	18
Sulfur-36	36	16	16	20

1.123 Sulfur-32 is the lightest isotope, because it has the fewest number of neutrons.

1.125 $^{16}_{8}\text{O}$, $^{17}_{8}\text{O}$, $^{18}_{8}\text{O}$

1.127 a. Chlorine-35 has mass number 35. Chlorine-37 has mass number 37.
b. The atomic number for both isotopes is 17.
c. Chlorine-35 has 18 neutrons, while chlorine-37 has 20 neutrons.

1.129 The physical properties of an element are the characteristics of the element such as color and consistency that it has on its own. The chemical properties of an element reflect the manner in which it interacts with other elements or substances. See page 21.

1.131 Elements in the first two and last six columns of the periodic table are the main group elements. The transition metals are located in columns 1B–8B.

1.133 The alkali metals are located in group 1A of the periodic table, while alkaline earth metals are located in group 2A of the periodic table.

1.135 a. physical properties   **b.** physical properties

1.137 a. nonmetal   **b.** metalloid   **c.** nonmetal

1.139 a. potassium   **b.** radon

1.141 the electron

1.143 a. shell $n = 1$: 2 electrons; shell $n = 2$: 3 electrons. Three valence electrons.
b. shell $n = 1$: 2 electrons; shell $n = 2$: 8 electrons; shell $n = 3$: 5 electrons. Five valence electrons.

1.145 The s orbital is spherical in shape and the p orbital has a two lobed shape, similar to a dumbbell.

1.147 group 4A

1.149 The macronutrients are Na, K, Mg, Ca, P, S, and Cl. The micronutrients are V, Cr, Mn, Fe, Co, Cu, Zn, Mo, Si, Se, F, and I. Macronutrients are required in our diet in large quantities—more than 100 mg a day, while micronutrients are required in quantities of less than 100 mg a day.

1.151 Your body needs micronutrients in trace quantities, less than 100 mg a day.

1.153 Iron is part of hemoglobin that transports oxygen in red blood cells. Iron is also found in many other oxygen transport molecules and enzymes that are involved in extracting energy from the foods you eat.

Chapter 2

2.39 An element is composed of one type of atom, defined by its atomic number. Compounds are composed of two or more different elements.

2.41 Covalent compounds are formed between *nonmetal* atoms.

2.43 (+ and +) and (− and −) are repulsive interactions.

2.45 a. Ca^{2+} **b.** Cr^{3+} and Cr^{6+} **c.** N^{3-} **d.** Ag^+

2.47 A cation is a positively charged atom: It has more protons than electrons. A cation is formed when an atom loses electrons. An anion is a negatively charged atom: It has more electrons than protons. An anion is formed when an atom gains electrons. Group 1A, 2A, and 3A metals and the transition elements form cations.

2.49 Ions are not formed from group 8A elements because these elements have a full outer shell of electrons.

2.51 H^+ or H^-. If hydrogen gains an electron, it will have a full electron shell, similar to helium. If hydrogen loses an electron, it will not have any electrons in its outer shell.

2.53 a. Magnesium cation: 12 protons, 10 electrons. It has a +2 charge because the ion lost two electrons to achieve a full outermost electron shell.
b. Mercury cation: 80 protons, 78 electrons. It has a +2 charge because it has two more protons than electrons.
c. Chlorine anion or chloride: 17 protons, 18 electrons. It has a −1 charge because the chlorine atom gained an electron to achieve a complete shell of electrons.
d. Fluorine anion or fluoride: 9 protons, 10 electrons. It has a −1 charge because the fluorine atom gained an electron to achieve a complete shell of electrons.

e. Oxygen anion or oxide: 8 protons, 10 electrons. It has a −2 charge because oxygen gained 2 electrons to achieve a complete electron shell.

2.55 An ionic compound is formed by the mutual attraction of cations and anions. The ions are held together by electrostatic attraction.

2.57 When the ionic lattice of NaCl is dissolved in water, the lattice structure falls apart. The sodium ions (Na^+) and chloride ions (Cl^-) separate, and each ion is surrounded by water molecules.

2.59 a. potassium ion, K^+, and bromide, Br^-
b. magnesium ion, Mg^{2+}, and chloride, Cl^-
c. potassium ion, K^+, and iodide, I^-
d. barium ion, Ba^{2+}, and chloride, Cl^-
e. sodium ion, Na^+, and fluoride, F^-

2.61 a. LiI; lithium (cation), iodide (anion)
b. RbF; rubidium (cation), fluoride (anion)
c. $CaBr_2$; calcium (cation), bromide (anion)
d. BaI_2; barium (cation), iodide (anion)
e. FeS; iron (cation), sulfide (anion)
f. Al_2O_3; aluminum (cation) oxide (anion)

2.63 a. strontium oxide   **b.** potassium iodide   **c.** gallium oxide   **d.** lithium fluoride   **e.** sodium iodide   **f.** iron(III) oxide

2.65 Na^+

2.67 a. Zn^{2+}, OH^- **b.** Cu^+, $CH_3CO_2^-$ **c.** Sn^{4+}, Cl^- **d.** V^{5+}, O^{2-} **e.** Cr^{6+}, O^{2-}

2.69 The ions in ionic compounds are held together by electrostatic forces, the attraction between oppositely charged species. The atoms in covalent compounds are held together by covalent bonds. In a covalent bond, two atoms share two valence electrons.

2.71 The diatomic elements are hydrogen, nitrogen, oxygen, fluorine, chlorine, bromine, and iodine.

2.73 a. $\cdot\ddot{C}\cdot$ **b.** $H\cdot$ **c.** $:\ddot{O}\cdot$ **d.** $\cdot\ddot{P}\cdot$

2.75 a. The carbon atom contains eight bonding electrons and zero nonbonding electrons.
b. The nitrogen atom contains six bonding electrons and two nonbonding electrons.
c. The carbon and nitrogen atoms each have an octet of electrons; the hydrogen atom has a duet of electrons.
d. There is a single bond between hydrogen and carbon. There is a triple bond between carbon and nitrogen.
e. Six electrons are shared between the carbon atom and the nitrogen atom in the triple bond.

2.77 $:\ddot{Br} - \ddot{Br}:$ This molecule is an element.

2.79 a.

2.81 The phosphorus atoms have an expanded octet. Each atom has ten valence electrons.

2.83 SO_3 is sulfur trioxide.

2.85 PCl_3 is phosphorus trichloride. PCl_5 is phosphorus pentachloride. The names differ in the prefix attached to indicate the number of chlorine atoms.

2.87 a. hydrogen carbonate (or bicarbonate) ion
b. acetate ion
c. hydroxide ion

2.89 a. NH_4^+ **b.** CO_3^{2-} **c.** HPO_4^{2-}

2.91 a. The overall charge is -1.
b. Hydrogen carbonate is a polyatomic atom because it is charged and composed of several atoms joined by covalent bonds.
c. $NaHCO_3$ **d.** $Ca(HCO_3)_2$

2.93 a. The overall charge is -2.
b. This ion is composed of several atoms joined by covalent bonds.
c. K_2HPO_4 **d.** $MgHPO_4$

2.95 a. Na_3PO_4 **b.** NH_4Cl **c.** $Mg(OH)_2$

2.97 $Ca_3(PO_4)_2$

2.99 Avogadro's number is the number items in one mole of items. It is 6.02×10^{23}.

2.101 58.69 g

2.103 a mole of zirconium

2.105 58.12 amu

2.107 0.0180 mol

2.109 0.012 mol

2.111 240 g (There are only two significant figures.)

2.113 a. The molar mass of aluminum is 26.98 g/mol.
b. The molar mass of hydrogen is 2.016 g/mol.
c. The molar mass of calcium is 40.08 g/mol.
d. The molar mass of nitrogen is 28.02 g/mol.

2.115 a. 73.89 g/mol **b.** 212.3 g/mol.

2.117 albumin

2.119 The molecular formula of urea is CH_4N_2O. The molar mass of urea is 60.06 g/mol. Urea is a covalent compound because there are only covalent bonds between the atoms in the molecule. There are from 1×10^{-4} mol to 3×10^{-4} mol of urea in 1 dL of serum.

Chapter 3

3.21 a.

The electron geometry is tetrahedral. The molecular shape is tetrahedral. The bond angles are 109.5°.

b.

The electron geometry is tetrahedral. The molecular shape is trigonal pyramidal. The bond angles are approximately 109.5°.

c.

The electron geometry is trigonal planar. The molecular shape is bent. The bond angles are 120°.

3.23 The molecule is linear.

3.25 The electron geometry is tetrahedral. The possible molecular shapes are tetrahedral, trigonal pyramidal, and bent. The bond angles for all three molecular shapes are 109.5°.

3.27 The trigonal pyramidal shape is three-dimensional.

3.29 The bond angles are 180°.

3.31 The bond angles are 109.5°.

3.33 a. The molecular geometry is linear. The bond angle around the carbon atom is 180°.
b. The molecular geometry is trigonal planar. The bond angles around the carbon atom are 120°.

3.35 The geometry is tetrahedral. The O—S—O bond angle is 109.5°.

3.37 The bond angle around the central atom decreases as the number of electron groups surrounding the central atom increases because there are more groups to distribute in the same amount of space. The greatest bond angle is 180°; the smallest bond angle typically is 109.5°.

3.39 a.

The molecular shape is trigonal planar around each carbon atom. The bond angles are all 120°.

b.

The molecular shape is tetrahedral. The bond angles are 109.5°.

c.

The molecular shape is tetrahedral around each carbon atom. The bond angles are 109.5°.

d. $H—C \equiv C—H$

The molecular shape around each carbon atom is linear. The bond angles are 180°.

3.41 a.

b. The bond angle around each carbon atom is 109.5°.
c. The bond angle around the oxygen atom is approximately 109.5°.
d. The molecular shape around each carbon atom is tetrahedral.
e. The molecular shape around the oxygen atom is bent.

3.43 The geometry is trigonal planar when there are three bonding groups and no nonbonding groups surrounding a central atom. The bond angles are 120°. The geometry is trigonal pyramidal when there are three bonding groups and one nonbonding group surrounding a central atom. The bond angles are 109.5°.

3.45 The geometry is linear for two atoms because two points define a line and no other geometry is possible.

3.47 **2.**

$$\overset{\cdot\cdot}{\underset{\|}{O}}$$
$$H-C-H$$

    4. There are three electron groups surrounding the central carbon atom.

    5. There are three groups of bonding electrons.

    6. There are no nonbonding electrons on the central atom.

    7. The electron geometry is trigonal planar.

    8. The molecular shape of formaldehyde is trigonal planar.

    9. This molecule has a two-dimensional, flat shape.

    10. The H—C—O bond angle is 120°. All the bond angles are the same.

    11. The trigonal planar shape allows all three groups of electrons to be spaced the farthest possible distance from each other. In the T-shape, two groups are farther apart, but two groups are closer together.

3.49 Ball-and-stick models show the bond angles present in a molecule. When the size of the atoms is important, a space-filling model is used.

3.51 Electronegativity is a measure of the ability of an atom to attract electrons to itself in a covalent bond.

3.53 **a.** Oxygen is the more electronegative element.
    b. Oxygen is the more electronegative element.
    c. Fluorine is the more electronegative element.

3.55 Electronegativity increases as you move up in a group of elements.

3.57 The halogens are the most electronegative.

3.59 If the electronegativities of the two atoms in a covalent bond are similar, then the bond will be nonpolar. If the electronegativities are different, then the bond will be a polar covalent bond.

3.61 **a.** Water is polar.

    b. Ethanol is polar.

    c. C_2H_4 is not polar.

3.63 **a.** I_2 is nonpolar. The two atoms in the covalent bond are the same.
    b. CH_4 is nonpolar. Carbon and hydrogen have similar electronegativities.
    c. H—Br is polar. Hydrogen and bromine have different electronegativities.

3.65 A covalent bond occurs between two atoms within the same molecule and is much stronger than an intermolecular force of attraction, the force of attraction between atoms in different molecules.

3.67 The three types of intermolecular forces of attraction are: dispersion forces, dipole–dipole interactions, and hydrogen bonds. The strongest intermolecular forces of attraction are hydrogen bonds. The weakest forces of attraction are dispersion forces.

3.69 Dispersion forces are the weakest of the intermolecular forces because they are created from temporary dipoles.

3.71

H	Br	H	Br
Br	H	Br	H
H	Br	H	Br
Br	H	Br	H

The H—Br molecules exhibit dipole–dipole interactions. They are lined up so that the partial positive charge (H) lies next to the partial negative charge (Br) of an adjacent molecule.

3.73 **a.** C_5H_{12} is nonpolar, so it should only exhibit dispersion forces.
    b. Acetone is polar, so it should exhibit dispersion forces and dipole–dipole interactions.
    c. Water is polar and has O—H bonds, so it should exhibit hydrogen bonding as well as dispersion forces, and dipole–dipole interactions.

3.75 Ice floats on water because hydrogen bonding causes the collection of molecules to occupy a greater volume in the solid state than in the liquid state, so the mass per volume is less for the solid state than the liquid state.

3.77

The dashed lines represent the hydrogen bond between the oxygen atom of one molecule and the hydrogen atom of another molecule.

3.79 Hydrogen bonding gives DNA a helical shape.

3.81 Estrogen binds to the estrogen receptor and activates several genes. This gene activation also stimulates the proliferation of breast cancer cells.

3.83 The estrogen receptor is a large protein molecule. Estradiol fits the estrogen binding site perfectly because it has a complimentary shape to the binding site—a cavity within the receptor.

3.85 When Tamoxifen binds to the receptor, it changes the shape of the receptor, preventing gene activation.

Chapter 4

4.53 The three physical states of matter are the solid state, the liquid state, and the gas state.

4.55 The solid state and the liquid state have fixed volumes compared to the container. A gas will expand to fill the container completely.

4.57 A rock rolling down a hill is in motion, so it has kinetic energy.

4.59 **a.** Kinetic energy (the biker is in motion).
    b. Potential energy (the hiker has stored energy).
    c. Kinetic energy (the helium atoms are in motion).
    d. Potential energy (the chemical bonds in the molecules of wax have stored energy).

4.61 Heavy molecules have more kinetic energy, if they have the same average speed.

4.63 In the solid state, molecules have the least amount of kinetic energy.

4.65 Steam molecules have the most kinetic energy because the water molecules are in the gas phase.

4.67 88 °F; 304 K. You would be wearing summer clothes.

4.69 37.0 °C

4.71 −270 °C; −450 °F

4.73 Intermolecular forces are greatest in the solid state.

4.75 A block of aluminum would sink in a container of molten aluminum; the atoms are much more closely packed together in the solid state.

4.77 Only intermolecular forces of attraction are affected when a change of state occurs.

4.79 Energy needs to be removed to achieve freezing, condensation, or deposition.

4.81 **a.** Liquid → gas **b.** Liquid → solid
c. Liquid → solid **d.** Liquid → gas **e.** Gas → liquid

4.83 Steam causes burns because of the change of state that occurs when steam comes in contact with your skin. Steam condenses when it comes in contact with the skin, which requires heat to be removed from the steam by an amount equivalent to the heat of vaporization of water. The heat is removed from the steam and transferred to your skin. Additional heat is transferred to your skin as the liquid water cools from 100°C to 37°C.

4.85 Ethanol can form hydrogen bonds, while carbon dioxide cannot. In order for ethanol to enter the gas phase, more energy must be supplied to break the hydrogen bonds holding the ethanol molecules together.

4.87 **a.** For brick: 59 cal **b.** for ethanol: 170 cal **c.** for wood: 29 cal. Ethanol requires the greatest input of heat to warm because ethanol has the largest heat capacity, 0.58 cal/g °C.

4.89 **a.** Amount of heat transferred by the water cooling: 1575 cal, rounded to 1600 cal.
b. Amount of heat transferred to condense steam: 13,500 cal, rounded to 14,000 cal
The total amount of heat released is 14,000 cal + 1600 cal = 15,600 cal.

4.91 Atmospheres are a unit of pressure. Atmospheric pressure is the pressure caused by the molecules of air pressing down on us as a result of gravity.

4.93 **a.** 3.1 psi **b.** 110 torr

4.95 0.6 psi

4.97 Systolic pressure: 170 mmHg; diastolic pressure: 112 mmHg. The patient's blood pressure is $\frac{175}{112}$. This patient has hypertension.

4.99 The vapor pressure of acetone is lower, so acetone has a higher boiling point than methylene chloride.

4.101 The boiling point of water should be lower at the base camp for Mt. Everest because the atmospheric pressure is significantly lower there.

4.103 One mole of carbon dioxide occupies 22.4 L at STP.

4.105 92 L

4.107 9.8 mol

4.109 At the lower elevation in Denver, the atmospheric pressure is higher compared to the mountains where you capped the bottle. As the atmospheric pressure increases as you come down from the mountains, it forces the volume of gas in the water bottle to decrease.

4.111 380 L

4.113 0.6 L

4.115 Upon exhalation, the pressure of the lungs *increases* as the volume of the lungs *decreases*.

4.117 Gay-Lussac's law states that the pressure of a gas is directly proportional to the temperature of a gas. If you heat the can of unopened beans directly on the stove, the temperature of the gas inside the can will increase. As the temperature of the gas increases, the pressure of the gas will increase, which could cause the can of beans to explode.

4.119 Temperatures are higher in the summer, and pressure is directly proportional to temperature, so the pressure inside the tires will be higher in summer.

4.121 31.7 psi

4.123 Charles' law states that temperature and volume are directly proportional to each other. As you heat the cake in the oven, the carbon dioxide in the cake heats up; therefore, the volume of the carbon dioxide increases and the cake rises.

4.125 9.4 L

4.127 As the air inside of the balloon is heated up, the volume of the air increases (Charles' law). The number of molecules of air in the balloon stays the same, but the density (mass/volume) of the heated air decreases. The less dense air inside the balloon will cause the balloon to float.

4.129 5.9 L

4.131 Dalton's law states that each gas in a mixture of gases will exert a pressure independent of the other gases present; and each gas will behave as if it alone occupied the total volume. The sum of the partial pressures of each gas present is equal to the total pressure.

4.133 0.61 atm

4.135 The person who is scuba diving is under greater pressure and therefore will have a higher concentration of oxygen in their blood.

4.137 Desflurane has a larger Henry's constant than diethyl ether. The concentration of desflurane will be lower in the blood; therefore, the patient would regain consciousness more quickly.

4.139 The increase in pressure within the chamber causes the nitrogen bubbles to redissolve. The redissolved nitrogen circulates to the lungs where it can safely be exhaled. Henry's law shows that there is a direct relationship between the pressure and the concentration of a gas in solution. As the pressure increases, the concentration of nitrogen in the blood should increase.

4.141 Carbon monoxide is poisonous because it binds to hemoglobin more strongly than oxygen. When carbon monoxide enters the bloodstream, it binds with hemoglobin, replacing the oxygen, and the level of oxygen available to tissues drops to dangerous levels.

4.143 a. 0.059 mL
  b. 0.026 mL
4.145 7.6×10^{-3} mol

Chapter 5

5.41 The two components of a solution are the solute and the solvent. The solvent is present in the greater amount.

5.43 a. solute: tin and sometimes phosphorus; solvent: copper.
  b. solute: carbon dioxide, sugar; flavorings; solvent: water.
  c. solute: carbon dioxide, ethanol; solvent: water
  d. solute: isopropyl alcohol; solvent: water.

5.45 Concentration is a measure of how much solute is dissolved in a given quantity of solution.

5.47 1.25 mg/L. Iron is an electrolyte.

5.49 140 mg/mL

5.51 1.5 g

5.53 1.2×10^{-2} mol/L

5.55 1.2×10^{-10} mol/L

5.57 2.67×10^4 mmol/L

5.59 4.0 meq/L

5.61 The IU takes into account the biological effect of the solute.

5.63 8 mL

5.65 0.01 mL/min

5.67 12 mg/hr, rounded to 10 mg/hr

5.69 $\dfrac{35 \text{ mg erythromycin}}{\text{kg} \times \text{day}}$ Yes, it is within the recommended range.

5.71 0.18 mol/L

5.73 1×10^{-4} mol/L

5.75 In a suspension, the particles are analogous to the solute of a solution. The dispersion medium is analogous to the solvent of a solution.

5.77 a. solution   **b.** colloid   **c.** solution and colloid
  d. solution   **e.** solution   **f.** colloid

5.79 The solutes in whole blood are less than 1 nm in size and include ions such as Ca^{2+}, molecules such as glucose, and gases such as oxygen. The colloidal particles in whole blood range in size from 1 nm to 100 nm and include proteins and starch. The suspended particles in whole blood are greater than 100 nm in size and include red blood cells, white blood cells, and platelets.

5.81 A semipermeable membrane allows the passage of small molecules and ions across the membrane while preventing the passage of larger molecules and ions.

5.83 Simple diffusion is the spontaneous movement of a molecule or ion from a region of higher concentration to a region of lower concentration.

5.85 a. Water will flow from solution B to solution A.
  b. Water will flow from solution A to solution B.
  c. The solutions are isotonic. No net flow of water will occur between the solutions.

5.87 In osmosis, water flows from a hypotonic solution to a hypertonic solution.

5.89 No, you should not expect dialysis to occur. With isotonic solutions, there is no difference in concentration between the two solutions and so no net crossing of the membrane from either side will occur.

5.91 Creatine will diffuse from solution A to pure water because that is the direction of lower solute concentration. Solution B can be periodically replaced with pure water to encourage further diffusion of creatine from solution A. Eventually, almost all of the creatine will be separated from solution A. The globulin remains in solution A because it is too large a molecule to cross the semipermeable membrane.

5.93 The effective concentration of ions in Solution A is 0.078 M $\times$ 2, or 0.16 M. The effective concentration of ions in Solution B is 4.8×10^{-4} M $\times$ 3, or 0.0014 M. Solution B has the lower concentration of ions, so water will flow from compartment B to compartment A.

5.95 The walls of the arterial capillaries in the kidneys are membranes that selectively allow waste products to diffuse from the bloodstream into the kidneys. Urea, ions, glucose, and amino acids are small enough to pass through these membranes.

5.97 In hemodialysis, ions and other solutes, such as urea and creatine, flow out of the blood into the dialysate. The concentrations of ions, urea, and creatine are lower in the dialysate than in the blood. These solutes will diffuse from a region of higher concentration to a region of lower concentration.

5.99 The two types of kidney dialysis are hemodialysis and peritoneal dialysis. Hemodialysis needs to be performed at a medical center and uses a dialyzer to remove the waste products and excess water from the patient's blood. Peritoneal dialysis can be performed at home and uses the patient's abdominal cavity as a filter to remove the waste products.

5.101 0.02 mol/L

5.103 The effective concentration of ions in Solution A is 0.25 M $\times$ 2, or 0.50 M. The effective concentration of ions in Solution B is 0.25 M $\times$ 3, or 0.75 M. Solution A has the lower concentration of ions, so water will flow from compartment A to compartment B.

Chapter 6

6.45 Organic compounds contain carbon atoms. Inorganic compounds do not contain carbon.

6.47 The four types of hydrocarbons are alkanes, alkenes, alkynes, and aromatic hydrocarbons.

6.49 a. False   **b.** true   **c.** true   **d.** true

6.51 a. hydrophobic   **d.** insoluble in water
  e. soluble in other hydrocarbons

6.53 Water has a higher boiling point than methane. Water molecules form hydrogen bonds with other water molecules. Hydrogen bonds are stronger than the dispersion forces holding the methane molecules together, so more heat is needed to separate water molecules and form a gas.

6.55 Alkanes are hydrocarbons that contain only carbon–carbon single bonds and carbon–hydrogen bonds. They contain no multiple bonds. They are classified as saturated hydrocarbons because they contain the maximum number of hydrogen atoms for a given number of carbon atoms.

6.57 Every atom in an alkane must have a tetrahedral geometry; therefore, the overall shape of the molecule takes on a zigzag appearance when the chain has three or more carbon atoms. Yes, it can have other shapes due to the different conformations that arise from C—C bond rotation.

6.59 The geometry of a carbon atom in an alkane is tetrahedral.

6.61 The geometry of a carbon atom in a triple bond of an alkyne is linear.

6.63 a. Different compounds because they have different formulas. The two compounds are C_3H_8 and C_2H_6.
b. Different compounds because they are structural isomers.
c. Different compounds because they have different formulas. The two compounds are C_4H_{10} and C_5H_{12}.
d. Identical compounds, but different conformations.

6.65 The structure on the left has five carbons in a straight chain. The structure on the right has three carbons in the main chain and two CH_3 groups branching off the middle carbon. Structural isomers have the same chemical formula.

6.67 a.

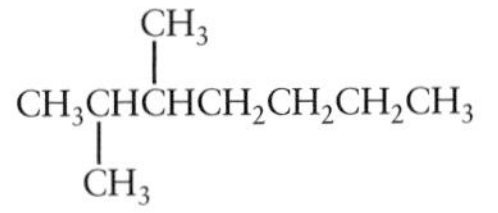

b. 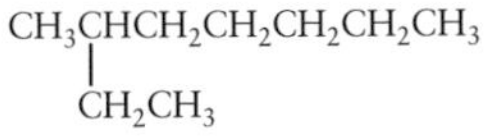 $CH_3CHCH_2CH_2CH_2CH_2CH_3$
       CH_2CH_3

c. 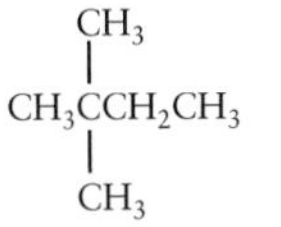
       CH_3
$CH_3CCH_2CH_3$
       CH_3

d. $CH_3CH_2CH_2CH_3$

e. 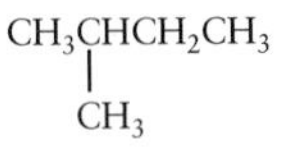 $CH_3CHCH_2CH_3$
       CH_3

f. $CH_3CH_2CH_2CH_2CH_2CH_2CH_3$

6.69 a.

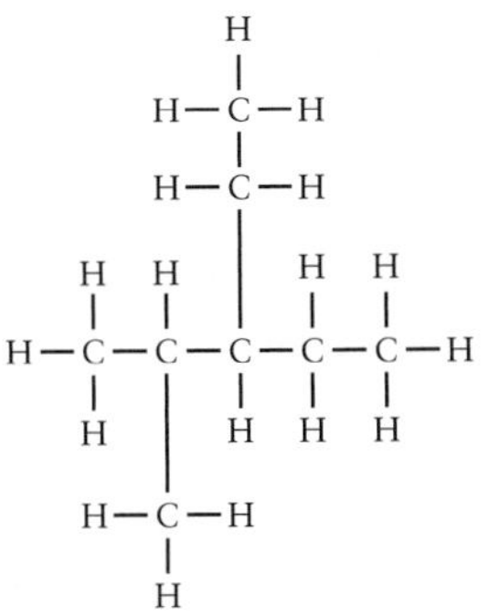

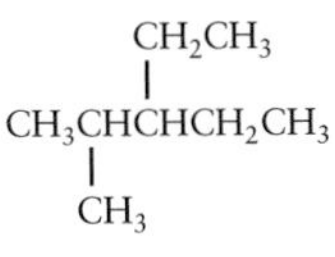

b.

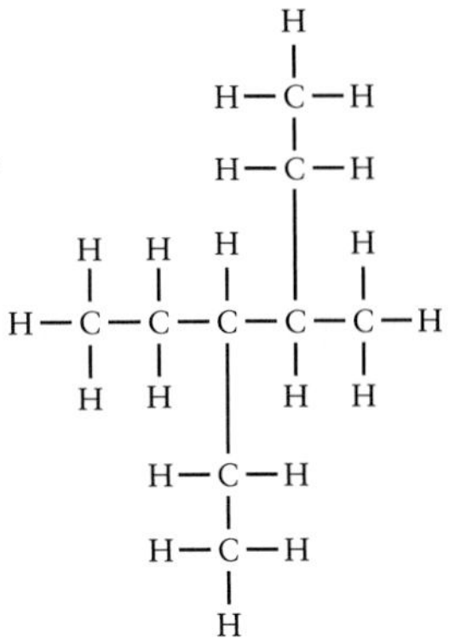

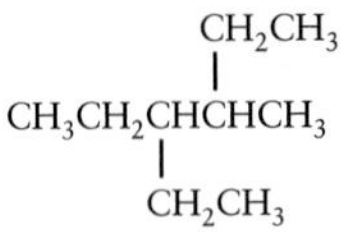

$CH_3CH_2CHCHCH_3$
    CH_2CH_3

c.

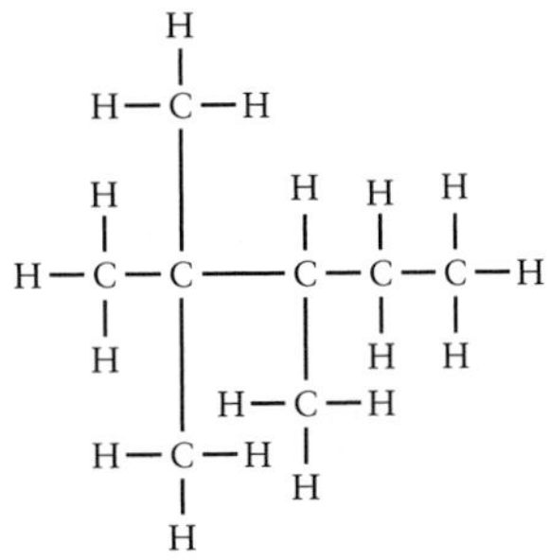

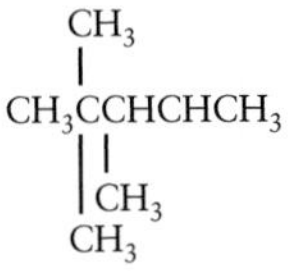

$CH_3CCHCHCH_3$
   CH_3
   CH_3

d.

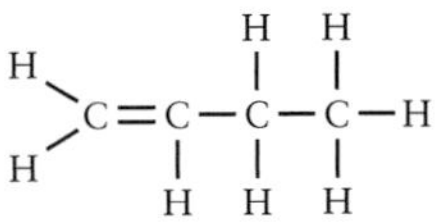

$CH_2CHCH_2CH_3$

6.71 a.

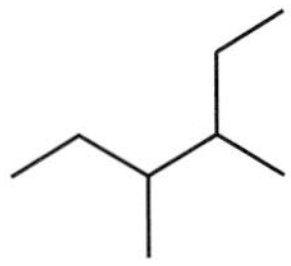

b.

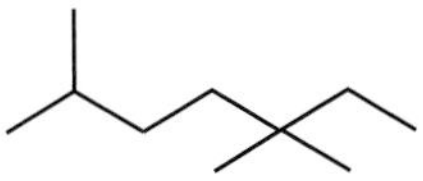

c.

6.73 Three carbons in a ring:  . Five carbons in a ring: 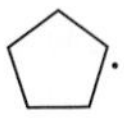 . The ring with five carbons is more common in nature. The three-carbon ring is strained; the bond angles are forced to be 60° when they should be 109.5°.

6.75

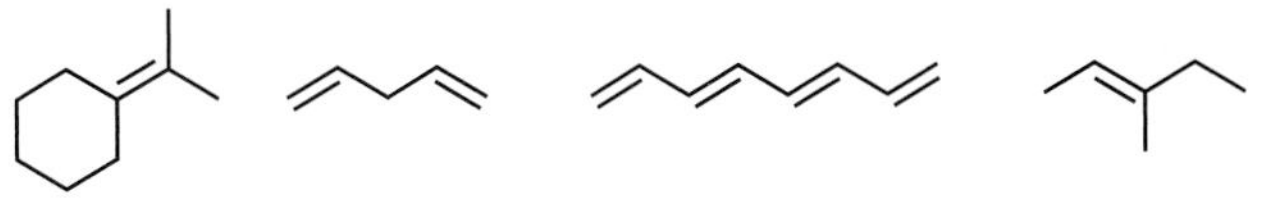

Simple alkene   Diene   Polyene   Simple alkene

6.77

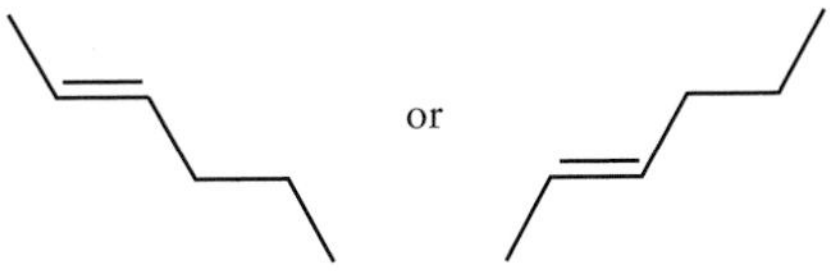

or

6.79 Yes, the rotational freedom around a carbon–carbon double bond is different from that around a carbon–carbon single bond. There is no rotation about a carbon–carbon double bond, while there is rotation about a carbon–carbon single bond.

6.81 The C—C—C bond angle is 180° because the carbon atoms are part of a carbon–carbon triple bond.

6.83 The three parts of an IUPAC name are the prefix, the root, and the suffix.

6.85 a. Heptane **b.** butane **c.** propane

6.87

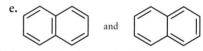

Hexane   2-Methylpentane   3-Methylpentane

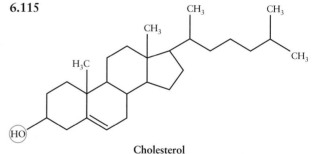

2,2-Dimethylbutane   2,3-Dimethylbutane

If two compounds end up having the same name, then they are the same compound.

6.89 a. 3-methylhexane **b.** 4-propylheptane
c. 2,3,3-trimethylhexane

6.91 a. Cyclooctane **b.** cycloheptane **c.** cyclopentane

6.93 a. *trans*-2-Pentene **b.** 1-hexene **c.** 2,3-dimethyl-2-pentene **d.** 3-ethyl-2-hexene **e.** cyclopentene **f.** cycloheptene

6.95 a. **b.** **c.**

d.

6.97 a. 3-Ethyl-4-octyne **b.** 1-heptyne **c.** 3-hexyne
d. 2-methyl-5-decyne **e.** 2,5,6-trimethyl-3-heptyne

6.99 Benzene is classified as an unsaturated hydrocarbon because it has less than the maximum number of hydrogen atoms for six carbon atoms.

6.101 Benzene and cyclohexane both have six carbon atoms in a ring. Benzene has six hydrogen atoms and three carbon–carbon double bonds and is more stable due to the delocalization of the electrons. The carbon atoms in benzene are trigonal planar. Cyclohexane has 12 hydrogen atoms and only carbon–carbon single bonds. In cyclohexane the carbon atoms are tetrahedral.

6.103

6.105 a. Propylbenzene **b.** 1,3-dimethylbenzene
c. 1-ethyl-2-methylbenzene

6.107 a.

PABA

The "benz" part of the root name appears in the IUPAC name for PABA.
b. The substitution on the aromatic ring is 1,4-.

6.109 a. Vitamins E and K_1 contain aromatic rings. Vitamins A and D_3 can be classified as polyenes.
b. There are 13 carbons in the long hydrocarbon chain. It is a branched hydrocarbon.
c. The two heteroatoms in vitamin E are oxygen atoms.
d. These vitamins are hydrophobic. They contain mostly carbon and hydrogen atoms and thus are similar to fat molecules.
e. Since these vitamins are hydrophobic, they are not soluble in water (the main component of urine). Therefore, they would not be readily excreted in urine.
f. Vitamin A plays a role in the chemistry of vision.
g. There are six substituents on the aromatic ring in vitamin E.
h. Vitamin D_3 and cholesterol are very similar in structure. Vitamin D_3 is missing one C—C bond in the second six-carbon ring and has three more C—C double bonds than cholesterol. Yes, it seems plausible that the body can synthesize cholesterol out of vitamin D_3.

6.111 a. Yes
b. No
c. Yes. Naphthalene and anthracene should be flat molecules because all of the carbon atoms are trigonal planar.
d. The bond angles are 120°.
e.

and

6.113 A carcinogen is a molecule or compound that causes cancer. Benzene is a carcinogen.

6.115

Cholesterol

6.117 Cholesterol is used to build cell membranes and make bile, vitamin D, and many important steroid hormones.

6.119 Plaque contains cholesterol and fats and can build up along the artery walls, causing restricted blood flow.

6.121 The carotid artery is most likely blocked.

6.123 The liver synthesizes cholesterol.

Chapter 7

7.31 a. 1-Hexanol. Primary alcohol. $CH_3(CH_2)_4CH_2OH$
b. 2-Pentanol. Secondary alcohol.

$$CH_3CH_2CH_2CHCH_3$$
(OH)

c. Cyclopentanol. Secondary alcohol.
d. Cyclobutanol. Secondary alcohol.
e. 2-Methyl-3-pentanol. Secondary alcohol.

7.33

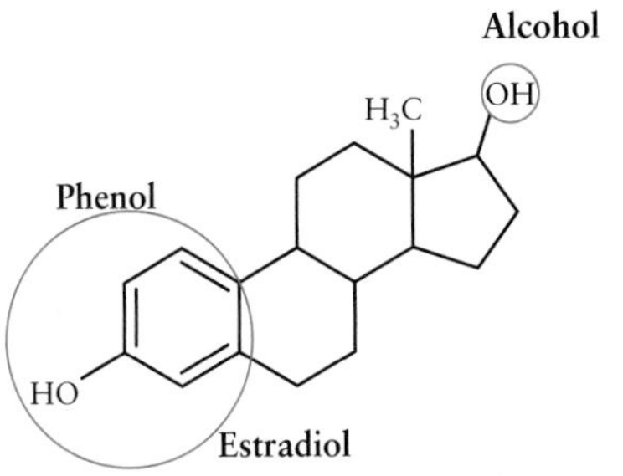

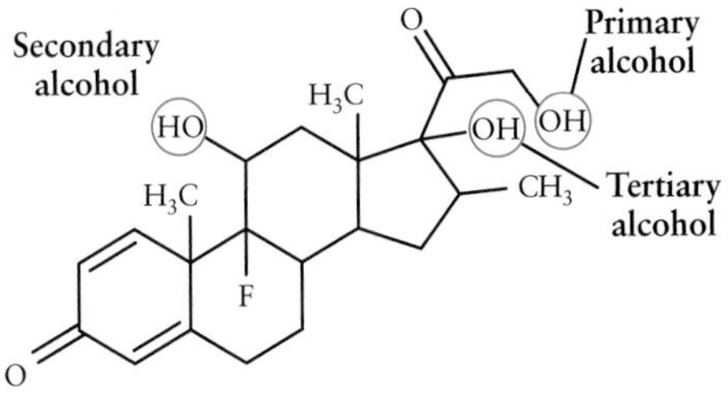

Phenol

Estradiol

Alcohol

H₃C

OH

HO

Secondary alcohol

Primary alcohol

Tertiary alcohol

H₃C

HO

H₃C

O

OH OH

CH₃

F

O

Betamethasone

7.35 The alcohol 2-methyl-1-butanol will be more soluble in water. The OH group of the alcohol can hydrogen bond with water. The hydrocarbon cannot hydrogen bond with water.

7.37 An ether has two carbon atoms (R groups) attached to an oxygen atom. An alcohol has one carbon atom (R group) and a hydrogen atom attached to an oxygen atom.

7.39

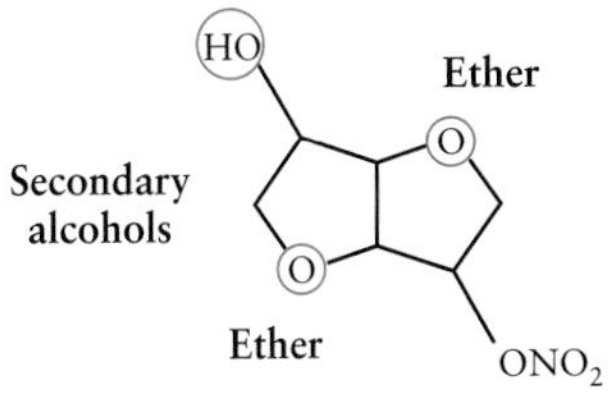

HO

Ether

Secondary alcohols

O

O

Ether

ONO₂

7.41 a. Propanal **b.** pentanal **c.** 2-hexanone
d. 3-pentanone **e.** cyclohexanone

7.43 a. and **b.**

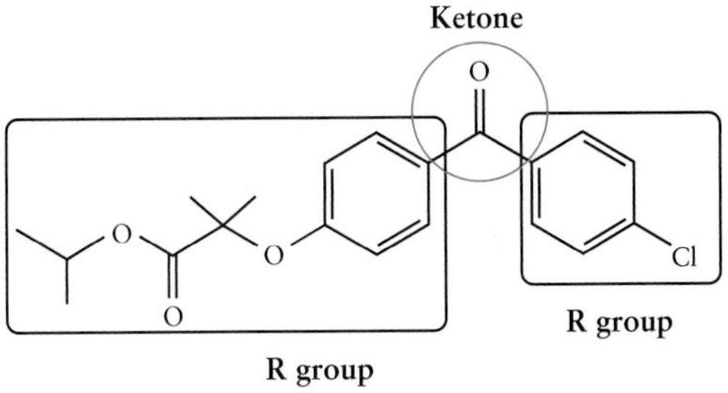

Ketone

O

O

O

Cl

R group

R group

c.

Aromatic ring

Aromatic ring

O

O

O

Cl

Ether

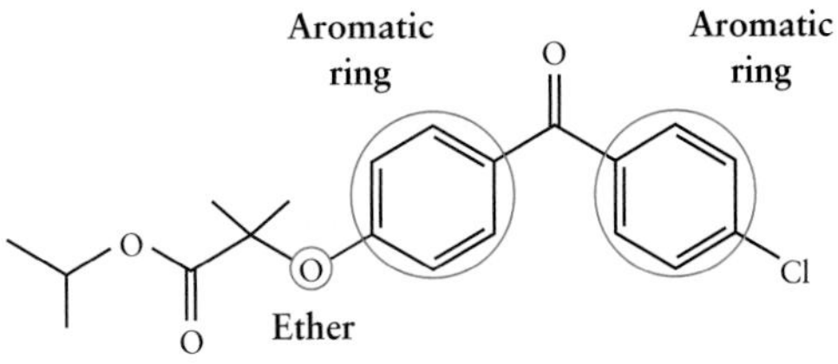

7.45 a. A carboxylic acid has an R group and an OH group attached to a carbonyl carbon; an ester has an R group and an OR group attached to a carbonyl carbon.

b. A carboxylic acid has an R group and an OH group attached to a carbonyl carbon; a thioester has an R group and an SR group attached to a carbonyl carbon.

c. An ester has an oxygen atom attached to a carbonyl group and an R group; a thioester has a sulfur atom attached to a carbonyl group and an R group.

d. A carboxylic acid has an R group and an OH group attached to a carbonyl carbon; an amide has an R group and a nitrogen atom attached to the carbonyl carbon. The nitrogen atom in an amide may be attached to one or more hydrogen atoms or one or more R groups.

e. An amide has a nitrogen atom attached to a carbonyl carbon and may be attached to one or more hydrogen atoms or one or more R groups; an amine does not have a carbonyl carbon. An amine only has one or more hydrogen atoms or one or more R groups attached to a nitrogen atom.

f. A carboxylic acid has an R group and an OH group attached to a carbonyl carbon; an alcohol has an OH group, but does not contain a carbonyl group.

7.47 a. Carboxylic acid

O

OH

OH

OH

HO

OH

O

Carboxylic acid

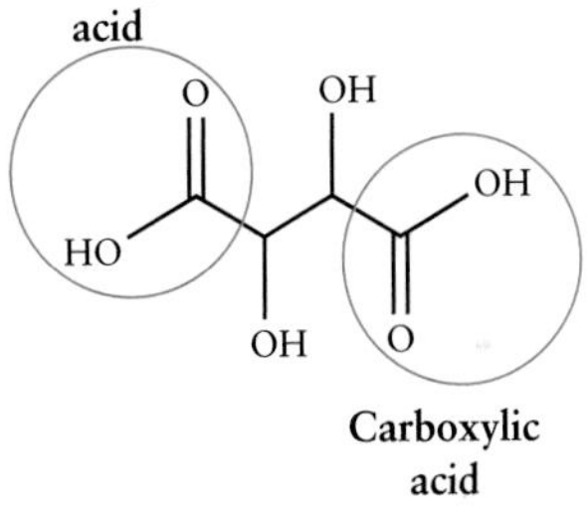

b. Alcohol

O

OH

OH

HO

OH

O

Alcohol

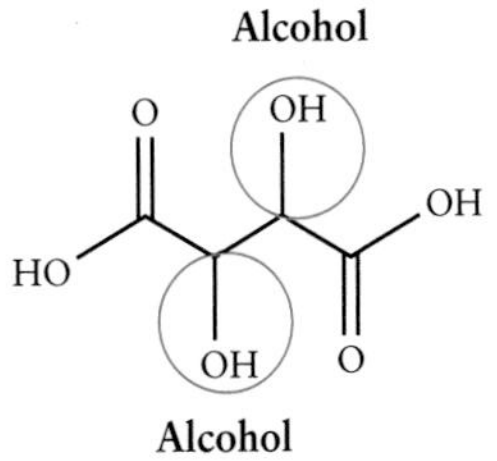

c. Yes, tartaric acid should be soluble in water. The alcohol groups and the carboxylic acid groups can hydrogen bond with water molecules.

7.49 a.

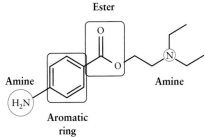

Ester

Amine

H₂N

Amine

Aromatic ring

b. Both amine groups are in their neutral form.
c. The amine on the left is attached to one R group. The amine on the right is attached to three R groups.

7.51 a.

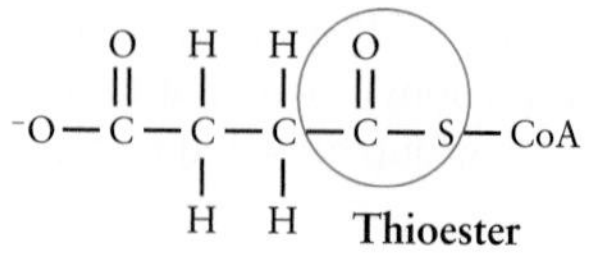

O H H O
‖ │ │ ‖
⁻O—C—C—C—C—S—CoA
│ │
H H   **Thioester**

b.

O H H O
‖ │ │ ‖
⁻O—C—C—C—C—S—CoA
│ │
H H

Carboxylate ion

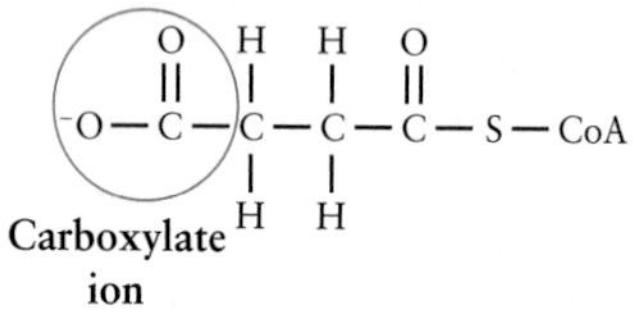

This functional group is in its ionic form.

7.53

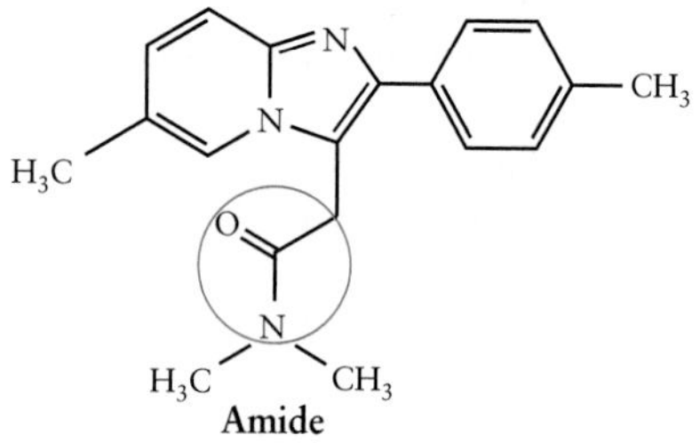

Amide

There are two methyl groups attached to the nitrogen atom.

7.55 a.

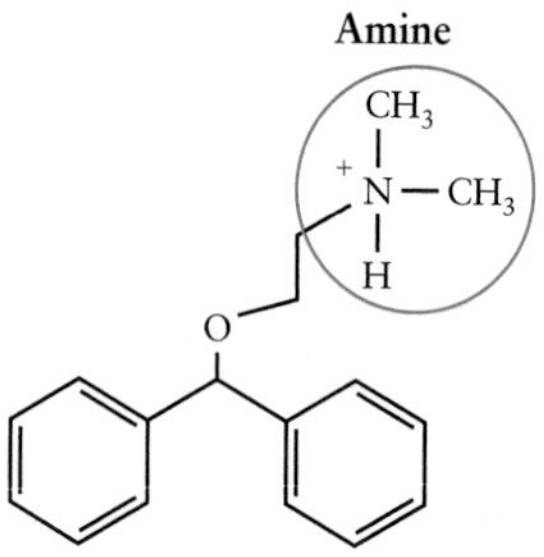

Amine

b. The amine in Benadryl is in its ionic form.

c.

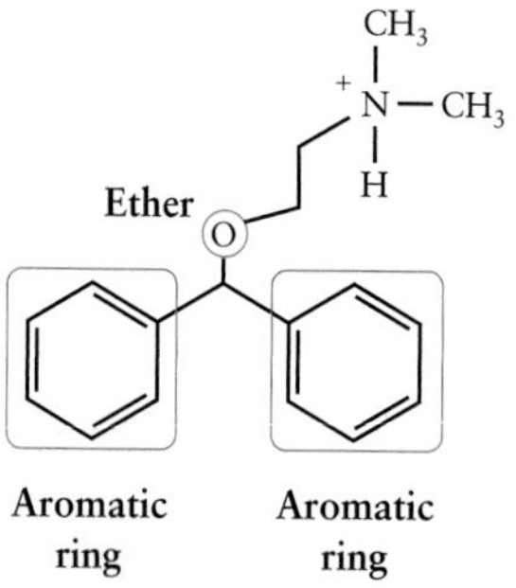

Ether

Aromatic ring   **Aromatic ring**

7.57

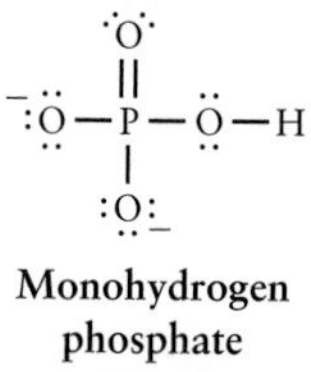

$\ddot{O}$
‖
⁻$\ddot{O}$—P—$\ddot{O}$—H
│
:$\ddot{O}$:⁻

Monohydrogen phosphate

It is often called inorganic phosphate and abbreviated P_i.

7.59 A person who has schizophrenia has excess dopamine in the brain; a person who has Parkinson's disease has a decreased amount of dopamine in the brain.

7.61 neurotransmitters; amine.

7.63 Dopamine cannot pass through the blood–brain barrier to reach the brain where it is needed.

7.65 Removal of a carboxylic acid functional group converts L-dopa into dopamine.

7.67 Parkinson's disease involves loss of dopamine-producing neurons.

7.69 a.–d.

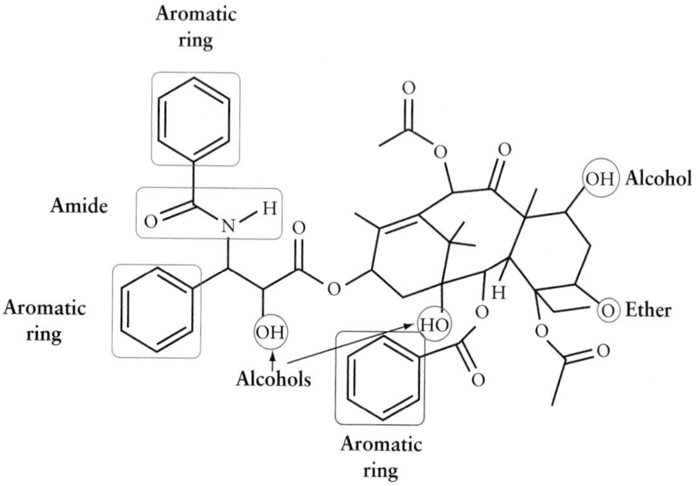

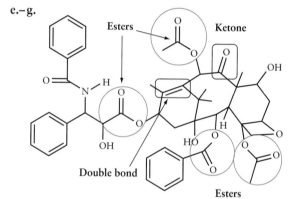

e.–g.

Chapter 8

8.21 The abbreviation (s) indicates that the substance is in the solid state. The abbreviation (g) indicates that the substance is in the gaseous state. The abbreviation (aq) indicates that the substance is dissolved in water.

8.23 conservation of matter

8.25 a. C_2H_5OH (l) + 3 O_2 (g) → 2 CO_2 (g) + 3 H_2O (l)
  b. 2 C_6H_{14} (l) + 19 O_2 (g) → 12 CO_2 (g) + 14 H_2O (l)
  c. CH_3OCH_3 (g) + 3 O_2 (g) → 2 CO_2 (g) + 3 H_2O (l)
  d. 2 C_3H_6 (g) + 9 O_2 (g) → 6 CO_2 (g) + 6 H_2O (l)

8.27 15.0 g of glucose produces 22.0 g of carbon dioxide.

8.29 2 C_4H_{10} (g) + 13 O_2 (g) → 8 CO_2 (g) + 10 H_2O (l)
  1.05 g of butane produces 1.63 g of water.

8.31 There are 1000 "small c" calories in one "capital C" Calorie. "Capital C" Calories are normally reported on nutritional food labels.

8.33 a. 234 cal **b.** 99.1 cal **c.** 2.07 × 10⁴ cal
  d. 3.52 × 10⁵ cal

8.35 ΔH is the enthalpy change of a reaction, or the amount of heat energy transferred in a reaction.

8.37 In an exothermic reaction, the products are lower in energy than the reactants. If the reaction is reversed, the products become the reactants and the reactants the products. In the reverse reaction, the products are higher in energy than the reactants; therefore, the reverse reaction is an endothermic reaction.

8.39 a. Exothermic. Heat is a product of the reaction.
 b. Exothermic. Heat is a product of the reaction.
 c. Endothermic. Heat needs to be added to the reaction.
 d. Endothermic. Heat needs to be added to the reaction.

8.41 A calorimeter is the name of the instrument that measures the heat content of various substances.

8.43 a. Total Calories supplied by almonds: 859 Cal or 860 Cal.
 b. Total Calories supplied by the banana: 121 Cal or 120 Cal.
 c. Total Calories supplied by the cheddar cheese: 109 Cal or 110 Cal.
 d. Total Calories supplied by the glazed donut: 190 Cal.
 e. Total Calories supplied by the swordfish: 124 Cal or 120 Cal.

8.45 Adenosine triphosphate, ATP, serves as the important energy carrier molecule between catabolic and anabolic reactions.

8.47 carbohydrates, fats, and proteins

8.49 The synthesis of proteins is an anabolic pathway.

8.51 Chemical kinetics is the study of reaction rates, how fast reactants in a chemical reaction are converted into products.

8.53 a. The reaction shown is an endothermic reaction.
 b.

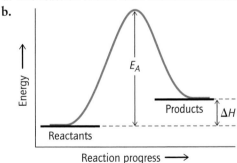

 c.

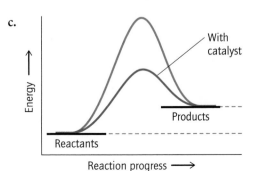

8.55 A catalyst does not affect the value of ΔH. A catalyst affects the value of E_A by making it smaller.

8.57 In the cell, chemical reactions occur at normal body temperature and at a relatively constant concentration. Therefore, to increase the rate of a biochemical reaction, enzymes are used. Enzymes reduce the freedom of motion available to reactants; they lower the activation energy by forcing reactants into a spatial orientation conducive to reaction.

8.59

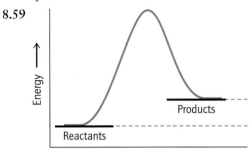

Reaction progress without a catalyst ⟶

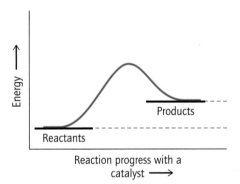

Reaction progress with a catalyst ⟶

The reaction without the catalyst has the higher activation energy.

8.61 800 Cal

8.63 6.95×10^3 Cal

8.65 a. Condensation is an exothermic physical process. Heat must be removed.
 b. Deposition is an exothermic physical process. Heat must be removed.
 c. The temperature in the surrounding air would rise because heat energy is removed from the molecules undergoing condensation or deposition and transferred to the surrounding air molecules.

8.67 Heat + $ZnCO_3 \rightarrow ZnO + CO_2$

8.69 Yes, the products are the same. Combustion of food in a calorimeter and by the human body both produce carbon dioxide, water, and energy.

8.71 Spirometry measures a patient's oxygen uptake. A patient breathes in 100% oxygen from a prefilled spirometer. The patient continues to re-breathe oxygen. Gases exhaled from the patient include carbon dioxide and unused oxygen. The exhaled carbon dioxide is removed, and the amount of oxygen uptake is determined by the decrease in volume of the spirometer.

Chapter 9

9.25

$$H\text{—}C\text{—}C\text{—}C\text{—}O\text{—}H \quad + \quad H_2O \quad \rightleftharpoons$$

Lactic acid   Water

$$H_3O^+ \quad + \quad H\text{—}C\text{—}C\text{—}C\text{—}O^-$$

Hydronium   Lactate ion
ion

9.27

$$C\text{—}C\text{—}\overset{..}{N}\text{—}CH_3 \quad + \quad H_2O \quad \rightleftharpoons$$

$$C\text{—}C\text{—}\overset{+}{N}\text{—}CH_3 \quad + \quad OH^-$$

9.29 a. H_3O^+ **b.** H_2O **c.** HI
d. $CH_3CH_2CH_2CH_2COOH$

9.31 $HClO_4 + H_2O \rightarrow ClO_4^- + H_3O^+$

9.33 $H_2SO_4 + H_2O \rightarrow HSO_4^- + H_3O^+$
$HSO_4^- + H_2O \rightarrow SO_4^{2-} + H_3O^+$

9.35 For weak bases, most of the base (reactant) is present in solution; only a small percentage of the base accepts a proton from water.

9.37 a. Weak acid; most of the acid is still present at equilibrium.
b. Weak base
c. Strong acid; it completely dissociates.

9.39 a.

$$H\text{—}O\text{—}C\text{—}C\text{—}C\text{—}O\text{—}H$$

b.

$$H\text{—}O\text{—}C\text{—}C\text{—}C\text{—}O\text{—}H \quad + \quad H_2O \quad \rightleftharpoons$$

$$H_3O^+ \quad + \quad H\text{—}O\text{—}C\text{—}C\text{—}C\text{—}O^-$$

$$H\text{—}O\text{—}C\text{—}C\text{—}C\text{—}O^- \quad + \quad H_2O \quad \rightleftharpoons$$

$$H_3O^+ \quad + \quad {}^-O\text{—}C\text{—}C\text{—}C\text{—}O^-$$

9.41 a. Propanoic acid, water, propanoate, and hydronium ion are present at equilibrium.
b. The concentrations of propanoic acid and propanoate are constant at equilibrium.

c. The two opposing arrows indicate that both the forward and reverse reactions occur simultaneously.

9.43 Le Châtelier's principle states that, when a reaction at equilibrium is disturbed, the reaction responds by shifting in the direction that restores equilibrium: either the forward direction (shift to the right) or the reverse direction (shift to the left).

9.45 a. If more ammonia is added, the reaction will shift to the left to consume the excess NH_3 present.
b. If more NH_4^+ is added, the reaction will shift to the right to consume the excess NH_4^+.

9.47 Hemoglobin will react with the excess oxygen to form the O_2–hemoglobin complex (shifting to the left) to restore the equilibrium. The removal of hemoglobin then prevents the reaction to the right that forms CO-hemoglobin and favors the reaction to the left to make more O_2-hemoglobin.

9.49 $2 HCl + Ba(OH)_2 \rightarrow 2 H_2O + BaCl_2$

9.51 $3 HCl + Al(OH)_3 \rightarrow 3 H_2O + AlCl_3$

9.53 a. Basic, pH > 7   **b.** Acidic, pH < 7
c. Neutral, pH = 7   **d.** Acidic, pH < 7
e. Basic, pH > 7   **f.** Basic, pH > 7
g. Acidic, pH < 7

9.55

$[H_3O^+]$	$[OH^-]$	Is the Solution Acidic, Neutral, or Basic?
1.0×10^{-3}	1.0×10^{-11}	Acidic pH = 3
1.0×10^{-12}	1.0×10^{-2}	Basic pH = 12
1.0×10^{-7}	1.0×10^{-7}	Neutral pH = 7
1.0×10^{-5}	1.0×10^{-9}	Acidic pH = 5
1.0×10^{-9}	1.0×10^{-5}	Basic pH = 9

9.57 The pH of apple juice is 3.50. It is acidic.
$[OH^-] = 3.1 \times 10^{-11}$ M

9.59 The pH of milk is 6.49. Milk is acidic.
$[OH^-] = 3.1 \times 10^{-8}$ M

9.61 A buffer is a solution that resists changes in pH upon addition of small amounts of acid or base.

9.63 When a small amount of acid enters the bloodstream, the bicarbonate ion (HCO_3^-) present reacts with the acid, H_3O^+, forming water and carbonic acid (H_2CO_3). The products do not increase the concentration of H_3O^+, and therefore the pH does not change.

9.65 a. NaOH, a strong base, would react with carbonic acid.
b. HCl, a strong acid, would react with bicarbonate.
c. NH_3, a weak base, would react with carbonic acid.
d. CH_3COO^-, a weak base, would react with carbonic acid.

9.67 The carbonic acid part of the buffer should be given to the patient to decrease the pH.

9.69 Proteins degrade naturally in the gastrointestinal tract due to the reaction of acids and enzymes with proteins. Also, most intact proteins cannot cross the membrane barrier in the small intestine to proceed to the circulatory system. If the protein could pass through the membrane barrier in the small intestine, it might damage the epithelial cells lining the intestine.

9.71 No, it does not. Protons bind to certain functional groups in a protein, altering its shape.

9.73 Enteric coatings are stable in acidic environment; therefore, the enteric coating controls where in the digestive tract the medication is absorbed.

9.75 The enteric coating is acidic because it reacts in an alkaline environment—a neutralization reaction.

Chapter 10

10.23 Alcohols, thiols, amines, carboxylic acids, esters, thioesters, and amides are involved in acyl transfers.

10.25 a. A ketone has a carbonyl group attached to two R groups.
  b. An ester has a carbonyl group attached to an R group and an OR group.
  c. A thioester has a carbonyl group attached to an R group and an SR group.

10.27 oxidation

10.29 An organic reactant that gains hydrogen atoms has undergone reduction.

10.31 An organic reactant that gains oxygen atoms has undergone oxidation.

10.33 a. Oxidation: $Zn: \rightarrow Zn^{2+} + 2e^-$
    Reduction: $2H^+ + 2e^- \rightarrow H{-}H$
    Cl^- is a spectator ion.
  b. Oxidation: $2Na\cdot \rightarrow 2Na^+ + 2e^-$
    Reduction: $:\ddot{C}l{-}\ddot{C}l: + 2e^- \longrightarrow 2\,:\ddot{C}l:^-$

10.35 a. Acetaldehyde has undergone an oxidation reaction; it has gained oxygen atoms. NAD^+ has undergone a reduction reaction; it has gained a hydrogen atom.
  b. The reactant has undergone an oxidation reaction: The product has lost hydrogen atoms. FAD has undergone a reduction reaction: it has gained hydrogen atoms.

10.37

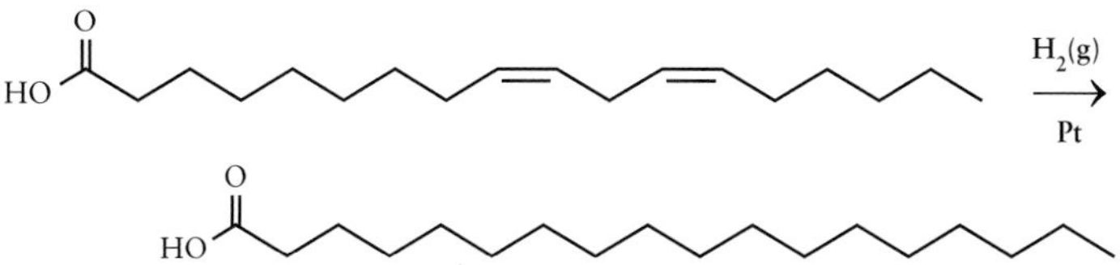

This reaction is a type of reduction reaction known as a catalytic hydrogenation. The product has gained hydrogen atoms.

10.39 An unsaturated fat is more likely to be a liquid.

10.41 $FADH_2$ is most likely used in the biochemical reduction of an alkene.

10.43 a.

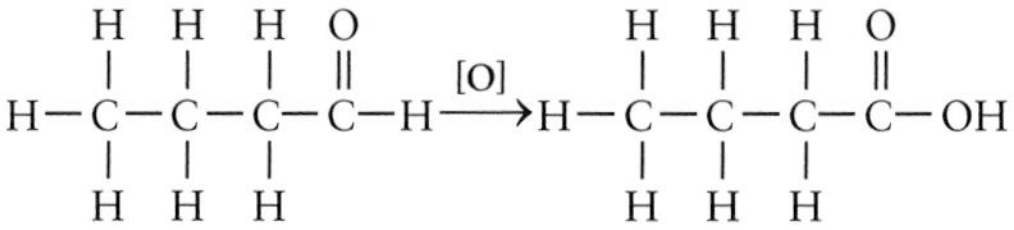

  b.

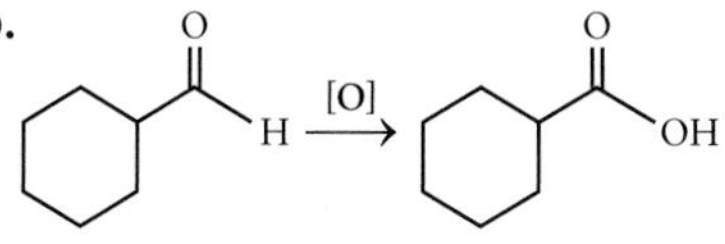

  c.

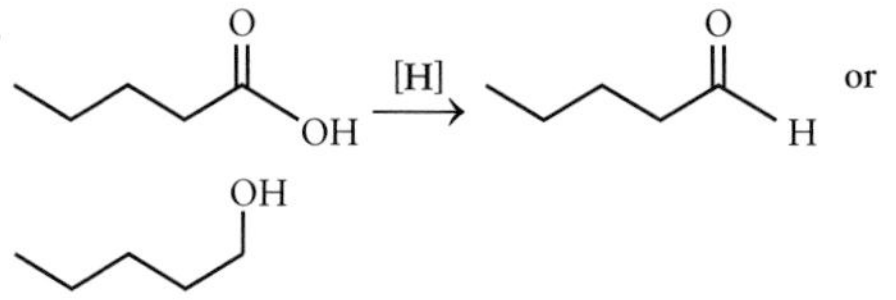

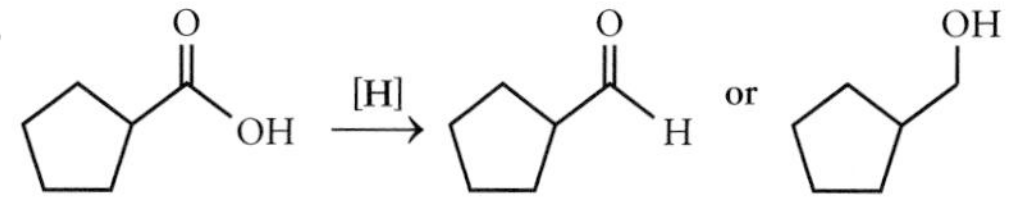

10.45 a. $C_3H_8 + 5\,O_2 \rightarrow 3\,CO_2 + 4\,H_2O$
  b.

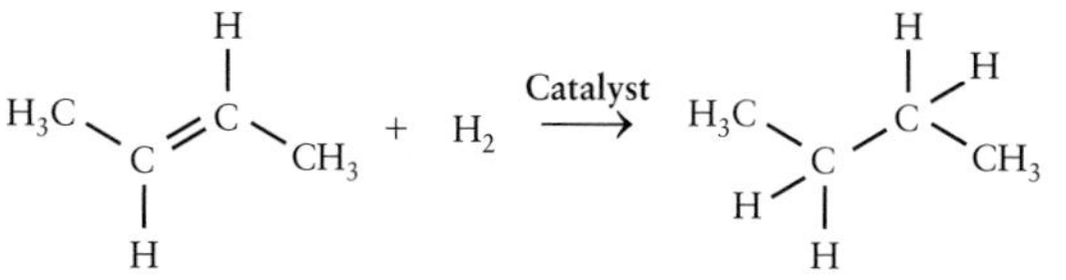

  c.

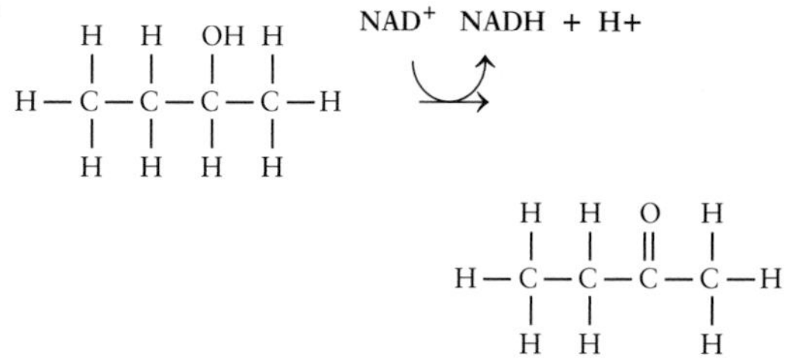

  d. $FADH_2 + NAD^+ \rightarrow FAD + NADH + H^+$

10.47 a.

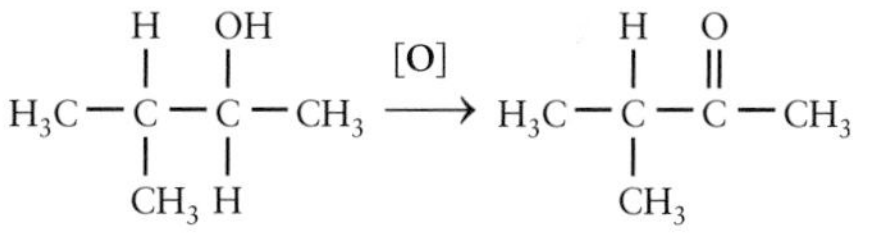

  b.

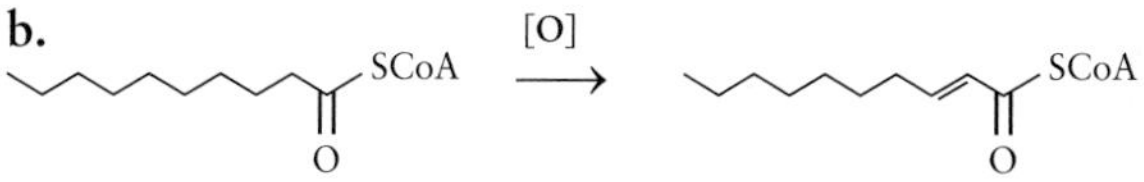

  c. $FADH_2$
  d.

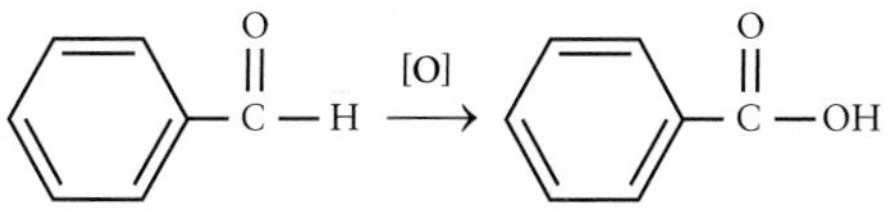

10.49 d. A free radical does not contain an octet of electrons.

10.51 a. hydration

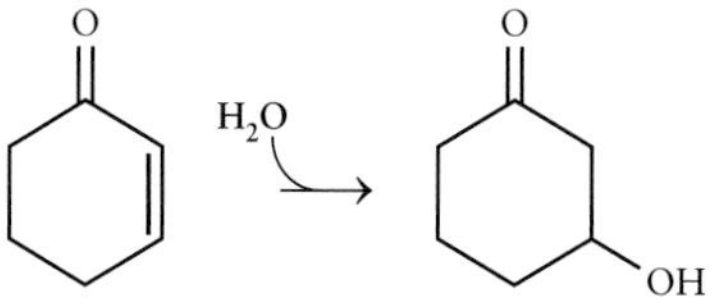

  b. dehydration

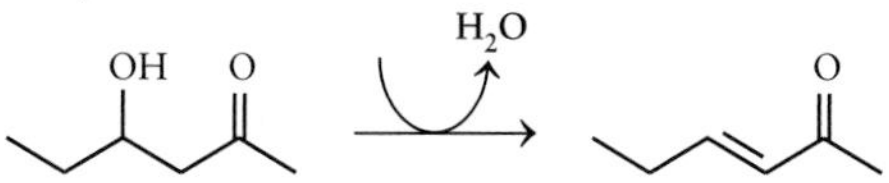

  c. hydration

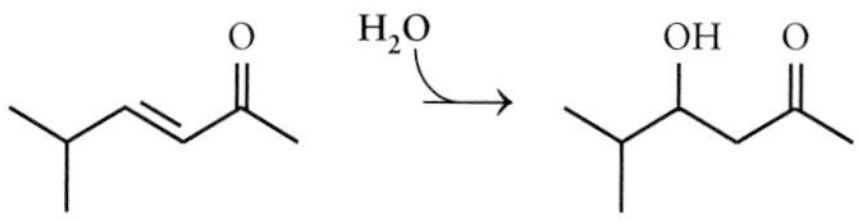

10.53 a.

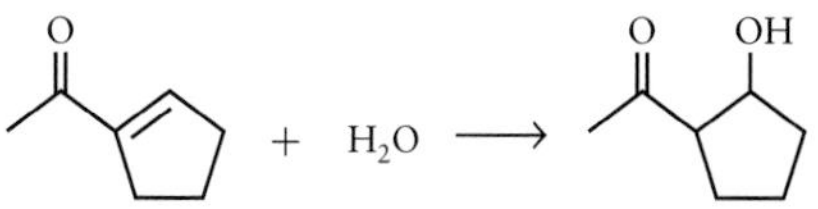

  b.

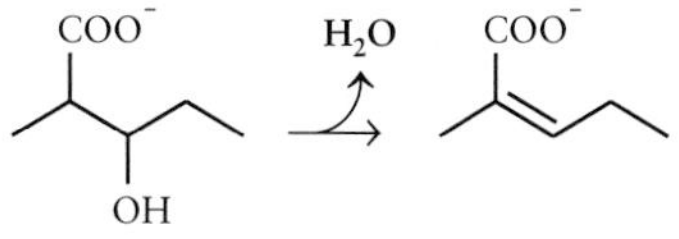

10.55 a.

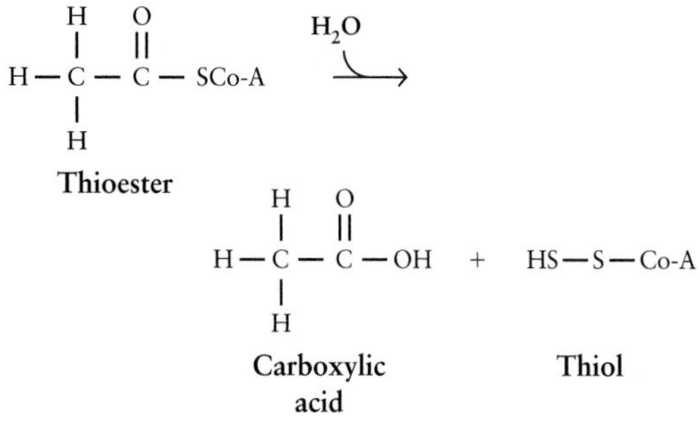

Thioester

$\xrightarrow{H_2O}$

Carboxylic acid + Thiol

b.

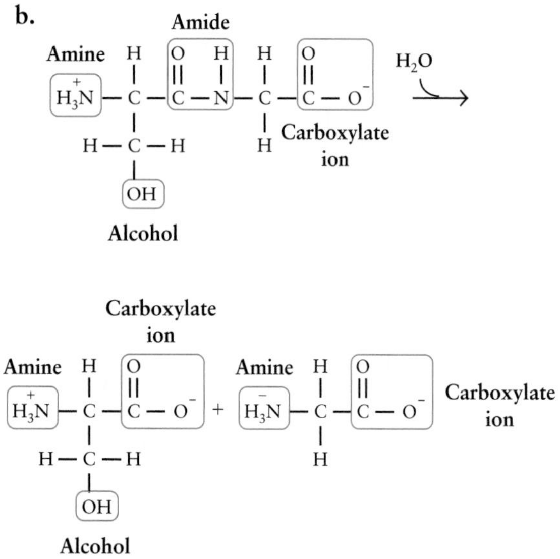

Amide

Amine — Carboxylate ion

Alcohol

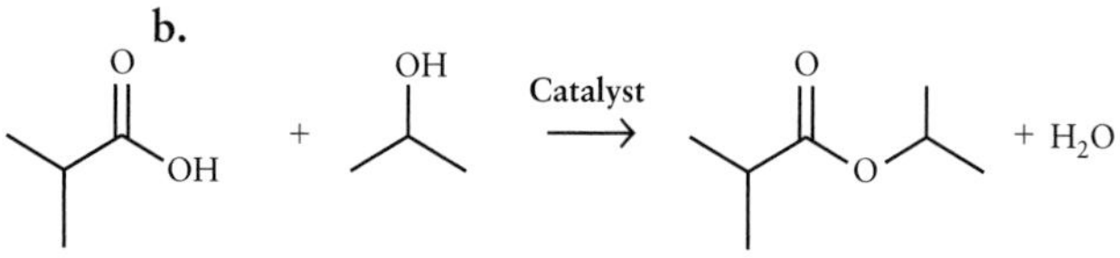

$\xrightarrow{H_2O}$

Carboxylate ion

Amine + Amine — Carboxylate ion

Alcohol

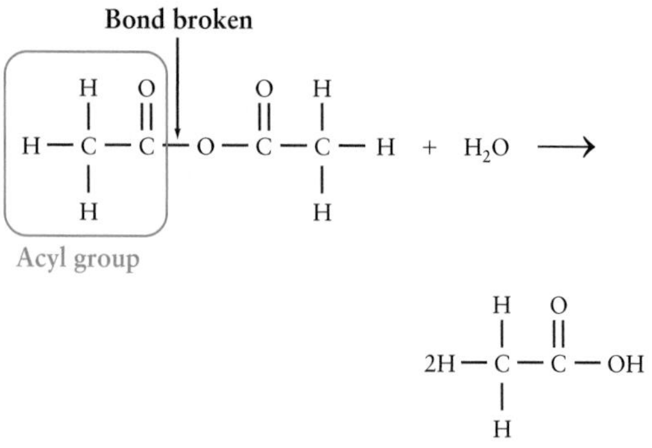

10.57 a.

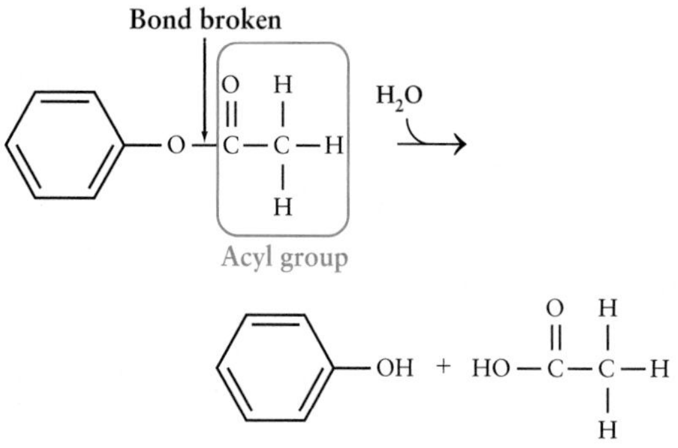

Bond broken

Acyl group

$\xrightarrow{H_2O}$

$\longrightarrow$ OH + HO—C—C—H

b.

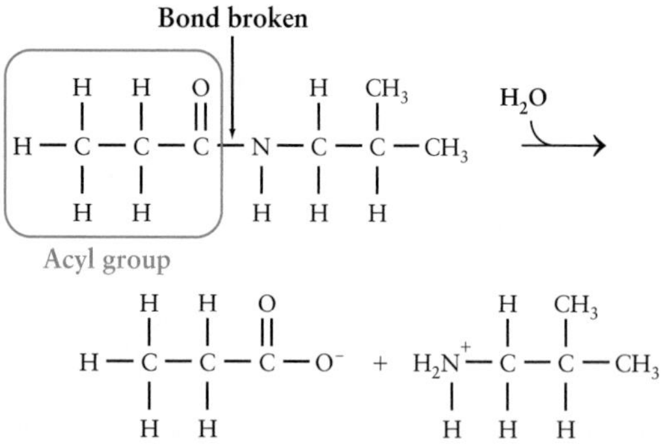

Bond broken

Acyl group

$\xrightarrow{H_2O}$

c.

Bond broken

Acyl group

$\xrightarrow{H_2O}$

d.

Bond broken

Acyl group + H_2O $\longrightarrow$

$2H—C—C—OH$

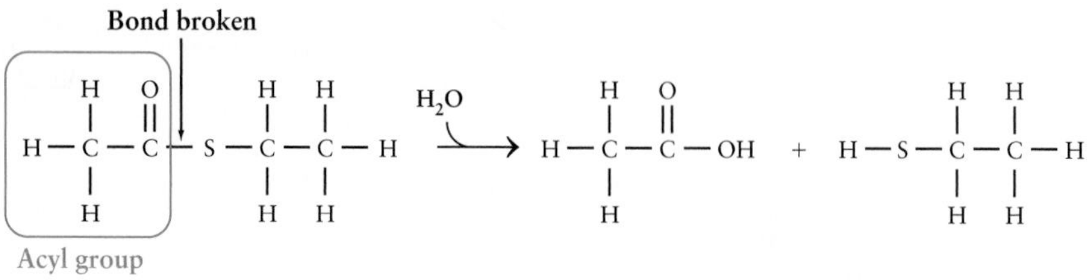

10.59 a.

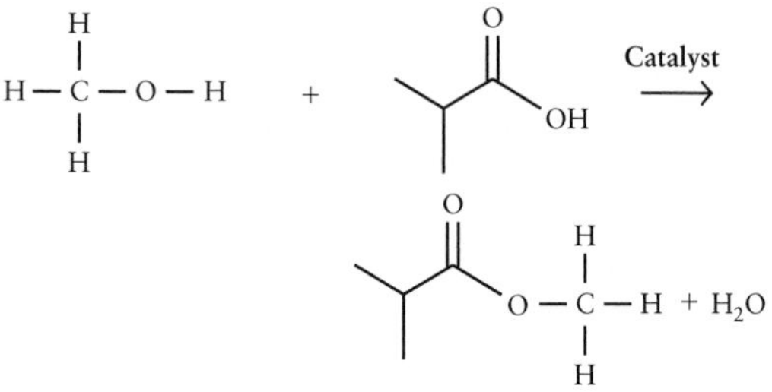

$H—C—O—H$ + $\xrightarrow{Catalyst}$ + H_2O

b.

$\xrightarrow{Catalyst}$ + H_2O

c.

$\xrightarrow{Catalyst}$

+ H_2O

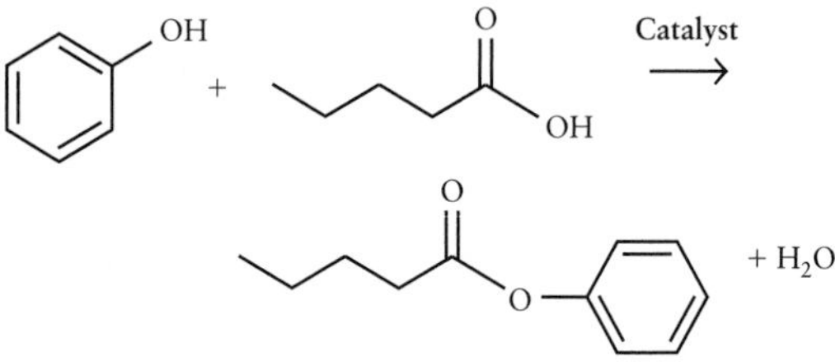

10.61 A phosphoryl group can be transferred to the oxygen atom of water or the oxygen atom of an alcohol group, just like an acyl group.

10.63 a.

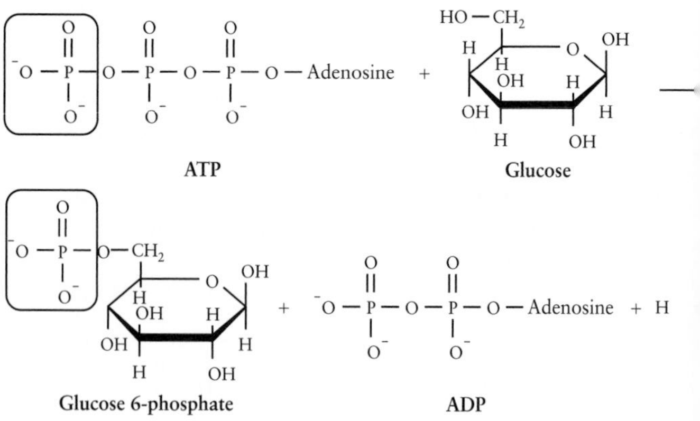

ATP + Glucose

Glucose 6-phosphate + ADP + H

b. The phosphate group is being transferred from a triphosphate.

c. The phosphate group is being transferred to an alcohol group.

10.65 a. hydrolysis, a type of acyl group transfer reaction

b. oxidation–reduction

c. hydration reaction

10.67 a.

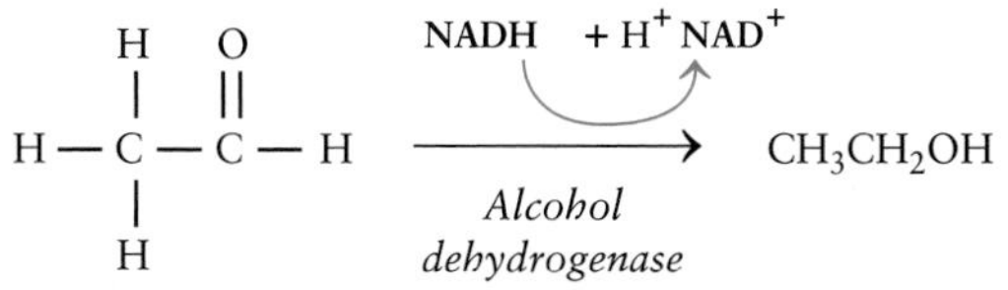

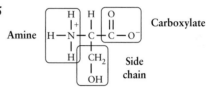

b. In the reaction that occurs in yeast, NADH is the reactant.

c. In the reaction that occurs in yeast, acetaldehyde is reduced.

10.69 The reaction is a reduction reaction. FAD gains electrons when it gains two hydrogen atoms. FAD is derived from riboflavin, vitamin B_2.

10.71 The four fat-soluble vitamins are vitamin A, vitamin D, vitamin E, and vitamin K. These vitamins are soluble in fat and not in water because they are hydrophobic, and they are stored in fat tissue in the body.

10.73 A vitamin deficiency occurs when the body does not have enough of a particular vitamin. It can be caused by insufficient vitamin intake in the diet or poor vitamin absorption. The symptoms of riboflavin deficiency include inflammation of the tongue and skin, lesions in the mouth, and cataracts.

Chapter 11

11.45 a. Glycine

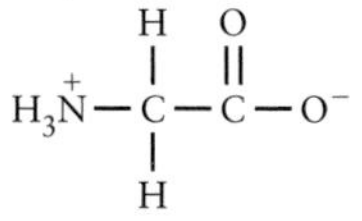

$$H_3\overset{+}{N}-\underset{\underset{H}{|}}{\overset{\overset{H}{|}}{C}}-\overset{\overset{O}{||}}{C}-O^-$$

b. Aspartic acid

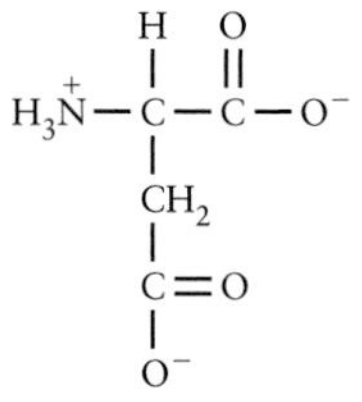

$$H_3\overset{+}{N}-\underset{\underset{\underset{\underset{O^-}{|}}{C=O}}{\overset{|}{CH_2}}}{\overset{\overset{H}{|}}{C}}-\overset{\overset{O}{||}}{C}-O^-$$

c. Tyrosine

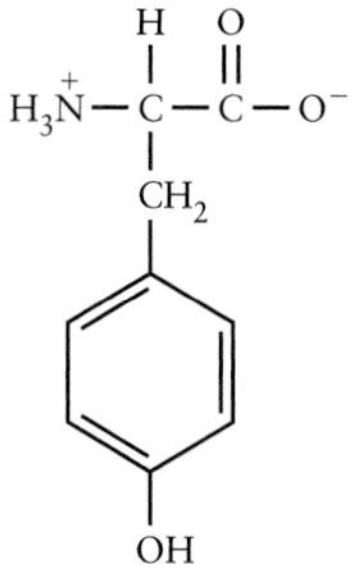

$$H_3\overset{+}{N}-\underset{\overset{|}{CH_2}}{\overset{\overset{H}{|}}{C}}-\overset{\overset{O}{||}}{C}-O^-$$

(with benzene ring and OH)

d. Cysteine

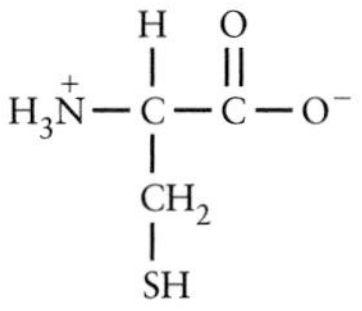

$$H_3\overset{+}{N}-\underset{\underset{\underset{SH}{|}}{CH_2}}{\overset{\overset{H}{|}}{C}}-\overset{\overset{O}{||}}{C}-O^-$$

e. Glutamic acid

$$H_3\overset{+}{N}-\underset{\underset{\underset{\underset{\underset{O^-}{|}}{C=O}}{\overset{|}{CH_2}}}{\overset{|}{CH_2}}}{\overset{\overset{H}{|}}{C}}-\overset{\overset{O}{||}}{C}-O^-$$

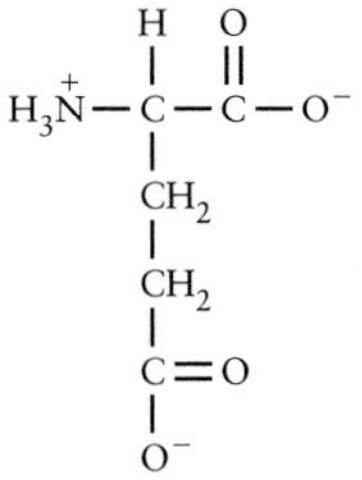

11.47 The amino acids with nonpolar side chains are glycine, alanine, valine, leucine, isoleucine, proline, tryptophan, phenylalanine, and methionine. The amino acids with basic side chains are lysine, arginine, and histidine. The amino acids with acidic side chains are aspartic acid and glutamic acid. The amino acids with polar but neutral side chains are serine, threonine, tyrosine, cysteine, glutamine, and asparagine.

11.49 Essential amino acids are amino acids that are not synthesized by the body and must be supplied through the diet.

11.51 a. cysteine **b.** asparagine and glutamine

11.53 Cysteine has a thiol functional group.

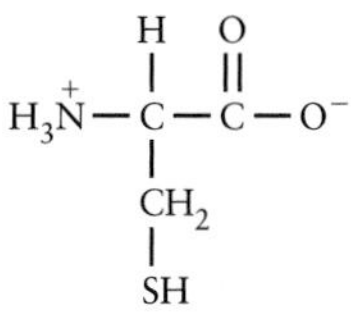

$$H_3\overset{+}{N}-\underset{\underset{\underset{SH}{|}}{CH_2}}{\overset{\overset{H}{|}}{C}}-\overset{\overset{O}{||}}{C}-O^-$$

11.55

Amine $H-\underset{\underset{H}{|}}{\overset{\overset{H}{|+}}{N}}-\underset{\underset{\underset{OH}{|}}{CH_2}}{\overset{\overset{H}{|}}{C}}-\overset{\overset{O}{||}}{C}-O^-$ Carboxylate / Side chain

a. The amine is in its ionized form.

b. The proton has been lost from the carboxylic acid, forming a carboxylate ion.

c. Yes, the amino acid shown is a zwitterion. At approximately physiological pH, most amino acids exist as a zwitterion.

d. This amino acid is serine.

e. The side chain is polar.

f. At pH = 12, serine has the structure shown below. The net charge is −2.

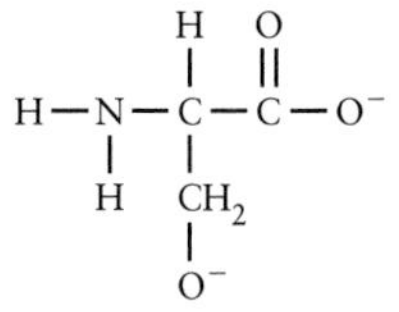

$$H-\underset{\underset{H}{|}}{\overset{\overset{H}{|}}{N}}-\underset{\underset{\underset{O^-}{|}}{CH_2}}{\overset{\overset{H}{|}}{C}}-\overset{\overset{O}{||}}{C}-O^-$$

11.57

Amino Acid	Non-polar	Polar	Acidic Side Chain	Basic Side Chain	Neutral Side Chain	Essential Amino Acid
Phe	x					x
Asn		x			x	
Met	x					x

11.59 Meat contains all the essential amino acids.

11.61 a. A corkscrew is chiral.

b. An orange is achiral.

c. A car is chiral.

d. A nail is achiral.

11.63 A racemic mixture of alanine would contain a 50:50 mixture of L-alanine and D-alanine.

11.65 a.

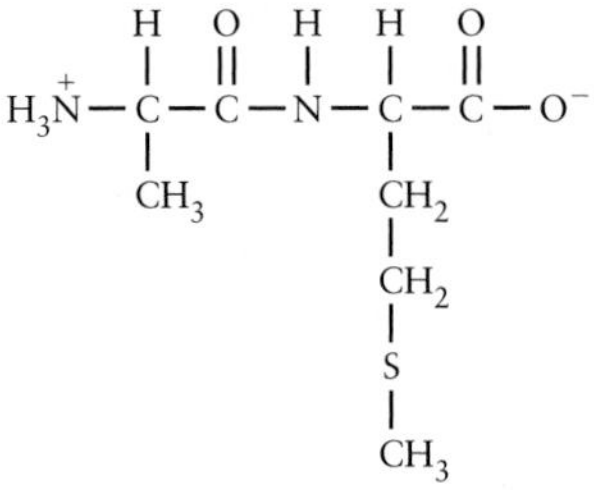

Ala-Met 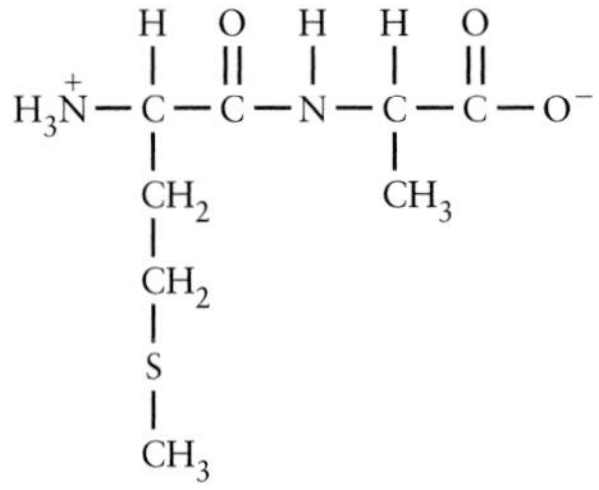   Met-Ala

b.

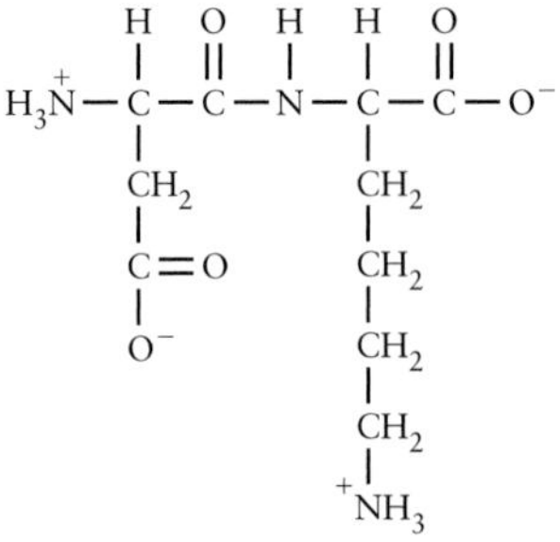

Asp-Lys 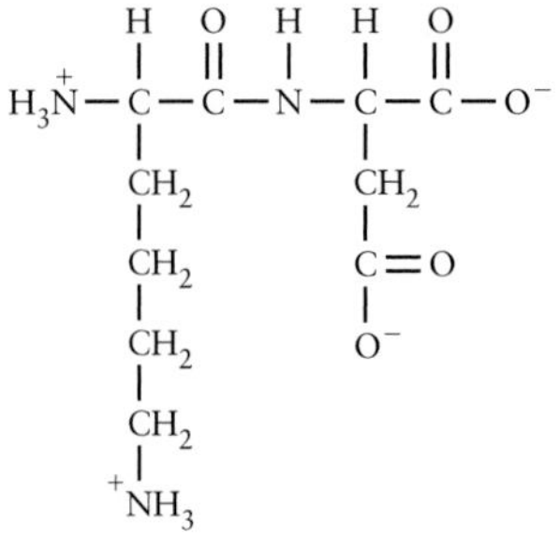   Lys-Asp

c.

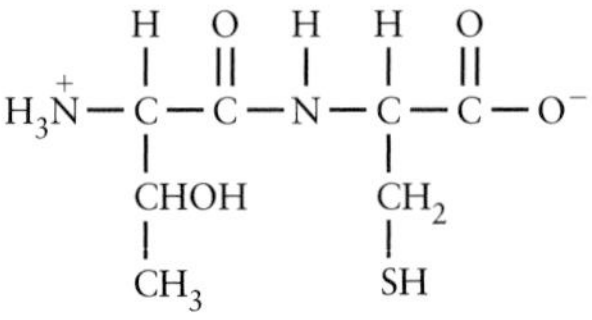

Thr-Cys 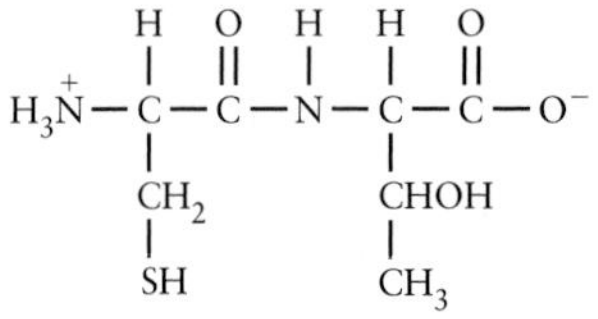   Cys-Thr

d.

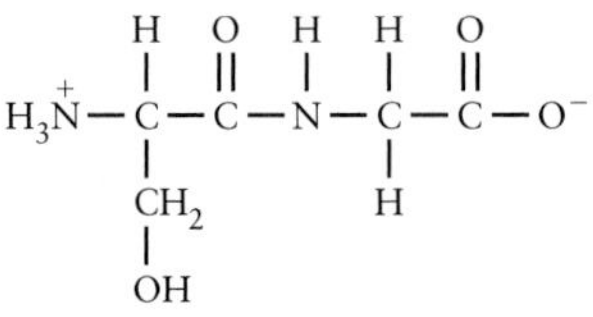

Ser-Gly 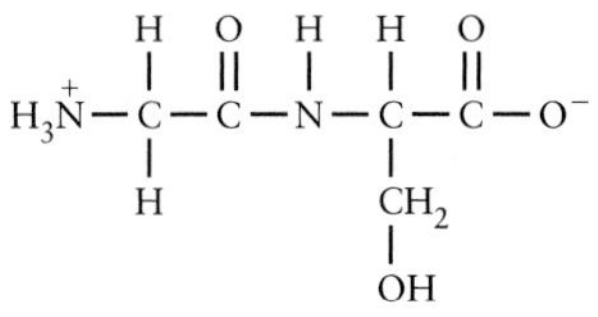   Gly-Ser

11.67 Leu-Glu-His

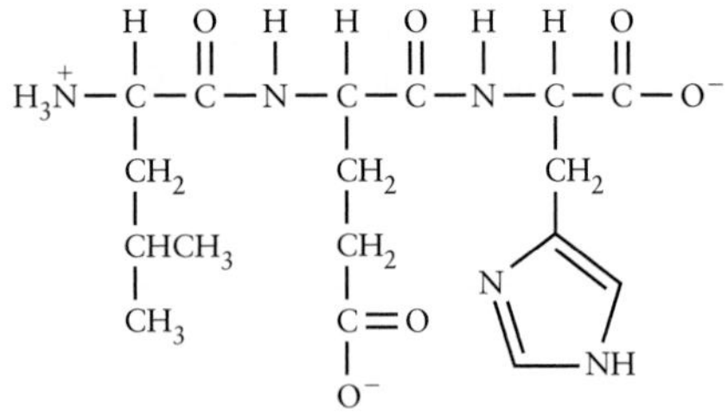

11.69 a. Lys-Thr-Thr-Lys-Ser
 b. The N-terminal amino acid is lysine.
 c. The C-terminal amino acid is serine.
 d. Lysine has an amine functional group, and serine and threonine have alcohol functional groups.

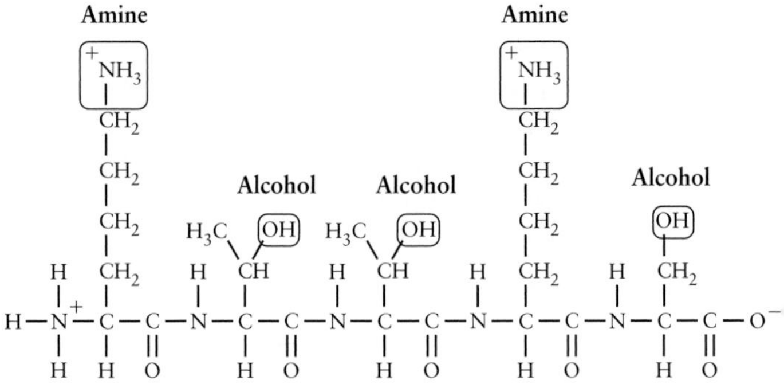

11.71 Vasopressin should be more polar because it contains Arg instead of Leu, a polar amino acid.

11.73 The DNA of the cell contains the code for the amino acid sequence.

11.75 The two most common forms of secondary structure are the α-helix and the β-pleated sheet. Secondary structure is formed by hydrogen bonding between the amide carbonyl and the amide nitrogen of the peptide backbone.

11.77 The four types of electrostatic interactions that are responsible for the tertiary structure of a protein are: disulfide bridges, salt bridges, hydrogen bonding, and dispersion forces. A disulfide bridge:

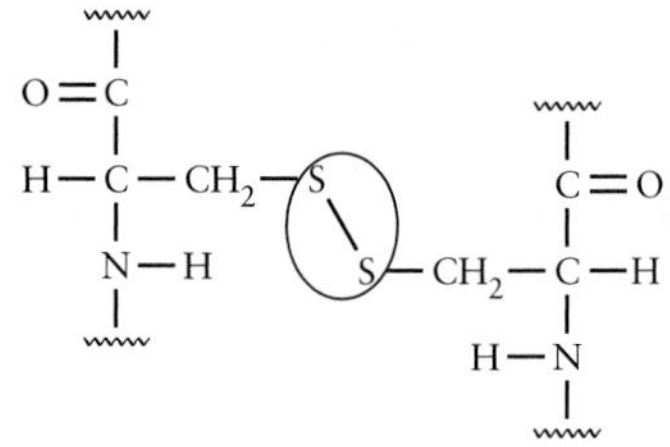

Salt bridge:

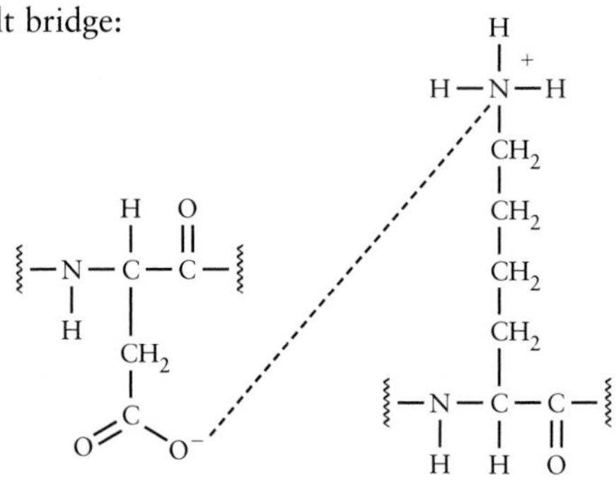

Hydrogen bonding:

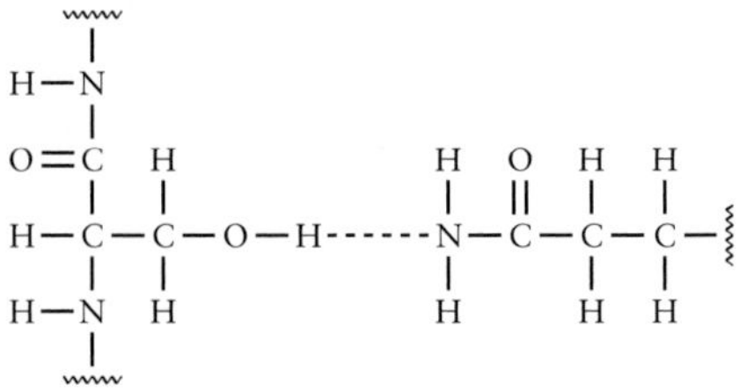

Dispersion forces:

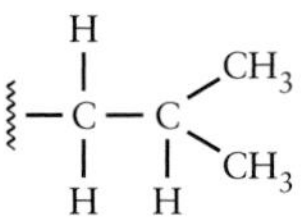

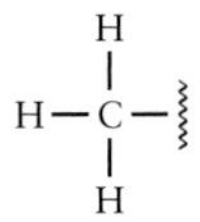

11.79 Aspartic acid, glutamic acid, lysine, arginine, and histidine

11.81 The three general classes of proteins distinguished by their quaternary structure and solubility are fibrous proteins, globular proteins, and membrane proteins.

11.83

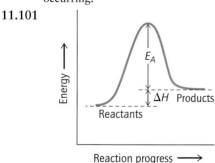

11.85 Most fibrous proteins provide structure to tissues and are found in hair, skin, fingernails (keratins); skin, blood vessels, heart, lung, intestines, tendons, ligaments (elastins); and connective tissue (collagens). The quaternary structure provides elasticity and rigidity of connective tissue.

11.87 Membrane proteins are found in the cell membrane.

11.89 Heat, pH changes, mechanical agitation, detergents, and some metals may denature a protein. When a protein is denatured it loses its secondary, tertiary, and quaternary structure. Since the denatured protein has lost its shape, it has also lost its function.

11.91 $E + S \rightleftharpoons ES \rightleftharpoons E + P$
E is the enzyme, S is the substrate, ES is the enzyme–substrate complex, and P is the product.

11.93 Some cofactors include Fe^{2+}, Mg^{2+}, and Zn^{2+}.

11.95 The substrate binds to the active site through ionic interactions and intermolecular forces of attraction (dispersion forces, dipole–dipole forces, and hydrogen bonding).

11.97 The activation energy, E_A, is lower for the enzyme-catalyzed reaction compared to the uncatalyzed one.

11.99 A competitive inhibitor competes with the substrate for the active site of the enzyme because it too has a structure that is complementary to the active site. By binding to the active site, the competitive inhibitor blocks the substrate from binding to the enzyme and thus prevents the reaction from occurring.

11.101

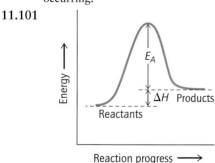

11.103 Angiotension II initiates a number of physiological events that raise blood pressure. Therefore, stopping the production of angiotension II makes it less likely that blood pressure will rise.

11.105 The *angiotension converting enzyme* breaks the peptide bond between Phe and His in angiotension I.

11.107 When valine replaces glutamic acid in hemoglobin, an acidic polar side chain has been replaced with a nonpolar side chain and a potential salt bridge has been removed, changing the shape of the protein. The tertiary and quaternary shape of the protein has been affected.

11.109 At low pH, the carboxylic acids are in their un-ionized form. They cannot form salt bridges with positively charged side chains. Therefore, the tertiary and quaternary structure of the protein is disrupted.

Chapter 12

12.35 Starch provides energy for the plant: Cellulose provides structure for the plant.

12.37 a. The Fischer projections shown are D-sugars. The OH group that is farthest away from the CHO group is pointing to the right.

b. L-arabinose

c. D-arabinose and D-xylose are diastereomers. The OH group and H groups are reversed at C-2 and C-3.

12.39 In the α anomer, the OH group on the anomeric carbon is pointing down. In the β anomer, the OH group on the anomeric carbon is pointing up.

12.41 a. Furanose, five membered ring, β anomer, OH group is pointing up.

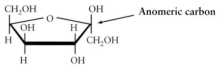

b. Pyranose, six membered ring, β anomer, OH group is pointing up.

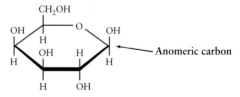

12.43 a.

$$\begin{array}{c}
^1CHO \\
H \!-\!^2\!-\! OH \\
HO \!-\!^3\!-\! H \\
H \!-\!^4\!-\! OH \\
HO \!-\!^5\!-\! H \\
^6CH_2OH
\end{array}$$

L-Idose

b. The OH group on carbon 5 determines that it is an L-sugar.

c. D-idose and L-idose are enantiomers, nonsuperposable mirror images.

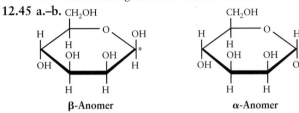

D-**Idose**

d. D-idose and D-glucose are diastereomers.

12.45 a.–b.

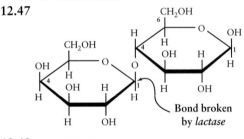

β-**Anomer**   α-**Anomer**

c. These two structures are diastereomers.

d. These two sugars interconvert through a process called mutarotation. The ring of one anomer opens to give the open-chain form and then recloses to give the other anomer.

12.47

Bond broken by *lactase*

12.49

Isomaltose

12.51 Giraffes have bacteria in their digestive tracts that produce β-*glycosidases*. These β-*glycosidases* can hydrolyze the β(1→4) glycosidic bonds in cellulose and produce glucose.

12.53 In D-glucosamine, the OH group on C2 has been replaced by an amine group.

12.55 a. Maltulose is a disaccharide since two monosaccharrides make up maltulose.

b.–c.

Glycosidic bond

d.

Glucose   Fructose

Hydrolysis produces glucose and fructose.

12.57 Glycolysis is a process that converts *glucose* into *pyruvate*.

12.59 Every intermediate in glycolysis contains a *phosphoryl* group. The 2− charge on the phosphoryl group prevents the intermediate from diffusing out of the cell.

12.61 The primary role of carbohydrates in the body is to produce energy.

12.63 Step 4 converts fructose-1, 6-bisphosphate, a molecule containing six carbon atoms, into glyceraldehyde-3-phosphate and dihydroxyacetone, two molecules that contain three carbon atoms each.

12.65 In step 2, a pyranose ring, glucose-6-phosphate, is isomerized to a furanose ring, fructose-6-phosphate.

12.67 People with AB blood are called universal recipients because they can receive any type of blood: A, B, AB, or O. Type A and type B blood components are found in type AB blood, so it is recognized as part of the type AB recipient's blood. Type O blood has the trisaccharide component of type AB blood, so it is also recognized by the recipient.

12.69 a. He or she is hypoglycemic.
b. Glycogenolysis, the conversion of glycogen into glucose, is activated when glucose levels fall.
c. Glucose levels typically fall in between meals and during exercise.
d. Glucagon is secreted when glucose levels fall.

12.71 Normal fasting glucose levels are 70–110 mg/dL.

12.73 Someone with Type I diabetes must manage his or her diet carefully, controlling the amount of carbohydrates consumed, so that glucose levels do not rise significantly.

12.75 The hormone glucagon has a complementary role to insulin.

Chapter 13

13.33 Lipids are insoluble in water and soluble in nonpolar organic solvents.

13.35 Lipids have an extensive hydrocarbon structure; therefore, they are nonpolar and insoluble in water.

13.37 Oleic acid is a Δ^9 fatty acid and ω-9 fatty acid.

13.39 Linolenic acid is an ω-3 fatty acid.

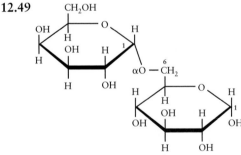

13.41 The presence of a cis double bond lowers the melting point of a fatty acid. The bend created by the double bond reduces the number of points of contact between the fatty acid molecules, leading to fewer dispersion forces.

13.43 Fats are derived primarily from saturated fatty acids. Saturated fatty acids pack together tightly, causing greater dispersion forces and, hence, higher melting points.

13.45 Coconut oil should be a solid at room temperature because it consists mostly of saturated fats, which pack together tightly.

13.47

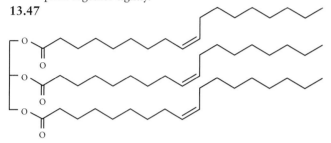

Olive oil is considered a healthy fat because it consists mostly of unsaturated fatty acids.

13.49 A triglyceride would not be soluble in water. It contains mostly nonpolar hydrogen and carbon atoms.

13.51 Safflower oil is lower in saturated fats.

13.53 Olive oil is a healthier food because it contains fewer saturated fats.

13.55 Unsaturated trans fats should pack more closely. The trans double bond does not cause a bend in the chain, whereas the cis double bond causes a bend. The unsaturated trans fat would have a higher melting point because these fats can pack together tightly and exhibit more dispersion forces leading to a higher melting point.

13.57 Lipids are the main component of the cell membrane.

13.59 Glycerphospholipids and sphingomyelins contain a phosphate group.

13.61 The amine and the phosphate form the polar head of the molecule.

13.63 **a.** It is a glycolipid. It has a carbohydrate attached.
b. It is a sphingolipid. It has a sphingosine backbone.

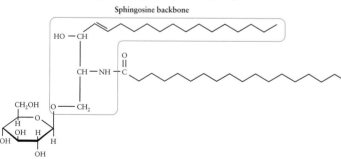

c. It was derived from the fatty acid stearic acid. The fatty acid component is saturated. An amide functional group connects the fatty acid to the backbone.
d. This lipid is a cerebroside. It has a monosaccharide as the carbohydrate component.

e. It contains a glucose derivative. The glycosidic linkage is β: The oxygen atom at the anomeric carbon bond is above the ring.
f.

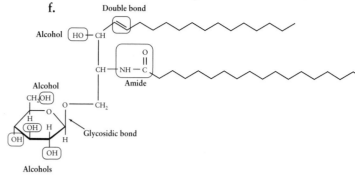

13.65 Dispersion forces and hydrogen bonding are responsible for the way membrane lipids organize to form a cell membrane.

13.67 Cholesterol adds rigidity to the cell membrane.

13.69 Small nonpolar molecules such as oxygen and carbon dioxide pass freely through the cell membrane, as does water.

13.71 The body obtains fatty acids from dietary fat and body fat.

13.73 Triglycerides are packaged into chylomicrons to be transported through the lymph system and the bloodstream.

13.75 The triglycerides are found on the interior of the lipoprotein since triglycerides are nonpolar and the interior of the lipoprotein is hydrophobic.

13.77 146 ATP

13.79 The first and third steps of β-oxidation are oxidation steps.

13.81 Cholesterol is similar to a fatty acid in that both molecules are amphipathic. They both have a polar head and a nonpolar tail.

13.83 Glucocortoids regulate carbohydrate, protein, and lipid metabolism. They also have powerful anti-inflammatory and immunosuppressant activity.

13.85 The six-membered ring attached to an OH group has been converted into an aromatic ring. The double bond in the adjacent six-membered ring has been reduced to a single bond. The hydrocarbon chain attached to the five-membered ring has been converted into an OH group.

13.87 Eicosanoids are a class of lipids derived from arachidonic acid (eicosanoic acid) that serve as signaling molecules to stimulate inflammation.

13.89 The symptoms of inflammation are redness, swelling, pain, and warmth.

13.91 **a.** Leukotriene A_4 will not be produced when a *lipoxygenase* inhibitor is present.
b. Prostaglandin H_2, thromboxane A_2, and prostaglandin E_2 will not be produced when a *cycloxygenase* inhibitor is present.

13.93 Aspirin inhibits *cyclooxygenase*, the key enzyme involved in the formation of prostaglandin H_2.

13.95 An NSAID is a **n**onsteroidal **a**nti-**i**nflammatory **d**rug. NSAIDs include aspirin, ibuprofen (Motrin and Advil), and naproxen (Aleve).

Chapter 14

14.23 Catabolic pathways break down larger molecules into smaller molecules and produce energy. Anabolic pathways build larger molecules from smaller molecules and consume energy.

14.25 Glycolysis, β-oxidation of fatty acids, and the degradation of several amino acids produce acetyl CoA.

14.27 The citric acid cycle begins with the reaction of acetyl CoA with oxaloacetate. The last step of the biochemical pathway regenerates oxaloacetate.

14.29 a.

$$
\begin{array}{c}
COOH \\
| \\
CH_2 \\
| \\
H-C-COOH \\
| \\
HO-C-H \\
| \\
COOH
\end{array}
$$

b. The difference between the two isomers is the location of the hydroxyl group.

c. They are structural isomers.

14.31 After the first half of the citric acid cycle, two carbon atoms have been removed from citrate. Thus carbon dioxide molecules are expelled to remove two carbon atoms. Two molecules of NADH are produced in the first half of the citric acid cycle.

14.33 The citric acid cycle produces *three* NADH molecules. Every NADH molecule produces 3 ATP molecules.

14.35 Steps 3 and 4 produce the carbon dioxide that we exhale. The carbon atom in carbon dioxide comes from oxaloacetate.

14.37 A new carbon–carbon bond is formed. Oxaloacetate and acetyl CoA react to form this new bond.

14.39 a. An oxidation–reduction reaction.

b. Succinate is oxidized.

c. FAD is reduced.

d. A carbon–carbon single bond is oxidized to a carbon–carbon double bond.

14.41 a. An oxidation–reduction reaction.

b. A secondary alcohol is converted into a ketone and NAD$^+$ is converted into NADH.

c. L-malate is oxidized.

d. The oxaloacetate produced in this step can react with another molecule of acetyl CoA, starting another round of the citric acid cycle.

14.43 Glycolysis—the conversion of glucose to pyruvate—occurs in the cytoplasm.

14.45 Ions can pass easily through the outer membrane.

14.47 The matrix has a higher pH.

14.49 The potential energy stored in the unequal distribution of protons is called the proton-motive force.

14.51 Cyanide poisoning is lethal because cyanide is an irreversible inhibitor of *cytochrome c oxidase,* a key enzyme in the last step of the electron transport chain. Cyanide brings the electron-transport chain to a halt, by preventing phosphorylation of ADP and depriving the cell of the energy needed to function.

14.53 *ATP synthase* uses the proton-motive force to drive phosphorylation of ADP to ATP.

14.55 A spontaneous reaction continues on its own once it has started, whereas energy needs to be continuously supplied to a nonspontaneous reaction for it to proceed.

14.57 b. and **c.** A plant growing in sunlight needs the constant energy from the sun to keep growing. Swinging on a swing requires constant energy to keep the swing in motion.

14.59 b., d., and **e.** represent an increase in entropy. When a reaction produces more product molecules than reactant molecules, the product molecules can move in more ways than the reactant molecules. Molecules in liquid water have more freedom of motion than those in snow. Molecules in steam have more freedom of motion than those in liquid water.

14.61 A decrease in entropy. The enzyme–substrate complex has less freedom of motion than the individual enzyme and substrate. The entropy of the surroundings increases more than the E-S entropy decreases; therefore, the entropy of the universe increases.

14.63 a. spontaneous, ΔG is negative

b. nonspontaneous, ΔG is positive

c. spontaneous, ΔG is negative

d. spontaneous, ΔG is negative

14.65 The reaction is spontaneous, ΔG is negative.

14.67 a. Spontaneous. There is an increase in entropy because there are more product molecules than reactant molecules. $\Delta S > 0$ and $\Delta H < 0$, therefore $\Delta G < 0$.

b. Nonspontaneous. $\Delta H > 0$, $\Delta S < 0$, therefore $\Delta G > 0$.

c. Spontaneous. $\Delta H < 0$, $\Delta S > 0$, therefore, $\Delta G < 0$.

14.69 Temperature plays a key role in determining ΔG in two cases: (1) when ΔH is negative and ΔS is negative and (2) when ΔH is positive and ΔS is positive. Temperature is multiplied by ΔS in the equation for free energy; therefore, entropy has a greater influence at higher temperatures.

14.71 a. The phosphorylation of glucose is endergonic.

b. The overall coupled reaction is exergonic.

c. $\Delta G_{overall} = -3.5$ kcal/mol

14.73 a. The formation of pyruvate is exergonic, ΔG is negative.

b. The overall coupled reaction is exergonic.

c. $\Delta G_{overall} = -7.5$ kcal/mol

d. The reaction of phosphoenolpyruvate to form pyruvate provided the energy to drive the phosphorylation of ADP.

14.75 If left untreated, PKU leads to mental retardation and regular seizures.

14.77 In someone who does not have PKU, phenylalanine is either converted into tyrosine or incorporated into polypeptide chains.

14.79 Because PAH is a protein, it will be broken down into amino acids in the stomach during digestion if taken in pill form.

Chapter 15

15.29 A gene is a section of DNA that contains the instructions for making a protein. A mutation is a permanent alteration in one or more nucleotides of a gene, which in turn produces an altered protein.

15.31 A nucleotide contains a nitrogen-containing ring called a base, a monosaccharide, and a phosphate group. A nucleoside is a nucleotide without the phosphate group.

15.33 a.–f.

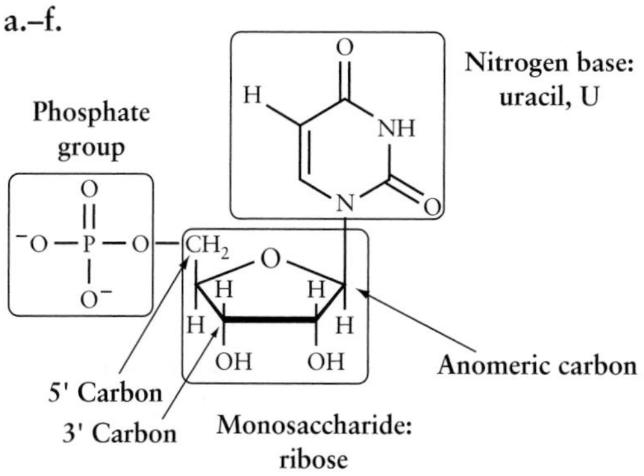

g. The nucleotide would be found in RNA because the monosaccharide is D-ribose, not 2-deoxyribose.

15.35

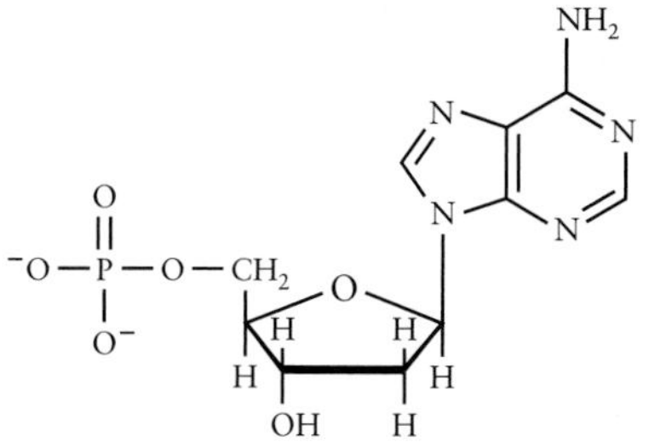

15.37 When two nucleotides are joined together to form a dinucleotide, the new bond is formed between the alcohol on the 3' carbon of one nucleotide and the phosphate group on the 5' carbon of the other nucleotide. A phosphate ester bond is produced.

15.39 4096

15.41

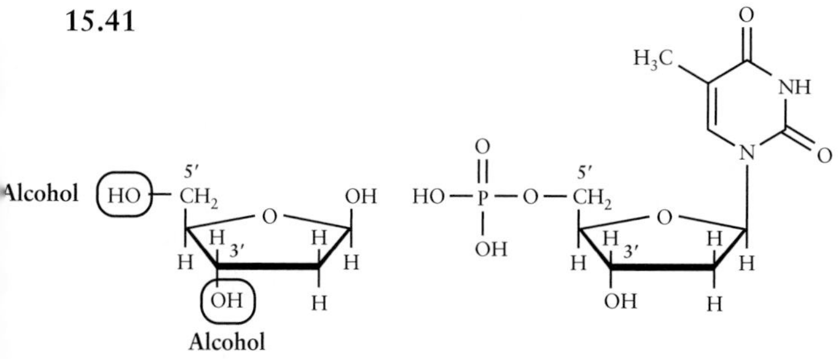

The nucleotide would be found in DNA because the monosaccharide is 2-deoxyribose.

15.43

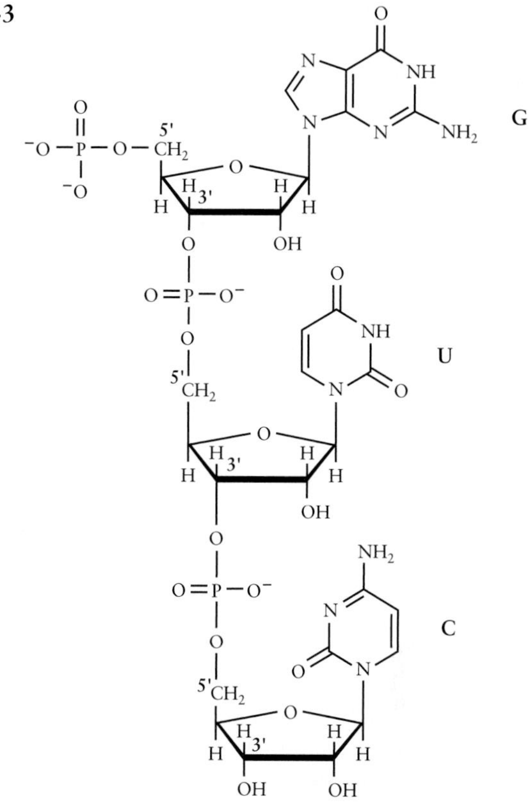

15.45 The monosaccharide and phosphate groups are located on the outside of the three-dimensional DNA structure. These groups are hydrophilic and can interact with the polar aqueous environment.

15.47 Hydrogen bonds exist between the base pairs.

15.49 A C—G base pair is linked by three hydrogen bonds.

15.51 The complementary sequence is dGCTATC. 15 hydrogen bonds.

15.53 a. dATACGG   **b.** dTTGGAC
c. dGGATAA   **d.** dCATAGG

15.55 Most human cells contain 46 chromosomes.

15.57 See Figure 15-12. For example, diabetes mellitus and schizophrenia.

15.59 DNA replication begins with the unraveling of the supercoiled DNA to expose the double helix.

15.61

	DNA 1	DNA 2
Parent:	dGGTACGCTT	dCCATGCGAA
Daughter:	dCCATGCGAA	dGGTACGCTT

15.63 Hydrogen bonds

15.65 When DNA polymerase detects an error in the daughter strand, it signals other enzymes to replace and repair incorrectly placed nucleotides. The mistake rate is less than one in one billion molecules after proofreading.

15.67 a. dCGTACTGGA: Parent strand
    dCCATGAACT: Daughter strand (Errors in daughter strand are in red.)
    dGCATGACCT: Corrected daughter strand
  b. dGAGTATCT: Parent strand
    dCTCCTCGA: Daughter strand (Errors in daughter strand are in red.)
    dCTCATAGA: Corrected daughter strand

15.69 RNA is single-stranded and contains D-ribose. It also contains uracil, U, instead of thymine, T.

15.71 *RNA polymerase* catalyzes the synthesis of a new mRNA molecule. It copies the sequences on the template strand of DNA from the 3' end to the 5' end.

15.73 a. dCCGGAATATA: DNA strand
   GGCCUUAUAU: mRNA strand
   b. dAAGGCCAATT: DNA strand
   UUCCGGUUAA: mRNA strand
   c. dGTACACGTCG: DNA strand
   CAUGUGCAGC: mRNA strand

15.75 The two important regions of tRNA are the anticodon loop and the 3' end covalently bound to one particular amino acid.

15.77 The anticodon AAA will correspond to the codon UUU; therefore, the amino acid will be phenylalanine.

15.79 No more amino acids will be added to the growing peptide chain when a stop codon is reached. The stop codons are UAA, UAG, and UGA.

15.81 The codons for alanine are GCU, GCC, GCA, and GCG.

15.83 a. mRNA: GUUGCUCGU
   Anticodons on tRNA: CAA CGA GCA
   Tripeptide: Val-Ala-Arg
   b. mRNA: CUACGCGGU
   Anticodons on tRNA: GAU GCG CCA
   Tripeptide: Leu-Arg-Gly
   c. mRNA: AGUAACUCG
   Anticodons on tRNA: UCA UUG AGC
   Tripeptide: Ser-Asn-Ser
   d. mRNA: UAUGAUACC
   Anticodons on tRNA: AUA CUA UGG
   Tripeptide: Tyr-Asp-Thr

15.85 a. DNA strand: dTTACCCGACGGC
   mRNA strand: AAUGGGCUGCCG
   anticodons on tRNA: UUA CCC GAC GGC
   amino acid sequence: Asn-Gly-Leu-Pro
   b. DNA strand: dGAAGCAACCATA
   mRNA strand: CUUCGUUGGUAU
   anticodons on tRNA: GAA GCA ACC AUA
   amino acid sequence: Leu-Arg-Trp-Tyr
   c. DNA strand: dATGACCACAGAA
   mRNA strand: UACUGGUGUCUU
   anticodons on tRNA: AUG ACC ACA GAA
   amino acid sequence: Tyr-Trp-Cys-Leu
   d. DNA strand: dCGACATCCTCTA
   mRNA strand: GCUGUAGGAGAU
   anticodons on tRNA: CGA CAU CCU CUA
   amino acid sequence: Ala-Val-Gly-Asp

15.87 DNA: dAGTAAA
   mRNA: UCAUUU
   dipeptide: Ser-Phe
   a. DNA mutation: dAG**C**AAA
   mRNA: UC**G**UUU
   dipeptide: Ser-Phe
   The dipeptide would be the same.

b. DNA mutation: dAGTA**A**T
   mRNA: UCAU**U**A
   dipeptide: Ser-Leu
   The dipeptide would be different.
   c. DNA mutation: dAT**A**AA
   mRNA: UA**U**UU
   peptide: Tyr-
   The dipeptide would be different.
   d. DNA mutation: dAGTAA
   mRNA: UCAUU
   peptide: Ser-
   If the next nucleotide in the mRNA sequence is U or C, then the dipeptide will be the same. If the next nucleotide in the mRNA sequence is A or G, then the dipeptide will be a different dipeptide, Ser-Leu.

15.89 The HIV virus contains RNA.

15.91 The HIV virus uses *reverse transcriptase*, *protease*, *integrase*, and *ribonuclease* to replicate.

15.93 HIV infects T-lymphocytes.

15.95 HIV uses its genomic RNA to code for DNA.

15.97 The HIV virus rapidly mutates and becomes resistant to any one type of drug therapy; therefore, the best treatment often is to use combination drug therapies.

15.99 *Protease* inhibitors prevent the release of essential enzymes required for the virus to replicate.

Chapter 16

16.37 The atomic number represents the number of protons in an isotope. The mass number represents the sum of the protons and neutrons in the nucleus.

16.39

Radioisotope	Atomic number	Mass number	Number of neutrons	Number of protons
Thallium-201	81	201	120	81
Selenium-75	34	75	41	34
Cobalt-60	27	60	33	27
Ba-131	56	131	75	56

16.41 Electromagnetic radiation is a form of energy that travels through space as a wave at the speed of light.

16.43 Radio waves have insufficient energy to damage biological tissues, while x-rays can cause significant damage to biological tissue.

16.45 a. Radio-wave radiation has the longer wavelength.
   b. X-ray radiation has the longer wavelength.
   c. Visible radiation has the longer wavelength.

16.47 a. Gamma-rays are more damaging to biological tissue because they are higher in energy.
   b. Ultraviolet radiation is more damaging to biological tissue because it is higher in energy.

16.49 $^{241}_{95}\text{Am} \rightarrow {}^{237}_{93}\text{Np} + {}^{4}_{2}\alpha$

16.51 Both positrons and β-particles are high-speed high-energy particles. The difference between the two is the charge. The positron has a charge of +1, while a β-particle has a charge of −1.

16.53 The nuclear equation is $^{12}_{5}B \rightarrow ^{12}_{6}C + ^{0}_{-1}\beta$. The daughter nuclide is carbon-12.

16.55 a. $^{26}_{11}Na \rightarrow ^{26}_{12}Mg + ^{0}_{-1}\beta$

b. $^{210}_{86}Rn \rightarrow ^{206}_{84}Po + ^{4}_{2}\alpha$

c. $^{52}_{26}Fe \rightarrow ^{52}_{25}Mn + ^{0}_{+1}\beta$

16.57 1.1 g of iodine-131 is left after 32 days. Thirty-two days represents four half-lives.

16.59 The half-life of Tc-99m is only 6 hours, so the patient's exposure to radiation can be kept to a minimum because Tc-99m in a patient's body decays quickly.

16.61 Half of a half-life has elapsed, or 33 hours.

16.63 Artificial radioisotopes are created by bombarding nuclei with high-energy particles generated in an accelerator.

16.65 $^{1}_{0}n + ^{59}_{27}Co \rightarrow ^{60}_{27}Co$ Cobalt-59 is used to produce cobalt-60.

16.67 $^{238}_{92}U + ^{4}_{2}\alpha \rightarrow ^{239}_{94}Pu + 3 ^{1}_{0}n$ U-238 is bombarded with an α-particle to produce Pu-239 and three neutrons.

16.69 particle accelerator; period 7; the actinoids

16.71 Ca^{2+}

16.73 A mutation occurs when radiation breaks phosphodiester bonds sufficiently close to each other on opposite strands of DNA. These mutations are passed on when the cell reproduces, possibly causing cancer.

16.75 X-rays do not have enough penetrating power to pass through the lead apron.

16.77 a. β-particle **b.** gamma-ray **c.** gamma-ray

16.79 β-Particles and positrons have the same penetrating power and the same energy.

16.81 The bequerel and curie measure the same property of radiation: the rate of radioactive emissions from a sample. The abbreviations are Bq for bequerel and Ci for curie. A millicurie would be abbreviated mCi.

16.83 An absorbed dose measures the energy of radiation absorbed per mass of tissue, but does not take into account the penetrating power of the radiation. The effective dose takes into account both the penetrating power of radiation and the amount of energy to give a biological effect. The units for absorbed dose are the Gray and the Rad. The effective dose is measured in sieverts and rem.

16.85 An LD_{50} indicates a level of exposure that would result in death in 50% of the population in 30 days.

16.87 The CT scan

16.89 The quarter has a higher density than the tissue in the esophagus. The quarter absorbed more x-rays and is lighter in color than the tissue of the esophagus.

16.91 The MRI is best for imaging soft tissue areas of the body. The technique uses low-energy radiowaves that do not pose a risk of biological damage. An MRI is not ideal for imaging denser tissues in the body such as bones and joints.

16.93 1.3 s

16.95 a. $^{192}_{77}Ir \rightarrow ^{192}_{78}Pt + ^{0}_{-1}\beta + \delta$. The daughter nuclide is Pt-192.

b. A thick lead apron should be used to protect the healthy tissues from the β and γ radiation.

Glossary/Index

NOTE: Page numbers followed by f indicate figures; those followed by t indicate tables.

A blood type, 458–459, 459t
AB blood type, 458–459, 459t
Abbreviations, for dosages, 16
Absolute temperature scale The Kelvin scale. This temperature scale assigns a temperature of zero to the theoretical condition in which all molecular motion has stopped, 114, 114t, 115f
Absorbed dose The energy of radiation absorbed per mass of tissue, 582, 582t
Accelerator A sophisticated instrument that creates high-energy, fast moving particles, 579
ACE inhibitors, 420–421
Acetaldehyde, 241–242, 241f
  in alcohol metabolism, 342, 342f, 346
Acetamide, 252, 252f
Acetaminophen, naming of, 215, 216
Acetic acid, 244f, 245, 245f, 310t, 313
  in acetyl CoA, 504, 504f
  in alcohol metabolism, 342, 342f
  in esterification, 367
Acetone, vapor pressure of, 126t
Acetyl CoA A central molecule in metabolism used to transfer acyl groups in acyl transfer reactions. Contains a thioester functional group, 249, 249f, 367
  in β-oxidation, 488–489, 488f, 490f, 504
  in carbohydrate catabolism, 433–434, 434f, 451, 456, 504
  in citric acid cycle, 503–510, 506f
  in fatty acid catabolism, 487–489, 488f, 490f, 504
  hydrolysis of, 504
  pyruvate oxidation to, 456
  structure of, 504, 504f
Acetylene, 203
Achiral Superposable on its mirror image, and therefore the same as its mirror image, 394
Acid(s) A substance that produces protons (H+) when dissolved in water; a proton donor, 308–323
  antacids and, 322–323, 322t
  in aqueous solutions, 309, 310–312, 314–316, 314f, 316f
  in biochemistry and medicine, 310t

biological roles of, 310t
conjugate, 311–313
definitions of, 308, 309
fatty. *See* Fatty acids
hydronium ions and, 309
overview of, 308–309
as proton donors, 309, 311–312
strong, 314–315, 314f, 315t, 316f
structure of, 310t
water as, 313
weak, 315–316, 316f
Acid reflux, 314
Acid-base homeostasis The body's maintenance of the proper pH, 319–321, 320f, 320t, 321f, 333
Acid-base reactions
  buffers and, 330–333, 330f, 331f
  in digestion, 334–335
  equilibrium in, 319–321, 320f, 320t, 321f, 333
  Le Châtelier's principle and, 319–321, 321f, 439
  neutralization, 322–323
  pH and, 324–329
Acidosis A serious medical condition diagnosed when the pH of the blood falls below the normal range, 333
Aconitase, in citric acid cycle, 506f, 510t
Acquired immunodeficiency syndrome (AIDS), 559–561
Actinides, 26, 26f
Activation energy (E_A) The minimum amount of energy that must be attained by the reactants for a reaction to proceed, 294–295, 295f, 298, 415
  catalysts and, 298, 298f, 415
Active site The pocket or cleft within a protein where the chemical reaction takes place. Also known as the binding site, 416
Acyl group, 358
Acyl group transfer reaction(s) A type of reaction that interconverts carboxylic acids and their derivatives: esters, thioesters, and amides. RC=O is an acyl group, 358–369, 358f, 373t
  in citric acid cycle, 506f, 508, 510t
  coenzyme A in, 504

esterification, 365–367, 373t
hydrolysis, 359–363, 359f, 369, 373t
Addition, significant figures in, 12
Adenine, 312t
  in acetyl CoA, 504, 504f
  complementary base for, 546, 546f
  in DNA, 538, 538f
  in RNA, 538, 539f
Adenosine, 262, 539, 539f
Adenosine diphosphate (ADP). *See* ADP (adenosine diphosphate)
Adenosine monophosphate. *See* AMP (adenosine monophosphate)
Adenosine triphosphate. *See* ATP (adenosine triphosphate)
Adipocytes Fat cells, 469
ADP (adenosine diphosphate)
  in phosphoryl transfer reactions, 369–371
  phosphorylation of, 448, 503, 510–516. *See also* Oxidative phosphorylation
  ATP synthase in, 516, 516f, 517f
  in citric acid cycle, 505, 506f, 510t
  electron-transport chain and, 451–454, 453f
  in glycolysis, 451–454, 453f
Adrenaline, 257, 312t
Aerobic conditions Conditions in which oxygen is present, 456
Aggregates, 163
AIDS, 559–561
Alanine
  chirality of, 395, 395f
  Fischer projections of, 397, 397f
  structure of, 317, 317f, 391f
Albumin, 72, 72f
Alcohol(s) A functional group having the form R–OH, 232, 232f, 233–238, 263t, 264t. *See also* Ethanol
  classification of, 233–234, 234f
  in esterification, 365–367, 365t
  from hydration reactions, 356
  hydrogen bonds in, 236
  from hydrolysis, 359–363, 359f
  hydroxyl groups in, 234, 234f
  intermolecular forces of attraction in, 236–237
  in membrane lipids, 477–478, 477f, 478f
  naming of, 235, 237–238

Amylose The minor constituent of starch; it is an unbranched polysaccharide with $\alpha(1\rightarrow4)$ linkages, 445–447, 446t
hydrolysis of, 448–449, 449f

Anabolic reactions Biochemical reactions that convert smaller molecules into larger molecules such as proteins and DNA. Anabolic reactions consume energy overall, 293–294, 293f, 343f, 502
energy for, 526

Anabolic steroids, 468–469, 468f, 492

Anaerobic conditions Conditions in which oxygen is not present, 456

Analgesics Pain medications, 232
naming of, 204t, 215, 216
narcotic, 231f, 232

Analogs, functional groups and, 232

Analytical balance, 9f

Androgenic effects, of steroids, 468

Androgens, 242, 242f, 468–469, 468f, 492, 492f

Anemia, 2
sickle-cell, 388, 411, 411f, 557

Anesthesia, Henry's Law and, 136

Angioplasty, 219, 220f

Angiotensin, 420

Angiotensin-converting enzyme (ACE) inhibitors, 420–421

Angular molecular geometry, 85, 85f

Aniline, 256, 256f

Anions Negatively charged ions resulting from a nonmetal atom gaining electrons, so that the number of electrons is greater than the number of protons, 44. *See also* Ion(s)
in formula unit, 48–50, 61
names of, 46, 50

Anomeric carbon The carbon atom in the ring form of a monosaccharide that is bonded to the ring oxygen atom and bears a hydroxyl group, 436–437, 436f, 437f
in disaccharides, 442, 442f, 444t
mutarotation and, 438–439, 439f
in nucleosides, 538–539

Antacids, 322–323, 322t

Antibiotics, amides in, 252, 253f

Anticodon loop The three nucleotides on tRNA that correspond to the complementary codon on mRNA, 554–556, 555f

Antiestrogens, for breast cancer, 80, 102–103, 102f, 103f

Antifreeze poisoning, 308

Antihypertensives, 420–421

Anti-inflammatory agents, 420, 494

Antioxidants Natural substances found in fruits, vegetables, and tea that prevent the damaging effects of oxidation by reducing harmful oxidizing agents, 355

Antipsychotics, 266

Antiretroviral drugs, 561

Aqueous solutions Homogeneous mixtures where water is the solvent, 150. *See also* Solutions; Water
acids in, 309, 310–312, 314–316, 314f
bases in, 309, 310–312, 315–317, 315f, 316f
pH of, 324–329

Arachidonic acid, 470, 470t
in inflammation, 493–494, 493f

Arginine
side chains of, 390f, 392
structure of, 391f

Aromatic compounds, 213, 263t

Aromatic hydrocarbons Six-membered rings written as alternating double and single bonds, although six of these electrons are actually distributed evenly over the six carbon atoms in the ring, 181, 181f, 213–217. *See also* Benzene

Aromatic rings, 344t

Arrhenius definition, 309

Artificial radioisotopes Man-made radioisotopes, 569, 579–580

Ascorbic acid, 374, 374t

Asparagine, structure of, 391f

Aspartame, in phenylketonuria, 528

Aspartic acid
side chains of, 390f, 392
structure of, 391f

Aspirin, functional groups in, 249, 249f
naming of, 215, 216

Atherosclerosis, 218–219

Atmospheres, 122t, 123

Atmospheric pressure The pressure exerted by the weight of air at any given place on the earth, 122–123, 123f

Atom(s) The smallest component of an element that still displays the characteristics of the element, 18–21
in compounds, 43. *See also* Compounds
conservation in reactions, 278, 279, 280–281
electronegativity of, 90–91, 91f
nucleus of, 19–20
structure of, 18–20, 19f, 19t

Atomic mass, 23
average, 23
in periodic table, 23–24, 24f
vs. atomic number, 24
formula mass and, 64
molar mass and, 65
molecular mass and, 63
in periodic table, 23–24, 24f

Atomic mass units (amu), 20, 64

Atomic number The number of protons in an atom; it defines the element, 20, 21, 21t, 22
in periodic table, 23–24, 24f
vs. average atomic mass, 24

Atomic scale, 3, 3f, 4t

ATP (adenosine triphosphate), 262, 262f, 294
from β-oxidation, 488–489, 488f, 490f
in citric acid cycle, 503, 505, 506f
in fatty acid oxidation, 487–489, 488f
functions of, 518
in glycolysis, 453f, 454
hydrolysis of, free energy change in, 525–526
in membrane transport, 480
in phosphoryl transfer reactions, 369–371
synthesis of
ATP synthase in, 516
in citric acid cycle, 505, 506f, 510t
electron-transport chain in, 451–454, 453f
energy source for, 516
in glycolysis, 451–454, 453f
oxidative phosphorylation in, 510–518

ATP synthase, 516, 516f

Atypical antipsychotics, 266

Autoionization of water The reaction of water molecules with one another to produce a small amount of hydronium ions (H_3O^+) and hydroxide ions (OH^-) in pure water, 311

Average atomic mass A weighted average of the mass of the isotopes of an element based on the natural abundance of each isotope and its mass in the periodic table, 23
in periodic table, 23–24, 24f
vs. atomic number, 24

Avogadro's number, 65
in mass-particle conversion, 69

Axons, 265, 265f

Background radiation Environmental radiation from natural sources, 569

Carboxylic acid derivatives, 244, 245f, 254t, 263t
  hydrolysis of, 359–363, 359f
Cardiac disease
  cholesterol and, 218–219, 218f–220f
  coronary, 218–219, 218f–220f
Cardiovascular disease, cholesterol and, 218–219, 218f–220f
Caries, dental, 432, 456
β-Carotene, 197, 197f, 202
Catabolic reactions Reactions that convert large molecules, such as carbohydrates, proteins and fats, into smaller molecules. Catabolic reactions release energy overall, 293–294, 293f, 343f. See also Metabolism
  of carbohydrates, 433, 434f, 448–458, 517f. See also Carbohydrates, catabolism of
  citric acid cycle in, 504–510
  electron-transport chain in, 505, 512–515, 512f–514f
  energy for, 519–527. See also Bioenergetics
  of fatty acids, 483–489, 517f. See also Fatty acids, catabolism of
  glycolysis in, 448–457
  in mitochondria, 510–511
  overview of, 517–518, 517f
  oxidative phosphorylation in, 506f, 509, 510–516, 510t. See also Oxidative phosphorylation
  in phosphoryl group transfer, 371
  of proteins, 334–335, 360, 517f
  sites of, 510–511
Catalysts Substances that increase the rate of reaction by lowering the activation energy for the reaction, 298, 298f, 414–421, 415t. See also Enzyme(s)
  activation energy and, 298, 298f, 415
  pH and, 417, 417f
  temperature and, 417, 417f
Catalytic hydrogenation The reduction of carbon-carbon double bonds to carbon-carbon single bonds using hydrogen gas (H_2) in the presence of a metal catalyst (Pd, Pt, etc.), 349
Cations Positively charged ions resulting from the loss of electrons from a metal atom, so that the number of electrons is less that the number of protons, 44. See also Ion(s)
  in formula unit, 48–50, 61
  naming of, 46, 50
Cavities, dental, 432, 456
CD4 receptors, 559–561

Conversion Factors

Length
1 inch = 2.54 cm (exact)
1 m = 39.37 inches

Mass
1 lb = 453.5 g
1 kg = 2.205 lb

Volume
1 L = 1.057 quarts
1 cm^3 = 1 mL
1 dL = 100 mL

Temperature
$°F = (\frac{9}{5} \times °C) + 32$
$°C = \frac{5}{9}(°F - 32)$
$K = °C + 273$
$°C = K - 273$

Energy
1 cal = 4.184 J (exact)
1 Cal = 1 kcal
1 kcal = 10^3 cal

Pressure
1 atm = 760 mmHg (exact)
= 760 torr (exact)
= 14.70 psi
= 1.013 × 10^5 Pa

Number of atoms or molecules
1 mole = 6.02 × 10^{23}
atoms, molecules, or
formula units

Common Metric Prefixes

Scale	Prefix	Symbol	Factor	Factor in Scientific Notation
Macroscale	giga	G	1 000 000 000	10^9
	mega	M	1 000 000	10^6
	kilo	k	1 000	10^3
	Base unit		1	1
	deci	d	0.1	10^{-1}
	centi	c	0.01	10^{-2}
	milli	m	0.001	10^{-3}
Microscale	micro	μ	0.000 001	10^{-6}
Nanoscale	nano	n	0.000 000 001	10^{-9}
Atomic	pico	p	0.000 000 000 001	10^{-12}
Subatomic	femto	f	0.000 000 000 000 001	10^{-15}

Common Transition Metal and Polyatomic Ions

Common transition metal ions

Period 4	V^{5+}	Cr^{2+}, Cr^{3+}	Mn^{2+}	Fe^{2+}, Fe^{3+}	Co^{2+}, Co^{3+}	Cu$^+$, Cu^{2+}	Zn^{2+}
Period 5		Mo^{6+}				Ag$^+$	
Period 6							Hg$^+$, Hg^{2+}

Common polyatomic anions

Name	Formula	Name	Formula	Name	Formula
Acetate	CH$_3$CO$_2^-$	Hydroxide	OH$^-$	Phosphate	PO$_4^{3-}$
Hydrogen carbonate (bicarbonate)	HCO$_3^-$	Hypochlorite	OCl$^-$	Hydrogen phosphate	HPO$_4^{2-}$
Carbonate	CO$_3^{2-}$	Nitrate	NO$_3^-$	Sulfate	SO$_4^{2-}$
Cyanide	CN$^-$	Nitrite	NO$_2^-$	Sulfite	SO$_3^{2-}$
Dihydrogen phosphate	H$_2$PO$_4^-$				

Common polyatomic cations

Name	Formula
Hydronium	H$_3$O$^+$
Ammonium	NH$_4^+$